ENCYCLOPEDIA OF STATISTICAL SCIENCES

Volume 1

A to Buys–Ballot Table

ENCYCLOPEDIA OF STATISTICAL SCIENCES

Second Edition

Volume 1

A to Buys–Ballot Table

The *Encyclopedia of Statistical Sciences* is available Online at
http://www.mrw.interscience.wiley.com/ess

 WILEY-INTERSCIENCE

A John Wiley & Sons, Inc., Publication

Published by John Wiley & Sons, Inc., Hoboken, New Jersey
Published simultaneously in Canada

For general information on our other products and services, please contact our Customer Care Department within the United States at (800) 762-2974, outside the United States at (317) 572-3993 or fax (317) 572-4002.

Wiley also publishes its books in a variety of electronic formats. Some content that appears in print may not be available in electronic format. For more information about Wiley products, visit our web site at www.wiley.com.

Library of Congress Cataloging-in-Publication Data:

Encyclopedia of statistical sciences / edited by N. Balakrishnan ... [et al.]—2nd ed.
      p. cm.
    "A Wiley-Interscience publication."
    Includes bibliographical references and index.
    ISBN-13: 978-0-471-15044-2 (set)
    ISBN-10: 0-471-15044-4 (set: cloth)
    ISBN 0-471-74391-7 (v. 1)
     1. Statistics—Encyclopedias. I. Balakrishnan, N., 1956–

  QA276.14.E5 2005
  519.5'03—dc22

                                                          2004063821

Printed in the United States of America

10 9 8 7 6 5 4 3 2 1

NORMAN LLOYD JOHNSON, 1917–2004

We dedicate this Second Edition of the Encyclopedia of Statistical Sciences in honor of Norman Johnson, who, along with Sam Kotz, was its Founding Editor-in-Chief. A biographical entry appears in Volume 6.

CONTRIBUTORS

H. Ahrens, *Zentralinstitut fur Mathematiks, Berlin, Germany. Biometrical Journal*

Sergei A. Aivazian, *The Russian Academy of Sciences, Moscow, Russia.* Bol'shev, Login Nikolaevich

James Algina, *University of Florida, Gainesville, FL.* Behrens-Fisher Problem—II

Frank B. Alt, *University of Maryland, College Park, MD.* Bonferroni Inequalities and Intervals; Autoregressive Error, Hildreth–Lu Scanning Method

Torben G. Andersen, *Northwestern University, Evanston, IL.* ARCH and GARCH Models

O. D. Anderson, *Civil Service College, London, United Kingdom.* Box–Jenkins Model

D. F. Andrews, *University of Toronto, Toronto, ON, Canada.* Andrews Function Plots

Barry C. Arnold, *University of California, Riverside, CA.* Bivariate Distributions, Specification of

S. Asmussen, *Aarhus Universitet, Aarhus, Denmark.* Branching Processes

V. Bagdonavicius, *Université Victor Segalen Bordeaux, Bordeaux, France.* Accelerated Life Testing

Charles R. Baker, *University of North Carolina, Chapel Hill, NC.* Absolute Continuity

David L. Banks, *Institute of Statistics and Decision Science, Durham, NC.* Bootstrapping—II

D. Basu, *Florida State University, Tallahasee, FL.* Basu Theorems

Niels G. Becker, *Australian National University, Canberra, Australia.* Back-Projection, Method of

James O. Berger, *Duke University, Durham, NC.* Bayes Factors

G. K. Bhattacharyya, *University of Wisconsin, Madison, WI.* Bivariate Sign Test, Bennett's

Tim Bollerslev, *Duke University, Durham, NC.* ARCH and GARCH Models

K. O. Bowman, *Oak Ridge National Laboratory, Oak Ridge, TN.* Approximations to Distributions

David R. Brillinger, *University of California, Berkeley, CA.* Anscombe, Francis John

Lyle D. Broemeling, *University of Texas M.D. Anderson Cancer Center, Houston, TX.* Box–Cox Transformation—I

Caitlin E. Buck, *University of Oxford, Oxford, United Kingdom.* Archaeology, Statistics in—II

Foster B. Cady, *ASA, Alexandria, VA.* Biometrics

L. Le Cam, *University of California, Berkeley, CA.* Asymptotic Normality of Experiments

Enrique Castillo, *Universidad de Cantabria, Santander, Spain.* Bayesian Networks

G. J. Chaitin, *IBM Research Division, New York, NY.* Algorithmic Information Theory

John M. Chambers, *Bell Labs—Lucent Technologies, Murray Hill, NJ.* Algorithms, Statistical

Shein-Chung Chow, *National Health Research Institutes, Zhunan, Taiwan.* Bioavailability and Bioequivalence

Gauss M. Cordeiro, *Universidade Federal da Bahiam, Recife, Brazil. Brazilian Journal of Probability and Statistics*

D. R. Cox, *University of Oxford, Oxford, United Kingdom.* Bartlett Adjustment—I

Sarah C. Darby, *University of Oxford, Oxford, United Kingdom.* Berkson Error Model

Anirban Dasgupta, *Purdue University, West Lafayette, IN.* Binomial and Multinomial Parameters, Inference On

Somnath Datta, *University of Georgia, Athens, GA.* Bootstrap

F. Downton, *The University, Birmingham, United Kingdom.* Betting, Labouchère Systems

N. R. Draper, *University of Wisconsin, Madison, WI.* Backward Elimination Selection Procedure

Paul S. Dwyer, *University of Michigan, Ann Arbor, MI. Annals of Applied Probability*

Jean Edmiston, *Greenhitche, Kent, United Kingdom.* Annals of Human Genetics

W. J. Ewens, *University of Pennsylvania, Philadelphia, PA.* Ascertainment Sampling

Tom Fearn, *University College London, London, United Kingdom.* Berkson Error Model

Eric D. Feigelson, *Pennsylvania State University, University Park, PA.* Astronomy, Statistics in

Martin R. Frankel, *CUNY-Baruch College, Cos Cob, CT.* Balanced Repeated Replications

D. A. S. Fraser, *Univeristy of Toronto, Toronto, ON, Canada.* Ancillary Statistics, First Derivative

G. H. Freeman, *University of Warwick, Coventry, United Kingdom.* Agriculture, Statistics in

Wayne A. Fuller, *Iowa State University, Ames, IA.* Area Sampling

K. Ruben Gabriel, *University of Rochester, Rochester, NY.* Biplots

J. Gani, *Australian National University, Canberra, Australia.* Applied Probability; Applied Probability Journals; Bartlett, Maurice Stevenson

Alan E. Gelfand, *Duke University, Durham, NC.* Archaeology, Statistics in—I

Edward I. George, *University of Pennsylvania, Philadelphia, PA.* Bayesian Model Selection

Jayanta Ghosh, *Indian Statistical Institute, Calcutta, India.* Basu, Debabrata

A. Goldman, *University of Nevada, Las Vegas, NV.* Blackjack

Michael Goldstein, *University of Durham, Durham, United Kingdom.* Bayes Linear Analysis

W. A. Golomski, *W.A. Golomski & Associates, Chicago, IL.* American Society for Quality (ASQ)

I. J. Good, *Virginia Polytechnic Institute & State University, Blacksburg, VA.* Axioms of Probability; Belief, Degrees of

Bernard G. Greenberg, *University of North Carolina, Chapel Hill, NC.* Biostatistics, Classical

Nels Grevstad, *Metropolitan State College, Denver, CO.* Binomial Distribution: Sample Size Estimation

Peter C. Gøtzsche, *The Nordic Cochrane Centre, Copenhagen, Denmark.* Assessment Bias

S. J. Haberman, *Center for Statistical Theory and Practice Education Testing Service, Princeton, NJ.* Association, Measures of

Ali S. Hadi, *The American University in Cairo, Cairo, Egypt.* Bayesian Networks

Peter Hall, *Australian National University, Canberra, Australia.* Binning; Block Bootstrap

David J. Hand, *Imperial College, London, United Kingdom.* Banking, Statistics in

D. M. Hawkins, *University of Minnesota, Minneapolis, MN.* Branch-and-Bound Method

Juha Heikkinen, *The Finnish Forest Research Institute Vantaa Research Centre, Unioninkatu, Helsinki.* Biogeography, Statistics in

C. C. Heyde, *Australian National University, Canberra, Australia.* Australian and New Zealand Journal of Statistics

M. Hills, *British Museum, London, United Kingdom.* Allometry; Anthropology, Statistics in

W. G. S. Hines, *University of Guelph, Guelph, ON, Canada.* Box–Cox Transformations: Selecting for Symmetry

Klaus Hinkelmann, *Virginia Polytechnic Institute & State University, Blacksburg, VA.* Aitken Equations

D. V. Hinkley, *Univeristy of California, Santa Barbara, CA.* Annals of Statistics; Angular Transformation

W. Hoeffding, *University of North Carolina, Chapel Hill, NC.* Asymptotic Normality

Robert V. Hogg, *University of Iowa, Iowa, IA.* Adaptive Methods

Jason C. Hsu, *Ohio State University, Columbus, OH.* Bioequivalence Confidence Intervals

Xiaoming Huo, *Georgia Institute of Technology, Atlanta, GA.* Beamlets and Multiscale Modeling

Harri Högmander, *University of Jyvaskyla, Jyvaskyla, Finland.* Biogeography, Statistics in

G. M. Jenkins, *Gwilym Jenkins & Partners, Ltd., Lancaster, United Kingdom.* Autoregressive–Integrated Moving Average (ARIMA) Models; Autoregressive–Moving Average (ARMA) Models

Wesley O. Johnson, *University of California, Davis, CA.* Binary Data, Multivariate

V. M. Joshi, *University of Western Ontario, London, ON, Canada.* Admissibility

John D. Kalbfleisch, *University of Michigan, Ann Arbor, MI.* Ancillary Statistics—I

G. V. Kass, *University of Witwatersrand, Witwatersrand, South Africa.* Automatic Interaction Detection (AID) Techniques

Ravindra Khattree, *Michigan State University, East Lansing, MI.* Antieigenvalues and Antieigenvectors

Karl Christoph Klauer, *Psychometric Society, Research Triangle Park, NC.* Agreement Analysis, Basic

Stuart Klugman, *Drake University, Des Moines, IA.* Actuarial Science

Kathleen Kocherlakota, *University of Manitoba, Winnipeg, MB, Canada.* Bivariate Discrete Distributions

Subrahmaniam Kocherlakota, *University of Manitoba, Winnipeg, MB, Canada.* Bivariate Discrete Distributions

John E. Kolassa, *Rutgers University, Piscataway, NJ.* Asymptotics, Higher Order

Alex J. Koning, *Erasmus University Rotterdam, Rotterdam, The Netherlands*. Bahadur Efficiency, Approximate

Helena Chmura Kraemer, *Stanford University, Stanford, CA*. Agreement, Measures of; Biserial Correlation

K. Krickeberg, *University Rene Descartes, Paris, France*. Bernoulli Society

Austin F. S. Lee, *Boston University Metropolitan College, Boston, MA*. Behrens—Fisher Problem, Lee—Gurland Test

Fred C. Leone, *American Statistical Association, Alexandria, VA*. American Statistical Association

Edward A. Lew, *Association of Life Insurance Medical Directors, Punta Gorda, FL*. Actuarial Health Studies

Dennis V. Lindley, *Woodstock, Somerset, United Kingdom*. Bayesian Inference; Assessment of Probabilities; Basu's Elephant

H. Li, *Washington State University, Pullman, WA*. Aging First-Passage Times

Christopher J. Lloyd, *Australian Graduate School of Management, Sydney, Australia*. Ancillary Statistics—II

Wei-Yin Loh, *University of Wisconsin, Madison, WI*. Box—Cox Transformations—II

Bryan F. J. Manly, *West Inc., Laramie, WY*. Animal Populations, Manly—Parr Estimators

Barry H. Margolin, *University of North Carolina, Chapel Hill, NC*. Blocks, Randomized Complete; Blocks, Balanced Incomplete

J. S. Marron, *University of North Carolina, Chapel Hill, NC*. Bandwidth Selection

Ian W. McKeague, *Columbia University, New York, NY*. Additive Risk Model, Aalen's

Piotr W. Mikulski, *University of Maryland, College Park, MD*. Bonferroni, Carlo Emilio

John Neter, *University of Georgia, Athens, GA*. Auditing, Statistics in; *American Statistician, The*

Paul Newbold, *University of Nottingham, Nottingham, United Kingdom*. Business Forecasting Methods

M. Nikulin, *Steklov Mathematical Institute, St. Petersburg, Russia*. Accelerated Life Testing

R. M. Norton, *College of Charleston, Charleston, SC*. Arc-Sine Distribution

J. Keith Ord, *Georgetown University, Washington, DC*. Aggregation

S. C. Pearce, *University of Kent at Canterbury, Kent, United Kingdom*. Analysis of Covariance; Analysis of Variance

S. K. Perng, *Kansas State University, Manhattan, KS*. Bahadur Efficiency

Campbell B. Read, *Southern Methodist University, Dallas, TX*. Accuracy and Precision; Bivariate Normal Distribution, Fieller's Theorem

Nancy Reid, *University of Toronto, Toronto, ON, Canada*. Ancillary Statistics, First Derivative; Asymptotic Expansions—II

G. K. Robinson, *The Commonwealth Scientific and Industrial Research Organization, South Clayton, Australia*. Behrens–Fisher Problem—I

Robert N. Rodriguez, *SAS Institute Inc., Cary, NC*. Burr Distributions

Peter J. Rousseeuw, *Renaissance Technologies Corporation, Suffolk, NY*. Boxplot, Bivariate

Ida Ruts, *University of Antwerp, Antwerp, Belgium*. Boxplot, Bivariate

Shinichi Sakata, *University of British Columbia, Vancouver, BC, Canada*. Breakdown Point

Edward G. Schilling, *Rochester Institute of Technology, Rochester, NY*. Acceptance Sampling

H. T. Schreuder, *USDA Forest Service Rocky Mountain Research Station, Fort Collins, CO*. Bias Reduction, Quenouille's Method for

David W. Scott, *Rice University, Houston, TX*. Averaged Shifted Histogram

George A. F. Seber, *University of Auckland, Auckland, New Zealand*. Adaptive Sampling

A. R. Sen, *Oakland University, Rochester, MI*. Animal Science, Statistics in

P. K. Sen, *University of North Carolina, Chapel Hill, NC*. Approgression; Antiranks

E. Seneta, *University of Sydney, Sydney, Australia*. Boltzmann, Ludwig Edward; Bernstein, Sergei Natanovich; Bienaymé, Irenée-Jules; Abbe, Ernst; Boscovich, Ruggiero Giuseppe

R. J. Serfling, *University of Texas at Dallas, Plano, TX*. Asymptotic Expansions—I

G. Shafer, *Rutgers University, Princeton, NJ*. Belief Functions; Bernoullis, The

M. Shaked, *University of Arizona, Tucson, AZ*. Aging First-Passage Times

L. R. Shenton, *University of Georgia, Athens, GA*. Approximations to Distributions

O. Sheynin, *Berlin, Germany*. Achenwall, Gottfried

E. Shoesmith, *University of Buckingham, Buckingham, United Kingdom*. Arbuthnot, John

R. R. Sitter, *Simon Fraser University, Burnaby, BC, Canada*. Balanced Resampling Using Orthogonal Multiarrays

Walter L. Smith, *University of North Carolina, Chapel Hill, NC*. Birth-and-Death Processes

Andrew Solow, *Woods Hole Oceanographic Institution, Woods Hole, MA.* Atmospheric Statistics

Brenda Sowan, *Imperial College, London, United Kingdom.* Biometrika

Stephen M. Stigler, *University of Chicago, Chicago, IL.* Arithmetic Mean; Bahadur, Raghu Raj

Lajos Takács, *Case Western Reserve University, Cleveland, OH.* Ballot Problems

Wai-Yuan Tan, *Memphis State University, Memphis, TN.* Aids Stochastic Models

H. Taylor, *Towson University, Towson, MD.* Brownian Motion

Peter Thompson, *Statistics Research Associates Ltd., Wellington, New Zealand.* Bayes p-Values

Blaza Toman, *National Institute of Standards and Technology, Gaithersburg, MD.* Bayesian Experimental Design

D. S. Tracy, *University of Windsor, Windsor, ON, Canada.* Angle Brackets

R. L. Trader, *University of Maryland, College Park, MD.* Bayes, Thomas

Robert K. Tsutakawa, *University of Missouri, Columbia, MO.* Bioassay, Statistical Methods in

Frederic A. Vogel (Retired), *National Agricultural Statistics Service, Washington, DC.* Agricultural Surveys

Larry Wasserman, *Carnegie Mellon University, Pittsburgh, PA.* Bayesian Robustness

G. S. Watson, *Princeton University, Princeton, NJ.* Aitken, Alexander Craig

Mike West, *Duke University, Durham, NC.* Bayesian Forecasting

Halbert White, *University of California, La Jolla, CA.* Breakdown Point

Grace L. Yang, *University of Maryland, College Park, MD.* Biometric Functions

Sandy L. Zabell, *Northwestern University, Evanston, IL.* Bortkiewicz, Ladislaus Von

Tonglin Zhang, *Purdue University, West Lafayette, IN.* Binomial and Multinomial Parameters, Inference On

PREFACE TO SECOND EDITION

This Second Edition of the *Encyclopedia of Statistical Sciences* (ESS2) is accompanied by the important and significant step of presenting it online. The First Edition (ESS1) consisted of nine volumes with entries appearing in sequential alphabetic order plus a Supplement, published between 1982 and 1989, and three Update volumes published between 1997 and 1999.

The purposes of the Encyclopedia remain largely as they were in the First Edition, and they cannot be expressed any better than through the words of the Editors in Prefaces written for ESS1 Volume 1 in 1982 and for ESS Update Volume 1 written in 1997.

However, the pace of research in Statistics and burgeoning methodologies in its applications continue to be so rapid that the Editors and the publisher considered it to be a priority to find a way to incorporate timely revisions of material in many ESS1 entries as well as to introduce new ones. We have deleted a few ESS1 entries, replacing them with fresh contributions. Many entries clearly needed to be updated, but we frequently chose to leave some ESS1 entries unchanged while adding to each a second companion contribution that complements the original and brings it up to date.

The Editors have invited contributions on new important subject areas, representing the state of the art at the beginning of the twentyfirst century of the Common Era. These include data mining, Bayesian networks, beamlets and multiscale modeling, false discovery rate, agricultural surveys, image processing, radar detection, and cancer stochastic models, to name a few. There are new or additional entries on such subjects as medical statistics, forestry, loglinear models in contingency tables, government statistics, and survey methodology. For what has already been named the Century of Genetics, there is an updated entry on statistics in genetics, and new entries on bioinformatics and microarray analysis.

The Editors have consolidated and added new unsigned entries (those contributed by former and current ESS Editors). We became aware that the titles of many ESS1 entries put them at risk of being hidden from readers who might want to locate them, and have tried to make ESS2 more user-friendly by inserting blind entries or rewording the titles. For example, the blind entry "CORRELATION COEFFICIENTS, GLAHN AND HOOPER *See* GLAHN AND HOOPER CORRELATION COEFFICIENTS" should make readers aware of the existence of the referred entry.

With the help of the Internet we have aimed at updating all the ESS1 entries covering statistical and related journals and societies, and at ensuring that all those that were founded by 1950 have been included. A post-1950 explosion in new societies and journals is referred to in the short entries JOURNALS, STATISTICAL and SOCIETIES AND ORGANIZATIONS, STATISTICAL, and readers can find listed there several websites linked to societies and journals that are not featured in these pages.

It is with deep regret that we announce the death on November 18, 2004, of Norman Lloyd Johnson, Founding Editor of the *Encyclopedia* and Chairman of the Advisory Board (see the biographical entry JOHNSON, NORMAN LLOYD for an appreciation of his career). To the end of his life he maintained a keen interest in the Encyclopedia and in the preparation of this Second Edition, providing the Editors with many suggested improvements to the unsigned entries, so many indeed that some of them have yet to receive our full attention.

We are very fortunate that N. Balakrishnan and Brani Vidakovic joined the team of Editors-in-Chief for the Second Edition. We are very grateful to Sam Kotz for his many new contributions and for his many worthwhile suggestions. Our thanks go to the Advisory Board for many worthwhile suggestions and for valuable advice, to our Editorial Assistant Sheila Crain, without whose patience and hard work we might never have made our deadlines, to Steve Quigley, Surlan Murrell, Shirley Thomas at John Wiley and Sons, and above all to our many contributors, whose work forms the backbone of this project. Special thanks go to Anestis Antoniadis, Barry Arnold, Adelchi Azzalini, David Banks, Sir David Cox, Erich Lehmann, Eugene Seneta, Alex Shapiro, Jef Teugels, Mike Titterington, and Edward Waymire.

N. BALAKRISHNAN
McMaster University,
Hamilton, Ontario

CAMPBELL B. READ
Southern Methodist University,
Dallas, Texas

BRANI VIDAKOVIC
Georgia Institute of Technology,
Atlanta, Georgia

PREFACE TO FIRST EDITION

The purpose of this encyclopedia is to provide information about an extensive selection of topics concerned with statistical theory and the applications of statistical methods in various more or less scientific fields of activity. This information is intended primarily to be of value to readers who do not have detailed information about the topics but have encountered references (either by field or by use of specific terminology) that they wish to understand. The entries are not intended as condensed treatises containing all available knowledge on each topic. Indeed, we are on guard against placing too much emphasis on currently fashionable, but possibly ephemeral, matters. The selection of topics is also based on these principles. Nevertheless, the encyclopedia was planned on a broad basis—eight volumes, each of approximately 550 pages—so that it is possible to give attention to nearly all the many fields of inquiry in which statistical methods play a valuable (although not usually, or necessarily, a predominant) role.

Beyond the primary purpose of providing information, we endeavored to obtain articles that are pleasant and interesting to read and encourage browsing through the volumes. There are many contributors, for whose cooperation we are grateful, and a correspondingly wide range of styles of presentation, but we hope that each is attractive in its own way. There is also, naturally and inevitably, a good deal of variation among the (mathematical and technical-scientific) levels of the entries. For some topics, considerable mathematical sophistication is needed

for adequate treatment; for others, it is possible to avoid heavy reliance on mathematical formulation.

We realize that even an eight-volume compendium cannot incorporate all of the terms, notions, and procedures that have appeared in statistical literature during the last century. There are also contributions by scientists who paved the way, as early as the seventeenth century, toward the statistical sciences as they are known today. We endeavored to include historical background and perspective when these seem important to the development of statistical methods and ideas.

It is to be expected that most readers will disagree with the relative amount of emphasis accorded to certain fields, and will find that some topics of considerable interest have been omitted. While this may reflect a lack of judgment or knowledge (or both) on our part, it is inevitable, because each person has a specific, idiosyncratic viewpoint on statistical matters (as on others). Our intention is to mirror the state of the art in the last quarter of the twentieth century, including terms (in particular mathematical) that found a place in the language of statistical methodology during its formative years.

We have two ways of cross-referencing: First, when a possibly unfamiliar term appears in an entry, reference to another entry is indicated by an asterisk, or by direct reference (e.g., *See* HISTOGRAMS). An asterisk sometimes refers to the preceding word but quite frequently to the preceding phrase. For example, "... random variable*" refers to the

xiii

entry on random variables rather than on variables. We feel that this notation is the simplest possible. Second, most articles conclude with a list of related entries of potential interest for further reading. These two sets of cross-references may overlap but are usually not identical. The starred items are for utility, whereas the list is more for interest. Neither set is exhaustive and we encourage individual initiative in searching out further related entries.

Since our primary purpose is to provide information, we neither avoid controversial topics nor encourage purely polemic writing. We endeavor to give fair representation to different viewpoints but cannot even hope to approximate a just balance (if such a thing exists).

In accordance with this primary purpose, we believe that the imposition of specific rules of style and format, and levels of presentation, must be subordinate to the presentation of adequate and clear information. Also, in regard to notation, references, and similar minutiae, we did not insist on absolute uniformity although we tried to discourage very peculiar deviations that might confuse readers.

The encyclopedia is arranged lexicographically in order to entry titles. There are some inconsistencies; for example, we have "CHEMISTRY, STATISTICS IN" but "STATISTICS IN ASTRONOMY." This simply reflects the fact that the encyclopedia is being published serially, and the second of these entries was not available when the first volume was in production. (This volume does, however, contain the "dummy" entry "ASTRONOMY, STATISTICS IN *See* STATISTICS IN ASTRONOMY.")

We are indeed fortunate that Professor Campbell B. Read joined us as Associate Editor on October 1, 1980. Professor Read's active participation in the editorial process and the numerous improvements he contributed to this project have been invaluable. The Co-Editors-in-Chief express their sincerest appreciation of his expertise.

We also express our thanks to the members of the Advisory Board for their valuable advice and expert suggestions; to the Editorial Assistant, Ms. June Maxwell, for her devotion and for contributions to the project far beyond the call of duty; and last, but certainly not least, to all the contributors, who responded enthusiastically to our call for partnership in this undertaking.

Unsigned entries are contributed by the Editors—Samuel Kotz, Norman L. Johnson, and Campbell B. Read—either jointly or individually.

SAMUEL KOTZ
College Park, Maryland

NORMAN L. JOHNSON
Chapel Hill, North Carolina

January 1982

ABOUT THE EDITORS

N. BALAKRISHNAN, PhD, is a Professor in the Department of Mathematics and Statistics at McMaster University, Hamilton, Ontario, Canada. He has published widely in different areas of statistics including distribution theory, order statistics and reliability. He has authored a number of books including four volumes in "Distributions in Statistics Series" of Wiley, coauthored with N. L. Johnson and S. Kotz. He is a Fellow of the American Statistical Association and an Elected Member of the International Statistical Institute.

CAMPBELL B. READ, PhD, is Professor Emeritus of Statistical Science at the Institute for the Study of Earth and Man at Southern Methodist University in Dallas, Texas. He studied mathematics at the University of Cambridge in England, and obtained a PhD in mathematical statistics from the University of North Carolina at Chapel Hill. He is the author of several research papers in sequential analysis, properties of statistical distributions, and contingency table analysis. He is the author of several biographies appearing in "Leading Personalities in Statistical Sciences" and of various entries in this Encyclopedia, and is a coauthor of "Handbook of the Normal Distribution". He is an elected member of the International Statistical institute, has served on the faculty of the American University of Beirut, Lebanon, and was a Senior Research Scholar in 1987 at Corpus Christi College, University of Cambridge.

BRANI VIDAKOVIC, PhD, is Professor of Statistics at The Wallace H. Coulter Department of Biomedical Engineering at Georgia Institute of Technology, Atlanta, Georgia.

He obtained a BS and MS in mathematics at the University of Belgrade, Serbia, and a PhD in statistics at Purdue University, West Lafayette, Indiana. He is the author or coauthor of several books and numerous research papers on minimax theory, wavelets and computational and applied statistics. Dr. Vidakovic is a member of the Institute of Mathematical Statistics, American Statistical Association, International Society for Bayesian Analysis, and Bernoulli Society, and an elected member of the International Statistical Institute.

ENCYCLOPEDIA OF STATISTICAL SCIENCES

Volume 1

A to Buys–Ballot Table

A

AALEN'S ADDITIVE RISK MODEL.
See ADDITIVE RISK MODEL, AALEN'S

ABAC

A graph from which numerical values may be read off, usually by means of a grid of lines corresponding to argument values.

See also NOMOGRAMS.

ABACUS

A simple instrument to facilitate numerical computation. There are several forms of abacus. The one in most common use at present is represented diagramatically in Fig. 1. It consists of a rectangular framework *ABCD* with a cross-piece *PQ* parallel to the longer sides, *AB* and *CD*, of the rectangle. There are a number (at least eight, often more) of thin rods or wire inserted in the framework and passing through *PQ*, parallel to the shorter sides, *AD* and *BC*. On each rod there are threaded four beads between *CD* and *PQ*, and one bead between *PQ* and *AB*.

Analogously to the meaning of position in our number system, the extreme right-hand rod corresponds to units; the next to the left, tens; the next to the left, hundreds; and so on. Each bead in the lower rectangle (*PQCD*) counts for 1, when moved up, and each bead in the upper rectangle (*ABQP* counts for 5. The number shown in Fig. 2 would be 852 if beads on all rods except the three extreme right-hand ones are as shown for the three extreme left-hand rods (corresponding to "zero").

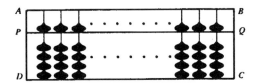

Figure 1. Diagrammatic representation of the form of abacus presently in common use.

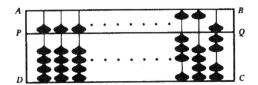

Figure 2. Abacus that would be showing the number 852 if beads on all rods except the three extreme right-hand ones are as shown for the three extreme left-hand rods (corresponding to "zero").

The Roman abacus consisted of a metal plate with two sets of parallel grooves, the lower containing four pebbles and the upper one pebble (with a value five times that of each pebble in the corresponding groove of the lower set). The Japanese and Chinese abacus (still in use) consists of a frame with beads on wires. The Russian abacus, which originated in the sixteenth century (the modern version in the eighteenth century), is also still in use.

BIBLIOGRAPHY

Dilson, J. (1968). *The Abacus: A Pocket Computer.* St. Martin's Press, New York.

Gardner, M. (1979). *Mathematical Circus.* Alfred A. Knopf, New York, Chap. 18.

Pullan, J. (1969). *The History of the Abacus.* F. A. Praeger, New York.

ABBE, ERNST

> ***Born:*** January 23, 1840, in Eisenach, Germany.
>
> ***Died:*** January 14, 1905, in Jena, Germany.
>
> ***Contributed to:*** theoretical and applied optics, astronomy, mathematical statistics.

The recognition of Abbe's academic talent by those in contact with him overcame a childhood of privation and a financially precarious situation very early in his academic career, when he completed "On the Law of Distribution of Errors in Observation Series," his inaugural dissertation for attaining a lecture-ship at Jena University at the age of 23 [1]. This dissertation, partly motivated by the work of C. F. Gauss*, seems to contain his only contributions to the probability analysis of observations subject to error. These contributions constitute a remarkable anticipation of later work in distribution theory and time-series* analysis, but they were overlooked until the late 1960s [5,8], and almost none of the early bibliographies on probability and statistics (a notable exception being ref. 10) mention this work. In 1866, Abbe was approached by Carl Zeiss, who asked him to establish a scientific basis for the construction of microscopes; this was the beginning of a relationship that lasted throughout his life, and from this period on his main field of activity was optics [9] and astronomy*.

Abbe shows, first, that the quantity $\Delta = \sum_{i=1}^{n} Z_i^2$, where Z_i, $i = 1, \ldots, n$, are n independently and identically distributed $N(0, 1)$ random variables, is described by a chi-square* density with n degrees of freedom [5,8], although this discovery should perhaps be attributed to I. J. Bienaymé* [4]. Second, again initially by means of a "discontinuity factor" and then by complex variable methods, Abbe obtains the distribution of $\Theta = \sum_{j=1}^{n}(Z_j - Z_{j+1})^2$, where $Z_{n+1} = Z_1$, and ultimately that of Θ/Δ, a ratio of quadratic forms* in $Z_1, \ldots, Z_n$ very close in nature to the definition of what is now called the first circular serial correlation* coefficient, and whose distribution under the present conditions is essentially that used to test the null hypothesis of Gaussian white noise* against a first-order autoregression alternative, in time-series* analysis [3]. (The distribution under such a null hypothesis was obtained by R. L. Anderson in 1942.) Knopf [6] expresses Abbe's intention in his dissertation as being to seek a numerically expressible criterion to determine when differences between observed and sought values in a series of observations are due to chance alone.

REFERENCES

1. Abbe, E. (1863). *Über die Gesetzmässigkeit der Vertheilung der Fehler bei Beobachtungsreihen*. Hab. schrift., Jena (Reprinted as pp. 55–81 of ref. 2.)

2. Abbe, E. (1906). *Gesammelte Abhandlungen*, Vol. 2. G. Fischer, Jena.

3. Hannan, E. J. (1960). *Time Series Analysis*. Methuen, London, pp. 84–86.

4. Heyde, C. C. and Seneta, E. (1977). *I. J. Bienaymé: Statistical Theory Anticipated*. Springer-Verlag, New York, p. 69.

5. Kendall, M. G. (1971). *Biometrika*, **58**, 369–373. (Sketches Abbe's mathematical reasoning in relation to the contributions to mathematical statistics.)

6. Knopf, O. (1905). *Jahresber. Dtsch. Math.-Ver.*, 14, 217–230. [One of several obituaries by his associates; nonmathematical, with a photograph. Another is by S. Czapski (1905), in *Verh. Dtsch. Phys. Ges.*, 7, 89–121.]

7. Rohr, L. O. M. von (1940). *Ernst Abbe*. G. Fischer, Jena. (Not seen.)

8. Sheynin, O. B. (1966). *Nature (Lond.)*, **211**, 1003–1004. (Notes Abbe's derivation of the chi-square density.)

9. Volkman, H. (1966). *Appl. Opt.*, **5**, 1720–1731. (An English-language account of Abbe's life and contributions to pure and applied optics; contains two photographs of Abbe, and further bibliography.)

10. Wölffing, E. (1899). *Math. Naturwiss. Ver. Württemberg* [Stuttgart], *Mitt.*, (2) **1**, 76–84. [Supplements the comprehensive bibliography given by E. Czuber (1899), in *Jahresber. Dtsch. Math.-Ver.*, **7** (2nd part), 1–279.]

See also Chi-Square Distribution; Quadratic Forms; Serial Correlation; and Time Series Analysis and Forecasting Society.

E. Seneta

ABEL'S FORMULA

(Also known as the Abel identity.) If each term of a sequence of real numbers $\{a_i\}$ can be represented in the form $a_i = b_i c_i, i = 1, \ldots, n$, then $a_1 + a_2 + \cdots + a_n$ can be expressed as

$$s_1(b_1 - b_2) + s_2(b_2 - b_3) + \cdots$$
$$+ s_{n-1}(b_{n-1} - b_n),$$

where $s_i = c_1 + \cdots + c_i$. Equivalently,

$$\sum_{k=n}^{m} b_k c_k = B_m c_{m+1} - B_{n-1} c_n$$
$$+ \sum_{k=n}^{m} B_k(c_k - c_{k+1}),$$

where $B_k = \sum_{l=1}^{k} b_l$.

This representation is usually referred to as Abel's formula, due to Norwegian mathematician Niels Henrik Abel (1802–1829). (The continuous analog of this formula is the formula of integration by parts.) It is useful for manipulations with finite sums.

BIBLIOGRAPHY

Knopp, K. (1951). *Theory and Application of Infinite Series*, 2nd ed. Blackie, London/Dover, New York.

ABSOLUTE ASYMPTOTIC EFFICIENCY (AAE). See ESTIMATION, CLASSICAL

ABSOLUTE CONTINUITY

Absolute continuity of measures, the Radon–Nikodym theorem*, and the Radon–Nikodym derivative* are subjects properly included in any basic text on measure and integration. However, both the mathematical theory and the range of applications can best be appreciated when the measures are defined on an infinite-dimensional linear topological space. For example, this setting is generally necessary if one wishes to discuss hypothesis testing* for stochastic processes with infinite parameter set. In this article we first define basic concepts in the area of absolute continuity, state general conditions for absolute continuity to hold, and then specialize to the case where the two measures are defined on either a separable Hilbert space or on an appropriate space of functions. Particular attention is paid to Gaussian measures.

The following basic material is discussed in many texts on measure theory*; see, e.g., ref. 23. Suppose that (Ω, β) is a measurable space, and that μ_1 and μ_2 are two probability measures on (Ω, β). μ_1 is said to be *absolutely continuous* with respect to μ_2 ($\mu_1 \ll \mu_2$) if A in β and $\mu_2(A) = 0$ imply that $\mu_1(A) = 0$. This is equivalent to the following: $\mu_1 \ll \mu_2$ if and only if for every $\epsilon > 0$ there exists $\delta > 0$ such that $\mu_2(A) < \delta$ implies that $\mu_1(A) \leqslant \epsilon$. Similar definitions of absolute continuity can be given for nonfinite signed measures; this article, however, is restricted to probability measures. When $\mu_1 \ll \mu_2$, the Radon–Nikodym theorem* states that there exists a real-valued β-measurable function f such that $\mu_1(A) = \int_A f \, d\mu_2$ for all A in β. The function f, which belongs to $L_1[\Omega, \beta, \mu_2]$ and is unique up to μ_2-equivalence, is called the Radon–Nikodym derivative of μ_1 with respect to μ_2, and is commonly denoted by $d\mu_1/d\mu_2$. In statistical and engineering applications $d\mu_1/d\mu_2$ is usually called the likelihood ratio*, a term that has its genesis in maximum likelihood estimation*.

Absolute continuity and the Radon–Nikodym derivative have important applications in statistics. For example, suppose that $X : \Omega \to \mathbb{R}^N$ is a random vector. Suppose also that under hypothesis H_1 the distribution function of X is given by $F_1 = \mu_1 \circ X^{-1}[F_1(x) = \mu_1\{\omega : X(\omega) \leqslant x\}]$, whereas under H_2, X has the distribution function $F_2 = \mu_2 \circ X^{-1}$. F_i defines a Borel measure on $\mathbb{R}^1$; one says that F_i is induced from μ_i by X. A statistician observes one realization (sample path) of X, and wishes to design a statistical test to optimally decide in favor of H_1 or H_2. Then, under any of several classical decision criteria of mathematical statistics (e.g., Bayes risk, Neyman–Pearson*, minimum probability of error), an optimum decision procedure when $\mu_1 \ll \mu_2$ is to form the test statistic* $\Lambda(X) = [dF_1/dF_2](X)$ and compare its value with some constant, C_0; the decision is then to accept H_2 if $\Lambda(X) \leqslant C_0$, accept H_1 if $\Lambda(X) > C_0$. The value of C_0 will depend on the properties of F_1 and F_2 and on the optimality criterion. For more details, see HYPOTHESIS TESTING*.[1]

Two probability measures μ_1 and μ_2 on (Ω, β) are said to be equivalent ($\mu_1 \sim \mu_2$) if $\mu_1 \ll \mu_2$ and $\mu_2 \ll \mu_1$. They are orthogonal, or extreme singular ($\mu_1 \perp \mu_2$) if there exists a set A in β such that $\mu_2(A) = 0$ and $\mu_1(A) = 1$. For the hypothesis-testing problem discussed above, orthogonal induced measures permit one to discriminate perfectly between H_1 and H_2. In many practical applications, physical considerations rule out perfect discrimination. The study of conditions for absolute continuity then becomes important from the aspect of verifying that the mathematical model is valid.

In the framework described, the random vector has range in $\mathbb{R}^N$. However, absolute continuity, the Radon–Nikodym derivative, and their application to hypothesis-testing problems are not limited to such finite-dimensional cases. In fact, the brief comments above on hypothesis testing apply equally well when X takes its value in an infinite-dimensional linear topological space, as when $X(\omega)$ represents a sample path* from a stochastic process* (X_t), $t \in [a, b]$. (The infinite-dimensional case does introduce interesting mathematical complexities that are not present in the finite-dimensional case.)

GENERAL CONDITIONS FOR ABSOLUTE CONTINUITY

We shall see later that special conditions for absolute continuity can be given when the two measures involved have certain specialized properties, e.g., when they are both Gaussian. However, necessary and sufficient conditions for absolute continuity can be given that apply to any pair of probability measures on any measurable space (Ω, β). Further, if (Ω, β) consists of a linear topological space Ω and the smallest σ-field β containing all the open sets (the Borel σ-field), then additional conditions for absolute continuity can be obtained that apply to any pair of probability measures on (Ω, β). Here we give one well-known set of general necessary and sufficient conditions. First, recall that if (Ω, β, P) is a probability space and F a collection of real random variables on (Ω, β), then F is said to be uniformly integrable with respect to P [23] if the integrals $\int_{\{\omega : |f(\omega)| \geqslant c\}} |f(\omega)| dP(\omega), c > 0, f$ in F, tend uniformly to zero as $c \to \infty$. An equivalent statement is the following: F is uniformly integrable (P) if and only if

(a) $$\sup_F \int_\Omega |f(\omega)| dP(\omega) < \infty$$

and

(b) For every $\epsilon > 0$ there exists $\delta > 0$ such that $P(A) < \delta$ implies that

$$\sup_F \int_A |f(\omega)| dP(\omega) \leqslant \epsilon.$$

Theorem 1. Suppose that μ_1 and μ_2 are two probability measures on a measurable space (Ω, β). Suppose that $\{\mathscr{F}_n, n \geqslant 1\}$ is an increasing family of sub-σ-fields of β such that β is the smallest σ-field containing $\cup_n \mathscr{F}_n$. Let μ_i^n be the restriction of μ_i to $\mathscr{F}_n$. Then $\mu_1 \ll \mu_2$ if and only if

(a) $$\mu_1^n \ll \mu_2^n \quad \text{for all } n \geqslant 1,$$

and

(b) $$\{d\mu_1^n/d\mu_2^n, n \geqslant 1\}$$

is uniformly integrable (μ_2).

When $\mu_1 \ll \mu_2$, then $d\mu_1/d\mu_2 = \lim_n d\mu_1^n/d\mu_2^n$ almost everywhere (a.e.) $d\mu_2$.

Condition (a) of Theorem 1 is obviously necessary. The necessity of (b) follows from the fact that $\{d\mu_1^n/d\mu_2^n, \mathscr{F}_n : n \geqslant 1\}$ is a martingale* with respect to μ_2. This property, and the martingale convergence theorem, yield the result that $d\mu_1/d\mu_2 = \lim_n d\mu_1^n/d\mu_2^n$ a.e. $d\mu_2$. Sufficiency of (a) and (b) follows from the second definition of uniform integrability given above and the assumption that β is the smallest σ-field containing $\cup_n \mathscr{F}_n$.

Conditions (a) and (b) of Theorem 1 are also necessary and sufficient for $\mu_1 \ll \mu_2$ when the family of increasing σ-fields $(\mathscr{F}_t)$ has any directed index set.

A number of results frequently used to analyze absolute continuity can be obtained from Theorem 1. This includes, for example, Hájek's divergence criterion [20] and Kakutani's theorem on equivalence of infinite product measures [29] (a fundamental result in its own right).

The conditions of Theorem 1 are very general. However, in one respect they are somewhat unsatisfactory. They usually require that one specify an infinite sequence of Radon–Nikodym derivatives $\{d\mu_1^n/d\mu_2^n, n \geqslant 1\}$. It would be preferable to have a more direct method of determining if absolute continuity holds. One possible alternative when the measures are defined on a separable metric space involves the use of characteristic functions*. The characteristic function of a probability measure defined on the Borel σ-field of a separable metric space completely and uniquely specifies the measure [38]. Thus in such a setting, two characteristic functions contain all the information required to determine whether absolute continuity exists between the associated pair of measures. The use of characteristic functions offers a method for attacking the following problem. For a given measure μ on (Ω, β) determine the set $\mathscr{P}_\mu$ of all probability measures on (Ω, β) such that $\nu \ll \mu$ for all ν in $\mathscr{P}_\mu$. Some results on this problem are contained in ref. 3; further progress, especially detailed results for the case of a Gaussian measure μ on Hilbert space, would be useful in several important applications areas (detection of signals in noise, stochastic filtering, information theory*).

PROBABILITY MEASURES ON HILBERT SPACES

There has been much activity in the study of probability measures on Banach spaces [1,4,5,31]. Here we restrict attention to the case of probabilities on Hilbert spaces; this is the most important class of Banach spaces for applications, and the theory is relatively well developed in this setting.

Let $\mathbf{H}$ be a real separable Hilbert space with inner product $\langle \cdot, \cdot \rangle$ and Borel σ-field Γ (*see* SEPARABLE SPACE). Let μ be a probability measure on Γ. For any element y in $\mathbf{H}$, define the distribution function F_y by $F_y(a) = \mu\{x : \langle y, x \rangle \leqslant a\}$, a in $(-\infty, \infty)$. μ is said to be *Gaussian* if F_y is Gaussian for all y in $\mathbf{H}$. It can be shown that for every Gaussian μ there exists a self-adjoint trace-class nonnegative linear operator R_μ in $\mathbf{H}$ and an element m_μ in $\mathbf{H}$ such that

$$\langle y, m_\mu \rangle = \int_{\mathbf{H}} \langle y, x \rangle d\mu(x) \qquad (1)$$

and

$$\langle R_\mu, v \rangle = \int_{\mathbf{H}} \langle y - m_\mu, x \rangle \langle v - m_\mu, x \rangle \, d\mu(x) \quad (2)$$

for all y and v in $\mathbf{H}$. R_μ is called the *covariance (operator)* of μ, and m_μ is the *mean (element)*. Conversely, to every self-adjoint nonnegative trace-class operator R_μ and element m in $\mathbf{H}$ there corresponds a unique Gaussian measure μ such that relations (1) and (2) are satisfied. Non-Gaussian measures μ may also have a covariance operator R_μ and mean element m_μ satisfying (1) and

(2); however, the covariance R_μ need not be trace-class. For more details on probability measures on Hilbert space, see refs. 17, 38, and 53.

Elegant solutions to many problems of classical probability theory (and applications) have been obtained in the Hilbert space framework, with methods frequently making use of the rich structure of the theory of linear operators. Examples of such problems include Sazanov's solution to obtaining necessary and sufficient conditions for a complex-valued function on $\mathbf{H}$ to be a characteristic function* [49]; Prohorov's conditions for weak compactness of families of probability measures, with applications to convergence of stochastic processes [43]; the results of Mourier on laws of large numbers* [34]; the results of Fortét and Mourier on the central limit theorem* [15,34]; and conditions for absolute continuity of Gaussian measures. The latter problem is examined in some detail in the following section. The study of probability theory in a Hilbert space framework received much of its impetus from the pioneering work of Fortét and Mourier (see refs. 15 and 34, and the references cited in those papers). Their work led not only to the solution of many interesting problems set in Hilbert space, but also to extensions to Banach spaces and more general linear topological spaces [1,4,5,15,31,34].

The infinite-dimensional Hilbert spaces $\mathbf{H}$ most frequently encountered in applications are $L_2[0, T]$ ($T < \infty$) and l_2. For a discussion of how Hilbert spaces frequently arise in engineering applications, *see* COMMUNICATION THEORY, STATISTICAL. In particular, the interest in Gaussian measures on Hilbert space has much of its origin in hypothesis testing and estimation problems involving stochastic processes: detection and filtering of signals embedded in Gaussian noise. For many engineering applications, the noise can be realistically modeled as a Gaussian stochastic process with sample paths almost surely (a.s.) in $L_2[0, T]$ or a.s. in l_2. When $\mathbf{H}$ is $L_2[0, T]$, a trace-class covariance operator can be represented as an integral operator whose kernel is a covariance function. Thus suppose that $(X_t), t \in [0, T]$, is a measurable zero-mean stochastic process on (Ω, β, P), inducing the measure μ on the Borel σ-field of $L_2[0, T]$; $\mu(A) = P\{\omega : X(\omega) \in A\}$. Then $E \int_0^T X_t^2(\omega) dt < \infty$ if and only if μ has a trace-class covariance operator R_μ defined by $[R_\mu f](t) = \int_0^T R(t, s) f(s) ds, f$ in $L_2[0, T]$, where R is the covariance function of (X_t). If R_μ is trace-class, then $E \int_0^T X_t^2(\omega) dt = \text{trace } R_\mu$.

ABSOLUTE CONTINUITY OF PROBABILITY MEASURES ON HILBERT SPACE

If **H** is finite-dimensional and μ_1 and μ_2 are two zero-mean Gaussian measures on Γ, it is easy to see that μ_1 and μ_2 are equivalent if and only if their covariance matrices have the same range space. However, if **H** is infinite-dimensional, this condition (on the ranges of the covariance operators) is neither necessary nor sufficient for $\mu_1 \sim \mu_2$. The study of conditions for absolute continuity of two Gaussian measures on function space has a long and active history. Major early contributions were made by Cameron and Martin [6,7] and by Grenander [18]. The work of Cameron and Martin was concerned with the case when one measure is Wiener measure (the measure induced on $C[0, 1]$ by the Wiener process*) and the second measure is obtained from Wiener measure by an affine transformation. Grenander obtained conditions for absolute continuity of a Gaussian measure (induced by a stochastic process with continuous covariance) with respect to a translation. Segal [50] extended the work of Cameron and Martin to a more general class of affine transformations of Wiener measure. Segal also obtained [50] conditions for absolute continuity of Gaussian "weak distributions." These necessary and sufficient conditions can be readily applied to obtain sufficient conditions for equivalence of any pair of Gaussian measures on **H**; they can also be used to show that these same conditions are necessary. Complete and general solutions to the absolute continuity problem for Gaussian measures were obtained by Feldman [12] and Hájek [21]. Their methods are quite different. The main result, in each paper, consists of two parts: a "dichotomy theorem," which states that any two Gaussian measures are either equivalent or orthogonal; and conditions that are necessary and sufficient for equivalence.

The following theorem for Gaussian measures on Hilbert space is a modified version of Feldman's result [12]; several proofs have been independently obtained (Kallianpur and Oodaira [30], Rao and Varadarajan [44], Root [45]).

Theorem 2. Suppose that μ_1 and μ_2 are two Gaussian measures on Γ, and that μ_i has covariance operator R_i and mean $m_i, i = 1, 2$. Then:

1. either $\mu_1 \sim \mu_2$ or $\mu_1 \perp \mu_2$;
2. $\mu_1 \sim \mu_2$ if and only if all the following conditions are satisfied:
   (a) range $(R_1^{1/2}) = $ range $(R_2^{1/2})$;
   (b) $R_1 = R_2^{1/2}(I + T)R_2^{1/2}$, where I is the identity on **H** and T is a Hilbert–Schmidt operator in **H**:
   (c) $m_1 - m_2$ is in range $(R_1^{1/2})$.

Various specializations of Theorem 2 have been obtained; see the references in refs. 8 and 47. Two of the more interesting special cases, both extensively analyzed, are the following: (1) both measures induced by stationary Gaussian stochastic processes; (2) one of the measures is Wiener measure. In the former case, especially simple conditions can be given when the two processes have rational spectral densities; see the papers by Feldman [13], Hájek [22], and Pisarenko [40,41]. In this case, when the two measures have the same mean function, $\mu_1 \sim \mu_2$ if and only if $\lim_{|\lambda| \to \infty} f_1(\lambda)/f_2(\lambda) = 1$, where f_i is the spectral density* of the Gaussian process inducing μ_i. Moreover, this occurs if and only if the operator T appearing in Theorem 2 is also trace-class [22]. For the case where one of the measures is Wiener measure, see the papers by Shepp [51], Varberg [54,55], and Hitsuda [24].

The problem of determining the Radon–Nikodym derivative for two equivalent Gaussian measures on a Hilbert space has been studied, especially by Rao and Varadarajan [44]. For convenience, we use the notation of Theorem 2 and assume now that all covariance operators are strictly positive. In the case where the Hilbert space is finite-dimensional, the log of the Radon–Nikodym derivative $d\mu_1/d\mu_2$ (log-likelihood ratio*) is easily seen to be a quadratic-linear form; that is, $\log \Lambda(X) = \langle x, Wx \rangle + \langle x, b \rangle + \text{constant}$, where the linear operator $W = \frac{1}{2}(R_2^{-1} - R_1^{-1})$, $b = R_1^{-1}m_1 - R_2^{-1}m_2$, and $\log \equiv \log_e$. However, when $\mathbf{H}$ is infinite-dimensional, the log-likelihood ratio need not be a quadratic-linear form defined by a bounded linear operator. This holds true even if the operator T of Theorem 2 is not only Hilbert–Schmidt, but is also trace class. However, when T is Hilbert–Schmidt, one can always express the log of the Radon–Nikodym derivative as an almost surely convergent series [44]. The essential difficulty in characterizing the likelihood ratio for infinite-dimensional Hilbert space is that the operators R_1 and R_2 cannot have *bounded* inverses and these two inverses need not have the same domain of definition. Even if range $(R_1) = $ range (R_2), so that $R_2^{-1} - R_1^{-1}$ is defined on range (R_1), it is not necessary that $R_2^{-1} - R_1^{-1}$ be bounded on range (R_1).

In the finite-dimensional case, if $R_1 = R_2$, then $\log \Lambda(X) = \langle x, b \rangle + $ constant, with b defined as above, so that the log-likelihood ratio is a bounded linear form. This need not be the case for infinite-dimensional Hilbert space; in general, $\log \Lambda(X)$ will be a bounded linear form (when $R_1 = R_2$) if and only if $m_1 - m_2$ is in the range of R_1. As can be seen from Theorem 1, this condition is strictly stronger than the necessary and sufficient condition for $\mu_1 \sim \mu_2$, which (with $R_1 = R_2$) is that $m_1 - m_2$ be in range $(R_1^{1/2})$.

If the two measures are induced by stationary Gaussian processes with rational spectral densities, expressions for the likelihood ratio can be given in terms of the spectral densities; see the papers by Pisarenko [41] and Hájek [22].

In many applications, only one of the two measures can be considered to be Gaussian. For this case, a useful sufficient condition for absolute continuity is given in ref. 2. This condition can be applied when the two measures are induced by stochastic processes (X_t) and (Y_t), where (Y_t) is a function of (X_t) and a process (Z_t) that is independent of (X_t). In particular, if (X_t) is Gaussian and $(Y_t) = (X_t + Z_t)$, then conditions for absolute continuity can be stated in terms of sample path properties of the (Z_t) process (absolute continuity, differentiability, etc.). Such

conditions can often be verified in physical models by knowledge of the mechanisms generating the observed data, when the distributional properties of the (Z_t) process are unknown. When (X_t) is the Wiener process on $[0, T]$, conditions for absolute continuity of the induced measures on $L_2[0, T]$ can be obtained from the results of refs. 10, 27, and 28. Some of these results do not require independence of (X_t) and (Z_t).

Other results on absolute continuity of measures on Hilbert space have been obtained for infinitely divisible measures [16], measures induced by stochastic processes with independent increments [16], admissible translations of measures [42,52], and for a fixed measure and a second measure obtained from the first measure by a nonlinear transformation [16]. With respect to admissible translates, Rao and Varadarajan [44] have shown that if μ is a zero-mean measure having a trace-class covariance operator, R, then the translate of μ by an element y is orthogonal to μ if y is not in range $(R^{1/2})$. A number of these results are collected in the book by Gihman and Skorohod [17], which also contains much material on basic properties of probability measures on Hilbert space, and on weak convergence*. The book by Kuo [33] contains not only basic material on probability measures on Hilbert spaces (including absolute continuity), but also an introduction to some topics in probability on Banach spaces.

ABSOLUTE CONTINUITY OF MEASURES INDUCED BY STOCHASTIC PROCESSES

Many problems involving stochastic processes are adequately modeled in the framework of probability measures on Hilbert space, provided that the sample paths of each process of interest belong almost surely to some separable Hilbert space. However, this condition is not always satisfied; even when it is satisfied, one may prefer conditions for absolute continuity stated in terms of measures on $\mathbb{R}^T$ (the space of real-valued functions on T), where T is the parameter set of the process. For example, a class of stochastic processes frequently considered are those having almost all paths in $D\,[0, 1]$. $D\,[0, 1]$ is the set of all real-valued functions having limits from both left and right existing at all points of (0, 1), with either left-continuity or right-continuity at each point of (0, 1), and with a limit from the left (right) existing at 1(0). D [0, 1] is a linear metric space* under the Skorohod metric [38], but this metric space is not a Hilbert space.

The general conditions for absolute continuity stated in Theorem 1 apply in any setting. Moreover, necessary and sufficient conditions for equivalence of measures (most frequently on $\mathbb{R}^T$) induced by two Gaussian stochastic processes can be stated in a number of ways: The reproducing kernel Hilbert space (r.k.H.s.) of the two covariance functions [30,37,39]; operators and elements in an L_2 space of real-valued random functions [12]; operators and elements in an L_2-space of random variables [46]; and tensor products [35]. Hájek's conditions for absolute continuity in terms of the divergence [21] apply to the general case. Sato [48] has stated conditions for absolute continuity in terms of a representation for all Gaussian processes whose induced measure on $\mathbb{R}^T$ is equivalent to the measure induced by a given Gaussian process. Several of these results are presented in [8]. Many other papers on absolute continuity for measures induced by two Gaussian processes have appeared; space does not permit an attempt at a complete bibliography.

Use of the r.k.H.s. approach to study linear statistical problems in stochastic processes was first explicitly and systematically employed by Parzen; the r.k.H.s. approach was also implicit in the work of Hájek (see the papers by Hájek [22] and Parzen [39] and their references).

For non-Gaussian processes, results on absolute continuity have been obtained for Markov processes* [16,32], diffusion processes* [36], locally infinitely divisible processes [16], semimartingales* [25], point processes* [26], and non-Gaussian processes equivalent to the Wiener process [9,10,27,28].

Dudley's result [9] is of particular interest to researchers interested in Gaussian measures. Suppose that (W_t) is the Wiener process on [0,1] with zero mean and unity variance parameter, and that $\beta(\cdot, \cdot)$ is a continuous real-valued function on $\mathbb{R} \times [0, 1]$. Let $Y_t = \beta(W_t, t)$. Dudley shows in ref. 9 that the measure on function space induced by

(Y_t) is absolutely continuous with respect to Wiener measure if and only if $\beta(u, t) = u + \phi(t)$ or $\beta(u, t) = -u + \phi(t)$, where ϕ is in the r.k.H.s. of the Wiener covariance min(t, s). The methods used to prove this result rely heavily on some of the special properties of the Wiener process, such as the fact that (W_t) has the strong Markov property, and laws of the iterated logarithm* for the Wiener process (obtained in ref. 9). A characterization of admissible β's for other Gaussian processes with continuous paths would be of much interest; such characterizations would necessarily require a different approach, and this problem is very much open at present.

The absolute continuity problem discussed in refs. 10, 27, and 28 has received much attention, partly because of its connection to signal detection* and nonlinear filtering*. One considers a measurable process (Y_t) defined by $Y_t = \int_0^t h_s\, ds + W_t, 0 \leqslant t \leqslant T$, where (W_t) is a zero-mean Wiener process and (h_s) is a stochastic process with sample paths a.s. in $L_1[0, T]$. Let μ_Y and μ_W be the measures induced by (Y_t) and (W_t) on the space of continuous functions on [0, 1]. Conditions for $\mu_Y \ll \mu_W, \mu_Y \sim \mu_W$, and results on the Radon–Nikodym derivative have been obtained in refs. 10, 27, and 28. In the special case where (h_s) is independent of (W_t), a sufficient condition for $\mu_Y \sim \mu_W$ is that $\int_0^T h_s^2 ds < \infty$ for almost all sample paths of (h_s). This condition is also sufficient for $\mu_Y \ll \mu_W$ if the process (h_s) is only assumed independent of future increments of (W_t).

Finally, we mention a result of Fortét [14], who has obtained a sufficient condition for orthogonality of two measures when one is Gaussian, expressed in terms of the r.k.H.s. of the two covariances. Suppose that μ_i is a probability measure on $\mathbb{R}^T, T = [0, 1]$, with r.k.H.s. H_i and mean function m_i. Then if μ_1 is Gaussian, μ_1 and μ_2 are orthogonal unless both the following conditions are satisfied: (a) $H_1 \subset H_2$; and (b) $m_1 - m_2 \in H_2$.

NOTE

1. The ubiquitous nature of the Radon–Nikodym derivative in various hypothesis-testing applications can be attributed to its being a necessary and sufficient statistic* [11].

REFERENCES

1. Aron, R. M. and Dineen, S., eds. (1978). Vector Space Measures and Applications, I. *Lect. Notes Math.*, 644. (Contains most of the papers on probability theory presented at the Conference on Vector Space Measures and Applications, Dublin, 1978.)

2. Baker, C. R. (1973). *Ann. Prob.*, **1**, 690–698.

3. Baker, C. R. (1979). *Lect. Notes Math.*, **709**, 33–44.

4. Beck, A., ed. (1976). Probability in Banach spaces. *Lect. Notes Math.*, 526. (Collection of papers on various topics; Proceedings of First International Conference on Probability in Banach Spaces, Oberwolfach, 1975.)

5. Beck, A., ed. (1979). Probability in Banach Spaces, II. *Lect. Notes Math.* 709. (Proceedings of Second International Conference on Probability in Banach Spaces, Oberwolfach, 1978.)

6. Cameron, R. H. and Martin, W. T. (1944). *Ann. Math.*, **45**, 386–396.

7. Cameron, R. H. and Martin, W. T. (1945). *Trans. Amer. Math. Soc.*, **58**, 184–219.

8. Chatterji, S. D. and Mandrekar, V. (1978). In *Probabilistic Analysis and Related Topics*, Vol. 1, A. T. Bharucha-Reid, ed. Academic Press, New York, pp. 169–197. (Extensive bibliography.)

9. Dudley, R. M. (1971). *Zeit. Wahrscheinlichkeitsth.*, **20** 249–258; correction: *ibid.*, **30** (1974), 357–358.

10. Duncan, T. E. (1970). *Inf. Control*, **16**, 303–310.

11. Dynkin, E. B. (1951). *Uspehi Mat. Nauk* (N.S.), **6**, 68–90; translation: *Select. Transl. Math. Statist. Prob.* **1**, 17–40, (1961).

12. Feldman, J. (1958). *Pacific J. Math.*, **8**, 699–708; correction: *ibid.*, **9**, 1295–1296 (1959).

13. Feldman, J. (1960). *Pacific J. Math.*, **10**, 1211–1220.

14. Fortét, R. (1973). *Ann. Inst. Henri Poincaré*, **9**, 41–58.

15. Fortét, R. and Mourier, E. (1955). *Studia Math.*, **15**, 62–79.

16. Gihman, I. I. and Skorohod, A. V. (1966). *Russ. Math. Surv. (Uspehi Mat. Nauk.)*, **21**, 83–156.

17. Gihman, I. I. and Skorohod, A. V. (1974). *The Theory of Stochastic Processes*, Vol. 1. Springer-Verlag, New York. (Basic material on probability on metric spaces; many results on absolute continuity; also has extensive treatment of weak convergence of probability measures, and applications to convergence of stochastic processes.)

18. Grenander, U. (1950). *Ark. Mat.*, **1**, 195–277.

19. Grigelonis, B. (1973). *Lect. Notes Math.*, **330**, 80–94.

20. Hájek, J. (1958). *Czech. Math. J.*, **8**, 460–463.

21. Hájek, J. (1958). *Czech. Math. J.*, **8**, 610–618.

22. Hájek, J. (1962). *Czech. Math. J.*, **12**, 404–444.

23. Hewitt, E. and Stromberg, K. (1965). *Real and Abstract Analysis*. Springer-Verlag, New York.

24. Hitsuda, M. (1968). *Osaka J. Math.*, **5**, 299–312.

25. Jacod, J. and Mémin, J. (1976). *Zeit. Wahrscheinlichkeitsth.*, **35**, 1–37.

26. Kabanov, Yu. M., Liptser, R. S., and Shiryayev, A. N. (1973). *Lect. Notes Math.*, **550**, 80–94.

27. Kadota, T. T. and Shepp, L. A. (1970). *Zeit. Wahrscheinlichkeitsth.*, **16**, 250–260.

28. Kailath, T. and Zakai, M. (1971). *Ann. Math. Statist.*, **42**, 130–140.

29. Kakutani, S. (1948). *Ann. Math.*, **49**, 214–224.

30. Kallianpur, G. and Oodaira, H. (1963). In *Time Series Analysis* (Proc. 1962 Symp.), M. Rosenblatt, ed. Wiley, New York, pp. 279–291.

31. Kuelbs, J., ed. (1978). *Probability on Banach Spaces*. Marcel Dekker, New York. (Five papers in areas of central limit theorem, Gaussian measures, martingales.)

32. Kunita, H. (1976). *Lect. Notes Math.* **511**, 44–77.

33. Kuo, H. -H. (1975). Gaussian Measures in Banach Spaces. *Lect. Notes Math.*, 463. (Basic material on Gaussian measures on Hilbert space, including absolute continuity; also contains results on abstract Wiener spaces for Gaussian measures on Banach space.)

34. Mourier, E. (1953). *Ann. Inst. Henri Poincaré*, **13**, 161–244.

35. Neveu, J. (1968). *Processus aléatoires gaussiens*. University of Montreal Press, Montreal. (Results on Gaussian processes, including absolute continuity. Extensive bibliography.)

36. Orey, S. (1974). *Trans. Amer. Math. Soc.*, **193**, 413–426.

37. Pan Yi-Min (1972). *Select. Transl. Math. Statist. Prob.*, **12**, 109–118; translation of *Shuxue Jinzhan*, **9**, 85–90 (1966).

38. Parthasarathy, K. R. (1967). *Probability Measures on Metric Spaces*. Academic Press, New York. (Excellent collection of results available

through the mid-1960s in various areas of probability on metric spaces (not including absolute continuity), plus foundations.)

39. Parzen, E. (1963). In *Time Series Analysis* (Proc. Symp. 1962), M. Rosenblatt, ed. Wiley, New York, pp. 155–169.

40. Pisarenko, V. (1961). *Radio Eng. Electron.*, **6**, 51–72.

41. Pisarenko, V. (1965). *Theory Prob. Appl.*, **10**, 299–303.

42. Pitcher, T. S. (1963). *Trans. Amer. Math. Soc.*, **108**, 538–546.

43. Prohorov, Yu. V. (1956). *Theory Prob. Appl.*, **1**, 157–214.

44. Rao, C. R. and Varadarajan, V. S. (1963). *Sankhya A*, **25**, 303–330.

45. Root, W. L. (1963). In *Time Series Analysis* (Proc. 1962 Symp.), M. Rosenblatt, ed. Wiley, New York, 292–346.

46. Rosanov, Yu. V. (1962). *Theory Prob. Appl.*, **7**, 82–87.

47. Rosanov, Yu. V. (1971). Infinite-Dimensional Gaussian Distributions. *Proc. Steklov Inst. Math.*, No. 108. [Monograph containing results on absolute continuity, measurable linear functionals (including a zero-one law for linear manifolds). Extensive bibliography of papers on absolute continuity published prior to 1968.]

48. Sato, H. (1967). *J. Math. Soc. Japan*, **19**, 159–172.

49. Sazanov, V. V. (1958). *Theory Prob. Appl.*, **3**, 201–205.

50. Segal, I. E. (1958). *Trans. Amer. Math. Soc.*, **88**, 12–41.

51. Shepp, L. A. (1966). *Ann. Math. Statist.*, **37**, 321–354.

52. Skorohod, A. V. (1970). *Theory Prob. Appl.*, **15**, 557–580.

53. Skorohod, A. V. (1974). *Integration in Hilbert Space*. Springer-Verlag, New York. (Basic results concerning measures on Hilbert space; many results on absolute continuity.)

54. Varberg, D. E. (1964). *Trans. Amer. Math. Soc.*, **113**, 262–273.

55. Varberg, D. E. (1966). *Notices Amer. Math. Soc.*, **13**, 254.

See also COMMUNICATION THEORY, STATISTICAL; GAUSSIAN PROCESSES; LIKELIHOOD RATIO TESTS; MARTINGALES; MEASURE THEORY IN PROBABILITY AND STATISTICS; RADON–NIKODYM THEOREM; SEPARABLE SPACE; and STOCHASTIC PROCESSES.

CHARLES R. BAKER

ABSOLUTE DEVIATION

The numerical value of the difference between two quantities regardless of its sign. If $\hat{\theta}$ is an estimate of θ, its absolute deviation from θ is $|\theta - \hat{\theta}|$.

See also CHEBYSHEV'S INEQUALITY; MEAN DEVIATION; and MOMENTS.

ABSOLUTE MOMENT

The expected value of the modulus (absolute value) of a random variable X, raised to power r is its rth absolute (crude) moment

$$\nu_r' = E[|X|^r].$$

The quantity

$$\nu_r = E[|X - E[X]|^r]$$

is the rth absolute central moment of X.
ν_1 is the mean deviation*.

ACCELERATED LIFE TESTING

NOTATIONS

Accelerated life models relate the lifetime distribution to the explanatory variable (stress, covariate). This distribution can be defined by the survival function. But the sense of accelerated life models is best seen if they are formulated in terms of the hazard rate function.

Suppose at first that the explanatory variable $x(\cdot) \in E$ is a deterministic time function:

$$x(\cdot) = (x_1(\cdot), \ldots, x_m(\cdot))^T : [0, \infty) \to B \in \mathbf{R}^m,$$

where E is a set of all possible stresses, $x_i(\cdot)$ is univariate explanatory variable. If $x(\cdot)$ is constant in time, $x(\cdot) \equiv x = const$, we shall write x instead of $x(\cdot)$ in all formulas. We note E_1 a set of all constant in time stresses, $E_1 \subset E$.

Denote informally by $T_{x(\cdot)}$ the failure time under $x(\cdot)$ and by

$$S_{x(\cdot)}(t) = \mathbf{P}\{T_{x(\cdot)} \geqslant t\}, \quad t > 0, \quad x(\cdot) \in E,$$

the *survival function* of $T_{x(\cdot)}$. Let $F_{x(\cdot)}(t) = 1 - S_{x(\cdot)}(t)$ be the *cumulative distribution function* of $T_{x(\cdot)}$. The *hazard rate function* of $T_{x(\cdot)}$ under $x(\cdot)$ is

$$\alpha_{x(\cdot)}(t) = \lim_{h \downarrow 0} \frac{1}{h} \mathbf{P}\{T_{x(\cdot)} \in [t, t+h) \mid T_{x(\cdot)} \geqslant t\}$$
$$= -\frac{S'_{x(\cdot)}(t)}{S_{x(\cdot)}(t)}.$$

Denote by

$$A_{x(\cdot)}(t) = \int_0^t \alpha_{x(\cdot)}(u)du = -ln\{S_{x(\cdot)}(t)\}$$

the *cumulative hazard function* of $T_{x(\cdot)}$.

Each specified accelerated life model relates the hazard rate (or other function) to the explanatory variable in some particular way. To be concise the word stress will be used here for explanatory variable.

We say that a stress $x_2(\cdot)$ is higher than a stress $x_1(\cdot)$ and we write $x_2(\cdot) > x_1(\cdot)$, if for any $t \geqslant 0$ the inequality $S_{x_1(\cdot)}(t) \geqslant S_{x_2(\cdot)}(t)$ holds and exists $t_0 > 0$ such that $S_{x_1(\cdot)}(t_0) > S_{x_2(\cdot)}(t_0)$. We say also that the stress $x_2(\cdot)$ is *accelerated* with respect to the stress $x_1(\cdot)$. It is evident that by the same way one can consider *decelerated* stresses.

At the end we note that if the stress is a stochastic process $X(t)$, $t \geqslant 0$, and $T_{X(\cdot)}$ is the failure time under $X(\cdot)$, then denote by

$$S_{x(\cdot)}(t) = \mathbf{P}\{T_{X(\cdot)} \geqslant t | X(s) = x(s), 0 \leqslant s \leqslant t\},$$

the conditional survival function. In this case the definitions of models should be understood in terms of these conditional function.

STRESSES IN ALT

In accelerated life testing (ALT) the most used types of stresses are: constant in time stresses, step-stresses, progressive (monotone) stresses, cyclic stresses and random stresses (see, for example, Duchesne & Lawless (2000, 2002), Duchesne & Rosenthal (2002), Duchesne (2000), Gertsbach & Kordonsky (1997), Miner (1945), Lawless (1982), Nelson (1990), Meeker & Escobar (1998), Nikulin & Solev (2000), Shaked & Singpurwalla (1983)).

The most common case is when the stress is unidimensional, for example, pressure, temperature, voltage, but then more then one accelerated stresses may be used. The monotone stresses are used, for example, to construct the so-called collapcible models, the accelerated degradation models, etc.

The mostly used time-varying stresses in ALT are step-stresses: units are placed on test at an initial low stress and if they do not fail in a predetermined time t_1, the stress is increased. If they do not fail in a predetermined time $t_2 > t_1$, the stress is increased once more, and so on. Thus step-stresses have the form

$$x(u) = \begin{cases} x_1, & 0 \leqslant u < t_1, \\ x_2, & t_1 \leqslant u < t_2, \\ \dots & \dots \\ x_m, & t_{m-1} \leqslant u < t_m, \end{cases} \quad (1)$$

where $x_1, \dots, x_m$ are constant stresses. Sets of step-stresses of the form (1) will be denoted by E_m, $E_m \subset E$. Let E_2, $E_2 \subset E$, be a set of step-stresses of the form

$$x(u) = \begin{cases} x_1, & 0 \leqslant u < t_1, \\ x_2, & u \geqslant t_1, \end{cases} \quad (2)$$

where $x_1, x_2 \in E_1$.

SEDYAKIN'S PRINCIPLE AND MODEL

Accelerated life models could be at first formulated for constant explanatory variables. Nevertheless, before formulating them, let us consider a method for generalizing such models to the case of time-varying stresses.

In 1966 N. Sedyakin formulated his famous *physical principle in reliability* which states that for two identical populations of units functioning under different stresses x_1 and x_2, two moments t_1 and t_2 are equivalent if the probabilities of survival until these moments are equal:

$$\mathbf{P}\{T_{x_1} \geqslant t_1\} = S_{x_1}(t_1) = S_{x_2}(t_2)$$
$$= \mathbf{P}\{T_{x_2} \geqslant t_2\}, \quad x_1, x_2 \in E_1.$$

If after these equivalent moments the units of both groups are observed under the same stress x_2, i.e. the first population is observed under the step-stress $x(\cdot) \in E_2$ of the form (2) and the second all time under the constant stress x_2, then for all $s > 0$

$$\alpha_{x(\cdot)}(t_1 + s) = \alpha_{x_2}(t_2 + s). \qquad (3)$$

Using this idea of Sedyakin, we considered some generalisation of the model of Sedyakin to the case of any time-varying stresses by supposing that the hazard rate $\alpha_{x(\cdot)}(t)$ at any moment t is a function of the value of the stress at this moment and of the probability of survival until this moment. It is formalized by the following definition.

Definition 1. We say that *Sedyakin's model (SM) holds on a set of stresses E if there exists on $E \times \mathbf{R}^+$ a positive function g such that for all $x(\cdot) \in E$*

$$\alpha_{x(\cdot)}(t) = g\left(x(t), A_{x(\cdot)}(t)\right). \qquad (4)$$

The fact that the SM does not give relations between the survival under different constant stresses is a cause of non-applicability of this model for estimation of reliability under the design (usual) stress from accelerated experiments. On the other hand, restrictions of this model when not only the rule (4) but also some relations between survival under different constant stresses are assumed, can be considered. These narrower models can be formulated by using models for constant stresses and the rule (4). For example, it can be shown that the well known and mostly used accelerated failure time model for time-varying stresses is a restriction of the SM when the survival functions under constant stresses differ only in scale.

MODEL OF SEDYAKIN FOR STEP-STRESSES

The mostly used time-varying stresses in accelerated life testing are step-stresses (2) or (1). Let us consider the meaning of the rule (4) for these step-stresses. Namely, we shall show that in the SM the survival function under the step-stress is obtained from the survival functions under constant stresses by the *rule of time-shift*.

Proposition 1. *If the SM holds on E_2 then the survival function and the hazard rate under the stress $x(\cdot) \in E_2$ satisfy the equalities*

$$S_{x(\cdot)}(t) = \begin{cases} S_{x_1}(t), & 0 \leqslant t < t_1, \\ S_{x_2}(t - t_1 + t_1^*), & t \geqslant t_1, \end{cases} \qquad (5)$$

and

$$\alpha_{x(\cdot)}(t) = \begin{cases} \alpha_{x_1}(t), & 0 \leqslant t < t_1, \\ \alpha_{x_2}(t - t_1 + t_1^*), & t \geqslant t_1, \end{cases} \qquad (6)$$

respectively; the moment t_1^* is determined by the equality $S_{x_1}(t_1) = S_{x_2}(t_1^*)$.

From this proposition it follows that for any $x(\cdot) \in E_2$ and for all $s \geqslant 0$

$$\alpha_{x(\cdot)}(t_1 + s) = \alpha_{x_2}(t_1^* + s). \qquad (7)$$

It is the model on E_2, proposed by Sedyakin (1966).

Let us consider a set E_m of stepwise stresses of the form (1). Set $t_0 = 0$. We shall show that the *rule of time-shift* holds for the SM on E_m.

Proposition 2. *If the SM holds on E_m then the survival function $S_{x(\cdot)}(t)$ satisfies the equalities:*

$$S_{x(\cdot)}(t) = S_{x_i}(t - t_{i-1} + t_{i-1}^*), \quad if$$
$$t \in [t_{i-1}, t_i), \ (i = 1, 2, \ldots, m), \qquad (8)$$

where t_i^* satisfy the equations

$$S_{x_1}(t_1) = S_{x_2}(t_1^*), \ldots, S_{x_i}(t_i - t_{i-1} + t_{i-1}^*)$$
$$= S_{x_{i+1}}(t_i^*), \ (i = 1, \ldots, m - 1). \qquad (9)$$

From this proposition it follows that for all $t \in [t_{j-1}, t_j)$ we have

$$A_{x(\cdot)}(t) = A_{x_j}(t - t_{j-1} + t_{j-1}^*).$$

In the literature on ALT (see Bhatta-charyya & Stoejoeti (1989), Nelson (1990), Meeker & Escobar (1998)) the model (8) is also called the *basic cumulative exposure model*.

We note here that the SM can be not appropriate in situations of periodic and quick change of the stress level or when switch-up of the stress from one level to the another can imply failures or shorten the life, see Bagdonavičius & Nikulin (2002).

ACCELERATED FAILURE TIME MODEL

Accelerated life models describing dependence of the lifetime distribution on the stresses will be considered here. The considered models are used in survival analysis and reliability theory analyzing results of accelerated life testing. A number of such models was proposed by engineers who considered physics of failure formation process of certain products or by statisticians (see, for example, Bagdonavičius, Gerville-Reache, Nikulin (2002), Cox & Oakes (1984), Chen (2001), Cooke & Bedford (1995), Crowder, Kimber, Smith & Sweeting (1991), Duchesne & Rosenthal (2002), Duchesne Lawless (2002), Gertsbach & Kordonskiy (1969, 1997), Gerville-Reache & Nikoulina (2002), Kartashov (1979), Lawless (2000), LuValle (2000), Meekers & Escobar (1998), Nelson (1990), Nelson & Meeker (1991), Singpurwalla & Wilson (1999), Viertl (1988), Xu & Harrington (2001),...) To introduce in this topic we consider at first the most famous Accelerated Failure Time (AFT) model. We start from the definition of this model for constant stresses.

Suppose that under different constant stresses the survival functions differ only in scale:

$$S_x(t) = S_0\{r(x)t\} \quad \text{for any} \quad x \in E_1, \quad (10)$$

where the survival function S_0, called the *baseline survival function*, does not depend on x. For any fixed x the value $r(x)$ can be interpreted as the *time-scale change constant* or the acceleration (deceleration) constant of survival functions.

Applicability of this model in accelerated life testing was first noted by Pieruschka (1961), see also Sedyakin (1966). It is the most simple and the most used model in failure time regression data analysis and ALT (see, Cooke & Bedford (1995), Nelson (1990), Meeker & Escobar (1998), Chen (2001), Hu & Harrington (2001), etc. ...)

Under the AFT model on E_1 the distribution of the random variable

$$R = r(x)T_x$$

does not depend on $x \in E_1$ and its survival function is S_0. Denote by m and σ^2 the mean and the variance of R, respectively. In this notations we have

$$\mathbf{E}(T_x) = m/r(x), \quad \mathbf{Var}(T_x) = \sigma^2/r^2(x),$$

and hence the coefficient of variation

$$\frac{\mathbf{E}(T_x)}{\sqrt{\mathbf{Var}(T_x)}} = \frac{m}{\sigma}$$

does not depend on x.

The survival functions under any $x_1, x_2 \in E_1$ are related in the following way:

$$S_{x_2}(t) = S_{x_1}\{\rho(x_1, x_2)t\},$$

where the function $\rho(x_1, x_2) = r(x_2)/r(x_1)$ shows the degree of scale variation. It is evident that $\rho(x, x) = 1$.

Now we consider the definition of the AFT model for time-varying stresses, using the Sedyakin's prolongation of the model (10) on E_1 to another model on E.

Definition 2. *The AFT model holds on E if there exists on E a positive function r and on $[0, \infty)$ a survival function S_0 such that*

$$S_{x(\cdot)}(t) = S_0 \left(\int_0^t r\{x(u)\} \, du \right)$$

$$\text{for any} \quad x(\cdot) \in E. \quad (11)$$

The most used way of application of AFT model is the following. The baseline survival function S_0 is taken from some class of parametric distributions, such as Weibull, lognormal, loglogistic, or from the class of regression models such as Cox proportional hazards model, frailty model, linear transformation model, additive hazards model, etc. ...

The AFT model in this form (11) was studied in Bagdonavičius (1990), Bagdonavičius & Nikulin (2002), Cox & Oakes (1984), Chen (2001), Meeker & Escobar (1998), Nelson (1990), Sedyakin (1966), Viertl (1988), Xu & Harrington (2001), etc.

AFT MODEL FOR STEP-STRESSES

The next two proposals give the forms of the survival functions in AFT model on E_m.

Proposition 3. *If the AFT model holds on E_2 then the survival function under any stress $x(\cdot) \in E_2$ of the form (2) verifies the equality*

$$S_{x(\cdot)}(t) = \begin{cases} S_{x_1}(t), & 0 \leqslant \tau < t_1, \\ S_{x_2}(t - t_1 + t_1^*), & \tau \geqslant t_1, \end{cases}$$

(12)

where

$$t_1^* = \frac{r(x_1)}{r(x_2)} t_1.$$

(13)

Proposition 4. *If the AFT model holds on E_m then the survival function $S_{x(\cdot)}(t)$ verifies the equalities:*

$$S_{x(\cdot)}(t) = S_0 \left\{ \sum_{j=1}^{i-1} r(x_j)(t_j - t_{j-1}) \right.$$

$$\left. + r(x_i)(t - t_{i-1}) \right\}$$

$$= S_{x_i} \left\{ t - t_{i-1} + \frac{1}{r(x_i)} \right.$$

$$\left. \times \sum_{j=1}^{i-1} r(x_j)(t_j - t_{j-1}) \right\},$$

$$t \in [t_{i-1}, t_i), \quad (i = 1, 2, \dots, m).$$

So the AFT model in the form (11) verifies the Sedyakin's principle on E_m. The proofs of these propositions one can find in Bagdonavičius and Nikulin (1995, 2002).

Remark 1. Suppose that $z_0(\cdot)$ is a fixed (for example, *usual* or *standard*) stress and $S_0 = S_{z_0(\cdot)}$, $z_0(\cdot) \in E$. In this case the AFT model is given by (11). Denote

$$f_{x(\cdot)}(t) = S_0^{-1}(S_{x(\cdot)}(t))$$

$$= \int_0^t r\{x(u)\} du, \quad x(\cdot) \in E,$$

$$\text{with} \quad f_{x(\cdot)}(0) = 0. \quad (14)$$

This relation shows that the moment t under any stress $x(\cdot) \in E$ is equivalent to the moment $f_{x(\cdot)}(t)$ under the usual stress $z_0(\cdot)$, therefore $f_{x(\cdot)}(t)$ is called *the resource used till the moment t under the stress $x(\cdot)$.*

PARAMETRIZATION OF THE AFT MODEL

Let $x(\cdot) = (x_0(\cdot), x_1(\cdot), \dots, x_m(\cdot))^T \in E$ be a possibly time-varying and multidimensional explanatory variable; here $x_0(t) \equiv 1$ and $x_1(\cdot), \dots, x_m(\cdot)$ are univariate explanatory variables.

Under the AFT model the survival function under $x(\cdot)$ is given by (11). Often the function r is parametrized in the following form:

$$r(x) = e^{-\beta^T x}, \quad (15)$$

where $\beta = (\beta_0, \dots, \beta_m)^T$ is a vector of unknown regression parameters. In this case the parametrized AFT model is given by the next formula:

$$S_{x(\cdot)}(t) = S_0 \left(\int_0^t e^{-\beta^T x(\tau)} d\tau \right), \quad x(\cdot) \in E.$$

(16)

Here $x_j(\cdot)$ $(j = 1, \dots, m)$ are not necessary the observed explanatory variables. They may be some specified functions $z_j(x)$. Nevertheless, we use the same notation x_j for $z_j(x)$.

If the explanatory variables are constant over time then the model (16), or (10), is written as

$$S_x(t) = S_0 \left(e^{-\beta^T x} t \right), \quad x \in E, \quad x_j \in E_1, \quad (17)$$

and the logarithm of the failure time T_x under x may be written as

$$\ln\{T_x\} = \beta^T x + \varepsilon, \quad (18)$$

where the survival function of the random variable ε is $S(t) = S_0(\ln t)$. It does not depend on x. Note that if the failure-time distribution is lognormal, then the distribution of the random variable ε is normal, and we have the *standard multiple linear regression model.*

Let us consider, following Nelson (1990), Meeker & Escobar (1998), Viertl (1988), some examples. For this we suppose at first that the explanatory variables are interval-valued (load, temperature, stress, voltage, pressure).

Suppose at first that x is one-dimensional, $x \in E_1$.

Example 1. Let

$$r(x) = e^{-\beta_0 - \beta_1 x}. \tag{19}$$

It is the *log-linear model*.

Example 2. Let

$$r(x) = e^{-\beta_0 - \beta_1 \log x} = \alpha_1 x^{\beta_1}. \tag{20}$$

It is the *power rule model*.

Example 3. Let

$$r(x) = e^{-\beta_0 - \beta_1/x} = \alpha_1 e^{-\beta_1/x}. \tag{21}$$

It is the *Arrhenius model*.

Example 4. Let

$$r(x) = e^{-\beta_0 - \beta_1 \ln \frac{x}{1-x}} = \alpha_1 \left(\frac{x}{1-x} \right)^{-\beta_1},$$

$$0 < x < 1. \tag{22}$$

It is the *Meeker-Luvalle* model (1995).

The Arrhenius model is used to model product life when the explanatory variable is the temperature, the power rule model—when the explanatory variable is voltage, mechanical loading, the log-linear model is applied in endurance and fatigue data analysis, testing various electronic components (see Nelson (1990)). The model of Meeker-Luvalle is used when x is the proportion of humidity.

The model (16) can be generalized. One can suppose that $\ln r(x)$ is a linear combination of some specified functions of the explanatory variable:

$$r(x) = \exp\left\{ -\beta_0 - \sum_{i=1}^{k} \beta_i z_i(x) \right\}, \tag{23}$$

where $z_i(x)$ are specified functions of the explanatory variable, $\beta_0, \ldots, \beta_k$ are unknown (possibly not all of them) parameters.

Example 5. Let

$$r(x) = e^{-\beta_0 - \beta_1 \log x - \beta_2/x} = \alpha_1 x e^{-\beta_2/x}, \tag{24}$$

where $\beta_1 = -1$. It is the *Eyring model*, applied when the explanatory variable x is the temperature.

INTERPRETATION OF THE REGRESSION COEFFICIENTS IN AFT MODEL

Suppose that the stresses are constant over time. Then under the AFT model (17) the p-quantile of the failure time T_x is

$$t_p(x) = e^{\beta^T x} S_0^{-1}(1-p), \quad x \in E_1, \tag{25}$$

so the logarithm

$$\ln\{t_p(x)\} = \beta^T x + c_p, \quad x \in E_1, \tag{26}$$

is a linear function of the regression parameters; here $c_p = \ln(S_0^{-1}(1-p))$.

Let $m(x) = \mathbf{E}\{T_x\}$ be the *mean life* of units under x. Then

$$m(x) = e^{\beta^T x} \int_0^\infty S_0(u) du, \quad x \in E_1, \tag{27}$$

and the logarithm

$$\ln\{m(x)\} = \beta^T x + c, \quad x \in E_1, \tag{28}$$

is also a linear function of the regression parameters; here

$$c = \ln\left\{ \int_0^\infty S_0(u) du \right\}.$$

Denote by

$$MR(x,y) = \frac{m(y)}{m(x)} \quad \text{and} \quad QR(x,y) = \frac{t_p(y)}{t_p(x)},$$

$$x, y \in E_1, \tag{29}$$

the ratio of means and quantiles, respectively. For the AFT model on E_1 we have

$$MR(x,y) = QR(x,y) = e^{\beta^T(y-x)}, \tag{30}$$

and hence $e^{\beta^T(y-x)}$ *is the ratio of means, corresponding to the stresses x and y.*

TIME-DEPENDENT REGRESSION COEFFICIENTS

The AFT model usually is parametrized in the following form (16). In this case the resource used till the moment t under stress $x(\cdot)$ is given by (14), from which it follows that at any moment t the resource usage rate

$$\frac{\partial}{\partial t} f_{x(\cdot)}(t) = e^{-\beta^T x(t)}, \quad x(\cdot) \in E,$$

depends only on the value of the stress $x(\cdot)$ at the moment t. More flexible models can be obtained by supposing that the coefficients β are time-dependent, i.e. taking

$$\frac{\partial}{\partial t} f_{x(\cdot)}(t) = e^{-\beta^T(t)x(t)} = e^{-\sum_{i=0}^m \beta_i(t)x_i(t)},$$

$$x(\cdot) \in E,$$

If the function $\beta_i(\cdot)$ is increasing or decreasing in time then the effect of ith component of the stress is increasing or decreasing in time.

So we have the model

$$S_{x(\cdot)}(t) = S_0 \left\{ \int_0^t e^{-\beta^T(u)x(u)} du \right\}. \quad (31)$$

It is the AFT model with time-dependent regression coefficients. We shall consider the coefficients $\beta_i(t)$ in the form

$$\beta_i(t) = \beta_i + \gamma_i g_i(t), \quad (i = 1, 2, \ldots, m),$$

where $g_i(t)$ are some specified deterministic functions or realizations of predictable processes. In such a case the AFT model with time-dependent coefficients and constant or time-dependent stresses can be written in the usual form (16) with different interpretation of the stresses. Indeed, set

$$\theta = (\theta_0, \theta_1, \ldots, \theta_{2m})^T$$
$$= (\beta_0, \beta_1, \ldots, \beta_m, \gamma_1, \ldots, \gamma_m)^T,$$
$$z(\cdot) = (z_0(\cdot), z_1(\cdot), \ldots, z_{2m}(\cdot))^T$$
$$= (1, x_1(\cdot), \ldots, x_m(\cdot), x_1(\cdot)g_1(\cdot), \ldots,$$
$$x_m(\cdot)g_m(\cdot))^T. \quad (32)$$

Then

$$\beta^T(u)x(u) = \beta_0 + \sum_{i=1}^m (\beta_i + \gamma_i g_i(t))x_i(t)$$
$$= \theta^T z(u).$$

So the AFT model with the time-dependent regression coefficients can be written in the standard form

$$S_{x(\cdot)} = S_0 \left\{ \int_0^t e^{-\theta^T z(u)} du \right\}, \quad (33)$$

where the unknown parameters and the explanatory variables are defined by (32).

PLANS OF EXPERIMENTS IN ALT

As it was said before the purpose of ALT is to give estimators of the main reliability characteristics: *the reliability function* $S_0 = S_{x^{(0)}}$, *the p-quantile* $t_p(x^{(0)})$ *and the mean value* $m(x_0)$ *under usual (design) stress* $x^{(0)}$, using data of accelerated experiments when units are tested at higher than usual stress conditions. Different plans of experiments are used in ALT with dynamic environment.

The first plan of experiments.

Denote by $x_0 = (x_{00}, x_{01}, \ldots, x_{0m})$, $x_{00} = 1$, the usual stress. Generally accelerated life testing experiments are done under an one-dimensional stress ($m = 1$), sometimes under two-dimensional ($m = 2$).

Let $x_1, \ldots, x_k$ be constant over time *accelerated stresses*:

$$x_0 < x_1 < \cdots < x_k;$$

here $x_i = (x_{i0}, x_{i1}, \ldots, x_{im}) \in E_m$, $x_{i0} = 1$. The usual stress x_0 *is not used* during experiments. According to the first plan of experiment k groups of units are tested. *The ith group of* n_i *units,* $\sum_{i=1}^k n_i = n$, *is tested under the stress* x_i. *The data can be complete or independently right censored.*

If the form of the function r is completely unknown and this plan of experiments is used, the function S_{x_0} can not be estimated even if it is supposed to know a parametric family to which belongs the distribution $S_{x_0}(t)$.

For example, if $S_0(t) = e^{-(t/\theta)^\alpha}$ then for constant stresses

$$S_x(t) = \exp\left\{-\left(\frac{r(x)}{\theta}t\right)^\alpha\right\}, \quad x \in E_1. \quad (34)$$

Under the given plan of experiments the parameters

$$\alpha, \frac{r(x_1)}{\theta}, \ldots, \frac{r(x_k)}{\theta}, \quad x_i \in E_1, \quad (35)$$

and the functions $S_{x_1}(t), \ldots, S_{x_k}(t)$ may be estimated. Nevertheless, the function $r(x)$ being completely unknown, the parameter $r(x_0)$ can not be written as a known function of these estimated parameters. So $r(x_0)$ and, consequently, $S_{x_0}(t)$ can not be estimated.

Thus, the function r must be chosen from some class of functions. Usually the model (17) is used.

The second plan of experiments. In step-stress accelerated life testing the second plan of experiments is as follows:

n units are placed on test at an initial low stress and if it does not fail in a predetermined time t_1, the stress is increased and so on. Thus, all units are tested under the step-stress $x(\cdot)$ of the form:

$$x(\tau) = \begin{cases} x_1, & 0 \leqslant \tau < t_1, \\ x_2, & t_1 \leqslant \tau < t_2, \\ \cdots & \cdots \\ x_k, & t_{k-1} \leqslant \tau < t_k; \end{cases} \quad (36)$$

where $x_j = (x_{j0}, \ldots, x_{jm})^T \in E_m$, $x_{j0} = 1$, $t_0 = 0, t_k = \infty$.

In this case the function $r(x)$ should be also parametrized because, even when the usual stress is used until the moment t_1, the data of failures occurring after this moment do not give any information about the reliability under the usual stress when the function $r(x)$ is unknown. Thus, the model (16) should be used. From the Proposition 2 it follows that for step-stresses the form (36) we have: if $t \in [t_{i-1}, t_i), i = 1, \ldots, k$

$$S_{x(\cdot)}(t) = S_0 \left\{ \mathbf{1}_{\{i>1\}} \sum_{j=1}^{i-1} e^{-\beta^T x_j}(t_j - t_{j-1}) \right.$$

$$\left. + e^{-\beta^T x_i}(t - t_{i-1}) \right\}. \quad (37)$$

Now we consider *the third plan of experiments.* Application of the first two plans may not give satisfactory results because assumptions on the form of the function $r(x)$ are done. These assumptions can not be statistically verified because of lack of experiments under the usual stress.

If the function $r(x)$ is completely unknown, and the coefficient of variation (defined as the ratio of the standard deviation and the mean) of failure times is not too large, the following plan of experiments may be used.

The third plan of experiments. Suppose that the failure time under the usual stress x_0 takes large values and most of the failures occur after the moment t_2 given for the experiment. According to this plan two groups of units are tested:

a) the first group of n_1 units under a constant accelerated stress x_1;

b) the second group of n_2 units under a step-stress: time t_1 under x_1, and after this moment under the usual stress x_0 until the moment t_2, $x_1, x_2 \in E_1$, i.e. under the stress $x_2(\cdot)$ from E_2:

$$x_2(\tau) = \begin{cases} x_1, & 0 \leqslant \tau \leqslant t_1, \\ x_0, & t_1 < \tau \leqslant t_2. \end{cases} \quad (38)$$

Units use much of their resources until the moment t_1 under the accelerated stress x_1, so after the switch-up failures occur in the interval $[t_1, t_2]$ even under usual stress. The AFT model implies that

$$S_{x_1}(u) = S_{x_0}(ru),$$

where $r = r(x_1)/r(x_0)$, and

$$S_{x_2(\cdot)}(u) = \begin{cases} S_{x_0}(ru), & 0 \leqslant u \leqslant t_1, \\ S_{x_0}(rt_1 + u - t_1), & t_1 < u \leqslant t_2, \end{cases}$$

or, shortly,

$$S_{x_2(\cdot)}(t) = S_{x_0}(r(u \wedge t_1) + (u - t_1) \vee 0),$$

$$x_2(\cdot) \in E_2, \quad (39)$$

with $a \wedge b = \min(a, b)$ and $a \vee b = \max(a, b)$.

It will be shown in the next section that if the third plan is used, and both functions S_{x_0} and $r(x)$ are completely unknown, semiparametric estimation of S_{x_0} is possible.

The third plan may be modified. The moment t_1 may be chosen as random. The most natural is to choose t_1 as the moment when the failures begin to occur.

At the end we consider the *fourth plan of experiment*, which is applied when the failure-time distribution under the usual stress is *exponential*. According to this plan k groups of units are observed. The i-th group of n_i units is tested under one-dimensional constant stress $x^{(i)}$ until the r_i-th failure ($r_i \leqslant n$), (type two censoring). *The failure moments of the i-th group are*

$$T_{(i1)} \leqslant T_{(i2)} \leqslant \cdots \leqslant T_{(ir_i)}.$$

DATA

We suppose that n units are observed. The ith unit is tested under the value $x^{(i)}(\cdot) = (x_1^{(i)}(\cdot), \ldots, x_m^{(i)}(\cdot))^T$ of a possibly time-varying and multi-dimensional explanatory variable $x(\cdot)$, according to the plan of experiment. The data are supposed to be independently right censored.

Let T_i and C_i be the failure and censoring times of the ith unit,

$$X_i = T_i \wedge C_i, \quad \delta_i = \mathbf{1}_{\{T_i \leqslant C_i\}}.$$

As it is well known the right censored data may be presented in the form

$$(X_1, \delta_1), \ldots, (X_n, \delta_n). \tag{40}$$

or

$$(N_1(t), Y_1(t), t \geqslant 0), \ldots, (N_n(t), Y_n(t), t \geqslant 0), \tag{41}$$

where

$$N_i(t) = \mathbf{1}_{\{X_i \leqslant t, \delta_i = 1\}}, \quad Y_i(t) = \mathbf{1}_{\{X_i \geqslant t\}}. \tag{42}$$

Here $N_i(t)$ is the number of observed failures of the ith unit in the interval $[0, t]$, and $Y_i(t)$ is the indicator of being at risk just prior to the moment t.

Using this presentation of data the statistical analysis of an appropriate accelerated life model can be done as is shown, for example, in Andersen, Borgan, Gill & Keiding (1993), Bagdonavičius and Nikulin (2002), Lawless (1982), Meeker & Escobar (1998).

ESTIMATION AND TESTING IN ALT

If the functions $r(\cdot)$ and $S_0(\cdot)$ are unknown we have a nonparametric model. The function $r(\cdot)$ can be parametrized. If the baseline function S_0 is completely unknown, in this case the we have a semiparametric model. Very often the baseline survival function S_0 is also taken from some class of parametric distributions, such as Weibull, lognormal, loglogistique, etc. In this case we have a parametric model and the maximum likelihood estimators of the parameters are obtained by almost standard way for any plans. Parametric case was studied by many people, see, for example, Bagdonavičius, Gerville-Réache, Nikoulina and Nikulin (2000), Gertsbakh & Kordonskiy (1969), Gerville-Réache & Nikoulina (2000), Glaser (1984), Hirose (1993), Iuculano & Zanini (1986), Lin & Ying (1995), LuValle (2000), Mazzuchi & Soyer (1992), Nelson (1990), Meeker & Escobar (1998), Sedyakin (1966), Sethuraman & Singpurwalla (1982), Schmoyer (1986), Shaked & Singpurwalla (1983), Viertl (1988). Nonparametric analysis of AFT model was considered by Basu & Ebrahimi (1982), Lin & Ying (1995), Lawless (1982), Robins & Tsiatis (1992), Schmoyer (1991), Ying (1993), Bagdonavičius and Nikulin (2000). Semiparametric case was considered by Tsiatis (1991), Lin & Ying (1995), Duchesne & Lawless (2002), Bagdonavičius and Nikulin (2002).

Tsiatis (1991), (constant stresses), Robins and Tsiatis (1992), Lin and Ying (1995) (time-dependent stresses) give asymptotic properties of the regression parameters for random right censored data. Lin and Ying (1995) give also semiparametric procedures for making inference about β (but not about the survival function and other reliability characteristics) which by pass the estimation of the covariance matrix of $\hat{\beta}$. All above mentioned papers boundedness of the density of the censoring variable is required. In the case of accelerated life testing when type one censoring is generally used, this condition is not true. In the case of accelerated life testing when type one censoring is generally used, this condition does not hold. Bagdonavičius and Nikulin (2002) give asymptotic properties of the estimators under the third plan of experiments. Using these properties Lin & Ying (1995) and Bagdonavičius & Nikulin (2002) studied tests for nullity of the regression coefficients, namely for testing the hypothesis

$$H_{k_1,k_2,\dots,k_l} : \beta_{k_1} = \cdots = \beta_{k_l} = 0,$$

$$(1 \leqslant k_1 < k_2 < \cdots < k_l). \tag{43}$$

MODELING IN ALT IN TERMS OF RESOURCE USAGE

The accelerated life (time transformation) models with dynamic environment give the possibility to include the effect of the usage history on the lifetime distributions of various units, subjects, items, populations, etc. Accelerated life models as time transformation models can be formulated using the notion of the resource introduced by Bagdonavičius and Nikulin (1995). This notion gives a general approach for construction time transformation models in terms of the rate of resource usage and it gives a simple physical interpretation of considered models.

Let Ω be a population of units and suppose that the failure-time of units under stress $x(\cdot)$ is a random variable $T_{x(\cdot)} = T_{x(\cdot)}(\omega), \omega \in \Omega$, with the survival function $S_{x(\cdot)}(t)$ and the cumulative distribution function $F_{x(\cdot)}(t)$. The moment of failure of a concrete item $\omega_0 \in \Omega$ is given by a nonnegative number $T_{x(\cdot)}(\omega_0)$.

The proportion $F_{x(\cdot)}(t)$ of units from Ω which fail until the moment t under the stress $x(\cdot)$ is also called the *uniform resource of population used until the moment t*. The same population of units Ω, observed under different stresses $x_1(\cdot)$ and $x_2(\cdot)$ use different resources until the same moment t if $F_{x_1(\cdot)}(t) \neq F_{x_2(\cdot)}(t)$. In sense of equality of used resource the moments t_1 and t_2 are equivalent if $F_{x_1(\cdot)}(t_1) = F_{x_2(\cdot)}(t_2)$.

The random variable

$$R^U = F_{x(\cdot)}(T_{x(\cdot)}) = 1 - S_{x(\cdot)}(T_{x(\cdot)})$$

is called the *uniform resource*. The distribution of the random variable R^U does not depend on $x(\cdot)$ and is uniform on $[0, 1)$. The uniform resource of any *concrete* item $\omega_0 \in \Omega$ is $R^U(\omega_0)$. It shows the proportion of the population Ω which fails until the moment of the unit's ω_0 failure $T_{x(\cdot)}(\omega_0)$.

The considered definition of the resource is not unique. Take any continuous survival function G such that the inverse $H = G^{-1}$ exists. In this case the distribution of the statistics $R^G = H(S_{x(\cdot)}(T_{x(\cdot)}))$ doesn't depend on $x(\cdot)$ and the survival function of R^G is G. The random variable R^G is called the *G-resource* and the number

$$f_{x(\cdot)}^G(t) = H(S_{x(\cdot)}(t)),$$

is called the *G-resource used until the moment t*. Note that in the case of the uniform resource $H(p) = 1 - p, p \in (0, 1]$.

Accelerated life models can be formulated specifying the way of resource usage, i.e. in terms of the rate of resource usage.

Note often the definitions of accelerated life models are formulated in terms of *exponential resource usage*, when $G(t) = e^{-t}, t \geqslant 0$, because *the exponential resource usage rate is nothing else but the hazard rate and the used resource is the cumulative hazard rate* respectively.

Let $\alpha_{x(\cdot)}(t)$ and $A_{x(\cdot)}(t)$ be the hazard rate and the cumulative hazard rate under $x(\cdot)$. The exponential resource is obtained by taking $G(t) = e^{-t}, t \geqslant 0$, and $H(p) = G^{-1}(p) = -\ln p$, so it is the random variable

$$R = A_{x(\cdot)}(T_{x(\cdot)})$$

with standard exponential distribution. For any t the number $A_{x(\cdot)}(t) \in [0, \infty)$ is the exponential resource used until the moment t under stress $x(\cdot)$. The rate of exponential resource usage is the hazard rate $\alpha_{x(\cdot)}(t)$.

All models can be formulated in terms of other resources than exponential. Let us consider at first one particular resource. Suppose that x_0 is a fixed (for example, usual) stress and $G = S_{x_0}$. For any $x(\cdot) \in E \supset E_1$ set

$$f_{x(\cdot)}(t) = S_{x_0}^{-1}(S_{x(\cdot)}(t)).$$

Then the moment t under any stress $x(\cdot) \in E$ is equivalent to the moment $f_{x(\cdot)}(t)$ under the usual stress x_0. The survival function of the resource R is S_{x_0}. As it was shown by (14) in the Remark 1 the AFT model (11) is determined by the S_{x_0}-resource usage rate by the next differential equation:

$$\frac{\partial}{\partial t} f_{x(\cdot)}(t) = r\{x(t)\}, \quad \text{for any} \quad x(\cdot) \in E,$$

$$\text{with} \quad f_{x(\cdot)}(0) = 0.$$

An natural generalization of the AFT model on E is obtained by changing, for example, the right part of this equation by the next way:

$$\frac{\partial f_{x(\cdot)}(t)}{\partial t} = r\{x(t)\} \, t^{\nu(x(t))-1} \quad \text{for any} \quad x(\cdot) \in E,$$

$$\text{with} \quad f_{x(\cdot)}(0) = 0, \qquad (44)$$

where ν is a positive function on E. This equality implies that

$$S_{x(\cdot)}(t) = S_{x_0} \left(\int_0^t r\{x(\tau)\} \tau^{\nu(x(\tau))-1} d\tau \right)$$

$$\text{for any} \quad x(\cdot) \in E. \qquad (45)$$

In this model variation of stress changes locally not only the scale but also the shape of distribution. This model is known as *the changing shape and scale model* (CHSS), see Bagdonavičius & Nikulin (2000).

To show the usefulness of the notion of the resource let us consider, following Bagdonavičius and Nikulin (1997), the so-called *generalized additive-multiplicative* (GAM) model on E, given in terms of the rate of resource usage by the next differential equation

$$\frac{\partial f_0^G(t)}{\partial t} = r[x(t)] \frac{\partial f_0^G(t)}{\partial t} + a(x(t)),$$

$$\text{with} \quad f_0^G(0) = f_{x(\cdot)}^G(0) = 0, \quad (46)$$

for some functions a and r (positive) on E, where $f_0^G(t) = H(S_0(t))$. In this model the stress influences the rate of resource using as multiplicatively as additively. The last equation implies that

$$S_{x(\cdot)}(t) = G \left\{ \int_0^t r[x(\tau)] dH(S_0(\tau)) \right.$$

$$\left. + \int_0^t a(x(\tau)) d\tau \right\}. \qquad (47)$$

Consider some particular cases.

1. Taking $G(t) = e^{-t} \mathbf{1}_{\{t \geqslant 0\}}$ we obtain:

$$\frac{\partial f_{x(\cdot)}^G(t)}{\partial t} = -\frac{S'_{x(\cdot)}(t)}{S_{x(\cdot)}(t)} = \alpha_{x(\cdot)}(t),$$

$$\frac{\partial f_0^G(t)}{\partial t} = -\frac{S'_0(t)}{S_0(t)} = \alpha_0(t),$$

where $\alpha_{x(\cdot)}(t)$ is the hazard rate under the covariate $x(\cdot)$, $\alpha_0(t)$ is the baseline hazard rate. So we obtain the model:

$$\alpha_{x(\cdot)}(t) = r[x(t)] \alpha_0(t) + a(x(t)).$$

It is the *additive-multiplicative semiparametric model* (Lin & Ying (1996). If $a(x(t)) \equiv 0$, we obtain the *proportional hazards model* or *Cox model*:

$$\alpha_{x(\cdot)}(t) = r[x(t)] \alpha_0(t).$$

If $r[x(t)] \equiv 1$, we obtain the *additive hazards model*:

$$\alpha_{x(\cdot)}(t) = \alpha_0(t) + a(x(t)).$$

2. Taking $G(t) = \exp\{-\exp\{t\}\}, \quad t \in R^1,$ we obtain

$$\frac{\partial f_{x(\cdot)}^G(t)}{\partial t} = \frac{\alpha_{x(\cdot)}(t)}{A_{x(\cdot)}(t)}, \quad \frac{\partial f_0^G(t)}{\partial t} = \frac{\alpha_0(t)}{A_0(t)},$$

where

$$A_{x(\cdot)}(t) = \int_0^t \alpha_{x(\cdot)}(\tau) d\tau, \quad A_0(t) = \int_0^t \alpha_0(\tau) d\tau$$

are the cumulated hazards rates. So we have the new model:

$$\frac{\alpha_{x(\cdot)}(t)}{A_{x(\cdot)}(t)} = r[x(t)] \frac{\alpha_0(t)}{A_0(t)} + a(x(t)).$$

3. Taking $G(t) = 1/(1+t), t \geqslant 0$, we obtain

$$\frac{\partial f_{x(\cdot)}^G(t)}{\partial t} = \frac{\alpha_{x(\cdot)}(t)}{S_{x(\cdot)}(t)}, \quad \frac{\partial f_0^G(t)}{\partial t} = \frac{\alpha_0(t)}{S_0(t)}.$$

So we have the *generalized logistic regression* model:

$$\frac{\alpha_{x(\cdot)}(t)}{S_{x(\cdot)}(t)} = r[x(t)] \frac{\alpha_0(t)}{S_0(t)} + a(x(t)).$$

If $a(x(t)) \equiv 0$, we obtain the *logistic regression model* since we have:

$$\frac{1}{S_{x(\cdot)}(t)} - 1 = r[x(t)] \left(\frac{1}{S_0(t)} - 1 \right).$$

4. Taking $G(t) = 1/(1 + e^t)$, $t \in R^1$, we obtain

$$\frac{\partial f_{x(\cdot)}^G(t)}{\partial t} = \frac{\alpha_{x(\cdot)}(t)}{1 - S_{x(\cdot)}(t)}, \quad \frac{\partial f_0^G(t)}{\partial t} = \frac{\alpha_0(t)}{1 - S_0(t)}.$$

So we have the new model:

$$\frac{\alpha_{x(\cdot)}(t)}{1 - S_{x(\cdot)}(t)} = r[x(t)] \frac{\alpha_0(t)}{1 - S_0(t)} + a(x(t)).$$

5. Take $G(t) = \Phi(\ln t)$, $t \geq 0$, where $\Phi(\cdot)$ is the distribution function of the standard normal law. If $a(x(t)) \equiv 0$, then in terms of survival functions we obtain the model:

$$\Phi^{-1}(1 - S_{x(\cdot)}(t)) = \ln r[x(t)] + \Phi^{-1}(1 - S_0(t)).$$

It is the *generalized probit model*.
6. Taking $G = S_0$, we obtain

$$S_{x(\cdot)}(t) = S_0 \left\{ \int_0^t \sigma[x(\tau)]d\tau \right\}.$$

where $\sigma(x(t)) = r[x(t)] + a(x(t))$. It is the *accelerated life model*, given by (11). As one can see, the GAM model contains many interesting sub-models, which are well adapted to solve the statistical problems in ALT, and the notions of the resource and the rate of resource usage give a power instrument for modeling in ALT.

REFERENCES

1. Andersen, P. K., Borgan, O., Gill, R. D. and Keiding, N. (1993). *Statistical Models Based on Counting Processes*. New York: Springer.
2. Bagdonavičius, V. (1990). Accelerated life models when the stress is not constant. *Kybernetika*, **26**, 289–295.
3. Bagdonavičius, V., Nikulin, M. (1995). *Semiparametric models in accelerated life testing*. Queen's Papers in Pure and Applied Mathematics, **98**. Kingston: Queen's University, Canada.
4. Bagdonavičius, V., Nikulin, M. (1997). Analysis of general semiparametric models with random covariates. *Revue Roumaine de Mathématiques Pures et Appliquées*, **42**, #5–6, 351–369.
5. Bagdonavičius, V., Nikulin, M. (1998). *Additive and multiplicative semiparametric models in accelerated life testing and survival analysis*. Queen's Papers in Pure and Applied Mathematics, **108**. Kingston: Queen's University, Canada.
6. Bagdonavičius, V., Gerville-Réache, L., Nikoulina, V., Nikulin, M. (2000). Expériences accélérées: analyse statistique du modèle standard de vie accélérée. *Revue de Statistique Appliquée*, XLVIII, 3, 5–38.
7. Cox, D. R. and Oakes, D. (1984). *Analysis of Survival Data*. New York: Methuen (Chapman and Hall).
8. Chen, Y. Q. (2001). Accelerated Hazards Regression Model and Its Adequacy for Censored Survival Data. *Biometrics*, **57**, 853–860.
9. Cooke R., Bedford, T. (1995). Analysis of Reliability Data Using Subsurvival Functions and Censoring Mosels. In: *Recent Advances in Life-Testing and Reliability*, (Ed. Balakrishnan, N.), Boca Raton: CRC Press, 12–41.
10. Duchesne, Th., Rosenthal, J. F. (2002). Stochastic Justification of Some Simple Reliability Models. Preprint of the Department of Statistics, University of Toronto, Toronto, ON, Canada, 18p.
11. Duchesne, Th., Lawless, J. (2002). Semiparametric Inference Methods for General Time Scale Models. *Lifetime Data Analysis*, **8**, 263–276.
12. Duchesne, T. (2000). Methods Common to Reliability and Survival Analysis. In: *Recent Advances in Reliability Theory. Methodology, Practice and Inference*, (Eds. N. Limnios and M. Nikulin), Boston: Birkhauser, 279–290.
13. Duchesne, T. and Lawless, J. (2000). Alternative Time Scale and Failure Time Models. *Lifetime Data Analysis*, **6**, 157–179.
14. Gertsbakh, L. B. and Kordonskiy, K. B. (1969). *Models of Failure*. New York: Springer-Verlag.
15. Gertsbakh, L. B. and Kordonskiy, K. B. (1997). Multiple time scale and the lifetime coefficient of variation: Engineering applications. *Lifetime Data Analysis*, **2**, 139–156.
16. Gerville-Réache, L. and Nikoulina, V. (1999). Analysis of Reliability Characteristics Estimators in Accelerated Life Testing, In: *Statistical and Probabilistic Models in Reliability*,

(Eds. D. Ionescu and N. Limnios), Boston: Birkhauser, 91–100.

17. Glaser, R. E. (1984). Estimation for a Weibull Accelerated Life Testing Model, *Naval Research Logistics Quarterly*, **31**, 4, 559–570.

18. Hirose, H. (1993). Estimation of Threshsold Stress in Accelerated Life-Testing, *IEEE Transaction on reliability*, **42**, 650–657.

19. Kalbfleisch, J. D. and Prentice, R. L. (1980). *The Statistical Analysis of Failure Time Data*. New York: John Wiley and Sons.

20. Kartashov, G. D. (1979). Methods of Forced (Augmented) Experiments (in Russian). Moscow: *Znaniye Press*.

21. Klein, J. P. and Basu, A. P. (1981). Weibull Accelerated Life Tests When There are Competing Causes of Failure, *Communications in Statistical Methods and Theory*, **10**, 2073–2100.

22. Klein, J. P. and Basu, A. P. (1982). Accelerated Life Testing under Competing Exponential Failure Distributions, *IAPQR Trans.*, **7**, 1–20.

23. Lawless, J. F. (1982). *Statistical Models and Methods for Lifetime Data*. New York: John Wiley and Sons.

24. Lawless, J. F. (2000). Dynamic Analysis of Failures in Repairable Systems and Software. In: *Recent Advances in Reliability Theory*. (Eds. Limnios, N., Nikulin, M), Boston: Birkhauser, 341–354.

25. Lin, D. Y. and Ying, Z. (1995). Semiparametric inference for accelerated life model with time dependent covariates. *Journal of Statist. Planning and Inference*, **44**, 47–63.

26. Lin, D. Y. and Ying, Z. (1996). Semiparametric analysis of the general additive-multiplicative hazards model for counting processes. *Ann. Statistics*, **23**, #5, 1712–1734.

27. LuValle, M. (2000). A Theoretical Framework for Accelerated Testing. In: *Recent Advances in Reliability Theory. Methodology, Practice and Inference*, (Eds. N. Limnios and M. Nikulin), Boston: Birkhauser, 419–434.

28. Miner, M. A. (1945). Cumulative Damage in Fatigue, *J. of Applied Mechanics*, **12**, A159–A164.

29. Meeker, W. Q., Escobar, L. A. (1998). *Statistical Methods for Reliability Data*. New York: John Wiley and Sons.

30. Nelson, W. (1990). *Accelerated Testing: Statistical Models, Test Plans, and Data Analysis*. New York: John Wiley and Sons.

31. Nelson, W. and Meeker, W. (1991). Accelerated Testing: Statistical models, test plans, and data analysis. *Technometrics*, **33**, 236–238.

32. Nikulin, M. S., and Solev, V. N. (2002). Testing Problem for Increasing Function in a Model with Infinite Dimensional Nuisance Parameter. In: *Goodness-of-fit Tests and Validity of Models*, (Eds. C. Huber, N. Balakrishnan, M. Nikulin and M. Mesbah), Boston: Birkhauser.

33. Sethuraman, J. and Singpurwalla, N. D. (1982). Testing of hypotheses for distributions in accelerated life tests. *JASA*, **77**, 204–208.

34. Singpurwalla, N. (1995). Survival in Dynamic Environments. *Statistical Sciences*, **1**, #10, 86–103.

35. Viertl, R. (1988). *Statistical Methods in Accelerated Life Testing*. Göttingen: Vandenhoeck and Ruprecht.

36. Tsiatis, A. A. (1990). Estimating regression parameters using linear rank tests for censored data. *Ann. Statist.*, **18**, 354–72.

37. Robins, J. M., and Tsiatis, A. A. (1992). Semiparametric estimation of an accelerated failure time model with time dependent covariates. *Biometrika*, **79**, #2, 311–319.

V. Bagdonavicius
M. Nikulin

ACCEPTABLE QUALITY LEVEL (AQL)

This is usually defined as the maximum percent defective (or the maximum number of defects per 100 units) that can be considered satisfactory for a process average.

BIBLIOGRAPHY

Brownlee, K. A. (1965). *Statistical Theory and Methodology in Science and Engineering*, 2nd ed. Wiley, New York.

Johnson, N. L. and Leone, F. (1977). *Statistics and Experimental Design in Engineering and the Physical Sciences*, 2nd ed., Vol. 1. Wiley, New York, Chap. 10.

Juran, J. M., ed. (1964). *Quality-Control Handbook*, 2nd ed. McGraw-Hill, New York (first ed., 1953).

See also ACCEPTANCE PROCESS ZONE and QUALITY CONTROL, STATISTICAL.

ACCEPTANCE ERROR

A term used in the theory of testing hypotheses* to denote a decision to accept a hypothesis H_0 when that hypothesis is not valid. It is also called a Type I error*.

See also HYPOTHESIS TESTING and LEVEL OF SIGNIFICANCE.

ACCEPTANCE NUMBER

Given a sampling plan*, the acceptance number c denotes the maximum number of defective items that can be found in the sample without leading to rejection of the lot.

BIBLIOGRAPHY

Duncan, A. J. (1974). *Quality Control and Industrial Statistics*, 4th ed. Richard D. Irwin, Homewood, Ill.

Standards Committee of ASQC* (1971). ASQC Standard A2 (rev. 1978).

See also ACCEPTABLE QUALITY LEVEL (AQL); QUALITY CONTROL, STATISTICAL; and SAMPLING PLANS.

ACCEPTANCE PROCESS ZONE

The acceptance process level (APL) is a fundamental notion in quality control. It is the process level most remote from the standard that still yields product quality that we are willing to accept with high probability. Since most specifications are two-sided (i.e., requiring characteristics to lie within a specified tolerance band), it is usually appropriate to specify both an upper and a lower APL.

The band around the nominal value between the upper and lower acceptance process level (APL) values is called the *acceptance process zone*.

FURTHER READING

Duncan, A. J. (1986). *Quality Control and Industrial Statistics* (4th ed.). Irwin, Homewood, Ill.

Montgomery, D. C. (2001). *Introduction to Statistical Quality Control* (3rd ed.). Wiley, New York/Chichester.

See also QUALITY CONTROL, STATISTICAL; REJECTABLE PROCESS LEVEL (RPL); and SAMPLING PLANS.

ACCEPTANCE REGION (IN TESTING HYPOTHESES)

A hypothesis test* divides the space T of a test statistic* T into two complementary regions, C (the critical region) and $T - C$. The region $T - C$ is called the *acceptance region or nonrejection region*. This region is characterized by the property that if the *value* of the test statistic falls into this region the hypothesis under test is accepted.

See also CRITICAL REGION and HYPOTHESIS TESTING.

ACCEPTANCE SAMPLING

The term "acceptance sampling" relates to the acceptance or rejection of a product or process on the basis of sampling inspection*. It has been pointed out that "sampling inspection is the process of evaluating the quality of material by inspecting some but not all of it" [4]. Its methods constitute decision rules for the disposition or sentencing of the product sampled. In this sense it may be contrasted with survey sampling*, the purpose of which is largely estimation*.

Sampling plans*, which specify sample size and acceptance criteria, are fundamental to acceptance sampling. Such plans may be based on a simple dichotomous classification of conformance or nonconformance of a quality characteristic to specified criteria (*attributes plans*) or on a comparison of statistics computed from quantitative measurements to numerical criteria developed from the specifications and from assumptions about the shape and nature of the distribution of individual measurements (*variables plans*). An example of the former is the attributes plan: sample 50 items and accept the lot of material from which the sample was taken if two or fewer items are found nonconforming; reject otherwise. An example of the latter is the variables plan: sample 12 items and accept the lot if the sample mean is more than 2 standard deviations above the lower specification limit; reject otherwise.

When a process parameter such as the mean* or standard deviation* is specified, the sampling plan resolves itself simply into a test of hypothesis. *See* HYPOTHESIS TESTING. That is, the sampling plan might be a *t*-test* if the plan is imposed to assure that the process mean conforms to a specified value. These tests are called variables plans for process parameter and are commonly used in the sampling of bulk materials. *See* BULK SAMPLING. More complicated situations arise when it is necessary to test the proportion of product beyond or between specification limits through the use of a measurement criterion. Such tests are referred to as variables plans for proportion nonconforming.

The operating characteristic (OC) curve (complement of the power* curve) is the primary measure of performance of a sampling plan. It shows the probability of acceptance as a function of the value of the quality characteristic. As such, it provides a description of the protection afforded by the plan as well as a vehicle for comparison of various acceptance sampling plans and procedures. Two types of OC curves are distinguished. The type A operating characteristic curve relates to the inspection of individual lots of product and shows lot probability of acceptance as a function of lot quality. In attributes inspection, it is computed from the hypergeometric distribution*. The type B operating characteristic curve relates to the process that produced the product to be inspected and shows the proportion of lots accepted in a continuing series as a function of the process average. In attributes inspection, it is computed either from the binomial distribution* when inspection is for proportion nonconforming, or from the Poisson distribution* when inspection is for nonconformities per unit. Details of the nature and construction of type A and type B operating characteristic curves are given in ref. 2.

Often the performance of a plan is characterized by two points on the OC curve: a producer's quality level with high probability of acceptance and a consumer's quality level with low probability of acceptance. The corresponding risks are called the *producer's risk* (of rejection) and the *consumer's risk* (of acceptance). (The producer's risk is conventionally taken to be 0.05 and the consumer's risk is 0.10.) The ratio of the consumer's quality level to the producer's quality level is called the *operating* (or *discrimination*) *ratio*. It describes the steepness of the OC curve and hence the capability of the plan to distinguish

between acceptable and unacceptable quality. So-called two-point plans can be derived from the producer's and consumer's quality levels through their operating ratios. A third point on the OC curve that is commonly referenced is the indifference quality level*, at which the probability of acceptance is 0.50. Sets of plans have been developed using the indifference quality level and the relative slope of the OC curve at that point.

Acceptance sampling procedures progress far beyond the simple single sampling plan to other, more complex procedures. Double-sampling* plans allow the possibility of two samples, a second sample being taken if the results of the first sample are not sufficiently definitive. This concept can be extended to multiple sampling plans, involving more than two samples. Sequential procedures* are applied in acceptance sampling to achieve the excellent discrimination and economy of sample size associated with such methods. These plans may be used in both attributes and variables inspection.

Various measures and associated curves have been developed to describe the properties of sampling plans. The *average sample number (ASN)* *curve* describes the average sample size for various quality levels when using double*, multiple*, sequential*, or other procedures. The *average outgoing quality (AOQ)* *curve* shows the average proportion of nonconforming product the consumer will receive if rejected lots are 100% inspected plotted against quality level. Its maximum is called the *average outgoing quality limit (AOQL)**. Under such a procedure, the average total inspection curve shows the total number of units inspected in both sampling and 100% inspection and can be used to estimate and compare inspection loads.

For a continuing sequence of lots sampling plans may be combined into sampling schemes consisting of two or more plans used together with switching rules that establish the procedure for moving from one of the plans to another. Schemes can be constructed to give protection for both the producer and the consumer which is superior to that of the constituent plans with a reduction of sample size for the protection afforded. Such schemes are usually specified by an *acceptable quality level (AQL)** which, when exceeded, will eventually lead to a switch to a tighter plan with consequent economic and psychological pressure on the supplier to improve the quality of the product. Sampling schemes have their own OC curves. Sometimes options for discontinuation of inspection are incorporated into the procedure. Sampling schemes may be combined into sampling systems that select specific schemes by prescribed rules. The most important sampling systems are military standards* MIL-STD-105D for attributes and MIL-STD-414 for variables. Their civilian counterparts are ANSI Z1.4 and ANSI Z1.9 in the United States and international standards ISO 2859 and ISO 3951, respectively.

The variety of approaches in acceptance sampling is almost limitless. Continuous sampling plans are used on streams of output where lots are difficult to define. Chain sampling* plans link the criteria for the immediate sampling plan to past results. Grand lot sampling procedures combine samples from lots that have been shown to be homogeneous, to achieve larger sample size and greater discrimination. Skip-lot plans* provide for the inspection of only a fraction of the lots submitted. Acceptance control charts* can be used to visually portray the results of inspection through the medium of the control chart. Bayesian plans introduce economic considerations and prior results and estimates into the sampling equation. Special plans have been developed for various areas of application, such as compliance testing, reliability and life testing*, and safety inspection. *See* SAMPLING PLANS.

The application and administration of sampling inspection demands a broad range of knowledge of statistical methodology, because the determination of what, where, when, how, and to what quality levels the inspection is to be carried out is largely empirical. Acceptance sampling procedures are an integral part of quality control* practice and serve to distinguish between quality levels as a vehicle for quality improvement. In application, however, they should, where possible, be supplemented and eventually supplanted by statistical techniques for process quality control (such as control charts*)

in an effort to *prevent* the occurrence of nonconforming material rather than to *detect* it after it has been produced.

The history and development of acceptance sampling is described in detail by Dodge [1], who originated and developed many acceptance sampling procedures. The statistical methodology of acceptance sampling has been treated specifically by Schilling [5] and in the context of industrial statistics as a whole by Duncan [3].

REFERENCES

1. Dodge, H. F. (1969). *J. Quality Tech.*, **1**: Part I, Apr., 77–88; Part II, July, 155–162; Part III, Oct., 225–232; **2**: Part IV, Jan., 1970, 1–8.

2. Dodge, H. F. and Romig, H. G. (1959). *Sampling Inspection Tables—Single and Double Sampling*, 2nd ed. Wiley, New York.

3. Duncan, A. J. (1986). *Quality Control and Industrial Statistics*, 4th ed. Richard D. Irwin, Homewood, Ill.

4. Freeman, H. A., Friedman, M., Mosteller, F., and Wallis, W. A., eds. (1948). *Sampling Inspection Principles: Procedures and Tables for Single, Double and Sequential Sampling in Acceptance Inspection and Quality Control.* McGraw-Hill, New York.

5. Schilling, E. G. (1981). *Acceptance Sampling in Quality Control*. Marcel Dekker, New York.

See also AVERAGE OUTGOING QUALITY LIMIT (AOQL); CONTROL CHARTS; LOT TOLERANCE TABLES, DODGE–ROMIG; QUALITY CONTROL, STATISTICAL; and SAMPLING PLANS.

EDWARD G. SCHILLING

ACCURACY AND PRECISION

The *accuracy* of an observation or a statistic derived from a number of observations has to do with how close the value of the statistic is to a supposed "true value".

In forecasting, accuracy is a measure of how close the forecast $\hat{Y}_t$ of an observation Y_t at time t is to Y_t; *see* PREDICTION AND FORECASTING. *See* also REGRESSION VARIABLES, SELECTION OF and FINAL PREDICTION ERROR CRITERIA, GENERALIZED for model choices in multiple regression aimed at reduction of error (and hence at improved accuracy).

In estimation theory accuracy measures how close an estimate $\hat{\theta}$ of a parameter θ is to the "true value" of θ. The accuracy of $\hat{\theta}$ can be measured, for example, in terms of the mean absolute error or of the mean squared error* (MSE) of $\hat{\theta}$.

Accuracy should be distinguished from *precision*. Precision of measurement indicates the resolving power of a measuring device and is frequently given by the number of decimal places reported in the measurements made with the device. The precision of an estimator $\hat{\theta}$, on the other hand, measures how tightly the distribution of $\hat{\theta}$ clusters about its center (say, its expected value) [1, Sec. 4.1]. One has

$$\text{MSE}(\hat{\theta}) = \{\text{Variance of } \hat{\theta}\} + (\text{bias of } \hat{\theta})^2;$$

Here the accuracy of $\hat{\theta}$ can be measured via $\text{MSE}(\hat{\theta})$ and its precision via $\text{Var}(\hat{\theta})$. Carl Friedrich Gauss's measure of precision is

$$1/\{\sqrt{2} \times (\text{standard deviation of } \hat{\theta})\}$$

where the quantity $\sqrt{2}\times$ (standard deviation) is known as the *modulus*. Gauss's measure satisfies the intuitive notion that the precision increases as the standard deviation decreases.

NEYMAN AND WOLFOWITZ ACCURACY

Let T be a statistic* based on a sample from a population having an unknown parameter θ, and let $(L_1(T), L_2(T))$ be a confidence interval for θ. If

$$Q(\theta_0) = \text{Pr}\,[L_1(T) \leqslant \theta_0 \leqslant L_2(T)|\theta],$$

then [2,4] $Q(\theta_0)$ is the *Neyman accuracy* of the confidence interval. It is a measure of the accuracy of $(L_1(T), L_2(T))$ in excluding the false value $\theta_0 \neq \theta$ of θ. The interval with the smaller Neyman accuracy is said to be *more selective* [5]; see CONFIDENCE INTERVALS AND REGIONS. If

$$W(a,b) = aE\{(L_1(T) - \theta)^2\} + bE\{(L_2(T) - \theta)^2\},$$

then [2] $W(\cdot, \cdot)$ is the *Wolfowitz accuracy* of the confidence interval [6], and measures how close the confidence limits L_1 and L_2 are to the true value of θ; see also [3], where the efficiency of competing confidence intervals is measured inter alia by the ratio of their Wolfowitz accuracies when $a = b = 1$.

REFERENCES

1. Bickel, P. J. and Doksum, K. A. (1977). *Mathematical Statistics: Basic Ideas and Selected Topics*, Holden-Day, San Francisco.

2. Ghosh, B. K. (1975). A two-stage procedure for the Behrens-Fisher problem, *J. Amer. Statist. Ass.*, **70**, 457–462.

3. Harter, H. L. (1964). Criteria for best substitute interval estimators, with an application to the normal distribution, *J. Amer. Statist. Ass.*, **59**, 1133–1140.

4. Neyman, J. (1937). Outline of a theory of statistical estimation based on the classical theory of probability, *Philos. Trans. Royal. Soc. A*, **236**, 333–380.

5. Stuart, A. and Ord, J. K. (1991). *Kendall's Advanced Theory of Statistics*, Vol. 2 (5th ed.). Oxford University Press, Oxford, U.K. (Secs. 20.14, 20.15).

6. Wolfowitz, J. (1950). Minimax estimates of the mean of a normal distribution with known variance. *Ann. Math. Statist.*, **21**, 218–230.

See also FINAL PREDICTION ERROR CRITERIA, GENERALIZED; MEAN DEVIATION; MEAN SQUARED ERROR; PREDICTION AND FORECASTING; and VARIANCE.

ACHENWALL, GOTTFRIED

Born: October 20, 1719, in Elbing, Germany.

Died: May 1, 1772, in Göttingen, Germany.

Contributed to: *Staatswissenschaft* ("university statistics").

Achenwall was born into the family of a merchant. In 1738–1740 he acquired a knowledge of philosophy, mathematics, physics, and history at Jena; then he moved to Halle, where, without abandoning history, he studied the law and *Staatswissenschaft* (the science of the state; also known as "university statistics"). Apparently in 1742 Achenwall returned for a short time to Jena, then continued his education in Leipzig. In 1746 he became Docent at Marburg, and in 1748, extraordinary professor at Göttingen (ordinary professor of law and of philosophy from 1753), creating there the Göttingen school of statistics. Its most eminent member was A. L. Schlözer (1735–1809). Achenwall married in 1752, but his wife died in 1754, and he had no children.

Achenwall followed up the work of Hermann Conring (1606–1681), the founder of *Staatswissenschaft*, and was the first to present systematically, and in German rather than in Latin, the Conring tradition. According to both Conring and Achenwall, the aim of statistics was to describe the climate, geographical position, political structure, and economics of a given state, to provide an estimate of its population, and to give information about its history; but discovering relations between quantitative variables was out of the question. For Achenwall [1, p. 1], "the so-called statistics" was the *Staatswissenschaft* of a given country.

Since 1741, "statisticians" have begun to describe states in a tabular form, which facilitate the use of numbers, a practice opposed by Achenwall. Even in 1806 and 1811 [5, p. 670] the use of tabular statistics was condemned because numbers were unable to describe the spirit of a nation.

Nevertheless, Achenwall [4, Chap. 12] referred to Süssmilch,* advised state measures fostering the multiplication of the population, recommended censuses, and even [4, p. 187] noted that its "probable estimate" can be gotten by means of "yearly lists of deaths, births, and marriages." The gulf between statistics (in the modern sense) and *Staatswissenschaft* was not as wide as it is usually supposed to have been. Leibniz's manuscripts, written in the 1680s, present a related case. First published in 1866 and reprinted in 1977, they testify that he was both a political arithmetician and an early advocate of tabular description (both with and without the use of numbers) of a given state; see [10,222–227,255].

REFERENCES

1. Achenwall, G. (1748). *Vorbereitung zur Staatswissenschaft*. Göttingen. (An abridged version of this was included in his next contribution.)

2. Achenwall, G. (1749). *Abriß der neuesten Staatswissenschaft der vornehmsten europäischen Reiche und Republicken zum Gebrauch in seinen academischen Vorlesungen*. Schmidt, Göttingen.

3. Achenwall, G. (1752). *Staatsverfassung der europäischen Reiche im Grundrisse*. Schmidt, Göttingen. This is the second edition of the *Abriß*. Later editions: 1756, 1762, 1767, 1768, 1781–1785, 1790–1798. By 1768 the title had changed to *Staatsverfassung der heutigen vornehmsten europäischen Reiche and Völker*, and the publisher was Witwe Wanderhoeck.

4. Achenwall, G. (1763). *Staatsklugheit nach ihren Grundsätzen*. Göttingen. (Fourth ed., 1779).

5. John, V. (1883). The term "Statistics." *J. R. Statist. Soc.* **46**, 656–679. (Originally published in German, also in 1883.)

6. John, V. (1884). *Geschichte der Statistik*. Encke, Stuttgart.

7. Lazarsfeld, P. (1961). Notes on the history of quantification in sociology—trends, sources and problems, *Isis*, **52**, 277–333. Reprinted (1977) in *Studies in the History of Statistics and Probability*, Sir Maurice Kendall and R. L. Plackett. eds. Vol. 2, pp. 213–269. Griffin, London and High Wycombe

8. Leibniz, G. W. (1886). *Sämmtliche Schriften und Briefe*. Reihe 4, Bd. 3. Deutsche Akad. Wiss., Berlin.

9. Schiefer, P. (1916). *Achenwall und seine Schule*. München: Schrödl. (A Dissertation.)

10. Sheynin, O. B. (1977). Early history of the theory of probability. *Arch. Hist. Ex. Sci.*, **17**, 201–259.

11. Solf, H. H. (1938). *G. Achenwall. Sein Leben und sein Werk, ein Beitrag zur Göttinger Gelehrtengeschichte*. Mauser, Forchheim. Oberfranken. (A dissertation.)

12. Westergaard, H. (1932). *Contributions to the History of Statistics*. King, London. (Reprinted, New York, 1968, and The Hague, 1969.)

13. Zahn, F. and Meier, F. (1953). Achenwall, *Neue deutsche Biogr.* 1, 32–33. Duncker and Humblot, Berlin.

O. SHEYNIN

ACTUARIAL HEALTH STUDIES

Medico-actuarial studies originated in the United States in the 1890s from concerted efforts to improve the underwriting of life insurance risks [15]. The mortality investigations undertaken were aimed to isolate and measure the effects of selected risk factors, such as occupational hazards, medical conditions, and build. The underlying hypothesis was that each of the factors (or certain combinations of factors) influencing mortality could be regarded as an independent variable and the total mortality risk could be treated as a linear compound of a number of independent elements.

The first comprehensive study, known as the *Specialized Mortality Investigation*, was carried out by the Actuarial Society of America and published in 1903 [1]. It covered the experience of 34 life insurance companies over a 30-year period and focused attention on the mortality in 35 selected occupations, 32 common medical conditions, and several other factors affecting mortality. It was followed in 1912 by the *Medico-Actuarial Mortality Investigation* [2], sponsored jointly by the Actuarial Society of America and the Association of Life Insurance Company Medical Directors, which included a much wider variety of occupations, medical conditions, and other factors, among them abuse of alcohol. This study laid the broad lines on which such investigations have been conducted since.

The assessment of the long-term risk in life insurance was seen as requiring analysis of mortality by sex, age at issue, and duration since issue of insurance for policies issued under similar underwriting rules. Cohorts of policyholders were followed over long periods of time. Attention was focused on the mortality in the years following issue of insurance in order to trace the effects on mortality of the selection exercised by insurance companies through medical examinations and other screening for insurance, as well as effects of antiselection by applicants for insurance who withheld information relevant to their health. The extent of class selection, that is, the reflection of the underlying mortality in the segments of the population from which the insured lives were drawn, was brought out in the mortality experienced among the insured lives under study after many years since issue of insurance had elapsed. Most important, however, the patterns of the mortality experienced over longer periods of time indicated the incidence of the extra mortality by duration, which permitted classifying the long-term risk as one of decreasing extra mortality, relatively level extra mortality, or increasing extra mortality.

Analyses of mortality by cause shed light on the causes mainly responsible for excess mortality and also on the causes of death which could be controlled to some degree by screening applicants for life insurance. Successive medico-actuarial investigations permitted some conclusions regarding the trends in mortality associated with different factors affecting mortality, notably occupational hazards, build, blood pressure, various medical conditions, and changing circumstances.

METHODOLOGY

In selecting a particular cohort of policyholders for study, because of interest in some particular factor influencing mortality, it was the practice in medico-actuarial investigations to exclude individuals who also presented other kinds of risks. Specifically, all individuals were excluded from the study if they were also subject to any other kind of risk that would have precluded issuing insurance at standard premium rates. Consequently, such extra mortality as was found in the study could properly be associated with the factor of interest rather than with the combined effects of this factor and other elements at risk.

The findings of medico-actuarial investigations have been customarily expressed as ratios of actual deaths in the cohort of policyholders under study to the expected deaths, which were calculated on the basis of contemporaneous death rates among otherwise similar life insurance risks accepted at standard premium rates. Such mortality ratios usually were computed by sex, age groups at issue of the insurance, duration since issue of the insurance, and causes of death. The calculation of expected deaths involves accurate estimates of the exposed to risk. Because of the varying forms of records, individual and grouped, different tabulating procedures have been employed, as a rule considering deaths within a unit age interval as having occurred at the end of that interval.

Ratios of actual to expected mortality provide very sensitive measures of mortality and therefore may fluctuate widely in finer subdivisions of the experience level. They have the merit, however, of revealing even small departures from expected mortality in broad groupings. In some circumstances the patterns of excess mortality are more clearly perceived from the extra deaths per 1000 than from corresponding mortality ratios; this often is the case during the period immediately following surgery for cancer. It is important to keep in mind that mortality ratios generally decrease with age, so that the mortality ratios for all ages combined can be materially affected by the age composition of a population.

Proportions surviving a specified period of time, even though they provide absolute measures of longevity in a population, have rarely been used in medico-actuarial investigations, because they are relatively insensitive to appreciable changes in mortality; small differences in proportions surviving may be difficult to assess. Relative proportions surviving have been used occasionally in medico-actuarial studies where very high mortality rates occur, as among cancer patients.

In all mortality comparisons, but particularly in comparisons of mortality ratios and relative proportions surviving, it is the suitability of the basis for calculating expected deaths that makes the figures meaningful. If such a basis is regarded as a fixed yardstick, then the reliability of comparisons based on small numbers of deaths can be tested by determining whether an observed deviation from this basis is or is not significant in probability terms; if it is significant, then what are the limits in probability terms within which the "true" value of the observed deviation can be expected to lie [14]?

In medico-actuarial mortality investigations the numbers of deaths in most classifications have usually been quite large. The mortality ratios shown for such classifications have therefore been taken as reasonably reliable estimates of the "true" values of the mortality ratios in the underlying population. As a rule of thumb, when the number of policies terminated by death was 35 or greater and some doubt attached to the significance of the mortality ratio, confidence limits were calculated at the 95% confidence level on the assumption of a normal distribution; when the number of policies terminated by death was less than 35, confidence limits have been calculated on the assumption of a Poisson distribution.

INTERPRETING THE FINDINGS

Medico-actuarial investigations have been based on the experience among men and women insured under ordinary life insurance policies. These insured lives have been drawn predominantly from the middle-class and better-off segments of the population and have passed the screening for life insurance which results in the rejection of about 2% of all applicants and the charging of extra premiums to about 6% of all applicants. Initially at least more than 9 out of 10 persons are accepted for life insurance at standard premium rates and those issued insurance at standard premium rates are in ostensibly good health. In recent years the death rates of men aged 25 or older insured under standard ordinary policies have ranged from 25 to 35% of the corresponding population death rates in the first two policy years, from 40 to 50% of the corresponding population death rates at policy year durations 3–5, and from 55 to 75% of the corresponding population death rates after 15 or more years have elapsed since issue of insurance. The corresponding figures for women insured under standard ordinary policies have been similar to those of male insured lives at ages over 50, but were closer to population death rates at the younger ages.

Inasmuch as the underwriting rules determine which applicants are accepted for standard insurance and which for insurance at extra premium rates, the mortality experience in medico-actuarial investigations has occasionally been affected to a significant degree by changes in underwriting practices to more lenient or stricter criteria.

The mortality findings in medico-actuarial investigations relate to the status of individual at time of issue of the insurance. The experience therefore reflects not only the effects of some individuals becoming poorer risks with the passage of time, but also of some individuals becoming better risks (e.g., leaving employment in a hazardous occupation or benefiting from medical or surgical treatment) and withdrawing from the experience. Where the effects of employment in hazardous occupations or of certain physical impairments are deferred, it is essential that the study cover a sufficiently long period of time for the deferred mortality to become manifest. This is particularly important in the case of overweight and hypertension [13].

The results of medico-actuarial investigations have been relatively free from bias arising from failure to trace the experience among those withdrawing. Considerable evidence has been accumulated to show that insured lives who cease paying premiums and thus automatically remove themselves from observation are as a group subject to somewhat lower mortality [12].

It should also be kept in mind that the mortality ratios shown in medico-actuarial studies were computed on the basis of the number of policies (or amounts of insurance) and not on the basis of lives. In classifications involving small numbers of policies terminated by death it has been necessary to look into the data to determine whether the results had been affected by the death of a single individual with several policies (or with large amounts of insurance). This has usually been noted in the descriptive text.

The data in medico-actuarial investigations are very accurate with respect to reported ages and remarkably complete in the follow-up*. The information obtained on applications for life insurance with respect to past medical histories requires some qualification. The great majority of applicants for life insurance admit some physical impairment or medical history; if the impairment or history appears to be significant, attention is focused on it in the medical examination for insurance and a statement from the attending physician may be obtained. The medical examination on modest amounts of life insurance is not as comprehensive as a diagnostic examination in clinical practice, where the physician is in position to study a patient for a longer period of time, more intensively, and with the patient's full cooperation. Applicants for insurance not infrequently forget or try to conceal unfavorable aspects of their personal or family medical histories, particularly with respect to questionable habits. Even when reasonably complete details are elicited, there are usually practical limits on the extent to which it is feasible to check up on indefinite statements and vague diagnoses reported on applications for life insurance. Only on applications for large amounts of

insurance would two or more medical examinations by different physicians be called for and intensive effort made to clarify obscure findings. Broadly speaking, the medical findings on life insurance medical examinations stand up very well, but the medical impairments studied in medico-actuarial mortality investigations often represent less differentiated conditions which cannot be characterized as precisely as is sometimes possible in clinical studies [13].

On the other hand, it has proved feasible on applications for life insurance to obtain fuller details of occupation and avocation (even approximate exposure to occupational hazards) than has been possible in many epidemiological* studies.

FINDINGS OF MEDICO-ACTUARIAL INVESTIGATIONS

The *Medico-Actuarial Mortality Investigation* of 1912 covered the period from 1885 to 1909 [2]. It produced tables of average weights for men and women by age and height which remained in general use as a weight standard until 1960. The mortality experienced according to variations in build indicated some extra mortality among underweights at ages under 35, associated with materially greater risk of tuberculosis and pneumonia, but at ages 35 and older the lowest mortality was found among those 5 to 10 pounds underweight. Overweight was found to be associated with increased death rates from heart disease, diabetes, and cerebral hemorrhage. The investigation also included 76 groups of medical impairments, 68 occupations, four categories of women studied according to marital status, and insured blacks and North American Indians.

The *Occupational Study 1926* dealt with some 200 occupations or groups of occupations, separately for those where occupational accidents were the dominant element of extra risk and those where nonoccupational accidents, pneumonia, cirrhosis of the liver, cancer, or other causes, were suspect as responsible for the extra risk [3].

The *Medical Impairment Study 1929* [4], which covered the period from 1909 to 1928, and its 1931 Supplement[11] broadly confirmed the findings of the Medico-Actuarial Investigation as to average weights by age and height at ages 25 and older and the effects on mortality of departures from average weight. The study centered on 122 groups of medical impairments, including a number of combinations of two impairments treated as a single element of risk. The more important findings related to the extra mortality on heart murmurs, elevated blood pressure, and albumin and sugar in the urine. The findings on elevated blood pressure indicated clearly that systolic blood pressures in excess of 140 mm were associated with significant extra mortality, which at the time was contrary to medical opinion [10].

Smaller investigations of the mortality among insured lives with various medical impairments followed, published under the titles *Impairment Study 1936* [6] and *Impairment Study 1938* [19]. Together they included 42 groups of medical impairments, among them persons with a history of cancer, gastric and duodenal ulcers, gall bladder disease, and kidney stone, including surgery for these conditions.

In 1938 an extensive investigation was also undertaken of the mortality according to variations in systolic and diastolic pressure. This study, known as the *Blood Pressure Study 1938*, covered the period from 1925 to 1938 [8,9]. It confirmed earlier findings that diastolic pressures in excess of 90 mm as well as systolic blood pressures in excess of 140 mm were associated with at least 25% extra mortality, and it brought out clearly that various minor impairments accompanying elevated blood pressure, notably overweight, increased the risk appreciably.

The *Occupational Study 1937* covered numerous occupations over the period 1925 to 1936 [7]. It developed the extra mortality among those employed in the manufacturing, distribution, and serving of alcoholic beverages. It also indicated some decline since the early 1920s in the accidental death rates in many occupations.

The *Impairment Study 1951*, which covered the period 1935 to 1950, reviewed the mortality experience for 132 medical impairment classifications on policies issued during the years 1935 through 1949 [16]. It showed lower extra mortality than that found in the

Medical Impairment Study 1929 for medical impairments due to infections, for conditions treated surgically, for diseases of the respiratory system, and for some women's diseases. Because of the inclusion in the study of smaller groups of lives with specific impairments, greater use was made of confidence intervals based on the Poisson distribution.

The *Build and Blood Pressure Study 1959* covered the experience on about 4,500,000 policies over the period 1935 to 1953 [17]. It focused on changes in the mortality experienced among underweights and overweights, elevated blood pressures, and combinations of overweight and hypertension with other impairments. New tables of average weights for men and women by age and height were developed, which showed that men had gained weight while women had reduced their average weights since the 1920s. Moderate underweights showed very favorable mortality, while marked overweights recorded somewhat higher relative mortality. The mortality on slight, moderate, and marked elevations in blood pressure registered distinctly higher mortality than found in earlier investigations.

The *Occupational Study 1967* covered the period from 1954 to 1964 and was limited to occupations believed to involve some extra mortality risks [18]. Only the following occupations, on which there was substantial experience, recorded a crude death rate in excess of 1.5 per 1000:

Lumberjacks
Mining operators
Explosive workers
Construction crane workers
Shipbuilding operators
Structural iron workers
Railroad trainmen and switchmen
Taxi drivers
Marine officers and crew
Guards and watchmen
Marshals and detectives
Sanitation workers
Porters
Elevator operators
Persons selling, delivering, or serving alcoholic beverage

The mortality in most occupations decreased from that reported in the *Occupational Study 1937*. Significant reductions occurred among mining officials and foremen, workers in metal industry, telecommunication linemen, longshoremen, firefighters, police officers, window cleaners, hotelkeepers, saloonkeepers and bartenders, and most laborers. Relative mortality increased for lumberjacks, railroad trainmen and switchmen, truck drivers, marine crew and guards, and watchmen.

The *Build Study 1979* [21] and the *Blood Pressure Study 1979* [22] each covered about 4,250,000 policies over the period 1954 to 1971. They showed that the average weights for men had continued to increase, as did the average weights for women under 30; women 30 and older registered decreases in average weights as compared with the *Build and Blood Pressure Study 1959*. The excess mortality among overweights was found to be substantially the same as in the earlier study, but somewhat higher mortality was recorded among moderate overweights. Nevertheless, the optimum weights (those associated with the lowest mortality) were again found to be in the range of 5 to 10% below average weight, even though the average weights for men had increased significantly. The excess mortality on elevated blood pressures was found to be distinctly lower than in the *Build and Blood Pressure Study 1959*. A cohort of 24,000 men who had been treated for hypertension exhibited virtually normal mortality among those whose blood pressures had been reduced to below 140 systolic and 90 diastolic after treatment. The study adduced the most convincing evidence thus far available that recent treatment for hypertension was highly effective for many years. In progress at this time is another medico-actuarial investigation of the mortality among insured lives with a wide variety of medical impairments, covering the period from 1955 through 1974.

REFERENCES

References 1, 3 to 10, and 16 to 21 are original reports.

1. Actuarial Society of America (1903). *Specialized Mortality Investigation*. New York.

2. Actuarial Society of America and Association of Life Insurance Medical Directors of America (1912–1914). *Medico-Actuarial Mortality Investigation*, 5 vols. New York. (Original reports; some of these cover basic design of studies.)

3. Actuarial Society of America and Association of Life Insurance Medical Directors of America (1926). *Occupational Study 1926*. New York.

4. Actuarial Society of America and Association of Life Insurance Medical Directors of America (1929). *Medical Impairment Study 1929*. New York.

5. Actuarial Society of America and Association of Life Insurance Medical Directors of America (1932). *Supplement to Medical Impairment Study 1929*. New York.

6. Actuarial Society of America and Association of Life Insurance Medical Directors of America (1936). *Impairment Study 1936*. New York.

7. Actuarial Society of America and Association of Life Insurance Medical Directors of America (1937). *Occupational Study 1937*. New York.

8. Actuarial Society of America and Association of Life Insurance Medical Directors of America (1939). *Blood Pressure Study 1939*. New York.

9. Actuarial Society of America and Association of Life Insurance Medical Directors of America (1940). *Supplement to Blood Pressure Study 1939*. New York.

10. Association of Life Insurance Medical Directors of America and the Actuarial Society of America (1925). *Blood Pressure Study*. New York.

11. Batten, R. W. (1978). *Mortality Table Construction*. Prentice-Hall, Englewood Cliffs, N.J.

12. Benjamin, B. and Haycocks, H. W. (1970). *The Analysis of Mortality and Other Actuarial Statistics*. Cambridge University Press, London.

13. Lew, E. A. (1954). *Amer. J. Publ. Health*, **44**, 641–654. (Practical considerations in interpretation of studies.)

14. Lew, E. A. (1976). In *Medical Risks*. R. B. Singer and L. Levinson, eds. Lexington Books, Lexington, Mass., Chap. 3. (Limitations of studies.)

15. Lew, E. A. (1977). *J. Inst. Actuaries*, **104**, 221–226. (A history of medico-actuarial studies.)

16. Society of Actuaries (1954). *Impairment Study 1951*. New York. (A highly readable original report.)

17. Society of Actuaries (1960). *Build and Blood Pressure Study 1959*. Chicago. (A highly readable original report.)

18. Society of Actuaries (1967). *Occupational Study 1967*. New York.

19. Society of Actuaries and Association of Life Insurance Medical Directors of America (1938). *Impairment Study 1938*. New York.

20. Society of Actuaries and Association of Life Insurance Medical Directors of America

(1980). *Blood Pressure Study 1979*. Chicago. (Very readable.)

21. Society of Actuaries and Association of Life Insurance Medical Directors of America (1980). *Build Study 1979*. Chicago. (Highly readable.)

See also Actuarial Science; Clinical Trials—II; Epidemiological Statistics—I; Follow-Up; Life Tables; Rates, Standardized; and Vital Statistics.

Edward A. Lew

ACTUARIAL SCIENCE

Actuarial science is an applied mathematical and statistical discipline in which data-driven models are constructed to quantify and manage financial risk. The term "actuarial statistics" is not in common use because a well-defined set of statistical techniques useful to actuaries has not been established. This topic could also be viewed as a discussion of the types of data (mortality, morbidity*, accident frequency, and severity) collected by actuaries, but that will not be the focus here. This entry will concentrate on two particular statistical endeavors in which actuaries have played a major role—construction of mortality tables and credibility theory.

CONSTRUCTION OF MORTALITY TABLES

From the 1600s, governments sold annuities based on individual's lifetimes. To be useful as a fund-raising mechanism, the cost of the annuity needed to be greater than the expected cost of the benefit. Although not the first mortality table (or life table*), the work of Halley [10] combined the construction of a mortality table with the concept of expected present value. From the life table, for a person of current age x, it is possible to get the probability distribution of the number of years remaining, that is,

$$_k|q_x = \text{Pr(death is between ages } x + k$$

$$\text{and } x + k + 1), \ k = 0, 1, \ldots.$$

In addition, if the annuity is to pay one monetary unit at the end of each year, provided the annuitant is alive, the expected present value is

$$a_x = \sum_{k=1}^{\infty} {}_k|q_x(v + v^2 + \cdots + v^k)$$

where $v = 1/(1 + i)$ and i is the rate of interest.

A few years later, de Moivre* [19] introduced an approximation based on linear interpolation* between values in the life table (his table did not have survival probabilities at each integral age). This approximation continues to be used today and is referred to as the *uniform distribution of deaths* assumption [4], Chap. 3.

Life tables for actuarial use were constructed on an ad-hoc basis until the middle of the twentieth century when the so-called "actuarial method" was developed. It is loosely based on an assumption put forth by Balducci [2], viz.,

Pr(a person age $x + t$ dies before age $x + 1$)

$$= (1 - t) \text{Pr(a person age } x$$

dies before age $x + 1$),

$0 < t < 1$ (*see* Life Tables, Balducci Hypothesis). The result is an exposure-based formula that estimates the key life-table quantity as

$q_x = $ Pr(a person age x dies before age $x + 1$)

$ = $ number of observed deaths / exposure.

For a life observed between ages x and $x + 1$, the exposure contribution is the portion of the year the life was observed, except for deaths, for which the exposure is the time from first observation to age $x + 1$.

This estimator is inconsistent [5]. However, it has one quality that made it extremely valuable. Given the types of records commonly kept by insurance companies, this formula was easy to implement by hand, or using mainframe computers prevalent through the 1980s. A good exposition of the actuarial method and its practical applications is Reference 3. Since then, actuaries have used the more accurate Kaplan-Meier* [13] and maximum likelihood estimation* procedures. These concepts are introduced in an actuarial setting in Reference 6.

Once the values of q_x have been obtained, a second actuarial contribution has been the smoothing of these values to conform with the a priori notion that from about age five onward the values should be smoothly increasing. The process of smoothing mortality rate estimates is called graduation*. An introduction to all of the commonly used methods is given in Reference 17. Two of the more commonly used methods, interpolation*, and Whittaker, will be discussed here. Both methods create the graduated rates as a linear combination of surrounding values.

The interpolation method requires that the observations be grouped in a manner that creates estimates of q_x at every k (often 5) years of age. This is done by first aggregating the deaths (say, d_x) and exposures (say, e_x) at the surrounding ages, to create, for example, with $k = 5$,

$$d_x^* = d_{x-2} + d_{x-1} + d_x + d_{x+1} + d_{x+2},$$

$$e_x^* = e_{x-2} + e_{x-1} + e_x + e_{x+1} + e_{x+2}.$$

Because these series are often convex, an improved aggregated value can be found from King's pivotal point formula [14]:

$$d_x^{**} = -0.008d_{x-5}^* + 0.216d_x^* - 0.008d_{x+5}^*$$

$$e_x^{**} = -0.008e_{x-5}^* + 0.216e_x^* - 0.008e_{x+5}^*.$$

Finally, the mortality rate at age x is given by $q_x^{**} = d_x^{**}/u_x^{**}$.

The most commonly used interpolation formula is the Karup-King formula

$$q_{x+j} = sq_{x+5}^{**} + 0.5s^2(s-1)\delta^2 q_{x+5}^{**}$$
$$+(1-s)q_x^{**} + 0.5(1-s)^2(-s)\delta^2 q_x^{**},$$
$$j = 0, 1, 2, 3, 4, 5,$$

where $s = j/5$ and $\delta^2 q_x^{**} = q_{x+5}^{**} - 2q_x^{**} + q_{x-5}^{**}$ is the second central difference (see FINITE DIFFERENCES, CALCULUS OF). This formula uses four mortality rates and has the property that if those rates lie on a quadratic curve, the interpolated values will reproduce that curve. In addition, the cubic curves that connect consecutive mortality rates will have identical first derivatives where they meet.

Another popular formula is due to Jenkins [12] (see Eq. 7 in the entry GRADUATION). It requires fourth central differences and thus involves six points. It reproduces third-degree polynomials and adjacent curves will have identical first and second derivatives. To achieve these goals, the formula does not match the original mortality rates. That is, $q_{x+0} \neq q_x^{**}$.

The Whittaker method [22] can be derived by a Bayesian argument or from arguments similar to those used in creating smoothing splines (see SPLINE FUNCTIONS). Let q_x, $x = 0, \ldots, n$, be the original estimates; let v_x, $x = 0, \ldots, n$, be the graduated values; and let w_x, $x = 0, \ldots, n$, be a series of weights. Then, the graduated values are those that minimize the expression

$$\sum_{x=0}^{n} w_x(v_x - q_x)^2 + h \sum_{x=0}^{n-z} (\Delta^z v_x)^2$$

(see GRADUATION, WHITTAKER–HENDERSON). The weights are often chosen as either the exposure (sample size) at each age or the exposure divided by $q_x(1 - q_x)$, which would approximate using the reciprocal of the variance as the weight. The value of z controls the type of smoothing to be effected. For example, $z = 3$ leads to graduated values that tend to follow a quadratic curve. The choice of h controls the balance between fit (having the graduated values be close to the original values) and smoothing (having the graduated values follow a polynomial).

CREDIBILITY

Credibility theory is used by actuaries in the setting of premiums based on prior or corollary information. Two common situations are experience rating and classification ratemaking. An example of the former is workers compensation insurance. Suppose a particular employer had been charged a standard rate on the basis of expecting $\$x$ of claim payments per thousand dollars of payroll. In the previous year, the employer had claims of $\$y$ per thousand dollars of payroll, where $y < x$. The employer believes that a reduction in premium is warranted, while the insurer may claim that the result was simply good fortune.

A credibility procedure will base the next premium on the value $zy + (1 - z)x$, where $0 \leqslant z \leqslant 1$ and z is called the *credibility factor*. The magnitude of z is likely to depend on the sample size that produced y, the variance of y, and, perhaps, some measure of the accuracy of x.

With regard to classification ratemaking, consider setting premiums for automobile insurance. Separate rates may be needed for various combinations of gender, age, location, and accident history. Let y be an estimate based on the data for a particular combination of factors and let x be an estimate based on all the data. Because some combinations may occur infrequently, the reliability of y may be low. A credibility estimate using $zy + (1 - z)x$ may be more accurate (though biased). Credibility analysis succeeds for just that reason. By applying the factor $1 - z$ to an estimator that is more stable, the reduction in variance may offset the effect of bias, producing a smaller mean square error.

Two approaches to credibility have evolved. One, usually credited to Mowbray [20], has been termed *limited fluctuation credibility*. The question reduces to determining the sample size needed so that the relative error when estimating the mean will be less than $k\%$ with probability at least $p\%$. A normal or Poisson approximation along with a variance estimate is usually sufficient to produce the answer. If the sample size exceeds this number, then $z = 1$ (full credibility) is used. If not, z is customarily set equal to the square root of the ratio of the actual sample size to that needed for full credibility. Assuming no error in the quantity being multiplied by $1 - z$, the effect is to reduce the variance to equal that which would have been obtained with the full credibility sample size. The simplicity of this method causes it to remain popular. Its drawback is that it does not allow for the increased bias as z decreases, nor does it allow for any error in the quantity being multiplied by $1 - z$.

The second method has been termed *greatest accuracy credibility* and bears a strong resemblance to Bayesian analysis. It was introduced by Whitney [21] with a more thorough derivation produced by Bailey [1] and a modern derivation by Bühlmann [7]. This approach begins by assuming that a sample of size n is obtained from an individual. The observations are independent realizations of the random variable X with a distribution function that depends on the vector parameter θ. Define

$$E(X|\theta) = \mu(\theta) \text{ and } \mathrm{Var}(X|\theta) = v(\theta).$$

Further, assume that θ is unknown, but has been drawn at random from a random variable Θ with distribution function $F_\Theta(\theta)$. Finally, assume that $\mu(\theta)$ is to be estimated by a linear function of the observations, that is,

$$\widehat{\mu(\theta)} = \alpha_0 + \alpha_1 X_1 + \cdots + \alpha_n X_n.$$

The objective is to minimize

$$E_{\Theta, X_1, \dots, X_n} \left\{ \left[\widehat{\mu(\Theta)} - \mu(\Theta) \right]^2 \right\}.$$

That is, the squared error should be minimized both over all possible observations and all possible parameter values. For a particular insured with a particular value of θ, the squared error may be larger than if the sample mean were used, but for others it will be smaller so that the overall error is reduced.

The solution is

$$\widehat{\mu(\theta)} = z\bar{x} + (1 - z)\mu; \;\; \mu = E[\mu(\Theta)],$$

$$z = \frac{n}{n + k}, \; k = \frac{E[v(\Theta)]}{\mathrm{Var}[\mu(\Theta)]}.$$

It turns out to be the Bayesian (posterior mean) solution for certain common cases such as normal-normal and Poisson-gamma.

In practice, the indicated quantities must usually be estimated. An approach given in Bühlmann and Straub [8] provides an empirical Bayes* estimate, derived by a method of moments* approach. This is not unreasonable, because the distribution of Θ is not an *a priori* opinion, but rather a real, if unobservable, distribution of how characteristics vary from policyholder to policyholder or group to group. With data on several policyholders or groups, it is possible to estimate the needed moments. A true Bayesian model can be constructed by placing a prior distribution on the parameters of the distribution of Θ.

This is done for the normal-normal model in Reference 15.

Textbooks that develop these credibility topics and more (all include an English language version of the Bühlmann-Straub formula) include references 9, 11, 16, Chap. 5; and 18. A comprehensive list of book and article abstracts through 1982 is found in reference 23.

REFERENCES

1. Bailey, A. (1950). Credibility procedures. *Proc. Casualty Actuarial Soc.*, **37**, 7–23, 94–115.

2. Balducci, G. (1921). Correspondence. *J. Inst. Actuaries*, **52**, 184.

3. Batten, R. (1978). *Mortality Table Construction*. Prentice Hall, Englewood Cliffs, N.J.

4. Bowers, N., Gerber H., Hickman, J., Jones, D., and Nesbitt, C. (1997). *Actuarial Mathematics*, 2nd ed. Society of Actuaries, Schaumburg, Ill.

5. Breslow, N. and Crowley, J. (1974). A large sample study of the life table and product limit estimates under random censorship. *Ann. Stat.*, **2**, 437–453.

6. Broffitt, J. (1984). Maximum likelihood alternatives to actuarial estimators of mortality rates. *Trans. Soc. Actuaries*, **36**, 77–142.

7. Bühlmann, H. (1967). Experience rating and credibility. *ASTIN Bull.*, **4**, 199–207.

8. Bühlmann, H. and Straub, E. (1970). Glaubwürdigkeit für Schadensätze (credibility for loss ratios). *Mitteilungen der Vereinigung Schweizerisher Versicherungs-Mathematiker*, **70**, 111–133. (English translation (1972) in *Actuarial Research Clearing House*).

9. Dannenburg, D., Kass, R. and Goovaerts, M. (1996). *Practical Actuarial Credibility Models*. Ceuterick, Leuven, Belgium.

10. Halley, E. (1694). An estimate of the degrees of the mortality of mankind, drawn from curious tables of births and funerals at the city of Breslau; with an attempt to ascertain the price of annuities on lives. *Philos. Trans.*, **17**, 596–610.

11. Herzog, T. (1996). *Introduction to Credibility Theory*. Actex, Winsted, Conn.

12. Jenkins, W. (1927). Graduation based on a modification of osculatory interpolation. *Trans. Am. Soc. Actuaries*, **28**, 198–215.

13. Kaplan, E. and Meier, P. (1958). Nonparametric estimation from incomplete observations. *J. Am. Stat. Assoc.*, **53**, 457–481.

14. King, G. (1887). Discussion: the graphic method of adjusting mortality tables (by T. Sprague). *J. Inst. Actuaries*, **26**, 114.

15. Klugman, S. (1987). Credibility for classification ratemaking via the hierarchical linear model. *Proc. Casualty Actuarial Soc.*, **74**, 272–321.

16. Klugman, S., Panjer, H. and Willmot, G. (1998). *Loss Models: From Data to Decisions*. Wiley, New York.

17. London, D. (1985). *Graduation: The Revision of Estimates*. Actex, Winsted, Conn.

18. Mahler, H. and Dean, C. (2001). *Credibility*. In *Foundations of Casualty Actuarial Science*, 4th ed. Casualty Actuarial Society, Arlington, Va.

19. de Moivre, A. (1725). *Annuities Upon Lives*. Fayram, Motte and Pearson, London.

20. Mowbray, A. (1914). How extensive a payroll exposure is necessary to give a dependable pure premium? *Proc. Casualty Actuarial Soc.*, **1**, 24–30.

21. Whitney, A. (1918). The theory of experience rating. *Proc. Casualty Actuarial Soc.*, **4**, 274–292.

22. Whittaker, E. and Robinson, G. (1924). *The Calculus of Observations*. Blackie and Sons, London.

23. de Wit, G., ed. (1986). Special issue on credibility theory. *Insurance Abstr. Rev.*, **2**(3).

See also ACTUARIAL HEALTH STUDIES; DEMOGRAPHY; GRADUATION; GRADUATION, WHITTAKER–HENDERSON; LIFE TABLES; LIFE TABLES, BALDUCCI HYPOTHESIS; MORBIDITY; MULTIPLE DECREMENT TABLES; POPULATION, MATHEMATICAL THEORY OF; POPULATION PROJECTION; RATES, STANDARDIZED; and VITAL STATISTICS.

STUART KLUGMAN

ACTUARIAL STATISTICS.   See ACTUARIAL SCIENCE

ADAPTIVE IMPORTANCE SAMPLING (AIS).   See IMPORTANCE SAMPLING

ADAPTIVE METHODS

In adaptive statistical inference, we use the sample to help us select the appropriate type of statistical procedure needed for the situation under consideration. For a simple illustration of this, say that we use the sample kurtosis* K as a selector statistic [3]. One adaptive point estimator, T, for the center of a distribution would be

$$T = \begin{cases} \text{midrange}^*, & K \leqslant 2, \\ \text{arithmetic mean}^*, & 2 < K < 5, \\ \text{median}^*, & 5 \leqslant K. \end{cases}$$

That is, if the sample looks as if it arises from a short-tailed distribution, the average of the largest and smallest items of the sample is used as our estimator. If it looks like a long-tailed situation, the median is used. Otherwise, our estimate is the arithmetic mean (average) $\bar{x}$.

To generalize this illustration somewhat, suppose that we have a whole family (not necessarily finite) of possible distributions. Within this family of distributions, take a few representative ones, say $F_1, F_2, \ldots, F_k$. Now, for each of these k distributions, suppose that we can find a good statistic to make the inference under consideration. Let us say that these respective statistics are $T_1, T_2, \ldots, T_k$. We observe the sample from a distribution; and with a selector statistic, say Q, we determine which one of $F_1, F_2, \ldots, F_k$ seems closest to the underlying distribution from which the sample arises. If Q suggests that we have been sampling from F_i, then we would use the statistic T_i; or if Q suggests that we might be someplace between F_i and F_j, then we could use a combination of T_i and T_j; or more generally, Q could dictate a statistic that is a linear combination of all the statistics, $T_1, T_2, \ldots, T_k$: let us say

$$T = \sum_{i=1}^{k} W_i T_i, \quad \sum_{i=1}^{k} W_i = 1,$$

where the weights $W_1, W_2, \ldots, W_k$ are functions of the statistic Q. If it looks more like the sample arises from F_i, then, of course, the weight W_i would be large.

Consider a very simple example in which we are trying to choose the best of three types of concrete [10]. The compression strengths were tested after bars of each type had been exposed to severe environmental conditions

Table 1.

Concrete	A	B	C
Ordered	5060	5625	4880
Observations	5398	6020	6030
	5820	6270	6290
	6131	6636	6372
	6400	6880	6920
	7527	7337	8320
	7560	8170	8581
Midrange	6310.0	6897.5	6730.5
Mean	6270.86	6705.43	6770.43
Modified median	6122.00	6609.86	6471.86

for a period of 1 year. Seven $(n = 7)$ observations were taken for each type of cement, where the observations are the breaking strengths of the bars measured in pounds per square inch. Let us denote the order statistics* of a sample by $y_1 \leqslant y_2 \leqslant \cdots \leqslant y_7$. However, since we do not know from what underlying distribution these arose, we choose three representative distributions: the short-tailed uniform* using the midrange $(y_1 + y_7)/2$ as an estimate of center, the normal* using the average $\bar{x}$ as the estimate, and the long-tailed double exponential* with a modified median $(3y_3 + 8y_4 + 3y_5)/14$ as the statistic. These statistics were computed for each of the three samples and are given in Table 1 together with the original data.

It is interesting to note that using the midrange or median statistics, concrete B looks to be the best, whereas $\bar{x}$ suggests concrete C. Accordingly, a selector statistic is needed, and we use

$$Q = \frac{(y_7 - y_1)/2}{\sum |y_i - M|/7},$$

the ratio of one-half of the range divided by the mean deviation* from the sample median M. (Q is defined somewhat differently when $n > 20$.) The average of the three Q values is computed to obtain $\overline{Q} = 1.876$. The midrange, average, or median is selected respectively, according to whether $\overline{Q}$ falls below, in between, or above

$$2.08 - (2/n) \quad \text{and} \quad 2.96 - (5.5/n);$$

that is, with $n = 7$, 1.794 and 2.174. (The formulas for these cutoffs have been determined empirically.) Since $\overline{Q} = 1.876$, it seems as if the distribution has fairly normal tails; thus the statistic $\overline{x}$ chooses concrete C as the best.

We must understand, however, that the inference under consideration is not necessarily a point estimate in the general situation. We could be considering a confidence interval* or a test of hypothesis*. Moreover, making an inference in this manner, that is, selecting the underlying distribution and then making the inference from the same data, can certainly destroy certain probabilities that are of interest in statistics. For example, if we are constructing a nominal 95% confidence interval, we can actually spoil the confidence coefficient* by such a procedure, so that it might actually be 0.80 or even 0.70. Or if we are making a test of a statistical hypothesis, the significant level might not be $\alpha = 0.05$, but 0.15 or 0.25. Despite this fact, however, the adaptive idea is useful in good data analysis; therefore, it is necessary for us to adjust our theories to the applications. That is, we want our theories to support the applications, not oppose them.

This forces us to look at some of the difficulties associated with the corresponding sampling distribution theory. Let us say that θ is the location parameter* and we are interested in testing the hypothesis $H_0 : \theta = \theta_0$ against the hypothesis $H_1 : \theta > \theta_0$. Again, suppose that we have a family of distributions for which θ is the location parameter of each member of the family. If we are sampling from F_i, say, we would reject the hypothesis H_0 and accept the alternative hypothesis H_1 if some statistic, say Z_i, was greater than or equal to c_i; i.e., $Z_i \geqslant c_i$, $i = 1, 2, \ldots, k$. Therefore, our adaptive test might be something like this: reject H_0 and accept H_1 if

$$Z = \sum_{i=1}^{k} W_i Z_i \geqslant c,$$

where $W_1, W_2, \ldots, W_k$ are functions of some selector statistic, say Q. The significance level of the test is then

$$\Pr\left[\sum_{i=1}^{k} W_i Z_i \geqslant c | H_0 \right].$$

This probability is difficult to compute, so let us first consider a special and easier situation

in which each of the W's is equal to 0 or 1. Of course, only one W_i can equal 1, and the rest must equal 0. Thus if Q suggests that F_i is the underlying distribution, then we will use Z_i. That is, if $Q \in R_i$, where $R_1, R_2, \ldots, R_k$ are appropriate mutually exclusive* and exhaustive* sets, we will select Z_i for the test statistic. Under these conditions, the significance level would be

$$\sum_{i=1}^{k} \Pr[Q \in R_i \quad \text{and} \quad Z_i \geqslant c_i | H_0].$$

If each of the individual tests is made at the 0.05 significance level, it has been observed in practice that this significance level is frequently somewhat larger than that nominal significance level of 0.05.

There is a certain desirable element of model building in this entire procedure; that is, we observe the data and select the model that seems appropriate, and then we make the statistical inference* for the situation under consideration. However, there can be some cheating in doing this; that is, if we construct the model from given data and then make a test of hypothesis using those data, our nominal significance level is not necessarily the correct one. Moreover, even some researchers carry this to an extreme by selecting the test procedure that favors what they want (usually rejection of the null hypothesis*). They might then quote a significance level of 0.05, while the real α, for the overall selection and testing procedure might be higher than 0.25.

There is a method, however, of "legalizing" this cheating. Suppose that the selector statistic Q and each Z_i are independent under the null hypothesis H_0. Then the significance level is

$$\sum_{i=1}^{k} \Pr[Q \in R_i \quad \text{and} \quad Z_i \geqslant c_i | H_0]$$

$$= \sum_{i=1}^{k} \Pr[Q \in R_i | H_0] \Pr[Z_i \geqslant c_i | H_0]$$

$$= \alpha \sum_{i=1}^{k} \Pr[Q \in R_i | H_0] = \alpha,$$

provided that each individual test is made at the nominal significance level α. That is, this common significance level α is exactly the same as the overall significance level. The important feature of this is to make certain that the selector statistic Q is independent of the test statistic. One elegant way of achieving this is through distribution-free (nonparametric) methods* [5].

To illustrate the beauty of the nonparametric methods in these situations, let us consider the two-sample problem. Suppose that we have two independent continuous-type distributions, F and G. The null hypothesis H_0 is the equality of the two corresponding functions. Say that the sample $X_1, X_2, \ldots, X_m$ arises from F, and the sample $Y_1, Y_2, \ldots, Y_n$ arises from G. We suggest three nonparametric statistics that can be used to test this null hypothesis [2]. The first, Tukey's quick test*, is used when the underlying distributions have short tails, like those of the uniform distribution*. Tukey's statistic is

$$T_1 = (\#Y's > \text{ largest } X)$$
$$+ (\#X's < \text{ smallest } Y).$$

A large T_1 would suggest the alternative hypothesis H_1 that the Y's tend to be larger than the X's. Thus we reject H_0 and accept H_1 if T_1 is greater than or equal to c_1, where

$$\Pr[T_1 \geqslant c_1|H_0] = \alpha.$$

The second statistic T_2 is that of Mann, Whitney, and Wilcoxon. This statistic is a good one in case the underlying distributions have middle tails, like those of the normal* or logistic* distributions. After combining the two samples, we determine the ranks of the Y's in the combined sample; say those ranks are $R_1, R_2, \ldots, R_n$. One form of the Mann-Whitney-Wilcoxon statistic* is

$$T_2 = \sum_{i=1}^{n} R_i.$$

Now we reject H_0 and accept H_1 if T_2 is greater than or equal to c_2, where

$$\Pr[T_2 \geqslant c_2|H_0] = \alpha.$$

The third statistic is that associated with the median test. It is $T_3 = \#Y's$ greater than the combined sample median. We reject H_0 if that statistic, T_3, is greater than or equal to c_3, where

$$\Pr[T_3 \geqslant c_3|H_0] = \alpha.$$

Each of the probabilities denoted by α in these three tests does not depend on the form of the underlying continuous distribution, and sometimes these tests are called distribution-free tests.

For an example of each of these statistics, refer to data on the three types of concrete, and let the samples from A and B represent, respectively, the X and Y values with $m = n = 7$. The computed statistics are $T_1 = 3$, $T_2 = 59$, with $T_3 = 4$. Let us now consider an adaptive procedure that selects one of these three statistics. Considering the combined sample (i.e., the X's and Y's together) use a selector statistic, say Q, and decide whether we have short-tailed distributions, in which case we use the T_1 test; middle-tailed distributions, in which case we use the T_2 test; or long-tailed distributions, in which case we use the T_3 test. It turns out that the overall (selecting and testing) significance level will also equal α because each of T_1, T_2, and T_3 is independent of Q. The reason we have this independence under H_0 is that the order statistics of the combined sample are complete, sufficient statistics for the underlying "parameter," the common distribution $F = G$. Moreover, it is well known that the complete, sufficient statistics for $F = G$ are then independent of statistics whose distributions do not depend upon $F = G$, such as T_1, T_2, and T_3. However, the selector statistic Q is a function of the complete, sufficient statistics, and thus it is also independent of each of the statistics T_1, T_2, and T_3, under H_0. Incidentally, in our example, using the Q and $\overline{Q}$ associated with the illustration about concrete, the statistics T_2 for middle-tailed distributions would be selected and $T_2 = 59$ has a p-value of 0.288. Thus the null hypothesis would not be rejected at the significance level of $\alpha = 0.05$.

Although these nonparametric methods can be generalized to multivariate situations such as regression*, many statisticians do not find them extremely satisfactory in data analysis. Possibly the newer robust statistics show more promise in adaptation; some of them are "almost distribution-free" and lend themselves better to data analysis. Although it is impossible to give many details on robustness in this short article, the idea is illustrated with the trimmed mean.

Suppose that we are attempting to make an inference about the center θ of a symmetric distribution. Let $X_1 \leqslant X_2 \leqslant \cdots \leqslant X_n$ represent the items of a random sample, ordered according to magnitude. The β-trimmed mean is

$$\overline{X}_\beta = \frac{1}{h} \sum_{i=g+1}^{n-g} X_i,$$

where β is usually selected so that $g = \eta\beta$ is an integer (otherwise, $g = [\eta\beta]$, the greatest integer in $\eta\beta$) and where $h = n - 2g$. Of course, $\overline{X}_{\beta=0} = \overline{X}$.

It is well known that

$$Z = \frac{\overline{X} - 0}{S/\sqrt{n-1}},$$

where $S^2 = \sum_{i=1}^n (X_i - \overline{X})^2/n$, has a t-distribution* with $n - 1$ degrees of freedom provided that the sample arises from a normal distribution. However, even though the underlying distribution is nonnormal (without really long tails), Z still has a distribution fairly close to this t-distribution. This is what we mean by "almost distribution-free." Now it is not so well known, but true [11], that

$$Z_\beta = \frac{\overline{X}_\beta - 0}{\sqrt{SS(\beta)/h(h-1)}},$$

where

$$SS(\beta) = (g+1)(X_{g+1} - \overline{X}_\beta)^2$$
$$+ (X_{g+2} - \overline{X})^2 + \cdots$$
$$+ (X_{n-g-1} - \overline{X})^2$$
$$+ (g+1)(X_{n-g} - \overline{X}_\beta)^2,$$

has an approximate t-distribution with $h - 1$ degrees of freedom for many underlying distributions, so that Z_β is almost distribution-free. Of course, $Z_{\beta=0} = Z$.

In an adaptive procedure using some Z_β to make an inference about θ, a selector statistic, such as the kurtosis K or Q, can be used to choose an appropriate β. This β will be larger for larger values of K and Q. In making inferences about θ based upon a selected Z_β, the overall confidence coefficient or the overall significance level will deviate somewhat from the nominal one. However, these deviations are not great; in many instances we have found that α equals something like 0.06 rather than the nominal $\alpha = 0.05$. Thus we can place great reliability on the level of the resulting inferences.

These adaptive and robust methods have been extended to multivariate situations and the interested reader is referred to some of the following articles and their references for further study. The future seems bright for adaptive methods, and these will bring applications and theory closer together.

REFERENCES

1. Andrews, D. F., Bickel, P. J., Hampel, F. R., Huber, P. J., Rogers, W. H., and Tukey, J. W. (1972). *Robust Estimates of Location.* Princeton University Press, Princeton, N. J.

2. Conover, W. F. (1971). *Practical Nonparametric Statistics.* Wiley, New York.

3. Hogg, R. V. (1967). *J. Amer. Statist. Ass.,* **62**, 1179–1186.

4. Hogg, R. V. (1974). *J. Amer. Statist. Ass.,* **69**, 909–927.

5. Hogg, R. V., Fisher, D. M., and Randles, R. H. (1975). *J. Amer. Statist. Ass.,* **70**, 656–661.

6. Hogg, R. V. (1979). *Amer. Statist.,* **33**, 108–115.

7. Huber, P. J. (1973). *Ann. Math. Statist.,* **43**, 1041–1067.

8. Huber, P. J. (1973). *Ann. Statist.,* **1**, 799–821.

9. Jaeckel, L. A. (1971). *Ann. Math. Statist.,* **42**, 1540–1552.

10. Randles, R. H., Ramberg, J. S., and Hogg, R. V. (1973). *Technometrics,* **15**, 769–778.

11. Tukey, J. W. and McLaughlin, D. H. (1963). *Sankhyā A,* **25**, 331–352.

See also DISTRIBUTION-FREE METHODS; EXPLORATORY DATA ANALYSIS; and ROBUST ESTIMATION.

ROBERT V. HOGG

ADAPTIVE SAMPLING

Adaptive sampling is a method of unequal probability sampling whereby the selection of sampling units at any stage of the sampling process depends on information from the units already selected. In general terms it means that if you find what you are looking for at a particular location, you sample in the vicinity of that location with the hope of obtaining even more information.

Methods of estimation were initially developed in the three pioneering papers of Thompson [23–25] and the sampling book by Thompson [26]. The material considered in this review is described briefly by Seber and Thompson [20], while full details are given in the book by Thompson and Seber [31].

ADAPTIVE CLUSTER SAMPLING

Suppose we have a population spread over a large area which is highly clumped but is generally sparse or empty between clumps. If one selects a simple random sample (without replacement) of units, then most of the units selected will be empty. Density estimation based on this meager information will then have poor precision. Further, if the population species is rare, we will get little physiological information about individuals. It would be better to begin with an initial sample and, if individuals are detected on one of the selected units, then sample the neighboring units of that unit as well. If further individuals are encountered on a unit in the neighborhood, then the neighborhood of that unit is also added to the sample, and so on, thus building up a cluster of units. We call this *adaptive cluster sampling**. If the initial sample includes a unit from a clump, then the rest of the clump will generally be sampled. Such an approach will give us a greater number of individuals.

As well as counting individuals, we may wish to measure some other characteristic of the unit, for example plant biomass or pollution level, or even just note the presence or absence of some characteristic using an indicator variable. In addition to rare-species and pollution studies, we can envisage a wide range of populations which would benefit from adaptive sampling, for example populations which form large aggregations such as fish, marine mammals, and shrimp. We can also add mineral deposits and rare infectious diseases in human populations (e.g., AIDS) to our list. Recently the method has been used in sampling houses for a rare characteristic [5] and in sampling animal habitats [15].

To set out the steps involved in adaptive cluster sampling we begin with a finite population of N units indexed by their "labels" $(1, 2, \ldots, N)$. With unit i is associated a variable of interest y_i for $i = 1, 2, \ldots, N$. The object is to select a sample, observe the y-values for the units in the sample, and then estimate some function of the population y-values such as the population total $\sum_{i=1}^{N} y_i = \tau$ or the population mean $\mu = \tau/N$.

The first step is to define, for each unit i, a neighborhood consisting of that unit and a set of "neighboring" units. For example, we could choose all the adjacent units with a common boundary, which, together with unit i, form a cross. Neighborhoods can be defined to have a variety of patterns; the units (plots) in a neighborhood do not have to be contiguous. However, they must have a *symmetry* property, that is, if unit j is in the neighborhood of unit i, then unit i is in the neighborhood of unit j. We assume, for the moment, that these neighborhoods do not depend on y_i.

The next step is to specify a condition C (for instance, $y > c$, where c is a specified constant). We now take an initial random sample of n_1 units selected with or without replacement from the N units in the population. Whenever the y-value of a unit i in the initial sample satisfies C, all units in the neighborhood of unit i are added to the sample. If in turn any of the added units satisfies the condition, still more units are added. The process is continued until a cluster of units is obtained which contains a "boundary" of units called *edge* units that do not satisfy C. If a unit selected in the initial sample does not satisfy C, then there is no augmentation and we have a cluster of size one. The process is demonstrated in Fig. 1, where the units are plots and the neighborhoods form a cross. Here y_i is the number of animals on plot i, and $c = 0$, so that a neighborhood is added every time animals are found. In Fig. 1a we see one of the initial plots which happens to contain one animal. As it is on the edge of

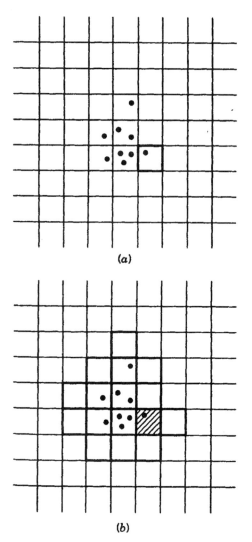

Figure 1. (a) Initial sample plot. (b) Cluster obtained by adding adaptively.

a "clump," we see that the adaptive process leads to the cluster of plots in Fig. 1b.

We note that even if the units in the initial sample are distinct, as in sampling without replacement, repeats can occur in the final sample, as clusters may overlap on their edge units or even coincide. For example, if two non-edge units in the same cluster are selected in the initial sample, then that whole cluster occurs twice in the final sample. The final sample then consists of n_1 (not necessarily distinct) clusters, one for each unit selected in the initial sample.

APPLICATIONS AND EXTENSIONS

In applications, other methods are sometimes used for obtaining the initial sample. For instance, in forestry the units are trees and these are usually selected by a method of unequal probability sampling*, where the probability of selecting a tree is proportional to the basal area of a tree (the cross-sectional area of a tree at the basal height—usually 4.5 feet in the USA). Roesch [16] described a number of estimators for this situation.

In ecology, larger sample units other than single plots are often used. For example, a common sampling unit is the strip transect, which we might call the primary unit. In its adaptive modification, the strip would be divided up into smaller secondary units, and if we found animals in a secondary unit, we would sample units on either side of that unit, with still further searching if additional animals are sighted while on this search. Strips are widely used in both aerial and ship surveys of animals and marine mammals. Here the aircraft or vessel travels down a line (called a *line transect**), and the area is surveyed on either side out to a given distance. Thompson [24] showed how the above theory can be applied to this sampling situation. He pointed out that a primary unit need not be a contiguous set of secondary units. For example, in some wildlife surveys the selection of sites chosen for observation is done systematically (with a random starting point), and a single systematic selection then forms the primary unit. We can then select several such primary units without replacement and add adaptively as before. Such a selection of secondary units will tend to give better coverage of the population then a simple random sample.

Clearly other ways of choosing a primary unit to give better coverage are possible. Munholland and Borkowski [13,14] suggest using a Latin square* + 1 design selected from a square grid of secondary units (plots). The Latin square gives a secondary unit in every row and column of the grid, and the extra (i.e. +1) unit ensures that any pair of units has a positive probability of being included in the initial sample. The latter requirement is needed for unbiased variance estimation.

In some situations it is hard to know what c should be for the condition $y > c$. If we choose c too low or too high, we end up with a feast or famine of extra plots. Thompson [28] suggested using the data themselves, in fact the order statistics*. For example, c could be the rth largest y-value in the initial sample statistic, so that the neighborhoods are now determined by the y-values. This method would be particularly useful in pollution studies, where the location of "hot spots" is important.

Another problem, regularly encountered with animal population studies, is that not all animals are detected. Thompson and Seber [30] developed tools for handling incomplete detectability for a wide variety of designs, including adaptive designs, thus extending the work of Steinhorst and Samuel [22].

Often we are in a multivariate situation where one needs to record several characteristics or measurements on each unit, e.g. the numbers of different species. Thompson [27] pointed out that any function of the variables can be used to define the criterion C, and obtained unbiased estimates of the mean vector and covariance matrix for these variables.

We can use any of the above methods in conjunction with stratification. If we don't allow the clusters to cross stratum boundaries, then individual stratum estimates are independent and can be combined in the usual fashion. Thompson [25] extended this theory to allow for the case where clusters do overlap. Such an approach makes more efficient use of sample information.

Finally, there are two further developments relating to design, namely, selecting networks without replacement and a two-stage sampling procedure [17,18].

UNBIASED ESTIMATION

Although the cluster is the natural sample group, it is not a convenient entity to use for theoretical developments, because of the double role that edge units can play. If an edge unit is selected in the initial sample, then it forms a cluster of size 1. If it is not selected in the initial sample, then it can still be selected by being a member of any cluster for which it is an edge unit. We therefore introduce the idea of the network A_i for unit i, defined to be the cluster generated by unit i but with its edge units removed. In Fig. 1(b) we get the sampled network by omitting the empty units from the sampled cluster. Here the selection of *any* unit in the network leads to the selection of *all* of the network. If unit i is the only unit in a cluster satisfying C, then A_i consists of just unit i and forms a network of size 1. We also define any unit which does not satisfy C to be a network of size 1, as its selection does not lead to the inclusion of any other units. This means that all clusters of size 1 are also networks of size 1. Thus any cluster consisting of more than one unit can be split into a network and further networks of size 1 (one for each edge unit). In contrast to having clusters which may overlap on their edge units, the distinct networks are *disjoint* and form a *partition* of the N units.

Since the probability of selecting a unit will depend on the size of the network it is in, we are in the situation of unequal-probability sampling and the usual estimates based on equal-probability sampling will be biased. However, we have the well-known Horvitz—Thompson* (HT) and Hansen–Hurwitz (HH) estimators (cf. refs. [8] and [9]) for this situation, the latter being used in sampling with replacement. These estimators, however, require knowing the probability of selection of each unit in the final sample. Unfortunately these probabilities are only known for units in networks selected by the initial sample and not for the edge units attached to these networks. Therefore, in what follows we ignore all edge units which are not in the initial sample and use only network information when it comes to computing the final estimators.

Motivated by the HT estimator for the population mean μ, we consider

$$\hat{\mu} = \frac{1}{N} \sum_{i=1}^{N} y_i \frac{I_i}{E[I_i]},$$

where I_i takes the value 1 if the initial sample intersects network A_i, and 0 otherwise; $\hat{\mu}$ is an unbiased estimator for sampling with or without replacement.

Another possible estimator (motivated by the HH estimator) which is also obviously unbiased for sampling with or without replacement, is

$$\tilde{\mu} = \frac{1}{N} \sum_{i=1}^{N} y_i \frac{f_i}{E[f_i]},$$

where f_i is the number of times that the ith unit in the final sample appears in the estimator, that is, the number of units in the initial sample which fall in (intersect) A_i determined by unit i; $f_i = 0$ if no units in the initial sample intersect A_i. It can be shown that

$$\tilde{\mu} = \frac{1}{n_1} \sum_{i=1}^{n_1} w_i = \overline{w}, \quad \text{say},$$

where w_i is the mean of the observations in A_i, i.e., $\overline{w}$ is the mean of the n_1 (not necessarily distinct) network means. *See also* NETWORK ANALYSIS.

ADAPTIVE ALLOCATION

There are other ways of adaptively adding to an initial sample. For instance, suppose the population is divided up into strata or primary units each consisting of secondary units. An initial sample of secondary units is taken in each primary unit. If some criterion is satisfied such as $\overline{y} > c$, then a further sample of units is taken from the *same* primary unit. Kremers [12] developed an unbiased estimator for this situation.

If the clumps tend to be big enough so that they are spread over several primary units, we could use what is found in a particular primary unit to determine the level of the sampling in the next. This is the basis for the theory developed by Thompson et al. [29]. Other forms of augmenting the initial sample which give biased estimates are described by Francis [6,7] and Jolly and Hampton [10,11]. This kind of adaptive sampling based on allocating more units rather than adding more neighborhoods is called *adaptive allocation*.

RAO—BLACKWELL MODIFICATION

An adaptive sample can be defined as one for which the probability of obtaining the sample depends only on the distinct unordered y-observations in the sample, and not on the y-values outside the sample. In this case d, the set of distinct unordered labels in the sample together with their associated y-values, is minimal sufficient for μ. This is proved for "conventional designs" by Cassel et al. [3] and Chaudhuri and Stenger [4], and their proofs readily extend to the case of adaptive designs. (This extension is implicit in Basu [1].) This means that an unbiased estimator which is not a function of d can be "improved" by taking the expectation of the estimator conditional on d to give an estimator with smaller variance. For example, consider three unbiased estimators of μ, namely $\overline{y}_1$ (the mean of the initial sample of n_1 units), $\hat{\mu}$, and $\tilde{\mu}$. Each of these depends on the order of selection, as they depend on which n_1 units are in the initial sample; $\tilde{\mu}$ also depends on repeat selections; and when the initial sample is selected with replacement, all three estimators depend on repeat selections. Since none of the three estimators is a function of the minimal sufficient statistic d, we can apply the Rao—Blackwell theorem*. If T is any one of the three estimators, then $E[T|d]$ will give a better unbiased estimate, i.e. one with smaller variance. We find that this estimator now uses all the units including the edge units.

Finally we mention the "model-based" or "superpopulation" approach (cf. Särndal et al. [19], for example). Here the population vector **y** of y-values is considered to be a realization of a random vector **Y** with some joint distribution F, which may depend on an unknown parameter ϕ. In a Bayesian framework ϕ will have a known prior distribution. For this model-based approach, Thompson and Seber [31] indicate which of the results for conventional designs carry over to adaptive designs and which do not. They also show in their Chapter 10 that optimal designs tend to be adaptive.

RELATIVE EFFICIENCY

An important question one might ask about adaptive sampling is "How does it compare with, say, simple random sampling?" This question is discussed by Thompson and Seber [31, Chapter 5], and some guidelines are given. Cost considerations are also important. Simple examples given by them throughout their book suggest that there are large gains in efficiency to be had with clustered populations. Two simulation studies which shed light on this are by Brown [2] and Smith et al. [21].

REFERENCES

1. Basu, D. (1969). Role of the sufficiency and likelihood principles in sample survey theory. *Sankhyā A*, **31**, 441–454.

2. Brown, J. A. (1994). The application of adaptive cluster sampling to ecological studies. In *Statistics in Ecology and Environmental Monitoring*, D. J., Fletcher and B. F. J., Manly. eds., University of Otago Press, Dunedin, New Zealand, pp. 86–97.

3. Cassel, C. M., Särndal, C. E., and Wretman, J. H. (1977). *Foundations of Inference in Survey Sampling*, Wiley, New York.

4. Chaudhuri, A. and Stenger, H. (1992). *Survey Sampling: Theory and Methods*. Marcel Dekker, New York.

5. Danaher, P. J. and King, M. (1994). Estimating rare household characteristics using adaptive sampling. *New Zealand Statist.*, **29**, 14–23.

6. Francis, R. I. C. C. (1984). An adaptive strategy for stratified random trawl surveys. *New Zealand J. Marine and Freshwater Res.*, **18**, 59–71.

7. Francis, R. I. C. C. (1991). Statistical properties of two-phase surveys: comment. *Can. J. Fish. Aquat. Sci.*, **48**, 1128.

8. Hansen, M. M. and Hurwitz, W. N. (1943). On the theory of sampling from finite populations. *Ann. Math. Statist.*, **14**, 333–362.

9. Horvitz, D. G. and Thompson, D. J. (1952). A generalization of sampling without replacement from a finite universe. *J. Amer. Statist. Assoc.*, **47**, 663–685.

10. Jolly, G. M. and Hampton, I. (1990). A stratified random transect design for acoustic surveys of fish stocks. *Can. J. Fish. Aquat. Sci.*, **47**, 1282–1291.

11. Jolly, G. M. and Hampton, I. (1991). Reply of comment by R. I. C. C. Francis. *Can. J. Fish. Aquat. Sci.*, **48**, 1128–1129.

12. Kremers, W. K. (1987). Adaptive Sampling to Account for Unknown Variability Among Strata. *Preprint No. 128*, Institut für Mathematik, Universität Augsburg, Germany.

13. Munholland, P. L. and Borkowski, J. J. (1993). Adaptive Latin Square Sampling +1 Designs. *Technical Report No. 3-23-93*, Department of Mathematical Sciences, Montana State University, Bozeman.

14. Munholland, P. L. and Borkowski, J. J. (1996). Latin square sampling +1 designs. *Biometrics*, **52**, 125–132.

15. Ramsey, F. L. and Sjamsoe'oed, R. (1994). Habitat association studies in conjunction with adaptive cluster samples. *J. Environmental Ecol. Statist.*, **1**, 121–132.

16. Roesch, F. A., Jr. (1993). Adaptive cluster sampling for forest inventories. *Forest Sci.*, **39**, 655–669.

17. Salehi, M. M. and Seber, G. A. F. (1997). Adaptive cluster sampling with networks selected without replacement. *Biometrika*, **84**, 209–219.

18. Salehi, M. M. and Seber, G. A. F. (1997). Two-stage adaptive cluster sampling. *Biometrics*, **53**, 959–970.

19. Särndal, C. E., Swensson, B., and Wretman, J. (1992). *Model Assisted Survey Sampling*. Springer-Verlag, New York.

20. Seber, G. A. F. and Thompson, S. K. (1994). Environmental adaptive sampling. In *Handbook of Statistics, Vol. 12 (Environmental Sampling)*, G. P. Patil and C. R. Rao, eds., New York, North Holland/Elsevier Science, pp. 201–220.

21. Smith, D. R., Conroy, M. J., and Brakhage, D. H. (1995). Efficiency of adaptive cluster sampling for estimating density of wintering waterfowl. *Biometrics*, **51**, 777–788.

22. Steinhorst, R. K. and Samuel, M. D. (1989). Sightability adjustment methods for aerial surveys of wildlife populations. *Biometrics*, **45**, 415–425.

23. Thompson, S. K. (1990). Adaptive cluster sampling. *J. Amer. Statist. Ass.*, **85**, 1050–1059.

24. Thompson, S. K. (1991). Adaptive cluster sampling: Designs with primary and secondary units. *Biometrics*, **47**, 1103–1115.

25. Thompson, S. K. (1991). Stratified adaptive cluster sampling. *Biometrika*, **78**, 389–397.

26. Thompson, S. K. (1992). *Sampling*. Wiley, New York.

27. Thompson, S. K. (1993). Multivariate aspects of adaptive cluster sampling. In *Multivariate Environmental Statistics*, G. P. Patil and C. R. Rao, eds., New York, North Holland/Elsevier Science, pp. 561–572.

28. Thompson, S. K. (1995). Adaptive cluster sampling based on order statistics. *Environmetrics*, **7**, 123–133.

29. Thompson, S. K., Ramsey, F. L., and Seber, G. A. F. (1992). An adaptive procedure for

sampling animal populations. *Biometrics*, **48**, 1195–1199.

30. Thompson, S. K. and Seber, G. A. F. (1994). Detectability in conventional and adaptive sampling. *Biometrics*, **50**, 712–724.

31. Thompson, S. K. and Seber, G. A. F. (1996). *Adaptive Sampling*. Wiley, New York.

See also ADAPTIVE METHODS; CLUSTER SAMPLING; LINE TRANSECT SAMPLING; POPULATION SIZE, HORVITZ-THOMPSON ESTIMATOR FOR; RAO–BLACKWELL THEOREM; TRANSECT METHODS; and UNEQUAL PROBABILITY SAMPLING.

GEORGE A. F. SEBER

ADDITION THEOREM

Let A_i and A_j be two events defined on a sample space. Then

$$\Pr[A_i \cup A_j] = \Pr[A_i] + \Pr[A_j] - \Pr[A_i \cap A_j],$$

where $\Pr[A_i \cup A_j]$ denotes the probability of A_i or A_j or both occurring, $\Pr[A_i]$ and $\Pr[A_j]$ denote respectively the probability of A_i and the probability of A_j, and $\Pr[A_i \cap A_j]$ denotes the probability of both A_i and A_j occurring.

The theorem is extended for the general case of n events as follows:

$$\Pr[A_1 \cup \cdots \cup A_n] = \sum_{i=1}^{n} \Pr[A_i]$$
$$- \sum_{i_1 \, <i_2}^{n-1 \quad n} \Pr[A_{i_1} \cap A_{i_2}]$$
$$+ \sum_{i_1 \, <i_2 \, <i_3}^{n-2 \; n-1 \quad n} \Pr[A_{i_1} \cap A_{i_2} \cap A_{i_3}]$$
$$- \cdots + (-1)^{n+1} \Pr[\cap_{i=1}^{n} A_i].$$

It is also called *Waring's theorem*.

See also BONFERRONI INEQUALITIES AND INTERVALS; BOOLE'S INEQUALITY; and INCLUSION-EXCLUSION METHOD.

ADDITIVE RISK MODEL, AALEN'S

THE MODEL

In medical statistics and survival analysis*, it is important to assess the association between risk factors and disease occurrence or mortality. Underlying disease mechanisms are invariably complex, so the idea is to simplify the relationship between survival patterns and covariates in such a way that only essential features are brought out. Aalen's (1980) additive risk model [1] is one of three well-developed approaches to this problem, the others being the popular proportional hazards model introduced by D. R. Cox in 1972 (*see* PROPORTIONAL HAZARDS MODEL, COX'S), and the accelerated failure-time model, which is a linear regression model with unknown error distribution, introduced in the context of right-censored survival data by R. G. Miller in 1976.

Aalen's model expresses the conditional hazard function $\lambda(t|\mathbf{z})$ of a survival time T as a linear function of a p-dimensional covariate vector $\mathbf{z}$:

$$\lambda(t|\mathbf{z}) = \boldsymbol{\alpha}(t)'\mathbf{z} = \sum_{j=1}^{p} \alpha_j(t)z_j, \qquad (1)$$

where $\boldsymbol{\alpha}(t)$ is a nonparametric p-vector of regression functions [constrained by $\lambda(t|\mathbf{z}) \geqslant 0$] and $\mathbf{z} = (z_1, \ldots, z_p)'$. Some authors refer to (1) as the linear hazard model.

As a function of the covariates $z_1, \ldots, z_p$, the *additive* form of Aalen's model contrasts with the *multiplicative* form of Cox's model:

$$\lambda(t|\mathbf{z}) = \lambda_0(t)\exp\{\boldsymbol{\beta}'\mathbf{z}\}$$
$$= \lambda_0(t)\prod_{j=1}^{p}\exp\{\beta_j z_j\},$$

where $\lambda_0(t)$ is a nonparametric baseline hazard function and $\boldsymbol{\beta}$ is a vector of regression parameters. Aalen's model has the feature that the influence of each covariate can vary separately and nonparametrically through time, unlike Cox's model or the accelerated failure-time model. This feature can be desirable in some applications, especially when there are a small number of covariates.

Consider the following simple example with three covariates: T is the age at which an individual contracts melanoma (if at all), $z_1 =$ indicator male, $z_2 =$ indicator female, and $z_3 =$ number of serious sunburns as a child. Then the corresponding regression functions, α_1, α_2, and α_3, can be interpreted as the (age-specific) background rates of melanoma for males and females and as the excess rate of melanoma due to serious sunburns is childhood, respectively.

Aalen's model is expected to provide a reasonable fit to data, since the first step of a Taylor series expansion of a general conditional hazard function about the zero of the covariate vector can be expressed in the form (1). It is somewhat more flexible than Cox's model and can be especially helpful for exploratory data analysis*. A rough justification for the additive form can be given in terms of p independent competing risks*, since the hazard function of the minimum of p independent random variables is the sum of their individual hazard functions.

It is generally sensible to include a nonparametric baseline function in the model, by augmenting $\mathbf{z}$ with a component that is set to 1. Also, it is often natural to center the covariates in some fashion, so the baseline can be interpreted as the "hazard" function for an "average" individual. In some cases, however, a baseline hazard is already implicit in the model and it is not necessary to center the covariates, as in the melanoma example above.

Aalen originally proposed his model in a counting process* setting, which allows time-dependent covariates and general patterns of censorship, and which can be studied using powerful continuous-time martingale* techniques. In a typical application the observed survival times are subject to right censorship, and it is customary to assume that the censoring time, C say, is conditionally independent of T given $\mathbf{z}$. One observes $(X, \delta, \mathbf{z})$, where $X = T \wedge C$ and $\delta = I\{X = T\}$. Aalen's model (1) is now equivalent to specifying that the counting process $N(t) = I(X \leqslant t, \delta = 1)$, which indicates an uncensored failure by time t, has intensity process

$$\lambda(t) = \boldsymbol{\alpha}(t)'\mathbf{y}(t),$$

where $\mathbf{y}(t) = \mathbf{z}I\{X \geqslant t\}$ is a covariate process.

MODEL FITTING

To fit Aalen's model one first estimates the p-vector of integrated regression functions $\mathbf{A}(t) = \int_0^t \boldsymbol{\alpha}(s)\,ds$. Denote by $(t_i, \delta_i, \mathbf{z}_i)$ the possibly right-censored failure time t_i, indicator of noncensorship δ_i, and covariate vector $\mathbf{z}_i$ for n individuals. Let $\mathbf{N} = (N_1, \ldots, N_n)'$ and $\mathbf{Z} = (\mathbf{y}_1, \ldots, \mathbf{y}_n)'$, where N_i is the counting process and $\mathbf{y}_i$ is the associated covariate process for individual i.

Aalen [1] introduced an ordinary least squares (OLS) type estimator of $\mathbf{A}(t)$ given by

$$\hat{\mathbf{A}}(t) = \int_0^t (\mathbf{Z}'\mathbf{Z})^{-1}\mathbf{Z}'\,d\mathbf{N},$$

where the matrix inverse is assumed to exist; $\hat{\mathbf{A}}$ is a step function, constant between uncensored failures, and with jump

$$\boldsymbol{\Delta}_i = \left(\sum_{t_k \geqslant t_i} \mathbf{z}_k \mathbf{z}_k' \right)^{-1} \mathbf{z}_i \qquad (2)$$

at an uncensored failure time t_i. The matrix inverse exists unless there is collinearity* between the covariates or there are insufficiently many individuals at risk at time t_i. A heuristic motivation for $\hat{\mathbf{A}}$ comes from applying the method of least squares to increments of the multivariate counting process $\mathbf{N}$. The estimator is consistent and asymptotically normal [14,9]. The covariance matrix of $\hat{\mathbf{A}}(t)$ can be estimated [1,2] by $\hat{\mathbf{V}}(t) = \sum_{t_i \leqslant t} \delta_i \boldsymbol{\Delta}_i \boldsymbol{\Delta}_i'$.

Plots of the components of $\hat{\mathbf{A}}(t)$ against t, known as *Aalen plots*, are a useful graphical diagnostic tool for studying time-varying covariate effects [2,3,4,7,9,12,13]. Mau [12] coined the term Aalen plots and made a strong case for their importance in survival analysis*. Roughly constant slopes in the plots indicate periods when a covariate has a non-time-dependent regression coefficient; plateaus indicate times at which a covariate has no effect on the hazard. Interpretation of the plots is helped by the inclusion of pointwise or simultaneous confidence limits. An approximate pointwise $100(1 - \alpha)\%$ confidence interval for the jth component of $\mathbf{A}(t)$ is given by

$$\hat{\mathbf{A}}_j(t) \pm z_{\alpha/2}\hat{\mathbf{V}}_{jj}(t)^{1/2},$$

where $z_{\alpha/2}$ is the upper $\alpha/2$ quantile of the standard normal distribution and $\hat{\mathbf{V}}_{jj}(t)$ is the jth entry on the diagonal of $\hat{\mathbf{V}}(t)$. To avoid wild fluctuations in the plots (which occur when the size of the risk set is small), estimation should be restricted to time intervals over which the matrix inverse in (2) is numerically stable.

Figure 1 shows an Aalen plot based on survival data for 495 myelamatosis patients [17]. The plot gives the estimated integrated regression function for one particular covariate, serum β_2-microglobulin, which was log-transformed to adjust for skewness. Pointwise 95% confidence limits are also shown. Serum β_2-microglobulin is seen to have a strong effect on survival during the first two years of follow-up.

The vector of regression functions α can be estimated by smoothing the increments of $\hat{\mathbf{A}}$. One approach is to extend Ramlau-Hansen's kernel estimator [19] to the additive-risk-model setting [3,9,14]. For a kernel function K that integrates to 1 and some bandwidth $b > 0$,

$$\hat{\alpha}(t) = b^{-1} \sum_{i=1}^{n} K\left(\frac{t - t_i}{b}\right) \Delta_i$$

consistently estimates α provided the bandwidth tends to zero at a suitable rate with increasing sample size. Plots of the regression function estimates in some real and simulated data examples have been given by Aalen [3].

Huffer and McKeague [9] introduced a weighted least squares* (WLS) estimator of $\mathbf{A}$; see Fig. 2 for a comparison with the OLS estimator. The weights consistently estimate $[\lambda(t|\mathbf{z}_i)]^{-1}$ and are obtained by plugging $\hat{\alpha}(t)$ and $\mathbf{z} = \mathbf{z}_i$ into (1). The WLS estimator is an approximate maximum-likelihood estimator and an approximate solution to the score equations [20]. It is consistent and asymptotically normal provided $\lambda(t|\mathbf{z})$ is bounded away from zero [9,14]. Furthermore, the WLS

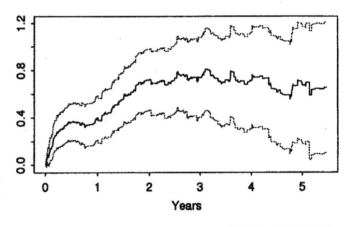

Figure 1. An Aalen plot with 95% confidence limits for the myelamatosis data.

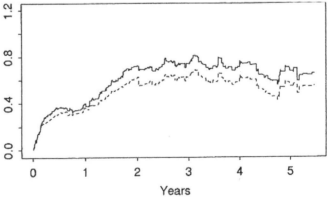

Figure 2. Comparison of Aalen plots of the WLS estimates (dashed line) and OLS (solid line) estimates for the myelamatosis data.

estimator is asymptotically efficient in the sense of having minimal asymptotic variance [6,20,4]. Simulation studies [9] show that significant variance reductions are possible using WLS compared with OLS estimators, especially in large samples where the weights are more stable. When there are no covariates ($p = 1$ and $\mathbf{z}_i = 1$), the OLS and WLS estimators reduce to the Nelson—Aalen estimator of the baseline cumulative hazard function. Simultaneous confidence bands for $\mathbf{A}$ based on OLS and WLS estimators, for continuous or grouped data, can be found in refs. 9, 15.

Tests of whether a specific covariate (say the jth component of $\mathbf{z}$) has any effect on survival can be carried out within the Aalen-model setting. The idea is to test the null hypothesis $H_0 : \mathbf{A}_j(t) = 0$ over the follow-up period. This can be done [2] using a test statistic of the form $\sum_{i=1}^n w(t_i)\boldsymbol{\Delta}_{ij}$ for a suitable weight function w. Kolmogorov—Smirnov-type tests* are also available [9]; such tests are equivalent to checking whether the confidence band for the jth component of $\mathbf{A}$ contains the zero function.

To predict survival under Aalen's model, one estimates the conditional survival probability $P(T > t|\mathbf{z}) = \exp\{-\mathbf{A}(t)'\mathbf{z}\}$. This can be done using the product-limit estimator

$$\hat{P}(T > t|\mathbf{z}) = \prod_{t_i \leqslant t}(1 - \boldsymbol{\Delta}_i'\mathbf{z}),$$

or by plugging $\hat{\mathbf{A}}(t)$ into $P(T > t|\mathbf{z})$ in place of the unknown $\mathbf{A}(t)$. When there are no covariates, $\hat{P}(T > t|\mathbf{z})$ reduces to the Kaplan–Meier estimator* of the survival function corresponding to the baseline hazard.

MODEL DIAGNOSTICS

Some goodness-of-fit checking procedures are available for additive risk models. Aalen [2,3] suggested making plots against t of sums of the martingale residual processes $\hat{M}_i(t) = \delta_i I(t_i \leqslant t) - \hat{\mathbf{A}}(t_i \wedge t)'\mathbf{z}_i$ over groups of individuals. If the model fits well, then the plots would be expected to fluctuate around the zero line. McKeague and Utikal [16] suggested the use of a standardized residual process plotted against t and $\mathbf{z}$, and developed a formal goodness-of-fit test for Aalen's model.

Outlier detection has been studied by Henderson and Oman [8], who considered the effects on $\hat{\mathbf{A}}(t)$ of deletion of an observation from a data set. They show that unusual or influential observations* can be detected quickly and easily. They note that Aalen's model has an advantage over Cox's model in this regard, because closed-form expressions for the estimators are available, leading to exact measures of the effects of case deletion.

Mau [12] noticed that Aalen plots are useful for diagnosing time-dependent covariate effects in the Cox model. To aid interpretation of the plots in that case, Henderson and Milner [7] suggested that an estimate of the shape of the curve expected under proportional hazards be included.

RELATED MODELS

In recent years a number of variations on the additive structure of Aalen's model have been introduced. McKeague and Sasieni [17] considered a partly parametric additive risk model in which the influence of only a subset of the covariates varies nonparametrically over time, and that of the remaining covariates is constant:

$$\lambda(t|\mathbf{x}, \mathbf{z}) = \boldsymbol{\alpha}(t)'\mathbf{x} + \boldsymbol{\beta}'\mathbf{z}, \qquad (3)$$

where $\mathbf{x}$ and $\mathbf{z}$ are covariate vectors and $\boldsymbol{\alpha}(t)$ and $\boldsymbol{\beta}$ are unknown. This model may be more appropriate than (1) when there are a large number of covariates and it is known that the influence of only a few of the covariates is time-dependent. Lin and Ying [10] studied an additive analogue of Cox's proportional hazards model that arises as a special case of (3):

$$\lambda(t|\mathbf{z}) = \alpha_0(t) + \boldsymbol{\beta}'\mathbf{z}. \qquad (4)$$

Efficient WLS-type estimators for fitting (3) and (4) have been developed.

A variation in the direction of Cox's proportional hazards model [5] has been studied by Sasieni [21,22]: the *proportional excess hazards* model

$$\lambda(t|\mathbf{x}, \mathbf{z}) = \alpha_0(t|\mathbf{x}) + \lambda_0(t)\exp\{\boldsymbol{\beta}'\mathbf{z}\}, \qquad (5)$$

where $\alpha_0(t|\mathbf{x})$ is a known background hazard (available from national mortality statistics say) and $\lambda_0(t)$ and β are unknown. A further variation in this direction is due to Lin and Ying [11], who considered an *additive–multiplicative hazards* model that includes

$$\lambda(t|\mathbf{x}, \mathbf{z}) = \boldsymbol{\gamma}'\mathbf{x} + \lambda_0(t) \exp\{\boldsymbol{\beta}'\mathbf{z}\}, \qquad (6)$$

where $\boldsymbol{\gamma}$, $\boldsymbol{\beta}$, and $\lambda_0(t)$ are unknown. Finding efficient procedures for fitting the models (5) and (6) involves a combination of Cox partial likelihood* techniques and the estimation of efficient weights similar to those needed for the standard additive risk model (1).

CONCLUSION

Despite the attractive features of Aalen's model as an alternative to Cox's model in many application, it has received relatively little attention from practitioners or researchers. Cox's model has been perceived to be adequate for most applications, but it can lead to serious bias when the influence of covariates is time-dependent. Fitting separate Cox models over disjoint time intervals (years, say) is an ad hoc way around this problem. Aalen's model, however, provides a more effective approach. Interest in it, and especially in Aalen plots, is expected to increase in the future.

Acknowledgment
This research was partially supported by NSF Grant ATM-9417528.

REFERENCES

1. Aalen, O. O. (1980). A model for nonparametric regression analysis of counting processes. *Lecture Notes Statist.*, **2**, 1–25. Springer-Verlag, New York. (Aalen originally proposed his model at a conference in Poland; this paper appeared in the conference proceedings.)

2. Aalen, O. O. (1989). A linear regression model for the analysis of life times. *Statist. Med.*, **8**, 907–925. (A readable introduction to additive risk models with an emphasis on graphical techniques. Results based on the Aalen and Cox models are compared.)

3. Aalen, O. O. (1993). Further results on the nonparametric linear regression model in survival analysis. *Statist. Med.*, **12**, 1569–1588. (Studies the use of martingale residuals for assessing goodness-of-fit of additive risk models.)

4. Andersen, P. K., Borgan, Ø., Gill, R. D., and Keiding, N. (1993). *Statistical Models Based on Counting Processes*. Springer-Verlag, New York. (A comprehensive and in-depth survey of the counting process approach to survival analysis, including the additive risk model and its associated estimators.)

5. Cox, D. R. (1972). Regression models and life tables (with discussion). *J. Roy. Statist. Soc. B*, **34**, 187–220.

6. Greenwood, P. E. and Wefelmeyer, W. (1991). Efficient estimating equations for nonparametric filtered models. In *Statistical Inference in Stochastic Processes*, N. U. Prabhu and I. V. Basawa, eds., pp. 107–141. Marcel Dekker, New York. (Shows that the weighted least-squares estimator is an approximate maximum-likelihood estimator and that this implies asymptotic efficiency.)

7. Henderson, R. and Milner, A. (1991). Aalen plots under proportional hazards. *Appl. Statist.*, **40**, 401–410. (Introduces a modification to Aalen plots designed to detect time-dependent covariate effects in Cox's proportional-hazards model.)

8. Henderson, R. and Oman, P. (1993). Influence in linear hazard models. *Scand. J. Statist.*, **20**, 195–212. (Shows how to detect unusual or influential observations under Aalen's model.)

9. Huffer, F. W. and McKeague, I. W. (1991). Weighted least squares estimation for Aalen's additive risk model. *J. Amer. Statist. Ass.*, **86**, 114–129.

10. Lin, D. Y. and Ying, Z. (1994). Semiparametric analysis of the additive risk model. *Biometrika*, **81**, 61–71.

11. Lin, D. Y. and Ying, Z. (1995). Semiparametric analysis of general additive–multiplicative hazard models for counting processes. *Ann. Statist.*, **23**, 1712–1734.

12. Mau, J. (1986). On a graphical method for the detection of time-dependent effects of covariates in survival data. *Appl. Statist.*, **35**, 245–255. (Makes a strong case that Aalen plots can provide important information that might be missed when only Cox's proportional-hazards model is applied.)

13. Mau, J. (1988). A comparison of counting process models for complicated life histories. *Appl. Stoch. Models Data Anal.*, **4**, 283–298.

(Aalen's model is applied in the context of intermittent exposure in which individuals alternate between being at risk and not.)

14. McKeague, I. W. (1988). Asymptotic theory for weighted least squares estimators in Aalen's additive risk model. In *Statistical Inference from Stochastic Processes*, N. U. Prabhu, ed., *Contemp. Math.*, 80, 139–152. American Mathematical Society, Providence. (Studies the asymptotic properties of estimators in the counting process version of Aalen's model.)

15. McKeague, I. W. (1988). A counting process approach to the regression analysis of grouped survival data. *Stoch. Process. Appl.*, **28**, 221–239. (Studies the asymptotic properties of grouped data based estimators for Aalen's model.)

16. McKeague, I. W. and Utikal, K. J. (1991). Goodness-of-fit tests for additive hazards and proportional hazards models. *Scand. J. Statist.*, **18**, 177–195.

17. McKeague, I. W. and Sasieni, P. D. (1994). A partly parametric additive risk model. *Biometrika*, **81**, 501–514.

18. Miller, R. G. (1976). Least squares regression with censored data. *Biometrika*, **63**, 449–464.

19. Ramlau-Hansen, H. (1983). Smoothing counting process intensities by means of kernel functions. *Ann. Statist.*, **11**, 453–466.

20. Sasieni, P. D. (1992). Information bounds for the additive and multiplicative intensity models. In *Survival Analysis: State of the Art*, J. P. Klein and P. K. Goel, eds., pp. 249–265. Kluwer, Dordrecht. (Shows asymptotic efficiency of the weighted least-squares estimators in the survival analysis setting.)

21. Sasieni, P. D. (1995). Efficiently weighted estimating equations with application to proportional excess hazards. *Lifetime Data Analysis*, in press.

22. Sasieni, P. D. (1996). Proportional excess hazards. *Biometrika*, **83**, 127–141.

See also COUNTING PROCESSES; PROPORTIONAL HAZARDS MODEL, COX'S; SEMIPARAMETRICS; and SURVIVAL ANALYSIS.

IAN W. MCKEAGUE

ADMISSIBILITY

DEFINITION

Admissibility is a very general concept that is applicable to any procedure of statistical inference*. The statistical literature contains discussions of admissibility of estimators*, confidence intervals*, confidence sets, tests of hypotheses, sampling designs in survey sampling*, and so on. For each class of procedures, there is formulated a definition of admissibility which is appropriate for that class only. But all such definitions are based on a common underlying notion—that a procedure is admissible if and only if there does not exist within that class of procedures another one which performs uniformly at least as well as the procedure in question and performs better than it in at least one case. Here "uniformly" always means for all values of the parameter* (or parameters) that determines the (joint) probability distribution* of the random variables under investigation. It thus remains only to define how the condition of "performing as well as" is interpreted in each case. All such definitions are based closely on that of the admissibility of a decision rule* formulated in Abraham Wald's theory of statistical decision functions or decision theory*, as it is briefly called. In fact, the importance of the notion of admissibility in statistical theory rests on the adoption of the decision-theoretic approach to statistical problems formulated in Wald's theory.

DECISION THEORY

Wald's theory formulates the following general model for statistical decision making. Let $\mathscr{S}$ denote the sample space of the random variables under investigation, of which the true (joint) probability distribution is unknown, it being known only to belong to a family $\mathscr{P} = (P_\theta, \theta \in \Omega)$. Depending upon the object of the investigation (e.g., point or interval estimation or hypothesis testing*, etc.), there is a specific set $\mathscr{a}$ of all possible decisions a which the statistician may make. A decision rule δ is a function which prescribes for each sample point z how the decision would be made, i.e., some specific a chosen from $\mathscr{a}$. (δ may either assign a unique a to each z—such a rule is called a nonrandomized rule—or it may assign to each z a probability distribution δ_z on $\mathscr{a}$, the choice of a specific a being made according to that probability distribution by an independent random experiment.) Consequences of wrong decisions are allowed for by assuming a suitable nonnegative loss function* $\mathscr{L}(a, \theta)$. When the experiment, i.e., the taking of observations on the random variables, is repeated a large number of times under identical conditions, the average long-run loss converges to its "expectation" or "mean value" $\mathscr{R}(\theta, \delta)$ which is called the risk function*. Admissibility of decision rules is defined in terms of the risk function as follows:

A decision rule δ_2 is better than a rule δ_1 if

$$\mathscr{R}(\theta, \delta_2) \leqslant \mathscr{R}(\theta, \delta_1) \quad \text{for all } \theta \in \Omega \quad (1)$$

and the strict inequality in (1) holds for at least one $\theta \in \Omega$.

A decision rule δ is admissible if there exists no rule δ_1 which is better than δ. See DECISION THEORY for further details.

A LIMITATION

A limitation of the admissibility principle may be noted here. The central problem of decision theory is: Under the conditions of a given decision problem, how should the choice of a decision rule be effected from the class of all decision rules? The admissibility criterion requires that inadmissible rules be left out of consideration. This leaves the class of admissible rules, which, however, is generally very large, and admissibility says nothing as to how a choice should be made from this large class. The choice is therefore made in practice by applying other statistical principles, such as unbiasedness*, minimaxity, and invariance*, or by taking into consideration the statistician's prior beliefs* regarding the weight to be attached to the different possible values of the parameter. In the last-mentioned case, if the prior beliefs are expressed as a probability distribution τ on Ω, the risk function $\mathscr{R}(\theta, \delta)$ integrated with respect to τ gives the Bayes risk $r(\tau, \delta)$. The appropriate decision rule, called the Bayes rule, is then the one that minimizes $r(\tau, \delta)$ for given τ. This is the Bayesian mode of inference. In Bayesian inference*, there is thus an optimum decision rule which often is unique, determined by the prior beliefs, and hence the concept of admissibility has

less importance. In Wald's theory, however, the approach is non-Bayesian and based on the long-run frequency. Thus the risk function $\mathscr{R}(\theta, \delta)$ represents the average loss in the long run when the experiment is repeated a large number of times under identical conditions. There was the same approach in the earlier Neyman–Pearson* theories of interval estimation and hypothesis testing, which are included in Wald's theory as particular cases. This approach is often referred to as the N-P-W approach. The importance of the criterion of admissibility is thus related to the N-P-W approach.

Another point to be noted is that if a decision rule is inadmissible, there exists another that should be used in preference to it. But this does not mean that every admissible rule is to be preferred over any inadmissible rule. It is easy to construct examples of rules that are admissible but which it would be absurd to use. An example in point estimation is an estimator that is equal to some constant k whatever be the observations. The risk function then vanishes for $\theta = k$ and the estimator is an admissible one. (See ref. 11 for a more sophisticated example.)

PARTICULAR PROCEDURES

Any statistical inference procedure such as point or interval estimation* corresponds simply to a class of decision rules. But the definition of admissibility considered appropriate for a particular procedure may not always be equivalent to the decision theory definition in terms of a risk function. For example, consider interval estimation of the parameter θ in a probability density function $f(x, \theta)$ on the basis of independent observations. Let x denote collectively the observations and T_1, T_2 the statistics (functions of x) defining the confidence interval. [The pair (T_1, T_2) constitutes in this case the decision rule.] Admissibility for confidence intervals is defined as follows. "A set of confidence intervals $\{T_1, T_2\}$ is admissible if there exists no other set $\{T_1^*, T_2^*\}$ such that

(a) $T_2^*(\mathbf{x}) - T_1^*(\mathbf{x}) \leqslant T_2(\mathbf{x}) - T_1(\mathbf{x})$

      for all $\mathbf{x}$,

(b) $P_\theta\{T_1^*(\mathbf{x}) \leqslant \theta \leqslant T_2^*(\mathbf{x})\}$

$$\geqslant P_\theta\{T_1(\mathbf{x}) \leqslant \theta \leqslant T_2(\mathbf{x})\}$$

for all θ,

the strict inequality in (b) holding for at least one θ. [The probabilities in (b) are the inclusion probabilities.] This definition is obviously not reducible to one based on the risk function, as there are two inequalities; moreover, the first of these is required to hold at each sample point $\mathbf{x}$, there being no averaging over the sample space, which is an essential ingredient in the definition of a risk function.

In the case of point estimation*, on the other hand, the definition of admissibility is identical with that for decision rules. A loss function that is found reasonable and mathematically convenient, particularly in point estimation problems, is the squared error, i.e., $L(t, \theta) = c(t - \theta)^2$, where t is the estimate and c is any positive constant. Further if, as often is the case, the estimators are restricted to the class of unbiased* estimators, then the admissibility criterion reduces to one based on the variance or equivalently the efficiency* of the estimator.

In the case of hypothesis testing* of the null hypothesis $\theta = \theta_0$, the decision-theoretic definition of admissibility reduces to one based on the power function* in the Neyman–Pearson theory by a suitable choice of the loss function, namely, by putting $\mathscr{L}(\theta, a_1) = 0$ and $\mathscr{L}(\theta, a_2) = 1$, where for the value θ, a_1 is the correct decision and a_2 the incorrect one. The set α consists in this case of only two points, corresponding to the rejection or nonrejection of the null hypothesis. (See ref. 2 for a paper dealing with admissibility of tests.)

SPECIAL TYPES OF ADMISSIBILITY

These are extensions of the basic notion of admissibility.

Strong and Weak Admissibility

In some cases it is found necessary to introduce weak and strong versions of admissibility, strong admissibility being based on a more stringent criterion. For example, in the problem of interval estimation (see the section "Particular Procedures"), if condition (a) is replaced by

(a*) $E_\theta\{T_2^* - T_1^*\} \leqslant E_\theta\{T_2 - T_1\}$ for all θ,

and the sign of strict inequality required to hold for at least one θ in either (a*) or in (b), we obtain a more stringent criterion as the set of alternatives is enlarged. (See ref. 8 for weak and strong admissibility of confidence sets.)

ϵ-Admissibility

A decision rule δ_0 is said to be ϵ-admissible if there exists no other decision rule δ_1 such that

$$\mathscr{R}(\theta, \delta_1) < \mathscr{R}(\theta, \delta_0) - \epsilon \quad \text{for all } \theta \in \Omega.$$

(See "Decision Theory" section of this entry for definitions of these terms.)

ϵ-admissibility provides a measure of the extent by which an inadmissible rule falls short of being admissible. (See ref. 10 for an application.)

Uniform Admissibility

This term is special to survey sampling* theory. Earlier investigations had related mostly to the admissibility of a particular estimator e_1 under a given sampling design p_1. But in survey sampling, the choice of the sampling design is generally, subject to certain limitations of cost and time, within the statistician's control. This leads to the notion of the joint admissibility of the pair (e_1, p_1) within a class of pairs (e, p). (Such a pair is now called a *sampling strategy* or, more simply, a *strategy*.) It would be pointless to consider the joint admissibility within the class of all possible pairs (e, p), as then the only admissible sampling design would be that in which the whole population is observed with probability 1. It is therefore necessary to place a restriction on the class e of designs. The restrictions usually assumed are that the expected sample size or the expected sampling cost under p should not exceed certain limits, as these are the restraints that generally apply in practice. Of course, the particular sampling design p_1

must also satisfy the restriction. The term "uniform admissibility" denotes just this joint admissibility of an estimator e_1 and a sampling design p_1, defined as usual, within a class of pairs (e, p) such that p belongs to a specified class e of designs. Note that the uniform admissibility of (e_1, p_1) is a stronger property than the admissibility of e_1 under the given design p_1, as the former implies the latter. (See refs. 14 and 15 for some recent related results.)

Hyperadmissibility

This notion is special to survey sampling theory* and denotes broadly that an estimator is admissible for the population as a whole and also for every possible subpopulation of that population (see ref. 6 for more details).

Admissibility within a Restricted Class

It is often necessary to consider the admissibility of a procedure within a restricted class of procedures. For example, in the case of point estimation, an unbiased estimator T_0 of a parameter θ is said to be admissible within the unbiased class if there exists no other unbiased estimator T_1 of θ that is "better" than T_0.

RELATIONS WITH COMPLETENESS* AND EFFICIENCY*

In Wald's theory, the notion of admissibility is intimately related to that of completeness. The theory requires that the statistician should restrict the choice of a decision rule to the class of all admissible rules. But in general, there is no simple characteristic that distinguishes the class of all admissible rules from that of all inadmissible ones. This leads to the notion of a complete class that contains all the admissible rules: "A class e of decision rules is said to be *complete* if given any rule δ_1 not in e, there exists at least one rule δ_0 in e that is better than δ_1." (See the "Decision Theory" section for definition of betterness.) Hence if a class of rules is known to be complete, the statistician may validly restrict the choice of decision rule to such class as all the excluded rules are necessarily inadmissible. Of course, after choosing a particular

rule from a complete class, it would have to be tested for its admissibility. It is further shown in the theory that there exists a simply characterized class of decision rules (the class of generalized Bayes rules) which under very general conditions forms a complete class.

Essential completeness is a sharpening of the notion of completeness "A class e of decision rules is said to be *essentially complete* if given any rule δ_1 not in e, there exists at least one rule δ in e which is as good as δ_1, i.e., such that $\mathscr{R}(\theta, \delta_0) \leqslant \mathscr{R}(\theta, \delta_1)$ for all $\theta \in \Omega$." Clearly, a statistician may validly restrict the choice of a decision rule to an essentially complete class if it exists. It is shown in the theory that "the class of decision rules based on a sufficient statistic is always essentially complete." This proposition provides the decision-theoretic justification for the sufficiency principle* in statistics. For some related propositions such as the Rao–Blackwell theorem*, see ref. 4.

STEIN'S RESULT

A notable result relating to admissibility is that of Stein [16]: For k independent normal variables, the sample means are jointly inadmissible for the population means with the squared errors as loss function if $k \geqslant 3$. The theoretical and practical implications of Stein's results have been a matter of debate [1,1]; see JAMES–STEIN ESTIMATORS and SHRINKAGE ESTIMATORS.

SURVEY SAMPLING*

Survey sampling essentially involves no new point of principle. The commonly considered estimation problem is to estimate the population total. If the squared error is taken as the loss function, as is often the case, the admissibility of an estimator is defined as follows: An estimator $e(s, \mathbf{x})$ is admissible if there does not exist any other estimator $e'(s, \mathbf{x})$ such that

$$\sum_s p(s)[e'(s, \mathbf{x}) - T(\mathbf{x})]^2$$

$$\leqslant \sum_s p(s)[e(s, \mathbf{x}) - T(\mathbf{x})]^2 \quad \text{for all } \mathbf{x}$$

and the strict inequality holds for at least one **x**. Here $\mathbf{x} = (x_1, x_2, \ldots, x_N)$ denotes the population vector and is the parameter, $T(\mathbf{x}) = \sum_{i=1}^{N} x_i$, $p(s)$ is the probability of the sample s under the chosen sampling design, and N denotes the number of units in the population. The estimator $e(s, \mathbf{x})$ must, of course, depend only on the x_i observed in the sample. See Survey Sampling for further details.

The following important general results have been proved recently. (The lateness of the results is a consequence of the fact the correct model for survey sampling was developed only after 1950; see Survey Sampling.)

1. The sample mean is admissible as the estimator of the population mean in the entire class of all estimators whatever the sampling design, and for a very wide class of loss functions [7].
2. The Horwitz–Thompson estimator* is always admissible in the restricted class of unbiased estimators [5].

Suppose that samples are taken independently from k different finite populations. Are the sample means together jointly admissible for the population means with squared error as loss function? It is found that they are. Thus in the case of finite populations, an effect corresponding to Stein's result for the multivariate normal population does not occur. This is a very recent result [9].

REFERENCES

1. Alam, K. (1975). *J. Multivariate Anal.*, **5**, 83–95.
2. Brown, L. D., Cohen, A., and Strawderman, W. E. (1979). *Ann. Statist.*, **3**, 569–578.
3. Efron, B. and Morris, C. (1973). *J. R. Statist. Soc. B*, **35**, 379–421.
4. Ferguson, T. S. (1967). *Mathematical Statistics: A Decision-Theoretical Approach*. Academic Press, New York.
5. Godambe, V. P. and Joshi, V. M. (1965). *Ann. Math. Statist.*, **36**, 1707–1722.
6. Hanurav, T. V. (1968). *Ann. Math. Statist.*, **39**, 621–641.
7. Joshi, V. M. (1968). *Ann. Math. Statist.*, **39**, 606–620.
8. Joshi, V. M. (1969). *Ann. Math. Statist.*, **40**, 1042–1067.
9. Joshi, V. M. (1979). *Ann. Statist.*, **7**, 995–1002.
10. Kagan, A. M. (1970). *Sankhyā A*, **32**, 37–40.
11. Makani, S. M. (1977). *Ann. Statist.*, **5**, 544–546.
12. Neyman, J. (1937). *Philos. Trans. R. Soc. Lond. A*, **236**, 333–380.
13. Neyman, J. and Pearson, E. S. (1933). *Philos. Trans. R. Soc. Lond. A*, **231**, 289–337.
14. Scott, A. J. (1975). *Ann. Statist.*, **3**, 489–491.
15. Sekkappan, R. M. and Thompson, M. E. (1975). *Ann. Statist.*, **3**, 492–499.
16. Stein, C. (1956). *Proc. 3rd Berkeley Symp. Math. Stat. Prob.*, Vol. 1. University of California Press, Berkeley, Calif., pp. 197–206.
17. Wald, A. (1950). *Statistical Decision Functions*. Wiley, New York.

See also Bayesian Inference; Decision Theory; Estimation, Classical; Hypothesis Testing; Inference, Statistical; James–Stein Estimators; Shrinkage Estimators; and Survey Sampling.

V. M. Joshi

ADVANCES IN APPLIED PROBABILITY. See Applied Probability Journals

AFFLUENCE AND POVERTY INDEXES.
See Indexes, Affluence and Poverty

AGGREGATE

The word "aggregate" has several meanings. As a verb, it means putting together, or combining, elements that usually differ in some notable respect. As a noun, it is used to describe the result of this process. The word is also sometimes used as a synonym for "total," as in "aggregate production" and "aggregate debt."

In geology, and especially in mining engineering, the word is specifically applied to collections of samples of ore.

See also Arithmetic Mean and Geology, Statistics in.

AGGREGATE INDEX NUMBERS. See Index Numbers

AGGREGATION

Aggregation may be a phenomenon of direct interest, as in the study of biological populations, or it may reflect the necessary reduction of primary data to produce a usable statistical summary, as in the construction of index numbers*. Since the two topics are quite distinct, we consider them separately.

AGGREGATION AS AN OBSERVABLE PHENOMENON

It is often true that events (individuals) cluster in time or space or both (e.g., larvae hatching from eggs laid in a mass, aftershocks of an earthquake). Thus if the random variable of interest is the number of events occurring in an interval of time (or in a selected area), the clustering is manifested in a greater probability of extreme events (large groups) than would be expected otherwise. Alternatively, individual members of a population may be in close proximity because of environmental conditions. In either case, the population is said to be aggregated.

A standard initial assumption (corresponding to the absence of aggregation) is that the random variable follows a Poisson distribution*, and various indices have been proposed to detect departures from the Poisson process*. These methods are based upon data collected either as quadrat counts or as measurements of distance (or time) from randomly selected individuals (or points) to the nearest individual, known as nearest-neighbor distances. For example, the index of dispersion is defined as $I = s^2/m$, where m and s^2 denote the sample mean and variance. For the Poisson process, $E(I|m > 0) \doteq 1$; values of I significantly greater than 1 suggest aggregation; $I < 1$ is indicative of regular spacing of the individuals [5, Chap. 4]. Other measures, based upon both quadrat counts and distances, are summarized in Pielou [15, Chaps. 8 and 10] and Cormack [6]. When different kinds of individual (e.g., species) have different aggregation patterns, this make inferences about population characteristics such as diversity* much more difficult.

If the Poisson process is used to describe parents (or centers), each parent may give rise to offspring (or satellites). If these clusters are independent but have identical size distributions, the resulting distribution for the total count is a (Poisson) randomly stopped sum distribution*. If environmental heterogeneity is postulated, a compound distribution*, usually based on the Poisson, is appropriate. For both classes of Poisson-based distributions, $I > 1$. These standard distributions lack an explicit spatial or temporal dimension, for which a dispersal mechanism must be incorporated. The resulting model, known as a center-satellite process, has three components: a Poisson process for locating the cluster center, a distribution to generate the number of satellites, and a dispersal distribution to describe displacements from the center. This class of processes was introduced by Neyman and Scott [14] and is mathematically equivalent to the class of doubly stochastic Poisson processes defined for heterogeneity (See Bartlett [3, Chap. 1]).

A more empirical approach to aggregation is that of Taylor [18], who suggests that the population mean, μ, and variance, σ^2, are related by the power law:

$$\sigma^2 = A\mu^b, \qquad A > 0, \quad b > 0.$$

It is argued that values of b greater than 1 (the Poisson value) reflect density dependence in the spatial pattern of individuals. Although this view has been contested (see the discussion in Taylor [18]), a substantial body of empirical evidence has been presented in its support [19].

Knox [12] developed a test to detect the clustering of individuals in space and time, which may be formulated as follows. Suppose that n individuals (e.g., cases of a disease) are observed in an area during a time period. If cases i and j are less than a specified critical distance from one another, set the indicator variable $w_{ij} = 1$; otherwise, set $w_{ij} = 0$. Similarly, if i and j occur within a specified time of one another, set $y_{ij} = 1$; otherwise, set $y_{ij} = 0$. Then the space-time interaction coefficient is

$$\text{STI} = \sum_{i \neq j} \sum w_{ij} y_{ij}$$

For example, for a disease such as measles, we might consider cases within 1 mile of each other occurring 10 days or less apart (the length of the latent period). If n_S and n_T denote the number of adjacent pairs in space and time, respectively, and both are small relative to n, then the conditional distribution of STI given n_S, n_T, and n is approximately Poisson with expected value $n_S n_T / n$. The test has been extended to several spatial and temporal scales by Mantel [13]. For further details, see Cliff and Ord [5, Chaps. 1 and 2].

[*Editor's Addendum.* A population that has its individuals evenly distributed is frequently called *overdispersed*, while an aggregated population with its individuals clustered in groups can be described as *underdispersed*; see OVERDISPERSION.]

AGGREGATION AS A STATISTICAL METHOD

Aggregation in this sense involves the compounding of primary data in order to express them in summary form. Also, such an exercise is necessary when a model is specified at the micro (or individual) level but the usable data refer to aggregates. Then the question that arises is whether the equations of the micro model can be combined in such a way as to be consistent with the macro (or aggregate) model to which the data refer.

We may wish to compound individual data records, such as consumers' expenditures, or to combine results over time and/or space. The different cases are described in turn.

Combining Individual Records

Consider a population of N individuals in which the ith individual ($i = 1, \ldots, N$) has response Y_i to input x_i of the form

$$Y_i = f(x_i, \beta_i) + \epsilon_i$$

where β_i denotes a (vector of) parameter(s) specific to the ith individual and ϵ_i denotes a random-error term. For example, the equation may represent the consumer's level of expenditure on a commodity given its price. Then the total expenditure is $Y = \sum Y_i$ (summed over $i = 1, \ldots, N$), and the average input is $x = \sum x_i / N$.

In general, it is not possible to infer an exact relationship between Y and x from the micro relations. The few results available refer to the linear aggregation of linear equations. Theil [20] showed that when f denotes a linear function so that

$$Y_i = \alpha_i + \beta_i x_i + \epsilon_i,$$

perfect aggregation is possible, in that we may consider a macro relation of the form

$$Y = \alpha + \beta x^* + \epsilon,$$

where $\alpha = \sum \alpha_i$, $\beta = \sum \beta_i$, $x^* = \sum \beta_i x_i / \beta$, and $\epsilon = \sum \epsilon_i$. That is, we must use the weighted average x^* rather than the natural average x. Further, a different aggregation procedure is required for each regressor variable and for the same regressor variable with different response variables [2, Chap. 20; 20, Chap. 2]. If we use the natural average, the macro relation is

$$Y = \alpha + \beta x + N \operatorname{cov}(x_i, \beta_i) + \epsilon,$$

where the covariance is evaluated over the N members of the population and represents the aggregation bias. This bias is small, for example, when x is the price variable in a consumer demand equation, but may be much more substantial when x denotes consumers' income in such a relationship. When the micro relationship is a nonlinear function of x_i, the nonlinearity will generate a further source of aggregation bias. It must be concluded that exact aggregation is rarely possible, although the bias may be small in many cases. For further discussion and recent developments of the theory, see Ijiri [11].

Aggregation of Groups

Instead of forming a macro relation from a known group of individuals, we may wish to identify suitable groups from a finer classification of individuals. This is a necessary step in the construction of broad industrial classifications for use in input-output systems. See ECONOMETRICS. Blin and Cohen [4] propose a method of cluster analysis* for solving this problem.

Temporal Aggregation

Variates may be continuous or summed over a unit time interval, although the variate is recorded only as an aggregate over periods of r units duration. For a model that is linear in the regressor variables and has time-invariant parameters, aggregation is straightforward provided that there are no lagged variables. However, if

$$Y_t = \alpha + \beta x_{t-k} + \epsilon_t$$

for some $k > 0$ and k not a multiple of r, exact aggregation is not possible; any aggregated model will involve x-values for two or more time periods [20, Chap. 4]. Such models are often formulated using distributed lags*. Also, Granger and Morris [9] show that the autoregressive–moving average (ARMA) models* are often appropriate in this case.

The aggregation of time series* exhibiting positive autocorrelation tends to increase the values of the various test statistics and thereby give rise to overoptimistic assessments of the model [17]. However, Tiao and Wei [21] have shown that whereas aggregation can considerably reduce the efficiency of parameter estimators*, it has much less effect upon prediction efficiency. Indeed, it has been shown that there are circumstances where forecasts from aggregate relations may be more accurate than an aggregate of forecasts from micro relations [1,10].

The discussion so far has assumed that β does not vary over time. For a discussion of aggregation when there is a change of regime (time-varying parameters), see Goldfeld and Quandt [8, Chap. 4].

Spatial Aggregation

Many problems in spatial aggregation are similar to those of time series, but they are further compounded by the (sometimes necessary) use of areas of irregular shape and different size. Yule and Kendall [22] first showed how different aggregations of spatial units affect estimates of correlation*. Cliff and Ord [5, Chap. 5] give a general review of methods for estimating the autocovariance and cross-correlation functions using a nested hierarchy of areal sampling units. The estimation of these functions for irregular areas depends upon making rather restrictive assumptions about the nature of interaction between areas.

Cliff and Ord considered data from the *London Atlas*; a 24×24 lattice of squares of side 500 meters was laid over the Greater London area and the percentage of land used for commercial (X), industrial (Y), office (Z), and other purposes was recorded for each square. The correlation between each X and Y for different combinations of grid squares were as follows:

	Size of Spatial Unit			
	1×1	2×2	4×4	8×8
corr(X, Z)	0.19	0.36	0.67	0.71
corr(Y, Z)	0.09	0.16	0.33	0.34

The correlation functions exhibit an element of mutual exclusion at the smallest levels, and the positive correlation for larger spatial units indicates the general effects of areas zoned for housing and nonhousing purposes. Here, as in time series, we must think in terms of a distance or time-dependent correlation function and not a unique "correlation" between variables.

A PERSPECTIVE

Aggregation appears both as a phenomenon of interest in its own right and as a necessary evil in modeling complex processes. The center-satellite models have proved useful in astronomy [14], in ecology [3,15], in geography [5], and several other disciplines. At the present time, such processes offer a flexible tool for simulation work, although further work on the theory and analysis of such processes is desirable; the data analytic distance methods of Ripley [16] represent a useful step in the right direction. In epidemiology [12,13] and hydrology (models of storms, etc.) the development of clusters in both space and time is important, although relatively little work exists to date. The work of Taylor [18] represents a challenge to the theoretician, as useful models generating the empirical regularities observed by Taylor are still lacking.

Econometricians seem to be turning away from the view that an aggregate model is the sum of its parts and placing more emphasis upon aggregated models per se. The nonlinearities of the aggregation procedure, combined with the complexities of the underlying processes [8], suggest that aggregated models with time-dependent parameters are likely to play an increasing role in economics and other social sciences.

Where the level of aggregation is open to choice [4], further work is needed to identify suitable procedures for combining finer units into coarser ones. Similar problems arise in quadrat sampling* [15, p. 222].

The use of a sample to estimate the mean over an area or volume is of interest in the geosciences (e.g., drillings in an oil field). Estimators for such aggregates are based upon a variant of generalized least squares* known as "kriging"*; see Delfiner and Delhomme [7] and the papers of Matheron cited therein for further details.

In all these areas, much remains to be discussed about the statistical properties of the estimators currently used, and there is still plenty of scope for the development of improved methods.

REFERENCES

1. Aigner, D. J., and Goldfeld, S. M. (1974). *Econometrica*, **42**, 113–134.

2. Allen, R. G. D. (1959). *Mathematical Economics*. Macmillan, London. (Outlines the aggregation problem in econometric modeling and sets it in the context of other aspects of mathematical economics; written at an intermediate mathematical level.)

3. Bartlett, M. S. (1975). *The Statistical Analysis of Spatial Pattern*. Chapman & Hall, London. (A concise introduction to the theory of spatial point processes and lattice processes with a variety of applications in ecology.)

4. Blin, J. M. and Cohen, C. (1977). *Rev. Econ. Statist.*, **52**, 82–91. (Provides a review and references on earlier attempts to form viable aggregates as well as new suggestions.)

5. Cliff, A. D. and Ord, J. K. (1981). *Spatial Processes: Models, Inference and Applications*. Pion, London. (Discusses aggregation problems in the context of spatial patterns with examples drawn from ecology and geography; written at an intermediate mathematical level.)

6. Cormack, R. M. (1979). In *Spatial and Temporal Processes in Ecology*, R. M. Cormack and J. K. Ord, eds. International Co-operative Publishing House, Fairland, Md., pp. 151–211. (An up-to-date review of spatial interaction models with an extensive bibliography. Other papers in the volume cover related aspects.)

7. Delfiner, P. and Delhomme, J. P. (1975). In *Display and Analysis of Spatial Data*, J. C. Davis and J. C. McCullagh, eds. Wiley, New York; pp. 96–114. (This volume contains several other papers of general interest on spatial processes.)

8. Goldfeld, M. and Quandt, R. E. (1976). *Studies in Nonlinear Estimation*. Cambridge, Mass.: Ballinger.

9. Granger, C. W. J. and Morris, J. J. (1976). *J. R. Statist. Soc. A*, **139**, 246–257.

10. Grunfeld, Y. and Griliches, Z. (1960). *Rev. Econ. Statist.*, **42**, 1–13.

11. Ijiri, Y. (1971). *J. Amer. Statist. Ass.*, **66**, 766–782. (A broad review of aggregation for economic models and an extensive bibliography.)

12. Knox, E. G. (1964). *Appl. Statist.*, **13**, 25–29.

13. Mantel, N. (1967). *Cancer Res.*, **27**, 209–220.

14. Neyman, J. and Scott, E. L. (1958). *J. R. Statist. Soc. B*, **20**, 1–43. (The seminal paper on clustering processes.)

15. Pielou, E. C. (1977). *Mathematical Ecology*. Wiley, New York. (Describes methods for the measurement of aggregation among individuals; extensive bibliography.)

16. Ripley (1977).

17. Rowe, R. D. (1976). *Int. Econ. Rev.*, **17**, 751–757.

18. Taylor, L. R. (1971). In *Statistical Ecology*, Vol. 1, G. P. Patil et al., eds. Pennsylvania State University Press, University Park, Pa., pp. 357–377.

19. Taylor, L. R. and Taylor, R. A. J. (1977). *Nature, (Lond.)*, **265**, 415–421.

20. Theil, H. (1954). *Linear Aggregates of Economic Relations*. North-Holland, Amsterdam. (The definitive work on aggregation for economic models.)

21. Tiao, G. C. and Wei, W. S. (1976). *Biometrika*, **63**, 513–524.

22. Yule, G. U. and Kendall, M. G. (1965). *An Introduction to the Theory of Statistics*. Charles Griffin, London.

See also DISPERSION THEORY, HISTORICAL DEVELOPMENT OF; DIVERSITY INDICES; ECOLOGICAL STATISTICS; ECONOMETRICS; and OVERDISPERSION.

J. KEITH ORD

AGING FIRST-PASSAGE TIMES

First-passage times of appropriate stochastic processes* have often been used to represent times to failure of devices or systems subjected to shocks and wear, random repair times, and random interruptions during their operations. Therefore the aging properties of such first-passage times have been widely investigated in the reliability* and maintenance literature. In this entry we overview some basic results regarding first-passage times that have aging properties that are related as follows (the definitions of these notions are given later):

$$PF_2 \Rightarrow IFR \Rightarrow DMRL \Rightarrow NBUE$$
$$\Downarrow \qquad\qquad \Uparrow$$
$$IFRA \Rightarrow NBU \Rightarrow NBUC$$

Some terminology and notation that are used throughout the entry are described below. Let $\{X_n, n \geq 0\}$ be a discrete-time stochastic process with state space $[0, \infty)$. We assume that the process starts at 0, that is, $P\{X_0 = 0\} = 1$. For every $z \geq 0$ we denote by T_z the first time that the process crosses the threshold z, that is, $T_z \equiv \inf\{n \geq 0 : X_n \geq z\}$ ($T_z = \infty$ if $X_n < z$ for all $n \geq 0$). If, for example, T_z is new better than used (NBU) [increasing failure rate (IFR), increasing failure-rate average (IFRA), etc.] for any $z \geq 0$ (see HAZARD RATE AND OTHER CLASSIFICATIONS OF DISTRIBUTIONS) then the process $\{X_n, n \geq 0\}$ is called an NBU [IFR, IFRA, etc.] process. In a similar manner one defines an NBU [IFR, IFRA, etc.] continuous-time process. In this entry we point out many instances of NBU, IFRA, IFR, and other processes that have first-passage times with similar aging properties. We do not consider here antiaging properties such as NWU (new worse than used), DFR (decreasing failure rate), DFRA (decreasing failure-rate average), etc. Some antiaging properties of first-passage times can be found in the references.

MARKOV PROCESSES

Consider a discrete-time Markov process* $\{X_n, n \geq 0\}$ with the discrete state space $\mathbb{N}_+ = \{0, 1, \ldots\}$. Denote by $\boldsymbol{P} = \{p_{ij}\}_{i \in \mathbb{N}_+, j \in \mathbb{N}_+}$

the transition matrix of the process. Keilson, in a pioneering work [16], obtained many distributional properties, such as complete monotonicity, for various first-passage times of such processes. The strongest aging property that we consider here is that of log-concavity. A nonnegative random variable is said to have the *Pólya frequency of order* 2 (PF$_2$) property if its (discrete or continuous) probability density is log-concave. *See* PÓLYA TYPE 2 FREQUENCY (PF$_2$) DISTRIBUTIONS. Assaf et al. [4] have shown the following result:

PF$_2$ Theorem. If the transition matrix $\boldsymbol{P}$ is totally positive of order 2 (TP$_2$) (that is, $p_{ij}p_{i'j'} \geq p_{i'j}p_{ij'}$ whenever $i \leq i'$ and $j \leq j'$), then T_z has a log-concave density for all $z \geq 0$, that is, $\{X_n, n \geq 0\}$ is a PF$_2$ process.

They also extended this result to some continuous-time Markov processes with discrete state space. Shaked and Shanthikumar [35] extended it to continuous-time pure jump Markov processes with continuous state space. *See* also TOTAL POSITIVITY.

An aging notion that is weaker than the PF$_2$ property is the notion of IFR (*increasing failure rate*). A discrete nonnegative random variable T is said to have this property if $P\{T \geq n\}$ is log-concave on $\mathbb{N}_+$, or, equivalently, if its discrete hazard rate* function, defined by $P\{T = n\}/P\{T \geq n\}$, is nondecreasing on $\mathbb{N}_+$. If $\boldsymbol{P}$ is the transition matrix of the discrete-time Markov process $\{X_n, n \geq 0\}$, then let $\boldsymbol{Q}$ denote the matrix of left partial sums of $\boldsymbol{P}$. Formally, the i, jth element of $\boldsymbol{Q}$, denoted by q_{ij}, is defined by $q_{ij} = \sum_{k=0}^{j} p_{ik}$. Durham et al. [10] essentially proved the following result.

IFR Theorem. If $\boldsymbol{Q}$ is TP$_2$, then T_z is IFR for all $z \geq 0$, that is, $\{X_n, n \geq 0\}$ is an IFR process.

This strengthens previous results of Esary et al. [13] and of Brown and Chaganty [8]. Using ideas of the latter, it can be extended to some continuous-time Markov processes with discrete state space. Shaked and Shanthikumar [35] have extended this result to continuous-time pure jump Markov processes with continuous state space; see also refs. 17, 38.

One application of the IFR result described above is given in Kijima and Nakagawa [18]. They considered a Markov process $\{X_n, n \geq 0\}$ defined by $X_n = bX_{n-1} + D_n, n = 1, 2, \ldots (X_0 \equiv 0)$, where $0 \leq b \leq 1$, and the D_n's are independent nonnegative random variables. Such processes arise in studies of imperfect preventive maintenance policies, where each maintenance action reduces the current damage by $100(1 - b)\%$, and D_n is the total damage incurred between the $(n - 1)$st and the nth maintenance action. Let G_n denote the distribution function of D_n. They showed that if $G_n(x)$ is TP$_2$ in n and x, and if $G_n(x)$ is log-concave in x for all n, then $\{X_n, n \geq 0\}$ is an IFR process.

Since the IFR property is weaker than the PF$_2$ property, one would expect that a condition weaker than the assumption that P is TP$_2$ would suffice to guarantee that $\{X_n, n \geq 0\}$ is an IFR process. Indeed, the assumption that Q is TP$_2$ is weaker than the assumption that P is TP$_2$.

Sometimes the time that the maximal increment of a Markov process passes a certain critical value is of interest. Thus, Li and Shaked [21] studied first-passage times of the form $T_{z,u} \equiv \inf\{n \geq 1 : X_n \geq z \text{ or } X_n - X_{n-1} \geq u\}[T_{z,u} = \infty \text{ if } X_n < z \text{ and } X_n - X_{n-1} < u \text{ for all } n]$. The process $\{X_n, n \geq 0\}$ is said to have a *convex transition kernel* if $P\{X_{n+1} > x + y | X_n = x\}$ is nondecreasing in x for all y. They have shown the following result:

Increment IFR Theorem. If $\{X_n, n \geq 0\}$ has monotone sample paths, and a convex transition kernel, and if P is TP$_2$, then $T_{z,u}$ is IFR for all z and u.

They also obtained a version of this result for some continuous-time Markov processes with discrete state space.

An aging notion that is weaker than the IFR property is the notion of IFRA (*increasing failure-rate average*). A discrete nonnegative random variable T is said to have this property if either $P\{T = 0\} = 1$ or $P\{T = 0\} = 0$ and $[P\{T > n\}]^{1/n}$ is nonincreasing in $n = 1, 2, \ldots$. The process $\{X_n, n \geq 0\}$ is said to be *stochastically monotone* if $P\{X_{n+1} > x | X_n = y\}$ is nondecreasing in y for every x. Equivalently, $\{X_n, n \geq 0\}$ is said to be stochastically

monotone if q_{ij} is nonincreasing in i for all j, where Q is the matrix of left partial sums of P. From a general result of Shaked and Shanthikumar [34] we obtain the following result.

IFRA Theorem. If $\{X_n, n \geq 0\}$ is stochastically monotone and if it has nondecreasing sample paths, then T_z is IFRA for all $z \geq 0$, that is, $\{X_n, n \geq 0\}$ is an IFRA process.

This result strengthens previous results of Esary et al. [13] and of Brown and Chaganty [8]. Using ideas of the latter, it can be extended to some continuous-time Markov processes with discrete state space. Drosen [9] and Shaked and Shanthikumar [35] have extended it to continuous-time pure jump Markov processes with continuous state space. Özekici and Günlük [28] have used this result to show that some interesting Markov processes that arise in the study of some maintenance policies are IFRA.

Since the IFRA property is weaker than the IFR property, one would expect that a condition weaker than the assumption that Q is TP$_2$ would suffice to guarantee that $\{X_n, n \geq 0\}$ is an IFRA process. Indeed, the assumption that $\{X_n, n \geq 0\}$ is stochastically monotone is weaker than the assumption that Q is TP$_2$. However, for the IFRA result we need to assume that $\{X_n, n \geq 0\}$ has nondecreasing sample paths, whereas there is no such assumption in the IFR result.

When the time that the maximal increment of a Markov process passes a certain critical value is of interest, then one studies $T_{z,u}$. Li and Shaked [21] have shown the following result.

Increment IFRA Theorem. If $\{X_n, n \geq 0\}$ has monotone convex sample paths and a convex transition kernel, then $T_{z,u}$ is IFRA for all z and u.

They also obtained a version of this result for some continuous-time Markov processes with discrete state space.

An aging notion that is weaker than the IFRA property is the notion of NBU (*new better than used*). A discrete nonnegative random variable T is said to have this property if $P\{T \geq n\} \geq P\{T - m \geq n | T \geq m\}$ for all $n \geq 0$ and $m \geq 0$. Brown and Chaganty [8] proved the following result.

NBU Theorem. If $\{X_n, n \geqslant 0\}$ is stochastically monotone, then T_z is NBU for all $z \geqslant 0$, that is, $\{X_n, n \geqslant 0\}$ is an NBU process.

This strengthens previous results of Esary et al. [13]. Brown and Chaganty [8] extended it to some continuous-time Markov processes with discrete state space.

Since the NBU property is weaker than the IFRA property, it is not surprising that the condition that suffices to imply that $\{X_n, n \geqslant 0\}$ is NBU is weaker than the conditions that imply that $\{X_n, n \geqslant 0\}$ is IFRA.

Marshall and Shaked [25] have identified a condition that is different than stochastic monotonicity, and that still yields that $\{X_n, n \geqslant 0\}$ is an NBU process. Explicitly, they showed that if the strong Markov process $\{X_n, n \geqslant 0\}$ starts at 0, and is free of positive jumps (that is, it can jump up only at most one unit at a time), then it is an NBU process. They obtained a similar result for continuous-time strong Markov processes with discrete or continuous state space. For example, a Wiener process which starts at 0 is an NBU process. Also, if $\{(X_1(t), X_2(t), \ldots, X_m(t)), t \geqslant 0\}$ is a Brownian motion* in $\mathbb{R}^m$, and $Y(t), \equiv [\sum_{i=1}^{m} X_i^2(t)]^{1/2}$, then the Bessel process $\{Y(t), t \geqslant 0\}$ is NBU.

Li and Shaked [21] studied the NBU property of $T_{z,u}$, the time that the maximal increment of a Markov process passes a certain critical value, and obtained the following result.

Increment NBU Theorem. If $\{X_n, n \geqslant 0\}$ has a convex transzition kernel, then $T_{z,u}$ is NBU for all z and u.

They also obtained a version of this result for some continuous-time Markov processes with discrete state space.

An aging notion that is weaker than the NBU property is the notion of NBUC (*new better than used in convex ordering*). A nonnegative random variable T is said to have this property if $E[\phi(T)] \geqslant E[\phi(T - m)|T \geqslant m]$ for all $m \geqslant 0$ and for all nondecreasing convex functions ϕ for which the expectations are defined. If $\boldsymbol{P}$ is the transition matrix of $\{X_n, n \geqslant 0\}$, then the potential matrix of $\{X_n, n \geqslant 0\}$, which we denote by $\boldsymbol{R}$, is defined by $\boldsymbol{R} \equiv \sum_{n=0}^{\infty} \boldsymbol{P}^n$. Let $\overline{\boldsymbol{R}}$ denote the matrix of

left partial sums of $\boldsymbol{R}$, that is, if r_{ij} denotes the ijth element of $\boldsymbol{R}$ and $\bar{r}_{ij}$ denotes the ijth element of $\overline{\boldsymbol{R}}$, then $r_{ij} \equiv \sum_{k=0}^{j} r_{ik}$. Also, let us define the following matrix: $\boldsymbol{R}_m \equiv \sum_{n=m}^{\infty} \boldsymbol{P}^n$, and let $\overline{\boldsymbol{R}}_m$ denote the matrix of left partial sums of $\boldsymbol{R}_m$, that is, if $r_{m,ij}$ denotes the ijth element of $\boldsymbol{R}_m$, and $\bar{r}_{m,ij}$ denotes the ijth element of $\overline{\boldsymbol{R}}_m$, then $\bar{r}_{m,ij} \equiv \sum_{k=0}^{j} r_{m,ik}$. Ocón and Pérez [27] have shown the following result.

NBUC Theorem. If $\bar{r}_{m,ij}$ is nonincreasing in i for all j and m, then T_z is NBUC for all $z \geqslant 0$; that is, $\{X_n, n \geqslant 0\}$ is an NBUC process.

In addition to the condition given in the NBUC Theorem, they also assumed that $\{X_n, n \geqslant 0\}$ has nondecreasing sample paths, but if, for a fixed z, one modifies $\{X_n, n \geqslant 0\}$ so that z is an absorbing state, then the almost sure monotonicity of the sample paths is not required for the conclusion that $\{X_n, n \geqslant 0\}$ is NBUC. They have also extended this result to some continuous-time Markov processes with discrete state space.

Since the NBUC property is weaker than the NBU property, one would expect that a condition weaker than stochastic monotonicity should suffice to imply that $\{X_n, n \geqslant 0\}$ is NBUC. Indeed, it can be shown that if $\{X_n, n \geqslant 0\}$ is stochastically monotone then $\bar{r}_{m,ij}$ is nonincreasing in i for all j and m.

An aging notion that is weaker than the NBUC property is the notion of NBUE (*new better than used in expectation*). A nonnegative random variable T is said to have this property if $E[T] \geqslant E[T - m|T \geqslant m]$ and all $m \geqslant 0$. Karasu and Özekici [15] obtained the following result.

NBUE Theorem. If $\bar{r}_{ij}$ is nonincreasing in i for all j, then T_z is NBUE for all $z \geqslant 0$, that is, $\{X_n, n \geqslant 0\}$ is an NBUE process.

In addition to the condition given in the NBUE Theorem, they also assumed that $\{X_n, n \geqslant 0\}$ has nondecreasing sample paths, but, again, if for a fixed z one modifies $\{X_n, n \geqslant 0\}$ so that z is an absorbing state, then the almost sure monotonicity of the sample paths is not required for the conclusion that $\{X_n, n \geqslant 0\}$ is NBUE. They have also extended this result to some continuous-time Markov processes with discrete state space.

Since the NBUE property is weaker than the NBUC property, it is not surprising that the condition that suffices to imply that $\{X_n, n \geqslant 0\}$ is NBUE is weaker than the condition that implies that $\{X_n, n \geqslant 0\}$ is NBUC.

We close this section with an aging notion that is weaker than the IFR but stronger than the NBUE notion. A nonnegative random variable T is said to have the DMRL (*decreasing mean residual life*) property if $E[T - m | T \geqslant m]$ is nonincreasing in $m \geqslant 0$. If $\boldsymbol{P}$ is the transition matrix of the discrete-time Markov process $\{X_n, n \geqslant 0\}$, then let $\boldsymbol{Q}_m$ denote the matrix of left partial sums of $\boldsymbol{P}^m$. Denote the ijth element of $\boldsymbol{Q}_m$ by $q_{m,ij}$. Ocón and Pérez [27] have shown the following result.

DMRL Theorem. If $\bar{r}_{m,ij}/q_{m,ij}$ is nonincreasing in i for all j and m, then T_z is DMRL for all $z \geqslant 0$, that is, $\{X_n, n \geqslant 0\}$ is an DMRL process.

In addition to the condition given in the DMRL Theorem, they also assumed that $\{X_n, n \geqslant 0\}$ has nondecreasing sample paths, but, again, if for a fixed z one modifies $\{X_n, n \geqslant 0\}$ so that z is an absorbing state, then the almost sure monotonicity of the sample paths is not required for the conclusion that $\{X_n, n \geqslant 0\}$ is DMRL. They have also extended this result to some continuous-time Markov processes with discrete state space.

CUMULATIVE DAMAGE PROCESSES

Suppose that an item is subjected to shocks occurring randomly in (continuous) time according to a counting process* $\{N(t), t \geqslant 0\}$. Suppose that the ith shock causes a nonnegative random damage X_i, and that damages accumulate additively. Thus, the damage accumulated by the item at time t is $Y(t) = \sum_{i=1}^{N(t)} X_i$. We assume that the damage at time 0 is 0. The process $\{Y(t), t \geqslant 0\}$ is called a *cumulative damage* * *shock process*. Suppose that the item fails when the cumulative damage exceeds a threshold z.

If $\{N(t), t \geqslant 0\}$ is a Poisson process*, and if the damages X_i are independent and identically distributed and are independent of $\{N(t), t \geqslant 0\}$, then $\{Y(t), t \geqslant 0\}$ is a Markov process. The results described in the preceding section can then be used to derive aging properties of the first-passage time T_z. For example, the process $\{Y(t), t \geqslant 0\}$ clearly has monotone sample paths, and is stochastically monotone. Therefore it is IFRA. Esary et al. [13] noticed that if the X_i's are not necessarily identically distributed, but are merely stochastically increasing (that is, $P\{X_i > x\}$ is nondecreasing in i for all x), then the process $\{Y(t), t \geqslant 0\}$ is still IFRA. In fact, they identified even weaker conditions on the X_i's that ensure that $\{Y(t), t \geqslant 0\}$ is IFRA. As another example of the application of the results of the preceding section to the process $\{Y(t), t \geqslant 0\}$, suppose that the damages X_i are independent and identically distributed with a common log-concave distribution function; then the process $\{Y(t), t \geqslant 0\}$ is IFR. In fact, Esary et al. [13] and Shaked and Shanthikumar [35] have identified even weaker conditions on the X_i's that ensure that $\{Y(t), t \geqslant 0\}$ is IFR.

If $\{N(t), t \geqslant 0\}$ is a nonhomogeneous (rather than homogeneous) Poisson process, then $\{Y(t), t \geqslant 0\}$ is not a Markov process, even if the damages X_i are independent and identically distributed, and are independent of $\{N(t), t \geqslant 0\}$. Let $\Lambda(t), t \geqslant 0$, be the mean function of the process $\{N(t), t \geqslant 0\}$. From results of A-Hameed and Proschan [3] it follows that the IFRA results mentioned in the previous paragraph still hold provided Λ is star-shaped. It also follows that the IFR results mentioned in the previous paragraph still hold provided Λ is convex.

Sumita and Shanthikumar [40] have studied a cumulative damage wear process* in which $\{N(t), t \geqslant 0\}$ is a general renewal* process. Let the interarrivals of $\{N(t), t \geqslant 0\}$ be denoted by $U_i, i = 1, 2, \ldots$. Thus, $(U_i, X_i), i = 1, 2, \ldots$, are independent and identically distributed pairs of nonnegative random variables. It is not assumed that, for each i, U_i and X_i are independent. In fact, the model is of particular interest when, for each i, U_i and X_i are not independent. If we define $Y(t) \equiv \sum_{i=1}^{N(t)} X_i$, then in general $\{Y(t), t \geqslant 0\}$ is not a Markov process. They showed that if the U_i's are NBU, and if the pairs (U_i, X_i) possess some positive dependence properties, then $\{Y(t), t \geqslant 0\}$ is an NBU process. They also showed that if the U_i's are NBUE, and if the pairs (U_i, X_i) possess some other positive dependence properties, then $\{Y(t), t \geqslant 0\}$

is an NBUE process. Furthermore, they also showed that if the U_i's are HNBUE (*harmonic new better than used in expectation*) (that is, $\int_t^\infty P\{U_i > x\}\,dx \leqslant \mu \exp(-t/\mu)$ for $t \geqslant 0$, where $\mu = E[U_i]$), if the X_i's are exponential random variables, and if the pairs (U_i, X_i) possess some positive dependence properties, then $\{Y(t), t \geqslant 0\}$ is an HNBUE process. They obtained similar results for the model in which the nth interarrival depends on the $(n-1)$st jump X_{n-1} (rather than on the nth jump). In ref. 39 they considered a wear process that, at time t, is equal to $\max_{0 \leqslant n < N(t)}\{X_n\}$. Again, they did not assume that, for each i, U_i and X_i are independent. they identified conditions under which this process is NBU, NBUE, or HNBUE.

In the preceding paragraphs it has been assumed that the threshold z is fixed. But in many applications it is reasonable to allow it to be random, in which case denote the threshold by Z, and the first-passage time to Z by T_Z. For this model Esary et al. [13] have shown that if Z is IFRA [respectively, NBU] then T_Z is IFRA [respectively, NBU]. They have also shown that if the identically distributed random damages X_i have the PF$_2$ property, and if Z has the PF$_2$ property, then T_Z has the PF$_2$ property.

A-Hameed and Proschan [3] considered a random threshold cumulative damage model in which the damages X_i are still independent, but are not necessarily identically distributed, and the counting process $\{N(t), t \geqslant 0\}$ is a nonhomogeneous Poisson process with mean function Λ. In fact, they assumed that the ith random damage has the gamma distribution* with shape parameter a_i and rate parameter b. Let $A_k \equiv \sum_{i=1}^k a_i, k = 1, 2, \ldots$. They showed that if A_k is convex [respectively, star-shaped, superadditive] in k, if Λ is convex [respectively, star-shaped, super-additive], and if Z is IFR [respectively, IFRA, NBU]. For the special case when all the a_i's are equal, they showed that if Λ is convex [star-shaped] and if Z is DMRL [NBUE], then T_Z is DMRL [NBUE]. For this special case, Klefsjö [19] showed that if Λ is star-shaped and if Z is HNBUE, then T_Z is HNBUE. Abdel-Hameed [1,2] and Drosen [9] have extended some of these results to pure jump wear processes.

The reader is referred to Shaked [32] for more details and further references on cumulative damage processes.

NON-MARKOVIAN PROCESSES

In many applications in reliability theory*, the underlying wear process is non-Markovian. Various researchers have tried to obtain aging properties of first-passage times for some such processes. We describe some fruits of their efforts.

IFRA Closure Theorem [29]. If $\{X_i(t), t \geqslant 0\}, i = 1, 2, \ldots, n$ are independent IFRA processes, each with nondecreasing sample paths, then $\{\phi(X_1(t), X_2(t), \ldots, X_n(t)), t \geqslant 0\}$ is also an IFRA process whenever ϕ is continuous and componentwise nondecreasing.

The following result follows from refs. 11, 26.

NBU Closure Theorem. If $\{X_i(t), t \geqslant 0\}, i = 1, 2, \ldots, n$, are independent NBU processes, each with nondecreasing sample paths, then $\{\phi(X_1(t), X_2(t), \ldots, X_n(t)), t \geqslant 0\}$ is also an NBU process whenever ϕ is continuous and componentwise nondecreasing.

More general results are described in the next section.

Marshall and Shaked [25] and Shanthikumar [37] have identified a host of non-Markovian processes that are NBU. We will try to describe some of these processes in plain words. One class of NBU wear processes that Marshall and Shaked [25] have identified is the following, with shocks and recovery: Shocks occur according to a renewal process with NBU interarrivals. Each shock causes the wear to experience a random jump, where the jumps are independent and identically distributed and are independent of the underlying renewal process. These jumps may be negative as long as the wear stays nonnegative. Between shocks the wear changes in some deterministic manner which depends on the previous history of the process. This deterministic change may correspond to a partial recovery of the underlying device.

A second class of NBU wear processes that Marshall and Shaked [25] have identified is the following, with random repair times: The process starts at 0, and before the first shock it increases in some deterministic manner. Shocks occur according to a Poisson process. Each shock causes the wear to experience a random jump (usually a negative jump, but the process is set equal to 0 if such a jump would carry it below 0), where the jumps are independent and identically distributed, and are independent of the underlying Poisson process. Between shocks the wear increases in some deterministic manner where the rate of increase depends only on the current height of the process. This deterministic change may correspond to a continuous wear of the underlying device, and the jumps correspond to repairs that reduce the wear.

A third class of NBU wear processes that Marshall and Shaked [25] have identified is that of Gaver–Miller [14] processes. These have continuous sample paths that alternately increase and decrease in a deterministic fashion where the rate of increase or decrease depends only on the current height of the process. The random durations of increase are independent and identically distributed exponential random variables, and the random durations of decrease are independent and identically distributed NBU random variables.

Shanthikumar [37] has generalized the first two kinds of processes mentioned above. In particular, he allowed the times between jumps and the magnitude of jumps to be dependent, and he still was able to prove, under some conditions, that the resulting wear processes are NBU. His results also extend the NBU results of Sumita and Shanthikumar [40]. Lam [20] has gone even further and identified a class of stochastic processes that are even more general than those of Shanthikumar [37]. She showed that the processes in that class are NBUE. Marshall and Shaked [26] extended the NBU results that are described above to processes with state space $[0, \infty)^m$.

Semi-Markov processes* are more general than Markov processes in the sense that the sojourn time of the process in each state has a general distribution rather than being exponential. Using coupling arguments, Shanthikumar [37] was able to formulate a set of conditions under which semi-Markov processes

are NBU. Lam [20] obtained conditions under which semi-Markov processes are NBUE.

For some more details on the aging properties of first-passage times of non-Markovian processes, and for further references, see the review by Shaked [32].

PROCESSES WITH STATE SPACE $\mathbb{R}^M$

Let $\{\mathbf{X}(t), t \geqslant 0\} = \{(X_1(t), X_2(t), \ldots, X_m(t)), t \geqslant 0\}$ be a stochastic process on $\mathbb{R}_+^m \equiv [0, \infty)^m$. A set $U \subseteq \mathbb{R}_+^m$ is an *upper set* if $\mathbf{x} \in U$ and $\mathbf{y} \geqslant \mathbf{x}$ implies that $\mathbf{y} \in U$. The first-passage time of the process $\{\mathbf{X}(t), t \geqslant 0\}$ to an upper set U is defined by $T_U \equiv \inf\{t \geqslant 0 : \mathbf{X}(t) \in U\}$ [$T_U = \infty$ if $\mathbf{X}(t) \notin U$ for all $t \geqslant 0$]. The process $\{\mathbf{X}(t), t \geqslant 0\}$ is called an IFRA [NBU] process if T_U is IFRA [NBU] for all closed upper sets $U \subseteq \mathbb{R}_+^m$. Clearly, every component $\{X_i(t), t \geqslant 0\}$ of an IFRA [NBU] process $\{\mathbf{X}(t), t \geqslant 0\}$ is an IFRA [NBU] process on $\mathbb{R}_+$. In this section we consider only processes that start at $\mathbf{0}$. The following characterizations of IFRA and NBU processes are taken from refs. 26, 36.

IFRA and NBU Characterization Theorem

(i) The process $\{\mathbf{X}(t), t \geqslant 0\}$, with nondecreasing sample paths, is IFRA if, and only if, for every choice of closed upper sets $U_1, U_2, \ldots, U_n$, the random variables $T_{U_1}, T_{U_2}, \ldots, T_{U_n}$ satisfy that $\tau(T_{U_1}, T_{U_2}, \ldots, T_{U_n})$ is IFRA for every coherent life function τ. (For a definition of coherent life functions see, e.g., Esary and Marshall [12] or Barlow and Proschan [5].)

(ii) The process $\{\mathbf{X}(t), t \geqslant 0\}$, with nondecreasing sample paths, is NBU if, and only if, for every choice of closed upper sets $U_1, U_2, \ldots, U_n$, the random variables $T_{U_1}, T_{U_2}, \ldots, T_{U_n}$ satisfy that $\tau(T_{U_1}, T_{U_2}, \ldots, T_{U_n})$ is NBU for every coherent life function τ.

The following IFRA closure properties can be derived from results in Marshall [22].

General IFRA Closures Theorem

(i) Let $\{\mathbf{X}_i(t), t \geqslant 0\}$ be an IFRA process on $\mathbb{R}_+^{m_i}$ with nondecreasing sample paths, $i = 1, 2, \ldots, n$. If these n processes are independent, then $\{(\boldsymbol{X}_1(t), \boldsymbol{X}_2(t), \ldots, \boldsymbol{X}_n(t)), t \geqslant 0\}$ is an IFRA process on $\mathbb{R}_+^{\sum_{i=1}^n m_i}$.

(ii) Let $\{\boldsymbol{X}(t), t \geqslant 0\}$ be an IFRA process on $\mathbb{R}_+^{m_1}$, and let $\boldsymbol{\psi} : \mathbb{R}_+^{m_1} \to \mathbb{R}_+^{m_2}$ be a nondecreasing continuous function such that $\boldsymbol{\psi}(\boldsymbol{0}) = \boldsymbol{0}$. Then $\{\boldsymbol{\psi}(\boldsymbol{X}(t)), t \geqslant 0\}$ is an IFRA process on $\mathbb{R}_+^{m_2}$.

The following NBU closure properties can be derived from results in Marshall and Shaked [26].

General NBU Closures Theorem

(i) Let $\{\boldsymbol{X}_i(t), t \geqslant 0\}$ be an NBU process on $\mathbb{R}_+^{m_i}$ with nondecreasing sample paths, $i = 1, 2, \ldots, n$. If these n processes are independent, then $\{(\boldsymbol{X}_1(t), \boldsymbol{X}_2(t), \ldots, \boldsymbol{X}_n(t)), t \geqslant 0\}$ is an NBU process on $\mathbb{R}_+^{\sum_{i=1}^n m_i}$.

(ii) Let $\{\boldsymbol{X}(t), t \geqslant 0\}$ be an NBU process on $\mathbb{R}_+^{m_1}$, and let $\boldsymbol{\psi} : \mathbb{R}_+^{m_1} \to \mathbb{R}_+^{m_2}$ be a nondecreasing continuous function such that $\boldsymbol{\psi}(\boldsymbol{0}) = \boldsymbol{0}$. Then $\{\boldsymbol{\psi}(\boldsymbol{X}(t)), t \geqslant 0\}$ is an NBU process on $\mathbb{R}_+^{m_2}$.

A discrete-time Markov process $\{\boldsymbol{X}_n, n \geqslant 0\}$ with state space $\mathbb{R}^m$ is said to be *stochastically monotone* if $P\{\boldsymbol{X}_{n+1} \in U | \boldsymbol{X}_n = \boldsymbol{x}\}$ is nondecreasing in $\boldsymbol{x}$ for all upper sets $U \subseteq \mathbb{R}^m$. Brown and Chaganty [8] have shown that if such a stochastically monotone process, with state space $\mathbb{R}_+^m$, starts at $\boldsymbol{0}$, then it is an NBU process. They also showed that some continuous-time Markov processes are NBU. Brown and Chaganty [8] and Shaked and Shanthikumar [34] have shown that if such a stochastically monotone process, with state space $\mathbb{R}_+^m$, starts at $\boldsymbol{0}$ and has nondecreasing sample paths, then it is an IFRA process. They also showed that some continuous-time Markov processes are IFRA. (In fact, they as well as Marshall [22] and Marshall and Shaked [26], have considered processes with state spaces that are much more general than $\mathbb{R}_+^m$.)

To see an application of their IFRA results, consider the following model of Ross [30]. Suppose that shocks hit an item according to a nonhomogeneous Poisson process $\{N(t), t \geqslant 0\}$ with mean function Λ. The ith

shock inflicts a nonnegative random damage X_i. The X_i's are assumed to be independent and identically distributed, and are also assumed to be independent of the underlying nonhomogeneous Poisson process. Suppose that there is a function D such that the total damage after n shocks is $D(X_1, X_2, \ldots, X_n, 0, 0, 0 \ldots)$, where D is a nonnegative function whose domain is $\{(x_1, x_2, \ldots), x_i \geqslant 0, i = 1, 2, \ldots\}$. Define $Y(t) \equiv D(X_1, X_2, \ldots, X_{N(t)}, 0, 0, 0, \ldots), t \geqslant 0$. If $\Lambda(t)/t$ is nondecreasing in $t > 0$, if D is nondecreasing in each of its arguments, and if $D(x_1, x_2, \ldots, x_n, 0, 0, 0, \ldots)$ is permutation symmetric in $x_1, x_2, \ldots, x_n$, for all n, then $\{Y(t), t \geqslant 0\}$ is an IFRA process.

A function $\phi : \mathbb{R}_+^m \to \mathbb{R}_+$ is said to be *subhomogeneous* if $\alpha\phi(\boldsymbol{x}) \leqslant \phi(\alpha\boldsymbol{x})$ for all $\alpha \in [0, 1]$ and all $\boldsymbol{x}$. Note that every coherent life function τ is a nondecreasing subhomogeneous function. A vector $(S_1, S_2, \ldots, S_n)$ of nonnegative random variables is said to be MIFRA (*multivariate increasing failure-rate average*), in the sense of Block and Savits [7], if $\phi(S_1, S_2, \ldots, S_n)$ is IFRA for any nondecreasing subhomogeneous function ϕ (see Marshall and Shaked [24] for this interpretation of the MIFRA property). In a similar manner Marshall and Shaked [24] have defined the notion of MNBU (*multivariate new better than used*). According to Block and Savits [7], a stochastic process $\{\boldsymbol{X}(t), t \geqslant 0\}$ on $\mathbb{R}_+^m$ is said to be a MIFRA process if, for every finite collection of closed upper sets $U_1, U_2, \ldots, U_n$ in $\mathbb{R}_+^m$, the vector $(T_{U_1}, T_{U_2}, \ldots, T_{U_n})$ is MIFRA. Clearly, every MIFRA process is an IFRA process. Block and Savits [7] showed that there exist IFRA processes that are not MIFRA. In a similar manner one can define MNBU processes. Clearly, every MIFRA process is also an MNBU process. It may be of interest to compare the definition of MIFRA and MNBU processes to the characterizations given in the IFRA and NBU Characterization Theorem above. Some multivariate cumulative damage wear processes that are MIFRA will be described now.

Consider m items that are subjected to shocks that occur according to (one) Poisson process $\{N(t), t \geqslant 0\}$. Let X_{ij} denote the damage inflicted by the ith shock on the jth item, $i = 1, 2, \ldots, j = 1, 2, \ldots, m$. Suppose that the vectors $\boldsymbol{X}_i = (X_{i1}, X_{i2}, \ldots, X_{im}), i = 1, 2, \ldots$, are independent. Assume that the damages accumulate additively. Thus, the wear process $\{Y(t), t \geqslant 0\} = \{(Y_1(t), Y_2(t), \ldots, Y_m(t)), t \geqslant 0\}$ here has state space $\mathbb{R}_+^m$, where $Y_j(t) = \sum_{i=1}^{N(t)} X_{ij}, t \geqslant 0, j = 1, 2, \ldots, m$.

IFRA Shock Model Theorem [31]. If $\boldsymbol{X}_i \leqslant_{\text{st}} \boldsymbol{X}_{i+1}$ (that is, $E[\psi(\boldsymbol{X}_i)] \leqslant E[\psi(\boldsymbol{X}_{i+1})]$) for all nondecreasing functions ψ for which the expectations are well defined), $i = 1, 2, \ldots$, then $\{\boldsymbol{Y}(t), t \geqslant 0\}$ is an IFRA process.

Thus, if the jth item is associated with a fixed threshold z_j (that is, the item fails once the accumulated damage of item j crosses the threshold z_j), $j = 1, 2, \ldots, m$, then, by the IFRA Characterization Theorem, the vector of the lifetimes of the items $(T_{z_1}, T_{z_2}, \ldots, T_{z_m})$ satisfies that $\tau(T_{z_1}, T_{z_2}, \ldots, T_{z_m})$ is IFRA for every coherent life function τ.

MIFRA Shock Model Theorem [31]. If $\boldsymbol{X}_i =_{\text{st}} \boldsymbol{X}_{i+1}$ (that is, the $\boldsymbol{X}_i$'s are identically distributed) then $\{\boldsymbol{Y}(t), t \geqslant 0\}$ is a MIFRA process.

Thus, the vector of the lifetimes of the items $(T_{z_1}, T_{z_2}, \ldots, T_{z_m})$ is such that $\phi(T_{z_1}, T_{z_2}, \ldots, T_{z_m})$ is IFRA for every nondecreasing subhomogeneous function ϕ.

Shaked and Shanthikumar [36] have extended the above results to processes with state spaces more general than $\mathbb{R}^m$. They also showed that these results still hold if the shocks occur according to a birth process with nondecreasing birth rates (rather than a homogeneous Poisson process). Marshall and Shaked [23] obtained some multivariate NBU properties for the model described above.

For additional details regarding multivariate IFRA and NBU processes see refs. 32, 33.

REFERENCES

1. Abdel-Hameed, M. (1984). Life distribution properties of devices subject to a Lévy wear process. *Math. Oper. Res.*, **9**, 606–614.

2. Abdel-Hameed, M. (1984). Life distribution properties of devices subject to a pure jump damage process. *J. Appl. Probab.*, **21**, 816–825.

3. A-Hameed, M. S. and Proschan, F. (1973). Nonstationary shock models. *Stochastic Process. Appl.*, **1**, 383–404.

4. Assaf, D., Shaked, M., and Shanthikumar, J. G. (1985). First-passage times with PF_r densities. *J. Appl. Probab.*, **22**, 185–196. (The PF_r property for Markov processes with discrete state space is established in this paper.)

5. Barlow, R. E. and Proschan, F. (1975). *Statistical Theory of Reliability and Life Testing: Probability Models*. Holt, Rinehart and Winston. (This is a comprehensive book that discusses various life distributions and many useful probabilistic models in reliability theory.)

6. Block, H. W. and Savits, T. H. (1980). Multivariate increasing failure rate average distributions. *Ann. Probab.*, **8**, 793–801.

7. Block, H. W. and Savits, T. H. (1981). Multidimensional IFRA processes. *Ann. Probab.*, **9**, 162–166. (Processes with multivariate IFRA first-passage times are considered in this paper.)

8. Brown, M. and Chaganty, N. R. (1983). On the first passage time distribution for a class of Markov chains. *Ann. Probab.*, **11**, 1000–1008. (The IFR, IFRA, and NBU properties for Markov processes with monotone transition matrices are presented in this paper.)

9. Drosen, J. W. (1986). Pure jump models in reliability theory. *Adv. Appl. Probab.*, **18**, 423–440. (The IFR, IFRA, and NBU properties for Markov pure jump processes with continuous state space are discussed in this paper.)

10. Durham, S., Lynch, J., and Padgett, W. J. (1990). TP₂-orderings and the IFR property with applications. *Probab. Eng. Inform. Sci.*, **4**, 73–88. (The IFR property for Markov processes with discrete state space is established in this paper.)

11. El-Neweihi, E., Proschan, F., and Sethuraman, J. (1978). Multistate coherent systems. *J. Appl. Probab.*, **15**, 675–688.

12. Esary, J. D. and Marshall. A. W. (1970). Coherent life functions. *SIAM J. Appl. Math.*, **18**, 810–814.

13. Esary, J. D., Marshall, A. W., and Proschan, F. (1973). Shock models and wear processes. *Ann. Probab.*, **1**, 627–649. (This paper is a pioneering work on aging properties of first-passage times of cumulative damage wear processes.)

14. Gaver, D. P. and Miller, R. G. (1962). Limiting distributions for some storage problems. In *Studies in Applied Probability and Management Science*, K. J. Arrow, S. Karlin, and H. Scarf, eds. Stanford University Press, pp. 110–126.

15. Karasu, I. and Özekici, S. (1989). NBUE and NWUE properties of increasing Markov processes. *J. Appl. Probab.*, **26**, 827–834.

16. Keilson, J. (1979). *Markov Chains Models—Rarity and Exponentiality*. Springer-Verlag, New York. (Early results on the complete monotonicity of some first-passage times of Markov processes with discrete state space are described in this book.)

17. Kijima, M. (1989). Uniform monotonicity of Markov processes and its related properties. *J. Oper. Res. Soc. Japan*, **32**, 475–490.

18. Kijima, M. and Nakagawa, T. (1991). A cumulative damage shock model with imperfect preventive maintenance. *Naval Res. Logist.*, **38**, 145–156.

19. Klefsjö, B. (1981). HNBUE survival under some shock models. *Scand. J. Statist.*, **8**, 39–47.

20. Lam, C. Y. T. (1992). New better than used in expectation processes. *J. Appl. Probab.*, **29**, 116–128. (The NBUE property for semi-Markov processes and some other non-Markovian processes is given in this paper.)

21. Li, H. and Shaked, M. (1995). On the first passage times for Markov processes with monotone convex transition kernels. *Stochastic Process. Appl.*, **58**, 205–216. (The aging properties for the increment processes of some Markov processes are considered in this paper.)

22. Marshall, A. W. (1994). A system model for reliability studies. *Statist. Sinica*, **4**, 549–565.

23. Marshall, A. W. and Shaked, M. (1979). Multivariate shock models for distributions with increasing hazard rate average. *Ann. Probab.*, **7**, 343–358.

24. Marshall, A. W. and Shaked, M. (1982). A class of multivariate new better than used distributions. *Ann. Probab.*, **10**, 259–264.

25. Marshall, A. W. and Shaked, M. (1983). New better than used processes. *Adv. Appl. Probab.*, **15**, 601–615. (This paper describes a host of non-Markovian processes with NBU first-passage times.)

26. Marshall, A. W. and Shaked, M. (1986). NBU processes with general state space. *Math. Oper. Res.*, **11**, 95–109. (This paper discusses processes with multivariate NBU first-passage times.)

27. Ocón, R. P. and Pérez, M. L. G. (1994). On First-Passage Times in Increasing Markov Processes. *Technical Report*, Department of Statistics and Operations Research, University of Granada, Spain.

28. Özekici, S. and Günlük, N. O. (1992). Maintenance of a device with age-dependent exponential failures. *Naval Res. Logist.*, **39**, 699–714.

29. Ross, S. M. (1979). Multivalued state component systems. *Ann. Probab.*, **7**, 379–383.

30. Ross, S. M. (1981). Generalized Poisson shock models. *Ann. Probab.*, **9**, 896–898.

31. Savits, T. H. and Shaked, M. (1981). Shock models and the MIFRA property. *Stochastic Process. Appl.*, **11**, 273–283.

32. Shaked, M. (1984). Wear and damage processes from shock models in reliability theory. In *Reliability Theory and Models: Stochastic Failure Models, Optimal Maintenance Policies, Life Testing, and Structures*, M. S. Abdel-Hameed, E. Çinlar, and J. Queen, eds. Academic Press, pp. 43–64. (This is a survey paper on wear processes and shock models with a list of references up to 1983.)

33. Shaked, M. and Shanthikumar, J. G. (1986). IFRA processes. In *Reliability and Quality Control*, A. P. Basu, ed., Elsevier Science, pp. 345–352.

34. Shaked, M. and Shanthikumar, J. G. (1987). IFRA properties of some Markov jump processes with general state space. *Math. Oper. Res.*, **12**, 562–568. (The IFRA property is extended to Markov processes with a general state space in this paper.)

35. Shaked, M. and Shanthikumar, J. G. (1988). On the first-passage times of pure jump processes. *J. Appl. Probab.*, **25**, 501–509.

36. Shaked, M. and Shanthikumar, J. G. (1991). Shock models with MIFRA time to failure distributions. *J. Statist. Plann. Inference*, **29**, 157–169.

37. Shanthikumar, J. G. (1984). Processes with new better than used first passage times. *Adv. Appl. Probab.*, **16**, 667–686. (The NBU property for semi-Markov processes and some other non-Markovian processes is discussed in this paper.)

38. Shanthikumar, J. G. (1988). DFR property of first-passage times and its preservation under geometric compounding. *Ann. Probab.*, **16**, 397–406.

39. Shanthikumar, J. G. and Sumita, U. (1984). Distribution properties of the system failure time in a general shock model. *Adv. Appl. Probab.*, **16**, 363–377.

40. Sumita, U. and Shanthikumar, J. G. (1985). A class of correlated cumulative shock models. *Adv. Appl. Probab.*, **17**, 347–366.

See also Cumulative Damage Models; Hazard Rate and Other Classifications of Distributions; Jump Processes; Markov Processes; Reliability, Probabilistic; Shock Models; and Wear Processes.

M. Shaked

H. Li

AGREEMENT ANALYSIS, BASIC

Basic agreement analysis [4] is a model-based approach to analyzing subjective categorical data*. Two or more raters place objects into K unordered, mutually exclusive, and exhaustive categories. Basic agreement analysis provides a principled measure of the amount of inter-rater agreement as well as estimates of rater bias and the true probabilities $\tau(i)$ of the categories $i, i = 1, \ldots, K$. The approach thereby controls for possible confounding effects of systematic rater bias in the analysis of subjective categorical data.

Each rater's judgment process is modeled as a mixture of two components: an error process that is unique for the rater in question, and an agreement process that operationalizes the true values of the objects to be classified. The probability $\Pr[X_r = i]$ that rater r places a randomly selected object into category i is given by

$$\Pr[X_r = i] = \lambda \tau(i) + (1 - \lambda)\epsilon_r(i),$$

where $\epsilon_r(i)$ is the probability of i under Rater r's error process, and λ, the percentage of judgments governed by the agreement process, is assumed to be the same for all raters in the simplest model. The coefficient λ quantifies the amount of agreement between raters, and is closely related to the well-known kappa index of Cohen [1]; see KAPPA COEFFICIENT. In fact, basic agreement analysis can be considered as a systematization of Cohen's idea to correct for agreement by chance in the analysis of subjective categorical data.

For two raters,

$$\Pr[X_r = i, X_2 = j] = \Pr[X_1 = i]\Pr[X_2 = j]$$
$$+ \begin{cases} \lambda^2 \tau(i)[1 - \tau(i)] & \text{if } i = j, \\ -\lambda^2 \tau(i)\tau(j) & \text{if } i \neq j; \end{cases}$$

model parameters can be estimated on the basis of cross-tabulation of the raters' judgments in *agreement matrices* [4]. For the special case of $K = 2$, see [3]. The model is a member of the class of *general processing tree models* [2].

The basic agreement model is a measurement error model that allows more focused analyses of experiments employing subjective categorical data from several raters, for whom ratings have measurement error distributions that can induce bias in the evaluation of scientific hypotheses of interest.

REFERENCES

1. Cohen, J. (1960). A coefficient of agreement for nominal scales. *Educ. and Psych. Meas.*, **20**, 37–46.
2. Hu, X. and Batchelder, W. H. (1994). The statistical analysis of general processing tree models with the EM algorithm. *Psychometrika*, **59**, 21–47.
3. Klauer, K. C. (1996). Urteilerübereinstimmung für dichotome Kategoriensysteme. *Diagnostica*, **42**, 101–118.
4. Klauer, K. C. and Batchelder, W. H. (1996). Structural analysis of subjective categorical data. *Psychometrika*, **61**, 199–240.

See also AGREEMENT, MEASURES OF; CATEGORICAL DATA, SUBJECTIVE; and KAPPA COEFFICIENT.

KARL CHRISTOPH KLAUER

AGREEMENT, MEASURES OF

Measures of agreement are special cases of measures of association* or of correlation* that are designed to be sensitive not merely to deviations from independence, but specifically to deviations indicating agreement. These are measures most commonly used to assess reliability (*see* GROUP TESTING) or reproducibility of observations. In this context, it is usually assured that one has better than chance agreement. Consequently, the statistical problems of interest revolve not around the issue of testing the null hypothesis of independence but around estimation of the population measure of agreement.

Typically, one samples n subjects and has m observations on each, say $X_{i1}, X_{i2}, \ldots, X_{im}$ $(i = 1, 2, \ldots, n)$, where the marginal distributions of the observations are the same for all $j = 1, 2, \ldots, m$. Typically, a measure of agreement is zero when all ratings are independent and identically distributed, and 1.0 if $\Pr\{X_{ij} = X_{ij}'\} = 1$ for all i and $j \neq j'$.

Controversies as to the *validity* of certain measures of agreement can be generated if the assumption of equal marginal distributions is violated [1]. This assumption, however, imposes no major limitation. One need only randomly assign the m observations for each subject to positions $1, 2, \ldots, m$ prior to analysis.

Suppose that

$$X_{ij} = \mu + \xi_i + \epsilon_{ij},$$

where $\xi_i \sim N(0, \sigma_\xi^2), \epsilon_{ij} \sim N(0, \sigma_\epsilon^2), \rho = \sigma_\xi^2 / (\sigma_\xi^2 + \sigma_\epsilon^2)$, and the "true" values of ξ_i and "errors" ϵ_{ij} are independent for all i and j. For this type of interval or ratio-level data, the intraclass correlation coefficient* is one such measure of agreement. This measure is most readily computed by applying a oneway analysis of variance* with each subject constituting a group. The intraclass correlation coefficient then is

$$r_I = \frac{F - 1}{F + m - 1},$$

where F is the F-statistic* to test for subject differences.

Since

$$F \sim \frac{(m - 1)\rho + 1}{1 - \rho} F_{n-1, n(m-1)},$$

nonnull tests and confidence intervals for the parameter ρ can be structured on this basis [2].

A nonparametric analogue of the same procedure is based on the use of coefficient of concordance*. In this case the n observations for each value of j are rank ordered $1, 2, \ldots, n$, with ties given the average of ranks that would have been assigned had there been no ties. One may then calculate the intraclass correlation coefficient based on the ranks r_S. This statistic, r_S, is the average Spearman rank correlation coefficient* between pairs of ratings and is related to the coefficient of concordance W by the relationship

$$r_S = \frac{mW - 1}{m - 1}.$$

The distribution of r_S is approximately of the same form as that of r_I [3].

These two measures, r_I and r_S, are designed to measure agreement for measurements taken on ordinal, interval, or ratio scales. For measurements taken on the nominal scale, the kappa coefficient* performs the same function [4]; its nonnull distribution is not known theoretically, and jack-knife* or bootstrap* methods are suggested for estimation and testing of this measure [5].

To illustrate these methods we use scores on a test of memory for 11 subjects ($n = 11$) each tested three times ($m = 3$). The three scores per subject are listed in random time order in Table 1. The intraclass correlation coefficient was found to be $r_I = 0.59$. The rank orders for each rating appear as superscripts in the table. The average Spearman coefficient based on these data was $r_S = 0.45$.

Finally, one may dichotomize (or classify) the scores in innumerable ways. For illustration we defined a "positive" test as a score of 50 or above, a "negative" test as a score below 50. The kappa coefficient for this dichotomization was $k = 0.05$.

This example illustrates an important point in evaluating magnitudes of measures of agreement. The measure of agreement reflects the nature of the population sampled (i.e., σ_ξ^2), the accuracy of the observation (i.e., σ_ϵ^2), and the nature of the observation itself (interval vs. ordinal vs. nominal). Consequently, poor measures of agreement are obtained if the observation is insensitive to the variations inherent in the population either because of an intrinsic scaling problem or because of inaccuracy of measurement.

Further, there are many more measures of agreement than these common ones proposed in the literature, because there are many ways of conceptualizing what constitutes agreement and how to measure disagreement. To see this, let $D(X_{ij}, X_{rs})$ be any metric reflecting agreement between two observations X_{ij} and X_{rs} (even if the X's are multivariate) with $D(X_{ij}, X_{rs}) = 1$ if and only if $X_{ij} \equiv X_{rs}$. If D_w is the mean of D's between all $n\binom{m}{2}$ pairs of observations within subjects, and D_t the mean of D's between all $\binom{nm}{2}$ pairs of observations, then a measure of agreement is

$$\frac{D_w - D_t}{1 - D_t}.$$

For example, when X_{ij} is a rank order vector, one might propose that $D(X_{ij}, X_{rs})$ be

Table 1.

Subject	Rating		
	1	2	3
1	44^4	48^7	54^5
2	57^8	45^6	$40^{2.5}$
3	$46^{5.5}$	32^3	58^7
4	66^9	50^8	55^6
5	$46^{5.5}$	43^4	50^4
6	72^{11}	83^{11}	70^{11}
7	35^3	28^2	$64^{9.5}$
8	67^{10}	66^{10}	63^8
9	26^2	44^5	$64^{9.5}$
10	16^1	21^1	19^1
11	48^7	69^9	$40^{2.5}$

a rank correlation coefficient between such vectors. Such a measure has been proposed as an extension of kappa [5] for multiple choices of categories, or as a measure of intergroup rank concordance [6].

As a result, the magnitude of a measure of agreement is determined not only by the nature of the population, the accuracy of the observation, and the nature of the observation itself, but, finally, by the metric of agreement between observations selected as the basis of the measure of agreement. *See also* SIMILARITY, DISSIMILARITY AND DISTANCE, MEASURES OF.

REFERENCES

1. Bartko, J. J. (1976). *Psychol. Bull.*, **83**, 762–765.

2. Bartko, J. J. (1966). *Psychol. Rep.*, **19**, 3–11.

3. Fleiss, J. L. and Cuzick, J. (1979). *Appl. Psychol. Meas.*, **3**, 537–542.

4. Kraemer, H. C. (1981). *Biometrika*, **68**, 641–646.

5. Kraemer, H. C. (1980). *Biometrics*, **36**, 207–216.

6. Kraemer, H. C. (1976). *J. Amer. Statist. Ass.*, **71**, 608–613.

See also ASSOCIATION, MEASURES OF; CONCORDANCE, COEFFICIENT OF; INTRACLASS CORRELATION COEFFICIENT; KAPPA COEFFICIENT; PSYCHOLOGICAL TESTING THEORY; SIMILARITY, DISSIMILARITY AND DISTANCE, MEASURES OF; SPEARMAN RANK CORRELATION COEFFICIENT; and VARIANCE COMPONENTS.

HELENA CHMURA KRAEMER

AGRICULTURAL SURVEYS

The purpose of this paper is to describe the statistical methodologies that have been developed over time that are somewhat unique to the field of agriculture. Agricultural Statisticians throughout history have been innovative in the use of available statistical theory and adapting it to their needs.

From a statistical point of view, the characteristics of agricultural surveys are similar to surveys associated with other social or physical phenomena. First, one must define the population and then prepare a frame that represents it. Then, unless one is to do a complete census*, sampling theory is used to select a sample from the frame. Estimators that reflect the sampling design are developed and include methodologies to impute for missing data (*see* IMPUTATION) and to estimate for nonresponse*. The statistics literature is rich with theory and methods to support these issues.

Agriculture has several features that pose methodological problems. One first needs to consider the definition of agriculture. Agriculture involves the use of land, the culture or raising of a living organism through different life cycles, and the concept of ownership. For example, the use of land, culture, and ownership as criteria separate fishing from aquaculture and tree farming from forestry. This definition easily points to land as the population. For example, agricultural land is simply all land used for the culture or raising of plants or animals under some form of private ownership. One of the first difficulties is that it is also necessary to describe agriculture as a business, with the need to provide measures of the economic or demographic situation of farms and the people operating the farms. This suggests that the sampling frame* consists of a list of farms or farm operators that account for the land associated with agriculture.

The need to account for both land and the farm operators has dominated the statistical thinking of agricultural statisticians and this led to the development of multiple frame sampling that will be described in greater detail.

The other unique feature of agriculture is that it provides the food and fiber that feeds and clothes a country's population. Food security becomes an issue, which means that timely and accurate forecasts of future supplies are needed for policy purposes and to ensure that markets operate efficiently. Agriculture is seasonal—especially crop production—which means that all of the supply becomes available in a short period of time and is held in storage and used until the next harvest period.

This paper will provide an overview of the history of the development of statistical procedures to forecast and measure agricultural production. The examples will mostly apply to the United States but, where appropriate, methods used in other countries will also be presented. The material will draw heavily from Reference 20, which traces the history of agricultural statistics as developed by the United States Department of Agriculture (USDA).

BEFORE PROBABILITY SAMPLING

The basic method to obtain agricultural statistics in many developing and developed countries was to use the administrative levels of government. The basic unit was an administrative unit such as a village or small agricultural region. The village or local area administrators provided subjective measures of areas planted and production and other agricultural measures. No sampling was involved as village totals were sent up the administrative line. This practice is still widespread in many developing countries, for several reasons. First, their farmers may not be educated enough to understand the concept of area or production or are unwilling to answer questions. Cost is also an issue. However, these administrative data are subject to manipulation by the various levels of government through which they pass, which means that these countries are facing the need to modernize their methodology. The administrative system used in China is typical of procedures used in many countries [21].

Basic statistics in the United States were initially provided by periodic censuses of agriculture, with surveys of voluntary reporters in intervening years to measure change from the census benchmark.

Prior to the 1880 agricultural census in the United States, only information about total crop production and livestock inventories was obtained. The 1880 census also obtained information about crop acreages. These census enumerations of acreage provided benchmarks for estimating crop acreages for years between census years. This was the beginning of the procedure that is still used to forecast and estimate crop production. Basically, it calculates crop production as the product of the two separate estimates of acreage and yield per acre. Once planted, crop averages usually do not change very much between planting and harvest. There is also less year-to-year variability between crop acres than there is between yield per acre. In general, the estimates through the nineteenth century were linked to the decennial Census of Agriculture conducted by the Bureau of the Census*. The USDA relied upon correspondents reporting their assessment of year-to-year changes in their locality to make the annual estimates. As might be suspected, small year-to-year biases in the measures of change linked to a census could grow to a widening gap over the years between the USDA estimates and the next census benchmark level. This problem led to the development of improved methodology.

The most important statistics produced were and still are the forecasts of the production of crops such as wheat, corn, soybeans, and cotton, followed by end-of-season estimates of actual production. For reasons given above, the forecasts and estimates of production were determined by separately estimating or forecasting acreage planted and average yields per acre. There was no "sampling frame" of farms; there were only lists of correspondents who would voluntarily respond to a mailed inquiry.

In the absence of probability sampling theory, much effort went into improving estimating procedures to measure crop acreages and to forecast crop yields. These procedures are discussed below in chronological order and are described more thoroughly in Reference 3.

Starting in 1912, the *Par Method* was adopted to translate farmer reported crop condition values early in the crop season into a probable yield per acre that would be realized at harvest. The par method to forecast yield (y) consisted of the following components:

$\bar{y} = CY_m/C_m$, where

C_m = the previous 10-yr average condition for the given month,

Y_m = the previous 10-yr average yield per acre realized at the end of the season,

C = current condition for the given month.

The forecasting model was simply a line passing through the origin and the point (C, $\bar{y}$). A separate par yield ($\bar{y}$) was established for each state, crop, and month. In actual practice, subjective modification of the means was considered necessary to remove the effects of atypical conditions. For example, a drought that may occur only once every 10 or 15 year would greatly affect the 10-yr average conditions and yield. To aid in these adjustments, 5- and 10-yr moving averages* were computed to identify unusual situations or trends, and if necessary, exclude the atypical observations.

The development of simple graphic solutions prior to the use of regression and correlation theory was a major breakthrough as a practical means to forecast crop yields, and this approach was implemented in the late 1920s. Data for a sufficient number of years had accumulated, so that final end-of-season estimates of yields could be plotted against averages of condition reports from farmers for each crop in each State.

The condition was reflected on the x-axis and end-of-season yields on the y-axis. The plotted points depicted a linear description of early season crop conditions compared to end-of-season yields. The yield forecast for the current system was obtained by entering the current condition on the chart and finding the best fit with historic yields subjectively.

Graphical regression techniques provided a consistent method to translate survey data into estimates, which in effect adjusted for persistent bias in the data caused by the purposive sampling procedures. This method quickly replaced the par method and was adopted rapidly.

The following discussion describes early attempts to estimate the acreage to be harvested.

Because the ideas of probability sampling had not yet been formed, the procedures used to estimate acres for harvest were more difficult than those to estimate average yields. The USDA used its state offices to enlarge the lists of farm operators, but there was no complete list of farms that could be used for survey purposes. Therefore, the estimating procedures relied upon establishing a base from the most recent Census of Agriculture and estimating the percent change from year to year. As the country developed with a road system and a postal service, a common data collection* approach was the Rural Carrier Survey. The postal carriers would drop a survey form in each mailbox along their route. A common procedure during that time was to include two columns in the questionnaire when asking the farmer questions about acreage planted to each crop. During the current survey, the farmer was asked to report the number of acres planted this year and the number of acres planted the previous year in each crop. This method was subject to several reporting biases, including memory bias, and it led to matching "identical" reports from year to year to remove the memory bias. The matching of identical reports did improve the estimates, but was considerably more labor intensive because the name matching had to be done by hand. The process was also complicated by problems with operations changing in size, and it was inherently biased because it did not account for new entrants to agriculture.

This methodology was subject to a potentially serious bias, caused by the selective or purposive nature of the sample. In an effort to make an allowance for this bias, a relative indicator of acreage was developed in 1922; it became known as the *ratio relative* and contained the following components:

R_1 = Ratio of the acreage of a given crop to the acreage of all land in farms (or crops) for the current year as reported by the sample of farm operators;

R_2 = Same as R_1 but for the previous year;

$\hat{y} = (R_1/R_2) \times$ (Estimated total area of the crop the previous year).

The ratio R_1/R_2 was essentially a measure of change in the crop area from the previous year. The assumption was that the previous year's area was known without error. The belief was that this ratio held the bias resulting from the purposive sampling constant from one year to the next. A reported limitation was the extreme variability in the acreage ratios between the sample units. This was countered by increasing the "number" of farms surveyed and weighting the results by size of farm.

By 1928, matched farming units that reported in both years were used to compute the ratio relative. This reduced the influence of the variability between sample units. When looking back at the variance of the estimate from a current perspective, one may examine the components (also assuming probability sampling).

$$\text{Var}(\bar{y}) = \text{Var}(R_1) + \text{Var}(R_2) - 2\,\text{cov}(R_1, R_2).$$

This shows why the use of matching reports improved the ratio relative estimator. However, this did not solve the problem, because by using matching reports, farms going into or out of production of a particular crop were not properly represented. Therefore, statisticians continued their efforts to develop a more objective method of gathering and summarizing survey data.

Some statisticians in the early 1920s would travel a defined route on the rural roads or via railway routes and record the number of telephone or telegraph poles opposite fields planted to each crop. The relative change in the pole count for each crop from year to year provided a measure of the average change in crop acreage. This method was generally unsatisfactory, because large portions of the United States still did not have telephone service; the pole count method was therefore not widely used.

A more refined method of estimating acreage was developed in the mid-1920s. A *crop meter* was developed and attached to an automobile speedometer to measure the linear frontage of crops along a specified route. The same routes were covered each year. This made possible a direct comparison of the number of feet in various crops along identical routes from the current year and the previous year. The properties of the estimator are described in Reference 12.

THE TWENTIETH CENTURY AFTER PROBABILITY SAMPLING

A milestone in the evolution of statistical methodology for agriculture was the development of the master sample of agriculture [13,14]. This was a cooperative project involving Iowa State University, the US Department of Agriculture, and the US Bureau of the Census. This area-sampling* frame demonstrated the advantages of probability sampling. The entire landmass of the United States was subdivided into area-sampling units using maps and aerial photographs. The sampling units had identifiable boundaries for enumeration purposes. The area-sampling frame had several features that were extremely powerful for agricultural surveys.

By design, it was complete in that every acre of land had a known probability of being selected. Using rules of association to be described presently, crops and livestock associated with the land could also be measured with known probabilities. The Master Sample of Agriculture was based on a stratified design—the strata defined to reflect the frequency of occurrence of farmsteads. Area-sampling units varied in size in different areas of the country to roughly equalize the number of farm households in each area-sampling unit.

The master sample was used for many probability surveys, but not on a recurring basis, because of added costs arising from the area samples having to be enumerated in person. The panel surveys of farm operators, while not selected using probability theory, were very much cheaper to conduct, because the collection was done by mail. It was not until 1961 that pressures to improve the precision of the official estimates resulted in the US Congress appropriating funds for a national level area frame survey on an annual recurring basis. During the early 1960s, the Master Sample of Agriculture was being replaced by a new area frame that was stratified via land use categories on the basis of the intensity of cultivation of crops. This methodology is still used. The process of developing the area frame is now much more sophisticated, relying upon satellite imagery and computer-aided stratification [18]. The method of area frame sampling described here is generally referred to as *Area Frame of Segments*.

An alternative is to select an *Area Frame of Points*. The usual practice is to divide the landmass into grids, and within each grid selecting a sample of points. Then, data collection involves identifying what is on each point or identifying the farm associated with the point; Reference 9 describes a sample design starting with 1800×1800 m grids. Each grid contained a total of 36 points—each point having a 3×3 m dimension. A random sample of 3 points was selected from each grid.

This method of sampling is easier to implement than segment sampling and is being used in Europe. However, data collected from point samples are less suitable for matching with satellite images or data. Data coming from segments are based on a drawing of the fields on aerial photographs. The use of the area frame led to the development of some new estimators, described below.

The sampling unit for the area sample frame is a segment of land—usually identified on an aerial photograph for enumeration. The segment size generally ranged from 0.5 to 2 sq mi, depending upon the availability of suitable boundaries for enumeration and the density of the farms. The basic area frame estimator was the design-based unbiased estimate of the total,

$$y_a = \sum_h \sum_i e_{hi} \bullet y_{hi}$$

where y'_{hi} was the ith segment total for an item in the hth stratum and e_{hi} was the reciprocal of the probability of selecting the ith segment in the hth stratum.

During the frame development process, the segment boundaries are determined without knowledge of farm or field boundaries. Therefore, an early (and continuing) difficulty

was how to associate farms with sample segments during data collection. Three methods have evolved, which are referred to both as methods of association* and as estimators. Let y_{hil} be the value of the survey item on the lth farm having all or a portion of its land in the ith sample segment. Then different estimators arise, depending on how survey items on farms are associated with the sample segments. We present these next.

Farm (Open):

The criterion for determining whether a farm is in a sample or not is whether its headquarters are located within the boundaries of the sample segment. This estimator was most practicable when farm operations were generally homogeneous, that is, they produced a wide variety of items, some of which may not have appeared in the segment. This estimator was also useful for items such as number of hired workers and of animals born that are difficult to associate with a parcel of land. The extreme variation in size of farms and the complex rules needed to determine if the farm headquarters were in the segment resulted in large sampling and nonsampling errors,

$$y'_{hi} = \sum_l F_{hil} y_{hil},$$

where $F_{hil} = 1$, if the operator of farm l lives in the segment; 0, otherwise.

Tract (Closed):

The tract estimator is based on a rigorous accounting of all land, livestock, crops, and so on, within the segment boundaries, regardless of what part of a farm may be located within the boundaries of the segment. The method offered a significant reduction in both sample and nonsampling errors over the farm method, because reported acreages could be verified by map or photograph. The estimator is robust in that the maximum amount that can be reported for a segment is limited by its size. The estimator is especially useful for measuring acres in specific crops,

$$y'_{hi} = \sum_l T_{hil} y_{hil},$$

where

$$T_{hil} = \frac{\text{Amount of item on farm } l \text{ in segment } i}{\text{Total amount of item on farm } l}.$$

Weighted:

The difficulty with the tract estimate was that some types of information, such as economic, could only be reported on a whole-farm basis. This led to the development of the weighted procedure, in which data are obtained on a whole-farm basis, for each farm with a portion of its land inside a sample segment. The whole farm data are prorated to the segment on the basis of the proportion of each farm's land that is inside the segment. This estimator provides the advantage of a smaller sampling error than either the farm or tract procedures. On the minus side, data collection costs increased 15 to 20% because of increased interviewing times, and intractable nonsampling errors are associated with determining the weights. This estimator is also used to estimate livestock inventories, number of farm workers, and production expenditures,

$$y'_{hi} = \sum_{l} W_{hil} y_{hil},$$

where

$$W_{hil} = \frac{\text{Acres of farm } l \text{ in segment } i}{\text{Total acres in farm } l}.$$

Ratio:

The area frame sample was designed so that 50 to 80% of the segments were in the sample from year to year. This allowed the computation of the usual ratio estimators* such as the year-to-year matched segment ratios of change.

While the area frame sampling and estimating procedures were being refined, this period also saw a rapid change in the structure of agriculture. Farms became more specialized and much larger. This introduced more variability that required much larger area frame sample sizes.

The proportion of farms having livestock was decreasing rapidly during this period. The variation in numbers of livestock on such farms also had increased dramatically.

The combination of these two factors meant that either resources for an extremely large area frame sample would be needed or alternative sampling frames were needed. In the early 1960s, H.O. Hartley* at Iowa State University was approached about this problem. The result was his 1962 paper laying out the basic theory of multiple frame sampling and estimation, and summarized in Reference 6. This was followed by Reference 5, which more fully developed the concepts of multiple frame sampling and estimation methodology and which also developed multiple frame estimators with reduced variances.

As implied by its name, multiple frame sampling involves the use of two or more sampling frames. If there are two frames, there are three possible poststrata or domains— sample units belonging only to frame A, sample units belonging only to frame B, and finally the domain containing sample units belonging to both frames A and B. As pointed out by Hartley, the sampling and estimation theory to be used depended on knowing in advance of sampling whether the domain and frame sizes were known. This determined whether theories applying to poststratification or domain estimation were to be used.

In the agricultural situation, the area-sampling frame provided 100% coverage of the farm population. There was also a partial list of farms, which could be stratified by size or item characteristic before sampling. Domain membership and sizes are unknown prior to sampling, thus sample allocation is by frame and domain estimation theories apply. The theory requires that after sampling, it is necessary to separate the sampled units into their proper domain. This meant area sample units had to be divided into two domains—farms not on the list, and farms on the list.

By definition, all farms represented by the list were also in the area frame. The Hartley estimator for this situation was

$$\hat{Y}_H = N_a(\bar{y}_a + P\bar{y}'_{ab}) + N_b Q\bar{y}''_{ab},$$

where $\bar{y}_a$ represents area sample units not in the list, $\bar{y}'_{ab}$ represents area sample units overlapping the list frame, $\bar{y}''_{ab}$ represents the list frame, and $P + Q = 1$.

The weights P and Q were to be determined to minimize var $(\hat{Y}_H)$. This sampling and estimation theory was used for surveys to measure farm labor numbers and wage rates, livestock inventories, and farm production expenditure costs. Because of the considerable variation in the sizes of farms and the sampling efficiencies that occurred from the stratification in the list frame, the majority of the weight went to the list frame portion of the estimator; that is, P was small and Q was large.

An alternative estimator is suggested in Reference 10. With it, units in the list frame that are in the area frame sample are screened out of the area frame portion of the survey. In other words, $P = 0$ and

$$\hat{Y}_H = N_a \bar{y}_a + N_b \bar{y}_{ab}''.$$

Additional analysis [5] suggested that, for a fixed cost, the screening estimator would have the lower variance whenever the cost of sampling from the list frame is less than the difference between the cost of sampling from the area frame and the cost of screening the area frame sample to identify those also in the list frame.

For those reasons, the screening estimator is used exclusively. The increased use of telephone enumeration for the list sample reflects personal to telephone enumeration cost ratios of 1 to 15 in some cases. The area frame sample is surveyed in its entirety in June each year. Farms that overlap the list frame are screened out and the area domain representing the list incompleteness is defined. During the next 12-month period, a series of multiple frame quarterly surveys are conducted to measure livestock inventories, crop acreages and production, and grain in storage. Other multiple frame surveys during the year cover farm labor and production expenditures. Each survey relies upon the multiple frame-screening estimator.

Multiple frame is still the dominant sampling methodology used in the United States. Its use has also spread to many countries [23]. The methods used in other countries differ only by their choice of area frame sampling, that is, point sampling versus square segments.

As the structure of agriculture became widely diverse in the United States, the basic use of stratification of the list frame of farms

and estimators such as direct and ratio estimators were becoming increasingly inefficient.

The sampling methods now used to select farms from the list frame are described in Reference 15. This procedure was named *Multiple Probability Proportional to Size*[*] (Multiple PPS), because in effect multiple samples are selected. The frame is optimized for a particular characteristic, and a PPS sample is selected using a measure of size representing the characteristic. This process can be repeated multiple times for each variable of interest to be included in the combined sample.

The next step is to combine the samples into one overall sample and recalculate the sample weights.

The probability of selecting the ith farm is $m = \max(\pi_{i1}, \ldots, \pi_{iM})$, where from 1 to M samples have been selected, and the individual probabilities of selecting a given farm from each sample is noted. The maximum probability of selecting each farm from across the individual samples becomes the farm's sample weight.

The next step is to calibrate the sample weights so that the final estimation becomes model unbiased.

The use of Poisson Permanent Random Number (PPRN) sampling, as described in Reference 17, is used to select the PPS samples.

In these designs, every population is assigned a permanent random number between 0 and 1. Unit l is selected if its random number is less than its maximum probability of being selected.

PPRN sampling furthermore allows one to think of a sample drawn with inclusion probabilities as the union of M PPRN samples, each drawn using the same permanent random number and individual probabilities. This is convenient when one is interested in estimates of different combinations of target variables in different surveys.

For example, the USDA makes estimates for potatoes in June and December, row crops (e.g., soybeans and corn) in March, June, and December, and small grains (e.g., wheat and barley) in March, July, September, and December. It wants to contact the same farms throughout the year, but has little interest in sampling a farm for the September survey if it has not historically had small grains. Thus, PPRN samples of farms using the same permanent random number can be drawn for potatoes, row crops, and small grains, each with its own selection probabilities. The union of all three is the overall sample in June. Similarly, the union of the row-crops and small-grains samples is the overall sample in March. The use of Poisson sampling is discussed in greater detail in Reference 2.

The domain determination has been the most difficult operational aspect to tackle in developing, implementing, and using multiple frame methodology [19]. As the structure of farms becomes more complicated with complex corporate and partnership arrangements, the survey procedures require a substantial effort to minimize nonsampling errors associated with domain determination.

Since the first crop report was issued in 1863, the early season forecasts of crop production continued to be some of the most critical and market sensitive information prepared by the USDA. The development of probability sampling theory and the area-sampling frame provided a foundation upon which to replace judgement-based estimates of locality conditions to forecast yields per acre. In 1954, research was initiated to develop forecasting techniques based on objective counts and measurements that would be independent of judgment-based estimates. The use of nonrepresentative samples of farmers continued to be used to report conditions in their locality and individual farms during this period, however.

Research on the use of corn and cotton objective methods began in 1954 followed by work on wheat and soybeans in 1955 and sorghum in 1958. Early results showed that a crop-cutting survey at harvest time based on a probability sample of fields would provide estimates of yield per acre with good precision. Countries such as India and China have also used this methodology, estimating final yields and production with pioneering work described in Reference 16. The methods used in China are described in Reference 21. There were two difficulties when attempting to forecast yields. One difficulty is to forecast yield before the crop is mature, and it is even

more difficult to do so before the plants have set fruit.

A two-step sampling procedure is used. First, a sample of fields is selected from those identified during the annual area frame survey as having the crop of interest. Self-weighting samples are selected. Observations within fields are made in two randomly located plots with each selected field. Selected plots for most crops include two adjacent rows of predetermined length. The probable yield per acre is a function of the number of plants, the number of fruits per plant, and the size or weight of the fruit. Early in the crop season, the number of plants is used to forecast the number of fruits, with historical averages used for fruit weights. After fruit is present, several measurements are obtained to project final fruit weight. For example, the length and diameter of corn ears are obtained from ears within the sample plots. When the crop is mature, the sample plots are harvested, and the fruit counted and weighed for the final yield estimate. The early season counts and measurements from within the sample plots are combined with the data from the harvested fruit, and become part of a database that is used to develop forecasting models in subsequent years. After the farmer harvests the sample field, another set of sample plots is located and grain left on the ground is gleaned and sent to a laboratory where it is weighed and used to measure harvest loss. During the forecast season, historical averages are used to estimate harvest losses.

Simple linear and multiple regression models are used to describe past relationships between the prediction variables and the final observations at maturity. Typically, early season counts and end-of-season harvest weights and counts from within each unit are used. They are first screened statistically for outlier* and leverage* points [4]. Once these atypical data are identified and removed, the remaining data are used to create current forecast equations.

The basic forecast models for all crops are essentially the same, in that they consist of three components: the number of fruits, average fruit weight, and harvest loss.

The net yield per acre as forecast for each sample plot is computed as follows:

$$y_i = (F_i C_i W_i) - L_i,$$

where

F_i = Number of fruit harvested or forecast to be harvested in the ith sample plot,

C_i = Conversion factor using the row space measurement to inflate the plot counts to a per acre basis,

W_i = Average weight of fruit harvested or forecast to be harvested,

L_i = Harvest loss as measured from postharvest gleanings (the historical average is used during the forecast, season),

Yield forecasts = $\sum_i (y_i/n)$ for the n sample fields.

Separate models are used to forecast the number of fruits (F_i) to be harvested and the final weights (W_i). The variables used in each model vary over the season, depending upon the growth stage at the time of each survey. At the end of the crop season, F_i and W_i are actual counts and weights of fruit for harvest.

The major contributor to forecast error is the difficulty of forecasting fruit weight early in the season. Many factors such as planting date, soil moisture, and temperatures at pollination time crucially affect a plant's potential to produce fruit. While the fruit can be counted early in the season, the plant does not always display characteristics that provide an indication of final fruit weight. While each plant's potential to produce fruit is affected by previous circumstances, that information is locked inside the plant—often until fruit maturity.

Over the years, the USDA has conducted extensive research to improve the basic yield forecast models. Examples of this work appear in Reference 1. Models using weather data were continuously being developed and compared against the traditional objective yield models, but always fell short. The plant measurements reflected the impacts of weather, and the use of weather data does not add to the precision. Another effort involved an attempt to model the plant growth and to use these models, known as *plant process models*, for yield forecasting. They did not prove to be feasible in a sample survey environment.

USE OF SATELLITE IMAGERY AND DATA IN AGRICULTURAL STATISTICS

The first satellite designed to monitor land use was the land observatory (Landsat) satellite launched in 1972. Several generations of Landsat satellites have since been launched and placed in orbit, for example in 1999. The satellites are designed to travel in almost perfectly circular, near-polar orbit passes over the sunlit side of the planet several times daily. The orbit shifts westward so that every part of the surface of the earth is imaged every 16 days. The satellite contains a sensor referred to as the Thematic Mapper (TM). This camera-like device divides the images into picture elements (pixels) and measures the brightness of each pixel in seven portions of the electronic spectrum. The TM scanner pixel size is 30 m sq, for which there are measures of light reflectance for seven bands of the electromagnetic spectrum. The French government launched the SPOT satellite in 1986, which contains a sensor that provides a 20-m resolution. The sensor is pointable, which allows the satellite to observe the same area on the ground several days in a row. Data from both LANDSAT and SPOT satellites are available in either photographic or digital form. The net result is large amounts of data about the land and the vegetation it carries.

The set of measurements for each pixel, its signature, can be used to separate crop areas by type or by different types of land use. It is only to the degree that the spectral signatures for different crops and land uses can be separated that satellite data become useful.

It soon became evident that the accuracy of the crop and land use classifications as derived from the satellite data would be greatly improved by using ground truth data. The methodology developed to use data from the area frame surveys as ground truth to improve the accuracy of the classification of pixels by land use or crop cover is described in Reference 11. This process first involves obtaining ground-to-pixel registration. Discriminant functions are developed from pixels matched to ground truth data. The discriminant functions are then used to classify all elements in a satellite scene. In other words, every pixel is assigned to a target crop or land use. Regression estimators are used to estimate the population parameters.

This complete classification of pixels by crop or land use in effect provides complete coverage of a given land area. A popular product is the set of cropland data layers prepared for entire states. Since each pixel is georeferenced, these cropland data in a geographic information system can be linked to transportation corridors, watershed boundaries, or any other georeferenced data.

The National Oceanic and Atmospheric Administration (NOAA) has launched a series of weather satellites that also carry an imaging system, the Advanced Very High Resolution Radiometer (AVHRR). This imaging has a pixel size of 1.1 km versus the 30-m pixel size of the TM. A key result, however, is that the entire globe can be imaged daily instead of once every 16 days. The AVHRR images provide both data and images on vegetation conditions [22]. The daily images or weekly composites are used by governmental agencies and marketing boards to monitor crop conditions over large areas to make decisions for agricultural marketing and for early warning of food aid requirements.

Satellite data classified by land use categories are used extensively to design and prepare area-sampling frames [7]. The satellite data are also spatially referenced using latitude/longitude coordinates. Therefore, they can be used along with mapping products showing natural boundaries such as roads and rivers.

THE FUTURE

The future holds many challenges. There is a growing need to understand and measure agriculture's affect on the environment. There is a related need for policy makers and others to know how their decisions about trade and the environment affect the production decisions made by farmers and their resulting economic situation. There is a growing concern about the demographics of farm operations as they shrink in number with the increasing size of farms. The interrelationship between these variables will need to be measured, pointing to an integration of sampling methods with the use of satellite data. Statistical theory will pave the way.

REFERENCES

1. Arkin, G. F., Vanderlip, R. L., and Ritchie, J. T. (1976). A dynamic Grain sorghum growth model. *Trans. Am. Soc. Agricult. Engrs.*, **19**(4), 622–630.

2. Bailey, J. and Kott, P. (1997). An application of multiple frame sampling for multipurpose surveys, *Proceedings of American Statistical Association*.

3. Becker, J. A. and Harlan, C. L. (1939). Developments in the crop and livestock reporting service since 1920. *J. Farm Econ.*, **21**, 799–827.

4. Beckman, R. J. and Cook, R. D. (1983). Outliers. *Technometrics*, **25**, 119–149.

5. Cochran, R. S. (1965). *Theory and Application of Multiple Frame Sampling*. Ph.D. Dissertation, Iowa State University, Ames, Iowa.

6. Cochran, W. G. (1977). *Sampling Techniques*, 3rd ed. John Wiley, New York.

7. Cotter, J. and Tomczac, G., (1994). An image analysis system to develop area sampling frames for agricultural surveys. *Photogrammetr. Eng. Remote Sens.*, **60**(3), 299–306.

8. Food and Agricultural Organization of the United Nations. (1998). *Multiple Frame Agricultural Surveys*. FAO Statistical Development Series, Vol. 2.

9. Fuller, W. A. and Burnmeister, L. F. (1972). "Estimators for Samples Selected from Two Overlapping Frames". *Proceedings of the Social Statistics Section*. American Statistical Association, pp. 245–249.

10. Gallego, F. J. (1995). Sampling Frames of Square Segments. Report EUR 16317, Office for Official Publications of the European Communities, Luxembourg, ISBN 92-827-5106-6.

11. Hartley, H. O. (1962). "Multiple Frame Surveys". *Proceedings of the Social Statistics Section*. American Statistical Association, pp. 203–206.

12. Hanuschak, G. R., Sigman, R., Craig, M., Ozga, M., Luebbe, R., Cook, P., Kleweno, D., and Miller, C. (1979). *Obtaining Timely Crop Area Estimates Using Ground-Gathered and Landsat Data*. Technical Bulletin No. 1609, USDA, Washington, D.C.

13. Hendricks, W. A. (1942). *Theoretical Aspects of the Use of the Crop Meter*. Agricultural Marketing Service, USDA.

14. King, A. J. and Jessen, R. J. (1945). The master sample of agriculture. *J. Am. Stat. Assoc.*, **4D**, 38–56.

15. King, A. J. and Simpson, G. D. (1940). New developments in agricultural sampling. *J. Farm Econ.*, **22**, 341–349.

16. Kott, P. S. and Bailey, J .T. (2000). The theory and practice of maximal Brewer selection with Poisson PRN sampling. *International Conference on Establishment Surveys II*, Buffalo, New York.

17. Mahalanobis, P. C. (1946). Sample surveys of crop yields in India. *Sankhya*, 269–280.

18. Ohlsson, E. (1995). "Coordination of Samples Using Permanent Random Numbers". *Business Survey Methods*. Wiley, New York, pp. 153–169.

19. Tortora, R. and Hanusachak, G., (1988). Agricultural surveys and technology. *Proceedings of 1998 American Statistical Association Meeting*.

20. Vogel, F. A. (1975). "Surveys with Overlapping Frames—Problems in Application". *Proceedings of the Social Statistics Section*. American Statistical Association, pp. 694–699.

21. Vogel, F. A. (1995). The evolution and development of agricultural statistics at the United States Department of Agriculture. *J. Off. Stat.*, **11**(2), 161–180.

22. Vogel, F. A. (1999). *Review of Chinese Crop Production Forecasting and Estimation Methodology*. U.S. Department of Agriculture Miscellaneous Publication No. 1556.

23. Wade, G., Mueller, R., Cook, P., and Doraiswamy, T. (1994). AVHRR map products for crop condition assessment: a geographic system approach. *Photogrammetr. Eng. Remote Sens.*, **60**(9), 1145–1150.

See also AGRICULTURE, STATISTICS IN and CROP AREA ESTIMATION, LANDSAT DATA ANALYSIS IN.

FREDERIC A. VOGEL (Retired)

AGRICULTURE, STATISTICS IN

The area covered by this topic is so vast that whole volumes have been written on it, e.g., ref. 8—and that is only an introduction. The use of statistical techniques in agricultural research goes back many years, and indeed agriculture was one of the areas in which modern analytical techniques were first devised. The interchange of ideas between statistical and agricultural science has been of mutual benefit to both subjects. This continues to the present day, and all that is done here is to point out particular topics of joint importance to the two sciences.

HISTORY

The earliest paper describing what may be thought of as a statistically designed agricultural experiment appears to be that of Cretté de Palluel [4]. This concerned an experiment on the fattening of sheep in which 16 animals, four each of four different breeds, were fed on four diets, one of each breed per diet. The animals were killed at four monthly intervals so that the experiment could be regarded, in modern terms, either as a $\frac{1}{4}$ replicate* of a 4^3 factorial* or a 4×4 Latin square*. This experiment, which antedates the founding of modern agricultural research stations by more than half a century, shows in a simple form the principles of good experimental design* and analysis.

Agricultural journals have been in existence in their present form since the early years of this century, and many now have statisticians on their editorial boards. Thus the *Journal of Agricultural Science* has been published in Cambridge since 1905 and deals with many branches of agriculture and animal husbandry. The *Journal of Agricultural Research* was founded in Washington in 1913 and changed to *Agronomy Journal* in 1949, reflecting its prime concern with crops. *Tropical Agriculture* has been published since 1924 in Trinidad, and the *Indian Journal of Agricultural Science* since 1931 in New Delhi. These two deal primarily with tropical agriculture, as does *Experimental Agriculture*, which started in 1930 in Oxford as the *Empire Journal of Experimental Agriculture* and dropped its imperial connections in 1965. All these journals have a long and honorable history of statistical writing, from early papers on the methodology of the analysis of field data in the *Journal of Agricultural Science* in the 1920s to several papers on the techniques of intercropping trials in *Experimental Agriculture* in the late 1970s.

Courses on statistical methods applied to agriculture have been taught for many years, one of the first being those by G. W. Snedecor* at Iowa State College as early as 1915. However, the first statistician appointed to work at an agricultural research station was R. A. Fisher*, who went to Rothamsted Experimental Station in 1919. Within a few years Fisher had developed the technique of analysis of variance* for use in analyzing the results of agricultural experiments; he was also quick to emphasize the importance of replication and randomization in field trials and introduced the randomized block design*. A good summary of Fisher's early work is given by Yates [12].

PRESENT POSITION

From the 1930s onward, statistical methods for agricultural use have been greatly extended, both by the introduction of new techniques and by their use in agricultural research throughout the world. Thus, at Rothamsted, Fisher's colleague and successor F. Yates introduced more complex experimental designs. Among others, Yates recognized the importance of extensive experimentation: Crowther and Yates [5] gave a comprehensive summary of fertilizer trials in northern Europe from 1900 to that time. Yates also used statistical methods in surveys of agricultural practice, from 1944 onward [14]. Again, these statistical techniques were initially employed in agronomy and crop husbandry, but similar principles were soon applied to experiments with animals, despite their often greater expense and difficulty. A comprehensive statement of the part statistics, and statisticians, can play in planning field experiments was given by Finney [7], and the position since then has changed only in detail, not in broad outline.

METHODS OF EXPERIMENTAL DESIGN AND ANALYSIS

The main techniques used in practice for design and analysis of agricultural experiments continue to be based largely on Fisherian principles. Thus, since all agricultural work is subject to biological variability, treatments in comparative experiments are replicated in space, and sometimes in time also. Further, the application of any treatment to a particular set of plants or animals, or piece of ground, is usually randomized, possibly with some restrictions, although systematic designs* are sometimes used for particular purposes. These same principles are used, to a lesser degree, in the design of surveys, the random element occurring in the selection of units to be sampled.

The most commonly used experimental design for field trials is the randomized block* design, in which the area of land available for experimentation is divided into *blocks*, within which it is hoped that soil conditions are reasonably uniform; the blocks are subdivided into *plots* to which treatments* are applied. (The names "block" and "plot," now widely used in experimental design, reflect the agricultural context in which they were first applied.) There are three main lines of development of practical designs, in the directions of factorial experimentation*, incomplete block designs*, and row and column designs*. Full details are given in the relevant articles elsewhere, but there are whole books devoted to the topic of experimental design, e.g., Cochran and Cox [2] for ways of allocating treatments to plots and Cox [3] for other aspects of the planning of practical experiments.

The standard technique for analyzing the results of agricultural experiments is the analysis of variance*. Although this has its critics and is certainly not universally applicable, it remains the usual method for assessing whether the variation among a group of treatments is greater than would occur if all the observed effects were due to chance. However, this technique occupies only the middle range of the examination of experimental results: it is first necessary to summarize observed data to see whether they have any meaning at all, and it is frequently desirable to synthesize the observed results into more formal models, which may advance agricultural theory as well as practice.

Since it is common to take many records on an agricultural crop or animal, the first task is to sort out those on which to conduct a formal statistical analysis. For example, if a crop is harvested over a long period of time (e.g., tomatoes or coffee), does one wish to analyze total yield, or early yield, or indeed the proportion of the total yield in a specified time? Again, there may be derived variables of interest: in experiments with animals, it could be the digestibility of the feed or the butterfat content of the milk. In pest and disease control trials it is often far more important to determine the damage on a crop than to assess the total yield, damaged and undamaged together. All these preliminaries are a vital part of the statistical assessment of a trial*; also, noting apparently anomalous values may help to pinpoint errors in recording, or alternatively, lead to the discovery of quite unsuspected effects.

Formal statistical analysis is not always necessary when the main purpose of a trial is just to obtain preliminary information for use in a further trial, for example at an early stage in a plant breeding project. However, it is common to conduct analyses of variance on trial results, if only to provide an assessment of residual variation* after allowing for treatment effects*. Some trials have treatments that are quantitative in nature, and the technique of regression* as well as analysis of variance will be useful at this formal stage. With two variables, judicious use of analysis of covariance* permits the effect of one variable on another to be assessed and allowed for. When, as is common, many variables have been recorded, multivariate methods of analysis (see MULTIVARIATE ANALYSIS OF VARIANCE (MANOVA)) may be used as an alternative to the separate analysis of each record.

Although analysis of variance and its derivatives are undoubtedly the methods most commonly used for data analysis, they are not the only ones; many other techniques may be used to supplement or replace them. Thus an important area for studying experimental techniques is the investigation of best plot sizes. An early study here was that by Fairfield Smith [6] of the relation between plot size and variability. Subsequent work has shown that it is also often necessary to take account of possible variation due to individual plants as well as the environment, while there are many nonstatistical factors that have to be considered in practice. Studies of animal breeding trials and components of variance (see VARIANCE COMPONENTS) have proceeded together ever since the work of Henderson [9] dealing with a nonorthogonal set of data on dairy cows. Many agricultural experiments are now conducted to provide data for testing a mathematical model, and there are biologically important models that do not fall conveniently into the linear form suitable for analysis-of-variance techniques. One example among many is the set of models describing the relations between crop yield and plant density, work on which is conveniently summarized by Willey and Health [11]. There is now much interest in plant disease epidemiology, and although the earlier theoretical work, both biological and mathematical, was not relevant to practical agriculture, some of the more recent studies are, e.g., ref. 1. Finally, the design and analysis of series of trials often present problems different in kind from those for a single trial: for trials of crop varieties, references range in time from 1938 [13] to 1980 [10].

APPLICATION AREAS

There is now scarcely an agricultural experimental station anywhere in the world that does not use statistical techniques of the types outlined here; indeed, many have their own statisticians. This is true not only of the United States and the United Kingdom, where these methods started, but of other countries in the English-speaking world. The language barrier has proved no impediment, and striking advances have been made in many European countries, including the Netherlands, East Germany, and Poland. Further, the methods, although originating largely in the more developed countries with a temperate climate, have been used in tropical developing countries, such as India, Israel, and others in Asia, together with those in Africa and Latin America.

Experiments on many crops now use statistical methods; these include a wide range of temperate cereals, fruit, vegetables, and forage crops, and an even wider range of tropical cereals and plantation crops. Experiments in the area of animal husbandry and disease control also use statistical techniques (although the methods used on large and expensive long-lived animals cannot be identical with those on short-term annual crops). Surveys using statistical methods have been conducted on an equally wide range of temperate and tropical practices in agriculture and animal husbandry. Indeed, the use of statistical methods now permeates the whole of research and development in agriculture and related disciplines throughout the world.

REFERENCES

1. Butt, D. J. and Royle, D. J. (1974). In *Epidemics of Plant Diseases*, J. Kranz, ed., pp. 78–114.

2. Cochran, W. G. and Cox, G. M. (1957). *Experimental Designs*, 2nd ed. Wiley, New York.

3. Cox, D. R. (1958). *Planning of Experiments*. Wiley, New York.

4. Cretté de Palluel (1788; English version by A. Young, 1790). *Ann. Agric.*, **14**, 133–139.

5. Crowther, E. M. and Yates, F. (1941). *Emp. J. Exper. Agric.*, **9**, 77–97.

6. Fairfield Smith, H. (1938). *J. Agric. Sci. Camb.*, **28**, 1–23.

7. Finney, D. J. (1956). *J. R. Statist. Soc. A*, **119**, 1–27.

8. Finney, D. J. (1972). *An Introduction to Statistical Science in Agriculture*, 4th ed. Blackwell, Oxford.

9. Henderson, C. R. (1953). *Biometrics*, **9**, 226–252.

10. Patterson, H. D. and Silvey, V. (1980). *J. R. Statist. Soc. A*, **143**, 219–240.

11. Willey, R. W. and Heath, S. B. (1969). *Adv. Agron.*, **21**, 281–321.

12. Yates, F. (1964). *Biometrics*, **20**, 307–321.

13. Yates, F. and Cochran, W. G. (1938). *J. Agric. Sci. Camb.*, **28**, 556–580.

14. Yates, F., Boyd, D. A., and Mathison, I. (1944). *Emp. J. Exp. Agric.*, **12**, 164–176.

See also ANALYSIS OF COVARIANCE; ANALYSIS OF VARIANCE; and FISHER, RONALD AYLMER.

G. H. FREEMAN

AIC. See MODEL SELECTION: AKAIKE'S INFORMATION CRITERION

AIDS STOCHASTIC MODELS

AIDS is an infectious disease caused by a retrovirus called human immunodeficiency virus (HIV) [17]. The first AIDS case was diagnosed in Los Angeles, CA, USA in 1980 [14]. In a very short period, the AIDS epidemic has grown into dangerous proportions. For example, the World Health Organization (WHO) and the United Nation AIDS Program (UNAIDS) estimated that 5 million had acquired HIV in 2003, and about 40 million people are currently living with AIDS. To control AIDS, in the past 10 years, significant advances had been made in treating AIDS patients by antiviral drugs through cocktail treatment protocol [10]. However, it is still far from cure and the disease is still spreading, especially in Africa and Asia. For preventing the spread of HIV, for controlling AIDS, and for understanding the HIV epidemic, mathematical models that take into account the dynamic of the HIV epidemic and the HIV biology are definitely needed. From this perspective, many mathematical models have been developed [1,8,11,22,24–26,29,42,55]. Most of these models are deterministic models in which the state variables (i.e., the numbers of susceptible people, HIV-infected people, and AIDS cases) are assumed as deterministic functions, ignoring completely the random nature of these variables. Because the HIV epidemic is basically a stochastic process, many stochastic models have been developed [5–7,18,33–37,39–44,46–49,53–54,57–64, 66–68,73–76]. This is necessary because as shown by Isham [23], Mode et al. [40,41], Tan [48,54], Tan and Tang [62], and Tan et al. [63], in some cases, the difference between the mean numbers of the stochastic models and the results of deterministic models could be very substantial; it follows that in these cases, the deterministic models would provide very poor approximation to the corresponding mean numbers of the stochastic models, leading to misleading and sometimes confusing results.

STOCHASTIC TRANSMISSION MODELS OF HIV EPIDEMIC IN POPULATIONS

Stochastic transmission models for the spread of HIV in populations were first developed by Mode et al. [40,41] and by Tan and Hsu [59,60] in homosexual populations; see also References 5–7, 18, 47–49, 54, 57, 66–68, 73, and 74. These models have been extended to IV drug populations [19,53,62] and to heterosexual populations [39,46,75–76]. Many of these models have been summarized in books by Tan [55] and by Mode and Sleeman [42]. Some applications of these models have been illustrated in Reference 55.

To illustrate the basic procedures for deriving stochastic transmission models for the HIV epidemic, consider a large population at risk for AIDS. This population may be a population of homosexual men, or a population of IV drug users, or a population of single males, single females, and married couples, or a mixture of these populations. In the presence of HIV epidemic, then there are three types of people in the population: S people (susceptible people), I people (infective people), and A people (AIDS patients). S people are healthy people but can contract HIV to become I people through sexual contact and/or IV drug contact with I people or A people or through contact with HIV-contaminated blood. I people are people who have contracted HIV and can pass the HIV to S people through sexual contact or IV drug contact with S people. According to the 1993 AIDS case definition [15] by the Center of Disease Control (CDC) at Atlanta, GA, USA, an I person will be classified as a clinical AIDS patient (A person) when this person develops at least one of the AIDS symptoms specified in Reference 15 and/or when his/her $CD4^{(+)}$ T-cell counts fall below $200/mm^3$. Then, in this population, one is dealing with a high-dimensional stochastic process involving the numbers of S people, I people and AIDS cases.

To develop realistic stochastic models for this process, it is necessary to incorporate many important risk variables and social and behavior factors into the model, and to account for the dynamics of the epidemic process. Important risk variables that have significant impacts on the HIV epidemic are age, race, sex, and sexual (or IV drug use) activity levels defined by the average number of sexual (or IV drug use) partners per unit time; the important social and behavior factors are the IV drug use, the mixing patterns between partners, and the condom use that may reduce the probability of transmission of the HIV viruses. To account for these important risk variables and for IV drug use, the population is further stratified into subpopulations.

Given that the population has been stratified into sub populations by many risk factors, the stochastic modeling procedures of the HIV epidemic essentially boils down to two steps. The first step involves modeling the transmission of HIV from HIV carriers to susceptible people. This step would transform S people to I people ($S \longrightarrow I$). This step is the dynamic part of the HIV epidemic process and is influenced significantly by age, race, social behavior, sexual level, and many other factors. This step is referred to as the transmission step. The next step is the modeling of HIV progression until the development of clinical AIDS and death in people who have already contracted HIV. This step is basically the step transforming I people into A people ($I \longrightarrow A$) and is influenced significantly by the genetic makeup of the individual and by the person's infection duration that is defined by the time period elapsed since he/she first contracts the HIV. This step is referred to as the HIV progression step. By using these two steps, the numbers of S people, I people, and AIDS cases are generated stochastically at any time given the numbers at the previous time. These models have been referred by Tan and his associates [48–49,54,57,61,63,66–68,73–76] as chain multinomial models since the principle of random sampling dictates that aside from the distributions of recruitment and immigration, the conditional probability distributions of the numbers of S people, I people, and AIDS cases at any time, given the numbers at the previous time, are related to multinomial distributions.

THE TRANSMISSION STEP: THE PROBABILITY OF HIV TRANSMISSION

The major task in this step is to construct the probabilities of HIV transmission from infective people or AIDS cases to S people by taking into account the dynamic aspects of the HIV epidemic. These probabilities are functions of the mixing pattern that describes how people from different risk groups mix together and the probabilities of transmission of HIV from HIV carriers to susceptible people given contacts between these people. Let $S_i(t)$ and $I_i(u,t)(u = 0,\ldots,t)$ denote the numbers of S people and I people with infection duration u at time t in the ith risk group respectively. Let the time unit be a month and denoted by $p_i(t; S)$, the conditional probability that an S person in the ith risk group contracts HIV during the tth month given $\{S(t), I(u,t), u = 0, 1, \ldots, t\}$. Let $\rho_{ij}(t)$ denote the probability that a person in the ith risk group selects a person in the jth risk group as a partner at the tth month and $\alpha_{ij}(u,t)$ the probability of HIV transmission from an infective person with infection duration u in the jth risk group to the susceptible person in the ith risk group given contacts between them during the tth month. Assume that because of the awareness of AIDS, there are no contacts between S people and AIDS cases and that there are n risk groups or subpopulations. Then $p_i(t; S)$ is given by:

$$p_i(t; S) = 1 - \{1 - \psi_i(t)\}^{X_i(t)} \qquad (1)$$

where $X_i(t)$ is the number of partners of the S person in the ith risk group during the tth month and $\psi_i(t) = \sum_{j=1}^{n} \rho_{ij}(t)\{I_j(t)/T_j(t)\}\overline{\alpha}_{ij}(t)$ with $T_j(t) = S_j(t) + \sum_{u=0}^{t} I_j(u,t)$ and $\overline{\alpha}_{ij}(t) = \frac{1}{I_j(t)} \sum_{u=0}^{t} I_j(u,t)\alpha_{ij}(u,t)$ with $I_j(t) = \sum_{u=0}^{t} I_j(u,t)$.

If the $\alpha_{ij}(u,t)$ are small, then $\{1 - \psi_i(t)\}^{X_i(t)} \cong \{1 - X_i(t)\psi_i(t)\}$ so that

$$p_i(t; S) \cong X_i(t)\psi_i(t)$$

$$= X_i(t) \sum_{j=1}^{n} \rho_{i,j}(t)\frac{I_j(t)}{T_j(t)}\overline{\alpha}_{ij}(t). \qquad (2)$$

Notice that in Equations 1 and 2, the $p_i(t; S)$ are functions of $\{S_i(t), I_i(u,t), u = 0, 1, \ldots, t\}$ and hence in general are random variables. However, some computer simulation studies by Tan and Byer [57] have indicated that if the $S_i(t)$ are very large, one may practically assume $p_i(t; S)$ as deterministic functions of time t and the HIV dynamic.

THE PROGRESSION STEP: THE PROGRESSION OF HIV IN HIV-INFECTED INDIVIDUALS

The progression of HIV inside the human body involves interactions between $CD4^{(+)}T$ cells, $CD8^{(+)}T$ cells, free HIV, HIV antibodies, and other elements in the immune system, which will eventually lead to the development of AIDS symptoms as time increases. It is influenced significantly by the dynamics of the interactions between different types of cells and HIV in the immune system, treatment by antiviral drugs, and other risk factors that affect the speed of HIV progression. Thus, it is expected that the progression of I to A depends not only on the calendar time t but also on the infection duration u of the I people as well as the genetic makeup of the I people. This implies that the transition rate of $I \longrightarrow A$ at time t for I people with infection duration u is in general a function of both u and t; this rate will be denoted by $\gamma(u,t)$. Let T_{inc} denote the time from HIV infection to the onset of AIDS. Given $\gamma(u,t)$, the probability distribution of T_{inc} can readily be derived [55, chapter. 4]. In the AIDS literature, T_{inc} has been referred to as the HIV incubation period and the probability distribution of T_{inc} the HIV incubation distribution. These distributions have been derived by Bachetti [4], Longini et al. [33], and by Tan and his associates [50–53,57,61,64–65] under various conditions. These probability distributions together with many other distributions, which have been used in the literature have been tabulated and summarized in Reference 55, chapter 4.

STOCHASTIC MODELING OF HIV TRANSMISSION BY STATISTICAL APPROACH

To develop stochastic models of HIV transmission, it is necessary to take into account the dynamic of the HIV epidemic to construct the probabilities $p_S(i,t) = \beta_i(t)\Delta t$ of HIV transmission. To avoid the dynamic aspect, statisticians assume $p_i(t; S)$ as deterministic functions of i and t and proceed to estimate these probabilities. This is a nonparametric procedure, which has ignored all information about the dynamics of the HIV epidemic; on the other hand, it has minimized the misclassification or misspecification of the dynamic of the HIV epidemic. In the literature, these probabilities are referred to as the infection incidence.

To illustrate the statistical approach, let T_I denote the time to infection of S people and $f_i(t)$ the probability density of T_I in the ith risk group. Then, $f_i(t) = \beta_i(t) exp\{-\int_0^t \beta_i(x)dx\}$. The $f_i(t)$ have been referred by statisticians as the HIV infection distributions.

To model the HIV progression, let T_A denote the time to AIDS of S people and $h_i(t)$ the probability density of T_A in the ith risk group. Then $T_A = T_I + T_{inc}$, where T_{inc} is the HIV incubation period. If T_I and T_{inc} are independently distributed of each other, then

$$h_i(t) = \int_0^t f_i(x)g_i(x,t)dx, \qquad (3)$$

where $g_i(s,t)$ is the density of the HIV incubation distribution given HIV infection at time s in the ith risk group. Notice that since the transition rates of the infective stages are usually independent of the risk group, $g_i(s,t) = g(s,t)$ are independent of i. In what follows, it is thus assumed that $g_i(s,t) = g(s,t)$ unless otherwise stated.

Let ω_i denote the proportion of the ith risk group in the population. For an individual taken randomly from the population, the density of T_A is given by:

$$h(t) = \sum_{i=1}^{n} w_i h_i(t) = \int_0^t \left\{ \sum_{i=1}^{n} w_i f_i(x)g_i(x,t) \right\} dx$$

$$= \int_0^t f(x)g(x,t)dx, \qquad (4)$$

where $\sum_{i=1}^{n} \omega_i f_i(t) = f(t)$ is the density of the HIV infection distribution for people taken randomly from the population.

Equation 4 is the basic equation for the backcalculation method. By using this equation and by interchanging the order of summation and integration, it can easily be shown that the probability that a S person at time 0 taken randomly from the population will become an AIDS case for the first time during $(t_{j-1}, t_j]$ is

$$P(t_{j-1}, t_j) = \int_{t_{j-1}}^{t_j} \int_0^t f(u)g(u,t) \, du \, dt$$

$$= \left\{ \int_0^{t_j} - \int_0^{t_{j-1}} \right\} \int_0^t f(u)g(u,t) \, du \, dt$$

$$= \int_0^{t_j} f(u) \left\{ G(u,t_j) - G(u,t_{j-1}) \right\} , du, \qquad (5)$$

where $G(u,t) = \int_u^t g(u,x) \, dx$ is the cumulative distribution function (cdf) of the HIV incubation period, given HIV infection at time u.

Equation 5 is the basic formula by means of which statisticians tried to estimate the HIV infection or the HIV incubation based on AIDS incidence data [8,55]. This has been illustrated in detail in Reference 8 and in Reference 55. There are two major difficulties in this approach, however. One difficulty is that the problem is not identifiable in the sense that one cannot estimate simultaneously the HIV infection distribution and the HIV incubation distribution. Thus, one has to assume the HIV incubation distribution as known if one wants to estimate the HIV infection distribution; similarly, one has to assume the HIV infection distribution as known if one wants to estimate the HIV incubation distribution. Another difficulty is that one has to assume that there are no immigration, no competing death, and no other disturbing factors for Equation 5 to hold; see Reference 55, chapter 5. These difficulties can readily be resolved by introducing state space models; see References 55, 56, 73, and 74.

STOCHASTIC TRANSMISSION MODELS OF HIV EPIDEMIC IN HOMOSEXUAL POPULATIONS

In the United States, Canada, Australia, New Zealand, and Western European countries, AIDS cases have been found predominantly amongst the homosexual, bisexual, and intravenous drug-user community with only a small percentage of cases being due to heterosexual contact. Thus, most of the stochastic models for HIV spread were first developed in homosexual populations [5–7, 40–41,47,49,57,59–60,62,66–68,73–74].

To illustrate how to develop stochastic models for the HIV epidemic, consider a large homosexual population at risk for AIDS that has been stratified into n risk groups of different sexual activity levels. Denote by $A_i(t)$ the number of new AIDS cases developed during the tth month in the ith risk group and let the time unit be a month. Then, one is entertaining a high-dimensional discrete-time stochastic process $\underset{\sim}{U}(t) = \{S_i(t), I_i(u, t), u = 0, \ldots, t, A_i(t), i = 1, \ldots, n\}$. To derive basic results for this process, a convenient approach is by way of stochastic equations. This is the approach proposed by Tan and his associates in modeling the HIV epidemic [49,54–57,63–64,66–68,73–76].

THE STOCHASTIC DIFFERENCE EQUATIONS AND THE CHAIN MULTINOMIAL MODEL

To develop the stochastic model for the above process, let $\{\Lambda_i(S, t), \mu_i(S, t)\}$ and $\{\Lambda_i(u, t), \mu_i(u, t)\}$ denote the recruitment and immigration rate, and the death and migration rate of S people and $I(u)$ people at the tth month in the ith risk group respectively. Then, given the probability $p_i(t; S)$ that the S person in the ith risk group would contract HIV during the tth month, one may readily obtain $\underset{\sim}{U}(t + 1)$ from $\underset{\sim}{U}(t)$ by using multinomial distributions, for $t = 0, 1, \ldots,$. This procedure provides stochastic difference equations for the numbers of S people, $I(u)$ people, and the number of new A people at time t. These models have been referred to by Tan [48,49,54,55] and by Tan and his coworkers [57,61,63,64, 66–68,73,74] as chain multinomial models.

To illustrate, let $R_i(S, t), F_i(S, t)$, and $D_i(S, t)$ denote respectively the number of recruitment and immigrants of S people, the number of $S \to I(0)$ and the total number of death of S people during the tth month in the ith risk group. Similarly, for $u = 0, 1, \ldots, t$, let $R_i(u, t), F_i(u, t)$ and $D_i(u, t)$ denote respectively the number of recruitment and immigrants of $I(u)$ people, the number of $I(u) \to A$ and the total number of death of $I(u)$ people during the tth month in the ith risk group. Then, the conditional probability distribution of $\{F_i(S, t), D_i(S, t)\}$ given $S_i(t)$ is multinomial with parameters $\{S_i(t), p_i(t; S), \mu_i(S, t)\}$ for all $i = 1, \ldots, n$; similarly, the conditional

probability distribution of $\{F_i(u, t), D_i(u, t)\}$ given $I_i(u, t)$ is multinomial with parameters $\{I_i(u, t), \gamma_i(u, t), \mu_i(u, t)\}$, independently of $\{F_i(S, t), D_i(S, t)\}$ for all $\{i = 1, \ldots, n, u = 0, 1, \ldots, t\}$. Assume that $E\{R_i(S, t)|S(t)\} = S_i(t)\Lambda_i(S, t)$ and $E\{R_i(u, t)|I_i(u, t)\} = I_i(u, t)\Lambda_i(u, t)$. Then, one has the following stochastic difference equations for $\{S_i(t), I_i(u, t), i = 1, \ldots, n\}$:

$$S_i(t + 1) = S_i(t) + R_i(S, t)$$
$$-F_i(S, t) - D_i(S, t)$$
$$= S_i(t)\{1 + \Lambda_i(S; t) - p_i(t; S)$$
$$-\mu_i(S, t)\} + \epsilon_i(S, t + 1), \quad (6)$$

$$I_i(0, t + 1) = F_i(S, t) = S_i(t)p_i(t; S)$$
$$+\epsilon_i(0, t + 1), \quad (7)$$

$$I_i(u + 1, t + 1) = I_i(u, t) + R_i(u, t)$$
$$-F_i(u, t) - D_i(u, t)$$
$$= I_i(u, t)\{1 + \Lambda_i(u, t)$$
$$-\gamma_i(u, t) - \mu_i(S, t)\}$$
$$+\epsilon_i(u + 1, t + 1), \quad (8)$$

$$A_i(t + 1) = \sum_{u=0}^{t} F_i(u, t)$$
$$= \sum_{u=0}^{t} I_i(u, t)\gamma_i(u, t) + \epsilon_i(A, t). \quad (9)$$

In Equations 6 to 9, the random noises $\{\epsilon_i(S, t), \epsilon_i(u, t), u = 0, 1, \ldots, t, \epsilon_i(A, t)\}$ are derived by subtracting the conditional mean numbers from the corresponding random variables. It can easily be shown that these random noises have expected values 0 and are uncorrelated with the state variables.

Using the above stochastic equations, one can readily study the stochastic behaviors of the HIV epidemic in homosexual populations and assess effects of various risk factors on the HIV epidemic and on some intervention procedures. Such attempts have been made by Tan and Hsu [59,60], Tan [48,54], and Tan, Tang and Lee [63] by using some simplified models. For example, Tan and Hsu [59–60] have shown that the effects of intervention by decreasing sexual contact rates depend heavily on the initial number of infected people; when the initial number is small, say 10, then the effect is quite significant. On the other hand, when the initial number is large, say 10000, then the effect of decreasing sexual contact rates is very small. The Monte Carlo studies by Tan and Hsu [59,60] have also revealed some effects of "regression on the mean" in the sense that the variances are linear functions of the expected numbers. Thus, although the variances are much larger than their respective mean numbers, effects of risk factors on the variance curves are quite similar to those of these risk factors on the mean numbers.

By using the above equations, one may also assess the usefulness of deterministic models in which $S_i(t), I_i(r,t), r = 1, \cdots, k, A_i(t)$ are assumed as deterministic functions of i and t, ignoring completely randomness of the HIV epidemic process. The system of equations defining the deterministic models is derived by ignoring the random noises from Equations 6 to 9. Thus, one may assume that the deterministic models are special cases of the stochastic models. However, the above equations for $S_i(t)$ and $I_i(0,t)$ are not the same as the equations for the mean numbers of $S_i(t)$ and $I_i(0,t)$ respectively. This follows from the observation that since the $p_S(i,t)$ are functions of $S_i(t)$ and $I_i(u,t), u = 1, \ldots, k, E[S_i(t)p_S(i,t)] \neq E[S_i(t)] \times E[p_S(i,t)]$. As shown in Reference 55, the equations for the mean numbers of $S_i(t)$ and $I_i(0,t)$ differ from the corresponding ones of the deterministic model in that the equations for the means of $S_i(t)$ and $I_i(0,t)$ contain additional terms involving covariances $Cov\{S_i(t), p_S(i,t)\}$ between $S_i(t)$ and $p_S(i,t)$. Thus, unless these covariances are negligible, results of the deterministic models of the HIV epidemic would in general be very different from the corresponding mean numbers of the stochastic models of the HIV epidemic. As shown by Isham [23], Mode et al. [40,41], Tan [48,54], Tan and Tang [62], and Tan, Tang and Lee [63], the difference between results of deterministic models and the mean numbers of the stochastic models could be very substantial. The general picture appears to be that the stochastic variation would in general speed up the HIV epidemic. Further, the numbers of I people and A people computed by the deterministic model would underestimate the true numbers in the short run but

overestimate the true numbers in the long run. These results imply that, in some cases, results of the deterministic model may lead to misleading and confusing results.

THE PROBABILITY DISTRIBUTION OF THE STATE VARIABLES

Let $\mathbf{X} = \{\mathbf{X}(0), \ldots, \mathbf{X}(t_M)\}$ with $\mathbf{X}(t) = \{S_i(t), I_i(u,t), u = 0, 1, \ldots, t, i = 1, \ldots, n\}$. To estimate the unknown parameters and to assess stochastic behaviors of the HIV epidemic, it is of considerable interest to derive the probability density $P(\mathbf{X}|\Theta)$ of $\mathbf{X}$. By using multinomial distributions for $\{R_i(S,t), F_i(S,T)\}$ and for $\{R_i(u,t), F_i(u,T)\}$ as above, this probability density can readily be derived. Indeed, denoting by $g_{i,s}\{j; t|\mathbf{X}(t)\}$ and $g_{i,u}\{j; t|\mathbf{X}(t)\}$ the conditional densities of $R_i(S,t)$ given $\mathbf{X}(t)$ and of $R_i(u,t)$ given $B\mathbf{X}(t)$ respectively, one has

$$P(\mathbf{X}|\Theta) = P\{\mathbf{X}(0)|\Theta\} \prod_{t=1}^{t_M} P\{\mathbf{X}(t)|\mathbf{X}(t-1)\}$$

$$= P\{\mathbf{X}(0)|\Theta\} \prod_{t=1}^{t_M} \prod_{i=1}^{n} P\{S_i(t)|\mathbf{X}(t-1)\}$$

$$\times \left\{ \prod_{u=0}^{t} P[I_i(u,t)|\mathbf{X}(t-1)] \right\}. \quad (10)$$

In Equation 10, the $P\{S_i(t+1)|\mathbf{X}(t)\}$ and the $P\{I_i(u+1,t+1)|\mathbf{X}(t)\}$ are given respectively by

$$P\{S_i(t+1)|\mathbf{X}(t)\}$$

$$= \binom{S_i(t)}{I_i(0,t+1)} [p_i(S,t)]^{I_i(0,t+1)} h_i(t|S), \quad (11)$$

$$P\{I_i(u+1,t+1)|\mathbf{X}(t)\}$$

$$= \sum_{r=0}^{I_i(u,t)} \binom{I_i(u,t)}{r} [\gamma_i(u,t)]^r h_{i,r}(u,t|I)$$
$$\text{for } u = 0, 1, \ldots, t, \quad (12)$$

where the $h_i(t|S)$ and the $h_{i,r}(u,t|I)$ are given respectively by:

$$h_i(t|S) = \sum_{j=0}^{S_i(t+1)-S_i(t)+I_i(0,t+1)} g_{i,S}\{j,t|\boldsymbol{X}(t)\}$$

$$\times \binom{S_i(t) - I_i(0,t+1)}{a_{i,S}(j,t)} [d_i(S,t)]^{a_{i,S}(j,t)}$$

$$\times \{1 - p_i(S,t) - d_i(S,t)\}^{b_{i,S}(j,t)},$$

with $a_{i,S}(j,t) = S_i(t) - S_i(t+1) - I_i(0,t+1) + j$ and $b_{i,S}(j,t) = S_i(t+1) - j$, and

$$h_{i,r}(u,t|I) = \sum_{j=0}^{I_i(u+1,t+1)-I_i(u,t)+r} g_{i,u}\{j,t|\boldsymbol{X}(t)\}$$

$$\times \binom{I_i(u,t) - r}{a_{i,u}(r,j,t)} [d_i(u,t)]^{a_{i,u}(r,j,t)}$$

$$\times \{1 - \gamma_i(u,t) - d_i(u,t)\}^{b_{i,u}(r,j,t)},$$

with $a_{i,u}(r,j,t) = I_i(u,t) - I_i(u+1,t+1) - r + j$ and $b_{i,u}(r,j,t) = I_i(u+1,t+1) + r - 2j$.

In the above equations, notice that aside from the immigration and recruitment, the distribution of $\boldsymbol{X}$ is basically a product of multinomial distributions. Hence, the above model has been referred to as a chain multinomial model; see References 48, 49, 54, 57, 61, 63, 64, 66–68, 73, and 74. Tan and Ye [74] have used the above distribution to estimate both the unknown parameters and the state variables via state space models.

THE STAGED MODEL OF HIV EPIDEMIC IN HOMOSEXUAL POPULATIONS

In the above model, the number of state variables $I_i(u,t)$ and hence the dimension of the state space increases as time increases; that is, if the infection duration is taken into account, then the size of the dimension of the state space increases as time increases. To simplify matters, an alternative approach is to partition the infective stage into a finite number of substages with stochastic transition between the substages and assume that within the substage, the effects of duration is the same. This is the approach proposed by Longini and his associates [33–37]. In the literature, such staging is usually achieved by using the number of CD4$^{(+)}$ T-cell counts per mm^3 blood. The staging system used by Satten and Longini [43,44] is I_1, CD4 counts $\geqslant 900/mm^3$; I_2, $900/mm^3 >$

CD4 counts $\geqslant 700/mm^3$; I_3, $700/mm^3 >$ CD4 counts $\geqslant 500/mm^3$; I_4, $500/mm^3 >$ CD4 counts $\geqslant 350/mm^3$; I_5, $350/mm^3 >$ CD4 counts $\geqslant 200/mm^3$; I_6, $200/mm^3 >$ CD4 counts. (Because of the 1993 AIDS definition by CDC [15], the I_6 stage is merged with the AIDS stage (A stage).)

The staging of the infective people results in a staged model for the HIV epidemic. Comparing these staged models with the previous model, the following differences are observed: (i) Because of the infection duration, the nonstaged model is in general not Markov. On the other hand, if one assumes that the transition rates of the substages are independent of the time at which the substage were generated, the staged models are Markov. (ii) For the nonstaged model, the number of different type of infectives always increase nonstochastically as time increases. These are referred to as expanding models by Liu and Chen [32]. On the other hand, the number of substages of the infective stage in the staged model is a fixed number independent of time. (iii) For the nonstaged model, the infective people increase its infection duration nonstochastically, always increasing by one-time unit with each increase of one-time unit. However, they transit directly to AIDS stochastically. On the other hand, for the staged model, the transition from one substage to another substage is stochastic and can either be forward or backward, or transit directly to AIDS. Because of the random transition between the infective substages, one would expect that the staging has introduced more randomness into the model than the nonstaged model.

Assuming Markov, then one may use some standard results in Markov chain theory to study the HIV epidemic in staged models. This has been done by Longini and his associates [33–37]. Alternatively, by using exactly the same procedures as in the previous model, one may derive the stochastic equation for the state variables as well as the probability distributions of the state variables. By using these equations and the probability distributions, one can then study the stochastic behaviors and to assess effects of risk variables and the impact of some intervention procedures. This has been done by Tan and his associates [48–49,54–55,57,63–64,66–68]. By using the San Francisco homosexual population as an example, they have shown that the staged model gave similar results as the previous model and hence the same conclusions. These results indicates that the errors of approximation and the additional variations due to stochastic transition between the substages imposed by the staging system are in general quite small for the HIV epidemic in homosexual populations. On the other hand, because of the existence of a long asymptomatic infective period with low infectivity, it is expected that the staged model would provide a closer approximation to the real world situations then the nonstaged model.

Another problem in the staged model is the impacts of measurement error as the CD4 T-cell counts are subject to considerable measurement error. To take this into account, Satten and Longini [44] have proposed a hidden Markov model by assuming the measurement errors as Gaussian variables. The calculations done by them did not reveal a significant impact of these errors on the HIV epidemic, however, indicating that the effects of measurement errors on the HIV epidemic of the staged model is not very significant.

STOCHASTIC MODELS OF HIV TRANSMISSION IN COMPLEX SITUATIONS

In Africa, Asia, and many south American countries, although homosexual contact and IV drug use may also be important avenues, most of the HIV epidemic are developed through heterosexual contacts and prostitutes [38]. The dynamics of HIV epidemic in these countries are therefore very different from those in the United States, Canada, and the western countries, where the major avenues of HIV transmission are homosexual contact and sharing needles and IV drug use. It has been documented that even in homosexual populations, race, age, and risk behaviors as well as many other risk variables would significantly affect the HIV epidemic [66–67,75–76]. To account for effects of many risk factors such as sex, race, and age, the above simple stochastic model has been extended into models under complex situations [19,39,46,53,66–67,75–76].

To develop stochastic models of HIV transmission in complex situations, the basic procedures are again the same two steps as described above: (i) Stratifying the population into subpopulations by sex, race, and risk factors, derive the probabilities of HIV transmission from infective people to susceptible people in each subpopulation. These probabilities usually involve interactions between people from different risk groups and the structure of the epidemic. (ii) Develop steps for HIV progression within each subpopulation. It appears that because the dynamics of HIV epidemic are different under different situations, the first step to derive the probabilities of HIV transmission varies from population to population depending on different situations. Given that the probabilities of HIV transmission from infective people to susceptible people have been derived for each subpopulation, the second step is similar to the progression step of the procedures described above. That is, the only major difference between different models lies in the derivation of the probabilities $p_i(t; S)$ for the ith subpopulation. Assuming that there are no sexual contacts with AIDS cases, the general form of $p_i(t; S)$ is

$$p_i(t; S) = X_i(t) \sum_j \rho_{ij}(t) \frac{I_j(t)}{T_j(t)} \overline{\alpha}_{ij}(t) \qquad (13)$$

Notice that Equation 13 is exactly of the same form as Equation 2; yet, because of the different dynamics in different models, the $\rho_{ij}(t)$'s are very different between different models. In populations of IV drug users, HIV spread mainly through sharing IV needles in small parallel groups. In these populations, therefore, $\rho_{ij}(t)$ is derived by first forming small groups and then spread HIV by sharing needles between members within the group. This is the basic formulation by means of which Capasso et al. [9], Gani and Yakowitz [19], Kaplan [28], and Kaplan et al. [30] derived the probabilities of HIV infection of S people by infective people. Along this line, Tan [55, Chap. 4] has formulated a general procedure to derive this probability.

In populations stratified by race, sex, age, sexual activity levels, and risk behaviors and involving married couples, to derive $\rho_{ij}(t)$ one needs to take into account some realistic preference patterns. These realities include *(i)* People tend to mix more often with people of the same race, same age group, and same sexual activity level; *(ii)* people with high sexual activity levels and/or old age tend to select sexual partners indiscriminately; *(iii)* if the age difference between the two partners is less than five years, then age is not an important factor in selecting sexual partners, and *(vi)* race and sexual activity level may interact with each other to affect the selection of sexual partners; *(v)* a happy marriage would reduce external marital relationship. Taking many of these factors into account and assuming that members of small populations select members from larger populations, Tan and Xiang [66,67] and Tan and Zhu [75,76] have proposed a selective mixing pattern through the construction of acceptance probabilities. Intuitively, this mixing can be expressed as a product of two probability measures: The first is the probability of selecting members from subpopulations with larger effective population size via acceptance probabilities; the second is the conditional probability of selecting members of infective people with different infection duration from the selected population.

By using the selective mixing pattern, Tan and Xiang [66,67] have developed stochastic models for the HIV epidemic in homosexual populations taking into account race, age, and sexual activity levels. Their results indicate that race and age affect the HIV epidemic mainly through the numbers of different sexual partners per partner per month and their interactions with the mixing pattern. Increasing the transition rates of infective people by race and/or by age seems to have some impact on the HIV progression but the effects are much smaller than those from the average numbers of different sexual partners per partner per month and the mixing patterns. Thus, the observed result that there is a much larger proportion of AIDS cases from black people than from white people in the US population ([16]) is a consequence of the following observations: *(i)* Black people in general have larger number of sexual partners per unit time than white people. *(ii)* There is a large proportion of restricted mixing pattern and mixing patterns other than

proportional mixing while under these mixing patterns, black people appear to contract HIV much faster than white people.

By using selective mixing pattern, Tan and Zhu [75,76] have developed stochastic models involving single males, single females, married couples, and prostitutes. Their results indicate that the prostitute factor may be the main reason for the rapid growth of the HIV epidemic in some Asian countries such as Thailand and India, which have a large prostitute population. Their Monte Carlo studies also suggest that rapid growth of the HIV epidemic in some Asian countries may be arrested or controlled by promoting extensive use of condoms by prostitutes combined with a campaign of AIDS awareness in the younger and sexually active populations.

STATE SPACE MODELS OF THE HIV EPIDEMIC

State space models of stochastic systems are stochastic models consisting of two submodels: The stochastic system model, which is the stochastic model of the system, and the observation model, which is a statistical model based on available observed data from the system. That is, the state space model adds one more dimension to the stochastic model and to the statistical model by combining both of these models into one model. This is a convenient and efficient approach to combine information from both stochastic models and statistical models. It takes into account the basic mechanisms of the system and the random variation of the system through its stochastic system model and incorporate all these into the observed data from the system; and it validates and upgrades the stochastic model through its observation model and the observed data of the system and the estimates of the state variables. It is advantageous over both the stochastic model and the statistical model when used alone since it combines information and advantages from both of these models. Specifically, one notes that *(i)* Because of additional information, many of the identifiability problems in statistical analysis are nonexistent in state space models; see References 73 and 74 for some examples. *(ii)* It provides an optimal procedure to update the model by new data that may become available in the future. This is the smoothing step of the state space models; see References 2, 12, and 20. *(iii)* It provides an optimal procedure via Gibbs sampling to estimate simultaneously the unknown parameters and the state variables of interest; see References 55, chapter 6; 56, chapter 9; 73; and 74. *(iv)* It provides a general procedure to link molecular events to critical events in population and at the cellular level; see Reference 58.

The state space model (Kalman-filter model) was originally proposed by Kalman and his associates in the early 60s for engineering control and communication [27]. Since then it has been successfully used as a powerful tool in aerospace research, satellite research, and military missile research. It has also been used by economists in econometrics research [21] and by mathematician and statisticians in time series research [3] for solving many difficult problems that appear to be extremely difficult from other approaches. In 1995, Wu and Tan [78,79] had attempted to apply the state space model and method to AIDS research. Since then many papers have been published to develop state space models for the HIV epidemic and the HIV pathogenesis; see References 13, 68–73, 77, and 78. Alternatively, by combining the Markov staged model with Gaussian measurement errors for the CD4 T-cell counts, Satten and Longini [44] have proposed a Hidden Markov model for the HIV epidemic; however, it is shown by Tan [56] that this is a special case of the state space models.

Although Tan and Ye [74] have applied the state space model for the HIV epidemic in the Swiss population of IV drug users, to date, the state space models for HIV epidemic are primarily developed in homosexual populations. To illustrate how to develop state space models for the HIV epidemic, we will thus use the San Francisco homosexual population as an example, although the general results apply to other populations as well.

A STATE SPACE MODEL FOR THE HIV EPIDEMIC IN THE SAN FRANCISCO HOMOSEXUAL POPULATION

Consider the San Francisco homosexual population, in which HIV spread primarily by sexual contact [16]. For this population, Tan and Ye [73] have developed a state space model for the HIV epidemic. For this state space model, the stochastic system model was represented by stochastic difference equations. The observation model of this state space model is based on the monthly AIDS incidence data (i.e., data of new AIDS cases developed during a month period). This data is available from the gofer network of CDC. This is a statistics model used by statistician through the backcalculation method. Combining these two models into a state space model, Tan and Ye [73] have developed a general Bayesian procedure to estimate the HIV infection, the HIV incubation, as well as the numbers of susceptible people, infective people, and AIDS cases. Notice that this is not possible by using the stochastic model alone or by using the statistic model alone because of the identifiability problem.

THE STOCHASTIC SYSTEM MODEL

To develop a stochastic model for the San Francisco population, Tan and Ye [74] have made two simplifying assumptions: *(i)* By visualizing the infection incidence and hence the infection distribution as a mixture of several sexual levels, one sexual activity level may be assumed. *(ii)* Because the population size of the city of San Francisco changes very little, for the S people and I people it is assumed that the number of immigration and recruitment is about the same as the number of death and migration. Tan and Ye [74] have shown that these assumptions have little impacts on the probability distributions of the HIV infection and the HIV incubation. On the basis of these assumptions, then the state variables are $U(t) = \{S(t), I(u,t), u = 0, 1, \ldots, t, A(t)\}$.

Assuming that the probability $p_S(t)$ of HIV infection of S people and the probability $\gamma(s,t) = \gamma(t-s)$ of $I(u) \rightarrow AIDS$ as deterministic functions of time, then the parameters are $\theta(t) = \{p_S(t), \gamma(u,t) = \gamma(t-u), u = 0, 1, \ldots, t\}'$. The densities of the HIV infection and the HIV incubation are given by $f_I(t) = p_S(t)\Pi_{i=0}^{t-1}[1 - p_S(i)] = G_S(t-1)p_S(t), t = 1, \ldots, \infty$ and $g(t) = \gamma(t)\Pi_{j=0}^{t-1}[1 - \gamma(j)] = G_I(t-1)\gamma(t), t = 1, \ldots, \infty$ respectively.

Let $F_S(t)$ be the number of S people who contract HIV during the t-th month and $F_I(u, t)$ the number of $I(u, t) \rightarrow A$ during the tth month. Then,

$$S(t + 1) = S(t) - F_S(t), \qquad (14)$$

$$I(0, t + 1) = F_S(t), \qquad (15)$$

$$I(u + 1, t + 1) = I(u, t) - F_I(u, t), \qquad (16)$$

$$u = 0, \ldots, t,$$

where $F_S(t)|S(t) \sim B\{S(t), p_S(t)\}$ and $F_I(u, t)|$ $I(ut) \sim B\{I(u, t), \gamma(u)\}, u = 0, 1, \ldots, t$.

Put $\Theta = \{\theta(t), t = 1, \ldots, t_M\}, \underset{\sim}{X}(t) = \{S(t),$ $I(u, t), u = 0, 1, \ldots t\}$ and $\boldsymbol{X} = \{\underset{\sim}{X}(1), \ldots,$ $\underset{\sim}{X}(t_M)\}$, where t_M is the last time point. Then, the probability distribution of $\boldsymbol{X}$ given Θ and given $\underset{\sim}{X}(0)$ is

$$P\{\boldsymbol{X} | \underset{\sim}{X}(0)\} = \prod_{j=0}^{t_M-1} P\{\underset{\sim}{X}(j + 1) | \underset{\sim}{X}(j), \Theta\}$$

where

$$Pr\{\underset{\sim}{X}(j + 1) | \underset{\sim}{X}(j), \Theta\}$$

$$= \binom{S(t)}{I(0, t + 1)} [p_S(t)]^{I(0, t+1)}$$

$$\times [1 - p_S(t)]^{S(t) - I(0, t+1)}$$

$$\times \prod_{u=0}^{t} \binom{I(u, t)}{I(u, t) - I(u + 1, t + 1)}$$

$$\times [\gamma(u)]^{I(u,t) - I(u+1,t+1)}$$

$$\times [1 - \gamma(u)]^{I(u+1,t+1)}. \qquad (17)$$

Notice that the above density is a product of binomial densities and hence has been referred to as the chain binomial distribution.

For the HIV epidemic in the San Francisco homosexual population, Tan and Ye [73] have assumed January 1, 1970, as $t = 0$ since the first AIDS case in San Francisco appeared in 1981 and since the average incubation period for HIV is about 10 years. It is also assumed that, in 1970, there are no infective people but to start the HIV epidemic, some HIV were introduced into the population in 1970. Thus, one may take $I(0, 0) = 36$ because this is the number of AIDS in San Francisco in 1981. Tan and Ye [73] have assumed the size

of the San Francisco homosexual population in 1970 as 50000 because with a 1% increase in population size per year by the US census survey [77], the estimate of the size of the San Francisco homosexual population is 58,048 = 50,000 × (1.01)[15] in 1985, which is very close to the estimate 58,500 of the size of the San Francisco homosexual population in 1985 by Lemp et al. [31].

THE OBSERVATION MODEL.

Let $y(j + 1)$ be the observed AIDS incidence during the jth month and $A(t + 1)$ the number of new AIDS cases developed during the tth month. Then the stochastic equation for the observation model is

$$y(j + 1) = A(j + 1) + \xi(j + 1)$$

$$= \sum_{u=0}^{j} F_I(u, j) + \xi(j + 1)$$

$$= \sum_{u=0}^{j} [I(u, t) - I(u + 1, t + 1)]$$

$$+ \xi(t + 1)$$

$$= \sum_{u=0}^{j} I(u, j)\gamma(u) + \epsilon_A(j + 1) + \xi(j + 1)$$

$$= \sum_{u=0}^{j} I(u, j)\gamma(u) + e(j + 1), \qquad (18)$$

where $\xi(t + 1)$ is the random measurement error associated with observing $y(j + 1)$ and $\epsilon_A(j + 1) = [F_s(t) - S(t)p_s(t)] + \sum_{u=1}^{j} [F_I(u, t) - I(u, t)\gamma(u)]$.

Put $\mathbf{Y} = \{y(j), j = 1, \ldots, t_M\}$. Assuming that the $\xi(j)$ are independently distributed as normal with means 0 and variance σ_j^2, then the likelihood function $P\{\mathbf{Y}|\mathbf{X}, \Theta\} = L(\Theta|\mathbf{Y}, \mathbf{X})$ given the state variables is

$$P\{\mathbf{Y}|\mathbf{X}, \Theta\}$$

$$\propto \prod_{j=1}^{t_M} \left(\sigma_j^{-1} \exp\left\{ -\frac{1}{2\sigma_j^2} [y(j) - A(j)]^2 \right\} \right). \qquad (19)$$

Notice that under the assumption that $\{p_s(t), \gamma(t)\}$ are deterministic functions of t,

$$E[S(t+1)] = E[S(t)][1 - p_s(t)]$$

$$= E[S(t-1)] \prod_{i=t-1}^{t} \{1 - p_s(i)\}$$

$$= E[S(0)] \prod_{i=0}^{t} \{1 - p_s(i)\}$$

$$= E[S(0)]G_s(t), \tag{20}$$

$$E[F_s(t)] = E[S(t)]p_s(t)$$

$$= E[S(0)]G_s(t-1)p_s(t)$$

$$= E[S(0)]f_I(t), \tag{21}$$

$$E[I(u+1, t+1)] = E[I(u,t)][1 - \gamma(u)]$$

$$= E[I(0, t-u)] \prod_{j=0}^{u} [1 - \gamma(j)]$$

$$= E[I(0, t-u)]G_I(u)$$

$$= E[S(0)]f_I(t-u)G_I(u). \tag{22}$$

Hence,

$$E[I(u,t) - I(u+1, t+1)]$$

$$= E[S(0)]f_I(t-u)\{G_I(u-1) - G_I(u)\}$$

$$= E[S(0)]f_I(t-u)G_I(u-1)\gamma(u)$$

$$= E[S(0)]f_I(t-u)g(u). \tag{23}$$

It follows that

$$\sum_{u=0}^{t} E[I(u,t) - I(u+1, t+1)]$$

$$= E[S(0)] \sum_{u=0}^{t} f_I(t-u)g(u)$$

$$= E[S(0)] \sum_{u=0}^{t} f_I(u)g(t-u), \tag{24}$$

so that

$$y(j+1) = E[S(0)] \sum_{u=0}^{t} f_I(u)g(t-u) + e(j+1). \tag{25}$$

Notice that Equation 25 is the convolution formula used in the backcalculation method [4,55]. This implies that the backcalculation method is the observation model in the state space model. The backcalculation method is not identifiable because using Equation 25 alone and ignoring information from the stochastic system model, the information is not sufficient for estimating all the parameters.

A GENERAL BAYESIAN PROCEDURE FOR ESTIMATING THE UNKNOWN PARAMETERS AND THE STATE VARIABLES

By using the state space model, Tan and Ye [73,74] have developed a generalized Bayesian approach to estimate the unknown parameters and the state variables. This approach will combine information from three sources: *(i)* Previous information and experiences about the parameters in terms of the prior distribution of the parameters, *(ii)* biological information via the stochastic system equations of the stochastic system, and *(iii)* information from observed data via the statistical model from the system.

To illustrate the basic principle of this method, let $P(\Theta)$ be the prior distribution of Θ. Then, the joint distribution of $\{\Theta, X, Y\}$ is given by $P(\Theta, X, Y) = P(\Theta)P(X|\Theta)P(Y|X, \Theta)$. From this, the conditional distribution $P(X|\Theta, Y)$ of X given (Θ, Y) and the conditional posterior distribution $P(\Theta|X, Y)$ of Θ given (X, Y) are given respectively by

(A): $P(X|\Theta, Y) \propto P(X|\Theta)P(Y|X, \Theta)$

(B): $P(\Theta|X, Y) \propto P(\Theta)P(X|\Theta)P(Y|X, \Theta)$

Given these probability densities, one may use the multilevel Gibbs sampling method to derive estimates of Θ and X given Y [45]. This is a Monte Carlo sequential procedure alternating between two steps until convergence: *(i)* Given $\{\Theta, Y\}$, one generates X by using $P(X|\Theta, Y)$ from (A). These are the Kalman-filter estimates. *(ii)* Using the Kalman-filter estimates of X from (A) and given Y, one generates values of Θ by using $P(\Theta|\tilde{X}, Y)$ from (B). Iterating between these two steps until convergence, one then generates random samples from the conditional probability distribution $P(X|Y)$ independently of Θ, and from the posterior distribution $P(\Theta|Y)$ independently of X, respectively. This provides the Bayesian estimates of Θ given data and the Bayesian estimates of X given data, respectively. The proof of the convergence can be developed by using basic theory of stationary distributions in irreducible and aperiodic Markov chains; see Reference 56, chapter 3.

Using the above approach, one can readily estimate simultaneously the numbers of S people, I people, and AIDS cases as well as the parameters $\{p_S(t), \gamma(t)\}$. With the estimation of $\{p_S(t), \gamma(t)\}$, one may then estimate the HIV infection distribution $f_I(t)$ and the HIV incubation distribution $g(t)$. For the San

Francisco homosexual population, the estimates of $f_I(t)$ and $g(t)$ are plotted in Figs. 1 and 2. Given below are some basic findings by Tan and Ye [73]:

(a) From Fig 1; the estimated density of the HIV infection clearly showed a mixture of distributions with two obvious peaks. The first peak (the higher peak) occurs around January 1980 and the second peak around March 1992.

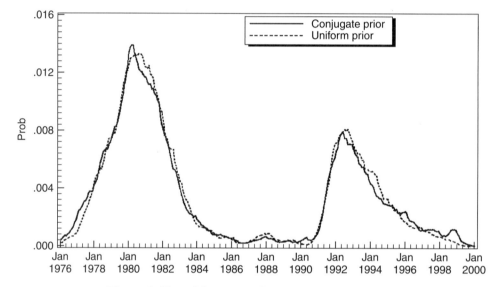

Figure 1. Plots of the estimated HIV infection distribution

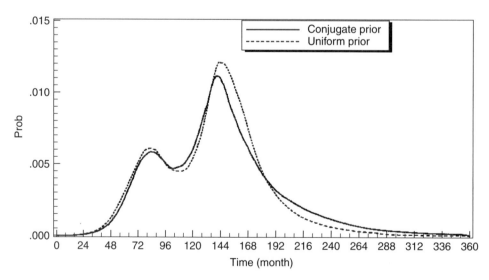

Figure 2. Plots of the estimated HIV incubation distribution

The mixture nature of this density implies that there are more than one sexual activity levels with a high proportion of restricted mixing (like-with-like) mixing. The second peak also implies a second wave of infection although the infection intensity is much smaller than the first wave.

(b) From Fig 2, the estimated density of the HIV incubation distribution also appeared to be a mixture of distributions with two obvious peaks. The first peak is around 75 months after infection and is much lower than the second peak which occurs around 140 months after infection. This result suggests a multistage nature of the HIV incubation.

(c) The Kalman-filter estimates of the AIDS incidence by the Gibbs sampler are almost identical to the corresponding observed AIDS incidence respectively. This result indicates that the Kalman-filter estimates can trace the observed numbers very closely if observed numbers are available.

(d) Figuring a 1% increase in the population size of San Francisco yearly, Tan and Ye [73] have also estimated the number of S people and the I people in the San Francisco population. The estimates showed that the total number of S people before January 1978 were always above 50,000 and were between 31,000 and 32,000 during January 1983 and January 1993. The total number of people who do not have AIDS were estimated around 50,000 before January 1993. It appeared that the total number of infected people reached a peak around the middle of 1985 and then decreased gradually to the lowest level around 1992. The estimates also showed that the number of infected people had two peaks, with the higher peak around the middle of 1985, the second peak around the year 2000 and with the lowest level around 1992.

Extending the above state space model to include immigration and death, Tan and

Ye [74] have also analyzed the data of the Swiss population of homosexual men and IV drug users by applying the above generalized Bayesian method. The estimated density of the HIV infection in the Swiss homosexual population is quite similar to that of the San Francisco population except that in the Swiss population, the first peak appears about 6 months earlier and the second peak about 2 years earlier. Similarly, the estimated density of the HIV incubation distribution in the Swiss population is a mixture of distributions with two obvious peaks. The higher peak occurs around 320 months after infection and the lower peak occurs around 232 months after infection. In the Swiss population, the estimates of the immigration and recruitment rates are about 10 times greater than those of the estimates of the death and retirement rates of the I people, suggesting that the size of the Swiss homosexual and bisexual population is increasing with time. Another interesting point is that, in the Swiss population, the estimates of the death and retirement rates of infective people were much greater (at least 100 times greater) than those of S people, suggesting that HIV infection may have increased the death and retirement rates of HIV infected people.

REFERENCES

1. Anderson, R. M. and May, R. M. (1992). *Infectious Diseases of Humans: Dynamics and Control*. Oxford University Press, Oxford, U.K.

2. Anderson, B. D. O. and Moore, J. B. (1979). *Optimal Filtering*. Prentice-Hall, Englewood Cliffs, N.J.

3. Aoki, M. (1990). *State Space Modeling of Time Series*, 2nd ed. Spring-Verlag, New York.

4. Bacchetti, P. R. (1990). Estimating the incubation period of AIDS comparing population infection and diagnosis pattern. *J. Amer. Statist. Assoc.*, **85**, 1002–1008.

5. Billard, L. and Zhao, Z. (1991). Three-stage stochastic epidemic model: an application to AIDS. *Math. Biosci.*, **107**, 431–450.

6. Billard, L. and Zhao, Z. (1994). Multi-stage non-homogeneous Markov models for the acquired immune deficiency syndrome epidemic. *J. R. Stat. Soc. B*, **56**, 673–686.

7. Blanchard, P., Bolz, G. F., and Krüger, T. (1990). "Modelling AIDS-Epidemics or Any

Veneral Disease on Random Graphs". In *Stochastic Processes in Epidemic Theory*, Lecture Notes in Biomathematics, No. 86, eds. J. P. Gabriel, C. Lefévre, and P. Picard, eds. Springer-Verlag, New York, pp. 104–117.

8. Brookmeyer, R. and Gail, M. H. (1994). *AIDS Epidemiology: A Quantitative Approach.* Oxford University Press, Oxford, U.K.

9. Capassor, V., Di Somma, M., Villa, M., Nicolosi, A., and Sicurello, F. (1997). "Multistage Models of HIV Transmission Among Injecting Drug Users via Shared Injection Equipment". In *Advances in Mathematical Population Dynamics- Molecules, Cells and Man, Part II. Population Dynamics in Diseases in Man*, O. Arino, D. Axelrod, and M. Kimmel, eds. World Scientific, Singapore, Chapter 8, pp. 511–528.

10. Carpenter, C. C. J., Fischl, M. A., Hammer, M. D., Hirsch, M. S., Jacobsen, D. M., Katzenstein, D. A., Montaner, J. S. G., Richman, D. D., Saag, M. S., Schooley, R. T., Thompson, M. A., Vella, S., Yeni, P. G., and Volberding, P. A. (1996). Antiretroviral therapy for the HIV infection in 1996. *J. Amer. Med. Ass.*, **276**, 146–154.

11. Castillo-Chavez, C., ed. (1989). *Mathematical and Statistical Approaches to AIDS Epidemiology*, Lecture Notes in Biomathematics 83. Springer-Verlag, New York.

12. Catlin, D. E. (1989). *Estimation, Control and Discrete Kalman Filter.* Springer-Verlag, New York.

13. Cazelles, B. and Chau, N. P. (1997). Using the Kalman filter and dynamic models to assess the changing HIV/AIDS epidemic. *Math. Biosci.*, **140**, 131–154.

14. CDC. (1981). Pneumocystis pneumonia - Los Angeles. *MMWR*, **30**, 250–252.

15. CDC, (1992). Revised classification system for HIV infection and expanded surveillance case definition for AIDS among adolescents and adults. *MMWR*, **41**(RR-17), 1–19.

16. CDC. (1997). *Surveillance Report of HIV/AIDS*. Atlanta, Ga., June 1997.

17. Coffin, J., Haase, J., Levy, J. A., Montagnier, L., Oroszlan, S., Teich, N., Temin, H., Toyoshima, K., Varmus, H., Vogt, P., and Weiss, R. (1986). Human immunodeficiency viruses. *Science*, **232**, 697.

18. Gani, J. (1990). "Approaches to the Modelling of AIDS". In *Stochastic Processes in Epidemic Theory*, Lecture Notes in Biomathematics No. 86, J. P. Gabriel, C. Leférre, and P. Picards, eds. Springer-Verlag, Berlin, pp. 145–154.

19. Gani, J. and Yakowitz, S. (1993). Modeling the spread of HIV among intravenous drug users. *IMA J. Math. Appl. Med. Biol.*, **10**, 51–65.

20. Gelb, A. (1974). *Applied Optimal Estimation.* MIT Press, Cambridge, Mass.

21. Harvey, A. C. (1994). *Forcasting, Structural Time Series Models and the Kalman Filter.* Cambridge University Press, Cambridge, U.K.

22. Hethcote, H. W. and Van Ark, J. M. (1992). *Modeling HIV Transmission and AIDS in the United States*, Lecture Notes in Biomathematics 95. Springer-Verlag, New York.

23. Isham, V. (1991). Assessing the variability of stochastic epidemics. *Math. Biosci.*, **107**, 209–224.

24. Isham, V. and Medley, G., eds. (1996). *Models for Infectious Human Diseases: Their Structure and Relation to Data.* Cambridge University Press, Cambridge, U.K.

25. Jager, J. C. and Ruittenberg, E. J., eds. (1988). *Statistical Analysis and Mathematical Modelling of AIDS.* Oxford University Press, Oxford, U.K.

26. Jewell, N. P., Dietz, K., and Farewell, V. T. (1992). *AIDS Epidemiology: Methodological Issues.* Birkhäuser, Basel.

27. Kalman, R. E. (1960). A new approach to linear filter and prediction problems. *J. Basic Eng.*, **82**, 35–45.

28. Kaplan, E. H. (1989). Needles that kill: modeling human immuno-deficiency virus transmission via shared drug injection equipment in shooting galleries. *Rev. Infect. Dis.*, **11**, 289–298.

29. Kaplan, E. H. and Brandeau, M. L., eds. (1994). *Modeling the AIDS Epidemic.* Raven Press, New York.

30. Kaplan, E. H., Cramton, P. C., and Paltiel, A. D. (1989). "Nonrandom Mixing Models of HIV Transmission". In *Mathematical and Statistical Approaches to AIDS Epidemiology*, Lecture Notes in Biomathematics 83, C. Castillo-Chavez, ed. Springer-Verlag, Berlin, pp. 218–241.

31. Lemp, G. F., Payne, S. F., Rutherford, G. W., Hessol, N. A., Winkelstein, W. Jr., Wiley, J. A., Moss, A. R., Chaisson, R. E., Chen, R. T., Feigal, D. W. Jr., Thomas, P. A., and Werdegar, D. (1990). Projections of AIDS morbidity and mortality in San Francisco. *J. Am. Med. Assoc.*, **263**, 1497–1501.

32. Liu, J. S. and Chen, R. (1998). Sequential Monte Carlo method for dynamic systems. *J. Am. Stat. Assoc.*, **93**, 1032–1044.

33. Longini Ira, M. Jr., Byers, R. H., Hessol, N. A., and Tan, W. Y. (1992). Estimation of the stage-specific numbers of HIV infections via a Markov model and backcalculation. *Stat. Med.*, **11**, 831–843.

34. Longini Ira, M. Jr., Clark, W. S., Byers, R. H., Ward, J. W., Darrow, W. W., Lemp, G. F. and Hethcote, H. W. (1989). Statistical analysis of the stages of HIV infection using a Markov model. *Stat. Med.*, **8**, 831–843.

35. Longini Ira, M. Jr., Clark, W. S., Gardner, L. I., and Brundage, J. F. (1991). The dynamics of CD4+ T-lymphocyte decline in HIV-infected individuals: a Markov modeling approach. *J. AIDS*, **4**, 1141–1147.

36. Longini Ira, M. Jr., Clark, W. S., and Karon, J. (1993). Effects of routine use of therapy in slowing the clinical course of human immunodeficiency virus (HIV) infection in a population based cohort. *Am. J. Epidemiol.*, **137**, 1229–1240.

37. Longini Ira, M. Jr., Clark, W. S., Satten, G. A., Byers, R. B., and Karon, J. M., (1996). "Staged Markov Models Based on CD4$^{(+)}$ T-Lymphocytes for the Natural History of HIV Infection. In *Models for Infectious Human Diseases: Their Structure and Relation to Data*, V. Isham and G. Medley, eds. Cambridge University Press, Cambridge, U.K., pp. 439–459.

38. Mann, J. M. and Tarantola, D. J. M. (1998). HIV 1998: The global picture. *Sci. Am.*, **279**, 82–83.

39. Mode, C. J. (1991). A stochastic model for the development of an AIDS epidemic in a heterosexual population. *Math. Biosci.*, **107**, 491–520.

40. Mode, C. J., Gollwitzer, H. E., and Herrmann, N. (1988). A methodological study of a stochastic model of an AIDS epidemic. *Math. Biosci.*, **92**, 201–229.

41. Mode, C. J., Gollwitzer, H. E., Salsburc, M. A., and Sleeman, C. K. (1989). A methodological study of a nonlinear stochastic model of an AIDS epidemic with recruitment. *IMA J. Math. Appl. Med. Biol.*, **6**, 179–203.

42. Mode, C. J. and Sleeman, C. K. (2000). *Stochastic Processes in Epidemiology: HIV/AIDS, Other Infectious Diseases and Computers*. World Scientific, River Edge, N.J.

43. Satten, G. A. and Longini Ira, M. Jr. (1994). Estimation of incidence of HIV infection using cross-sectional marker survey. *Biometrics*, 50, 675–68.

44. Satten, G. A. and Longini Ira, M. Jr. (1996). Markov chain with measurement error: estimaxting the 'True' course of marker of the progression of human immunodeficiency virus disease. *Appl. Stat.*, **45**, 275–309.

45. Shephard, N. (1994). Partial non-Gaussian state space. *Biometrika*, **81**, 115–131.

46. Tan, W. Y. (1990). A Stochastic model for AIDS in complex situations. *Math. Comput. Modell.*, **14**, 644–648.

47. Tan, W. Y. (1991). Stochastic models for the spread of AIDS and some simulation results. *Math. Comput. Modell.*, **15**, 19–39.

48. Tan, W. Y. (1991). "On the Chain Multinomial Model of HIV Epidemic in Homosexual Populations and Effects of Randomness of Risk Factors". In *Mathematical Population Dynamics 3*, O. Arino, D. E. Axelrod, and M. Kimmel, eds. Wuerz Publishing, Winnepeg, Manitoba, Canada, pp. 331–353.

49. Tan, W. Y. (1993). The chain multinomial models of the HIV epidemiology in homosexual populations. *Math. Comput. Modell.*, **18**, 29–72.

50. Tan, W. Y. (1994). On the HIV Incubation Distribution under non-Markovian Models. *Stat. Probab. Lett.*, **21**, 49–57.

51. Tan, W. Y. (1994). On first passage probability in Markov models and the HIV incubation distribution under drug treatment. *Math. Comput. Modell.*, **19**, 53–66.

52. Tan, W. Y. (1995). On the HIV incubation distribution under AZT treatment. *Biometric. J.*, **37**, 318–338.

53. Tan, W. Y. (1995). A stochastic model for drug resistance and the HIV incubation distribution. *Stat. Probab. Lett.*, **25**, 289–299.

54. Tan, W. Y. (1995). "On the Chain Multinomial Model of HIV Epidemic in Homosexual Populations and Effects of Randomness of Risk Factors". In *Mathematical Population Dynamics 3*, O. Arino, D. E. Axelrod, and M. Kimmel, eds. Wuerz Publishing, Winnepeg, Manitoba, Canada, pp. 331–356.

55. Tan, W. Y. (2000). *Stochastic Modeling of AIDS Epidemiology and HIV Pathogenesis*. World Scientific, River Edge, N.J.

56. Tan, W. Y. (2002). *Stochastic Models With Applications to genetics, Cancers, AIDS and Other Biomedical Systems*. World Scientific, River Edge, N.J.

57. Tan, W. Y. and Byers, R. H. (1993). A stochastic model of the HIV epidemic and the HIV infection distribution in a homosexual population. *Math. Biosci.*, **113**, 115–143.

58. Tan, W. Y., Chen, C. W., and Wang, W. (2000). *A generalized state space model of carcinogenesis*. Paper Presented at the *2000 International*

Biometric Conference at UC. Berkeley, Calif., July 2–7, 2000.

59. Tan, W. Y. and Hsu, H. (1989). Some stochastic models of AIDS spread. *Stat. Med.*, **8**, 121–136.

60. Tan, W. Y. and Hsu, H. (1991). "A Stochastic Model for the AIDS Epidemic in a Homosexual Population". In *"Mathematical Population Dynamics*, eds. O. Arion, D. E. Axelrod, and M. Kimmel, eds. Marcel Dekker, New York, Chapter 24, pp. 347–368.

61. Tan, W. Y, Lee, S. R., and Tang, S. C. (1995). Characterization of HIV infection and seroconversion by a stochastic model of HIV epidemic. *Math. Biosci.*, **126**, 81–123.

62. Tan, W. Y. and Tang, S. C. (1993). Stochastic models of HIV epidemics in homosexual populations involving both sexual contact and IV drug use. *Math. Comput. Modell.*, **17**, 31–57.

63. Tan, W. Y., Tang, S. C., and Lee, S. R. (1995). Effects of Randomness of risk factors on the HIV epidemic in homo-sexual populations. *SIAM J. Appl. Math.*, **55**, 1697–1723.

64. Tan, W. Y., Tang, S. C., and Lee, S. R. (1996). Characterization of the HIV incubation distributions and some comparative studies. *Stat. Med.*, **15**, 197–220.

65. Tan, W. Y., Tang, S. C., and Lee, S. R. (1998). Estimation of HIV seroconversion and effects of age in San Francisco homosexual populations. *J. Appl. Stat.*, **25**, 85–102.

66. Tan, W. Y. and Xiang, Z. H. (1996). A stochastic model for the HIV epidemic and effects of age and race. *Math. Comput. Modell.*, **24**, 67–105.

67. Tan, W. Y. and Xiang, Z. H.(1997). "A Stochastic Model for the HIV Epidemic in Homosexual Populations and Effects of Age and Race on the HIV Infection". In *Advances in Mathematical Population Dynamics- Molecules, Cells and Man". Part II. Population Dynamics in Diseases in Man*, O. Arino, D. Axelrod, and M. Kimmel, eds. World Scientific, River Edge, N.J, Chapter 8, pp. 425–452.

68. Tan, W. Y. and Xiang, Z. H. (1998). State space models of the HIV epidemic in homosexual populations and some applications. *Math. Biosci.*, **152**, 29–61.

69. Tan, W. Y. and Xiang, Z. H. (1998). "Estimating and Predicting the numbers of T Cells and Free HIV by Non-Linear Kalman Filter". In *Artificial Immune Systems and Their Applications*, D. DasGupta, ed. Springer-Verlag, Berlin, pp. 115–138.

70. Tan, W. Y. and Xiang, Z. H.(1998). "State Space Models for the HIV Pathogenesis". In *"Mathematical Models in Medicine and Health Sciences*, M. A. Horn, G. Simonett, and G. Webb, eds. Vanderbilt University Press, Nashville, Tenn., pp. 351–368.

71. Tan, W. Y. and Xiang, Z. H. (1999). Modeling the HIV epidemic with variable infection in homosexual populations by state space models. *J. Stat. Inference Plann.*, **78**, 71–87.

72. Tan, W. Y. and Xiang, Z. H. (1999). A state space model of HIV pathogenesis under treatment by anti-viral drugs in HIV-infected individuals. *Math. Biosci.*, **156**, 69–94.

73. Tan, W. Y. and Ye, Z. Z. (2000). Estimation of HIV infection and HIV incubation via state space models. *Math. Biosci.*, **167**, 31–50.

74. Tan, W. Y. and Ye, Z. Z. (2000). Simultaneous estimation of HIV infection, HIV incubation, immigration rates and death rates as well as the numbers of susceptible people, infected people and AIDS cases. *Commun. Stat. (Theory Methods)*, **29**, 1059–1088.

75. Tan, W. Y. and Zhu, S. F. (1996). A stochastic model for the HIV infection by heterosexual transmission involving married couples and prostitutes. *Math. Comput. Modell.*, **24**, 47–107.

76. Tan, W. Y. and Zhu, S. F. (1996). A stochastic model for the HIV epidemic and effects of age and race. *Math. Comput. Modell.*, **24**, 67–105.

77. U.S. Bureau of the Census. (1987). *Statistical Abstract of the United States: 1980*, 108th ed. Washington, D.C.

78. Wu, H. and Tan, W. Y. (1995). "Modeling the HIV Epidemic: A State Space Approach". *ASA 1995 Proceedings of the epidemiology Section*. ASA, Alexdria, Va., pp. 66–71.

79. Wu, H. and Tan, W. Y. (2000). Modeling the HIV epidemic: a state space approach. *Math. Comput. Modell.*, **32**, 197–215.

See also CLINICAL TRIALS; EPIDEMICS; and EPIDEMIOLOGICAL STATISTICS—I.

WAI-YUAN TAN

AITCHISON DISTRIBUTIONS

These form a class of multivariate distributions with density functions:

$$f_X(\mathbf{X}|\alpha, \beta) \propto \left[\prod_{i=1}^{p} x_i^{\alpha_i - 1} \right]$$

$$\times \exp\left[-\frac{1}{2} \sum_{i<j}^{p} \sum \beta_{ij} (\log x_i - \log x_j)^2 \right],$$

$$0 < x_i, \quad \sum_{i=1}^{p} x_i = 1,$$

$$\alpha_i \geqslant 0, \quad \beta \text{ nonnegative definite.}$$

Although there are p variables $x_1, \ldots, x_p$, the distribution is confined to the $(p - 1)$-dimensional simplex: $0 \leqslant x_i, \sum_{i=1}^{p} x_i = 1$.

If $\beta = 0$ we have a *Dirichlet distribution**; if $\alpha_i = 0$ for all i we have a *multivariate logistic-normal distribution*.

Methods of estimating the parameters (α and β) are described by Aitchison [1, pp. 310–313].

REFERENCE

1. Aitchison, J. (1986). *Statistical Analysis of Compositional Data*. Chapman and Hall, London and New York.

See also COMPOSITIONAL DATA; DIRICHLET DISTRIBUTION; FREQUENCY SURFACES, SYSTEMS OF; and LOGISTIC-NORMAL DISTRIBUTION.

AITKEN, ALEXANDER CRAIG

Alexander Craig Aitken was born April 1, 1885 in Dunedin, New Zealand, and died in Edinburgh, Scotland, on November 3, 1967, where he had spent his working life. Dunedin is a rather Scottish community on the southern tip of New Zealand. His father was a farmer. Aitken made important contributions to statistics, numerical analysis, and algebra. His extraordinary memory, musical gift, attractive personality, work, and teaching talents are well described in the obituary articles [1,2].

After attending Otago Boys High School, he studied at Otago University classical languages for two years. In April, 1915 he enlisted in the New Zealand Infantry and served in Gallipoli, Egypt, and France in World War I—experiences movingly described in a manuscript (written while he was recovering from wounds in 1917), but not turned into a book [3] until 1962—his last publication. Aitken had total recall of his past, and this section of it always gave him great pain. His platoon was all but wiped out in the battle of the Somme along with all records. He was able to remember and write down all the information in the records of all these men. Upon recovery, he returned to Otago University. He could not study mathematics there, though it was realized that he had a gift for it. He then taught languages in his old school for three years. Upon graduating in 1920, he married a fellow student, and later they had a son and a daughter.

Fortunately, in 1923 he was given a scholarship to study mathematics under E. T. Whittaker in Edinburgh. The first edition of Whittaker and Robinson's *The Calculus of Observations* [4] (W&R) appeared in 1924. Whittaker had earlier considered *graduation** or *smoothing of data* as a statistical problem; his motivation was mainly actuarial. So arose—almost—what we now call splines. The function chosen to be minimized was the sum of the squares of the differences on the observed u_n and "true" values u'_n plus a multiple of the sum of squares of the *third* differences of the "true" values. How to execute this was Aitken's Ph.D. problem. Whittaker was so pleased with Aitken's results that he was awarded a D.Sc. instead and a staff appointment. His method is given in W&R. In a preceding section in W&R, they speak of the *method of interlaced parabolas*, in which they fit a cubic polynomial to each successive four graduated values u'_n—this allows interpolation. Had the second term in the minimand been the integral of the square of the third derivative of the interpolating function, they would have invented the modern method of getting a spline.

The Mathematics Department in Edinburgh had over many years very broad interests that spanned all of applied mathematics, and Whittaker was a towering figure in many fields. In particular, it was then the only place in Britain that taught determinants and matrices, and these immediately appealed to Aitken. In 1932 he published with H. W. Turnbull, *An Introduction of the Theory of Canonical Matrices* [5]. Its last chapter gives, among other applications, some to statistics. So by then he had shown his real interests—algebra, numerical analysis, and statistics. Aitken succeeded Whittaker in the Chair in Mathematics in 1946, holding it until he retired in 1965.

Aitken was a renowned teacher. In 1939 he published the first two volumes in the Oliver & Boyd series (of which he was a joint editor with D. E. Rutherford) *University Mathematical Texts*. They were *Statistical Mathematics* [6] and *Determinants and Matrices* [7]. These two books were my first acquaintance with his work. When I was an undergraduate there were very few books on these topics with any style to them.

Around 1949, as a graduate student, I became aware of his research, largely written in the preceding ten years, in statistics, especially of his matrix treatment of least squares (see e.g. ref. [8])—idempotents like $X(X^tX)^{-1}X^t$, the use of the trace operator (e.g. $Ex^tAx = \text{Tr } EAxx^t$), etc. This has now become standard. Indeed, with the linear model $y = X\beta + u$ where the random error vector has mean 0 and nonsingular covariance matrix V, the estimating equations $X^tV^{-1}y = X^tV^{-1}X\hat{\beta}$ are known as the *Aitken equations**. But he never mentioned vector spaces, although he would have been aware of the background geometry. He wrote many papers about least squares and especially about the fitting of polynomials and on other statistical topics. We have only singled out here three areas to mention.

In the late thirties he gave a student from New Zealand, H. Silverstone, a Ph.D. topic—the optimal estimation of statistical parameters—which he had apparently already worked out for a scalar parameter. (See Aitken & Silverstone (1941) [9].) In a 1947 paper [10] Aitken completed the work for many parameters. But he has never received credit for this work, which seems unfair to me. *For the time at which it was written*, the formulation is correct, as are the answers. But instead of a direct proof of a minimum, they simply give the "Euler equation" that the calculus of variations (C of V) throws up, though with the full knowledge that that method has difficulties. This was very natural, as the study of minimal surfaces was then of great interest, and much

use was made of the C of V in mathematical physics.

To take the simpler problem with his notation, let $\Phi(x, \theta)$ be the density of the vector x, which ranges over a region which is the same for all θ, and be uniformly differentiable with respect to θ. Suppose there exists $t(x)$, a minimum-variance estimator of θ. Then $\int t\partial\Phi/\partial\theta \, dx = 1$, and $\int(t - \theta)^2\Phi dx$ must be a minimum. Writing $I(t) = \int(t - \theta)^2\Phi dx - 2\lambda \int t\partial\Phi/\partial\theta dx$, consider any other estimator $t + h$. Then

$$I(t + h) = I(t) + \int h\left(2(t - \theta)\Phi - 2\lambda\frac{\partial\Phi}{\partial\theta}\right) dx + \int h^2\Phi \, dx \geqslant I(t)$$

if and only if $t - \theta = \lambda\partial\Phi/\partial\theta$, where λ may be a function of θ. This is the "Euler" equation they give (and, I would guess, the proof they had), and from which they correctly draw all the now well-known conclusions. It took however many other statisticians many years to clarify these questions and to find what other assumptions are necessary. M. S. Bartlett has published some correspondence [11] with Aitken.

Several other features of Aitken's life are his love of music, his awesome memory, and his arithmetic ability. He played the violin well and for a time was leader of the Edinburgh University Musical Society Orchestra, which was sometimes conducted by his close friend Sir Donald Tovey. His violin was particularly important to him when in the Army, and it now is displayed in his old school. He said that 75% of the time his thoughts were musical. On occasion he would demonstrate his arithmetic feats. He wrote a book [12] against the decimalization of the English coinage—the use of twelve, which has so many factors, appealed to him. He was able to dictate rapidly the first 707 digits of π.

Among his honors, he was a Fellow of the Royal Societies of Edinburgh and London and of the Royal Society of Literature.

REFERENCES

1. Copson, E. T., Kendall, D. G., Miller, J. C. P., and Ledermann, W. (1968). Obituary articles. *Proc. Edinburgh Math. Soc.*, **16**, 151–176.

2. Whittaker, J. M. and Bartlett, M. H. (1968). Alexander Craig Aitken. *Biogr. Mem. Fell. R. Soc. Lond.*, **14**, 1–14.

3. Aitken, A. C. (1962). *Gallipoli to the Somme*. Oxford University Press, London. 177 pp.

4. Whittaker, E. T. and Robinson, G. (1924). *The Calculus of Observations*. Blackie & Son, London and Glasgow (6th impression, 1937).

5. Turnbull, H. W. and Aitken, A. C. (1932). *An Introduction to the Theory of Canonical Matrices*. Blackie & Son, London and Glasgow. 192 pp.

6. Aitken, A. C. (1939). *Statistical Mathematics*. Oliver & Boyd, Edinburgh. 153 pp.

7. Aitken, A. C. (1939). *Determinants and Matrices*. Oliver & Boyd, Edinburgh. 144 pp.

8. Aitken, A. C. (1934–35). On least squares and linear combinations of observations. *Proc. Roy. Soc. Edinburgh A*, **55**, 42–48.

9. Aitken, A. C. and Silverstone, H. (1941). On the estimation of statistical parameters. *Proc. Roy. Soc. Edinburgh A*, **61**, 186–194.

10. Aitken, A. C. (1947). On the estimation of many statistical parameters. *Proc. Roy. Soc. Edinburgh A*, **62**, 369–370.

11. Bartlett, M. S. (1994). Some pre-war statistical correspondence. In *Probability, Statistics and Optimization*, F. P. Kelly, ed. Wiley, Chichester and New York, pp. 297–413.

12. Aitken, A. C. (1962). *The Case against Decimalization*. Oliver & Boyd, Edinburgh, 22 pp.

G. S. WATSON

AITKEN EQUATIONS

One of the most important results in the theory of linear models is the Gauss—Markov theorem*. It establishes that for the linear model $\mathbf{Y} = \mathbf{X}\beta + \mathbf{e}$ with

$$E(\mathbf{e}) = \phi \quad \text{and} \quad E(\mathbf{ee}^T) = \sigma^2\mathbf{I} \quad (1)$$

the minimum variance linear unbiased estimator for an estimable function $\lambda^T\beta$ is given by $\lambda^T\hat{\beta}$, where $\hat{\beta}$ is a solution of the normal equations*

$$\mathbf{X}^T\mathbf{X}\hat{\beta} = \mathbf{X}^T\mathbf{y}. \quad (2)$$

The case of equal variances and zero correlations for the elements of $\mathbf{e}$, as given in (1), was generalized first by Aitken* [1] to the case where the observations and hence the elements of $\mathbf{e}$ have different variances and/or are correlated, i.e., (1) is replaced by

$$E(\mathbf{e}) = \phi, E(\mathbf{ee}^T) = \Sigma = \sigma^2\mathbf{V}. \quad (3)$$

In (3), Σ represents a known (positive definite) variance—covariance matrix, but since it needs to be known only up to a constant, Σ is often rewritten as $\sigma^2\mathbf{V}$, where $\mathbf{V}$ represents a known matrix (note that for the case of equal variances σ^2, $\mathbf{V}$ is a correlation matrix). As a consequence of the variance—covariance structure (3) the equations (2) are replaced by the Aitken equations

$$\mathbf{X}^T\mathbf{V}^{-1}\mathbf{X}\hat{\beta} = \mathbf{X}^T\mathbf{V}^{-1}\mathbf{y}, \quad (4)$$

which are obtained by minimizing the expression (see [1])

$$(\mathbf{y} - \mathbf{X}\beta)^T\mathbf{V}^{-1}(\mathbf{y} - \mathbf{X}\beta).$$

It is for this reason that the method is also referred to as weighted least squares* or generalized least squares, as compared to ordinary least squares (OLS), which leads to (2).

With the conditions (3) the minimum-variance linear unbiased estimator for an estimable function $\lambda^T\beta$ is given by $\lambda^T\hat{\beta}$, where $\hat{\beta}$ is a solution to the Aitken equations (4). A proof can be found in ref. [2].

An interesting and important question is: For $\mathbf{V} \neq \mathbf{I}$. when are the OLS and the weighted least squares estimators for $\lambda^T\beta$ the same? The answer is (see e.g. [2]), when $\mathbf{VX} = \mathbf{XQ}$ for some matrix $\mathbf{Q}$. This condition holds, for example, for mixed linear models with balanced data.

REFERENCES

1. Aitken, A. C. (1934/35). On least squares and linear combination of observations. *Proc. Roy. Soc. Edinburgh*, **55**, 42–48.

2. Hinkelmann, K. and Kempthorne, O. (1994). *Design and Analysis of Experiments*, Vol. 1. Wiley, New York.

See also GENERAL LINEAR MODEL and WEIGHTED LEAST SQUARES.

KLAUS HINKELMANN

AKAIKE'S INFORMATION CRITE-RION.

See MODEL SELECTION: AKAIKE'S INFORMATION CRITERION

ALEATORY VARIABLE

An obsolete term for *random variable*.

ALGEBRA OF EVENTS

Let Ω be a space whose points correspond to the possible outcomes of a random experiment. Certain subsets of Ω are called *events*, and *probability* is assigned to these subsets. A collection $\mathscr{F}$ of subsets of Ω is called an *algebra* (the term *field* is also used) if the following conditions are satisfied:

(a) The space Ω belongs to $\mathscr{F}$ ($\Omega \in \mathscr{F}$).

(b) The collection $\mathscr{F}$ is closed under complementation and finite union. Formally:

b₁: If $A \in \mathscr{F}$, then the complement $\overline{A}$ (also denoted as A^c) belongs to $\mathscr{F}$.

b₂: If $A_1, \ldots, A_n \in \mathscr{F}$, then union $\cup_{i=1}^n A_i$ (also denoted as $A_1 \cup \ldots \cup A_n) \in \mathscr{F}$.

[Since $\overline{(\cup_{i=1}^n \overline{A_i})} = \cap_{i=1}^n A_i$, b_1 and b_2 imply that an algebra is also closed under finite intersection.]

If in place of b_2 we require $\mathscr{F}$ to be closed under *countable* union, namely, if $A_1, A_2, \ldots \in \mathscr{F}$, then $\cup_{i=1}^\infty A_i \in \mathscr{F}$, the collection $\mathscr{F}$ is called a σ-*algebra* (or σ-*field*). The notion of the σ-algebra of events is a basic concept for theoretical probability theory.

See also AXIOMS OF PROBABILITY.

ALGORITHM

An algorithm is a rule for performing a calculation—usually, although not necessarily, numerical. For example, one might have algorithms for classificatory purposes, as well as for evaluation of roots of determinantal equations. Algorithms do not provide any background for the calculations to which they refer, either in terms of motivation or justification.

Algorithms for specific purposes are described in separate entries, in particular in the article ALGORITHMS, STATISTICAL.

ALGORITHMIC INDEPENDENCE.

See ALGORITHMIC INFORMATION THEORY

ALGORITHMIC INFORMATION THEORY

The Shannon entropy* concept of classical information theory* [9] is an ensemble notion; it is a measure of the degree of ignorance concerning which possibility holds in an ensemble with a given a priori probability distribution*

$$H(p_1, \ldots, p_n) \equiv -\sum_{k=1}^n p_k \log_2 p_k.$$

In algorithmic information theory the primary concept is that of the *information content* of an individual object, which is a measure of how difficult it is to specify or describe how to construct or calculate that object. This notion is also known as *information-theoretic complexity*. For introductory expositions, see refs. 1, 4, and 6. For the necessary background on computability theory and mathematical logic, see refs. 3, 7, and 8. For a more technical survey of algorithmic information theory and a more complete bibliography, see ref. 2. See also ref. 5.

The original formulation of the concept of algorithmic information is independently due to R. J. Solomonoff [22], A. N. Kolmogorov* [19], and G. J. Chaitin [10]. The information content $I(x)$ of a binary string x is defined to be the size in bits (binary digits) of the smallest program for a canonical universal computer U to calculate x. (That the computer U is universal means that for any other computer M there is a prefix μ such that the program μp makes U do exactly the same computation that the program p makes M do.) The *joint information* $I(x, y)$ of two strings is defined to be the size of the smallest program that makes U calculate both of them. And the *conditional* or *relative information* $I(x|y)$ of x

given y is defined to be the size of the smallest program for U to calculate x from y. The choice of the standard computer U introduces at most an $O(1)$ uncertainty in the numerical value of these concepts. [$O(f)$ is read "order of f" and denotes a function whose absolute value is bounded by a constant times f.]

With the original formulation of these definitions, for most x one has

$$I(x) = |x| + O(1) \qquad (1)$$

(here $|x|$ denotes the length or size of the string x, in bits), but unfortunately

$$I(x,y) \leqslant I(x) + I(y) + O(1) \qquad (2)$$

holds only if one replaces the $O(1)$ error estimate by $O(\log I(x)I(y))$.

Chaitin [12] and L. A. Levin [20] independently discovered how to reformulate these definitions so that the subadditivity property (2) holds. The change is to require that the set of meaningful computer programs be an instantaneous code, i.e., that no program be a prefix of another. With this modification, (2) now holds, but instead of (1) most x satisfy

$$I(x) = |x| + I(|x|) + O(1)$$
$$= |x| + O(\log |x|).$$

Moreover, in this theory the decomposition of the joint information of two objects into the sum of the information content of the first object added to the relative information of the second one given the first has a different form than in classical information theory. In fact, instead of

$$I(x,y) = I(x) + I(y|x) + O(1), \qquad (3)$$

one has

$$I(x,y) = I(x) + I(y|x, I(x)) + O(1). \qquad (4)$$

That (3) is false follows from the fact that $I(x, I(x)) = I(x) + O(1)$ and $I(I(x)|x)$ is unbounded. This was noted by Chaitin [12] and studied more precisely by Solovay [12, p. 339] and Gač [17].

Two other concepts of algorithmic information theory are *mutual* or *common information* and *algorithmic independence*. Their

importance has been emphasized by Fine [5, p. 141]. The mutual information content of two strings is defined as follows:

$$I(x : y) \equiv I(x) + I(y) - I(x,y).$$

In other words, the mutual information* of two strings is the extent to which it is more economical to calculate them together than to calculate them separately. And x and y are said to be algorithmically independent if their mutual information $I(x : y)$ is essentially zero, i.e., if $I(x,y)$ is approximately equal to $I(x) + I(y)$. Mutual information is symmetrical, i.e., $I(x : y) = I(y : x) + O(1)$. More important, from the decomposition (4) one obtains the following two alternative expressions for mutual information:

$$I(x : y) = I(x) - I(x|y, I(y)) + O(1)$$
$$= I(y) - I(y|x, I(x)) + O(1).$$

Thus this notion of mutual information, although it applies to individual objects rather than to ensembles, shares many of the formal properties of the classical version of this concept.

Up until now there have been two principal applications of algorithmic information theory: (a) to provide a new conceptual foundation for probability theory and statistics by making it possible to rigorously define the notion of a *random sequence**, and (b) to provide an information-theoretic approach to metamathematics and the limitative theorems of mathematical logic. A possible application to theoretical mathematical biology is also mentioned below.

A random or patternless binary sequence x_n of length n may be defined to be one of maximal or near-maximal complexity, i.e., one whose complexity $I(x_n)$ is not much less than n. Similarly, an infinite binary sequence x may be defined to be random if its initial segments x_n are all random finite binary sequences. More precisely, x is random if and only if

$$\exists c \forall n [I(x_n) > n - c]. \qquad (5)$$

In other words, the infinite sequence x is random if and only if there exists a c such that for all positive integers n, the algorithmic information content of the string consisting of the first n bits of the sequence x, is bounded from below by $n - c$. Similarly, a *random real number* may be defined to be one having the property that the base 2 expansion of its fractional part is a random infinite binary sequence.

These definitions are intended to capture the intuitive notion of a lawless, chaotic, unstructured sequence. Sequences certified as random in this sense would be ideal for use in Monte Carlo* calculations [14], and they would also be ideal as one-time pads for Vernam ciphers or as encryption keys [16]. Unfortunately, as we shall see below, it is a variant of Göel's famous incompleteness theorem that such certification is impossible. It is a corollary that no pseudorandom number* generator can satisfy these definitions. Indeed, consider a real number x, such as $\sqrt{2}$, π, or e, which has the property that it is possible to compute the successive binary digits of its base 2 expansion. Such x satisfy

$$I(x_n) = I(n) + O(1) = O(\log n)$$

and are therefore maximally nonrandom. Nevertheless, most real numbers are random. In fact, if each bit of an infinite binary sequence is produced by an independent toss of an unbiased coin, then the probability that it will satisfy (5) is 1. We consider next a particularly interesting random real number, Ω, discovered by Chaitin [12, p. 336].

A. M. Turing's theorem that the halting problem is unsolvable is a fundamental result of the theory of algorithms [4]. Turing's theorem states that there is no mechanical procedure for deciding whether or not an arbitrary program p eventually comes to a halt when run on the universal computer U. Let Ω be the probability that the standard computer U eventually halts if each bit of its program p is produced by an independent toss of an unbiased coin. The unsolvability of the halting problem is intimately connected to the fact that the halting probability Ω is a random real number, i.e., its base 2 expansion is a random infinite binary sequence in the very strong sense (5) defined above. From (5) it follows that Ω is normal (a notion due to E. Borel [18]), that Ω is a Kollectiv* with respect to

all computable place selection rules (a concept due to R. von Mises and A. Church [15]), and it also follows that Ω satisfies all computable statistical tests of randomness* (this notion being due to P. Martin-Löf [21]). An essay by C. H. Bennett on other remarkable properties of Ω, including its immunity to computable gambling schemes, is contained in ref. 6.

K. Gödel established his famous incompleteness theorem by modifying the paradox of the liar; instead of "This statement is false" he considers "This statement is unprovable." The latter statement is true if and only if it is unprovable; it follows that not all true statements are theorems and thus that any formalization of mathematical logic is incomplete [3,7,8]. More relevant to algorithmic information theory is the paradox of "the smallest positive integer that cannot be specified in less than a billion words." The contradiction is that the phrase in quotes only has 14 words, even though at least 1 billion should be necessary. This is a version of the Berry paradox, first published by Russell [7, p. 153]. To obtain a theorem rather than a contradiction, one considers instead "the binary string s which has the shortest proof that its complexity $I(s)$ is greater than 1 billion." The point is that this string s cannot exist. This leads one to the metatheorem that although most bit strings are random and have information content approximately equal to their lengths, it is impossible to prove that a specific string has information content greater than n unless one is using at least n bits of axioms. See ref. 4 for a more complete exposition of this information-theoretic version of Gödel's incompleteness theorem, which was first presented in ref. 11. It can also be shown that n bits of assumptions or postulates are needed to be able to determine the first n bits of the base 2 expansion of the real number Ω.

Finally, it should be pointed out that these concepts are potentially relevant to biology. The algorithmic approach is closer to the intuitive notion of the information content of a biological organism than is the classical ensemble viewpoint, for the role of a computer program and of deoxyribonucleic acid (DNA) are roughly analogous. Reference 13 discusses possible applications of the concept of mutual algorithmic information to theoretical biology; it is suggested that a living organism might be defined as a highly correlated region, one whose parts have high mutual information.

GENERAL REFERENCES

1. Chaitin, G. J. (1975). *Sci. Amer.*, **232** (5), 47–52. (An introduction to algorithmic information theory emphasizing the meaning of the basic concepts.)

2. Chaitin, G. J. (1977). *IBM J. Res. Dev.*, **21**, 350–359, 496. (A survey of algorithmic information theory.)

3. Davis, M., ed. (1965). *The Undecidable—Basic Papers on Undecidable Propositions, Unsolvable Problems and Computable Functions.* Raven Press, New York.

4. Davis, M. (1978). In *Mathematics Today: Twelve Informal Essays*, L. A. Steen, ed. Springer-Verlag, New York, pp. 241–267. (An introduction to algorithmic information theory largely devoted to a detailed presentation of the relevant background in computability theory and mathematical logic.)

5. Fine, T. L. (1973). *Theories of Probability: An Examination of Foundations.* Academic Press, New York. (A survey of the remarkably diverse proposals that have been made for formulating probability mathematically. Caution: The material on algorithmic information theory contains some inaccuracies, and it is also somewhat dated as a result of recent rapid progress in this field.)

6. Gardner, M. (1979). *Sci. Amer.*, **241** (5), 20–34. (An introduction to algorithmic information theory emphasizing the fundamental role played by Ω.)

7. Heijenoort, J. van, ed. (1977). *From Frege to Gödel: A Source Book in Mathematical Logic, 1879–1931.* Harvard University Press, Cambridge, Mass. (This book and ref. 3 comprise a stimulating collection of all the classic papers on computability theory and mathematical logic.)

8. Hofstadter, D. R. (1979). *Gödel, Escher, Bach: An Eternal Golden Braid.* Basic Books, New York. (The longest and most lucid introduction to computability theory and mathematical logic.)

9. Shannon, C. E. and Weaver, W. (1949). *The Mathematical Theory of Communication.* University of Illinois Press, Urbana, Ill. (The first and still one of the very best books on classical information theory.)

ADDITIONAL REFERENCES

10. Chaitin, G. J. (1966). *J. ACM*, **13**, 547–569; **16**, 145–159 (1969).

11. Chaitin, G. J. (1974). *IEEE Trans. Inf. Theory*, **IT-20**, 10–15.

12. Chaitin, G. J. (1975). *J. ACM*, **22**, 329–340.

13. Chaitin, G. J. (1979). In *The Maximum Entropy Formalism*, R. D. Levine and M. Tribus, eds. MIT Press, Cambridge, Mass., pp. 477–498.

14. Chaitin, G. J. and Schwartz, J. T. (1978). *Commun. Pure Appl. Math.*, **31**, 521–527.

15. Church, A. (1940). *Bull. AMS*, **46**, 130–135.

16. Feistel, H. (1973). *Sci. Amer.*, **228** (5), 15–23.

17. Gač, P. (1974). *Sov. Math. Dokl.*, **15**, 1477–1480.

18. Kac, M. (1959). *Statistical Independence in Probability, Analysis and Number Theory.* Mathematical Association of America, Washington, D.C.

19. Kolmogorov, A. N. (1965). *Problems of Inf. Transmission*, **1**, 1–7.

20. Levin, L. A. (1974). *Problems of Inf. Transmission*, **10**, 206–210.

21. Martin-Löf, P. (1966). *Inf. Control*, **9**, 602–619.

22. Solomonoff, R. J. (1964). *Inf. Control*, **7**, 1–22, 224–254.

See also ENTROPY; INFORMATION THEORY AND CODING THEORY; MARTINGALES; MONTE CARLO METHODS; PSEUDO-RANDOM NUMBER GENERATORS; STATISTICAL INDEPENDENCE; and RANDOMNESS, TESTS OF.

G. J. CHAITIN

ALGORITHMS, STATISTICAL

Traditionally, in mathematics, the term "algorithm"* means "some special process for solving a certain type of problem" [3].[1] With the advent of automatic computing, the term was adopted to refer to the description of a process in a form suitable for implementation on a computer. Intuitively, an algorithm is useful in mathematics or in computing if the "type of problem" is well defined and if the "special process" can be used effectively for these problems. A reasonable definition of the term for our purposes is:

An algorithm is a process for the solution of a type of problem, such that the process can be implemented computationally without significant difficulty and that the class of problems treated is computationally specific and well understood.

Statistical algorithms are those algorithms having useful application to problems encountered in statistics. They are not, it should be emphasized, restricted to algorithms written by or specifically for statisticians. Such a restriction would exclude a wide range of useful work and, unfortunately, would still include a number of inferior approaches to some problems.

In the general process of using computers to assist in statistical analysis of data, three aspects are frequently important: the recognition of the need for an algorithm (more generally, the role of algorithms in the overall approach); the attempt to find or implement a suitable algorithm; and judgments about the quality of an algorithm. Let us consider each of these questions in turn.

ALGORITHMS AND STATISTICAL COMPUTING

The importance of good algorithms derives from their role as building blocks supporting reliable, flexible computing. Statisticians (and equally, physicists, chemists, engineers, and other users of computers) have tended to plunge in with ad hoc attacks on specific computing problems, with relatively little use of existing algorithms or research in computing. Many arguments, some of them quite sound, support this approach. The end user is interested in the "answer" (in our case, the statistical analysis), not in the process that produces it. Particularly at early stages of statistical computing, the statistician was often not familiar with computing, either in the sense of a user or in the more important sense of understanding some of the basic principles of computation. The problem to be solved often appeared straightforward, with a solution that was qualitatively obvious to the statistician. In this case, finding or developing an algorithm seems a waste of valuable time. Furthermore, it may not be at all obvious how a statistical problem can be formulated in appropriate terms for

algorithms which frequently were devised for other problem areas.

Paradoxically, some of the statistical systems and packages developed to assist statisticians aggravate the tendency to take ad hoc rather than algorithmic approaches. The many conveniences of using high-level systems make it tempting to rig an intuitively plausible solution within the system rather than reach outside to find a high-quality algorithm for the problem. In many systems, the process of integrating such an algorithm into the system may require a high degree of programming skill and knowledge of the system's inner workings.

Although arguments for casual solutions to statistical computing problems have some force, there are stronger arguments that statisticians should try to integrate high-quality algorithms into their computing. Two arguments are particularly important. First, the use of good computational algorithms generally improves the use of our own time, in spite of the widely held intuition to the contrary. Second, the quality and the defensibility of the statistical analysis of data is eventually inseparable from the quality of the underlying computations.

Support for the first argument is that well-chosen algorithms will not only increase the chance that a particular statistical computation succeeds relatively quickly, but will usually greatly simplify the (inevitable) process of adapting the computation to new data or to a change in the analysis. As for the second argument, this is asserting both that the statistician should understand what an analysis has produced, in clear and precise terms, and also that the operational steps should be communicable and independently reproducible by others. Well-defined and correct computations are needed if statistical analysis is to satisfy fundamental criteria of scientific validity. This requires, in turn, that computations in statistical systems and specially programmed data analysis be based on algorithms that are accepted as correct implementations of valid computational methods.

As computing evolves, statisticians should be able to combine the convenience of statistical systems with the use of high-quality algorithms. Statistical systems increasingly incorporate good algorithms for the common operations. Advances in techniques of language design and implementation can simplify the process of integrating new algorithms into such systems, gradually merging the process of user programming and system extension [2].

SPECIFYING ALGORITHMS

(This section is directed largely to persons writing algorithms.) Our definition of an algorithm requires that it be suitable for implementation on a computer but deliberately does not restrict the form in which the algorithm is specified. The specification may be translatable mechanically into computer steps, i.e., may be given in a programming language. Alternatively, the specification may be instructions that someone familiar with a programming language can understand and implement. An important goal of computer science is to merge the two forms by improving programming languages and the art of algorithm design to the point that algorithms can be presented in a computer-readable form which is at the same time comprehensible to reasonably well informed human beings. Steps toward this goal, such as techniques of structured programming, are of value in that they increase the chance that one can understand what an algorithm does and hence the degree of confidence in its correctness.

The most convenient specification of an algorithm would intuitively seem to be in a programming language that is locally available, so that a running program could in principle be generated directly. This convenience has to be tempered by the need to understand the algorithm and occasionally by the unsuitability of the common programming languages (e.g., FORTRAN) to handle certain problems (e.g., random number generation*). Clear verbal descriptions are still important as supplements to program code. In some cases, semiverbal presentations can be used either as supplements or as replacement for actual code. Two styles of semiverbal description are used: natural language statements organized into numbered steps, usually with iteration among the steps;

and "pidgin" programming languages, with most of the description identical to some language, but with natural language inserted where the actual code would be harder to understand and with details omitted.

Given that an algorithm is to be presented in a programming language, which one will be most helpful? The overwhelming majority of published and otherwise generally circulated algorithms are written in FORTRAN, at least for scientific computing. Presenting an algorithm in this language is then likely to make it implementable widely (at least on the larger computers) and allow it to be used with many existing programs. Other, older languages are less frequently used. ALGOL60 was designed specifically to bridge the previously mentioned gap between readability and implementability; however, at the time of its design, it could take only a partial step in this direction. Although many early published algorithms were written in ALGOL60 (for a time the only accepted language for algorithm sections), FORTRAN has largely taken over, in spite of its deficiencies in generality and readability. Many of the ALGOL60 algorithms were subsequently translated into FORTRAN (some examples will be mentioned in the section "Finding Algorithms"). Other languages, such as PL-1 and COBOL, have at most marginal relevance to algorithms for statistics.

Three other languages do, however, need to be considered: APL, BASIC, and PASCAL. These are all important for interactive computing, particularly on the smaller machines. APL is widely used in statistical analysis; its advantages are its interactive nature and a general, convenient approach to arrays*. Some existing algorithms written in APL have been ad hoc and poorly designed. Nevertheless, there is a large community of users. Also, improvements in APL have made the description of some calculations more attractive (in particular, the inclusion in APL of some key operators to support the kind of calculations done in regression* and multivariate analysis*).

BASIC is also interactive, but in appearance is a (simplified) language of the FORTRAN family. It shares the advantages of availability on small computers, relative ease of initial learning, and a sizable user community. As with APL, the language has suffered at times from algorithms written without enough understanding of the problem. Both languages have been somewhat neglected by the computer-science community involved in developing high-quality algorithms. The neglect is a combination of professional isolation and some intrinsic flaws in the languages themselves. For example, both languages are rather clumsy for expressing the iterative calculations that most algorithms involve. BASIC, in addition, may make the process of separate definition of algorithms difficult.

PASCAL is again oriented to the use of small, interactive computers and can be learned fairly easily. It derives, however, from the ALGOL family of languages. PASCAL is a simple, structured language, well adapted to writing many types of algorithms in a clear and readable form. One of its attractions, in fact, is to the portion of the computer-science community interested in writing programs whose correctness can be formally verified. For these reasons, PASCAL is perhaps the most attractive new language for the specification of algorithms. At the time of writing, however, its applications to statistical computing are minimal. Applications of PASCAL are mostly to non-numerical problems. Its importance as a vehicle for statistical algorithms is largely in the future.

FINDING ALGORITHMS

There are several sources for statistical algorithms, with no simple process for searching them all. Algorithms from the various sources will tend to differ in reliability and in the convenience of implementation. Roughly in descending order of overall reliability, the major sources are:

Published Algorithm Sections

Several computing journals have published algorithms in one of a set of accepted programming languages (typically FORTRAN and ALGOL60). These algorithms have been independently referred and (in principle) tested. They should conform to specified requirements for quality (see the next section) established by journal policy. The journal *Applied Statistics** publishes such an algorithm section specifically for statistical computing. Some statistical algorithms have also appeared in *Communications in Statistics** (B). Major general algorithm sections appear in *Transactions on Mathematical Software* and *The Computer Journal*. The publication *Collected Algorithms of the Association of Computing Machinery* reprints the former set of general algorithms and contains an important cumulative index, covering most published algorithm sections as well as many algorithms published separately in scientific journals.

General Algorithm Libraries

These are collections of algorithms, usually distributed in machine-readable form, for a wide range of problems. Although the algorithms are often the work of many people, the libraries usually exert some central editorial control over the code. As a result, from the user's viewpoint, greater uniformity and simplicity can be achieved. However, the distributors may not be as disinterested judges of the library contents as are editors of algorithm sections. Confidence in the quality of the library rests to a large extent on evaluation of the organization distributing it. The International Mathematical and Statistical Library (IMSL), specifically oriented to statistical algorithms, is distributed by an independent organization in suitable FORTRAN source for many computer systems. The National Algorithm Group (NAG) is a publicly sponsored British organization designed to coordinate the distribution and development of algorithms. In this work it has had the cooperation of a number of professional groups, including the Royal Statistical Society*. A number of scientific laboratories also maintain and distribute general algorithm libraries, e.g., the PORT library (Bell Laboratories), Harwell Laboratory (U.K. Atomic Energy Research Establishment), and the National Physical Laboratory.

Specialized Algorithm Packages

These are less general collections of algorithms than the previous. They provide a range of solutions to a set of related problems, frequently in greater detail than that provided by general libraries. In addition, they attack some problem areas that tend to be ignored by published algorithms, such as graphics*. Questions of reliability will be similar to the general algorithm libraries. A series of specialized packages has been developed with the cooperation of Argonne National Laboratories, covering topics such as eigenvalue problems, linear equations, and function approximation. Graphics packages include the GR-Z package (Bell Laboratories) for data analysis and the DISSPLA package (a general-purpose system distributed commercially).

Scientific Journals

In addition to published algorithm sections, many published papers contain algorithm descriptions, either in one of the semiverbal forms or in an actual programming language. A qualified referee should have examined the paper, but unless given an explicit statement, it is probably unwise to assume that the algorithm has been independently implemented and tested. Nevertheless, there are a number of problems for which the only satisfactory published algorithms are of this form (e.g., some random number generation* techniques).

Unpublished Papers; Program Sharing

These categories are perhaps last resorts— least in average quality but certainly not least in quantity. It may be that more algorithms exist in these forms than in all other categories combined. They are usually not referred, except unintentionally by users, and one should expect to spend time testing them before putting them into regular use. Simply finding out about the algorithms requires considerable effort. Of most help are library search techniques and centralized clearinghouses for technical reports (such as the National Technical Information Service in the United States).

With increased familiarity, the process of searching the various sources will become more straightforward. Services provided by technical libraries, such as literature searches (now often computerized and relatively inexpensive) and centralized listings of papers, books, and memoranda, are

extremely valuable. Modern library personnel are often very knowledgeable and helpful in searching through the jungle of technical literature. Of course, once one or more algorithms have been found, there remains the question of whether they are adequate and, if not, what steps can be taken to improve or replace them.

THE QUALITY OF ALGORITHMS

The problem of evaluating algorithms has no simple solution. For most statistical applications, a sensible judgment about algorithms requires some understanding of the computational methods being used. The discussion in Chambers [1] and in the further references cited there provides background to some of the computational methods important for statistics. Although it is tempting to hope that some mechanical evaluation of algorithms could resolve their quality thoroughly, this is rarely the case. Most problems are too complex for an evaluation that treats the algorithm as a black box*; i.e., as a phenomenon to be judged only by its empirical performance, without regard for the techniques used. A tendency to use only this approach to evaluate statistical software is regrettable, particularly since it reinforces the overall ad hoc approach which has been detrimental to statistical computing in the past.

In the process of evaluating algorithms, both empirically and in terms of the method used, one may apply some general guidelines. Four helpful classes of questions are the following.

Is The Algorithm Useful? Does it solve the problem at hand? Is it general enough for all the cases likely to be encountered? Will it adapt to similar problems to be encountered later, or will a new algorithm have to be found essentially from scratch?

Is The Algorithm Correct? Will it run successfully on all the cases? If not, will it detect and clearly indicate any failures? For numerical calculations, what guarantees of accuracy are available? (If not theoretical estimates beforehand, are there at least reliable measures of accuracy after the fact?) We emphasize again that such judgments require understanding of what numerical methods can do to solve the problem.

How Hard Is It To Implement And Use? Does the form of the algorithm require considerable local effort (e.g., because the algorithm is written in English or in a programming language not locally available)? Does the algorithm as implemented make inconvenient assumptions (such as limits on the size of problem that can be handled)? Is it written portably, or are there features that will need to be changed locally? Most important, is the algorithm comprehensible, so that there is some hope of fixing problems or making modifications after one is committed to its use?

Is The Algorithm Efficient? Will its requirements for storage space, running time, or other computer resources be modest enough to make it practical for the problems at hand? Are there convenient, general estimates of these requirements? Issues of efficiency are often overemphasized, in the sense that the human costs involved in the previous questions are far more important in most applications. Nevertheless, we can still encounter problems that exceed the capacity of current computing, and it is good to be careful of such situations. As with accuracy, it is important to understand what computing science can do for the problem. Both theoretical estimates (of the order of difficulty, frequently) and empirical estimates are helpful.

NOTE

1. For the general reader, the complete Oxford English Dictionary gives algorithm (preferably algorism) as meaning the arabic numerals, with the chastening added meaning of a cipher or nonentity.

REFERENCES

1. Chambers, J. M. (1977). *Computational Methods for Data Analysis*. Wiley, New York.
2. Chamber, J. M. (1980). *Amer. Statist.*, 34, 238–243.
3. James, G. and James, R. C. (1959). *Mathematics Dictionary*. D. Van Nostrand, Princeton, N.J.

BIBLIOGRAPHY

The following is a selection of references, by subject, to some of the more useful algorithms for statistical applications. In most cases, the algorithms are presented in the form of subprograms or procedures in some programming language. A few are (reasonably precise) verbal descriptions to be followed by the reader in implementing the procedure. There is, of course, no assertion that these are "best" algorithms. Most of them do present a good combination of reliability, generality, and simplicity.

In addition to these references, several organizations provide algorithm libraries. Two sources that should be mentioned specifically for statistical and scientific computing are:

International Mathematical and Statistical Libraries, Inc. (IMSL), 7500 Bellaire Boulevard, Houston, Tex. 77036, USA.

The Numerical Algorithms Group (NAG), 7 Banbury Road, Oxford OX2 6NN, England.

Fourier Transforms

Singleton, R. C. (1968). *Commun. ACM*, **11**, 773–779.

Singleton, R. C. (1969). *IEEE Trans. Audio Electroacoust.*, **17**, 93–103.

Graphics

Akima, H. (1978). *ACM Trans. Math. Software*, **4**, 148–159.

Becker, R. A. and Chambers, J. M. (1977). *The GR-Z System of Graphical Subroutines for Data Analysis*. Write to Computer Information Library, Bell Laboratories, Murray Hill, N. J. 07974.

Crane, C. M. (1972). *Computer J.*, **15**, 382–384.

Doane, D. P. (1976). *Amer. Statist.*, **30**, 181–183.

Lewart, C. R. (1973). *Commun. ACM*, **16**, 639–640.

Newman, W. M. and Sproull, R. F. (1979). *Principles of Interactive Computer Graphics*, 2nd ed. McGraw-Hill, New York.

Scott, W. (1979). *Biometrika*, **66**, 605–610.

Nonlinear Models

Brent, R. P. (1973). *Algorithms for Minimization without Derivatives*. Prentice-Hall, Englewood Cliffs, N. J.

Gill, P. E. and Murray, W. (1970). *A Numerically Stable Form of the Simplex Algorithm*. Rep. Math. No. 87, National Physical Laboratory, Teddington, England.

Lill, A. S. (1970). *Computer J.*, **13**, 111–113; also, *ibid.*, **14**, 106, 214 (1971).

O'Neill, R. (1971). *Appl. Statist.*, **20**, 338–345.

Shanno, D. F. and Phua, K. H. (1976). *ACM Trans. Math. Software*, **2**, 87–94.

Numerical Approximation

Cody, W. J., Fraser, W., and Hart, J. F. (1968). *Numer. Math.*, **12**, 242–251.

Hart, J. F., Cheney, E. W., Lawson, C. L., Maehly, H. J., Mesztenyi, C. K., Rice, J. R., Thacher, H. C., and Witzgall, C. (1968). *Computer Approximations*. Wiley, New York.

Numerical Integration

Blue, J. L. (1975). *Automatic Numerical Quadrature: DQUAD. Comp. Sci. Tech. Rep. No. 25*, Bell Laboratories, Murray Hill, N. J.

Gentleman, W. M. (1972). *Commun. ACM*, **15**, 353–355.

Numerical Linear Algebra

Businger, P. A. and Golub, G. H. (1969). *Commun. ACM*, **12**, 564–565.

Wilkinson, J. H. and Reinsch, C. eds. (1971). *Handbook for Automatic Computation* Vol. 2: *Linear Algebra*. Springer-Verlag, Berlin. (Contains a wide selection of algorithms, many subsequently used in the EISPAK package developed at Argonne Laboratories.)

Programming

Ryder, B. G. (1974). *Software—Pract. Exper.*, **4**, 359–377.

Sande, G. (1975). *Proc. 8th Comp. Sci. Statist. Interface Symp.* Health Sciences Computing Facility, UCLA, Los Angeles, Calif., pp. 325–326.

Random Numbers

Ahrens, J. H. and Dieter, U. (1972). *Commun. ACM*, **15**, 873–882.

Ahrens, J. H. and Dieter, U. (1974). *Computing*, **12**, 223–246.

Ahrens, J. H. and Dieter, U. (1974). *Acceptance-Rejection Techniques for Sampling from the Gamma and Beta Distributions. Tech. Rep. No. AD*-782478, Stanford University, Stanford, Calif. (Available from National Technical Information Service.)

Chambers, J. M., Mallows, C. L., and Stuck, B. W. (1976). *J. Amer. Statist. Ass.*, **71**, 340–344.

Kinderman, A. J. and Monahan, J. F. (1977). *ACM Trans. Math. Software*, **3**, 257–560.

Regression

Barrodale, I. and Roberts, F. D. K. (1974). *Commun. ACM*, **17**, 319–320.

Chambers, J. M. (1971). *J. Amer. Statist. Ass.*, **66**, 744–748.

Daniel, J. W., Gragg, W. B., Kaufman, L., and Stewart, G. W. (1976). *Math. Comp.*, **30**, 772–795.

Gentleman, W. M. (1974). *Appl. Statist.*, **23**, 448–454.

Wampler, R. H. (1979). *ACM Trans. Math. Software*, **5**, 457–465.

Sorting

Brent, R. P. (1973). *Commun. ACM*, **16**, 105–109.

Chambers, J. M. (1971). *Commun. ACM*, **14**, 357–358.

Loesser, R. (1976). *ACM Trans. Math. Software*, **2**, 290–299.

Singleton, R. C. (1969). *Commun. ACM*, **12**, 185–186.

Utilities

Fox, P., Hall, A. D., and Schryer, N. L. (1978). The PORT Mathematical Subroutine Library. *ACM. Trans. Math. Software*, **4**, 104–126. (Also Bell Laboratories, Murray Hill, N. J., *Computing Sci. Tech. Rep. No. 47*.)

Kernighan, B. W. and Plauger, P. J. (1976). *Software Tools*. Addison-Wesley, Reading, Mass.

See also COMPUTERS AND STATISTICS; STATISTICAL PACKAGES; and STATISTICAL SOFTWARE.

JOHN M. CHAMBERS

ALIAS

When two (or more) parameters affect the distribution of a test statistic* in similar ways, each is said to be an alias of the other(s). The term is especially associated with fractional factorial designs*, in the analysis of which certain sums of squares have distributions that can reflect the existence of any one, or some, of a number of different effects.

See also CONFOUNDING and FRACTIONAL FACTORIAL DESIGNS.

ALIAS GROUP. See Fractional Factorial Designs

ALIASING. See Confounding

ALIAS MATRIX. See Fractional Factorial Designs

ALLAN VARIANCE. See Successive Differences

ALLGEMEINES STATISTISCHES ARCHIV

The *Allgemeines Statistisches Archiv* is the Journal of the Deutsche Statistische Gesellschaft* (German Statistical Society) and began publication in 1890.

The journal provides an international forum for researchers and users from all branches of statistics. The first part (Articles) contains contributions to statistical theory, methods, and applications. There is a focus on statistical problems arising in the analysis of economic and social phenomena. All papers in this part are refereed. In order to be acceptable, a paper must either present a novel methodological approach or a result, obtained by substantial use of statistical methods, which has a significant scientific or societal impact. For further information, readers are referred to the parent society website, www.dstatg.de.

See also Statistische Gesellschaft, Deutsche.

ALLOKURTIC CURVE

An allokurtic curve is one with "unequal" curvature, or a skewed*, as distinguished from an isokurtic* curve (which has equal curvature and is symmetrical). This term is seldom used in modern statistical literature.

See also Kurtosis and Skewness: Concepts and Measures.

ALLOMETRY

It is rare, in nature, to observe variation in size without a corresponding variation in shape. This is true during the growth of an organism when radical shape changes are commonplace; when comparing different species from the same family; and even when comparing mature individuals from the same species. The quantitative study of this relationship between size and shape is known loosely as *allometry* and the main tool is the log-log plot. If X and Y are two dimensions that change with size, then the way each changes relative to the other is best studied by plotting $\log X$ vs. $\log Y$. In the past this was usually done on special log-log graph paper, but calculators have rendered such devices obsolete. Natural algorithms will be used in this article (and are recommended).

Some examples are shown in Figs. 1 to 3. In Fig. 1 the points represent different individuals, each measured at one point during growth. In Fig. 2 the points refer to mature individuals. In Fig. 3 the points refer to different species, and X and Y now refer to mean values (or some other typical values) for the species. The value of the log-log plot is that it provides a simple summary of departures from *isometric** size variation, i.e., variation in which geometric similarity is maintained. If X and Y are both linear dimensions, then isometric variation corresponds to a constant ratio Y/X, which in turn corresponds to a line

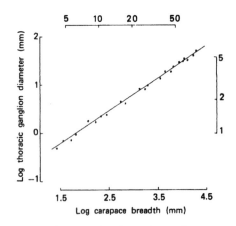

Figure 1. Growth of crabs (redrawn from Huxley [10]). The slope of the line is approximately 0.6.

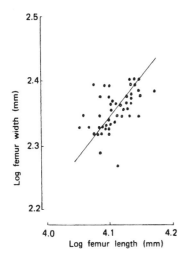

Figure 2. Variation between mature pine martens (redrawn from Jolicoeur [11]). The slope of the line is approximately 1.3.

is the natural scale on which to study departures from isometric size variation. Linear plots using original units have also been used, but these have the disadvantage that isometry now corresponds to a line passing through the origin rather than a line of slope 1 (in log units), and there are usually no measurements in the region of the origin.

The log-log plot has also been widely used when comparing species in the general study of the effect of scale on form and function. Once again a linear relationship often proves adequate, but it is not only departures from isometry that are important but also deviations of individual species from the allometric line.

Gould [6] should be consulted for further details about the history of allometry and for a review of applications.

STATISTICAL METHODS FOR ESTIMATING A STRAIGHT LINE

The central problem is to estimate the straight-line relationship displayed in the log-log plot. To avoid too many "logs," the values of $\log X$ and $\log Y$ will be referred

of slope 1 on the log-log plot. If Y is linear but X is a volume (or weight), then isometric variation corresponds to a constant value for $Y/X^{1/3}$, i.e., a line of slope 0.33 in the log-log plot. In a similar fashion the slope is 0.50 when Y is linear and X an area and 0.67 when Y is an area and X a volume. Slopes of X on Y are the reciprocals of the slopes of Y on X.

The log-log plot was first used systematically by Huxley [9,10] and Teissier [23]. They found that for growth studies, using a wide variety of dimensions and organisms, the plot could often be adequately summarized by a straight line, i.e., a power law of the type $Y = bX^\alpha$ in original units. The coefficient α is the slope in the log-log plot and $\log b$ is the intercept: $\log Y = \alpha \log X + \log b$. For linear dimensions $\alpha = 1$ corresponds to isometry* and $\alpha \neq 1$ is referred to as allometric growth: positive if $\alpha > 1$ and negative if $\alpha < 1$. Departures from a simple straight-line relationship are not uncommon, the best known being where there is a sudden change in relative growth rate during development.

Some importance was initially attached to providing a theoretical basis for the power law, but today the relationship is recognized as being purely empirical. It is the "roughly linear relationship" so widely used in biology, but it happens to be in log units because this

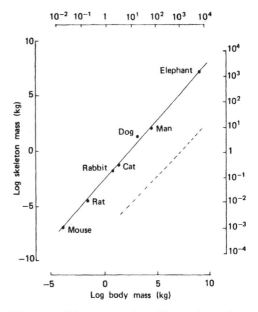

Figure 3. "Mouse-to-elephant" line (redrawn from Schmidt-Nielsen, p. 5, in Pedley [21]). The dashed line has slope 1.

to as x and y. The equation of a straight line passing through (x_0, y_0) with slope α(y on x) is $y - y_0 = \alpha(x - x_0)$. This corresponds to the power law $Y = bx^\alpha$, where $\log b = y_0 - \alpha x_0$. Figure 4 shows four different approaches to fitting a line. In the first two the lines are chosen to minimize the sum of squares of deviations* $\sum d^2$ and correspond to regression* of y on x, and x on y. In the third the line minimizing $\sum d^2$ is called the major axis and in the fourth the value of $\sum d_1 d_2$ is minimized to produce the reduced major axis [14] or D-line [25]. All four lines pass through the centroid of points $(\bar{x}, \bar{y})$ and differ only in their values for α the slope of y on x. All the estimates may be expressed in terms of r, S_x, and S_y, the sample correlation* and standard deviations* of x and y. They are:

1. $\hat{\alpha} = rS_y/S_x$ (regression of y on x)
2. $\hat{\alpha} = r^{-1}S_y/S_x$ (regression of x on y)
3. $\hat{\alpha} = $ slope of the major axis (given below)
4. $\hat{\alpha} = S_y/S_x$ with sign the same as that of r.

The two regression estimates differ from one another and there is no natural way of resolving the question of which to choose. For this reason regression theory is not as helpful in allometry as in other branches of applied statistics. The slope of the major axis is found by first obtaining the positive root of the equation in t,

$$t/(1 - t^2) = r\lambda/(1 - \lambda^2), \qquad (1)$$

where $\lambda = S_y/S_x$ or S_x/S_y, whichever is less than 1. If $\lambda = S_y/S_x < 1$, then t is the slope in

units of y on x. If $\lambda = S_x/S_y < 1$, then t is the slope in units of x on y and t^{-1} is the slope in units of y on x. When r is negative, the same numerical value is used for t but the sign is now negative.

Of the last two estimates the most popular has been the reduced major axis. The major axis has been criticized because it depends on the ratio $\lambda = S_y/S_x$ in a nonlinear way so that a change of units for x and y does not affect t in the same way as it affects λ. This has no validity when x and y refer to log measurements because a change of units for the measurements X and Y leaves r, S_x, and S_y unchanged. The major axis has also been criticized as being more difficult to compute than the others, but in fact (1) is easy to solve either as a quadrate or graphically. For values of λ in the range 0.2 to 0.8 a good approximation is $t \simeq \lambda + 0.3 \log r$. As λ approaches 1, the right-hand side of (1) tends to ∞, so t tends to 1. There is no general agreement as to which of these two estimates is to be preferred. Fortunately, when r is high, they give very similar answers.

Statistical sampling properties of the estimates are based on the bivariate normal distribution*. Approximate large sample standard errors* are given by Kermack and Haldane [14]:

$$SE(\hat{\alpha}) \simeq \sqrt{\left(\frac{1 - r^2}{r^2 n}\right)} t$$

for the major axis

$$SE(\hat{\alpha}) \simeq \sqrt{\left(\frac{1 - r^2}{n}\right)} \frac{S_y}{S_x}$$

for the reduced major axis.

Since $t > S_y/S_x$, the standard error for the major axis is always greater than that for the reduced major axis (considerably greater if r is low). This is an argument in favor of the reduced major axis, but such an estimate would make little sense if r were too low. When $S_x = S_y$ and $r = 0$, it amounts to drawing a line of slope 1 through a circular cloud of points. At least the large sampling error of the major axis serves as a warning that the line is difficult to determine when r is low. As a numerical example, consider the case $r = 0.8$, $n = 20$, $\lambda = S_y/S_x =$

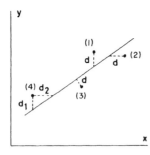

Figure 4. Four methods of fitting a straight line.

0.6. The slope of the major axis is found from $t/(1 - t^2) = 0.75$, a quadratic equation yielding $t = 0.5352$ ($t = \tan\{\frac{1}{2}\tan^{-1}(2 \times 0.75)\}$ on a calculator). The slope of the reduced major axis is $\lambda = 0.6$ and the approximation for t is $\lambda + 0.3\log 0.8 = 0.5331$. The two estimates for α are 0.54 ± 0.09 (major axis) and 0.60 ± 0.08 (reduced major axis). When $r = 0.4$ they become 0.33 ± 0.17 and 0.60 ± 0.12, respectively.

An alternative to basing estimates directly on the bivariate normal distribution* is to assume that x and y would lie exactly on a straight line if it were not for "errors." The errors are due to biological variation rather than to measurement errors* in this context. The model is called the linear functional relationship model. It is possible to estimate α provided that the ratio of error variances is assumed known. In fact, if they are assumed equal, the estimate turns out to be the major axis. However, no sampling theory based on the model is available and the approach has not proved very fruitful. Further details are given in Sprent [22] and Kendall and Stuart [13]. Bartlett's estimate of slope also belongs to this class [2]. It is an example of the use of an instrumental variate* (see Kendall and Stuart [13]). the estimate has enjoyed some popularity in allometry, but unfortunately the assumptions underlying its use are not usually fulfilled unless the scatter* is low. A detailed numerical example of the use of Bartlett's estimate has been given by Simpson et al. [21].

STATISTICAL TESTS

The most commonly required test is for departure from isometry. For both the major axis and reduced major axis, $\alpha = 1$ implies that $\sigma_y = \sigma_x$ (in terms of population parameters), and this is most conveniently tested for by computing $z = y - x$ and $w = y + x$ for each individual. Since $\text{cov}(z, w) = \sigma_y^2 - \sigma_x^2$, the test for $\sigma_y = \sigma_x$ is equivalent to testing for zero correlation* between z and w. If the isometric value of α is 0.33 (for example) rather than 1, then $z = y - 0.33x$ and $w = y + 0.33x$ are used to test whether $\sigma_y = 0.33\sigma_x$ in the same way.

Confidence intervals* for α may be obtained using the approximate large sample standard errors for $\log\hat\alpha$. These are easily derived from those given earlier and are

$$SE(\log\hat\alpha) \simeq \sqrt{\left(\frac{1 - r^2}{r^2 n}\right)}$$

for the major axis

$$SE(\log\hat\alpha) \simeq \sqrt{\left(\frac{1 - r^2}{n}\right)}$$

for the reduced major axis.

There is some advantage to using the SE of $\log\hat\alpha$ rather than $\hat\alpha$ since the formulas do not involve the population value α. A better approximation for the reduced major axis has been given by Clarke [4]. See also Jolicoeur [12].

An important question in allometry is whether or not one group of individuals or species is an allometric extension of another. Figure 5 illustrates a case where this is so and also shows why the use of different lines can give different answers to this question. Group A is a linear extension of group B only along the major axis, not along the regression line of y on x. There seems to be no generally recommended statistical test for this situation. Analysis of covariance* would be used with regression lines,* and this suggests a rough test for use with the major axis. First obtain a pooled estimate* of the common slope of the axis in the two groups as $\hat\alpha = \frac{1}{2}(\hat\alpha_1 + \hat\alpha_2)$ and then test for differences between groups in a direction perpendicular to this slope, ignoring any sampling error in $\hat\alpha$. This is equivalent to calculating $z = y - \hat\alpha x$ for each individual (or species) in each group and then testing whether the mean of z differs between groups.

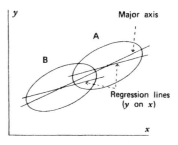

Figure 5. Allometric extension.

Statistical tests are often omitted in allometry and there are a number of reasons for this. One is that in many growth studies the random error* is small compared to the systematic change and the conclusions are apparent. Another is that where points represent species, which cannot be thought of as randomly sampled from a population, the relevance of statistical sampling theory is slight. Even in growth studies the individuals are rarely a proper random sample from a defined population, so that significance tests* play a less important role than they do in randomized experiments*. Finally, the inadequacy of statistical methods based on the bivariate normal distribution or the linear functional relationship model is an important factor.

MULTIVARIATE ALLOMETRY

When more than two measurements of an organism are studied, an overall view of the joint variation is desirable. For p measurements $X_1, \ldots, X_p$ the log-log plot generalizes to a plot of $x_1 = \log X_1, \ldots, x_p = \log X_p$ in p dimensions, a useful concept even though it is not possible to actually plot the points. Isometry corresponds to the direction vector $\boldsymbol{\alpha}_0 = (1/\sqrt{p}, \ldots, 1/\sqrt{p})$ with the usual provision that if X_i represents an area or volume it is replaced by $X_i^{1/2}$ or $X_i^{1/3}$, respectively. Allometry again corresponds to departures from isometry, but clearly the possibilities are considerably wider in p dimensions than in two dimensions. A further problem is that there is no longer a large body of empirical evidence to suggest that departures from isometry are usually adequately summarized by a straight line. However, such a departure is a sensible starting point, and it may be summarized by a direction vector $\boldsymbol{\alpha} = (\alpha_1, \ldots, \alpha_p)$, where $|\boldsymbol{\alpha}| = 1$ (i.e., $\sum \alpha_i^2 = 1$). The angle between $\boldsymbol{\alpha}$ and $\boldsymbol{\alpha}_0$ is given by $\cos\theta = \sum \alpha_i/\sqrt{p}$.

The direction $\boldsymbol{\alpha}$ may be estimated using either the major or the reduced major axis. In the latter $\hat{\alpha}$ is taken to be the direction vector proportional to $(S_1, \ldots, S_p)$ where S_i is the sample standard deviation of x_i. Using the major axis, $\hat{\alpha}$ is taken equal to the direction of the first eigenvector* of the covariance matrix* of $x_1, \ldots, x_p$. This eigenvector is often referred to as the first principal component*.

Corruccini and Henderson [5] give a good example of the use of the major axis. The linear functional relationship model has also been generalized to many variables [8,22], but there are few examples of its actual use.

The statistical sampling theory for many variables encounters a major difficulty that does not exist in the bivariate case. This is the need to specify the error structure about the allometric line. In the bivariate case variation perpendicular to the line is in one dimension only and may be summarized by a standard deviation. In p dimensions the space perpendicular to the allometric line has $p - 1$ dimensions, so that covariation as well as variation must be specified. The simplest thing is to assume zero covariance perpendicular to the line, which implies that all eigenvalues* of the covariance matrix apart from the first are equal. A fairly straightforward test for this hypothesis has been described by Morrison [16, p. 250]. The eigenvector corresponding to the first eigenvalue is the allometric direction, and for this to be a sensible summary of the data the first eigenvalue must be considerably greater than the $p - 1$ (equal) eigenvalues. This is the same qualitative judgment that has to be made in the bivariate case. The first eigenvalue is the variance in the allometric direction and the others are the variances in directions perpendicular to the line. Variation along the line should be much greater than that about the line. Isometry corresponds to a further specialization of this situation to the case where the first eigenvector is $(1/\sqrt{p}, \ldots, 1/\sqrt{p})$. Kshirsagar [15] gives an overall χ^2 test* for both the isometric direction and independent variation about the line and also shows how to partition the χ^2 into its two components. Morrison [16] describes a test for the isometric direction which does not rest on the assumption that all other eigenvalues are equal. Anderson [1] should be consulted for the detailed derivation of the tests described in Morrison's book.

Mosimann [17,18] has proposed a more general approach which avoids the very restrictive assumption of independent variation orthogonal to $\boldsymbol{\alpha}_0$ or $\boldsymbol{\alpha}$. He defines size as a function G of the original measurements $\mathbf{X}$, with the property that $G(\mathbf{X}) > 0$ and $G(a\mathbf{X}) = aG(\mathbf{X})$ for $a > 0$. A shape vector $\mathbf{Z}(\mathbf{X})$

is defined to be any dimensionless vector with $p - 1$ components. Isometry is then defined to be statistical independence* of shape and size, and it is shown that for a given choice of size, either all shape vectors are independent of size or none are. Mosimann also shows that if shape is independent of one measure of size, then it cannot be independent of any other measure of size.

A test for isometry of shape with respect to a given size can be carried out on a log scale using multiple regression* techniques to test independence*. For example, if $(x_1, \ldots, x_p)$ are the measurements on a log scale, then a common measure of size is $\bar{x} = \sum x_i / p$ and a possible shape vector is $(x_1 - \bar{x}, \ldots, x_{p-1} - \bar{x})$. If R^2 is the multiple correlation* between $\bar{x}$ and $(x_1 - \bar{x}, \ldots, x_{p-1} - \bar{x})$ based on n individuals, then $\{R^2(n - p)\}/\{(1 - R^2)(p - 1)\}$ may be used to test the independence of shape and size. Provided that $(x_1, \ldots, x_p)$ have a multivariate normal distribution*, the statistic has the F-distribution* with $p - 1$, $n - p$ degrees of freedom when the null hypothesis is true. Mosimann and James [19] give an example of this test, but no attempt is made to quantify departures from isometry, and this seems to be a weakness of the approach so far.

The problem of whether one group of points is an allometric extension of another in p dimensions does not seem to have been dealt with explicitly. There have been some advances in the wider problem of making comparisons between groups, ignoring variation in certain given directions (e.g., the isometric direction). Burnaby [3] has extended discriminant analysis* to deal with this situation and Gower [7] has extended Burnaby's work to cover the case where the direction to be ignored is an estimated allometric direction.

REFERENCES

1. Anderson, T. W. (1963). *Ann. Math. Statist.*, **34**, 122–148.

2. Bartlett, M. S. (1949). *Biometrics*, **5**, 207–212.

3. Burnaby, T. P. (1966). *Biometrics*, **22**, 96–110.

4. Clarke, M. R. B. (1980). *Biometrika*, **67**, 441–446.

5. Corruccini, R. S. and Henderson, A. M. (1978). *Amer. J. Phys. Anthropol.*, **48**, 203–208.

6. Gould, S. J. (1966). *Biol. Rev.*, **41**, 587–640.

7. Gower, J. C. (1976). *Bull. Geol. Inst. Univ. Upps.* (N.S.), **7**, 1–10.

8. Hopkins, J. W. (1966). *Biometrics*, **22**, 747–760.

9. Huxley, J. S. (1924). *Nature (Lond.)*, **114**, 895–896.

10. Huxley, J. S. (1932). *Problems of Relative Growth*. Methuen, London.

11. Jolicoeur, P. (1963). *Growth*, **27**, 1–27.

12. Jolicoeur, P. (1968). *Biometrics*, **24**, 679–682.

13. Kendall, M. G. and Stuart, A. (1967). *The Advanced Theory of Statistics*, Vol. 2., 2nd ed., Charles Griffin, London.

14. Kermack, K. A. and Haldane, J. B. S. (1950). *Biometrika*, **37**, 30–41.

15. Kshirsagar, A. M. (1961). *Biometrika*, **48**, 397–407.

16. Morrison, D. F. (1976). *Multivariate Statistical Methods* (2nd ed.), McGraw-Hill, New York.

17. Mosimann, J. E. (1970). *J. Amer. Statist. Ass.*, **65**, 930–945.

18. Mosimann, J. E. (1975). Statistical problems of size and shape, I, pp. 187–217. Statistical problems of size and shape, II, pp. 219–239. In *Statistical Distributions in Scientific Work*, Vol. 2, G. P. Patil, S. Kotz, and J. K. Ord, eds. D. Reidel, Dordrecht.

19. Mosimann, J. E. and James, F. C. (1979). *Evolution*, **33**, 444–459.

20. Pedley, T. J. (1977). *Scale Effects in Animal Locomotion*. Academic Press, New York.

21. Simpson, G. G., Roe, A., and Lewontin, R. C. (1960). *Quantitative Zoology* (rev. ed.). Harcourt Brace, New York.

22. Sprent, P. (1969). *Models in Regression and Related Topics*. Methuen, London.

23. Teissier, G. (1931). *Trav. Stn. Biol. Roscoff*, **9**, 227–238.

24. Teissier, G. (1948). *Biometrics*, **4**, 14–53.

FURTHER READING

D'Arcy Thompson's book *Growth and Form* (Cambridge University Press, Cambridge, 1917, 1942) provides the standard introduction to problems of scale in form and function. An account of some recent work is given in *Scale Effects in Animal Locomotion* by T. J. Pedley (Academic Press, New York, 1977). S. J. Gould's article "Allometry and size in ontogeny and phylogeny" (*Biol. Rev.*, **41**, 587–640) gives a comprehensive survey of the uses of allometry, and Sprent's 1972 article "The mathematics of size and shape" (*Biometrics*, **28**, 23–37) reviews the mathematical and statistical side. D. L. Pilbeam and S. J. Gould, in their 1974 contribution "Size and scaling in human evolution" (*Science*, **186**, 892–901), give an interesting account of the use of allometry to investigate evolutionary relationships. A variety of different allometric studies are reported in *Problems of Relative Growth* by J. S. Huxley (Methuen, London, 1932), and a good discussion of some of the problems with allometry is given in *Essays on Growth and Form Presented to D'Arcy Wentworth Thompson*, edited by W. E. Le Gros Clark and P. B. Medewar (Clarendon Press, Oxford, 1945).

See also Anthropology, Statistics in; Principal Component Analysis, Generalized; Regression (Various); and Size and Shape Analysis.

M. Hills

ALMOST CERTAIN CONVERGENCE.

See Convergence of Sequences of Random Variables

ALMOST-LACK-OF-MEMORY (ALM) DISTRIBUTIONS

A random variable X is defined to have an almost-lack-of-memory (ALM) distribution [1] if the Cauchy functional equation

$$\Pr(X - b < x | X \geqslant b) = \Pr(X < x) > 0,$$

here given in probabilistic form, holds for infinitely many values of $b, b > 0$. Continuous distributions with this property possess a density $f(\cdot)$, almost surely, that satisfies (see Ref. 1)

$$f_X(nc + x) = \alpha^n(1 - \alpha)f(x), \quad 0 < x < c, \quad (1)$$

$0 < \alpha < 1, n = 0, 1, 2 \ldots$, for any arbitrary positive integer c.

Let $S(\cdot)$ be a survival function defined in $[0, \infty)$, so that $S(x) = 1 - F(x)$ for a cumulative distribution function (cdf) $F(\cdot)$, and for c, a positive integer, the functional equation

$$S(nc + x) = S(nc)S(x), \quad x \geqslant 0, \quad (2)$$

$n = 1, 2, \ldots$, is a particular case of the ALM property (1). If (2) holds for a nonnegative random variable with cdf $F(x) = \Pr(X \leqslant x)$, $x > 0$, then [2] $F(\cdot)$ is of the form

$$F(x) = 1 - \alpha^{[x/c]} + \alpha^{[x/c]}F(x - [x/c]c),$$

$$x \geqslant c > 0, \quad (3)$$

where $\alpha = S(c) < 1$ and $[x]$ is the smallest integer less than or equal to x. Hence, each cdf $F(\cdot)$ satisfying Equation 1 is uniquely determined by the values of $F(x)$ for $0 \leqslant x < c$.

As an application, consider a queueing system with instantaneous repairs after any failure of a constant-lifetime server. Then each cdf satisfying Equation 3 is the cdf of the "blocking time," that is, the total time taken by a customer, and conversely [1,2].

REFERENCES

1. Chukova, S. and Dimitrov, B. (1992). On distributions having the almost-lack-of-memory property. *J. Appl. Probab.*, **29**, 691–698.
2. Lin, G. D. (1999). Letter to the editor. *J. Appl. Probab.*, **31**, 595–605.

See also Exponential Distribution and Queueing Theory.

α-LAPLACE DISTRIBUTION.   See Linnik Distribution

ALTERNATIVE HYPOTHESIS

A hypothesis that differs from the hypothesis being tested is an alternative hypothesis, usually one to which it is hoped that the test used will be sensitive. Alternative hypotheses should be chosen having regard to (1) what situation(s) are likely to arise if the hypothesis tested is not valid and (2) which ones, among these situations, it is of importance to detect.

Usually, a whole class of alternative hypotheses, rather than a single one, is used.

See also Critical Region; Hypothesis Testing; Level of Significance; Null Hypothesis; and Power.

AMERICAN JOURNAL OF HUMAN GENETICS

The American Journal of Human Genetics (AJHG) was founded in 1948, and is published monthly by the University of Chicago Press in two volumes per year, six issues per volume. The website for the journal is www.journals.uchicago.edu/AJHG

As stated on the website:

> "*AJHG* is [sm1]a record of research and review relating to heredity in humans and to the application of genetic principles in medicine, psychology, anthropology, and social services, as well as in related areas of molecular and cell biology. Topics explored by *AJHG* include behavioral genetics, biochemical genetics, clinical genetics, cytogenetics, dysmorphology, genetic counseling, immunogenetics, and population genetics and epidemiology".

AJHG is the official scientific publication of the American Society of Human Genetics.

AMERICAN SOCIETY FOR QUALITY (ASQ)

[This entry has been updated by the Editors.]

The United States is generally considered the country that founded modern quality control. The work of Walter Shewhart* of the American Telephone and Telegraph Company and his associates George Edwards, Harold Dodge, and Harry Romig forms the nucleus around which the movement grew. However, it was the crisis of World War II that gave impetus to the field. Business managers realized that government-mandated quality control programs for defense products had an equally important application in civilian products.

During World War II a series of short courses were conducted throughout the country on statistical quality control*. Those who attended were encouraged to get together to exchange ideas and to reinforce their new-found knowledge. A series of local societies and several regional ones were founded throughout the country. In 1946, they formed a confederation called the American Society for Quality Control, which acts as the primary professional society for the United States, Mexico, and Canada. The headquarters office, initially in New York, was later transferred to Milwaukee, Wisconsin, where it is now located.

In 1997 the society dropped 'Control' from its title, and, as the American Society for Quality (ASQ), adopted a new mission of promoting performance excellence worldwide.

The ASQ has encouraged community colleges and universities to offer courses in the field. Despite the demand, most institutions of higher education choose not to have separate curricula; rather, courses are woven into other curricula.

The society is organized in modified matrix fashion. There are in 2004 more than 250 local geographic sections throughout major industrial areas of the United States, Canada and Mexico. Some members also choose to join one or more of the 27 industrial or subdiscipline-oriented divisions. They usually hold one or more national conferences per year. The divisions are listed on the ASQ's website, www.asq.org.

By 1991 the membership of ASQ exceeded 100,000, and in 2001 membership extended to 122 countries. Members are classified as Members, Senior Members, or Fellows.

Honors are presented by the society at the national, divisional, and sectional levels. Best known are the Shewhart Medal, the Edwards Medal, the Grant Award, and the Brumbaugh Award.

The initial journal of ASQC was *Industrial Quality Control*, later renamed *Quality Progress*. The *Journal of Quality Technology** has been published since 1969. *Technometrics** is a journal published jointly with the American Statistical Association. In addition, the *Transactions* of the Annual Technical Conference have been found to be a useful source of practical information. *Quality Management Journal* and *Software Quality Professional* began publication in 1993 and 1998, respectively. Many other publications are published centrally and by the divisions, conference boards, and local sections.

In the early days of the ASQC the emphasis was primarily on the statistical basis of sampling schemes* for raw materials coming into a plant or warehouse. There was also interest in reducing the cost of final inspection. The early emphasis was on detection of nonconforming products. A bit later the emphasis changed to the prevention of defects. Shewhart's work in the economic control of quality* in manufacturing provided the basis for this work. Today, managerial, motivational, and engineering aspects get a more balanced hearing. The ASQ is the primary vehicle for teaching these concepts. It has an important educational arm which conducts courses in quality and reliability throughout North America. In addition, educational materials are available for persons wishing to conduct courses under local auspices. Examinations for certification as Certified Quality Engineer (CQE), Certified Reliability Engineer (CRE), Certified Quality Technician (CQT), Certified Quality Manager (CQM), and Certified Reliability Technician (CRT) are available throughout the world for both members and nonmembers.

The savings due to reliable quality control programs initiated since World War II approached $1 trillion worldwide.

In later years ASQ became involved in coordinating national quality standards on behalf of the American National Standards Institute. These have worldwide impact.

The society maintains close relations with societies associated with quality in various parts of the world. These include the European Organization for Quality, the Japanese Union of Scientists and Engineers, the New Zealand Organization for Quality Assurance, the Australian Organization for Quality Control, and a variety of others.

See also JOURNAL OF QUALITY TECHNOLOGY; QUALITY CONTROL, STATISTICAL; and TECHNOMETRICS

W. A. GOLOMSKI

AMERICAN SOCIETY FOR QUALITY CONTROL (ASQC). See AMERICAN SOCIETY FOR QUALITY (ASQ)

AMERICAN STATISTICAL ASSOCIATION

[This entry has been updated by the Editors.]

The American Statistical Association, founded in 1839 as a nonprofit corporation, has as its purpose "to foster, in the broadest manner, statistics and its applications, to promote unity and effectiveness of effort among all concerned with statistical problems, and to increase the contribution of statistics to human welfare." It is a professional association whose membership is open to individuals with interest and background in the development and use of statistics in both methodology and application.

The Association is governed by an elected board of directors, which represents geographical areas, subject-matter areas, and the offices of President, Past President, President-elect, Secretary-treasurer, and three Vice-presidents. There is also a council made up of representation from each of the chapters of the association as well as the sections, as noted below. There are presently (in 2004) 78 chapters. Membership at the end of 2003 exceeded 17,000 in the U.S., Canada and other countries.

A central office is maintained at 1429 Duke Street, Alexandria, VA - 22314-3415; tel. (703) 684-1221. The staff includes an Executive Director, Director of Programs, Director of Operations, and support staff. Technical editing for some journals is maintained at this office.

The chapters of the ASA have individual programs throughout the year. These vary from one or two to as many as 20 meetings in a single year. Each chapter is autonomous in its program of activities. Chapters vary in size from 25 to 1800 members. A chapter may sponsor a seminar of one or more days or a short course in a subject-matter area or a set of methodological techniques. Chapters often collaborate with local or regional units of other professional associations.

Within ASA there are 21 sections. These represent different areas of activity and include Bayesian Statistical Science, Government Statistics, Health Policy Statistics, Nonparametric Statistics, Quality and Productivity, Risk Analysis, Statistical Consulting, Statistical Graphics, Statistics and the Environment, Statistics in Defense and National Security, Statistics in Epidemiology, Statistics in Sports, Business and Economics, Social Statistics, Statistical Education, Physical and Engineering Sciences, Biometrics, Statistical Computing,

Survey Research Methods, Biopharmaceutics, Teaching of Statistics in the Health Sciences, and Statistics in Marketing. The sections develop various activities which are useful to statistical practitioners and researchers. These include (1) cosponsorship of regional meetings with other professional associations, (2) review of statistical computer packages, (3) development of visual-aids materials, (4) appraisal of surveys, and (5) recommendations of statistical curricula in health sciences, computer sciences, industrial statistics, etc. Sections cosponsor symposia, workshops, and special topic meetings. A number of the sections participate in advisory capacities in such areas as national and international standards, educational programs, and federal statistical activities.

One of the strengths of the association is its committees, which vary from advisory boards to government agencies to joint committees with other professional associations. There are approximately 90 committees within the association. Although some are the usual "in-house" committees and some are short-term ad hoc committees, 23 relate to activities and programs outside the statistical profession, as well as to key issues and concerns on national and international levels. For example, the Committee on Law and Justice Statistics reviews the programs and structure of statistical data collection and analysis in the U.S. Department of Justice. Advisory committees serve the U.S. Bureau of the Census* and the Energy Information Administration of the Department of Energy. Possibilities for similar committees are continually being studied. The areas of privacy and confidentiality, as well as the statistical components of legislative action, are serviced by active committees. For the latter, a consortium of 10 professional associations was formed. A joint committee of the American Statistical Association and the National Council of Teachers of Mathematics works on developing the teaching of Statistics and Probability at every level of school education from grades K-12. Some activities are in close communication with similar groups in other countries.

The association holds an annual meeting. Often, this is in cooperation with other statistical and related societies, e.g., the International Biometric Society*, Eastern and Western North American Regions, and the Institute of Mathematical Statistics*. This is usually preceded by one- or two-day short courses in statistical methodology. There is also joint sponsorship in national and regional activities with such professional organizations as the American Society for Quality (ASQ)*, the Society of Actuaries, etc.

The association engages in programs of research and development in various areas of statistics in the types of activities that would not ordinarily be considered appropriate for a particular university or research institution. At times, the association engages in national and international projects in an attempt to make statistics even more useful to the profession and to the public at large. An example of this was the international seminar on the Transfer of Methodology between Academic and Government Statisticians.

Some of the programs in which the ASA has been engaged have far-reaching consequences in the areas of social science, survey research methods, physical and engineering sciences, and health sciences, to state a few. For example, beginning in 1977, the association engaged in a joint venture with the U.S. Bureau of the Census, "Research to Improve the Social Science Data Base." This included research fellows and trainees in time series*, demography*, computer software, editing* of large data sets, and a number of related areas. Another research program concerns the quality of surveys.

Educational programs, often in conjunction with other professional associations, include the development of resource material and visual aids for the teaching of statistics*, a visiting lecturer program, development and testing of a statistical curriculum for secondary schools, and short courses. The association also sponsors research fellowships and traineeships in which an individual spends a major portion of a year at a federal statistical agency. In the area of continuing education, videotapes of short courses and special lectures are made available to a wide segment of professional statisticians, not only in ASA chapters, but at universities and in industrial organizations.

In 1976, in collaboration with ASA members in several Latin American countries, the association initiated a program of symposia and site visits to stimulate the development of statistics in these countries. The result has been increased professional activity and the development of new statistical associations in some of the host countries.

The American Statistical Association is affiliated with a number of national and international organizations. It is one of approximately 50 national statistical associations affiliated with the International Statistical Institute*. Furthermore, it maintains representation in sections of the American Association for the Advancement of Science (AAAS). Through its representatives the ASA is active in some AAAS programs, such as scientific freedom and human rights and the international consortium of professional associations. The ASA maintains representation on the councils of the American Federation of Information Processing Societies, the Conference Board of the Mathematical Sciences, the Social Science Research Council, and the National Bureau of Economic Research.

The Association publishes seven journals; *Journal of the American Statistical Association**, *The American Statistician**, *Technometrics** (jointly with the American Society for Quality Control), the *Journal of Agricultural, Biological and Environmental Statistics** (jointly with the International Biometric Society*), the *Journal of Business and Economic Statistics**, the *Journal of Computational and Graphical Statistics* (jointly with the Institute of Mathematical Statistics* and Interface Foundation of North America), and the *Journal of Educational and Behavioral Statistics** (jointly with the American Educational Research Association). ASA also publishes *Current Index to Statistics** (with the Institute of Mathematical Statistics*) and the three magazines *Amstat News, Chance* and *Stats,* the last-named reaching more than 3,000 student members of ASA. Beyond this, there are proceedings of its annual meetings by sections and various reports on conferences, symposia, and research programs.

The American Statistical Association is dedicated to be of service to its members abroad as well as its members within the United States and Canada, and has from time to time assisted other associations in the development of programs. Its services extend to the entire profession and to society in general. New programs are developed as needs become evident. Occasionally, the Association is called upon to provide an independent review or appraisal of statistical programs or the statistical content of a critical program in science or technology.

The website for ASA and its publications is www.amstat.org.

FRED C. LEONE

AMERICAN STATISTICIAN, THE

[This entry has been updated by the Editors.]

The American Statistician is one of the principal publications of the American Statistical Association* (ASA). The *Journal of the American Statistical Association** (*JASA*) is devoted largely to new developments in statistical theory and its extensions to a wide variety of fields of application. *The American Statistician* emphasizes the professional development of ASA members by publishing articles that will keep persons apprised of new developments in statistical methodology through expository and tutorial papers dealing with subjects of widespread interest to statistical practitioners and teachers. It is published quarterly.

The origins of *The American Statistician* can be traced to the *American Statistical Association Bulletin,* which, from 1935 until 1947, served as the organ of information relating to ASA chapter activities, members, annual meetings, and employment opportunities for statisticians. At the end of World War II, the need was recognized for a more ambitious publication, providing expanded news coverage as well as serving the needs for professional development, and publication of *The American Statistician* was authorized. The first issue appeared in August 1947.

The contents of the first issue provide a preview of the major emphases that prevailed for many years. A major portion of the issue (11 of a total of 25 pages) was devoted to news items, including information about ASA programs, government statistics, other statistical societies, chapter activities, and news about members. The articles dealt with statistical applications in engineering and process control ("Statistical Engineering" by Tumbleson), electronic computing ("New High-Speed Computing Devices" by Alt), and the teaching of statistics ("A Well-Rounded Curriculum in Statistics" by Neiswanger and Allen). In addition, a note by Haemer on graphic presentation, the first in a series, appeared, and a "Questions and Answers" department to serve as a consulting forum, edited by Mosteller, was begun. The early emphasis on teaching of statistics continued throughout the period 1947–1979. Similarly, the early recognition of electronic computation was matched by continuing attention to statistical computing, with particular emphasis on this area found since the mid-1970s. The current "Statistical Computing and Graphics" section dates back to 1974.

Teaching of statistics has always been a major focus of *The American Statistician*. The section "The Teacher's Corner" was begun in 1962 and is still current in 2004. Other sections are "Statistical Practice", "General", "Reviews of Books and Teaching Materials", "Statistical Computing and Graphics", and "Letters".

The professional interests of statisticians were the focus of several early papers. In the 1970s, renewed interest in statistical consulting led to more frequent articles on this subject. It is interesting to note that an article "The Outlook for Women is Statistics," by Zapoleon, appeared in 1948.

Occasionally, special symposia papers have been published, such as a series of papers on "Reliability and Usability of Soviet Statistics" (1953) and the papers of a symposium on unemployment statistics (1955). During 1981, the "Proceedings of the Sixth Symposium on Statistics and the Environment" was published as a special issue of *The American Statistician*.

The year 1974 represented a major turning point for *The American Statistician*. Beginning in that year, all news items were moved to the new ASA publication *Amstat News*. In 1973, these items accounted for about one-third of the journal's pages, and a drop in the number of pages published from 1973 to 1974 corresponds largely to this shift. Since 1974, *The American Statistician* has devoted its pages exclusively to papers and notes on statistics.

The special departments of *The American Statistician* varied during the period 1947–1980. For example, the two initial departments "Questions and Answers" (a statistical consulting* forum) and "Hold That Line" (concerned with graphic presentation*) were terminated in 1953 and 1951 respectively. The department "Questions and Answers" was revived in 1954, but was modified from a consulting forum to a section containing essays on a variety of statistical subjects. The new "Questions and Answers" department was edited by Ernest Rubin from 1954 until 1973, when the department was terminated. Rubin wrote many of the essays published during this period.

In 1948, an editorial committee of six persons was organized to assist the editor. In 1954, this committee gave way to seven associate editors, a number that grew to 31 by the end of 2004.

The editorial review process for manuscripts is similar to that used by many other professional journals. All manuscripts that are potentially suitable for the journal are assigned to an associate editor for review.

The publication policy for *The American Statistician* is developed by the ASA Committee for Publications. The policy in effect during the late 1970s called for articles of general interest on (i) important current national and international statistical problems and programs, (ii) public policy matters of interest to the statistical profession, (iii) training of statisticians, (iv) statistical practice, (v) the history of statistics*, (vi) the teaching of statistics, and (vii) statistical computing. In addition, expository and tutorial papers on subjects of widespread interest to statistical practitioners and teachers are strongly encouraged.

The website for the journal can be accessed via that for the ASA, www.amstat.org

JOHN NETER

AMSTAT NEWS. See AMERICAN STATISTI-
CIAN, THE

ANALYSIS OF COVARIANCE

The analysis of covariance is a special form of
the analysis of variance* and mathematically
need not be distinguished from it, although
there are differences in utilization. (Any reader who is unfamiliar with the analysis of
variance is advised to read the article on that
topic before proceeding.) Using the analysis
of covariance, an experiment or other investigation is planned and the analysis of variance
is sketched out, based upon the model

$$\mathbf{y} = \mathbf{M}\theta + \eta,$$

where $\mathbf{y}$ is the vector of n data and η of
n independent residuals*, θ is a vector of p
parameters, and $\mathbf{M}$ is an $n \times p$ matrix relating to the data to the parameters. It is then
realized that there are q variates that could
be measured which might explain some of
the variation in $\mathbf{y}$, so the model is extended
to read

$$\mathbf{y} = \mathbf{M}\theta + \mathbf{D}\beta + \eta = \left(\mathbf{M} \dot{\vdots} \mathbf{D}\right)\begin{pmatrix}\theta \\ \cdots \\ \beta\end{pmatrix} + \eta,$$

where $\mathbf{D}$ is an $n \times q$ matrix of supplementary data and β is a vector of q regression
coefficients*, one appropriate to each variate.
By this extension it is hoped to improve the
estimate of θ. Following the nomenclature
usual in correlation* and regression*, the values of y make up the dependent variate and
those in the columns of $\mathbf{D}$ the independent
variates.

As has been said, that is not different
in its essentials from an ordinary analysis of
variance. Thus there is nothing novel in introducing parameters that are not themselves
under study but might serve to explain irrelevant variation. The blocks of an experimental
design* will serve as an example. Further,
the effect of blocks can be removed in either
of two ways. If there were three of them,
three block parameters could be introduced
in θ, probably with some implied constraint

to reduce them effectively to two. Alternatively, two independent variates, x_1 and x_2,
could be introduced, such that x_1 was equal
to $+1$ in block I, to -1 in block II, and to
0 in block III, while x_2 took the values $+1$,
$+1$, and -2 in the three blocks, respectively.
The outcome would be the same. Where a
variate is thus derived from characteristics
of the design rather than from measurement
it is called a "pseudo-variate"*. The device is
one that links the analyses of variance and
covariance as a single technique.

Nevertheless, the user will continue to see
them as different, usually thinking of the
analysis of variance as the form visualized at
the inception of the investigation and of the
analysis of covariance as a means of coping
with accidents and afterthoughts.

HISTORY AND DEVELOPMENT

The idea of allowing for an independent
variate originated with Fisher* [4], who
unfortunately did not appreciate that a
covariance adjustment necessarily introduces nonorthogonality. The derivation of
standard errors* of means is due to Wishart
[6]. Bartlett [1] considerably extended the
usefulness by introducing pseudo-variates for
incomplete data. Later developments have
tended to assimilate the method to the analysis of variance.

FORM OF THE CALCULATIONS

Nowadays, there are numerous computer
packages able to carry out the necessary calculations. Nevertheless, some more detailed
understanding of them can be helpful.

To take first the case of only one independent variate, the analysis of variance
for its data, x, will give a sum of squared
deviations* for error that has a quadratic
form, i.e., it can be written as $\mathbf{x}'\mathbf{Hx}$, where $\mathbf{H}$
is some positive semidefinite matrix derived
from the design. A corresponding quantity,
$\mathbf{y}'\mathbf{Hy}$, exists for the dependent variate, $\mathbf{y}$. It
will also be necessary to know $\mathbf{y}'\mathbf{Hx} = \mathbf{x}'\mathbf{Hy}$.
Then, for a single independent variate, β,
the regression coefficient of y on x, equals
$\mathbf{y}'\mathbf{Hx}/\mathbf{x}'\mathbf{Hx}$. Also, the sum of squared deviations is reduced from $\mathbf{y}'\mathbf{Hy}$ with f degrees
of freedom to $\mathbf{y}'\mathbf{Hy} - (\mathbf{y}'\mathbf{Hx})^2/(\mathbf{x}'\mathbf{Hx})$ with

$(f - 1)$ when all values of y are adjusted to a standard value of x. This new mean-squared deviation* will be written as σ^2.

In the analysis of covariance the variation in x is regarded as a nuisance because it disturbs the values of y, the variate actually under study. Accordingly, any mean of y, e.g., a treatment mean*, is adjusted to a standard value of x. If the corresponding mean of x differs from this standard by d, the mean of y needs to be adjusted by βd. Similarly, if a difference of means of y is under study and the corresponding means of x differ by d, the same adjustment needs to be applied to make the y-means comparable.

An adjustment of βd will have a variance of $\sigma^2 d^2/(\mathbf{x}'\mathbf{H}\mathbf{x})$. If no adjustment had taken place, the variance of the y-mean (or difference of y-means) would have been, say, $A(\mathbf{y}'\mathbf{H}\mathbf{y})/f$, where A is a constant derived from the design. After adjustment the corresponding figure is $[A + d^2/(\mathbf{x}'\mathbf{H}\mathbf{x})]\sigma^2$, which is not necessarily a reduction, although sometimes the advantage will be considerable.

These results are readily generalized to cover p independent variates. Let $\mathbf{C}$ be a $(p + 1) \times (p + 1)$ matrix; the first row and the first column relate to the dependent variate and the others to the independent variates taken in some standard order. The element in the row for variate u and the column for variate v is $\mathbf{u}'\mathbf{H}\mathbf{v}$. Then writing $\mathbf{C}$ in partitioned form,

$$\mathbf{C} = \begin{pmatrix} \mathbf{Y} & \mathbf{P}' \\ \mathbf{P} & \mathbf{X} \end{pmatrix},$$

the new error sum of squared deviations is $Y - \mathbf{P}'\mathbf{X}^{-1}\mathbf{P}$ with $(f - p)$ degrees of freedom, thus giving σ^2, and the vector of regression coefficients, β, is $\mathbf{X}^{-1}\mathbf{P}$. If an adjustment of $\beta'\mathbf{d}$ is applied to a mean, it will have a variance of $\mathbf{d}'\mathbf{X}^{-1}\mathbf{d}\sigma^2$.

Some special points need attention. For example, in the analysis of variance some lines, e.g., the treatment line in the analysis of data from an experiment in randomized blocks*, can be obtained directly without a second minimization. This is not so when covariance adjustments are introduced; it is necessary first to find $E = \mathbf{y}'\mathbf{H}\mathbf{y} - (\mathbf{x}'\mathbf{H}\mathbf{y})^2/\mathbf{x}'\mathbf{H}\mathbf{x}$ and then to ignore treatments in order to find $\mathbf{y}\mathbf{H}'_0\mathbf{y}$, $\mathbf{x}'\mathbf{H}_0\mathbf{y}$, and $\mathbf{x}'\mathbf{H}_0\mathbf{x}$,

where $\mathbf{H}_0$ is some other matrix, and to attribute $(E_0 - E)$ to treatments, where $E_0 = \mathbf{y}'\mathbf{H}_0\mathbf{y} - (\mathbf{x}'\mathbf{H}_0\mathbf{Y})^2/\mathbf{x}'\mathbf{H}_0\mathbf{x}$; i.e., it is necessary to allow for the possibility that β_0, the regression coefficient when treatments are included with error, will be different from β, the regression coefficient* based on error alone. Some have argued against this complication on the grounds that the two cannot really be different, but reflection shows that they could be. In an agricultural field experiment, for example, the error may derive chiefly from differences in available nutrient from one part of the field to another. If treatments are the exaggeration of such differences by fertilizer applications, all may be well, but if they are something entirely different, such as pruning or a change of variety, it is not reasonable to expect that β_0 will be the same as β. Now that computation is so easy, the complication should be accepted at all times. It is, in any case, required by the mathematics.

Similar complications arise in split-plot* situations, because the two regression coefficients derived from the two error lines, one for main plots and the other from subplots, are often different. Since the two errors can be made up from quite different sources, i.e., the balance of different kinds of uncontrolled variation depends on the plot size, a comparison of the two regression coefficients can be illuminating. The difficulties are chiefly those of presentation and are much relieved if an intelligible explanation can be given for why adjustments, which may necessarily depend upon different regressions, behave as they do.

NUMERICAL EXAMPLE

The experiment described in the article on the analysis of variance has an available independent variate, namely x, the number of boxes of fruit, measured to the nearest tenth of a box, for the four seasons previous to the application of treatments. Full data are set out in Table 1. There need be no difficulty about the sums of squared deviations. The sums of products of deviations are here found simply by multiplying corresponding deviations from each plot and adding. In general, wherever in the calculation of sums of squared deviations a function of x or y

Table 1. Yields from a Soil Management Trial on Apple Trees

Treatment	Block[a]							
	I		II		III		IV	
	x	y	x	y	x	y	x	y
A	8.2	287	9.4	290	7.7	254	8.5	307
B	8.2	271	6.0	209	9.1	243	10.1	348
C	6.8	234	7.0	210	9.7	286	9.9	371
D	5.7	189	5.5	205	10.2	312	10.3	375
E	6.1	210	7.0	276	8.7	279	8.1	344
S	7.6	222	10.1	301	9.0	238	10.5	357

[a]x represents boxes of fruit per plot in the 4 years preceding the application of treatments; y, the crop weight in pounds during the 4 years following.
Source: These data were first presented by S. C. Pearce [5] and have been considered in some detail by D. R. Cox [3] and C. I. Bliss [2].

is squared, the sum of products of deviations is found by multiplying the function of x by the corresponding function of y. To cope with the plot feigned to be missing, it will be convenient to use a pseudo-variate, w, that has the value 1 for treatment A in block 1, and 0 elsewhere. Once that is done it does not matter what values for x and y are assigned to the missing plot*. Here, where it is intended to calculate the analysis with the plot first included and later excluded, it will be convenient to use the actual values throughout—8.2 and 287, respectively.

Allowing only for blocks, sums of squares and products are:

	y	w	x
y	24,182	51.50	710.1
w	51.50	0.8333	1.100
x	710.1	1.100	31.63

Allowing for treatments as well, they are:

	y	w	x
y	23,432	42.75	688.3
w	42.75	0.6250	0.958
x	688.3	0.958	24.23

Ignoring w and the missing plot for the moment, the sum of squared deviations allowing only for blocks is

$$24{,}182 - (710.1)^2/31.63 = 8240$$

with 19 degrees of freedom. Allowing for treatments also, it is

$$23{,}432 - (688.3)^2/24.23 = 3880$$

with 14 degrees of freedom. The analysis of variance now reads

Source	d.f.	Sum of Squares	Mean Square	F
Treatments	5	4360	872	3.15
Error	14	3880		
Treatments + error	19	8240		

The new picture is very different. The F-value* of 3.15 is significant ($P < 0.05$). Clearly, the independent variate has effected an improvement. If there is any doubt, an analysis of variance establishes it, namely,

Source	d.f.	Sum of Squares	Mean Square	F
Regression	1	19,552	19,552	70.58
Error	14	3,880	277	
Regression + error	15	23,432		

The regression coefficient is $688.3/24.23 = 28.41$. Adjusting treatment means for y to a standard value of $x = 8.308$ (its mean), that for A is

$$274.5 - 28.41(8.450 - 8.308) = 280.5$$

and for **S**

$$279.5 - 28.41(9.300 - 8.308) = 251.3.$$

The last figure shows a large effect of the adjustment. It appears that the randomization* had assigned treatment **S** to some heavily cropping trees, as the values of x show, and it had therefore shown up better than it should. To take the difference between the adjusted means of treatments **A** and **S**, i.e., $29.2 = 280.5 - 251.3$, the standard error is

$$\sqrt{277 \left\{ \frac{1}{4} + \frac{1}{4} + \frac{(9.300 - 8.450)^2}{24.23} \right\}}$$

$$= 12.1,$$

which suggests that A is, in fact, a more fruitful treatment than S. The other treatments can be investigated in the same way. The covariance adjustment has had two beneficial effects. It has markedly reduced the error variance and it has given better-based treatment means.

The case of the "missing plot" could have been dealt with in the same way. Adjusting y by w instead of x, the error sum of squared deviations would have been $23,432 - (42.75)^2 / 0.6250 = 20,508$ with 14 degrees of freedom, as before. For treatments and error together, the corresponding figure would have been $24,182 - (51.50)^2 / 0.8333 = 20,999$ with 19 degrees of freedom, which leaves 491 with 5 degrees of freedom for treatments. The error variance is now $1465 = (20,508/14)$. The method has the advantage of giving easily a figure for the standard error of the difference between treatment means. Thus, that for treatment A and any other treatment is

$$\sqrt{1465 \left\{ \frac{1}{4} + \frac{1}{4} + \frac{(0.2500 - 0.0000)^2}{0.6250} \right\}}$$

$$= 29.6$$

There is, however, no objection to using two independent variates, e.g., both w and x. In that case the new sum of squared deviations ignoring treatments is

$$24,182 - (51.50 \quad 710.1) \begin{pmatrix} 0.8333 & 1.100 \\ 1.100 & 31.63 \end{pmatrix}^{-1}$$

$$\times \begin{pmatrix} 51.50 \\ 710.1 \end{pmatrix}$$

$$= 7295 \text{ with 18 degrees of freedom.}$$

For the error alone the corresponding figure is 3,470 with 13 degrees of freedom. That leads to the following analysis of variance:

Source	d.f.	Sum of Squares	Mean Square	F
Treatments	5	3825	765	2.87
Error	13	3470	267	
Treatments + error	18	7295		

CHOICE OF INDEPENDENT VARIATES

Modern computer packages permit the simultaneous use of several independent variates, and thus they extend the usefulness of the techniques. Nevertheless, the temptation to introduce every available independent variate is to be resisted.

The original use of covariance was to effect adjustments where some disturbing factor had not been controlled. That remains the most common application. For example, a suspicion may arise that a thermostat is not functioning properly. Pending its replacement, the investigator may introduce periodic measurements of temperature, which are then used as an independent variate in the analysis of data. Assuming that there is a straight-line relationship between the data, y, and the temperature, the outcome might be much the same apart from the loss of a degree of freedom from the error and an arbitrary loss of regularity in the standard errors*. Again, in an agricultural context it may be found that an experiment has been laid down on variable soil, a crisis that could be resolved by a covariance adjustment on measurements of soil texture or soil acidity. Sometimes, too, there are quantities that cannot be controlled precisely. It is not possible, for example, to grow test animals of exactly the same body weight or to find patients with identical blood counts. In these instances covariance adjustments may be used as a matter of course.

At one time the technique had a bad reputation, arising, it is said, from one much-publicized example. The story goes that a field experiment was conducted on the yield of barley, a covariance adjustment being made on germination rates. As a result the error sum of squared deviations* was gratifyingly reduced, but the treatment effects also disappeared. An obvious explanation is that the treatments had affected yield by way of the germination rates and in no other manner. If so, the conclusion could have been of some importance. The story is not a warning against using covariance adjustments at all; indeed, it shows their value in revealing mechanisms, but it does warn against facile interpretations, especially when the independent variate does not justify its name but is dependent upon the treatments.

If the covariate is measured before the application of treatments, which are then allocated at random, no question need arise. Nor does any arise when the aim is to find what the treatments do to the dependent variate apart from the obvious indirect effect through the independent. Sometimes, however, the user cannot be sure whether the treatments do or do not affect the independent variate*. It is wise in such cases to be very cautious. It is true that subjecting the figures to an analysis of variance may decide the matter. If, however, the proposed independent variate is an array of ones and zeros, indicating the presence or absence of some feature, the user is unlikely to obtain any useful guidance. Even a more amenable variate may give an inconclusive response.

CONSTANCY OF THE REGRESSION COEFFICIENT

It is by no means obvious that the regression coefficients will be independent of the treatments, and in some instances the assumption may verge on the absurd. Occasionally, the difficulty can be met by transformation of the variates. Where that is not feasible, an algebraic solution by least squares is usually not possible. Thus, with data from a designed experiment it is simple to fit separate regression coefficients if the design is completely randomized but not if there are blocks, as usually there will be. Although with the aid of a computer some kind of minimization can often be achieved, it is open to question whether a constant block effect regardless of treatments in conjunction with variable regression coefficients makes a convincing and realistic model.

ACCIDENTS AND MISHAPS

It has been recognized for a long time that the analysis of covariance provides a theoretically sound way of dealing with data from damaged experiments. For example, suppose that an experiment has been designed in such a way that a program is available for calculating an analysis of variance, but m plots (i.e., units) have been lost. (It must be reasonable to assume that the loss is not a result of the treatments that have been applied.) Each gap in the data is filled with a convenient value such as zero, the treatment mean, or even an arbitrary number. It is now required to estimate the deviation between the value for the missing plot given by the method of least squares and the value that has been assigned. That is done by writing down a pseudo-variate for each missing value. It equals zero for all plots except the one to which it refers, when it equals 1. A covariance adjustment on the pseudo-variates will give a correct analysis, the regression coefficient of the dependent variate on any pseudo-variate being minus the required deviation for the plot. The method has several advantages. For one thing, unlike many methods for dealing with incomplete data, it gives a correct F-value for any effect. For another, it gives correct standard errors for treatment contrasts*. Also, it obtains degrees of freedom without special adjustment.

A similar problem can arise when the data for any plot (or unit) is the sum of an unspecified number of observations, such as weighings. If someone makes a mistake, there can be doubt whether a certain observation belongs to this plot or that. The difficulty can sometimes be resolved by attributing it first to one and then to the other. If residuals look extraordinary in one case but are unremarkable in the other, the difficulty is over, but sometimes doubt will remain. A similar problem arises when samples are taken, say for chemical analyses, and some labels are lost. The samples can still be analyzed and a total found. In both examples it is possible to state the total for the two plots without knowing how to apportion it between them. It suffices to attribute the total to one plot and zero to the other and to adjust by a pseudo-variate equal to +1 and −1 for the two plots and to zero for all others. If three plots are involved in the muddle, two pseudo-variates are required. The total is attributed to one plot and zero to the others. The first pseudo-variate equals $+2$ for the plot with the total, -1 for the other two involved, and zero for the others. It therefore serves to apportion a correct amount to the plot credited with the total. A second pseudo-variate apportions the rest between the other two affected plots, being equal to $+1$ and -1 for those plots and to zero for all others. The method can be extended easily to more complicated cases.

Adjustments of this sort can be made in conjunction. Thus provided that the program can manage so many, it is permissible to have some independent variates for adjustment by related quantities, others to allow for missing values, and still others to apportion mixed-up values. All these adjustments can be made correctly and simultaneously.

TRANSFORMATION OF VARIATES

It should be noted that the independent variates, despite their name, play a role in the model analogous to that of **M** rather than that of y, i.e., no assumptions are involved about their distributions, which are perhaps known, but nothing depends upon them. Accordingly, there is no need to seek variance-stabilizing transformations* for them. It is, however, still necessary to consider if y needs one, since not only must the elements of η be distributed independently but also they should have equal variances. In the case of the independent variates, the need is for them to be linearly related to y (or to the transformation of y) and that may call for a transformation of a different kind. Alternatively, it may be desirable to introduce the same variate twice but in different forms. Thus, it has already been mentioned that a field experiment on variable land might be improved by an adjustment on soil acidity. However, unless the species is one that favors extremely acid or extremely alkaline soils, there will almost certainly be an optimal value somewhere in the middle of the range of acidity and it would be sensible to introduce both soil pH and its square to allow the fitting of a parabola.

The correct choice of transformation for the independent variate is especially important if the intention is to enquire how far treatments affect the dependent variate other than through the independent. It is then essential to fit the right relationship; an inept choice of transformations can do harm.

An additional variate does little harm but it is not wise to load an analysis with adjustments in the hope of something emerging. A further independent variate should be included only for good reason, but it should not be omitted if there are good reasons for regarding it as relevant. If it does nothing, there should be reserve about taking it out again. Clearly, if variates are included when they reduce the error variance and excluded if they do not, bias must result. Also, in presenting results it is more convincing to report that some quantity or characteristic was allowed for but had in fact made little difference than to ignore it.

LEVEL OF ADJUSTMENT OF AN INDEPENDENT VARIATE

Basically, the method consists of estimating the partial regression coefficients* of the dependent variate upon each of the independent variates and then adjusting the dependent variate to correspond to standard values of the independent variates. What these standard values should be requires some thought.

First, as long as only linear relations are in question, it does not much matter, because differences in adjusted means of the dependent variate will be unaffected, although not the means themselves. If, however, the first independent variate is some measured quantity, x, and the second is x^2 introduced to allow for curvature, a computer package, which does not know that the two are related, will adjust the first to $\bar{x}$, the mean of x, and the second to $(\overline{x^2})$, which will not be the same as $(\bar{x})^2$. Probably little harm will have been done, but the point needs to be noted.

Pseudo-variates do not usually cause much trouble with their standard values as long as only differences are in question. When they are used for missing plots, the adjustment should be to zero, corresponding to the presence of the plot and not to a mean value, if actual means of the dependent variate are to relate to those given by other methods.

INTERPRETATION OF AN ANALYSIS OF COVARIANCE

An analysis of covariance having been calculated, the first step usually is to look at the error-mean-squared deviation to see if it is as small as was hoped. If it is not, there are two possibilities. One is that, as with the analysis of variance, some important source of variation has been left in error; the other comes from the independent variates having had little effect. The latter case needs to be noted because a research team can go on using adjustments believing them to be a sovereign remedy, even though in fact they do no good. To test the matter formally, it is sufficient to carry out an F-test* using the two minimizations provided by the analysis of variance with and without the independent variates. Sometimes only one of the independent variates is in question; it may be helpful to repeat the calculations omitting that variate to see if it has really had a useful effect.

The testing of effects is the same as for the analysis of variance. They should be examined in logical order and tables prepared to show all effects of importance. Where there is any possibility of an independent variate having been affected by the treatments, it may be advisable to examine the position using the analysis of variance.

The standard error of a treatment mean depends party upon the design and the error variance, as in the analysis of variance, and partly on the magnitude of the adjustments that have been required. In one sense these modifications of standard errors are a help. Cases commonly dealt with using pseudovariates, missing plots, for example, require modifications that are not easy to make except by covariance and then they are made automatically. On the other hand, a measured independent variate will give arbitrary variation in the standard errors, which can be very awkward, for instance, in a multiple comparison test*. Incidentally, some computer packages disguise the situation by giving a common mean standard error for a set of quantities which, but for the adjustments, would all have been determined with the same precision. Although often convenient, the practice can also be misleading.

REFERENCES

1. Bartlett, M. S. (1937). *J. R. Statist. Soc. Suppl. 4*, 137–183.
2. Bliss, C. I. (1967). *Statistics in Biology*, Vol. 2. McGraw-Hill, New York, Chap. 20.
3. Cox, D. R. (1958). *Planning of Experiments.* Wiley, New York, Chap. 4.
4. Fisher, R. A. (1935). *The Design of Experiments.* Oliver & Boyd, Edinburgh. (A passage was added to later editions of *Statistical Methods for Research Workers* prior to the appearance of *The Design of Experiments*.)
5. Pearce, S. C. (1953). *Field Experimentation with Fruit Trees and Other Perennial Plants.* Commonwealth Bureau of Horticulture and Plantation Crops, Farnham Royal, Slough, England, App. IV.
6. Wishart, J. (1936). *J. R. Statist. Soc. Suppl. 3*, 79–82.

FURTHER READING

The general literature on the subject is rather scanty. *Biometrics** devoted Part 2 of Volume 13 (1957) to a series of papers on the analysis of covariance. Later *Communications in Statistics** similarly devoted Volume A8, Part 8 (1979) to the topic. There is a valuable account of the method in C. I. Bliss's *Statistics in Biology*, Volume II, Chapter 20 (McGraw-Hill, New York). Also, there are short but illuminating descriptions by D. J. Finney in *An Introduction to Statistical Science in Agriculture* (Munksgaard, Copenhagen, 1962) and by D. R. Cox in Chapter 4 of *Planning of Experiments* (Wiley, New York, 1958).

See also AGRICULTURE, STATISTICS IN; ANALYSIS OF VARIANCE; DESIGN OF EXPERIMENTS: INDUSTRIAL AND SCIENTIFIC APPLICATIONS; FACTOR ANALYSIS; GENERAL LINEAR MODEL; and REGRESSION (Various Entries).

S. C. PEARCE

ANALYSIS OF MEANS FOR RANKS (ANOMR)

The analysis of means for ranks (ANOMR) procedure, attributed to Bakir [1], assumes k independent samples of sizes n_i, $i = 1, \ldots, k$. Observations

$$X_{ij}, i = 1, \ldots, k; j = 1, \ldots, n_i,$$

are selected from k continuous populations that may differ only in their location parameters $\mu_i, i = 1, \ldots, k$. We test $H_0 : \mu_1 = \mu_2 = \cdots = \mu_k$ versus the alternative that not all μ_is are equal.

Replacing X_{ij} by its rank R_{ij} in the combined sample of size $N = \sum_{i=1}^{K} n_{i-1}$, we calculate

$$\overline{R}_i = \sum_{j=1}^{n_i} (R_{ij}/n_i).$$

The null hypothesis is rejected if for any i

$$|\overline{R}_i - \overline{R}| \geqslant C,$$

where $\overline{R} = \frac{1}{2}(N + 1)$ is the grand mean of all the ranks, and where the critical value C is a function of the size α of the test, of k, and of $n_1, \ldots, n_k$. Tables of values of C for selected combinations for $\alpha, k = 3, 4$, and for small values of n_i have been provided [1]. For equal sample sizes ($n_i = n, i = 1, \ldots, k$), a Bonferroni* approximation based on the Wilcoxon rank sum* statistic is available [1], which turns out to be satisfactory when α is in the vicinity of 0.072.

An asymptotically normal procedure is suggested [1]. Specifically, under H_0 and when the sample sizes are equal, each R_i has the expected value $(N + 1)/2$ and variance

$$
\begin{aligned}
\sigma^2 &= \frac{\text{Var}(R_{ij})}{n} \cdot \left(\frac{N - n}{N - 1} \right) \\
&= \frac{(N - n)(N + 1)}{12n} \\
&= \frac{(k - 1)(kn + 1)}{12},
\end{aligned}
$$

since $\text{Var}(R_{ij}) = (N^2 - 1)/12$. Note that H_0 here corresponds to the classical Kruskal–Wallis test* that, however, involves

more general alternatives than does the ANOMR procedure here.

The standard differences $W_i = |\overline{R}_i - \overline{R}|/\sigma$ define the random vector $\boldsymbol{W} = (W_1, \ldots, W_k)'$. As $N \to \infty$, the limiting distribution of $\boldsymbol{W}$ is a (singular) multivariate distribution with equicorrelated structure, given by

$$\text{Corr}(W_i, W_j) = 1/(k - 1).$$

The large-sample ANOMR test is based on the following rule: reject H_0 if for any i

$$|W_i| \geqslant w,$$

where the critical value w is obtained from the analysis of means (ANOM) table of Nelson [2] with appropriate values of k and infinite degrees of freedom. Further details are provided in Reference 3.

REFERENCES

1. Bakir, S. T. (1989). Analysis of means using ranks. *Commun. Stat. Simul. Comp.*, **18**, 757–775.

2. Nelson, L. S. (1983). Exact critical values for the analysis of means. *J. Qual. Tech.*, **15**, 40–44.

3. Wludka, P. S. and Nelson, P. R. (1999). Two non-parametric, analysis-of-means-type tests for homogeneity of variances. *J. Appl. Stat.*, **26**, 243–256.

See also KRUSKAL–WALLIS TEST.

ANALYSIS OF VARIANCE

The analysis of variance is best seen as a way of writing down calculations rather than as a technique in its own right. For example, a significance of a correlation coefficient*, r, based on n pairs of observations, can be tested in several ways. Using the analysis of variance the calculations are set out thus:

Source	Degrees of Freedom	Sum of Squared Deviations	Mean Squared Deviation
Correlation	1	Sr^2	Sr^2
Error	$n - 2$	$S(1 - r^2)$	$S(1 - r^2)/(n - 2)$
Total	$n - 1$	S	

The significance is then judged by an F-test*, i.e., from the ratio of the two mean squared deviations, which is $(n-2)r^2/(1-r^2)$ with one and $(n-2)$ degrees of freedom. By using a standard format, standard tests are available, thus obviating the need for numerous formulas and a range of tables, but in general the same conclusions will be reached however the calculations are set out.

HISTORY AND DEVELOPMENT

In its origins the method was devised by Fisher [1] for the study of data from agricultural experiments. At first the only designs, i.e., randomized blocks* and Latin squares*, were orthogonal, but later F. Yates [4] showed how to deal with those that were nonorthogonal, like balanced incomplete blocks*, and those with a factorial structure of treatments [5]. From this point development was rapid, so that the analysis of variance was being used for a wide range of problems that involved the studying of a linear hypothesis*. The original problem of analyzing data from block designs acquired a new dimension from the use of matrices, due to Tocher [3], and the solution of the normal equations in terms of generalized inverses*. *See* GENERAL LINEAR MODEL.

THE METHOD DESCRIBED

In essentials the method depends upon the partition of both degrees of freedom and the sums of squared deviations between a component called "error" and another, which may be termed the "effect," although generally it will have a more specific name. Thus, in the example above the effect was the correlation. The nomenclature should not mislead. The sum of squared deviations for the effect is influenced by error also, which is thought of as an all-pervading uncertainty or noise* distributed so that, in the absence of the effect i.e., on the null hypothesis*, the expectation of the two sums of squared deviations will be in the ratio of their respective degrees of freedom. Hence the mean squared deviations, i.e., the sums of squares divided by their degrees of freedom, should have similar expectations. If, however, the effect does exist, it will inflate its own mean squared

deviation but not that of the error, and if large enough, will lead to significance being shown by the F-test. In this context, F equals the ratio of the mean squared deviations for the effect and for error.

In practice, analyses of variance are usually more complicated. In the example, the total line was obtained by minimizing the sum of squared residuals*, $\sum_i \eta_i^2$, in

$$y_i = \alpha + \eta_i,$$

whereas that for error came from minimizing a similar quantity in

$$y_i = \alpha + \beta x_i + \eta_i.$$

In short, the test really investigated the existence or non-existence of β. In the example α is common to both minimizations, representing as it does a quantity needed to complete the model but not itself under test, whereas β was in question. That is the general pattern. Thus in a randomized block design the blocks form a convenient "garbage can" where an ingenious experimenter can dispose of unwanted effects such as spatial position, different observers, sources of material, and much else that would disturb the experiment if not controlled. Consequently, they must be allowed for, although no one is studying the contents of garbage cans. There will also be parameters for treatments*, which are under study. Minimization with respect to the block parameters alone gives a measure of the remaining variation, i.e., that due to treatments and uncontrollable error. A further minimization on block and treatments parameters together gives the error line, that for treatments being found by difference. (The fact that it can be found more easily by direct calculation obscures its real origins.) The block line, relating as it does to variation that has been eliminated, is not really relevant but is ordinarily included.

Such an analysis is called "intrablock"*, studying as it does variation within blocks and discarding any between them. In some instances it is possible to derive an "interblock"* analysis, in which the block parameters are regarded as random variables. The procedure then is to minimize the sum of their squares, both when the treatment parameters are included and when they are excluded.

The additional information can be worthwhile, but not necessarily so. For example, if each block is made up in the same way with respect to treatments, a study of the differences between blocks can provide no information about the effects of treatments. Also, unless the number of blocks appreciably exceeds that of treatments, there will not be enough degrees of freedom to determine the interblock error properly. Not least, if good use has been made of blocks as garbage cans, the distribution of their parameters must be regarded as arbitrary.

Complications arise when there are several effects. Here it is advisable to form the error allowing for them all, although that will be considered in more detail below. The problems arise rather in deciding the order of testing, but that is often a matter of logic rather than statistics. For example, with a factorial design* of treatments, if it appears that there is an interaction* of factors A and B, the conclusion should be that the response to the various levels of A depends on the level of B, and *vice versa*. If that is so, there is no point in examining the main effects of factors A and B, since each relates to the response to the levels of one factor when it has been averaged over levels of the other. The only true interpretation must rest upon a two-way table* of means. Again, if the example is extended to cover parabolic effects, i.e.,

$$y_i = \alpha + \beta x_i + \gamma x_i^2 + \eta_i,$$

and if it appears that γ should be included in the model, i.e., the relationship of x_i and y_i is not a straight line, there need be no detailed study of β, since it has no meaning except as the slope of such a line. However, it is always necessary to consider what really is under test. For example, it could be that

$$y_i = \alpha + \gamma x_i^2 + \eta_i$$

was the expected relationship and doubts had arisen whether βx_i was not needed as well. The analysis of variance is an approach of wonderful subtlety, capable of adaptation to a wide range of problems. It is used to best advantage when it reflects the thinking and questioning that led to the inception of the investigation in the first place. Consequently, each analysis should be individual and should

Table 1. Yields in Pounds per Plot over Four Seasons from an Experiment on Soil Management with Apple Trees

Treatment	Blocks				Totals
	I	II	III	IV	
A	287	290	254	307	1138
B	271	209	243	348	1071
C	234	210	286	371	1101
D	189	205	312	375	1081
E	210	276	279	344	1109
S	222	301	238	357	1118
Totals	1413	1491	1612	2102	6618

study each question in logical order. *There is no place for automated procedures, as if all research programs raised the same questions.* Also, although the analysis of variance had its origins in the testing of hypotheses*, there is no reason for leaving it there. It can shed light on the sources of experimental error*; it can suggest confidence limits* for means and differences of means and much else. In the hands of a thoughtful user it has unimagined potentiality; as an unthinking process it leads to few rewards.

NUMERICAL EXAMPLE

The data in Table 1 [2] represent yields per plot from an apple experiment in four randomized blocks, I through IV. There were six treatments. One of them, S, was the standard practice in English apple orchards of keeping the land clean during the summer, letting the weeds grow up in the fall, and turning them in for green manure in the spring. The rest, A through E, represented alternative methods in which the ground was kept covered with a permanent crop. The interest then lay in finding out if any of the other methods showed any improvement over S.

It is first necessary to find the sum of squared deviations ignoring treatments and considering only blocks. The estimated value for each plot, i, is

$$y_i = \beta_j + \eta_i,$$

where β_j is the parameter for the block, j, in which it finds itself. It will quickly appear that the sum of $\eta_i^2 = (y_j - \beta_j)^2$ is minimized where β_i is taken to be the appropriate block mean, i.e.,

$$\beta_1 = 235.50, \quad \beta_2 = 248.50,$$

$$\beta_3 = 268.67, \quad \beta_4 = 350.33.$$

With these values known it is possible to write down η_i for each plot; e.g., those for blocks I and II and treatments A and B are

$$
\begin{array}{ll}
51.50 & 41.50 \ \ldots \\
35.50 & -39.50 \ \ldots
\end{array}
$$

$$\vdots \quad \vdots$$

The sum of these quantities squared comes to 24,182; it has $20(= 24 - 4)$ degrees of freedom, since there are 24 data to which four independent parameters have been fitted. It is now required to fit treatment parameters as well, i.e., to write

$$y_i = \beta_j + \gamma_k + \eta_i.$$

In a randomized block design* in which all treatment totals are made up in the same way with respect to block parameters, i.e., the design is orthogonal*, it is sufficient to estimate a treatment parameter, γ_k, as the difference of the treatment mean from the general mean. The table of deviations, η_i, now starts

$$
\begin{array}{ll}
42.75 & 32.75 \ \ldots \\
43.50 & -31.50 \ \ldots
\end{array}
$$

$$\vdots \quad \vdots$$

The sum of these squares is now 23,432 with 15 degrees of freedom, because an additional five degrees of freedom have been used to estimate how the six treatment means diverge from the general mean. (Note that only five such quantities are independent. When five have been found, the sixth is known.) The analysis of variance is therefore

Source	d.f.	Sum of Squares	Mean Square
Treatment	5	750	150
Error	15	23,432	1562
Treatments + error	20	24,182	

There is no suggestion that the treatment mean square has been inflated relative to the error, and therefore no evidence that the treatments in general have had any effect. However, the study really relates to comparisons between A through E and S. In view of the orthogonality of the design and the fact that each treatment mean is based on four data, the variance of a difference of two means is $(\frac{1}{4} + \frac{1}{4})1562 = 781$, the standard error being the square root of that quantity, i.e., 27.9. There is no question of any other treatment being an improvement on the standard because all except A give smaller means. (However, the situation can be changed; *see* ANALYSIS OF COVARIANCE.)

The analysis above was for an orthogonal design. If, however, the datum for treatment A in block 1 had been missing, a more complicated situation would have arisen. (It is true that in practice a missing plot value would be fitted, but it is possible to carry out a valid analysis of variance without doing that.) The deviations allowing only for blocks start

$$
\begin{array}{ll}
— & 32.75 \ \ldots \\
45.80 & -47.50 \ \ldots
\end{array}
$$

$$\vdots \quad \vdots$$

the mean for block I being now 225.2. The sum of squared deviations is 20,999 with 19 degrees of freedom. The so-called normal equations*, derived from the block and treatment totals, are

$$
\begin{aligned}
1126 &= 5\beta_1 + \sum \gamma - \gamma_1 & 1138 &= \sum \beta - \beta_1 + 3\gamma_1 \\
1491 &= 6\beta_2 + \sum \gamma & 1071 &= \sum \beta \quad + 4\gamma_2 \\
1612 &= 6\beta_3 + \sum \gamma & 1101 &= \sum \beta \quad + 4\gamma_3 \\
2102 &= 6\beta_4 + \sum \gamma & 1081 &= \sum \beta \quad + 4\gamma_4 \\
& & 1109 &= \sum \beta \quad + 4\gamma_5 \\
& & 1118 &= \sum \beta \quad + 4\gamma_6,
\end{aligned}
$$

where $\sum \beta = \beta_1 + \beta_2 + \beta_3 + \beta_4$ and $\sum \gamma = \gamma_1 + \gamma_2 + \gamma_3 + \gamma_4 + \gamma_5 + \gamma_6$. At this point it is necessary to say that no one in practice solves the normal equations as they stand because better methods are available, e.g., the Kuiper–Corsten iteration* or the use of generalized inverses*. However, there is no objection in principle to a direct solution, the difficulty being that there are not enough equations for the parameters. An equation of constraint, e.g.,

$$3\gamma_1 + 4\gamma_2 + 4\gamma_3 + 4\gamma_4 + 4\gamma_5 + 4\gamma_6 = 0,$$

can always be used. Further, the one just suggested, which makes the treatment parameters sum to zero over the whole experiment, is very convenient. It follows that

$$\beta_1 = 224.34 \quad \gamma_1 = -5.74$$
$$\beta_2 = 248.74 \quad \gamma_2 = -5.39$$
$$\beta_3 = 268.91 \quad \gamma_3 = 2.11$$
$$\beta_4 = 350.57 \quad \gamma_4 = -2.89$$
$$\gamma_5 = 4.11$$
$$\gamma_6 = 6.36.$$

Hence subtracting the appropriate parameters from each datum, the values of η_i are

$$\begin{matrix} — & 47.00 & \dots \\ 52.05 & -34.35 & \dots \\ \vdots & \vdots & \end{matrix}$$

The sum of their squares is now 20,508 with 14 degrees of freedom, yielding the following analysis of variance:

Source	d.f.	Sum of Squares	Mean Square
Treatments	5	491	98
Error	14	20,508	1465
Treatments + error	19	20,999	

MULTIPLE COMPARISONS*

Over the years there has grown up an alternative approach to testing in the analysis of variance. As long as there are only two treatments or two levels of a factor, the F-test has a clear meaning but if there are three or more, questions arise as to where the differences are. Some care is needed here, because background knowledge is called for. If, for example, it had appeared that different varieties of wheat gave different percentages of a vegetable protein in their grain, the result would surprise no one and would merely indicate a need to assess each variety separately. At the other extreme, if the treatments formed a highly structured set, it might be quite obvious what should be investigated next. Thus if it had appeared that the outcome of a chemical reaction depended upon

the particular salt used to introduce a metallic element, the science of chemistry is so developed that a number of lines of advance could probably be suggested immediately, some of which might receive preliminary study from further partition of the treatment line. (Of course, no body of data can confirm a hypothesis suggested by itself.) Sometimes, however, the experimenter is in the position of suspecting that there could be a structure but he does not know what it is. It is then that multiple comparison tests have their place. Like much else in the analysis of variance, their unthinking use presents a danger, but that is not to deny them a useful role. Also, an experimenter confronted with a jumble of treatment means that make no sense could well adopt, at least provisionally, the treatment that gave the highest mean, but would still want to know how it stood in relation to its nearest rivals. It might be so much better that the others could be discarded, or it might be so little better that further study could show that it was not really to be preferred. Such a test can be useful. Like others, it can be abused.

COMPOSITION OF ERROR

In its original form the analysis of variance was applied to data from agricultural field experiments designed in randomized blocks. The error was then clearly identified as the interaction of treatments and blocks, i.e., the F-test investigated the extent to which treatment differences were consistent from block to block. If a difference of means was much the same in each block, it could clearly be relied upon; if it had varying values, the significance was less well established. In other instances the error may be the interaction of treatments and some other factor such as occasion or operator. In all such cases the purport of the test is clear.

Sometimes, however, the error is less open to interpretation, being little more than a measure of deviations from a hypothesis that is itself arbitrary. Thus, if the agricultural field experiment had been designed in a Latin square*, the error would have been made up of deviations from parameters for treatments added to those for an underlying fertility pattern, assumed itself to derive from additive effects of rows and columns. If, as is usually the case, there is no prior reason why the fertility pattern should have that form, there is a danger of the error sum of squared deviations being inflated by sources that will not affect the treatment line, thus reversing the usual position. In fact, it is not unknown for the error mean square deviation to be the larger, and sometimes there is good reason for it.

The subject of error is sometimes approached by distinguishing fixed effects* from random*. In the first, the levels are determinate and reproducible, like pressure or temperature, whereas in the latter they are to be regarded as a random selection of possible values, like the weather on the successive days of an investigation. For many purposes the distinction is helpful. The ideal, as has been suggested, is an error that represents the interaction between a fixed effect and a random effect of the conditions over which generalization is required, but other possibilities can be recognized. For example, an error made up of interactions between fixed effects is nearly useless unless it can be assumed that there will be no real interaction.

Further questions arise when confounding* is introduced. The experimenter may have decided that there can be no interaction of factors A, B, and C and would then be ready to have it confounded. The experimenter may get rid of it in that way if possible, but if it remains in the analysis it should be as part of error. However, if the experimenter believes that $A \times B \times C$ never exists, he or she cannot believe that its value depends upon the level of D, so $A \times B \times C \times D$ also should either be confounded or included in error. Such decisions can be made readily enough before analysis begins; they are more difficult afterward. Plainly, it would be wrong to start with an error that was acceptable and add to it those high order interactions* that were small and to exclude from it those that were large. The error that finally resulted would obviously be biased*. On the other hand, there are occasions when sight of the data convinces the experimenter that his or her preconceptions were wrong. In that case the experimenter usually does well to confine the analysis to an indubitable

error, e.g., the interaction of blocks and treatments, and to regard all else as subject to testing. As with the Latin square*, mistaken assumptions about the composition of error can lead to the inflation of its sum of squared deviations.

The question of what is and what is not error depends to some extent upon the randomization*. Let there be b blocks of xy treatments, made up factorially by x levels of factor X and y levels of factor Y. The obvious partition of the $(bxy - 1)$ degrees of freedom between the data is

Blocks	$(b - 1)$
X	$(x - 1)$
Y	$(y - 1)$
$X \times Y$	$(x - 1)(y - 1)$
Blocks $\times X(I)$	$(b - 1)(x - 1)$
Blocks $\times Y(II)$	$(b - 1)(y - 1)$
Blocks $\times X \times Y(III)$	$(b - 1)(x - 1)(y - 1)$.

No one need object if each of the effects X, Y and $X \times Y$ is compared with its own interaction with blocks, i.e., with I, II, and III, respectively. If, however, all treatment combinations have been subject to the same randomization procedure, the components I, II, and III can be merged to give a common error with $(b - 1)(xy - 1)$ degrees of freedom. In a split-plot* design, where X is allocated at random to the main plots, II and III can together form a subplot error, leaving I to form that for main plots. Given a strip-plot* (or crisscross) design, on the other hand, each component of error must be kept separate. The example illustrates the old adage: "As the randomization is, so is the analysis." Sometimes there are practical difficulties about randomization and they can raise problems in the analysis of data, but they are not necessarily insuperable ones. For example, in the split-plot case it may not be feasible to randomize the main plot factor, X, in which case it is vitiated and so is its interaction with blocks (component I), but it might still be permissible to use the subplot analysis*. Again, if the subplot factor, Y, has to be applied systematically (it might be different occasions on which the main plot was measured), component II may be vitiated, but $X \times Y$ can still be compared with component III.

SUITABILITY OF DATA

With so many computer packages available it is easy to calculate an analysis of variance that is little better than nonsense. Some thought therefore needs to be given to the data before entering the package.

Strictly, the data should be continuous. In fact, it is usually good enough that they spread themselves over 10 or more points of a scale. When they do not, e.g., as with a body of data that consists mostly of ones and zeros, the sum of squared deviations for error is inflated relative to those for effects with consequent loss of sensitivity. Discrete data often come from Poisson* or binomial* distributions and may call for one of the variance-stabilizing transformations* to be considered below.

It is also required that the residuals* (the η_i in the examples) should be distributed independently and with equal variance. Independence is most important. Where the data come from units with a spatial or temporal relationship, independence is commonly achieved by a randomization of the treatments. If that is impracticable, the analysis of variance in its usual forms is better avoided. Equality of variance is more problematical. Where treatments are equally replicated*, the F-test is fairly robust* against some treatments giving a higher variance than others. If, however, attention is directed to a subset of treatments, as happens with multiple comparisons* and can happen when the treatment line is partitioned, it is necessary to have a variance estimated specifically for that subset or serious bias could result. Here two main cases need to be distinguished. Some treatments may involve more operations than others and may therefore give rise to larger errors. For example, the injection of a plant or animal can itself give rise to variation that is absent when other methods of administration are adopted. In the other case, the variance for any treatment bears a functional relationship to the mean value for that treatment.

To take the first case, if the error is the interaction of treatment and blocks, it is an easy matter to partition the treatment line into three parts: (1) between the groups, one with higher and one with lower variance; (2) within one group; and (3) within the other. The error can then be partitioned accordingly. Depending upon circumstances, it may be easier to regard each component of treatments as having its own error or to concoct a variance for each contrast of interest between the treatment means. Alternatively, if the treatments of high and low variance are associated with two levels of a factor, e.g., administration by injection as compared with some other method, a better way and one that places fewer constraints on the nature of error may be to group the data into pairs according to the two levels, a pair being the same with respect to other factors, and to analyze separately the sum and difference of data from the pair. In effect, that is virtually the same as regarding the design as if it were in split plots, the factor associated with the divergent variances being the one in subplots.

In the other case, where the variance of any observation depends upon the mean, the usual solution is to find a variance-stabilizing transformation*. Thus, if the variance is proportionate to the mean, as in a Poisson distribution*, it may be better to analyze the square roots of the data rather than the data themselves. Such transformations can be useful, especially when they direct attention to some quantity more fundamental than that measured. Thus given the end product of a growth process, it is often more profitable to study the mean growth rate than the final size, because that is what the treatments have been affecting, and similarly the error has arisen because growth rates have not been completely determined. The approach can, however, cause problems when the transformation is no more than a statistical expedient, especially when it comes to the interpretation of interactions. Thus suppose that an untreated control has led to 90% of insects surviving, applications of insecticide A had given a survival rate of 60% while B had given 45%. A declaration that the two insecticides had not interacted would lead most people to suppose that A and B in combination had given a survival rate of 30% ($30 = 60 \times 45/90$). If the data are analyzed without transformation, a zero interaction would imply 15% for the combination (15 = 60 + 45 − 90). Using the appropriate angular transformation, the figures become (0) 71.6, (A) 50.8, (B) 42.1, leading to an expectation for (AB) of 21.3 for zero interaction. This last figure corresponds to 13.2%, which is even further from the ordinary meaning of independence*.

INTERPRETATION OF AN ANALYSIS OF VARIANCE

As has been said, the analysis of variance is a subtle approach that can be molded to many ways of thought. It is at its best when the questions implicit in the investigation are answered systematically and objectively. It is not a method of "data snooping," the treatment line being partitioned and repartitioned until something can be declared "significant." Nor is it rigid, as if there were some royal way to be followed that would ensure success. Its function is to assist a line of thought, so any unthinking procedure will be unavailing.

The first step is usually to look at the mean squared deviation for error, which sums up the uncontrolled sources of variation. An experienced person may note that its value is lower than usual, which could suggest that improved methods have paid off, or it could be so large as to show that something had gone wrong. If that is the case, an examination of residuals may show which observations are suspect and provide a clue for another time. It may even lead to a positive identification of the fault and the exclusion of some data. The fault, however, must be beyond doubt; *little credence attaches to conclusions based on data selected by the investigator to suit his own purposes.*

Next, it is wise to look at the line for blocks or whatever else corresponds to the control of extraneous variation. It is possible for a research team to fill their garbage cans again and again with things that need not have been discarded. If the block line rarely shows any sign of having been inflated, it could well be that a lot of trouble is being taken to control sources of variation that are of little importance anyway. Also, if the sources of variation are indeed so little understood, it could be that important components are being left in error.

All this is a preliminary to the comparison of treatment effects* and error. Here some thought is needed. First, the partition of the treatment line into individual effects may be conventional but irrelevant to immediate needs. For example, in a $2 \times p$ factorial set of treatments, there may be no interest in the main effects* and interactions*, the intention being to study the response of the factor with two levels in p different conditions. Even if the partition is suitable, the order in which the effects should be studied needs consideration. As has been explained, a positive response to one test may render others unnecessary.

However, the need may not be for testing at all. The data may give a set of means, and it is only necessary to know how well they have been estimated. Even if the need is for testing, there are occasions when an approach by multiple comparisons* is called for rather than by F-test. Also, there are occasions when a significance test* has proved negative, but interest centers on its power*; that is, the enquiry concerns the probability of the data having missed a difference of specified size, supposing that it does exist.

A difficult situation arises when a high-order interaction appears to be significant but without any support from the lower-order interactions contained in it. Thus if $A \times B \times C$ gives a large value of F while $B \times C$, $A \times C$, $A \times B$, A, B, and C all seem unimportant, it is always possible that the 1 : 20 chance, or whatever it may be, has come off. Before dismissing awkward effects, however, it is as well to look more closely at the data. The whole interaction could depend upon one observation that is obviously wrong. If, however, all data for a particular treatment combination go the same way, whether up or down, there is the possibility of some complicated and unsuspected phenomenon that requires further study.

Anyone who interprets an analysis of variance should watch out for the inflated error. If there is only one treatment effect and that is small, little can be inferred, but if several F-values for treatment effects are well below expectation, it is at least possible that the error has been badly conceived. For example, blocks may have been chosen so ineptly that, instead of bringing together similar plots or units, each is composed of dissimilar ones. Again, the distribution of the residuals may be far from normal. Randomization* may have been inadequate, or the error may represent deviations from an unlikely model. The matter should not be left. A valuable pointer might be obtained to the design of better investigations in the future.

REFERENCES

1. Fisher, R. A. (1925). *Statistical Methods for Research Workers*. Oliver & Boyd, Edinburgh.

2. Pearce, S. C. (1953). *Field Experimentation with Fruit Trees and Other Perennial Plants*. Tech. Commun. Bur. Hort. Plantation Crops, Farnham Royal, Slough, England, **23**. App IV.

3. Tocher, K. D. (1952). *J. R. Statist. Soc. B*, **14**, 45–100.

4. Yates, F. (1933). *J. Agric. Sci.*, **23**, 108–145.

5. Yates, F. (1937). The Design and Analysis of Factorial Experiments. *Tech. Commun. Bur. Soil Sci. Rothamsted*, **35**.

FURTHER READING

A standard work is H. Scheffé's book *The Analysis of Variance* (Wiley, New York, 1959). Other useful books are *Statistical and Experimental Design in Engineering and the Physical Sciences* by N. L. Johnson and F. C. Leone (Wiley, New York, 1966), especially Volume 2, and *The Linear Hypothesis: A General Theory* by G. A. F. Seber (Charles Griffin, London, 1966). In *Experiments: Design and Analysis* (Charles Griffin, London, 1977), J. A. John and M. H. Quenouille discuss the analysis of variance for many standard designs, and G. B. Wetherill's *Intermediate Statistical Methods* (Chapman & Hall, London, 1980) looks with some care at various models. In Chapter 9 of their book, *Applied Regression Analysis* (Wiley, New York, 1966), N. R. Draper and H. Smith show the relationship of the analysis of variance to regression methods. Other useful references are Chapter 3 of C. R. Rao's *Advanced Statistical Methods in Biometric Research*, and Chapter 4 of G. A. F. Seber's *Linear Regression Analysis*, both published by Wiley, New York, the first in 1952 and the second in 1976.

See also Agriculture, Statistics in; Analysis of Covariance; Confounding; Design of Experiments; F-Tests; General Linear Model; Multiple Comparisons—I; and Regression (Various).

S. C. Pearce

ANALYSIS OF VARIANCE, WILK–KEMPTHORNE FORMULAS

For ANOVA* of data from cross-classified experiment designs with unequal but proportionate class frequencies, Wilk and Kempthorne [3] have developed general formulas for the expected values of mean squares (EMSs) in the ANOVA table. We will give appropriate formulas for an $a \times b$ two-way cross-classification.

If the number of observations for the factor level combination (i, j) is n_{ij}, then proportionate class frequencies require

$$n_{ij} = n u_i v_j, \quad i = 1, \dots, a, \quad j = 1, \dots, b.$$

In this case the Wilk and Kempthorne formulas, as presented by Snedecor and Cochran [2] for a variance components* model, are: For main effect* factor A,

$$\text{EMS} = \sigma^2 + \frac{nuv(1 - u^*)}{a - 1} \\ \times \{(v^* - B^{-1})\sigma_{AB}^2 + \sigma_A^2\};$$

for main effect of factor B,

$$\text{EMS} = \sigma^2 + \frac{nuv(1 - v^*)}{b - 1} \\ \times \{(u^* - A^{-1})\sigma_{AB}^2 + \sigma_B^2\};$$

for main interaction* $A \times B$,

$$\text{EMS} = \sigma^2 + \frac{nuv(1 - u^*)(1 - v^*)}{(a - 1)(b - 1)} \sigma_{AB}^2,$$

where

$$u = \sum_{i=1}^{a} u_i, \quad u^* = \left(\sum_{i=1}^{a} u_i^2 \right) \Big/ u^2,$$

$$v = \sum_{j=1}^{b} v_j, \quad v^* = \left(\sum_{j=1}^{b} v_j^2 \right) \Big/ v^2,$$

$\sigma_A^2, \sigma_B^2, \sigma_{AB}^2$ are the variances of the (random) terms representing main effects of A and B and the $A \times B$ interactions in the model, respectively, and σ^2 is the (common) variance of the residual terms.

Detailed numerical examples and discussions are available, for example in Bancroft [1].

REFERENCES

1. Bancroft, T. A. (1968). *Topics in Intermediate Statistical Methods*, Vol. 1. Iowa State University Press, Ames, IA.

2. Snedecor, G. W. and Cochran, W. G. (1967). *Statistical Methods*, 7th ed. Iowa State University Press, Ames, IA.

3. Wilk, M. B. and Kempthorne, O. (1955). *J. Amer. Statist. Ass.*, **50**, 1144–1167.

See also ANALYSIS OF VARIANCE; FACTORIAL EXPERIMENTS; INTERACTION; MAIN EFFECTS; and VARIANCE COMPONENTS.

ANCILLARY STATISTICS—I

Let X be an observable vector of random variables with probability density function* (PDF) $f_X(x; \theta)$, where θ is an unknown parameter taking values over a space Λ. If the distribution of the statistic, or vector of statistics, $A = a(X)$ is not dependent upon θ, then A is said to be ancillary for (the estimation of) θ. Suppose now that X is transformed in a $1:1$ manner to the pair (S, A). The joint PDF of S, A can be written

$$f_{S|A}(s; \theta|a)f_A(a), \qquad (1)$$

where the second term is free of θ. As the examples below will make apparent, an ancillary statistic often indexes the precision of an experiment; certain values of A indicate that the experiment has been relatively informative about θ; others indicate a less informative result. For this reason, it is often argued that procedures of inference should be based on the conditional distribution* of the data given the observed value of the ancillary. For example, estimation* and hypothesis-testing* procedures are based on the first term of (1), and their frequency properties

are evaluated over the reference set in which $A = a$ is held fixed at its observed value.

In the simplest context, ancillary statistics arise in experiments with random sample size.

Example 1. Contingent upon the outcome of a toss of an unbiased coin, either 1 or 10^4 observations are taken on a random variable Y which has a $N(\theta, 1)$ distribution. The sample size N is ancillary and, in estimating or testing hypotheses about θ, it seems imperative that N be regarded as fixed at its observed value. For example, a size α test of $\theta = 0$ versus the one-sided alternative $\theta > 0$ has critical region* $\overline{Y} > z_\alpha$ if $N = 1$ or $\overline{Y} > 10^{-2}z_\alpha$ if $N = 10^4$, where $\overline{Y}$ is the sample mean and z_α is the upper α point of a $N(0, 1)$ distribution. Conditional on the observed value of N, this test is uniformly most powerful* (UMPT), and since this test is conditionally of size α for each N, it is also unconditionally of size α. It is discomforting, however, that power considerations when applied to the unconditional experiment do not lead to this test [5].

A second example illustrates further the role of ancillary statistics.

Example 2. Let $X_{1:n}, \ldots, X_{n:n}$ be the order statistics* of a random sample from the density $f(x - \theta)$, where f is of specified form and $-\infty < \theta < \infty$. It is easily seen from considerations of the group of location transformations that the statistics $A_i = X_{i:n} - X_{1:n}$, $i = 2, \ldots, n$, are jointly ancillary for θ. R. A. Fisher* [7] describes these statistics as giving the "configuration" of the sample and their observed values can be viewed as being descriptive of the observed likelihood* function. For example, if $n = 2$ and $f(z) = \pi^{-1}(1 + z^2)^{-1}$, $-\infty < z < \infty$, the likelihood is unimodal if $A_2 \leqslant 1$ and bimodal if $A_2 > 1$, while $\overline{X} = (X_1 + X_2)/2$ is the maximum likelihood estimator* in the first case and a local minimum in the second. Here again, a conditional approach to the inference problem is suggested.

By an easy computation, the conditional density of $M = X_{1:n}$ given $A_2 = a_2, \ldots, A_n = a_n$ is

$$c \prod_1^n f(m + a_i - \theta), \qquad (2)$$

where $a_1 = 0$ and c is a constant of integration. The choice of the minimum M is arbitrary; the maximum likelihood estimator T or any other statistic that measures location may be used in its place. All such choices will lead to equivalent conditional inferences*. The essential point is that inferences can be based on the conditional distribution* (2).

Both of the foregoing examples suggest that, at least in certain problems, a conditionality principle is needed to supplement other statistical principles. The approach then seems clear: In evaluating the statistical evidence, repeated sampling criteria should be applied to the conditional experiment defined by setting ancillary statistics at their observed values. As will be discussed in the section "Nonuniqueness and Other Problems," however, there are some difficulties in applying this directive.

It should be noted that some authors require that an ancillary A be a function of the minimal sufficient statistic* for θ. This is discussed further in the section just mentioned and in "Conditionality and the Likelihood Principle."

RECOVERY OF INFORMATION

Ancillary statistics were first defined and discussed by Fisher [7], who viewed their recognition and use as a step toward the completion of his theory of exhaustive estimation [8, pp. 158 ff.]. For simplicity, we assume that θ is a scalar parameter and that the usual regularity conditions apply so that the Fisher information* may be written

$$I_X(\theta) = E[\partial \log f_X(x; \theta)/\partial\theta]^2$$
$$= -E[\partial^2 \log f_X(x; \theta)/\partial\theta^2].$$

If the maximum likelihood estimator $T = \hat{\theta}(X)$ is sufficient* for θ, then $I_T(\theta) = I_X(\theta)$ for all $\theta \in \Lambda$. Fisher calls T exhaustive because all information is retained in reducing the data to this scalar summary.

It often happens, however, that T is not sufficient and that its sole use for the estimation of θ entails an information loss measured by $I_X(\theta) - I_T(\theta)$, which is nonnegative for all θ and positive for some θ. Suppose, however, that T can be supplemented with a set of ancillary statistics A such that (T, A) are jointly sufficient* for θ. The conditional information in T given $A = a$ is defined as

$$I_{T|A} = a^{(\theta)}$$
$$= -E[\partial^2 \log f_{T|A}(t; \theta|a)/\partial\theta^2|A = a]$$
$$= -E[\partial^2 \log f_{T,A}(t, a; \theta)/\partial\theta^2|A = a],$$

since $f_A(a)$ is free of θ. Thus since T, A are jointly sufficient,

$$E[I_{T|A}(\theta)] = I_{T,A}(\theta) = I_X(\theta).$$

The average information in the conditional distribution of T given A is the whole of the information in the sample. The use of the ancillary A has allowed for the total recovery of the information on θ. Depending on the particular observed outcome $A = a$, however, the conditional information $I_{T|A} = a^{(\theta)}$ may be greater or smaller than the expected information $I_X(\theta)$.

Viewed in this way, an ancillary statistic A quite generally specifies the informativeness of the particular outcome actually observed. To some extent, the usefulness of an ancillary is measured by the variation in $I_{T|A}(\theta)$.

Although only a scalar parameter θ has been considered above, the same general results hold also for vector parameters. In this case, $I_X(\theta)$ is the Fisher information matrix* and $I_X(\theta) - I_T(\theta)$ is nonnegative definite. If T is the vector of maximum likelihood estimators, A is ancillary and T, A are jointly sufficient, the conditional information matrix, $I_{T|A}(\theta)$, has expectation $I_X(\theta)$ as above.

NONUNIQUENESS AND OTHER PROBLEMS

Several difficulties arise in attempting to apply the directive to condition on the observed values of ancillary statistics.

1. There are no general constructive techniques for determining ancillary statistics.

2. Ancillaries sometimes exist which are not functions of the minimal sufficient statistic*, and conditioning upon their observed values can lead to procedures

that are incompatible with the sufficiency principle.

3. There is, in general, no maximal ancillary.

In this section we look at problems 2 and 3. It should be noted that in certain problems (e.g., Example 2), group invariance* arguments provide a partial solution to problem 1. Even in such problems, however, there can be ancillaries present which are not invariant. An interesting example is given by Padmanabhan [10]. In the context of Example 2 with $f(\cdot)$ a standard normal density* and $n = 2$, he defines the statistic

$$B = \begin{cases} X_1 - X_2 & \text{if } X_1 + X_2 \geqslant 1, \\ X_2 - X_1 & \text{if } X_1 + X_2 < 1, \end{cases}$$

and shows that B is ancillary but not invariant.

Basu [2] has given several examples of nonunique ancillary statistics. The first of these concerns independent bivariate normal* variates (X_i, Y_i), $i = 1, \ldots, n$, with means 0, variances 1, and correlation ρ. In this example, $(X_1, \ldots, X_n)$ and $(Y_1, \ldots, Y_n)$ are each ancillary and conditional inference would clearly lead to different inferences on ρ. this is an example of problem 2 above, and to avoid this difficulty many authors require that the ancillary be a function of the minimal sufficient statistic (e.g., ref. 5). If an initial reduction to the minimal sufficient set, $\sum X_i^2 + \sum Y_i^2$ and $\sum X_i Y_i$, is made, there appears to be no ancillary present.

Not all examples of nonuniqueness are resolved by this requirement. Cox [4] gives the following example, which derives from another example of Basu.

Example 3. Consider a multinomial distribution* on four cells with respective probabilities $(1 - \theta)/6$, $(1 + \theta)/6$, $(2 - \theta)/6$, and $(2 + \theta)/6$, where $|\theta| < 1$. Let $X_1, \ldots, X_4$ represent the frequencies in a sample of size n. Each of the statistics

$$A_1 = X_1 + X_2 \qquad A_2 = X_1 + X_4$$

is ancillary for θ, but they are not jointly ancillary.

If conditional inference* is to be useful in such problems, methods for selecting from among competing ancillaries are needed. Cox [4] notes that the usefulness of an ancillary is related to the variation in $I_{T|A}(\theta)$ (see the preceding section) and suggests (again with scalar θ) that the ancillary be chosen to maximize

$$\text{var}\{I_{T|A}(\theta)\}.$$

In general, this choice may depend on θ. In the example above, however, Cox shows that A_1 is preferable to A_2 for all θ. The choice of variance as a measure of variation is, of course, arbitrary.

Barnard and Sprott [1] argue that the ancillary's role is to define the shape of the likelihood function, and that, in some problems, invariance* considerations lead to a straightforward selection between competing ancillaries. In the example above, the estimation problem is invariant under reflections with $\theta \leftrightarrow -\theta$ and $X_1 \leftrightarrow X_2$, $X_3 \leftrightarrow X_4$. Under this transformation, the ancillary A_1 is invariant while $A_2 \leftrightarrow n - A_2$. Thus under a natural group of transformations, the ancillary A_1 and not A_2 is indicated. This type of argument suggests that *invariance**, and not ancillarity, is the key concept.

CONDITIONALITY AND THE LIKELIHOOD PRINCIPLE*

In a fundamental paper, Birnbaum [3] formulates principles of sufficiency, conditionality, and likelihood and shows that the sufficiency* and conditionality principles are jointly equivalent to the likelihood principle. In this section we outline Birnbaum's arguments and some of the subsequent work in this area.

Birnbaum introduces the concept of the "total evidential meaning" (about θ) of an experiment E with outcome x and writes $\mathscr{E} \downarrow (E, x)$. Total evidential meaning is left undefined but the principles are formulated with reference to it, as follows.

The Sufficiency Principle (S). Let E be an experiment with outcomes x, y and t be a sufficient statistic. If $t(x) = t(y)$, then

$$\mathscr{E} \downarrow (E, x) = \mathscr{E} \downarrow (E, y).$$

This principle (S) is almost universally accepted by statisticians, although some would limit the types of experiments E to which the principle applies.

Let $L(\theta; x, E)$ denote the likelihood* function of θ on the data x from experiment E.

The Likelihood Principle* (L). Let E_1 and E_2 be experiments with outcomes x_1 and x_2, respectively. Suppose further that

$$L(\theta; x_1, E_1) \propto L(\theta; x_2, E_2).$$

Then $\mathscr{E} \downarrow (E_1, x_1) = \mathscr{E} \downarrow (E_2, x_2)$.

This principle (L) (sometimes called the strong likelihood principle) asserts that only the observed likelihood is relevant in assessing the evidence. It is in conflict with methods of significance testing*, confidence interval procedures, or indeed any methods that are based on repeated sampling*. The sufficiency principle is sometimes called the weak likelihood principle since, by the sufficiency of the likelihood function, it is equivalent to the application of the likelihood principle to a single experiment.

An experiment E is said to be a mixture experiment* with components E_h if, after relabeling of the sample points, E may be thought of as arising in two stages. First, an observation h is made on a random variable H with known distribution, and then x_h is observed from the component experiment E_h. The statistic H is an ancillary statistic.

The Conditionality Principle* (C). Let E be a mixture experiment with components E_h. Then

$$\mathscr{E} \downarrow (E, (h, x_h)) = \mathscr{E} \downarrow (E_h, x_h).$$

This principle asserts that the inference we should draw in the mixture experiment with outcome (h, x_h) should be the same as that drawn from the simpler component experiment E_h when x_h is observed.

Birnbaum's Theorem. $(S) + (C) \Leftrightarrow (L)$.

Proof. It follows immediately that (L) implies (C). Also (L) $\Rightarrow$ (S) since, by the factorization theorem*, two outcomes giving the same value of a sufficient statistic yield proportional likelihood functions. To show that (C) and (S) together imply (L), let E_1 and E_2 be two experiments with outcomes x_1 and

x_2, respectively, such that $L(\theta; x_1, E_1) \propto L(\theta; x_2, E_2)$. Let $\Pr[H = 1] = 1 - \Pr[H = 2] = p$ be a specified nonzero probability. In the mixture experiment E with components E_h, $h = 1, 2$, the outcomes $(H = 1, x_1)$ and $(H = 2, x_2)$ give rise to proportional likelihoods. Since the likelihood function is itself minimally sufficient, (S) implies that

$$\mathscr{E} \downarrow (E, (H = 1, x_1)) = \mathscr{E} \downarrow (E, (H = 2, x_2)).$$

On the other hand, (C) implies that

$$\mathscr{E} \downarrow (E, (H = h, x_h)) = \mathscr{E} \downarrow (E_h, x_h), \quad h = 1, 2,$$

and hence $\mathscr{E} \downarrow (E_1, x_1) = \mathscr{E} \downarrow (E_2, x_2)$. Thus $(S) + (C) \Rightarrow (L)$.

Since almost all frequency-based methods of inference contradict the likelihood principle, this result would seem to suggest that sufficiency and conditionality procedures are jointly incompatible with a frequency theory. It should be noted, however, that the foregoing argument pertains to the symmetric use of sufficiency and conditionality arguments. The likelihood principle would appear to follow only under such conditions.

Durbin [6] restricted the conditionality principle to apply only after an initial reduction to the minimal sufficient statistic. He defined a reduced experiment E' in which only T is observed and considered a revised conditionality principle (C') which applied to this reduced experiment. In essence, this restricts attention to ancillary statistics that are functions of the minimal sufficient statistic. This change apparently obviates the possibility of deducing (L).

A second approach [9] classifies ancillaries as being experimental or mathematical. The former are ancillary by virtue of the experimental design and the purpose of the investigation. They are ancillary statistics regardless of the parametric model chosen for the chance setup being investigated. Mathematical ancillaries, on the other hand, are ancillary because of the particular parametric model assumed. In the examples given above, N is an experimental ancillary in Example 1 while the ancillaries A_1 and A_2 in Example 3 are mathematical. Example 2 may be interpreted in two ways. If this is a measurement model* whereby the response

X arises as a sum, $X = \theta + e$, e being a real physical entity with known distribution, then the ancillaries $A_2, \ldots, A_n$ are experimental. (The purpose of the experiment is to determine the physical constant θ.) More usually, an experiment with data of this type is designed to determine the distribution of X and the model $f(x - \theta)$ is a preliminary specification. In this case, the ancillaries are mathematical. The primary principle is taken to be the experimental conditionality principle and other principles (e.g., sufficiency) are applied only after conditioning on any experimental ancillaries present.

REFERENCES

1. Barnard, G. A. and Sprott, D. A. (1971). In *Waterloo Symposium on Foundations of Statistical Inference*. Holt, Rinehart and Winston, Toronto, pp. 176–196. (Investigates relationships between ancillaries and the likelihood function.)

2. Basu, D. (1964). *Sankhyā A*, **26**, 3–16. (Discusses the problem of nonuniqueness and gives several interesting examples.)

3. Birnbaum, A. (1962). *J. Amer. Statist. Ass.*, **57**, 269–306. (A fundamental paper dealing with the formalization of inference procedures and principles.)

4. Cox, D. R. (1971). *J. R. Statist. Soc. B*, **33**, 251–255.

5. Cox, D. R. and Hinkley, D. V. (1974). *Theoretical Statistics*, Chapman & Hall, London. (Probably the most complete text reference on ancillary statistics.)

6. Durbin, J. (1970). *J. Amer. Statist. Ass.*, **65**, 395–398.

7. Fisher, R. A. (1936). *Proc. Amer. Acad. Arts Sci.*, **71**, 245–258. (First discussion of the existence and use of ancillary statistics.)

8. Fisher, R. A. (1973). *Statistical Methods and Scientific Inference*, 3rd ed. Oliver & Boyd, Edinburgh. (A discussion of the recovery of information and the use of ancillaries.)

9. Kalbfleisch, J. D. (1975). *Biometrika*, **62**, 251–268.

10. Padmanabhan, A. R. (1977). *Amer. Statist.*, **31**, 124.

See also Ancillary Statistics, First Derivative; Basu Theorems; Conditional Inference; Fiducial Inference; Inference, Statistical; Likelihood; and Sufficient Statistics.

JOHN D. KALBFLEISCH

ANCILLARY STATISTICS—II

Since the original entry "Ancillary Statistics" appeared in *ESS* Vol. 1 in 1982, further study has been devoted to the properties and applications of ancillary statistics. This entry will emphasize some of the practical developments, such as methods of approximating conditional distributions and of construction of ancillaries, as well as further problems of nonuniqueness. We also address the role of ancillaries when a nuisance parameter is present, an area not covered in the original article.

ANCILLARIES WITH NO NUISANCE PARAMETERS

The first mention of the term "ancillary" occurs in Fisher [12], where it is pointed out that, when the maximum-likelihood estimator is not fully efficient, the information it provides may be enhanced by the adjunction as an ancillary statistic of the second derivative of the likelihood function, giving the weight to be attached to the value of the estimator. In Fisher [13], he also requires that the distribution of an ancillary statistic be free of the unknown parameter. This requirement is now generally recognized as the characteristic property. In fact the second derivative of the likelihood function, more commonly called the observed information, is not generally ancillary; however, Efron and Hinkley [11] showed that, under certain conditions, it is a better estimator of the conditional variance of the estimator than is the expected Fisher information*.

Useful surveys of the properties of ancillary statistics have been given by Buehler [6] and by Lehmann and Scholz [15]. These surveys discuss, among other things, the sense in which one type of ancillary may function as an index of precision of the sample. There is an important distinction to be made between those ancillary statistics that are part of the minimal sufficient statistic (for the parameter of interest) and those that are not. Fisher's applications refer to the former cases of which he states: "The function of the ancillary statistic is analogous to providing a true, in place of an approximate, weight for the value of the estimate." Ancillaries that are a function of the minimal sufficient statistic we term *internal*, and those that are not we term *external*.

External ancillaries are of interest primarily when nuisance parameters are present, a discussion of which forms the second part of our contribution. Internal ancillaries are available for location and scale families of distributions, and indeed to all transformation models (see below). For such families Fisher [13] provided the initial analysis, which was developed and clarified by Pitman [22]. Pitman estimators have optimal properties within the conditional framework of evaluation. These were covered in the original article.

Approximate Ancillaries

In the first section of our discussion we will understand the term ancillary to mean internal ancillary. For many statistical models ancillary statistics do not exist. Indeed, formulation of precise conditions under which an ancillary does exist remains an open problem. Certainly, a model which does admit an ancillary statistic may not admit one upon minor modification. If such minor modifications are not to make a radical difference to the mode of inference, then we are driven to condition on statistics that are approximately ancillary in an appropriate sense.

This problem was pursued by Cox [7]. Concentrating on the problem of testing $\theta = \theta_0$, we argue that, whatever test statistic is to be used, its distribution should be computed conditional on a statistic A, possibly depending on θ_0, chosen so that its distribution depends on θ as little as possible, locally near θ_0. Assuming appropriate regularity, consider the Taylor expansion

$$E(A; \theta) \approx E(A; \theta_0) + c_1(\theta - \theta_0)$$
$$+ c_2(\theta - \theta_0)^2$$

of the mean of A about θ_0. The statistic A is *first-order local ancillary* if the coefficient $c_1 = 0$. In this case, for values of θ close to θ_0, the mean of A depends little on θ. For instance, at $\theta = \theta_0 + \delta/\sqrt{n}$, close to θ_0, where n is the sample size, the error in the above approximation is $O(n^{-3/2})$ and dependence on

θ occurs in the $O(n^{-1})$ term, assuming that A is scaled to have asymptotic unit variance. The statistic A is *second-order local ancillary* if both coefficients c_1 and c_2 are zero as well as the first coefficient in an expansion of the second moment. In many cases, such as for curved exponential families*, local ancillaries are simply constructed from linear or quadratic forms in the canonical sufficient statistics. Barndorff-Nielsen [2] uses a statistic A which, for fixed $\hat\theta$, is an affine (i.e., linear) function of the canonical sufficient statistics and is approximately free of θ in its mean and variance.

Example. Let (X, Y) be bivariate normal with standard mean and variance, and correlation θ. From an identical independent sample, the sufficient statistic for θ is

$$S = \sum_{i=1}^{n}(X_i Y_i, X_i^2 + Y_i^2).$$

In this case S_2 has mean $2n$, entirely free of θ, but has variance $4n(1 + \theta^2)$. Then

$$A = \frac{S_2 - 2n}{\sqrt{4n(1 + \theta_0^2)}}$$

is first-order local ancillary for θ near θ_0. A locally conditional confidence region for θ can be given approximately by inverting an Edge-worth expansion for the conditional distribution of S_1 given A; a general expression is given by Cox [7].

Information Recovery

When the sufficient statistic S is expressed as (T, A), then, using standard notation,

$$E\{I_{T|A}(\theta)\} = I_S(\theta) - I_A(\theta).$$

Thus, conditioning on approximate ancillaries involves loss of information, on average equal to the amount of information in the conditioning variable. On the other hand, the conditional analysis delivers a more relevant evaluation of the precision of estimation, as discussed earlier.

An alternative requirement of an approximate ancillary A is hence that the Fisher information* in it should be small. Statistics typically have information of order n in the sample size. A first-order approximate ancillary has information $O(1)$. This can be achieved by requiring that the derivative of $E(A)$ with respect to θ should be $O(n^{-1/2})$ (Lloyd [17]). By arranging a special condition on the variance and skewness of A and their derivatives, the information can be reduced to $O(n^{-1})$. Where a statistic is first-order ancillary not only locally but globally, it will be first-order approximate ancillary. Conversely, a statistic that is first-order approximate ancillary locally will be first-order local ancillary.

Balancing Information Loss Against Relevance

Cox's criterion judges the effectiveness of A as a conditioning variable by how large $\mathrm{Var}(I_{T|A})$ is. On the other hand, information I_A is lost on average. An exact ancillary may be quite ineffective while another statistic, which is only approximately ancillary, is much more effective. The statistic most "θ-free" in its distribution is not necessarily the best conditioning variable.

It is not clear how to balance the effectiveness against the information loss. Some progress has been made (Lloyd [18]); however, at present there seems to be no entirely objective way of making this tradeoff.

Nonuniqueness of Conditional Inference*

McCullagh [20] has given an example where the minimal sufficient statistic* may be expressed either as $(\hat{\theta}, A)$ or as $(\hat{\theta}, A^*)$, where A and A^* have exactly the same distributions and where $\mathrm{var}(I_{T|A})$ is identical for both A and A^*. Thus there is no hope of deciding which to condition on from consideration of marginal or conditional distributions. Further, (A, A^*) together are not ancillary.

The example is beautifully simple. An independent sample is taken from the Cauchy distribution with location θ_1 and scale θ_2. The entire parameter $\theta = (\theta_1, \theta_2)$ is of interest. The configuration ancillary A is the n-vector with ith component $(x_i - \hat{\theta}_1)/\hat{\theta}_2$. Fisher [13] and Pitman [22] recommend using the distribution of $\hat{\theta}$ conditional on the configuration A. Now for any real numbers a, b, c, d the transformed data

$$X_i^* = \frac{aX_i + b}{cX_i + d}$$

is also Cauchy in distribution with parameters θ^*, a simple transformation of θ. The X_i^* data are equivalent to the original data; it makes no difference which we use. However the configuration A^* of the X_i^* data is not the same as A. Which density do we use to assess $\hat{\theta}$, the one conditional on A or on A^*?

McCullagh shows that for large deviations, i.e., when $\hat{\theta} - \theta$ is $O(1)$, the two conditional densities differs relatively by $O(n^{-1/2})$, which is of the same magnitude as the difference between conditional and unconditional densities. When the observed value of the configuration statistic is atypically extreme, this difference is even larger. The conclusion is that the general recommendation to condition on exact ancillaries is ambiguous.

The Likelihood-Ratio Statistic

Let $L(\theta)$ be the likelihood function maximized at $\hat{\theta}$. Then the likelihood-ratio statistic is

$$w(\theta_0) = 2\log\{L(\hat{\theta})/L(\theta_0)\}.$$

Confidence intervals are set by collecting values of θ_0 for which $w(\theta_0)$ is less than a quantile of its sampling distribution which is approximately χ_p^2, where p is the dimension of θ. The approximation improves as cumulants of derivatives of the log-likelihood (usually proportional to sample size n) diverge. A slight adjustment to $w(\theta_0)$, called the Bartlett correction (see BARTLETT ADJUSTMENT—I), reduces the error of this distributional approximation to $O(n^{-2})$.

Such intervals are attractive because they produce sensible confidence sets even for anomalous likelihood functions, such as those with multiple maxima, divergent maximum, or regions of zero likelihood. Classical confidence regions based on the estimator and standard error are, in contrast, always elliptical in shape. Thus, likelihood-based intervals seem, in an informal sense, to better summarize what the data say about the parameter. Making inference more relevant to the particular data set is a fundamental reason for conditioning.

It has been shown that to high order, $w(\theta_0)$ is independent of the local ancillary $A(\theta_0)$, and when there are several choices for $A(\theta_0)$ it is approximately independent of all of them; see Efron and Hinkley [11], Cox [7], McCullagh [19]. Intervals based directly on the likelihood-ratio statistic are therefore automatically approximately conditional on the local ancillary, whereas alternative approaches via the score or Wald statistics are not. Seen another way, conditioning on the local ancillary leads to inference that agrees qualitatively with the likelihood function.

Approximate Conditional Distributions

Once an appropriate conditioning statistic is identified, there remains the problem of computing the conditional distribution of the information-carrying statistic. A remarkable amount such conditional distributions in more recent years. Among several remarkable properties of these approximations, the most important is that the conditioning statistic need not be explicitly identified.

Let A be an approximately ancillary statistic, as yet unspecified explicitly. Barndorff-Nielsen [2,3] gives the approximation

$$P^*_{\hat{\theta}|A=a}(t) = c|\hat{J}|^{1/2}\frac{L(t)}{L(\hat{\theta})} \qquad (1)$$

to the density function of $\hat{\theta}$ given $A = a$, where $\hat{J}$ is minus the second derivative matrix of the log-likelihood function evaluated at the maximum, and $|\hat{J}|$ denotes the determinant of this nonnegative definite matrix. The constant $c(a, t)$ is a norming constant, although in important cases it does not depend on the argument t. The norming constant possibly excepted, the formula is well suited to practical use, involving only a knowledge of the likelihood function.

The p^*-formula* is a synthesis and extension of the results of Fisher [13] and Pitman [22] for location and scale parameters. They showed that the distribution of the estimator conditional on the configuration ancillary was essentially the likelihood function itself, renormalized to integrate to 1. The p^*-formula, while approximate in general, establishes a similar link between conditional inference* and the shape of the likelihood function. The presence of $|\hat{J}|$ allows for the particular parametrization.

Transformation models are models that are closed under the action of a group of transformations, and that induce the same transformation on the parameter. An example is a location model where translation of the data translates the mean by the same amount. For transformation models an exact analogue of the configuration ancillary exists and the p^*-formula gives the exact conditional distribution. A curved exponential family is a multidimensional exponential family with a smooth relation between the canonical parameters. In this case the p^*-formula gives the distribution of the maximum-likelihood estimator conditional on the affine ancillary with accuracy $O(n^{-3/2})$ and the norming constant is a simple power of 2π. This has been established by relating the formula to the so-called saddle-point approximation*.

ANCILLARIES WITH NUISANCE PARAMETERS*

Let λ now represent the unknown parameter vector for a set of data $\{x\}$ with density function $f(x; \lambda)$. Only rarely are all components of λ of equal interest; we are typically interested in separately summarizing what the data say about specific components of interest.

Consider the case where $\lambda = (\theta, \phi)$. We take θ to be the parameter of interest and ϕ a nuisance (sometimes also called accessory) parameter. An inferential statement about θ is desired without regard to the value of ϕ. For instance we require a confidence region for θ whose coverage is at least close to nominal not only for all θ but for all ϕ.

A host of methods exist for achieving this under specific circumstances and a wide ranging survey was given by Basu [5]. The use of ancillary statistics, defined in an extended and appropriate sense, is central to many of these methods.

Pivotals and External Ancillaries

If R is a $100(1 - \alpha)\%$ confidence region for θ, then the indicator function of the event $\theta \in R$ is a binary random variable taking value 1 with probability $1 - \alpha$. It thus has a distribution free of both θ and ϕ. A function $P(\{x\}, \theta)$ which has a distribution free of λ is called a *pivotal*. When θ is known, P is an ancillary statistic in the model $f(x; \theta, \phi)$ with θ known, i.e. ancillary with respect to ϕ. Thus a necessary and sufficient condition for construction of a confidence interval for θ is that for any θ one can construct an ancillary with respect to ϕ.

Note that the pivotal is used directly for inference on θ, for instance by collecting all those values of θ for which $P(\{x\}, \theta)$ lies within the central part of its known distribution. There is no suggestion that it be conditioned on, and the rationale for its construction is quite distinct from arguments given earlier.

While such an ancillary should be a function of the minimal sufficient statistic for λ, the sufficiency principle would not require it to be a function of the minimal sufficient statistic for ϕ. External ancillary statistics therefore have some relevance to the elimination of nuisance parameters from inference. In a wide class of cases, external ancillaries will actually be independent of the sufficient statistic for ϕ; see Basu [4].

For continuous distributions admitting a complete sufficient statistic, an external ancillary may always be constructed as the distribution function of the data, conditional on the sufficient statistic. This construction is discussed by Rosenblatt [23] and extends to the construction of a pivotal by considering the interest parameter θ known. The pivotal is the distribution function of the data conditional on a complete sufficient statistic for ϕ. This "statistic" may also depend on θ; see Lloyd [16] for theory and for the following example.

Example. Let X, Y be independently distributed exponentially with means ϕ^{-1} and $(\theta + \phi)^{-1}$ respectively. From a sample of n independent pairs, inferences about θ are required, free of the nuisance parameter ϕ. The sufficient statistic for ϕ, whose distribution also involves θ, is $S = \Sigma(X_i + Y_i)$. (In other examples, S may even involve θ in its definition.) The conditional distribution function of the remaining element of the minimal

sufficient statistics $T = \Sigma X_i$ is

$$\int_{s-x}^{s} g(u, \theta)\, du \Big/ \int_{0}^{s} g(u, \theta)\, du,$$

where $g(u, \theta) = [u(s - u)]^{n-1} e^{-\theta u}$. For $n = 1$ the expression simplifies to

$$P(X, \theta) = \frac{1 - e^{-\theta X}}{1 - e^{-\theta(X+Y)}},$$

a probability that gives directly the significance of any proposed value of θ.

For the normal and gamma distributions, which admit complete sufficient statistics for location and scale respectively, the independence of external ancillaries (location- and scale-free statistics, respectively) are particularly important and useful, as the following examples show.

Example. A sample of size n is drawn from a population of unknown mean μ and standard deviation σ and is to be tested for normality. The mean $\overline{X}$ and standard deviation S of the sample are calculated. Then a sample of n from a standard normal population is drawn, and its mean $\overline{X}^*$ and standard deviation S^* calculated. The standardized members of the original sample are

$$\frac{X_i - \mu}{\sigma} = \frac{X_i - \overline{X}}{\sigma} + \frac{\overline{X} - \mu}{\sigma}.$$

Now $\overline{X}$ is complete sufficient for μ, and the external ancillary comprises the statistics $X_i - \overline{X}$, which are independent of $\overline{X}$ and therefore of the last term above. Thus the distribution is unchanged if we replace it with the equidistributed $\overline{X}^*$. Further,

$$\frac{X_i - \overline{X}}{\sigma} = \frac{X_i - \overline{X}}{S}\frac{S}{\sigma}.$$

The statistics $A_i = (X_i - \overline{X})/S$ are external ancillary for (μ, σ) and independent of both $\overline{X}$ and S, and so we may replace S/σ by the equidistributed S^*. Hence finally we have

$$\frac{X_i - \mu}{\sigma} \stackrel{d}{=} A_i S^* + \overline{X}^*,$$

and these statistics are available for testing normality in the absence of knowledge of the unknown parameters. This device is due to Durbin [10], who used a different method of proof.

Example. In some sampling situations the probability of an item appearing in the sample is inversely proportional to its measured values (such as lifetime), and a weighted mean $W = S_2/S_1$, where $S_1 = \Sigma X_i$ and $S_2 = \Sigma X_i^2$, is a reasonable estimator. The expectation and other moments of this estimator are readily determined if the X_i are $\Gamma(m)$-distributed, where the scale parameter is suppressed. The effect of the method of sampling is to reduce the shape parameter to $m - 1$. For a sample of n, the total S_1 is $\Gamma(n(m - 1))$-distributed, is sufficient for the scale parameter, and is independent of any scale-free statistic, such as $W/S_1 = S_2/S_1^2$. Hence $E(W) = E(W/S_1)E(S_1)$ and $E(S_2) = E(W/S_1)E(S_1^2)$, so that, in terms of the moments,

$$E(W) = \frac{E(S_1)S(S_2)}{E(S_1^2)} = \frac{nm(m - 1)}{nm - n + 1}.$$

Thus W has a small negative bias, $O(n^{-1})$.

Ancillarity in the Presence of a Nuisance Parameter

Suppose we have the following factorization of the density function:

$$f(x; \lambda) = f(x; a, \theta)f(a; \theta; \phi)$$

for some statistic A. The first factor is the density conditional on $A = a$, and the second the marginal distribution of A; the salient feature is that the first factor depends on λ only through θ.

It is generally agreed that we must have this feature present in the factorization of $f(x; \lambda)$ before we can say that A is ancillary for θ in any sense (Barndorff-Nielsen [1]). An alternative description is that A is "sufficient" for ϕ. However, the requirement that the marginal distribution of A be completely free of θ seems too strong. Even when the distribution of A depends on θ, A may be incapable of providing information (in the common usage of the term) about θ by itself,

because information concerning θ cannot be separated from ϕ in the experiment described by $f(a; \theta, \phi)$. One definition that makes this notion precise requires

1. that for every pair θ_1, θ_2 and for every $a, f(a; \theta_1, \phi)/f(a; \theta_2, \phi)$ run through all positive values as ϕ varies,
2. that given possible values θ_1, θ_2, ϕ, and a, there exist possible values θ, ϕ_1, ϕ_2 such that

$$\frac{f(a; \theta_1, \phi)}{f(a; \theta_2, \phi)} = \frac{f(a; \theta, \phi_1)}{f(a; \theta, \phi_2)}.$$

In this case we say that A is ancillary for θ and inference should be carried out using the distribution conditional on $A = a$ (see ref. 1). Under slightly different conditions, Godambe [14] has shown that the conditional likelihood leads to an estimating function* that is optimal amongst those not depending on ϕ. Yip [24,25] provides examples of this type where the information loss is negligible.

Approximate Conditional Likelihood

Maximum-likelihood estimates are not always consistent. The first example was given by Neyman and Scott [21]. Another simple example involves pairs of twins, one of each pair being given a treatment that increases the odds of a binary response by θ. The underlying rate of response ϕ_1 is specific to each pair. As more twins are collected, $\hat{\theta}$ converges to θ^2 rather than θ. This example is not artificial. Moreover, maximum-likelihood estimation is generally poor whenever a model involves a large number of nuisance parameters.

Let $L(\theta, \phi)$ be the likelihood function maximized at $(\hat{\theta}, \hat{\phi})$, and let $\hat{\phi}_0$ partially maximize L with respect to ϕ for fixed θ. The *profile likelihood* function

$$L_p(\theta) = L(\theta, \hat{\phi}_\theta)$$

shares many logical properties of likelihood and is maximized at $\theta = \hat{\theta}$. However, we have seen that it may perform very poorly. When a sufficient statistic A for ϕ exists, the density function $f(x; a, \theta)$ conditional on $A = a$ is free of ϕ. The *conditional likelihood* $L_c(\theta)$ is just this conditional density considered as a function of θ. The conditional likelihood avoids the bias and consistency problems of the unconditional likelihood. Thus, in the presence of nuisance parameters, another role of conditioning on the ancillary A is to reduce the bias of estimation. To perform this alternative function A need not be "free of θ" in any sense.

When the model $f(x; \lambda)$ is an exponential family with canonical parameter θ, such a conditional likelihood is always obtained simply by conditioning on $\hat{\phi}$. The resulting conditional likelihood can be treated much as an ordinary likelihood, and standard asymptotic theory often applies, conditional on any value $\hat{\phi}$. Davison [9] has applied these ideas to generalized linear models* as well as the approximation to be described below. Appropriately conditional inferences are easily computed within a standard computer package.

More generally, when a sufficient statistic $A(\theta_0)$ for θ exists for fixed θ_0, the conditional likelihood may be calculated. Barndorff-Nielsen [3] gives an approximation, called *modified profile likelihood*, which is

$$L_M(\theta) \approx \left| \frac{\partial \hat{\phi}}{\partial \hat{\phi}_\theta} \right| |\hat{J}_\theta|^{-1/2} L_p(\theta),$$

where $\hat{J}_\theta$ is the observed information for ϕ when θ is known. The only difficult factor to compute is the first, which can be made to approximately equal unity if ϕ is chosen to be orthogonal to θ; see Cox and Reid [8]. The factor involving j_θ penalizes values of θ that give relatively high information about ϕ. In examples where the estimate $\hat{\theta}$ is inconsistent, use of $L_M(\theta)$ reduces the inconsistency but may not completely eliminate it. The attraction is that conditioning has implicitly been performed merely by directly using a likelihood function, albeit approximately.

It is unclear what the conceptual and practical properties of $L_M(\theta)$ are when no sufficient statistics for ϕ exist (even when allowed to be functions of θ). There are also cases where conditioning produces a degenerate distribution, primarily when the data are discrete. A simple example is a logistic regression* where interest is in the intercept. The use of modified profile likelihood in such cases has not yet been fully justified.

REFERENCES

1. Barndorff-Nielsen, O. (1978). *Information and Exponential Families*. Wiley, New York.

2. Barndorff-Nielsen, O. (1980). Conditionality resolutions. *Biometrika*, **67**, 293–310.

3. Barndorff-Nielsen, O. (1983). On a formula for the distribution of the maximum likelihood estimator. *Biometrika*, **70**, 343–365.

4. Basu, D. (1955). On statistics independent of a complete sufficient statistic. *Sankhya*, **15**, 377–380.

5. Basu, D. (1977). On the elimination of nuisance parameters. *J. Amer. Statist. Ass.*, **72**, 355–366.

6. Buehler, R. J. (1982). Some ancillary statistics and their properties (with discussion). *J. Amer. Statist. Ass.*, **77**, 582–594.

7. Cox, D. R. (1980). Local ancillarity. *Biometrika*, **67**, 279–286.

8. Cox, D. R. and Reid, N. (1987). Parameter orthogonality and approximate conditional inference (with discussion). *J. Roy. Statist. Soc. B*, **49**, 1–39.

9. Davison, A. (1988). Approximate conditional inference in generalised linear models. *J. Roy. Statist. Soc. B*, **50**, 445–461.

10. Durbin, J. (1961). Some methods of constructing exact tests. *Biometrika*, **48**, 41–55.

11. Efron, B. and Hinkley, D. V. (1978). Assessing the accuracy of the maximum likelihood estimator: observed versus expected Fisher information (with discussion). *Biometrika*, **65**, 457–487.

12. Fisher, R. A. (1925). Theory of statistical estimation. *Proc. Cambridge Philos. Soc.*, **22**, 700–725.

13. Fisher, R. A. (1934). Two new properties of mathematical likelihood. *Proc. Roy. Soc. A*, **144**, 285–307.

14. Godambe, V. P. (1976). Conditional likelihood and unconditional optimal estimating equations. *Biometrika*, **63**, 277–284.

15. Lehmann, E. L. and Scholz, F. W. (1991). Ancillarity. In *Current Issues in Statistical Inference: Essays in honour of D. Basu*, M. Ghosh and P. K. Pathak, eds., IMS Lecture Notes—Monograph Series, pp. 32–51.

16. Lloyd, C. J. (1985). On external ancillarity. *Austral. J. Statist.*, **27**, 202–220.

17. Lloyd, C. J. (1991). Asymptotic expansions of the Fisher information in a sample mean. *Statist. Probab. Lett.*, **11**, 133–137.

18. Lloyd, C. J. (1992). Effective conditioning. *Austral. J. Statist.*, **34**(2), 241–260.

19. McCullagh, P. (1984). Local sufficiency. *Biometrika*, **71**, 233–244.

20. McCullagh, P. (1991). *On the Choice of Ancillary in the Cauchy Location Scale Problem.* Tech. Rep. 311, University of Chicago.

21. Neyman, J. and Scott, E. L. (1948). Consistent estimates based on partially consistent observations. *Econometrica*, **16**, 1–32.

22. Pitman, E. J. G. (1939). The estimation of the location and scale parameter of a continuous population of any given form. *Biometrika*, **30**, 391–421.

23. Rosenblatt, M. (1952). Remarks on a multivariate transformation. *Ann. Math. Statist.*, **23**, 470–472.

24. Yip, P. (1988). Inference about the mean of a Poisson distribution in the presence of a nuisance parameter. *Austral. J. Statist.*, **30**, 299–306.

25. Yip, P. (1991). Conditional inference on a mixture model for the analysis of count data. *Commun. Statist. Theory Methods*, **20**, 2045–2057.

See also ANCILLARY STATISTICS, FIRST DERIVATIVE; CONDITIONAL INFERENCE; NUISANCE PARAMETERS; and p^*-FORMULA.

CHRISTOPHER J. LLOYD
EVAN J. WILLIAMS
PAUL S. F. YIP

ANCILLARY STATISTICS, FIRST DERIVATIVE

The usual definition of an ancillary statistic* $a(y)$ for a statistical model $\{f(y; \theta) : \theta \in \Omega\}$ requires that the distribution of $a(y)$ be fixed and thus free of the parameter θ in Ω. A first derivative ancillary (at θ_0) requires, however, just that the first derivative at θ_0 of the distribution of $a(y)$ be zero; in an intuitive sense this requires ancillarity for θ restricted to a small neighborhood $(\theta_0 \pm \delta)$ of θ_0 for δ sufficiently small. The notion of a first derivative ancillary was developed in Fraser [4] and named in Fraser and Reid [6].

Fisher's discussion and examples of ancillaries [3] indicate that he had more in mind than the fixed θ-free distribution property, but this was never revealed to most readers' satisfaction. One direction was to require that y in some sense measure θ, as in the location model (*see* ANCILLARY STATISTICS—I); this notion was generalized in terms of structural models* which yield a well-defined ancillary and a well-defined conditional distribution to be used for conditional inference*. In a related manner the usefulness of first derivative ancillaries arises in a context where the variable y in a sense measures θ, locally at a value θ_0, and particularly in some asymptotic results that will be described briefly below.

First we give a simple although somewhat contrived example of a first derivative ancillary. Consider a sample $(y_1, \ldots, y_n)$ from the model $\varphi(y - \theta)(1 + y^2\theta^2)/(1 + \theta^2 + \theta^4)$, where φ denotes the standard normal density. Then $(y_1 - \bar{y}, \ldots, y_n - \bar{y})$ is first derivative ancillary at $\theta = 0$, as this model agrees with the model $\varphi(y - \theta)$ at $\theta = 0$, and the first derivatives with respect to θ of the two models are also identical at $\theta = 0$. Also $(y_1 - \bar{y}, \ldots, y_n - \bar{y})$ is the standard configuration ancillary for the model $\varphi(y - \theta)$; *see* ANCILLARY STATISTICS—I.

There are a number of likelihood-based methods that lead to highly accurate test procedures for scalar parameters. The typical context has a continuous model $f(y; \theta)$, where y has dimension n and θ has dimension p, and interest centers on a component scalar parameter ψ, where $\theta = (\lambda, \psi)$, say.

In cases where the dimension of the minimal sufficient statistic is the same as the dimension of θ, as in an exponential model or via conditionality with a location or transformation model, there are now well-developed, very accurate methods for obtaining a p-value, $p(\psi_0)$, for testing the value ψ_0 for the parameter ψ; see, for example, Barndorff-Nielsen [1], Fraser and Reid [5,6], and Reid [8].

Extension to the $n > p$ case requires a dimension reduction by sufficiency or conditionality. In this context, sufficiency turns out to be too specialized and conditionality requires an ancillary or approximate ancillary statistic. While suitable approximate ancillary statistics have long been known to exist, there seems not to be any generally applicable construction method. Particular examples are discussed in Barndorff-Nielsen and Cox [2]. The first derivative ancillary [6] provides the ingredients for a simple construction method.

Consider an independent coordinate y of a model with scalar parameter θ, and suppose the distribution function $F(y; \theta)$ is stochastically monotone at θ_0. Then a transformation $x = x(y)$ to a new variable with distribution function $G(x; 0)$ exists such that $G(x; \theta)$ and $G(x - (\theta - \theta_0); \theta_0)$ agree up to their first derivatives at $\theta = \theta_0$. If this property holds for each coordinate, then $(x_1 - \bar{x}, \ldots, x_n - \bar{x})$ is first derivative ancillary at $\theta = \theta_0$, as in the simple example above.

The conditional distribution given this ancillary has tangent direction $(1, \ldots, 1)'$ in terms of the x-coordinates and tangent direction $\upsilon = (\upsilon_1, \ldots, \upsilon_n)'$ in terms of the y-coordinates, where

$$\upsilon_i = \frac{\partial F_i(y_i; \theta)}{\partial \theta} \bigg/ \frac{\partial F_i(y_i; \theta)}{\partial y_i} \bigg|_{(y^0, \hat{\theta}^0)};$$

the calculation is at the observed data y^0 with corresponding maximum likelihood estimate $\hat{\theta}^0$, and the subscript i is to designate the ith coordinate distribution function. More concisely, we can write $\upsilon = \partial y / \partial \theta|_{(\bar{y}^0, \hat{\theta}^0)}$, where the differentiation is for a fixed value of the distribution function. In most problems this is easy to calculate.

An interesting feature of the above type of first derivative ancillary is that it can be adjusted to give an approximate ancillary statistic to third order without altering its tangent directions at the data point. (A third-order approximate ancillary statistic is a statistic whose distribution is free of θ in a specific sense: see Reid [7].) It turns out that for approximation of p-values to third order [meaning with relative error $O(n^{-3/2})$], the only needed information concerning the ancillary is information on the tangent directions at the data point. As a result the first derivative ancillary at the maximum likelihood value provides a means to generalize third-order likelihood asymptotics from the case $n = p$ to the typical general case with $n > p$ [6].

REFERENCES

1. Barndorff-Nielsen, O. E. (1986). Inference on full or partial parameters based on the standardized signed log-likelihood ratio. *Biometrika*, **73**, 307–322.

2. Barndorff-Nielsen, O. E. and Cox, D. R. (1994). *Inference and Asymptotics*. Chapman and Hall, London.

3. Fisher, R. A. (1973). *Statistical Methods and Scientific Inference*, 3rd ed. Oliver & Boyd, Edinburgh. (Includes a discussion of ancillary statistics with many examples.)

4. Fraser, D. A. S. (1964). Local conditional sufficiency. *J. R. Statist. Soc. B*, **26**, 52–62.

5. Fraser, D. A. S. and Reid, N. (1993). Simple asymptotic connections between densities and cumulant generating function leading to accurate approximations for distribution functions. *Statist. Sinica*, **3**, 67–82.

6. Fraser, D. A. S. and Reid, N. (1995). Ancillaries and third-order significance. *Utilitas Math.*, **47**, 33–53.

7. Reid, N. (1995). The roles of conditioning in inference. *Statist. Sci.*, **10**, 138–157.

8. Reid, N. (1996). Asymptotic expansions. In *Encyclopedia of Statistical Sciences Update*, vol. 1, S. Kotz, C. B. Read, and D. L. Banks, eds. Wiley, New York, pp. 32–39.

See also ANCILLARY STATISTICS—I; ASYMPTOTIC EXPANSIONS—II; CONDITIONAL INFERENCE; STRUCTURAL MODELS; and SUFFICIENT STATISTICS.

D. A. S. FRASER
N. REID

ANDERSON–DARLING TEST FOR GOODNESS OF FIT. See GOODNESS OF FIT, ANDERSON–DARLING TEST OF

ANDERSON, OSKAR NIKOLAEVICH

Born: August 2, 1887, in Minsk, Russia.

Died: February 12, 1960, in Munich, Federal Republic of Germany.

Contributed to: correlation analysis, index numbers, quantitative economics, sample surveys, time-series analysis, nonparametric methods, foundations of probability, applications in sociology.

The work and life of Oskar Anderson received a great deal of attention in the periodical statistical literature following his death in 1960. In addition to the customary obituary in the *Journal of the Royal Statistical Society (Series A)*, there was a relatively rare long appreciative article—written by the famous statistician and econometrist H. O. A. Wold,* whose biography also appears in this volume—in the *Annals of Mathematical Statistics*, with an extensive bibliography, and a remarkable obituary and survey of his activities in the *Journal of the American Statistical Association*, written by the well-known statistician G. Tintner, also containing a bibliography. Furthermore there was detailed obituary in *Econometrica*. The first entry in the *International Statistical Encyclopedia* (J. M. Tanur and W. H. Kruskal, eds.) contains a rather detailed biography and an analysis of Anderson's contributions.

Some of this special interest may be associated with unusual and tragic events the first 40 years of Anderson's life, as aptly noted by Wold (1961):

> The course of outer events in Oskar Anderson's life reflects the turbulence and agonies of a Europe torn by wars and revolutions. [In fact Fels (1961) noted that "a daughter perished when the Andersons were refugees; a son a little later. Another son fell in the Second World War."]

Both German and Russian economists and statisticians compete to claim Oskar Anderson as their own. His father became a professor of Finno-Ugric languages at the University of Kazan (the famous Russian mathematician I. N. Lobachevsky, who was born in Kazan, was also a professor at the University). The Andersons were ethnically German. Oskar studied mathematics at the University of Kazan for a year, after graduating (with a gold medal) from gymnasium in that city in 1906. In 1907, he entered the Economics Department of the Polytechnical Institute at St. Petersburg. From 1907 to 1915 he was a assistant to A. A. Chuprov* at the Institute, and a librarian of the Statistical-Geographical "Cabinet" attached to it. He proved himself an outstanding student of Chuprov's, whose influence on Anderson persisted throughout his life. Also, during the years 1912–1917, he was a lecturer at a "commercial gymnasium" in St. Petersburg, and managed to obtain a law degree. Among his other activities at that time, he organized and participated in an expedition in 1915 to Turkestan to carry out an agricultural survey in the area around the Syr Darya river. This survey was on a large scale, and possessed a representativity ahead of contemporary surveys in Europe and the USA. In 1917 he worked as a research economist for a large cooperative society in southern Russia.

In 1917, Anderson moved to Kiev and trained at the Commercial Institute in that city, becoming a docent, while simultaneously holding a job in the Demographic Institute of the Kiev Academy of Sciences, in association with E. Slutskii.*

In 1920, he and his family left Russia, although it was said that Lenin had offered him a very high position in the economic administration of the country. It is possible that his feelings of loyalty to colleagues who were in disfavor with the authorities influenced this decision. For a few years he worked as a high-school principal in Budapest, and then for many years (1924–1943) except for a two-year gap he lived in Bulgaria, being a professor at the Commercial Institute in Varna from 1924 to 1933, and holding a similar position at the University of Sofia from 1935 to 1942. (In the period 1933–1935 he was a Rockefeller Fellow in England and Germany, and his first textbook on mathematical statistics was published.) While in Bulgaria he was very active in various sample surveys and censuses, utilizing, from time to time, the methodology of purposive sampling. In 1940 he was sent to Germany by the Bulgarian government to study rationing, and in 1942, in the midst of World War II, he accepted an appointment at the University of Kiel.

After the war, in 1947, Anderson became a professor of statistics in the Economics Department of the University of Munich, and he remained there till his death in 1960. His son Oskar Jr. served as a professor of economics in the University of Mannheim.

Anderson was a cofounder—with Irving Fisher and Ragnar Frisch—of the Econometric Society. He was also a coeditor of the journal *Metrika*. In the years 1930–1960 he was one of the most widely known statisticians in Central and Western Europe, serving as a link between the Russian and Anglo-American schools in statistics, while working within the German tradition exemplified by such statisticians as Lexis* and von

Bortkiewicz.* He contributed substantially to the statistical training of economists in German universities. His main strengths lay in systematic coordination of statistical theory and practice; he had good intuition and insight, and a superb understanding of statistical problems. His second textbook, *Probleme der Statistischen Methodenlehre*, published in 1954, went through three editions in his lifetime; a fourth edition appeared posthumously in 1962. He was awarded honorary doctorates from the Universities of Vienna and Mannheim and was an honorary Fellow of the Royal Statistical Society.

His dissertation in St. Petersburg, "On application of correlation coefficients in dynamic series," was a development of Chuprov's ideas on correlation. He later published a paper on this topic in *Biometrika*, and a monograph in 1929. While in Varna, he published a monograph in 1928 (reprinted in Bonn in 1929), criticizing the Harvard method of time-series analysis, and developed his well-known method of "variate differences" (concurrently with and independently of W. S. Gosset*). This method compares the estimated variances of different orders of differences in a time series to attempt to estimate the appropriate degree of a polynomial for a local fit. In 1947, Anderson published a long paper in the *Schweizerische Zeitschrift für Volkswirtschaft und Statistik*, devoted to the use of prior and posterior probabilities in statistics, aiming at unifying mathematical statistics with the practices of statistical investigators. Anderson was against abstract mathematical studies in economics, and often criticized the so-called "Anglo-American school," claiming that the Econometric Society had abandoned the goals originally envisioned by its founders.

During the last period of his life, he turned to nonparametric methods, advocating, *inter alia*, the use of Chebyshev-type inequalities, as opposed to the "sigma rule" based on assumptions of normality. Some of his endeavors were well ahead of his time, in particular, his emphasis on causal analysis of nonexperimental data, which was developed later by H. Wold,* and his emphasis on the importance of elimination of systematic errors in sample surveys. Although he had severed physical contact with the land of his birth as early as 1920, he followed with close attention the development of statistics in the Soviet Union in the thirties and forties, subjecting it to harsh criticism in several scorching reviews of the Marxist orientation of books on statistics published in the USSR.

REFERENCES

1. Anderson, O. N. (1963). *Ausgewählte Schriften*, H. Strecker and H. Kellerer, eds. Mohr Verlag, Tübingen, Germany. 2 vols. (These collected works, in German, of O. N. Anderson include some 150 articles, originally written in Russian, Bulgarian, and English as well as German. They are supplemented by a biography.)

2. Fels, E. (1961). Oskar Anderson, 1887–1960. *Econometrica*, **29**, 74–79.

3. Fels, E. (1968). Anderson, Oskar N. In *International Encyclopedia of Statistics*, J. M. Tanur and W. H. Kruskal, eds. Free Press, New York.

4. Rabikin, V. I. (1972). O. Anderson—a student of A. A. Chuprov, *Uchenye Zapiski po Statistike*, **18**, 161–174.

5. Sagoroff, S. (1960). Obituary: Oskar Anderson, 1887–1960, *J. R. Statist. Soc. A.* **123**, 518–519.

6. Tintner, G. (1961). The statistical work of Oscar Anderson. *J. Amer. Statist. Ass.*, **56**, 273–280.

7. Wold, H. (1961). Oskar Anderson: 1887–1960, *Ann. Math. Statist.*, **32**, 651–660.

ANDREWS FUNCTION PLOTS

Function plots are displays of multivariate data in which all dimensions of the data are displayed. Each observation is displayed as a line or function running across the display. The plots are useful in detecting and assessing clusters* and outliers*. Statistical properties of the plots permit tests of significance* to be made directly from the plot.

The display of data of more than two dimensions requires special techniques. Symbols may be designed to represent simultaneously several dimensions of the data. These may be small symbols used in a scatter plot with two dimensions of the data giving the location of the symbol of the page. Anderson [1] gives examples of such glyphs*. Patterns involving one or more of the plotting dimensions are most easily detected.

Alternatively, these may be larger symbols displayed separately. Chernoff faces* are an example of this type. Although no two dimensions have special status as plotting coordinates, the detection of patterns is more awkward. Function plots are a method of displaying large (page size) symbols simultaneously.

CONSTRUCTION OF PLOTS

Although only few statisticians have experience in displaying items of more than three dimensions, all statisticians are familiar with displays of functions $f(t)$. These may be considered as infinite dimensional. This suggests a mapping of multivariate data, observation by observation, into functions and then displaying the functions. Many such mappings are possible, but the mapping proposed here has many convenient statistical properties.

For each observation involving k dimensions $\mathbf{x}' = (x_1, \ldots, x_k)$, consider the function

$$f_x(t) = x_1\sqrt{2} + x_2 \sin t + x_3 \cos t$$
$$+ x_4 \sin 2t + x_5 \cos 2t + \cdots$$

plotted for values $-\pi < t < \pi$. Each observation contributes one line running across the display. The completed display consists of several such lines.

STATISTICAL PROPERTIES

The mapping $x \to f_x(t)$ preserves distances. For two points x, y the equation

$$\sum_{i=1}^{k} (x_i - y_i)^2 = \pi^{-1} \int_{-\pi}^{\pi} \left[f_x(t) - f_y(t)\right]^2 dt$$

implies that two functions will appear close *if and only if* the corresponding points are close.

The mapping is linear. This implies that

$$\bar{f}_x(t) = f_{\bar{x}}(t).$$

If the data have been scaled so that the variates are approximately independent with the same variance σ^2, then the variance of $f_x(t)$ is constant, independent of t, or almost constant. Since

$$\mathrm{var}(f_x(t)) = \sigma^2(\tfrac{1}{2}\pi + \sin^2 t + \cos^2 t$$
$$+ \sin^2 2t + \cos^2 2t + \ldots)$$
$$= \sigma^2(\tfrac{1}{2} + k/2 + R),$$

where $R = 0$ if k is odd and $\sin^2[(k + 1)/2]$ if k is even. This relation may be used to produce and display confidence bands* and

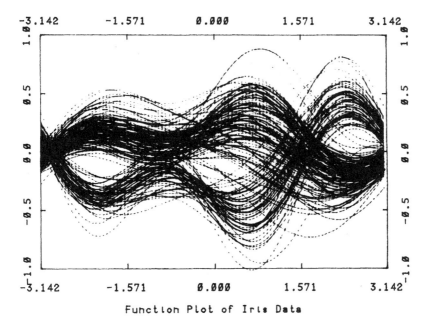

Function Plot of Iris Data

Figure 1. All species—150 observations.

tests for outliers*. These tests may be made for preselected values of t or marginally for all values of t. Scheffé's method of multiple comparison* may be used here.

EXAMPLE

Figure 1 is a plot of the Fisher iris data. These data consist of observations of four variables (log units) on 150 iris flowers. The example is commonly used to demonstrate multivariate techniques. Figure 1 clearly demonstrates the separation of one group. This group consists of one species. This is verified in Fig. 2, which is the plot of this species alone. Note the presence of two "outliers" represented by two straggling lines.

FURTHER NOTES

In some applications, with large data sets, the data may be summarized for each value of t by selected order statistics* of the values $f_x(t)$. Thus a complex plot may be reduced to a plot of the median, the quartiles, the 10% points, and the outlying observations. The order statistics were chosen so that the lines will be almost equally spaced for Gaussian (normal) data.

The order of the variables included in the specification of the function has no effect on the mathematical of statistical properties, although it does affect the visual appearance of the display. Some experience suggests that dominant variables should be associated with the lower frequencies.

GENERAL REFERENCE

1. Anderson, E.   (1960).   *Technometrics*,   **2**, 387–392.

BIBLIOGRAPHY

Andrews, D. F. (1972). *Biometrics*, **28**, 125–136.

Chernoff, H. (1973). *J. Amer. Statist. Ass.*, **68**, 361–368.

Fisher, R. A. (1936). *Ann. Eugen. (Lond.)*, **7** (Pt. II), 179–188.

Gnanadesikan, R. (1977). *Methods for Statistical Data Analysis of Multivariate Observations*. Wiley, New York.

See also CHERNOFF FACES; GRAPHICAL REPRESENTATION OF DATA; and MULTIVARIATE ANALYSIS.

D. F. ANDREWS

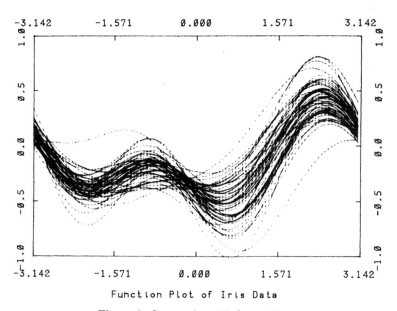

Function Plot of Iris Data

Figure 2. One species—50 observations.

ANGLE BRACKETS

Tukey [4] used angle brackets for symmetric means. These are power product sum* (augmented monomial symmetric functions) divided by the number of terms forming the sum, thus giving the mean power product [5, p. 38].

If observations in a sample are $x_1, \ldots, x_n$, the power product sum $[P] = [p_1 \ldots p_\pi] = \sum_{\neq}^n x_i^{p_1} x_j^{p_2} \ldots x_l^{p_\pi}$, where the sum is over all permutations of the subscripts, and no two subscripts are equal. The number of terms forming the sum is $n(n-1) \ldots (n - \pi + 1) = n^{(\pi)}$. The symmetric mean or angle bracket is then $\langle P \rangle = [P]/n^{(\pi)}$. Thus

$$\langle 1 \rangle = \frac{1}{n}[1] = \frac{1}{n} \sum_1^n x_i = \bar{x},$$

$$\langle r \rangle = \frac{1}{n}[r] = \frac{1}{n} \sum_1^n x_i^r = m_r',$$

$$\langle rs \rangle = \frac{1}{n^{(2)}}[rs] = \frac{1}{n(n-1)} \sum_{i \neq j}^n x_i^r x_j^s,$$

$$\langle 11 \rangle = \frac{1}{n^{(2)}}[11] = \frac{1}{n(n-1)} \sum_{i \neq j}^n x_i x_j.$$

But

$$[1]^2 = [2] + [11]$$

as

$$\left(\sum_1^n x_i \right)^2 = \sum_1^n x_i^2 + \sum_{i \neq j}^n x_i x_j.$$

Hence

$$\langle 11 \rangle = \frac{1}{n-1} \{ n \langle 1 \rangle^2 - \langle 2 \rangle \}.$$

Tukey [4] and Schaeffer and Dwyer [3] give such recursion formulas for the computation of angle brackets. Elementary examples of computing angle brackets are given in Keeping [2]. Two angle brackets may be multiplied by the rule

$$\langle r \rangle \langle s \rangle = \frac{n-1}{n} \langle rs \rangle + \frac{1}{n} \langle r+s \rangle.$$

A similar symmetric mean may be defined for the population $x_1, \ldots, x_N$, and denoted by $\langle P \rangle_N = [P]_N / N^{(\pi)}$, where $[P]_N$ denotes the power product sum $\sum_{\neq}^N$ for the population. Then, if E_N denotes the expected value for the finite population, it follows from an argument of symmetry [3,4] that $E_N \langle P \rangle = \langle P \rangle_N$. Tukey [4] calls this property "inheritance on the average." It makes angle brackets attractive in the theory of sampling from finite populations*, as sample brackets are unbiased estimates* of corresponding population brackets.

Every expression that is (1) a polynomial, (2) symmetric, and (3) inherited on the average can be written as a linear combination of angle brackets with coefficients that do not depend on the size of the set of numbers concerned [4].

Since Fisher's k-statistic* k_p is defined as $k_p = \sum_p (-1)^{\pi-l}(\pi-1)! C(P) \langle P \rangle$ and the finite population K-parameter is $K_p = \sum_p (-1)^{\pi-l}(\pi-1)! C(P) \langle P \rangle_N$, it directly follows that $E_N(k_p) = K_p$. For infinite populations, $E \langle P \rangle = \mu_{p_1}' \ldots \mu_{p_\pi}'$, and hence $E(k_p) = \kappa_p$. *See* FISHER'S k-STATISTICS for more details.

Tukey [5] defines polykays* by a symbolic multiplication of the k-statistics written as linear combinations of angle brackets. The symbolic product of the brackets is a bracket containing the elements in the brackets multiplied, i.e., $\langle p_1 \ldots p_\pi \rangle \circ \langle q_1 \ldots q_x \rangle = \langle p_1 \ldots p_\pi q_1 \ldots q_x \rangle$. Thus $k_{21} = k_2 \circ k_1 = \{ \langle 2 \rangle - \langle 11 \rangle \} \circ \langle 1 \rangle = \langle 21 \rangle - \langle 111 \rangle$. Tukey [4] uses angle brackets in the consideration of randomized (or random) sums. Hooke [1] extends them to generalized symmetric means for a matrix.

REFERENCES

1. Hooke, R. (1956). *Ann. Math. Statist.*, **27**, 55–79.

2. Keeping, E. S. (1962). *Introduction to Statistical Inference.* D. Van Nostrand, Princeton, N. J.

3. Schaeffer, E., and Dwyer, P. S. (1963). *J. Amer. Statist. Ass.*, **58**, 120–151.

4. Tukey, J. W. (1950). *J. Amer. Statist. Ass.*, **45**, 501–519.

5. Tukey, J. W. (1956). *Ann. Math. Statist.*, **27**, 37–54.

See also FISHER'S *k*-STATISTICS; POLYKAYS; and POWER PRODUCT SUMS.

D. S. TRACY

ANGULAR TRANSFORMATION. See
VARIANCE STABILIZATION

ANIMAL POPULATIONS, MANLY–PARR ESTIMATORS

The Manly—Parr estimator of population size can be calculated from data obtained by the capture-recapture method* of sampling an animal population. The assumption made by Manly and Parr [5] is that an animal population is sampled on a series of occasions in such a way that on the *i*th occasion all the N_i animals then in the population have the same probability p_i of being captured. In that case the expected value of n_i, the number of captures, is $E(n_i) = N_i p_i$, so that

$$N_i = E(n_i)/p_i. \qquad (1)$$

Manly and Parr proposed that p_i be estimated by the proportion of the animals known to be in the population at the time of the *i*th sample that are actually captured at that time. For example, in an open population (where animals are entering through births and leaving permanently through deaths and emigration) any animal seen before the time of the *i*th sample and also after that time was certainly in the population at that time. If there are C_i individuals of this type, of which c_i are captured in the *i*th sample, then $\hat{p}_i = c_i/C_i$ is an unbiased estimator of p_i. Using equation (1) the Manly—Parr estimator of N_i is then

$$\hat{N}_i = n_i/\hat{p}_i = n_i C_i/c_i. \qquad (2)$$

Based upon a particular multinomial* model for the capture process, Manly [4] gives the approximate variance

$$\mathrm{var}(\hat{N}_i) \simeq N_i(1 - p_i)(1 - \theta_i)/(p_i \theta_i), \qquad (3)$$

where θ_i is the probability of one of the N_i animals being included in the class of C_i animals known to certainly be in the population

at the time of the *i*th sample. This variance can be estimated by

$$\mathrm{V\hat{a}r}(\hat{N}_i) = \hat{N}_i(C_i - c_i)(n_i - c_i)/c_i^2. \qquad (4)$$

Manly and Parr also proposed estimators for survival rates and birth numbers in the open population situation. Let r_i denote the number of animals that are captured in both the *i*th and the $(i + 1)$th samples. The expected value of this will be $E(r_i) = N_i s_i p_i p_{i+1}$, where s_i is the survival rate over the period between the two samples for the population as a whole. This relationship together with equation (1) suggests the Manly—Parr survival rate estimator

$$\hat{s}_i = r_i/(n_i \hat{p}_{i+1}), \qquad (5)$$

where $\hat{p}_{i+1} = c_{i+1}/C_{i+1}$.

At the time of the $(i + 1)$th sample the population size N_{i+1} will be made up of the survivors from the previous sample time, $N_i s_i$, plus the number of new entries B_i to the population (the births). On this basis the Manly—Parr estimator of the number of births is

$$\hat{B}_i = \hat{N}_{i+1} - \hat{s}_i \hat{N}_i. \qquad (6)$$

As an example of the use of the Manly—Parr equations consider the data shown in Table 1. The values of n_i, c_i, C_i, and r_i that are needed for equations (2), (4), (5), and (6) are shown at the foot of the table. Using these, eq. (2) produces the population size estimates $\hat{N}_2 = 94.5$ and $\hat{N}_3 = 82.9$. The square roots of the estimated variances from equation (4) then give the estimated standard errors $S\hat{e}(\hat{N}_2) = 16.9$ and $S\hat{e}(\hat{N}_3) = 16.7$. Eq. (5) gives the estimated survival rates $\hat{s}_1 = 0.812$ and $\hat{s}_2 = 0.625$. Finally, Eq. (6) produces the estimated birth number $\hat{B}_2 = 23.8$. The data in this example are part of the illustrative data used by Manly and Parr [5].

Seber [7], Southwood [8], and Begon [1] discuss the Manly—Parr method in the context of capture—recapture methods in general. There are two principal competitors for the analysis of data from open populations, the Jolly—Seber method [3,6] and the Fisher—Ford method [2]. The main theoretical advantage of the Manly—Parr approach is that it does not require the assumption that the probability of survival is the same for animals of all ages.

REFERENCES

1. Begon, M. (1979). *Investigating Animal Abundance*. Edward Arnold, London. (An introduction to capture—recapture methods aimed mainly at biologists. There is a lengthy discussion on the relative merits of different methods of analysis for capture—recapture data.)

2. Fisher, R. A. and Ford, E. B. (1947). *Heredity*, **1**, 143–174.

3. Jolly, G. M. (1965). *Biometrika*, **52**, 225–247.

4. Manly, B. F. J. (1969). *Biometrika*, **56**, 407–410.

5. Manly, B. F. J. and Parr, M. J. (1968). *Trans. Soc. Brit. Ent.*, **18**, 81–89.

6. Seber, G. A. F. (1965). *Biometrika*, **52**, 249–259.

7. Seber, G. A. F. (1982). *The Estimation of Animal Abundance and Related Parameters*, 2nd. ed. Charles Griffin, London. (This is the standard reference for statistical and mathematical aspects of capture—recapture methods.)

8. Southwood, T. R. E. (1978). *Ecological Methods*. Chapman and Hall, London. (This is a standard text for ecologists. One chapter gives a good survey of both statistical and practical aspects of capture—recapture methods.)

See also Animal Science, Statistics in; Capture–Recapture Methods—I; Ecological Statistics; and Wildlife Sampling.

BRYAN F. J. MANLY

ANIMAL SCIENCE, STATISTICS IN

INTRODUCTION AND HISTORY

The introduction of modern statistical methods has been a slower process in animal science than in agriculture*. In designing experiments there have been limitations in the maintenance costs and duration of some of the experiments. Sampling methods for estimating animal numbers have been difficult to adopt, because animals, being mobile, may be counted repeatedly.

The first application of statistical methods to animal populations was in 1890 when Weldon [44] showed that the distributions of different measurements (ratios to total length) made on four local races of shrimp (*Crangon vulgaris*) closely followed the normal law. Weldon [45] found that the frontal breadths of Naples crabs formed into asymmetric curves with a double hump. It was this problem of dissection of a frequency distribution into two normal components that led to Karl Pearson's [27] first statistical memoir. The capture—mark—recapture method was first used by Petersen [28] in his studies of the European plaice and the earliest beginnings in the quantitative study of bird populations were made by Jones [21].

Some developments in the planning and analysis of animal experiments and in sampling methods for estimating animal numbers will be reviewed in the following sections.

DESIGN OF EXPERIMENTS

Choice of Experimental Unit

In feeding trials the experimental unit may be a pen. Often one pen of animals is fed on each treatment. The animals are assigned at random to each pen and the treatments are assigned at random to the pens. With a single pen of animals receiving a treatment, the effect of factors such as breed and age cannot be separated from treatment differences to provide a valid estimate of error [19,23].

Change-over Designs

One way of controlling the variation among animals in experiments is to adopt change-over designs* in which different treatments are tested on the same animal, such that each animal receives in sequence two or more treatments in successive experimental periods. The analysis of switchback trials for more than two treatments is dealt with in ref. 24. Incomplete block designs*, where the number of treatments exceeds the number of experimental units per block, are given in refs. 25 and 26.

Complete Block Designs

Randomized block designs (RBD) and Latin squares are common in animal experiments. A series of Latin squares is recommended for use in large experiments on rodents [39].

Alternatively, analysis of covariance* has been used in which a concomitant observation is taken for eliminating the effect of variations arising out of the peculiarities of individual animals. Thus, to test the effect of treatments on milk production of cows, the yield in the previous lactation is used as a concomitant variate to reduce the error in the experimental yield. The method assumes that (i) the regression of y on the concomitant variate x is significant and (ii) x does not vary significantly among treatments.

Scientists often object to use of RBDs or Latin squares*, which make dosing and sampling procedures difficult to operate in practice. In such cases, a standard order of groups may be used within each block, e.g., use of a standard Latin square within a battery of cages in rodent studies.

Split-Plot Designs*

Incomplete blocks or fractional replication designs may be used to eliminate the influence of litter differences from treatment effects when the number of treatments is large. An experiment on mice in which larger units are split into smaller ones is described in ref. 8 to provide increased precision on the more interesting comparisons.

Repeated Measurements* Experiments

A RBD in which the experimental units are animals that receive each treatment in turn is a repeated measurement experiment. This is generally analyzed as a split-plot design, which assumes equal correlation for every pair of subplot treatments. Multivariate methods that do not require this assumption are given in ref. 5.

Regression*

A common problem in animal studies is the estimation of x from the regression of y on x when y is easy to measure but x is difficult and expensive. Thus y and x may represent the length and age of a fish and we may want to estimate the age composition to predict the status of the stock in future years. Other examples are given in ref. 16. A polynomial regression* of average daily gains of each of several Holstein calves on time was fitted in ref. 1 and analysis of variance* (ANOVA)

done on the regression coefficients. Nonlinear regressions for describing weight—age relationship in cattle are compared in ref. 3. Fitzhugh [14] reviewed analysis of growth curves* and strategies for altering their shape.

Anova Models

Sometimes observations may be lost due to death or premature drying of lactation. The estimation of variance components* in mixed models with unbalanced data are dealt with in ref. 17 and illustrations from animal experiments are given in ref. 2. Analysis of balanced experiments for linear and normal models is discussed in ref. 18.

Transformations

Often the data violate basic assumptions of the methods of analysis. For violations due to heterogeneity, transformation of the individual observations into another scale may reduce heterogeneity, increase precision in treatment comparisons, and reduce nonadditivity [41]. Illustrations from animal experiments are given in ref. 16.

In management trials with pigs a score of x (out of 100) is assigned to each pig for treatment comparisons. Treatment effects will not be additive and the transformation $\log[(x + \frac{1}{2})/(100 + \frac{1}{2} - x)]$ given in ref. 8 would rectify this.

Categorical Data

Log-linear* and generalized least-squares approaches have been used for categorical data* on the incidence of pregnancy in ewes and of buller steers [33]. Log-linear models for capture—recapture* data from closed populations are mainly discussed in refs. 6 and 34; extensions to open populations are in ref. 7.

ESTIMATION OF ANIMAL NUMBERS OR DENSITY

Two important reasons for using sampling methods for estimation of animal numbers or density are (1) limitation of funds, time, and resources and (2) lesser disturbance of the population and its environment than by total count operations. Some important methods for estimating animal numbers will be summarized in the following sections.

Strips, Transects, Or Quadrats

A common method is to count the number within a long rectangle of known area called a *transect*, a short rectangle called a *strip*, or a square called a *quadrat*. The average density per unit area is estimated by taking total counts over randomly chosen areas. The total population is estimated by multiplying average density by the total area of the population. Quadrat sampling* is preferred for some big game species and its size and shape will depend upon the habitat, abundance, and mobility of the species. When a population is randomly distributed, the number(s) of quadrats to be sampled in an area is given by $s = S/(1 + NC^2)$, where S is the total number of quadrats, N is the size of the population being sampled, and C is the coefficient of variation* of $\hat{N}$ [34].

When the population density varies over large areas, stratified sampling with optimum allocation* is recommended. Siniff and Skoog [40] allocated a sampling effort for six caribou strata on the basis of preliminary population estimates of big game in aerial surveys of Alaska caribou conducted in 22,000 square miles of the Nelchina area. The method was based on rectangular sample plots 4 square miles in area on which observers attempted to count all animals. Two important advantages of the method were (i) reduction of the observer bias in sighting by using large transects and (ii) the use of an efficient design through optimum allocation of effort based on available information on caribou distribution. Sampling effort allocated to strata reduced the variance by more than half over simple random sampling*. Other examples are given in refs. 13 and 36. Where it is difficult or time consuming to count all animals in all the sampled quadrats, two-phase sampling using ratio or regression methods may be adopted. Somewhat similar to quadrat sampling in birds and mammals is the use of sampling in estimating fish population in a section of a stream. Sample areas are selected and fish captured by seining or electrical shocking. The areas are screened to prevent the dispersal of the fish while they are being caught.

Stratification in weirs or seines [20] is common to the Atlantic herring or sardine fishing off the New England and New Brunswick coasts.

Strip Sampling. Parallel lines one strip width ($2W$, say) apart determine the population of strips. All the n animals observed within the sampled strips are counted. The estimate of total number of animals (N) is given by $A \cdot n/(2LW)$, where L and A represent the length and population area of the strip. Strip sampling involves less risk of repeated counting and is generally recommended in aerial surveys [42,43].

Line Transect Sampling. An observer walks a fixed distance L along a transect or set of transects and records for each of the n sighted animals, its right angle distance $y_i(i = 1, 2, \ldots, n)$ from the transect or its distance r_i from the observer or both. The technique is most useful when the animals move so fast that they can only be seen when they are flushed into the open. Various parametric estimators have been suggested in refs. 15 and CAPTURE–RECAPTURE METHODS—I. Nonparametric estimators are given in refs. 4 and 10. References 15 and 34 deal with consequences for departures from the assumptions. All the estimates are of the form $A \cdot n/(2LW)$, where W is some measure of one-half the "effective width" of the strip covered by the observer as he moves down the transect. The transect method can also be used in aerial surveys. The procedure is the same as for walking or driving and is likewise subject to error. *See also* LINE INTERCEPT SAMPLING and LINE INTERSECT SAMPLING.

Capture—Mark—Recapture (CMR) Methods

A number of M animals from a closed population are caught, marked, and released. On a second occasion, a sample of n individuals are captured. If m is the number of marked animals in the sample, a biased estimate of N and its variance are

$$\hat{N} = \frac{n}{m}M,$$

$$v(\hat{N}) = \frac{\hat{N}^2(\hat{N} - m)(\hat{N} - n)}{Mn(\hat{N} - 1)}.$$

This was first given in ref. 28 using tagged plaice. Because the coefficient of variation of

$\hat{N}$ is approximately given by $1/m^{1/2}$, it follows that for $\hat{N}$ to be efficient, we should have sufficient recaptures in the second sample. For closed populations $\hat{N}$ appears to be the most useful estimate of N if the basic assumptions [34] underlying the method are met. CAPTURE–RECAPTURE METHODS—I contains recent developments in estimating N for "closed" and "open" populations when marking is carried over a period of time. A capture—recapture design robust to unequal probability of capture is given in ref. 29. The linear logistic binary regression model has been used in ref. 30 to relate the probability of capture to auxiliary variables for closed populations. With tag losses due to nonreporting or misreporting of tag returns by hunters, $\hat{N}$ will be an overestimate of N [35]. A treatment of response and nonresponse errors in Canadian waterfowl surveys is given in refs. 37 and 38; response errors were found larger than the sum of sampling and nonresponse errors.

Change-In-Ratio and Catch-Effort Methods. The Change-In-Ratio method estimates the population size by removing individual animals if the change in ratio of some attribute of the animal, e.g., age or sex composition, is known. For a closed population the maximum-likelihood estimator (MLE) of the population total $N_t(t = 1, 2)$, based on samples n_t at the beginning and end of the "harvested period," is given by

$$\hat{N}_t = \left(R_m - R\hat{P}_t\right) / \left(\hat{P}_1 - \hat{P}_2\right),$$

where R_m and $R_f(R = R_m + R_f)$ are, respectively, the males and females caught during the harvested period $\hat{P}_t = m_t/n_t (t = 1, 2)$, where m_t is the number of males at time t. The method assumes (i) a closed population, (ii) $\hat{P}_t = P = $ const for all t, and (iii) R_m and R_f are known exactly. A detailed discussion when the assumptions are violated is given in ref. 34. In fisheries attempts have been made to estimate populations using harvest and age structure data [32].

In the catch-effort method, one unit of sampling effort is assumed to catch a fixed proportion of the population. It is shown in refs. 9 and 22 that

$$E(C_t|k_t) = K(N - k_t),$$

where C_t is the catch per unit effort in the tth period, k_t is the cumulative catch through time period $(t - 1)$, K is the catchability coefficient, and N is the initial population size. C_t plotted against k_t provides estimates of the intercept KN and the slope K, whence N can be estimated.

Indices

Indices are estimates of animal population derived from counts of animal signs, e.g., pellet group, roadside count of breeding birds, nest counts, etc. The investigator records the presence or absence of a species in a quadrat and the percentage of quadrats in which the species is observed, giving an indication of its relative importance. Stratified sample surveys are conducted annually in U.S.A. and Canada for measuring changes in abundance of nongame breeding birds during the breeding season [12,31].

Big game populations are estimated by counting pellet groups in sample plots or transects. A method for calibrating an index by using removal data is given in ref. 11.

SUMMARY

A basic problem in experimental design with animals is substantial variation among animals owing to too few animals per unit. Hence, the need for concomitant variables* and their relationship with the observations for reducing the effect of variations due to individual animals in a unit. The alternative approach, to reduce this variation by picking experimental animals from a uniform population is not satisfactory, since the results may not necessarily apply to populations with much inherent variability.

Animal scientists are often not convinced of the importance of good experimentation, so that laboratory experiments are not necessarily the best to detect small differences. Closer cooperation is needed between the animal scientist and the statistician to yield the best results. Where the underlying assumptions are false, it is important to investigate whether this would invalidate the conclusions. Animal scientists should be alert to the severity of bias in the use of regression for calibration of one variable as an indicator for another when the problem is an inverse one and the relation between the variables is not very close.

In experiments with repeated observations, with successive observations correlated on the same animal, the animal scientist should consult the statistician in the use of the most appropriate method of analysis.

Sampling techniques for estimating the size of animal populations are indeed difficult to implement since animals often hide from us or move so fast that the same animal may be counted more than once. The choice of method would depend on the nature of the population, its distribution, and the method of sampling. Where possible, the design should be flexible enough to enable the use of more than one method of estimation.

In the past, the Petersen method has been used with too few captures and recaptures, leading to estimates with low precision. As far as possible, more animals should be marked and recaptured to ensure higher precision. Planning for a known precision has been difficult owing to lack of simple expressions for errors. Where the total sample can be split into a number of interpenetrating subsamples*, separate estimates of population size can be formed, resulting in simpler expressions for overall error. This point needs examination for population estimates.

The CMR method assumes that both mortality and recruitment are negligible during the period of data collection*. Another assumption underlying the method is that marked and unmarked animals have the same probability of being caught in the second sample. When these assumptions are violated, it is useful to compare the results with other estimating procedures and, if possible, test on a known population. The latter is recommended for use in animal populations when the methods are likely to vary widely in accuracy.

Nonresponse and response errors often form a high proportion of the total error in estimation of animal numbers or their density. As an example of such errors, visibility bias of the observer in aerial surveys, which results in a proportion of the animal being overlooked, may be cited. In such cases, the population total or its density should be estimated by different groups of experienced observers, either through the use of interpenetrating subsamples or bias corrected by use of alternative methods, e.g., air—ground comparisons.

REFERENCES

1. Allen, O. B., Burton, J. H., and Holt, J. D. (1983). *J. Animal Sci.*, **55**, 765–770.
2. Amble, V. M. (1975). *Statistical Methods in Animal Sciences*. Indian Society of Agricultural Statistics, New Delhi, India.
3. Brown, J. E., Fitzhugh, H. A., and Cartwright, T. C. (1976). *J. Animal Sci.*, **42**, 810–818.
4. Burnham, K. P. and Anderson, D. R. (1976). *Biometrics*, **32**, 325–336.
5. Cole, J. W. L. and Grizzle, J. E. (1966). *Biometrics*, **22**, 810–828.
6. Cormack, R. M. (1979). In *Sampling Biological Populations*, R. M. Cormack, G. P. Patil, and D. S. Robson, eds. Satellite Program in Statistical Ecology, International Cooperative Publishing House, Fairland, MD, pp. 217–255.
7. Cormack, R. M. (1984). *Proc. XIIth International Biom. Conf.*, pp. 177–186.
8. Cox, D. R. (1958). *Planning of Experiments*. Wiley, New York.
9. De Lury, D. B. (1947). *Biometrics*, **3**, 145–167.
10. Eberhardt, L. L. (1978). *J. Wildl. Manag.*, **42**, 1–31.
11. Eberhardt, L. L. (1982). *J. Wildl. Manag.*, **46**, 734–740.
12. Erskine, A. J. (1973). *Canad. Wildl. Serv. Prog. Note*, **32**, 1–15.
13. Evans, C. D., Troyer, W. A., and Lensink, C. J. (1966). *J. Wildl. Manag.*, **30**, 767–776.
14. Fitzhugh, H. A. (1976). *J. Animal Sci.*, **42**, 1036–1051.
15. Gates, C. E. (1969). *Biometrics*, **25**, 317–328.
16. Gill, J. L. (1981). *J. Dairy Sci.*, **64**, 1494–1519.
17. Henderson, C. R. (1953). *Biometrics*, **9**, 226–252.
18. Henderson, C. R. (1969). In *Techniques and Procedures in Animal Science Research*. American Society for Animal Sciences, pp. 1–35.
19. Homeyer, P. G. (1954). *Statistics and Mathematics in Biology*. O. Kempthorne et al., eds. Iowa State College Press, Ames, IA, pp. 399–406.
20. Johnson, W. H. (1940). *J. Fish. Res. Board Canad.*, **4**, 349–354.

21. Jones, L. (1898). *Wils. Orn. Bull.*, **18**, 5–9.

22. Leslie, P. H. and Davis, D. H. S. (1939). *J. Animal Ecol.*, **8**, 94–113.

23. Lucas, H. L. (1948). *Proc. Auburn Conference on Applied Statistics*, p. 77.

24. Lucas, H. L. (1956). *J. Dairy Sci.*, **39**, 146–154.

25. Patterson, H. D. (1952). *Biometrika*, **39**, 32–48.

26. Patterson, H. D., and Lucas, H. L. (1962). Tech. Bull. No. 147, N. C. Agric. Exper. Stn.

27. Pearson K. (1894). *Philos. Trans. R. Soc. Lond., A*, **185**, 71–110.

28. Petersen, C. E. J. (1896). *Rep. Dan. Biol. Stat.*, **6**, 1–48.

29. Pollock, K. H. and Otto, M. C. (1983). *Biometrics*, **39**, 1035–1049.

30. Pollock, K. H., Hines, J. E., and Nichols, J. D. (1984). *Biometrics*, **40**, 329–340.

31. Robbins, C. S. and Vanvelzen, W. T. (1967). *Spec. Sci. Rep. Wildl. No. 102*, U.S. Fish Wildl. Serv.

32. Ricker, W. E. (1958). *Bull. Fish. Board Canad.*, **119**, 1–300.

33. Rutledge, J. J. and Gunsett, F. C. (1982). *J. Animal Sci.*, **54**, 1072–1078.

34. Seber, G. A. F. (1980). *The Estimation of Animal Abundance and Related Parameters*, 2nd ed. Griffin, London, England.

35. Seber, G. A. F. and Felton, R. (1981). *Biometrika*, **68**, 211–219.

36. Sen, A. R. (1970). *Biometrics*, **26**, 315–326.

37. Sen, A. R. (1972). *J. Wildl. Manag.*, **36**, 951–954.

38. Sen, A. R. (1973). *J. Wildl. Manag.*, **37**, 485–491.

39. Shirley, E. (1981). *Bias*, **8**, 82–90.

40. Siniff, D. B. and Skoog, R. O. (1964). *J. Wildl. Manag.*, **28**, 391–401.

41. Tukey, J. W. (1949). *Biometrics*, **5**, 232–242.

42. Watson, R. M., Parker, I. S. C., and Allan, T. (1969). *E. Afr. Wildl. J.*, **7**, 11–26.

43. Watson, R. M., Graham, A. D., and Parker, I. S. C. (1969). *E. Afr. Wildl. J.*, **7**, 43–59.

44. Weldon, W. F. R. (1890). *Proc. R. Soc. Lond.*, **47**, 445–453.

45. Weldon, W. F. R. (1893). *Proc. R. Soc. Lond.*, **54**, 318–329.

BIBLIOGRAPHY

Seber, G. A. F. (1986). *Biometrics*, **42**, 267–292. (Reviews animal abundance estimation

techniques developed between 1979 and 1985, some of which supplant previous methods. This is a very thorough survey, with over 300 references.)

See also CAPTURE–RECAPTURE METHODS—I; ECOLOGICAL STATISTICS; LINE TRANSECT SAMPLING; QUADRAT SAMPLING; and WILDLIFE SAMPLING.

A. R. SEN

ANNALES DE L'INSTITUT HENRI POINCARÉ (B)

The *Annales de l'Institut Henri Poincaré, Section B, Probabilités et Statistiques* publishes articles in French and English, and is an international journal covering all aspects of modern probability theory and mathematical statistics, and their applications.

The journal is published by Elsevier; website links are www.elsevier.com/locate/anihpb and www.sciencedirect.com/science/journal/02460203 . The Editor-in-Chief works with an international Editorial Board of twelve or so members.

ANNALS OF APPLIED PROBABILITY

[This entry has been updated by the Editors.]

The *Annals of Applied Probability (AAP)* is the youngest of the *"Annals"* series to be created by the Institute of Mathematical Statistics* (IMS).

The website for *AAP* is www.imstat.org/aap/.

In 1973 the original *Annals of Mathematical Statistics* was split into the *Annals of Statistics** and the *Annals of Probability**. Over the next two decades both of these expanded noticeably in size: the 1973–1979 volumes of *Annals of Probability* contained around 1100 pages, but the 1993 volume more than doubled this.

By the late 1980s, the Council of the IMS had decided to further split its publications in probability theory and its applications into two journals—the continuing *Annals of Probability* and the *Annals of Applied Probability*. There were factors other than increasing size involved in this decision. Paramount amongst them was the feeling that the *Annals of*

Probability was failing to attract outstanding papers in applied probability with the ease that it was attracting such papers in the theory of probability. The IMS felt that this was partly because the *Annals of Probability* was seen as being aimed at an audience and an authorship interested in the purer aspects of probability theory, despite the efforts of successive editors to solicit papers with applications.

The first volume of the *Annals of Applied Probability* appeared in 1991, and *AAP* quickly established itself as one of the leading journals in the area, seamlessly carrying on the tradition of the other *Annals* published by the IMS.

EDITORIAL POLICY

The *Annals of Applied Probability* has two overriding criteria for accepting papers, other than formal correctness and coherence. These are

1. that the results in the paper should be genuinely applied or applicable; and
2. that the paper should make a serious contribution to the mathematical theory of probability, or in some other sense carry a substantial level of probabilistic innovation, or be likely to stimulate such innovation.

The first criterion in particular can be hard to define, and in some cases it has a broad interpretation. But in differentiating this journal from its sibling journal, the *Annals of Probability*, it is a criterion that is applied with some care. The Editorial Board has rejected a number of excellent papers because the authors did not make a convincing case for applicability, and indeed several of these have been forwarded (with the author's permission) to the other journal, and some have been accepted there.

The second, more mathematical criterion is also taken seriously, and in this the *Annals of Applied Probability* follows the tradition of the original *Annals of Mathematical Statistics*, which was set up to provide an outlet for the more mathematical papers written by members of the American Statistical Association. Thus the *Annals of Applied Probability* has rejected a number of excellent papers in, say, queueing theory, or in the theory of biological models, where the applications were indeed of interest but where the contribution to the mathematical or probabilistic theory was felt to be limited.

The editorial policy of *AAP* is stated on the website exhibited above. In addition to the two criteria just discussed, it states:

> "The *Annals of Applied Probability* aims to publish research of the highest quality reflecting the varied facets of contemporary Applied Probability. Primary emphasis is placed on importance and originality ... .
> "Mathematical depth and difficulty will not be the sole criteria with respect to which submissions are evaluated. Fundamentally, we seek a broad spectrum of papers which enrich our profession intellectually, and which illustrate the role of probabilistic thinking in the solution of applied problems (where the term "applied" is often interpreted in a side sense)."
> "Most papers should contain an Introduction which presents a discussion of the context and importance of the issues they address and a clear, non-technical description of the main results. The Introduction should be accessible to a wide range of readers. Thus, for example, it may be appropriate in some papers to present special cases or examples prior to general, abstract formulations. In other papers a discussion of the general scientific context of a problem might be a helpful prelude to the main body of the paper. In all cases, motivation and exposition should be clear".

All papers are refereed.

TYPES OF PAPERS CARRIED

Stating such general criteria for publication is one thing; seeing them in operation in practice is another. The style of a journal is always best described in terms of its actual content, and although this is by no means stationary in time, the following areas are ones in which the *Annals of Applied Probability* carried very significant contributions in its first years:

1. Queueing networks, with the Lanchester Prize-winning paper on loss networks by Frank Kelly, papers on unexpectedly unstable networks, and much

of the seminal literature on fluid model approximations;

2. Stochastic models in finance*, with applications to investment strategies, options analysis, arbitrage models, and more;

3. Rates of convergence papers for Markov chains*, with contributions ranging from finite to general space chains;

4. Foundational work on Markov-chain Monte Carlo* models (*see* also GIBBS SAMPLING);

5. Development of stochastic algorithms in computing, and analysis of their properties.

This list is by no means exhaustive; the behavior of diffusion processes*, branching processes*, and the like is of course widely studied in applied probability, and excellent papers in these areas have appeared in the *Annals of Applied Probability*. However, to be encouraged by the Editors, there needs to be more than formal technical novelty in such papers, and authors are encouraged to consider this carefully when submitting to the journal.

STRUCTURE AND ACCEPTANCE PROCESS

Currently, the *Annals of Applied Probability* has an Editorial Board, consisting of the Editor (appointed by the IMS Council for a three-year term), a Managing Editor (with shared responsibilities for the *Annals of Probability*) and 25 or so Associate Editors from around the world. Past Editors of the *Annals of Probability* are:

J. Michael Steele, 1991–1993,
Richard L. Tweedie, 1994–1996,
Richard T. Durrett, 1997–1999,
Søren Asmussen, 2000–2002,
Robert Adler, 2003–.

FUTURE PLANNING

As with many journals, the *Annals of Applied Probability* is moving towards an era of improved publication speed using electronic means.

See also ANNALS OF PROBABILITY; ANNALS OF STATISTICS; and INSTITUTE OF MATHEMATICAL STATISTICS.

PAUL S. DWYER

ANNALS OF EUGENICS.   See ANNALS OF HUMAN GENETICS

ANNALS OF HUMAN GENETICS

The *Annals of Human Genetics* was one of the earliest and has remained one of the foremost journals concerned with research into genetics*; it is specifically concerned with human genetics*. The home page for the journal is www.gene.ucl.ac.uk/anhumgen.

The original title was *Annals of Eugenics*, subtitled "A Journal for the scientific study of racial problems." The word "Eugenics" had been coined by Sir Francis Galton*, who defined it in 1904 as "the science which deals with all the influences that improve the inborn qualities of a race; also those that develop them to the utmost advantage." In his foreword to the first volume of the *Eugenics Review* in 1909 he said that "the foundation of Eugenics is laid by applying mathematical statistical treatment to a large collection of facts." He was a man of wide scientific interests, which included stockbreeding, psychology, and the use of fingerprints for identification, and he was a cousin of Charles Darwin. He was also one of the early white explorers of Africa and a prolific writer on these and many other subjects.

The journal was founded in 1925 by Karl Pearson*, who was head of the Department of Applied Statistics and of the Galton Laboratory at University College; it was printed by Cambridge University Press. The Galton Laboratory had been founded under Galton's will, and Karl Pearson was the first occupant of the Galton Chair of Eugenics. He had been an intimate friend and disciple of Galton and remained faithful to many of his ideas. The journal's aims were set out in a foreword to the first volume, where eugenics is defined as "the study of agencies under social control that may improve or impair the racial qualities of future generations either physically or mentally." The quotations from Galton and Darwin still retained on the cover of the

Annals indicate its commitment to mathematical and statistical techniques. Until the arrival of the computer and the resulting enormous increase in the amount of information collected, the journal emphasized the necessity of publishing data with papers; this is now most often deposited in record offices when it is extensive.

Karl Pearson retired in 1933 and was succeeded as editor in 1934 by Ronald A. Fisher*, who was also keenly interested in the development of statistical techniques, but was critical of some of Pearson's statistical methods. He became Galton Professor in 1934, and the editorship of the *Annals* went with the Chair then as now.

Fisher innovated many well-known statistical techniques and showed how they could be applied to genetical problems. He changed the subtitle of the *Annals* to "A Journal devoted to the genetic study of human populations." He co-opted several members of the Eugenics Society on to the editorial board. This society had been founded independently in 1908 as the Eugenics Education Society; its purpose was to propagate Galton's ideas and the work of the Laboratory, and Galton had accepted the presidency. Ronald Fisher had been an active member from its early days. This partnership seems only to have lasted until the outbreak of war, when Fisher returned to Rothamsted Experimental Station. He stayed there until 1943, when he accepted the Chair of Genetics at Cambridge and his editorship ended. In the foreword to the first volume of the *Annals* which he edited, Vol. 6, he announces its policy to be consistent with the aims of its founder:

> The contents of the journal will continue to be representative of the researches of the Laboratory and of kindred work, contributing to the further study and elucidation of the genetic situation in man, which is attracting increasing attention from students elsewhere. The two primary disciplines which contribute to this study are genetics and mathematical studies.

In 1945, Lionel S. Penrose succeeded him as editor of the *Annals*. He was a distinguished medical man and alienist, and under him the journal became more medical in content. Some of his papers on Down's anomaly,

a permanent interest of his, and other aspects of inherited mental illness appeared in the journal. A feature of it in his time was the printing of pedigrees of inherited diseases covering several generations. He was responsible for changing the title from *Annals of Eugenics* to *Annals of Human Genetics*, a change for which it was necessary for an act of Parliament to be passed. The subtitle was also changed again to "A Journal of Human Genetics." He retired in 1965, and so did M. N. Karn, who had been the assistant editor since prewar days. Penrose was editor for a longer period than either of his predecessors and under his guidance the journal broadened its coverage and drew its contributions from a wider field.

Harry Harris, also a medical man and biochemist, succeeded to the Galton Chair and the editorship in 1965, and coopted C. A. B. Smith, mathematician and biometrician, who had been on the editorial board since 1955, to be coeditor. Professor Harris was also head of a Medical Research Council Unit of Biochemical Genetics which became associated with the Galton Laboratory. Reflecting the editors' interests, the contents of the *Annals* inevitably became more concerned with biochemical genetics and statistics; *Annals of Eugenics* was dropped from the title page.

In 1975, Harris accepted the Chair of Human Genetics at the University of Pennsylvania, and for the next two years Cedric Smith was virtually the sole editor, as the Galton Chair remained vacant. In 1978, Elizabeth B. Robson was appointed to it, and she and C. A. B. Smith became the editors. The journal has always been edited at the Galton Laboratory.

The journal is a specialized one dealing with human genetics, but has changed with the progress of research and changing methods, currently publishing material related directly to human genetics or to scientific aspects of human inheritance.

As the website for the journal points out, during the latter part of the twentieth century it became clear that our understanding of variation in the human genome could be enhanced by studying the interaction between the fields of population genetics and molecular pathology. Accordingly, contemporary topics included in the *Annals of Human*

Genetics include human genome variation, human population genetics, statistical genetics, the genetics of multifactorial diseases, Mendelian disorders and their molecular pathology, and pharmacogenetics. Animal as well as human models may be considered in some of these areas.

Most articles appearing in the journal report full-length research studies, but many issues include at least one review paper. Shorter communications may also be published.

For many years the journal was published by the Galton Laboratory under the ownership of University College London. Since 2003 it is published by Blackwell. It has an Editor-in-Chief, a Managing Editor, five or so Senior Editors, a Reviews Editor and an international Editorial Board of 30 or so members. All papers are rigorously referred.

See also ENGLISH BIOMETRIC SCHOOL; FISHER, RONALD AYLMER; GALTON, FRANCIS; HUMAN GENETICS, STATISTICS IN—II; and PEARSON, KARL—I.

JEAN EDMISTON
The Editors

ANNALS OF MATHEMATICAL STATISTICS. See ANNALS OF STATISTICS

ANNALS OF PROBABILITY

[This entry has been updated by the Editors.]

The *Annals of Probability (AP)* was one of two journals that evolved from the *Annals of Mathematical Statistics (AMS)* in 1973, the other being the *Annals of Statistics* (AS)*. All three journals are (or were) official publications of the Institute of Mathematical Statistics* (IMS).

Readers are referred to the entries ANNALS OF STATISTICS and INSTITUTE OF MATHEMATICAL STATISTICS for the evolution both of *AP* and *AS* out of *AMS*.

The *Annals of Probability* is published bimonthly, each volume consisting in the six issues of each calendar year. Due to the expansion of *AP* in the 1970s and 1980s, the IMS Council decided to split *AP* into two journals; *Annals of Probability* continued publication, but with a theoretical focus, and *Annals of Applied Probability* (AAP)*

Table 1. Editors, The Annals of Probability

Ronald Pyke,	Burgess Davis,
1972–1975	1991–1993
Patrick Billingsley,	Jim Pitman,
1976–1978	1994–1996
R. M. Dudley,	S. R. S. Varadhan,
1979–1981	1996–1999
Harry Kesten,	Thomas K. Kurtz,
1982–1984	2000–2002
Thomas M. Liggett,	Steve Lalley,
1985–1987	2003–2005
Peter Ney,	
1988–1990	

with a focus on applications began publication in 1991. See the entry on *AAP* for further discussion on the rationale for the split.

The editorship of AP has been held roughly for three-year periods; see Table 1. In the first issue (Feb. 1973) appeared the policy statement:

"The main aim of the *Annals of Probability* and the *Annals of Statistics* is to publish original contributions related to the theory of statistics and probability. The emphasis is on quality, importance and interest; formal novelty and mathematical correctness alone are not sufficient. Particularly appropriate for the *Annals* are important theoretical papers and applied contributions which either stimulate theoretical interest in important new problems or make substantial progress in existing applied areas. Of special interest are authoritative expository or survey articles, whether on theoretical areas of vigorous recent development, or on specific applications. All papers are referred."

The current editorial policy of AP appears on the journal website www.imstat.org/aop/:

"The *Annals of Probability* publishes research papers in modern probability theory, its relation to other areas of mathematics, and its applications in the physical and biological sciences. Emphasis is on importance, interest, and originality—formal novelty and correctness are not sufficient for publication. The *Annals* will also publish authoritative review papers and surveys of areas in vigorous development."

Currently the Editorial Board is comprised of the Editor, a Managing Editor, and 25 or so Associate Editors from around the world.

See also ANNALS OF APPLIED PROBABILITY; ANNALS OF STATISTICS; and INSTITUTE OF MATHEMATICAL STATISTICS.

ANNALS OF STATISTICS

[This entry has been updated by the Editors.]

The *Annals of Statistics (AS)*, first published in 1973, is one of the two journals resulting from a split of the old *Annals of Mathematical Statistics*, the other journal being the *Annals of Probability**. Three current *Annals* are official publications of the Institute of Mathematical Statistics* (IMS); in the late 1980s the IMS Council decided to split its publications on probability theory and its applications further; *see* ANNALS OF APPLIED PROBABILITY.

The website for the *Annals of Statistics* is www.imstat.org/aos/.

The original *Annals of Mathematical Statistics (AMS)* started before the Institute existed, and in fact was originally published by the American Statistical Association* in 1930. At that time it had become apparent that the *Journal of the American Statistical Association* (JASA)* could not adequately represent the interests of mathematically inclined statisticians, who were beginning to do important research. Willford King, writing in a prefatory statement to the first issue of *AMS*, said:

> The mathematicians are, of course, interested in articles of a type which are not intelligible to the non-mathematical readers of our Journal. The Editor of our Journal [JASA] has, then, found it a puzzling problem to satisfy both classes of readers.
>
> Now a happy solution has appeared. The Association at this time has the pleasure of presenting to its mathematically inclined members the first issue of the *ANNALS OF STATISTICS*, edited by Prof. Harry C. Carver of the University of Michigan. This Journal will deal not only with the mathematical technique of statistics, but also with the applications of such technique to the fields of astronomy, physics, psychology, biology, medicine, education, business, and economics. At present, mathematical articles along these lines are scattered through a great variety of publications. It is hoped that in the future they will be gathered together in the *Annals*.

The seven articles that followed in that first issue covered a very wide range indeed, as their titles suggest:

Remarks on Regression

Synopsis of Elementary Mathematical Statistics

Bayes Theorem

A Mathematical Theory of Seasonal Indices

Stieltjes Integrals in Mathematical Statistics

Simultaneous Treatment of Discrete and Continuous Probability by Use of Stieltjes Integrals

Fundamentals of Sampling Theory

Harry Carver, the founding editor, took on sole responsibility for the young journal when in 1934 the *ASA* stopped its financial support. He continued to publish privately until 1938 when the IMS took over the financial responsibility. Actually, the IMS had come into existence in 1935, and the *Annals* had been its official publication from the start. But Carver, a prime mover in starting the IMS, insisted that it not be tied down by support of the journal.

After the crucial period of Carver's editorship, S. S. Wilks was appointed editor, a post he held from 1938 until 1949. Since that time the editorship has been held for 3-year periods, initially by Wilks's appointed successors (Table 1).

Each editor has headed a distinguished editorial board, but none compares to the illustrious board established in 1938: Wilks, Craig, Neyman (co-editors), Carver, Cramér, Deming, Darmois, R. A. Fisher*, Fry, Hotelling*, von Mises, Pearson, Rietz, and Shewhart.

With this auspicious start, the *AMS* became the focal point for developments in theoretical statistics, particularly those developments associated with the general mathematical theories of estimation, testing, distribution theory, and design of experiments. The full impact of the *Annals*, past and present, is clearly seen in the bibliographies of most books on theoretical statistics. Many of the newer statistical methods can be traced back to pioneering research papers in *AMS*.

Table 1.

Editors of *Annals of Mathematical Statistics*

H. C. Carver (1930–1938)	William H. Kruskal (1958–1961)
S. S. Wilks (1938–1949)	J. L. Hodges, Jr. (1961–1964)
T. W. Anderson (1950–1952)	D. L. Burkholder (1964–1967)
E. L. Lehmann (1953–1955)	Z. W. Birnbaum (1967–1970)
T. E. Harris (1955–1958)	Ingram Olkin (1970–1972)

Editors of *Annals of Statistics*

Ingram Olkin (1972–1973)	Lawrence D. Brown and
I. R. Savage (1974–1976)	John A. Rice, 1995–1997
R. G. Miller, Jr. (1977–1979)	James O. Berger and
David Hinkley (1980–1982)	Hans R. Kunsch, 1998–2000
Michael D. Perlman, 1983–1985	John I. Marden and
Willem R. Van Zwet, 1986–1988	Jon A. Wellner, 2001–2003
Arthur Cohen, 1989–1991	Morris L. Eaton and
Michael Woodroofe, 1992–1994	Jianqing Fan, 2004–2006

After some 40 years of growing strength and size, the *AMS* was split in 1972 during the editorship of I. Olkin, who continued as first editor of *AS*. Each volume of *AS* is devoted entirely to research articles. (The news items and notices that used to appear in *AMS* until 1972 are published in the *IMS Bulletin*, issued bimonthly.) The journal does not publish book reviews. A volume currently consists of six bimonthly issues. All papers are referred under the general guidelines of editorial policy.

EDITORIAL POLICY

The following statement was made in the 1938 volume of *JASA*:

The *Annals* will continue to be devoted largely to original research papers dealing with topics in the mathematical theory of statistics, together with such examples as may be useful in illustrating or experimentally verifying the theory. However, in view of the purpose of the Institute of Mathematical Statistics which, interpreted broadly, is to stimulate research in the mathematical theory of statistics and to promote cooperation between the field of pure research and fields of application, plans are being made to extend the scope of the *Annals* to include expository articles from time to time on various fundamental notions, principles, and techniques in statistics. Recognizing that many theoretical statistical problems have their origin in various fields of pure and applied science

and technology, papers and shorter notes dealing with theoretical aspects of statistical problems arising in such fields will be welcomed by the editors.

The current editorial policy of the journal is stated on its website www.imstat.org/aos/, as follows:

"*The Annals of Statistics* aims to publish research papers of highest quality, reflecting the many facets of contemporary statistics. Primary emphasis is placed on importance and originality, not on formalism.

The discipline of statistics has deep roots in both mathematics and in substantive scientific fields. Mathematics provides the language in which models and the properties of statistical methods are formulated. It is essential for rigor, coherence, clarity and understanding. Consequently, our policy is to continue to play a special role in presenting research at the forefront of mathematical statistics, especially theoretical advances that are likely to have a significant impact on statistical methodology or understanding. Substantive fields are essential for continue vitality of statistics, since they provide the motivation and direction for most of the future developments in statistics. We thus intend to also publish papers relating to the role of statistics in inter-disciplinary investigations in all fields of natural, technical and social sciences. A third force that is reshaping statistics is the computational revolution, and the *Annals* will also welcome developments in this area. Submissions in these two latter categories will be evaluated primarily by the relevance of the issues addressed and the creativity of the proposed solutions.

"Lucidity and conciseness of presentation are important elements in the evaluation of submissions. The introduction of each paper should be accessible to a wide range of readers. It should thus discuss the context and importance of the issues addressed and give a clear, nontechnical description of the main results. In some papers it may, for example, be appropriate to present special cases or specific examples prior to general, abstract formulations, while in other papers discussion of the general scientific context of a problem might be a helpful prelude to the body of the paper."

Currently two Editors, a Managing Editor, and 40 or so Associate Editors in many countries serve on the Editorial Board.

AS continues to recognize its singular and historic role as publisher of general theory, while reflecting the impact that theory does and should have on practical problems of current interest. For that reason, relevance and novelty are at least as important as mathematical correctness.

See also ANNALS OF APPLIED PROBABILITY; ANNALS OF PROBABILITY; and INSTITUTE OF MATHEMATICAL STATISTICS.

D. V. HINKLEY

ANNALS OF THE INSTITUTE OF STATISTICAL MATHEMATICS

The first issue of this journal appeared in 1949. It is published in English by Kluwer.

The aims and scope of *AISM* are presented at the journal's website www.kluweronline .com/issn/0020-3157, as follows:

"*Annals of the Institute of Statistical Mathematics* (AISM) provides an international forum for open communication among statisticians and researchers working with the common purpose of advancing human knowledge through the development of the science and technology of statistics.
AISM will publish broadest possible coverage of statistical papers of the highest quality. The emphasis will be placed on the publication of papers related to: (a) the establishment of new areas of application; (b) the development of new procedures and algorithms; (c) the development of unifying theories; (d) the analysis and improvement of existing procedures and theories; and the communication of empirical findings supported by real data.
"In addition to papers by professional statisticians, contributions are also published by authors working in various fields of application. Authors discussing applications are encouraged to contribute a complete set of data used in their papers to the AISM Data Library. The Institute of Statistical Mathematics will distribute it upon request from readers (see p. 405 and 606, Vol. 43, No. 3, 1991). The final objective of AISM is to contribute to the advancement of statistics as the science of human handling of information to cope with uncertainties. Special emphasis will thus be placed on the publication of papers that will eventually lead to significant improvements in the practice of statistics."

AISM currently has an Editor-in-Chief, six Editors and 40 Associate Editors. All papers published in the journal are refereed.

ANOCOVA   TABLE. See ANALYSIS   OF COVARIANCE; ANOVA TABLE

ANOVA TABLE

An ANOVA (analysis of variance) table is a conventional way of presenting the results of an analysis of variance*. There are usually four columns, headed

1. Source (of variation)
2. Degrees of freedom*
3. Sum of squares
4. Mean square*

Columns 1 and 2 reflect the size and pattern of the data being analyzed and the model being used. Column 4 is obtained by dividing the entry (in the same row) in column 3 by that in column 2.

Sometimes there is a fifth column, giving the ratios of mean squares to a residual mean square* (or mean squares). These statistics are used in applying the standard F-test* used in the analysis of variance.

The value of the ANOVA table is not only in its convenient and tidy presentation of the quantities used in applying analysis-of-variance tests. The juxtaposition of all the quantities used for a number of different tests (or the mean squares for many different sources of variation) can provide valuable insight into the overall structure of variation. For example, in the analysis of a factorial experiment*, the groups of interactions* of specified order can provide evidence of relatively great variation arising when a particular factor (or group of factors) is involved.

The term "ANOCOVA table" is also used (although rather infrequently) to describe similar tables relevant to the analysis of covariance*.

See also ANALYSIS OF VARIANCE.

ANSARI—BRADLEY W-STATISTICS.

See SCALE TESTS, ANSARI—BRADLEY

ANSCOMBE DATA SETS

A celebrated classical example of role of residual analysis and statistical graphics in statistical modeling was created by Anscombe [1]. He constructed four different data sets (X_i, Y_i), $i = 1, \ldots, 11$ that share the same descriptive statistics $(\overline{X}\,\overline{Y}, \hat{\beta}_0, \hat{\beta}_1, MSE, R^2, F)$ necessary to establish linear regression fit $\hat{Y} = \hat{\beta}_0 + \hat{\beta}_1 X$.

The following statistics are common for the four data sets:

Sample size N	11
Mean of $X(\overline{X})$	9
Mean of $Y(\overline{Y})$	7.5
Intercept $(\hat{\beta}_0)$	3
Slope $(\hat{\beta}_1)$	0.5
Estimator of σ, (s)	1.2366
Correlation $r_{X,Y}$	0.816

A linear model is appropriate for Data Set 1; the scatterplots and residual analysis suggest that the Data Sets 2–4 are not amenable to linear modeling.

REFERENCE

1. Anscombe, F. (1973). Graphs in Statistical Analysis, *American Statistician*, **27** [February 1973], 17–21.

FURTHER READING

Tufte, E. R., *The Visual Display of Quantitative Information*, Graphic Press, 1983.

ANSCOMBE, FRANCIS JOHN

Frank Anscombe was born in Hertfordshire and grew up in Hove, England. His parents were Francis Champion Anscombe (1870–1942) and Honoria Constance Fallowfield Anscombe (1888–1974). His father worked for a pharmaceutical company in London and his mother was an early female graduate of the University of Manchester. Frank attended Trinity College, Cambridge, on a merit scholarship. He obtained a B.A. in

						Set 1					
X	10	8	13	9	11	14	6	4	12	7	5
Y	8.04	6.95	7.58	8.81	8.33	9.96	7.24	4.26	10.84	4.82	5.68
						Set 2					
X	10	8	13	9	11	14	6	4	12	7	5
Y	9.14	8.14	8.74	8.77	9.26	8.10	6.13	3.10	9.13	7.26	4.74
						Set 3					
X	10	8	13	9	11	14	6	4	12	7	5
Y	7.46	6.77	12.74	7.11	7.81	8.84	6.08	5.39	8.15	6.42	5.73
						Set 4					
X	8	8	8	8	8	8	8	19	8	8	8
Y	6.58	5.76	7.71	8.84	8.47	7.04	5.25	12.50	5.56	7.91	6.89

Mathematics in 1939 with first class honors, and an M.A. in 1943. During the war years, he was with the British Ministry of Supply, concerned with assessment of weapons and quality control of munitions production. From 1945 to 1948, he was in the statistics department at Rothamsted Experimental Station, Hertfordshire. In 1954, he married John Tukey's sister-in-law, Phyllis Elaine Rapp. They had four children (Francis, Anthony, Frederick, and Elizabeth). Anscombe died in 2001 after a long illness.

ACADEMIC CAREER

From 1948 to 1956, Frank Anscombe was lecturer in mathematics at the Statistical Laboratory, the University of Cambridge in England. In 1956, he moved to the mathematics department of Princeton University, as associate and then full professor. He left Princeton in 1963 to found the statistics department at Yale University. He chaired that department for six years, developing a graduate program. The department became known for its careful balance of theory and applications. He was a member of important advisory and evaluation committees. He retired in 1988.

STATISTICAL CONTRIBUTIONS

Anscombe made important contributions to the British World War II effort. In particular, he was concerned with the deployment of weapons, the aiming of anti-aircraft rockets, and the development of strategies for massing guns. After the war, he worked at Rothamsted on the applications of statistics to agriculture. During these years, he published papers in *Nature, Biometrika*, and *Biometrics,* and had a discussion paper in the *Journal of the Royal Statistical Society* [1–3,12]. He often wrote on the problems of sampling inspection and sequential estimation. One in particular [5] is a discussion paper on sequential analysis invited by the Royal Statistical Society. Later, however [7], he wrote, "Sequential analysis is a hoax."

The next decade saw Anscombe evolving into a subjective Bayesian on the one hand, but on the other delving directly into data analysis concerned with uses of residuals*

and into tricky problems created by outliers*. The Fourth Berkeley Symposium paper [6] and the *Technometrics* paper with John Tukey [13] were landmarks in the history of residual analysis.

The final stages of Anscombe's research career saw a move into computing [8,9]. In the Preface of the book [9], he writes that the work is "a festivity in ... honor" of J. W. Tukey and K. E. Iverson. Anscombe's last published paper [10] concerned testing in clinical trials*.

Two of Anscombe's theoretical papers are often referred to. Reference 4 presents Anscombe's Theorem, which provides conditions for replacing a fixed sample size by a random stopping time. The result was later extended by Rényi [14]. Reference 11 contains the Anscombe–Auman work showing that Savage's derivation of expected utility* can be considerably simplified.

CONCLUDING REMARKS

Anscombe was an elected member of the International Statistical Institute* and a charter member of the Connecticut Academy of Science and Engineering. He was the R. A. Fisher Lecturer in 1982. Throughout his career, he prepared pithy book reviews and contributed to discussions in a lively manner.

Anscombe was concerned with the popularization and simplification of statistics. For example, he had papers in *The American Statistician* and once wore a sombrero at an Institute of Mathematical Statistics meeting in an attempt to enliven it. He had important interests outside of statistics, including classical music, poetry, art, and hiking.

REFERENCES

1. Anscombe, F. J. (1948). The transformation of Poisson, binomial and negative-binomial data. *Biometrika*, **35**, 246–254.

2. Anscombe, F. J. (1948). The validity of comparative experiments (with discussion). *J. R. Stat. Soc. A*, **111**, 181–211.

3. Anscombe, F. J. (1949). The statistical analysis of insect counts based on the negative binomial distribution. *Biometrics*, **5**, 165–173.

4. Anscombe, F. J. (1952). Large-sample theory of sequential estimation. *Proc. Cambridge Philos. Soc.*, **48**, 600–607.

5. Anscombe, F. J. (1953). Sequential estimation (with discussion). *J. R. Stat. Soc. B*, **15**, 1–29.

6. Anscombe, F. J. (1961). Examination of residuals. *Proc. Fourth Berkeley Symp. Math. Statist. Probab.*, **1**, 1–36.

7. Anscombe, F. J. (1963). Sequential medical trials. *J. Am. Stat. Assoc.*, **58**, 365–383.

8. Anscombe, F. J. (1968). "Regression Analysis in the Computer Age". *Proc. Thirteenth Conf. Design Expts. Army Res. Dev. Testing.* U.S. Army Research Office, Durham, NC, pp. 1–13.

9. Anscombe, J. J. (1981). *Computing in Statistical Science Through APL*. Springer-Verlag, New York.

10. Anscombe, F. J. (1990). The summarizing of clinical experiments by significance levels. *Stat. Med.*, **9**, 703–708.

11. Anscombe, F. J. and Auman, J. (1963). A definition of subjective probability. *Ann. Math. Stat.*, **34**, 199–205.

12. Anscombe, F. J. and Singh, B. N. (1948). Limitation of bacteria by micro-predators in soil. *Nature*, **161**, 140–141.

13. Anscombe, F. J. and Tukey, J. W. (1963). The examination and analysis of residuals. *Technometrics*, **5**, 141–160.

14. Rényi, A. (1960). On the central limit theorem for the sum of a random number of independent random variables. *Acta Math. Acad. Sci. Hung.*, **11**, 97–102.

See also BAYESIAN INFERENCE; OUTLIERS; RESIDUALS; and SEQUENTIAL ESTIMATION.

DAVID R. BRILLINGER

ANTHROPOLOGY, STATISTICS IN

Statistical methods were introduced into physical anthropology by Adolphe Quetelet* and Francis Galton* during the nineteenth century. From their work grew the "Biometrics School," headed by Pearson* and Karl Weldon W.F.R.*, whose members studied the variation between local races in an attempt to clarify the processes of inheritance and evolution. Human skulls, in particular, were intensively studied because of the availability of historical material. The statistical treatment of data from skulls raised many new problems, and Pearson [23] took one of the first steps in multivariate analysis* by introducing the coefficient of racial likeness*, a statistic based on all measurements and used to assess the significance of differences between groups. There is less emphasis on craniometry today than there was in Pearson's time, but the basic statistical problem is still with us. How should variation in the shape of complex objects such as bones be described?

Pearson believed passionately in measurement* as the basis of all science, although he recognized that most advances in the study of shape had actually relied on visual comparisons. Unfortunately, the measurement of shape is very difficult. In the majority of published examples the procedure has been to identify common landmarks on the objects and then to measure angles and linear distances. The hope is that a statistical summary of these data will embody a summary of shape. The early biometricians were restricted to simple statistical summaries by primitive computing equipment, but techniques of multivariate analysis are now commonly used to attempt a more explicit description of shape variation within a group and to make comparisons between groups.

Landmarks are hard to find on some objects, and an alternative approach is to record the coordinates of a large number of points on the object. Measures of shape must then be based on geometric properties of the surfaces containing the points, and should be independent of which points are chosen. The technique has so far been limited to outlines from sections of the original objects for which curves rather than surfaces are relevant. This has been for practical rather than theoretical reasons. A plot of radial distance versus angle has been used to summarize outlines, but this has the disadvantage of depending on the choice of origin. An alternative is a plot of tangent direction versus distance round the outline [27]. In both cases the plots may be described quantitatively using Fourier series*. The geometric approach is well reviewed in Bookstein [5].

SHAPE AND SIZE

The special problem of determining the extent to which shape is related to size is referred to as allometry*. Apart from this concern, size variation is usually of little interest in studies of shape. However, linear distances inevitably reflect the size of an object so that size variation can be a nuisance. Since only the relative magnitude of distances is important for shape, the distances are often replaced by ratios of one distance to another. If shape is related to size, such ratios still might well be related to size, but the degree of relationship will be much less than for the original distances. It is usual to choose one distance that is strongly influenced by size and to use this as the denominator when expressing other distances as ratios. Mosimann [18] and Corruccini [7] have suggested the use of a symmetric function of all the size-dependent variables as the denominator.

VARIATION WITHIN A GROUP

Measurements made on an object are regarded as a vector of observations x on a vector of variables X. The individual variables in X are referred to as $X_1, \ldots, X_v$. Data from a group of n objects consists of n vectors, $x_1, \ldots, x_n$, which together form a $n \times v$ data matrix. The rows of this matrix, which are the vectors x_i, may be represented as n points in v-space and the columns as v points in n-space. The two representations are sometimes referred to as Q and R, respectively. The Euclidean metric is used in both spaces so that in row space (Q) the distance between two objects is, in matrix notation, $(\mathbf{x} - \mathbf{y})^T(\mathbf{x} - \mathbf{y})$. Thus two objects that are close in row space are similar in respect of the v measurements.

The usual statistical summary is based on the mean* and standard deviation* of each variable, together with correlations* between pairs of variables. This depends on the distribution being roughly multivariate normal*, an assumption that may be partially tested by inspecting the distribution of each variable separately (which should be normal) and each possible bivariate plot (which should be linear). If there is no correlation, then the variation in shape is uninteresting: objects vary, but not in any consistent way. Suppose now that all objects are roughly the same size. Then a high correlation (positive or negative) is regarded as evidence that the two variables are constrained by the necessity for the object to stay in the same class of

shapes and that jointly they are measuring a single aspect of shape. A negative correlation can be converted to a positive one by using the reciprocal of a measurement or ratio, or the complement of an angle, and this is usually done to ensure positive correlations as far as possible. If a group of variables has high positive intercorrelations, then the group is taken to be measuring a single aspect of shape. Different aspects of shape will have relatively low correlation, by definition. Statistically, this amounts to grouping the variables on the basis of their correlations. It may be done by eye for a small number of variables or by extracting principal components for a larger number $(v > 10)$. If the objects do not have the same size, then the interpretation of correlations depends on the nature of the variables. Correlation among linear distances will almost certainly be partly, perhaps largely, due to size variation. Correlation among angles and ratios will generally indicate constraints of shape.

Principal component analysis* extracts the components $(Z_1, \ldots, Z_v)$ from the covariances* between the original variables. Each component is a linear combination of the variables $(X_1, \ldots, X_v)$. If $Z_1 = a_1 X_1 + \cdots + a_v X_v$, then a_i is called the loading* of X_i on the first component and is proportional to the covariance between Z_1 and X_i. When the data have been standardized by reducing each variable by its mean and scaling each to have unit standard deviation, then covariance equals correlation. In this case the loadings are used to group variables according to their correlations with the first few components and hence with each other. Sometimes the procedure is reversed and the first few components are "named" according to the variables they are associated with. This is logically equivalent to grouping the variables*. Howells [14] gives a good example.

If the observed values of X for an object are substituted in the expression for Z_1, the result is a *score* for that object on Z_1. The scores on Z_1 and Z_2 may be plotted, using rectangular axes, to give an approximation to the representation of objects in row space. The quality of the approximation depends on how much of the overall variability between objects has been reproduced by Z_1 and Z_2. The plot is useful for spotting any lack of homogeneity* in the group of objects. (Rao [25] gives a very detailed account of the different uses of principal components.)

VARIATION BETWEEN GROUPS

If the comparison of several groups is to be meaningful, each must have a representative shape. In other words, the groups must be homogeneous, displaying some variation, but not too much. We shall assume that there are k groups in all, distinguishing between them by using different letters $(x, y, z, \ldots)$ to refer to a typical vector of measurements in each group. The vector of means for each group is regarded as representing a mean shape. For example, McLearn et al. [16] reconstruct a typical profile from the mean of measurements taken from a large number of individual profiles of the human face. In this example the result still looked like a human face, i.e., $\bar{x}$ satisfied the same geometric constraints as each x_i in $\bar{x} = \sum x_i/n$, but since these constraints are in general nonlinear, this will not always be the case.

When comparing groups there are two kinds of questions: comparison between pairs of groups and an overall comparison. The latter requires some metric* enabling one to decide whether group x is closer to y than to z, and a wide variety have been proposed. A good review is given in Weiner [26]. The most commonly used metric is now $(\mathbf{x} - \mathbf{y})^T \Sigma^{-1}(\mathbf{x} - \mathbf{y})$, where Σ is a positive definite $p \times p$ matrix. This may be interpreted as follows. If a multivariate normal density with covariance Σ is centered at x, then all points y that are equiprobable in this density are equidistant from x in this metric. If $\Sigma = \mathbf{I}$, the contours of equal probability are spheres; otherwise, they are ellipsoids. The metric is satisfactory only for groups with similar patterns of covariance, in which case Σ is taken equal to $\mathbf{S}$, the pooled covariance matrix within groups. This leads to D^2, equal to $(\bar{\mathbf{x}} - \bar{\mathbf{y}})^T \mathbf{S}^{-1}(\bar{\mathbf{x}} - \bar{\mathbf{y}})$, as the measure of distance between groups. It is clear from the derivation that the comparison of two large values of D^2 is unlikely to be satisfactory. In fact, although D^2 takes account of correlations within groups, it is, like all metrics, rather a blunt instrument when used for assessing affinity.

With a lot of groups the pairs of D^2 values can be confusing, so a visual overall picture of the interrelationships between groups is produced using principal components*. A $k \times v$ data matrix in which the rows are now the group means is used and principal component analysis is carried out in the D^2 metric [12]. The resulting components are called canonical variates or sometimes discriminant functions*. Scores for each group on the first two or three canonical variates are plotted using rectangular axes. Good examples of the use of canonical variates are those of Ashton, et al. [2], Day and Wood [9], and Oxnard [19].

If something is known about the function of the objects, then it may be possible to order the groups according to this function, at least roughly, e.g., low, medium, and high. We assume that function is not quantifiable, so that only a rough ordering is possible. The plot will show whether or not this ordering is associated with the major dimensions of shape variation. Of course, there may well be other measures of shape more highly associated with function than the first few canonical variates. Aspects of shape that are highly variable are not necessarily those most highly associated with function.

The D^2-metric itself is not without disadvantages. First the assumption that the pattern of variation is constant for all groups is inherently unlikely with shape studies. Second, when a large number of variables is used to ensure that shape is adequately described, they often contain redundancies which lead to nearly singular matrices $\mathbf{S}$ and hence to very large and meaningless D^2-values. Some progress has been made in overcoming these and other difficulties. To avoid distortion from studying scores on just the first two or three canonical variates, Andrews [1] has suggested representing each group by a weighted combination of trigonometric functions of θ with weights equal to the scores on all v canonical variates. As θ varies, each group is represented by a smooth curve. Burnaby [6] has shown how the D^2-metric may be adjusted to measure change only in certain directions in row space. This can be useful when it is required to avoid a direction corresponding to growth or size change. Penrose [24] showed how the adjusted D^2-metric becomes equal to the Euclidean metric when all correlations are equal. To avoid D^2 altogether, some authors have pooled the groups and studied the extent to which the individual objects can be clustered, either visually using principal components and the Euclidean metric, or automatically using various clustering* algorithms [20]. The techniques of multidimensional scaling* [15] and principal coordinates* [11] are also relevant here.

FOSSILS

The stimulus to study variation in modern groups of man often comes from a particularly important fossil find. If canonical variates for the modern groups have been evaluated, then scores on these variates can be obtained for the fossil, and it can be placed on the plot of modern group means relative to the first two or three canonical variates. This plot indicates its position relative to the modern groups, but the result must be treated with caution. Both the modern groups and the fossil should fit well into the two- or three-dimensional plot, for otherwise their relative positions will be distorted. If the fit is poor, then the actual D^2 distances of the fossil from the modern groups should be compared, but if these are all large, then the assessment of affinity is bound to be unsatisfactory. Day and Wood [9], after associating a canonical variate with function, used it to predict function for a fossil that was very different from the modern groups used to derive the canonical variate. This situation is in some ways analogous to that encountered when using a regression line* to predict values outside the range on which the line was based.

PREVALENCE OF ATTRIBUTES*

Consider v attributes of an object measured by $X_1, \ldots, X_v$, where these now take only two possible values (presence/absence). The comparison between groups rests on a comparison of the prevalence for each attribute. If $p(X_i)$ is the prevalence for X_i, then the difference between two groups is measured on the transformed scale $\theta = \sin^{-1} \sqrt{p}$. On this scale the standard deviation of θ is approximately $1/(4n)$. *See* VARIANCE STABILIZATION. If the v attributes are independent, then v differences, $\theta_i - \theta_i'$, may be combined to provide an overall distance $d^2 = \sum (\theta_i - \theta_i')^2$. Berry and Berry [4] give an example based on attributes of the skull. Edwards [10] has generalized this to cover attributes with more than two states, such as blood groups.

STATISTICAL TESTS AND PREDICTION

Since there is rarely any element of randomization* in data collection for physical anthropology, the role of significance tests* is less important than that of description. Three tests are commonly performed: equality of covariance matrices between groups, difference between two groups based on D^2, and zero intercorrelation within a set of variables. All three are based on the multivariate normal distribution*. Full details are given in Morrison [17].

A vector **x** which is incomplete cannot be used in multivariate analysis* without the missing values being replaced by some estimates. The group means for the relevant variables are sometimes used, but a better method is to use the set of complete vectors to predict the missing values using multiple regression* [3].

Predicting the sex of bones can be dealt with statistically if reference groups of known sex are available. Each unknown bone is allocated to the closer of the male and female reference groups using the D^2-metric. Day and Pitcher–Wilmott [8] give an example. The maturity of bones is assessed from characteristics that can be ordered with respect to maturity. Scaling techniques* are used to make the assessment quantitative [13].

REFERENCES

1. Andrews, D. F. (1972). *Biometrics*, **28**, 125–136.

2. Ashton, E. H., Healy, M. J. R., and Lipton, S. (1957). *Proc. R. Soc. Lond. B*, **146**, 552–572.

3. Beale, E. M. L. and Little, R. J. A. (1975). *J. R. Statist. Soc. B*, **37**, 129–145.

4. Berry, A. C. and Berry, R. J. (1967). *J. Anat.*, **101**, 361–379.

5. Bookstein, F. L. (1978). The Measurement of Biological Shape and Shape Change. *Lect. Notes Biomath.*, 24. Springer-Verlag, Berlin.

6. Burnaby, T. P. (1966). *Biometrics*, **22**, 96–110.

7. Corruccini, R. S. (1973). *Amer. J. Phys. Anthropol.*, **38**, 743–754.

8. Day, M. H. and Pitcher-Wilmott, R. W. (1975). *Ann. Hum. Biol.*, **2**, 143–151.

9. Day, M. H. and Wood, B. A. (1968). *Man*, **3**, 440–455.

10. Edwards, A. W. F. (1971). *Biometrics*, **27**, 873–881.

11. Gower, J. C. (1966). *Biometrika*, **53**, 325–338.

12. Gower, J. C. (1966). *Biometrika*, **53**, 588–590.

13. Healy, M. J. R. and Goldstein, H. (1976). *Biometrika*, **63**, 219–229.

14. Howells, W. W. (1972). In *The Functional and Evolutionary Biology of Primates: Methods of Study and Recent Advances*, R. H. Tuttle, ed. Aldine-Atherton, Chicago, pp. 123–151.

15. Kruskal, J. B. (1964). *Psychometrika*, **29**, 1–27.

16. McLearn, I., Morant, G. M., and Pearson, K. (1928). *Biometrika*, **20B**, 389–400.

17. Morrison, D. F. (1967). *Multivariate Statistical Methods*. McGraw-Hill, New York.

18. Mosimann, J. E. (1970). *J. Amer. Statist. Ass.*, **65**, 930–945.

19. Oxnard, C. E. (1973). *Form and Pattern in Human Evolution*. University of Chicago Press, Chicago.

20. Oxnard, C. E. and Neely, P. M. (1969). *J. Morphol.*, **129**, 1–22.

21. Pearson, E. S. (1936). *Biometrika*, **28**, 193–257.

22. Pearson, E. S. (1938). *Biometrika*, **29**, 161–248.

23. Pearson, K. (1926). *Biometrika*, **18**, 105–117.

24. Penrose, L. S. (1954). *Ann. Eugen. (Lond.)*, **18**, 337–343.

25. Rao, C. R. (1964). *Sankhyā A*, **26**, 329–358.

26. Weiner, J. S. (1972). *The Assessment of Population Affinities*, J. S. Weiner, ed. Clarendon Press, Oxford.

27. Zahn, C. T. and Roskies, R. Z. (1972). *IEEE Trans. Computers*, **C-21**, 269–281.

FURTHER READING

The book by Oxnard [19] provides the best available general introduction to the subject, and contains a good bibliography. The series of papers by Howells (1951 onward; see the bibliography in ref. [19]) contain good examples of most of the important methodological contributions made during this period. The use of canonical variates is well explained in Ashton et al. [2]. Bookstein's monograph [5] provides a useful antidote to the uncritical use of interlandmark distances; although more mathematical than the other references cited it is well worth the effort. Technical details about principal components and canonical variates are best obtained from Rao [25], Gower [11,12], and Morrison [17]. For historical details the early volumes of *Biometrika* should be consulted, particularly Volume 1 and the account of Karl Pearson's life and work given by E. S. Pearson [21,22].

See also ALLOMETRY; CLUSTER ANALYSIS; CORRELATION; DISCRIMINANT ANALYSIS; MULTIDIMENSIONAL SCALING; MULTIVARIATE ANALYSIS; *PATTERN RECOGNITION*; PRINCIPAL COMPONENT ANALYSIS, GENERALIZED; and REGRESSION.

M. HILLS

ANTIEIGENVALUES AND ANTIEIGENVECTORS

INTRODUCTION

It follows from the celebrated Cauchy–Schwarz* inequality that for a real symmetric positive definite matrix $\mathbf{A}$ of order $p \times p$, $(\mathbf{x}'\mathbf{A}\mathbf{x})^2 \leqslant \mathbf{x}'\mathbf{A}^2\mathbf{x} \cdot \mathbf{x}'\mathbf{x}$, with equality if and only if $\mathbf{A}\mathbf{x}$ is proportional to $\mathbf{x}$. Thus, the optimization problem

$$\max_{\mathbf{x} \neq 0} \frac{\mathbf{x}'\mathbf{A}\mathbf{x}}{\sqrt{\mathbf{x}'\mathbf{A}^2\mathbf{x} \cdot \mathbf{x}'\mathbf{x}}}, \tag{1}$$

has its optimum value at 1 and it is attained when $\mathbf{x}$ is an eigenvector* of $\mathbf{A}$. Thus, if $\{\boldsymbol{\gamma}_1, \boldsymbol{\gamma}_2, \ldots, \boldsymbol{\gamma}_p\}$ is the set of orthogonal eigenvectors of $\mathbf{A}$ corresponding to the eigenvalues* $\lambda_1 \geqslant \lambda_2 \geqslant \ldots \geqslant \lambda_p$, then any of these eigenvectors will solve Equation 1. The corresponding minimization problem

$$\min_{\mathbf{x} \neq 0} \frac{\mathbf{x}'\mathbf{A}\mathbf{x}}{\sqrt{\mathbf{x}'\mathbf{A}^2\mathbf{x} \cdot \mathbf{x}'\mathbf{x}}}, \tag{2}$$

however, is solved by $\left(\frac{\lambda_p}{\lambda_1 + \lambda_p}\right)^{1/2} \boldsymbol{\gamma}_1 \pm \left(\frac{\lambda_1}{\lambda_1 + \lambda_p}\right)^{1/2} \boldsymbol{\gamma}_p$ and the corresponding minimum value is called the *Kantorovich bound* [17] (*see* KANTOROVICH INEQUALITY), namely, $\frac{2\sqrt{\lambda_1\lambda_p}}{\lambda_1+\lambda_p}$. Since this corresponds to the minimization problem rather than maximization, it may be appropriate to call the quantity

$$\mu_1 = \frac{2\sqrt{\lambda_1\lambda_p}}{\lambda_1+\lambda_p}$$

an *antieigenvalue* and the solution vectors $\left(\frac{\lambda_p}{\lambda_1+\lambda_p}\right)^{1/2}\boldsymbol{\gamma}_1 \pm \left(\frac{\lambda_1}{\lambda_1+\lambda_p}\right)^{1/2}\boldsymbol{\gamma}_p$ the corresponding *antieigenvectors*. In an attempt to imitate the theory of eigenvalues and eigenvectors, we may call μ_1 the smallest antieigenvalue and the corresponding antieigenvector may be denoted by η_1 in a generic way. The next antieigenvalue μ_2 and corresponding antieigenvector η_2 are defined as the optimized value and the solution to the optimization problem

$$\min_{\mathbf{x}\neq 0, \mathbf{x}\perp\eta_1} \frac{\mathbf{x}'\mathbf{A}\mathbf{x}}{\sqrt{\mathbf{x}'\mathbf{A}^2\mathbf{x}\cdot\mathbf{x}'\mathbf{x}}}. \qquad (3)$$

The other antieigenvalues and the corresponding antieigenvectors are similarly defined by requiring that the corresponding antieigenvector be orthogonal to all those previously obtained. A set of these orthogonal antieigenvectors is given by $\{\eta_1, \eta_2, \ldots, \eta_k\}$, where $k = [\frac{p}{2}]$ and

$$\eta_i = \left(\frac{\lambda_{p-i+1}}{\lambda_i+\lambda_{p-i+1}}\right)^{1/2}\boldsymbol{\gamma}_i$$
$$+ \left(\frac{\lambda_i}{\lambda_i+\lambda_{p-i+1}}\right)^{1/2}\boldsymbol{\gamma}_{p-i+1},$$

$i = 1, 2, \ldots, k$. The corresponding antieigenvalues are given by

$$\mu_i = \frac{2\sqrt{\lambda_i\lambda_{p-i+1}}}{\lambda_i+\lambda_{p-i+1}}.$$

Clearly, because of the arbitrariness of the signs of the eigenvectors, there are 2^k such sets for a given set $\{\boldsymbol{\gamma}_1, \boldsymbol{\gamma}_2, \ldots, \boldsymbol{\gamma}_p\}$ of eigenvectors. Further, if there were repeated eigenvalues, say for example, if $\lambda_i = \lambda_{p-i+1}$ for some i, then any vector in the plane generated by $\boldsymbol{\gamma}_i$ and $\boldsymbol{\gamma}_{p-i+1}$ is also an eigenvector. Consequently, in this case, η_i are eigenvectors as well as antieigenvectors of $\mathbf{A}$. The corresponding antieigenvalue μ_i is equal to 1, which is the value of the maximum as well as of the minimum. From a statistical point of view, these antieigenvalues are uninteresting.

The development of the mathematical theory of antieigenvalues and antieigenvectors can be traced to Gustafson [6] and Davis [3]. Notable contributions were made by Gustafson and his associates in a series of papers [7–16]. Khattree [20–22] provides some statistical interpretations and computational details involving antieigenvalues. Khattree [21] also defines what he terms as the *generalized antieigenvalue of order r* of a symmetric positive definite matrix $\mathbf{A}$ and the corresponding *antieigenmatrix* through the optimization problem

$$\min_{\mathbf{X}, \mathbf{X}'\mathbf{X}=\mathbf{I}_r} \frac{|\mathbf{X}'\mathbf{A}\mathbf{X}|}{\sqrt{|\mathbf{X}'\mathbf{A}^2\mathbf{X}|}}, \qquad (4)$$

where the minimized value $\mu_{[r]}$ is the generalized antieigenvalue of order $r(\leqslant [\frac{p}{2}])$, and the corresponding $p \times r$ suborthogonal matrix $\mathbf{X}$ solving the above problem is the antieigenmatrix. Clearly, $\mathbf{X}$ is not unique; for a given $\mathbf{X}$, $\mathbf{PX}$, where $\mathbf{P}$ is orthogonal, also solves Equation 4. In fact [21] one choice of $\mathbf{X}$ is the matrix whose columns are $\eta_1, \eta_2, \ldots, \eta_r$. Correspondingly,

$$\mu_{[r]} = \prod_{i=1}^{r} \frac{2\sqrt{\lambda_i\lambda_{p-i+1}}}{\lambda_i+\lambda_{p-i+1}} = \mu_1\mu_2\ldots\mu_r. \qquad (5)$$

A paper by Drury et al. [4] nicely connects antieigenvalues and the generalized antieigenvalue stated in Equation 5 with many matrix inequalities. In that context, although these authors also talk about the quantity given in Equation 5 above, their usage of the term "generalized antieigenvalue" should be contrasted with our definition. In fact, their generalization, as stated in their Theorem 1, is toward replacing the positive definiteness of $\mathbf{A}$ by nonnegative definiteness, and in the context of Löwner ordering.

Gustafson [11] provides an interesting interpretation of $\cos^{-1}(\mu_1)$ as the largest angle by which $\mathbf{A}$ is capable of turning any vector $\mathbf{x}$. The quantities $\cos^{-1}(\mu_i), i = 2, 3, \ldots$ can be similarly defined, subject to the orthogonality conditions as indicated in Equation 3.

STATISTICAL APPLICATIONS

The concepts of antieigenvalues and antiei-genvectors were independently used in statistics about the same time as these were discovered in mathematics, although without any such nomenclatures. It is appropriate to interpret these quantities statistically and provide a reference to the contexts in which they naturally arise.

Let $\mathbf{z}$ be a $p \times 1$ random vector with variance-covariance matrix* proportional to the identity matrix $\mathbf{I}_p$. The problem of finding a nonnull vector $\mathbf{x}$, such that the correlation* between the linear combinations $\mathbf{z}'\mathbf{x}$ and $\mathbf{z}'\mathbf{A}\mathbf{x}$ (with $\mathbf{A}$ symmetric positive definite) is minimum, is essentially the problem stated in Equation 2 [22]. Under certain conditions on the correlation structure, it is also the minimum possible value of the parent-offspring correlation [22].

Venables [27], Eaton [5], and Schuene-meyer and Bargman [25] have shown that a lower bound on certain canonical correlation* is given by $1 - \mu_1^2$. Furthermore, this bound is sharp. Bartmann and Bloomfield [1] generalize this result by stating that

$$\prod_{i=1}^{r}(1 - \rho_i^2) \geqslant \mu_{[r]}, \qquad (6)$$

where

$$\Sigma = \begin{bmatrix} \Sigma_{11} & \Sigma_{12} \\ \Sigma_{21} & \Sigma_{22} \end{bmatrix},$$

with Σ, Σ_{11}, and Σ_{22} as symmetric positive definite variance-covariance matrices of order $p \times p, r \times r$ and $(p - r) \times (p - r)$, respectively, and the canonical correlations $\rho_i, i = 1, 2, \ldots, r$ are defined by $\rho_i^2 = i^{th}$ eigenvalue of $\Sigma_{11}^{-1}\Sigma_{12}\Sigma_{22}^{-1}\Sigma_{21}$. Without any specific reference to the term antieigenvalue, their problem of minimizing the left-hand side of Equation 6 essentially boils down to that stated in Equation 4.

Bloomfield and Watson [2] and Knott [23] in two back to back articles in *Biometrika** consider the problem of determining the efficiency of least square* estimator of β in the linear model $\mathbf{y} = \mathbf{X}\beta + \varepsilon, \varepsilon \sim (\mathbf{0}, \mathbf{A})$ set up. Let $\mathbf{X}$ be of order $n \times k$ with $k \leqslant [\frac{n}{2}]$. If the least squares estimator $\hat{\beta} = (\mathbf{X}'\mathbf{X})^{-1}\mathbf{X}'\mathbf{y}$ is used instead of the general least squares estimator

$$\tilde{\beta} = (\mathbf{X}'\mathbf{A}^{-1}\mathbf{X})^{-1}\mathbf{X}'\mathbf{A}^{-1}\mathbf{y},$$

then the relative efficiency e of $\hat{\beta}$ relative to $\tilde{\beta}$ may be defined as the ratio of the generalized variances

$$e = \frac{|D(\tilde{\beta})|}{|D(\hat{\beta})|} = \frac{|\mathbf{X}'\mathbf{X}|^2}{|\mathbf{X}'\mathbf{A}^{-1}\mathbf{X}||\mathbf{X}'\mathbf{A}\mathbf{X}|},$$

where $D(.)$ stands for the variance-covariance matrix. The above-mentioned authors were then interested in finding a lower bound on the above, which if attained, would represent the worst-case scenario. This problem can be reduced to Equation 4 with the transformation $\mathbf{Z} = \mathbf{A}^{-\frac{1}{2}}\mathbf{X}$. Thus, the worst case occurs when the columns of $\mathbf{Z}$ (or $\mathbf{X}$) span the subspace generated by the antieigenvectors $\{\boldsymbol{\eta}_1, \boldsymbol{\eta}_2, \ldots, \boldsymbol{\eta}_k\}$. The minimum value of the relative efficiency is given by $\mu_{[k]}$ as in Equation 5.

Venables [27] has considered the problem of testing sphericity* ($H_0 : \Sigma = \sigma^2\mathbf{I}$) of a variance-covariance matrix Σ. The sample variance-covariance matrix $\mathbf{S}$ is given as data and $f\mathbf{S} \sim W_p(f, \Sigma)$, the Wishart distribution* with f degrees of freedom along with $E(\mathbf{S}) = \Sigma$. He argues that the null hypothesis is true if and only if any arbitrary r-dimensional subspace ($r \leqslant p$) of Σ is an invariant subspace. Thus, the null hypothesis is the intersection of all such subhypotheses stating the invariance, and the intersection is taken over all such subspaces. The likelihood ratio test for all such subhypotheses can be derived, and with an argument that one accepts H_0 if and only if all such subhypotheses are accepted, a test statistic for H_0 can be obtained by taking the maximum of such test statistics. This maximization problem turns out to be equivalent to Equation 4 and hence the test statistic for testing H_0 is given by $\mu_{[r]}$ computed from the matrix $\mathbf{S}$. The result can be intuitively justified from the fact that each product term in Equation 5 provides a measure of the eccentricity of the ellipsoids defined by $\mathbf{S}$. Another article [18] and its correction in Reference 19 deal with the same problem but using a different mathematical argument, and still arrives at $\mu_{[r]}$ as the test statistic. Also see Reference 26.

Since the Kantorovich inequality plays an important role in determining the rates of convergence [24] in many mathematical optimization problems, the smallest antieigenvalue

$$\mu_1 = \frac{2\sqrt{\lambda_1 \lambda_p}}{\lambda_1 + \lambda_p} = 2\left[\sqrt{\frac{\lambda_1}{\lambda_p}} + \sqrt{\frac{\lambda_p}{\lambda_1}}\right]^{-1},$$

has a one-to-one correspondence with the condition number λ_1/λ_p. Thus, μ_1 and, more generally, $\mu_1, \mu_2 \ldots$ as well as $\mu_{[r]}$ can be viewed as the condition indices. In linear models, their applications hold promise in assessing multicollinearity* problems.

The topic of antieigenvalues and antieigenvectors may not be very well known among mathematicians. Perhaps statisticians have been slightly ahead of mathematicians in using the concepts implicit in the definitions of the antieigenvalues and antieigenvectors. It is anticipated that some of the mathematical research done on this topic will find further applications in statistics and will provide insight into many other statistical problems.

Acknowledgment

The author's work was partially sponsored by an Oakland University Research fellowship.

REFERENCES

1. Bartmann, F. C. and Bloomfield, P. (1981). Inefficiency and correlation. *Biometrika*, **68**, 67–71.

2. Bloomfield, P. and Watson, G. S. (1975). The inefficiency of least squares. *Biometrika*, **62**, 121–128.

3. Davis, C. (1980). Extending the Kantorovich inequalities to normal matrices. *Lin. Alg. Appl.*, **31**, 173–177.

4. Drury, S. W., Liu, S., Lu, C.-Y, Puntanen, S., and Styan, G. P. H. (2002). *Some Comments on Several Matrix Inequalities with Applications to Canonical Correlations: Historical Background And Recent Developments.* Sankhyā, A64, 453–507.

5. Eaton, M. L. (1976). A maximization problem and its application to canonical correlation. *J. Multivariate Anal.*, **6**, 422–425.

6. Gustafson, K. (1972). "Antieigenvalue Inequalitities in Operator Theory". In *Inequalities III, Proceedings of the Los Angeles Symposium 1969*, O. Shisha, ed. Academic Press, New York, pp. 115–119.

7. Gustafson, K. (1994a). Operator trigonometry. *Lin. Mult. Alg.*, **37**, 139–159.

8. Gustafson, K. (1994b). Antieigenvalues. *Lin. Alg. Appl.*, **208/209**, 437–454.

9. Gustafson, K. (1995). Matrix trigonometry. *Lin. Alg. Appl.*, **217**, 117–140.

10. Gustafson, K. (1996). *Operator Angles (Gustafson), Matrix Singular Angles (Wielandt), Operator Deviations (Krein), Collected Works of Helmug Wielandt II,* B. Huppert and H. Schneider, eds. De Gruyters, Berlin, Germany.

11. Gustafson, K. (1999). The geometrical meaning of the Kantorovich-Wielandt inequalities. *Lin. Alg. Appl.*, **296**, 143–151.

12. Gustafson, K. (2000). An extended operator trigonometry. *Lin. Alg. Appl.*, **319**, 117–135.

13. Gustafson, K. and Rao, D. K. M. (1977a). Numerical range and accretivity of operator products. *J. Math. Anal. Appl.*, **60**, 693–702.

14. Gustafson, K. and Rao, D. K. M. (1997b). *Numerical Range: The Field of Values of Linear Operators and Matrices.* Springer-Verlag, New York.

15. Gustafson, K. and Seddighin, M. (1989). Antieigenvalue bounds. *J. Math. Anal. Appl.*, **143**, 327–340.

16. Gustafson, K. and Seddighin, M. (1993). A note on total antieigenvectors. *J. Math. Anal. Appl.*, **178**, 603–611.

17. Kantorovich, L. B. (1948). Functional analysis and applied mathematics. *Uspehi. Mat. Nauk.*, **3**, 89–185 (Translated from the Russian by Benster, C. D. (1952), Report 1509, National Bureau of Standards, Washington, D.C).

18. Khatri, C. G. (1978). Some optimization problems with applications to canonical correlations and sphericity tests, *J. Multivariate Anal.*, **8**, 453–467 (Corrected by Khatri [19]).

19. Khatri, C. G. (1982). *J. Multivariate Anal.*, **12**, 612; Erratum (to Khatri [18]).

20. Khattree, R. (2001). On calculation of antieigenvalues and antieigenvectors. *J. Interdiscip. Math.*, **4**, 195–199.

21. Khattree, R. (2002). Generalized antieigenvalues of order *r*. *Am. J. Math. Manage. Sci.*, **22**, 89–98.

22. Khattree, R. (2003). Antieigenvalues and antieigenvectors in statistics. *J. Stat. Plann. Inference* (Special issue in celebration of C.R. Rao's 80th birthday), **114**, 131–144.

23. Knott, M. (1975). On the minimum efficiency of least squares. *Biometrika*, **62**, 129–132.

24. Luenberger, D. (1984). *Linear and Nonlinear Programming*, 2nd ed. Addison-Wesley, Reading, Mass.

25. Schuenemeyer, J. H. and Bargman, R. E. (1978). Maximum eccentricity as a union-intersection test statistic in multivariate analysis. *J. Multivariate Anal.*, **8**, 268–273.

26. Srivastava, M. S. and Khatri, C. G. (1978). *An Introduction to Multivariate Statistics*, North Holland, New York.

27. Venables, W. (1976). Some implications of the union-intersection principle for tests of sphericity, *J. Multivariate Anal.*, **6**, 185–190.

See also CANONICAL ANALYSIS; KANTOROVICH INEQUALITY; LINEAR ALGEBRA, COMPUTATIONAL; and SPHERICITY, TESTS OF.

RAVINDRA KHATTREE

ANTIMODE

An antimode is the opposite of a mode* in the sense that it corresponds to a (local) minimum frequency. As with the mode, it is sometimes desired that the name should be applied only to global, and not to local minima. The more common use, however, includes local minima.

Note that whereas $x = 1$ is a mode of the PDF*,

$$f_X(x) = \begin{cases} 2x & 0 \leqslant x \leqslant 1, \\ 0 & \text{elsewhere,} \end{cases}$$

$x = 0$ is not an antimode of this PDF. On the other hand,

$$f_X(x) = \begin{cases} |x| & -1 \leqslant x \leqslant 1, \\ 0 & \text{elsewhere,} \end{cases}$$

has an antimode at $x = 0$ (and modes at $x = -1$ and $x = 1$).

The antimode itself refers to the frequency (or PDF) at the antimodal value of the argument.

See also MEAN, MEDIAN, AND MODE.

ANTIRANKS

Nonparametric tests and estimates are generally based on certain statistics which depend on the sample observations $X_1, \ldots, X_n$ (real-valued) only through their ranks* $R_1, \ldots, R_n$, where

$$R_i = \text{number of indices } r$$
$$(1 \leqslant r \leqslant n) : X_r \leqslant X_i, \qquad (1)$$

for $i = 1, \ldots, n$. If $X_{n:1} \leqslant \cdots \leqslant X_{n:n}$ stand for the sample order statistics*, then we have

$$X_i = X_{n:R_i}, \quad i = 1, \ldots, n. \qquad (2)$$

Adjustment for ties can be made by dividing equally the total rank of the tied observations among themselves. Thus if $X_{n:k} < X_{n:k+1} = \cdots = X_{n:k+q} < X_{n:k+q+1}$, for some $k, 0 \leqslant k \leqslant n - 1$, and $q \geqslant 1$ (where $X_{n:0} = -\infty$ and $X_{n:n+1} = +\infty$), then for the q tied observations (with the common value $X_{n:k+1}$), we have the *midrank* $k + (q + 1)/2$.

Let us now look at (2) from an opposite angle: For which index (S_k), is X_{S_k} equal to $X_{n:k}$? This leads us to define the *antiranks* $S_1, \ldots, S_n$ by

$$X_{n:i} = X_{S_i}, \quad \text{for } i = 1, \ldots, n. \qquad (3)$$

Note the inverse operations in (2) and (3) as depicted below:

$$X_1, \ldots, X_i, \ldots, X_{S_i}, \ldots, X_n$$
$$X_{n:1}, \ldots, X_{n:i}, \ldots, X_{n:R_i}, \ldots, X_{n:n}, \qquad (4)$$

so that, we have

$$R_{S_i} = S_{R_i} = i, \quad \text{for } i = 1, \ldots, n, \qquad (5)$$

and this justifies the terminology: Antiranks.

Under the null hypothesis that $X_1, \ldots, X_n$ are exchangeable random variables, $\mathbf{R} = (R_1, \ldots, R_n)$, the vector of ranks, takes on each permutation of $(1, \ldots, n)$ with the common probability $(n!)^{-1}$. By virtue of (5), we obtain that under the same null hypothesis, $\mathbf{S} = (S_1, \ldots, S_n)$, the vector of antiranks, has the same (discrete) uniform permutation distribution. In general, for the case of ties neglected, (5) can be used to obtain the distribution of **S** from that of **R** (or vice versa), although when the X_i are not exchangeable, this distribution may become quite cumbrous. Under the null hypothesis of exchangeability*, for suitable functions of **R** (i.e. rank- statistics), *permutational central limit theorems** provide asymptotic solutions, and by virtue of (5), these remain applicable to antirank statistics as well.

For mathematical manipulations, often **S** may have some advantage over **R**. To illustrate this point, consider a typical *linear rank statistic** (T_n) of the form $\sum_{i=1}^{n} c_i a_n(R_i)$, where the c_i are given constants and $a_n(1), \ldots, a_n(n)$ are suitable scores. By (5), we have

$$T_n = \sum_{i=1}^{n} c_{S_i} a_n(i), \qquad (6)$$

and this particular form is more amenable to censoring schemes (*see* PROGRESSIVE CENSORING SCHEMES). If we have a type II censoring (at the kth failure), then the censored version of T_n in (6) is given by

$$T_{nk} = \sum_{i=1}^{k} (c_{S_i} - \bar{c}_n)[a_n(i) - a_n^*(k)], \quad k \geqslant 0, \qquad (7)$$

where

$$\bar{c}_n = n^{-1} \sum_{i=1}^{n} c_i,$$

$$a_n^*(k) = (n - k)^{-1} \sum_{j=k+1}^{n} a_{n(j)}, \, k < n$$

and

$$a_n^*(n) = 0.$$

For the classical *Kolmogorov–Smirnov tests**, these antiranks may be used to express the statistics in neat forms and to study their distributions in simpler manners; we may refer to Hájek and Šidák [1] for a nice account of these.

In life-testing* problems and clinical trials*, for rank procedures in a time-sequential setup, the antiranks play a vital role; for some details, see Sen [2,3].

REFERENCES

1. Hájek, J. and Šidák, Z. (1967). *Theory of Rank Tests*. Academic, New York.
2. Sen, P. K. (1981). *Sequential Nonparametrics*. Wiley, New York.
3. Sen, P. K. (1985). *Theory and Application of Sequential Nonparametrics*. SIAM, Philadelphia.

See also EMPIRICAL DISTRIBUTION FUNCTION (EDF) STATISTICS; KOLMOGOROV–SMIRNOV STATISTICS; LIMIT THEOREM, CENTRAL; LINEAR RANK TESTS; ORDER STATISTICS; PERMUTATIONAL CENTRAL LIMIT THEOREMS; PROGRESSIVE CENSORING SCHEMES; and TIME-SEQUENTIAL INFERENCE.

P. K. SEN

ANTISMOOTHING. See FOURIER SERIES CURVE ESTIMATION AND ANTISMOOTHING

ANTITHETIC VARIATES

If T_1 and T_2 are unbiased estimators of a parameter θ, then $T = \frac{1}{2}(T_1 + T_2)$ is also unbiased and has variance $\frac{1}{4}\{\text{var}(T_1) + \text{var}(T_2) + 2\,\text{cov}(T_1, T_2)\}$. This variance is reduced (for fixed $\text{var}(T_1)$ and $\text{var}(T_2)$) by reducing $\text{cov}(T_1, T_2)$ and making the correlation between T_1 and T_2 negative and as large, numerically, as possible. Pairs of variates constructed with this aim in view are called *antithetic variates*.

The concept arose in connection with estimation of integrals by simulation experiments [2,3] (*see* MONTE CARLO METHODS). The following example [1, p. 61] may help to clarify ideas. If X is uniformly distributed* between 0 and 1, then for any function $g(x)$,

$$E[g(X)] = \int_0^1 g(x)\,dx,$$

so $g(x)$ is an unbiased estimator* of $\int_0^1 g(x)dx$. So is $g(1 - X)$, since $(1 - X)$ is also uniformly distributed between 0 and 1. If $g(x)$ is a monotonic function of x, $g(x)$ and $g(1 - X)$ are negatively correlated and are "antithetic variates." In particular,

$$\text{var}(g(X) + g(1 - X)) \leqslant \frac{1}{2}\,\text{var}(g(X)).$$

The construction of a variate antithetic to $g(X)$ can be extended in a simple way to cases when the function $g(x)$ is not monotonic, but the interval 0 to 1 can be split into a finite number of intervals in each of which $g(x)$ is monotonic.

The method can also be applied to estimation of multivariate integrals by a straightforward extension. Use of antithetic variables can be a powerful method of increasing accuracy of estimation from simulation in appropriate situations.

REFERENCES

1. Hammersley, J. M. and Handscomb, D. C. (1964). *Monte Carlo Methods*. Methuen, London.

2. Hammersley, J. M. and Mauldon, J. G. (1956). *Proc. Camb. Philos. Soc.*, **52**, 476–481.

3. Hammersley, J. M. and Morton, K. W. (1956). *Proc. Camb. Philos. Soc.*, **52**, 449–475.

4. Tukey, J. W. (1957). *Proc. Camb. Philos. Soc.*, **53**, 923–924.

See also MONTE CARLO METHODS and NUMERICAL INTEGRATION.

APL.   See STATISTICAL SOFTWARE

APPLIED PROBABILITY

Applied probability is that field of mathematical research and scholarship in which the theory and calculus of probability are applied to real-life phenomena with a random component. Such applications encompass a broad range of problems originating in the biological, physical, and social sciences, as well as engineering and technology.

The term "applied probability" first appeared as the title of the proceedings of a symposium on the subject, held by the American Mathematical Society in 1955 [3]. It became popular through its use by the *Methuen Monographs in Applied Probability and Statistics*, edited from 1959 by M. S. Bartlett. The two fields are closely related: applied probability is concerned primarily with modeling random phenomena* (*see* STOCHASTIC PROCESSES), while statistics serves to estimate parameters* and test the goodness of fit* of models to observed data. Bartlett has expressed the opinion that neither field could exist without the other; this is a viewpoint shared by many applied probabilists.

There are currently several periodicals publishing material in applied probability; the principal ones in order of their dates of first publication are *Teoriya Veroyatnostei i ee Primeneniya** (1956), (English translation: *Theory of Probability and Its Applications**) *Zeitschrift für Wahrscheinlichkeitstheorie* (1962), the *Journal of Applied Probability** (1964), *Advances in Applied Probability** (1969), *Stochastic Processes and Their Applications** (1973), and *Annals of Probability** (1973). Other mathematical or statistical journals publish the occasional paper in applied probability, and there is considerable discussion of applied probability models in journals of biology, physics, psychology, operations research, and engineering. Among these are *Theoretical Population Biology*, the *Journal of Statistical Physics**, *Psychometrika*, *Operations Research**, the *Journal of Hydrology*, and the *Journal of the Institute of Electrical Engineers*.

It is impossible to give a comprehensive description of current work in applied probability; perhaps a few illustrative examples selected at random from the recent literature in each of the biological, physical, social, and technological areas will serve to indicate the breadth of the field.

BIOLOGICAL SCIENCES: BIRD NAVIGATION; OPTIMAL HUNTING

David Kendall [2] investigated some interesting models of bird navigation. Ornithologists have surmised that birds navigate instinctively by reference to the sun and stars; Kendall constructed two models to simulate such navigation realistically.

The first is referred to as the Manx model, after the Manx shearwater, which flies across the Atlantic to its European breeding grounds. In this model the bird flies laps of approximately 20 miles each, at a speed of roughly 40 mph. At the end of each lap it redirects itself toward its goal but commits an angular error, having the von Mises or wrapped normal distribution*. When it arrives within a radius of approximately 10 miles of its home, it recognizes its destination and heads directly for it. The second Bessel model allows both the lap length and the deflection from the correct direction to be random. Under appropriate conditions, the two models converge to Brownian motion*.

In his paper, Kendall analyzes bird data on which to base his models and tests these by repeated simulations*, leading to graphical representations of Manx and Bessel flights. He compares these models theoretically and numerically, carries out some diffusion* approximations, and concludes with a lengthy study of the hitting times to the circumference of the homing target. Kendall's work is a model of what is most illuminating in the applied probabilist's approach to a scientific problem. Practical data are carefully considered, and a suitable model is found to fit it. The apprentice applied probabilist could well base his methods and style on Kendall's work. See also ORNITHOLOGY, STATISTICS IN.

Another of the many interesting problems which arise in a biological context is that of determining optimal hunting or harvesting policies for animal populations. Abakuks [1] has discussed such a policy for a population growing according to a stochastic logistic* scheme, subject to natural mortality. The object is to maximize the long-term average number of animals hunted or harvested per unit time. It is shown that there is a critical population size x_c such that hunting or harvesting is optimal if and only if the population is greater or equal to this number.

PHYSICAL SCIENCES: ISING LATTICES

In statistical mechanics, one may need to determine the partition function for large lattices of points representing crystals, particles, or atoms. The Ising model, which helps to characterize qualitative changes at critical parameter values, is therefore important in theoretical physics (see LATTICE SYSTEMS).

Consider a rectangular lattice of $N = m \times n$ points in a plane; these points are labeled 1 to N. At each site i, the random variable X_i may take the values ± 1, where $+1$ may,

for example, correspond to the formation of a crystal. The joint distribution of the site variables is given by

$$P\{\mathbf{X} = \mathbf{x}\} = k^{-1}(a) \exp\left(a \sum x_i x_j\right),$$

where $k(a) = \sum_{\mathbf{x}} \exp\left(a \sum x_i x_j\right)$ is the partition function, and the $\sum x_i x_j$ is taken over all nearest-neighbor* pairs; $\mathbf{X}$ is a simple Markov random field*.

Pickard [4] has recently obtained some limit theorems for the sample correlation* between nearest neighbors in an Ising lattice for the noncritical case. This provides a model for asymptotic testing and estimation of the correlation between nearest neighbors, based on experimental data.

SOCIAL SCIENCES: MANPOWER SYSTEMS; ECONOMIC OPTIMIZATION

A firm or company is a hierarchically graded manpower system, usually modeled by a Markov chain. The probabilities p_{ij} of this chain denote annual promotion rates from grade i to grade j of the firm; the state of the graded manpower system is described from year to year by the numbers of individuals in each grade (see MANPOWER PLANNING).

The simpler manpower models are mainly linear, but Vassiliou [7] considered a high-order nonlinear Markovian model for promotion based on three principles. The first is the ecological principle that promotions should be proportional to suitable staff available for them, as well as to vacancies for promotion. The second is that resignations from different grades are different, and the third is an inertia principle which prescribes that when there is a reduced number of vacancies, promotions may still be made faster than the ecological principle suggests.

Difference equations* for the mean number of individuals in each grade are obtained, and the model is used to provide detailed numerical forecasts for probabilities of promotion in organizations with five grades. A comparison is made between predictions based on an earlier linear model*, and the present nonlinear model*; actual data from a large British firm are found to be adequately described by the latter.

Optimization may be important in a variety of sociological contexts, some purely economic, others more technological. Vered and Yechiali [8] studied the optimization of a power system for a private automatic telephone exchange (PABX). Several maintenance policies are considered, which depend on the set of parameters (m, u, μ, υ) of the PABX system, where n is the number of independent rectifiers in parallel, m the minimal number of units that will keep the system operative, μ their mean time to failure* and υ the period of time between regular maintenance visits. Repairs are carried out every υ units of time, or when the system fails. The authors determine the optimal (m, n, μ, υ) to minimize costs for a required level of reliability* of the PABX system, and provide tables of numerical results for the optimal parameters in both the cases of periodic and emergency maintenance.

TECHNOLOGICAL SCIENCES: RELIABILITY

All engineering components have a failure point; it is therefore of importance to study the reliability of mechanical or electronic parts to determine the probability of their failure times. In this context, two recent problems, the first studied by Szász [5] concerning two lifts, and the second by Taylor [6] on the failure of cables subjected to random loads, will be of interest.

Szász [5] examines a building that has two lifts working independently of each other. The functioning of each lift forms an alternating renewal process* with working-time distribution F and repair-time distribution G. Suppose that this latter distribution $G = G(x, \epsilon)$ depends on $\epsilon > 0$ in such a way that its mean $\int x \, dG(x, \epsilon)$ tends to zero as $\epsilon \to 0$. The author sets out to find the asymptotic distribution of the first instant τ^ϵ at which both lifts are simultaneously out of order, as $\epsilon \to 0$.

It is shown that under certain conditions, as $\epsilon \to 0$, the normalized point process W^ϵ: $w_1^\epsilon < w_2^\epsilon < \cdots$, where $w_k^\epsilon = \epsilon \tau_k^\epsilon$, tends to a Poisson process* with parameter $2\lambda^{-2}$, where $\lambda = \int x \, dF(x)$. Thus subject to certain very general conditions, one must expect breakdowns of both lifts to occur according to a Poisson process. For the favorable case in which λ is very large, this process will have a very small mean.

Taylor [6] is concerned with the reliability of deep-sea cables made up of fiber bundles, and the effect on them of random loads generated by waves that rock the ocean vessels deploying them. A simple model with random loads is studied, subject to power-law breakdown, such that the failure time T under constant load L follows the negative exponential distribution*

$$\Pr[T > x] = \exp(-KL^\rho x) \qquad (x \geqslant 0),$$

where $K > 0$ and $\rho \geqslant 1$.

The asymptotic distribution of T under random loads is derived and Taylor shows that random loads have a significant effect on the lifetime of a cable. The loss in mean lifetime cannot be predicted from the first few moments of the load process; it depends on the entire marginal probability distribution of the load, as well as the power-law exponent ρ. It is shown that the asymptotic variance of the lifetime has two components, the first due to the variation of individual fibers, and the second to the variation of the load.

EXTENT AND FUTURE OF THE FIELD

It is of interest to know the methods of probability theory that are most used in attacking problems of applied probability. The most commonly applied areas are found to be Markov chains and processes (including diffusion processes*), branching processes* and other stochastic processes* (mostly stationary), limit theorems*, distribution theory and characteristic functions*, methods of geometrical probability*, stopping times, and other miscellaneous methods.

To list the subsections of the main categories given in the first four sections exhaustively would be impossible, but in the biological sciences one finds that population processes, mathematical genetics, epidemic theory, and virology are the major subfields. In the technological sciences, operations research*, queueing theory*, storage (see DAM THEORY), and traffic theory are possibly the most active areas.

Applied probability is a very broad subject; as we have seen, it encompasses real-life problems in a variety of scientific and other fields. Although the subject feeds on practical problems, it requires a very high level of theoretical competence in probability. In solving these problems, every approach that proves successful is a useful one. Classical mathematical analysis, numerical analysis, statistical calculations, limit theorems*, simulation, and every other branch of mathematics are legitimate weapons in the search for a solution. Applied probability, although a relatively small branch of mathematics, relies on the resources of the entire subject. It maintains itself successfully without specific affiliation to any particular school or tradition, whether it be British, French, Russian, or North American, while drawing on the best aspects of them all. Its strength lies in the universality of its traditions and the versatility of its mathematical methods.

A further point of importance is the delicate interrelation of theory and practice in applied probability. Without practice (which involves computation and statistics), applied probability is trivial; without theory, it becomes shallow. Close contact is required with experiment and reality for the healthy development of the subject. The collection and analysis of data cannot be avoided, and a certain amount of numerical work will always prove necessary. In attacking problems of applied probability there is a complete cycle from the examination of data to the development of a theoretical model, followed by the statistical verification of the model and its subsequent refinement in the light of its goodness of fit*.

It seems possible that too much effort has been diverted into model building for its own sake, as well as in following through the mathematical refinements of new models. The further development of applied probability requires consideration of real-life problems and the validation of models for these based on observed data. Research workers in the field are aware that only by paying close attention to data and considering genuine problems can their contributions to the subject achieve full scientific stature. For interested readers, a selected bibliography is appended.

REFERENCES

Note: The following references are all highly technical.

1. Abakuks, A. (1979). *J. Appl. Prob.*, **16**, 319–331.

2. Kendall, D. G. (1974). *J. R. Statist. Soc. B*, **36**, 365–402.

3. McColl, L. A., ed. (1957). *Applied Probability* [Proc. Symp. Appl. Math., 7 (1955)]. McGraw-Hill, New York (for the American Mathematical Society.)

4. Pickard, D. (1976). *J. Appl. Prob.*, **13**, 486–497.

5. Szász, D. (1977). *Ann. Prob.*, **5**, 550–559.

6. Taylor, H. M. (1979). *Adv. Appl. Prob.*, **11**, 527–541.

7. Vassiliou, P.-C. G. (1978). *J. R. Statist. Soc. A*, **141**, 86–94.

8. Vered, G. and Yechiali, U. (1979). *Operat. Res.*, **27**, 37–47.

BIBLIOGRAPHY

Bailey, N. T. J. (1975). *The Mathematical Theory of Infectious Diseases and Its Applications.* Charles Griffin, London. (Detailed survey of modeling of infectious diseases.)

Barlow, R. E. and Proschan, F. (1965). *Mathematical Theory of Reliability.* Wiley, New York. (Basic principles of reliability.)

Cohen, J. W. (1969). *The Single Server Queue.* North-Holland, Amsterdam. (Comprehensive treatise on queueing.)

Ewens, W. J. (1969). *Population Genetics.* Methuen, London. (Short monograph on mathematical genetics.)

Iosifescu, M. and Tautu, P. (1973). *Stochastic Processes and Applications in Biology and Medicine*, 2 vols. Springer-Verlag, Berlin. (Compendium of stochastic methods in biology and medicine.)

Keyfitz, N. (1968). *Introduction to Mathematics of Population.* Addison-Wesley, Reading, Mass. (Basic principles of demography.)

Kimura, M. and Ohta, T. (1971). *Theoretical Aspects of Population Genetics.* Monographs in Population Biology 4. Princeton University Press, Princeton, N.J. (Monograph on population genetics with emphasis on diffusion.)

Newell, G. F. (1971). *Applications of Queueing Theory.* Chapman & Hall, London. (Short practical account of queueing applications.)

Pielou, E. C. (1969). *An Introduction to Mathematical Ecology.* Wiley-Interscience, New York. (Basic principles of mathematical ecology.)

Pollard, J. H. (1973). *Mathematical Models for the Growth of Human Populations*. Cambridge University Press, Cambridge. (Good introduction to human population models.)

Ross, S. M. (1970). *Applied Probability Models with Optimization Applications*. Holden-Day, San Francisco. (Elementary text with broad range of examples.)

Syski, R. (1960). *Introduction to Congestion Theory in Telephone Systems*. Oliver & Boyd, Edinburgh. (Stochastic processes in telephone traffic.)

Takács, L. (1962). *Introduction to the Theory of Queues*. Oxford University Press, New York. (Introductory treatment of queueing theory.)

Thompson, C. J. (1972). *Mathematical Statistical Mechanics*. Macmillan, New York. (Introduction to statistical mechanics.)

See also CRYSTALLOGRAPHY, STATISTICS IN; DAMAGE MODELS; DAM THEORY; ECOLOGICAL STATISTICS; QUEUEING THEORY; RELIABILITY; RENEWAL THEORY; and STATISTICAL GENETICS; STOCHASTIC PROCESSES.

J. GANI

APPLIED PROBABILITY JOURNALS

[*Journal of Applied Probability (JAP); Advances in Applied Probability (AAP)*]

[This entry has been updated by the Editors.]

JAP is an international journal which first appeared in June 1964; it is published by the Applied Probability Trust, based in the University of Sheffield, England. For some years it was published also in association with the London Mathematical Society. The journal provides a forum for research papers and notes on applications of probability theory to biological, physical, social, and technological problems. A volume of approximately 900 to 1,250 pages is published each year, consisting of four issues, which appear in March, June, September, and December. *JAP* considers research papers not exceeding 20 printed pages in length, and short communications in the nature of notes or brief accounts of work in progress, and pertinent letters to the editor.

AAP is a companion publication of the Applied Probability Trust, launched in 1969. It publishes review and expository papers in applied probability, as well as mathematical and scientific papers of interest to probabilists. A volume of approximately 900 to 1,200 pages is published each year; it also consists of four issues appearing in March, June, September, and December. *AAP* considers review papers; longer papers in applied probability which may include expository material; expository papers on branches of mathematics of interest to probabilists; papers outlining areas in the biological, physical, social, and technological sciences in which probability models can be usefully developed; and papers in applied probability presented at conferences that do not publish their proceedings. In addition, *AAP* has a section featuring contributions relating to stochastic geometry and statistical applications (*SGSA*). Occasionally, a special *AAP* supplement is published to record papers presented at a conference of particular interest.

JAP and *AAP* have a wide audience with leading researchers in the many fields where stochastic models are used, including operations research, telecommunications, computer engineering, epidemiology, financial mathematics, information systems and traffic management.

EARLY HISTORY

In 1962, the editor-in-chief of the two journals, J. Gani, made his first attempts to launch the *Journal of Applied Probability*. At that time, a large number of papers on applications of probability theory were being published in diverse journals dealing with general science, statistics, physics, applied mathematics, economics, and electrical engineering, among other topics. There were then only two probability journals, the Russian *Teoriya Veroyatnostei* (English translation: *Theory of Probability and Its Applications**) and the German *Zeitschrift für Wahrscheinlichkeitstheorie**. Neither of these specialized in applications of probability theory, although papers in applied probability occasionally appeared in them.

Having assembled an editorial board in 1962, Gani attempted to launch the *Journal of Applied Probability* with assistance from the Australian National University and the Australian Academy of Science. He was not successful in this, and it was only after raising private contributions from himself, Norma McArthur, and E. J. Hannan*, both of the Australian National University, that he was able to provide half the finance necessary for the publication of *JAP*. In May 1963, with the support of D. G. Kendall of the University of Cambridge, he was able to persuade the London Mathematical Society to donate the remaining half of the funds required, and thus collaborate in the publication of the journal.

In February 1964, agreement was reached that the ownership of *JAP* should be vested in the Applied Probability Trust, a non-profit-making organization for the advancement of research and publication in probability, and more generally mathematics. The four trustees were to include the three Australian sponsors of the journal and one trustee nominated by the London Mathematical Society. The agreement was ratified legally on June 1, 1964, although de facto collaboration had already begun several months before.

The first issue of *JAP* appeared in June 1964. Every effort was made to prevent the time lag between submission and publication of a paper exceeding 15 months; this became an established policy for both *JAP* and *AAP*.

ORGANIZATION

The office of the Applied Probability Trust is located at the University of Sheffield, England. The website for the Trust, and a link to the journals, is www.shef.ac.uk/uni/companies/apt.

Each of the two journals has an international Editorial Board consisting of an Editor-in-Chief, two (*JAP*) or three (*AAP*) Coordinating Editors and between 20 and 30 Editors, some of whom serve on the Board for both journals.

It is impossible for the *JAP* and *AAP* to collect together every diverse strand of the subject, but probabilists can now find most of the applied material they require in a few periodicals. Among these are the two mentioned earlier, as well as the more recent *Stochastic Processes and Their Applications**, and the Institute of Mathematical Statistics'

Annals of Probability*, both first published in 1973, also the *Annals of Applied Probability**, launched by the IMS in 1991, and the journal *Bernoulli*, since 1995 a publication of the International Statistical Institute*. It is the policy of *JAP* and *AAP* to contribute to future development.

J. GANI

The Editors

APPLIED STATISTICS. See *JOURNAL OF THE ROYAL STATISTICAL SOCIETY*

APPROGRESSION

Approgression is the use of regression functions (usually linear) to approximate the truth, for simplicity and predictive efficiency. Some *optimality* results are available.

The term seems to be due to H. Bunke in

> Bunke, H. (1973). Approximation of regression functions. *Math. Operat. Statist.*, **4**, 314–325.

A forerunner of the concept is the idea of an *inadequate* regression model in

> Box, G. E. P. and Draper, N. R. (1959). A basis for the selection of a response surface design. *J. Amer. Statist. Ass.*, **54**, 622–654.

The paper:

> Bandemer, H. and Näther, W. (1978). On adequateness of regression setups. *Biom. J.*, **20**, 123–132

helps one to follow some of the treatment in Section 2.7 of

> Bunke, H. and Bunke, O., eds. (1986). *Statistical Inference in Linear Models*. Wiley, New York,

which is exceedingly general and notationally opaque. The Bunke and Bunke account has about 12 relevant references. More recent work is to be found in

> Zwanzig, S. (1980). The choice of appropriate models in non-linear regression. *Math. Operat. Statist. Ser. Statist.*, **11**, 23–47

and in

Bunke, H and Schmidt, W. H. (1980). Asymptotic results on non-linear approximation of regression functions and weighted least-squares. *Math. Operat. Statist. Ser. Statist.*, **11**, 3–22,

which pursue the concept into nonlinearity. A nice overview is now available in

Linhart, H. and Zucchini, W. (1986) *Model Selection*. Wiley, New York.

See also LINEAR MODEL SELECTION; REGRESSION (VARIOUS ENTRIES); and RESPONSE SURFACE DESIGNS.

P.K. SEN

APPROXIMATING INTEGRALS, TEMME'S METHOD

Let $h_n(x)$ be a sequence of functions of a real variable x such that h_n and its derivatives are of order $O(1)$ as $n \to \infty$. (We shall suppress the subscript n in the sequel.) Consider the integral

$$\int_z^\infty h(x)\sqrt{n}\phi(\sqrt{n}x)dx, \qquad (1)$$

where $\phi(\cdot)$ is the standard normal density function. Let $g(x) = [h(x) - h(x)]/x$, $g_1(x) = [g'(x) - g'(0)]/x$, and $g_{j+1}(x) = [g'_j(x) - g'_j(0)]/x$.

Temme [1] approximates the "incomplete" integral (1) by

$$[1 - \Phi(\sqrt{n}z)] \int_{-\infty}^\infty h(x)\sqrt{n}\phi(\sqrt{n}x)dx$$
$$+ \frac{1}{n}\sqrt{n}\phi(\sqrt{n}z)[g(z) + R(z)],$$

where $\Phi(\cdot)$ denotes the standard normal cdf and

$$R(z) = \frac{1}{n}g_1(z) + \frac{1}{n^2}g_2(z) + \dots.$$

Let $\bar{c}$ be the constant such that

$$\bar{c}\int_\infty^\infty h(x)\sqrt{n}\phi(\sqrt{n}x)dx = 1.$$

Assuming that $\bar{c} = 1 + O(n^{-1})$, one can approximate

$$\int_z^\infty \bar{c}h(x)\sqrt{n}\phi(\sqrt{n}x)dx. \qquad (2)$$

by

$$[1 - \Phi(\sqrt{n}z)] + \frac{1}{n}\sqrt{n}\phi(\sqrt{n}z)\frac{h(z) - h(0)}{z}. \qquad (3)$$

Knowledge of the value of $\bar{c}$ is thus not needed for this approximation. The error of (3) as an approximation to (2) is $\phi(\sqrt{n}z)$ $O(n^{-\frac{3}{2}})$.

For fixed z, the *relative* error of the approximation is $O(n^{-1})$.

These formulas provide approximations to tail probabilities, which, unlike those derived via the saddle-point* method, can be integrated analytically.

REFERENCE

1. Temme, N. M. (1982). The uniform asymptotic expansion of a class of integrals related to cumulative distribution functions. *SIAM J. Math. Anal.*, **13**, 239–253.

APPROXIMATIONS TO DISTRIBUTIONS

HISTORICAL DEVELOPMENTS

The main currents of thought in the development of the subject can be traced to the works of Laplace* (1749–1827), Gauss* (1777–1855), and other scientists of a century or so ago. The normal* density played an important role because of its alleged relation to the distribution of errors of measurement, and it is said that Gauss was a strong "normalist" to the extent that departures were due to lack of data. Natural and social phenomena called for study, including the search for stability, causality, and the semblance of order. Quetelet* (1796–1874) exploited the binomial* and normal* distributions, paying particular attention to the former in his investigations of data reflecting numbers of events. Being well versed in mathematics and the natural sciences (he was a contemporary of Poisson*, Laplace, and Fourier), he searched for order in social problems, such as comparative crime rates and human characteristics, posing a dilemma for the current notions of free will. His contributions to social physics, including the concepts for stability and stereotypes, paved the way for modern sociology. Lexis (1837–1914) opened up fresh avenues of research with his sequences of dependent trials, his aim being to explain departures from the binomial exhibited by many demographic* studies.

Urn problems*, involving drawing of balls of different colors, with and without replacement, have played an important part in distributional models. Karl Pearson* [58] interested himself in a typical discrete distribution (the hypergeometric*) originating from urn sampling, and was led to the formulation of his system of distributions by a consideration of the ratio of the slope to ordinate of the frequency polygon*. Pearson's system (see the section "The Pearson System"), including a dozen distinct types, was to play a dominant role in subsequent developments.

The effect of transformation, with natural examples such as the links between distribution of lengths, areas, and volumes, also received attention; for example, Kapteyn [40] discussed the behavior of a nonlinear mapping of the normal.

Which models have proved useful and enlightening? It is impossible to answer this question purely objectively, for utility often depends on available facilities. Statistical development over this century has reacted specifically, first to all applied mathematics, and second, to the influence of the great digital computer invasion. So models exploiting mathematical asymptotics have slowly given way to insights provided by massive numerical soundings.

For example, the Pearson system probability integrals, involving awkward quadratures* for implementation, have recently [56] been tabulated and computerized. Its usefulness has become obvious in recent times. Similarly, the Johnson [35] translation system*, at one time prohibitive in numerical demands, now has extensive tabulated solutions, which, however, present no problem on a small computer.

Questions of validation of solutions are still mostly undecided. As in numerical quadrature, different appropriate formulas on differential grids form the basis for error analysis, and so for distributional approximation, we must use several approaches for comparison; a last resort, depending on the circumstances, would be simulation* studies.

In fitting models by moments it should be kept in mind that even an infinite set of moments may not determine a density uniquely; it turns out that there are strange nonnull functions, mathematical skeletons, for which all moments are zero. This aspect of the problem, usually referred to as the problem of moments, has been discussed by Shohat and Tamarkin [66].

On the other hand, from Chebyshev-type inequalities*, distributions having the same first r moments, cannot be too discrepant in their probability levels. For the four-moment case, the subject has been studied by Simpson and Welch [67]. They consider $\Pr[x < y] = \alpha$ and the problem of bounds for y given α. Also, it should be kept in mind that moments themselves are subject to constraints; thus for central moments, $\mu_4 \mu_2 - \mu_3^2 \geqslant \mu_2^3$.

TEST STATISTICS IN DISTRIBUTION

The distribution of test statistics (Student's t^*, the standard deviation*, the coefficient of variation*, sample skewness*, and kurtosis*) has proved a problem area and excited interest over many decades. Here we are brought face to face with the fact that normality rarely occurs, and in the interpretation of empirical data conservatism loosened its grip slowly and reluctantly.

Outstanding problems are many. For example, consider statistics which give information over and above that supplied by measures of scale and location. Skewness and kurtosis are simple illustrations, the former being assessed by the third central moment, and the latter by the fourth. The exact distributions of these measures under the null hypothesis, normal universe sampled, are still unknown, although approximations are available (see "Illustrations"). Similarly, Student's t and Fisher's F^*, although known distributionally under normality, are in general beyond reach under alternatives. These problems direct attention to approximating distributions by mathematical models.

In passing, we note that whereas large sample assumptions usually slant the desired distribution toward a near neighborhood of normality, small-sample assumptions bring out the amazing intricacies and richness of distributional forms. McKay [44] showed the density of the skewness in samples of four from a normal universe to be a complete elliptic integral; Geary [30] shows the density for samples of five and six to be a spline function*, and remarkably, smoothness of density becomes evident for larger samples. Hoq et al. [33] have derived the exact density of Student's ratio, a noncentral version, for samples of three and four in sampling from an exponential distribution*. Again, spline functions* appear, and it becomes evident that the problem presents insuperable difficulties for larger samples.

In this article we discuss the following systems:

1. K. Pearson's unimodal curves based on a differential equation.
2. Translation systems* based on a mapping of a normal, chi-squared*, or other basic density.
3. Perturbation models based on the normal density, arising out of studies of the central limit theorem*.
4. Multivariate models.
5. Discrete distributions*.

Literature

Johnson and Kotz [37] give an up-to-date comprehensive treatment, the first of its kind, with many references. Discrete, continuous univariate, and multivariate distributions are discussed. A comprehensive account of distributions, with many illuminating examples and exercises, is provided by Kendall and Stuart [42]. Patil and Joshi [53] give a comprehensive summary with most important properties. Bhattacharya and Ranga Rao [4] give a theoretical and mathematical account of approximations related to the central limit theorem*. Some specialized approaches are given in Crain [16] and Dupuy [22].

THE PEARSON SYSTEM

The Model

The model (*see* PEARSON SYSTEM OF DISTRIBUTIONS) is defined by solutions of the differential equation

$$y' = -(x + a)y/(Ax^2 + Bx + C), \qquad (1)$$

$y(\cdot)$ being the density. We may take here (but not necessarily in the following discussion) $E(x) = 0$, $\mathrm{var}(x) = 1$, so that $E(x^3) = \sqrt{\beta_1}$, $E(x^4) = \beta_2$, where the skewness $\beta_1 = \mu_3^2/\mu_2^3$ and the kurtosis $\beta_2 = \mu_4/\mu_2^2$. Arranging (1) as

$$x^s(Ax^2 + Bx + C)y' + x^s(x + a)y = 0 \qquad (2)$$

and integrating, using $s = 0, 1, 2, 3$, we find

$$A = (2\beta_2 - 3\beta_1 - 6)/\Delta,$$

$$B = \sqrt{\beta_2}(\beta_2 + 3)/\Delta,$$

$$C = (4\beta_2 - 3\beta_1)/\Delta, \qquad a = B,$$

$$\Delta = 10\beta_2 - 12\beta_1 - 18.$$

[If $\Delta = 0$, then define $A\Delta = \alpha, B\Delta = \beta$, $C\Delta = \gamma$, so that (1) becomes

$$y' = 2y\sqrt{\beta_1}/(x^2 - 2x\sqrt{\beta_1} - 3),$$

leading to a special case of type 1.]

Note that $\sqrt{\beta_1}$, β_2 uniquely determine a Pearson density. Moreover, it is evident that solutions of (1) depend on the zeros of the quadratic denominator. Although Karl Pearson* identified some dozen types, it is sufficient here, in view of the subsequent usage of the system, to describe the main densities. We quote noncentral (μ_s') or central (μ_s) moments for convenience, and expressions for $\sqrt{\beta_1}$ and β_2.

Normal:

$$y(x) = (2\pi)^{-1/2} \exp(-\tfrac{1}{2}x^2) \qquad (x^2 < \infty). \quad (3a)$$

$$\mu_1' = 0, \qquad \mu_2 = 1,$$

$$\sqrt{\beta_1} = 0, \qquad \beta_2 = 3.$$

Type 1 (or Beta*):

$$y(x) = \Gamma(a+b)x^{a-1}(1-x)^{b-1}/(\Gamma(a)\Gamma(b))$$

$$(0 < x < 1; a, b > 0). \tag{3b}$$

$$\mu_1' = a/\alpha_0, \qquad \mu_2 = ab/(\alpha_0^2\alpha_1),$$

$$\mu_3 = 2ab(b-a)/(\alpha_0^3\alpha_1\alpha_2),$$

$$\mu_4 = 3ab(ab(\alpha_0 - 6) + 2\alpha_0^2)/(\alpha_0^4\alpha_1\alpha_2\alpha_3);$$

$$\alpha_s = a + b + s.$$

Gradient at $x = 0$ is finite if $a > 1$.

Gradient at $x = 1$ is finite if $b > 1$.

$$\sqrt{\beta} \geqslant 0 \qquad \text{if } b \geqslant a.$$

Type I may be U-, Π-, or J-shaped according to the values of a and b.

Type III (Gamma*, Chi-Squared*):

$$y(x) = (x/a)^{\rho-1}\exp(-x/a)/(a\Gamma(\rho))$$

$$(0 < x < \infty; a, \rho > 0). \tag{3c}$$

$$\mu_1' = a\rho, \qquad \mu_2 = a^2\rho,$$

$$\sqrt{\beta_1} = 2/\sqrt{\rho}, \qquad \beta_2 = 3 + 6/\rho.$$

If the density is zero for $x < -s$, then Type III becomes

$$y(x) = ((x+s)/a)^{\rho-1}$$

$$\cdot \exp(-(x+s)/a)/(a\Gamma(\rho))$$

$$(-s < x < \infty),$$

and the only modification in the moments is that $\mu_1' = s + a\rho$. For other types, see Elderton and Johnson [26].

Recurrence for Moments

For the standardized central moments $\nu_s = \mu_s/\sigma^s(\mu_2 = \sigma^2)$, we have from (1),

$$\nu_{s+1} = \frac{s}{D_s}\left\{(\beta_2 + 3)\nu_s\sqrt{\beta_1}\right.$$

$$\left. + (4\beta_2 - 3\beta_1)\nu_{s-1}\right\}$$

$$(s = 1, 2, \ldots; \nu_0 = 1, \nu_1 = 0), \tag{4}$$

where $D_s = 6(\beta_2 - \beta_1 - 1) - s(2\beta_2 - 3\beta_1 - 6)$. Note that if $2\beta_2 - 3\beta_1 - 6 < 0$, then $D_s > 0$ since $\beta_2 - \beta_1 - 1 > 0$. Thus in the Type I

region of the (β_1, β_2) plane, all moments exist, whereas below the Type III line, $2\beta_2 - 3\beta_1 - 6 > 0$, only a finite number of moments exists; in this case the highest moment ν_s occurs when $s = [x] + 1$, where $x = 6(\beta_2 - \beta_1 - 1)/(2\beta_2 - 3\beta_1 - 6)$ and $[x]$ refers to the integer part of x.

Evaluation of Percentage Points

First, for a given probability level α, we may seek t_α, where

$$\int_{t_\alpha}^{\infty} y(x)\,dx = \alpha \tag{5}$$

and $y(\cdot)$ is a solution of the Pearson differential equation, with $\alpha = 1$ when $t_\alpha = -\infty$. It is possible to solve (5) by fragmentation, employing tailor-made procedures for each subregion of the $(\sqrt{\beta_1}, \beta_2)$ space. An alternative, encompassing the whole system, is to use (5) and (1) as simultaneous equations for the determination of t_α. Computer programs have been constructed by Amos and Daniel [1] and Bouver and Bargmann [5].

The converse problem of finding α given t_α has also been solved by the two approaches.

This problem of the relation between t_α and α, as one might expect, transparently reflects the fashions and facilities available over its history of a century or so. Computerized numerical analysis*, in modern times, has dwarfed the quadrature* and inverse interpolation* problems involved, and directed attention away from applied mathematical expertise. Nonetheless, there is a rich literature on probability integral problems, much of it relating to various aspects of closest approximation.

The Johnson et al. [39] tables have been available for two decades and give lower and upper points at the percent levels 0.1, 0.25, 0.5, 1.0, 2.5, 5.0, 10.0, 25.0, and 50.0 in terms of (β_1, β_2); the tables are given with additions in the Pearson and Hartley tables [56]. Interpolation is frequently necessary, however. Approximation formulas for standard percent levels 1.0, 5.0, etc., have been given by Bowman and Shenton [8,9].

Illustrations

Pearson curves have been fitted to distributions of the following statistics.

Skewness Statistic. $\sqrt{b_1}$, defined as $m_3/m_2^{3/2}$, where for the random sample $X_1, X_2, \ldots, X_n$, $m_i = \sum (X_j - \overline{X})^i/n$, and $\overline{X}$ is the sample mean. For sampling from the normal, Fisher [28] demonstrated the independence of $\sqrt{b_1}$ and m_2, from which exact moments can be derived. E. S. Pearson [55] gives comparisons of approximations, including Pearson Type VII, for $n = 25(5)40(10)60$ and probability levels $\alpha = 0.05$ and 0.01.

D'Agostino and Tietjen [20] also give comparisons for $n = 7, 8, 15, 25, 35$ at $\alpha = 0.1$, 0.05, 0.025, 0.01, 0.005, and 0.001.

Mulholland [40] in a remarkable study of $\sqrt{b_1}$ has developed approximations to its density for samples $4 \leqslant n \leqslant 25$. The work uses an iterative integral process based on examination of the density discontinuities and to a certain extent follows earlier work of this kind by Geary [30].

When nonnormal sampling is involved, independence property of $\overline{X}$ and S^2 breaks down, and a Taylor series* for $\sqrt{b_1}$ and its powers can be set up leading to an asymptotic series in n^{-1}. Examples of Pearson approximations have been given by Bowman and Shenton [9] and Shenton and Bowman [65].

Kurtosis Statistic b_2. This is defined as m_4/m_2^2, and the independence of $\overline{X}$ and S^2 in normal sampling enables exact moments to be evaluated. Pearson approximations are given in Pearson [54].

Noncentral χ^2. For independent $X_i \in N(0, 1)$ define

$$\chi^{'2} = \sum (a_i + X_i)^2.$$

Pearson approximations are given in Pearson and Hartley [56, pp. 10–11, 53–56]; see also Solomon and Stephens [68].

Watson's U_N^{2*}. For approximations to this goodness-of-fit statistic, see Pearson and Hartley [56,77].

Miscellaneous. Bowman et al. [11] have studied Pearson approximations in the case of Student's t under nonnormality.

Four moments of Geary's ratio* of the mean deviation to the standard deviation in normal samples show the density to be near the normal for $n \geqslant 5$; Pearson curves give an excellent fit [7].

Discussion

In approximating a theoretical density by a four-moment Pearson density, we must remember that the general characteristics of the one must be reflected in the other. Bimodel and multimodal densities, for example, should in general be excluded. Also, for $\sqrt{b_1}$ and $n = 4$ under normality, the true distribution consists of back-to-back J-shapes. However, the Pearson curve is $\cap$-shaped (see McKay [44], and Karl Pearson's remarks). A comparison is given in Table 1.

Since the four-moment fit ignores end points of a statistic (at least they are not fitted in the model), approximations to extreme percentage points will deteriorate sooner or later. Thus, the maximum value of $\sqrt{b_1}$ is known [61] to be $(n - 2)/\sqrt{n - 1}$, so a Type 1 model will either over or under estimate this end point (see Table 1). Similarly, an associated matter concerns tail abruptness; Pearson models usually deteriorate at and near the blunt tail.

Again, exact moments of a statistic may be intractable. Approximations therefore induce further errors, but usually percentage points are not finely tuned to the measures of skewness and kurtosis.

Finally, in general a unimodal density near the normal ($\sqrt{\beta_1}$ small, $\beta_2 = 3$ approximately) should be well approximated by a Pearson density.

Literature

The classical treatise is Elderton's *Frequency Curves and Correlation* [25]. This has now been revised and appears as *Systems of Frequency Curves* [26] by W. P. Elderton and N. L. Johnson (Cambridge University Press). It contains a complete guide to fitting Pearson curves to empirical data, with comments on other approximation systems.

The basic papers by Karl Pearson are those of 1894 [57], 1895 [58], and 1901 [59]. They are of considerable interest in showing the part played by the normal law of errors and modifications of it; for example, Pearson argues that data might be affected by a second normal component to produce the appearance of nonnormality, and solves the two-component normal mixture problem completely. It has since been shown [10]

Table 1. Pearson and Johnson Approximations

Population	Statistic	Sample Size	A[a]	$\alpha = 0.01$	0.05	0.10	0.90	0.95	0.99
Exponential	$\sqrt{m_2}$	$n = 3$[b]	E	0.045	0.109	0.163	1.293	1.610	2.340
			P	0.057	0.113	0.164	1.296	1.612	2.337
			P*	0.368	0.368	0.369	1.290	1.632	2.374
			S_B	0.019	0.102	0.164	1.289	1.609	2.358
		$n = 4$	E	0.095	0.180	0.245	1.342	1.632	2.303
			P	0.108	0.184	0.245	1.343	1.633	2.293
			P*	0.397	0.397	0.399	1.347	1.654	2.322
			S_B	0.083	0.178	0.246	1.340	1.632	2.308
	$v = (s_x/m_1')$	$n = 10$	M	0.481	0.587	0.648	1.231	1.351	1.617
			P	0.499	0.591	0.646	1.237	1.353	1.598
			P*	0.531	0.600	0.650	1.238	1.356	1.601
			S_B^*	0.509	0.596	0.650	1.237	1.355	1.604

				$\alpha = 0.90$	0.950	0.975	0.990	0.995	0.999
Normal	$\sqrt{b_1}$	$n = 4$[c]	E	0.831	0.987	1.070	1.120	1.137	1.151
			P	0.793	0.958	1.074	1.178	1.231	1.306
			S_B	0.791	0.955	1.071	1.179	1.237	1.327
		$n = 8$	E	0.765	0.998	1.208	1.452	1.606	1.866
			P	0.767	0.990	1.187	1.421	1.583	1.929
			S_U	0.767	0.990	1.187	1.421	1.583	1.929

				$\alpha = 0.01$	0.05	0.10	0.90	0.95	0.99
	b_2	$n = 10$	M	1.424	1.564	1.681	3.455	3.938	4.996
			P	1.495	1.589	1.675	3.471	3.931	4.921
			S_B	1.442	1.577	1.677	3.456	3.921	4.933
			D	1.39	1.56	1.68	3.53	3.95	5.00
			A_G	1.287	1.508	1.649	3.424	3.855	4.898

[a] E, exact; P, four-moment Pearson on statistic; P*, four-moment Pearson on square of statistic; M, Monte Carlo of 100,000 runs; A_G, Anscombe and Glynn [2]; D, D'Agostino and Tietjen [19]; E for $\sqrt{m_2}$ derived by Lam [43]; S_U, S_B, Johnson; S_B^*, Johnson on square of statistic.

[b] For $n = 3$, P* has density $c_0(m_2 - 0.136)^\alpha/(m_2 + 13.168)^\beta$, $\alpha = -0.740$, $\beta = 7.770$; P has density $c_1(\sqrt{m_2} - 0.024)^\alpha/(\sqrt{m_2} + 13.076)^\beta$, $\alpha = 0.738$, $\beta = 39.078$.

[c] For $n = 4$, P is the density $c_2(1.896^2 - b_1)^{3.26}$; from theory $\sqrt{b_1}$ max. is 1.155.

that if exact moments are available (five are needed) then there may be one or two solutions to the problem, or quite possibly none at all. Pearson's examples note these cases.

Again, Pearson from time to time expresses his displeasure at the sloppy tabulation and abbreviation of data. One example he considers concerns the distribution of 8,689 cases of enteric fever received into the Metropolitan Asylums Board Fever Hospitals, 1871–1893. There are 266 cases reported for "under age 5" and 13 for "over age 60." Fitting types I and III to the data by moments, he notes that both models suggest an unlikely age for first onset of the disease (-1.353 and -2.838 years, respectively) and similarly an unlikely upper limit

to the duration of life (385 years). (It is quite possible that he was well aware of the small chances involved, and we should keep in mind the point that to arrive at bounds of any kind for the data was something not supplied by conventional normal theory modeling.)

It should be emphasized that a model for a distribution of experimental data encounters problems that are different from those of the corresponding situation for a theoretical statistic. In the latter, errors are due solely to choice of model, although different solution procedures are possible; for example, parameters may be determined by moments, or by percentiles, and also in part from a knowledge of end points. However, in the case of experimental data, parameters are estimated by different procedures (least squares*, maximum likelihood*, moments*, etc.) and there is the question of biases*, variances, etc., and more detailed knowledge of the sampling distribution of the estimates.

Since higher moments of data are subject to considerable sampling errors, a recent [48] method of fitting is recommended when one (or both) end point(s) is known. The procedure calculates a modified kurtosis using the end point and four moments.

TRANSLATION SYSTEMS

Early Ideas

Suppose that the berries of a fruit have radii (measured in some decidable fashion) normally distributed $N(R, \sigma^2)$. Then surface area $S = \pi r^2$ will no longer be normal and indeed will be skewed distributionally. This idea was developed a little later than the introduction of the Pearson system by Kapteyn [40] in his treatment of skew frequency curves in biology. From a different point of view and using a hypothesis relating to elementary errors, Wicksell [71] traced the relation between certain transformation of a variate and a genetic theory of frequency.

Kapteyn considered the transformations

$$y = (X + h)^q - m \qquad [y \in N(0, 1)]$$

with $-\infty < q < \infty$, the special case $q \to 0$ leading to a logarithmic transformation.

The development of the subject was perhaps retarded because of overanxiety to trace a transformation's relation to natural phenomena and mathematical difficulties. Moreover, the Kapteyn mapping proved intractable mathematically in most cases.

Johnson's Systems

Johnson [35] introduced two transformations of the normal density (*see* JOHNSON'S SYSTEM OF DISTRIBUTIONS). His S_U system relates to a hyperbolic sine function, has doubly infinite range, and

$$y = \sinh\left(\frac{X - \gamma}{\delta}\right) \tag{6}$$
$$(-\infty < y < \infty, \delta > 0),$$

where $X \in N(0, 1)$.

Similarly, the S_B system relates to densities with bounded range, and

$$y = 1/(1 + e^{(\gamma - X)/\delta}) \qquad (0 < y < 1, \delta > 0), \tag{7}$$

where again $X \in N(0, 1)$.

These transformations are both one-to-one, and the S_U case readily yields to moment evaluation, whereas S_B does not.

For S_U, the first four moments are

$$\mu'_1(y) = -\sqrt{\omega} \sinh \Omega,$$
$$\mu_2(y) = (\omega - 1)(\omega \cosh(2\Omega) + 1)/2,$$
$$\mu_3(y) = -(\omega - 1)^2 \sqrt{\omega}\{(\omega^2 + 2\omega) \sinh(3\Omega) + 3 \sinh \Omega\}/4, \tag{8}$$
$$\mu_4(y) = (\omega - 1)^2\{d_4 \cosh(4\Omega) + d_2 \cosh(2\Omega) + d_0\},$$

where

$$d_4 = \omega^2(\omega^4 + 2\omega^3 + 3\omega^2 - 3)/8,$$
$$d_2 = \tfrac{1}{2}\omega^2(\omega + 2),$$
$$d_0 = 3(2\omega + 1)/8, \qquad \text{and}$$
$$\ln \omega = 1/\delta^2, \quad \Omega = \gamma/\delta.$$

Note that since $\omega > 1$, $\mu_3(y)$ has the sign of $(-\Omega)$, and unlike the structure of the Pearson system, here all moments are functions of ω and Ω. (For the Pearson system, mean and variance are not affected by $\sqrt{\beta_1}$ and β_2, whereas they are for S_U.)

If Y is a variate, then set

$$Y = (y + p)/q \tag{9a}$$

so that

$$\begin{aligned} v'_1 &= E(Y) = (\mu'_1(y) + p)/q \\ v_2 &= \operatorname{var}(Y) = (\mu_2(y))/q^2 \end{aligned} \tag{9b}$$

Determine p and q. The values of $\mu'_1(y)$ and $\mu_2(y)$ are set by equating the skewness ($\sqrt{\beta_1}$), and kurtosis (β_2) of Y to those of y; assistance here is given in Tables 34 and 35 in Pearson and Hartley [56].

Johnson's S_U system [35] immediately provides percentiles from those of the normal, and also an equivalent normal variate,

$$X = \gamma + \delta \sinh^{-1} y.$$

However, one must keep in mind that the S_U density is still only a four-moment approximation.

To fix the domain of validity of S_U, let $\Omega = -k$ and $k \to \infty$, so that the mean of y tends to ∞ and $\sigma^2 \sim \omega(\omega - 1)e^{2k}/4$. Then if $t = y/\sigma$ from (6), $X = c + \ln t$, which corresponds to a log-normal transformation, and from (8)

$$\sqrt{\beta_1} = (\omega + 2)\sqrt{\omega - 1},$$
$$\beta_2 = \omega^4 + 2\omega^3 + 3\omega^2 - 3, \qquad (10)$$

the parametric form of the boundary in the $(\sqrt{\beta_1}, \beta_2)$ plane.

Examples of S_U and S_B are given in Table 1.

Literature

An adequate description of S_U and S_B is given by Pearson and Hartley [56, pp. 80–87], including tables to facilitate fitting. Moreover, the iterative scheme for the evaluation of the parameters of S_U given by Johnson [36] is readily programmed for a small computer. A rational fraction solution for ω, leading also to a value of Ω, has been developed by Bowman and Shenton [10]; a similar scheme is also available for the S_B-system [13].

SERIES DEVELOPMENTS

These originated in the nineteenth century, and are related to procedures to sharpen the central limit theorem*. For large n, for $z_1, z_2, \ldots, z_n$ mutually independent variates with common standard deviation σ, the distribution of $s = (\sum z)/(\sigma \sqrt{n})$ is approximately normal. A better approximation appears from Charlier's A-series [14],

$$\phi(x) + (a_3/3!)\phi^{(3)}(x)$$
$$+ (a_4/4!)\phi^{(4)}(x) + \cdots \qquad (11)$$

where $\phi(x) = (2\pi)^{-1/2} \exp\left(-\frac{1}{2}x^2\right)$ is the standard normal density. [This approximation involves derivatives of $\phi(x)$.] Cramér [17,18] has proved that certain asymptotic properties of (11) hold. Another version of the series

development (11) takes the form $\Phi(d/dx)\phi(x)$, where

$$\Phi(t) \equiv \exp \sum \epsilon_j(-t)^j/j!, \qquad (12)$$

the operator changing the cumulants* (κ_r) of $\phi(\cdot)$ to $(\kappa_r + \epsilon_r)$. This, from Cramér's work (see his note and references in ref. 18), has similar asymptotic properties with respect to the central limit theorem*.

Since the derivatives of the normal density are related to Hermite polynomials* (see CHEBYSHEV–HERMITE POLYNOMIALS), (11) may be written

$$[1 - (a_3/3!)H_3(x)$$
$$+ (a_4/4!)H_4(x) - \cdots]\phi(x), \qquad (13)$$

and if this approximates a density $f(x)$, then using orthogonality,

$$a_s = (-1)^s E[H_s(x)],$$

where the expectation* operator refers to $f(x)$. Thus in terms of cumulants,

$$a_3 = -\kappa_3, \quad a_4 = \kappa_4,$$
$$a_5 = -\kappa_5, \quad a_6 = \kappa_6 + 10\kappa_3^2,$$

etc., the coefficients a_1, a_2 being zero because of the use of the standard variate x.

If empirical data are being considered, only the first five or six terms of (1) can be contemplated because of large sampling errors of the moments involved. However, this problem does not arise in the approximation of theoretical structures, and cases are on record (for example, ref. 47) in which the twentieth polynomial has been included; quite frequently the density terms turn out to have irregular sign and magnitude patterns.

The Edgeworth* form [23,24],

$$\phi(x)\{1 + (\kappa_3/3!)H_3(x) + (\kappa_4/4!)H_4(x)$$
$$+ (\kappa_5/5!)H_5(x)$$
$$+ (\kappa_6 + 10\kappa_3^2)H_6(x)/6! + \cdots\}$$

has one advantage over (1), in that when applied to certain statistics, the standardized cumulants κ_r may be shown to be of order $1/n^{(1/2)r-1}$, where n is the sample size, so that if carried far enough the coefficients will tend to exhibit a regular magnitude pattern.

Nonnormal Kernels

The basic function $\phi(\cdot)$ need not be normal, and Romanowsky [62] introduced generalizations such as gamma- and beta*-type densities, with associated Laguerre* and Jacobi* orthogonal polynomials. The normal density is, however, generally favored because of the simplicity of the Chebyshev-Hermite system of polynomials.

Cornish–Fisher Expansions

Using a differential series, such as (12) with normal kernel, Cornish and Fisher [15] derived a series for the probability integral of a variate whose cumulants are known (*see* CORNISH–FISHER AND EDGEWORTH EXPANSIONS). By inversion of series, they derived a series for the deviate at a given probability level in terms of polynomials in the corresponding normal deviate. Fisher and Cornish [29] extended the earlier study to include terms of order n^{-3}, these involving the eighth cumulant and polynomials of degree seven. Further generalizations are due to Finney [27], who treated the case of several variates; Hill and Davis [32], who gave a rigorous treatment indicating the procedure for the derivation of a general term; and Draper and Tierney [21], who tabulated the basic polynomials involved in the terms to order n^{-4} and cumulants up to κ_{10}; unpublished results of Hill and Davis for the higher-order polynomials agreed with those found by Draper and Tierney.

Some Applications

In his paper on testing for normality* (*see* DEPARTURES FROM NORMALITY, TESTS FOR), Geary [31] considered the distribution of t under nonnormality, basing it on what he called a differential series [expression (12)] with kernel the density of t^* under normality, i.e., $c_n(1 + t^2/(n-1))^{-n/2}$. He derived a series to order n^{-2} for the probability integral, using it cautiously for a sample of 10 under only moderate nonnormality. In an earlier study [30] searching for precise forms for the density of the sample skewness under normality, he used the Cornish-Fisher series to assess probability levels using cumulants up to the eighth. Mulholland [47] developed the subject further, providing a recursive scheme for

evaluating at least theoretically any moment of $\sqrt{b_1}$, and a Charlier series for the probability integral up to polynomials of degree 20.

Series involving Laguerre polynomials and a gamma density have been exploited by Tiku. For example, his study [69] of the variance ratio* and Student's t [70], both under nonnormality, involve terms up to order n^{-2}.

Discussion

In approximating distributions, convergence* questions, and even to a less extent, asymptotic properties are irrelevant, for we are concerned with a finite number of terms, since very rarely can general terms be found. Thus questions of nonnegativity of the density arise [3,64], and internal checks for closeness of approximation via convergence or otherwise are not available. At best the Charlier and Edgeworth series can only serve as backup models.

MULTIVARIATE DENSITIES

The Pearson system defined in (1) becomes, in bivariate form,

$$\frac{1}{y}\frac{\partial y}{\partial x_i} = \frac{P_i(x_1, x_2)}{Q_i(x_1, x_2)} \quad (i = 1, 2),$$

where P_i and Q_i are functions of x_1 and x_2 of degrees 1 and 2, respectively. There are again several types, including the bivariate normal* with density

$$\phi(x_1, x_2) = (1 - \rho^2)^{-1/2}(2\pi)^{-1}$$

$$\times \exp\left\{\frac{-\frac{1}{2}x^2 - \rho xy - \frac{1}{2}y^2}{1 - \rho^2}\right\}, \rho^2 < 1,$$

in standard form, which reduces to a product form when $\rho = 0$ (i.e., when the variates are uncorrelated).

Another well-known type is the Dirichlet* density,

$$y(x_1, x_2) = cx_1^{a-1}x_2^{b-1}(1 - x_1 - x_2)^{c-1}$$
$$(a, b, c > 0)$$

with domain $x_1, x_2 > 0, x_1 + x_2 \leqslant 1$.

Series development based on the Charlier model take the form

$$\{1 + a_{10}\partial_1 + a_{01}\partial_2 + (a_{20}/2!)\partial_1^2 + a_{11}\partial_1\partial_2$$
$$+ (a_{02}/2!)\partial_2^2 + \cdots\}\phi(x_1, x_2),$$

where $\partial_i \equiv \partial/\partial x_i$.

Similarly, there are multivariate translation systems, following the Johnson S_U and S_B models.

The Pearson system, for which marginal distributions are of Pearson form, can be fitted by moments, a new feature being the necessity to use product moments*, the simplest being related to the correlation between the variates. A similar situation holds for Charlier multivariate developments, and again, as with the single-variate case, negative frequencies can be a problem.

Literature

A sustained study of empirical bivariate data, at least of historical interest, is given in K. Pearson's 1925 paper [60].

A brief discussion of frequency surfaces* will be found in Elderton and Johnson [26]. A more comprehensive treatment, including modern developments, is that of N. L. Johnson and S. Kotz [37]. Mardia's work [45] is a handy reference.

DISCRETE DISTRIBUTIONS

There is now a richness of variety of discrete distributions, and the time is long passed when there were possibly only three or so choices, the binomial*, the Poisson*, and the hypergeometric*; the geometric appearing infrequently in applications.

Just as a differential equation is used to define the Pearson system of curves, so analogs in finite differences* may be used to generate discrete density functions [41,49, 58]. Again, if $G_i(t)$, $i = 1, 2, \ldots$, are the probability generating functions* of discrete variates, then new distributions arise from G_1 $(G_2(t))$ and similar structures; for example, Neyman's contagious distributions* are related to choosing G_1 and G_2 as Poisson generating functions.

There are many applications where distribution approximation models are required for random count data* [51,52]. We also mention occupancy problems* [38], meteorological phenomena relating to drought frequency, storm frequency, degree-days, etc. However, the need for approximating discrete distributions is not common, especially with the advent of computer facilities. A classical exception is the normal approximation to the binomial distribution when the index n is large; in this case, if the random variate is x, then with probability parameter p, we consider the approximation $(x - np)/\sqrt{npq}$ to be nearly normal (for refinements, see, e.g., Molenaar [46]).

Again the Poisson or binomial density functions may be used as the basis for Charlier-type approximation expansions. Thus for the Poisson function $\psi(x) = e^{-m}m^x/x!$, we consider

$$f(x) = \{1 + \alpha_2 \nabla_x^2/2! + \alpha_3 \nabla_x^3/3!t$$
$$+ \cdots\}\psi(x),$$

where $\nabla g(x) \equiv g(x) - g(x - 1)$ is a backward difference*. This type of approximation may be considered for random variates taking the values $x = 0, 1, \ldots$, when moments exist and the density to be approximated is complicated.

Literature

Historical information has been given by Särndal [63], and comprehensive accounts are those of Johnson and Kotz [37] and Patil and Joshi [53]. A short account, with new material, is given by J. K. Ord [50].

Acknowledgment

This research was operated by Union Carbide Corporation under Contract W-7405-eng-26 with the U.S. Department of Energy.

REFERENCES

1. Amos, D. E. and Daniel, S. L. (1971). *Tables of Percentage Points of Standardized Pearson Distributions. Rep. No. SC-RR-71 0348*, Sandia Laboratories, Albuquerque, N.M.

2. Anscombe, F. and Glynn, W. J. (1975). *Distribution of the Kurtosis Statistics b_2 for Normal Samples. Tech. Rep. 37*, Dept. of Statistics, Yale University, New Haven, Conn.

3. Barton, D. E. and Dennis, K. E. (1952). *Biometrika*, **39**, 425–427.

4. Bhattacharya, R. N. and Ranga Rao, R. (1976). *Normal Approximations and Asymptotic Expansions*. Wiley, New York.

5. Bouver, H. and Bargmann, R. (1976). *Amer. Stat. Ass.: Proc. Statist. Computing Sect.*, pp. 116–120.

6. Bowman, K. O., Beauchamp, J. J., and Shenton, L. R. (1977). *Int. Statist. Rev.*, **45**, 233–242.

7. Bowman, K. O., Lam, H. K., and Shenton, L. R. (1980). *Reports of Statistical Application Res., Un. Japanese Scientists and Engineers*, **27**, 1–15.

8. Bowman, K. O., Serbin, C. A., and Shenton, L. R. (1981). *Commun. Statist. B*, **10**(1), 1–15.

9. Bowman, K. O. and Shenton, L. R. (1973). *Biometrika*, **60**, 155–167.

10. Bowman, K. O. and Shenton, L. R. (1973). *Biometrika*, **60**, 629–636.

11. Bowman, K. O. and Shenton, L. R. (1979). *Biometrika*, **66**, 147–151.

12. Bowman, K. O. and Shenton, L. R. (1979). *Commun. Statist. B*, **8**(3), 231–244.

13. Bowman, K. O. and Shenton, L. R. (1980). *Commun. Statist. B*, **9**(2), 127–132.

14. Charlier, C. V. L. (1905). *Ark. Mat. Astron. Fys.*, **2**(15).

15. Cornish, E. A. and Fisher, R. A. (1937). *Rev. Inst. Int. Statist.*, **4**, 1–14.

16. Crain, B. R. (1977). *Siam J. Appl. Math.*, **32**, 339–346.

17. Cramér, H. (1928). *Skand. Aktuarietidskr*, **11**, 13–74, 141180.

18. Cramér, H. (1972). *Biometrika*, **59**, 205–207.

19. D'Agostino, R. B. and Tietjen, G. L. (1971). *Biometrika*, **58**, 669–672.

20. D'Agostino, R. B. and Tietjen, G. L. (1973). *Biometrika*, **60**, 169–173.

21. Draper, N. R. and Tierney, D. E. (1973). *Commun. Statist.*, **1**(6), 495–524.

22. Dupuy, M. (1974). *Int. J. Computer Math. B*, **4**, 121–142.

23. Edgeworth, F. Y. (1905). *Camb. Philos. Trans.*, **20**, 36–66, 113141.

24. Edgeworth, F. Y. (1907). *J. R. Statist. Soc.*, **70**, 102–106.

25. Elderton, W. P. (1960). *Frequency Curves and Correlation*. Cambridge University Press, Cambridge.

26. Elderton, W. P. and Johnson, N. L. (1969). *Systems of Frequency Curves*. Cambridge University Press, Cambridge.

27. Finney, D. J. (1963). *Technometrics*, **5**, 63–69.

28. Fisher, R. A. (1928). *Proc. Lond. Math. Soc.*, **2**, 30, 199–238.

29. Fisher, R. A. and Cornish, E. A. (1960). *Technometrics*, **2**, 209–225.

30. Geary, R. C. (1947). *Biometrika*, **34**, 68–97.

31. Geary, R. C. (1947). *Biometrika*, **34**, 209–242.

32. Hill, G. W. and Davis, A. W. (1968). *Ann. Math. Statist.*, **39**, 1264–1273.

33. Hoq, A. K. M. S., Ali, M. M., and Templeton, J. G. (1977). Distribution of Student's Ratio Based on the Exponential Distribution. *Working Paper No. 77-001*, Dept. of Industrial Engineering, University of Toronto, Toronto.

34. Hotelling, H. (1961). *Proc. 4th Berkeley Symp. Math. Stat. Prob.*, Vol. 1. University of California Press, Berkeley, Calif., pp. 319–359.

35. Johnson, N. L. (1949). *Biometrika*, **36**, 149–176.

36. Johnson, N. L. (1965). *Biometrika*, **52**, 547–558.

37. Johnson, N. L. and Kotz, S. (1972). *Distributions in Statistics: Continuous Multivariate Distributions*. Wiley, New York.

38. Johnson, N. L. and Kotz, S. (1977). *Urn Models and Their Application*. Wiley, New York.

39. Johnson, N. L., Nixon, E., Amos, D. E., and Pearson, E. S. (1963). *Biometrika*, **50**, 459–498.

40. Kapteyn, J. C. (1903). *Skew Frequency Curves in Biology and Statistics*. Noordhoff, Groningen.

41. Katz, L. (1963). *Proc. Int. Symp. Discrete Distrib.*, Montreal, pp. 175–182.

42. Kendall, M. G. and Stuart, A. (1969). *The Advanced Theory of Statistics*, 3rd ed., Vol. 1: Distribution Theory. Charles Griffin, London/Hafner Press, New York.

43. Lam, H. K. (1978). The Distribution of the Standard Deviation and Student's *t* from Nonnormal Universes. Ph.D. dissertation, University of Georgia.

44. McKay, A. T. (1933). *Biometrika*, **25**, 204–210.

45. Mardia, K. V. (1970). Families of Bivariate Distributions. *Griffin's Statist. Monogr. No. 20*.

46. Molenaar, W. (1970). Approximation to the Poisson, Binomial, and Hypergeometric Distribution Functions. *Math. Centre Tracts No. 31*. Mathematisch Centrum, Amsterdam.

47. Mulholland, H. P. (1977). *Biometrika*, **64**, 401–409.

48. Müller, P. -H. and Vahl, H. (1976). *Biometrika*, **63**, 191–194.

49. Ord, J. K. (1967). *J. R. Statist. Soc. A*, **130**, 232–238.

50. Ord, J. K. (1975). *Families of Frequency Distributions. Griffin's Statist. Monogr. No. 30.*

51. Patil, G. P. (1965). *Classical and Contagious Discrete Distributions* (Proc. Int. Symp., Montreal). Pergamon Press, New York.

52. Patil, G. P. ed. (1970). *Random Counts in Physical Science, Geological Science, and Business*, Vols. 1–3. The Penn State Statistics Series. Pennsylvania State University Press, University Park, Pa.

53. Patil, G. P. and Joshi, S. W. (1968). *A Dictionary and Bibliography of Discrete Distributions*. Oliver & Boyd, Edinburgh.

54. Pearson, E. S. (1963). *Biometrika*, **50**, 95–111.

55. Pearson, E. S. (1965). *Biometrika*, **52**, 282–285.

56. Pearson, E. S. and Hartley, H. O. (1972). *Biometrika Tables for Statisticians*, Vol. 2. Cambridge University Press, Cambridge.

57. Pearson, K. (1894). *Philos. Trans. R. Soc. Lond. A*, **185**, 71–110.

58. Pearson, K. (1895). *Philos. Trans. R. Soc. Lond. A*, **186**, 343–414.

59. Pearson, K. (1901). *Philos. Trans. R. Soc. Lond. A*, **197**, 443–459.

60. Pearson, K. (1925). *Biometrika*, **17**, 268–313.

61. Pearson, K. (1933). *Biometrika*, **25**, 210–213.

62. Romanowsky, V. (1925). *Biometrika*, **16**, 106–116.

63. Särndal, C.-E. (1971). *Biometrika*, **58**, 375–391.

64. Shenton, L. R. (1951). *Biometrika*, **38**, 58–73.

65. Shenton, L. R. and Bowman, K. O. (1975). *J. Amer. Statist. Ass.*, **70**, 220–229, 349.

66. Shohat, J. A. and Tamarkin, J. E. (1943). *The Problem of Moments*. American Mathematical Society, New York.

67. Simpson, J. A. and Welch, B. L. (1960). *Biometrika*, **47**, 399–410.

68. Solomon, H. and Stephens, M. A. (1978). *J. Amer. Statist. Ass.*, **73**, 153–160.

69. Tiku, M. L. (1964). *Biometrika*, **51**, 83–95.

70. Tiku, M. L. (1971). *Aust. J. Statist.*, **13**(3), 142–148.

71. Wicksell, S. D. (1917). *Ark. Mat. Astron. Fys.*, **12**(20).

See also ASYMPTOTIC EXPANSIONS—I; CORNISH–FISHER AND EDGEWORTH EXPANSIONS; FREQUENCY CURVES, SYSTEMS OF; FREQUENCY SURFACES, SYSTEMS OF; GRAM–CHARLIER SERIES; JOHNSON'S SYSTEM OF DISTRIBUTIONS; and PEARSON SYSTEM OF DISTRIBUTIONS.

K. O. BOWMAN
L. R. SHENTON

APPROXIMATIONS TO FUNCTIONS.
See FUNCTIONS, APPROXIMATIONS TO

A PRIORI DISTRIBUTION

The term *a priori distribution* is used to describe a distribution ascribed to a parameter in a model. It occurs most commonly in the application of Bayesian methods. The a priori distribution is usually supposed to be known exactly—and not to depend on unknown parameters of its own.

See also BAYESIAN INFERENCE.

ARBITRARY ORIGIN.   See CODED DATA

ARBITRARY SCALE.   See CODED DATA

ARBUTHNOT, JOHN

Born: April 29, 1667, in Arbuthnott, Kincardineshire, Scotland.

Died: February 27, 1735, in London, England.

Contributed to: early applications of probability, foundations of statistical inference, eighteenth-century political satire, maintaining the health of Queen Anne.

John Arbuthnot was the eldest son of Alexander, parson of the village of Arbuthnott, near the east coast of Scotland. They were part of the Aberdeenshire branch of the Arbuthnot family, whose precise relationship to the Viscounts Arbuthnott of Kincardineshire, and to the ancient lairds of the family, is not known. In later years, John invariably signed his name "Arbuthnott," yet "Arbuthnot" consistently appears on his printed works. The latter is evidently the more ancient form, to which the Kincardineshire branch added a second "t" some time in the seventeenth century.

John entered Marischal College, Aberdeen, at the age of fourteen, and graduated with an M.A. in medicine in 1685. Following the Glorious Revolution of 1688–1689, John's father was deprived of his living in 1689, because he would not accept Presbyterianism. He died two years later, and soon after that John left Scotland to settle in London, where initially he made his living teaching mathematics. It was during this period that he produced one of the two pieces of work that have merited him a place in the history of probability and statistics. He translated, from Latin into English, Huygens' *De Ratiociniis in Ludo Aleae*, the first probability text [7]. Arbuthnot's English edition of 1692, *Of the Laws of Chance* [3], was not simply a translation. He began with an introduction written in his usual witty and robust style, gave solutions to problems Huygens* had posed, and added further sections of his own about gaming with dice and cards.

In 1694, he was enrolled at University College, Oxford as a "fellow commoner," acting as companion to Edward Jeffreys, the eldest son of the Member of Parliament for Brecon. Arbuthnot developed lasting friendships with Arthur Charlett, the Master of University College, and with David Gregory, the Savilian Professor of Astronomy at Oxford. It seems, though, that Edward Jeffreys did not use his time at Oxford well, and in 1696, Arbuthnot became "resolv'd on some other course of life." He moved briefly to St. Andrews and there presented theses which earned him a doctor's degree in medicine in September 1696.

Having returned to London, he quickly earned himself a reputation as a skilled physician and a man of learning. Two of his early scientific publications were *An Examination of Dr. Woodward's Account of the Deluge* (1697; John Woodward was Professor of Physic at Gresham College) [3] and *An Essay on the Usefulness of Mathematical Learning* (1701) [1]. He became a Fellow of both the Royal Society (1704) and the Royal College of Physicians (1710), and served on two prestigious committees of the former. The first, which included Sir Isaac Newton* (then President of the Royal Society), Sir Christopher Wren, and David Gregory, was set up in 1705 to oversee the publication of the astronomical observations of the Astronomer Royal, John Flamsteed. Arbuthnot was directly involved in trying to secure Flamsteed's cooperation in carrying the venture through to completion, in the face of hostility between Flamsteed and Newton, and accusations on each side that the other was obstructing the project. After long delays, the observations were finally published in 1712. Arbuthnot was less active in the second of the committees, appointed in 1712 to deliberate on the rival claims of Newton and Leibniz to invention of the "method of fluxions" (differential calculus).

In 1705, Arbuthnot became Physician Extraordinary to Queen Anne, and in 1709 Physician in Ordinary. Until Queen Anne's death in 1714, he enjoyed a favored position at the Court. In 1711, he held office in the Customs, and in 1713 he was appointed Physician at Chelsea College.

Outside the statistical community, he is most widely remembered for his comic satirical writings. He was the creator of the figure John Bull, who has become a symbol of the English character. John Bull featured in a series of pamphlets written by Arbuthnot in 1713; they contain a witty allegorical account of the negotiations taking place towards the settlement of the War of the Spanish Succession and were, in due course, put together as *The History of John Bull* [1]. Arbuthnot was a close colleague of Jonathan Swift, Alexander Pope, and John Gay, and a founder member of the Scriblerus Club, an association of scholars formed around 1714 with the aim of ridiculing pedantry and poor scholarship. It is thought that much of the *Memoirs of the Extraordinary Life, Works, and Discoveries of Martinus Scriblerus* [1] came from Arbuthnot's pen. His writings ranged widely over aspects of science, mathematics, medicine, politics and philosophy. Often he was modest or careless in claiming authorship, particularly of his political satire.

The paper he presented to the Royal Society of London on April 19, 1711 [2] has attracted most attention from historians of statistics and probability (it appeared in a volume of the *Philosophical Transactions* dated 1710, but published late). Arbuthnot's paper was "An Argument for Divine Providence, taken from the constant Regularity observ'd in the Births of both Sexes." In it, he maintained that the guiding hand of a divine being was to be discerned in the nearly constant ratio of male to female christenings recorded annually in London over the years 1629 to 1710. Part of his reasoning is recognizable as what we would now call a sign test, so Arbuthnot has gone down in statistical history as a progenitor of significance testing*.

The data he presented showed that in each of the 82 years 1629–1710, the annual number of male christenings had been consistently higher than the number of female christenings, but never very much higher. Arbuthnot argued that this remarkable regularity could not be attributed to chance*, and must therefore be an indication of divine providence. It was an example of the "argument from design," a thesis of considerable theological and scientific influence during the closing decades of the seventeenth century and much of the next. Its supporters held that natural phenomena of many kinds showed evidence of careful and beneficent design, and were therefore indicative of the existence of a supreme being.

Arbuthnot's representation of "chance" determination of sex at birth was the toss of a fair two-sided die, with one face marked M and the other marked F. From there, he argued on two fronts: "chance" could not explain the very close limits within which the annual ratios of male to female christenings had been observed to fall, neither could it explain the numerical dominance, year after year, of male over female christenings.

He pursued the first argument by indicating how the middle term of the binomial expansion, for even values of the size parameter n, becomes very small as n gets large. Though he acknowledged that in practice the balance between male and female births in any one year was not exact, he regarded his mathematical demonstration as evidence that, "if mere Chance govern'd," there would be years when the balance was not well maintained.

The second strand of his argument, concerning the persistent yearly excess of male over female christenings, was the one which ultimately caught the attention of historians of statistics. He calculated that the probability of 82 consecutive years in which male exceeded female christenings in number, under the supposition that "chance" determined sex, was very small indeed. This he took as weighty evidence against the hypothesis of chance, and in favor of his alternative of divine providence. He argued that if births were generated according to his representation of chance, as a fair two-sided die, the probability of observing an excess of male over female births in any one year would be no higher than one-half. Therefore the probability of observing 82 successive "male years" was no higher than $(\frac{1}{2})^{82}$ (a number of the order of 10^{-25} or 10^{-26}). The probability of observing the data given the "model," as we might now say, was very small indeed, casting severe doubt on the notion that chance determined sex at birth. Arbuthnot proceeded to a number of conclusions of a religious or philosophical nature, including the observation that his arguments vindicated the undesirability of polygamy in a civilized society.

We can see in Arbuthnot's probabilistic reasoning some of the features of the modern hypothesis test. He defined a null hypothesis ("chance" determination of sex at birth) and an alternative (divine providence). He calculated, under the assumption that the null hypothesis was true, a probability defined by reference to the observed data. Finally, he argued that the extremely low probability he obtained cast doubt on the null hypothesis and offered support for his alternative.

Arbuthnot's reasoning has been thoroughly examined by modern statisticians and logicians, most notably by Hacking [5, 6]. We have, of course, the benefit of more than 250 years of hindsight and statistical development. The probability of $(\frac{1}{2})^{82}$, on which hinged Arbuthnot's dismissal of the "chance" hypothesis, was one of a well-defined reference set, the binomial distribution with parameters 82 and one-half. It was the lowest and most extreme probability in this reference set, and hence also in effect a tail-area probability. And it was an *extremely low* probability. Arbuthnot made only the last of these points explicit.

Arbuthnot's advancement of an argument from design did not single him out from his contemporaries. Nor were his observations on the relative constancy of the male to female birth ratio radical. What was novel was his attempt to provide a statistical "proof" of his assertions, based on a quantitative concept of chance, explicitly expressed and concluded in numerical terms.

An unpublished manuscript in the Gregory collection at the University of Edinburgh indicates that Arbuthnot had been flirting with ideas of probabilistic proof well before 1711, possibly as early as 1694 [4]. In his 10-page "treatise on chance" is an anticipation of his 1711 argument concerning the middle term of the binomial as n gets large, as well as two other statistical "proto-tests" concerning the lengths of reign of the Roman and Scottish kings. The chronology of the first seven kings of Rome was suspect, he suggested, because they appeared to have survived far longer on average than might reasonably be expected from Edmund Halley's life table, based on the mortality bills of Breslau. In the case of the Scottish kings, on the other hand, the evidence seemed to indicate that mortality amongst them was higher than might be expected from Halley's table. However, neither of the calculations Arbuthnot outlined had the clarity of statistical modeling evident in his 1711 paper, nor did they culminate in a specific probability level quantifying the evidence.

Arbuthnot's 1711 paper sparked off a debate which involved, at various times, William 'sGravesande* (a Dutch scientist who later became Professor of Mathematics, Astronomy and Philosophy at the University of Leiden), Bernard Nieuwentijt (a Dutch physician and mathematician), Nicholas Bernoulli*, and Abraham de Moivre*. 'sGravesande developed Arbuthnot's test further, attempting to take into account the close limits within which the male-to-female birth ratio fell year after year. Bernoulli, on the other hand, questioned Arbuthnot's interpretation of "chance." He proposed that the fair two-sided die could be replaced by a multifaceted die, with 18 sides marked M and 17 marked F. If tossed a large number of times, Bernoulli maintained, such a die would yield ratios of M's to F's with similar variability to the London christenings data. Certain aspects of the exchanges between the participants in the debate can be seen as attempts to emulate and develop Arbuthnot's mode of statistical reasoning, but have not proved as amenable to reinterpretation within modern frameworks of statistical logic.

Though Arbuthnot's 1711 argument tends now to be regarded as the first recognizable statistical significance test, it is doubtful whether his contribution, and the debate it provoked, provided any immediate stimulus to ideas of statistical significance testing. The obvious impact was to fuel interest in the "argument from design," in the stability of statistical ratios, and in the interplay of one with the other.

An oil painting of Arbuthnot hangs in the Scottish National Portrait Gallery, Edinburgh. By all accounts, he was a charitable and benevolent man. In a letter to Pope, Swift said of him: "Our doctor hath every quality in the world that can make a man amiable and useful; but alas! he hath a sort of slouch in his walk."

REFERENCES

1. Aitken, G. A. (1892). *The Life and Works of John Arbuthnot*. Clarendon Press, Oxford.

2. Arbuthnot, J. (1710). An argument for divine providence, taken from the constant regularity observ'd in the births of both sexes. *Phil. R. Soc. London, Trans.*, 27, 186–190. Reprinted in (1977). *Studies in the History of Statistics and Probability*, M. G. Kendall and R. L. Plackett, eds., Griffin, London, **vol. 2**, pp. 30–34.

3. Arbuthnot, J. (1751). *Miscellaneous works of the late Dr. Arbuthnot, with Supplement*. Glasgow. [Arbuthnot's *Of the Laws of Chance* (1692) was included as *Huygens' de Ratiociniis in Ludo Aleae*: translated into English by Dr. Arbuthnot.]

4. Bellhouse, D. R. (1989). A manuscript on chance written by John Arbuthnot. *Int. Statist. Rev.*, **57**, 249–259.

5. Hacking, I. (1965). *Logic of Statistical Inference*. Cambridge University Press, pp. 75–81.

6. Hacking, I. (1975). *The Emergence of Probability*. Cambridge University Press, pp. 166–171.

7. Huygens, C. (1657). De ratiociniis in ludo aleae. In *Exercitationum mathematicarum libri quinque*, F van Schooten, ed. Amsterdam.

BIBLIOGRAPHY

Beattie, L. M. (1935). *John Arbuthnot: Mathematician and Satirist*. Harvard Studies in English, vol. XVI. Harvard University Press, Cambridge, Massachusetts.

Hald, A. (1990). *A History of Probability and Statistics and their Applications before 1750*. Wiley, New York.

Shoesmith, E. (1987). The continental controversy over Arbuthnot's argument for divine providence. *Historia Math.* **14**, 133–146.

Stephen, L. and Lee, S. (eds.) (1917). John Arbuthnot. In *Dictionary of National Biography*, vol. I. Oxford University Press.

E. Shoesmith

ARCH AND GARCH MODELS

Many time series* display time-varying dispersion, or uncertainty, in the sense that large (small) absolute innovations* tend to be followed by other large (small) absolute innovations. A natural way to model this phenomenon is to allow the variance to change through time in response to current developments of the system. Specifically, let $\{y_t\}$ denote the observable univariate discrete-time stochastic process of interest. Denote the corresponding innovation process by $\{\epsilon_t\}$, where $\epsilon_t \equiv y_t - E_{t-1}(y_t)$, and $E_{t-1}(\cdot)$ refers to the expectation conditional on time-$(t-1)$ information. A general specification for the innovation process that takes account of the time-varying uncertainty would then be given by

$$\epsilon_t = z_t \sigma_t, \qquad (1)$$

where $\{z_t\}$ is an i.i.d. mean-zero, unit-variance stochastic process*, and σ_t represents the time-t latent volatility; i.e., $E(\epsilon_t^2 | \sigma_t) = \sigma_t^2$. Model specifications in which σ_t in (1) depends non-trivially on the past

innovations and/or some other latent variables are referred to as *stochastic volatility* (SV) models. The historically first, and often most convenient, SV representations are the *autoregressive conditionally heteroscedastic* (ARCH) models pioneered by Engle [21]. Formally the ARCH class of models are defined by (1), with the additional restriction that σ_t must be measurable with respect to the time-$(t-1)$ observable information set. Thus, in the ARCH class of models $\text{var}_{t-1}(y_t) \equiv E_{t-1}(\epsilon_t^2) = \sigma_t^2$ is predetermined as of time $t-1$.

VOLATILITY CLUSTERING

The ARCH model was originally introduced for modeling inflationary uncertainty, but has subsequently found especially wide use in the analysis of financial time series. To illustrate, consider the plots in Figs. 1 and 2 for the daily Deutsche-mark—U.S. dollar (DM/$) exchange rate and the Standard and Poor's 500 composite stock-market index (S&P 500) from October 1, 1979, through September 30, 1993. It is evident from panel (a) of the figures that both series display the long-run swings or trending behavior that are characteristic of unit-root*, or $I(1)$, nonstationary processes. On the other hand, the two return series, $r_t = 100 \ln(P_t/P_{t-1})$, in panel (b) appear to be covariance-stationary. However, the tendency for large (and for small) absolute returns to cluster in time is clear.

Many other economic and financial time series exhibit analogous volatility clustering features. This observation, together with the fact that modern theories of price determination typically rely on some form of a risk—reward tradeoff relationship, underlies the very widespread applications of the ARCH class of time series models in economics and finance* over the past decade. Simply treating the temporal dependencies in σ_t as a nuisance would be inconsistent with the trust of the pertinent theories. Similarly, when evaluating economic and financial time series forecasts it is equally important that the temporal variation in the forecast error uncertainty be taken into account.

The next section details some of the most important developments along these lines.

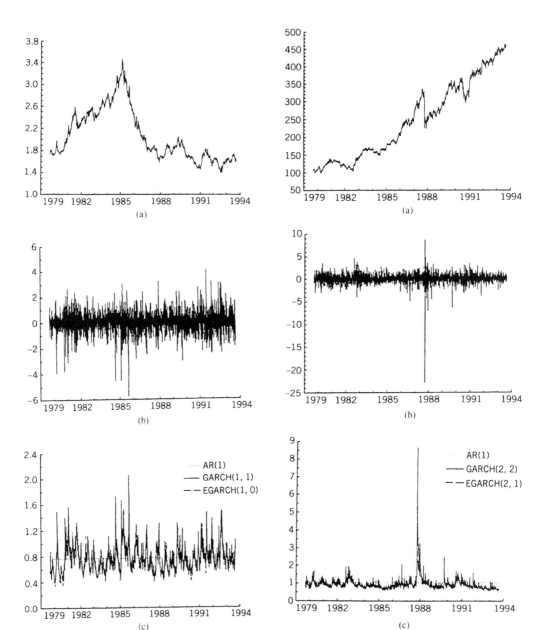

Figure 1. Daily deutsche-mark—U.S. dollar exchange rate. Panel (a) displays daily observations on the DM/U.S. $ exchange rate, s_t, over the sample period October 1, 1979 through September 30, 1993. Panel (b) graphs the associated daily percentage appreciation of the U.S. dollar, calculated as $r_t \equiv 100 \ln(s_t/s_{t-1})$. Panel (c) depicts the conditional standard-deviation estimates of the daily percentage appreciation rate for the U.S. dollar implied by each of the three volatility model estimates reported in Table 1.

Figure 2. Daily S&P 500 stock-market index. Panel (a) displays daily observations on the value of the S&P 500 stock-market index, P_t, over the sample period October 1, 1979 through September 30, 1993. Panel (b) graphs the associated daily percentage appreciation of the S&P 500 stock index excluding dividends, calculated as $r_t \equiv 100 \ln(P_t/P_{t-1})$. Panel (c) depicts the conditional standard-deviation estimates of the daily percentage appreciation rate for the S&P 500 stock-market index implied by each of the three volatility-model estimates reported in Table 2.

For notational convenience, we shall assume that the $\{\epsilon_t\}$ process is directly observable. However, all of the main ideas extend directly to the empirically more relevant situation in which ϵ_t denotes the time-t innovation of another stochastic process, y_t, as defined above. We shall restrict discussion to the univariate case; most multivariate generalizations follow by straightforward analogy.

GARCH

The definition of the ARCH class of models in (1) is extremely general, and does not lend itself to empirical investigation without additional assumptions on the functional form, or smoothness, of σ_t. Arguably, the two most successful parameterizations have been the *generalized ARCH*, or GARCH (pq), model of Bollerslev [7] and the *exponential GARCH*, or EGARCH (p, q), model of Nelson [46]. In the GARCH (p, q) model, the conditional variance is parametrized as a distributed lag of past squared innovations and past conditional variances,

$$\sigma_t^2 = \omega + \sum_{i=1}^{q} \alpha_i \epsilon_{t-i}^2 + \sum_{j=1}^{p} \beta_j \sigma_{t-j}^2$$
$$\equiv \omega + \alpha(B)\epsilon_t^2 + \beta(B)\sigma_t^2, \qquad (2)$$

where B denotes the backshift (lag) operator; i.e., $B^i y_t \equiv y_{t-i}$. For $\alpha_i > 0$, this parametrization directly captures the tendency for large (small) ϵ_{t-i}^2's to be followed by other large (small) squared innovations. Of course, for the conditional variance in (2) to be positive almost surely, and the process well defined, the coefficients in the corresponding infinite ARCH representation for σ_t^2, expressed in terms of $\{\epsilon_{t-i}^2\}_{i=1}^{\infty}$, must all be nonnegative, i.e., $[1 - \beta(B)]^{-1}\alpha(B)$, where all of the roots of $1 - \beta(x) = 0$ are assumed to be outside the unit circle.

On rearranging the terms in (2), we obtain

$$[1 - \alpha(B) - \beta(B)]\epsilon_t^2 = \omega + [1 - \beta(B)]v_t, \quad (3)$$

where $v_t \equiv \epsilon_t^2 - \sigma_t^2$. Since $E_{t-1}(v_t) = 0$, the GARCH(p, q) formulation in (3) is readily interpreted as an ARMA(max$\{p, q\}, p$) model for the squared innovation process $\{\epsilon_t^2\}$; see

Milhøj [43] and Bollerslev [9]. Thus, if the roots of $1 - \alpha(x) - \beta(x) = 0$ lie outside the unit circle, then the GARCH(p, q) process for $\{\epsilon_t\}$ is covariance-stationary, and the unconditional variance equals $\sigma^2 = \omega[1 - \alpha(1) - \beta(1)]^{-1}$. Furthermore, standard ARMA*-based identification and inference procedures may be directly applied to the process in (3), although the heteroscedasticity in the innovations, $\{v_t\}$, renders such an approach inefficient.

In analogy to the improved forecast accuracy obtained in traditional time-series analysis by utilizing the conditional as opposed to the unconditional mean of the process, ARCH models allow for similar improvements when modeling second moments. To illustrate, consider the s-step-ahead ($s \geqslant 2$) minimum mean square error* forecast for the conditional variance* in the simple GARCH(1, 1) model,

$$E_t(\epsilon_{t+s}^2) = E_t(\sigma_{t+s}^2)$$
$$= \omega \sum_{i=0}^{s-2} (\alpha_1 + \beta_1)^i + (\alpha_1 + \beta_1)^{s-1}\sigma_{t+1}^2.$$
$$(4)$$

If the process is covariance-stationary, i.e. $\alpha_1 + \beta_1 < 1$, it follows that $E_t(\sigma_{t+s}^2) = \sigma^2 + (\alpha_1 + \beta_1)^{s-1}(\sigma_{t+1}^2 - \sigma^2)$. Thus, if the current conditional variance is large (small) relative to the unconditional variance, the multistep forecast is also predicted to be above (below) σ^2, but converges to σ^2 at an exponential rate as the forecast horizon lengthens. Higher-order covariance-stationary models display more complicated decay patterns [3].

IGARCH

The assumption of covariance stationarity has been questioned by numerous studies which find that the largest root in the estimated lag polynomial $1 - \hat{\alpha}(x) - \hat{\beta}(x) = 0$ is statistically indistinguishable from unity. Motivated by this stylized fact, Engle and Bollerslev [22] proposed the so-called *integrated GARCH*, or IGARCH(p, q), process, in which the autoregressive polynomial in (3) has one unit root; i.e., $1 - \alpha(B) - \beta(B) \equiv (1 - B)\phi(B)$, where $\phi(x) \neq 0$ for $|x| \leqslant 1$. However, the notion of a unit root* is intrinsically a linear concept, and considerable care should

Table 1. Daily Deutsche-Mark—U.S. Dollar Exchange-Rate Appreciation

---

AR(1):(a)

$$r_t = \begin{matrix} -0.002 \\ (0.013) \\ [0.013] \end{matrix} \quad \begin{matrix} -0.033 \cdot r_{t-1} + \epsilon_t \\ (0.017) \\ [0.019] \end{matrix}$$

$$\sigma_t^2 = \begin{matrix} 0.585 \\ (0.014) \\ [0.022] \end{matrix}$$

Log1 $= -4043.3$, $b_3 = -0.25$, $b_4 = 5.88$, $Q_{20} = 19.69$, $Q_{20}^2 = 231.17$

AR(1)—GARCH(1, 1): (b)

$$r_t = \begin{matrix} -0.001 \\ (0.012) \\ [0.012] \end{matrix} \quad \begin{matrix} -0.035 \cdot r_{t-1} \\ (0.018) \\ [0.019] \end{matrix} \quad +\epsilon_t$$

$$\sigma_t^2 = \begin{matrix} 0.019 \\ (0.004) \\ [0.004] \end{matrix} \quad \begin{matrix} +0.103 \cdot \epsilon_{t-1}^2 \\ (0.011) \\ [0.015] \end{matrix} \quad \begin{matrix} +0.870 \cdot \sigma_{t-1}^2 \\ (0.012) \\ [0.015] \end{matrix}$$

Log1 $= -3878.8, b_3 = -0.10, b_4 = 4.67, Q_{20} = 32.48, Q_{20}^2 = 22.45$

AR(1)—EGARCH(1, 0): (c)

$$r_t = \begin{matrix} 0.005 \\ (0.012) \\ [0.012] \end{matrix} \quad \begin{matrix} -0.034 \cdot r_{t-1} + \epsilon_t \\ (0.018) \\ [0.017] \end{matrix}$$

$$\ln \sigma_t^2 = \begin{matrix} -0.447 \\ (0.081) \\ [0.108] \end{matrix} \quad \begin{matrix} +[0.030 \cdot z_{t-1} \\ (0.009) \\ [0.013] \end{matrix} \quad \begin{matrix} +0.208 \cdot (|z_{t-1}| - \sqrt{2/\pi})] \\ (0.019) \\ [0.022] \end{matrix} \quad \begin{matrix} +0.960 \cdot \ln(\sigma_{t-1}^2) \\ (0.004) \\ [0.008] \end{matrix}$$

Log1 $= -3870.8$, $b_3 = -0.16$, $b_4 = 4.54$, $Q_{20} = 33.61$, $Q_{20}^2 = 24.22$

Notes: All the model estimates are obtained under the assumption of conditional normality; i.e., $z_t \equiv \epsilon_t \hat{\sigma}_t^{-1}$ i.i.d. $N(0, 1)$. Conventional asymptotic standard errors based on the inverse of Fisher's information matrix are given in parentheses, while the numbers in square brackets represent the corresponding robust standard errors as described in the text. The maximized value of the pseudo-log-likelihood function is denoted Log1. The skewness and kurtosis of the standardized residuals, $\hat{z}_t = \hat{\epsilon}_t \hat{\sigma}_t^{-1}$, are given by b_3 and b_4, respectively. Q_{20} and Q_{20}^2 refer to the Ljung—Box portmanteau test for up to 20th-order serial correlation in $\hat{z}_t$ and $\hat{z}_t^2$, respectively.

be exercised in interpreting persistence in nonlinear models. For example, from (4), the IGARCH(1, 1) model with $\alpha_1 + \beta_1 = 1$ behaves like a random walk*, or an $I(1)$ process, for forecasting purposes. Nonetheless, by repeated substitution, the GARCH(1, 1) model may be written as

$$\sigma_t^2 = \sigma_0^2 \prod_{i=1}^{t} (\alpha_1 z_{t-i}^2 + \beta_1)$$

$$+ \omega \left(1 + \sum_{j=1}^{t-1} \prod_{i=1}^{j} (\alpha_1 z_{t-i}^2 + \beta_1)\right).$$

Thus, as Nelson [44] shows, strict stationarity and ergodicity of the GARCH(1, 1) model requires only geometric convergence of $\{\alpha_1 z_t^2 + \beta_1\}$, or $E[\ln(\alpha_1 z_t^2 + \beta_1)] < 0$, a weaker condition than arithmetic convergence, or $E(\alpha_1 z_t^2 + \beta_1) = \alpha_1 + \beta_1 < 1$, which is required for covariance stationarity. This also helps to explain why standard maximum likelihood* based inference procedures, discussed below, still apply in the IGARCH context [39,42,51].

EGARCH

While the GARCH(p, q) model conveniently captures the volatility clustering phenomenon, it does not allow for asymmetric effects in the evolution of the volatility process. In the EGARCH(p, q) model of Nelson [46], the logarithm of the conditional variance is given as an ARMA(p, q) model in both the absolute size and the sign of the

Table 2. Daily S&P 500 Stock-Market Index Returns

AR(1):(a)

$$r_t \;=\; \underset{(0.017)}{\underset{[0.018]}{0.039}} \;+\; \underset{(0.017)}{\underset{[0.056]}{0.055 \cdot r_{t-1}}} \;+\; \epsilon_t$$

$$\sigma_t^2 \;=\; \underset{(0.025)}{\underset{[0.151]}{1.044}}$$

Log1 $= -5126.7$, $b_3 = -3.20$, $b_4 = 75.37$, $Q_{20} = 37.12$, $Q_{20}^2 = 257.72$

AR(1)—GARCH(2, 2): (b)

$$r_t \;=\; \underset{(0.014)}{\underset{[0.015]}{0.049}} \;+\; \underset{(0.018)}{\underset{[0.019]}{0.058 \cdot r_{t-1}}} \;+\; \epsilon_t$$

$$\sigma_t^2 \;=\; \underset{(0.004)}{\underset{[0.008]}{0.014}} \;+\; \underset{(0.018)}{\underset{[0.078]}{0.143 \cdot \epsilon_{t-1}^2}} \;-\; 0.103 \cdot \epsilon_{t-2}^2 \;+\; \underset{(0.021)}{\underset{[0.077]}{0.885 \cdot \sigma_{t-1}^2}} \;+\; \underset{(0.101)}{\underset{[0.098]}{0.060 \cdot \sigma_{t-2}^2}} \;\;\underset{(0.092)}{\underset{[0.084]}{}}$$

Log1 $= -4658.0$, $b_3 = -0.58$, $b_4 = 8.82$, $Q_{20} = 11.79$, $Q_{20}^2 = 8.45$

AR(1)—EGARCH(2, 1): (c)

$$r_t \;=\; \underset{(0.014)}{\underset{[0.015]}{0.023}} \;+\; \underset{(0.018)}{\underset{[0.017]}{0.059 \cdot r_{t-1}}} \;+\; \epsilon_t$$

$$\ln \sigma_t^2 \;=\; \underset{(0.175)}{\underset{[0.333]}{0.281}} + (1 \;\; \underset{(0.031)}{\underset{[0.046]}{-0.927 \cdot B}})[\;\; \underset{(0.014)}{\underset{[0.046]}{-0.093 \cdot z_{t-1}}} + \; \underset{(0.018)}{\underset{[0.058]}{0.173 \cdot (|z_{t-1}| - \sqrt{2/\pi})}}]$$

$$+ \underset{(0.062)}{\underset{[0.113]}{1.813 \ln \sigma_{t-1}^2}} \;\; \underset{(0.061)}{\underset{[0.112]}{-0.815 \ln \sigma_{t-2}^2}}$$

Log1 $= -4643.4$, $b_3 = -0.60$, $b_4 = 9.06$, $Q_{20} = 8.93$, $Q_{20}^2 = 9.37$

Notes: See Table 1.

lagged innovations,

$$\ln \sigma_t^2 = \omega + \sum_{i=1}^{p} \varphi_i \ln \sigma_{t-i}^2 + \sum_{j=0}^{q} \psi_j g(z_{t-1-j})$$

$$\equiv \omega + \varphi(B) \ln \sigma_t^2 + \psi(B) g(z_t), \qquad (5)$$

$$g(z_t) = \theta z_t + \gamma [|z_t| - E(|z_t|)], \qquad (6)$$

along with the normalization $\psi_0 \equiv 1$. By definition, the news impact function $g(\cdot)$ satisfies $E_{t-1}[g(z_t)] = 0$. When actually estimating EGARCH models the numerical stability of the optimization procedure is often enhanced by approximating $g(z_t)$ by a smooth function that is differentiable at zero. Bollerslev et al. [12] also propose a richer parametrization for this function that downweighs the influence of large absolute innovations. Note that the EGARCH model still predicts that large (absolute) innovations follow other large innovations, but if $\theta < 0$ the effect is accentuated for negative ϵ_t's. Following Black [6],

this stylized feature of equity returns is often referred to as the "leverage effect."

ALTERNATIVE PARAMETRIZATIONS

In addition to GARCH, IGARCH, and EGARCH, numerous alternative univariate parametrizations have been suggested. An incomplete listing includes: *ARCH-in-mean*, or ARCH-M [25], which allows the conditional variance to enter directly into the equation for the conditional mean of the process; *nonlinear augmented ARCH*, or NAARCH [37], *structural ARCH*, or STARCH [35]; *qualitative threshold ARCH*, or QTARCH [31]; *asymmetric power ARCH*, or AP-ARCH [19]; *switching ARCH*, or SWARCH [16,34]; *periodic GARCH*, or PGARCH [14]; and *fractionally integrated GARCH*, or FIGARCH [4]. Additionally, several authors have proposed the inclusion of various asymmetric terms in the conditional-variance equation to better capture the aforementioned leverage* effect; see e.g., refs. 17, 26, 30.

TIME-VARYING PARAMETER AND BILINEAR MODELS

There is a close relation between ARCH models and the widely-used time-varying parameter class of models. To illustrate, consider the simple ARCH(q) model in (2), i.e., $\sigma_t^2 = \omega + \alpha_1 \epsilon_{t-1}^2 + \cdots + \alpha_q \epsilon_{t-q}^2$. This model is observationally equivalent to the process defined by

$$\epsilon_t = w_t + \sum_{i=1}^{q} a_i \epsilon_{t-i},$$

where $w_t, a_1, \ldots, a_q$ are i.i.d. random variables with mean zero and variances $\omega, \alpha_1, \ldots, \alpha_q$, respectively; see Tsay [54] and Bera et al. [5] for further discussion. Similarly, the class of bilinear time series models discussed by Granger and Anderson [32] provides an alternative approach for modeling nonlinearities; see Weiss [56] and Granger and Teräsvirta [33] for a more formal comparison of ARCH and bilinear models. However, while time-varying parameter and bilinear models may conveniently allow for heteroskedasticity* and/or nonlinear dependencies through a set of nuisance parameters, in applications in economics and finance the temporal dependencies in σ_t are often of primary interest. ARCH models have a distinct advantage in such situations by directly parametrizing this conditional variance.

ESTIMATION AND INFERENCE

ARCH models are most commonly estimated via maximum likelihood. Let the density for the i.i.d. process z_t be denoted by $f(z_t; v)$, where v represents a vector of nuisance parameters. Since σ_t is measurable with respect to the time-$(t-1)$ observable information set, it follows by a standard prediction-error decomposition argument that, apart from initial conditions, the log-likelihood* function for $\epsilon_T \equiv \{\epsilon_1, \epsilon_2, \ldots, \epsilon_T\}$ equals

$$\log L(\epsilon_T; \xi, v) = \sum_{t=1}^{T} \left[\ln f(\epsilon_t \sigma_t^{-1}; v) - \tfrac{1}{2} \ln \sigma_t^2 \right],$$

(7)

where ξ denotes the vector of unknown parameters in the parametrization for σ_t. Under conditional normality,

$$f(z_t; v) = (2\pi)^{-1/2} \exp\left(-\tfrac{1}{2} z_t^2\right). \quad (8)$$

By Jensen's inequality*, $E(\epsilon_t^4) = E(z_t^4) \times E(\sigma_t^4) \geqslant E(z_t^4) E(\sigma_t^2)^2 = E(z_t^4) E(\epsilon_t^2)^2$. Thus, even with conditionally normal* innovations, the unconditional distribution for ϵ_t is leptokurtic. Nonetheless, the conditional normal distribution often does not account for all the leptokurtosis in the data, so that alternative distributional assumptions have been employed; parametric examples include the $t-$distribution* in Bollerslev [8] and the generalized error distribution (GED) in Nelson [46], while Engle and Gonz'alez-Rivera [23] suggest a nonparametric* approach. However, if the conditional variance is correctly specified, the normal quasiscore vector based on (7) and (8) is a martingale* difference sequence when evaluated at the true parameters, ξ_0; i.e., $E_{t-1}[\tfrac{1}{2}(\nabla_\xi \sigma_t^2)\sigma_t^{-2}(\epsilon_t^2 \sigma_t^{-2} - 1)] = 0$. Thus, the corresponding quasi-maximum-likelihood* estimate (QMLE), $\hat{\xi}$, generally remains consistent, and asymptotically valid inference may be conducted using an estimate of a robustified version of the asymptotic covariance matrix, $\mathbf{A}(\xi_0)^{-1}\mathbf{B}(\xi_0)\mathbf{A}(\xi_0)^{-1}$, where $\mathbf{A}(\xi_0)$ and $\mathbf{B}(\xi_0)$ denote the Hessian* and the outer product of the gradients respectively [55]. A convenient form of $\mathbf{A}(\hat{\xi})$ with first derivatives only is provided in Bollerslev and Wooldridge [15].

Many of the standard mainframe and PC computer-based packages now contain ARCH estimation procedures. These include E-VIEW, RATS, SAS, TSP, and a special set of time series libraries for the GAUSS computer language.

TESTING

Conditional moment (CM) based misspecification tests are easily implemented in the ARCH context via simple auxiliary regressions [50, 53, 57, 58]. Specifically, following Wooldridge [58], the moment condition

$$E_{t-1}[(\lambda_t \sigma_t^{-2})(\epsilon_t^2 - \sigma_t^2)\sigma_t^{-2}] = 0 \quad (9)$$

(evaluated at the true parameter ξ_0) provides a robust test in the direction indicated by the vector λ_t of misspecification indicators. By selecting these indicators as appropriate functions of the time-$(t-1)$ information set, the test may be designed to have asymptotically optimal power* against a specific alternative; e.g., the conditional variance specification may be tested for goodness of fit over subsamples by letting λ_t be the relevant indicator function, or for asymmetric effects by letting $\lambda_t \equiv \epsilon_{t-1} I\{\epsilon_{t-1} < 0\}$, where $I\{\cdot\}$ denotes the indicator function for $\epsilon_{t-1} < 0$. Lagrange-multiplier-type* tests that explicitly recognize the one-sided nature of the alternative when testing for the presence of ARCH have been developed by Lee and King [40].

EMPIRICAL EXAMPLE

As previously discussed, the two time-series plots for the DM/\$ exchange rate and the S&P 500 stock market index in Figs. 1 and 2 both show a clear tendency for large (and for small) absolute returns to cluster in time. This is also borne out by the highly significant Ljung—Box [41] portmanteau tests* for up to 20th-order serial correlation* in the squared residuals from the estimated AR(1) models, denoted by Q_{20}^2 in panel (a) of Tables 1 and 2. To accommodate this effect for the DM/\$ returns, Panel (b) of Table 1 reports the estimates from an AR(1)—GARCH(1, 1) model. The estimated ARCH coefficients are overwhelmingly significant, and, judged by the Ljung—Box test, this simple model captures the serial dependence in the squared returns remarkably well. Note also that $\hat{\alpha}_1 + \hat{\beta}_1$ is close to unity, indicative of IGARCH-type behavior. Although the estimates for the corresponding AR(1)—EGARCH(1, 0) model in panel (c) show that the asymmetry coefficient θ is significant at the 5% level, the fit of the EGARCH model is comparable to that of the GARCH specification. This is also evident from the plot of the estimated volatility processes in panel (c) of Fig. 1.

The results of the symmetric AR(1)—GARCH(2, 2) specification for the S&P 500 series reported in Table 2 again suggest a very high degree of volatility persistence.

The largest inverse root of the autoregressive polynomial in (3) equals $\frac{1}{2}\{\hat{\alpha}_1 + \hat{\beta}_1 + [(\hat{\alpha}_1 + \hat{\beta}_1)^2 + 4(\hat{\alpha}_2 + \hat{\beta}_2)]^{1/2}\} = 0.984$, which corresponds to a half-life of 43.0, or approximately two months. The large differences between the conventional standard errors* reported in parentheses and their robust counterparts in square brackets highlight the importance of the robust inference procedures with conditionally nonnormal innovations. The two individual robust standard errors for α_2 and β_2 suggest that a GARCH(1, 1) specification may be sufficient, although previous studies covering longer time spans have argued for higher-order models [27,52]. This is consistent with the results for the EGARCH(2, 1) model reported in panel (c), where both lags of $g(z_t)$ and $\ln \sigma_t^2$ are highly significant. On factorizing the autoregressive polynomial for $\ln \sigma_t^2$, the two inverse roots equal 0.989 and 0.824. Also, the EGARCH model points to potentially important asymmetric effects in the volatility process. In summary, the GARCH and EGARCH volatility estimates depicted in panel (c) of Fig. 2 both do a good job of tracking and identifying periods of high and low volatility in the U.S. equity market.

FUTURE DEVELOPMENTS

We have provided a very partial introduction to the vast ARCH literature. In many applications a multivariate extension is called for; see refs. 13, 18, 10, 11, 24, 48 for various parsimonious multivariate parametrizations. Important issues related to the temporal aggregation of ARCH models are addressed by Drost and Nijman [20]. Rather than directly parametrizing the functional form for σ_t in (1), Gallant and Tauchen [29], and Gallant et al. [28] have developed flexible nonparametric techniques for analysis of data with ARCH features. Much recent research has focused on the estimation of stochastic volatility models in which the process for σ_t is treated as a latent variable* [1,36,38]. For a more detailed discussion of all of these ideas, see the many surveys listed in the Bibliography below.

A conceptually important issue concerns the rationale behind the widespread empirical findings of IGARCH-type behavior, as exemplified by the two time series analyzed above. One possible explanation is provided by the continuous record asymptotics developed in a series of papers by Nelson [45,47] and Nelson and Foster [49]. Specifically, suppose that the discretely sampled observed process is generated by a continuous-time diffusion, so that the sample path for the latent instantaneous volatility process $\{\sigma_t^2\}$ is continuous almost surely. Then one can show that any consistent ARCH filter must approach an IGARCH model in the limit as the sampling frequency increases. The empirical implications of these theoretical results should not be carried too far, however. For instance, while daily GARCH(1, 1) estimates typically suggest $\hat{\alpha}_1 + \hat{\beta}_1 \approx 1$, on estimating GARCH models for financial returns at intraday frequencies, Andersen and Bollerslev [2] document large and systematic deviations from the theoretical predictions of approximate IGARCH behavior.

This breakdown of the most popular ARCH parametrizations at the very high intraday frequencies has a parallel at the lowest frequencies. Recent evidence suggests that the exponential decay of volatility shocks in covariance-stationary GARCH and EGARCH parametrizations results in too high a dissipation rate at long horizons, whereas the infinite persistence implied by IGARCH-type formulations is too restrictive. The fractionally integrated GRACH, or FIGARCH, class of models [4] explicitly recognizes this by allowing for a low hyperbolic rate of decay in the conditional variance function. However, a reconciliation of the empirical findings at the very high and low sampling frequencies within a single consistent modeling framework remains an important challenge for future work in the ARCH area.

REFERENCES

1. Andersen, T. G. (1996). Return volatility and trading volume: an information flow interpretation of stochastic volatility. *J. Finance*, **51**, 169–204.

2. Andersen, T. G. and Bollerslev, T. (1997). Intraday periodicity and volatility persistence in financial markets. *J. Empirical Finance*, **4**, 2–3.

3. Baillie, R. T. and Bollerslev, T. (1992). Prediction in dynamic models with time-dependent conditional variances. *J. Econometrics*, **52**, 91–113.

4. Baillie, R. T., Bollerslev, T., and Mikkelsen, H. O. (1996). Fractional integrated generalized autoregressive conditional heteroskedasticity. *J. Econometrics*, **74**, 3–30.

5. Bera, A. K., Higgins, M. L., and Lee, S. (1993). Interaction between autocorrelation and conditional heteroskedasticity: a random coefficients approach. *J. Bus. and Econ. Statist.*, **10**, 133–142.

6. Black, F. (1976). Studies of stock market volatility changes. *Proc. Amer. Statist. Assoc.*, Business and Economic Statistics Section, 177–181.

7. Bollerslev, T. (1986). Generalized autoregressive conditional heteroskedasticity. *J. Econometrics*, **31**, 307–327.

8. Bollerslev, T. (1987). A conditional heteroskedastic time series model for speculative prices and rates of return. *Rev. Econ. and Statist.*, **69**, 542–547.

9. Bollerslev, T. (1988). On the correlation structure for the generalized autoregressive conditional heteroskedastic process. *J. Time Ser. Anal.*, **9**, 121–131.

10. Bollerslev, T. (1990). Modelling the coherence in short-run nominal exchange rates: a multivariate generalized ARCH approach. *Rev. Econ. and Statist.*, **72**, 498–505.

11. Bollerslev, T. and Engle, R. F. (1993). Common persistence in conditional variance. *Econometrica*, **61**, 166–187.

12. Bollerslev, T., Engle, R. F., and Nelson, D. B., (1994). ARCH models. In *Handbook of Econometrics*, vol. **4**, R. F. Engle and D. McFadden, eds., Elsevier Science–North Holland, Amsterdam.

13. Bollerslev, T., Engle, R. F., and Wooldridge, J. M. (1988). A capital asset pricing model with time varying covariances. *J. Polit. Econ.*, **96**, 116–131.

14. Bollerslev, T. and Ghysels, E. (1996). Periodic autoregressive conditional heteroskedasticity. *J. Bus. and Econ. Statist.*, **14**, 139–151.

15. Bollerslev, T. and Wooldridge, J. M. (1992). Quasi-maximum-likelihood estimation and inference in dynamic models with time varying covariances. *Econometric Rev.*, **11**, 143–172.

16. Cai. J. (1994). A Markov model switching regime ARCH. *J. Bus. and Econ. Statist.*, **12**, 309–316.

17. Campbell, J. Y. and Hentschel, L. (1992). No news is good news: an asymmetric model of changing volatility in stock returns. *J. Financial Econ.*, **31**, 281–318.

18. Diebold, F. X. and Nerlove, M. (1989). The dynamics of exchange rate volatility: a multivariate latent factor ARCH model. *J. Appl. Econometrics*, **4**, 1–21.

19. Ding, Z., Granger, C. W. J., and Engle, R. F. (1993). A long memory property of stock market returns and a new model. *J. Empirical Finance*, **1**, 83–106.

20. Drost, F. C. and Nijman, T. E. (1993). Temporal aggregation of GARCH processes. *Econometrica*, **61**, 909–928.

21. Engle, R. F. (1982). Autoregressive conditional heteroskedasticity with estimates of the variance of U.K. inflation. *Econometrica*, **50**, 987–1008.

22. Engle, R. F. and Bollerslev, T. (1986). Modelling the persistence of conditional variances. *Econometric Rev.*, **5**, 1–50, 8187.

23. Engle, R. F. and Gonzalez-Rivera, G. (1991). Semiparametric ARCH models, *J. Bus. and Econ. Statist.*, **9**, 345–359.

24. Engle, R. F. and Kroner, K. F. (1995). Multivariate simultaneous generalized ARCH. *Econometric Theory*, **11**, 122–150.

25. Engle, R. F., Lilien, D. M., and Robins, R. P. (1987). Estimating time varying risk premia in the term structure: the ARCH-M model. *Econmetrica*, **55**, 391–407.

26. Engle, R. F., and Ng, V. (1993). Measuring and testing the impact of news on volatility. *J. Finance*, **48**, 1749–1778.

27. French, K. R., Schwert, G. W., and Stambaugh, R. F. (1987). Expected stock returns and volatility. *J. Financial Econ.*, **19**, 3–30.

28. Gallant, A. R., Rossi, P. E. and Tauchen, G. (1993). Nonlinear dynamic structures. *Econometrica*, **61**, 871–907.

29. Gallant, A. R. and Tauchen, G. (1989). Semi non-parametric estimation of conditionally constrained heterogeneous processes: asset pricing applications. *Econometrica*, **57**, 1091–1120.

30. Glosten, L. R., Jagannathan, R., and Runkle, D. E. (1993). On the relation between the expected value and the volatility of the nominal excess return on stocks. *J. Finance*, **48**, 1779–1801.

31. Gourieroux, C. and Monfort, A. (1992). Qualitative threshold ARCH models. *J. Econometrics*, **52**, 159–199.

32. Granger, C. W. J. and Andersen, A. P. (1978). *An Introduction to Bilinear Time Series Models*. Vandenhoech and Ruprecht, Göttingen.

33. Granger, C. W. J. and Teräsvirta, T. (1993). *Modelling Nonlinear Economic Relationships*. Oxford University Press, Oxford, England.

34. Hamilton, J. D. and Susmel, R. (1994). Autoregressive conditional heteroskedasticity and changes in regime. *J. Econometrics*, **64** 307–333.

35. Harvey, A. C., Ruiz, E., and Sentana, E. (1992). Unobserved component time series models with ARCH disturbances. *J. Econometrics*, **52**, 129–158.

36. Harvey, A. C., Ruiz, E., and Shephard, N. (1994). Multivariate stochastic variance models. *Rev. Econ. Stud.*, **61**, 247–264.

37. Higgins, M. L. and Bera, A. K. (1992). A class of nonlinear ARCH models. *Int. Econ. Rev.*, **33**, 137–158.

38. Jacquier, E., Polson, N. G., and Rossi, P. E. (1994). Bayesian analysis of stochastic volatility models. *J. Bus. and Econ. Statist.*, **12**, 371–417.

39. Lee, S. W. and Hansen, B. E. (1994). Asymptotic theory for the GARCH(1, 1) quasi-maximum-likelihood estimator. *Econometric Theory*, **10**, 29–52.

40. Lee, J. H. H. and King, M. L. (1993). A locally most mean powerful based score test for ARCH and GARCH regression disturbances. *J. Bus. and Econ. Statist.*, **7**, 259–279.

41. Ljung, G. M. and Box, G. E. P. (1978). On a measure of lack of fit in time series models. *Biometrika*, **65**, 297–303.

42. Lumsdaine, R. L. (1996). Consistency and asymptotic normality of the quasi-maximum likelihood estimator in IGARCH(1, 1) and covariance stationary GARCH(1, 1) models. *Econometrica*.

43. Milhøj, A. (1985). The moment structure of ARCH processes. *Scand. J. Statist.*, **12**, 281–292.

44. Nelson, D. B. (1990). Stationarity and persistence in the GARCH(1, 1) model. *Econometric Theory*, **6**, 318–334.

45. Nelson, D. B. (1990). ARCH models as diffusion approximations. *J. Econometrics*, **45**, 7–38.

46. Nelson, D. B. (1991). Conditional heteroskedasticity in asset returns: a new approach. *Econometrica*, **59**, 347–370.

47. Nelson, D. B. (1992). Filtering and forecasting with misspecified ARCH models I: getting

the right variance with the wrong model. *J. Econometrics*, **52**, 61–90.

48. Nelson, D. B., Braun, P. A., and Sunier, A. M. (1995). Good news, bad news, volatility, and betas, *J. Finance*, **50**, 1575–1603.

49. Nelson, D. B. and Foster, D. (1995). Filtering and forecasting with misspecified ARCH models II: making the right forecast with the wrong model. *J. Econometrics*, **67**, 303–335.

50. Newey, W. K. (1985). Maximum likelihood specification testing and conditional moment tests. *Econometrica*, **53**, 1047–1070.

51. Pagan, A. (1996). The econometrics of financial markets. *J. Empirical Finance*, **3**, 15–102.

52. Pagan, A. R. and Schwert, G. W., (1990). Alternative models for conditional stock volatility. *J. Econometrics*, **45**, 267–290.

53. Tauchen, G. (1985). Diagnostic testing and evaluation of maximum likelihood models. *J. Econometrics*, **30**, 415–443.

54. Tsay, R. S. (1987). Conditional heteroskedastic time series models. *J. Amer. Statist. Ass.*, **82**, 590–604.

55. Weiss, A. A. (1986). Asymptotic theory for ARCH models: estimation and testing. *Econometric Theory*, **2**, 107–131.

56. Weiss, A. A. (1986). ARCH and bilinear time series models: comparison and combination. *J. Bus. and Econ. Statist.*, **4**, 59–70.

57. White, H. (1987). Specification testing in dynamic models. In *Advances in Econometrics: Fifth World Congress*, vol. I, T. F. Bewley, ed. Cambridge University Press, Cambridge, England.

58. Wooldridge, J. M. (1990). A unified approach to robust regression based specification tests. *Econometric Theory*, **6**, 17–43.

BIBLIOGRAPHY

Bera, A. K. and Higgins, M. L. (1993). ARCH models: properties, estimation and testing. *J. Econ. Surv.*, **7**, 305–366. (This article discusses the properties of the most commonly used univariate and multivariate ARCH and GARCH models.)

Bollerslev, T., Chou, R. Y., and Kroner, K. F. (1992). ARCH modelling in finance: a review of theory and empirical evidence. *J. Econometrics*, **52**, 5–59. (Chronicles more than two hundred of the earliest empirical applications in the ARCH literature.)

Bollerslev, T., Engle, R. F., and Nelson, D. B. (1994). ARCH models. In *Handbook of Econometrics*, vol. 4, R. F. Engle and D. McFadden, eds. Elsevier Science–North Holland, Amsterdam. (This chapter contains a comprehensive discussion of the most important theoretical developments in the ARCH literature.)

Brock, W. A., Hsieh, D. A., and LeBaron, B. (1991). *Nonlinear Dynamics, Chaos and Instability: Statistical Theory and Economic Evidence*. MIT Press, Cambridge, MA. (Explores the use of ARCH models for explaining nonlinear, possibly deterministic, dependencies in economic data, along with the implications for statistical tests for chaos.)

Diebold, F. X. and Lopez, J. A. (1996). Modeling volatility dynamics. In *Macroeconomics: Developments, Tensions and Prospects*, K. Hoover, ed. Kluwer, Amsterdam. (An easily accessible survey, including some recent results on volatility prediction.)

Enders, W. (1995). *Applied Econometric Time Series*. Wiley, New York. (An easy accessible review of some of the most recent advances in time-series analysis, including ARCH models, and their empirical applications in economics.)

Engle, R. F. (1995). *ARCH: Selected Readings*. Oxford University Press, Oxford, England. (A collection of some of the most important readings on ARCH models, along with introductory comments putting the various papers in perspective.)

Granger, C. W. J. and Teräsvirta, T. (1993). *Modelling Nonlinear Economic Relationships*. Oxford University Press, Oxford, England. (Explores recent theoretical and practical developments in the econometric modeling of nonlinear relationships among economic time series.)

Hamilton, J. D. (1994). *Time Series Analysis*. Princeton University Press, Princeton. (A lucid and very comprehensive treatment of the most important developments in applied time-series analysis over the past decade, including a separate chapter on ARCH models.)

Mills, T. C. (1993). *The Econometric Modelling of Financial Time Series*. Cambridge University Press, Cambridge, England. (Discusses a wide variety of the statistical models that are currently being used in the empirical analysis of financial time series, with much of the coverage devoted to ARCH models.)

Nijman, T. E. and Palm, F. C. (1993). GARCH modelling of volatility: an introduction to theory and applications. In *Advanced Lectures in Economics*, vol. II, A. J. De Zeeuw, ed. Academic Press, San Diego, California. (Discusses the statistical properties and estimation of GARCH models motivated by actual empirical applications in finance and international economics.)

Pagan, A. (1996). The econometrics of financial markets, J. Empirical Finance. (Contains a discussion of the time-series techniques most commonly applied in the analysis of financial time series, including ARCH models.)

Shephard, N. (1996). Statistical aspects of ARCH and stochastic volatility. In *Likelihood, Time Series with Econometric and Other Applications*. D. R. Cox, D. V. Hinkley, and O. E. Barndorff-Nielsen, eds. Chapman and Hall, London. (A discussion and comparison of the statistical properties of ARCH and stochastic volatility models, with an emphasis on recently developed computational procedures.)

Taylor, S. J. (1986). *Modelling Financial Time Series*. Wiley, Chichester, England. (Early treatment of both ARCH and stochastic-volatility-type models for financial time series, with an extensive discussion of diagnostics.)

Taylor, S. J. (1994). Modeling stochastic volatility. *Math. Finance*, **4**, 183–204. (An empirical driven comparison of ARCH and stochastic volatility models.)

See also AUTOREGRESSIVE–MOVING AVERAGE (ARMA) MODELS; DTARCH MODELS; FINANCE, STATISTICS IN; HETEROSCEDASTICITY; and TIME SERIES.

TORBEN G. ANDERSEN

TIM BOLLERSLEV

ARCHAEOLOGY, STATISTICS IN—I

The application of statistical thought to archaeology has been a slow process. This reluctance arises because (1) archaeological data rarely can be gathered in a well-designed statistical experiment; (2) describing empirical findings requires an expertise that is not easily modeled.

In recent years, however, the central problem of archaeology, generally labeled "typology," and the important related problem of "seriation" have received considerable mathematical and statistical attention, which we discuss herewith. The advent of the high-speed computer has made feasible analyses of large sets of archaeological data which were previously impracticable. The application of routine statistical methodology has been infrequent and nonsystematic (see ref. 5, Chap. 13). A recent article by Mueller [23]

is noteworthy, illustrating the use of sampling schemes in archaeological survey. The important question of how to locate artifact sites in a region that cannot be totally surveyed is examined. Simple random sampling and stratified sampling are compared in conjunction with an empirical study.

The artifact provides the class of entities with which archaeology is concerned. Typology is concerned with the definition of artifact types. Since mode of usage is unobservable, definition arises from an assortment of qualitative and quantitative variables (e.g., shape, color, weight, length) yielding a list of attributes for each artifact. Artifact types are then defined in terms of "tight clusters of attributes" [19]. The definition of types is usually called taxonomy* and the methods of numerical taxonomy have come to be employed in archaeological typology. (One does find in the literature references to taxonomy for archaeological sites, an isomorphic problem to that of typology for artifacts.)

"Typological debate" [11] has run several decades, resulting in a voluminous literature (see ref. 5). The issues of contention include such matters as whether types are "real" or "invented" to suit the researcher's purposes, whether there is a "best" classification of a body of materials, whether types can be standardized, whether types represent "basic" data, and whether there is a need for more or fewer types. Statistical issues arise in the construction of a typology.

Krieger's effort is a benchmark in unifying the typological concept. An earlier article in this spirit by Gozodrov [10] is deservedly characterized by Krieger as "tentative and fumbling" [19, p. 271]. Krieger reveals the variance in published thought on the classification* issue through examination of the work on pottery description and on projectile style. He articulates the "typological method" and cites Rouse [28] as a good illustration. The earliest quantitative work is by Spaulding [30]. A previous paper by Kroeber [20] concerned itself solely with relating pairs of attributes. The usual χ^2 statistic* (*see* CHI-SQUARE TEST—I) as well as other measures of association* were studied for 2×2 presence–absence attribute tables. From this lead Spaulding suggests, given attribute lists for each artifact in the collection, the preparation of cross-tabulations of all attributes (attribute categories are no longer restricted to presence-absence). Two-way tables are χ^2-tested, leading to the clustering of nonrandomly associated attributes. Artifact types are then defined by identifying classes of artifacts that exhibit sets of associated attributes.

Attempting to formalize Spaulding's technique led archaeologists into the realm of cluster analysis*. A basic decision in cluster analysis is whether the items or the components of the item data vectors are to be clustered. Specifically, are we clustering artifacts (Q-mode* of analysis, as it has been called) or attributes (R-mode* of analysis)? In defining a typology a Q-mode of analysis is appropriate, but the unfortunate use of the phrase "cluster of attributes" by both Krieger and Spaulding has resulted in a persistent confusion in the literature. Factor* and principal components* analyses have occasionally been employed as R-mode analyses to group attributes by significant dimensions interpretable as underlying features of the data.

Typology, then, involves discerning clusters of artifacts on the basis of similarity of their attributes. Similarity between artifacts is measured by similarity functions that arise in other (e.g., psychological and sociological) settings as well. Nonmathematically, a similarity function between two vectors reflects the closeness between components of the vectors as an inverse "distance." For a set of vectors the similarities between all pairs are arranged in a "similarity" matrix*. Beginning with such a matrix, clustering procedures are usually effected with the assistance of a computer.

The earliest computer typologies were done by single-link (nearest-neighbor) clustering [29, p. 180]. Links are assigned from largest similarities and clusters are derived from linked units. Unwelcome "chaining" often occurs, whence average linkage (weighted and unweighted) and complete linkage procedures have been suggested [29, p. 181]. Their sensitivity to spurious large similarities led Jardine and Sibson [15] to formulate double-link cluster analysis, but now "chaining" returns. Thus Hodson [13] proposes a K-means cluster analysis approach.

The total collection is partitioned into a predetermined number of clusters. Rules are defined for transferring artifacts from one cluster to another until a "best" clustering is obtained. The procedure is repeated for differing initial numbers, with expertise determining the final number of types defined. The approach can accommodate very large collections.

Similarities implicitly treat attributes in a hierarchical manner in defining types. Whallon [34] suggests that often a hierarchy of importance among attributes exists and that this is how archaeologists feel their way to defining types. Attribute trees are defined where presence or absence of an attribute creates a branch. Employing χ^2 values computed over appropriate 2×2 tables, the sequence of attributes forming the tree is achieved and also the definition of types from such sequences. Cell frequency problems plague many of the χ^2 values. Read [25] formalizes hierarchical classification in terms of partitions of a set of items and allows both discrete and continuous attributes. Clark [3] extends these ideas, assuming discrete attributes and setting them in appropriate higher-order contingency tables* to which log-linear models* are fitted.

In summary, then, a typology is usually obtained by clustering* (through an appropriate procedure) artifacts having similar attributes and defining types through these attributes. Recent hierarchical classification approaches show promise for data sets exhibiting weak clustering.

The next natural step in archaeological enterprise is the comparison of artifact collections, the process called seriation. In broadest terms, seriation consists of arranging a set of collections in a series with respect to similarity of the component artifacts to infer ordering in some nonobservable (usually time) dimension. The collections will typically be grave lots or assemblages. The chronological inference is drawn by the assumption that the degree of similarity between two collections varies inversely with separation in time. Such an assumption implicitly requires a "good" typology for the collections. Of course, other dimensions (e.g., geographic, cultural) may also affect the degree of similarity between collections and confound a "time" ordering. Techniques such as stratigraphy*, dated inscriptions, cross-ties with established sequences, or radiocarbon dating, if available, would thus preempt seriation. If, in addition to order, one wishes relative distance (in units of time) between the collections, we have a scaling problem as well. The seriation literature is quite extensive. Sterud [33] and Cowgill [6] provide good bibliographies.

The general principles of sequence dating originate with Flinders Petrie [24]. Brainerd [2] and Robinson [27] in companion papers set forth the first formalized mathematical seriation procedure. Robinson offers the methodology with examples and is credited as first to have linked similarities with sequencing. Brainerd provides the archaeological support for the method as well as the interpretation of its results. Some earlier formal attempts in the literature include Spier [31], Driver and Kroeber [8], and Rouse [28]. An assortment of earlier ad hoc seriations are noted by Brainerd [2, p. 304], who comments that they were often qualified as provisional pending stratigraphic support. Kendall [17], making more rigorous the ideas of Petrie, sets the problem as one of estimation. The observed collections Y_i are assumed independent with independent components Y_{ij} indicating the number of occurrences of the jth artifact type in collection i. Each Y_{ij} is assumed Poisson* distributed with mean μ_{ij}, a function of parameters reflecting abundance, centrality, and dispersion. P, the permutation of the Y_i's yielding the true temporal order, is also a parameter. A maximum likelihood* approach enables maximization over the μ_{ij}'s independently of P and yields a scoring function $S(P)$ to be maximized over all permutations. But for as few as 15 collections, exhaustive search for the maximum is not feasible.

Similarities help again the similarity matrix now being between collections, described by vectors, with components noting incidence or abundance of artifact types. Using similarities, an order is specified up to reversibility, with expertise then directing it. Labeling the similarity matrix by F, the objective of a seriation is to find a permutation of the rows and columns to achieve a matrix A with elements a_{ij} such that a a_{ij} increases in j for $j < i$; a_{ij} decreases in j for

$j > i$. A similarity matrix having this form has been called a Robinson matrix; the process of manipulating F to this form has been called petrifying. A permutation achieving this form must be taken as ideal under our assumptions but one need not exist. Practically, the goal is to get "close" (in some sense) to a Robinson form.

Taking Robinson's lead, archaeologists such as Hole and Shaw [14], Kuzara et al. [21], Ascher and Ascher [1], and Craytor and Johnson [7], studying large numbers of collections, develop orderings with elaborate computer search procedures (e.g., rules to restrict the search, sampling from all permutations, trial-and-error manipulations). Kendall [18], making these approaches more sophisticated, develops the "horseshoe method," based upon a multidimensional scaling* program in two dimensions. Both theory and examples suggest that with repeated iterations of such a program, a two-dimensional figure in the shape of a horseshoe may be expected if the data are amenable to a seriation. The horseshoe is then unfolded to give a one-dimensional order. Kadane [16] also suggests a computer-based approach by relating the problem of finding a "best" permutation to the traveling salesman problem*. In both cases one seeks a minimum-path-length permutation for a set of points, a problem for which effective computer solutions exist.

Sternin [32] takes a more mathematical tact. He sets the model $F = PAP^T + E$ where P is an unknown permutation matrix and E an error matrix accounting for the possible inability to restore F to exactly a Robinson form. Sternin argues that for certain types of Robinson matrices (e.g., exponential, Green's, and Toeplitz matrices*), the components of eigenvectors corresponding to the two largest eigenvalues will exhibit recognizable patterns. With $E = 0$, F and A have the same eigenvalues*, so Sternin suggests rearranging the components of the corresponding eigenvectors* of F to these patterns.

Gelfand [9] presents two "quick and dirty" techniques. Both methods guarantee the obtaining of the ideal P, if one exists. The better method takes each collection in turn as a reference unit, sequencing all other collections about it. After orienting each of these sequences in the same direction, a final order is obtained by "averaging" these sequences. The averaging should reduce the effect of E and yield a sequence invariant to the original order. An index of fit for a permutation similar to the stress measure used in multidimensional scaling is given, enabling comparison of orders. If an ideal permutation exists, it will minimize this index. Renfrew and Sterud [26] describe a "double-link" method analogous to double-link clustering.

In summary, if an ideal seriation exists, it can be found. If not, but if the data are sufficiently "one-dimensional," the foregoing techniques yield orders from which, with minor modifications suggested by expertise or index of fit, a "best" sequence can be produced.

We now turn to brief discussion of an example. The La Tène Cemetery at Munsingen-Rain near Berne, Switzerland, has proved a rich source of archaeological evidence and has been discussed in numerous articles over the past 15 years. (Hodson [12] provides the definitive work.) The excavation consists of 59 "closed-find" graves. Within these graves were found considerable numbers of fibulae, anklets, bracelets, etc. These ornamental items are typical of the more complex kinds of archaeological material in providing a wide range of detail that allows almost infinite variation within the basic range. A typology for these items was developed employing single-link cluster analysis, average-link cluster analysis, and a principal components analysis*. As a result, some 70 varieties or "types" were defined. A 59×70 incidence matrix of types within graves was created and converted to a 59×59 similarity matrix between graves. This matrix has been seriated using both the Kendall horseshoe method and Gelfand's technique. The unusual, almost linear form of the cemetery implies a geographical sequencing, which enabled Hodson to establish a very satisfactory seriation. The serial orders obtained by Kendall and by Gelfand in the absence of this information are both in good agreement with Hodson's.

In conclusion, the two books by Clarke [4,5] provide the best current picture of quantitative work in archaeology. Specifically, the articles by Hill and Evans [11] and by Cowgill [6] in the earlier book present excellent synopses on typology and seriation, respectively. The article by Hodson [13] is delightful in bringing some very sophisticated statistical thought to these problems. Finally, the volume from the conference in Mamaia [22] documents a very significant dialogue between archaeologists and statisticians and mathematicians. It bodes well for future analytic work in archaeology.

REFERENCES

1. Ascher, M. and Ascher, R. (1963). *Amer. Anthropol.*, **65**, 1045–1052.

2. Brainerd, G. W. (1951). *Amer. Antiq.*, **16**, 301–313.

3. Clark, G. A. (1976). *Amer. Antiq.*, **41**, 259–273.

4. Clarke, D. L. (1972). *Models in Archaeology*. Methuen, London.

5. Clarke, D. L. (1978). *Analytical Archaeology*, 2nd ed. Methuen, London.

6. Cowgill, G. L. (1972). In *Models for Archaeology*, D. L. Clarke, ed. Methuen, London, pp. 381–424.

7. Craytor, W. B. and Johnson, L. (1968). *Refinements in Computerized Item Seriation*. Mus. Nat. History, Univ. Oreg. Bull. 10.

8. Driver, H. E. and Kroeber, A. L. (1932). *Univ. Calif. Publ. Amer. Archaeol. Ethnol.*, **31**, 211–256.

9. Gelfand, A. E. (1971). *Amer. Antiq.*, **36**, 263–274.

10. Gozodrov, V. A. (1933). *Amer. Anthropol.*, **35**, 95–103.

11. Hill, J. N. and Evans, R. K. (1972). In *Models in Archaeology*, D. L. Clarke, ed. Methuen, London, pp. 231–274.

12. Hodson, F. R. (1968). *The LaTène Cemetery at Munsingen-Rain*. Stämpfi, Berne.

13. Hodson, F. R. (1970). *World Archaeol.*, **1**, 299–320.

14. Hole, F. and Shaw, M. (1967). *Computer Analysis of Chronological Seriation. Rice Uuiv. Stud. No. 53(3)*, Houston.

15. Jardine, N. and Sibson, R. (1968). *Computer J.*, **11**, 177.

16. Kadane, J. B. (1972). *Chronological Ordering of Archaeological Deposits by the Minimum Path Length Method. Carnegie-Mellon Univ. Rep. No. 58* Carnegie-Mellon University, Dept. of Statistics, Pittsburgh, Pa.

17. Kendall, D. G. (1963). *Bull. I.S.I.*, **40**, 657–680.

18. Kendall, D. G. (1971). In *Mathematics in the Archaeological and Historical Sciences*. Edinburgh Press, Edinburgh, pp. 215–252.

19. Krieger, A. D. (1944). *Amer. Antiq.*, **9**, 271–288.

20. Kroeber, A. L. (1940). *Amer. Antiq.*, **6**, 29–44.

21. Kuzara, R. S., Mead, G. R., and Dixon, K. A. (1966). *Amer. Anthropol.*, **68**, 1442–1455.

22. *Mathematics in the Archaeological and Historical Sciences* (1971). Edinburgh Press, Edinburgh.

23. Mueller, J. W. (1974). *The Use of Sampling in Archaeology Survey. Amer. Antiq. Mem. No. 28.*

24. Petrie, W. M. Flinders (1899). *J. Anthropol. Inst.*, **29**, 295–301.

25. Read, D. W. (1974). *Amer. Antiq.*, **39**, 216–242.

26. Renfrew, C. and Sterud, G. (1969). *Amer. Antiq.*, **34**, 265–277.

27. Robinson, W. S. (1951). *Amer. Antiq.*, **16**, 293–301.

28. Rouse, I. (1939). *Prehistory in Haiti, A Study in Method. Yale Univ. Publ. Anthropol. No. 21.*

29. Sokal, R. R. and Sneath, H. A. (1963). *Principles of Numerical Taxonomy.* W. H. Freeman, San Francisco.

30. Spaulding, A. C. (1953). *Amer. Antiq.*, **18**, 305–313.

31. Spier, L. (1917). *An Outline for a Chronology of Zuni Ruins. Anthropol. Papers Amer. Mus. Nat. History*, **18**, Pt. 3.

32. Sternin, H. (1965). *Statistical Methods of Time Sequencing. Stanford Univ. Rep. No. 112*, Dept. of Statistics, Stanford University, Stanford, Calif.

33. Sterud, G. (1967). *Seriation Techniques in Archaeology.* Unpublished M.S. thesis, University of California at Los Angeles.

34. Whallon, R. (1972). *Amer. Antiq.*, **37**, 13–33.

See also CLUSTER ANALYSIS; MULTIDIMENSIONAL SCALING; SIMILARITY, DISSIMILARITY AND DISTANCE, MEASURES OF; and TRAVELING-SALESMAN PROBLEM.

ALAN E. GELFAND

ARCHAEOLOGY, STATISTICS IN—II

Applications of statistics to archaeological data interpretation are widespread and can be divided broadly into two groups: those which are descriptive in nature (used primarily to reduce large and/or complex data sets to a more manageable size) and those which are model-based (used to make inferences about the underlying processes that gave rise to the data we observe). Approaches of the first type are most commonly adopted and, in general, are appropriately used and well understood by members of the archaeological profession. Model-based approaches are less widely used and usually rely upon collaboration with a professional statistician.

In the preceding entry Gelfand has provided an excellent survey of the application of statistics to archaeology up to and including the late 1970s. This entry supplements the earlier one, and the emphasis is on work undertaken since that time. Even so, this entry is not exhaustive, and readers are also encouraged to consult the review article of Fieller [15]. Statistics forms an increasingly important part of both undergraduate and graduate courses in archaeology, and several modern textbooks exist. At an introductory level, Shennan [28] assumes very little background knowledge and introduces the reader to both descriptive and model-based approaches. Baxter [1] concentrates on the interpretation of multivariate data in archaeology. The focus is on exploratory rather than model-based approaches, since this has been the primary approach to multivariate data adopted by the archaeological community. Buck et al. [4] take up where the other authors leave off. Their work uses a range of case studies that require model-based approaches and advocates a Bayesian approach so that all prior information is included in the data interpretation process.

We turn first to the uses archaeologists make of simple descriptive statistics. Most modern archaeological field work (and much undertaken in the laboratory) results in the collection of enormous quantities of numeric data. These might take the form of length and breadth measurements used to characterize particular types of artifacts (for example, human and animal bones, pottery vessels, or metal artifacts such as swords or knives) or counts of finds in particular locations (for example, pottery fragments observed on the surface of a site prior to excavation, grave goods deposited with the body at time of burial, or different types of pollen grains obtained by coring different parts of an archaeological landscape). Although such data can represent an enormous range of different archaeological phenomena, the same kinds of statistical approaches are likely to be used to compress the information to a manageable size for presentation and interpretation. Most common are means and standard deviations, percentages, scatter plots, bar charts (2-D and 3-D), and line graphs.

Most archaeological site reports contain a selection of these types of data presentation. In a recent example Cunliffe [14] reports on 25 years of excavation of the Iron Age hill-fort at Danebury in Hampshire, UK. This report provides examples of all the descriptive statistical techniques outlined above and some model-based ones too.

Model-based approaches to archaeological data interpretation have been rather slow to take off, since very few "off the peg" approaches are suitable. Nonetheless, some professional statisticians have shown an interest in helping to interpret archaeological data, and a range of subject-specific model-based approaches have been developed; the most famous is probably the approach used in an attempt to order chronologically (or seriate) archaeological deposits on the basis of the artifact types found within them. A good example might be the desire to establish the order in which bodies were placed in a cemetery on the basis of the grave goods found with them. The basic model is that objects come into use (or "fashion") and then after a period of time go out of use again, but never come back. This model is used not because archaeologists believe that it completely represents the past, but because it adequately reflects the nature of human activity and is not so sophisticated that it cannot be easily adopted in practice. (Gelfand made it the center of the preceding entry.) Development and formalization of the basic model can be attributed to Robinson [27], but see also refs. 6, 19, 21. The early works assumed that archaeological data would conform to the model exactly; Laxton and Restorick [21] noted that there was great potential for stochastic components in archaeological data and modeled this

into their approach; and Buck and Litton [6] adopted the same model, but suggested a Bayesian approach so that prior information could also be explicitly modeled into the interpretation process.

Other areas of archaeology that have benefited from adoption of model-based approaches include:

interpretation of soil particle size data in an attempt to understand the nature of the climate and landscape that gave rise to currently observed deposits [16],

consideration of the minimum numbers of individuals represented within assemblages of archaeological artifacts such as disarticulated skeletons [17,18,32],

interpretation of radiocarbon dates in the light of any available relative chronological information from excavation or literary sources [22,5,9], interpretation of data from site surveys; for example soil phosphate analysis or soil resistance measurements [10,3],

identifying the optimum duration and digging strategies for archaeological excavations [23], formalizing descriptions of the shapes and structural mechanics of prehistoric vaulted structures in Europe [11,12,13,20], and interpretation of multivariate chemical compositional data from archaeological artifacts (such as ceramics or glass) collected in an attempt to identify the geological source or site of manufacture [1,8].

Many of these applications are very new and have arisen from recent developments in statistics rather than from any specific changes in archaeological practice. Let us consider the interpretation of radiocarbon dates (for detailed information on radiocarbon dating in archaeology see Bowman [2]) and in so doing return to the report of the Danebury excavations [14]. The first discussion of radiocarbon dating at Danebury (Orton [25]) suggested a mechanism whereby radiocarbon data and archaeological information could be combined within an explicit mathematical framework. The work was completed, however, before the radiocarbon community had adopted a suitable calibration* curve from the many that had been published and before suitable statistical procedures had been developed to allow anything more than point estimates (rather than full distributional information) to be computed. But between 1983 and 1990 dramatic changes took place which we briefly document here.

First, we develop some notation. Suppose that we wish to obtain an estimate for the date of death (θ) of an organic sample found during excavation of the hill-fort at Danebury. This sample is sent for analysis at a radiocarbon dating laboratory, which returns an estimate of the radiocarbon age and an associated estimate of the laboratory error, represented by $y \pm \sigma$. Now the amount of radioactive carbon in the atmosphere has not been constant over time—indeed, it has varied considerably—and as a result a calibration curve is required to map radiocarbon age onto the calendar time scale. The first internationally agreed version of such a curve was published in 1986 [26,29]. It takes the form of bidecadal data that provide a nonmonotonic piecewise-linear calibration curve, which we represent by $\mu(\theta)$. By convention y is then modeled as normally distributed with mean $\mu(\theta)$ and standard deviation σ. This means that for any single radiocarbon determination $y \pm \sigma$ the (posterior) probability distribution of the calendar date θ can fairly readily be computed. Several specialist computer programs exist to do this (for example, CALIB [30,31]). However, because the calibration curve is nonmonotonic and because, in practice, the laboratory errors are often quite large, one radiocarbon determination often does not provide us with much information about the calendar date of interest. Indeed, posterior distributions* commonly have a range of several hundred years.

In an attempt to improve on this, archaeologists soon realized that groups of related determinations would be much more likely to provide precise information than would single ones. This was the approach adopted at Danebury. Cunliffe [14, Table 40, p. 132] reports a total of 60 determinations, all collected with the aim of refining the chronology at Danebury. At the outset Cunliffe realized that sophisticated statistical investigation

would be required to make the most of the data available, and his collaboration with Orton began. Between them Orton and Cunliffe developed a model that reflected Cunliffe's beliefs about the relative chronological information at the site.

Naylor and Smith [24] took the story one step further by determining not only that there was a model to be built and that the calibration curve must be allowed for in the interpretation process, but also that the relative chronological information (provided by archaeological stratigraphy) represented prior information, and that the whole problem could be represented using extremely elegant mathematical models.

On the basis of pottery evidence, Cunliffe divided the chronology of the site into four distinct phases. Initially (but see below) he saw these phases as following one another in a strict sequence. A major reason for taking so many radiocarbon samples at Danebury was to learn about the calendar dates of the phase boundaries of the four abutting phases. Consequently, the archaeologists carefully ascribed each of the organic samples to one (and only one) of the four ceramic phases.

In order to explain the statistical approach, we label the calendar dates associated with the 60 samples $\boldsymbol{\theta} = \{\theta_1, \theta_2, \ldots, \theta_{60}\}$. We then label the calendar dates of the phase boundaries so that the calendar date of the start of ceramic phase 1 is event 1 and the calendar date of the end of the same phase as event 2. In the same manner the calendar date of the start of phase 2 is event 3 and the calendar date of the end of the phase is event 4. Thus, in order to bound four ceramic phases, we need to define eight events. We then represent these events using the notation $\boldsymbol{\Psi} = \{\Psi_1, \Psi_2, \ldots, \Psi_8\}$. Since the four phases are modeled as abutting, however, we have $\Psi_2 = \Psi_3$, $\Psi_4 = \Psi_5$, and $\Psi_6 = \Psi_7$, and [since calibrated radiocarbon determinations are conventionally reported "before present" (BP), where "present" is 1950]

$$\Psi_1 > \Psi_3 > \Psi_5 > \Psi_7 > \Psi_8. \tag{1}$$

Since the beginning of each phase is always before its end, it can also be stated that

$$\Psi_{2j-1} > \Psi_{2j} \quad (j = 1, 2, 3, 4). \tag{2}$$

It was $\Psi_1, \Psi_3, \Psi_5, \Psi_7$, and Ψ_8 for which Naylor and Smith [24] provided estimates, using computer software based on quadrature methods. Their methodology was coherent, concise, and elegant, but unfortunately they did not work directly with archaeologists. As a result they made one or two fundamental archaeological errors.

With the benefit of hindsight, it may be better that things happened this way, since dramatic changes were taking place in applied statistics that altered the whole approach to calculating Bayesian posterior estimates and revolutionized the way archaeologists think about radiocarbon calibration. The advances in the modeling of archaeological phenomena had taken place against the development of Markov chain Monte Carlo* methods for computing Bayesian posteriors. Realistic models could be used for both likelihoods and priors, since posteriors would simply be simulated from their conditionals. All that is required is large amounts of computer power. Since applied statisticians usually have access to powerful computer workstations, fruitful collaborations were undertaken; Buck et al. [9] is one example.

To understand the approach, note first that, if D is taken to represent the set of all dates for the phase boundaries that satisfy the two sets of inequalities in (1) and (2), the joint prior density for Ψ is

$$\Pr(\Psi) = \begin{cases} c, & \Psi \in D, \\ 0, & \text{otherwise,} \end{cases}$$

where c is a constant. We also need to model the distribution of the θ's within each phase. In the absence of firm archaeological information on deposition rates of the relevant organic samples, it was assumed that if the ith radiocarbon sample is associated with phase j, then its calendar date θ_{ij} (where $j = 1, 2, 3, 4$) is uniformly distributed over the interval Ψ_{2j-1}, Ψ_{2j}, i.e., a uniform deposition rate is assumed:

$$\Pr(\theta_{ij}|\Psi_{2j-1}, \Psi_{2j})$$
$$= \begin{cases} (\Psi_{2j-1} - \Psi_{2j})^{-1}, & \Psi_{2j-1} > \theta_{ij} > \Psi_{2j}, \\ 0, & \text{otherwise.} \end{cases}$$

Then assuming that, conditional on the Ψ_k's ($k = 1, 2, \ldots, 8$), the θ_{ij}'s are independent, we have

$$\Pr(\theta|\Psi) = \prod_{j=1}^{4}\prod_{i=1}^{n_j} \Pr(\theta_{ij}|\Psi_{2j-1}, \Psi_{2j}),$$

where n_j is the number of samples in phase j. If x_{ij} represents the radiocarbon determination with standard deviation σ_{ij} which corresponds to the sample with calendar date θ_{ij}, then x_{ij} is a realization of a random variable X_{ij} with mean $\mu(\theta_{ij})$ and variance σ_{ij}^2. Since the calibration curve is piecewise linear,

$$\mu(\theta) = \begin{cases} a_1 + b_1\theta & (\theta \leqslant t_0), \\ a_l + b_l\theta & (t_{l-1} < \theta \leqslant t_l, l = 1, 2, \ldots, L), \\ a_L + b_L\theta & (\theta > t_L), \end{cases}$$

where the t_l are the knots of the calibration curve, $L + 1$ is the number of knots, and a_l and b_l are known constants.

Using Stuiver and Pearson's [29,26] calibration curves and the relative chronological information described above, Buck et al. [9] calibrated the 60 radiocarbon determinations to obtain posterior probability distributions for $\Psi_1, \Psi_3, \Psi_5, \Psi_7$ and Ψ_8 via Markov chain Monte Carlo simulation. They took the joint posterior density to be

$$\Pr(\theta, \Psi|x, \sigma^2) \sim \Pr(x|\theta, \Psi, \sigma^2)\Pr(\theta|\Psi)\Pr(\Psi),$$

where

$$\Pr(x|\Psi, \theta, \sigma^2) = \prod_{j=1}^{4}\prod_{i=1}^{n_j} \Pr(x_{ij}|\theta_{ij}, \sigma_{ij}^2),$$

and the likelihood is given by

$$\Pr(x_{ij}|\theta_{ij}, \sigma_{ij}^2)$$
$$= (2\pi\sigma_{ij}^2)^{-1/2}\exp\left(-\frac{[x_{ij} - \mu(\theta_{ij})]^2}{2\sigma_{ij}^2}\right);$$

the priors, $\Pr(\theta|\Psi)$ and $\Pr(\Psi)$, are given above. Originally developed to solve a particular problem, this methodology is fairly general and was published in an archaeological journal. As a result, after about ten years of developmental work and collaboration between more than half a dozen individuals, a model-based, fully coherent approach was available in a forum accessible to archaeological researchers.

This was not the end of the story. Between 1992 and 1995 Cunliffe undertook a reassessment of the data from Danebury and decided that his initial interpretation of the ceramic phases was not entirely appropriate; a more appropriate assessment was that phases 1, 2 and 3 abut one another, but that phase 4 may overlap phase 3. In addition, phase 4 definitely began after the end of phase 2. This required a restatement of relationships between boundary parameters. The new information is that $\Psi_1 > \Psi_2 = \Psi_3 > \Psi_4 = \Psi_5 > \Psi_6$ and $\Psi_5 \geqslant \Psi_7 > \Psi_8$. Having developed a general approach to the work reported in Buck et al. [9], it was possible to alter the model restrictions and recompute the posterior probability distributions for the six parameters now of interest [7].

A large number of researchers (archaeologists and statisticians) have been involved in aiding the interpretation of the radiocarbon determinations from Danebury. As a result there is now a well-developed and widely tested model-based framework in which radiocarbon calibration and interpretation can take place. In general, in order to obtain access to powerful computers, archaeologists currently need to collaborate with statisticians; some see this as a great drawback. In the foreseeable future this is likely to change, but for the moment there are some benefits: applied model-based statistics is not treated as a black box technology, and the applied statistician working on the project ensures that each application is approached afresh, the models are tailor-made for the problem under study, and no presumptions are made that might color judgment about the available prior information.

In summary, statistics is an essential tool for the investigation and interpretation of a wide range of archaeological data types. Descriptive statistics are widely used to allow archaeologists to summarize and display large amounts of otherwise uninterpretable data. Model-based statistics are increasingly widely used, but continue to be most commonly adopted by teams of archaeologists and statisticians working in collaboration.

Most model-based statistical interpretations are still seen to be at the cutting edge of archaeological research.

REFERENCES

1. Baxter, M. J. (1994). *Exploratory Multivariate Analysis in Archaeology*. Edinburgh University Press, Edinburgh.

2. Bowman, S. (1990). *Interpreting the Past: Radiocarbon Dating*. British Museum Publications, London.

3. Buck, C. E., Cavanagh, W. G., and Litton, C. D. (1988). The spatial analysis of site phosphate data. In *Computer Applications and Quantitative Methods in Archaeology*, S. P. Q. Rahtz, ed. British Archaeological Reports, Oxford, pp. 151–160.

4. Buck, C. E., Cavanagh, W. G., and Litton, C. D. (1996). *The Bayesian Approach to Interpreting Archaeological Data*. Wiley, Chichester.

5. Buck, C. E., Kenworthy, J. B., Litton, C. D., and Smith, A. F. M. (1991). Combining archaeological and radiocarbon information: a Bayesian approach to calibration. *Antiquity*, **65**(249), 808–821.

6. Buck, C. E. and Litton, C. D. (1991). A computational Bayes approach to some common archaeological problems. In *Computer Applications and Quantitative Methods in Archaeology 1990*, K. Lockyear and S. P. Q. Rahtz, eds. Tempus Reparatum, British Archaeological Reports, Oxford, pp. 93–99.

7. Buck, C. E. and Litton, C. D. (1995). The radio-carbon chronology: further consideration of the Danebury dataset. In *Danebury: An Iron Age Hill-fort in Hampshire, vol. 6, a Hill-fort Community in Hampshire*, Report 102. Council for British Archaeology, pp. 130–136.

8. Buck, C. E. and Litton, C. D. (1996). Mixtures, Bayes and archaeology. In *Bayesian Statistics 5*, A. P. Dawid, J. M. Bernardo, J. Berger, and A. F. M. Smith, eds. Clarendon Press, Oxford, pp. 499–506.

9. Buck, C. E., Litton, C. D., and Smith, A. F. M. (1992). Calibration of radiocarbon results pertaining to related archaeological events. *J. Archaeol. Sci.*, **19**, 497–512.

10. Cavanagh, W. G., Hirst, S., and Litton, C. D. (1988). Soil phosphate, site boundaries and change-point analysis. *J. Field Archaeol.*, **15**(1), 67–83.

11. Cavanagh, W. G. and Laxton, R. R. (1981). The structural mechanics of the Mycenaean

tholos tombs. *Ann. Brit. School Archaeol. Athens*, **76**, 109–140.

12. Cavanagh, W. G. and Laxton, R. R. (1990). Vaulted construction in French Megalithic tombs. *Oxford J. Archaeol.*, **9**, 141–167.

13. Cavanagh, W. G., Laxton, R. R., and Litton, C. D. (1985). An application of change-point analysis to the shape of prehistoric corbelled domes; i. The maximum likelihood method. *PACT*, **11**(III.4), 191–203.

14. Cunliffe, B. (1995). *Danebury: an Iron Age Hill-fort in Hampshire, vol. 6*, a Hill-fort Community in Hampshire, Report 102. Council for British Archaeology.

15. Fieller, N. R. J. (1993). Archaeostatistics: old statistics in ancient contexts. *Statistician*, **42**, 279–295.

16. Fieller, N. R. J. and Flenley, E. (1988). Statistical analysis of particle size and sediments. In *Computer Applications and Quantitative Methods in Archaeology*, C. L. N. Ruggles and S. P. Q. Rahtz, eds. British Archaeological Reports, Oxford, pp. 79–94.

17. Fieller, N. R. J. and Turner, A. (1982). Number estimation in vertebrate samples. *J. Archaeol. Sci.*, **9**(1), 49–62.

18. Horton, D. R. (1984). Minimum numbers: a consideration. *J. Archaeol. Sci.*, **11**(3), 255–271.

19. Kendall, D. G. (1971). Abundance matrices and seriation in archaeology. *Z. Wahrsch. Verw. Geb.*, **17**, 104–112.

20. Laxton, R. R. Cavanagh, W. G., Litton, C. D., Buck, R., and Blair, C. E. (1994). The Bayesian approach to archaeological data analysis: an application of change-point analysis for prehistoric domes. *Archeol. e Calcolatori*, **5**, 53–68.

21. Laxton, R. R. and Restorick, J. (1989). Seriation by similarity and consistency. In *Computer Applications and Quantitative Methods in Archaeology 1989*. British Archaeological Reports, Oxford, pp. 229–240.

22. Litton, C. D. and Leese, M. N. (1991). Some statistical problems arising in radiocarbon calibration. In *Computer Applications and Quantitative Methods in Archaeology 1990*, K. Lockyear and S. P. Q. Rahtz, eds. Tempus Reparatum, Oxford, pp. 101–109.

23. Nakai, T. (1991). An optimal stopping problem in the excavation of archaeological remains. *J. Appl. Statist.*, **28**, 924–929.

24. Naylor, J. C. and Smith, A. F. M. (1988). An archaeological inference problem. *J. Amer. Statist. Ass.*, **83**(403), 588–595.

25. Orton, C. R. (1983). A statistical technique for integrating C-14 dates with other forms of dating evidence. In *Computer Applications and Quantitative Methods in Archaeology*. J. Haigh, ed. School of Archaeological Science, University of Bradford, pp. 115–124.

26. Pearson, G. W. and Stuiver, M. (1986). High-precision calibration of the radiocarbon time scale, 500–2500 BC. *Radiocarbon*, **28**(2B), 839–862.

27. Robinson, W. S. (1951). A method for chronologically ordering archaeological deposits. *Amer. Antiquity*, **16**, 293–301.

28. Shennan, S. (1997). *Quantifying Archaeology*, 2nd edition. Edinburgh University Press, Edinburgh.

29. Stuiver, M. and Pearson, G. W. (1986). High-precision calibration of the radiocarbon time scale, AD 1950–500 BC. *Radiocarbon*, **28**(2B), 805–838.

30. Stuiver, M. and Reimer, P. (1986). A computer program for radiocarbon age calibration. *Radiocarbon*, **28**(2B), 1022–1030.

31. Stuiver, M. and Reimer, P. (1993). Extended ^{14}C data base and revised CALIB 3.0 ^{14}C age calibration program. *Radiocarbon*, **35**(1), 215–230.

32. Turner, A. and Fieller, N. R. J. (1985). Considerations of minimum numbers: a response to Horton. *J. Archaeol. Sci.*, **12**(6), 477–483.

See also ALGORITHMS, STATISTICAL; CALIBRATION—I; and MARKOV CHAIN MONTE CARLO ALGORITHMS.

CAITLIN E. BUCK

ARC-SINE DISTRIBUTION

The arc-sine distribution is a name attributed to a discrete and several continuous probability distributions. The discrete and one of the continuous distributions are principally noted for their applications to fluctuations in random walks*. In particular, the discrete distribution describes the percentage of time spent "ahead of the game" in a fair coin tossing contest, while one of the continuous distributions has applications in the study of waiting times*. The distribution most appropriately termed "arc-sine" describes the location, velocity, and related attributes at random time of a particle in simple harmonic motion. Here "random time" means that the time of observation is independent of the initial phase angle, $0 \leqslant \theta_0 < 2\pi$.

The arc-sine distribution with parameter $b > 0$ has support $[-b, b]$ and PDF $\pi^{-1}(b^2 - x^2)^{-1/2}$ for $-b < x < b$. The position at random time of a particle engaged in simple harmonic motion with amplitude $b > 0$ has the arc-sine (b) distribution.

If X is an arc-sine (1) random variable (RV) and $b \neq 0$, then the RV $Y = bX$ has arc-sine ($|b|$) distribution. Salient features of this distribution are:

$$\text{moments:} \begin{cases} EX^{2k} = 2^{-2k} \begin{pmatrix} 2k \\ k \end{pmatrix} \\ EX^{2k+1} = 0 \end{cases}$$

$$(k = 0, 1, 2, \ldots)$$

CDF: $(\sin^{-1} x + \pi/2)/\pi \quad (-1 < x < 1)$
characteristic function: $Ee^{itX} = J_0(t)$,

where $J_0(t)$ is the Bessel function* of the first kind, of order 0, $\sum_{k=0}^{\infty} (-1)^k (t/2)^{2k}/(k!)^2$.

Let $\sim$ denote "is distributed as." In ref. 6, Norton showed that if X_1 and X_2 are independent arc-sine(b) RVs, then $b(X_1 + X_2)/2 \sim X_1 X_2$, and in ref. 7 made the following conjecture. Let X_1 and X_2 be independent identically distributed RV's having all moments, and let F denote the common CDF. Then the only nondiscrete F for which $b(X_1 + X_2)/2 \sim X_1 X_2$ is the arc-sine(b) distribution. This conjecture was proved by Shantaram [8].

Arnold and Groeneveld [1] proved several results. Let X be a symmetric RV. Then $X^2 \sim (1 + X)/2$ if and only if $X \sim$ arc-sine(1). If X is symmetric and $X^2 \sim 1 - X^2$, then $X \sim 2X\sqrt{1 - X^2}$ if and only if $X \sim$ arc-sine(1). If X_1 and X_2 are symmetric independent identically distributed RVs with $X_i^2 \sim 1 - X_i^2$, then $X_1^2 - X_2^2 \sim X_1 X_2$ if and only if $X_i \sim$ arc-sine(1).

Feller [3] discusses distributions that have acquired the arc-sine name. Set $u_{2k} = \begin{pmatrix} 2k \\ k \end{pmatrix} 2^{-2k}$, $k = 0, 1, 2, \ldots (u_0 = 1)$. Let X_k equal ± 1 according to the kth outcome in a fair coin tossing game, and let $S_n = \sum_{k=1}^{n} X_k$ denote the net winnings of a player through epoch $n (S_0 = 0)$. From epochs 0 through $2n$, let Z_{2n} denote that epoch at which the last visit to the origin occurs. Then Z_{2n} necessarily assumes only even values and $\Pr[Z_{2n} = 2k] = u_{2k} u_{2n-2k} = \begin{pmatrix} 2k \\ k \end{pmatrix} \begin{pmatrix} 2n - 2k \\ n - k \end{pmatrix} 2^{-2n}$, $k =$

$0, 1, \ldots, n$. The probability distribution of the RV Z_{2n} is called the *discrete arc-sine distribution of order n*. Set $\Pr[Z_{2n} = 2k] = p_{2k,2n}$. The probability that in the time interval from 0 to $2n$ the S_j's are positive (the player is ahead) during exactly $2k$ epochs is $p_{2k,2n}$. This result is readily rephrased in terms of $x = k/n$, the proportion of the time the player is ahead. If $0 < x < 1$, the probability that at most $x(100)\%$ of the S_j's are positive tends to $2\pi^{-1} \sin^{-1} \sqrt{x}$ as $n \to \infty$. The corresponding PDF is $\pi^{-1}[x(1-x)]^{-1/2}, 0 < x < 1$, which has acquired the name "arc-sine density." Consideration of $p_{2k,2n}$ or the PDF shows that in a fair coin tossing game, being ahead one-half the time is the least likely possibility, and being ahead 0% or 100% of the time are the most likely possibilities. The probability that the first visit to the terminal value S_{2n} occurs at epoch $2k$ (or $2n - 2k$) is $p_{2k,2n}$. In a game of $2n$ tosses the probability that a player's maximum net gain occurs for the first time at epoch k, where $k = 2r$ or $k = 2r + 1$, is $\frac{1}{2} p_{2r,2n}$ for $0 < k < 2n$, u_{2n} for $k = 0$, and $\frac{1}{2} u_{2n}$ for $k = 2n$.

Feller also notes related results in other settings. Let $X_1, X_2, \ldots$ be independent symmetric RVs with common continuous CDF F. Let K_n denote the epoch (index) at which the maximum of $S_0, S_1, \ldots, S_n$ is first attained. Then $\Pr[K_n = k] = p_{2k,2n}$ and, for fixed $0 < \alpha < 1$, as $n \to \infty \Pr[K_n < n\alpha] \to 2\pi^{-1} \sin^{-1} \sqrt{\alpha}$. The number of strictly positive terms among $S_1, \ldots, S_n$ has the same distribution as K_n.

Standard beta* densities with support $[0, 1]$ and having form $f_\alpha(x) = [B(1 - \alpha, \alpha)]^{-1} x^{-\alpha} (1 - x)^{\alpha - 1}$, $0 < \alpha < 1$, are called *generalized arc-sine densities*. When $\alpha = \frac{1}{2}$, f_α is the "arc-sine density" $\pi^{-1}[x(1-x)]^{-1/2}$, $0 < x < 1$, mentioned earlier. Such PDFs play a role in the study of waiting times*. For example, let $X_1, X_2, \ldots$ be positive independent RVs with common CDF F and $S_n = \sum_{k=1}^{n} X_k$. Let N_t denote the random index for which $S_{N_t} \leq t < S_{N_t+1}$. Define $Y_t = t - S_{N_t}$. A result of Dynkin [2] is that if $0 < \alpha < 1$ and $1 - F(x) = x^{-\alpha} L(x)$, where $L(tx)/L(t) \to 1$ as $t \to \infty$, then the variable Y_t/t has limiting distribution with PDF f_α. Horowitz [4] extended Dynkin's result to semilinear Markov processes*. Imhof [5] considers the case in which t denotes time and

$\{X(t) : 0 \leqslant t \leqslant T\}$ is a stochastic process* satisfying certain conditions. If V denotes the elapsed time until the process reaches a maximum, then

$$\Pr[V < \alpha T] = 2\pi^{-1}\sin^{-1}\sqrt{\alpha}.$$

REFERENCES

1. Arnold, B. and Groeneveld, R. (1980). *J. Amer. Statist. Ass.*, **75**, 173–175. (Treats some characterizations of the distribution.)

2. Dynkin, E. B. (1961). *Select. Transl. Math. Statist. Probl.*, **1**, 171–189. (Requires probability theory.)

3. Feller, W. (1966). *An Introduction to Probability Theory and Its Applications.* Wiley, New York. (Provides a good survey of the arc-sine distributions.)

4. Horowitz, J. (1971). *Ann. Math. Statist.*, **42**, 1068–1074. (Requires elementary stochastic processes.)

5. Imhof, J. P. (1968). *Ann. Math. Statist.*, **39**, 258–260. (Requires probability theory.)

6. Norton, R. M. (1975). *Sankhyā, A*, **37**, 306–308. (Gives a characterization.)

7. Norton, R. M. (1978). *Sankhyā, A*, **40**, 192–198. (Treats primarily moment properties of discrete RVs.)

8. Shantaram, R. (1978). *Sankhyā, A*, **40**, 199–207. (Uses combinatorial identities.)

See also CHARACTERIZATIONS OF DISTRIBUTIONS; RANDOM WALKS; and TAKÁCS PROCESS.

R. M. NORTON

ARC-SINE TRANSFORMATION. See VARIANCE STABILIZATION

AREA SAMPLING

An area sample is a sample with primary sampling units* that are well-defined fractions of the earth's surface. The sampling frame* can be visualized as a map that has been subdivided into N nonoverlapping subareas that exhaust the total area of interest. The N distinct subareas are the primary sampling units. See SURVEY SAMPLING for a discussion of the meanings of primary sampling unit and sampling frame. The sampling units in area sampling are often called *segments* or *area segments*. The list of all area segments is the *area frame*.

Area samples are used to study characteristics of the land, such as the number of acres in specific crops, the number of acres under urban development, the number of acres covered with forest, or the fraction of cultivated acres subject to severe water erosion. Area sampling is an integral part of the U.S. Department of Agriculture's method of estimating acreages and yields of farm crops. *See* AGRICULTURE, STATISTICS IN.

An example of a recent large-scale area sample is the study of the potential for increasing the cropland of the United States conducted by the U.S. Soil Conservation Service. See Dideriksen et al. [1] and Goebel [2]. Area sampling is used heavily in forestry*. See Husch et al. [6] and Labau [13].

Area samples are also used when the observation units are persons or institutions for which a list is not available. For example, area frames are used in studies of the general population in the United States and in other countries where current lists of residents are not maintained. The Current Population Survey of the U.S. Bureau of the Census*, from which statistics on unemployment are obtained, is an area sample of the population of the United States.

Area sampling developed apace with probability sampling*. Mahalanobis* [14] in a discussion of area sampling described the contribution of Hubbock [5], who was responsible for a 1923 study that specified methods of locating a random sample* of areas used in estimating the yield of rice. King [12] cites a number of European studies of the 1920s and 1930s that used a type of area sampling.

In 1943, a large project was undertaken by the Statistical Laboratory of Iowa State College in cooperation with the Bureau of Agricultural Economics, U.S. Department of Agriculture, to design a national area sample of farms in the United States. The name *Master Sample* was applied to the project. The Bureau of the Census* also cooperated in the project and developed an area sample of cities, which, together with the Master Sample of rural areas, was used as a sample of the entire population. The materials developed in the Iowa State project, updated for changes in culture, are still used in the creation of area samples of the rural part of the United States. Stephan [18] provides an excellent review of sampling history.

The basic idea of area sampling is relatively simple, but the efficient implementation of the method requires some sophistication. It must be possible for the field worker (enumerator) to identify the boundaries of each area segment. Thus roads, streets, streams, fences, and other "natural boundaries" are used to define the segments, whenever possible. Aerial photographs, county highway maps, and street maps are materials commonly used in area sampling. Aerial photographs are particularly useful for studies outside heavily urbanized areas. *see* CENSUS.

A precise set of rules associating the elements of the population with the area segments must be developed when the population of interest is not a characteristic of the land itself. For a study of households, the households whose residence is on a given area segment are associated with that segment. Even for this relatively simple rule, problems arise. One problem is the definition of a household. There is also the problem of defining the primary residence in the case of multiple residences. For samples of farms or other businesses, it is common practice to associate the business with the segment on which the "headquarters" is located. See, for example, Jabine [7] for a set of rules of association.

In studies of the characteristics of the land itself, the definition of boundaries of the area segments is very important. A phenomenon called "edge bias" has been identified in empirical studies of crop yields. It has been observed that field workers tend to include plants near the boundary of a plot. Therefore, yields based on small areas are often biased upward. See Sukhatme [19] and Masuyama [16].

Any subdivision of the study area into segments will, theoretically, provide an unbiased estimator* of the population total. But the variances of the estimator obtained for two different partitions of the study area may be very different. Therefore, the design of efficient area samples requires that the area segments be as nearly equal in "size"

as is possible. In this context, size is a measure assigned to the area segments that is correlated with the characteristic of interest. An example of a measure of size is the number of households reported in the most recent census. The measure of size used in the Master Sample of Agriculture was the number of dwelling units indicated on county highway maps. This number was correlated with the number of farm headquarters.

In area samples used to study populations over time the size of the area segment will change. Gray and Platek [3] discuss methods of modifying the design in such situations.

Part of the cost of designing area samples is the cost of obtaining information on the estimated size of the area units from recent photos, maps, censuses, city directories, field visits, etc. In designing an area sample, the practitioner must balance the cost of obtaining additional information, the smaller variance of estimates obtained from units of nearly equal estimated size, and the practical requirement for boundaries that can be identified by the field worker.

REFERENCES

1. Dideriksen, R. I.,   Hidlebaugh, A. R.,   and Schmude, K. O. (1977). *Potential Cropland Study. Statist. Bull. No. 578*, U.S. Dept. of Agriculture.

2. Goebel, J. J. (1967). *Proc. Social Statist. Sect. Amer. Statist. Ass.*, 1976, pp. 350–354.

3. Gray, C. B. and Platek, R. (1968). *J. Amer. Statist. Ass.*, **63**, 1280–1297.

4. Houseman, E. E. (1975). *Area Frame Sampling in Agriculture. U.S. Dept. Agric. Bull. SRS No. 20*.

5. Hubbock, J. A. (1927). *Sampling for Rice Yield in Bihar and Orissa. Bull. No. 166*, Agricultural Research Institute, Pusa, Government of India. Reprinted in *Sankhyā*, **7**, 281–294 (1947).

6. Husch, B.,   Miller, C. I.,   and   Beers, T. W. (1972). *Forest Mensuration*. Ronald Press, New York.

7. Jabine, T. B. (1965). In *Estimation of Areas in Agricultural Statistics*. Food and Agriculture Organization of the United Nations, Rome, pp. 136–175.

8. Jessen, R. J. (1942). *Statistical Investigation of a Sample Survey for Obtaining Farm Facts*.

Res. Bull. 304, Iowa Agric. Exper. Stat., Iowa State College, Ames, Iowa.

9. Jessen, R. J. (1945). *J. Amer. Statist. Ass.*, **40**, 45–56.

10. Jessen, R. J. (1947). *J. Farm Econ.*, **29**, 531–540.

11. Jessen, R. J. (1978). *Statistical Survey Techniques*. Wiley, New York.

12. King, A. J. (1945). *J. Amer. Statist. Ass.*, **40**, 38–45.

13. Labau, V. J. (1967). *Literature on the Bitterlich Method of Forest Cruising*. U.S. Forest Serv. Res. Paper PNW-47, U.S. Dept. of Agriculture.

14. Mahalanobis, P. C. (1944). *Philos. Trans. Ser. B*, **231**, 329–451.

15. Mahalanobis, P. C. (1947). *Sankhyā*, **7**, 269–280.

16. Masuyama, M. (1954). *Sankhyā*, **14**, 181–186.

17. Monroe, J. and Finkner, A. L. (1959). *Handbook of Area Sampling*. Chilton, New York.

18. Stephan, F. F. (1948). *J. Amer. Statist. Ass.*, **43**, 12–39.

19. Sukhatme, P. V. (1947). *J. Amer. Statist. Ass.*, **42**, 297–310.

See also AGRICULTURE, STATISTICS IN; FORESTRY, STATISTICS IN; SMALL AREA ESTIMATION; SURVEY SAMPLING; and U.S. BUREAU OF THE CENSUS.

WAYNE A. FULLER

ARES PLOTS

A procedure suggested by Cook and Weisberg [1] and originally called "an animated plot for adding regression variables smoothly" was designed to show the impact of adding a set of predictors to a linear model by providing an "animated" plot. The term ARES is an acronym for "adding regressors smoothly" [1]. The basic idea is to display a smooth transition between the fit of a smaller model and that of a larger one. In the case of the general linear model* we could start with the fit of the subset model

$$Y = X_1 \beta_1 + \epsilon \qquad (1)$$

and then smoothly add X_2 according to some control parameter $\lambda \in [0, 1]$ with $\lambda = 0$ corresponding to (1) and $\lambda = 1$ corresponding to the full model:

$$Y = X_1 \beta_1 + X_2 \beta_2 + \epsilon. \qquad (2)$$

The procedure consists in plotting

$$\{\hat{Y}(\lambda), e(\lambda)\},$$

where $\hat{Y}(\lambda)$ are the fitted values and $e(\lambda)$ are the corresponding residuals obtained when the control parameter is equal to λ. A similar device of plotting $\{e(\lambda), \lambda\}$ and $\{\hat{Y}(\lambda), \lambda\}$ was suggested by Pregibon [4], who call it a *traceplot* or λ-*trace*.

Cook and Weisberg [2] extend ARES for generalized linear models* (see, e.g., McCullagh and Nelder [3]) and provide details on the available software in the LISP-STAT code.

REFERENCES

1. Cook, R. D. and Weisberg, S. (1989). Regression diagnostics with dynamic graphs, Response. *Technometrics*, **31**, 309–311. (Original text, *ibid.*, 273–290.)

2. Cook, R. D. and Weisberg, S. (1994). ARES plots for generalized linear models. *Comput. Statist. and Data Anal.*, **17**, 303–315.

3. McCullagh, P. and Nelder, J. A. (1989). *Generalized Linear Models*, 2nd ed. Chapman and Hall, London.

4. Pregibon, D. (1989). Discussion of Cook, D. R. and Weisberg, S. (1989). *Technometrics*, **31**, 297–301.

See also GRAPHICAL REPRESENTATION OF DATA and REGRESSION DIAGNOSTICS.

ARFWEDSON DISTRIBUTION

This is the distribution of the number (M_0, say) of zero values among k random variables $N_1, N_2, \ldots, N_k$ having a joint equiprobable multinomial distribution*. If the sum of the N's is n, then

$$\Pr[M_0 = m] = \binom{k}{m} \sum_{i=0}^{m} (-1)^i \binom{m}{i} \left(\frac{m-i}{k}\right)^n$$

$$= k^{-n} \binom{k}{m} \Delta^m \mathbf{0}^n$$

$$(m = 0, 1, \ldots, k-1).$$

It is a special occupancy distribution*, being the number of cells remaining empty after n balls have been distributed randomly among k equiprobable cells.

FURTHER READING

Arfwedson, G. (1951). *Skand. Aktuarietidskr,* **34**, 121–132.

Johnson, N. L., Kotz, S. (1969). *Discrete Distributions.* Wiley, New York, p. 251.

Johnson, N. L., Kotz, S. and Kemp, A. W. (1992). *Discrete Distributions.* (2nd ed.) Wiley, New York, p. 414.

See also DIFFERENCE OF ZERO; MULTINOMIAL DISTRIBUTIONS; and OCCUPANCY PROBLEMS.

ARIMA MODELS. See AUTOREGRESSIVE–MOVING AVERAGE (ARMA) MODELS; BOX–JENKINS MODEL

ARITHMETIC MEAN

The arithmetic mean of n quantities $X_1, X_2, \ldots, X_n$ is the sum of the quantities divided by their number n. It is commonly denoted by $\overline{X}$, or when it is desirable to make the dependence upon the sample size explicit, by $\overline{X}_n$. Symbolically,

$$\overline{X} = (1/n) \sum_{i=1}^{n} X_i.$$

An alternative form of the definition, useful for iterative calculations, is

$$\overline{X}_{n+1} = \overline{X}_n + (X_{n+1} - \overline{X}_n)/(n+1);$$

$$\overline{X}_0 = 0.$$

Historically, the arithmetic mean is one of the oldest algorithmic methods for combining discordant measurements in order to produce a single value, although even so, few well-documented uses date back to before the seventeenth century [3,6,8]. Today it is the most widely used and best understood data summary in all of statistics. It is included as a standard function on all but the simplest hand-held calculators, and it enjoys the dual distinction of being the optimal method of combining measurements from several points of view, and being the least robust* such method according to others. Our discussion begins with the consideration of its distributional properties.

DISTRIBUTION

Suppose that the X_i are independent, identically distributed (i.i.d.) with CDF $F_X(x)$, mean μ, and variance σ^2. The distribution of $\overline{X}$ may be quite complicated, depending upon F_X, but it will always be true that

$$E[\overline{X}] = \mu, \tag{1}$$

$$\mathrm{var}(\overline{X}) = \sigma^2/n, \tag{2}$$

whenever these moments exist. For some distributions $F_X, \overline{X}$ possesses a distribution of simple form. If F_X is $N(\mu, \sigma^2)$, then $\overline{X}$ has a normal $N(\mu, \sigma^2/n)$ distribution*. If F_X is a Bernoulli (p) distribution*, then $n\overline{X}$ has a binomial bin(n, p) distribution*. If F_X is a Poisson (λ) distribution*, $n\overline{X}$ has a Poisson $(n\lambda)$ distribution. If F_X is a Cauchy distribution*, then $\overline{X}$ has the same Cauchy distribution. If F_X is a gamma distribution* or a chi-squared distribution*, then $\overline{X}$ has a gamma distribution*. The exact density of $\overline{X}$ when F_X is a uniform distribution* was derived as long ago as 1757 by T. Simpson [6,8]; it is a complicated case of a B-spline*. For further information about the distribution of $\overline{X}$ for different parametric families F_X, see the entries under those distributions.

Since $\overline{X}$ is a sum of independent random variables, many aspects of its distribution are amenable to study by using generating functions*. In particular, the characteristic function* of $\overline{X}$ can be given in terms of the characteristic function $\phi(t)$ of F_X as $\phi_{\overline{X}}(t) = [\phi(n^{-1}t)]^n$.

Much is known about the asymptotic behavior of $\overline{X}_n$ for large n. See LAWS OF LARGE NUMBERS; LIMIT THEOREM, CENTRAL. For example, the Kolmogorov strong law of large numbers* states that $\overline{X}_n \overset{a.s.}{\to} \mu$ as $n \to \infty$ if and only if $E|X_i| < \infty$ and $EX_i = \mu$. The classical central limit theorem states that if $\sigma^2 < \infty$, $n^{1/2}(\overline{X}_n - \mu)$ is asymptotically distributed as $N(0, \sigma^2)$. The degree of approximation that can be expected from this asymptotic result has received considerable study, although even the strongest available results are usually too pessimistic to be practically useful. See e.g., ASYMPTOTIC NORMALITY.

Various refinements to the normal approximation are also available. See CORNISH–FISHER AND EDGEWORTH EXPANSIONS.

If the measurements X_i are not independent, then of course the distribution of $\overline{X}$ may be more complicated. If the X_i form a stationary* sequence with $E[X_i] = \mu$ and $\text{var}(X_i) = \sigma^2$, then $E[\overline{X}] = \mu$, but $\text{var}(\overline{X})$ may be either larger or smaller than in the independent case. For example, if the X_i follow a first-order moving average process [with $X_i = \mu + a_i + \theta a_{i-1}$, where the a_i are i.i.d. with $E[a_i] = 0$, $\text{var}(a_i) = \sigma_a^2$, and $\sigma^2 = (1 + \theta^2)\sigma_a^2$], then $\rho = \text{corr}(X_i, X_{i+1}) = \theta/(1 + \theta^2)$ varies from -0.5 to 0.5, and

$$\text{var}(\overline{X}) = (\sigma^2/n)[1 + 2(1 - (1/n))\rho], \qquad (3)$$

which varies from σ^2/n^2 to $2\sigma^2/n - \sigma^2/n^2$. See TIME SERIES. If the measurements X_i are determined by sampling at random* from a finite population* of size N without replacement, then

$$\text{var}(\overline{X}) = \frac{\sigma^2}{n}\left(\frac{N - n}{N - 1}\right), \qquad (4)$$

where σ^2 is the population variance. See FINITE POPULATIONS, SAMPLING FROM.

STATISTICAL PROPERTIES

The arithmetic mean $\overline{X}$ is usually considered as an estimator of a population mean μ: if the X_i are i.i.d. with CDF $F_X(x)$ and finite mean μ, then $\overline{X}$ is an unbiased estimator* of μ regardless of F_X (in this sense it is a nonparametric estimator of μ). The same is true if the X_i are sampled at random without replacement from a finite population with mean μ. Chebyshev's inequality* tells us $\Pr[|\overline{X} - \mu| > \epsilon] \leqslant \text{var}(\overline{X})/\epsilon^2$, so $\overline{X}$ will in addition be a consistent estimator* of μ as long as $\text{var}(\overline{X}) \to 0$ as $n \to \infty$, which will

hold, for example, if $\sigma^2 < \infty$ in (2), (3), and (4). Laws of large numbers* provide several stronger consistency results. For the case of i.i.d. X_i with finite variances, (2) can be interpreted as stating that the precision of $\overline{X}$ as an estimator of μ increases as the square root of the sample size. In the i.i.d. case, the nonparametric unbiased estimator of $\text{var}(\overline{X})$ is s^2/n, where s^2 is the sample variance* $(n - 1)^{-1}\sum(X_i - \overline{X})^2$.

The arithmetic mean enjoys several optimality properties beyond unbiasedness and consistency. It is a special case of a least-squares estimator*; $\sum(X_i - c)^2$ is minimized by $c = \overline{X}$. As such, $\overline{X}$ has all the properties of least-squares estimators: The Gauss–Markov theorem* ensures that $\delta = \overline{X}$ minimizes $E(\delta - \mu)^2$ within the class of all linear unbiased estimators; when F_X is normal $N(\mu, \sigma^2)$, $\overline{X}$ is the maximum-likelihood estimator of μ; and from a Bayesian* perspective, $\overline{X}$ is at the maximum of the posterior distribution* of μ for a uniform prior distribution*. (In fact, C. F. Gauss* proved in 1809 that this later property of $\overline{X}$ characterized the normal within the class of all location parameter families.)

The optimality of $\overline{X}$ as an estimator of a parameter of course depends upon the parametric family in question, but in the i.i.d. case there are several examples of F_X [including $N(\mu, \sigma_0^2)$, σ_0^2 known; Poisson (μ); Bernoulli (μ); the one-parameter exponential (μ)*], where $\overline{X}$ is the maximum-likelihood estimator and a minimal sufficient* statistic. For a simple example of how the optimality of $\overline{X}$ depends upon the distribution and the criterion, however, see ref. 10.

Much attention to the arithmetic mean in recent years has focused upon its lack of robustness*, in particular, its sensitivity to aberrant measurements such as are likely to occur when sampling from heavy-tailed distributions. The most commonly noted example of this was noted as early as 1824 by Poisson: if F_X is a Cauchy distribution*, then $\overline{X}$ has the same Cauchy distribution (and thus no mean or variance no matter how large n is). Estimators such as the sample median* perform much more efficiently in this case. Indeed, even a small amount of heavy-tailed contamination* can in principle drastically effect the efficiency of $\overline{X}$ as an estimator of μ.

Opinion is divided on the question of whether such contamination occurs in practice with a severity or frequency to dictate drastic remedy; see refs. 1, 5, and 8 for an airing of these and related issues. Meanwhile, a vast array of estimators have been devised that are less sensitive than $\overline{X}$ to extreme measurements; the simplest of these (the Winsorized mean* and the trimmed mean*) are, in fact, equivalent to the calculation of the arithmetic mean of a modified sample.

The arithmetic mean is frequently used as a test statistic for testing hypotheses about the mean μ, often in the form of Student's t statistic*, $t = \sqrt{n}(\overline{X} - \mu_0)/s$, and as the basis of confidence intervals* for μ.

RELATIONSHIP TO OTHER MEANS

Two classical means, the geometric* and the harmonic*, are related simply to the arithmetic mean. If $Y_i = \ln X_i$ and $Z_i = X_i^{-1}$, the geometric mean of the X_i is given by $(\prod X_i)^{1/n} = \exp(\overline{Y})$ and the harmonic mean is $(\overline{Z})^{-1}$. Hardy et al. [5] discuss inequality relationships between these and more general mean functions, the simplest being that if all X_i are positive, then $(\overline{Z})^{-1} \leqslant \exp(\overline{Y}) \leqslant \overline{X}$. See GEOMETRIC MEAN; HARMONIC MEAN. Many other means have been related to the arithmetic mean in less mathematically precise ways. The best known such relationship is that between the arithmetic mean $\overline{X}$, the median m, and the mode* M for empirical distributions that are unimodal and skewed to the right; it is frequently true that $M \leqslant m \leqslant \overline{X}$. Furthermore, a rough rule of thumb that goes back at least to 1917 (see refs. 2, 4, 7 and 12) observes that these means often satisfy, approximately, the relationship $(\overline{X} - M) = 3(\overline{X} - m)$. The arithmetic mean may also be viewed as the expected valued or mean of the empirical distribution* which places mass $1/n$ at each X_i, a fact that points to several other characterizations of $\overline{X}$: it is the center of gravity of the X_i; it is value such that the sum of the residuals about that value is zero [$\sum(X_i - \overline{X}) = 0$]; it is a functional of the empirical* cumulative distribution function F_n,

$$\overline{X} = \int x \, dF_n(x).$$

For other definitions for which the arithmetic mean is a special case, *see* INDEX NUMBERS; ORDER STATISTICS; and ROBUST ESTIMATION.

REFERENCES

1. Andrews, D. F., Bickel, P. J., Hampel, F. R., Huber, P. J., Rogers, W. H. and Tukey, J. W. (1972). *Robust Estimates of Location: Survey and Advances.* Princeton University Press, Princeton, N.J. (This book presents the results of a Monte Carlo study of a large number of estimates, including the arithmetic mean.)

2. Doodson, A. T. (1917). *Biometrika,* **11,** 425–429.

3. Eisenhart, C. *The Development of the Concept of the Best Mean of a Set of Measurements from Antiquity to the Present Day.* ASA Presidential Address, 1971, to appear. (The early history of the arithmetic mean.)

4. Groeneveld, R. A. and Meeden, G. (1977). *Amer. Statist.,* **31,** 120–121.

5. Hardy, G. H., Littlewood, J. E. and Polya, G. (1964). *Inequalities.* Cambridge University Press, Cambridge.

6. Plackett, R. L. (1958). *Biometrika,* 45, 130–135. Reprinted in *Studies in the History of Statistics and Probability,* E. S. Pearson and M. G. Kendall, eds. Charles Griffin, London, 1970. (The early history of the arithmetic mean.)

7. Runnenburg, J. T. (1978). *Statist. Neerlandica,* **32,** 73–79.

8. Seal, H. L. (1977). *Studies in the History of Statistics and Probability,* Vol. 2, M. G. Kendall and R. L. Plackett, eds. Charles Griffin, London, Chap. 10. (Originally published in 1949.)

9. Stigler, S. M. (1977). *Ann. Statist.,* **5,** 1055–1098. (A Study of the performance of several estimates, including the arithmetic mean, on real data sets. Much of the discussion focuses on the characteristics of real data.)

10. Stigler, S. M. (1980). *Ann. Statist.,* **8,** 931–934. (An early example is discussed which shows that the arithmetic mean need not be optimal, even for samples of size 2, and even if all moments exist.)

11. Tukey, J. W. (1960). In *Contributions to Probability and Statistics,* I. Olkin et al., eds. Stanford University Press, Stanford, Calif., pp. 448–485.

12. Zwet W. R. van (1979). *Statist. Neerlandica*, **33**, 1–5.

STEPHEN M. STIGLER

ARITHMETIC PROGRESSION

This is a sequence of numbers with constant difference between successive numbers. The mth member of the sequence a_m can be expressed as

$$a_m = a_1 + (m - 1)d,$$

where d is the constant difference.

See also GEOMETRIC DISTRIBUTION.

ARITHMETIC TRIANGLE. See
COMBINATORICS

ARMA MODELS. See
AUTOREGRESSIVE–MOVING AVERAGE (ARMA) MODELS

ARRAY

This term is applied to the distribution of sample values of a variable Y, for a fixed value of another variable X. It refers especially to the frequency distribution (*see* GRAPHICAL REPRESENTATION OF DATA) formed by such values when set out in the form of a contingency table*. Such an array is formed only when the data are discrete or grouped.

See also ORTHOGONAL ARRAYS AND APPLICATIONS.

ARRAY MEAN

An array mean is the arithmetic mean* of the values of a variable Y in a group defined by limits on the values of variables $X_1, \ldots, X_k$ (an array*). It is an estimate of the regression function* of Y on $X_1, \ldots, X_k$.

See also ARRAY; LOG-LINEAR MODELS IN CONTINGENCY TABLES; and REGRESSION, POLYCHOTOMOUS.

ARS CONJECTANDI. See BERNOULLIS, THE

ARTIFICIAL INTELLIGENCE. See
STATISTICS AND ARTIFICIAL INTELLIGENCE

ASCERTAINMENT SAMPLING

Ascertainment sampling is used frequently by scientists interested in establishing the genetic basis of some disease. Because most genetically based diseases are rare, simple random sampling* will usually not provide sample sizes of affected individuals sufficiently large for a productive statistical analysis. Under ascertainment sampling, the entire family (or some other well-defined set of relatives) of a proband (i.e., an individual reporting with the disease) is sampled: we say that the family has been *ascertained through a proband*. If the disease does have a genetic basis, other family members have a much higher probability of being affected than individuals taken at random, so that by using data from such families we may obtain samples of a size sufficient to draw useful inferences. Additional benefits of sampling from families are that it yields linkage* information not available from unrelated individuals and that an allocation of genetic and environmental effects can be attempted.

The fact that only families with at least one affected individual can enter the sample implies that a conditional probability must be used in the data analysis, a fact recognized as early as 1912 by Weinberg [5]. However, the conditioning event is not that there is at least one affected individual in the family, but that the family is ascertained. These two are usually quite different, essentially because of the nature of the ascertainment process.

If the data in any family are denoted by D and the event that the family is ascertained by A, the ascertainment sampling likelihood* contribution from this family is the conditional likelihood $P(DA)/P(A)$, where both numerator and denominator probabilities depend on the number of children in the family. The major problem in practice with ascertainment sampling is that the precise nature of the sampling procedure must be known in order to calculate both numerator and denominator probabilities. In practice this procedure is often not well known, and this leads to potential biases in the estimation of genetic parameters, since while the numerator in the above probability can be written as the product P(D)P(A|D) of genetic and ascertainment parameters, the denominator cannot, thus confounding estimation of the genetic parameters with the properties of the ascertainment process.

This is illustrated by considering two commonly discussed ascertainment sampling procedures. The first is that of *complete ascertainment*, arising for example from the use of a registry of families, and sampling only from those families in the registry with at least one affected child. Here the probability of ascertainment is independent of the number of affected children. The second procedure is that of *single ascertainment*; here the probability that a family is sampled is proportional to the number of affected children. There are many practical situations where this second form of sampling arises—for example, if we ascertain families by sampling all eighth-grade students in a certain city, a family with three affected children is essentially three times as likely to be ascertained as a family with only one affected child.

These two procedures require different ascertainment corrections. For example, if the children in a family are independently affected with a certain disease, each child having probability p of being affected, the probability of ascertainment of a family with s children under complete ascertainment is $1 - (1 - p)^s$ and is proportional to p under single ascertainment. The difference between the two likelihoods caused by these different denominators can lead to significant bias in estimation of p if one form of ascertainment is assumed when the other is appropriate.

In practice the description of the ascertainment process is usually far more difficult than this, since the actual form of sampling used is seldom clear-cut (and may well be neither complete nor single ascertainment). For example, age effects are often important (an older child is more likely to exhibit a disease than a younger child, and thus more likely to lead to ascertainment of the family), different population groups may have different social customs with respect to disease reporting, the relative role of parents and children in disease reporting is often not clear-cut (and depends on the age of onset of the disease), and the most frequent method of obtaining data (from families using a clinic in which a physician happens to be collecting disease data) may not be described well by any obvious sampling procedure.

Fisher [3] attempted to overcome these problems by introducing a model in which complete and single ascertainment are special cases. In his model it is assumed that any affected child is a proband with probability π and that children act independently with respect to reporting behavior. Here π is taken as an unknown parameter; the probability that a family with i affected children is ascertained is $1 - (1 - \pi)^i$, and the two respective limits $\pi = 1$ and $\pi \to 0$ correspond to complete and single ascertainment, respectively. However, the assumptions made in this model are often unrealistic: children in the same family will seldom act independently in reporting a disease, and the value of π will vary from family to family and will usually depend on the birth order of the child. Further, under the model, estimation of genetic parameters cannot be separated from estimation of π, so that any error in the ascertainment model will imply biases in the estimation of genetic parameters.

Given these difficulties, estimation of genetic parameters using data from an ascertainment sampling procedure can become a significant practical problem. An *ascertainment-assumption-free* approach which largely minimizes these difficulties is the following. No specific assumption is made about the probability $\alpha(i)$ of ascertaining a family having i affected children. [For complete ascertainment, $\alpha(i)$ is assumed to be independent of i and, for single ascertainment, to be proportional to i—but we now regard $\alpha(i)$ as an unknown parameter.] The denominator in the likelihood* contribution from any ascertained family is thus $\sum_i p_i \alpha(i)$, where p_i, the probability that a family has i affected children, is a function only of genetic parameters. The numerator in the likelihood contribution is $P(D) = \alpha(i)$. The likelihood is now maximized jointly with respect to the genetic parameters and the $\alpha(i)$.

When this is done, it is found that estimation of the genetic parameters separates out from estimation of the $\alpha(i)$, and that the former can be estimated directly by using as the likelihood contribution $P(D)/P(i)$ from a family having i affected children. Estimation of ascertainment parameters is not necessary, and the procedure focuses entirely on genetic parameters, being unaffected by the nature of the ascertainment process.

More generally, the data D in any family can be written in the form $D = \{D_1, D_2\}$, where it is assumed that only D_1 affects the probability of ascertainment. Then the likelihood contribution used for such a family is $P(D_1, D_2)/P(D_1)$. This procedure [1] gives asymptotically unbiased parameter estimators no matter what the ascertainment process—all that is required is that the data D_1 that is "relevant to ascertainment" can be correctly defined.

These estimators have higher standard error* than those arising if the true ascertainment procedure were known and used in the likelihood leading to the estimate, since when the true ascertainment procedure is known, this procedure conditions on more data than necessary, leaving less data available for estimation. The increase in standard error can be quantified using information concepts. In practice, the geneticist must choose between a procedure giving potentially biased estimators by using an incorrect ascertainment assumption and the increase in standard error in using the ascertainment-assumption-free method.

Conditioning not only on D_1 but on further parts of the data does not lead to bias in the estimation procedure, but will lead to increased standard errors of the estimate by an amount which can be again quantified by information concepts. Further conditioning of this type sometimes arises with continuous data. For such data the parallel with the dichotomy "affected/not affected" might be "blood pressure not exceeding T/blood pressure exceeding T," for some well-defined threshold T, so that only individuals having blood pressure exceeding T can be probands. To simplify the discussion, suppose that only the oldest child in any family can be a proband. Should the likelihood be conditioned by the probability element $f(x)$ of the observed blood pressure x of this child, or should it be conditioned by the probability $P(X \geqslant T)$ that his blood pressure exceeds T? The correct ascertainment correction is always the probability that the family is ascertained, so the latter probability is correct. In using the former, one conditions not only on the event that the family is ascertained, but on the further event that the blood pressure is x. Thus no bias arises in this case from using the probability element $f(x)$, but conditioning on further information (the actual value x) will increase the standard error of parameter estimates.

The above example is unrealistically simple. In more realistic cases conditioning on the observed value (or values) often introduces a bias, since when any affected child can be a proband, $f(x)/P(X \geqslant T)$ is a density function only under single ascertainment. Thus both to eliminate bias and to decrease standard errors, the appropriate conditioning event is that the family is ascertained.

A further range of problems frequently arises when large pedigrees, rather than families, are ascertained. For example if, as above, sampling is through all eighth-grade students in a certain city, and if any such student in the pedigree (affected or otherwise) is not observed, usually by being in a part of the pedigree remote from the proband(s), then bias in parameter estimation will, in general, occur. In theory this problem can be overcome by an exhaustive sampling procedure, but in practice this is usually impossible. This matter is discussed in detail in ref. 4. A general description of ascertainment sampling procedures is given in ref. 2.

REFERENCES

1. Ewens, W. J. and Shute, N. C. E. (1986). A resolution of the ascertainment sampling problem I: theory. *Theor. Pop. Biol.*, **30**, 388–412.
2. Ewens, W. J. (1991). Ascertainment biases and their resolution in biological surveys. In *Handbook of Statistics*, vol. 8, C. R. Rao and R. Chakraborty, eds. North-Holland, New York, pp. 29–61.
3. Fisher, R. A. (1934). The effects of methods of ascertainment upon the estimation of frequencies. *Ann. Eugen.*, **6**, 13–25.

4. Vieland, V. J. and Hodge, S. E. (1994). Inherent intractability of the ascertainment problem for pedigree data: a general likelihood framework. *Amer. J. Human Genet.*, **56**, 33–43.

5. Weinberg, W. (1912). Further contributions to the theory of heredity. Part 4: on methods and sources of error in studies on Mendelian ratios in man. (In German.) *Arch. Rassen- u. Gesellschaftsbiol.*, **9**, 165–174.

See also Human Genetics, Statistics in—I; Likelihood; Linkage, Genetic; and Statistical Genetics.

W. J. E. Wens

ASIMOV'S GRAND TOUR

This is a representation of multivariate data by showing a sequence of bivariate projections of the data.

It is a technique of exploratory projection pursuit*, based on the Cramér–Wold* theorem, which asserts that the distribution of a p-dimensional random vector X is determined by the set of all one-dimensional distributions of the linear combinations $\alpha'X$; here $\alpha \epsilon R^p$ ranges through all "fixed" p-dimensional vectors. The underlying setting for construction of the Asimov grand tour [1] is the Grassmannian manifold $G_{2,p}$. This is the space of all unoriented planes in p-dimensional space. The sequence of projections should become rapidly and uniformly dense in $G_{2,p}$.

Three methods for choosing a path through $G_{2,p}$ are advocated by Asimov. These are the torus method, the at-random method, and the mixture of these two methods called the at-random walk*. The disadvantage of the grand tour is that it necessitates reviewing a large number of planes in order to find any structure. See also ref. [2].

REFERENCES

1. Asimov, D. (1985). The grand tour: A tool for viewing multi-dimensional data. *SIAM J. Sci. Stat. Comput.*, **6**(1), 128–143.

2. Buja, A., Cook, D., Asimov, D., and Harley, C. (1996). Theory and computational methods for dynamic projections in high-dimensional data visualization. *Tech. Memorandum*, Bellcore, NJ.

ASSESSMENT BIAS

LACK OF BLINDING

One of the most important and most obvious causes of assessment bias is lack of blinding. In empirical studies, lack of blinding has been shown to exaggerate the estimated effect by 14%, on average, measured as odds ratio [10]. These studies have dealt with a variety of outcomes, some of which are objective and would not be expected to be influenced by lack of blinding, for example, total mortality.

When patient reported outcomes are assessed, lack of blinding can lead to far greater bias than the empirical average. An example of a highly subjective outcome is the duration of an episode of the common cold. A cold doesn't stop suddenly and awareness of the treatment received could therefore bias the evaluation. In a placebo-controlled trial of vitamin C, the duration seemed to be shorter when active drug was given, but many participants had guessed they received the vitamin because of its taste [12]. When the analysis was restricted to those who could not guess what they had received, the duration was not shorter in the active group.

Assessments by physicians are also vulnerable to bias. In a trial in multiple sclerosis, neurologists found an effect of the treatment when they assessed the effect openly but not when they assessed the effect blindly in the same patients [14].

Some outcomes can only be meaningfully evaluated by the patients, for example, pain and well being. Unfortunately, blinding patients effectively can be very difficult, which is why active placebos are sometimes used. The idea behind an active placebo is that patients should experience side effects of a similar nature as when they receive the active drug, while it contains so little of a drug that it can hardly cause any therapeutic effect.

Since lack of blinding can lead to substantial bias, it is important in blinded trials to test whether the blinding has been compromised. Unfortunately, this is rarely done (Asbjørn Hróbjartsson, unpublished observations), and in many cases, double-blinding is little more than window dressing.

Some outcome assessments are not made until the analysis stage of the trial (see below). Blinding should, therefore, be used also during data analysis, and it should ideally be preserved until two versions of the manuscript—written under different assumptions, which of the treatments is experimental and which is control—have been approved by all the authors [8].

HARMLESS OR FALSE POSITIVE CASES OF DISEASE

Assessment bias can occur if increased diagnostic activity leads to increased diagnosis of true but harmless cases of disease. Many stomach ulcers are silent, that is, they come and go and give no symptoms. Such cases could be detected more frequently in patients who receive a drug that causes unspecific discomfort in the stomach.

Similarly, if a drug causes diarrhoea, this could lead to more digital, rectal examinations, and, therefore, also to the detection of more cases of prostatic cancer, most of which would be harmless, since many people die *with* prostatic cancer but rather few die *from* prostatic cancer.

Assessment bias can also be caused by differential detection of false positive cases of disease. There is often considerable observer variation with common diagnostic tests. For gastroscopy, for example, a kappa value of 0.54 has been reported for the interobserver variation in the diagnosis of duodenal ulcers [5]. This usually means that there are rather high rates of both false positive and false negative findings. If treatment with a drug leads to more gastroscopies because ulcers are suspected, one would therefore expect to find more (false) ulcers in patients receiving that drug. A drug that causes unspecific, nonulcer discomfort in the stomach could, therefore, falsely be described as an ulcer-inducing drug.

The risk of bias can be reduced by limiting the analysis to serious cases that would almost always become known, for example, cases of severely bleeding ulcers requiring hospital admission or leading to death.

DISEASE SPECIFIC MORTALITY

Disease specific mortality is very often used as the main outcome in trials without any discussion how reliable it is, even in trials of severely ill patients where it can be difficult to ascribe particular causes for the deaths with acceptable error.

Disease specific mortality can be highly misleading if a treatment has adverse effects that increases mortality from other causes. It is only to be expected that aggressive treatments can have such effects. Complications to cancer treatment, for example, cause mortality that is often ascribed to other causes although these deaths should have been added to the cancer deaths. A study found that deaths from other causes than cancer were 37% higher than expected and that most this excess occurred shortly after diagnosis, suggesting that many of the deaths were attributable to treatment [1].

The use of blinded endpoint committees can reduce the magnitude of misclassification bias, but cannot be expected to remove it. Radiotherapy for breast cancer, for example, continues to cause cardiovascular deaths even 20 years after treatment [2], and it is not possible to distinguish these deaths from cardiovascular deaths from other causes. Furthermore, to work in an unbiased way, death certificates and other important documents must have been completed and patients and documents selected for review without awareness of status, and it should not be possible to break the masking during any of these processes, including review of causes of death. This seems difficult to obtain, in particular, since those who prepare excerpts of the data should be kept blind to the research hypothesis [3].

Fungal infections in cancer patients with neutropenia after chemotherapy or bone-marrow transplantation is another example of bias in severely ill patients. Not only is it difficult to establish with certainty that a patient has a fungal infection and what was the cause of death; there is also evidence that some of the drugs (azole antifungal agents) may increase the incidence of bacteriaemias [9]. In the largest placebo-controlled trial of fluconazole, more deaths were reported on drug than on placebo (55 vs 46 deaths), but the authors also reported that fewer deaths were ascribed to acute systemic fungal infections (1 vs 10 patients, $P = 0.01$) [6]. However, if this subgroup result is to be believed, it would mean that fluconazole increased mortality from other causes (54 vs 36 patients, $P = 0.04$).

Bias related to classification of deaths can also occur within the same disease. After publication of positive results from a trial in patients with myocardial infarction [16], researchers at the US Food and Drug Administration found that the cause-of-death classification was "hopelessly unreliable" [15]. Cardiac deaths were classified into three groups: sudden deaths, myocardial infarction, or other cardiac event. The errors in assigning cause of death, nearly all, favoured the conclusion that sulfinpyrazone decreased sudden death, the major finding of the trial.

COMPOSITE OUTCOMES

Composite outcomes are vulnerable to bias when they contain a mix of objective and subjective components. A survey of trials with composite outcomes found that when they included clinician-driven outcomes, such as hospitalization and initiation of new antibiotics, in addition to objective outcomes such as death, it was twice as likely that the trial reported a statistically significant effect [4].

COMPETING RISKS

Composite outcomes can also lead to bias because of competing risks [13], for example, if an outcome includes death as well as hospital admission. A patient who dies cannot later be admitted to hospital. This bias can also occur in trials with simple outcomes. If one of the outcomes is length of hospital stay, a treatment that increases mortality among the weakest patients who would have had long hospital stays may spuriously appear to be beneficial.

TIMING OF OUTCOMES

Timing of outcomes can have profound effects on the estimated result, and the selection of time points for reporting of the results is often not made until the analysis stage of the trials, when possible treatment codes have been broken. A trial report of the antiarthritic drug, celecoxib, gave the impression that it was better tolerated than its comparators, but the published data referred to 6 months of follow-up, and not to 12 and 15 months, as planned, when there was little difference; in addition, the definition of the outcome had changed, compared to what was stated in the trial protocol [11].

Trials conducted in intensive care units are vulnerable to this type of bias. For example, the main outcome in such trials can be total mortality during the stay in the unit, but if the surviving patients die later, during their subsequent stay at the referring department, little may be gained by a proven mortality reduction while the patients were sedated. A more relevant outcome would be the fraction of patients who leave the hospital alive.

ASSESSMENT OF HARMS

Bias in assessment of harms is common. Even when elaborate, pretested forms have been used for registration of harms during a trial, and guidelines for their reporting have been given in the protocol, the conversion of these data into publishable bits of information can be difficult and often involves subjective judgments.

Particularly vulnerable to assessment bias is exclusion of reported effects because they are not felt to be important, or not felt to be related to the treatment. Trials that have been published more than once illustrate how subjective and biased assessment of harms can be. Both number of adverse effects and number of patients affected can vary from report to report, although no additional inclusion of patients or follow-up have occurred, and these reinterpretations or reclassifications sometimes change a nonsignificant difference into a significant difference in favor of the new treatment [7].

REFERENCES

1. Brown, B. W., Brauner, C., and Minnotte, M. C. (1993). Noncancer deaths in white adult cancer patients. *J. Natl. Cancer Inst.*, **85**, 979–987.

2. Early Breast Cancer Trialists' Collaborative Group. (2000). Favourable and unfavourable effects on long-term survival of radiotherapy for early breast cancer: an overview of the randomised trials. *Lancet*, **355**, 1757–1770.

3. Feinstein, A. R. (1985). *Clinical Epidemiology*. W. B. Saunders, Philadelphia, Pa.

4. Freemantle, N., Calvert, M., Wood, J., Eastaugh, J., and Griffin, C. (2003). Composite outcomes in randomized trials: greater precision but with greater uncertainty? *JAMA*, **289**, 2554–2559.

5. Gjørup, T., Agner, E., Jensen, L. B, Jensen, A. M., and Møllmann, K. M. (1986). The endoscopic diagnosis of duodenal ulcer disease. A randomized clinical trial of bias and interobserver variation. *Scand. J. Gastroenterol.*, **21**, 561–567.

6. Goodman, J. L., Winston, D. J., Greenfield, R. A., Chandrasekar, P. H., Fox, B., Kaizer, H., et al. (1992). A controlled trial of fluconazole to prevent fungal infections in patients undergoing bone marrow transplantation. *N. Engl. J. Med.*, **326**, 845–851.

7. Gøtzsche, P. C. (1989). Multiple publication in reports of drug trials. *Eur. J. Clin. Pharmacol.*, **36**, 429–432.

8. Gøtzsche, P. C. (1996). Blinding during data analysis and writing of manuscripts. *Control. Clin. Trials*, **17**, 285–290.

9. Gøtzsche, P. C and Johansen, H. K. (2003). Routine versus selective antifungal administration for control of fungal infections in patients with cancer (Cochrane Review). *The Cochrane Library*, Issue 3. Update Software, Oxford.

10. Jüni, P., Altman, D. G., and Egger, M. (2001). Systematic reviews in health care: assessing the quality of controlled clinical trials. *Br. Med. J.*, **323**, 42–46.

11. Jüni, P., Rutjes, A. W., and Dieppe, P. A. (2002). Are selective COX 2 inhibitors superior to traditional non steroidal anti-inflammatory drugs? *Br. Med. J.*, **324**, 1287–1288.

12. Karlowski, T. R., Chalmers, T. C., Frenkel, L. D., Kapikian, A. Z., Lewis, T. L., and Lynch, J. M. (1975). Ascorbic acid for the common cold: a prophylactic and therapeutic trial. *JAMA*, **231**, 1038–1042.

13. Lauer, M. S. and Topol, E. J. (2003). Clinical trials—multiple treatments, multiple end points, and multiple lessons. *JAMA*, **289**, 2575–2577.

14. Noseworthy, J. H., Ebers, G. C., Vandervoort, M. K., Farquhar, R. E., Yetisir, E., and Roberts, R. (1994). The impact of blinding on the results of a randomized, placebo-controlled multiple sclerosis clinical trial. *Neurology*, **44**, 16–20.

15. Temple, R. and Pledger, G. W. (1980). The FDA's critique of the anturane reinfarction trial. *N. Engl. J. Med.*, **303**, 1488–1492.

16. The Anturane Reinfarction Trial Research Group. (1980). Sulfinpyrazone in the prevention of sudden death after myocardial infarction. *N. Engl. J. Med.*, **302**, 250–256.

See also CLINICAL TRIALS and MEDICINE, STATISTICS IN.

PETER C. GØTZSCHE

ASSESSMENT OF PROBABILITIES

NORMATIVE AND DESCRIPTIVE VIEWS

Central to this entry is a person, conveniently referred to as 'you', who is contemplating a set of *propositions* $A, B, C, \ldots$. You are *uncertain* about some of them; that is, you do not know, in your current state of knowledge $\mathscr{K}$, whether they are true or false. (An alternative form of language is often employed, in which $A, B, C, \ldots$ are *events* and their *occurrence* is in doubt for you. The linkage between the two forms is provided by propositions of the form 'A has occurred'.) You are convinced that the only logical way to treat your uncertainty, when your knowledge is $\mathscr{K}$, is to assign to each proposition, or combination of propositions, a probability $\Pr[A|\mathscr{K}]$ that A is true, given $\mathscr{K}$. This probability measures the strength of your *belief* in the truth of A. The task for you is that of *assessing* your probabilities; that is, of providing numerical values. That task is the subject of this article but, before discussing it, some side issues need to be clarified.

When it is said that you think that uncertainty is properly described by probability, you do not merely contemplate assigning numbers lying between 0 and 1, the convexity rule. Rather, you wish to assign numbers that obey all three rules of the probability calculus; convexity, addition, and multiplication (*see* PROBABILITY, FOUNDATIONS OF—I). This is often expressed as saying that your beliefs must *cohere* (*see* COHERENCE—I). It is therefore clear that, in order to perform the assessment, you must understand the rules of probability and their implications. You must be familiar with the calculus. In a sense, you have set yourself a *standard*, of coherence, and wish to adhere to that standard, or norm. This is often called the *normative* view of probability. A surveyor uses the normative theory of Euclidean geometry.

In contrast to the normative is the *descriptive* view, which aims to provide a description of how people in general attempt to deal with their uncertainties. In other words, it studies how people assess the truth of propositions when they do not have the deep acquaintance with the probability calculus demanded of the normative approach, nor feel the necessity of using that calculus as the correct, logical tool. There are several types of people involved. At one extreme are children making the acquaintance of uncertainty for the first time. At the other extreme are sophisticated people who employ different rules from those of the probability calculus; for example, the rules of fuzzy logic. This entry is *not* concerned with the descriptive concept. The seminal work on that subject is Kahneman et al. [3]. A recent collection of essays is that of Wright and Ayton [8].

Knowledge gained in descriptive studies may be of value in the normative view. For example, the former have exposed a phenomenon called *anchoring*, where a subject, having assessed a probability in one state of knowledge, may remain unduly anchored to that value when additional knowledge is acquired, and not change sufficiently. The coherent subject will update probabilities by Bayes' rule of the probability calculus (*see* BAYES' THEOREM). Nevertheless, an appreciation of the dangers of anchoring may help in the avoidance of pitfalls in the assessment of the numerical values to use in the rule. In view of the central role played by Bayes' rule, the coherent approach is sometimes called *Bayesian*, at least in some contexts.

Texts on the probability calculus do not include material on the assignment of numerical values, just as texts on geometry do not discuss mensuration or surveying. Assessment is an adjunct to the calculus, as surveying is to Euclidean geometry. Yet the calculus loses a lot of its value without the numbers.

SUBJECTIVE PROBABILITY*

In the context adopted here of your contemplating uncertain propositions or events and adhering to the probability calculus, the form of probability employed is usually termed *subjective* or *personal*. Your appreciation of uncertainty may be different from mine. Subjective ideas have been around as long as probability, but it is only in the second half of the twentieth century that they have attained the force of logic. This is largely due to the work of Ramsey [6] (whose work lay for long unappreciated), Savage [7], and De Finetti [2], amongst others. In various ways, these writers showed that, starting from axioms held to be self-evident truths, the existence of numbers describing your uncertainty in a given state of knowledge could be established, and that these numbers obeyed the rules of the probability calculus. The coherent, normative view thus follows from the axioms. It is important that although the axioms adopted by these writers differ—some treating uncertainty directly, others incorporating action in the face of uncertainty—they all lead to the same conclusion of "the inevitability of probability." Logic deals with truth and falsity. Probability is an extension of logic to include uncertainty, truth and falsity being the extremes, 1 and 0.

With these preliminaries out of the way, let us turn to the central problem: how can you, adhering to the normative view, and wishing to be coherent, assign numerical values to your uncertainties?

FREQUENCY DATA

There is one situation, of common occurrence, where the assessment of a probability is rather simple. This arises when, in addition to the event A about which you are uncertain, there is a sequence $A_1, A_2, \ldots, A_n$ of similar events where the outcome is known to you. If just r of these events are true, then it seems reasonable to suppose your probability of A is about r/n. The classic example is that of n tosses of a coin, yielding r heads and $n - r$ tails, where you are uncertain about the outcome of the next toss. In these cases the frequency r/n is equated with belief*. The situation is of such common occurrence that probability has often been identified with frequency, leading to a *frequentist* view of probability. According to the logical, subjective view this is erroneous. It is the data which are frequentist, not the probability. For the near-identification of frequency and belief within the normative approach, some conditions have to be imposed on your belief structure. These have been given by De Finetti in the concept of exchangeability* amongst all the events considered. The complete identification of frequency and belief is unreasonable, as can be seen by taking the case $n = 1$, when the identification would only leave 0 and 1 as possible values of the probability for the second toss.

CLASSICAL VIEW

A second common situation where probabilities are easily assessed is where the possible outcomes of an uncertain situation, N in all, are believed by you to be equally probable. Thus your belief that a stranger was born on a specific date is about 1/365; here $N = 365$. This is the *classical* view of probability. Applied to the coin-tossing example of the last paragraph, the probability of heads is $\frac{1}{2}$. Again, the connection between the normative and classical views requires a belief judgment on your part, namely of equal belief in all N possibilities. Notice that in the birthday example, you, understanding the calculus, can evaluate your belief that, amongst 23 independent strangers, at least two share the same birthday; the probability is about $\frac{1}{2}$. Experience shows that this is typically not true in the description of people's behavior.

There remain many situations in which neither are frequency data available, nor is the classical approach applicable, and yet you have to express your belief in an uncertain proposition. An important example occurs in courts of law where the jury, acting as 'you', has to express its belief in the defendant's guilt. A lot of information is available, in the form of the evidence produced in court, but it is rarely of the type that leads to a frequentist or classical analysis. So we now turn to other methods of assessment.

SCORING RULES

One assessment tool is the scoring rule*. This is derived from a method used as an axiom system by De Finetti. If you express your belief in the occurrence of an event E by a number x, then when the status of E becomes known, the rule assigns a penalty score dependent on x and the status. A popular rule is the quadratic one $(E - x)^2$, where E is also the indicator function of the event: $E = 1(0)$ if E is true (false). With some pathological exceptions, if you attempt to minimize your total expected penalty score, you will make assertions x which are such that they, or some function of them, obey the laws of probability. With the quadratic rule, x obeys the laws. Another rule yields log odds, rather than probabilities. The classic exposition is by Savage [7]. It is possible to train people to make probability statements using scoring rules, and the method has been applied by meteorologists issuing probability weather forecasts.

COHERENCE

One thing is clear about probability assessment: it is best to study beliefs about several related events, rather than to study an event in isolation. The reason for this is that it gives an opportunity for the rules of probability to be invoked. For a single event, only the convexity rule, that probability lies in the unit interval, is relevant. The simplest case of two events, E and F, illustrates "the point." You might assess $\Pr[E], \Pr[F|E], \Pr[F|E^c]$ as three numbers, unrestricted except that they each lie in the unit interval. (E^c denotes the complement of E, and a fixed knowledge base is assumed.) The rules then imply values for $\Pr[F], \Pr[E|F]$, etc. You would then consider whether these implications are in reasonable agreement with your beliefs. If they are not, then some adjustment will need to be made to the three original values. With more events, more checks are available and therefore better assessments possible. The object is to reach a set of probabilities that are coherent. Guidance is also available on which probabilities are best to assess. For example, under some conditions, it is better to assess $\Pr[E|F]$ directly, rather than $\Pr[EF]$ and $\Pr[F]$, using their ratio as an indirect assessment [4]. Computer programs can be written wherein some probability assessments may be fed in and others either calculated or limits given to where they must lie. By De Finetti's fundamental theorem* [2], they must lie in an interval. If the interval is empty, the values fed in are incoherent and some reassessment is necessary. Since the normative view is founded on the concept of coherence, its use in numerical assessment seems essential. It is the preferred method for a jury, who should aim to be coherent with all the evidence presented to them.

ROBUSTNESS

There has been a great deal of study of the robustness of statistical procedures (see, e.g., ROBUST ESTIMATION). Suppose that you know that the procedure you are using is robust to one of the probabilities. That is, the final conclusions are not sensitive to the actual numerical value assigned to that probability. Then this helps in the assessment, because the evaluation need not be done with great care. On the other hand, if it is known that the procedure is sensitive to a value, great care should be devoted to its determination, for example, by checking its coherence with other numerical probabilities. Knowledge of robustness shows where the work on assessment should be concentrated.

BETTING

Another way to perform the assessment is for you to consider bets (see GAMBLING, STATISTICS IN). At what odds will you accept a bet on the event E? When this is repeated for several events, either coherence results, or a Dutch book*, in which you will lose money for sure, can be made against you. Implicit in this method is the assumption that you are invoking only monetary considerations and that your utility* for money is locally linear. The latter restriction might be achieved by making the amounts of money involved small. The

former is more difficult to attain. People commonly overreact by perceiving even a modest loss as a disaster or an embarrassment, and likewise are elated by a small win. A gain of $100 in a gamble is viewed differently from a gain of the same amount resulting from honest toil. Decision analysis, based on personalistic ideas, demonstrates that, to make a decision, only products of a probability and a utility matter. No decision changes if one is multiplied by c and the other divided by c. Probability and utility seem inseparable. Yet belief, as a basic concept, seems genuinely to exist in your mind, irrespective of any action you might take as a result of that belief. This conundrum awaits resolution before betting, or similar devices based on action, can be used with complete confidence for assessment.

CALIBRATION*

Suppose that you provide probabilities for a long sequence of events and, for a selected number p, we select all the events to which you assigned probability p. For realism, suppose your probabilities are only to one place of decimals. It seems reasonable to expect that a proportion p of the selected events should subsequently turn out to be true and the remainder false. If this happens for all p, you are said to be *well calibrated*. Generally, a curve of the proportion true against the stated value p is called a calibration curve. An example is again provided by meteorologists forecasting tomorrow's weather. Some studies show they are well calibrated [5].

EXPERT OPINION

Calibration may be a consideration when you need to use an expert's probability for your own purposes. For example, you have to decide whether to go on a picnic tomorrow, and you consult an expert weather forecaster, who says that there is a 90% probability that tomorrow will be fine. If the expert is thought by you to be well calibrated, then you may reasonably assess your probability of fine weather also to be 90%. But you could equally well proceed if you knew the meteorologist's calibration curve. For instance, if the meteorologist is found to be an optimist and only 75% of their predicted days at 90% turn out to be fine, you might assess your probability for tomorrow being fine as $\frac{3}{4}$.

Expert opinion has been studied in a different way by other researchers. Suppose an expert provides probability p for an event E of interest to you. How should this provision affect your probability for E? Bayes' theorem* in odds form says

$$\mathrm{Od}[E|p] = \frac{\Pr[p|E]}{\Pr[p|E^c]}\mathrm{Od}[E],$$

where Od means odds on and $\Pr[p|E]$ is your probability that, if E is true, the expert will announce p, and similarly for E false, E^c true. In the usual way, your odds are updated by multiplying by your likelihood ratio for the data p. The latter necessitates your assessment of your probabilities that the expert will announce p when the event is true, and also when it is false. Notice that this is not the calibration curve, which fixes p and examines a frequency.

Unexpected results follow from this analysis of expert opinion. Suppose several experts announce that an event has, for them, probability 0.7. Suppose further (a severe assumption) you feel the experts are independent; this is a probability judgment by you. Then calculation, following the probability calculus, suggests that your probability for the event should exceed 0.7. This is supported by the intuitive consideration that if the event were false, it would be astonishing that every expert consulted should think it more likely to be true than not. Contrast this with the agreement amongst several experts that two cities are 70 miles apart. You would unhesitatingly accept 70 as also your opinion. Probability behaves differently. A convenient reference is Cooke [1].

FUTURE WORK

It is remarkable how little attention has been paid to the measurement of probabilities. The sophisticated calculus applies to numbers that, outside the limited frequency and classical domains, have not been assessed satisfactorily. And those that have, have rarely been tested. For example, many research studies throughout science quote a significance level, stating that the null hypothesis was "significant at 5%." How many of such hypotheses have subsequently turned out to be false? Calibration might require 95%, but the few such efforts to assess accuracy in this way have been recent, and mainly appear in the medical literature. The personal view of probability would condemn them as not being sound statements of belief in the null hypotheses.

Euclidean geometry predates good mensuration by several centuries. It is to be hoped that it will not be necessary to wait that long before good assessment of beliefs fits with the other branch of mathematics, probability.

REFERENCES

1. Cooke, R. M. (1991). *Experts in Uncertainty: Opinions and Subjective Probability in Science.* Oxford University Press, Oxford.

2. De Finetti, B. (1974/5). *Theory of Probability: A Critical Introductory Treatment.* Wiley, London. (A translation from the Italian of the seminal text on the personal view of probability. Difficult, but rewarding, reading.)

3. Kahneman, D., Slovic, P., Tversky, A., eds. (1982). *Judgment under Uncertainty: Heuristics and Biases.* Cambridge University Press, Cambridge, United Kingdom.

4. Lindley, D. V. (1990). The 1988 Wald memorial lectures: the present position in Bayesian statistics. *Statist. Sci.*, **5**, 44–89. (With discussion.)

5. Murphy, A. H. and Winkler, R. L. (1984). Probability forecasting in meteorology. *J. Amer. Statist. Ass.*, **79**, 489–500.

6. Ramsey, F. P. (1926). Truth and probability. In *The Foundations of Mathematics and Other Logical Essays.* Routledge & Kegan Paul, London.

7. Savage, L. J. (1971). Elicitation of personal probabilities and expectations. *J. Amer. Statist. Ass.*, **66**, 783–801.

8. Wright, G. and Ayton, P., eds. (1994). *Subjective Probability.* Wiley, Chichester. (A valuable collection of essays on many aspects, normative and descriptive, of subjective probability.)

See also BAYESIAN INFERENCE; BAYES' THEOREM; BELIEF
FUNCTIONS; CALIBRATION—I; COHERENCE—I;
ELICITATION; SCORING RULES; SUBJECTIVE PROBABILITIES;
and SUBJECTIVE RANDOMNESS.

D. V. LINDLEY

ASSIGNABLE CAUSE

In model building, effects of certain factors
("causes") are allowed for in construction of
the model. Ideally, all causes likely to have
noticeable effects should be so represented.
Such causes are often called "assignable
causes." A better term might be "recognized
causes."

Usually, there are, in fact, effects arising from causes that are not allowed for
("assigned") in the model. It is hoped that
these will not be seriously large; they are supposed to be represented by random variation*
included in the model.

Note that not all assignable causes may
be actually used ("assigned") in the model.
In the interests of simplicity, causes with
recognized potential for effect may be omitted
if the magnitudes of the effects are judged
likely to be small.

See also STATISTICAL MODELING.

ASSOCIATION, MEASURES OF

Measures of association are numerical assessments of the strength of the statistical dependence of two or more qualitative variables.
The common measures can be divided into
measures for nominal polytomous variables*
and measures for ordinal polytomous variables*.

MEASURES FOR NOMINAL VARIABLES

The most common measures of association for
nominal variables are measures of prediction
analogous in concept to the multiple correlation coefficient* of regression analysis*.

Consider two polytomous random variables X and Y with respective finite ranges
I and J. A measure of prediction $\phi_{Y \cdot X}$ of
Y given X depends on a measure Δ of the
dispersion of a polytomous random variable.

Such a measure is always nonnegative, with
$\Delta_Y = 0$ if and only if Y is essentially constant; i.e., $\Delta_Y = 0$ if and only if for some
$j \in J$, $p_{\cdot j} = \Pr[Y = j] = 1$. The measure does
not depend on the labeling* of elements. For-
mally, one may require that $\Delta_Y = \Delta_{\sigma(Y)}$ if σ
is a one-to-one transformation from J into
a finite set J'. The added requirement is
imposed that the conditional dispersion* $\Delta_{Y \cdot X}$
of Y given X not to exceed the unconditional
dispersion Δ_Y of Y. Here $\Delta_{Y \cdot X}$ is the expected
value of $\Delta_{Y \cdot X}(X)$, and $\Delta_{Y \cdot X}(i)$ is the dispersion of Y given that $X = i \in I$. [The definition
of $\Delta_{Y \cdot X}(i)$ when $p_{i \cdot} = \Pr[X = i] = 0$ does not
matter.] The measure of prediction

$$\phi_{Y \cdot X} = 1 - \Delta_{Y \cdot X} / \Delta_Y$$

compares the conditional dispersion of Y
given X to the unconditional dispersion of Y,
just as the multiple correlation coefficient*
compares the expected conditional variance
of the dependent variable to its unconditional
variance. The measure $\phi_{Y \cdot X}$ is well defined if
Y is not essentially constant. When $\phi_{Y \cdot X}$ is
defined, $0 \leqslant \phi_{Y \cdot X} \leqslant 1$, with $\phi_{Y \cdot X} = 0$ if X and
Y are independently distributed and $\phi_{Y \cdot X} = 1$
if X is an essentially perfect predictor* of
Y, i.e., if for some function k from I to J,
$\Pr[Y = k(X)] = 1$.

Two common examples of such measures
of prediction are the λ coefficient of Guttman*
[6] and of Goodman and Kruskal [1] and the
τ-coefficient of Goodman and Kruskal* [1].
Let $p_{ij} = \Pr[X = i, Y = j]$, $i \in I, j \in J$. Then

$$\lambda_{Y \cdot X} = \frac{\left(\sum_{i \in I} \max_{j \in J} p_{ij} - \max_{j \in J} p_{\cdot j} \right)}{(1 - \max_{j \in J} p_{\cdot j})},$$

$$\tau_{Y \cdot X} = \frac{\left(\sum_{i \in I} \sum_{j \in J} p_{ij}^2 / p_{i \cdot} - \sum_{j \in J} p_{\cdot j}^2 \right)}{(1 - \sum_{j \in J} p_{\cdot j}^2)}.$$

In the last formula, 0/0 is defined as 0 to
ensure that $p_{ij}^2 / p_{i \cdot}$ is always defined.

In the case of $\lambda_{Y \cdot X}$, the measure $\Delta_Y = 1 - \max_{j \in J} p_{\cdot j}$ is the minimum probability of
error from a prediction that Y is a con-
stant k, while $\Delta_{Y \cdot X} = 1 - \sum_{i \in I} \max_{j \in J} p_{ij}$ is
the minimum probability of error from a pre-
diction that Y is a function $k(X)$ of X. In the
case of $\tau_{Y \cdot X}$, $\Delta_Y = 1 - \sum_{j \in J} p_{\cdot j}^2$ is the probability that the random variable Y' does not
equal Y, where Y' and Y are independent

and identically distributed (i.i.d.). Similarly, $\Delta_{Y \cdot X} = 1 - \sum_{i \in I} \sum_{j \in J} p_{ij}^2 / p_{i \cdot}$ is the probability that $Y' \neq Y$, where given X, Y and Y' are conditionally i.i.d.

Other measures of this type are also available. For example, Theil [18] has considered the measure

$$\eta_{Y \cdot X}$$

$$= -\sum_{i \in I} \sum_{j \in J} p_{ij} \log \left(\frac{p_{ij}}{p_{i \cdot} p_{\cdot j}} \right) \bigg/ \sum_{j \in J} p_{\cdot j} \log p_{\cdot j},$$

based on the entropy* measure $\Delta_Y = -\sum_{j \in J} p_{\cdot j} \log p_{\cdot j}$ and the conditional entropy measure

$$\Delta_{Y \cdot X} = -\sum_{i \in I} \sum_{j \in J} p_{ij} \log(p_{ij}/p_{i \cdot}).$$

In these measures, $0/0 = 0$ and $0 \log 0 = 0$.

The coefficient $\lambda_{Y \cdot X}$ has an attractively simple interpretation in terms of prediction*; however, $\lambda_{Y \cdot X}$ has the possible disadvantage that $\lambda_{Y \cdot X}$ may be 0 even if X and Y are dependent. In contrast, $\eta_{Y \cdot X}$ and $\tau_{Y \cdot X}$ are only 0 if X and Y are independent.

Partial and Multiple Association

As in Goodman and Kruskal [1,3], generalizations to cases involving three or more polytomous variables are straightforward. Consider a new polytomous variable W with finite range H. If $\Delta_{Y \cdot WX}$ denotes the conditional dispersion of Y given the polytomous vector (W, X), then the multiple association coefficient $\phi_{Y \cdot WX}$ may be defined as $1 - \Delta_{Y \cdot WX}/\Delta_Y$. The partial association of Y and W given X may then be defined as

$$\phi_{Y \cdot W \mid X} = 1 - \Delta_{Y \cdot WX}/\Delta_{Y \cdot X}.$$

Thus $\phi_{Y \cdot W \mid X}$ measures the additional predictive power of W given that X has already been used as a predictor of Y. If W and Y are conditionally independent given X, then $\phi_{Y \cdot W \mid X} = 0$. If X is not an essentially perfect predictor of Y but (W, X) is an essentially perfect predictor of Y, then $\phi_{Y \cdot W \mid X} = 1$. In general, if X is not an essentially perfect predictor of Y, one has $0 \leqslant \phi_{Y \cdot W \mid X} \leqslant 1$ and

$$1 - \phi_{Y \cdot WX} = (1 - \phi_{Y \cdot X})(1 - \phi_{Y \cdot Z \mid X}).$$

Symmetric Measures

A measure of prediction $\phi_{Y \cdot X}$ of Y given X is not generally equal to the corresponding measure $\phi_{X \cdot Y}$ for prediction of X by Y. This behavior can be contrasted with the square ρ^2 of the correlation coefficient* of two continuous random variables U and V. In the continuous case, ρ^2 measures the power of U as a predictor of V and the power of V as a predictor of U. In cases in which a symmetric measure is desired, Goodman and Kruskal [1] propose measures of the form

$$\phi_{XY} = 1 - (\Delta_{Y \cdot X} + \Delta_{X \cdot Y})/(\Delta_Y + \Delta_X).$$

For example,

$$\lambda_{XY} = \left(\sum_{i \in I} \max_{j \in J} p_{ij} + \sum_{j \in J} \max_{i \in I} p_{ij} \right.$$
$$\left. - \max_{j \in J} p_{\cdot j} - \max_{i \in I} p_{i \cdot} \right)$$
$$\cdot \left(2 - \max_{j \in J} p_{\cdot j} - \max_{i \in I} p_{i \cdot} \right)^{-1}.$$

The measure ϕ_{XY} is defined if either X or Y is not essentially constant. The coefficient ϕ_{XY} ranges between 0 and 1, with $\phi_{XY} = 0$ if X and Y are independent and $\phi_{XY} = 1$ if and only if for some functions k from I to J and m from J to I, $\Pr[Y = k(X)] = \Pr[X = m(Y)] = 1$.

Standardization

In some cases, it is desirable to standardize the marginal distributions* of X and Y before computation of a measure of association. For example, one may wish to find $\phi_{Y' \cdot X'}$, where X' has some standard reference distribution such as the uniform distribution* of I and the conditional distribution of Y' given that X' is identical to the conditional distribution of Y given X. If $p'_{i \cdot} = \Pr[X' = i]$, $i \in I$, and $p'_{\cdot j} = \sum_{i \in I} (p'_{i \cdot}/p_{i \cdot}) p_{ij}, j \in J$, where $p_{i \cdot} = 0$ only when $p'_{i \cdot} = 0$, then

$$\lambda_{Y' \cdot X'} = \frac{\left(\sum_{i \in I} (\max_{j \in J} p_{ij}) p'_{i \cdot}/p_{i \cdot} - \max_{j \in J} p'_{\cdot j} \right)}{(1 - \max_{j \in J} p'_{\cdot j})}.$$

Similarly, one may consider a measure $\phi_{Y^* \cdot X^*}$, where Y^* has the same standard marginal distributions and the conditional distribution of X^* given Y^* is the same as the conditional distribution of X given Y. More thorough standardization is also possible, as in Mosteller [13]. One may consider $\phi_{U\cdot V}$, where U and V have standard marginal distributions and $\Pr[U = i, V = j] = s_i t_j p_{ij}$, $i \in I$, $j \in J$, for some s_i, $i \in I$, and t_j, $j \in J$.

Optimal Prediction

Measures such as $\phi_{Y\cdot X}$ always have interpretations in terms of optimal prediction in the following sense. Some nonnegative and possibly infinite function $A_J(j, \mathbf{q})$ is defined for $j \in J$ and $\mathbf{q}$ in the simplex* S_J of vectors $\mathbf{q} = \langle q_j : j \in J \rangle$ with nonnegative coordinates with sum $\sum_{j \in J} q_j = 1$. This function represents a loss incurred if a probabilistic prediction $\mathbf{q}$ is made for Y and $Y = j$. The function is such that

$$d_J(\mathbf{q}) = \sum_{j \in J} q_j A_J(j, \mathbf{q}) \leqslant \sum_{j \in J} q_j A_J(j, \mathbf{q}')$$

$$(\mathbf{q}, \mathbf{q}' \in S). \quad (1)$$

The dispersion Δ_Y of Y is $d_J(\mathbf{p}_Y)$, where $\mathbf{P}_Y = \langle p_{\cdot j} : j \in J \rangle$. Thus Δ_Y is the minimum expected loss achievable by prediction of Y without knowledge of X. Similarly, $\Delta_{Y\cdot X}(i)$ is $d_J(\mathbf{p}_{Y\cdot X}(i))$, where $\mathbf{p}_{Y\cdot X}(i) = \langle p_{ij}/p_{i\cdot} : j \in J \rangle$. Thus $\Delta_{Y\cdot X}(i)$ is the minimum expected loss achievable in prediction of Y, given that it is known that $X = i$, and $\Delta_{Y\cdot X}$ is the minimum expected loss achievable in prediction of Y given that X is known.

In the case of $\lambda_{Y\cdot X}$, one may define

$$A_J(j, \mathbf{q}) = \begin{cases} 1, & j \notin B(\mathbf{q}), \\ 1 - 1/m(\mathbf{q}), & j \in B(\mathbf{q}), \end{cases}$$

where $j \in B(\mathbf{q})$ if $q_j = \max_{k \in J} q_k$ and $m(\mathbf{q})$ is the number of elements in $B(\mathbf{q})$. In the typical case in which q_j has a unique maximum at a coordinate $j = \rho(\mathbf{q})$, the penalty $A_J(j, \mathbf{q})$ is 1 for $j \neq \rho(\mathbf{q})$ and 0 otherwise. In the case of $\tau_{Y\cdot X}$, $A_J(j, \mathbf{q})$ is the squared distance $\sum_{k \in K}(\delta_{kj} - q_k)^2$, where δ_{kj} is 1 for $k = j$ and δ_{kj} is 0 for $k \neq j$, while for $\eta_{Y\cdot X}$, one has

$$A_J(j, \mathbf{q}) = -\log q_j.$$

The loss function $A_J(j, \mathbf{q})$ is almost uniquely determined by the dispersion measure d_J. If $d_J(\alpha \mathbf{q})$ is defined to be $\alpha d_J(\mathbf{q})$ for $\alpha \geqslant 0$

and $\mathbf{q} \in S_J$, then d_J is a concave function on the set O_J of vectors $\mathbf{q} = \langle q_j : j \in J \rangle$ with all coordinates nonnegative. As in Savage [16], it follows that (1) is satisfied by $A_J(j, \mathbf{q})$, $j \in J$, $\mathbf{q} \in S_J$, if and only if for all $\mathbf{q} \in S_J$ and $\mathbf{q}' \in O_J$,

$$d_J(\mathbf{q}') \leqslant d_J(\mathbf{q}) + \sum_{j \in J} A_J(j, \mathbf{q})(q_j' - q_j).$$

As in Rockafellar [15], the vector $A_J(\mathbf{q}) = \langle A_J(j, \mathbf{q}) : j \in J \rangle$ is called a supergradient of d_J at $\mathbf{q}$. Some $A_J(\mathbf{q})$ exists at each $\mathbf{q} \in S_J$. If $\mathbf{q}$ is an element in the simplex S_J with all coordinates positive and if d_J is differentiable at $\mathbf{q}$, then $A_J(j, \mathbf{q})$ must be the partial derivative at $\mathbf{q}$ of d_J with respect to q_j. Thus the $A_J(j, \mathbf{q})$, $j \in J$, are uniquely determined and continuous at almost every point $\mathbf{q}$ on the simplex S_J. For example, in the case of $\eta_{Y\cdot X}$,

$$d_j(\mathbf{q}) = -\sum_{j \in J} q_j \log q_j$$

$$+ \left(\sum_{j \in J} q_j \right) \log \left(\sum_{j \in J} q_j \right)$$

for $\mathbf{q} \in O_J$. If all q_j are positive and $\sum_{j \in J} q_j = 1$, then $A(j, \mathbf{q}) = -\log q_j$ is the partial derivative at $\mathbf{q}$ of d_J with respect to q_j.

Estimation of Measures of Prediction

Typically bivariate measures of prediction in which standardization is not involved are estimated on the basis of a contingency table* $\mathbf{n} = \langle n_{ij} : i \in I, j \in J \rangle$ with a multinomial distribution of sample size N and probabilities $\mathbf{p} = \langle p_{ij} : i \in I, j \in J \rangle$. The estimate $\hat{\mathbf{p}} = N^{-1}\mathbf{n}$ of $\mathbf{p}$ is substituted for $\mathbf{p}$ in the formulas for $\phi_{Y\cdot X}$ and ϕ_{XY}. If $n_{\cdot j} = \sum_{i \in I} n_{ij}$, $j \in J$, $n_{i\cdot} = \sum_{j \in J} n_{ij}$, $i \in I$, $\hat{\mathbf{p}}_X = \langle N^{-1}n_{i\cdot} : i \in I \rangle$, $\hat{\mathbf{p}}_Y = \langle N^{-1}n_{\cdot j} : j \in J \rangle$, $\hat{\mathbf{p}}_{Y\cdot X}(i) = \langle n_{ij}/n_{i\cdot} : j \in J \rangle$, $i \in I$, and $\hat{\mathbf{p}}_{X\cdot Y}(j) = \langle n_{ij}/n_{\cdot j} : i \in I \rangle$, $j \in J$, then

$$\hat{\phi}_{Y\cdot X} = 1 - N^{-1} \sum_{i \in I} n_{i\cdot} d_J(\hat{\mathbf{p}}_{Y\cdot X}(i))/d_J(\hat{\mathbf{p}}_Y),$$

$\hat{\phi}_{XY} = 1 - N^{-1}$

$$\times \frac{\left[\sum_{i \in I} n_{i\cdot} d_J(\hat{\mathbf{p}}_{Y\cdot X}(i)) + \sum_{j \in J} n_{\cdot j} d_I(\hat{\mathbf{p}}_{X\cdot Y}(j))\right]}{d_J(\hat{\mathbf{p}}_Y) + d_I(\hat{\mathbf{p}}_X)}.$$

For example,

$\hat{\lambda}_{Y\cdot X} =$

$$\left(\sum_{i \in I} \max_{j \in J} n_{ij} - \max_{j \in J} n_{\cdot j}\right) \Big/ \left(N - \max_{j \in J} n_{\cdot j}\right).$$

Extensions to multivariate measures are straightforward.

Normal Approximation

Normal approximations* for distributions of measures such as $\hat{\phi}_{Y\cdot X}$ and $\hat{\phi}_{XY}$ are readily obtained as in Goodman and Kruskal [3,4]. Assume that d_J is differentiable at $\mathbf{p}_Y$ and at $\mathbf{p}_{Y\cdot X}(i)$ for $i \in I' = \{i \in I : p_{i\cdot} > 0\}$. [Alternatively, it suffices if $A(j, \cdot)$ is continuous at $\mathbf{p}_Y$ whenever $p_{\cdot j} > 0$ and $A(j, \cdot)$ is continuous at $p_{Y\cdot X}(i)$, $i \in I$, whenever $p_{ij} > 0$.] Assume that Y is not essentially constant. Then $N^{1/2}(\hat{\phi}_{Y\cdot X} - \phi_{Y\cdot X})$ has the large-sample distribution $N(0, \sigma^2(\phi_{Y\cdot X}))$, where $\sigma^2(\phi_{Y\cdot X})$ is the variance of the random variable

$$H(Y|X) = [(1 - \phi_{Y\cdot X}) d_J(Y, \mathbf{p}_Y)$$
$$- d_J(Y, \mathbf{p}_{Y\cdot X}(X))]/d_J(\mathbf{p}_Y).$$

Since

$$E[H(Y|X)] = 0, \ \sigma^2(\phi_{Y\cdot X}) = \sum_{i \in I} \sum_{j \in J} p_{ij}[H(j|i)]^2.$$

For examples of formulas, see Goodman and Kruskal [3,4]. Assume, in addition, that d_I is differentiable at $\mathbf{p}_X$ and at $\mathbf{p}_{X\cdot Y}(j)$ for j such that $p_{\cdot j} > 0$. Thus $N^{1/2}(\hat{\phi}_{XY} - \phi_{XY})$ has large-sample distribution $N(0, \sigma^2(\phi_{XY}))$, where $\sigma^2(\phi_{XY})$ is the variance of

$$H(X, Y) = \{(1 - \phi_{XY})[d_I(X, \mathbf{p}_X) + d_J(Y, \mathbf{p}_Y)]$$
$$- d_I(X, \mathbf{p}_{X\cdot Y}(J)) - d_J(Y, \mathbf{p}_{Y\cdot X}(I))\}$$
$$\cdot [d_I(\mathbf{p}_X) + d_J(\mathbf{p}_Y)]^{-1}.$$

Again $E[H(X, Y)] = 0$ and $\sigma^2(\phi_{XY}) = \sum_{i \in I} \sum_{j \in J} p_{ij}[H(i, j)]^2$. Since differentiability implies continuous differentiability

in concave functions, $\sigma^2(\phi_{Y\cdot X})$ and $\sigma^2(\phi_{XY})$ possess consistent* estimates $\hat{\sigma}^2_{Y\cdot X}$ and $\hat{\sigma}^2_{XY}$ obtained by replacing p_{ij} by $\hat{p}_{ij} = N^{-1} n_{ij}$ in the relevant formulas. If $\sigma^2(\phi_{Y\cdot X}) > 0$, $0 < \alpha < 1$, and z_α is the upper $(\alpha/2)$-point of the $N(0, 1)$ distribution, then an approximate confidence interval* for $\phi_{Y\cdot X}$ of level α is

$$[\hat{\phi}_{Y\cdot X} - z_\alpha \hat{\sigma}(\phi_{Y\cdot X})/N^{1/2}, \hat{\phi}_{Y\cdot X}$$
$$+ z_\alpha \hat{\sigma}(\phi_{Y\cdot X})/N^{1/2}].$$

A similar argument applies to ϕ_{XY}.

Since $0 \leqslant \hat{\phi}_{Y\cdot X} \leqslant 1$, $\sigma(\phi_{Y\cdot X})$ must be 0 if a normal approximation applies and $\phi_{Y\cdot X}$ is 0 or 1. If all p_{ij} are positive and $\sigma(\phi_{Y\cdot X})$ is 0, then $\phi_{Y\cdot X} = 0$, for $\sigma_{Y\cdot X} = 0$ implies that $H(j|i)$ is always 0, so that

$$\Delta_Y = \sum_{j \in J} p_{\cdot j} A(j, \mathbf{p}_Y)$$
$$\leqslant \sum_{j \in J} p_{\cdot j} A(j, \mathbf{p}_{Y\cdot X}(i))$$
$$= (1 - \phi_{Y\cdot X}) \sum_{j \in J} p_{\cdot j} A(j, p_{\cdot j})$$
$$= (1 - \phi_{Y\cdot X}) \Delta_Y.$$

In the special case of $\hat{\lambda}_{Y\cdot X}$, Goodman and Kruskal [3,4] note that $\sigma^2(\lambda_{Y\cdot X})$ is defined whenever $m(\mathbf{p}_Y)$ and $m(\mathbf{p}_{Y\cdot X}(i))$, $i \in I$, are all 1, and $\sigma^2(\lambda_{Y\cdot X}) = 0$ only if $\lambda_{Y\cdot X}$ is 0 or 1. The normal approximation always applies to the estimate $\hat{\tau}_{Y\cdot X}$ and $\hat{\eta}_{Y\cdot X}$; however, a simple necessary and sufficient condition for $\sigma^2(\tau_{Y\cdot X})$ or $\sigma^2(\eta_{Y\cdot X})$ to be 0 appears difficult to find.

Sampling by Rows

An alternative sampling problem has also been considered in Goodman and Kruskal [3,4], which is particularly appropriate for a standardized measure $\phi_{Y'\cdot X'}$ in which the conditional distribution of Y' given $X' = i \in \boldsymbol{I}$ is the same as the conditional distribution of Y given X and X' has a known marginal distribution with $p'_i = \Pr[X' = i]$. Let each row $\langle n_{ij} : j \in \boldsymbol{J} \rangle$, $i \in \boldsymbol{I}$, have an independent multinomial* distribution with sample size $N_i > 0$ and probabilities $\langle p_{ij}/p_{i\cdot} : j \in \boldsymbol{J} \rangle$. For simplicity, assume that each $p_{i\cdot}$ is positive. Let $N = \sum N_i$, let N_i/N approach $p_{i\cdot}$, and let

N approach infinity. Consider the standardized estimate

$$\hat{\phi}_{Y'.X'} = 1 - \sum_{i \in I} p'_{i.} d_J(\hat{\mathbf{p}}_{Y.X}(i)) \Big/ d_J(\hat{\mathbf{p}}'_Y),$$

where $\hat{\mathbf{p}}'_Y = \langle p'_{.j} : j \in \boldsymbol{J} \rangle$, $\hat{p}'_{.j} = \sum_{i \in I} \cdot p'_{i.} n_{ij}/N_i$, $j \in \boldsymbol{J}$, and $\hat{\mathbf{p}}_{Y.X}(i) = \langle n_{ij}/N_i : j \in \boldsymbol{J} \rangle$ for $i \in \boldsymbol{I}$. Assume that d_J is differentiable at $\mathbf{p}'_Y = \langle p'_{.j} : j \in \boldsymbol{J} \rangle$ and at $\mathbf{p}_{Y.X}(i), i \in \boldsymbol{I}$. Then $N^{1/2}(\hat{\phi}_{Y'.X'} - \phi_{Y'.X'})$ has an approximate $N(0, \sigma^2(\phi_{Y'.X'}))$ distribution, where $\sigma^2(\phi_{Y'.X'})$ is the expected conditional variance

$$\sum_{i \in I} \sum_{j \in J} p_{ij} [H'(j|i)]^2 - \sum_{i \in I} \left[\sum_{j \in J} H'(j|i) p_{ij} \right]^2 \Big/ p_{i.}$$

of $H'(Y|X)$ given X. Here for $i \in I, j \in J$,

$$H'(j|i) = (p'_{i.}/p_{i.})[(1 - \phi_{Y'.X'}) A_J(Y, \mathbf{p}'_Y)$$
$$- A_J(Y, \mathbf{p}_{Y.X}(i))].$$

In the special case $p'_{i.} = p_{i.}$, one has $\phi_{Y'.X'}$ equal to the unstandardized coefficient $\phi_{Y.X}$, $H'(j|i) = H(j,i)$, and $\sigma^2(\phi_{Y'.X'}) \leqslant \sigma^2(\phi_{Y.X})$.

Clearly, $\sigma^2(\phi_{Y'.X'}) = 0$ if $\phi_{Y'.X'}$ is 0 or 1. In the case of $\lambda_{Y'.X'}$, $\sigma^2(\lambda_{Y'.X'}) = 0$ if and only if $\phi_{Y'.X'}$ is 0 or 1, as noted in Goodman and Kruskal [4]. More generally, $\sigma^2(\phi_{Y'.X'}) = 0$ implies that $\phi_{Y'.X'} = 0$ if all probabilities p_{ij} are positive. The proof is only slightly changed from the corresponding proof for $\sigma_{Y.X}$.

Older Measures

Numerous older measures of association between nominal variables are reviewed by Goodman and Kruskal [1,2] and by Kendall and Stuart [10, pp. 556–561]. The most common are based on the chi-square* statistic. They include the mean square contingency

$$\phi^2 = \sum_{i \in I} \sum_{j \in J} (p_{ij} - p_{i.} p_{.j})^2 / (p_{i.} p_{.j})$$

and the coefficient of contingency $C = [\phi^2/(1 + \phi^2)]^{1/2}$ of Pearson [14] and Tschuprow's [19] coefficient $T = [\phi^2/v^{1/2}]^{1/2}$. In the last expression $v = (r - 1)(s - 1)$, I has r elements, and J has s elements. These measures lack the functional interpretations available in the case of $\phi_{Y.X}$ or ϕ_{XY}.

MEASURES OF ASSOCIATION FOR ORDINAL POLYTOMOUS VARIABLES

The most commonly used measure of association for ordinal polytomous variables is the γ coefficient of Goodman and Kruskal [1–5]. Assume that the ranges I of X and J of Y are well ordered, so that if i and i' are in I, then $i < i'$, $i = i'$, or $i > i'$ and if j and j' are in J, then $j < j'$, $j = j'$, or $j > j'$. Let (X_1, Y_1) and (X_2, Y_2) be independently distributed pairs with the same distribution as (X, Y). Let $C = 2 \Pr[X_1 > X_2 \text{ and } Y_1 > Y_2]$ be the probability that either $X_1 > X_2$ and $Y_1 > Y_2$ or $X_1 < X_2$ and $Y_1 < Y_2$, so that (X_1, Y_1) and (X_2, Y_2) are concordant. Let $2D = 2 \Pr[X_1 > X_2 \text{ and } Y_2 > Y_1]$ be the probability that either $X_1 > X_2$ and $Y_1 < Y_2$ or $X_1 < X_2$ and $Y_1 > Y_2$, so that (X_1, Y_1) and (X_2, Y_2) are discordant. Then

$$\gamma_{XY} = \frac{C - D}{C + D}.$$

The coefficient is defined if $p_{ij} > 0$ and $p_{i'j'} > 0$ for some i, $i' \in I$ and j, $j' \in J$ with $i \neq i'$ and $j \neq j'$. One has $-1 \leqslant \gamma_{XY} \leqslant 1$, with $\gamma_{XY} = 0$ under independence of X and Y. For γ_{XY} to be 1, the nonzero p_{ij} must have an ascending staircase pattern, so that if $i < i'$, $p_{ij} > 0$, and $p_{i'j'} > 0$, then $j \leqslant j'$, while if $j < j'$, $p_{ij} > 0$, and $p_{i'j'} > 0$, then $i \leqslant i'$. Similarly, γ_{XY} can only be -1 if the nonzero p_{ij} have a descending staircase pattern. In the special case in which I and J have the two elements 1 and 2, γ_{XY} is the *coefficient of association* $(p_{11}p_{22} - p_{12}p_{21})/(p_{11}p_{22} + p_{12}p_{21})$ of *Yule* [20]. In this special case, $\gamma_{XY} = 0$ if and only if X and Y are independent, $\gamma_{XY} = 1$ only if $\Pr[X = Y] = 1$, and $\gamma_{XY} = -1$ only if $\Pr[X = Y] = 0$.

The measure γ_{XY} only considers pairs (X_1, Y_1) and (X_2, Y_2) in which $X_1 \neq X_2$ and $Y_1 \neq Y_2$. An alternative approach by Somers [17] considers all pairs with just $X_1 \neq X_2$. One obtains the asymmetric coefficient

$$\gamma_{Y.X} = (C - D)/\Pr[X_1 \neq X_2].$$

Again $-1 \leqslant \gamma_{Y.X} \leqslant 1$, with $\gamma_{Y.X} = 0$ if X and Y are independent. The coefficient $\gamma_{Y.X}$ can only be -1 (or 1) if γ_{XY} is -1 (or 1) and if for each $j \in J$, p_{ij} is positive for no more than one $i \in \boldsymbol{I}$. For further variants on γ_{XY}, see Kendall and Stuart [10, pp. 561–565].

Estimation of γ_{XY} and $\gamma_{Y \cdot X}$ is straightforward given a table $\mathbf{n} = \langle n_{ij} : i \in I, j \in J \rangle$ with a multinomial distribution with sample size N and probabilities $\mathbf{p} = \langle p_{ij} : i \in I, j \in J \rangle$. Let $\hat{C}$ be the sum of all products $2n_{ij}n_{i'j'}/N^2$ such that $i < i'$ and $j < j'$, and let $\hat{D}$ be the sum of all products $2n_{ij}n_{i'j'}/N^2$ such that $i < i'$ and $j < j'$. Then $\hat{\gamma}_{XY} = (\hat{C} - \hat{D})/(\hat{C} + \hat{D})$ and $\hat{\gamma}_{Y \cdot X} = (\hat{C} - \hat{D})/[1 - \sum_{i \in I}(n_{i \cdot}/N)^2]$. As noted in Goodman and Kruskal [3,4] $N^{1/2}(\hat{\gamma}_{XY} - \gamma_{XY})$ has an approximate $N(0, \sigma^2(\phi_{XY}))$ distribution, with

$$\sigma^2(\phi_{XY}) = \frac{16}{(C+D)^4} \sum_{i \in I} \sum_{j \in J} p_{ij}(CS_{ij} - DR_{ij})^2$$

$$\leqslant 2(1 - \gamma_{XY}^2)/(C+D),$$

$$S_{ij} = \Pr[X > i \text{ and } Y < j]$$
$$+ \Pr[X < i \text{ and } Y > j],$$

$$R_{ij} = \Pr[X > i \text{ and } Y > j]$$
$$+ \Pr[X < i \text{ and } Y < j].$$

Similarly, $N^{1/2}(\hat{\gamma}_{Y \cdot X} - \gamma_{Y \cdot X})$ has an approximate $N(0, \sigma^2(\gamma_{Y \cdot X}))$ distribution with $E = \Pr[X_1 \neq X_2, Y_1 \neq Y_2]$ and

$$\sigma^2(\gamma_{Y \cdot X}) = \frac{4}{(C+D+E)^4} \sum_{i \in I} \sum_{j \in J} p_{ij}$$
$$\cdot [(C-D)(1-p_{i \cdot})$$
$$- (C+D+E)(S_{ij} - R_{ij})]^2.$$

One has $\sigma^2(\gamma_{XY}) = 0$ if $|\gamma_{XY}| = 1$ and $\sigma^2(\gamma_{Y \cdot X}) = 0$ if $|\gamma_{Y \cdot X}| = 1$. If all p_{ij} are positive, then $\sigma^2(\gamma_{XY}) > 0$. In the special case of $\gamma_{XY} = 0$, one has $\gamma_{Y \cdot X} = 0$,

$$\sigma^2(\gamma_{XY}) = \frac{4}{(C+D)^2} \sum_{i \in I} \sum_{j \in J} p_{ij}(S_{ij} - R_{ij})^2,$$

$$\sigma^2(\gamma_{Y \cdot X}) = \frac{4}{(C+D+E)^2} \sum_{i \in I} \sum_{j \in J} p_{ij}$$
$$\times (S_{ij} - R_{ij})^2.$$

Kendall's τ^*, Spearman's ρ_S^*, and Goodman and Kruskal's γ^*

In contrast to the nominal measures of association, Goodman and Kruskal's γ coefficient remains well-defined if the respective ranges

I and J of X and Y are infinite and $\Pr[X = i] = \Pr[Y = j] = 0$ for any $i \in I$ and $j \in J$. The coefficient γ_{XY} is then Kendall's [8] τ measure $\tau_k = C - D$. It remains true that $-1 \leqslant \tau_k \leqslant 1$, with $\tau_k = 0$ when X and Y are independent. Estimation of τ_k is, however, best described in terms of independent pairs $(X_t, Y_t), 1 \leqslant t \leqslant N$, with common distribution (X, Y). Then *Kendall's τ statistic $\hat{\tau}_k$* is $2(N_c - N_d)/[N(N-1)]$, where there are N_c s and t with $1 \leqslant s < t \leqslant N$ such that (X_s, Y_s) and (X_t, Y_t) are concordant $(X_s < X_t$ and $Y_s < Y_t$ or $X_t < X_s$ and $Y_t < Y_s)$ and there are N_d s and t with $1 \leqslant s < t \leqslant N$ such that (X_s, Y_s) and (X_t, Y_t) are discordant $(X_s < X_t$ and $Y_s > Y_t$ or $X_t < X_s$ and $Y_t > Y_s)$. As N becomes large $N^{1/2}(\hat{\tau}_k - \tau_k)$ has an approximate $N(0, \sigma^2(\tau_k))$ distribution. *See* KENDALL'S TAU. Here $\sigma^2(\tau_k) = 16(F - C^2)$ and

$$F = \Pr[X_1 > X_2, X_1 > X_3, Y_1 > Y_2, Y_1 > Y_3]$$
$$+ \Pr[X_1 < X_2, X_1 < X_3, Y_1 < Y_2, Y_1 < Y_3]$$

is the probability that both (X_2, Y_2) and (X_3, Y_3) are concordant with (X_1, Y_1). If X and Y are independent, then $\sigma^2(\tau_k) = 4/9$. See Hoeffding [7] and Kruskal [12] for details.

Closely related to Kendall's τ is the *Spearman rank correlation coefficient* r_s. Assume that all X_t are distinct and all Y_t are distinct. Let R_t be the number of s with $X_s \leqslant X_t$, and let S_t be the number of t with $Y_s \leqslant Y_t$. Then r_s is the sample correlation of the pairs $(R_t, S_t), 1 \leqslant t \leqslant N$. The statistic r_s provides a measure

$$\rho_s = 6\{\Pr[X_1 > X_2, Y_1 > Y_3] - 1\}$$

of the probability that (X_1, Y_1) and (X_2, Y_3) are concordant. An alternative unbiased estimate of ρ_s is $\hat{\rho}_s = [(n+1)/(n-2)]r_s - [3/(n-2)]\hat{\tau}_k$, which has been termed the *unbiased grade coefficient* by Konijn [11]. One has $-1 \leqslant \rho_s \leqslant 1$, with $\rho_s = 0$ under independence of X and Y. Both $N^{1/2}(\hat{\rho}_s - \rho_s)$ and $N^{1/2}(r_s - \rho_s)$ have limiting distribution $N(0, \sigma^2(\rho_s))$ as N becomes large. The formula for $\sigma^2(\rho_s)$ has been given by Hoeffding [7]. Since it is somewhat complicated, it will be omitted here in the general case. Under independence of X and Y, $\sigma^2(\rho_s) = 1$.

For further details concerning these and related measures, see Kendall [9] and

Kruskal [12]. (*See* also KENDALL'S TAU—I, SPEARMAN RANK CORRELATION COEFFICIENT, and GOODMAN–KRUSKAL TAU AND GAMMA)

NUMERICAL EXAMPLE

To illustrate results, consider Table 1, which can be found in Goodman and Kruskal [1], among many other references. Let X refer to eye color and let Y refer to hair color.

Some resulting estimated measures of association are listed in Table 2. In the case of the measures for ordered variables, eye color is ordered from blue to brown and hair color is ordered from fair to black.

Asymptotic standard deviations are based on the assumption that the counts in Table 1 have a multinomial distribution. The asymmetry in X and Y and the large variations in the sizes of measures are to be expected. As noted as early as Goodman and Kruskal [1], instincts developed from regression analysis* are not necessarily appropriate for assessment of the size of measures of association for ordinal or nominal polytomous variables.

REFERENCES

1. Goodman, L. A. and Kruskal, W. H. (1954). *J. Amer. Statist. Ass.*, **49**, 732–764. (Goodman and Kruskal's λ, τ, and γ are defined and the basic criteria for measures of association are presented.)

2. Goodman, L. A. and Kruskal, W. H. (1959). *J. Amer. Statist. Ass.*, **54**, 123–163. (A valuable historical review and bibliography are provided.)

3. Goodman, L. A. and Kruskal, W. H. (1963). *J. Amer. Statist. Ass.*, **58**, 310–364. (Asymptotic distributions are obtained for estimates of Goodman and Kruskal's λ, τ, and γ. Further results appear in their 1972 paper.)

4. Goodman, L. A. and Kruskal, W. H. (1972). *J. Amer. Statist. Ass.*, **67**, 415–421.

5. Goodman, L. A. and Kruskal, W. H. (1979). *Measures of Association for Cross Classifications*. Springer-Verlag, New York. (A volume consisting of Goodman and Kruskal's four papers on measures of association, indicated above.)

6. Guttman, L. (1941). In *The Prediction of Personal Adjustment* (Bull. 48), P. Horst et al., eds. Social Science Research Council, New York, pp. 253–318.

7. Hoeffding, W. (1948). *Ann. Math. Statist.*, **19**, 293–325. (The methods presented here apply to numerous problems involving ordinal data.)

8. Kendall, M. G. (1938). *Biometrika*, **30**, 277–283.

9. Kendall, M. G. (1962). *Rank Correlation Methods*, 3rd ed. Charles Griffin, London. (A standard text on rank correlation statistics.)

10. Kendall, M. G. and Stuart, A. (1967). *The Advanced Theory of Statistics*, 2nd ed. Vol. 2. Charles Griffin, London.

Table 2. Estimated Measures of Association for Table 1

Measure	Estimate	Estimated Asymptotic Standard Deviation of Estimate
$\lambda_{Y \cdot X}$	0.192	0.012
$\lambda_{X \cdot Y}$	0.224	0.013
λ_{XY}	0.208	0.010
$\tau_{Y \cdot X}$	0.081	0.005
$\tau_{X \cdot Y}$	0.089	0.005
τ_{XY}	0.085	0.005
$\eta_{Y \cdot X}$	0.075	0.004
$\eta_{X \cdot Y}$	0.085	0.005
η_{XY}	0.080	0.004
γ_{XY}	0.547	0.013
$\gamma_{X \cdot Y}$	0.346	0.009
$\gamma_{Y \cdot X}$	0.371	0.010

Table 1.

Eye Color Group	Hair Color Group				Total
	Fair	Red	Brown	Black	
Blue	1768	47	807	189	2811
Gray or green	946	53	1387	746	3132
Brown	115	16	438	288	857
Total	2829	116	2632	1223	6800

11. Konijn, H. S. (1956). *Ann. Math. Statist.*, **27**, 300–323.

12. Kruskal, W. H. (1958). *J. Amer. Statist. Ass.*, **53**, 814–861. (A very helpful review.)

13. Mosteller, F. (1968). *J. Amer. Statist. Ass.*, **63**, 1–28.

14. Pearson, K. (1904). Mathematical Contributions to the Theory of Evolution. XIII. On the Theory of Contingency and Its Relation to Association and Normal Correlation. *Drapers' Company Res. Mem., Biometric Ser. 1.* (Pearson introduces classical association measures related to the chi-square statistic* and relates them to the bivariate normal distribution.*)

15. Rockafellar, R. T. (1970). *Convex Analysis.* Princeton University Press, Princeton, N.J.

16. Savage, L. J. (1971). *J. Amer. Statist. Ass.*, **66**, 783–801.

17. Somers, R. H. (1962). *Amer. Soc. Rev.*, **27**, 799–811.

18. Theil, H. (1970). *Amer. J. Sociol.*, **76**, 103–154.

19. Tschuprow, A. A. (1925). *Grundbegriffe und Grundprobleme der Korrelationstheorie.* Teubner, Leipzig.

20. Yule, G. U. (1900). *Philos. Trans. R. Soc. Lond. A*, **194**, 257–319. (A valuable early paper on measurement of association of dichotomous variables.)

See also CATEGORICAL DATA; CORRELATION; DEPENDENCE, MEASURES AND INDICES OF; GOODMAN–KRUSKAL TAU AND GAMMA; and LOG-LINEAR MODELS IN CONTINGENCY TABLES.

S. J. HABERMAN

ASTRONOMY, STATISTICS IN

Perhaps more than other physical sciences, astronomy is frequently statistical in nature. The objects under study are inaccessible to direct manipulation in the laboratory. The astronomer is restricted to observing a few external characteristics of objects populating the Universe, and inferring from these data their properties and underlying physics. From the seventeenth through nineteenth centuries, European astronomers were engaged in the application of Newtonian theory to the motions of bodies in the solar system. This led to discussions of the statistical treatment of scientific data, and played a critical role in the development of statistical theory. The twentieth century has seen

remarkable success in the applications of electromagnetism and quantum mechanics* to heavenly bodies, leading to a deep understanding of the nature and evolution of stars, and some progress in understanding galaxies and various interstellar and intergalactic gaseous media. Statistical theory has played a less important role in these advances of modern astrophysics. However, the last few years have seen some reemergence of interest in statistical methodology to deal with some challenging data analysis problems. Some examples of these contemporary issues are presented.

EARLY HISTORY

Celestial mechanics in the eighteenth century, in which Newton's law of gravity was found to explain even the subtlest motions of heavenly bodies, required the derivation of a few interesting quantities from numerous inaccurate observations. As described in detail by Stigler [40], this required advances in the understanding of statistical inference* and error distributions. Mayer, in his 1750 study of lunar librations, suggested a procedure of reconciling a system of 27 inconsistent linear equations in three unknowns by solving the equations in groups. Laplace*, in a 1787 analysis of the influence of Jupiter's gravity on Saturn's motion, suggested a more unified approach that led to Legendre's invention of the least-squares method in an 1805 study of cometary orbits. Shortly thereafter, in an 1809 monograph on the mathematics of planetary orbits, Gauss* first presented the normal (or Gaussian) distribution of errors in overdetermined systems of equations using a form of Bayes' theorem*, though the actual derivation was flawed.

Many other individuals also contributed substantially to both astronomy and statistics [40,14]. Galileo* gave an early discussion of observational errors concerning the distance to the supernova of 1572. Halley, famous for his early contributions to celestial mechanics, laid important foundations to mathematical demography and actuarial science. Bessel, codiscoverer of stellar parallaxes and the binary companion of Sirius, introduce the notion of probable error*. Adolphe Quetelet*, founder of the Belgian

Royal Observatory, led the application of probability theory to the social sciences. Airy, a Royal astronomer, is known both for his text on statistical errors and his study of telescope optics. *See also* LAWS OF ERROR—I, II, III.

STATISTICAL ASTRONOMY

As comprehensive all-sky catalogs of star positions were compiled and astronomical photography permitted faint stars to be located, a field known as "statistical astronomy" rose to importance in the first half of this century. It is concerned with various collective properties of stars including their luminosity and mass distribution functions, their spatial distribution in the Milky Way, their distances from us and the related problem of light absorption in the interstellar medium, and their motions with respect to the Sun and to the center of the galaxy. The principal results of these studies are discussed in the monumental 1953 monograph of Trumpler and Weaver [41]. Prephotographic statistical discussions are reviewed in ref. 37 and more recent findings can be found in refs. 38 and 12.

From these studies we have learned that the Sun resides about 25 thousand light-years off-center in a disk of a differentially rotating spiral galaxy, with stars of increasing ages occupying the galaxy's spiral arms, smooth disk, and halo. By comparing the galactic mass inferred from star counts with that inferred from their rotational velocities around the galactic center, the existence of a "dark matter" component in the outer regions of the galaxy is inferred. Dynamical studies of other galaxies confirm that the mass in visible stars and gas is dwarfed by the dark matter in galaxy halos; yet astronomers do not know whether the matter is in the form of planets, elementary particles, black holes, or some more exotic form.

We give two modern examples of studies in galactic astronomy. Murray [25] derives the joint densities of the observed parallaxes, proper motions, and brightnesses for 6,125 stars, and computes the luminosity function, scale height in the galactic disk, streaming motions, and outliers with high velocities. A maximum-likelihood technique is used; the principal limitation is that parametric (e.g., Gaussian luminosity functions and observational errors, exponential scale heights for stars in the disk) forms are assumed throughout. Caldwell and Ostriker [8] seek fits of a three-component model of the mass distribution of the galaxy to 14 observationally derived quantities constraining the size, density, and motions in the galaxy. A nonlinear least-squares minimization algorithm is used to find the minimum chi-squared* solution.

Perhaps the central problem in statistical astronomy is the derivation of the *intrinsic* luminosity distribution function of a class of stars from a survey of the stars with the greatest *apparent* brightnesses (usually called a magnitude- or flux-limited survey). The observed population contains an excess of high luminosity stars, which can be seen out to large distances, and a deficit of low luminosity stars, which are bright enough to appear in the sample only if they lie close to us. Intrinsic or experimental errors scatter faint stars preferentially into flux-limited samples. These and related problems are called the "Malmquist effects," after the Swedish astronomer who derived a correction in 1920 for the bias for Gaussian luminosity distributions.

Interest in luminosity functions reemerged during the last decade with attempts to understand the phenomenon of quasar evolution [42]. The density of quasars, which are galaxies possessing extremely violent activity in their nuclei probably due to an accreting massive black hole, per unit volume was thousands of times higher when the Universe was young than it is today, indicating evolution of the shape and/or amplitude of the luminosity function. In 1971, Lynden-Bell [21] derived a remarkably simple nonparametric maximum likelihood* procedure for extrapolating from a flux-limited data set (assumed to be randomly truncated) of quasar counts to obtain a complete luminosity function. The method was largely unused by astronomers, and unknown to statisticians until its reexamination in 1985 by Woodroofe [43]. A similar method was independently developed by Nicoll and Segal [30] in support of Segal's proposed chronometric cosmology. A more common approach to the quasar evolution problem is to fit the data to parametric evolution formulas using least-squares* or maximum-likelihood criteria; see, for example, Marshall [23].

STATISTICAL ANALYSIS OF MODERN ASTRONOMICAL DATA

The modern observational astronomer is typically schooled only in elementary techniques such as the chi-squared test* and least-squares regression, using computer software such as that given by Bevington [6]. Some nonparametric methods such as Kolmogorov–Smirnov* and rank correlation* tests have come into frequent use, and computer codes distributed by Press et al. [32] are likely to bring other methods into the astronomer's repertoire. Unfortunately, very few astronomers are familiar with the major statistical software* packages such as SAS or BMDP.

Interest in more sophisticated and specialized statistical techniques of data analysis has emerged during the last decade. Murtagh and Heck [26] have written a monograph on multivariate techniques for astronomers, with a thorough bibliography of astronomical applications. The proceedings of a 1983 conference devoted specifically to the subject of statistical methods in astronomy is available [39]. An informal international Working Group for Modern Astronomical Methodology has formed, with a newsletter and announcements of workshops published in ref. 7. Articles in the statistical literature include those of Kendall [19], Scott [36], and Narlikar [28]. Following are brief descriptions of a few topics of current interest.

*Censored data**. A common experiment in astronomy entails the observations of a preselected sample of objects at a new spectral band, but some of the objects are not detected. For example, only about 10% of optically selected quasars are detected with the most sensitive radio telescopes and about 50% with the most sensitive satellite-borne X-ray telescopes, unless unacceptably long exposure times are devoted to the experiment. The data thus suffer type I left-censoring in apparent brightness, and a quasi-random censoring in intrinsic luminosities because the quasars lie at different distances. The studies seek to measure the mean radio or X-ray luminosities of the sample, differences in the luminosity functions between subsamples, correlations and linear regressions between luminosities in the various spectral bands, and so forth.

Until recently astronomers were unaware that survival analysis* statistical methods used by biostatisticians and others were available that provide many solutions to these questions. Avni, an astrophysicist, independently derived the "redistribute-to-the-right" formulation of the Kaplan–Meier product-limit estimator* [3], and a maximum-likelihood linear regression* assuming normally distributed residuals [2]. Schmitt [35] has developed a linear regression procedure, based on the two-dimensional Kaplan–Meier estimator and bootstrap error analysis, which can be applied to data censored in both the dependent and independent variables. Schmitt, Isobe, and Feigelson have brought previously known survival analysis methods into use for astronomical applications [35,10,18,17]. The principal difficulties in adapting standard survival analysis to astronomical situations are that the censoring levels are usually not known precisely, and the censoring patterns in flux-limited data are usually not completely random. These problems have yet to be addressed in detail.

Spatial analysis of galaxy clustering. Our Milky Way is but one of 10^{11} detectable galaxies, each containing up to $\sim 10^{11}$ luminous stars and a greater but uncertain amount of nonluminous matter. The galaxies appear to be rushing away from each other in a universal expansion that started 10–20 billion years ago. Correct modeling of the spatial and velocity distribution of the visible galaxies is critical for understanding the history of the Universe, and for determining whether or not the Universe will cease expanding and recollapse. (Poisson point process*). Other statistical analyses followed including the nearest-neighbor* distance distribution, the multiplicity function, and, most extensively, the two- and three-point correlation functions [31]. The power law two-point correlation function found for galaxies could be explained as the result of simple gravitational interactions of matter in an initially homogeneous Universe with an appropriate spectrum of density fluctuations.

Galaxies are not distributed randomly but are strongly clustered on many spatial scales. Neyman and Scott [29,36] were among the first to study this during the 1950s, using a hierarchy of clusters distributed as a uniform Poisson point process*.

By the middle 1980s, however, surveys of thousands of galaxy recessional velocities had been completed, revealing an unexpectedly anisotropic clustering pattern, as illustrated in Fig. 1 [9]. The galaxies seem to be concentrated along the edges of giant shells. The effect is sometimes referred to as "filaments" or "sheets" surrounding "voids," with a "spongelike" topology. The largest structures may exceed one-tenth the size of the observable Universe. In addition, large regions of the Universe may be moving in bulk with respect to other regions [22]. None of these phenomena can be easily

explained by simple gravitational effects in an initially homogeneous Universe. Statistical modeling of these data has just begun. Suggested methods include minimal spanning trees* [5], random walks* [20], ridge-finding algorithms [24], measures of topological curvature [13], and a quadrupole elongation statistic [11]. The study of Lynden-Bell et al. [22] is also of methodological interest, illustrating how sophisticated maximum-likelihood models of galaxy location and motions have become.

Analysis of periodic time series. Stellar systems often exhibit periodic behavior, from vibrations or rotations of single stars to orbits of two or more bodies in mutual gravitational orbits. Time scales run from milliseconds to hundreds of years, and the data can involve any portion of the electromagnetic spectrum. For example, time series of X-ray emission from binary star systems where one companion of an accreting neutron star or

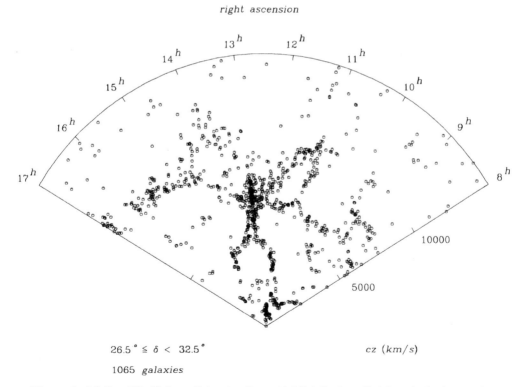

right ascension

$26.5° \leq \delta < 32.5°$ $cz \ (km/s)$

1065 galaxies

Figure 1. A "slice of the Universe" showing the spatial distribution of bright galaxies in a portion of the sky [9]. It shows the strongly anisotropic clustering pattern of "filaments" surrounding "voids." Courtesy of M. Geller, Center for Astrophysics.

black hole are quite interesting. Some systems show one or more strict periodicities, others random shot noise or white noise, others occasional sudden bursts of x-rays, and yet others combinations of $1/f$ noise and quasi-periodic oscillations. Certain of these behaviors are understood as rotational or orbital effects, whereas others are still mysterious.

A time-series* problem that has attracted recent methodological interest is the difficulty of establishing the existence of binary star orbits from optical photometric or velocity time series, which often consist of sparse, unevenly spaced, and noisy data. The issue is important because as many as half of the "stars" one sees at night may in fact be wide binaries. Classical Fourier analysis*, which assumes evenly spaced data points, is not an adequate method, and a variety of alternatives have been proposed. One is the periodogram*, which can be equivalent to least-squares fitting of sine curves to the data [34,16]. Several nonparametric methods have also been evaluated [15]. The principal difficulty with all methods is usually not finding the highest peak in the spectrum, but in evaluating its statistical significance and eliminating false alarms. Both analytical and bootstrap methods for measuring confidence limits have been suggested, but no consensus has emerged on the best treatment of the problem.

Image restoration techniques. Many classes of objects in the sky have complicated morphologies, and much effort is devoted to imaging them accurately. This entails compensation for the distortions caused by imperfect optics of the telescope or turbulence in the Earth's atmosphere, and nonlinearities in detectors such as photographic plates. Perhaps the greatest need for sophisticated image restoration* is in "aperture synthesis" interferometry, a technique developed in radio astronomy that combines the signals from separated individual radio antennas to produce a single high-resolution image [33,27]. The data consist of the two-dimensional Fourier transform of the brightness distribution in the sky, but the coverage in the Fourier plane is incomplete. Simple Fourier transformation thus gives an image contaminated by strong side lobes. Two restoration methods are commonly used: The "CLEAN" algorithm, which gives a least-squares fit to a superposition of many point sources; and the maximum entropy method*, which gives the most probable nonnegative smooth image consistent with the data. Both require prior knowledge of the coverage in the Fourier plane. The resulting CLEANed map can then be used as an improved model of the sky distribution to "self-calibrate" unpredicted instrumental or atmospheric disturbances. After many iterations, which can take hundreds of CPU-hours on large computers, images with extraordinarily high fidelity have been achieved with dynamic range (brightest spot in the image divided by the root mean square noise level) of order $10^5 : 1$. Maximum entropy image enhancement techniques are sometimes also used in optical and X-ray astronomy as well as radio interferometry.

Statistics of very few events. Important branches of astronomy have emerged in the last few decades from experimental physics that involve the detection of small numbers of discrete events. These include cosmic ray, X-ray, gamma-ray, and neutrino astronomy. With the proliferation of photon-counting detectors like charged-coupled devices, such data are becoming increasingly common in optical astronomy as well. The statistical procedures for interpreting these data are traditionally based on the Poisson distribution, though use of nonparametric and maximum likelihood techniques is also appearing.

A premier example of this problem was the detection of neutrinos from the supernova SN1987A, initiated by the gravitational collapse of a star in a satellite galaxy of our Milky Way. The detection represented the first astronomical object other than the Sun ever detected in neutrinos, and provides a unique opportunity to test a host of models of high-energy and nuclear astrophysics. The data consist of 11 events in 13 seconds in the Kamiokande-II underground detector, and 8 events in 6 seconds in the IMB detector. The timing of the events constrain the mass of the neutrino, which could be the most massive component of the Universe [4,1]. However, different treatments of the same data give considerably different neutrino mass limits.

SUMMARY

Many aspects of modern astronomy are statistical in character, and demand sophisticated statistical methodology. In light of this situation, and the rich history of the interaction between astronomy and statistics through the nineteenth century, it is surprising that the two communities have been so isolated from each other in recent decades. Astronomers dealing with censored data, for example, were unaware of the relevant progress in biostatistics* and industrial reliability applications until recently. Most are not familiar with multivariate techniques extensively used in the social sciences. Conversely, statisticians were unaware of Lynden-Bell's maximum-likelihood density estimation technique for a truncated distribution. There is little communication between astronomers analyzing the spatial distribution of galaxies (Fig. 1) and statisticians involved with point spatial processes* arising in other fields. Improved interactions between the two fields is clearly needed. It would give astronomers more effective techniques for understanding the Universe, and statisticians challenging and important problems to address.

REFERENCES

1. Arnett, W. D. and Rosner, J. L. (1987). *Phys. Rev. Lett.*, **58**, 1906.

2. Avni, Y. and Tananbaum, H. (1986). *Astrophys. J.*, **305**, 57.

3. Avni, Y., Soltan, A., Tananbaum, H., and Zamorani, G. (1980). *Astrophys. J.*, **235**, 694.

4. Bahcall, J. N. and Glashow, S. L. (1987). *Nature*, **326**, 476.

5. Barrow, J. D., Bhavsar, S. P., and Sonoda, D. H. (1985). *Monthly Notices R. Astron. Soc.*, **216**, 17.

6. Bevington, P. R. (1969). *Data Reduction and Error Analysis for the Physical Sciences.* McGraw-Hill, New York.

7. *Bulletin d'Information du Centre de Données Stellaires*, Observatoire de Strasbourg, 11 Rue de l'Université, 67000 Strasbourg, France.

8. Caldwell, J. A. and Ostriker, J. P. (1981). *Astrophys. J.*, **251**, 61.

9. de Lapparent, V., Geller, M. J., and Huchra, J. P. (1986). *Astrophys. J. Lett.*, **302**, L1.

10. Feigelson, E. D. and Nelson, P. I. (1985). *Astrophys. J.*, **293**, 192.

11. Fry, J. N. (1986). *Astrophys. J.*, **306**, 366.

12. Gilmore, G. and Carswell, B., eds. (1987). *The Galaxy.* Reidel, Dordrecht, Netherlands.

13. Gott, J. R., Melott, A. L., and Dikinson, M. (1986). *Astrophys. J.*, **306**, 341.

14. Hald, A. (1986). *Int. Statist. Rev.*, **54**, 211–220.

15. Heck, A., Manfroid, J., and Mersch, G. (1985). *Astron. Astrophys. Suppl.*, **59**, 63.

16. Horne, J. H. and Baliunas, S. L. (1986). *Astrophys. J.*, **302**, 757.

17. Isobe, T. and Feigelson, E. D. (1987). *ASURV: Astronomy Survival Analysis, software package.*

18. Isobe, T., Feigelson, E. D., and Nelson, P. I. (1986). *Astrophys J.*, **306**, 490.

19. Kendall, D. G. (1985). In *Celebration of Statistics*, A. C. Atkinson and S. E. Fienberg, eds. Springer Verlag, Berlin.

20. Kuhn, J. R. and Uson, J. M. (1982). *Astrophys. J. Lett.*, **263**, L47.

21. Lynden-Bell, D. (1971). *Monthly Notices R. Astron. Soc.*, **136**, 219.

22. Lynden-Bell, D., Faber, S. M., Burstein, D., Davies, R. L., Dressler, A., Terlevich, R. J., and Wegner, G. (1988). *Astrophys. J.*, **326**, 19.

23. Marshall, H. L. (1987). *Astron. J.*, **94**, 620.

24. Moody, J. E., Turner, E. L., and Gott, J. R. (1983). *Astrophys. J.*, **273**, 16.

25. Murray, C. A. (1986). *Monthly Notices R. Astron. Soc.*, **223**, 649.

26. Murtagh, F. and Heck, A. (1987). *Multivariate Data Analysis, Astrophysics and Space Science Library*, **131**. Reidel, Dordrecht, Netherlands.

27. Narayan, R. and Nityananda, R. (1986). *Ann. Rev. Astron. Astrophys.*, **24**, 127.

28. Narlikar, J. V. (1982). *Sankhyā B*, **42**, 125.

29. Neyman, J. and Scott, E. L. (1952). *Astrophys J.*, **116**, 144.

30. Nicoll, J. F. and Segal, I. E. (1983). *Astron. Astrophys.*, **118**, 180.

31. Peebles, P. J. E. (1980). *The Large-Scale Structure of the Universe.* Princeton University Press, Princeton, NJ.

32. Press, W. H., Flannery, B. P., Teukolsky, S. A., and Vetterling, W. T. (1986). *Numerical Recipes: The Art of Scientific Computing*. Cambridge University Press, London.

33. Roberts, J. A., ed (1984). *Indirect Imaging*. Cambridge University Press, London.

34. Scargle, J. D. (1982). *Astrophys. J.*, **263**, 835.

35. Schmitt, J. H. (1985). *Astrophys. J.*, **293**, 178.

36. Scott, E. L. (1976). In *On the History of Statistics and Probability*, D. B. Owen, ed. Marcel Dekker, New York.

37. Sheynin, O. B. (1984). *Archive for History of Exact Sciences*, **29**, 151–199.

38. *Stars and Stellar Systems* (1962–75). University of Chicago Press, Chicago. 9 volumes.

39. *Statistical Methods in Astronomy* (1983). SP-201, European Space Agency, ESA Scientific and Technical Publications, c/o ESTEC, Noordwijk, Netherlands.

40. Stigler, S. M. (1986). *The History of Statistics: The Measurement of Uncertainty Before 1900*. Harvard University Press, Cambridge, MA.

41. Trumpler, R. J. and Weaver, H. F. (1953). *Statistical Astronomy*. University of California Press, Berkeley, CA.

42. Weedman, D. (1986). *Quasar Astronomy*. Reidel, Dordrecht, Netherlands.

43. Woodroofe, M. (1985). *Ann. Statist.*, **13**, 163.

BIBLIOGRAPHY

Di Gesù, V., Scarsi, L., Crane, P., Friedman, J. H., and Levaldi, S., eds. (1985; 1986). *Data Analysis in Astronomy I; II*. Plenum, New York. (Proceedings of International Workshops held at Erice, Sicily, Italy in 1984 and 1985.)

Maistrov, L. E. (1974). *Probability Theory: A Historical Sketch*, translated by S. Kotz. Academic, New York.

EDITORIAL NOTE

S. R. Searle [*Commun. Statist. A*, **17**, 935–968 (1988)] notes that the earliest appearances of variance components* are in two books on astronomy, namely:

> Airy, G. B. (1861). *On the Algebraical and Numerical Theory of Errors of Observations and the Combination of Observations*. Macmillan, London.

and

> Chauvenet, W. (1863). *A Manual of Spherical and Practical Astronomy*. Lippincott, Philadelphia.

Searle also draws attention to the fact that the development of the least-squares* estimation of parameters in linear models was presented in books on astronomy, namely:

> Gauss, K. F. (1809). *Theoria Motus Corporum Coelestium in Sectionibus Conics Solem Ambientium*. Perthes and Besser, Hamburg.

and

> Legendre, A. M. (1806). *Nouvelles Methodes pour Determination des Orbites des Comètes*. Courcier, Paris.

See also Directional Data Analysis; Discriminant Analysis; Gauss, Carl Friedrich; Hierarchical Cluster Analysis; Least Squares; Linear Regression; Multivariate Analysis; Newcomb, Simon; Outlier Rejection, Chauvenet's Criterion; and Statistics: An Overview.

Eric D. Feigelson

ASYMMETRIC POPULATION

A population or a distribution that is not symmetric is *asymmetric*. The property of asymmetry is also related to skewness. It should be noted, however, that measures of skewness* usually correspond to some particular feature of symmetry. The third central moment μ_3, for example, is indeed zero for symmetric populations, but it can also be zero for populations that are asymmetric.

It is better to limit the use of the adjective "asymmetric" to distributions, and not apply it to populations.

See also Mean, Median, and Mode; Skewness: Concepts and Measures; and Skewness, Measures of.

ASYMPTOTIC EXPANSIONS—I

Many test statistics and estimators have probability distributions that may be approximated quite well by a normal distribution* in the case of a sufficiently large sample size. This is of practical use, for example, in such problems as determining critical regions* for tests of specified sizes and determining confidence regions* with specified confidence coefficients.

Let T_n denote a test statistic based on a sample $X_1, \ldots, X_n$ from a distribution F, let a_n and b_n be suitable normalizing constants, and let G_n denote the distribution of the normed statistic $(T_n - a_n)/b_n$. The "normal approximation" is expressed by

$$\lim_{n \to \infty} G_n(t) = \Phi(t) \qquad (-\infty < t < \infty), \quad (1)$$

where

$$\Phi(t) = (2\pi)^{-1/2} \int_{-\infty}^{t} \exp(-x^2/2)dx,$$

the standard normal distribution* function. Often (1) can be established under moderate regularity assumptions on the distribution function F and the functional form of the statistic T_n. See ASYMPTOTIC NORMALITY.

Of fundamental importance is the question of the error of approximation in (1) for a particular value of n. One useful type of answer is supplied by a "Berry-Esséen" rate, namely an assertion of the form

$$\sup_{-\infty < t < \infty} |G_n(t) - \Phi(t)| = O(n^{-1/2}), \quad (2)$$

available under additional restrictions on F and the form of T_n (see CENTRAL LIMIT THEOREMS, CONVERGENCE RATES FOR). A more refined answer is given by an *expansion* of the error $G_n(t) - \Phi(t)$ in powers of $n^{-1/2}$. This requires additional restrictions on F and T_n. However, not only does it provide detailed information in (2), but it also supplies a way to replace $\Phi(t)$ by an improved approximation. Below we shall survey such expansions for the key special case that T_n is a *sum*, and then we shall comment briefly on other cases. For more details see the entries devoted to specific expansions.

Let $X_1, X_2, \ldots$ be independent and identically distributed random variables with distribution F, mean μ, variance σ^2, and characteristic function* ψ. Let G_n denote the distribution of the normed sum $(\sum_1^n X_i - n\mu)/(n^{1/2}\sigma)$. If Cramér's condition

$$(C) \qquad \lim_{|z| \to \infty} \sup |\psi(z)| < 1$$

is satisfied and the kth moment of F is finite, then

$$\left| G_n(t) - \Phi(t) - \sum_{j=1}^{k-3} P_{3j-1}(t) e^{-t^2/2} n^{-j/2} \right|$$
$$< Mn^{-(k-2)/2}, \quad (3)$$

where M is a constant depending on k and F but not on n or t, and $P_m(t)$ is a polynomial (essentially the *Chebyshev–Hermite* polynomial*) of degree m in t. Indeed, we have the expressions

$$P_2(t)e^{-t^2/2} = -(\lambda_3/3!)\Phi^{(3)}(t),$$
$$P_3(t)e^{-t^2/2} = (\lambda_4/4!)\Phi^{(4)}(t)$$
$$+ (10\lambda_3^2/6!)\Phi^{(6)}(t), \ldots,$$

where λ_j denotes the jth cumulant* of F [the coefficient of $(iz)^j/j!$ in the MacLaurin series expansion of $\log \psi(z)$]. We may express λ_j (essentially) as a polynomial in the moments of F, obtaining in particular

$$\lambda_3 = \frac{E[(X-\mu)^3]}{\sigma^3} = \gamma_1$$

and

$$\lambda_4 = \frac{E[(X-\mu)^4]}{\sigma^4} - 3 = \gamma_2,$$

known as the coefficients of skewness* and of kurtosis*, respectively. Therefore, up to terms of order n^{-1}, which usually suffices in practical applications, the approximation given by (3) may be written conveniently in the form

$$G_n(t) \doteq \Phi(t) - \frac{\gamma_1}{6}(t^2 - 1)\phi(t)n^{-1/2}$$

$$- \left[\frac{\gamma_2}{24}(t^3 - 3t) + \frac{\gamma_1^2}{72} \right.$$

$$\left. \times (t^5 - 10t^3 + 15t) \right] \phi(t)n^{-1}, \quad (4)$$

with error $O(n^{-3/2})$ uniformly in t.

The expansion (3) is called the *Edgeworth expansion* for G_n. Corresponding expansions for the density g_n follow by replacing all functions of t by their derivatives. The assumption (C) is always satisfied if the distribution F has an absolutely continuous component. Analogs of (3) hold under alternative conditions on F, e.g., the case of a lattice distribution*. Versions also have been developed allowing the X_i's to have differing distributions or to be stationary dependent. Furthermore, other metrics besides $\sup_t |G_n(t) - \Phi(t)|$ have been treated. For extensive treatments of (3) and these various ramifications, see [2,3,7,9,10,12].

An inverse problem related to (3) concerns the equation

$$G_n(t_p) = 1 - p,$$

where $0 < p < 1$. The solution t_p may be expressed asymptotically as

$$t_p \sim \mu + \sigma w, \quad (5)$$

where w is given by the *Cornish–Fisher expansion**, which like (3) involves the quantities $\{\lambda_i\}$ and the Chebyshev–Hermite polynomials.

For detailed numerical illustration of the effectiveness of the expansions (3) and (5), see Abramowitz and Stegun [1], pp. 935–936, 955, 958–959. As noted in connection with (4), the improvement of the Edgeworth approximation over simply the normal approximation can be attributed to use of the coefficients of skewness and kurtosis; this provides a convenient intuitive basis for assessing the potential degree of improvement. Numerical illustration related to (4) is provided by Bickel and Doksum [5].

Finally, let us consider statistics other than sums. For such cases the question of asymptotic normality*, (1), has received extensive treatment. Secondarily, the associated Berry–Esséen rates, (2), have received attention. Consequently, results of types (1) and (2) are now available for several important wide classes of statistics: U-statistics*; von Mises differentiable statistical functions*; linear functions of order statistics; M-estimates*; and rank statistics*. Detailed exposition may be found in Serfling [11]. *see also* ASYMPTOTIC NORMALITY. However, except for isolated results, the question of asymptotic expansions analogous to (3) has only very recently gained intensive interest and development. For comments on Edgeworth expansions for rank statistics, such as the two-sample Wilcoxon statistic, *see* Hájek and Šidák [8], Sec. IV.4.2. Multivariate Edgeworth-type expansions are discussed by Chambers [6]. For a review of recent activity, see Bickel [4].

REFERENCES

1. Abramowitz, M. and Stegun, I. A., eds. (1965, 1970). *Handbook of Mathematical Functions.* U.S. Government Printing Office, Washington, D.C.

2. Bhattacharya, R. N. (1977). *Ann. Prob.,* **5**, 1–27.

3. Bhattacharya, R. N. and Ranga Rao, R. (1976). *Normal Approximation and Asymptotic Expansions.* Wiley, New York.

4. Bickel, P. J. (1974). *Ann. Statist.,* **2**, 1–20.

5. Bickel, P. J. and Doksum, K. A. (1977). *Mathematical Statistics.* Holden-Day, San Francisco.

6. Chambers, J. M. (1967). *Biometrika,* **54**, 367–383.

7. Cramér, H. (1970). *Random Variables and Probability Distributions,* 3rd ed. Cambridge University Press, Cambridge.

8. Hájek, J. and Šidák, Z. (1967). *Theory of Rank Tests.* Academic Press, New York.

9. Ibragimov, I. A. and Linnik, Y. V. (1971). *Independent and Stationary Sequences of Random Variables.* Wolters-Noordhoff, Groningen.

10. Petrov, V. V. (1975). *Sums of Independent Random Variables.* Springer-Verlag, New York.

11. Serfling, R. J. (1980). *Approximation Theorems of Mathematical Statistics*. Wiley, New York.

12. Wallace, D. (1958). *Ann. Math. Statist.*, **29**, 635–654.

See also APPROXIMATIONS TO DISTRIBUTIONS; ASYMPTOTIC NORMALITY; CENTRAL LIMIT THEOREMS, CONVERGENCE RATES FOR; CORNISH–FISHER AND EDGEWORTH EXPANSIONS; and LIMIT THEOREM, CENTRAL.

R. J. SERFLING

ASYMPTOTIC EXPANSIONS—II

Asymptotic expansions of functions are useful in statistics in three main ways. Firstly, conventional asymptotic expansions of special functions are useful for approximate computation of integrals arising in statistical calculations. An example given below is the use of Stirling's approximation to the gamma function. Second, asymptotic expansions of density or distribution functions of estimators or test statistics can be used to give approximate confidence limits for a parameter of interest or p-values for a hypothesis test. Use of the leading term of the expansion as an approximation leads to confidence limits and p-values based on the limiting form of the distribution of the statistic, whereas use of further terms often results in more accurate inference. Third, asymptotic expansions for distributions of estimators or test statistics may be used to investigate properties such as the efficiency of an estimator or the power of a test. The first two of these are discussed in this entry, which is a continuation of Serfling [35]. The third is discussed in the entry ASYMPTOTICS, HIGHER ORDER*.

An asymptotic expansion of a function is a reexpression of the function as a sum of terms adjusting a base function, expressed as follows:

$$f_n(x) = f_{0n}(x)[1 + b_{1n}g_1(x) + b_{2n}g_2(x)$$
$$+ \cdots + b_{kn}g_k(x) + O(b_{k+1,n})]$$
$$(n \to \infty). \qquad (1)$$

The sequence $\{b_{kn}\} = \{1, b_{1n}, b_{2n}, \ldots\}$ determines the asymptotic behavior of the expansion: in particular how the reexpression approximates the original function. Usual choices of $\{b_{kn}\}$ are $\{1, n^{-1/2}, n^{-1}, \ldots\}$ or $\{1, n^{-1}, n^{-2}, \ldots\}$; in any case it is required that $b_{kn} = o(b_{k-1,n})$ as $n \to \infty$. For sequences of constants $\{a_n\}$, $\{b_n\}$, we write $a_n = o(b_n)$ if $a_n/b_n \to 0$ as $n \to \infty$, and $a_n = O(b_n)$ if a_n/b_n remains bounded as $n \to \infty$. The notation $o_p(\cdot)$, $O_p(\cdot)$ is useful for sequences of random variables $\{Y_n\}$: Y_n is $o_p(a_n)$ if Y_n/a_n converges in probability to 0 as $n \to \infty$, and is $O_p(a_n)$ if $|Y_n/a_n|$ is bounded in probability as $n \to \infty$.

Asymptotic expansions are used in many areas of mathematical analysis. Three helpful textbooks are Bleistein and Handelsman [9], Jeffreys [25], and DeBruijn [18]. An important feature of asymptotic expansions is that they are not in general convergent series, and taking successively more terms from the right-hand side of (1) is not guaranteed to improve the approximation to the left-hand side. In the study of asymptotic expansions in analysis, emphasis is typically on $f_n(x)$ as a function of n, with x treated as an additional parameter, and n considered as the argument of the function. In these treatments it is usual to let n be real or complex, and the notation $f(z; x)$ or $f(z)$ is more standard. The functions $g_j(\cdot)$ that we have used in (1) are then just constants (in z), and the sequence $\{b_{kn}\}$ is typically $\{z^{-k}\}$ if $z \to \infty$ or $\{z_k\}$ if $z \to 0$.

An asymptotic expansion is used to provide an approximation to the function $f_n(x)$ by taking the first few terms of the right-hand side of (1): for example, we might write

$$f_n(x) = f_{0n}(x)[1 + b_{1n}g_1(x)].$$

Although the approximation is guaranteed to be accurate only as $n \to \infty$, it is often quite accurate for rather small values of n. It will usually be of interest to investigate the accuracy of the approximation for various values of x as well, an important concern being the range of values for x over which the error in the approximation is uniform.

In statistics the function $f_n(x)$ is typically a density function, a distribution function, a moment or cumulant generating function, for a random variable computed from a sequence of random variables of length n. For example, $f_n(x)$ could be the density of the standardized mean $\overline{X}_n$, say, of n independent, identically distributed random variables $X_i : \overline{X}_n = \sum X_i/n$. In this case the function $f_{0n}(x)$ is the limiting density for the standardized version of $\overline{X}_n$, usually the normal density function $\phi(x) = (2\pi)^{-1/2} \exp(-x^2/2)$; see ASYMPTOTIC NORMALITY.

ASYMPTOTIC EXPANSIONS OF SPECIAL FUNCTIONS

A familiar asymptotic expansion is that of the gamma function $\Gamma(z) = \int_0^\infty t^{z-1}e^{-t}\,dt$ by Stirling's formula* given, for example, in Abramowitz and Stegun [1, Sec. 6.1.37]:

$$\Gamma(z) = e^{-z}z^{z-1/2}(2\pi)^{1/2}$$

$$\times \left(1 + \frac{1}{12z} + \frac{1}{288z^2} + O(z^{-3})\right)$$

$$(z \to \infty \quad \text{in } |\arg z| < \pi). \qquad (2)$$

The leading term of the right-hand side is Stirling's approximation to the gamma function. There are similar expansions given in Abramowitz and Stegun [1, Chap. 6] for log $\Gamma(z)$ and its first two derivatives, the digamma and trigamma functions. The approximations given by the first several terms in these expansions are used, for example, in computing the maximum-likelihood estimator and its asymptotic variance for a sample from a gamma density with unknown shape parameter.

Another example is the asymptotic expansion for the tail of the standard normal cumulative distribution function:

$$1 - \Phi(z) = \int_z^\infty (2\pi)^{-1/2} \exp\left(-\frac{x^2}{2}\right) dx$$

$$= \phi(z)z^{-1}\left(1 - \frac{1}{z^2} + \frac{3}{z^4} + O(z^{-6})\right). \qquad (3)$$

The quantity $[1 - \Phi(z)]/\phi(z)$ is often called Mills' ratio*.

The asymptotic expansions given above are examples of expansions obtained using Laplace's method*. Laplace's method is also very useful for deriving approximations to integrals arising in Bayesian inference.

EDGEWORTH AND SADDLEPOINT EXPANSIONS

For statistics that are asymptotically normally distributed, the *Edgeworth expansion* for density or distribution functions gives a useful and readily computed approximation. Such statistics are typically either sample means or smooth functions of sample means, and the Edgeworth expansion for the distribution function of a standardized sample mean is given in Serfling [35]. We assume for simplicity that $X_1, \ldots, X_n$ are independent and identically distributed. Define $S_n = n^{1/2}(\overline{X}_n - \mu)/\sigma$, where μ and σ^2 are the mean and variance of X_i. The Edgeworth expansion for the density of S_n is

$$f_n(s) = \phi(s)\left[1 + \frac{1}{n^{1/2}}\frac{\lambda_3}{6}h_3(s)\right.$$

$$+ \frac{1}{n}\left(\frac{\lambda_4}{24}h_4(s) + \frac{\lambda_3^2}{72}h_6(s)\right)$$

$$\left. + O(n^{-3/2})\right], \qquad (4)$$

where λ_3 and λ_4 are the third and fourth cumulants of $(X_i - \mu)/\sigma$, and, $h_j(s) = (-1)^j \phi^{(j)}(s)/\phi(s)$ is the jth Hermite polynomial.

Note that (4) suggests the use of the first three terms as an approximation to the exact density, with the remaining terms absorbed into the expression $O(n^{-3/2})$. The full expansion for the distribution function is given in Serfling [35]. The Edgeworth expansion is derived in many textbooks; cf. the references in Serfling [35], Feller [22, Chap. 16], McCullagh [30, Chap. 6], and Barndorff-Nielsen and Cox [5, Chap. 4]. *See also* CORNISH–FISHER AND EDGEWORTH EXPANSIONS.

The Edgeworth approximation is quite accurate near the center of the density. In particular, at $\overline{x} = \mu$ the relative error in using the normal approximation is $O(n^{-1})$, and that in using the approximation suggested by (4) is $O(n^{-2})$, because the odd-order Hermite polynomials are 0 at $s = 0$. For large values of $|s|$, though, the approximation is often inaccurate for fixed values of n, as the polynomials oscillate substantially as $|s| \to \infty$. A particular difficulty is that the approximation to $f_n(s)$ may in some cases take negative values.

A different type of asymptotic expansion for the density of a sample mean is given by the *saddlepoint expansion*. Let the cumulant generating function of X_i be denoted by $K(t)$. The saddlepoint expansion for the density of $\overline{X}$ is defined by

$$f_n(\overline{x}) = \frac{1}{\sqrt{2\pi}} \left(\frac{n}{|K''(\hat{z})|}\right)^{1/2} \exp\{n[K(\hat{z}) - \hat{z}\overline{x}]\}$$
$$\times \left(1 + \frac{3\lambda_4(\hat{z}) - 5\lambda_3^2(\hat{z})}{24n} + O(n^{-2})\right), (5)$$

where $\hat{z}$ is called the *saddlepoint* and is defined by $K'(\hat{z}) = \overline{x}$, and the rth cumulant function $\lambda_r(z) = K^{(r)}(z)/[K''(z)]^{r/2}$. Note that this is an asymptotic expansion in powers of n^{-1}. The next term in the expansion is a complicated expression involving cumulant functions up to order 6. The leading term of (5) is the saddlepoint approximation* to the density of $\overline{X}_n$. This expression is always positive, but will not usually integrate to exactly one, so in practice it is renormalized. The renormalization improves the order of the approximation:

$$f_n(\overline{x}) = c \left(\frac{n}{|K''(\hat{z})|}\right)^{1/2} \exp\{n[K(\hat{z}) - \hat{z}\overline{x}]\}$$
$$\times [1 + O(n^{-3/2})]. \qquad (6)$$

Evaluating the saddlepoint approximation requires knowledge of the cumulant generating function $K(z)$. Approximations based on estimating $K(z)$ by estimating the first four cumulants are discussed in Easton and Ronchetti [21], Wang [41], and Cheah et al. [10].

The saddlepoint approximation to the distribution function of $\overline{X}$ can be obtained by integrating (6) or by applying the saddlepoint technique; the result, due to Lugannani and Rice [29], is

$$F_n(\overline{x}) = \Phi(r) + \phi(r)\left(\frac{1}{r} - \frac{1}{q}\right), \qquad (7)$$

where

$$r = \text{sign}(q)[2n[K(\hat{z}) - \hat{z}\overline{x}]]^{1/2},$$
$$q = \hat{z}[K''(\hat{z})]^{1/2}.$$

The approximation (7) is often surprisingly accurate throughout the range of $\overline{x}$, except near $\overline{x} = \mu$ or $r = 0$, where it should be replaced by its limit as $r \to 0$:

$$F_n(\mu) = \tfrac{1}{2} - \tfrac{1}{6}\lambda_3(0)/\sqrt{2\pi n}.$$

The approximation (7) has relative error $O(n^{-1})$ for all $\overline{x}$ and $O(n^{-3/2})$ for the so-called moderate deviation region $\overline{x} - \mu = O(n^{-1/2})$. It can be expressed in an asymptotically equivalent form by defining $r^* = r + r^{-1}\log(q/r)$: the approximation

$$F_n(\overline{x}) \doteq \Phi(r^*), \qquad (8)$$

originally due to Barndorff-Nielsen [4], is asymptotically equivalent to (7).

The approximations (5) and (7) were derived in Daniels [13] and Lugannani and Rice [29], respectively, using the saddlepoint technique of asymptotic analysis. Daniels [15] exemplifies the derivation of (7). Both Kolassa [27] and Field and Ronchetti [23] provide rigorous derivations of (5) using the saddlepoint method. General discussions of the saddlepoint method can be found in Bleistein and Handelsman [9] or Courant and Hilbert [11]. The approximations can also be derived from the Edgeworth expansion; cf. Barndorff-Nielsen and Cox [5, Chap. 4], where (5) is called the tilted Edgeworth expansion.

The Edgeworth and saddlepoint approximations for distribution functions discussed here apply to continuous random variables, and adjustments to the approximations are needed in the case that the variables X_i takes values on a lattice. The details are provided in Kolassa [27, Chaps. 3, 5].

For vector X_i of length m, say, multivariate versions of the Edgeworth and saddlepoint density approximations are readily available. The multivariate Edgeworth approximation requires for its expression a definition of generalized Hermite polynomials, which are conceptually straightforward but notationally complex. A brief account is given in Reid [33], adapted from McCullagh [30, Chap. 5]. The multivariate version of the saddlepoint approximation is the same as (6), with $K(z) = \log E \exp(z^T x)$, $\hat{z}^T \overline{x}$ interpreted as a scalar product of the two vectors

$\hat{z}$ and $\bar{x}$, and $|K''(\hat{z})|$ interpreted as the determinant of the $p \times p$ matrix of second derivatives of the cumulant generating function. The distribution-function approximation (7) is only available for the univariate case, but an approximation to the conditional distribution function $\Pr(\bar{X}_{(1)} \leqslant \bar{x}_{(1)} | \bar{x}_{(2)})$ is derived in Skovgaard [36] and extended in Wang [39] and Kolassa [27, Chap. 7]. The form of this approximation has proved very useful for inference about scalar parameters in the presence of nuisance parameters*.

The Edgeworth and saddlepoint approximations are both based on a limiting normal distribution for the statistic in question. Some statistics may have a limiting distribution that is not normal; in particular, sample maxima or minima usually have limiting distributions of the extreme-value form. For such statistics a series expansion of the density in which the basic function corresponds to the limiting density may be of more practical interest. In principle this is straightforward, and for example McCullagh [30, Chap. 5] derives Edgeworth-type expansions using arbitrary basis functions and the associated orthogonal polynomials. Examples of saddlepoint approximations based on nonnormal limits are discussed in Jensen [26] and Wood et al. [42].

STOCHASTIC ASYMPTOTIC EXPANSIONS

It is often very convenient in deriving asymptotic results in statistics to use *stochastic asymptotic expansions*, which are analogues of (1) for random variables. For a sequence of random variables $\{Y_n\}$ a stochastic asymptotic expansion is expressed as

$$Y_n = X_0 + X_1 b_{1n} + \cdots + X_k b_{kn}$$
$$+ O_p(b_{k+1,n}), \qquad (9)$$

where $\{X_0, X_1, \ldots\}$ have a distribution not depending on n. Stochastic asymptotic expansions are discussed in Cox and Reid [12] and in Barndorff-Nielsen and Cox [5, Chap. 5]. As an example, Cox and Reid [12] show that if Y_n follows a chi-squared distribution with n degrees of freedom, then

$$\frac{Y_n - n}{\sqrt{2n}} = X_0 + \frac{1}{n^{1/2}} \frac{\sqrt{2}}{3}(X_0^2 - 1)$$
$$+ O_p(n^{-1}),$$

where X_0 is a standard normal random variable. The relationship between stochastic asymptotic expansions and expansions for the corresponding distribution functions is discussed in Cox and Reid [12]. Expansions similar to (9) where the distributions of $X_0, X_1, \ldots$ are only asymptotically free of n are very useful in computing asymptotic properties of likelihood-based statistics.

An expansion closely related to (9) but usually derived in the context of the Edgeworth expansion for the distribution function of the sample means is the Cornish—Fisher expansion* for the quantile of the distribution function. As described in Serfling [35], an expansion for the value s_α satisfying $F_n(s_\alpha) = 1 - \alpha$ can be obtained by a reversion of the Edgeworth expansion. The result is

$$s_\alpha = z_\alpha + \frac{1}{6\sqrt{n}}(z_\alpha^2 - 1)\lambda_3$$
$$+ \frac{1}{24n}(z_\alpha^3 - 3z_\alpha)\lambda_4$$
$$- \frac{1}{36n}(2z_\alpha^3 - 5z_\alpha)\lambda_3^2 + O(n^{-3/2}),$$

where z_α satisfies $\Phi(z_\alpha) = 1 - \alpha$. Other asymptotic expansions for $F_n(s)$ lead to alternative expansions for s_α, and in particular the r^*-approximation given at (8) can be derived from the saddlepoint expansion for the distribution function of $\bar{X}_n$.

APPLICATIONS TO PARAMETRIC INFERENCE

There has been considerable development of statistical theory based on the use of higher-order approximations derived from asymptotic expansions. Likelihood-based inference, or inference from parametric models, has developed particularly rapidly, although the approximations are very useful in other contexts as well. Some examples of this will now be sketched.

Suppose that $X = (X_1, \ldots, X_n)$ is a sample from a parametric model that is an exponential family*, i.e., is of the form

$$f(x; \theta) = \exp[\theta^T x - b(\theta) - d(x)], \qquad (10)$$

where X and θ take values in $\mathbb{R}^m$, say. The minimal sufficient statistic is $S = s(X) = \sum X_i$, and the maximum-likelihood estimator of θ is a one-to-one function of $S : c'(\hat{\theta}) = S/n$. Thus the saddlepoint approximation of S given in (7) can be used to give an approximation to the density for $\hat{\theta}$, which takes the form

$$f_n(\hat{\theta}; \theta) = c|b''(\hat{\theta})|^{1/2}$$
$$\times \exp[(\theta - \hat{\theta})'s - nb(\theta) + nb(\hat{\theta})].$$

If we denote the log-likelihood function for θ based on x by $\ell(\theta; x)$, and the observed Fisher information* function $-\partial^2 \ell(\theta)/\partial\theta \partial\theta^T$ by $j(\theta)$, we can reexpress this approximation as

$$f_n(\hat{\theta}; \theta) = c|j(\hat{\theta})|^{1/2} \exp[\ell(\theta; x) - \ell(\hat{\theta}; x)] \quad (11)$$

This approximation to the density for the maximum-likelihood estimator is usually known as *Barndorff-Nielsen's approximation*, or, following Barndorff-Nielsen, the p^* *approximation*. Although it has been used here to illustrate the saddlepoint approximation in an exponential family, the approximation (11) is valid quite generally. This was exemplified and illustrated in Barndorff-Nielsen [2, 3] and several subsequent papers. A review of the saddlepoint approximation and the literature on the p^*-formula* through 1987 is given in Reid [33]. A general proof and discussion of the interpretation of the p^* formula is given in Skovgaard [38]. Chapters 6 and 7 of Barndorff-Nielsen and Cox [6] provide an extensive discussion of the p^* formula and its applications in parametric inference. The validity of (11) in more general models requires the existence of a one-to-one transformation from the minimal sufficient statistic to $(\hat{\theta}, a)$, where a is a complementary statistic with a distribution either exactly or approximately (in a specific sense) free of θ; such statistics are called exact or approximate ancillary* statistics. The right-hand side of (11) then approximates the conditional distribution of $\hat{\theta}$, given a.

An illustration of the cumulative-distribution-function approximation (7) in the exponential family is also instructive. Suppose in (10) that $m = 1$. Then (7) provides an approximation to the distribution function for the

maximum-likelihood estimate which is simply

$$F_n(\hat{\theta}; \theta) = \Phi(r) \left(\frac{1}{r} - \frac{1}{q} \right)$$
$$= \Phi \left(r - r^{-1} \log \frac{r}{q} \right) = \Phi(r^*), \quad (12)$$

where

$$r = \text{sign}(\hat{\theta} - \theta)\{2[\ell(\hat{\theta}) - \ell(\theta)]\},$$
$$q = (\hat{\theta} - \theta)|j(\hat{\theta})|^{1/2}$$

are the signed square root of the log-likelihood ratio statistic and the standardized maximum-likelihood estimator, respectively, and $r^* = r + r^{-1} \log(q/r)$.

As with the p^* approximation, the approximation (12) holds much more generally, with q replaced by a sometimes complicated statistic that depends on the underlying model and in particular on the exact or approximate ancillary statistic required for the validity of (11) in general models. Furthermore, (12) can be applied to marginal and conditional distributions for the maximum-likelihood estimate of a parameter of interest, after nuisance parameters have been eliminated via a marginal or conditional likelihood. A recent accessible reference is Pierce and Peters [32]. The approximation due to Skovgaard [36] is an important ingredient in this development. The r^* approximation, which for general families is due to Barndorff-Nielsen [4], is discussed in Barndorff-Nielsen and Cox [6, Chap. 6].

As an illustration of stochastic asymptotic expansions in likelihood-based inference, consider Taylor series expansion of the score equation $\ell'(\hat{\theta}) = 0$ (assuming for simplicity that this uniquely defines the maximum-likelihood estimator):

$$0 = \ell'(\theta) + (\hat{\theta} - \theta)\ell''(\theta)$$
$$+ \tfrac{1}{2}(\hat{\theta} - \theta)^2 \ell'''(\theta) + \cdots. \quad (13)$$

Reversion of this expansion gives an expansion for the maximum-likelihood estimator that can be expressed as

$$\sqrt{n}(\hat{\theta} - \theta) = \frac{Z_1(\theta)}{i(\theta)} + \frac{1}{\sqrt{n}}\left[\frac{Z_2(\theta)Z_1(\theta)}{[i(\theta)]^2}\right.$$
$$\left. + \frac{Z_1^3(\theta)\rho_3(\theta)}{2[i(\theta)]^3}\right] + O_p(n^{-1}), \quad (14)$$

where $Z_1 = n^{-1/2}\ell'(\theta)$ and $Z_2 = n^{-1/2} \times [\ell''(\theta) - ni(\theta)]$, $i(\theta) = n^{-1}E[\ell'(\theta)]^2$, $\rho_3(\theta) = n^{-1}E[\ell'''(\theta)]$. The random variables Z_1 and Z_2 are $O_p(1)$ and have mean zero. In expansions of this type it is much easier to keep track of the orders of various terms using these standardized variables Z; this notation is originally due to D. R. Cox, and is extensively use as well in McCullagh [30, Chap. 7].

The expansion (14) is a type of stochastic asymptotic expansion, although strictly speaking the distributions for Z_1, Z_2 are only asymptotically free of n. The leading term of (14) gives the usual asymptotic normal approximation for the maximum-likelihood estimator, and the next-order term is useful for deriving refinements of this. For example, it is readily verified that the expected value of $\hat{\theta}$ has the expansion

$$E(\hat{\theta}) = \theta + n^{-1}\frac{i'(\theta) + \rho_3(\theta)/2}{[i(\theta)]^2} + O(n^{-2}),$$

and that $\text{var}(\hat{\theta}) = [ni(\theta)]^{-1} + O(n^{-2})$.

The multivariate version of (14) is given in McCullagh [30, Chap. 7], as are extensions to the nuisance-parameter case and several illustrations of the use of these expansions. One particularly relevant application is the substitution of (14) into a Taylor series expansion of the log-likelihood ratio statistic $w(\theta) = 2[\ell(\hat{\theta}) - \ell(\theta)]$ to obtain an expansion of both the density and the expected value of $w(\theta)$. These expansions lead to the results

$$Ew(\theta) = m\left(1 + \frac{b(\theta)}{n} + O(n^{-2})\right)$$

and

$$\frac{w(\theta)}{1 + b(\theta)/n} = X_m^2[1 + O(n^{-3/2})], \quad (15)$$

where m is the dimension of θ and X_m^2 is a random variable following a χ^2 distribution on m degrees of freedom. The improvement of the approximation to the distribution of the log-likelihood ratio statistic given by (15) is called the *Bartlett correction*, after Bartlett [7], where the correction was derived for testing the equality of several normal variances (*see* BARTLETT ADJUSTMENT—I). It is a multivariate analogue of the improvement of the normal approximation to the signed square root given in (7). The expansion (15) is originally due to Lawley [28]: for details of the derivation see McCullagh [30, Chap. 7], Barndorff-Nielsen and Cox [6, Chap. 6], Bickel and Ghosh [8], and DiCiccio and Martin [19].

Approximations using the Edgeworth and saddlepoint expansions are also useful for statistics that are not derived from a likelihood-based approach to inference. Edgeworth expansions for more general statistics are discussed in Serfling [35] and in considerable generality in Pfanzagl [31]. Skovgaard [38] considers formulations for the density of minimum-contrast estimators that lead to the p^* approximation. Saddlepoint approximation to the density of M-estimators* is discussed in Daniels [14] and Field and Ronchetti [23]. Application of the saddlepoint approximation to the bootstrap is introduced in Davison and Hinkley [17], and explored further in Daniels and Young [16], DiCiccio et al. [20], Wang [40], and Ronchetti and Welsh [34].

A somewhat different application of asymptotic expansions in parametric inference is the use of the techniques outlined here to obtain asymptotic expansions for the efficiency of estimators and the power function of test statistics. One purpose of this is to provide a means for choosing among various estimators or test statistics that have the same efficiency or power to first order of asymptotic theory. Helpful surveys of these types of results are given in Skovgaard [37] and Ghosh [24]: see also the entry ASYMPTOTIC NORMALITY.

REFERENCES

1. Abramowitz, M. and Stegun, I. A. (1965). *Handbook of Mathematical Functions*. Dover, New York.

2. Barndorff-Nielsen, O. E. (1980). Conditionality resolutions. *Biometrika*, **67**, 293–310.

(Introduction to the p^* formula and proof of exactness for transformation models.)

3. Barndorff-Nielsen, O. E. (1983). On a formula for the distribution of the maximum-likelihood estimator. *Biometrika*, **70**, 343–365. (Easier to read than [2]; includes derivation of p^* from the saddlepoint approximation.)

4. Barndorff-Nielsen, O. E. (1986). Inference on full or partial parameters based on the standardized signed log-likelihood ratio. *Biometrika*, **73**, 307–322. (First paper on the use of r^* approximation.)

5. Barndorff-Nielsen, O. E. and Cox, D. R. (1990). *Asymptotic Techniques for Use in Statistics*. Chapman and Hall, London. (Presents detailed development of many asymptotic techniques that have proved useful for obtaining higher-order approximations in statistics, and includes a great number of useful examples. A very valuable reference text for this area. This article has been very strongly influenced by this text.)

6. Barndorff-Nielsen, O. E. and Cox, D. R. (1994). *Inference and Asymptotics*. Chapman and Hall, London. (Follows on from ref. [5], but emphasizes the role of higher-order asymptotics in parametric inference, as outlined in the last section of this article.)

7. Bartlett, M. S. (1937). Properties of sufficiency and statistical tests. *Proc. Roy. Soc. A*, **160**, 268–282.

8. Bickel, P. J. and Ghosh, J. K. (1990). A decomposition for the likelihood ratio statistic and Bartlett correction—a Bayesian argument. *Ann. Statist.*, **18**, 1070–1090.

9. Bleistein, N. and Handelsman, R. A. (1975). *Asymptotic Expansions of Integrals*. Holt, Rinehart and Winston, New York.

10. Cheah, P. K., Fraser, D. A. S., and Reid, N. (1993). Some alternatives to Edgeworth. *Canad. J. Statist.*, **21**, 131–138.

11. Courant, R. and Hilbert, D. (1950). *Methods of Mathematical Physics*. Wiley, New York.

12. Cox, D. R. and Reid, N. (1987). Approximations to non-central distributions. *Canad. J. Statist.*, **15**, 105–114.

13. Daniels, H. E. (1954). Saddlepoint approximations in statistics. *Ann. Math. Statist.*, **25**, 631–650. (The first derivation of the saddlepoint approximation for the density of the sample mean in the statistical literature. Very clearly written and important for any study of saddlepoint expansions. A helpful longer treatment is given in ref. [27].)

14. Daniels, H. E. (1983). Saddlepoint approximations for estimating equations. *Biometrika*, **70**, 89–96.

15. Daniels, H. E. (1987). Tail probability approximations. *Int. Statist. Rev.*, **55**, 37–48. (Derivation of distribution-function approximations using the saddlepoint technique. Illustration of the Lugannani—Rice approximation and two closely related approximations. Very clearly written. A helpful longer treatment is given in ref. [27].)

16. Daniels, H. E. and Young, G. A. (1991). Saddlepoint approximation for the studentized mean, with an application to the bootstrap. *Biometrika*, **78**, 169–179.

17. Davison, A. C. and Hinkley, D. V. (1988). Saddlepoint approximations in resampling methods. *Biometrika*, **75**, 417–431.

18. DeBruijn, N. G. (1970). *Asymptotic Methods in Analysis*, 2nd ed. North-Holland, Amsterdam. Reprinted by Dover, 1981.

19. DiCiccio, T. J. and Martin, M. A. (1993). Simple modifications for signed roots of likelihood ratio statistics. *J. R. Statist. Soc. B*, **55**, 305–316.

20. DiCiccio, T. J., Martin, M. A., and Young, G. A. (1992). Fast and accurate approximation to double bootstrap confidence intervals. *Biometrika*, **79**, 285–296.

21. Easton, G. S. and Ronchetti, E. (1986). General saddlepoint approximations with applications to L-statistics. *J. Am. Statist. Assoc.*, **81**, 420–430.

22. Feller, W. (1970). *Introduction to Probability Theory and Applications*, Vol. II, Wiley, New York.

23. Field, C. A. and Ronchetti, E. (1990). *Small Sample Asymptotics*. Institute of Mathematical Statistics, Hayward. (Emphasizes the use of exponential tilting in constructing approximations for statistics derived from sample means, with special emphasis on robust estimators.)

24. Ghosh, J. K. (1994). *Higher Order Asymptotics*. Institute of Mathematical Statistics, Hayward.

25. Jeffreys, H. (1962). *Asymptotic Approximations*. Oxford University Press.

26. Jensen, J. L. (1986). Inference for the mean of a gamma distribution with unknown shape parameter. *Scand. J. Statist.*, **13**, 135–151.

27. Kolassa, J. E. (1994). *Series Approximation Methods in Statistics*. Springer-Verlag, New York. (A very detailed derivation of saddlepoint and Edgeworth expansions that includes a lot of background material. Very helpful for

understanding how higher-order approximations are derived.)

28. Lawley, D. (1956). A general method for approximating to the distribution of the likelihood ratio criterion. *Biometrika*, **43**, 295–303.

29. Lugannani, R. and Rice, S. O. (1980). Saddlepoint approximation for the distribution of the sum of independent random variables. *Adv. Appl. Probab.*, **12**, 475–490. [Derivation of the approximation given in (7).]

30. McCullagh, P. (1987). *Tensor Methods in Statistics*. Chapman and Hall, London. (Introduces tensor calculus for use in multivariate statistical calculations. Derives multivariate Edgeworth and saddlepoint expansions, and considers application of asymptotic expansions and stochastic asymptotic expansions in parametric statistical inference.)

31. Pfanzagl, J. (1985). *Asymptotic Expansions for General Statistical Models*. Springer-Verlag, New York. (Rigorous development of expansions for efficiency of statistical estimators and power of statistical tests. A comparison of this approach and the one emphasized in this article is given in ref. [37].)

32. Pierce, D. A. and Peters, D. (1992). Practical use of higher order asymptotics for multiparameter exponential families (with discussion). *J. R. Statist. Soc. B*, **54**, 701–738.

33. Reid, N. (1988). Saddlepoint methods and statistical inference (with discussion). *Statist. Sci.*, **3**, 213–238. (A review of the saddlepoint approximation and its relationship to the p^* formula, and the uses of the p^* formula in inference.)

34. Ronchetti, E. and Welsh, A. (1994). Empirical saddlepoint approximations for multivariate M-estimators. *J. R. Statist. Soc. B*, **56**, 313–326.

35. Serfling, R. J. (1982). Asymptotic expansion. In *Encyclopedia of Statistical Sciences*. Wiley, New York, Vol. 1, pp. 137–138.

36. Skovgaard, I. M. (1987). Saddlepoint expansions for conditional distributions. *J. Appl. Probab.*, **24**, 875–887.

37. Skovgaard, I. M. (1989). A review of higher order likelihood inference. *Bull. Int. Statist. Inst.*, **53**, 331–350.

38. Skovgaard, I. M. (1990). On the density of minimum contrast estimators. *Ann. Statist.*, **18**, 779–789. (A very helpful derivation of the p^* formula is given in Section 3.)

39. Wang, S. (1990). Saddlepoint approximation for bivariate distributions. *J. Appl. Probab.*, **27**, 586–597.

40. Wang, S. (1990). Saddlepoint approximations in resampling. *Ann. Inst. Statist. Math.*, **42**, 115–131.

41. Wang, S. (1992). General saddlepoint approximations in the bootstrap. *Statist. Probab. Lett.*, **13**, 61–66.

42. Wood, A. T. A., Booth, J. G., and Butler, R. W. (1993). Saddlepoint approximations to the cdf of some statistics with nonnormal limit distributions. *J. Amer. Statist. Ass.*, **88**, 680–686.

See also ASYMPTOTIC NORMALITY; ASYMPTOTICS, HIGHER ORDER; CORNISH–FISHER AND EDGEWORTH EXPANSIONS; LAPLACE'S METHOD; p^*-FORMULA; and SADDLE POINT APPROXIMATIONS.

NANCY REID

ASYMPTOTIC NORMALITY

The exact distribution of a statistic is usually highly complicated and difficult to work with. Hence the need to approximate the exact distribution by a distribution of a simpler form whose properties are more transparent. The limit theorems* of probability theory provide an important tool for such approximations. In particular, the classical central limit theorems* state that the sum of a large number of independent random variables is approximately normally distributed under general conditions (see the section "Central Limit Theorems for Sums of Independent Random Variables"). In fact, the normal distribution* plays a dominating role among the possible limit distributions. To quote from Gnedenko and Kolmogorov [18, Chap. 5]: "Whereas for the convergence of distribution functions of sums of independent variables to the normal law only restrictions of a very general kind, apart from that of being infinitesimal (or asymptotically constant), have to be imposed on the summands, for the convergence to another limit law some very special properties are required of the summands." Moreover, many statistics behave asymptotically like sums of independent random variables (see the fifth, sixth, and seventh sections). All of this helps to explain the importance of the normal distribution* as an asymptotic distribution.

Suppose that the statistics $T_n, n = 1, 2, \ldots$, when suitably normed, have the standard normal limit distribution; i.e., for some constants $b_n > 0$ and a_n and for every real x we have

$$\Pr[(T_n - a_n)/b_n \leqslant x] \to \Phi(x) \qquad \text{as } n \to \infty$$
$$(1)$$

where

$$\Phi(x) = (2\pi)^{-1/2} \int_{-\infty}^{x} e^{-y^2/2} dy.$$

Then we say that T_n is asymptotically normal with mean a_n and variance b_n^2, or asymptotically normal (a_n, b_n^2). [Note that a_n and b_n^2 need not be the mean and the variance of T_n; indeed, (1) may hold even when T_n has no finite moments.]

It can be shown that if (1) holds for every x, the convergence is uniform in x, so that

$$\sup_{-\infty < x < \infty} |\Pr[(T_n - a_n)/b_n \leqslant x] - \Phi(x)| \to 0$$

$$\text{as } n \to \infty. \quad (2)$$

[This is due to the continuity of $\Phi(x)$.]

The knowledge that (1) or (2) holds is not enough for most statistical applications. For one thing, the statistician wants to know how large n has to be in order that the limit distribution may serve as a satisfactory approximation. Also, if the distribution of T_n depends on unknown parameters, the statistician wants to know how the values of the parameters affect the speed of convergence to the limit. Both goals are met, to some extent, by the Berry–Esseen theorem* (see below) and related results discussed in "Remainder Term in the Central Limit Theorem."

When the approximation provided by the limit distribution* is unsatisfactory, asymptotic expansions*, treated in the third section, may prove more helpful.

Conditions for the convergence of the moments of a statistic to the corresponding moments of its limit distribution are briefly discussed in the fourth section. The fifth section deals with the distributions of functions of asymptotically normal random variables. The asymptotic normality of functions of independent random variables and of sums of dependent random variables is considered in the sixth and seventh sections,

respectively. The final section deals with functional central limit theorems, which are concerned with asymptotic distributions of random functions.

CENTRAL LIMIT THEOREMS* FOR SUMS OF INDEPENDENT RANDOM VARIABLES

The following classical central limit theorem for the partial sums of an infinite sequence of independent, identically distributed (i.i.d.) random variables is due to Lindeberg.

Theorem 1. Let $X_1, X_2, \ldots$ be an infinite sequence of i.i.d. random variables with finite mean a and positive and finite variance σ^2. Then, as $n \to \infty$, $X_1 + \cdots + X_n$ is asymptotically normal $(an, \sigma^2 n)$.

In the following theorem only finite sequences of independent (not necessarily identically distributed) random variables are involved, which makes it better adapted to most applications.

Theorem 2. For each $N = 1, 2, \ldots$ let $X_{N1}, X_{N2}, \ldots, X_{Nn}$ be $n = n(N)$ independent random variables with finite p-th moments, for some $p > 2$. Let $B_N = \sum_j \text{var}(X_{Nj})$ (the index j runs from 1 to n). If

$$B_N^{-p/2} \sum_j E|X_{Nj} - EX_{Nj}|^p \to 0 \qquad \text{as } N \to \infty,$$

$$(3)$$

then $\sum_j X_{Nj}$ is asymptotically normal $(\sum_j EX_{Nj}, B_N)$.

This theorem is due to Liapunov. Condition (3) may be replaced by a weaker condition, due to Lindeberg, which does not assume finite moments of order >2 (see ref. 32).

For a general central limit theorem for sums of independent random variables which assumes no finite moments and for other central limit theorems, see ref. 32, Chap. IV, Sec. 4.

Multidimensional central limit theorems give conditions for the convergence of the distribution of a sum of independent random vectors to a multivariate normal distribution*; see Cramér [10] and Uspensky [40].

If the sums of independent random variables have probability densities, the latter will converge, under certain conditions, to a normal probability density. For results of this type, known as local* central limit theorems, see ref. 32.

REMAINDER TERM IN THE CENTRAL LIMIT THEOREM

The following result, due to Esseen [17], gives an explicit upper bound for the difference between the distribution function of a sum of independent random variables and the normal distribution function.

Theorem 3. Let $X_1, \ldots, X_n$ be independent random variables,

$$EX_j = 0, \qquad E|X_j|^3 < \infty, \qquad (j = 1, \ldots, n),$$

and let

$$B_n = \sum_{j=1}^n EX_j^2 > 0, \qquad L_n = \sum_{j=1}^n E|X_j|^3/B_n^{3/2}.$$

Then

$$\left| \Pr\left[B_n^{-1/2} \sum_{j=1}^n X_j \leqslant x \right] - \Phi(x) \right| \leqslant CL_n$$

$$\text{for all } x, \qquad (4)$$

where C is a numerical constant.

The assumption $EX_j = 0$ is made merely to simplify the notation. If $EX_j = a_j$, replace X_j by $X_j - a_j$ in the statement of the theorem.

The least value of C for which (4) holds is not known. It is known [2] that (4) is true with $C = 0.7975$ and is not true with $C < 0.4097$.

Note that Theorem 3 involves only one finite sequence of independent random variables and is not a limit theorem. It easily implies Theorem 2 with $p = 3$.

Under the further assumption that $X_1, \ldots, X_n$ are identically distributed, with $EX_1 = 0$, $EX_1^2 = \sigma^2$, inequality (4) simplifies to

$$\left| \Pr\left[n^{-1/2}\sigma^{-1} \sum_{j=1}^n X_j \leqslant x \right] - \Phi(x) \right| \leqslant Cn^{-1/2}\sigma^{-3}E|X_1|^3. \qquad (5)$$

This inequality was also derived by Berry [4] and is known as the Berry–Esseen inequality.

The upper bounds in (4) and (5) do not depend on x. S. V. Nagaev has shown that inequality (5) is still true (except perhaps for the value of the constant C) if the right side is multiplied with $(1 + |x|)^{-3}$. For this and related results, see ref. 32.

For extensions of these results to sums of independent random vectors, see ref. 5.

ASYMPTOTIC EXPANSIONS*

Let $X_1, X_2, \ldots$ be i.i.d. random variables, $EX_1 = 0$, $0 < \sigma^2 = EX_1^2 < \infty$,

$$F_n(x) = \Pr\left[\sum_{j=1}^n X_j \leqslant x\sigma n^{1/2} \right].$$

By Theorem 1, $F_n(x) \to \Phi(x)$ as $n \to \infty$. However, the approximation of $F_n(x)$ by $\Phi(x)$ is often too crude to be useful. There are expansions of the difference $F_n(x) - \Phi(x)$ in powers of $n^{-1/2}$ that may provide more accurate approximations.

The form of the expansion depends on whether the random variable X_1 is lattice* or nonlattice. [A random variable X is called a lattice random variable if, for some numbers $h > 0$ and a, the values of $(X - a)/h$ are integers; the largest h with this property is called the maximum span. Otherwise, X is nonlattice.]

Theorem 4. If the random variables X_1, $X_2, \ldots$ are i.i.d., nonlattice, and have a finite third moment, then

$$F_n(x) = \Phi(x) + \Phi'(x)Q_1(x)n^{-1/2}$$

$$+ o(n^{-1/2}) \qquad (6)$$

uniformly in x. Here $\Phi'(x) = (2\pi)^{1/2} \exp \cdot (-x^2/2)$ is the standard normal density function and

$$Q_1(x) = \frac{1}{6}\frac{EX_1^3}{\sigma^3}(1 - x^2).$$

For a proof and for extensions of (6) involving higher powers of $n^{-1/2}$, see refs. 5 and 32.

Expansions of this type have been studied by Chebyshev*, Edgeworth, Cramér, Esseen, and others.

Theorem 5. If $X_1, X_2, \ldots$ are i.i.d. lattice random variables taking the values $a + kh$ ($k = 0, \pm 1, \pm 2, \ldots$), where h is the maximum span, and have a finite third moment, then

$$F_n(x) = \Phi(x) + \Phi'(x)(Q_1(x) + S_1(x))n^{-1/2}$$
$$+ o(n^{-1/2}) \tag{7}$$

uniformly in x. Here

$$S_1(x) = \frac{h}{\sigma} S\left(\frac{x\sigma n^{1/2} - an}{h}\right),$$
$$S(x) = [x] - x + \tfrac{1}{2},$$

and $[x]$ is the largest integer $\leqslant x$.

This theorem is due to Esseen [17]; see also ref. 18. For an extension of (7) that involves higher powers of $n^{-1/2}$, see refs. 5 and 32.

Asymptotic expansions of the distribution function and the probability density function of a sum of independent random variables that need not be identically distributed are also treated in ref. 32.

CONVERGENCE OF MOMENTS

If a statistic T_n has a normal limit distribution, its moments need not converge to the corresponding moments of the latter; in fact, T_n need not have any finite moments.

If the conditions of Theorem 2 with a fixed $p > 2$ are satisfied then for all positive integers $q \leqslant p$, the qth absolute moment of $\sum_j (X_{Nj} - EX_{Nj})/B_N^{1/2}$ converges to the corresponding moment of the standard normal distribution; see S. N. Bernstein* [3] and Hall [23]. A similar result is due to Zaremba [41]. Bounds for the remainder terms in such limit theorems for moments have been obtained by von Bahr [1] and Hall [23], among others. An interesting discussion of the convergence of moments of certain statistics can be found in Cramér [9, Chap. 27].

FUNCTIONS OF ASYMPTOTICALLY NORMAL RANDOM VARIABLES

We often encounter statistics that are functions of sample moments or of generalized sample moments of the form $M_n = n^{-1} \sum_{j=1}^{n} g(X_j)$. If the X_j are i.i.d., $Eg(X_1) = a$, $\text{var}\, g(X_1) = \sigma^2$ $(0 < \sigma^2 < \infty)$, then M_n is asymptotically normal $(a, \sigma^2/n)$.

Theorem 6. Let the random variables M_n, $n \geqslant 1$, be asymptotically normal $(a, \sigma^2/n)$. If $H(x)$ is a function of the real variable x whose derivative $H'(x)$ exists and is $\neq 0$ and continuous at $x = a$, then $H(M_n)$ is asymptotically normal $(H(a), H'(a)^2 \sigma^2/n)$.

This result can be extended to functions of k moment-like statistics which are asymptotically k-variate normal. We state the extension for $k = 2$.

Theorem 7. Let the random vectors (M_{1n}, M_{2n}), $n \geqslant 1$, be asymptotically bivariate normal* with mean (a_1, a_2) and covariances σ_{ij}/n, $i, j = 1, 2$. If $H(x, y)$ is a function of the real variables x and y whose partial derivatives at (a_1, a_2),

$$H_1 = \partial H(x,y)/\partial x|_{(a_1,a_2)},$$
$$H_2 = \partial H(x,y)/\partial y|_{(a_1,a_2)},$$

exist and are not both zero, and which has a total differential at (a_1, a_2), so that

$$H(x,y) = H(a_1, a_2) + H_1 x + H_2 y$$
$$+ x\epsilon_1(x,y) + y\epsilon_2(x,y)$$

where $\epsilon_i(x,y) \to 0$ as $(x,y) \to (a_1, a_2)(i = 1, 2)$, then $H(M_{1n}, M_{2n})$ is asymptotically normal with mean $H(a_1, a_2)$ and variance $(H_1^2 \sigma_{11} + 2H_1 H_2 \sigma_{12} + H_2^2 \sigma_{22})/n$.

Proofs of these or closely related results can be found in refs. 9 and 26.

Note that the conditions of Theorems 6 and 7 are such that $H(M_n)$ and $H(M_{1n}, M_{2n})$ can be approximated by the linear terms of their Taylor expansions*. If the linear terms vanish and they can be approximated by the quadratic terms, the asymptotic distribution will be that of a quadratic form* in normal random variables.

ASYMPTOTIC NORMALITY OF FUNCTIONS OF INDEPENDENT RANDOM VARIABLES

Let $T_n = T_n(X_1, \ldots, X_n)$ be a function of the independent random variables $X_1, \ldots, X_n$. Suppose that $ET_n^2 < \infty$.

*Hájek's projection lemma** approximates T_n by the statistic

$$\hat{T}_n = \sum_{j=1}^{n} E[T_n | X_j] - (n-1)ET_n,$$

which is a sum of independent random variables. By the corollary of that entry we have:

Theorem 8. Let the stated assumptions be satisfied for all n. Suppose that $\hat{T}_n$ is asymptotically normal $(E\hat{T}_n, \text{var } \hat{T}_n)$ and that

$$(\text{var } \hat{T}_n)/\text{var}(T_n) \to 1 \qquad \text{as } n \to \infty.$$

Then T_n is asymptotically normal $(ET_n, \text{var } T_n)$.

Hájek [21] and Dupač and Hájek [14] used the projection lemma to prove the asymptotic normality of a simple linear rank statistic*,

$$\sum_{j=1}^{n} a_n(j) b_n(R_{nj}),$$

where $a_n(j)$, $b_n(j)$ are constants, $R_{n1}, \ldots, R_{nn}$ are the respective ranks* of $X_{n1}, \ldots, X_{nn}$, and, for each n, $X_{n1}, \ldots, X_{nn}$ are mutually independent, continuously distributed random variables. (For details, see the papers cited.) On the asymptotic normality of linear ranks statistics*, see also refs. 20 and 22. Compare also the end of the following section. Related results on multivariate linear rank statistics have been obtained in Ruymgaart and van Zuijlen [37] and the papers there cited.

Another class of statistics whose asymptotic normality can be proved with the help of Hájek's lemma are the U-statistics*,

$$U_n = \frac{1}{\binom{n}{m}} \sum_{1 \leqslant j_1 < \cdots < j_m \leqslant n} f(X_{j1}, \ldots, X_{jm}),$$

where m is a fixed integer, $n \geqslant m$, the X_j are mutually independent random variables, and

f is a real-valued function, symmetric in its m arguments; see ref. 24. A Berry–Esseen type bound for U-statistics* is derived in ref. 7.

Linear combinations of functions of order statistics* are asymptotically normal under general conditions; see Ruymgaart and van Zuijlen [36] and other work there cited.

There are statistics T_n which satisfy the conditions of Hájek's lemma and are asymptotically normally distributed, but whose asymptotic normality cannot be established by means of Theorem 8. A simple example is

$$T_n = X_1 X_2 + X_2 X_3 + \cdots + X_{n-1} X_n,$$

where the X_j are i.i.d. with a finite second moment. This is a special case of a sum of 1-dependent random variables; see the following section.

On the asymptotic normality of the sum of a random number of independent random variables (which is of interest in sequential analysis*), see ref. 8.

ASYMPTOTIC NORMALITY OF SUMS OF DEPENDENT RANDOM VARIABLES

Sums of independent random variables are asymptotically normal under general conditions. We may expect that the asymptotic normality will be preserved if the summands are allowed to be weakly dependent* in a suitable sense.

One way of expressing weak dependence is in terms of conditional expectations. For example, let $X_1, X_2, \ldots$ be a sequence of (possibly dependent) random variables and let $S_n = X_1 + \cdots + X_n$. Suppose that the Liapunov condition (3) with $N = n$, $X_{Nj} = X_j$ is satisfied and that $EX_j = 0$. In ref. 29, Sec. 31, it is shown that if, in addition, the conditional moments $E[X_j | S_{j-1}]$ and $E[X_j^2 | S_{j-1}]$ differ sufficiently little (in a specified sense) from the corresponding unconditional moments, then S_n is asymptotically normal. For other, related results, see ref. 29.

Dvoretzky [15] has shown that sums of dependent random variables are asymptotically normal under conditions such as those in the central limit theorems for sums of independent random variables (e.g., Theorem 2), except that quantities such as means, and the like, are replaced by conditional means, and the like, the conditioning being relative to the preceding sum.

Another notion of weak dependence that has proved fruitful is the following. For simplicity we restrict ourselves to stationary sequences (X_n), so that, for all n, the joint distribution of $X_{h+1}, \ldots, X_{h+n}$ does not depend on h. A stationary sequence (X_n) is said to satisfy the strong mixing condition if there are numbers $d(r)$ converging to 0 as $r \to \infty$ such that

$$| \Pr[A \cap B] - \Pr[A]\Pr[B]| \leqslant d(n - m)$$

for any events A and B determined by conditions on the random variables X_k, $k \leqslant m$ and X_k, $k \geqslant n$, respectively, and for all m, n ($m < n$).

Rosenblatt [34] has shown that the partial sums $X_1 + \cdots + X_n$ of a stationary sequence satisfying the strong mixing condition are asymptotically normal under conditions on some of their moments. For other sufficient conditions, see ref. 27.

A simple example of a sequence satisfying the strong mixing condition is an m-dependent sequence. The sequence $(X_n, n \geqslant 1)$ is said to be m-dependent if for all integers $1 \leqslant r \leqslant s < t \leqslant u$ the random vectors $(X_r, \ldots, X_s)$ and $(X_t, \ldots, X_u)$ are independent whenever $t - s > m$. A central limit theorem for sums of m-dependent random variables was proved in ref. 26. An improved version is due to Orey [30]. On Berry–Esseen type bounds for sums of m-dependent random variables, see Shergin [38].

Sums of m-dependent random variables and U-statistics have the feature in common that some subsets of the summands are mutually independent. A central limit theorem for more general sums of this type is due to Godwin and Zaremba [19].

For a central limit theorem for Markov chains* under conditions related to strong mixing, see Rosenblatt [35]. Sums of martingale differences are asymptotically normal under appropriate conditions; see, e.g., ref. 8.

Finally, we mention some of the so-called combinatorial central limit theorems, which have uses in sampling from a finite population* and in rank statistics. Let the random vector $(R_{n1}, \ldots, R_{nn})$ be uniformly

distributed on the $n!$ permutations of the integers $1, \ldots, n$, and let $a_n(j), b_n(j), j = 1, \ldots, n,$ be real numbers. Then the sums

$$\sum_{j=1}^{n} a_n(j)b_n(R_{nj})$$

are asymptotically normal under certain conditions on the $a_n(j), b_n(j)$; see ref. 20. A similar result [25] holds for sums of the form

$$\sum_{j=1}^{n} a_n(j, R_{nj}).$$

FUNCTIONAL CENTRAL LIMIT THEOREMS

Functional central limit theorems form a far-reaching extension of the classical central limit theorems. We confine ourselves to a brief description of some typical results in this area.

Let $X_1, X_2, \ldots$ be i.i.d. random variables with mean 0 and variance 1. Let $S_0 = 0, S_n = X_1 + \cdots + X_n, n \geqslant 1$, and define for $0 \leqslant t \leqslant 1$

$$Y_n(t) = n^{-1/2}S_{[nt]}$$
$$+ n^{-1/2}(nt - [nt])X_{[nt]+1},$$

where $[nt]$ is the largest integer $\leqslant nt$. Thus for given values of $n, X_1, \ldots, X_n, Y_n(t)$ is a continuous, piecewise linear function of t such that $Y_n(j/n) = n^{-1/2}S_j$ for $j = 0, 1, \ldots, n$.

Now let $W(t), 0 \leqslant t \leqslant 1$, be the standard Brownian motion* process (Wiener process*) on [0,1]. Thus for each fixed $t \in (0, 1]$ the random variable $W(t)$ is normally distributed with mean 0 and variance t ($W(0) = 0$), and for any finitely many points $t_1 < t_2 < \cdots < t_k$ in [0,1] the increments $W(t_2) - W(t_1), W(t_3) - W(t_2), \ldots, W(t_k) - W(t_{k-1})$ are mutually independent. Each increment $W(t_j) - W(t_{j-1})$ is normally distributed with mean 0 and variance $t_j - t_{j-1}$. These facts determine the joint (normal) distribution of $W(t_1), \ldots, W(t_k)$. It is known that the random function $W(t)$, $0 \leqslant t \leqslant 1$, is continuous with probability 1.

By a theorem of Donsker [11] the random functions Y_n converge in distribution*, as $n \to \infty$, to the random function W. The exact meaning of this statement is explained, e.g., in Billingsley [6a]. *See also* CONVERGENCE OF

SEQUENCES OF RANDOM VARIABLES. It has the important implication that for a large class of functionals $h(f)$ of a continuous function $f(t)$, $0 \leqslant t \leqslant 1$, the distributions of the random variables $h(Y_n)$ converge to that of $h(W)$. A trivial example is $h(f) = f(1)$. The implication that the distributions of $Y_n(1) = n^{-1/2} S_n$ converge to that of $W(1)$ is essentially equivalent to the central limit theorem, Theorem 1. A more interesting functional to which Donsker's theorem applies is $h(f) = \max_{0 \leqslant t \leqslant 1} f(t)$. Since $\max_{0 \leqslant t \leqslant 1} Y_n(t) = n^{-1/2} \max(0, S_1, \ldots, S_n)$, Donsker's theorem implies that

$$\lim_{n \to \infty} \Pr[n^{-1/2} \max(0, S_1, \ldots, S_n) \leqslant x]$$
$$= \Pr[\max_{0 \leqslant t \leqslant 1} W(t) \leqslant x]. \qquad (8)$$

A proof of Donsker's theorem can be found in Billingsley [6a], where also other applications of the theorem are discussed and other similar theorems are proved.

Donsker's theorem and theorems of a similar type are called functional central limit theorems.

For the limit in (8) we have

$$\Pr[\max_{0 \leqslant t \leqslant 1} W(t) \leqslant x] = \max(2\Phi(x) - 1, 0). \quad (9)$$

This can be proved from the properties of the Wiener process, or by applying the so-called invariance principle (not to be confused with the invariance principle* in statistical inference*). Donsker's theorem, just as the central limit theorem with $a = 0$ and $\sigma^2 = 1$, assumes only that the X_n are i.i.d. with mean 0 and variance 1. Thus in either theorem the limit is invariant in this class of distributions of X_n. Once it is known that the limit (8) exists, it can be evaluated directly by choosing the distribution of X_n in a convenient way; for details, see Billingsley [6a]. The idea of the invariance principle was first conceived by Erdös and Kac [16].

We conclude with another functional central limit theorem. Let $X_1, X_2, \ldots$ be i.i.d. random variables with common distribution function $F(t)$. Let $F_n(t)$ be the empirical distribution function* corresponding to the sample $X_1, \ldots, X_n$. Define the random function $Z_n(t)$,

t real, by

$$Z_n(t) = n^{1/2}(F_n(t) - F(t)),$$

First suppose that the X_n are uniformly distributed* with $F(t) = t$, $0 \leqslant t \leqslant 1$. In this case $F_n(t) - F(t) = 0$ outside of [0, 1], and we may restrict t to the interval [0, 1].

Let $W^0(t)$, $0 \leqslant t \leqslant 1$, be the Brownian* bridge process. This is the Gaussian process* on [0,1] whose distribution is specified by the requirements

$$EW^0(t) = 0,$$
$$EW^0(s)W^0(t) = \min(s, t) - st.$$

In the present case (F uniform) the random functions Z_n converge in distribution to the random function W^0, in a similar sense as the convergence of Y_n to W; see Donsker [12] or Billingsley [6a]. (An important difference is that the functions Z_n are not continuous as the Y_n are.) One implication is that

$$\lim_{n \to \infty} \Pr[\sup_t n^{1/2} |F_n(t) - F(t)| \leqslant x]$$
$$= \Pr[\sup_t |W^0(t)| \leqslant x]. \qquad (10)$$

Earlier, Kolmogorov [28] proved, by a different method, that the limit in (10) equals

$$1 - 2 \sum_{k=1}^{\infty} (-1)^{k+1} e^{-2k^2 x^2} \qquad (11)$$

for $x > 0$. Thus the probability on the right of (10) is equal to (11). The present approach to deriving results such as (10) was heuristically described by Doob [13] and made rigorous by Donsker [12].

The case where $F(t)$ is an arbitrary continuous distribution function can be reduced to the uniform case (by noting that $X_n' = F(X_n)$ is uniformly distributed on [0, 1]), and (10) remains valid. For the case where $F(t)$ is any distribution function on [0, 1], see Billingsley [6a].

The general foundations underlying the functional central limit theorems were laid by Prohorov [33] and Skorohod [39] and are expounded in the books of Billingsley [6a,6b]. *See also* Parthasarathy [31] and Loève [29, Chap. 13].

REFERENCES

1. Bahr, B. von (1965). *Ann. Math. Statist.*, **36**, 808–818.

2. Beek, P. van (1972). *Zeit. Wahrscheinlichkeitsth.*, **23**, 187–196.

3. Bernstein, S. N. (1939). *Dokl. Akad. Nauk SSSR* (Compt. rend.), **24**, 3–8.

4. Berry, A. C. (1941). *Trans. Amer. Math. Soc.*, **49**, 122–136.

5. Bhattacharya, R. N. and Ranga Rao, R. (1976). *Normal Approximation and Asymptotic Expansions.* Wiley, New York. (Thorough treatment of normal approximations and asymptotic expansions of distributions of sums of independent random variables and random vectors, with emphasis on error bounds.)

6a. Billingsley, P. (1968). *Convergence of Probability Measures.* Wiley, New York.

6b. Billingsley, P. (1971). *Weak Convergence of Measures.* SIAM, Philadelphia.

7. Callaert, H. and Janssen, P. (1978). *Ann. Statist.*, **6**, 417–421.

8. Chow, Y. -S. and Teicher, H. (1978). *Probability Theory: Independence, Interchangeability, Martingales.* Springer-Verlag, New York. (Includes careful proofs of representative results on asymptotic normality.)

9. Cramér, H. (1946). *Mathematical Methods of Statistics.* Princeton University Press, Princeton, N.J. (A classic text, still unsurpassed.)

10. Cramér, H. (1970). *Random Variables and Probability Distributions*, 3rd ed. Cambridge Tracts in Mathematics and Mathematical Physics No. 36. Cambridge University Press, Cambridge (first ed., 1937). (Concise monograph in the classical vein.)

11. Donsker, M. (1951). An Invariance Principle for Certain Probability Limit Theorems. *Mem. Amer. Math. Soc. No. 6.*

12. Donsker, M. (1952). *Ann. Math. Statist.*, **23**, 277–281.

13. Doob, J. L. (1949). *Ann. Math. Statist.*, **20**, 393–403.

14. Dupač, V. and Hájek, J. (1969). *Ann. Math. Statist.*, **40**, 1992–2017.

15. Dvoretzky, A. (1972). *Proc. 6th Berkeley Symp. Math. Statist. Prob.*, Vol. 2. University of California Press, Berkeley, Calif., pp. 513–535.

16. Erdös, P. and Kac, M. (1946). *Bull. Amer. Math. Soc.*, **52**, 292–302.

17. Esseen, C. -G. (1945). *Acta Math.*, **77**, 1–125. (A fundamental paper on asymptotic normality.)

18. Gnedenko, B. V. and Kolmogorov, A. N. (1968). *Limit Distributions for Sums of Independent Random Variables*, rev. ed. (Translated from the Russian by K. L. Chung.) Addison-Wesley, Reading, Mass. (A classic monograph on the subject of the title.)

19. Godwin, H. J. and Zaremba, S. K. (1961). *Ann. Math. Statist.*, **32**, 677–686.

20. Hájek, J. (1961). *Ann. Math. Statist.*, **32**, 501–523.

21. Hájek, J. (1968). *Ann. Math. Statist.*, **39**, 325–346.

22. Hájek, J. and Šidák, Z. (1967). *Theory of Rank Tests.* Academic Press, New York. (Includes results on asymptotic normality of rank statistics.)

23. Hall, P. (1978). *J. Aust. Math. Soc. A*, **25**, 250–256.

24. Hoeffding, W. (1948). *Ann. Math. Statist.*, **19**, 293–325. (On U-statistics.)

25. Hoeffding, W. (1951). *Ann. Math. Statist.*, **22**, 558–566.

26. Hoeffding, W. and Robbins, R. (1948). *Duke Math. J.*, **15**, 773–780. (On sums of m-dependent random variables.)

27. Ibragimov, I. A. and Linnik, Yu. V. (1971). *Independent and Stationary Sequences of Random Variables.* (Translated from the Russian and edited by J. F. C. Kingman.) Wolters-Noordhoff, Groningen. (Includes central limit theorems for sums of stationary random variables.)

28. Kolmogorov, A. N. (1933). *G. Ist. Att.*, **4**, 83–91.

29. Loève, M. (1977). *Probability Theory*, 4th ed., Vols. 1 and 2. Springer-Verlag, New York. (Includes treatment of sums of independent and dependent random variables.)

30. Orey, S. (1958). *Duke Math. J.*, **25**, 543–546.

31. Parthasarathy, K. R. (1967). *Probability Measures on Metric Spaces.* Academic Press, New York.

32. Petrov, V. V. (1975). *Sums of Independent Random Variables.* (Translated from the Russian by A. A. Brown.) Springer-Verlag, New York. (Contains a wealth of information on the subject of the title, with proofs of the more important results.)

33. Prohorov, Yu. V. (1956). *Theory Prob. Appl.*, **1**, 157–214.

34. Rosenblatt, M. (1956). *Proc. Natl. Acad. Sci. USA*, **42**, 43–47.

35. Rosenblatt, M. (1971). *Markov Processes: Structure and Asymptotic Behavior.* Springer-Verlag, New York. (Contains a chapter

on the central limit theorem for Markov processes.)

36. Ruymgaart, F. H. and van Zuijlen, M. C. A. (1977). *Ned. Akad. Wetensch. Proc. A*, **80**, 432–447.

37. Ruymgaart, F. H. and van Zuijlen, M. C. A. (1978). *Ann. Statist.*, **6**, 588–602.

38. Shergin, V. V. (1979). *Teor. Veroyatn. Ee Primen.*, **24**, 781–794. (In Russian. English translation to appear in *Theory Prob. Appl.*, **24**, 782–796.)

39. Skorohod, A. V. (1956). *Theory Prob. Appl.*, **1**, 261–290.

40. Uspensky, J. V. (1937). *Introduction to Mathematical Probability*. McGraw-Hill, New York. (This early text includes much material not found in other books.)

41. Zaremba, S. K. (1958). *Math. Zeit.*, **69**, 295–298.

See also CENTRAL LIMIT THEOREMS, CONVERGENCE RATES FOR; CONVERGENCE OF SEQUENCES OF RANDOM VARIABLES; ERGODIC THEOREMS; LIMIT THEOREMS; and LIMIT THEOREM, CENTRAL.

W. HOEFFDING

ASYMPTOTIC NORMALITY OF EXPERIMENTS

Asymptotic normality* is a feature of many probabilistic or statistical studies. It is often connected with the central limit theorem* for sums of independent variables, for martingales* [13], or for empirical processes* [9,29]. In contrast to these essentially probabilistic results, we consider a more statistical aspect of the situation.

Our framework is that of Wald's theory of *statistical decision functions** [37]. As in that book and in Blackwell [3], we abstract the idea of a statistical experiment by a mathematical structure consisting of a set Θ, a σ-field $\mathscr{A}$ carried by a space χ, and a family $\mathscr{E} = \{P_\theta : \theta \in \Theta\}$ of probability measures on $\mathscr{A}$ (*see* MEASURE THEORY IN PROBABILITY AND STATISTICS). (This is for one-stage experiments or for sequential ones where the stopping rule* is prescribed in advance; otherwise, additional structure is needed.) We don't care how the family $\mathscr{E}$ was created, whether by observing independent random variables or stochastic processes*, but only

study how the likelihood ratios* depend on the parameter $\theta \in \Theta$.

The set Θ is usually called the *set of states of nature* or the *parameter set*. It need not have any special structure, though in many cases it is a Euclidean space.

To further specify a statistical problem, Wald defines a set D of possible decisions and a loss function W on $\Theta \times D$ (*see* DECISION THEORY). A decision procedure ρ is a rule that attaches to each observable point $x \in \chi$ a probability measure ρ_x on D. Having observed x, one selects a $d \in D$ according to ρ_x. Then the statistician suffers a loss $W(\theta, d)$ if θ is the true state of nature. The risk $R(\theta, \rho)$ of the procedure ρ is the expected value under P_θ of the sustained loss.

We can now define Gaussian shift experiments, first in the familiar case where Θ is a Euclidean space, and then, to cope with signal detection with noise* or nonparametric* situations, in a more general setup that covers infinite-dimensional parameter sets.

Also, we will introduce a concept of distance between two experiments $\mathscr{E}$ and $\mathscr{F}$ having the same parameter set Θ. The distance Δ is defined through comparison of the risk functions available on $\mathscr{E}$ and $\mathscr{F}$ respectively. Combining this with the Gaussian shift experiments and introducing a variable n that tends to infinity, one obtains definitions of asymptotically Gaussian (or normal) sequences of experiments.

We then look at the classical local asymptotic normality (LAN) conditions for this framework, and discuss a concept of weak convergence* of experiments by using the distance Δ taken on finite subsets of Θ. Two of the most valuable results of that theory, the Hájek—Le Cam asymptotic minimax and convolution theorems, employ weak convergence as a basic assumption.

Finally, we revisit the distance Δ on experiments, showing that for a number of cases it allows automatic transfer, in asymptotic situations, of results known in the Gaussian case. This is illustrated by recent results of Nussbaum [27] on density estimation*. He shows that most nonparametric density estimation problems are asymptotically equivalent to estimation of a signal in a Gaussian-white-noise problem. The use of the distance Δ

provides the statistician with a firm grip on possible rates of convergence of estimates.

GAUSSIAN SHIFT EXPERIMENTS

Every statistician is familiar with some form of what we call Gaussian (or, for emphasis, Gaussian shift) experiments. A common example is one where the observable variable is a random vector X taking values in some Euclidean space with norm $\| \cdot \|$. One has a function $m(\theta)$ from the parameter space Θ to the same Euclidean space, and the density of X (with respect to the Lebesgue measure) is proportional to $\exp[-\frac{1}{2} \| x - m(\theta) \|^2]$.

The square norm $\| x \|^2$ is often written $\| x \|^2 = x'Mx$, where M is a positive definite matrix. These norms are characterized by the median* equality

$$\| (x + y)/2 - \|^2 + \| (x - y)/2 \|^2$$
$$= \tfrac{1}{2}[\| x \|^2 + \| y \|^2].$$

When a vector space is complete with respect to such a norm, it is a Hilbert space. Hilbert spaces are like Euclidean spaces except that they may be infinite-dimensional. They provide the basic framework for nonparametric estimation problems.

On Euclidean spaces there are more general experiments called heteroskedastic* Gaussian experiments. Their densities are proportional to

$$[\det M(\theta)]^2$$
$$\times \exp\{-\tfrac{1}{2}[X - m(\theta)]'M(\theta)[X - m(\theta)]\},$$

where M depends on θ. We shall not consider them, but concentrate instead on shift (or homoskedastic) experiments.

The name "Gaussian" refers to a theorem of C. F. Gauss [10]. He proved, in the one-dimensional case, that Gaussian experiments are the only shift families (see LOCATION-SCALE PARAMETER) where the average of the observations is the maximum-likelihood estimate* (see GAUSS, CARL FRIEDRICH).

There are Gaussian experiments where the observable entity is not a finite-dimensional vector X but an infinite-dimensional one or a stochastic process.

For instance, the theory of signal detection (in noise) introduces situations where the observable entity is a stochastic process $\{X(t) : t \in T\}$ consisting of a "signal" $\theta(t)$ and a "noise" $W(t)$ so that $X(t) = \theta(t) + W(t)$. Here θ is an unknown nonrandom function, and W is a Gaussian process* in the sense that for every finite set $\{t_1, t_2, \ldots, t_k\}$ the vector $\{X(t_1), X(t_2), \ldots, X(t_k)\}$ has a normal distribution* (cf. Ibragimov and Has'minskii [14, p. 321]). The study of nonparametric or semiparametric* inference quickly introduces infinite-dimensional Gaussian experiments (cf. Bickel et al. [1]).

Let us take a closer look at the experiment $\mathscr{G} = \{G_\theta : \theta \in \Theta\}$ whose densities are proportional to $\exp[-\frac{1}{2} \| X - m(\theta) \|^2]$ on a Euclidean space. Consider the log likelihood $\Lambda(u; s) = \ln dG_u/dG_s$. If u runs through Θ and X has the distribution G_s, this gives a stochastic process. Assume, for convenience, that $m(s) = 0$. Then

$$\Lambda(u; s) = \langle m(u), X \rangle - \tfrac{1}{2} \| m(u) \|^2,$$

where $\langle x, y \rangle$ is the inner product associated with $\| \cdot \|$; i.e.,

$$\langle x, y \rangle = \| (x + y)/2 \|^2 - \| (x - y)/2 \|^2 .$$

Observe the following properties:

1. The process $u \to \Lambda(u; s)$ is a Gaussian process.

2. It stays Gaussian if, instead of G_s inducing the distribution of X, one uses G_θ, $\theta \in \Theta$. The covariance* kernel $\text{Cov}[\Lambda(u; s), \Lambda(v; s)]$ is independent of the particular G_θ used.

3. Changing s to θ to induce the distributions of $\Lambda(u; s)$ adds $\text{Cov}[\Lambda(u; s), \Lambda(\theta; s)]$ to the expectation of $\Lambda(u; s)$.

The first property suggests the following definition:

Definition 1. Let $\mathscr{G} = \{G_\theta : \theta \in \Theta\}$ be an experiment with parameter set Θ. It is called *Gaussian* if:

(i) the G_θ are mutually absolutely continuous* and

(ii) there is an $s \in \Theta$ such that $\Lambda(\cdot; s)$ is a Gaussian process under G_s.

It is easy to show that when (i) holds, if (ii) holds for some $s_0 \in \Theta$, then (ii) holds for all $s \in \Theta$. Furthermore, properties 2 and 3 above are automatic. The covariance kernel of the process is independent of s. Shifting the distributions to those induced by G_θ adds $\text{Cov}[\Lambda(u; s), \Lambda(\theta; s)]$ to the expectation of $\Lambda(\theta; s)$ [23, p. 23].

Our Euclidean example had expectations $\mathbf{m}(\theta)$ in the same Euclidean space as $\mathbf{X}$. For the general case one needs Hilbert spaces. What takes the place of $\mathbf{X}$ will be a linear process.

Consider the space $\mathcal{M}_0$ of finite signed measures μ with finite support* on Θ and such that $\mu(\Theta) = 0$. The integrals $\int \Lambda(u; s)\mu (du)$ are finite linear combinations, with Gaussian distributions. We shall write them as $\langle \mu, Z \rangle$, where $\langle \cdot, \cdot \rangle$ is an inner product attached to the (Hilbert) square norm $\| \mu \|^2$, which is the variance of $\int \Lambda(u; s)\mu(du)$. To embed $\mathcal{M}_0$ into a true Hilbert space $\mathcal{H}$, equate to zero all μ such that $\| \mu \|^2 = 0$ and complete the resulting linear space.

Our likelihood ratio process becomes

$$\Lambda(u; s) = \langle (\delta_u - \delta_s), Z \rangle - \tfrac{1}{2} \| \delta_u - \delta_0 \|^2,$$

where δ_u is the probability measure that gives mass one to u. This can be extended to all of $\mathcal{H}$, giving new Gaussian measures $G(\mu)$ with log likelihood

$$\ln \frac{dG(\mu)}{dG(0)} = \langle \mu, Z \rangle - \tfrac{1}{2} \| \mu \|^2 .$$

If $\theta' \neq \theta''$ implies $G_{\theta'} \neq G_{\theta''}$, then the map $\theta \leadsto \delta_\theta - \delta_s$ is an embedding of Θ onto a certain subset Θ^* of $\mathcal{H}$.

The map $\mu \leadsto \langle \mu, Z \rangle$ is often called the canonical Gaussian process of $\mathcal{H}$. It is characterized (up to equivalence) by the property that it is linear, that the variable $\langle \mu, Z \rangle$ is an ordinary random variable with a normal distribution, and that under G_s, $\langle \mu, Z \rangle$ has expectation zero and variance $\| \mu \|^2$. (As before, variances and covariances do not depend on which $G(\upsilon), \upsilon \in \mathcal{H}$, is used to induce the distributions.)

Assuming that $\theta' \neq \theta''$ implies $G_{\theta'} \neq G_{\theta''}$, the statistical properties of the original experiment $\mathcal{H}$ depend only on the geometric properties of the image Θ^* of Θ in $\mathcal{H}$. More specifically, let $\mathcal{G}_1 = \{G_u : u \in \Theta_1\}$ and $\mathcal{G}_2 = \{G_\upsilon : \upsilon \in \Theta_2\}$ be two Gaussian experiments with parameter spaces Θ_1 and Θ_2 respectively. Introduce the square Hellinger distance*

$$H^2(u, \upsilon) = \tfrac{1}{2} \int (\sqrt{dG_u} - \sqrt{dG_\upsilon})^2.$$

A simple computation shows that the variance of $\Lambda(u; s) - \Lambda(\upsilon; s)$ is given by $-8 \ln[1 - H^2(u, \upsilon)]$. Thus an isometry (distance-preserving transformation) of the Θ_i^* corresponds to an isometry of the initial spaces Θ_i for H.

More concretely, take a point $s_1 \in \Theta_1$, and let s_2 be the point of Θ_2 that corresponds to s_1 in the isometry for H. Carry out the procedure described above for the log likelihoods $\Lambda_i(u; s_i)$. This will give spaces $\mathcal{H}_i$ and processes Z_i. The isometry between the Θ_i^* then extends to a linear isometry of the spaces $\mathcal{H}_i$ (cf. Le Cam [22, p. 239]). The process Z_1 is carried by the isometry to a process Z_1' that has the same distributions as Z_2. Thus the experiments $\mathcal{G}_1$ and $\mathcal{G}_2$ differ only by a relabeling of their parameter sets. Such relabelings are used very frequently. For instance $\mathcal{G}_1$ may be parametrized by a set of square-integrable functions f on the line, but one can then reparametrize it by the Fourier transforms $\tilde{f}$ of f. Whether one represents a square-integrable f by its Fourier coefficients or by its coefficients in some other orthonormal basis does not change anything essential.

For a simple example, consider the signal-plus-noise problem $X(t) = \theta(t) + \tilde{W}(t)$, where $t \in [0, 1]$ and $\tilde{W}$ is the standard Wiener process (see BROWNIAN MOTION). Assume that $\theta(t) = \int_0^t f(c)dc$ for a square-integrable function f. Then the associated Hilbert space $\mathcal{H}$ is the space L_2 of equivalence classes of square-integrable functions with the squared norm $\| f \|^2 = \int_0^1 f^2(c)dc$. Let Z denote standard white noise (equal to the "derivative" of $\tilde{W}$), and then the inner product $< f, Z >$ is $\int f(t)Z(dt)$.

Actually the experiment obtained in that manner is the archetype of all separable

Gaussian experiments. [Here "separable" means that the set $\{G_\theta : \theta \in \Theta\}$ contains a countable dense set for the Hellinger distance*, or equivalently, for the L_1-norm $\| G_u - G_v \| = \sup_{|\varphi| \leq 1} \int \varphi(dG_u - dG_v)$.] Any separable Gaussian experiment can be represented by taking for Θ^* a suitable subset of the space L_2, with the prescription that the observable process is $dX(t) = f_\theta(t)dt + Z(dt)$.

A DISTANCE BETWEEN EXPERIMENTS

Let $\mathscr{E} = \{P_\theta : \theta \in \Theta\}, \mathscr{F} = \{Q_\theta : \theta \in \Theta\}$ be two experiments with the same parameter set Θ. They can have observables in different spaces, say $(\chi, \mathscr{A})$ for $\mathscr{E}$ and $(\mathscr{Y}, \mathscr{B})$ for $\mathscr{F}$. The two experiments can be considered close to each other in a statistical sense if for any arbitrary pairs (D, W) of decision spaces D with loss function W, a risk function available on one of the two experiments can be closely matched by a risk function of the other experiment.

The idea of such possible matchings goes back to Wald [36]. He considered i.i.d. observations and matched any measurable set in the space of n observations to one in the space of maximum-likelihood estimates. His matching is such that the probabilities of the two sets differ little, as $n \to \infty$, uniformly in θ. Wald's demands on this set-to-set relation are too strict; they can be satisfied only under special circumstances.

Consider arbitrary decision spaces D and arbitrary loss functions W on $\Theta \times D$, but *restrict attention to functions W such that* $0 \leq W(\Theta, d) \leq 1$. Let $\mathscr{R}(\mathscr{E}, W)$ be the set of functions from Θ to $[0, \infty]$ that are possible risk functions for $\mathscr{E}$ and W, or at least as large as such possible risk functions (the use of functions larger than actually possible risk functions is a technical convenience). Instead of working with $\mathscr{R}(\mathscr{E}, W)$ we shall work with its closure $\overline{\mathscr{R}}(\mathscr{E}, W)$ for pointwise convergence on Θ.

In most cases $\mathscr{R}$ itself is already closed and thus equal to $\overline{\mathscr{R}}$. This happens for instance if the family $\{P_\theta : \theta \in \theta\}$ is dominated and if the W are lower semicontinuous in d on a decision space D that is locally compact. We shall call $\overline{\mathscr{R}}(\mathscr{E}, W)$ the *augmented* space of risk functions for $\mathscr{E}$ and W.

Definition 2. For $\epsilon \in [0, 1]$, the deficiency $\delta(\mathscr{E}, \mathscr{F})$ of $\mathscr{E}$ with respect to $\mathscr{F}$ does not exceed ϵ if for every D, every W such that $0 \leq W \leq 1$, and every $g \in \mathscr{R}(\mathscr{F}, W)$ there is an $f \in \overline{\mathscr{R}}(\mathscr{E}, W)$ such that $f(\theta) \leq g(\theta) + \epsilon$ for all $\theta \in \Theta$. The *deficiency* $\delta(\mathscr{E}, \mathscr{F})$ is the minimum of the numbers ϵ with that property.

Definition 3. The *distance* $\Delta(\mathscr{E}, \mathscr{F})$ between $\mathscr{E}$ and $\mathscr{F}$ is the number

$$\Delta(\mathscr{E}, \mathscr{F}) = \max\{\delta, (\mathscr{E}, \mathscr{F}), \delta(\mathscr{F}, \mathscr{E})\}.$$

Actually, Δ is not a distance but a pseudometric. It satisfies the triangle inequality (for given Θ), but there are always pairs $(\mathscr{E}, \mathscr{F})$ such that $\mathscr{E}$ and $\mathscr{F}$ are different but $\Delta(\mathscr{E}, \mathscr{F}) = 0$. This happens for instance if $\mathscr{E}$ is an experiment with observations $X_1, X_2, \ldots, X_n$ and $\mathscr{F}$ uses only a sufficient statistic*$T(X_1, X_2, \ldots, X_n)$ instead of the whole set $\{X_1, X_2, \ldots, X_n\}$.

To reiterate, if $\Delta(\mathscr{E}, \mathscr{F}) \leq \epsilon$, this means that for all D and all loss functions W subject to $0 \leq W \leq 1$, every possible function in the (augmented) set of risk functions for one experiment can be matched within ϵ by a function in the (augmented) set of risk functions for the other experiment.

The introduction of "deficiencies" and distances occurs in Le Cam [19]. Blackwell [3] and Stein [31] had previously considered the case where $\delta(\mathscr{E}, \mathscr{F}) = 0$, in which case one says that $\mathscr{E}$ is "better" or "stronger" than $\mathscr{F}$. The deficiency can be computed in terms of Bayes' risks. It is also obtainable by a *randomization* criterion. Let T be a linear transformation that transforms positive measures into positive measures of the same mass. Then

$$\delta(\mathscr{E}, \mathscr{F}) = \inf_T \sup_\theta \tfrac{1}{2} \| Q_\theta - TP_\theta \|,$$

where $\| m \|$ is the L_1-norm $\| m \| = \sup_\varphi \{\int \varphi dm : |\varphi| \leq 1\}$. Those linear transformations are limits of transformations obtainable through randomizations by Markov kernels.

A *Markov kernel* from $(\chi, \mathscr{A})$ to $(\mathscr{Y}, \mathscr{B})$ is a function $x \leadsto \chi$ and a probability measure K_x on $\mathscr{B}$ such that the functions $x \leadsto K_x(B)$ are $\mathscr{A}$-measurable or at least equivalent to $\mathscr{A}$-measurable functions. The corresponding transformation T is given by $(TP)(B) = \int K_x(B)P(dx)$.

The distance Δ applies to experiments $\mathscr{E}$ and $\mathscr{F}$ with the same parameter set Θ, but one can use it to define other distances: If S is a subset of φ, let $\mathscr{E}_s = \{P_\theta : \theta \in S\}$ be the experiment $\mathscr{E}$ with parameter set restricted to S. One can compute a distance $\Delta(\mathscr{E}_s, \mathscr{F}_s)$.

For asymptotic purposes one treats two sequences $\{\mathscr{E}_n\}$ and $\{\mathscr{F}_n\}$, with parameter set Θ_n. These sequences are asymptotically equivalent if $\Delta(\mathscr{E}_n, \mathscr{F}_n) \to 0$ as $n \to \infty$. For instance one can say that the $\{\mathscr{E}_n\}$ are *asymptotically Gaussian* if there are Gaussian experiments $\mathscr{G}_n = \{G_{\theta,n} : \theta \in \Theta_n\}$ such that $\Delta(\mathscr{E}_n, \mathscr{G}_n) \to 0$ as $n \to \infty$. Le Cam and Yang [23] say that the $\mathscr{E}_n$ are *weakly asymptotically Gaussian* if there are Gaussian $\mathscr{G}_n$ such that $\Delta(\mathscr{E}_{n,S_n}, \mathscr{G}_{n,S_n}) \to 0$ as long as the cardinality of $S_n \subset \Theta_n$ remains bounded above independently of n.

There is also a topology of weak convergence of experiments.

Definition 4. Let $\{\mathscr{E}_n\}$ be a sequence of experiments with common parameter set Θ. Let $\mathscr{F} = \{Q_\theta : \theta \in \Theta\}$ be another experiment with the same parameter set. One says that $\mathscr{E}_n \to \mathscr{F}$ weakly if for every *finite* subset $S \subset \Theta$ the distance $\Delta(\mathscr{E}_{n,S}, \mathscr{F}_S)$ tends to zero.

Weak convergence of experiments is equivalent to convergence in the ordinary sense of distributions of likelihood ratios [24, pp. 10–15]. One takes a finite set $\{\theta_0, \theta_1, \ldots, \theta_m\}$, and looks at the vector $\{dP_{\theta_{i,n}}/dP_{\theta_0,h} : i = 0, 1, 2, \ldots, m\}$ with the distributions induced by $P_{\theta_0,n}$. These should converge in the ordinary sense to the distributions of $\{dQ_{\theta_i}/dQ_{\theta_0} : i = 0, 1, \ldots, m\}$ under Q_{θ_0}. (Note that θ_0 is an *arbitrary* point in an *arbitrary* finite set. So the distributions under $P_{\theta_{i,n}}$ also converge. At times it is enough to consider only one particular θ_0, for instance if the $\{P_{\theta_{i,n}}\}$ and $\{P_{\theta_0,n}\}$ are contiguous sequences.)

Besides Δ, several other distances have been used. Torgersen [33,34] has made a deep study of a distance Δ_k defined exactly like Δ, but restricting the decision space D to have at most k elements (k decision problems) or even two elements (test problems). As k increases, so do the Δ_k; and Δ itself is $\sup_k \Delta_k$.

Convergence for the distance Δ is also linked to convergence in distribution of stochastic processes as defined by Prohorov [30]. Suppose $\mathscr{E} = \{P_\theta : \theta \in \Theta\}$ and

$\mathscr{F} = \{Q_\theta : \theta \in \Theta\}$ can be "coupled" to be on the same space and that they are dominated by a suitable probability measure μ. Let X be the process $X(\tau) = dP_\tau/d_\mu$ and let $Y(\tau) = dQ_\tau/d_\mu$ for $\tau \in \Theta$. Suppose that the two experiments are not only coupled on the same space but that they are coupled so that

$$\tfrac{1}{2} \sup_\tau E_\mu |X(\tau) - Y(\tau)| \leqslant \epsilon.$$

Then $\Delta(\mathscr{E}, \mathscr{F}) \leqslant \epsilon$. (An example of coupling occurred in our discussion of isometries for the parameter set of Gaussian experiments.)

By comparison, using the Skorohod embedding theorem (*see* SKOROHOD EMBEDDINGS) or Strassen's theorems [32], convergence of stochastic processes (for the uniform norm) would correspond to convergence to zero of $E_\mu \sup_\tau \{|X(\tau) - Y(\tau)| \wedge 1\}$, a much more delicate affair, since $E_\mu \sup_\tau$ is larger than $\sup_\tau E_\mu$.

LOCAL ASYMPTOTIC NORMALITY: THE LAN CONDITIONS

There are many situations in which one fixes a parameter value θ_0 and studies the behavior of experiments $\mathscr{E}_n = \{P_{\theta,n} : \theta \in \Theta\}$ in small shrinking neighborhoods of θ_0. For i.i.d. data, a famous example is Cramér's work on the roots of maximum-likelihood equations [4, p. 500]. Another example, for Markov processes*, is given by Billingsley [2].

Cramér imposed differentiability conditions on $f(x, \theta)$. An alternative approach, taken by Hájek and Le Cam, assumes that the densities $f(x, \theta)$ are defined with respect to some measure μ and requires the existence of a derivative in μ-quadratic mean of $\sqrt{f(x, \theta)}$. That is, one requires the existence of random vectors $Y(\theta_0)$ such that

$$\int \left(\frac{1}{|t|} \left| \sqrt{f(x, \theta_0 + t)} - \sqrt{f(x, \theta_0)} - \tfrac{1}{2} t' Y(\theta_0) \right| \right)^2 d\mu$$

tends to zero as $|t| \to 0$. The function $Y(\theta_0)$ is then square-integrable, and $\int |t' Y(\theta_0)| d\mu$ is equal to $t' J(\theta_0) t$, where $J(\theta_0)$ is the Fisher information* matrix $\int Y(\theta_0) Y'(\theta_0) d\mu$.

One relies upon local expansions of the log likelihood to study the behavior of these experiments. More precisely, assume that Θ is a subset of a vector space V with norm $|\cdot|$. One selects a sequence $\{\delta_n\}$ of linear maps from V to V and looks at the experiments

$$\mathscr{F}_{\theta_0,n} = \{P_{\theta_0+\delta_n t}, n : t \in V, \theta_0 + \delta_n t \in \Theta\}.$$

It is usually assumed that $\delta_n t \to 0$ as $n \to \infty$ for fixed $t \in V$. In the following, we always require that $\theta_0 + \delta_n t \in \Theta$.

For the Cramér conditions, or Hájek and Le Cam's condition of differentiability in quadratic mean, the space V is Euclidean and the maps δ_n are simply multiplications by $1/\sqrt{n}$, so that $\delta_n t = t/\sqrt{n}$. In contrast, the LAN* conditions of Le Cam [18] do not refer to i.i.d. or Markov processes. They ignore how the $P_{\theta,n}$ were arrived at and focus on the logarithm of likelihood ratios

$$\wedge_n(t) = \ln \frac{dP_{\theta_0} + \delta_n t, n}{dP_{\theta_0,n}}.$$

To state the conditions we shall use a particular Hilbertian or Euclidean norm $\|\cdot\|$ with its associated inner product $<\cdot,\cdot>$. We shall also need a contiguity* condition: Two sequences $\{P_n\}$ and $\{Q_n\}$ of probability measures on σ-fields $\mathscr{A}_n$ are called *contiguous* if sequences $\{X_n\}$ that tend to zero in probability for one of the sequences of measures also tend to zero for the other sequence.

The LAN conditions are as follows, with V an arbitrary vector space with norm $|\cdot|$:

LAN (1). There are random variables X_n and Hilbertian norms $\|\cdot\|$ such that whenever the nonrandom sequence $\{t_n\}$ tends to a limit t in the sense that $|t_n - t| \to 0$, the difference

$$\wedge_n(t_n) - \langle t_n, X_n \rangle + \tfrac{1}{2}\| t_n \|^2$$

tends to zero in $P_{\theta_0,n}$ probability.

LAN (2). If $|t_n - t| \to 0$, then the sequences $\{P_{\theta_0} + \delta_n t_n, n\}$ and $\{P_{\theta_0},n\}$ are contiguous.

One doesn't want the sets $\{t : t \in V, \theta_0 + \delta_n t \in \Theta\}$ to be excessively small. To prevent this, assume:

LAN (3). If V is finite-dimensional, the limit points $t = \lim_n t_n$ with $\theta_0 + \delta_n t_n \in \Theta$ are dense in an open subset of V. If V is infinite-dimensional, the same holds for every finite-dimensional subspace of V.

In these conditions the random X_n, the maps δ_n, and the norms $|\cdot|$ and $\|\cdot\|$ can depend on the particular θ_0 selected for attention. For instance, in the Hájek-Le Cam approach one can take $\| t \|^2 = t'J(\theta_0)t$ or one can take $\| t \|^2 = t't$ but make $\delta_n = \delta_n(\theta_0)$ so that $[\delta_n(\theta)_0 \delta'_n(\theta_0)]^{-1} = nJ(\theta_0)$.

Le Cam [18] contains a fourth, nonlocal requirement for estimates that converge at a suitable speed. If for each $\theta \in \Theta$ one has selected a sequence $\delta_n(\theta)$ of linear maps, the condition is as follows:

LAN (4). There exist estimates $\tilde{\theta}_n$ such that for each $\theta \in \Theta$ the norms $|\delta_n^{-1}(\theta)(\tilde{\theta}_n - \theta)|$ stay bounded in $P_{\theta,n}$ probability.

This permits the construction of asymptotically sufficient estimates (see Le Cam and Yang [24, §5.3]).

Together, these four conditions force Θ to be finite-dimensional. But although they were originally designed for Euclidean subsets Θ, the first three LAN conditions can be used for the infinite-dimensional spaces V needed in the study of signal-plus-noise processes or nonparametric investigations.

Let K be a compact subset of $(V, |\cdot|)$. Let $\mathscr{F}_{\theta_0,n}(K) = \{P_{\theta_0+\delta_n t,n} : t \in K, \theta_0 + \delta_n t \in \Theta\}$. Then the conditions LAN (1), (2), (3) imply the following:

(A) There are Gaussian experiments

$$\mathscr{G}_{\theta_0,n}(K) = \{G_{t,n} : t \in K, \theta_0 + \delta_n t \in \Theta\}$$

such that $\Delta[\mathscr{F}_{\theta_0,n}(K), \mathscr{G}_{\theta_0,n}(K)] \to 0$ as $n \to \infty$ for the distance Δ of Definition 3.

(B) For the norm $\|\cdot\|$ of LAN (1) and suitable random variables W_n one has

$$\ln \frac{dG_{t,n}}{dG_{0,n}} = \langle t, W_n \rangle - \tfrac{1}{2}\| t \|^2$$

with $\langle t, W_n \rangle$ distributed as $N(0, \| t \|^2)$. Thus the $\mathscr{G}_{\theta_0,n}(K)$ reflect the linear structure of V.

Conversely, if (A) and (B) hold, so do LAN (1) and LAN (2). But LAN (3) and LAN (4) cannot be consequences of (A) and (B).

For an example in the nonparametric i.i.d. case we work with densities f defined with respect to Lebesgue measure. For any integer $n \geqslant 1$ the $\sqrt{f}$ can be written in the form

$$\sqrt{f} = 1 - c\left(\frac{\| v_n \|}{2\sqrt{n}}\right) + \frac{v_n}{2\sqrt{n}},$$

where v_n is in the space $L_{2,0}$ of functions v such that $\int v d\lambda = 0$ and $\| v \|^2 = \int v^2 d\lambda < \infty$. Here c is a function from $[0, 1]$ to $[0, 1]$ defined by $[1 - c(z)]^2 = 1 - z^2$; it is used to ensure $\int f d\lambda = 1$. We restrict attention to the subset $L_{2,0}(n)$ of $L_{2,0}$ where

$$1 - c(\| v_n \| /2\sqrt{n}) + v_n/2\sqrt{n} \geqslant 0$$

and $\| v_n \|^2 \leqslant 4n$.

Consider subsets Θ_n of $L_{2,0}(n)$, and take n independent observations from the density

$$\left[1 - c\left(\frac{\| v \|}{2\sqrt{n}}\right) + \frac{v}{2\sqrt{n}}\right]^2.$$

It can be shown that such experiments satisfy LAN (1), (2) and (A), (B) for the squared norm $\| v \|^2 = \int v^2 d\lambda$. The notation has already rescaled the densities, so that $|\cdot|$ and δ_n do not appear, or one can take $|\cdot| = \| \cdot \|$ and δ_n to be the identity. This is all that is needed in most of the arguments and examples of Bickel et al. [1].

THE ASYMPTOTIC MINIMAX AND CONVOLUTION THEOREMS

Hájek [11,12] weakened LAN (1) and LAN (2) by letting $t_n = t$, independent of n. This necessitated the following strengthening of LAN (3).

1. There is a dense subset D of the metric space $(V, |\cdot|)$ such that for $t \in D$ one has $\theta_0 + \delta_n t \in \Theta$ for all n larger than some $n(t)$.

These weakened LAN (1), (2) conditions, together with (H), will be called the Hájek conditions. They do not imply statement (A) of the previous section, but they do imply the following:

Proposition. Under Hájek's conditions the experiments

$$\mathscr{F}_{\theta_0,n}(D) = \{P_{\theta_0 + \delta_n + t, n} : t \in D, \theta_0 + \delta_n t \in \Theta\}$$

converge weakly to a Gaussian experiment $\mathscr{G} = \{G_t : t \in D\}$ with log likelihood

$$L(t) = \langle t, Z \rangle - \tfrac{1}{2} \| t \|^2.$$

The weak convergence is that in Definition 4.

Such a weak convergence already has some interesting statistical consequences. One of them is as follows:

Asymptotic Minimax Theorem. Let $\mathscr{F}_n = \{Q_{t,n} : t \in D\}$ be experiments that converge weakly to a limit $\mathscr{G} = \{G_t : t \in D\}$, not necessarily Gaussian. Fix a loss function W that is bounded below for each fixed t. Let $\mathscr{R}(\mathscr{F}_n, W)$ be the (augmented) set of risk functions introduced in Definition 2, and let r be a function that *does not* belong to $\overline{\mathscr{R}}(\mathscr{G}, W)$. Then there is a finite set $F \subset D$, an $\alpha > 0$, and $N < \infty$ such that if $n \geqslant N$, the function $r + \alpha$ does not belong to $\overline{\mathscr{R}}(\mathscr{F}_{n,F}, W)$, where $\mathscr{F}_{n,F} = \{Q_{t,n} : t \in F\}$.

This result leads directly to Hájek's version of the asymptotic minimax theorem [12]. A stronger form of it is proved in Le Cam [20; 22; pp. 109–110]; for the i.i.d case, under Cramér-type conditions, see [17]. The theorem in [20] relies entirely on weak convergence of experiments and does not use the fact that the limit is Gaussian.

By contrast, Hájek's version of the convolution theorem relies on the fact that the limit a Gaussian shift experiment:

Convolution Theorem. Let V be a Euclidean space, and let Hájek's conditions be satisfied. Let $\mathscr{F}_n = \{Q_{t,n} : t \in g\}$ converge weakly to a Gaussian experiment with log-likelihood ratios

$$\ln \frac{dG_t}{dG_0} = \langle t, Z \rangle - \tfrac{1}{2} \| t \|^2.$$

Consider statistics T_n defined on $\mathscr{F}_n$ and such that $\mathscr{L}[T_n - t | Q_{t,n}]$ has a limit M independent of t for all $t \in S$. Then there is a probability measure π such that M is the convolution $\pi * G$, where $G = \mathscr{L}(Z)$.

This was proved by Hájek [11]; *see also* HÁJEK–INAGAKI CONVOLUTION THEOREM. It also applies to linear functions AT_n and AZ; the limiting distribution of $\mathscr{L}(AT_n)$ is a convolution of $\mathscr{L}(AZ)$. The theorem has been extended to infinite-dimensional (nonparametric) cases where V is a Hilbert space, with the norm $\| \cdot \|$ that occurs in LAN (1) (see Moussatat [26], Millar [25], and van der Vaart [35]).

The convolution theorem has been taken by many as a definition of optimality of estimates. If the T_n are such that $M = \pi * G$ with π concentrated at a point, they are called *optimal*, as other possible limiting distributions are more dispersed.

The Hájek-Le Cam asymptotic minimax and convolution theorems are widely used in asymptotic statistics. But they don't tell the whole story, as is shown in the following section. One of the reasons they cannot tell the whole story is that these theorems can yield only *lower* bounds for the risk of estimates. They cannot yield upper bounds. This is because they are really "local" theorems, and cannot imply such things as the global condition LAN (4). But they do imply, for example, that in the nonparametric i.i.d. case there are no estimates $\tilde{f}_n$ that can identify the square root of densities in such a way that $(n/2)\int(\sqrt{\tilde{f}_n} - \sqrt{f})^2 d\lambda$ stays bounded in probability for all densities f. We shall elaborate on this in the next section.

SOME GLOBAL ASYMPTOTIC NORMALITY SITUATIONS

The Hájek-Le Cam theorems yield lower bounds for the risk of estimates. Convergence in the strong sense of the distance Δ can yield both lower and upper bounds. In principle, this is possible only for bounded loss functions W; however, for unbounded loss functions (such as ordinary quadratic loss), if $\Delta(\mathscr{E}_n, \mathscr{F}_n) \to 0$ then one can truncate W to W_n such that $0 \leqslant W_n \leqslant b_n$ with $b_n\Delta(\mathscr{E}_n, \mathscr{F}_n)$ tending to zero. This is usually sufficient for practical purposes.

As an example of the need for strong convergence, consider the i.i.d. case where $P_{\theta,n}$ is the joint distribution of n variables $X_1, X_2, \ldots, X_n$. Suppose each X_j is Cauchy*

with center θ. Let $\mathscr{E}_n = \{P_{\theta,n} : \theta \in \mathbb{R}\}$ and $\mathscr{G}_n = \{G_{\theta,n} : \theta \in \mathbb{R}\}$, where $G_{\theta,n}$ is the distribution of Y, a $N(\theta, 2/n)$ random variable.

The $\mathscr{E}_n$ converge weakly to a limit where for $\theta \neq \theta'$ the corresponding measures are disjoint. This gives little information. For a fixed θ_0, say $\theta_0 = 0$, the rescaled experiments $\mathscr{F}_n = \{P_{\delta_n t, n} : t \in \mathbb{R}\}$ with $\delta_n t = t/\sqrt{n}$ satisfy the conditions of LAN (1), (2), and (3). By itself, this does not imply LAN (4), but it is sufficient to yield lower bounds on the asymptotic risk of estimates.

In this case one can prove that $\Delta(\mathscr{E}_n, \mathscr{G}_n) \to 0$ as $n \to \infty$. In fact one can prove that $\Delta(\mathscr{E}_n, \mathscr{G}_n) \leqslant C/\sqrt{n}$ for a suitable constant C. This says that the risk of estimates for $\mathscr{E}_n$ will behave like estimates for the Gaussian $\mathscr{G}_n$, except for a term of the type

$$C \parallel W_n \parallel /\sqrt{n},$$

where $\parallel W_n \parallel = \sup_{\theta,d} |W_n(\theta, d)|$. A similar assertion can be made for the general i.i.d. location family case, provided the density $f(x - \theta)$ has finite Fisher information.

As a more interesting problem, Nussbaum [27] considers a nonparametric situation where the variables are i.i.d. with densities f on $[0,1]$. Let Θ be the space of densities such that (1) for a fixed constant K and some fixed $\alpha > \frac{1}{2}$,

$$|f(x) - f(y)| \leqslant K|x - y|^\alpha,$$

and (2) there exists $\epsilon_0 > 0$ such that $f(x) \geqslant \epsilon_0$ for all $x \in [0, 1)$. Let $\mathscr{E}_n = \{P_{f,n} : f \in \theta\}$ be the experiment where $P_{f,n}$ is the joint distribution of n independent observations from $f \in \Theta$. Consider the Gaussian experiment $\mathscr{G}_n = \{G_{f,n} : f \in \Theta\}$ where the observable element is a process $Y(\tau)$ for $\tau \in [0, 1]$ such that

$$dY(\tau) = \sqrt{f(\tau)}d\tau + \frac{1}{2\sqrt{n}}Z(d\tau),$$

where Z is the standard Gaussian white noise* on $[0, 1]$. Nussbaum's result is as follows:

Theorem. As $n \to \infty$ the distance $\Delta(\mathscr{E}_n, \mathscr{G}_n)$ tends to zero.

Note that the shift parameter of $\mathscr{G}_n$ is the square root of the density. As in the nonparametric example of the previous section, this ensures homoskedasticity of $\mathscr{G}_n$.

People know a lot about estimation in the Gaussian case (see Donoho et al. [8], Donoho and Liu [7], Donoho and Johnstone [6], and the references therein). Using Nussbaum's theorem, one can readily transfer all of these Gaussian results to the case of density estimation in Θ, at least for bounded loss functions (and for unbounded ones, by truncation). It follows for instance that the speed of convergence of estimates will be the same in $\mathscr{E}_n$ as in $\mathscr{G}_n$.

Nussbaum gives many applications; we only cite one. Let $W_2^m(K)$ be the Sobolev ball of functions g where for the standard Fourier orthonormal basis on $[0, 1]$ the Fourier coefficients g_j satisfy $\sum (2nj)^{2m} g_j^2 \leqslant K$. Consider a truncated Hellinger loss

$$L(\hat{f}, f, n, c) = \min\{c, n^{1-r} \int (\hat{f}^{1/2} - f^{1/2})^2 d\lambda\},$$

where $r = (2m + 1)^{-1}$.

Let $\mathscr{F}(m, k, \epsilon)$ be the set of densities f such that $f \geqslant \epsilon > 0$ and $f^{1/2} \in W_2^m(K)$. Then for each $\epsilon > 0$ and for $m \geqslant 4$ there is a sequence c_n tending to infinity such that the minimax risk on $\mathscr{F}(m, k, \epsilon)$ for $L(\hat{f}, f, n, c_n)$ tends to $2^{2(r-1)} K^r \gamma(m)$, where $\gamma(m)$ is the Pinsker constant [28] relative to the Gaussian case. This result had been guessed but not proved before Nussbaum's theorem.

RELATED RESULTS

There are situations where approximability by Gaussian shift experiments is not possible, but where the Gaussian results are still useful. For inference on stochastic processes or time series, the LAN conditions may apply, but the locally asymptotically mixed normal (LAMN) conditions apply more generally (cf. [5,15,16]). Here the log-likelihood ratios have an expansion of the type

$$\wedge_n(\mathbf{t}) = \langle \mathbf{t}, S_n \rangle - \tfrac{1}{2} \mathbf{t}' \Gamma_n \mathbf{t} + \epsilon_n,$$

where ϵ_n tends to zero in probability but where the matrices, or linear maps, Γ_n stay random. They are not approximable by nonrandom maps. Nonetheless, a large number of Gaussian results can be applied *conditionally* given the Γ_n.

Several bounds on risks can be obtained through the study of Bayes' procedures when the posterior distributions are approximately Gaussian (see Le Cam and Yang [24, §5.4]). These results stay valid in the LAMN case. For a discussion of such Gaussian approximations to posterior distributions, see Jeganathan [15,16] and Le Cam [22, Chap. 12].

REFERENCES

1. Bickel, P., Klaassen, C., Ritov, Y., and Wellner, J. (1993). *Efficient and Adaptive Estimation for Semiparametric Models*. John Hopkins Series in the Mathematic Sciences.

2. Billingsley, P. (1961). *Statistical Inference for Markov Processes*. University of Chicago Press, Chicago.

3. Blackwell, D. (1951). Comparison of experiments. *Proc. 2nd Berkeley Symp. Math. Statist. Probab.* J. Neyman, ed. University of California Press, Berkeley, pp. 93–102.

4. Cramér, H. (1946). *Mathematical Methods of Statistics*. Princeton University Press.

5. Davies, R. (1985). Asymptotic inference when the amount of information is random. *Proc. Berkeley Conf. in Honor of Jerzy Neyman and Jack Kiefer*, vol. II, L. Le Cam and R. Olshen, eds. Wiley, New York, pp. 241–264.

6. Donoho, D. L. and Johnstone, I. M. (1998). Minimax estimation via wavelet shrinkage. *Ann Statist.*, **26**, 879–921.

7. Donoho, D. L. and Liu, R. (1991). Geometrizing rates of convergence, II and III. *Ann. Statist.*, **19**, 633–701.

8. Donoho, D. L., Liu, R., and MacGibbon, B. (1990). Minimax risk over hyperrectangles and implications. *Ann. Statist.* **18**, 1416–1437.

9. Dudley, R. (1978). Central limit theorems for empirical measures. *Ann. Probab.*, **6**, 899–929.

10. Gauss, C. F. (1809). *The Heavenly Bodies Moving about the Sun in Conic Sections (reprint)*. Dover, New York, 1963.

11. Hájek, J. (1970). A characterization of limiting distributions of regular estimates. *Z. Wahrsch. u. Verw. Geb.* **14**, 323–330.

12. Háek, J. (1972). Local asymptotic minimax and admissibility in estimation. *Proc. 6th Berkeley Symp. Math. Statist. Probab.*, vol. 1, L. Le Cam, J. Neyman, and E. Scott, eds. University of California Press, pp. 175–194.

13. Hall, P. and Heyde, C. (1980). *Martingale Limit Theory and Its Application*. Academic Press, New York.

14. Ibragimov, I. A. and Has'minskii, R. Z. (1981). *Statistical Estimation. Asymptotic Theory*. Springer-Verlag, New York.

15. Jeganathan, P. (1982). On the asymptotic theory of estimation when the limit of the log likelihood is mixed normal. *Sankhyā Ser. A*, **44, (part 2)**, 173–212.

16. Jeganathan, P. (1983). Some asymptotic properties of risk functions when the limit of the experiments is mixed normal. *Sankhyā Ser. A*, **45**, 66–87.

17. Le Cam, L. (1953). On some asymptotic properties of maximum-likelihood and related Bayes' estimates. *Univ. Calif. Publ. Statist.*, **1**, 277–330.

18. Le Cam, L. (1960). Locally asymptotically normal families of distributions. *Univ. Calif. Publ. Statist.*, **3**, 37–98.

19. Le Cam, L. (1964). Sufficiency and approximate sufficiency. *Ann. Math. Statist.*, **35**, 1419–1455.

20. Le Cam, L. (1979). On a theorem of J. Hájek. In *Contribution to Statistics: J. Hájek Memorial Volume*, J. Jurečková, ed. Akademia Praha, Czechoslovakia, pp. 119–137.

21. Le Cam, L. (1985). Sur l'approximation de familles de mesures par des familles gaussiennes. *Ann. Inst. Henri Poincaré, Probab. Statist.*, **21**, 225–287.

22. Le Cam. L. (1986). *Asymptotic Methods in Statistical Decision Theory*. Springer-Verlag, New York.

23. Le Cam, L. and Yang, G. L. (1988). On the preservation of local asymptotic normality under information loss. *Ann. Statist.*, **16**, 483–520.

24. Le Cam, L. and Yang, G. L. (1990). *Asymptotics in Statistics*, Springer-Verlag, New York.

25. Millar, P. W. (1985). Nonparametric applications of an infinite dimensional convolution theorem. *Z. Wahrsch. u. Verw. Geb.*, **68**, 545–556.

26. Moussatat, W. (1976). *On the asymptotic theory of statistical experiments and some of its applications*. Ph.D. thesis, University of California, Berkeley.

27. Nussbaum, M. (1996). Asymptotic equivalence of density estimation and white noise. *Ann. Statist.*, **24**, 2399–2430.

28. Pinsker, M. S. (1980). Optimal filtering of square integrable signals in Gaussian white noise. *Prob. Inf. Transmission*, **1**, 120–123.

29. Pollard, D. (1984). *Convergence of Stochastic Processes*. Springer-Verlag, New York.

30. Prohorov, Yu. V. (1956). Convergence of random processes and limit theorems in probability. *Theory Probab. Appl.*, **1**, 157–214.

31. Stein, C. (1951). *Notes on the Comparison of Experiments*, University of Chicago Press.

32. Strassen, V. (1965). The existence of probability measures with given marginals. *Ann. Math. Statist.*, **36**, 423–439.

33. Torgersen, E. N. (1970). Comparison of experiments when the parameter space is finite. *Z. Wahrsch. u. Verw. Geb.*, **16**, 219–249.

34. Torgersen, E. N. (1991). *Comparison of Statistical Experiments*. Cambridge University Press.

35. van der Vaart, A. (1988). *Statistical Estimation in Large Parameter Spaces*. Tract 44, Centrum voor Wiskunde en Informatica, Amsterdam.

36. Wald, A. (1943). Test of statistical hypotheses concerning several parameters when the number of observations is large. *Trans. Amer. Math. Soc.*, **54**, 426–482.

37. Wald, A. (1950). *Statistical Decision Functions*. Wiley, New York.

See also ABSOLUTE CONTINUITY; ASYMPTOTIC EXPANSIONS—II; CONTIGUITY; DECISION THEORY; GAUSSIAN PROCESSES; HÁJEK–INAGAKI CONVOLUTION THEOREM; MEASURE THEORY IN PROBABILITY AND STATISTICS; and MINIMAX DECISION RULES.

L. LE CAM

ASYMPTOTICS, HIGHER ORDER

Statisticians often seek to approximate quantities, such as the density of a test statistic evaluated at a fixed ordinate, that depend on a known parameter, such as the sample size, in a very complicated way. The resulting approximation should be tractable analytically or numerically. Asymptotic expansions* are approximations that can be expressed as a sum of a small number of terms, each of which is the product of a factor that is a simple function of the parameter, and a coefficient that does not depend on the parameter. Asymptotic expansions are distinguished from other potential approximations in that their accuracy is assessed by examining the limiting behavior of errors as the parameter approaches some limiting value. This limiting value is usually infinity in the very common case in which the known parameter is a measure of sample size.

Specifically, we will consider asymptotic approximations to a quantity f_n, depending on n, of the form

$$f_n = \sum_{j=1}^{r} g_j b_{jn} + R_{r,n}, \qquad (1)$$

with attention paid to the size of the remainder term $R_{r,n}$ as $n \to \infty$. Usually the coefficients b_{jn} are quantities that decrease as n moves toward its limiting value, and decrease more quickly for larger j. Typical choices for b_{jn} are inverse powers of n or its square root.

Frequently the quantity to be approximated is a function of another variable. For instance, when f_n is a function of x, (1) becomes

$$f_n(x) = \sum_{j=1}^{r} g_j(x) b_{jn} + R_{r,n}(x); \qquad (2)$$

the coefficients g_j and the error term are allowed to depend on x, and the factors b_{jn} and the error term are allowed to depend on n.

Methods with $r = 1$, using $g_1 b_{1n}$ to approximate f_n, are generally described by the term *first-order asymptotic*, and when additional terms are included one calls r defined in (1) the *order* of asymptotic approximation applied. Higher-order asymptotic methods, then, are defined here to be applications of (1) or (2) with $r > 1$; this includes second-order asymptotic methods.

We will consider innovative approaches to using terms in the series beyond the initial term $g_1 b_{1n}$ for assessing powers of tests and efficiencies of estimators, both asymptotically as n becomes large, and for finite n. Details may be omitted in discussing the applications of higher-order asymptotics, in order to avoid duplication with other *ESS* entries.

EFFICIENCY IN GENERAL

Much of the material in this section also appears in the entry EFFICIENCY, SECOND-ORDER. Common definitions of the relative efficiency of two statistical procedures involve comparisons of sample sizes necessary to provide equivalent precision. Specifically, suppose that k_n observations are required to give the second procedure the same precision as is realized with n observations from the first procedure; then the relative efficiency of these two procedures is k_n/n, and the *asymptotic relative efficiency* is ARE $= \lim_{n \to \infty} k_n/n$. Placing this in the context of (1),

$$\frac{k_n}{n} = \text{ARE} + R_{1,n},$$

where $\lim_{n \to \infty} R_{1,n} = 0$. \qquad (3)

Here equality of precision may refer to equality of mean square errors* of two estimates, or powers of two tests for a similar alternative; various precise definitions of equality of precision are discussed in the next section. The first procedure is preferable if ARE > 1, and the second procedure is preferable if ARE < 1.

Fisher [7] considered discrimination between estimation procedures when the asymptotic relative efficiency is unity, and explored situations in which a second-order version of (3) exists. Suppose that ARE $= 1$ and

$$\frac{k_n}{n} = 1 + \frac{d}{n} + R_{2,n},$$

where $\lim_{n \to \infty} \frac{R_{2,n}}{n} = 0$. \qquad (4)

The first procedure is preferable if $d > 0$, and the second procedure is preferable if $d < 0$. The relation (4) implies that

$$d = \lim_{n \to \infty} (k_n - n).$$

Hodges and Lehmann [9] define the *deficiency* of the two procedures to be $k_n - n$; the asymptotic deficiency of the second procedure relative to the first is then d, when this limit exists. A simple example is presented in the next section.

Hodges and Lehmann [9] give an example involving estimation in which the limit d is infinite, and so an expansion as above need not always exist. Pfanzagl [14] and Ghosh [8] present regularity conditions under which such an expansion will exist when comparing the higher-order efficiencies of first-order efficient tests; see the review article by Skovgaard [19].

COMPARISONS OF ESTIMATORS

Consider assessing an asymptotically normal and asymptotically unbiased estimator of a parameter. One might take as a definition of the efficiency of this estimator the ratio of the Cramér—Rao* lower bound for its variance to the actual achieved variance. One might also define the estimator's asymptotic efficiency as the limit of this ratio as some parameter, usually reflecting the sample size, increases [2, pp. 137 f.; 4, pp. 304 ff.]. Estimates with efficiency closer to one are preferable to those with lower efficiency. Estimators whose asymptotic efficiency is unity are *first-order efficient*.

When one considers collections of estimators one often is interested in their behavior relative to one another. The relative efficiency for two estimators is the inverse of the ratio of their variances, and the asymptotic relative efficiency is the limit of this inverse ratio. When the two estimators have variances approximately proportional to n in large samples, this definition of asymptotic relative efficiency coincides with the definition in terms of relative sample sizes needed to give equivalent precision. This section will place the differentiation between estimators in the context of (1).

As a simple example, consider estimating a mean of a population with finite variance, using a sample of n independent observations. If one procedure estimates the mean as the sample mean, and the second procedure estimates the mean as the sample mean with the first observation ignored, then $k_n = n + 1$, the relative efficiency is $(n + 1)/n$, and the asymptotic relative efficiency is 1. The deficiency is then 1 for all values of n, and so the asymptotic deficiency is also 1.

Fisher [7] argues heuristically that maximum-likelihood estimation* produces estimators that are first-order efficient, with variances, to first order, given by the Fisher information* in the whole sample, and that the loss in efficiency incurred by other estimators might be measured by the correlation of these other estimators with the maximum-likelihood estimator, or alternately by the differences between the whole sample information and the information in the sampling distribution of the estimator. Other authors have made these claims rigorous, and some of these results will be reviewed below. Wong [23] presents a more thorough rigorous review. The correlation between estimators may be used to build a definition for second-order efficiency which is equivalent to the Fisher information difference, under certain regularity conditions [17,18].

Higher-order asymptotic expansions for the mean square error for the maximum-likelihood estimator can be generated. Since expansions for the mean square error are related to expansions for the information content of the maximum-likelihood estimator, and the information expansion is simpler, the information expansion will be considered first. Efron [6] uses methods from differential geometry to define the statistical curvature* γ_θ of an inference problem, and relates it to the loss of efficiency when inference procedures designed for local alternatives are applied globally.

Consider a family of distributions on a sample space χ parametrized by θ taking values in $\Theta \subset R$, and suppose that $X \in \chi^n$ is a vector of n independent and identically distributed variables X_j. Let $l = l(\theta; X)$ be the log likelihood for X. Let $\ddot{l}(\theta, X)$ be the second derivative of the log likelihood with respect to θ. Let $i_n(\theta) = -E[\ddot{l}(\theta, X); \theta]$ be the Fisher information in the sample X, let $i_n^{\hat{\theta}}(\theta)$ be the Fisher information for the sampling distribution of $\hat{\theta}$, the maximum-likelihood estimator for θ, and let $i_1(\theta)$ be the Fisher information for X_1. If γ_θ^1 is the curvature defined for the distribution of a random variable X_1, and γ_θ^n is the curvature calculated for the distribution of X, then $\gamma_\theta^1 = \gamma_\theta^n / \sqrt{n}$. One may show that $\lim_{n \to \infty} [i_n(\theta) - i_n^{\hat{\theta}}(\theta)] = i_1(\theta)(\gamma_\theta^1)^2$, and hence $i_n^{\hat{\theta}}(\theta)/n = i_1(\theta) - i_1(\theta)\gamma_\theta^1/n + R_{2,n}$, where $\lim_{n \to \infty} nR_{2,n} = 0$, giving an asymptotic expansion* for the average information contained in a maximum-likelihood estimator. Efron [6] also produced an asymptotic expansion for the variance of the maximum-likelihood estimator at a parameter value θ_0, which contains the statistical curvature and additional terms involving the curvature of the bias of the result of one scoring iteration, and the bias of the maximum-likelihood estimator at θ_0. These terms are all of size $O(n^{-2})$, which means that after dividing by n^{-2} they remain bounded, and the error is of size $o(n^{-2})$, which means that after dividing by n^{-2} the error converges to zero as $n \to \infty$. For more details see EFFICIENCY, SECOND-ORDER. ·

COMPARISONS OF TESTS

Asymptotic comparisons of powers* of families of tests having exactly or approximately the same significance level have been examined by many authors. Generally their investigations have considered the problem of testing a null hypothesis that a statistical parameter θ takes a null value, of the form $H_0 : \theta = \theta_0$, using two competing tests $T_{1,n}$ and $T_{2,n}$, indexed by a parameter n, generally indicating the sample size. Their critical values $t_{1,n}$ and $t_{2,n}$ satisfy

$$P[T_{i,n} \geqslant t_{i,n}; H_0] = \alpha \quad \text{for} \quad i = 1, 2. \quad (5)$$

Measures of global asymptotic efficiency of two competing tests such as Bahadur efficiency* are generally functions of the parameter determining the distribution under an alternative, and since these functions generally do not coincide for competing tests, higher-order terms in sample size are generally not needed for choosing among tests of identical first-order efficiency.

Single-number measures of efficiency often times compare powers of tests whose sizes, exactly or approximately, are fixed and identical. For consistent tests and a fixed alternative hypothesis, distribution functions for the test statistics (or asymptotic approximations to these distribution functions) indicate an identical first-order asymptotic power of unity. Distinguishing between such tests, then, requires a local measure of relative efficiency such as Pitman efficiency*, which is the ratio of sample sizes necessary to give the same power against a local alternative. That is, alternatives of the form $H_A : \theta = \theta_0 + \epsilon/c_n$, where $c_n \to \infty$, are considered, and the limit $\lim_n (k_n/n)$ is desired, where

$$P[T_{1,n} \geqslant t_{1,n}; H_A] = P[T_{2,k_n} \geqslant t_{2,n}; H_A]. \quad (6)$$

Often competing tests can be found whose measures of asymptotic relative efficiency against local alternatives is unity. One frequently wishes to discriminate between two

such tests. Hodges and Lehmann [9] apply their concept of deficiency to the problem of comparing tests in cases in which sizes can be calculated exactly.

Often exact expressions for the probabilities in (5) and (6) are unavailable. In such cases the critical value, as well as the power usually must be approximated. Pfanzagl [14] notes that asymptotic comparisons of power are only interesting when significance levels of the tests agree to the same asymptotic order, and achieves this equality of size through a process of Studentization* in the presence of nuisance parameters*. Such equality of size might be obtained using a Cornish—Fisher* expansion to calculate the critical value for the test (see ASYMPTOTIC EXPANSIONS—II). Albers et al. [1] apply Edgeworth series to calculate the powers of nonparametric tests. The primary difficulty in such cases arises from the discrete nature of the distributions to be approximated. Pfaff and Pfanzagl [15] present applications of Edgeworth and saddlepoint* expansions to the problem of approximating power functions for test statistics with continuous distributions; they find that the Edgeworth series is more useful for analytic comparisons of power, while saddlepoint methods give more accurate numerical results. Applications to tests for sphericity*, and for the significance of a common cumulative logit in a model for ordered categorical data*, are presented by Chan and Srivastava [3] and Kolassa [10], respectively.

Asymptotic expansions of cumulative distribution functions can also be used to calculate asymptotic relative efficiencies and deficiencies. Those applications discussed here are under local alternatives. First-order approximations are generally sufficient to calculate asymptotic relative efficiencies; deficiency calculations generally require that second-order terms be included as well. Peers [13] uses approximations of the form (1) to approximate the cumulative distribution function of the likelihood-ratio statistics, Wald statistics*, and score statistics* as a mixture of noncentral χ^2 distributions. Taniguchi [21] calculates these terms to examine cases when the deficiency is zero and a third-order counterpart of (4) is required; comparisons may then be based on the final coefficient.

Many authors use higher-order asymptotic methods to aid in the construction of hypothesis tests; the most classical uses involve deriving the Bartlett correction to likelihood-ratio tests; *see also* BARTLETT ADJUSTMENT—I.

AN EXAMPLE

As an example, consider calculations of power in tests of equality of distribution for two populations classified into ordered categories. The Mann—Whitney—Wilcoxon statistic* is the score statistic for testing this null hypothesis against the alternative that the data are generated by the constant-cumulative-odds model of McCullagh [12]. Whitehead [22] proposes a method for approximating the power of the resulting test. Critical values are derived using a normal approximation with an approximate variance. Powers are calculated by approximating the alternative distribution as normal with the mean approximated by the parameter value times an approximation to the expected information. These first-order asymptotic methods will be compared with higher-order Edgeworth series methods.

One-sided tests of size $\alpha = 0.025$ for 2×4 contingency tables* with row totals both equal to 30 were examined. The column probabilities are those investigated by Whitehead [22], (0.289, 0.486, 0.153, 0.072).

For a variety of model parameters the first-order, higher-order Edgeworth, and Monte Carlo* approximations to these powers are presented. The Edgeworth series used was the distribution function corresponding to the density (4) from the entry on asymptotic approximations*:

$$F_n(s) = \Phi(s) - \phi(s)\left[\frac{\lambda_3}{6}h_2(s)\right.$$
$$\left. +\frac{\lambda_4}{24}h_3(s) + \frac{\lambda_3^2}{72}h_5(s)\right],$$

where the cumulants λ_k contain inverse powers of the row totals and are calculated by Kolassa [10], and the test statistic has been standardized to have zero mean and unit variance. All Monte Carlo approximations involve 1,000,000 samples, and so have standard errors no larger than 0.0005. Figure 1 contains the results. The Edgeworth approximations to the power are visually indistinguishable from the Monte Carlo approximations; the first-order approximations are not so close. Kolassa [10] shows that actual and nominal test sizes correspond more closely than the corresponding power values. Because of discreteness in the distribution of the test statistics, the first- and higher-order size calculations coincide in many cases; where they disagree, the nominal levels generating the Cornish—Fisher expansion appear to be substantially more accurate than the first-order approximations.

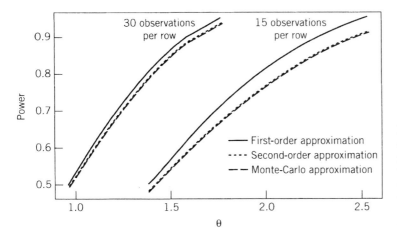

Figure 1. Comparisons of first- and second-order power approximations for ordered categorical data. (The second-order and Monte Carlo approximations are indistinguishable.)

OPEN QUESTIONS

Three topics among the open questions and ongoing difficulties in applying higher-order asymptotic methods will be reviewed here.

These are extending asymptotic methods to the more difficult situations involving statistics generated from nonindependent observations, simplifying regularity conditions, and simplifying the form of the resulting approximation.

Some results in the first area, particularly with references to ARMA processes, are presented by Swe and Taniguchi [20] and in the references they cite. The second area is perhaps not so important. Asymptotic methods, and saddlepoint methods in particular, are often remarkably accurate for a wide range of applications, and intuition often serves better than a careful evaluation of regularity conditions to indicate those contexts in which asymptotic methods are likely to be useful. Nevertheless, for certain applications careful examination of these conditions is sometimes required, and these conditions can often be quite complicated and not particularly intuitive [21].

Simplifying the form of asymptotic expansions is important for increasing their usefulness. Higher-order terms often are quite complicated [11,21]. Simplification might arise from choosing among various different asymptotic approximations accurate to the same order; see Daniels [5] and Pierce and Peters [16].

REFERENCES

1. Albers, W., Bickel, P. J., and van Zwet, W. R. (1976). Asymptotic expansions for the power of distribution free tests in the one-sample problem. *Ann. Statist.*, **4**, 108–156. (This paper presents Edgeworth series results for sums of discrete random variables, useful for implementing nonparametric tests.)

2. Bickel, P. J. and Doksum, K. A. (1977). *Mathematical Statistics: Basic Ideas and Selected Topics*. Holden-Day, Oakland, Calif. (A standard introductory text in theoretical statistics.)

3. Chan, Y. M. and Srivastava, M. S. (1988). Comparison of powers for the sphericity tests using both the asymptotic distribution and the bootstrap method, *Comm. Statist. Theory Methods*, **17**, 671–690.

4. Cox, D. R. and Hinkley, D. V. (1982). *Theoretical Statistics*. Chapman and Hall, London. (A standard text in theoretical statistics.)

5. Daniels, H. E. (1987). Tail probability approximations. *Internat. Statist. Rev.*, **55**, 37–46.

6. Efron, B. (1975). Defining the curvature of a statistical problem (with applications to second-order efficiency). *Ann. Statist.*, **3**, 1189–1242. (This paper presents a geometric approach to asymptotics for models that can be embedded in curved exponential families.)

7. Fisher, R. A. (1925). Theory of statistical estimation, *Proc. Cambridge Phil. Soc.*, **22**, 700–725.

8. Ghosh, J. K. (1991). Higher order asymptotics for the likelihood ratio, Rao's and Wald's tests. *Statist. Probab. Lett.*, **12**, 505–509.

9. Hodges, J. L., Jr., and Lehmann, E. L. (1970). Deficiency. *Ann. Math. Statist.*, **41**, 783–801. (This paper is an important work on higher-order asymptotic efficiency.)

10. Kolassa, J. E. (1995). A comparison of size and power calculations for the Wilcoxon statistic for ordered categorical data. *Statist. Med.*, **14**, 1577–1581.

11. Kolassa, J. E. (1996). Higher-order approximations to conditional distribution functions. *Ann. Statist.*, **24**, 353–364.

12. McCullagh, P. (1980). Regression models for ordinal data. *J. Roy. Statist. Soc. Ser. B*, **42**, 109–142.

13. Peers, H. W. (1971). Likelihood ratio and associated test criteria. *Biometrika*, **58**, 577–587. (This paper calculates approximations to statistical power at local alternatives in order to approximate efficiencies.)

14. Pfanzagl, J. (1980). Asymptotic expansions in parametric statistical theory. In *Developments in Statistics*, vol. 3, P. R. Krishnaiah, ed. Academic Press, New York, pp. 1–98. (This paper is an extended review article on the uses of asymptotics in statistical inference.)

15. Pfaff, T. and Pfanzagl, J. (1985). On the accuracy of asymptotic expansions for power functions. *J. Statist. Comput. Simul.*, **22**, 1–25.

16. Pierce, D. A. and Peters, D. (1992). Practical use of higher-order asymptotics for multiparameter exponential families. *J. Roy. Statist. Soc. Ser. B*, **54**, 701–737. (This paper reviews alternative forms of saddlepoint distribution function expansions.)

17. Rao, C. R. (1962). Efficient estimates and optimum inference procedures. *J. Roy, Statist. Soc. Ser. B*, **24**, 47–72.

18. Rao, C. R. (1963). Criteria of estimation in large samples. *Sankhyā*, **25**, 189–206.

19. Skovgaard, I. M. (1989). A review of higher-order likelihood methods. In *Bull. Int. Statist. Inst. Proc. Forty-seventh Session*. vol. III, International Statistical Institute, Paris, pp. 331–351. (This paper is a recent review article on higher-order asymptotic methods.)

20. Swe, M. and Taniguchi, M. (1991). Higher-order asymptotic properties of a weighted estimator for Gaussian ARMA processes. *J. Time Ser. Anal.*, **12**, 83–93.

21. Taniguchi, M. (1991). Third-order asymptotic properties of a class of test statistics under a local alternative. *J. Multivariate Anal.*, **37**, 223–238.

22. Whitehead, J. (1993). Sample size calculations for ordered categorical data. *Statist. Med.*, **12**, 2257–2271.

23. Wong, W. H. (1992). On asymptotic efficiency in estimation theory. *Statist. Sinica*, **2**, 47–68.

See also Asymptotic Expansions—II; Bahadur Efficiency; Efficiency, Second-Order; Fisher Information; Likelihood Ratio Tests; Log-linear Models in Contingency Tables; Mann–Whitney–Wilcoxon Statistic; Maximum Likelihood Estimation; Pitman Efficiency; Score Statistics; Statistical Curvature; and Wald's W-Statistics.

John E. Kolassa

ATMOSPHERIC STATISTICS

The term atmospheric statistics covers a large body of work. In broad terms, this work can be divided into two categories: *statistical atmospheric statistics* and *atmospheric atmospheric statistics*.

Although the distinction is not always clearcut, in statistical atmospheric statistics, the application of statistics to problems in atmospheric science is, to a large extent, incidental. The hallmark of statistical atmospheric statistics is that, with little modification, the same analysis could be applied to data from an entirely different field. For example, the possibility of weather modification* attracted a lot of interest in the atmospheric science community during the 1970s, and a number of cloud-seeding experiments were planned and performed. As in other fields, statistical issues arose over the design of these experiments and over the analysis and interpretation of their results. This work is reviewed in [5] and [12]; *see also* Weather Modification—I and Weather Modification—II.

A second example of statistical atmospheric statistics is the development of stochastic rainfall models [28,29]; *see also* Rainfall, Landforms, and Streamflow. Here, the problem is to construct a model of variations in rainfall intensity over time and space that can be used in the design of reservoirs and storm-sewer or flood control systems (*see* Dam Theory). Other examples of statistical atmospheric statistics are described in [21] and Meteorology, Statistics in.

In recent years, atmospheric issues with global environmental implications have received considerable attention. Some statistical work on two of these issues—stratospheric ozone depletion and global warming—is reviewed below.

STATISTICS IN WEATHER PREDICTION

In contrast to statistical atmospheric statistics, atmospheric statistics is characterized by a close connection to atmospheric science. Compared to other scientific fields, atmospheric science is organized to an unusual degree around a single practical problem: the prediction of the future state of the atmosphere. Bjerknes [1] referred to this as the ultimate problem in meteorology. Since the time of Bjerknes, atmospheric prediction has been viewed as an initial-value problem. That is, to predict the future state of the atmosphere it is necessary, first, to describe the current state of the atmosphere, and second, to predict the future state by applying the laws of atmospheric dynamics to the current state. The historical development and modern view of this two-step process are described in ref. 8.

The first step in this two-step process is the estimation of the current state of the atmosphere, as represented by one or more atmospheric variables, at a dense grid of locations (called *analysis points*) from relatively sparse observations made at irregularly spaced locations (called *observation points*). This is called *objective analysis*; the most prominent role for statistics in atmospheric prediction lies in it. Early methods for objective analysis were based on fitting polynomial or other parametric functions to observations distributed in space by weighted least squares* [14,23]. The weights in this fitting are used to allow for unequal variances in the observations, which may be from quite different types of instruments (surface stations, weather balloons, satellites, etc.). Both local fitting and global fitting have been used.

As noted, the state of the atmosphere is described by more than one variable (temperature, pressure, humidity, etc.). These variables are coupled through governing equations. Methods have been developed to incorporate the constraints imposed by these governing equations in *function fitting*. A popular approach was devised by Flattery [11] for the objective analysis of the geopotential and the two horizontal components of wind. Under this approach, the three variables are expressed through an expansion in Hough functions. Hough functions are the eigenfunctions of a linearized version of the Laplace tidal equations, a relatively simple model of atmospheric dynamics on the sphere [6]. Because the model is linearized, the constraint—in this case, the geostrophic relation under which the Coriolis forces are balanced by the atmospheric pressure gradient—is only approximately met.

During the early 1970s, the use of function fitting for objective analysis declined, due in part to numerical problems, and interest shifted to a method called *statistical interpolation*, which was introduced into atmospheric science by Gandin [13]. It is essentially equivalent to linear interpolation methods developed by Kolmogorov [18] and Wiener [40] and popularized by Matheron [20] under the name *kriging*.*

To begin with, consider the univariate case. Let $Y(x)$ be the value of the variable of interest at location x, let x_0 be an analysis point, and let $x_i, i = 1, 2, \ldots, n$, be nearby observation points. Under statistical interpolation, the estimate of $Y(x_0)$ is given by

$$Y^*(x_0) = \mu(x_0) + \sum_{i=1}^{n} w_i [Y(x_i) - \mu(x_i)],$$

where $\mu(x)$ is the mean of $Y(x)$. This can be rewritten in obvious matrix notation as

$$Y^*(x_0) = \mu(x_0) + \mathbf{w}'(\mathbf{Y} - \boldsymbol{\mu}).$$

The vector w of interpolation weights is chosen to minimize the variance of the interpolation error $Y(x_0) - Y^*(x_0)$. This vector is given by $\mathbf{w} = \mathbf{C}^{-1}\mathbf{c}$, where

$$\mathbf{C} = [\text{cov}(Y(x_i) - \mu(x_i), Y(x_j) - \mu(x_j))],$$

$$i, j = 1, 2, \ldots, n,$$

$$\mathbf{c} = [\text{cov}(Y(x_0) - \mu(x_0), Y(x_i) - \mu(x_i))],$$

$$i = 1, 2, \ldots, n.$$

The presence of measurement error* can also be incorporated into this approach.

To implement this approach, it is necessary to specify the mean field $\mu(x)$ and the spatial covariance function of the deviation field. In atmospheric science, $\mu(x)$ is called the *background*. In early work, the background was taken to be the average of historical measurements, and was called the *climatological background* or simply *climatology*. Currently, an iterative approach is used under which predictions from a numerical model based on the previous objective analysis are used as background for the current objective analysis [8]. The shift from climatology to forecast for background is a reflection of improved forecast accuracy. Numerical models generate predictions only at the analysis points and not at the observation points. Background at the observation points is interpolated from background at the analysis points. This is called *forward interpolation*.

In the simplest case, the spatial covariance of the deviation field is assumed to be separable into horizontal and vertical components; both of these components are assumed to be stationary, and the horizontal component is assumed to be isotropic. Under these assumptions, the covariance depends only on the distance between points. The deviations at the observation locations can be used to estimate spatial covariance at a number of distances and a parametric covariance function is fitted to these estimates. In some cases, the form of these parametric models is based on physical models [24]. Extensions of these methods have been made to accommodate anisotropic and nonstationary covariances. Statistical interpolation can be extended in a straightforward way to the multivariate case. Physical constraints (like geostrophic balance) can be imposed on the interpolated field, either exactly or approximately, by imposing constraints on the covariance structure of the multivariate field [19].

Research on objective analysis remains active. One area of current interest concerns the incorporation of asynoptic satellite data into objective analysis. This is called *continuous data assimilation*—the original idea was due to Charney et al. [7]. It involves what might be called *sequential objective analysis*, in which data are incorporated into an essentially continuous forward model integration. Another focus of current research involves the estimation of forecast errors in data-poor regions, such as over the oceans. Methods based on the Kalman filter* have shown promise here [9,38]. An area of research that may be of interest to statisticians concerns the use of splines* in objective analysis [19].

STATISTICS IN STRATOSPHERIC OZONE DEPLETION

Turning now to statistical atmospheric statistics, two problems of current interest in atmospheric science are the depletion of stratospheric ozone by chlorine-containing compounds such as chlorofluorocarbons (CFCs) and atmospheric warming due to increased atmospheric concentrations of carbon dioxide and other radiatively active gases. Statistical work on these issues, both of which have substantial environmental and economic implications, is reviewed in [32].

The potential depletion of stratospheric ozone, which shields the Earth's surface from ultraviolet radiation, is a serious concern. Some scientific aspects of ozone depletion are reviewed in [37]. While there is little doubt that significant depletion has occurred over

Antarctica, this is related to special atmospheric conditions (low temperatures and small atmospheric mixing), and questions remain about the rate of depletion at low latitudes. Total column ozone has been measured at 36 ground stations around the world since the 1960s. These stations are commonly grouped into seven regions. Because the measurements were made by a Dobson spectrophotometer, they are referred to as the *Dobson data*.

Statistical analyses of the Dobson data include refs. 4, 25, 26, and 36. The goal of these analyses was to estimate a global trend in stratospheric ozone concentration. One complicating factor in these analyses is that ozone concentration may be affected by such factors as solar variability, atmospheric nuclear testing, and volcanic eruptions, so that selection of variables becomes an issue. A second complicating factor concerns the form of the ozone depletion curve. Reinsel and Tiao [26] used a two-phase linear model, with slope 0 prior to 1970 and unknown slope ω thereafter. In contrast, Bloomfield et al. [4] used a depletion curve of the form ωm_t, where m_t was the depletion curve predicted by a photochemical model. For both models, $\omega = 0$ corresponds to the case of no depletion. For the specification adopted in [4], $\omega = 1$ corresponds to the case in which the predictions of the photochemical model are correct.

Reinsel and Tiao estimated depletion curves in each of the seven regions separately and combined these estimates through a random-effects model to estimate a global depletion rate. Bloomfield et al. used a more explicitly spatial model to estimate a global depletion rate. In neither case was significant global depletion found. However, using indirect measurements, Reinsel and Tiao identified significant ozone depletion of up to 4% per decade in the upper atmosphere where photochemical theory suggests that depletion should be greatest.

Since the late 1970s, the Total Ozone Mapping Spectrometer (TOMS) aboard the Nimbus 7 satellite has provided daily total column ozone data on a global $1°$ latitude by $1.25°$ longitude grid [17]. These data have recently been analyzed by Niu and Tiao [22]. The spatial coverage of these observations is much denser than of the ground stations, so more careful modeling of spatial correlation is needed. For a fixed latitude, let $Y_j(t)$ be the average ozone observation in month t at longitude j. Niu and Tiao adopted the additive model

$$Y_j(t) = s_j(t) + r_j(t) + \xi_j(t),$$

where $s_j(t)$ is a seasonal component and $r_j(t)$ is a linear trend component. The noise term $\xi_j(t)$ is assumed to follow a space—time autoregressive (STAR) model of the form

$$\xi_j(t) = \sum \alpha_k \xi_{j-k}(t) + \theta_k \xi_{j+k}(t)$$
$$+ \sum \phi_l \xi_j(t - l) + \epsilon_j(t),$$

where the $\epsilon_j(t)$ are independent, normal errors with mean 0 and variance that depends on t only through the month. Under this model, the error in period t at longitude j is related both to previous errors at the same longitude and to errors in the same period at different longitudes. The first summation in this expression runs from $k = 1$ to $k = q$, and the second from $l = 1$ to $l = p$. The parameters q and p define the spatial and temporal order of the STAR model, respectively. An interesting feature of this model is that longitude is a circular variable.

Niu and Tiao fitted this model to seven latitude bands in each hemisphere for the period November 1978 to May 1990. For the most part, a symmetric STAR (2, 1) error model was selected. This model includes two spatial lags and one temporal lag. The model is symmetric in the sense that $\alpha_k = \theta_k$. The results indicate significant reductions of up to several percent per decade at high latitudes. One possible extension of this analysis is to allow for latitudinal dependence in the error term $\xi_j(t)$.

STATISTICS IN GLOBAL WARMING

Concern over global warming stems from the observation that the atmospheric concentration of carbon dioxide and other radiatively active gases has increased significantly since the Industrial Revolution, due primarily to the consumption of fossil fuels and other human activities. These gases are responsible for the greenhouse effect that keeps the atmosphere warm enough to support life. An increase in their concentration effectively increases radiative forcing of the atmosphere. The direct effect of this increase in radiative forcing is atmospheric warming. However, the rate and magnitude of this warming is determined by certain feedbacks in the climate system, and there is considerable uncertainty about the ultimate atmospheric response. Scientific issues relating to global warming are reviewed in the reports of the Intergovernmental Panel on Climate Change [15,16].

Estimates of mean global surface temperature based on direct, instrumental measurements are available for the past 120 years. These estimates show an irregular warming trend amounting to around 0.5°C over the length of the record. Some statistical work in the area of global warming has involved efforts to assess the significance of this apparent trend. Let T_t be the observed increase in mean global surface temperature in year t over some baseline year. The basic model is

$$T_t = \beta_0 + \beta_1 t + \epsilon_t,$$

where ϵ_t represents variations in temperature around the linear trend. Interest centers on estimating β_1 and on testing the null hypothesis that $\beta_1 = 0$. A central problem in this work concerns the specification of the properties of the variability around the trend. In particular, failure to allow for positive serial dependence* in this variability will exaggerate the significance of the estimate of β_1. This issue was considered in generality in ref. 3. Woodward and Gray [42] focused on the case where ϵ_t follows an autoregressive moving-average (ARMA)* process, while Smith [30] considered a model with long-range dependence*. The results indicate that the estimate of β_1 (which is on the order of 0.004°C per year) is at least marginally significant. The one exception considered by Woodward and Gray is the case in which the process generating ϵ_t is a correlated random walk* (i.e., the ARMA model has a unit root*). However, this model is implausible on scientific grounds. In related work, Richards [27] used econometric methods to test a number of hypotheses about changes in global temperature.

This work on global warming uses only the data and exploits little if any scientific understanding of the climate system. Other statistical work in this area has focused on estimation and inference about parameters that are physically more meaningful than the slope of a linear trend. The sensitivity of the global surface temperature to changes in radiative forcing is summarized in a parameter called *temperature sensitivity*, commonly denoted by ΔT; it is defined as the equilibrium response of mean global surface temperature to an increase in radiative forcing of 4.4 W m^{-2} (corresponding to a doubling of the atmospheric concentration of carbon dioxide). A central problem in climatology is the estimation of ΔT.

It is possible to estimate the secular trend in the atmospheric concentrations of the principal greenhouse gases since 1860. These estimated trends can be converted into an estimated trend in overall radiative forcing over the same period. Radiative forcing has increased by around 2.2 W m^{-2} since 1860. The expected response of mean global surface temperature to this pattern of increase can be estimated using a simple climate model that includes ΔT as a parameter. As noted, annual estimates of the mean global surface temperature over this period show an overall warming of around 0.5°C.

The temperature sensitivity can be estimated by matching the model response to the observed warming. The basic model is

$$T_t = m_t(\Delta T) + \epsilon_t,$$

where $m_t(\Delta T)$ is the model response in year t to observed changes in radiative forcing for temperature sensitivity ΔT, and, as before, ϵ_t represents other temperature variations. It is again important in fitting this model—and crucial in constructing a confidence interval for ΔT—to recognize that the process ϵ_t is serially dependent. This process includes temperature responses to fluctuations in radiative forcing that are not included in the estimated secular trend. The effect of these fluctuations (which are due to events like volcanic eruptions and to other variations in solar activity, cloudiness, etc.) on temperature can persist for several years. This general approach was first applied in

ref. 41. In its most careful application, Bloomfield [2] adopted a fractionally integrated white-noise* model for ϵ_t. The corresponding point estimate of ΔT was around 1.4°C with an approximate 0.95 confidence interval of 0.7°C to 2.2°C.

The general approach to estimating ΔT outlined above depends on the accuracy of the estimated secular trend* in radiative forcing. For example, the radiative-forcing effects of sulfate aerosols were not included in Bloomfield's analysis. Sulfate aerosols, which are produced by volcanic eruptions and by burning coal, tend to have a cooling effect on the surface. This effect is due to the reflectivity of the aerosols and also to their role in providing condensation nuclei for clouds. Unfortunately, it is difficult to incorporate these effects in the kind of analysis undertaken by Bloomfield. Historical data on sulfate aerosols are incomplete. Unlike greenhouse gases, the cooling effect of sulfate aerosols tends to be localized, so that historical data on their regional distribution are needed. Finally, the radiative effects of sulfate aerosols are complicated, depending, for example, on whether the aerosol is sooty or not.

Motivated in part by the difficulties of incorporating sulfate aerosols into Bloomfield's approach, Solow and Patwardhan [35] developed an alternative approach to estimating ΔT that does not require knowledge of the secular trend in radiative forcing. This approach is based on the result that the statistical characteristics of the temperature response to short-term fluctuations in radiative forcing also depend on ΔT. Specifically, if ΔT is high, this response is large and persistent, while if ΔT is low, it is relatively small and transient. Briefly, Solow and Patwardhan estimated ΔT (via likelihood in the frequency domain) by fitting the spectrum of observed temperature variations around a smooth trend to a spectrum generated by model simulations with a fixed value of ΔT. In generating these model simulations, it was necessary to adopt a statistical model of short-term variations in radiative forcing. Following the literature, Solow and Patwardhan assumed that these variations followed a white-noise process with variance 1.0 W m^{-2}. This model is consistent with a short record of satellite measurements of the global radiation budget. The point estimate of ΔT found in this way was 1.4°C with an approximate 0.95 confidence interval of 0.9°C to 2.3°C. These results are virtually identical to those found by Bloomfield, suggesting that the cooling effect of sulfate aerosols has been minimal.

FUTURE DIRECTIONS

Unlike weather prediction, atmospheric environmental science has no real central problem. For this reason, directions in statistical work in this area are difficult to predict. Taking a somewhat personal view, one general area of future research concerns the use of observations to validate physical models that produce values of a suite of variables in two or three dimensions through time [31]. There is a conceptual difficulty here, because no physical model is truly correct and it is therefore only a matter of acquiring enough data to discover this. Apart from that, the problem is complicated by the need to describe the pattern of covariance of multivariate spatial—time series. On the subject-matter side, there is growing interest in understanding and predicting climate variability on the annual to decadal time scale. The most important component of climate variability on this time scale is the El Niño Southern Oscillation [10]. The best known manifestation of ENSO are so-called El Niño events, associated with regional changes in precipitation and other climate variables. An historical record of the timing and magnitude of these events was analyzed in refs. 33 and 34. However, this analysis just scratches the surface, and the area seems ripe for further statistical work.

REFERENCES

1. Bjerknes, V. (1911). *Dynamic Meteorology and Hydrography. Part II. Kinematics.* Carnegie Institute, New York.

2. Bloomfield, P. (1992). Trends in global temperature. *Climate Change*, **21**, 1–16.

3. Bloomfield, P. and Nychka, D. (1992). Climate spectra and detecting climate change. *Climatic Change*, **21**, 275–287.

4. Bloomfield, P., Oehlert, G., Thompson, M. L., and Zeger, S. (1984). A frequency domain analysis of trends in Dobson total ozone records. *J. Geophys. Res.*, **88**, 8512–8522.

5. Braham, J. J. (1979). Field experimentation in weather modification (with discussion). *J. Amer. Statist. Ass.*, **74**, 57–104.

6. Chapman, S. and Lindzen, R. (1970). *Atmospheric Tides*. Reidel, Hingham, Mass.

7. Charney, J., Halem, M., and Jastrow, R. (1969). Use of incomplete historical data to infer the present state of the atmosphere. *J. Atmos. Sci.*, **26**, 1160–1163.

8. Daley, R. (1991). *Atmospheric Data Analysis.* Cambridge University Press, Cambridge, UK.

9. Daley, R. (1992). Estimating model-error covariances for application to atmospheric data assimilation. *Monthly Weather Rev.*, **120**, 1735–1750.

10. Diaz, H. F. and Markgraf, V. (1992). *El Niño.* Cambridge University Press, Cambridge, UK.

11. Flattery, T. (1971). *Spectral models for global analysis and forecasting.* Proc. Sixth Air Weather Service Exchange Conf. Tech. Rep. 242, Air Weather Service, pp. 42–54.

12. Gabriel, K. R. (1979). Some issues in weather experimentation. *Commun. Statist. A*, **8**, 975–1015.

13. Gandin, L. (1965). *Objective Analysis of Meteorological Fields.* Israel Program for Scientific Translation, Jerusalem.

14. Gilchrist, B. and Cressman, G. (1954). An experiment in objective analysis. *Tellus*, **6**, 309–318.

15. IPCC (1990). *Climate Change: The IPCC Scientific Assessment.* Cambridge University Press, Cambridge, UK.

16. IPCC (1996). *Climate Change 1995.* Cambridge University Press, Cambridge, UK.

17. JGR (1984). Nimbus 7 scientific results. *J. Geophys. Res.*, **89**, 4967–5382.

18. Kolmogorov, A. (1941). Interpolated and extrapolated stationary random sequences. *Izv. Akad. Nauk SSSR Ser. Mat.*, **5**, 85–95.

19. Lorenc, A. (1981). A global three-dimensional multivariate statistical interpolation scheme. *Monthly Weather Rev.*, **109**, 701–721.

20. Matheron, G. (1971). *The Theory of Regionalized Variables and Its Application.* Ecole des Mines, Paris.

21. Neyman, J., Scott, E. L., and Wells, M. A. (1969). Statistics in meteorology. *Rev. Int. Statist. Inst.*, **37**, 119–148.

22. Niu, X. and Tiao, G. C. (1995). Modelling satellite ozone data. *J. Amer. Statist. Ass.*, **90**, 969–983.

23. Panofsky, H. (1949). Objective weather-map analysis. *J. Appl. Meteorol.*, **6**, 386–392.

24. Philips, N. (1986). The spatial statistics of random geostrophic modes and first-guess errors. *Tellus*, **A38**, 314–322.

25. Reinsel, G., Tiao, G. C., Wang, M. N., Lewis, R., and Nychka, D. (1981). Statistical analysis of stratospheric ozone data for detection of trend. *Atmos. Environment*, **15**, 1569–1577.

26. Reinsel, G. and Tiao, G. C. (1987). Impact of chlorofluoromethanes on stratospheric ozone. *J. Amer. Statist. Ass.*, **82**, 20–30.

27. Richards, G. R. (1993). Change in global temperature: a statistical analysis. *J. Climate*, **6**, 546–558.

28. Rodriguez-Iturbe, I., Cox, D. R., and Isham, V. (1987). Some models for rainfall based on stochastic point processes. *Proc. R. Soc. London A*, **410**, 269–288.

29. Rodriguez-Iturbe, I., Cox, D. R., and Isham, V. (1989). A point-process model for rainfall: further developments. *Proc. R. Soc. London A*, **417**, 283–298.

30. Smith, R. L. (1993). Longrange dependence and global warming. In *Statistics for the Environment*, V. Barnett and K. Turkman, eds. Wiley, Chichester.

31. Solow, A. R. (1991). On the statistical comparison of climate model output and climate data. In *Greenhouse-Gas-Induced Climatic Change*, M. Schlesinger, ed. Elsevier, Amsterdam.

32. Solow, A. R. (1994). Statistical methods in atmospheric science. In *Handbook of Statistics 12: Environmental Statistics*, G. P. Patil and C. R. Rao, eds. North-Holland, Amsterdam, pp. 717–734.

33. Solow, A. R. (1995). An exploratory analysis of a record of El Niño events, 1800–1987. *J. Amer. Statist. Ass.*, **90**, 72–79.

34. Solow, A. R. (1995). Testing for change in the frequency of El Niño events. *J. Climate*, **18**, 2563–2566.

35. Solow, A. R. and Patwardhan, A. (1994). Some model-based inference about global warming. *Environmetrics*, **5**, 273–279.

36. St. John, D., Bailey, W. H., Fellner, W. H., Minor, J. M., and Snee, R. D. (1982). Time series analysis of stratospheric ozone. *Commun. Statist. Theory and Methods*, **11**, 1293–1333.

37. Stolarski, R. S. (1982). Fluorocarbons and stratospheric ozone: a review of current knowledge. *Amer. Statist.*, **36**, 303–311.

38. Todling, R. and Cohn, S. E. (1994). Suboptimal schemes for atmospheric data assimilation based on the Kalman filter. *Monthly Weather Rev.*, **122**, 2530–2545.

39. Wahba, G. and Wendelberger, J. (1980). Some new mathematical methods for variational objective analysis using splines and cross-validation. *Monthly Weather Rev.*, **108**, 1122–1143.

40. Wiener, N. (1949). *Extrapolation, Interpolation, and Smoothing of Stationary Time Series*. Wiley, New York.

41. Wigley, T. M. L. and Raper, S. C. B. (1990). Natural variability of the climate system and detection of the greenhouse effect. *Nature*, **344**, 324–327.

42. Woodward, W. A. and Gray, H. L. (1993). Global warming and the problem of testing for trend in time series data. *J. Climate*, **6**, 953–962.

See also DAM THEORY; KRIGING; METEOROLOGY, STATISTICS IN; WEATHER FORECASTING, BRIER SCORE; WEATHER FORECASTING, EPSTEIN SCORING RULE IN; WEATHER MODIFICATION—I; and WEATHER MODIFICATION—II.

ANDREW SOLOW

ATTRIBUTE

In statistical usage this term often has the connotation "nonquantitative." It is applied to characteristics that are not easily expressed in numerical terms: e.g., temperament, taste, and species. The term *qualitative character* is used synonymously.

The *theory of attributes* (see Chapters 4 to 6 of Yule and Kendall [2]) is mainly concerned with analysis of contingency tables and categorical data*.

Attribute sampling, i.e., sampling "from a population whose members exhibit either an attribute, A, or its complement, $not - A$", includes repeated trials from a Bernoulli distribution [1, (Sec. 9.28)], leading to consideration of probabilities in binomial distributions (*see* BINOMIAL AND MULTINOMIAL PARAMETERS, INFERENCE ON and [1 (Secs. 9.28–9.33)]). However, attribute sampling may lead to consideration of probabilities in hypergeometric distributions* (when finite populations are involved), in negative binomial distributions* (for certain sequential sampling models) [1, (Sec. 9.37)] and in multinomial distributions*.

REFERENCES

1. Stuart, A. and Ord, J. K. (1987). *Kendall's Advanced Theory of Statistics*, Vol. 1: Distribution Theory (5th ed.). Oxford University Press, New York.
2. Yule, G. U. and Kendall, M. G. (1950). *Introduction to the Theory of Statistics*, 14th ed. Hafner, New York, Charles Griffin, London. (The early editions of this work, by G. U. Yule alone, contain a considerably longer discussion of attributes.)

AUDITING, STATISTICS IN

Auditing is a specialized area of accounting involving examination by independent auditors of the operation of an organization's system of internal control and of the financial statements in order to express an opinion on the financial statements prepared by the management of the organization. Internal auditors, who are members of the organization, also review the operation of the system of internal control to see if it is operating effectively.

Uses of statistical methods in auditing are a relatively recent phenomenon and are still evolving. Three major current uses are sampling of transactions to study the operation of the internal control system, sampling of accounts to study the correctness of recorded account balances, and regression analysis for analytical review.

STUDY OF INTERNAL CONTROL SYSTEM

Transactions, such as payments of bills received, are sampled to make inferences about the effectiveness of the internal control system. Usually with this type of sampling, auditors are concerned with a qualitative characteristic, namely, whether the internal control was operating improperly for the transaction. Inferences are made about the process proportion of transactions that are handled improperly by the internal control system. Typically, auditors desire assurance that the process proportion is reasonably small. Random samples of transactions are utilized and evaluated by standard statistical procedures based on the binomial* distribution or the Poisson* approximation. No special statistical problems are encountered in this application. Consequently, little research is currently being done in this area. A good summary is contained in Roberts [8].

STUDY OF ACCOUNT BALANCES

Accounts, such as accounts receivable and inventory, often consist of thousands of individual line items that may have a total value of millions of dollars. For such large accounts, it is not economical for the auditor to audit every line item. Consequently, auditors frequently select a sample of line items and audit these, on the basis of which inferences are made about the total error amount in the population. Thus, the characteristic of interest here is quantitative. Denoting the total amount recorded by the company for the account by Y and the total amount that the auditor would establish as correct if each line item in the account were audited by X, the total error amount in the account is $E = Y - X$.

Since the auditor usually knows the amounts recorded by the company for each line item in the population (called the *book amounts*), this information can be utilized in estimating the total error amount E. One estimator that incorporates this supplementary information is the difference estimator, which for simple random sampling of line items is

$$\hat{E} = \frac{N}{n} \sum_{i=1}^{n} d_i = N\overline{d},$$

where

$$d_i = y_i - x_i$$

is the difference between the book and audit amounts for the ith sample line item, $\overline{d}$ is the mean difference per sample line item, and N is the number of line items in the population.

Other estimators that incorporate the information on book amounts are the ratio and regression estimators. Each of these can be used with simple or stratified random sampling of line items. Still another means of incorporating the information about book amounts is through sampling with probability proportional to book amount and then utilizing an unbiased estimator.

Each of these procedures has some serious limitations in auditing applications. When the sample contains no errors, the estimated standard error of the estimator equals zero, as can be seen from the estimated variance of the difference estimator with simple random sampling of line items:

$$s^2(\hat{E}) = N^2 \frac{N-n}{Nn(n-1)} \sum_{i=1}^{n} (d_i - \overline{d})^2.$$

An estimated standard error of zero suggests perfect precision, which would be an unwarranted conclusion.

A second limitation of these supplementary information procedures is that a number of simulation* studies have found that large-sample confidence limits based on the normal distribution are frequently not applicable for sample sizes used in auditing. Neter and Loebbecke [5] studied four actual accounting populations and from these constructed a number of study populations with varying error rates. They utilized sample sizes of 100 and 200 with unstratified, stratified, and PPS sampling*. They found for the supplementary information procedures that the coverages for the large-sample upper confidence bounds for the total error amount (i.e., the proportion of times the bound is correct in repeated sampling) were frequently substantially below the nominal confidence level.

The search for alternative inference procedures that do not depend on large-sample theory has frequently involved monetary unit sampling, which involves the selection of individual monetary units from the population. Since the auditor cannot audit a single monetary unit but only the line item to which the unit pertains, any error found is then prorated to the monetary units belonging to the line item. The prorated errors are called *taints* in auditing, and are denoted

by $t_k = d_k/y_k$, where $d_k \neq 0$. An important case in auditing is when the errors in the population are all overstatement errors and the taints are restricted to positive values not exceeding 1.0. For this case, a conservative confidence bound for the total error amount can be constructed by assuming that all overstatement taints are at the maximum value of 1.0. The problem then becomes one of obtaining an upper confidence bound for the population proportion of monetary units that contain an error. Multiplying this bound by Y, the total number of monetary book units in the population, yields a conservative upper confidence bound for the total error amount E.

A difficulty with this conservative bound is that it is usually not tight enough. Stringer [9] developed a heuristic for reducing this bound when all observed taints are not at the maximum value of 1.0:

$$Y p_u(1 - \alpha; n, m) - Y \sum_{k=1}^{m} [p_u(1 - \alpha; n, k)$$
$$- p_u(1 - \alpha; n, k - 1)](1 - t_k),$$

where m is the observed number of errors in the sample of n monetary units, $t_1 \geqslant t_2 \geqslant \cdots \geqslant t_m$ are the observed taints, and $p_u(1 - \alpha; n, m)$ is the $1 - \alpha$ upper bound for a binomial proportion when m errors are observed in a sample of n.

Simulation studies (e.g., Reneau [7]) have shown that this Stringer bound has coverages always exceeding the nominal level and often close to 100%; also that the Stringer bound is not very tight and may involve substantial risks of making incorrect decisions when the bound is used for testing purposes. Leslie et al. [3] developed another heuristic bound, called a *cell bound*, that tends to be tighter than the Stringer bound and has good coverage characteristics for many accounting populations. The cell bound is based on cell sampling, where the frame of monetary units is divided into strata or cells of equal numbers of monetary units and one monetary unit is selected at random from each cell, the cell selections being made independently.

Fienberg et al. [2] developed a bound based on the multinomial distribution* by viewing monetary unit sampling in discretized form. The multinomial classes represent the different possible taints rounded to a specified degree. The procedure involves obtaining a joint confidence region for the multinomial parameters and then maximizing a linear function of the parameters, representing the total error amount in the population, over the confidence region. Because of the complexities of computation, Fienberg et al. utilized only a partial ordering of the sample outcomes for developing the joint confidence region. However, simulation studies (e.g., Plante et al. [6]) have shown that coverages for the multinomial bound are near to or above the nominal level for a variety of populations, particularly if cell sampling is employed.

Several approaches based on Bayesian* methodology have been proposed when monetary unit sampling is employed. Cox and Snell [1] proposed a Bayesian bound for which they assume that the population error rate for monetary units and the population mean taint for monetary units in error are independent parameters. These and other assumptions lead to a posterior distribution* of the total error amount that is a simple multiple of the F distribution*. Hence Bayesian bounds are very easy to obtain by this method. Neter and Godfrey [4] studied the behavior of the Cox and Snell bound in repeated sampling from a given population for sample size 100, and found that conservative prior parameter values exist so that the Cox and Snell bound has coverages near or above the Bayesian probability level for a wide range of populations. However, research in progress suggests that robustness may be a function of sample size, and not include all possible populations. Another recent proposal by Tsui et al. [11] is to combine the multinomial sampling model with a Dirichlet* prior distribution* to obtain Bayesian bounds for the total error amount.

ANALYTICAL REVIEW

Analytical review procedures are utilized by auditors to make internal and external consistency comparisons, such as comparing the current financial information with comparable information for preceding periods, or comparing operating data for the firm with data for the industry. Ratios have frequently been used in making these comparisons, such as comparing the ratio of cost of sales to sales for the current period with corresponding ratios for prior periods. Use of ratios for making comparisons assumes that the relationship between the two variables is linear through the origin. Often, this relationship may not be of this form. Also, use of ratio analysis does not permit the study of the simultaneous relationship of one variable with several others, such as when it is desired to examine the relationship of the revenue of a public utility to kilowatt hours, rate per kilowatt hour, and seasonal effects.

Regression analysis is now being used by some auditors for certain analytical review procedures. The data often are time series*, as when current periods are to be compared to prior periods. Sometimes, the data are cross-sectional, as when data for several hundred retail outlets of a firm are studied to identify ones that are outliers*. No special problems in applying regression analysis for analytical review have been encountered. The regression models employed have tended to be relatively simple ones. In using regression models, auditors are particularly concerned about identifying outliers that are worthy of further investigation. Much of the research to data has been concerned with developing rules that relate the identification of outliers with the amount of subsequent audit work that should be performed. Stringer and Stewart [10] have provided a comprehensive summary of the use of regression methods in auditing for analytical review.

REFERENCES

1. Cox, D. R. and Snell, E. J. (1979). *Biometrika*, **66**, 125–132.

2. Fienberg, S., Neter, J., and Leitch, R. A. (1977). *J. Amer. Statist. Ass.*, **72**, 295–302.

3. Leslie, D. A., Teitlebaum, A. D., and Anderson, R. J. (1979). *Dollar-Unit Sampling*. Copp, Clark, Pitman, Toronto, Canada. (This book provides a comprehensive description of the Stringer and cell bounds and their uses in auditing.)

4. Neter, J. and Godfrey, J. (1985). *J. R. Statist. Soc. C, (Appl. Statist.)*, **34**, 157–168.

5. Neter, J. and Loebbecke, J. K. (1975). *Behavior of Major Statistical Estimators in Sampling Accounting Populations*. American Institute of Certified Public Accountants, New York.

6. Plante, R., Neter, J., and Leitch, R. A. (1985). *Auditing*, **5**, 40–56.

7. Reneau, J. H. (1978). *Accounting Rev.*, **53**, 669–680.

8. Roberts, D. M. (1978). *Statistical Auditing*. American Institute of Certified Public Accountants, New York.

9. Stringer, K. W. (1963). *Proc. Bus. Econ. Statist. Sec., 1963*. American Statistical Association, pp. 405–411.

10. Stringer, K. W. and Stewart, T. R. (1985). *Statistical Techniques for Analytical Review in Auditing*. Deloitte, Haskins, and Sells, New York.

11. Tsui, K. W., Matsumura, E. M., and Tsui, K. L. (1985). *Accounting Rev.* **60**, 76–96.

See also CONTROL CHARTS; FINANCE, STATISTICS IN; and QUALITY CONTROL, STATISTICAL.

JOHN NETER

AUSTRALIAN AND NEW ZEALAND JOURNAL OF STATISTICS

[This entry has been updated by the Editors.]

The *Australian Journal of Statistics (AJS)* was founded by the (then) Statistical Society of New South Wales in 1959. Until 1997 it appeared three times a year, the three issues constituting a volume. The founding editor of the *AJS* was H. O. Lancaster, who served from 1959–1971. He was followed by C. R. Heathcote (1971–1973), C. C. Heyde (1973–1978), E. J. Williams (1978–1983), J. S. Maritz (1983–1989), C. A. McGilchrist (1989–1991), I. R. James (1991–1997) and S. J. Sheather (1997).

In 1998 *AJS* merged with *The New Zealand Statistician* as Volume 40 of *The Australian and New Zealand Journal of Statistics* (*see* also NEW ZEALAND STATISTICAL ASSOCIATION); the website is www.statsoc.org.au/Publications/ANZJS.htm, and the publisher is Blackwell.

AUSTRALIAN JOURNAL OF STATISTICS: HISTORY

The editorial policy of *AJS* aimed to achieve a balance between theoretical and applied articles in the following areas: (1) mathematical statistics, econometrics, and probability theory; (2) new applications of established statistical methods; (3) applications of newly developed methods; (4) case histories of interesting practical applications; (5) studies of concepts (particularly in economic and social fields) defined in terms suitable for statistical measurement; (6) sources and applications of Australian statistical data; and (7) matters of general interest, such as surveys of the applications of statistics in broad fields. No ranking is implied in this list. *AJS* also published critical book reviews and short book notices. A news and notes section regularly appeared until 1977, but this function was taken over by the *Statistical Society of Australia Newsletter*, which first appeared in May 1977.

An international perspective and coverage was intended for *AJS* and contributions from outside Australia were always welcomed; *see* STATISTICAL SOCIETY OF AUSTRALIA for further discussion. All papers were refereed.

At the time of establishment of *AJS*, the Statistical Society of New South Wales, based in Sydney and founded in 1947, was the only society of its kind in Australia. It assumed the responsibility for starting the journal, which, as its name implies, was intended to serve the statistical profession in Australia. The then president of the Statistical Society of New South Wales, P. C. Wickens, wrote in a foreword to the first issue of the journal: "It is hoped ... that it will not be long before statistical societies will be firmly established in other States, and when this occurs it will undoubtedly be necessary to reconsider the management of the Journal."

The hopes expressed in this statement were not long in coming to fruition. The Statistical Society of Canberra was formed in 1961. In October 1962 the Statistical Society of Australia was formed by the amalgamation of the New South Wales and Canberra Societies, which then became branches of the main society. At this stage the journal became the responsibility of the new society, but its editor and editorial policy remained unchanged. Further branches of the society were later formed in Victoria (1964), Western Australia (1964), and South Australia (1967). In 1976, responsibility for *AJS* was assumed by the Australian Statistical Publishing Association Inc., whose membership is coterminous with membership of the Central Council of the Statistical Society of Australia.

THE MERGER

Following the merger of *AJS* and *The New Zealand Statistician* in 1998, the combined *Australian and New Zealand Journal of Statistics (ANZJS)* has been published in four issues per year, under two Theory and Methods Editors (one of whom also serves as Managing Editor), an Applications Editors from New Zealand, a Book Review Editor, a Technical Editor, and an international Editorial Board (10 or so for Applications and 35 or so for Theory and Methods).

ANZJS publishes articles in four categories:

Applications: Papers demonstrate the application of statistical methods to problems faced by users of statistics. A particular focus is the application of newly developed statistical methodology to real data and the demonstration of better use of established statistical methodology in an area of application.

Theory & Methods: Papers make a substantive and original contribution to the theory and methodology of statistics, econometrics or probability. A special focus is given to papers motivated by, and illustrated with, real data.

Reviews: Papers give an overview of a current area of statistical research which consolidate and reconcile existing knowledge and make judgments about the most promising directions for future work.

Historical and General Interest: Papers discuss the history of statistics in Australia and New Zealand, the role of statistical organizations in private and government institutions and the analysis of datasets of general interest.

See also NEW ZEALAND STATISTICAL ASSOCIATION and STATISTICAL SOCIETY OF AUSTRALIA.

C. C. HEYDE
The Editors

AUSTRALIAN JOURNAL OF STATISTICS. See AUSTRALIAN AND NEW ZEALAND JOURNAL OF STATISTICS

AUTOCORRELATION FUNCTION.
See AUTOREGRESSIVE−INTEGRATED MOVING AVERAGE (ARIMA) MODELS

AUTOMATIC INTERACTION DETECTION (AID) TECHNIQUES

Automatic interaction detection (AID) is a technique for analyzing a large quantity of data (whose number of cases is typically hundreds or thousands) by subdividing it into disjoint exhaustive subsets so as best to "explain" a dependent variable on the basis of a given set of categorized predictors. Although first noted under a different name in 1959 [1], the current computer versions were introduced by a series of papers dating from 1963, starting with one by J. N. Morgan and J. A. Sonquist [8]. Since its inception AID has grown in popularity and found application (and misapplication) in many applied fields. It can be used as an end analysis in itself, or as a prior screening device to sift the variables and draw attention to certain interactions for specific inclusion in subsequent analysis by other methods.

All AID techniques (normally implemented as computer programs) operate in a stepwise manner, first subdividing the total data base according to one of the predictors* (chosen by the algorithm to maximize some given criterion), then reexamining separately and subdividing each of the groups formed by the initial subdivision, and continuing in a like manner on each new subgroup formed until some stopping criterion is reached. Some versions have a "look-ahead" feature which allows possible subdivisions of subdivisions to be examined before the primary subdivision is effected—a procedure that may be theoretically desirable but is currently not cost-effective or usually worthwhile for the majority of databases met in practice. This results in a tree-like structure called a dendrogram* in AID literature.

The "automatic" of AID refers to the computer program making all the decisions as to which predictor to use when, and how. This attribute can be countermanded by the user in some versions of AID. The "interaction" in AID refers to its design property that each subgroup formed in the analysis is treated *individually*. Thus the technique does not constrain itself to producing "additive" or "symmetric" models, although these may result anyway.

DEPENDENT VARIABLE

The various types of AID are distinguished by the nature of the dependent variable. Standard AID [10] operates on one that is an ordinal scalar, while the multivariate extension to an ordinal vector is handled by MAID [3]. A nominal scalar-dependent variable can be analyzed by "chi-squared* AID" CHAID [6] or "theta AID" THAID [7]. The theta statistic of the latter technique is the total proportion of observations that belong to a modal category in an AID subdivision. Thus, theta is bounded below by d^{-1} for a d-category dependent variable, and bounded above by unity.

The criterion for a "good" subdivision that "explains" the data depends on the nature of the dependent variable. Table 1 lists the possible types of dependent variable together with the appropriate classical-type criterion for each one. The criteria are all interrelated in that the suitable special case of each type

Table 1. Classical-Type Criteria for Various Dependent Variables

Type of Dependent Variable	Criterion
Ordinal scalar	t-test*, ANOVA*
Ordinal scalar and covariate	F-test*, ANOCOVA*
Nominal	Chi-square*
Ordinal vector	Hotelling's T^{2*}, MANOVA*
Ordinal vector and covariates	MANOCOVAR*

(e.g., ordinal vector of dimension 1, a dichotomous nominal variable) reduces to one of the others in the list. Other criteria have been proposed (e.g., THAID above), but as they are dissimilar to the classical criteria (even asymptotically), they are in need of theoretical backing.

Superficially, AID bears resemblance to stepwise regression* both in intent and procedure. A comparison of AID with this and other statistical techniques is given in ref. 10.

PREDICTORS

The categorized predictors are used to determine the possible subdivisions of the data at any one stage, and hence result in the subdivisions being meaningful in that a particular subgroup can be labeled according to which categories of which predictors define the subgroup. Anything from 2 to 100 predictors are typically used in a single analysis (depending also on the size of the data base), each predictor having from 2 up to about 10 categories (depending on type), although predictors with more categories have appeared in the literature.

The categories of each predictor are grouped by AID to define the subdivisions (in many versions this reduction is *always* to two subdivisions—i.e., a binary split), the allowable grouping determined by the type of predictor. These types, which carry names peculiar to AID, include monotonic (ordered categories in which only adjacent categories may be grouped), floating (monotonic plus an additional category that may be grouped with any of the others—usually used for missing information or "don't-knows"), free (nominal categories with no restriction on grouping), interleaved (a number of monotonic categories used for combining different ordered scales; special cases include the aforementioned types), and cyclic (monotonic with "wraparound," useful for U^*- or inverted U-distributions*). Modern theory has concentrated on the behavior of AID on monotonic and free predictors, where not unsurprisingly the Bonferroni inequality* applied to a dichotomous predictor produces reasonable conservative significance tests of subdivision differences for the multicategory case. The theory of piecewise-regression and cluster-ing are applicable to monotonic and free predictors, respectively, but unfortunately their asymptotic theories, which dwell on the increase in the number of categories (in our terminology), are not relevant, as the practical application of AID involves a small number of categories.

CRITICISM

Many criticisms have been leveled at AID-type techniques, the two major ones being that (1) AID produces too many subdivisions or idiosyncratic results (based on experi-ments with random data in which *no* sub-divisions should be produced), and (2) the interpretation of AID results are often fal-lacious. The first criticism is based on ear-lier versions of AID that contained no valid hypothesis test as to whether the subdivi-sions produced were statistically significant. Further, the older versions of AID made no distinction between the types of the predic-tors involved when choosing one on which to base the subdivisions—thus introducing a bias in favor of using multicategory free pre-dictors. Appropriate significance testing for standard AID has been developed in ref. 5. The latest versions of AID introduce testing as part and parcel of the technique, preferring to subdivide on the most "significant" rather than most "explanatory" predictor and thus remove both the aforementioned bias and the generation of nonsignificant splits. Ideally, a stringent significance level should be used to take into account the number of predictors present in the analysis, using e.g., Boole's (Waring) inequality*.

The second criticism is not so much an indictment of AID but rather of the ignorance displayed by some of its users. Added to this is the inadequate manner in which many current users present their AID results. Examples are completely omitting mention of sample sizes in the various subdivisions, the statistical significance* of these subdivi-sions, or appropriate auxiliary information that would allow a reader to repair this lack, at least in his or her own mind. The den-drogram resulting from an AID analysis is so appealing that a tendency has arisen for users to ignore the possible existence of com-peting predictors that would be revealed by an examination of the AID statistics produced for each predictor both before and after a subdivision. Certain apparent "interactions" quoted in the literature (see, e.g., the criti-cism [2] of the study [4] could well be spurious for this reason, and in certain circumstances, small changes in the data could cause dif-ferent predictors to be selected by AID, with consequent different conclusions reached by a naive user. This fault is probably magnified by the "automatic" nature of AID mentioned above).

Finally, among other possible limitations, the appropriateness of an AID analysis can be no better than the appropriateness of the splitting criterion as given in Table 1, and all the assumptions inherent in the theory behind these criteria and their properties naturally carry through to AID.

EXAMPLE

Figure 1 depicts the dendrogram* resulting from a CHAID analysis of 798 first-year com-merce students enrolled at the University of the Witwatersrand. The dependent vari-able is the student's midyear mark (June 1979) in the course Mathematics and Statis-tics IB (M & S1B), classified into the three groups: passed (>49%), borderline (40–49%), and failed (<40%).

Table 2 gives details of the first stage of the analysis and indicates the predictive power and optimal grouping of the categories of each predictor. Predictors 2 and 5 (whether a local matriculant, and sex) are immediately discarded since there is no significant differ-ence between the groups, as indicated by a significance level of $p = 1$. Predictors 3, 6, 7, and 8, while possessing an optimal grouping as shown on the right-hand side of the table, are nevertheless not considered significant. Note that predictor 6, which has nine cate-gories optimally grouped into two groups, is ostensibly "significant" since the 2×3 contin-gency table so formed has $p = 0.0014$; how-ever, taking into account the optimization that went into forming this table, a conser-vative estimate of the "significance" using a Bonferroni inequality* is $p = 0.35$—clearly

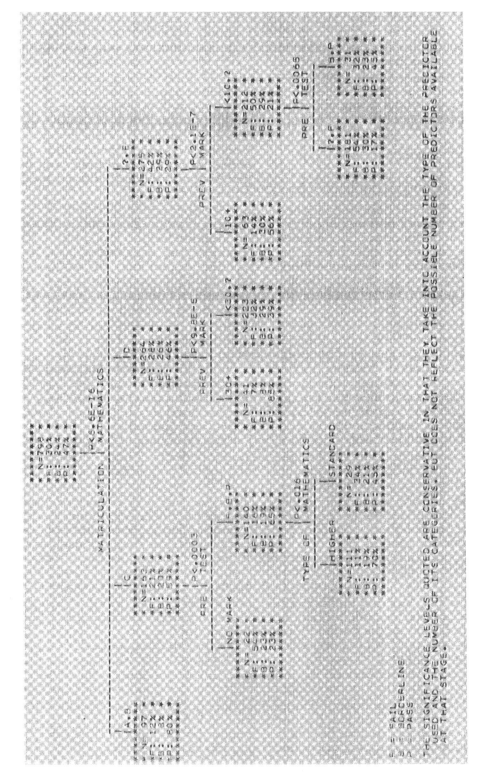

Figure 1. Dendrogram of CHAID analysis of first-year commerce students.

Table 2. First Stage of CHAID Analysis

Predictor	Number of Categories	Type	Ostensible Significance Level	Conservative Significance Level	Optimum Number of Groups	Optimum Grouping[a]
1. Matriculation mathematics (A, high mark; E, low)	6	Floating	3.0×10^{-17}	6.6×10^{-16}	4	AB C D ?E
2. Local matriculant? (Y, yes; N, no)	2	Monotonic	1.0	1.0	1	YN
3. Matriculation English (A, high mark; E, low)	6	Floating	0.018	0.40	3	?A B CDE
4. Year of matriculation (4, up to 1974; 5, 1975; 8, 1978)	6	Floating	4.1×10^{-5}	9.1×10^{-4}	3	8 7 ?654
5. Sex (M, male; F, female)	3	Floating	1.0	1.0	1	?MF
6. Course and rules registered under (G, general; BCom. full time; etc.)	9	Free	0.0014	0.35	2	G/7C9 LOPT
7. Year of study (1, first; . . . ; 5, fifth or more)	6	Floating	0.048	0.43	2	1 23?45
8. Type of matriculation mathematics (H, higher; S, standard)	3	Floating	0.036	0.11	2	?H S
9. Previous M & S1B mark (?, not a repeat student; 1, first; etc.)	9	Floating	1.1×10^{-9}	1.4×10^{-8}	2	23456 ?78
10. Pre-test (F, failed; B, borderline; P, passed)	4	Floating	8.7×10^{-11}	4.3×10^{-10}	2	?F BP
11. First midterm (April) test (F, failed; B, borderline; P, passed)	4	Floating	6.1×10^{-50}	3.0×10^{-49}	3	?F B P

[a]The symbol "?" refers to missing information and is the "floating" category.

not significant. Predictor 11 (midterm test mark) is clearly the best with conservative $p \leqslant 3.0 \times 10^{-49}$, but was included in the analysis for information only, and precluded from forming the basis of a subdivision of the data since the purpose of the analysis is to predict on the basis of information available before the commencement of the academic year. Predictor 1 is the best usable predictor ($p \leqslant 6.6 \times 10^{-16}$) and divides the data into four groups.

Figure 1 displays in detail the four-way subdivision of the total group from which it is clear that the pass rate declines from 80% for the students with high (A and B) matriculation mathematics marks, through 59 and 46% for those with intermediate marks (C and D) down to 29% for the lower marks (E). Those students for whom no mark is available (14 such students coded "?", mainly foreign or older students) are interestingly enough grouped with the poorest students.

The analysis then continued with each of the four subgroups. Information on each of the predictors similar to that in Table 2 was produced, from which further details are available. The students with mathematics symbol C are further divided in Fig. 1 according to their pre-test, where it is seen that those who did not attend the test (it was not compulsory, and implies that the students were skipping classes even at this early stage) perform worse than those who did—no matter what their mark! The groups with mathematics marks D and E were each further subdivided according to their mark in the same course last year (clearly only available for repeat students).

Finally, in the lowest level in Fig. 1, there are two further subdivisions. Although they are technically significant ($p \leqslant 0.016$ and $p \leqslant 0.0065$) they should be considered marginal, since these levels make no allowance for the number of predictors (effectively 10) examined. (They do, however, take into account the type and number of categories in the predictor used to define the subdivision.) Nevertheless, considering the pretest, it is comforting to note the similar poor performance of students who did not attend it, on the two occasions where this predictor was used.

This is merely a brief summary of some of the information and conclusions available from an AID-type analysis. The secondary details concerning the predictive power and optimum grouping of the categories within each predictor for each of the subdivisions provide valuable insight to the structure of the data and interrelationships of the predictors.

STATE OF THE ART AND FUTURE DEVELOPMENTS

The underlying theory behind valid hypothesis testing in AID is still embryonic. At present only standard AID has provision for a covariate; the other versions have yet to be so extended. Computer installations with powerful interactive terminals or personalized computer systems do not need the "automatic" decision making of AID. Instead, they could offer the researcher the opportunity to introduce additional background information and take various decisions dynamically along the lines of ref. 9. Such a feature is still to be implemented in AID.

REFERENCES

1. Belson, W. (1959). *Appl. Statist.*, **8**, 65–75.
2. Doyle, P. (1973). *Operat. Res. Quart.*, **24**, 465–467.
3. Gillo, M. W. and Shelly, M. W. (1974). *J. Amer. Statist. Ass.*, **69**, 646–653. (MAID.)
4. Heald, G. I. (1972). *Operat. Res. Quart.*, **23**, 445–457.
5. Kass, G. V. (1975). *Appl. Statist.*, **24**, 178–189. (Theory and additional references to AID.)
6. Kass, G. V. (1980). *Appl. Statist.*, **29**, 119–127. (CHAID.)
7. Morgan, J. N. and Messenger, R. C. (1973). *THAID—a sequential analysis program for the analysis of nominal scale dependent variables.* ISR, University of Michigan, Ann Arbor, Mich. (THAID.)
8. Morgan, J. N. and Sonquist, J. A. (1963). *J. Amer. Statist. Ass.*, **58**, 415–434.
9. Press, L. I., Rogers, M. S. and Shure, G. H. (1969). *Behav. Sci.*, **14**, 364–370.
10. Sonquist, J. A., Baker, E. L. and Morgan, J. N. (1971). *Search for Structure (Alias AID-III).* ISR, University of Michigan, Ann

Arbor, Mich. (Mainly a user manual, but contains background material, examples, and further references.)

See also CLASSIFICATION—I; COMPUTERS AND STATISTICS; PREDICTIVE ANALYSIS; and STEPWISE REGRESSION.

G. V. KASS

AUTOREGRESSIVE ERROR, HILDRETH–LU SCANNING METHOD

Consider the linear regression model

$$y_t = \beta_0 + \beta_1 X_{t1} + \cdots + \beta_{p-1} X_{t,p-1} + \epsilon_t,$$
$$t = 1, \ldots, N, \tag{1}$$

where y_t is the tth observation on the dependent variable, X_{tj} is the tth observation on the jth nonstochastic independent variable, and ϵ_t is the tth observation on the error term. This can be written in matrix form as $\mathbf{y} = \mathbf{X}\boldsymbol{\beta} + \boldsymbol{\epsilon}$, where y is $(N \times 1)$ in dimension, $\mathbf{X}$ is $(N \times p)$, $\boldsymbol{\beta}$ is $(p \times 1)$, and $\boldsymbol{\epsilon}$ is $(N \times 1)$. The usual assumptions on the error vector $\boldsymbol{\epsilon}$ are that $E(\boldsymbol{\epsilon}) = \mathbf{0}$ and $E(\boldsymbol{\epsilon}\boldsymbol{\epsilon}') = \sigma^2 \mathbf{I}$. In this case, the ordinary least-squares* estimator of β, denoted by $\hat{\boldsymbol{\beta}}_0$, is given by $\hat{\boldsymbol{\beta}}_0 = (\mathbf{X}'\mathbf{X})^{-1}\mathbf{X}'\mathbf{y}$ (see GENERAL LINEAR MODEL).

The assumption that the error terms are uncorrelated often breaks down in time-series* studies and sometimes in cross-sectional studies, in which case we state that the error terms are autocorrelated or serially correlated. We denote this by writing $E(\boldsymbol{\epsilon}\boldsymbol{\epsilon}') = \boldsymbol{\Sigma}$. Mixed autoregressive-moving average* processes are used to describe this serial correlation (see AUTOREGRESSIVE–MOVING AVERAGE (ARMA) MODELS). Specifically,

$$\epsilon_t = \phi_1 \epsilon_{t-1} + \cdots + \phi_p \epsilon_{t-p} + a_t$$
$$- \theta_1 a_{t-1} - \cdots - \theta_q a_{t-q}, \tag{2}$$

where $E(a_t) = 0$, $V(a_t) = \sigma_a^2$, and the a_t's are uncorrelated. To ensure stationarity, we require that the roots of $1 - \phi_1 x - \phi_2 x^2 - \cdots - \phi_p x^p = 0$ lie outside the unit circle. The case that has been considered most frequently in the econometrics* literature is when the error terms are AR (1). That is,

$$\epsilon_t = \phi_1 \epsilon_{t-1} + a_t. \tag{3}$$

In (3), it has become customary to replace ϕ_1 by ρ, where we require $|\rho| < 1$ for a stationary process*. We will focus our attention on the autocorrelation structure specified in (3).

For an AR (1) process, it is shown in Box and Jenkins [2] that $\sigma_\epsilon^2 = \sigma_a^2/(1 - \rho^2)$ and $E(\epsilon_t \epsilon_{t-k}) = \rho^k$. Thus the covariance matrix $\boldsymbol{\Sigma}$ associated with $(\epsilon_1, \epsilon_2, \ldots, \epsilon_N)$ is

$$\boldsymbol{\Sigma} = \sigma_\epsilon^2 \mathbf{A} = \sigma_\epsilon^2 \begin{bmatrix} 1 & \rho & \rho^2 & \cdots & \rho^{N-1} \\ \rho & 1 & \rho & \cdots & \rho^{N-2} \\ \rho^2 & \rho & 1 & \cdots & \rho^{N-3} \\ \vdots & \vdots & \vdots & & \vdots \\ \rho^{N-1} & \rho^{N-2} & \rho^{N-3} & \cdots & 1 \end{bmatrix}$$
$$= \sigma_a^2 \mathbf{B}. \tag{4}$$

When ρ is known, the generalized least-squares estimator of $\boldsymbol{\beta}$, denoted by $\hat{\boldsymbol{\beta}}_G$, is found by minimizing $(\mathbf{y} - \mathbf{X}\boldsymbol{\beta})'\mathbf{B}^{-1}(\mathbf{y} - \mathbf{X}\boldsymbol{\beta})$. Since $\mathbf{B}$ is positive definite, there is a nonsingular matrix $\mathbf{H}$ such that $\mathbf{B} = (\mathbf{H}'\mathbf{H})^{-1}$ and $\mathbf{B}^{-1} = \mathbf{H}'\mathbf{H}$. Thus minimizing $(\mathbf{y} - \mathbf{X}\boldsymbol{\beta})'\mathbf{B}^{-1}(\mathbf{y} - \mathbf{X}\boldsymbol{\beta})$ with respect to $\boldsymbol{\beta}$ is equivalent to minimizing $(\mathbf{y}^* - \mathbf{X}^*\boldsymbol{\beta})'(\mathbf{y}^* - \mathbf{X}^*\boldsymbol{\beta})$ via ordinary least-squares, where $\mathbf{y}^* = \mathbf{H}\mathbf{y}$ and $\mathbf{X}^* = \mathbf{H}\mathbf{X}$. It follows that

$$\hat{\boldsymbol{\beta}}_G = (\mathbf{X}^{*'}\mathbf{X}^*)^{-1}\mathbf{X}^{*'}\mathbf{y}^*$$
$$= (\mathbf{X}'\mathbf{B}^{-1}\mathbf{X})^{-1}\mathbf{X}'\mathbf{B}^{-1}\mathbf{y}. \tag{5}$$

For the AR(1) error structure, the transformation $\mathbf{H}$ that permits ordinary least-squares estimation is

$$\mathbf{H} = \begin{bmatrix} \sqrt{1 - \rho^2} & 0 & 0 & \cdots & 0 & 0 \\ -\rho & 1 & 0 & \cdots & 0 & 0 \\ 0 & -\rho & 1 & \cdots & 0 & 0 \\ \vdots & \vdots & \vdots & & \vdots & \vdots \\ 0 & 0 & 0 & \cdots & -\rho & 1 \end{bmatrix}. \tag{6}$$

In (5), one could have used $\mathbf{A}^{-1}$ or $\boldsymbol{\Sigma}^{-1}$ in place of $\mathbf{B}^{-1}$ since the scalars cancel out.

When ρ is not known, Judge et al. [7] point out that three procedures are available for parameter estimation: estimated generalized least-squares, nonlinear least-squares, and maximum likelihood*.

Let $\hat{\boldsymbol{\beta}}_E$ denote the estimated generalized least-squares estimator of $\boldsymbol{\beta}$. $\hat{\boldsymbol{\beta}}_E$ is obtained by using the estimator in (5) after estimating

ρ. Thus these procedures are called two-step procedures. Several methods are available for estimating ρ. These include:

1. The Cochrane–Orcutt procedure [3], where $\hat{\rho}_1 = \sum_{t=2}^{N} \hat{\epsilon}_t \hat{\epsilon}_{t-1} / \sum_{t=1}^{N} \hat{\epsilon}_t^2$ and the $\hat{\epsilon}_t$'s are obtained by using ordinary least-squares on $\mathbf{y} = \mathbf{X}\boldsymbol{\beta} + \boldsymbol{\epsilon}$. More precisely, this is termed the Prais–Winsten [10] procedure, since all N elements of $\mathbf{y}$ and all N rows of $\mathbf{X}$ were affected by the $\mathbf{H}$ transformation in obtaining $\mathbf{y}^*$ and $\mathbf{X}^*$. Theil [11] proposed a modification of $\hat{\rho}_1$: namely, $(N - p)/(N - 1)\hat{\rho}_1$.

2. The estimate of ρ obtained from the Durbin–Watson* statistic, D. Specifically, $\hat{\rho}_2 = 1 - D/2$. Theil and Nagar [12] give the following modification of $\hat{\rho}_2$: $(N^2 \hat{\rho}_2 + p^2)/(N^2 - p^2)$.

3. The estimate of ρ obtained from the Durbin procedure [4]. Let $\mathbf{H}_0$ denote the $(N - 1) \times N$ matrix obtained by deleting the first row of $\mathbf{H}$ in (6). Let $\hat{\rho}_3$ denote the estimated coefficient of y_{t-1} in the model: $\mathbf{H}_0 \mathbf{y} = \mathbf{H}_0 \mathbf{X}\boldsymbol{\beta} + \mathbf{H}_0 \boldsymbol{\epsilon}$. For the simple linear regression* model, we have

$$y_t = \beta_0(1 - \rho) + \rho y_{t-1}$$
$$+ \beta_1 x_t - \beta_1 \rho x_{t-1} + a_t, \quad t = 2, \ldots, N.$$

By using this method, Maddala [8] points out that one is ignoring the constraint that (coefficient of x_{t-1}) = −(coefficient of x_t). (coefficient of y_{t-1}).

In the nonlinear least-squares procedure, one needs to find those estimates of $\boldsymbol{\beta}$ and ρ that simultaneously minimize $(\mathbf{y}^* - \mathbf{X}^*\boldsymbol{\beta})'$ $(\mathbf{y}^* - \mathbf{X}^*\boldsymbol{\beta})$. Although nonlinear optimization algorithms can be used, Hildreth and Lu [6] suggested a search procedure. For values of ρ from -1.0 to 1.0 in increments of 0.1, calculate $\hat{\boldsymbol{\beta}}_G$ as stipulated in (5) and the corresponding sum of squares, $(\mathbf{y}^* - \mathbf{X}^*\hat{\boldsymbol{\beta}}_G)'(\mathbf{y}^* - \mathbf{X}^*\hat{\boldsymbol{\beta}}_G)$. Choose that value of ρ which minimizes this sum of squares. Higher decimal accuracy can be obtained by finding the sum of squares for several additional values of ρ near the minimizing value. Although the Hildreth–Lu method is not computationally efficient, the minimum sum of squares obtained should be global rather than local if some care

is exercised in the search procedure. Obviously, the value of ρ so obtained need not equal any of the values used in the estimated generalized least-squares procedure.

Under the assumption that the a_t's are normally distributed, the maximum-likelihood procedure can be used. Judge et al. [7] show that maximizing the concentrated likelihood function is equivalent to minimizing $(1 - \rho^2)^{-1/N}(\mathbf{y}^* - \mathbf{X}^*\boldsymbol{\beta})'(\mathbf{y}^* - \mathbf{X}^*\boldsymbol{\beta})$; this differs from the nonlinear least-squares procedure by the $(1 - \rho^2)^{-1/N}$ factor. Algorithms for maximizing the concentrated likelihood function are presented in Hildreth and Dent [5] and in Beach and MacKinnon [1], although a search procedure similar to the Hildreth–Lu method could be utilized.

Empirical results for some of the procedures discussed above are presented in Maddala [8]. For annual data from 1935 to 1954, and 10 different firms, Maddala regresses gross investment on two independent variables: value of the firm, and stock of plant and equipment. The results are presented in Table 1. Inspection of the entries in Table 1 reveals that the maximum-likelihood and Hildreth–Lu estimates are always in the same neighborhood, with Durbin's estimates differing substantially from these two.

Pindyck and Rubinfeld [9] also present two numerical examples using the Hildreth–Lu scanning method.

Although the name Hildreth–Lu has been reserved to refer to the search procedure for first-order autoregressive error, a similar search procedure could be employed for any ARMA error structure, as discussed in Judge et al. [7].

REFERENCES

1. Beach, C. M. and MacKinnon, J. G. (1978). *Econometrica*, **46**, 51–58.

2. Box, G. E. P. and Jenkins, G. M. (1970). *Time Series Analysis, Forecasting and Control*. Holden-Day, San Francisco.

3. Cochrane, D. and Orcutt, G. H. (1949). *J. Amer. Statist. Ass.*, **44**, 32–61.

4. Durbin, J. (1960). *J.R. Statist. Soc. B*, **22**, 139–153.

5. Hildreth C. and Dent W. (1974). In *Econometrics and Economic Theory: Essays in Honor of*

Table 1. Estimates of First-Order Autocorrelation Coefficient by Different Methods

Firm	Method			
	Cochrane–Orcutt	Durbin	Hildreth–Lu	Maximum Likelihood
GM	0.458	0.816	0.67	0.64
U.S. Steel	0.481	0.874	0.74	0.69
GE	0.461	1.061	0.50	0.47
Chrysler	−0.020	−0.346	−0.05	−0.04
Atlantic-Richfield	−0.236	−0.737	−0.22	−0.21
IBM	0.114	0.624	0.18	0.17
Union Oil	0.098	0.125	0.12	0.11
Westinghouse	0.241	0.297	0.30	0.28
Goodyear	0.246	0.706	0.39	0.36
Diamond Match	0.402	0.385	0.65	0.57

Jan Tinbergen, W. Sellekaert, ed. Macmillan, London, pp. 3–25.

6. Hildreth C. and Lu, J. Y. (1960). *Demand Relations with Autocorrelated Disturbances.* Mich. State Univ. Agric. Exp. Stn. Tech. Bull. 276, East Lansing, Mich.

7. Judge, G. G., Griffiths, W. E., Hill, R. C., and Lee, T. -C. (1980). *The Theory and Practice of Econometrics.* Wiley, New York.

8. Maddala, G. S. (1977). *Econometrics.* McGraw-Hill, New York.

9. Pindyck, R. S. and Rubinfeld, D. L. (1976). *Econometric Models and Economic Forecasts.* McGraw-Hill, New York.

10. Prais, S. J. and Winsten, C. B. (1954). *Trend Estimators and Serial Correlation.* Cowles Comm. Discuss. Paper No. 383, Chicago.

11. Theil, H. (1971). *Principles of Econometrics.* Wiley, New York.

12. Theil, H. and Nagar, A. L. (1961). *J. Amer. Statist. Ass.*, **56**, 793–806.

See also AUTOREGRESSIVE–MOVING AVERAGE (ARMA) MODELS; LEAST SQUARES; and SERIAL CORRELATION.

FRANK B. ALT

AUTOREGRESSIVE–INTEGRATED MOVING AVERAGE (ARIMA) MODELS

An important class of models for describing a single time series* z_t is the class of autoregressive–moving average models* referred to as ARMA(p, q)* models.

$$(z_t - \mu) = \phi_1(z_{t-1} - \mu) + \cdots + \phi_p(z_t - \mu)$$
$$+ a_t - \theta_1 a_{t-1} - \cdots - \theta_q a_{t-q}, \quad (1)$$

where the notation in (1) implies that (a) z_t is the original time series, or some suitable nonlinear transformation of it (such as a logarithm or a square root); (b) z_t is a stationary* time series with a fixed mean μ; (c) a_t is a random residual series, which can also be interpreted as the series of one-step-ahead forecast errors; and (d) $\phi_1, \ldots, \phi_p, \theta_1, \ldots, \theta_q, \mu$ are parameters to be estimated from the data. Alternatively, autoregressive–moving average (ARMA) models* may be written in terms of the backward-shift operator* B, such that $B^j z_t = z_{t-j}, B^j a_t = a_{t-j}$, as follows:

$$(z_t - \mu) = \frac{1 - \theta_1 B - \ldots - \theta_q B^q}{1 - \phi_1 B - \ldots - \phi_p B^p} a_t \quad (2)$$

$$= \frac{\theta(B)}{\phi(B)} a_t \quad (3)$$

Thus the ARMA(p, q) model represents the time series, or a suitable nonlinear transformation, as the output from a linear filter whose input is a random series and whose *transfer function** is a rational function of the backward-shift operator B.

The model (1) is not of immediate practical use because very few real-world time series are stationary time series in statistical equilibrium about a fixed mean μ. Instead, they are characterized by random changes in their level, slope, etc., and by the presence of seasonal patterns which also evolve with time. Traditional methods of handling such diverse behavior involve the decomposition of the time series into a "trend"*, a "seasonal component"*, and a "residual component." After removal of the trend and seasonal component, it is customary to describe the residual component by means of a stationary ARMA(p,q) model of the form (1). Such an approach suffers from the following disadvantages: (a) it is arbitrary as to what is called a trend and a seasonal component; (b) removal of the trend and seasonal component introduces additional autocorrelation into the residual component; and (c) the assumptions normally made about the behavior of the trend and seasonal component are unrealistic.

To overcome these difficulties a new class of models, called autoregressive–integrated moving average models, referred to as ARIMA models, has been developed to describe, *under the umbrella of one model*, trends, seasonality, and residual random behavior [1]. Moreover, such models for describing nonstationary* time series contain flexible structures which allow the trend and seasonal component to be nondeterministic, i.e., their statistical properties evolve in time. In addition, iterative methods have been developed [1] for identifying (or specifying), estimating* (or fitting*) and checking (or criticizing) such models given the data.

NONSEASONAL ARIMA MODELS

Consider the first-order autoregressive model

$$(z_t - \mu) = \phi(z_{t-1} - \mu) + a_t \qquad (4)$$

or, in backward-shift-operator notation $(B^j)z_t = z_{t-j}$,

$$(1 - \phi B)(z_t - \mu) = a_t,$$

which is stationary if $|\phi| < 1$. The solution of the difference equation* (4) may be written as the sum of the complementary function, i.e., the solution of $(1 - \phi B)(z_t - \mu) = 0$, and a particular integral, i.e., any function that satisfies (4). Relative to a time origin $t = 0$, the solution of (4) thus becomes

$$(z_t - \mu) = \phi^t(z_0 - \mu) + \sum_{j=1}^{t} \phi^{j-1} a_j. \qquad (5)$$

If $|\phi| > 1$, the first term in (5) dominates and the growth of the series is explosive. Although such explosive nonstationarity occurs in some situations (such as bacterial growth), for most practical situations it is convenient to work with a less severe form of nonstationarity. This can be achieved by setting $\phi = 1$ in (4) and (5), which then become

$$(1 - B)z_t = z_t - z_{t-1} = a_t, \qquad (6)$$

$$z_t = z_0 + \sum_{j=1}^{t} a_j, \qquad (7)$$

i.e., z_t is a random walk* model. More generally, we consider a nonstationary ARMA(p,q) model

$$\phi'(B)(z_t - \mu) = \theta(B)a_t, \qquad (8)$$

where $\phi'(B)$ is a nonstationary autoregressive operator. To prevent explosive nonstationarity, we impose the restriction that d of the factors of $\phi'(B)$ are unity, i.e., $\phi'(B) = \phi(B)(1 - B)^d$. Model (8) then becomes

$$\phi(B)(1 - B)^d z_t = \theta(B)a_t \qquad (9)$$

where $\phi(B)$ is a stationary autoregressive* operator and $\theta(B)$ is an invertible moving average* operator, as in a stationary autoregressive–moving average model. Since $(1 - B)z_t = z_t - z_{t-1} = \nabla z_t$, where ∇ is the backward difference operator, the model (9) can also be written

$$\phi(B)\nabla^d z_t = \theta(B)a_t \qquad (\nabla^0 = 1). \qquad (9)$$

Model (9) implies that whereas z_t is a nonstationary series, its dth difference $w_t = \nabla^d z_t$ is stationary and can be described by an autoregressive-moving average model. The model (9) is called an *autoregressive-integrated moving average model* or ARIMA(p,d,q) model, where p is the number of parameters in the autoregressive operator, d is the number of times that the series has to be differenced to induce stationarity, and q is the number of parameters in the moving average operator. Provided that the series does not contain seasonality, the ARIMA model (9) with small values of p, d, and q is capable of describing a wide range of practically occurring time series. When $d > 0$, the stationary series $w_t = \nabla^d z_t$ will usually have a zero mean. However, a useful generalization of (9) can be obtained by allowing w_t to have a nonzero mean, i.e.,

$$\phi(B)(\nabla z_t - \mu) = \theta(B)a_t. \qquad (10)$$

With $d > 0$, model (10) is capable of describing a *deterministic* polynomial trend of degree d as well as a stochastic nonstationary component. For example, when $d = 1$, the model is capable of describing nonstationary stochastic behavior over and above an underlying linear growth rate.

SPECIAL CASES OF NONSEASONAL ARIMA (P, D, Q) MODELS

With $p = 0$, $d = 1$, $q = 1$, $\mu = 0$, model (10) becomes

$$\nabla z_t = (1 - \theta B)a_t \qquad (11)$$

i.e., a nonstationary series whose first difference is stationary and can be represented as a first-order moving average model. Model (11) can be inverted to give

$$z_t = (1 - \theta)(z_{t-1} + \theta z_{t-2} + \theta^2 z_{t-3} + \cdots) + a_t. \qquad (12)$$

Thus the one-step-ahead forecast of z_t from origin $(t - 1)$ is an exponentially weighted moving average of past values of the series. By solving the difference equation, model (11) may also be written

$$z_t = a_t + (1 - \theta)(a_{t-1} + a_{t-2} + \cdots)$$
$$= a_t + l_{t-1} \qquad (13)$$

Although the series has no fixed mean, at a given time it has a local level l_{t-1} which is updated from time $(t - 1)$ to time t according to

$$l_t = l_{t-1} + (1 - \theta)a_t.$$

Thus when the random shock a_t occurs, a proportion $(1 - \theta)a_t$ of it is absorbed into the "level" of the series and the remaining proportion θa_t is "lost" from the system.

With $p = 0$, $d = 1$, $q = 1$, $\mu \neq 0$, model (10) becomes

$$\nabla z_t - \mu = (1 - \theta B)a_t. \qquad (14)$$

The solution of the difference equation (14) may be written as a complementary function (the solution of $\nabla z_t - \mu = 0$) and the particular integral (13), i.e.,

$$z_t = c + \mu t + a_t$$
$$+ (1 - \theta)(a_{t-1} + a_{t-2} + \cdots) \qquad (15)$$

and thus contains a "deterministic drift" term.

With $p = 0$, $d = 2$, $q = 2$, $\mu = 0$, model (10) becomes

$$\nabla^2 z_t = (1 - \theta_1 B - \theta_2 B^2)a_t. \qquad (16)$$

Thus z_t and ∇z_t are nonstationary series, and second-order differencing $\nabla^2 z_t$ is necessary to induce stationarity. It may be shown [1, p. 111] that model (16) implies that the series has a local "level" l_t and a local "slope" s_t which are updated from time $t - 1$ to time t by the new random shock a_t according to

$$l_t = l_{t-1} + s_{t-1} + (1 + \theta_2)a_t,$$
$$s_t = s_{t-1} + (1 - \theta_1 - \theta_2)a_t.$$

SEASONAL ARIMA MODELS

One of the deficiencies in handling seasonal time series in the past has been the absence of parametric models to describe seasonal behavior. A new class of models [3] for describing seasonality as well as nonstationary trends can be obtained by modification of the nonseasonal ARIMA model (10). Suppose that data become available at monthly intervals and that they are set out in the form of a two-way table in which the columns denote months and the rows denote years. The series for a particular column, say March, may contain a trend but is not seasonal in its behavior. Hence it is reasonable to link the observation for March in this year to observations in previous Marches by an ARIMA (P, D, Q) model of the form (9):

$$\Phi(B^S)\nabla_S^D z_t = \Theta(B^S)\alpha_t, \qquad (17)$$

where the autoregressive and moving average operators are now polynomials in B^s of degrees P and Q, respectively, and s is the seasonal period and equals 12 in this case. Also, the nonseasonal difference operator ∇ in (9) is replaced by the seasonal difference operator $\nabla_s z_t = z_t - z_{t-s}$ in (17) and ∇_s^D denotes the Dth seasonal difference. In general, the error terms α_t in models of the form (17) fitted to each month separately would not be random since the behavior of the series in March of this year will usually depend not only on what happened in previous Marches but also on the behavior of the series in February, January, etc., of this year. To describe this monthly dependence, we can use the nonseasonal ARIMA model

$$\phi(B)\nabla^d \alpha_t = \theta(B) a_t, \qquad (18)$$

where a_t is now a random series and $\phi(B)$ and $\theta(B)$ are polynomials in B of degrees p and q, respectively. Substituting (18) in (17), and allowing for the possibility that the differenced series may have a nonzero mean μ, we obtain the ARIMA $(p, d, q) \times (P, D, Q)$ multiplicative model

$$\begin{aligned} \phi_p(B)\Phi_P(B^s)(\nabla^d \nabla_s^D z_t - \mu) \\ = \theta_q(B)\Theta_Q(B^s) a_t, \end{aligned} \qquad (19)$$

where the subscripts on the operators denote the degrees of the polynomials involved.

In some cases, it may be better to work with a nonmultiplicative model in which the autoregressive operator, or the moving average operator, or both operators, cannot be factorized into a product of nonseasonal and seasonal operators, as in (19); for example, the right-hand side of (19) might take the form

$$(1 - \theta_1 B - \theta_{12} B^{12} - \theta_{13} B^{13}) a_t.$$

Seasonal models of the form (19) may be fitted to data with a range of seasonal periods: e.g., daily data $(s = 7)$, weekly data $(s = 12)$, and quarterly data $(s = 4)$. Moreover, several seasonal periods may occur simultaneously; e.g., hourly traffic data may display a cycle over a day $(s = 24)$ and a further cycle over a week $(s = 168)$. In such examples it may be necessary to add further seasonal autoregressive, moving average, and differencing operators to the model.

BUILDING ARIMA MODELS

Figure 1a shows part of a series consisting of the logarithms of the electricity consumption in one country. The series contains an upward trend and an annual cycle. ARIMA models of the form (19) may be fitted to data, using an iterative cycle of identification, estimation, and checking, as described below.

Initial analysis [4] suggested that to achieve a homoscedastic* distribution of the residuals a_t, it is necessary to apply a logarithmic transformation $\ln z_t$ to the data before fitting a model of the form (19). Alongside the plot of $\ln z_t$ shown in Fig. 1a is a plot of the autocorrelation function r_k of $\ln z_t$ as a function of the lag k. The autocorrelation function fails to damp out with the lag k and is indicative of nonstationarity [1]. Figure 1b shows the autocorrelation function of the nonseasonal difference $\nabla \ln z_t$. This autocorrelation function has peaks at $12, 24, 36, \ldots$, indicating nonstationarity with respect to the seasonal behavior and suggesting that further seasonal differencing is needed. Figure 1c shows the autocorrelation function of $\nabla \nabla_{12} \ln z_t$. This function contains no obvious trends, implying that the differenced and transformed series $w_t = \nabla \nabla_{12} \ln z_t$ is stationary. The next step is to arrive at an initial guess of the seasonal and nonseasonal autoregressive and moving average structure needed to explain the autocorrelation function of w_t. The autocorrelation functions of autoregressive-moving average models are characterized by a discrete number of spikes corresponding to the moving average part of the model and damped exponentials and/or damped sine waves corresponding to the autoregressive part of the model. The largest autocorrelations r_k in Figure 1c occurs at lags 1 and 12, suggesting an initial model of the form

$$\nabla \nabla_{12} \ln z_t = (1 - \theta B)(1 - \Theta B^{12}) a_t, \qquad (20)$$

where we may take as initial estimates of the parameters, $\theta = 0.30$ (based on $r_1 = -0.27$) and $\hat{\Theta} = 0.35$ (based on $r_{12} = -0.33$), using a procedure described in Box and Jenkins [1].

The initial model structure (20) may be fitted to the data by iterative calculation of the maximum-likelihood* estimates, starting

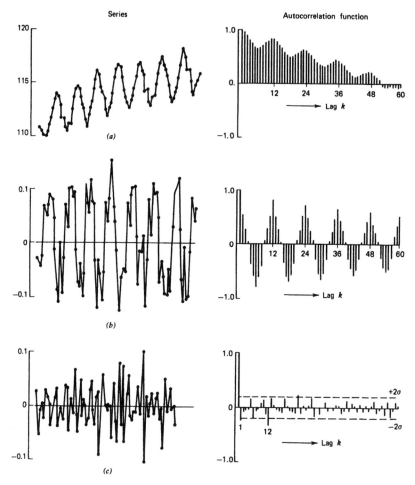

Figure 1. Various differences of the logarithms of national electricity consumption series, together with their corresponding autocorrelation functions: (a) $\ln Y_t$; (b) $\nabla \ln Y_t$; (c) $\nabla \nabla_{12} \ln Y_t$. Reproduced with the permission of GJP Publications from *Practical Experiences with Modeling and Forecasting Time Series* by Gwilym M. Jenkins.

from the initial values given above (see ref. 1, pp. 269–284, for the likelihood* function of autoregressive-moving average models). The fitted model, based on $N = 96$ observations, was

$$\nabla \nabla_{12} \ln Y_t = \frac{(1 - 0.73B)\,(1 - 0.83B^{12})}{\pm 0.08 \qquad a_t \pm 0.05} \qquad (21)$$

with estimated residual variance $\sigma_a^2 = 0.0006481(\sigma_a = 0.0255)$. The $\pm$ values underneath the estimated parameters denote the 1-standard-error* limits.

Examination of the residuals a_t in (21) showed that none of the residuals was large

compared with their standard deviation $\sigma_a = 0.0255$ and that 5 residuals out of 83 fell outside $\pm 2\sigma_a$, in reasonable accord with expectation. The largest autocorrelations $r_a(k)$ of the residuals a_t occurred at lags 6 and 24, suggesting some evidence of model inadequacy. However, further elaboration of the model revealed little improvement on model (21). Further details of how this model was built, and how it was elaborated to a transfer function* model relating electricity consumption to temperature, have been given by Jenkins [4].

Model (21) may be used to forecast future values z_{t+l} for each lead time $l = 1, 2, 3, \ldots$

from the current origin t by writing it at time $t + l$ in the form

$$\ln z_{t+l} - \ln z_{t+l-1} - \ln z_{t+l-12} + \ln z_{t+l-13}$$
$$= a_{t+l} - 0.73a_{t+l-1} - 0.83a_{t+l-12}$$
$$+ 0.61a_{t+l-13} \qquad (22)$$

and then taking conditional expectations at time t, bearing in mind that the conditional expectation of future values of the random series a_{t+l} for $l > 0$ are zero. For example, when $l = 1$, the one-step-ahead forecast $\hat{z}_t(1)$ can be calculated from

$$\ln \hat{z}_t(1) = \ln z_t + \ln z_{t-11} - \ln z_{t-12}$$
$$- 0.73a_t - 0.83a_{t-11} + 0.61a_{t-12}, \qquad (23)$$

where $a_{t-j} = \ln z_{t-j} - \ln \hat{z}_{t-j-1}(1)$ for $j \geqslant 0$. Thus the forecasts for each lead time l can be generated recursively, together with the standard deviations of the forecast errors $e_t(l) = \ln \hat{z}_{t+l} - \ln z(l)$ (see ref. 1). In the example above, further improvements in forecasting accuracy could be expected by introducing into the model other variables which are related with electricity consumption: e.g., temperature, industrial production, price. Such *transfer function* models are discussed in Box and Jenkins [1] and in Jenkins [4].

MULTIVARIATE ARIMA MODELS

Univariate ARIMA models may be generalized to deal with mutual interaction between several nonstationary time series. To illustrate the possibilities, consider two time series z_{1t} and z_{2t}. First, nonlinear transformation and nonseasonal differencing may be needed to produce stationary time series

$$w_{1t} = \nabla^{d_1} z_{1t}, \qquad w_{2t} = \nabla^{d_2} z_{2t}. \qquad (24)$$

Then it might be possible to describe the resulting stationary vector by a multivariate autoregressive–moving average model*

$$\begin{bmatrix} \phi_{11}(B) & \phi_{12}(B) \\ \phi_{21}(B) & \phi_{22}(B) \end{bmatrix} \begin{bmatrix} w_{1t} - \mu_1 \\ w_{2t} - \mu_2 \end{bmatrix} \qquad (25)$$
$$= \begin{bmatrix} \theta_{11}(B) & \theta_{12}(B) \\ \theta_{21}(B) & \theta_{22}(B) \end{bmatrix} \begin{bmatrix} a_{1t} \\ a_{2t} \end{bmatrix},$$

where a_{1t} and a_{2t} are the one-step-ahead forecast errors or residuals for z_{1t} and z_{2t}, respectively. If the forecasts are to be optimal, a_{1t} must be a random series, a_{2t} a random series, and a_{1t} and a_{2t} mutually uncorrelated series except possibly at simultaneous times. The model defined by (24) and (25) is an example of an ARIMA($\boldsymbol{P}, \boldsymbol{d}, \boldsymbol{Q}$) model, where $\boldsymbol{P}$ is a matrix whose elements (p_{ij}) define the degrees of the polynomials $\phi_{ij}(B)$ in the autoregressive matrix, the vector $\boldsymbol{d}$ has elements d_i which define the degrees of differencing needed to induce stationarity of the time series*, and $\boldsymbol{Q}$ is a matrix whose elements (q_{ij}) define the degrees of the polynomials $\theta_{ij}(B)$ in the moving average matrix. The foregoing models may also be generalized to deal with seasonality [2,4], and may be generalized by introducing explanatory variables* to explain the simultaneous behavior of the vector $\boldsymbol{z}_t$ of time series, leading to multiple output–multiple input transfer function models [4].

REFERENCES

1. Box, G. E. P. and Jenkins, G. M. (1970). *Time Series Analysis: Forecasting and Control*. Holden-Day, San Francisco (2nd ed., 1976).

2. Box, G. E. P., Hillmer, S. C., and Tiao, G. C. (1976). *NBER Census Conf. Seasonal Time Ser.*, Washington, D. C.

3. Box, G. E. P., Jenkins, G. M., and Bacon, D. W. (1967). In *Advanced Seminar on Spectral Analysis of Time Series, B. Harris*, ed., Wiley, New York, pp. 271–311.

4. Jenkins, G. M. (1979). *Practical Experiences with Modelling and Forecasting Time Series*. GJP Publications, St. Helier, N.J.

Further Reading

The basic theoretical properties of ARIMA models are given in *Time Series Analysis: Forecasting and Control* by G. E. P. Box and G. M. Jenkins (Holden-Day, San Francisco, 1970), together with practical procedures for identifying, fitting, and checking such models. Further accounts are given in *Applied Time Series Analysis for Managerial Forecasting* by C. R. Nelson (Holden-Day, San Francisco, 1973), *Time Series Analysis and Forecasting: The Box-Jenkins Approach* by O. D. Anderson (Butterworth, London, 1975), and *Forecasting Economic Time Series* by C. W. J. Granger and P. Newbold (Academic Press, New York, 1977). Theoretical and practical accounts of multivariate ARIMA models are given in *Practical Experiences with Modelling and Forecasting Time Series* by G. M. Jenkins (GJP Publications, St. Helier, N.J., 1979). Numerous papers in the field are published in *Statistical Literature* and are listed in *Current Index to Statistics** (CIS).

See also AUTOREGRESSIVE–MOVING AVERAGE (ARMA) MODELS; PREDICTION AND FORECASTING; SEASONALITY; TIME SERIES; and TRANSFER FUNCTION MODEL.

G. M. JENKINS

AUTOREGRESSIVE MODELS. See AUTOREGRESSIVE–MOVING AVERAGE (ARMA) MODELS

AUTOREGRESSIVE–MOVING AVERAGE (ARMA) MODELS

Data occurring in the form of time series* occur in many branches of the physical sciences, social sciences, and engineering. Figure 1 shows an example of a single time series* in which the value z_t of a certain variable (the average number of spots on the sun's surface) is plotted against time t. Earlier approaches to analyzing time series were based on decomposing the variance of the series into components associated with different frequencies, based on Fourier analysis* and leading more recently to methods based on spectral analysis*.

Alternative historical approaches were based on building a model for the time series in the time domain. The main motivation for building such a model was to forecast future values of the time series given its past history. However, such a model could be used (a) to gain a better understanding of the mechanisms generating the series, (b) to smooth the random variation in the series, and (c) to allow for dependence in the series when the data were used for other statistical purposes, such as testing the differences between the means of two sets of data or relating one time series to another as in some form of regression* analysis. The first practically

useful models for describing time series were the autoregressive models introduced by G. U. Yule and G. Walker (see below) and the moving average* models introduced by Slutsky, Wold, and others. Later, it was recognized that more general structures could be obtained by combining autoregressive and moving average models, leading to autoregressive-moving average (ARMA) models.

GENERAL LINEAR MODEL*

When forecasting* an observed time series z_t from a knowledge of its past history, it is natural to think of the forecast as being obtained by applying a set of weights to past values of the time series. Thus, the one-step-ahead forecast of z_t made from *origin* $(t-1)$ may be written

$$\text{forecast} = \pi_1 z_{t-1} + \pi_2 z_{t-2} + \pi_3 z_{t-3} + \cdots, \tag{1}$$

where π_j is the weight applied to the previous observation z_{t-j} in order to forecast z_t. When the future observation z_t comes to hand, it follows that

$$z_t = \text{forecast} + \text{forecast error} \tag{2}$$

Substituting for the forecast from (1) and denoting the forecast error by a_t, (2) becomes

$$z_t = \pi_1 z_{t-1} + \pi_2 z_{t-2} + \pi_3 z_{t-3} + \cdots + a_t. \tag{3}$$

If the forecast one step ahead of z_t is the best possible, then the forecast errors $a_t, a_{t-1}, a_{t-2}\ldots$ should be a *random series**, or *white noise**. If not, it should be possible to forecast the forecast errors and add this forecast to forecast (1) to obtain a better forecast. Model (3), with a_t a random series, is called a linear model*. In practice, the forecast errors a_t may depend on the level of the series, in which case a better representation is obtained by using a nonlinear transformation of z_t in (3), such as a log or square-root transformation*. From now on it will be assumed that the notation used in (3) denotes a representation for z_t or a suitable nonlinear transformation of z_t chosen so as to make the forecast errors a_t homoscedastic*. Although the representation (3) provides a useful general way

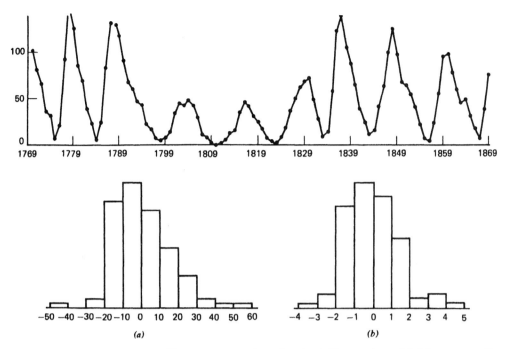

Figure 1. Plot of annual Wölfer sunspot numbers (1770–1869), together with histograms of residuals a_t from: (a) model $(z_t - 46.9) = 1.42(z_{t-1} - 46.9) - 0.73(z_{t-2} - 46.9) + a_t$; ($b$) model $(\sqrt{z_t} - 7.4) + 1.41(\sqrt{z_{t-1}} - 7.4) - 0.70(\sqrt{z_{t-2}} - 7.4) + a_t$.

of modeling a time series, it suffers from the disadvantage that it contains a large (potentially infinite) number of weights or parameters π_i. Since it would be impossible to estimate very accurately such a large number of weights, a practical solution to time-series problems requires a more parsimonious representation, containing as few parameters as possible. Such economy in parameterization can be achieved using autoregressive and moving average models.

PURE AUTOREGRESSIVE MODELS

From now on it is assumed that the time series is *stationary**, i.e., that it is in statistical equilibrium about a fixed mean μ and that it possesses, among other properties, a constant variance and a covariance structure which depends only on the difference k (or lag) between two time points. Suppose also that the weights π_i applied to past observations in the representation (3) are zero beyond a certain point p. Then, writing the series as a

deviation about its mean μ, (3) becomes

$$(z_t - \mu) = \phi_1(z_{t-1} - \mu) + \phi_2(z_{t-2} - \mu)$$
$$+ \cdots + \phi_p(z_{t-p} - \mu) + a_t, \quad (4)$$

where the finite set of weights or parameters ϕ_i may be estimated from the data. In words, (4) implies that the current deviation of the time series from its mean is a linear combination of the p previous deviations plus a random residual* a_t. The analogy between (4) and a multiple regression model* should be noted. Because the regressor variables in (4) are lagged values of the series itself and not distinct variables, (4) is called an *autoregressive* model of order p, or an AR(p). Model (4) also implies that the best forecast of z_t made from origin $(t - 1)$ is a linear combination of the past p-values of the series.

Introducing the backward-shift operator* B, such that $Bz_t = z_{t-1}$, $B^j z_t = z_{t-j}$, (4) may be written in the alternative form

$$(1 - \phi_1 B - \phi_2 B^2 - \cdots - \phi_p B^p)(z_t - \mu) = a_t. \quad (5)$$

Thus an AR(p) model is characterized by an operator

$$\phi(B) = (1 - \phi_1 B - \phi_2 B^2 - \cdots - \phi_p B^p), \quad (6)$$

which is a polynomial of degree p in the backward-shift operator B. The polynomial (6) may have real factors of the form $(1 - G_i B)$ or complex factors corresponding to complex roots of $\phi(B) = 0$. Complex factors indicate the presence of a quasi-cyclic* component in the data. Such cycles* do not have fixed periods, as in a sine wave, but are subject to random changes in amplitude, phase, and period. The fact that complex roots in (6) produce quasi-cyclical behavior in z_t may be seen by noting that if $p = 2$ and $a_t = 0$, the solution of the difference equation* (6) is a damped sine wave, as in the motion of a damped simple pendulum. When the zero on the right-hand side is replaced by a random series a_t, the sine wave is prevented from damping out by a series of random shocks, producing randomly disturbed sinusoidal behavior.

Autoregressive models were first suggested by G. U. Yule [9] who used a second-order model to describe the annual series of Wölfer sunspot numbers. Figure 1 shows a plot of this series, based on the annual average of daily readings, for the period 1770–1869. The fitted model* is

$$(z_t - 46.9) = \frac{1.42(z_{t-1} - 46.9)}{\pm 0.07}$$
$$- \frac{0.73(z_{t-2} - 46.9) + a_t}{\pm 0.07}, \quad (7)$$

where the $\pm$ values underneath the estimated parameters are their estimated standard error limits. The variance of the residuals a_t can be estimated together with the parameters μ, ϕ_1, and ϕ_2 [2] and was $\sigma_a^2 = 228.0$ in this example. The operator $(1 - 1.42B + 0.73B^2)$ corresponding to (7) has complex factors with a period p that can be calculated from

$$\cos \frac{2\pi}{p} = \frac{\phi_1}{2\sqrt{-\phi_2}} = \frac{1.42}{2\sqrt{0.73}}$$

and is 10.65 years. Figure 1 also shows the histogram* of the residuals a_t corresponding to model (7). The distribution is skew, suggesting that a transformation of

the data is needed before fitting a model. Using an approach due to Box and Cox [1] (see BOX–COX TRANSFORMATION—I), it may be shown that a better representation is obtained using the following model based on a square-root transformation*:

$$(\sqrt{z_t} - 7.4) = \frac{1.41}{\pm 0.07}(\sqrt{z_{t-1}} - 7.4)$$
$$- \frac{0.70}{\pm 0.07}(\sqrt{z_{t-2}} - 7.4) + a_t \quad (8)$$

with $\sigma_a^2 = 1.994$. Note that the parameter estimates are changed only very slightly by transformation. Its main affect is to shrink the peaks and stretch the troughs in the series, resulting in a more symmetric distribution* of the residuals, as shown in Fig. 1. The estimate of the average period corresponding to model (8) is 11.05 years, much closer than the previous value of 10.65 years to the period quoted by meteorologists for this series.

PURE MOVING AVERAGE MODELS

For autoregressive models the π-weights in the representation (8) have a cut-off after p, where p is the order of the model. In some situations it may be more appropriate to apply steadily declining weights to generate the forecasts rather than weights which have an abrupt cut-off. Such a pattern of weights may be obtained, e.g., by using a moving average model

$$z_t - \mu = a_t - \theta a_{t-1} = (1 - \theta B)a_t. \quad (9)$$

Inverting (9), we obtain

$$a_t = \frac{1}{1 - \theta B}(z_t - \mu)$$

and provided that $|\theta| < 1$ and $|B| < 1$, this expression may be expanded to give

$$a_t = (1 + \theta B + \theta^2 B^2 + \cdots)(z_t - \mu), \quad (10)$$

so that the π-weights decay exponentially. More generally, a moving average model of order q, or MA(q) is defined by

$$(z_t - \mu) = a_t - \theta_1 a_{t-1} - \theta_2 a_{t-2}$$
$$- \cdots - \theta_q a_{t-q} \quad (11)$$

where a_t is a random series. The model (11) contains q parameters $\theta_1, \theta_2, \ldots, \theta_q$, which can be estimated from the data. It implies that the current deviation of the series z_t from its mean μ is a linear combination of the current and q previous random shocks a_t (or one-step-ahead forecast errors) which have entered the system. In backward-shift notation, (11) may be written as $(z_t - \mu) = \theta(B)a_t$, where the moving average operator $\theta(B)$ is given by

$$\theta(B) = 1 - \theta_1 B - \theta_2 B^2 - \cdots - \theta_q B^q. \quad (12)$$

The model (11) has π-weights consisting of a mixture of real exponentials, corresponding to real factors of $\theta(B)$, and of damped sine waves, corresponding to complex factors of $\theta(B)$.

MIXED AUTOREGRESSIVE–MOVING AVERAGE MODELS

Result (10) shows that an MA(1) can be written as an autoregressive model of infinite order. If θ is small, say $\theta = 0.3$, then from a practical point of view the infinite series in (10) can be truncated after the term in B since it would require a long length of series to detect the parameter 0.09 in the next term $0.09B^2$ in the expansion. However, if θ is moderate or large, several terms would be needed in (10) to provide an adequate approximation to the single-parameter model (9). Thus if the moving average model were incorrectly specified as an autoregressive model, involving several parameters, the estimates of the parameters in the autoregressive representation would tend to have high standard errors and be highly correlated.

Conversely, the AR(1) model

$$(1 - \phi B)(z_t - \mu) = a_t \quad (13)$$

can be written as an infinite-order moving average model

$$(z_t - \mu) = a_t + \phi a_{t-1} + \phi^2 a_{t-2} + \cdots, \quad (14)$$

and hence estimation problems will be encountered if an autoregressive model is incorrectly specified as a moving average model. To achieve parsimony in parameterization in a given practical situation, it may be

necessary to include both autoregressive and moving average terms in the model. Thus the mixed autoregressive–moving average model, or ARMA(p, q), is defined by [8]

$$(z_t - \mu) = \phi_1(z_{t-1} - \mu) + \cdots$$
$$+ \phi_p(z_{t-p} - \mu) + a_t - \theta_1 a_{t-1}$$
$$- \cdots - \theta_q a_{t-q}. \quad (15)$$

Written in terms of the backward shift operator, (15) becomes

$$(z_t - \mu) = \frac{1 - \theta_1 B - \cdots - \theta_q B^q}{1 - \phi_1 B - \cdots - \phi_p B^p} a_t. \quad (16)$$

The form (15) represents the time series z_t (or an appropriate nonlinear transformation of z_t) as the output from a linear filter whose input is a random series and whose *transfer function** is a rational function of the backward-shift operator B. In words, (15) implies that the current deviation of the time series from its mean is a linear combination of the p previous deviations and of the current and q previous residuals a_t (or one-step-ahead forecast errors). The ARMA(p, q) model (15) is capable of generating π-weights in (1), the first p of which follow no fixed pattern and the remainder of which lie on a curve that is a mixture of damped exponentials and sine waves. Table 1 shows special cases of the general ARMA(p, q) model of the kind that frequently arise in practice.

STATIONARITY AND INVERTIBILITY CONDITIONS

The parameters in the ARMA(p, q) model (16) must satisfy the following two conditions. (a) For z_t to be written in the π-weight form (1), i.e.,

$$\left(1 - \sum_{j=1}^{\infty} \pi_j B^j \right)(z_t - \mu) = \theta^{-1}(B)\phi(B)z_t$$
$$= a_t$$
$$\text{for } |B| < 1,$$

the factors of $\theta(B)$ must be less than unity in modulus (the invertibility condition). This condition implies that the forecast weights

π_j die out; i.e., the forecast depends less on what has happened in the distant past than in the recent past. (b) For z_t to be written in the ψ-weight form

$$(z_t - \mu) = \left(1 + \sum_{j=1}^{\infty} \psi_j B^j\right) a_t = \phi^{-1}(B)\theta(B)a_t$$

$$\text{for } |B| < 1,$$

the factors of $\theta(B)$ must be less than unity in modulus (the stationarity condition). This condition implies that the series is stationary with finite variance. Table 2 shows the characteristic shapes of the π-weights and ψ-weights for the AR(p), MA(q), and AR-MA(p, q) models.

AUTOCORRELATION FUNCTIONS

The autocovariance functions $\gamma_k = E[(z_t - \mu)(z_{t+k} - \mu)]$ shows how the dependence between neighboring values of the series varies with the lag* k. It may be calculated from the autocovariance generating function

$$\gamma(B) = \sum_{k=-\infty}^{\infty} \gamma_k B^k$$
$$= \sigma_a^2 \phi^{-1}(B)\phi^{-1}(B^{-1})\theta(B)\theta(B^{-1}). \quad (17)$$

Table 2 shows the characteristic patterns of the autocorrelation functions $\rho_k = \gamma_k/\gamma_0$ of AR(p), MA(q), and ARMA (p, q) models. Such patterns may be used to provide an initial guess of the structure of an observed time series [2].

PARTIAL AUTOCORRELATION FUNCTIONS

A complementary tool to the autocorrelation function for identifying the structure of an ARMA(p, q) model is the partial autocorrelation function s_k [2]. The partial autocorrelation function may be estimated by fitting autoregressive models of orders $1, 2, 3, \ldots, k$ to a time series and picking out the estimates $s_1, s_2, \ldots, s_k$ of the *last* parameter in the model. Table 2 shows the partial autocorrelation function shapes corresponding to AR(p), MA(q), and ARMA(p, q) models. The duality in the properties of autoregressive and moving average models should be noted.

MULTIVARIATE AUTOREGRESSIVE–MOVING AVERAGE MODELS

If $\mathbf{z}_t$ denotes an m-vector of mutually interacting time series, the univariate ARMA(p, q) model (16) may be generalized to

$$\boldsymbol{\phi}(B)(\mathbf{z}_t - \boldsymbol{\mu}) = \boldsymbol{\theta}(B)\mathbf{a}_t, \quad (18)$$

where $\boldsymbol{\phi}(B)$ is an autoregressive matrix whose elements $\phi_{ij}(B)$ are autoregressive operators, $\boldsymbol{\mu}$ a vector of mean values, $\boldsymbol{\theta}(B)$ a moving average matrix with elements $\theta_{ij}(B)$, and $\mathbf{a}_t$ a vector of random series that are mutually uncorrelated. For further discussion of the properties of multivariate ARMA models, the reader is referred to Quenouille [7], Hannan [4], Box and Tiao [3], and Jenkins [5,6].

Since time series occurring in practice are rarely stationary with fixed means, ARMA models are of limited use in describing practical situations. The modifications necessary

Table 1. Some Simple Special Cases of the Autoregressive–Moving Average Model

(p, q)	Nature of Model	Mathematical Form of Model	Backward-Shift-Operator Form of Model
$(1, 0)$	First-order autoregressive	$z_t - \mu = \phi_1(z_{t-1} - \mu) + a_t$	$z_t - \mu = \dfrac{1}{1 - \phi_1 B}a_t$
$(2, 0)$	Second-order autoregressive	$z_t - \mu = \phi_1(z_{t-1} - \mu) + \phi_2(z_{t-2} - \mu) + a_t$	$z_t - \mu = \dfrac{1}{1 - \phi_1 B - \phi_2 B^2}a_t$
$(0, 1)$	First-order moving average	$z_t - \mu = a_t - \theta_1 a_{t-1}$	$z_t - \mu = (1 - \theta_1 B)a_t$
$(0, 2)$	Second-order moving average	$z_t - \mu = a_t - \theta_1 a_{t-1} - \theta_2 a_{t-2}$	$z_t - \mu = (1 - \theta_1 B - \theta_2 B^2)a_t$
$(1, 1)$	First-order autoregressive, first-order moving average	$z_t - \mu = \phi_1(z_{t-1} - \mu) + a_t - \theta_1 a_{t-1}$	$z_t - \mu = \dfrac{1 - \theta_1 B}{1 - \phi_1 B}a_t$

Table 2. Summary of Properties of AR, MA, and ARMA Models

	AR(p) Models	MA(q) Models	ARMA(p, q) Models
π-weights and partial autocorrelations function	Cutoff after p; follow no fixed pattern	Infinite; mixture of damped exponentials and sine waves	First p values follow no fixed pattern; thereafter, mixture of damped exponentials and sine waves
ψ-weights and autocorrelation function	Infinite; mixture of damped exponentials and sine waves	Cutoff after q; follow no fixed pattern	First q values follow no fixed pattern; thereafter, mixture of damped exponentials and sine waves

to make them practically useful are discussed under autoregressive-integrated moving average (ARIMA) models*.

REFERENCES

1. Box, G. E. P. and Cox, D. R. (1964). *J. Roy. Statist. Soc. Ser. B*, **26**, 211–252.

2. Box, G. E. P. and Jenkins, G. M., (1970). *Time Series Analysis; Forecasting and Control*. Holden-Day, San Francisco (2nd ed., 1976).

3. Box, G. E. P. and Tiao, G. C. (1977). *Biometrika*, **64**, 355–366.

4. Hannan, E. J. (1970). *Multiple Time Series*, Wiley, New York.

5. Jenkins, G. M. (1975). *Proc. 8th Int. Biometric Conf.*, Constanta, Romania, 1974. Editura Academici Republicii Socialists Romania, 53.

6. Jenkins, G. M. (1979). *Practical Experiences with Modelling and Forecasting Time Series*, GJP Publications, St. Helier, N. J.

7. Quenouille, M. H. (1957). *The Analysis of Multiple Time Series*, Charles Griffin, London.

8. Wold, H. (1938). *A Study in the Analysis of Stationary Time Series*, Almqvist & Wiksell, Stockholm (2nd ed., 1953).

9. Yule, G. U. (1927). *Philos. Trans. Roy. Soc. A.*, **226**, 267.

FURTHER READING

The first balanced account of theoretical and practical aspects of autoregressive–moving average models is given in the book by Wold [8]. Box and Jenkins [2] summarize the properties of these models and also give practical guidelines for identifying, fitting, and checking such models given the data. Pioneering work on multivariate models is to be found in Quenouille [7], and Hannan [4] discusses the theoretical background for both univariate and multivariate models. Practical guidelines for building multivariate autogressive-moving average models have been given by Jenkins [6].

See also AUTOREGRESSIVE–INTEGRATED MOVING AVERAGE (ARIMA) MODELS; MOVING AVERAGES; PREDICTION AND FORECASTING; TIME SERIES; and TRANSFER FUNCTION MODEL.

G. M. JENKINS

AVAILABILITY

Availability is a property of a system, defined as the proportion of time the system is functioning (properly). If failure and repair times are each distributed exponentially* with expected values θ and ϕ, then the availability is $\theta/(\theta + \phi)$. Sometimes the availability is defined generally in this way with $\theta = E$ [time to failure], $\phi = E$ [repair time].

BIBLIOGRAPHY

Gray, H. L. and Lewis, T. (1967). *Technometrics*, **9**, 465–471.

Nelson, W. (1968). *Technometrics*, **10**, 594–596.

See also QUALITY CONTROL, STATISTICAL and RELIABILITY, PROBABILISTIC.

AVERAGE CRITICAL VALUE METHOD

Geary's [1] average critical value (ACV) method applied to a statistic T used to test the hypothesis $H_0 : \theta = \theta_0$ against the alternative $H_1 : \theta = \theta_1$ determines what the difference between θ_0 to θ_1 should be for $E[T|\theta_1]$ to fall on the boundary of the critical region of the test. Test statistics with small differences correspond to tests with high efficiency*. The advantage of this method is that it requires only the calculation of the expected value $E[T|\theta_1]$ rather than the distribution of T. (The latter is required for power* function calculations.) Examples are given by Geary [1] and Stuart [2]. Stuart in 1967 showed that the ACV method of gauging the efficiency of tests can usually be represented as an approximation to the use of the asymptotic relative efficiency* of the tests.

REFERENCES

1. Geary, R. C. (1966). *Biometrika*, **53**, 109–119.
2. Stuart, A. (1967). *Biometrika*, **54**, 308–310.

See also EMPIRICAL DISTRIBUTION FUNCTION (EDF) STATISTICS.

AVERAGE EXTRA DEFECTIVE LIMIT (AEDL)

The average extra defective limit (AEDL) is a concept introduced by Hillier [1] as a measure of effectiveness of a continuous sampling plan* in adjusting to a process that has gone out of control.

Assume that a process has been operating in control at a quality level p_0 and then instantaneously deteriorates after the mth item to a level p_1, where $0 \leqslant p_0 < p_1 \leqslant 1$. Let D be the number of *uninspected* defectives among the next L items after the mth item is observed. The expected value of D, $E(D)$, is a well-defined quantity for a specific sampling plan. An average extra defective limit is the smallest number denoted by AEDL satisfying

$$E(D) \leqslant \text{AEDL} + L \times A$$

for all possible values of L, p_0, or p_1, where A is the average outgoing quality limit (AOQL)* of the plan.

Equivalently, let

$$X_m = \begin{cases} 1 & \text{if the } m\text{th item is defective} \\ & \text{but not inspected,} \\ 0 & \text{otherwise.} \end{cases}$$

Let the m_0th item be the last item before the shift in quality from p_0 to p_1; then

$$\text{AEDL} = \sup_{(p_0, p_1)} \sup_L \sum_{m=m_0+1}^{m_0+L} [E(X_m) - A].$$

Intuitively, AEDL is the upper limit to the expected number of "extra" defectives that will be left among outgoing items when the process goes out of control regardless of L, p_0, or p_1. Additional interpretations of AEDL, its uses, and methods of computation for continuous sampling plans have been discussed by Hillier [1].

REFERENCE

1. Hillier, F. S.   (1964).   *Technometrics*,   **6**, 161–178.

See also AVERAGE OUTGOING QUALITY LIMIT (AOQL); CONTINUOUS SAMPLING PLANS; and QUALITY CONTROL, STATISTICAL.

AVERAGE OUTGOING QUALITY (AOQ)

The definition of this concept suggested by the Standards Committee of ASQC [3] is: "the expected quality of the outgoing product following the use of an acceptance sampling plan* for a given value of incoming product quality." It is basically a ratio of defective items to total items, i.e., the total number of defectives in the lots accepted divided by the total number of items in those lots. Two other formal (but not equivalent) definitions are: (1) the average fraction defective in all lots after rejected lots have been sorted and cleared of defects—this is an average based on practically perfect lots (those sorted) and lots still with fraction of defectives approximately p (it is assumed that lots of stable quality are offered), and (2) the expected fraction of defectives, after substituting good items for bad ones in rejected lots, and in samples taken from accepted lots.

The AOQ serves as a performance measure associated with an (attribute) acceptance sampling plan* when the same sampling plan is used repeatedly.

Wortham and Mogg [4] present formulas for calculating AOQ for nine different ways of carrying out rectifying inspection*. For example, if the defective items are replaced by good ones (thereby returning the lot to its original size N), then

$$\text{AOQ} = \frac{P_a \times p \times (N - n)}{n},$$

where P_a is the probability of acceptance using the given acceptance plan, p the fraction defective, N the lot size, and n the sample size.

REFERENCES

1. Burr, I. W. (1953). *Engineering Statistics and Quality Control*, McGraw-Hill, New York.
2. Juran, J. M. ed. (1951). *Quality Control Handbook*. McGraw-Hill, New York.
3. Standards Committee of ASQC (1978). *Terms, Symbols and Definitions for Acceptance Sampling. ASQC Standard A3*.
4. Worthman, A. W. and Mogg, J. W. (1970). *J. Quality Tech.*, **2**(1), 30–31.

See also AVERAGE OUTGOING QUALITY LIMIT (AOQL) and QUALITY CONTROL, STATISTICAL.

AVERAGE OUTGOING QUALITY LIMIT (AOQL)

The current "official" definition of this concept as suggested by the Standards Committee of ASQC [3] reads: "For a given acceptance sampling plan* AOQL is the maximum AOQ over all possible levels of incoming quality."

Originally, AOQL was defined by Dodge [1] as the upper limit to the percent of defective units that remain in the output after inspection, given that the process is in statistical control (i.e., the proportion of defectives being produced is constant). In other words, it represents the worst average quality the consumer will accept under a particular rectifying inspection* scheme.

REFERENCES

1. Dodge, H. F. (1943). *Ann. Math. Statist.*, **14**, 264–279.
2. Sackrowitz, H. (1975). *J. Quality Tech.*, **7**, 77–80.
3. Standards Committee of ASQC (1978). *Terms, Symbols and Definitions for Acceptance Sampling, ASQC Standard A3*.
4. Wald, A. and Wolfowitz, J. (1945). *Ann. Math. Statist.*, **16**, 30–49.

See also AVERAGE OUTGOING QUALITY (AOQ) and SAMPLING PLANS.

AVERAGE RUN LENGTH (ARL)

The length of time the process must run, on the average, before a control chart* will indicate a shift in the process level* is called the average run length (ARL). It is, of course, desirable that the ARL should be long when no shift has occurred, but short when a shift has occurred.

The ARL is usually measured in terms of the number of consecutive points plotted on the control chart.

BIBLIOGRAPHY

Standards Committee of ASQC (1978). ASQC Standard AL.

See also CUMULATIVE SUM CONTROL CHARTS.

AVERAGE SAMPLE NUMBER (ASN)

In a sequential test* $\mathscr{S}$, the final size of the sample (N) required by the test is a random variable. If the sample sequential test $\mathscr{S}$ is carried out repeatedly, N will generally assume different values in successive repetitions of the test. The average amount of sampling per test that would result from the use of $\mathscr{S}$ is measured by the expected value of N and is called the average sampling number (ASN) of the test. If the test relates to the value of a parameter θ, we have formally

$$E[N; \theta] = \sum_{n=1}^{\infty} n p(n; \theta),$$

where $p(n; \theta) = \Pr[N = n|\theta]$ is the probability of reaching the terminal decision at sample size n. A graph showing $E[N; \theta]$ against various values of θ is called the ASN curve or ASN surface, according as θ is a scalar or a vector, respectively.

BIBLIOGRAPHY

Ghosh, B. K. (1970). *Sequential Tests of Statistical Hypotheses*. Addison-Wesley, Reading, Mass.

See also SEQUENTIAL ANALYSIS and STOPPING NUMBERS AND STOPPING TIMES.

AVERAGED SHIFTED HISTOGRAM

For a random univariate sample of size n, the classic histogram* takes the value

$$\hat{f}(x) = \frac{\text{bin count}}{nh},$$

where h is the width of a bin. Important details of proper histogram construction are usually overlooked. In practice, continuous data have a finite number of significant digits, with resultant accuracy $\pm\delta/2$. In this view, the raw data have been rounded and assume values at the midpoints of a finer histogram mesh with bin width δ. Let the kth such bin be denoted by $B_k = [t_k, t_{k+1})$, $-\infty < k < \infty$, where the bin edges $\{t_k\}$ satisfy $t_{k+1} - t_k = \delta$ for all k. If v_k is the bin count for B_k, then $\sum_k v_k = n$. Thus the classic histogram should obey two constraints: first, the bin width should be a multiple of δ, that is, $h = m\delta$ for some integer m; and second, the bins of the histogram should conform to the finer mesh, for example, $[t_k, t_{k+m})$. The bin count for this wider bin, $[t_k, t_{k+m})$, would be computed as $\sum_{i=0}^{m-1} v_{k+i}$.

If the point of estimation, x, falls in the (narrower) bin B_k, precisely which histogram with bin width h should be selected? Clearly, there are exactly m such *shifted histograms*, with explicit bin intervals ranging from $[t_{k-m+1}, t_{k+1})$ to $[t_k, t_{k+m})$. The ordinary *averaged shifted histogram* (ASH) estimates the density at $x \in B_k$ as the arithmetic mean of

these m shifted histogram estimates. A simple calculation shows that the ordinary ASH is given by

$$\hat{f}(x) = \frac{1}{m} \sum_{i=1-m}^{m-1} \frac{(m - |i|)v_{k+i}}{nh}$$

$$= \frac{1}{nh} \sum_{i=1-m}^{m-1} \left(1 - \frac{|i|}{m}\right) v_{k+i},$$

$$x \in B_k.$$

As $m \to \infty$, Scott [3] showed that $\hat{f}(x)$ converges to the kernel density estimator

$$\hat{f}(x) = \frac{1}{nh} \sum_{i=1}^{n} K\left(\frac{x - x_i}{h}\right),$$

with the triangle kernel $K(t) = 1 - |t|$ on $(-1, 1)$. The ASH may be generalized to mimic any kernel $K(t)$ defined on $(-1, 1)$ such that $K(\pm 1) = 0$ by

$$\hat{f}(x) = \frac{1}{n\delta} \sum_{i=1-m}^{m-1} w_m(i)v_{k+i}, \qquad x \in B_k,$$

by defining the weights

$$w_m(i) = \frac{K(i/m)}{\sum_{j=1-m}^{m-1} K(j/m)}.$$

In this form, the ASH is seen as a discrete convolution of adjacent bin counts. In practice, popular kernels are the biweight and triweight, which equal $\frac{15}{16}(1 - t^2)^2$ and $\frac{35}{32}(1 - t^2)^3$ on $(-1, 1)$, respectively. Silverman [6] provided a fast-Fourier-transform procedure for the Normal kernel, which does not have finite support. Fan and Marron [1] compare many algorithms.

MULTIVARIATE ASH

Binning* seems an effective device in dimensions up to four or five. For example, with bivariate data, (x_1, x_2), construct a fine mesh of size $\delta_1 \times \delta_2$. Then construct bivariate histograms with bins of size $h_1 \times h_2$, where $h_1 = m_1\delta_1$ and $h_2 = m_2\delta_2$. Then the bivariate

ASH is the average of the $m_1 \times m_2$ shifted histograms, and for $(x_1, x_2) \in B_{kl}$,

$$\hat{f}(x_1, x_2)$$

$$= \frac{1}{n\delta_1\delta_2} \sum_{i=1-m_1}^{m_1-1} \sum_{j=1-m_2}^{m_2-1} w_{m_1}(i)w_{m_2}(j)v_{k+i,l+j}$$

with obvious extension to more dimensions. Scott [4] discusses visualization of trivariate and quadrivariate densities with applications in clustering and discrimination.

COMPUTATION AND ASYMPTOTICS

The computational advantage of the ASH derives from the fact that the smoothing is applied to perhaps 50–500 bin counts, rather than to the raw data themselves. ASH software is available from *Statlib* or from ftp.stat.rice.edu by anonymous ftp.

From the point of view of statistical efficiency, the ordinary ASH represents a compromise between the histogram and triangle-kernel estimator*. For the univariate ASH, the asymptotic mean integrated squared error (AMISE) is

$$\text{AMISE}(h, m) = \frac{2}{3nh}\left(1 + \frac{1}{2m^2}\right) + \frac{h^2}{12m^2}R(f')$$

$$+ \frac{h^4}{144}\left(1 - \frac{2}{m^2} + \frac{3}{5m^2}\right)R(f''),$$

where $R(\phi) = \int \phi(x)^2 \, dx$. When $m = 1$, the bias has the usual $O(h^2)$ behavior of the histogram, whereas the bias is of order $O(h^4)$ as $m \to \infty$ as for the positive kernel estimator. If a piecewise linear interpolant of the ASH midpoints is used, then the bias is $O(h^4)$ for all m. Thus, linear interpolation is almost always recommended for the ASH.

The exact optimal choices of h and m depend on the unknown density through $R(f'')$. However, Terrell and Scott [7] provide useful upper bounds for h using any reasonable estimate of the standard deviation: for the ordinary ASH with $m = 1$, $h < 3.73\sigma n^{-1/3}$; for the ASH with $m \to \infty$, $h < 2.78\sigma n^{-1/5}$. These estimates converge at the rates $O(n^{-2/3})$ and $O(n^{-4/5})$, respectively. However, if the linear interpolant is used, the bound for the ASH with $m = 1$ is $h < 2.33\sigma n^{-1/5}$; the $m \to \infty$ bound is unchanged.

If δ is fixed, these formulas give upper bounds for m. A discussion of algorithms for choosing h may be found in Scott and Terrell [5] and Scott [4]. A simple table of factors to obtain smoothing parameters for use with other kernel weights is given on p. 142 in Scott [4].

ASH REGRESSION

An efficient nonparametric regression* estimate may be constructed by using the multivariate ASH. For example, with data (x, y), construct the bivariate ASH and then compute the conditional mean of the estimate. The result is particularly simple as $\delta_y \to 0$:

$$\hat{m}(x) = \frac{\sum_{i=1-m_1}^{m_1-1} w_{m_1}(i)v_{k+i}\bar{y}_{k+i}}{\sum_{i=1-m_1}^{m_1-1} w_{m_1}(i)v_{k+i}},$$

$$x \in B_k,$$

where $\bar{y}_k$ is the average of the y_i's for those points (x_i, y_i), where $x_i \in B_k$ and v_k is the univariate bin count. This estimator was introduced by Härdle and Scott [2]; two- and three-dimensional versions with application to spatial estimation from sample surveys of agriculture data are given in Whittaker and Scott [8].

EXAMPLES

Bradford Brown (see ref. [4], Appendix B.4) measured the thickness of 90 U.S. pennies to the nearest tenth of a mil. Two shifted histograms plus an ASH are shown in Fig. 1, all with smoothing parameter $h = 1.5$ mils. The visual impression of the two histograms is markedly different. The ASH essentially removes the effect of the choice of bin origin, which is a nuisance parameter. With a smaller smoothing parameter ($h = 0.9$), two extra modes appear (rather than only one) at 55.9 and 57.1 mils.

A fuller view of these data may be obtained by plotting them as a time series as in Fig. 2. Contours of the bivariate ASH, together with the regression ASH, are also displayed. Apparently, the thickness of pennies has changed more than once since World War II.

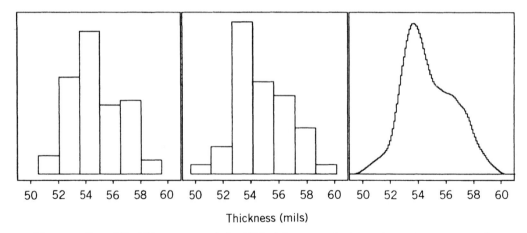

Figure 1. Two shifted histograms and the ASH of the penny thickness data. The bin origins for the histograms were 50.55 and 49.65. The triweight kernel was used for the ASH with $\delta = 0.1$ and $m = 1$.

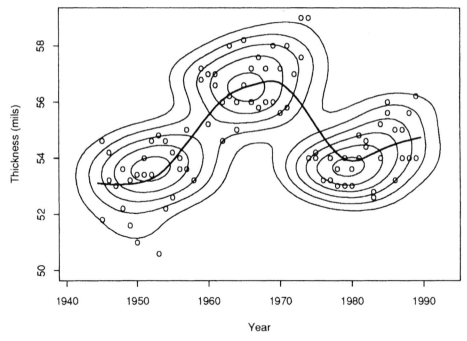

Figure 2. Bivariate ASH density and regression estimates of the penny data.

REFERENCES

1. Fan, J. and Marron, J. S. (1994). Fast implementations of nonparametric curve estimators. *J. Comput. Graph. Statist.*, **3**, 35–56.

2. Härdle, W. and Scott, D. W. (1992). Smoothing by weighted averaging of rounded points. *Comput. Statist.*, **7**, 97–128.

3. Scott, D. W. (1985). Averaged shifted histograms: effective nonparametric density estimators in several dimensions. *Ann. Statist.*, **13**, 1024–1040.

4. Scott, D. W. (1992). *Multivariate Density Estimation*. Wiley, New York.

5. Scott, D. W. and Terrell, G. R. (1987). Biased and unbiased cross-validation in density

estimation. *J. Amer. Statist. Ass.*, **82**, 1131–1146.

6. Silverman, B. W. (1982). Kernel density estimation using the fast Fourier transform, *J. R. Statist. Soc. B*, **31**, 93–97.

7. Terrell, G. R. and Scott, D. W. (1985). Oversmoothed nonparametric density estimates. *J. Amer. Statist. Ass.*, **80**, 209–214.

8. Whittaker, G. and Scott, D. W. (1994). Spatial estimation and presentation of regression surfaces in several variables via the averaged shifted histogram. *Comput. Sci. Statist.*, **26**, 8–17.

See also Graphical Representation of Data; Histograms; and Kernel Estimators.

David W. Scott

AXIAL DISTRIBUTIONS. See

Directional Distributions

AXIOMS OF PROBABILITY

THE AXIOMATIC METHOD

It is perhaps no coincidence that the axiomatic method in mathematics became prominent somewhat before the "use" theory of meaning became prominent in philosophy. An axiomatic system aims to capture and formalize some way of using language and it sidesteps the difficulties of explicit definitions.

The advantage of the axiomatic method is that theorems can in principle be deduced mathematically from the axioms without new assumptions creeping in surreptitiously, and without necessary philosophical commitment. In other words, the theorems can be proved rigorously according to the usual standards of pure mathematics without involvement in the controversial problems of application to the real world. In practice it is difficult to be totally rigorous, as has been found by philosophers of mathematics. An early example of the axiomatic method is in the geometry of Euclid, although his axioms do not satisfy most modern pure mathematicians. The value of a more precise axiomatic approach was emphasized by David Hilbert near the beginning of the twentieth century.

The approach has become a paradigm for pure mathematics, but less so for applied mathematics and for physics because it can lead to rigor mortis (to quote Henry Margenau's joke). Probability theory is both pure and applied*, so that different specialities put more or less emphasis on axiomatic systems.

NOTATION

Many theories of probability have been proposed, and many different notations have been used. In this article we use notations such as $P(E|F)$, which can be read as the probability of E given (or assuming, or conditional on) F. Here, depending on the theory or the context, E and F might denote propositions, events*, hypotheses*, scientific theories, or sets, or might even be abstract symbols, such as those in abstract algebra, without ordinary definitions but subject only to some axioms. We can regard $P(\cdot|\cdot)$ as a function of two variables, and the domains of E and F are not necessarily identical. The notation $P(E)$ is read "the probability of E" and is used either when F is taken for granted or, in some theories, not as a conditional probability* but as a so-called "absolute probability" in which officially nothing is "given" or "assumed" other than logic and mathematics (were that possible).

When a theory of probability is expressed axiomatically there will usually be axioms satisfied by such symbols as E and F and further axioms satisfied by the "probabilities" themselves. Some theories of probability are formulated as theories of rationality, and then the set of axioms needs to mention either decisions or "utilities" (="desirabilities"). *See* Decision Theory.

ARGUMENTS OF $P(E|F)$

A theory in which E and F denote sets or are abstract symbols can be regarded as a branch of pure mathematics, but propositions, events, and hypotheses are not purely mathematical concepts when they are interpreted as ordinary English words. We shall not try to define the English meanings of events, hypotheses, and theories, but the meaning of "proposition" is especially controversial: see, for example, Gale [11]. Perhaps the best definition is that a proposition is "the meaning of a clear statement." By ruling out unclear statements we are adequately justified in assuming that each proposition is capable of being either true or false, although really there are degrees of meaningfulness because statements can be more or less vague.

Moreover, a statement can be either empirically or mathematically meaningful, a point that is relevant to the choice of axioms. For example, to say that a measurement of a continuous variable lies between 5.25 and 5.35 inches is often empirically meaningful, whereas to say that it is exactly 5.30 inches, with no error at all, is at best mathematically meaningful within an idealized mathematical model. Again, to say that "the limit of the proportion of time that a coin comes up heads is approximately 0.5 in an infinite sequence of tosses" can be fully meaningful only within pure mathematics, because all sequences of tosses in the real world are of finite length.

AXIOMS FOR PROPOSITIONS AND SETS

The *conjunction* of propositions E and F is the proposition E & F and is denoted in this article by EF. The *disjunction* of E and F is denoted by $E \vee F$. This proposition asserts that E or F or both are true. The negation of E is denoted by $\tilde{E}$ or by $\sim E$. If E and F denote the same proposition, then we write $E = F$. (Other notations are in use.)

Some axioms for propositions are:

A1 If E is a proposition, then $\tilde{E}$ is also. This axiom might not be accepted by those who define a proposition as (scientifically) meaningful only if it is refutable if false. This first axiom, if applied to scientific propositions, forces us to the view that a proposition is also scientifically meaningful when it is confirmable if true. There are, however, degrees in these matters: see Good [15, pp. 492–494].

A2 $\sim (\sim E) = E$.

A3 If E and F are both propositions, then so is EF.

A4 *Commutative law.* $EF = FE$.

A5 *Associative law.* $E(FG) = (EF)G$.

A6 *De Morgan's law.* $\sim (EF) = \tilde{E} \vee \tilde{F}$. From this we can prove that the commutative and associative laws apply also to disjunctions.

A7 *Distributive laws*

$$E(F \vee G) = (EF) \vee (EG)$$

and

$$E \vee (FG) = (E \vee F)(E \vee G),$$

of which the second law can be inferred from the first by means of de Morgan's law*.

To these seven axioms, which are essentially the axioms of Boolean algebra, we can append the optional axiom (A8) and perhaps (A9):

A8 The conjunction and disjunction of a countably infinite number of propositions are propositions, with a corresponding "de Morgan law,"

$$\sim (E_1 E_2 E_3 \cdots) = \tilde{E}_1 \vee \tilde{E}_2 \vee \tilde{E}_3 \vee \cdots.$$

A9 The conjunction and disjunction of any infinite number of propositions are propositions, with yet another "de Morgan law": The negation of the conjunction is the disjunction of the negations.[1]

ORIGINS OF THE AXIOMS

In most theories of probability the probabilities lie in some sense between 0 and 1 and satisfy axioms somewhat resembling in appearance the addition and product axioms, namely:

$$P(A \vee B) = P(A) + P(B)$$

when A and B are mutually exclusive* and

$$P(AB) = P(A) \cdot P(B|A).$$

These comments will be clarified in what follows.

The addition and product axioms were known at least implicitly to Fermat* and Pascal* in 1654 and perhaps to Cardano* in the sixteenth century. But it is more convenient to express the axioms explicitly, formally, and completely.

Axioms are seldom slapped down arbitrarily; a system of axioms should be chosen either as a convenient form of other axioms or should be constructed to capture some intuitive ideas about the world or about mathematics, so that the system has some prior justification. The axioms can also be justified by their practical and philosophical implications; e.g., they should not be seen to lead to an irresolvable contradiction. Before the axioms can have practical meaning, some formal rules of application to the real world must be provided. Moreover, in practice a set of axioms and rules of application are still not sufficient: a theory needs to become to some extent a technique if it is to be useful. One needs informal suggestions of how to apply the theory, although these suggestions are not logically essential. In this article no more will be said about such practical suggestions because such matters belong properly to a discussion of the relationship between probability and statistics or between probability and practical decision making. *See* DECISION THEORY.

The prior justification of a set of axioms must depend on some concept of probability, however vague. One of the earliest concepts of probability was derived from games of chance*, such as those depending on coin spinning, dice throwing, and card drawing. In such games there are some symmetry properties that suggest that some outcomes are at least approximately equally probable, and this is so even for people who have not much idea of what probability means. In the context of such games one might be ready to accept Laplace's* definition of the probability of an event E as k/m, where m is the number of "equally possible" cases (meaning equally probable cases) that could occur, and k is the number of those cases that constitute E. When this "definition" is applicable, it leads to a familiar set of axioms. The main disadvantage of this approach is that a clearly exhaustive set of "equally probable cases" cannot usually be specified with reasonable

objectivity in scientific applications. Also, the definition is somewhat circular.

In many experiments or observational circumstances the kind of symmetry required for the direct application of the classical definition of probability is lacking. To get around this difficulty, or for other reasons, a definition in terms of long-run proportional frequency of "successes" was explicitly proposed by Leslie Ellis and Cournot* in 1843, and developed in much detail by Venn in 1866. (For these references and further history, see Keynes [21, pp. 92–93]). As usual with simple ideas, "frequentism" had been to some extent foreshadowed long before, e.g., by Aristotle, who said that the probable is what usually happens, or Greek words to that effect, but a self-respecting kudologist would not on that account attribute the theory to Aristotle alone. The frequentist definition is associated with physical probability rather than with logical or subjective (= personal) probability. For discussions of kinds of probability, *see* BELIEF, DEGREES OF and its references, and PROBABILITY, FOUNDATIONS OF—I.

It is by no means simple to construct a satisfactory definition of physical probability based on limiting frequencies. Consider, e.g., the following naive approach. By a "trial" we mean an experiment whose outcome is either some event E, or is the negation of E, a "failure" F. For example, a trial might be the tossing of a coin or the throw of a die and E might denote "heads" or "a six." Let an infinite sequence of such trials be performed under "essentially equivalent" conditions. Then the proportion of successes in the first n trials might tend to a limit p when $n \to \infty$. If so, then p might be called the probability of a success.

This naive definition of physical probability by long-run or limiting frequency* has some disadvantages. Even if we admit the possibility of an infinite sequence of trials in the real world, as some kind of approximation, the definition says nothing about whether the sequence of outcomes is in any sense random. A more sophisticated long-run-frequency definition of probability was proposed by von Mises [24] based on the prior notion of a random sequence or irregular Kollektiv*. This approach requires axioms for random sequences and again has severe mathematical and logical difficulties, although it can be presented fairly convincingly to the intuition in terms of generalized decimals [17, and references therein]. The theory of von Mises can be made logically rigorous and then leads to a familiar set of axioms for probability [6,23]. For a discussion of randomness, with further references, see also Coffa et al. [5] and the article on RANDOMNESS, BDS TEST FOR in the present encyclopedia.

An approach to the axioms via sharp absolute probabilities, when there are fewer than 26^M possible mutually exclusive propositions, where $M = 10^{1000}$, is to argue as follows. Suppose that the N mutually exclusive possible propositions $E_1, E_2, \ldots, E_N$ of interest have sharp probabilities approximated to ν places of decimals by $p_1, p_2, \ldots, p_N$, where ν is large, so that $p_i = m_i 10^{-\nu}$, where m_i is a positive integer, and $\sum m_i = 10^\nu$. For each i, take a well-shuffled pack of cards containing m_i equiprobable cards and use it to break E_i into m_i mutually exclusive propositions each of probability $10^{-\nu}$. This leads to 10^ν equally probable propositions and the classical definition can now be used to arrive at a familiar set of axioms. (Compare ref. 12, p. 33, where the argument was expressed somewhat differently.)

An approach that again assumes that probabilities mean something and can be expressed numerically was apparently first suggested by S. N. Bernstein* [3]. It depends on ideas such as that $P((E \vee F) \vee G)$ must equal $P(E \vee (F \vee G))$. It is assumed further that, when E and F are mutually exclusive, then $P(E \vee F)$ is some function of $P(E)$ and $P(F)$. The assumptions lead to functional equations that must be satisfied by probabilities, and these equations can be used to justify the axioms. The idea was developed independently by Schrödinger [32], Barnard and Good [12, pp. 107–108] and especially by R. T. Cox [7,8]. Cox's assumptions were weakened by Aczél [1]. The approach again leads to a familiar set of axioms, and seems to be the most convincing justification of these axioms for numerical subjective probability among those approaches that makes no reference to decisions or to gambles.

An advantage of bringing decisions or gambles into the discussion is that a prior intuitive concept of probability is then less necessary in arriving at axioms for subjective probabilities. We are then led to a behavioral approach that many people find more convincing than a more purely linguistic approach. With some ingenuity the behavioral approach can be developed to the point where no explicit definition of either probability or utility* is assumed, but only preferences between acts. This approach was adopted by F. P. Ramsey [27], B. de Finetti [9], and L. J. Savage [31]. They assumed that "your" preferences between acts can be completely ordered, and that the preferences satisfy desiderata for rationality that many people find compelling once the complete ordering is granted. These desiderata lead to the conclusion that if you were perfectly rational, you would behave as if you had a set of probabilities (degrees of belief*) satisfying familiar axioms, and a set of utilities, and that you would always prefer the act of maximum expected utility. In this approach the concepts of probability and utility are not separately defined, nor are they taken for granted. In fact, a perfectly rational person might not know the concepts of probability and utility, but these concepts can be used by some one else to describe the rational person. We can *imagine* a doctor or warrior, for example, who always makes the best decisions, although never having heard of probabilities.

THE PURELY MATHEMATICAL APPROACH

Since most of the philosophical approaches lead to somewhat similar formal theories, it is natural for a mathematician to choose a set of axioms based on earlier formalisms. By separating the symbols E, F, etc., from their concrete meanings, the mathematician can avoid philosophical controversies and get on with the job. This approach was adopted by A. N. Kolmogorov* [22], following some earlier writers. His axioms were expressed in the language of sets and measure theory. Borel–Lebesgue measure was introduced at the turn of the century: see, e.g., Carathéodory [4, p. 702]. Before 1890 set theory was not regarded as mathematically respectable, but by 1930 it was regarded as part of the foundation of pure mathematics.

Kolmogorov stated that he agreed with von Mises's frequency interpretation* of probability, but his axioms do not presuppose this interpretation. To begin with, he assumes that there is a set Ω of "elementary events" ω, but the concept of an elementary event requires no definition as long as we are concerned only with the mathematics, and each ω can instead be called a "point" if this helps the imagination. A class $\mathscr{S}$ of subsets of Ω are called "events," not yet to be interpreted in the ordinary sense; and it is assumed that if the subset S is an element of $\mathscr{S}$, then so is its complement $\Omega - S$. Furthermore, it is assumed that if S and T both belong to Ω, then so does the union of S and T. "Events" satisfying these assumptions are said to constitute an *algebra of events** or field of events. Note the similarity to the axioms for propositions given earlier. If the union of any countable infinity of events is also in $\mathscr{S}$, then $\mathscr{S}$ is unfortunately said to be a σ-algebra. We shall soon see why the condition of countability is assumed.

A symbol $P(S)$ is introduced and is called the (absolute) probability of S. It is assumed that $P(S)$ is a real number and lies in the closed interval [0,1], also that $\Omega \in \mathscr{S}$ and that $P(\Omega) = 1$. Finally, if a countable class of sets $S_1, S_2, S_3, \ldots$ are disjoint, then $P(S_1 \cup S_2 \cup \cdots) = P(S_1) + P(S_2) + \cdots$, where $\cup$ denotes union. This last assumption is called the axiom of complete additivity. A weaker axiom asserts the property only for two sets (which implies the property for any finite number of sets) instead of for a countable infinity of sets. The axiom of complete additivity is the main feature of Kolmogorov's system and makes his system highly reminiscent of Lebesgue measure.

The product axiom is introduced through the back door by defining the conditional probability $P(S|T)$ by the quotient $P(S \cap T)/P(T)$ when $P(T) \neq 0$, where $\cap$ denotes the intersection of sets.

The theory is applied by interpreting "events" as meaning physical events.

Kolmogorov's axioms, perhaps better called the measure-theoretic axioms, are the most popular among mathematical statisticians at present. He did not pretend that he had no predecessors, and Rényi [29, p. 55] cites Borel (1909), Lomnicki (1923), Lévy (1925), Steinhaus (1923), and Jordan (1925). As Rényi says, the measure-theoretic approach leads to a rigorous mathematical theory of stochastic processes.

The analogy with Lebesgue measure makes it clear why the axiom of complete additivity is stated only for a countable infinity of sets: the Lebesgue measure of a unit interval is unity, but this measure can hardly be expressed as the sum of the noncountable number of zero measures of the points in the interval. A similar objection has been raised by de Finetti [10, p. 124] against the axiom of complete additivity itself. For consider an infinite sequence in which it is known that there is precisely one "success" but where this success might be anywhere. In the von Mises theory the probability of a success would apparently be zero, and the axiom of complete additivity appears then to lead to the contradiction $1 = \sum 0 = 0$. Perhaps the resolution of this difficulty is to deny that the foregoing sequences count as Kollektivs and say that one models zero probability by a Kollektiv in which the limiting proportional frequency of "successes" is zero. Then there are a noncountable number of such Kollektivs and no paradox arises. (How you could recognize such Kollektivs in practice from finite initial segments is another problem.)

As indicated earlier, from a strictly practical point of view it makes no sense to choose a precise real number at random; in fact, to do so would be like selecting the infinite sequence of its decimal digits. When E and F denote meaningful practical propositions the measure-theoretic approach is not essential, but the approach compensates for this by its valuable mathematical convenience. Then again a price is paid because the mathematics becomes advanced.

The approach in terms of propositions appears more general than in terms of sets, because there need be no concept of an "elementary proposition." It would, however, be possible, at least if the total number of propositions if finite, to define an elementary proposition as a conjunction of propositions that is (a) not strictly impossible while (b) if it can be made any less probable by being conjoined with another proposition, then it becomes impossible. An elementary proposition would be an interpretation of an elementary event

ω. In this manner one might be able to subsume the propositional approach under the measure-theory umbrella, but if there were only a finite number of propositions, the axiom of complete additivity would be unnecessary.

AXIOMS EXPRESSED IN TERMS OF CONDITIONAL PROBABILITIES

Since probabilities in practice are always conditional, absolute probabilities do not capture anything concrete, so several writers have proposed sets of axioms stated directly in terms of conditional probabilities; e.g., Wrinch and Jeffreys [33], Keynes [21], Reichenbach [28], Good [12], Popper [26], and Rényi [29]. Some of these writers expressed the axioms in terms of the probabilities of propositions. We give here an example of such a system of axioms based on Good [12,13]. In these axioms, the symbol H does not necessarily denote a hypothesis, but in many applications it does.

A1 $P(E|F)$ *is a real number if F is not self-contradictory.* (A similar caveat applies in the remaining axioms.)

A2 $0 \leqslant P(E|F) \leqslant 1$.

A3 *If* $P(EF|H) = 0$, *then* $P(E \vee F|H) = P(E|H) + P(F|H)$.

A4 *If H logically implies E* (i.e., if $\overline{H} \vee E$ is a tautology), *then $P(E|H) = 1$* (but not conversely).

A5 (Axiom of equivalence.) *If neither HE nor HF is self-contradictory and HE implies F and HF implies E, then $P(E|H) = P(F|H)$.*

A6 $P(EF|H) = P(E|H)P(F|EH)$.

A7 $P(H^*|H^*) \neq 0$, *where H^* denotes the basic assumptions of logic and pure mathematics.*

A8 $P(E^*|H^*) = 0$ *for some proposition E^*.*

A9 (Complete additivity: optional.) *If* $P(E_iE_j|H) = 0 (i < j; i,j = 12,\ldots)$, *then* $P(E_1 \vee E_2 \vee \cdots |H) = \sum P(E_i|H)$.

A10 (The principle of cogent reason: optional: see Keynes [21, p. 56], Russell [30, p. 397], Good [12, p. 37].) *Let ϕ and ψ be propositional functions. Then, for*

all a and b for which the functions are defined, we have

$$P(\phi(a)|\psi(a)) = P(\phi(b)|\psi(b)).$$

For example, the probability of getting 7 hearts in a whist hand, given only that the pack contains 13 hearts, is the same if we change "hearts" to "diamonds."

In this theory, the main rule of application is obtained by thinking of the axioms as the workings of a black box* into which judgments of probability inequalities can be plugged and from which discernments of new inequalities can be read. This black-box theory is explained in more detail in the article BELIEF, DEGREES OF. Following Keynes [21], who, however, dealt with logical probabilities, Good assumes that (subjective) probabilities are only partially ordered and the use of axioms for sharp probabilities is only a device for expressing the theory in a highly intelligible form. From this form of the theory it is shown by Good [14] that one can derive axioms for the upper* and lower* probabilities themselves. For example, the product axiom splits into six axioms, one of which is

$$P_*(EF|H) \leqslant P^*(E|H) \cdot P_*(F|EH).$$

One can think of upper and lower probabilities as exterior and interior measures, and in a frequency theory they might correspond to upper and lower limits.

If one wishes to talk meaningfully about the probability of a mathematical theorem, as is desirable for the formalizing of "plausible reasoning" (see, e.g., Pólya [25]), then it is necessary as in Good [12, p. 49] to replace the axiom of equivalence by something like A5'. *If at time t you have seen that E and F are equivalent, then $P_t(E|H) = P_t(F|H)$ and $P_t(H|E) = P_t(H|F)$, where the subscript t is self-explanatory.* [Judgments are needed to decide whether $P_t(G|K) = P_s(G|K)$, where $t \neq s$.] This axiom allows subjective probabilities to vary as time passes, without changing ordinary empirical evidence. This is not, of course, the same as the elementary fact that $P(E|GH)$ is not in general equal to $P(E|H)$. By allowing probabilities to have the dynamic feature of varying as a consequence of calculations and thinking, such as the variables in FORTRAN, one can say meaningfully and quantitatively that a mathematical theorem conveys information, and one can also solve some otherwise intractable philosophical problems concerning scientific induction. (For example, see Good [16,18,19], where these varying probabilities are called "evolving" or "dynamic.")

The theory for partially ordered or comparative probability, as just discussed, extends immediately to a theory of rational behavior, by introducing utilities. For details, see Good [13].

A difficulty in theories of subjective probability, pointed out by Richard Jeffrey [20, p. 154], is that a subjective probability can change as a consequence of an experience that "you" cannot express in words. As a matter of fact, badly remembered experiences cause the same difficulty. Although you might have experiences that you cannot personally express fully in words, you can describe them as experiences that occurred at a certain time, and the meaning of this description can be regarded as a proposition in an extended sense. By allowing the meaning of "proposition" to be extended in this manner, the difficulty seems to be overcome without the need for any new axioms. The difficulty would not be overcome within a theory of logical rather than subjective probability.

A distinctive feature of Jeffrey [20, p. 83] is connected with utilities. Previous theories had led to the conclusion that if preference rankings* are sufficiently extensive, probabilities can be uniquely determined but utilities can be determined only up to linear transformations; that is, if a set of (expected) utilities is given, then each element u of this set can be replaced by $au + b$, where a and b are constants, and this substitution will have no effects on any recommended decisions. In Jeffrey's theory, in which both probabilities and utilities refer to propositions, the probabilities and utilities can undergo a class of transformations, the transformation of the utility being of the form $(au + b)/(cu + d)$. He attributes this result to independent personal communications from Kurt Gödel and Ethan Bolker.

In summary, distinct purposes and distinct philosophies can be associated with distinct systems of axioms of probability, although these systems fortunately have much in common.

NOTE

1. All these axioms and comments are applicable *mutatis mutandis* to sets as well as to propositions. For propositional functions, with "quantifiers" such as "for all" and "there exists," further axioms are necessary, but we shall not labor this point.

REFERENCES

1. Aczél, J. (1963). *Ann. Univ. Sci. Budap. Rolando Eötvös Nominatae Sect. Math.*, **6**, 3–11.
2. Barnard, G. A. (1949). *J. R. Statist. Soc. B*, **11**, 115–139.
3. Bernstein, S. N. (1917). An attempt at an axiomatic foundation for the calculus of probability (in Russian). *Khar'kov Univ. Kar'kovskoi mat. obshch. Soobshcheniia*, **15**, 209–274. Abstract in German by Bernstein in *Jb. Math.*, **48**, (1920–1921), 596–599.
4. Carathéodory, C. (1972). *Vorlesungen über Reelle Funktionen*, 2nd ed. Teubner, Leipzig. (Reprint: Chelsea, New York, 1948.)
5. Coffa, J. A., Good, I. J., and Kyburg, H. E. (1974). *PSA 1972* (Proc. 1972 Bienn. Meet. Philos. Sci. Ass.), K. F. Schaffner and R. S. Cohen, eds. D. Reidel, Dordrecht, pp. 103–149.
6. Copeland, A. H. (1937). *Trans. Amer. Math. Soc.*, **42**, 333–357.
7. Cox, R. T. (1946). *Amer. J. Phys.*, **14**, 1–13.
8. Cox, R. T. (1961). *The Algebra of Probable Inference*. Johns Hopkins University Press, Baltimore, Md.
9. Finetti, B. de (1937). *Ann. Inst. Henri Poincaré*, **7**, 1–68. English translation in *Studies in Subjective Probability*, H. E. Kyburg and H. E. Smokler, eds. Wiley, New York, 1964, pp. 95–158.
10. Finetti, B. de (1974). *Theory of Probability*, Vol. 1. Wiley, New York.
11. Gale, R. M. (1967). In *The Encyclopedia of Philosophy*, Vol. 5, Paul Edwards, ed. Macmillan/The Free Press, New York, pp. 494–505.

12. Good, I. J. (1950). *Probability and the Weighing of Evidence*. Charles Griffin, London/Hafner, New York.

13. Good, I. J. (1952). *J. R. Statist. Soc. B*, **14**, 107–114.

14. Good, I. J. (1962). In *Logic, Methodology, and Philosophy of Science*, E. Nagel, P. Suppes, and A. Tarski, eds. Stanford University Press, Stanford, Calif., pp. 319–329.

15. Good, I. J. (1962/1966). In *Theories of the Mind*, J. Scher, ed. Glencoe Free Press/Macmillan, New York, pp. 490–518. Misprints corrected in second edition.

16. Good, I. J. (1968). *Brit. J. Philos. Sci.*, **19**, 123–143.

17. Good, I. J. (1974). In the symposium cited in ref. 5, pp. 117–135.

18. Good, I. J. (1975). *Synthése*, **30**, 39–73.

19. Good, I. J. (1977). In *Machine Intelligence*, Vol. 8, E. W. Elcock and D. Michie, eds. Wiley, New York, pp. 139–150.

20. Jeffrey, R. (1965). *The Logic of Decision*. McGraw-Hill, New York.

21. Keynes, J. M. (1921). *A Treatise on Probability*. Macmillan, London (2nd ed., 1929).

22. Kolmogorov, A. N. (1933). *Grundbegriffe der Wahrscheinlichkeitsrechnung*. Springer-Verlag, Berlin (English translation: Chelsea, New York, 1950).

23. Martin-Löf, P. (1969). *Theoria*, **35**, 12–37.

24. Mises, R. von (1919). *Math. Zeit.*, **5**, 52–99.

25. Pólya, G. (1954). *Mathematics and Plausible Reasoning*, 2 vols. Princeton University Press, Princeton, N. J.

26. Popper, K. R. (1959). *The Logic of Scientific Discovery*. Hutchinson, London.

27. Ramsey, F. P. (1926/1931). *The Foundations of Mathematics and Other Logical Essays*. Kegan Paul, London; Harcourt Brace, New York.

28. Reichenbach, H. (1949). *The Theory of Probability*. University of California Press, Berkeley, Calif.

29. Rényi, A. (1970). *Foundations of Probability*. Holden-Day, San Francisco.

30. Russell, B. (1948). *Human Knowledge, Its Scope and Limitations*. Routledge & Kegan Paul, London.

31. Savage, L. J. (1954). *The Foundations of Statistics*. Wiley, New York. (2nd ed., Dover, New York).

32. Schrödinger, E. (1947). *Proc. R. Irish Acad.*, **51A**, 51–66, 141–146.

33. Wrinch, D. and Jeffreys, H. (1919). *Philos. Mag., 6th Ser.*, **38**, 715–731.

See also BELIEF, DEGREES OF; CHANCE—II; DECISION THEORY; PROBABILITY, FOUNDATIONS OF—I; PROBABILITY, HISTORY OF; and PROBABILITY THEORY: AN OUTLINE.

I. J. GOOD

AZZALINI'S FAMILY.   See SKEW-NORMAL FAMILY OF DISTRIBUTIONS

B

BACK-PROJECTION, METHOD OF

The term *back-projection* comes from the study of acquired immunodeficiency syndrome (AIDS) data, where the method has been applied extensively. The term *back-calculation* is also used, but back-projection seems to reflect better both what the method aims to do and that this task involves uncertainty. The method uses knowledge about the time delay between infection with the human immunodeficiency virus (HIV) and diagnosis with AIDS to project the AIDS incidence data backwards in time with the aim of reconstructing the incidence data for HIV infection. Reconstruction of the HIV infection curve is of considerable interest because it helps to predict future AIDS incidences and associated health-care costs. It also helps to study past trends in infection, giving it the potential to assess the effectiveness of past control strategies.

More generally, back-projection is relevant in the following situation. Consider a community of individuals who, over time, are at risk of having an event of type A occur to them. Any individual to whom A occurs is at risk of having an event of type B occur some time later. Suppose only occurrences of the event B are observed for a community of individuals. The method of back-projection addresses the task of constructing a curve, over time, that describes plausible incidences of the event A in this community. One example of two such events is initiation of a tumor on an internal organ (A) and its detection (B). In the HIV—AIDS context, infection with HIV is event A and diagnosis with AIDS is event B. In each case, occurrences of the event B are observed over time, but the times of occurrences of the event A are not observed. The HIV—AIDS setting is very natural for the method, and we therefore continue our discussion in that context. The paper by Bacchetti et al. [1], with its discussion by other authors, gives a useful review of the topic, and contains many references.

Here the discussion is in discrete time, because the method is usually applied to count data for months, quarters, or years. We adopt the quarter (three months) for the unit of time, because quarterly data are often used in applications.

As time origin take the time when the first HIV infection occurred in the community, or a time point prior to this. The quarters following this origin are labeled by $1, 2, \ldots, T$, where T is the most recent quarter for which a reliable AIDS incidence count is available. Let H_t and A_t denote the numbers of HIV infections and AIDS diagnoses in quarter t, respectively. The means of these two processes, namely the $\lambda_t = E(H_t)$ and the $\mu_t = E(A_t)$, are related by

$$\mu_t = \sum_{s=1}^{t} \lambda_s f_{t-s}, \quad t = 1, 2, \ldots,$$

where f_d is the probability that the duration between infection with HIV and AIDS diagnosis is d quarters. This equation is the basis of the method of back-projection. It is assumed that the $f_0, f_1, \ldots, f_{T-1}$ are known, i.e., estimated precisely from other studies. Most versions of the method of back-projection estimate the parameters $\lambda_1, \lambda_2, \ldots, \lambda_T$ and use them as the reconstruction of the $H_1, H_2, \ldots, H_T$.

The $\lambda_1, \lambda_2, \ldots, \lambda_T$ can be estimated by different approaches. It is tempting to leave them as separate parameters, but the estimates are then unstable, in the sense that the estimate $\hat{\lambda}_t$ may be small while the estimate $\hat{\lambda}_{t+1}$ is very high, and the nature of such fluctuations is very sensitive to the AIDS incidence data. This sensitivity is clearly undesirable, but so is the irregular nature of the estimated $\lambda_1, \lambda_2, \ldots, \lambda_T$. The force of infection responsible for the HIV infections stems from the total number of infective individuals in the community, which is usually large and fairly stable, so that very dramatic changes in the numbers infected in adjacent quarters are implausible. Furthermore, this instability makes it difficult to determine when the iterative methods used to obtain the parameter estimates have converged.

These two problems can be overcome by restricting the $\lambda_1, \lambda_2, \ldots, \lambda_T$ to be a smooth function over time. Some versions of the method of back-projection achieved this by assuming a smooth parametric form $\lambda_t = g(t; \theta), t = 1, 2, \ldots, T$, where the dimension p of the parameter $\theta = (\theta_1, \ldots, \theta_p)$ is much smaller than T; see Brookmeyer and Gail [6] and Rosenberg et al. [10] for two different forms of this approach. More recently there has been a preference [1,2,5,9] for keeping the $\lambda_1, \lambda_2, \ldots, \lambda_T$ distinct and estimating them in a way that ensures a smooth form over time. This approach is in the spirit of nonparametric density estimation* with kernal smoothing.

There are two reasons why the nonparametric approach seems preferable. Firstly, it does not rely on choosing an appropriate family of parametric forms; with a poor choice there could result considerable bias in the estimation of HIV incidence, particularly in the more recent past. The second reason concerns precision. When estimating the $\lambda_1, \lambda_2, \ldots, \lambda_T$ we are really estimating a function over time, and the achievable precision of estimates of λ_t depends on time t. In particular, AIDS incidence data contain relatively little information about recent HIV infection, because there is generally a long delay between HIV infection and AIDS diagnosis. In parametric models the precision is reflected through standard errors for the parameter estimates, and it is not clear that these errors are able to reflect the variation in the precision of estimates of HIV incidence over time in an appropriate manner.

A pragmatic choice for the probability model of HIV incidences is the Poisson distribution with the mean varying over time and assuming that the HIV incidences are independent. This assumption simplifies the analysis, because a sum of independent Poisson variates is also a Poisson variate. In particular, under this model the AIDS incidences are also independent Poisson variates. Support for the use of this model in the method of back-projection is given by Rosenberg and Gail [9], who show that for large epidemics* the use of this model gives results similar to those obtained by either using a quasi-likelihood* approach or an approach that conditions on the total number infected. Further support for the Poisson model is given by Becker and Xu [4], who demonstrate that for reconstructing the realized HIV incidence curve by back-projection the Poisson model is actually more appropriate than a transmission model.

With the Poisson assumption, the log likelihood corresponding to observed AIDS incidences $a_1, a_2, \ldots, a_T$, is

$$\ell(\lambda) = \sum_{t=1}^{T} a_t \log \left(\sum_{s=1}^{t} \lambda_s f_{t-s} \right)$$
$$- \sum_{s=1}^{T} \lambda_s F_{T-s} + \text{constant},$$

where $F_t = \sum_{d=0}^{t} f_d$, the cumulative distribution function of the incubation period for AIDS. Nonparametric back-projection requires the estimation of all the λ_t. Direct maximization of $\ell(\lambda)$ by the methods of calculus is usually troublesome, because the λ_t are constrained to be nonnegative. For example, if we want to fit this model by GLIM* [8], using a Poisson error and an identity link, then we are likely to have problems with the number of parameters being too large and obtaining negative estimates for nonnegative parameters.

The EM algorithm* (Dempster et al. [7]) conveniently overcomes the computational problems involved in maximizing $\ell(\lambda)$ with respect to λ. It produces estimates that are nonnegative, and its iterations involve only simple explicit formulas. Let n_{st} denote the number of individuals infected in quarter s and diagnosed with AIDS in quarter t. Adopting $\{n_{st}; 1 \leqslant s \leqslant t \leqslant T\}$ as the complete data set in the EM algorithm we find [2] that the E and M steps combined give the rth iteration of the estimate as

$$\lambda_s^{(r)} = \sum_{t=s}^{T} \hat{n}_{st}/F_{T-s}, \qquad (1)$$

where $\hat{n}_{st} = a_t \lambda_s^{(r-1)} f_{t-s} / \sum_{i=1}^{t} \lambda_i^{(r-1)} f_{t-i}$ "estimates" the unobserved n_{st}.

It can require a large number of iterations for convergence to be reached. This is often the case in applications of the EM algorithm, but in the present application this defect is exacerbated by the instability of the estimates $\hat{\lambda}_s$, making it unclear when convergence is reached. Smoothing overcomes this difficulty.

Several approaches have been used to obtain nonparametric estimates of the λ's that have a smooth form. Rosenberg and Gail [9] and Brookmeyer [5] use splines*, while Becker et al. [2] use the EMS algorithm, which adds a smoothing step after each iteration of the E and M steps of the EM algorithm. The latter approach is related to the use of a penalized likelihood*, which is proposed by Bacchetti et al. [1].

The approach using the EMS algorithm is the simplest to explain and to implement. It simply smoothes the updated estimates (1) by a moving average. That is, if we denote the EM-updated estimates (1) by $\lambda_s^{'(r)}$ then we use

$$\lambda_s^{(r)} = \sum_{i=1}^{k} w_i \lambda_{s+i-k/2}^{'(r)}$$

as the EMS-updated estimates, where k is an even integer and the weights w_i are positive, symmetric over $0, 1, \ldots, k$ and sum to 1. Iterations are continued until changes in the parameter estimates are minimal. For example, until $\sum_t (\lambda_t^{(t)} - \lambda_t^{(r-1)})^2 / (\lambda_t^{(r-1)})^2 < \varepsilon$, for a specified small value of ε.

Adding the smoothing step both generates a more plausible HIV infection curve and substantially reduces the number of iterations required to reach convergence.

The fitted AIDS incidence curve corresponding to the estimates $\hat{\lambda}_1, \hat{\lambda}_2, \ldots, \hat{\lambda}_T$ is given by

$$\hat{\mu}_t = \sum_{s=1}^{t} \hat{\lambda}_s f_{t-s}, \quad t = 1, 2, \ldots, T.$$

Short-term projections of AIDS incidences can be made by applying this equation with $t = T + 1, T + 2, \ldots$, which requires $\hat{\lambda}_s$ to be specified for $s = T + 1, T + 2, \ldots, t$. The choice of values for these unknown quantities is not crucial, because in the formula for making short-term projections of AIDS incidences these values are multiplied by very small probabilities. It is usually reasonable to set $\hat{\lambda}_s = \hat{\lambda}_T$ for $s = T + 1, T + 2, \ldots, t$.

The precision of short-term projections and of other estimates of interest can be quantified by simulation studies, which are easy to perform for the Poisson model. To simulate an AIDS incidence data set one simply generates realizations of independent Poisson variates having means $\mu_1, \mu_2, \ldots, \mu_T$, where estimates are used for the μ_t. Care is required when simulation is used to assess the precision of estimates derived by nonparametric back-projection with smoothing. As explained in ref. 3, it is necessary to simulate AIDS data sets using maximum likelihood estimates of the μ_t, as computed via the EM algorithm, rather than the estimates derived via the EMS algorithm. While appropriate estimates of the λ_s are obtained under the Poisson model, estimates of precision derived from simulations using the Poisson model tend to be optimistic. Standard errors derived from simulations based on the Poisson model tend to be smaller than those derived under simulations from transmission models [4].

A number of refinements are made to the method of back-projection in applications for the reconstruction of the HIV infection curve. For example, therapy with zidovudine, introduced in 1987, tends to increase the incubation period for those on therapy. This can be accommodated by allowing the probabilities f_d to depend appropriately on the calendar time of infection. Also, there tend to be random delays in surveillance AIDS data reaching the central registry. This can be accommodated by first adjusting the AIDS incidence data for reporting delays or by incorporating reporting delay data into the method of back-projection; see Bacchetti et al. [1], for example. It is also useful to try to enhance the precision of estimates by incorporating covariate data into the method of back-projection. In particular, age at diagnosis with AIDS is usually recorded, and the use of age as a covariate not only gives an estimate of an age-specific relative risk* of infection with HIV, but also improves the precision of back-projection estimates; see ref. 3.

REFERENCES

1. Bacchetti, P., Segal, M. R., and Jewell, N. P. (1993). Backcalculation of HIV infection rates (with discussion). *Statist. Sci.*, **8**, 82–119.

2. Becker, N. G., Watson, L. F., and Carlin, J. B. (1991). A method of nonparametric back-projection and its application to AIDS data. *Statist. Med.*, **10**, 1527–1542.

3. Becker, N. G. and Marschner, I. C. (1993). A method for estimating the age-specific relative risk of HIV infection from AIDS incidence data. *Biometrika*, **80**, 165–178.

4. Becker, N. G. and Xu Chao (1994). Dependent HIV incidences in back-projection of AIDS incidence data. *Statist. Med.*, **13**, 1945–1958.

5. Brookmeyer, R. (1991). Reconstruction and future trends of the AIDS epidemic in the United States, *Science* **253**, 37–42.

6. Brookmeyer, R. and Gail, M. H. (1988). A method for obtaining short-term projections and lower bounds on the size of the AIDS epidemic. *J. Amer. Statist. Ass.*, **83**, 301–308.

7. Dempster, A. P., Laird, N. M., and Rubin, D. B. (1977). Maximum likelihood for incomplete data via the EM algorithm (with discussion). *J. Roy. Statist. Soc. B*, **39**, 1–38.

8. Payne, C. D. (1985). *The GLIM System Release 3.77 Manual*. Numerical Algorithms Group, Oxford.

9. Rosenberg, P. S. and Gail, M. H. (1991). Backcalculation of flexible linear models of the human immunodeficiency virus infection curve. *App. Statist.*, **40**, 269–282.

10. Rosenberg, P. S., Gail, M. H., and Carroll, R. J. (1992). Estimating HIV prevalence and projecting AIDS incidence in the United States: a model that accounts for therapy and changes in the surveillance definition of AIDS. *Statist. Med.*, **11**, 1633–1655.

See also EM ALGORITHM; MEDICAL DIAGNOSIS, STATISTICS IN; MEDICINE, STATISTICS IN; PREDICTION AND FORECASTING; and SURVIVAL ANALYSIS.

NIELS G. BECKER

BACKWARD DIFFERENCE

A finite difference operator, ∇, defined by

$$\nabla f(x) = f(x) - f(x - 1).$$

Symbolically, $\nabla = 1 - E^{-1}$, where E is the displacement operator*. If the differencing interval is h, we define

$$\nabla^h f(x) = f(x) - f(x - h)$$

and

$$\nabla^h \equiv 1 - E^{-h}.$$

Powers of backward operators are defined recursively:

$$\nabla^2 f(x) = \nabla(\nabla f(x)) = \nabla f(x) - \nabla f(x - 1)$$
$$= f(x) - 2f(x - 1) + f(x - 2)$$

and

$$\nabla^n f(x) = \nabla(\nabla^{n-1} f(x)).$$

Newton's interpolation formula* (which is a finite difference analog of Taylor's series in differential calculus) allows us to evaluate the value of a function at any point x based on the value of the function at x_0 and the values of its successive differences at x_0, where $x = x_0 + uh$ and u is arbitrary. In terms of backward differences, Newton's interpolation formula is

$$f(x) = f(x_0) + \frac{x - x_0}{h} \nabla f(x_0)$$
$$+ \frac{(x - x_0)(x - x_{-1})}{2! h^2} \nabla^2 f(x_0)$$
$$+ \cdots,$$

where $x_i = x_0 + ih$ ($i = \pm 2, \ldots$).

BIBLIOGRAPHY

Johnson, N. L. and Kotz, S. (1977). *Urn Models and Their Application*. Wiley, New York.

Jordan, C. (1950). *Calculus of Finite Differences*, 2nd ed. Chelsea, New York.

See also FINITE DIFFERENCES, CALCULUS OF and FORWARD DIFFERENCE.

BACKWARD ELIMINATION SELECTION PROCEDURE

This is one of a number of alternative ways of fitting a "best" regression equation* using data on a response and a (usually fairly large) set of predictor* variables. Essentially, the method involves first fitting a least-squares* regression model using all the predictor variables and then testing the regression coefficient* of the predictor variable that provides the smallest "extra sum of squares"* when it is added to the model last. If this coefficient is statistically significant, the procedure is stopped and the full equation is used. Otherwise (and typically), this "worst predictor" is deleted, the equation is refitted by least squares, and the worst predictor of the new equation is checked, and eliminated if the corresponding regression coefficient is not statistically significant*. Predictor variables are deleted in this fashion until the worst predictor produces a statistically significant coefficient, when the procedure stops and the current equation is chosen. (*see* SELECTION PROCEDURES for details on various methods of variable selection.)

BIBLIOGRAPHY

Draper, N. R. and Smith, H. (1966). *Applied Regression Analysis*. Wiley, New York (2nd ed., 1981).

See also ELIMINATION OF VARIABLES and REGRESSION (Various Entries).

N. R. DRAPER

BACKWARD-SHIFT OPERATOR. See

BOX–JENKINS MODEL; GRADUATION

BAGPLOT. See BOXPLOT, BIVARIATE

BAHADUR EFFICIENCY

Both in theoretical and practical statistical work the choice of a good test procedure for a given hypothesis-testing* problem is of great importance. For finite sample size, the Neyman–Pearson theory* suggests that one select a test based on its power. Various optimality criteria that have been used in the finite-sample-size case include uniformly most powerful*, uniformly most powerful invariant*, uniformly most powerful unbiased*, etc. However, in many hypothesis-testing problems, an optimal *finite* sample test may not exist or may be difficult to determine. In this case an asymptotic approach is often quite useful. Among the asymptotic techniques used for comparing tests, the most frequently discussed approaches are Pitman efficiency* [11], Chernoff efficiency [7] (see below), Hodges–Lehmann efficiency* [9], and Bahadur efficiency [1–3].

Although Bahadur efficiency is a concept of theoretical nature, its basic idea is not difficult to understand. Let $s = (x_1, x_2, \ldots)$ be an infinite sequence of independent observations of a normal random variable X with mean θ and variance 1. Imagine that two statisticians A and B are supplied with the sequence $s = (x_1, x_2, \ldots)$ for testing $H_0 : \theta = 0$ against $H_1 : \theta = \theta_1(\theta_1 > 0)$. A uses $T_n^{(1)}$ and B uses $T_n^{(2)}$ as test statistics, where $T_n^{(i)}$ is a function of the first n observations $(x_1, x_2, \ldots, x_n)$. Suppose that in both cases, H_0 is rejected for large values of $T_n^{(i)}(i = 1, 2)$. The tail probability based on $T_n^{(i)}$ is defined as $P_{\theta=0}(T_n^{(i)} \geqslant T_n^{(i)}(s))$ for $i = 1, 2$, and in practical statistical work, tail probabilities are often computed for the observed values $x_1, \ldots, x_n$ in a given experiment. If the tail probability is less than or equal to level α, than H_0 is rejected. Suppose that $\theta = \theta_1$ and that data s is observed. Let N_i be the minimal sample size required for $T_n^{(i)}$ to reject H_0 at the significance level* α for $i = 1, 2$. $\{T_n^{(1)}\}$ is considered to be a better sequence of statistics than $\{T_n^{(2)}\}$ if $N_1 < N_2$, since the sample size required by $T_n^{(1)}$ to make the correct decision (rejection of H_0) is less than that of $T_n^{(2)}$. Therefore, it is reasonable to use the ratio N_2/N_1 as a measure of the relative efficiency of $T_n^{(1)}$ to $T_n^{(2)}$. However, the ratio is not very well suited for practical use because it depends on three arguments (α, s, θ_1). Moreover, it is often hard to compute. To avoid these difficulties, let us consider the limit of N_2/N_1 as $\alpha \to 0$. Assume that

$$\epsilon(s, \theta_1) = \lim_{\alpha \to 0} N_2/N_1$$

is a finite number. $\epsilon(s, \theta_1)$ can then be used as a measure of the asymptotic efficiency of

$\{T_n^{(1)}\}$ relative to $\{T_n^{(2)}\}$. Note that this method of comparison is stochastic since $\epsilon(s, \theta_1)$ is a random variable. At first sight it seems that $\epsilon(s, \theta_1)$ is not a very useful criterion for comparing tests since $\epsilon(s, \theta_1)$ still depends on two arguments. However, in many cases $\epsilon(s, \theta_1) = \epsilon^*(\theta_1)$ with probability 1. Loosely speaking, $\epsilon^*(\theta_1)$ is the Bahadur efficiency of $\{T_n^{(1)}\}$ relative to $\{T_n^{(2)}\}$ at $\theta = \theta_1$. If $\epsilon^*(\theta_1) > 1$, then $\{T_n^{(1)}\}$ is said to be more efficient. In the next paragraph we define Bahadur efficiency more precisely.

Let $X_1, X_2, \ldots$ be a sequence of independent, identically distributed (i.i.d.) random variables whose common probability distribution is indexed by a parameter θ ranging in a parameter space Θ. For every $1 \leqslant n \leqslant \infty$, let $P_\theta^{(n)}$ denote the probability distribution of $(X_1, \ldots, X_n)$. $P_\theta^{(n)}$ will often be abbreviated to P_θ to simplify the notation. Let $(\mathcal{S}, \mathcal{Q})$ be the sample space of the sequence of random variables $(X_1, X_2, \ldots)$. Every element s in $\mathcal{S}$ has the form $(x_1, x_2, \ldots)$ and $\mathcal{Q}$ is a σ-field of subsets of $\mathcal{S}$. We are interested in testing the null hypothesis* $H_0 : \theta \in \Theta_0$ against the alternative hypothesis* $H_1 : \theta \in \Theta_1$, where Θ_0 is a proper subset of Θ and $\Theta_1 = \Theta - \Theta_0$. For each $n = 1, 2, \ldots$, let $T_n^{(s)}$ be an extended real-valued function such that T_n is $\mathcal{Q}$-measurable and depends on s only through $(x_1, \ldots, x_n)$. T_n is considered to be a test statistic, large values of T_n being significant. Assume for simplicity, that there exists a distribution function $F_n(t)$ such that $P_\theta(T_n < t) = F_n(t)$ for all $\theta \in \Theta_0$ and for all t, $-\infty \leqslant t \leqslant +\infty$. For every s in $\mathcal{S}$ the tail probability of T_n is defined to be

$$L_n = L_n(s) = 1 - F_n(T_n(s)).$$

Note that L_n depends only on $(x_1, \ldots, x_n)$. Clearly, L_n is a random variable. In typical cases, if $\theta \in \Theta_1$, $L_n \to 0$ exponentially fast (with probability 1 $[P_\theta]$). We shall say that the sequence of test statistics $\{T_n\}$ has the (exact) Bahadur slope $c(\theta)$ if

$$\lim_{n \to \infty} (1/n) \log L_n(s) = -\tfrac{1}{2} c(\theta) \qquad (1)$$

with probability 1 $[P_\theta]$ when θ is the true parameter. Now for a given α, $0 < \alpha < 1$, and a given s, let $N = N(\alpha, s)$ be the smallest

integer m such that $L_n(s) < \alpha$ for all $n \geqslant m$ and let $N = \infty$ if no such m exists. Then N is the minimal sample size required for $\{T_n\}$ to reject H_0 when H_1 is true at level α. If (1) holds and $0 < c(\theta) < \infty$, then

$$\lim_{\alpha \to 0} N(\alpha, s) \left[\frac{c(\theta)}{2 \log(1/\alpha)} \right] = 1 \qquad (2)$$

with probability 1 $[P_\theta]$. To see this, choose a fixed θ and a fixed s such that

$$0 < c(\theta) < \infty$$

and

$$\lim_{n \to \infty} N^{-1} \log L_n(s) = -c(\theta)/2. \qquad (3)$$

Then $L_n(s) > 0$ for all sufficiently large n and $L_n \to 0$ as $n \to \infty$. It follows that $N < \infty$ for every $\alpha > 0$ and that $N \to \infty$ as $\alpha \to 0$. For all sufficient small α, say that $0 < \alpha < \alpha_1$. We have $L_N < \alpha \leqslant L_{N-1}$. Hence $N^{-1} \log L_N < N^{-1} \log \alpha \leqslant (N-1) N^{-1}$. $(N-1)^{-1} \log L_{N-1}$. It now follows from (3) that $N^{-1} \log \alpha \to -c(\theta)/2$ as $\alpha \to 0$. This proves (2).

From (2) we can say loosely that when $\theta \in \Theta_1$ the sample size required for $\{T_n\}$ to make a correct decision (rejection of H_0) at level α is approximately $2 \log(1/\alpha)(c(\theta))^{-1}$ for small α. The (exact) Bahadur slope $c(\theta)$ can thus be interpreted as a measure of the performance of $\{T_n\}$: For every $\theta \in \Theta_1$, the larger $c(\theta)$ is the faster $\{T_n\}$ makes the correct decision (rejection of H_0). If $\{T_n^{(i)}\}$, $i = 1, 2$, are two sequences of test statistics with (exact) Bahadur slope $0 < c_i(\theta) < \infty$ for all $\theta \in \Theta_1$, then (2) implies that

$$\frac{c_1(\theta)}{c_2(\theta)} = \lim_{\alpha \to 0} \frac{N_2(\alpha, s)}{N_1(\alpha, s)}$$

with probability 1 $[P_\theta]$, where $N_i(\alpha, s)$ is the sample size required for $T_n(i)$ to make the correct decision (rejection of H_0) at level α. This suggests $c_1(\theta)/c_2(\theta)$ as a measure of the asymptotic efficiency of $\{T_n^{(1)}\}$ relative to $\{T_n^{(2)}\}$, when θ is the true parameter. We will define, for every $\theta \in \Theta_1$,

$$\epsilon_{12}(\theta) = c_{1(\theta)}/c_{2(\theta)}$$

as the Bahadur efficiency of $\{T_n^{(1)}\}$ relative to $\{T_n^{(2)}\}$ at θ. If $\epsilon_{12}(\theta) > 1$, the sequence of test statistics $\{T_n^{(1)}\}$ is said to be more efficient than the sequence $\{T_n^{(2)}\}$ in the sense of Bahadur efficiency at θ.

To find the (exact) Bahadur slope of a sequence of test statistics $\{T_n\}$, the following result is often useful. Suppose that for every $\theta \in \Theta_1$

$$n^{-1/2}T_n \to b(\theta) \qquad (4)$$

with probability 1 $[P_\theta]$, where $-\infty < b(\theta) < +\infty$, and that

$$n^{-1}\log[1 - F_n(n^{1/2}t)] \to -f(t) \qquad (5)$$

for each t in an open interval I, where f is a continuous function on I, and $\{b(\theta) : \theta \in \Theta_1\} \subset I$. Then the (exact) Bahadur slope $c(\theta)$ exists and

$$c(\theta) = 2f(b(\theta)) \qquad (6)$$

for each $\theta \in \Theta_1$. To establish (6), choose a fixed $\theta \in \Theta_1$ and a fixed s such that

$$n^{-1/2}T_n(s) \to b(\theta)$$

as $n \to \infty$. Let $\epsilon > 0$ be so small that $b(\theta) + \epsilon$ and $b(\theta) - \epsilon$ are in I. Since $F_n(t)$ is nondecreasing in t, $L_n(s) = 1 - F_n(T_n(s))$ and $n^{1/2}(b(\theta) - \epsilon) \leqslant T_n(s) \leqslant n^{1/2}(b(\theta) + \epsilon)$ for sufficiently large n, it follows that $1 - F_n(n^{1/2}(b(\theta) + \epsilon)) \leqslant L_n(s) \leqslant 1 - F_n(n^{1/2} \cdot (b(\theta) - \epsilon))$ for sufficiently large n. From (5), we have $-f(b(\theta) + \epsilon) \leqslant \lim_{n\to\infty} \cdot \inf n^{-1}\log L_n(s) \leqslant \lim_{n\to\infty} \sup n^{-1}\log L_n(s) \leqslant -f(b(\theta) - \epsilon)$. Since f is continuous and ϵ is arbitrary, we conclude that $\lim_{n\to\infty} n^{-1} \cdot \log L_n(s) = -f(b(\theta))$. Hence $2^{-1}c(\theta) = f(b(\theta))$ for each $\theta \in \Theta_1$.

An example is given below to illustrate the computation of the (exact) Bahadur slopes of two sequences of test statistics. Let $X_1, X_2, \ldots$ be i.i.d. random variables with normal* $N(\theta, 1)$ distributions* and suppose that $H_0 : \theta = 0$ is to be tested against $H_1 : \theta > 0$. Consider $T_n(1) = n^{-1/2}\sum_{i=1}^{n}X_i$, $T_n^{(2)} = T_n^{(1)}/S_n$, where $S_n = [\sum_{i=1}^{n}(X_i - \overline{X})^2/(n-1)]^{1/2}$. H_0 is rejected for large values of $T_n^{(i)}$. Then $T_n^{(i)}$ satisfies (4) with $b_1(\theta) = \theta$ and $b_2(\theta) = \theta$ for every $\theta > 0$. Furthermore, $T_n^{(i)}$ satisfies (5) with $f_1(t) = t^2/2$ and $f_2(t) = (1/2)\log(1 + t^2)$. (For more details, see Bahadur [4].) It follows from (6) that $T_n^{(i)}$ has (exact) Bahadur slope $c_i(\theta)$ and $c_1(\theta) = \theta^2$ and

$c_2(\theta) = \log(1 + \theta^2)$ for every $\theta > 0$. Hence the Bahadur efficiency of $\{T_n^{(1)}\}$ relative to $\{T_n^{(2)}\}$ is

$$\epsilon_{12}(\theta) = \theta^2/\log(1 + \theta^2)$$

for every $\theta > 0$. Since $\epsilon_{12}(\theta) > 1$ for all $\theta > O\{T_n^{(1)}\}$ is considered to be a better sequence of test statistics in the sense of Bahadur efficiency.

Since the concepts of Chernoff efficiency and Hodges-Lehmann efficiency are closely related to the concept of Bahadur efficiency, we shall give a very brief introduction to these concepts here. If one wishes to compare statistics $T^{(1)}$ and $T^{(2)}$ for testing hypothesis $H_0 : \theta = \theta_0$, Let $N_i = N_i(\alpha, \beta, \theta_0, \theta), i = 1, 2$, denote sample sizes required for statistical tests based on statistic $T^{(i)}$ to have significance level α and power β for alternative θ. The ratio N_2/N_1 describes in general the relative efficiency of a test based on T_1 compared to T_2. Chernoff's approach is essentially equivalent to considering the limit of N_1/N_2 as α tends to zero and β to 1 at a controlled rate, i.e., $\lim_{\alpha\to0}(\alpha/(1-\beta)) = c$, where $0 < c < \infty$, while the Hodges-Lehmann approach takes the limit of N_1/N_2 as $\beta \to 1$.

A sequence of test statistics T_n is said to be asymptotically optimal in the sense of Bahadur efficiency for testing H_0 against H_1 if the (exact) Bahadur slope of $\{T_n\}$ is maximum among all sequences of test statistics for all $\theta \in \Theta_1$. A frequently used method of finding an asymptotically optimal sequence of test statistics is to find an upper bound for the (exact) Bahadur slope. A sequence $\{T_n\}$ is then clearly asymptotically optimal if its (exact) Bahadur slope is equal to this upper bound for every $\theta \in \Theta_1$. Suppose that for each θ the distribution of the single observation X admits a density function $f(x, \theta)$ with respect to a fixed measure μ. For any θ and θ_0 in Θ, let the Kullback-Liebler information* number K be defined by

$$K(\theta, \theta_0) = \int f(x, \theta)\log\frac{f(x, \theta)}{f(x, \theta_0)}\, d\mu.$$

Then $0 \leqslant K \leqslant \infty$ and $K = 0$ if and only if $P_\theta \equiv P_{\theta_0}$. For each $\theta \in \Theta$, let

$$J(\theta) = \inf\{K(\theta, \theta_0) : \theta_0 \in \Theta_0\}.$$

If $c(\theta)$ is the (exact) Bahadur slope of a sequence of test statistics $\{T_n\}$, then

$$c(\theta) \leqslant 2J(\theta)$$

for every $\theta \in \Theta_1$.

Under certain regularity conditions, the (exact) Bahadur slope of the sequence of likelihood ratio test* statistics exists and attains the upper bound $2J(\theta)$ for all $\theta \in \Theta_1$. Hence under these conditions, the likelihood ratio rests are asymptotically optimal in the sense of Bahadur efficiency.

An exceptional fine reference for Bahadur efficiency is the monograph written by Bahadur [4]. It contains an almost exhaustive bibliography on this subject (up to 1970). Groeneboom and Oosterhoff [8] wrote an interesting survey paper on Bahadur efficiency and probabilities of large deviations which contains more recent references. A much more general framework for obtaining asymptotical optimal tests in the sense of Bahadur efficiency was discussed in Bahadur and Raghavachari [5]. In Berk and Brown [6] the notion of Bahadur efficiency was extended to the sequential* case.

REFERENCES

1. Bahadur, R. R. (1960). *Ann. Math. Statist.*, **31**, 276–295.
2. Bahadur, R. R. (1960). *Contributions to Probability and Statistics.* Stanford University Press, Stanford, Calif., pp. 79–88.
3. Bahadur, R. R. (1960). *Sankhyā*, **22**, 229–252.
4. Bahadur, R. R. (1971). *Some Limit Theorems in Statistics.* SIAM, Philadelphia.
5. Bahadur, R. R. and Raghavachari M. (1972). *Proc. 6th Berkeley Symp. Math. Stat. Prob.*, Vol. 1. University of California Press, Berkeley, Calif., pp. 129–152.
6. Berk, R. H. and Brown, L. D. (1978). *Ann. Statist.*, **6**, 567–581.
7. Chernoff, H. (1952). *Ann. Math. Statist.*, **23**, 493–507.
8. Groeneboom, P. and Oosterhoff, J. (1977). *Statist. Neerlandica*, **31**, 1–24.
9. Hodges, J. L. and Lehmann, E. L. (1956). *Ann. Math. Statist.*, **27**, 324–335.
10. Pitman, E. J. G. (1948). *Nonparametric Statistical Inference.* Institute of Statistics, University of North Carolina, Chapel Hill, N. C. (unpublished).
11. Pitman, E. J. G. (1949). *Lecture Notes on Nonparametric Statistical Inferences.* Columbia University, New York (unpublished).

See also BAHADUR EFFICIENCY, APPROXIMATE; EFFICIENCY, SECOND-ORDER; HYPOTHESIS TESTING; and INFERENCE, STATISTICAL.

S. K. PERNG

BAHADUR EFFICIENCY, APPROXIMATE

As explained in the entry BAHADUR EFFICIENCY, the concept of exact Bahadur efficiency [1] requires the large deviation result

$$\lim_{n \to \infty} n^{-1} \log L_n(s) = -\tfrac{1}{2} c(\theta).$$

The derivation of this result often becomes the stumbling block in the application. As a "quick and dirty" variant Bahadur simultaneously proposed approximate Bahadur efficiency, valid for comparison of so-called standard sequences of test statistics. The sequence $\{T_n\}$ is *standard* if three conditions hold:

I. There exists a continuous probability function F such that

$$\lim_{n \to \infty} P_\theta(T_n < s) = F(s)$$

for each $\theta \in \Theta_0$ and $s \in \mathbb{R}$ (observe that F does not depend on θ).

II. There exists a positive constant a such that

$$\lim_{s \to \infty} s^{-2} \log[1 - F(s)] = -\tfrac{1}{2} a.$$

III. There exists a positive real-valued function b on Θ_1 such that

$$\lim_{n \to \infty} P_\theta(|n^{-1/2}T_n - b(\theta)| > s) = 0$$

for each $\theta \in \Theta_1$ and $s \in \mathbb{R}$.

Motivated by the fact that conditions I–III together imply

$$\lim_{n\to\infty} n^{-1} \log[1 - F(T_n(s))] = -\tfrac{1}{2} a[b(\theta)]^2,$$

the approximate Bahadur slope of a standard sequence $\{T_n\}$ is defined as $a[b(\theta)]^2$. Analogous to exact Bahadur efficiency, the approximate Bahadur efficiency of the sequence $\{T_{1n}\}$ with respect to another sequence $\{T_{2n}\}$ is defined as the ratio of their respective approximate Bahadur slopes $a_1[b_1(\theta)]^2$ and $a_2[b_2(\theta)]^2$. Since it is rather easy to verify whether a sequence of test statistics actually is standard, approximate Bahadur efficiency—in spite of its apparent shortcomings [1]—has become quite popular.

In its favor, Bahadur argued that for many well-known test statistics the limiting (as the alternative approaches the null hypothesis) approximate Bahadur efficiency is equal to the asymptotic relative Pitman efficiency*. Working with an extended version of the latter, Wieand [5] elaborated this point by presenting a condition under which the limiting approximate Bahadur efficiency coincided with the limiting (as the size of the test tends to zero) asymptotic relative Pitman efficiency:

III* There exists a constant $\epsilon^* > 0$ such that for every $\epsilon > 0$ and $\delta \in (0, 1)$ there exists a constant $C_{\epsilon,\delta}$ such that

$$P_\theta(|n^{-1/2}T_n - b(\theta)| \geqslant \epsilon b(\theta)) < \delta$$

for each $\theta \in \Theta_1$ and $n \in \mathbb{N}$ satisfying $\inf_{\theta_0 \in \Theta_0} d(\theta, \theta_0) < \epsilon^*$ and $n^{1/2}b(\theta) > C_{\epsilon,\delta}$.

Here d is an arbitrary metric on the parameter space Θ. Wieand pointed out that a standard sequence $\{T_n\}$ satisfies condition III* if there exists a nondefective cumulative distribution function Q such that

$$P_\theta(|T_n - n^{1/2}b(\theta)| > s) \leqslant 1 - Q(s) \quad (1)$$

for every $s > 0$ and $\theta \in \Theta_1$ satisfying $\inf_{\theta_0 \in \Theta_0} d(\theta, \theta_0) < \epsilon^*$ [choose $C_{\epsilon,\delta}$ so as to satisfy $Q(\epsilon C_{\epsilon,\delta}) > 1 - \delta$].

Evaluating limiting Pitman efficiency means that the alternative is sent to the null hypothesis, and, afterwards, the size α to zero (Fig. 1). The order of the operations is reversed for the limiting approximate Bahadur efficiency (Fig. 2). Under condition III* it holds for standard sequences of test statistics that both ways of approaching $(\Theta_0, 0)$ yield the same result. In fact, for many ways in which (θ, α) tends to $(\Theta_0, 0)$ the result is the same, provided that condition III* holds. Wieand's theorem states this coincidence for the simple null hypothesis, while requiring continuity and strict monotonicity of the tail of the asymptotic null distribution of the sequence of test statistics. In ref. 2 the theorem is extended to composite null hypotheses, and continuity and strict monotonicity conditions are discarded.

Condition III* only has meaning for sample sizes that are sufficiently large, and is therefore not capable of providing lower bounds for the smallest sample size such that the power at θ of the size-α test is at least β. Hence, Wieand's theorem should not be combined with alternative definitions of limiting Pitman efficiency that involve such a smallest sample size.

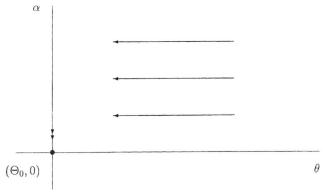

θ **Figure 1.** Limiting Pitman efficiency.

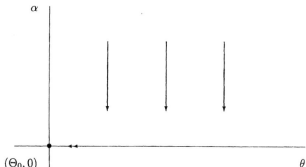

Figure 2. Limiting Bahadur efficiency. $(\Theta_0, 0)$

Condition III* may not hold when T_n shows sufficiently volatile behavior under the alternative hypothesis. To give an example, let $X_1, \ldots, X_n$ be independent nonnegative random variables following a common continuous cumulative distribution function $G(\cdot; \theta)$ under P_θ. Denote $-\log[1 - G(x; \theta)]$ by $\Lambda(x; \theta)$, and assume the existence of constants $v \in (0, 1)$ and $c_v < \infty$ such that

$$\int_{[0,t]} [1 - G(s; \theta)]^v d\Lambda(s; \theta_0) < c_v$$

$$\text{for every } \theta \in \Theta. \qquad (2)$$

The constant v indicates the effect the alternatives have on the tail of the distribution of the X_i's. We shall distinguish between moderate tail alternatives $0 < v < 1/2$ and heavy tail alternatives $1/2 < v < 1$. Define the empirical process by

$$U_n(x; \theta) = n^{-1/2} \sum_{i=1}^{n} [1_{\{X_i \leq x\}} - G(x; \theta)],$$

and introduce the test statistic

$$T_n = \sup_{x \in [0,\infty)} \int_{[0,x]} \frac{U_n(t-; \theta_0)}{1 - G(t; \theta_0)} dG(t; \theta_0)$$

$$= \sup_{x \in [0,\infty)} \int_{[0,x]} U_n(t-; \theta_0) d\Lambda(t; \theta_0).$$

We may write

$$|T_n - n^{1/2}b(\theta)| \leq c_v \sup_{x \in [0,\infty)} \frac{|U_n(x; \theta)|}{[1 - G(x; \theta)]^v},$$

where

$$b(\theta) = \sup_{x \in [0,\infty)} \int_{[0,x]} [G(t; \theta) - G(t; \theta_0)] d\Lambda(s; \theta_0)$$

[the existence of $b(\theta)$ is guaranteed by (2)]. Hence, (4.14) in Marcus and Zinn [3] yields

$$P_\theta(|T_n - n^{1/2}b(\theta)| > sn^{v-1/2})$$

$$\leq \frac{257}{2v - 1} \left(\frac{3c_v}{s}\right)^{1/v}$$

for heavy tail alternatives, and Lemma 2.2 in Pyke and Shorack [4] yields

$$P_\theta(|T_n - n^{1/2}b(\theta)| > s) \leq \frac{1}{1 - 2v} \left(\frac{c_v}{s}\right)^2$$

for moderate tail alternatives. Bahadur's condition III is always satisfied. Since the latter probability inequality is of the form (1), Wieand's condition III* is satisfied for moderate tail alternatives. However, for heavy tail alternatives condition III* is not satisfied.

In ref. 2 a weaker form of the condition is given, under which it still can be shown that we come close to the answer of the Bahadur approach for any trajectory in a restricted region, for instance trajectories as sketched in Fig. 3. The closer we want to be to the answer of the Bahadur approach, the smaller the area of admitted trajectories. This is reflected, e.g., by a lowering of the upper curve depicted in Fig. 3, and implies that we should restrict ourselves to tests of even smaller size. Thus, the quality of the approximation of the finite sample relative efficiency by the approximate Bahadur efficiency for alternatives close to the null hypothesis is only guaranteed in case of (very) small levels.

REFERENCES

1. Bahadur, R. R. (1960). Stochastic comparisons of tests. *Ann. Math. Statist.*, **31**, 276–295.

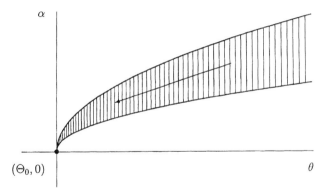

α

$(\Theta_0, 0)$

θ **Figure 3.** The restricted area of admitted trajectories.

2. Kallenberg, W. C. M. and Koning, A. J. (1995). On Wieand's theorem. *Statist. Probab. Lett.*, **25**, 121–132.

3. Marcus, M. B. and Zinn, J. (1985). The bounded law of the iterated logarithm for the weighted empirical distribution process in the non-i.i.d. case. *Ann. Probab.*, **12**, 335–360.

4. Pyke, R. and Shorack, G. R. (1968). Weak convergence of a two-sample empirical process and a new approach to Chernoff-Savage theorems. *Ann. Math. Statist.*, **39**, 755–771.

5. Wieand, H. S. (1976). A condition under which the Pitman and Bahadur approaches to efficiency coincide. *Ann. Statist.*, **4**, 1003–1011.

See also Bahadur Efficiency; Efficiency, Intermediate or Kallenberg and Pitman Efficiency.

Alex J. Koning

BAHADUR, RAGHU RAJ

Born: April 30, 1924, in Delhi, India

Died: June 7, 1997, in Chicago, Illinois, USA

Contributed to: Mathematical statistics, large deviation theory, maximum likelihood theory.

Raghu Raj Bahadur was born in Delhi, India, on April 30, 1924. At St. Stephen's College of Delhi University he excelled, graduating in 1943 with first class honors in mathematics. In 1944, he won a scholarship, and generously returned the money to the College to aid poor students. He continued on at Delhi, receiving a Masters degree in mathematics in 1945. After a year at the Indian Institute of Science in Bangalore, he was awarded a scholarship by the government of India for graduate studies, and in October 1947, after spending one year at the Indian Statistical Institute in Calcutta, Bahadur took an unusual and fateful step. While India was in the upheaval that followed partition and preceded independence, he traveled to Chapel Hill, North Carolina, to study mathematical statistics.

In barely over two years at the University of North Carolina, Bahadur completed his Ph.D. His dissertation focused on decision theoretic problems for k populations, a problem suggested by Harold Hotelling* (although Herbert Robbins served as his major professor). In a December 1949 letter of reference, Harold Hotelling wrote: "His thesis, which is now practically complete, includes for one thing a discussion of the following paradox: Two samples are known to be from Cauchy populations whose central values are known, but it is not known which is which. Probability of erroneous assignment of the samples to the two populations may be larger in some cases when the greater sample mean is ascribed to the greater population mean than when the opposite is done." His first paper, including this example, was published in the *Annals of Mathematical Statistics* [1].

At the winter statistical meetings in December 1949, W. Allen Wallis contacted him to sound him out –was he interested in joining the new group of statisticians being formed at the University of Chicago? He was interested, and Wallis arranged for Bahadur to start in the Spring Quarter of 1950. Bahadur's move to Chicago was to prove a pivotal event in his life. He left Chicago twice (in 1952 and in 1956), and he returned twice (in 1954 and in 1961). He never forgot his roots in India, and the pull of family and the intellectual community in Delhi caused him to return there time and again throughout his life, but Chicago had a special, irresistible allure for him. In the decade following 1948, Allen Wallis assembled an extraordinarily exciting and influential intellectual community. Starting with Jimmie Savage, William Kruskal, Leo Goodman, Charles Stein, and Raj Bahadur, he soon added David Wallace, Paul Meier, and Patrick Billingsley.

Bahadur thrived at Chicago, although sometimes the price was high. One of his great achievements was his 1954 paper on "Sufficiency and Statistical Decision Functions," [2] a monumental paper (it ran to 40 pages in the *Annals*) that is a masterpiece of both mathematics and statistics. The story of its publication tells much about the atmosphere in Chicago in those days. It was originally submitted in May of 1952, and, with Bahadur away in India, it was assigned to Jimmie Savage in Chicago as a referee. Savage was favorable and impressed, and fairly quick in his report (he took two months on what must have been a 100 page manuscript of dense mathematics), so why was there a two year delay in publication? It was not because of a backlog; the *Annals* was publishing with a three month delay in those days. Rather, it was the character of the report and the care of Bahadur's response. For, while Savage was favorable, his reports (eventually there were three) ran to 20 single-spaced pages, asking probing questions as well as listing over 60 points of linguistic and mathematical style. Somehow Bahadur survived this barrage, rewriting the paper completely, benefiting from the comments but keeping the work his own, and preserving over another referee's objections an expository style that explained the deep results both as mathematics and again as statistics.

From 1956 to 1961, Bahadur was again in India, this time as a Research Statistician at the Indian Statistical Institute, Calcutta, but in 1961, he returned to the University of Chicago to stay, except for two leaves that he took to come back to India. He retired in 1991 but continued to take vigorous part in the intellectual life of the Department as long as his increasingly frail health permitted. He died on June 7, 1997.

Bahadur's research in the 1950s and 1960s played a fundamental role in the development of mathematical statistics over that period. These works included a series of papers on sufficiency [2–4,16], investigations on the conditions under which maximum likelihood estimators will be consistent (including Bahadur's Example of Inconsistency) [6], new methods for the comparison of statistical tests (including the measure based upon the theory of large deviations now known as Bahadur Efficiency) [7,8], and an approach to the asymptotic theory of quantiles (now recognized as the Bahadur Representation of Sample Quantiles) [13]. C. R. Rao has written [18] "Bahadur's theorem [his 1957 converse to the Rao-Blackwell theorem [5]] is one of the most beautiful theorems of mathematical statistics." Other work included his approach to classification of responses from dichotomous questionnaires (including the Bahadur–Lazarsfeld Expansion) [9,10], and the asymptotic optimality of the likelihood ratio test in a large deviation sense [12]. Bahadur summarized his research in the theory of large deviations in an elegant short monograph, *Some Limit Theorems in Statistics* [14], and extended it with Zabell and Gupta [15,17]. Notes from Bahadur's 1985 lectures at the University of Chicago on the theory of estimation have been published [20].

Virtually everything Bahadur did was characterized by a singular depth and elegance. He took particular pleasure in showing how simplicity and greater generality could be allies rather than antagonists, as in his demonstration that LeCam's theorem on Fisher's bound for asymptotic variances could be derived from a clever appeal to the Neyman–Pearson Lemma [11]. He forever sought the "right" way of approaching a subject –a combination of concept and technique that not only yielded the result but also showed precisely how far analysis could go. Isaac Newton labored hard to draw the right diagram, to outline in simple steps a demonstration that made the most deep and subtle principles of celestial mechanics seem clear and unavoidably natural. Bahadur had a similar touch in mathematical statistics.

His own referee's reports were minor works of art; his papers often masterpieces.

Raj Bahadur was President of the IMS in 1974–1975, and he was the IMS's 1974 Wald Lecturer. He was honored by the Indian Society for Probability and Statistics in November 1987. In 1993, a Festschrift was published in his honor [19].

This account is an edited version of that in Reference 20, reprinted with the permission of the Institute of Mathematical Statistics.

REFERENCES

1. Bahadur, R. R. (1950). On a problem in the theory of k populations. *Ann. Math. Stat.*, **21**, 362–375.

2. Bahadur, R. R. (1954). Sufficiency and statistical decision functions. *Ann. Math. Stat.*, **25**, 423–462.

3. Bahadur, R. R. (1955). A characterization of sufficiency. *Ann. Math. Stat.*, **26**, 286–293.

4. Bahadur, R. R. (1955). Statistics and subfields. *Ann. Math. Stat.*, **26**, 490–497.

5. Bahadur, R. R. (1957). On unbiased estimates of uniformly minimum variance. *Sankhya*, **18**, 211–224.

6. Bahadur, R. R. (1958). Examples of inconsistency of maximum likelihood estimates. *Sankhya*, **20**, 207–210.

7. Bahadur, R. R. (1960). Stochastic comparison of tests. *Ann. Math. Stat.*, **31**, 276–295.

8. Bahadur, R. R. (1960). On the asymptotic efficiency of tests and estimates. *Sankhya*, **22**, 229–252.

9. Bahadur, R. R. (1961). "A Representation of the Joint Distribution of n Dichotomous Items". In *Studies in Item Analysis and Prediction*, H. Solomon, ed. Stanford University Press, Stanford, Calif., pp. 158–168.

10. Bahadur, R. R. (1961). "On Classification Based on Responses to n Dichotomous Items". In *Studies in Item Analysis and Prediction*, H. Solomon, ed. Stanford University Press, Stanford, Calif., pp. 169–176.

11. Bahadur, R. R. (1964). On Fisher's bound for asymptotic variances. *Ann. Math. Stat.*, **35**, 1545–1552.

12. Bahadur, R. R. (1965). An optimal property of the likelihood ratio statistic. *Proc. Fifth Berkeley Symp. Math. Stat. Prob.*, **1**, 13–26.

13. Bahadur, R. R. (1966). A note on quantiles in large samples. *Ann. Math. Stat.*, **37**, 577–580.

14. Bahadur, R. R. (1971). *Some Limit Theorems in Statistics*. NSF-CBMS Monograph No. 4. SIAM, Philadelphia.

15. Bahadur, R. R., Gupta, J. C. and Zabell, S. L. (1979). "Large Deviations, Tests, and Estimates". In *Asymptotic Theory of Statistical Tests and Estimation* (Hoeffding Volume). Academic Press, Orlando Fla, pp. 33–67.

16. Bahadur, R. R. and Lehmann, E. L. (1955). Two comments on sufficiency and statistical decision functions. *Ann. Math. Stat.*, **26**, 139–142.

17. Bahadur, R. R. and Zabell, S. L. (1979). On large deviations of the sample mean in general vector spaces. *Ann. Probab.*, **7**, 587–621.

18. Ghosh, J. K., Mitra, S. K., and Parthasarathy, K. R., eds. (1992). *Glimpses of India's Statistical Heritage*. Wiley Eastern, New York, Rao quote at p. 162.

19. Ghosh, J. K., Mitra, S. K., Parthasarathy, K. R., and Prakasa Rao, B. L. S., eds. (1993). *Statistics and Probability*. Wiley Eastern, New York.

20. Stigler, S. M., Wong, W. H., and Xu, D., eds. (2002). *R. R. Bahadur's Lecture Notes on the Theory of Estimation*, Vol. 38, *IMS Lecture Notes/Monograph Series*. Institute of Mathematical Statistics, Beechwood, Ohio. (Includes a full bibliography of Bahadur's published research, and two photographs of Bahadur.)

STEPHEN M. STIGLER

BAHADUR-LAZARSFELD EXPANSION

A reparameterization of the formula for a certain multinomial* probability, obtained by Bahadur [1]. Essentially, it is the identity

$$
P\left[\bigcap_{j=1}^{m}(X_j - x_j)\right] = \left\{\prod_{j=1}^{m} p_j^{x_j}(1 - p_j)^{1-x_j}\right\}
$$

$$
\times \left[1 + \sum_{j<k}\sum E[Z_j Z_k]z_j z_k \right.
$$

$$
+ \sum_{j<k<l}\sum\sum E[Z_j Z_k Z_l]z_j z_k z_l + \cdots
$$

$$
\left. + E[Z_1 Z_2 \cdots Z_m]z_1 z_2 \cdots z_m \right],
$$

where $X_1, \ldots, X_m$ are (not necessarily independent) Bernoulli variables (*see* BERNOULLI DISTRIBUTION), with $E[X_j] = p_j (j = 1, \ldots, m)$, $Z_j = (X_j - p_j)/(\sqrt{p_j(1 - p_j)})$ and $z_j = (x_j - p_j)/(\sqrt{p_j(1 - p_j)})$.

For one application, see Woodruff, et al. [2].

REFERENCES

1. Bahadur, R. R. (1961). In *Studies in Item Analysis and Production*, H. Solomon, ed. Stanford University Press, Stanford, Calif.

2. Woodruff, H. B., Ritter, G. L., Lowry, S. R., and Isenhour, T. L. (1975). *Technometrics*, **17**, 455–462.

See also LEXIAN DISTRIBUTION and MULTINOMIAL DISTRIBUTIONS.

BAIRE FUNCTION

This is a real-valued function belonging to the minimal closed class (in the sense of pointwise convergence) of functions containing all continuous functions.

These functions are often used when dealing with classical probabilistic laws related to limit theorems.

See also ZERO-ONE LAWS.

BALANCED INCOMPLETE BLOCK DESIGN. See BLOCKS, BALANCED INCOMPLETE; INCOMPLETE BLOCK DESIGNS

BALANCED REPEATED REPLICATIONS

Balanced repeated replications (BRR) is a technique that may be used for the estimation of the sampling errors* (variances) for complex statistics derived from probability sample designs which make use of both clustering (*see* CLUSTER SAMPLING)* and stratification. In such situations, traditional methods of sampling error estimation, based upon the Taylor expansion*, are often too complex for closed-form expression or too cumbersome for routine calculation.

The general approach of repeated replication was developed at the U.S. Census Bureau (*see* U.S. BUREAU OF THE CENSUS) [1] from basic replication concepts [4]. Orthogonal balancing was added later [2,3]. The BRR method can be briefly described as follows. Assume that we have a stratified design* with two primary selections (clusters) from each stratum. Let S denote the entire sample; let H_i denote the ith half-sample* formed by including one of the two primary selections from each of the strata; and let C_i denote the ith complement half-sample, formed by the primary selections in S not in H_i. If k repeated divisions of sample S, into half-samples and complement half-samples, satisfies an orthogonal design* (each stratum represented by a column, each half-sample by a row, the selection of a prespecified primary sampling unit* by $+1$ and the other by -1), the set of half and complement half-samples is said to be balanced.

For some statistic $g(S)$, the BRR estimate of variance is given by

$$\text{var}_{\text{BRR}-S}[g(S)] = (1/2k) \sum_{i=1}^{k} \left[(g(H_i) - g(S))^2 + (g(C_i) - g(S))^2 \right].$$

Depending upon the sample design, a finite population correction* term may be added to this formula.

If the amount of computation time required for both half-sample $g(H_i)$ and complement half-sample $g(C_i)$ is excessive, the following approximation may be used:

$$\text{var}_{\text{BRR}-H}[g(S)] = (1/k) \sum_{i=1}^{k} \left[g(H_i) - g(S) \right]^2.$$

In addition to their use for developing estimates of sampling variance from complex statistics (e.g., regression* and correlation* statistics), BRR methods have been employed for estimating variances of more simple means, proportions, and ratios when complex sample weighting procedures are in use. For example, BRR methods may be used when successive marginal weighting procedures have been applied. In this case, the weighting algorithm should be applied to each half- and complement half-sample prior to the development of $g(H_i)$ and $g(C_i)$.

When the number of strata is small, estimates of variance based upon fully balanced BRR should be treated as having degrees of freedom equal to the number of strata. This property may be derived in the case of simple linear estimators and has been demonstrated empirically in more complex situations.

REFERENCES

1. Deming, W. E. (1956). *J. Amer. Statist. Ass.*, **51**, 24–53.

2. Kish, L. and Frankel, M. R. (1970). *J. Amer. Statist. Ass.*, **65**, 1071–1094.

3. McCarthy, P. J. (1966). Replication: An Approach to the Analysis of Data from Complex Surveys. *Natl. Center Health Statist. Ser. 2*, **14**, (Washington, D. C.).

4. Mahalanobis, P. C. (1944). *Phil. Trans. R. Soc. Lond. B*, **231**, 329–451.

See also DESIGN OF EXPERIMENTS; HALF-SAMPLE TECHNIQUES; STRATIFIED DESIGNS; and SURVEY SAMPLING.

MARTIN R. FRANKEL

BALANCED RESAMPLING USING ORTHOGONAL MULTIARRAYS

Balanced resampling methods have long been used in surveys for variance estimation in stratified* multistage sampling (*see* BALANCED REPEATED REPLICATIONS and [5]). The technique uses orthogonal balancing to choose subset resamples. Its main advantage is that it yields consistent variance estimates for both smooth functions of means (e.g., the ratio and regression estimators [7]) and nonsmooth estimators (e.g., the median and the low-income proportion [9]), yet is not computer-intensive*.

Until recently, the main limitation on the application of balanced resampling methods was their inability to handle arbitrary within-stratum sample sizes, since the choice of stratum sample size is usually done to improve accuracy of the point estimator and not so as to allow ease of subsequent variance* estimation. Because of this, the method is usually used only when two primary sampling units are selected from each stratum (*see* BALANCED REPEATED REPLICATIONS). More recently, this limitation has been reduced through the use of mixed level orthogonal arrays (OAs) [4,13] and their generalization, balanced orthogonal multiarrays (BOMAs) [11].

Though these methods can be applied to stratified multistage sampling, we will restrict to stratified simple random sampling for simplicity. Stratified simple random sampling is done by partitioning the finite population of N units into L nonoverlapping strata of sizes N_h for $h = 1, \ldots, L$, so that $\sum_h N_h = N$. A simple random sample without replacement, s_h, of size n_h is then drawn independently from each stratum h. If y_{hi} denotes the value of interest for the i-th unit in the h-th stratum, an unbiased* estimate of the population mean, $\overline{Y}$, is $\overline{y}_{st} = \sum_h W_h \overline{y}_h$, where $\overline{y}_h = \sum_{i=1}^{n_h} y_{hi}/n_h$ and $W_h = N_h/N$. An unbiased estimate of the variance of $\overline{y}_{st}$ is given by $v(\overline{y}_{st}) = \sum_{h=1}^{L} P_h^2 n_h^{-1} \sum_{i=1}^{n_h} (y_{hi} - \overline{y}_h)^2$, where $P_h^2 = (1 - n_h/N_h)W_h^2/(n_h - 1)$. The population parameter of interest is often of the form $\theta = g(\overline{Y})$ and is estimated by $\hat{\theta} = g(\overline{y}_{st})$, where g is a smooth function and y may be vector-valued. For the purpose of illustrating the method, we will restrict consideration to this case with scalar y; however, the replication methods to follow extend to more general $\theta = T(F)$, where $F = \sum_h W_h F_h$, with $F_h(t) = N_h^{-1} \sum_i I_{(y_{hi} \leqslant t)}$ the population distribution function of stratum h, and T a smooth functional [10,13].

A balanced repeated replication (BRR) method forms R replicate estimates for $\hat{\theta} = g(\overline{y}_{st})$ by selecting R *balanced* subsets of the sample. These are used as follows: (here for an integer α we refer to a subset of size α as an α-subset, and r refers to the r-th replicate):

1. Let $\overline{y}_n^{(r)} = \alpha_{rh}^{-1} \sum_{i \in s_{rh}} y_{hi}^{(r)}$, the mean of y_{hi} for the units in the α_{rh}-subset s_{rh} of s_h.

2. Let $\bar{y}_{st}^{(r)} = \bar{y}_{st} + \sum_h P_{2h}^2(\bar{y}_h^{(r)} - \bar{y}_h)$, where $P_{2h}^2 = W_h^2/(n_h\tilde{\alpha}_h - 1)$, $\tilde{\alpha}_h = R^{-1}\sum_r 1/\alpha_{rh}$, and $\hat{\theta}^{(r)} = g(\bar{y}_{st}^{(r)})$.

3. A consistent variance estimator for $\hat{\theta}$ is given by

$$v_{\text{PBOMA}}(\hat{\theta}) = R^{-1}\sum_{r=1}^{R}(\hat{\theta}^{(r)} - \hat{\theta}^{(\cdot)})^2,$$

the sample variance of the R replicates of $\hat{\theta}$, where $\hat{\theta}^{(\cdot)} = R^{-1}\sum_r\hat{\theta}^{(r)}$.

In the above formulation it is assumed that the finite population corrections $(1 - n_h/N_h)$ are near 1, which is often the case. If not, one must replace P_{2h}^2 by $(1 - n_h/N_h)P_{2h}^2$. The term *balanced* refers to the method for selecting the α_{rh}-subsets s_{rh} of s_h for $h = 1,\ldots,L$ and $r = 1,\ldots,R$. These subsets, s_{rh}, are selected using the rh-th element of an $R \times L$ array or multiarray which satisfies some balance and orthogonality conditions. It is these conditions which ensure the validity of the resulting variance estimator.

An orthogonal array (*see* ORTHOGONAL ARRAYS AND APPLICATIONS) of strength 2, OA $(R, n_1 \times \cdots \times n_L)$, is an $R \times L$ matrix whose hth column has symbols $1,\ldots,n_h$ such that each symbol appears the same number of times within a column and for any pair of columns each possible combination of symbols appears equally often. If we associate each row with a replicate and each column with a stratum, this array can be used to obtain balanced replicates in a BRR method by letting $\alpha_{rh} = 1$ and s_{rh} be the 1-subset consisting of the y_{hi}, where i is given by the rhth element of the OA. For example, suppose we are in the simplest case, $n_h = 2$ (*see* BALANCED REPEATED REPLICATIONS) and $L = 7$. Associate y_{h1} and y_{h2} with the symbols $+$ and $-$, respectively, and let

$$B = \begin{pmatrix} + & + & + & + & + & + & + \\ - & + & - & + & - & + & - \\ - & - & + & + & - & - & + \\ + & - & - & + & + & - & - \\ + & + & + & - & - & - & - \\ - & + & - & - & + & - & + \\ - & - & + & - & + & + & - \\ + & - & - & - & - & + & + \end{pmatrix}, \quad (1)$$

obtained using the Hadamard* matrices given in ref. [12, p. 322]. Then the 4th replicate subset obtained from the 4th row of B, $(+,-,-,+,+,-,-)$ would be $(y_{11},y_{22},y_{32},y_{41},y_{51},y_{62},y_{72})$.

Viewing steps 1 and 2 in the BRR, we see that, using an OA when $n_h = 2$, we have $\bar{y}_h^{(r)} = y_{hi}^{(r)}$, $P_{2h} = W_h$, and thus $\bar{y}_{st}^{(r)} = \sum_h W_h^2\bar{y}_{hi}^{(r)}$, which is just the usual stratified mean applied to the subset data. This is the method described in BALANCED REPEATED REPLICATIONS.

In coding theory the definition of an OA is generalized to an orthogonal multiarray (OMA) by allowing each element to be a subset of the symbols which satisfy certain properties relevant to the construction of codes [1]. This general concept is used [11] to define a balanced orthogonal multiarray (BOMA), which includes OAs as a special case.

The general definition of a BOMA is quite technical and thus will be delayed. Instead, first consider a proper BOMA. This will allow consideration of some construction techniques.

Definition 1. A proper balanced orthogonal multiarray PBOMA $(R, n_1 \times \cdots \times n_L; \alpha_1,\ldots,\alpha_L)$ is an $R \times L$ array, $A = \{s_{rh}\}$, such that

(i) s_{rh} is an α_h-subset of the set $\{1,\ldots, n_h\}$ for $h = 1,\ldots,L$;

(ii) the number of rows in which a particular i in the stratum h appears is $R\alpha_h/n_h$;

(iii) the number of times i and i' appear together in the same α_h-subset is $R\alpha_h(\alpha_h - 1)/[n_h(n_h - 1)]$ for $i \neq i'$;

(iv) the number of rows in which i in column h and i' in column h' appear together for $h \neq h'$ is $R\alpha_h\alpha_h'/(n_hn_h')$.

One can then use the s_{rh} in the PBOMA to define the subsets in the BRR method. It is easily seen that OA $(R, n_1 \times \cdots \times n_L) =$ PBOMA $(R, n_1 \times \cdots \times n_L; 1^L)$ and that, using L columns of an $R \times R$ Hadamard matrix when $n_h = 2$ as in the above example, we obtain a case of a PBOMA$(R, 2^L; 1^L)$.

To illustrate some of the ideas used in construction of PBOMAs, we present a simple example, constructing the PBOMA$(24, 4^7; 2^7)$

Table 1. A PBOMA$(24, 4^7; 2^7)$

$h = 1$	2	3	4	5	6	7
(1,3)	(1,3)	(1,3)	(1,3)	(1,3)	(1,3)	(1,3)
(1,4)	(1,4)	(1,4)	(1,4)	(1,4)	(1,4)	(1,4)
(1,2)	(1,2)	(1,2)	(1,2)	(1,2)	(1,2)	(1,2)
(2,4)	(1,3)	(2,4)	(1,3)	(2,4)	(1,3)	(2,4)
(2,3)	(1,4)	(2,3)	(1,4)	(2,3)	(1,4)	(2,3)
(3,4)	(1,2)	(3,4)	(1,2)	(3,4)	(1,2)	(3,4)
(2,4)	(2,4)	(1,3)	(1,3)	(2,4)	(2,4)	(1,3)
(2,3)	(2,3)	(1,4)	(1,4)	(2,3)	(2,3)	(1,4)
(3,4)	(3,4)	(1,2)	(1,2)	(3,4)	(3,4)	(1,2)
(1,3)	(2,4)	(2,4)	(1,3)	(1,3)	(2,4)	(2,4)
(1,4)	(2,3)	(2,3)	(1,4)	(1,4)	(2,3)	(2,3)
(1,2)	(3,4)	(3,4)	(1,2)	(1,2)	(3,4)	(3,4)
(1,3)	(1,3)	(1,3)	(2,4)	(2,4)	(2,4)	(2,4)
(1,4)	(1,4)	(1,4)	(2,3)	(2,3)	(2,3)	(2,3)
(1,2)	(1,2)	(1,2)	(3,4)	(3,4)	(3,4)	(3,4)
(2,4)	(1,3)	(2,4)	(2,4)	(1,3)	(2,4)	(1,3)
(2,3)	(1,4)	(2,3)	(2,3)	(1,4)	(2,3)	(1,4)
(3,4)	(1,2)	(3,4)	(3,4)	(1,2)	(3,4)	(1,2)
(2,4)	(2,4)	(1,3)	(2,4)	(1,3)	(1,3)	(2,4)
(2,3)	(2,3)	(1,4)	(2,3)	(1,4)	(1,4)	(2,3)
(3,4)	(3,4)	(1,2)	(3,4)	(1,2)	(1,2)	(3,4)
(1,3)	(2,4)	(2,4)	(2,4)	(2,4)	(1,3)	(1,3)
(1,4)	(2,3)	(2,3)	(2,3)	(2,3)	(1,4)	(1,4)
(1,2)	(3,4)	(3,4)	(3,4)	(3,4)	(1,2)	(1,2)

in Table 1. Let $L = 7$ and $n_h = 4$ for $h = 1$, $\dots, 7$, with B given in (1) and

$$C = \begin{pmatrix} + & - & + & - \\ + & - & - & + \\ + & + & - & - \end{pmatrix}. \qquad (2)$$

Then $A = B \otimes C$, where $\otimes$ denotes the Kronecker* product, is obtained by replacing the $+$'s and $-$'s in B with $+C$ and $-C$. If we number the columns of $C1$–4, each row of C can be used to define a 2-subset of the four units by keeping the units with a $+$ sign in their column. So the three rows of C become the three subsets (1,3), (1,4), and (1,2), and similarly the three rows of $-C$ become the three subsets (2,4), (2,3), and (3,4). Doing this throughout A yields the PBOMA$(24, 4^7; 2^7)$ given in Table 1. This method generalizes easily to $n_h = p = 4m$ for arbitrary L, where m is an integer, to give a PBOMA$(R, p^L; (p/2)^L)$ with $(p - 1)(L + 1) \leqslant R \leqslant (p - 1)(L + 4)$.

The main difference in the general definition of a BOMA is the freedom of α_{rh} to vary over $r = 1, \dots, R$ for a given h.

Definition 2. A balanced orthogonal multiarray BOMA$(R, n_1 \times \cdots \times n_L; \alpha_1, \dots, \alpha_L)$ is an $R \times L$ array, $A = \{s_{rh}\}$, where $\alpha_h = (\alpha_{1h}, \dots, \alpha_{Rh})^T$, such that

(i) s_{rh} is an α_{rh}-subset of the set $\{1, \dots, n_h\}$ for $h = 1, \dots, L$.

(ii) (a) $\sum_r \delta_i(r, h)/\alpha_{rh} = R/n_h$ for all $1 \leqslant i \leqslant n_h$ and $h = 1, \dots, L$;

   (b) $\sum_r \delta_i(r, h)/\alpha_{rh}^2 = R\tilde{\alpha}_h/n_h$ for all $1 \leqslant i \leqslant n_h$ and $h = 1, \dots, L$, where $\tilde{\alpha}_h = R^{-1} \sum_r 1/\alpha_{rh}$;

   (c) $\sum_r \delta_i(r, h)\delta_i'(r, h)/\alpha_{rh}^2 = R(1 - \tilde{\alpha}_h)/[n_h(n_h - 1)]$ for all $i \neq i'$ and $h = 1, \dots, L$;

   (d) $\sum_r \delta_i(r, h)\delta_i'(r, h')/\alpha_{rh}\alpha_{rh}' = R/(n_h n_h')$ for all $1 \leqslant i \leqslant n_h$ and $1 \leqslant i' \leqslant n_h'$, and any pair $h \neq h'$.

Here

$$\delta_i(r, h) = \begin{cases} 1 & \text{if } i \in s_{rh}, \\ 0 & \text{otherwise} \end{cases}$$

for $i = 1, \dots, n_h, r = 1, \dots, R$, and $h = 1, \dots, L$.

Though this definition appears cumbersome, the general form allows greater flexibility for construction. In particular, one can use resolvable balanced incomplete block designs* [6, Chap. 11] to handle varying n_h-values [11]. Table 2 gives a portion of such a BOMA$(56, 8^4 \times 4^2 \times 2^1)$ (see ref. 11 for its construction).

Related to BRR and BOMAs is some work on balanced bootstrapping. The bootstrap* forms replicate estimates by appropriately taking random samples from the data. To obtain valid results the number of replicates must be large, in many cases much larger than necessary using BOMAs. The idea of balanced bootstrap of Davison et al. [2] and Graham et al. [3] is extended in ref. 8 to stratified sampling. A balanced bootstrap constructs balanced replicates, as opposed to randomly resampled replicates, in a similar fashion to using BOMAs, and it shares their properties. The main difference is that in the balanced bootstrap a unit may appear more than once in a given replicate.

Table 2. A BOMA($56, 8^4 \times 4^2 \times 2^1$)

r	$h = 1$	2	3	4	5	6	7
1	(1,3,5,7)	(1,3,5,7)	(1,3,5,7)	(1,3,5,7)	(1,3)	(1,3)	1
2	(2,4,6,8)	(1,3,5,7)	(2,4,6,8)	(1,3,5,7)	(2,4)	(1,3)	2
3	(2,4,6,8)	(2,4,6,8)	(1,3,5,7)	(1,3,5,7)	(2,4)	(2,4)	1
4	(1,3,5,7)	(2,4,6,8)	(2,4,6,8)	(1,3,5,7)	(1,3)	(2,4)	2
5	(1,3,5,7)	(1,3,5,7)	(1,3,5,7)	(2,4,6,8)	(2,4)	(2,4)	2
6	(2,4,6,8)	(1,3,5,7)	(2,4,6,8)	(2,4,6,8)	(1,3)	(2,4)	1
7	(2,4,6,8)	(2,4,6,8)	(1,3,5,7)	(2,4,6,8)	(1,3)	(1,3)	2
8	(1,3,5,7)	(2,4,6,8)	(2,4,6,8)	(2,4,6,8)	(2,4)	(1,3)	1
9	(1,2,5,6)	(1,2,5,6)	(1,2,5,6)	(1,2,5,6)	(1,4)	(1,4)	1
10	(3,4,7,8)	(1,2,5,6)	(3,4,7,8)	(1,2,5,6)	(2,3)	(1,4)	2
11	(3,4,7,8)	(3,4,7,8)	(1,2,5,6)	(1,2,5,6)	(2,3)	(2,3)	1
12	(1,2,5,6)	(3,4,7,8)	(3,4,7,8)	(1,2,5,6)	(1,4)	(2,3)	2
13	(1,2,5,6)	(1,2,5,6)	(1,2,5,6)	(3,4,7,8)	(2,3)	(2,3)	2
14	(3,4,7,8)	(1,2,5,6)	(3,4,7,8)	(3,4,7,8)	(1,4)	(2,3)	1
15	(3,4,7,8)	(3,4,7,8)	(1,2,5,6)	(3,4,7,8)	(1,4)	(1,4)	2
16	(1,2,5,6)	(3,4,7,8)	(3,4,7,8)	(3,4,7,8)	(2,3)	(1,4)	1
			$\vdots$				
49	(1,4,6,7)	(1,4,6,7)	(1,4,6,7)	(1,4,6,7)	(1,2,3,4)	(1,2,3,4)	1
50	(2,3,5,8)	(1,4,6,7)	(2,3,5,8)	(1,4,6,7)	(1,2,3,4)	(1,2,3,4)	2
51	(2,3,5,8)	(2,3,5,8)	(1,4,6,7)	(1,4,6,7)	(1,2,3,4)	(1,2,3,4)	1
52	(1,4,6,7)	(2,3,5,8)	(2,3,5,8)	(1,4,6,7)	(1,2,3,4)	(1,2,3,4)	2
53	(1,4,6,7)	(1,4,6,7)	(1,4,6,7)	(2,3,5,8)	(1,2,3,4)	(1,2,3,4)	2
54	(2,3,5,8)	(1,4,6,7)	(2,3,5,8)	(2,3,5,8)	(1,2,3,4)	(1,2,3,4)	1
55	(2,3,5,8)	(2,3,5,8)	(1,4,6,7)	(2,3,5,8)	(1,2,3,4)	(1,2,3,4)	2
56	(1,4,6,7)	(2,3,5,8)	(2,3,5,8)	(2,3,5,8)	(1,2,3,4)	(1,2,3,4)	1

REFERENCES

1. Brickell, E. F. (1984). A few results in message authentication. *Congr. Number.*, **43**, 141–154. (Provides a few results on authentication of codes, and the use of orthogonal arrays and orthogonal multiarrays in such.)

2. Davison, A. C., Hinkley, D. V., and Schechtman, E. (1986). Efficient bootstrap simulation, *Biometrika*, **73**, 555–566. (Provides a method for obtaining first-order balanced bootstrap replicates in simple random sampling which mainly affects bootstrap estimation of bias.)

3. Graham, R. L., Hinkley, D. V., John, P.W.M., and Shi, S. (1990). Balanced designs of bootstrap simulations. *J. R. Statist. Soc. B*, **52**, 185–202. (Provides a method for obtaining second-order balanced bootstrap replicates in simple random sampling which mainly affects bootstrap estimation of variance.)

4. Gupta, V. K. and Nigam, A. K. (1987). Mixed orthogonal arrays for variance estimation with unequal numbers of primary selections per stratum. *Biometrika*, **74**, 735–742. (Introduces the use of mixed orthogonal arrays with balanced repeated replications to produce variance estimates for means and totals in stratified sampling with unequal numbers of primary selections per stratum.)

5. Gurney, M. and Jewett, R. S. (1975). Constructing orthogonal replications for variance estimation. *J. Amer. Statist. Ass.*, **70**, 819–821. (Extends the balanced repeated replications method, which is applicable to stratified sampling with two primary selections per stratum, to situations with p primary selections per stratum, where p is a prime.)

6. John, P. W. M. (1971). *Statistical Design and Analysis of Experiments.* Macmillan, New York. (General techniques in statistical design of experiments.)

7. Krewski, D. and Rao, J. N. K. (1981). Inference from stratified samples, properties of linearization, jackknife, and balanced repeated replication methods. *Ann. Statist.*, **9**, 1010–1019. (Develops consistency results for the jackknife and balanced repeated replications variance estimators for smooth functions of means in stratified sampling as the number of strata goes to infinity.)

8. Nigam, A. K. and Rao, J. N. K. (1996). On balanced bootstrap for stratified multistage samples. *Statist. Sinica*, **6**, 199–214. (Develops the balanced bootstrap for arbitrary sample size in simple random sampling and for equal strata sample size in stratified sampling.)

9. Shao, J. and Rao, J. N. K. (1993). Standard errors for low income proportions estimated from stratified multi-stage samples. *Sankhyā A*, **55**, 393–414. (Obtains asymptotic distributions of two estimates of proportion of low-income economic families from stratified samples. Variance estimators based on balanced repeated replications and the delta method are proposed, and their consistency established.)

10. Shao, J. and Wu, C. F. J. (1992). Asymptotic properties of the balanced repeated replication method for sample quantiles. *Ann. Statist.*, **20**, 1571–1593. (Develops asymptotic results for balanced repeated replication methods for sample quantiles under quite broad assumptions. These results are applicable to balanced repeated replications using balanced orthogonal multiarrays.)

11. Sitter, R. R. (1993). Balanced repeated replications based on orthogonal multi-arrays. *Biometrika*, **80**, 211–221. (Extends balanced repeated replications to widely varying strata sample sizes through the use of balanced orthogonal multiarrays.)

12. Wolter, K. M. (1985). *Introduction to Variance Estimation*. Springer-Verlag, New York. (Variance estimation methods for complex survey data.)

13. Wu, C. F. J. (1991). Balanced repeated replications based on mixed orthogonal arrays, *Biometrika*, **78**, 181–188. (Develops the use of mixed orthogonal arrays with balanced repeated replications to produce variance estimates for functions of means and totals in stratified sampling with unequal numbers of primary selections per stratum through the use of rescaling.)

See also BALANCED REPEATED REPLICATIONS; ORTHOGONAL ARRAYS AND APPLICATIONS; and SAMPLING PLANS.

R. R. SITTER

BALANCING IN EXPERIMENTAL DESIGN

Treatment combinations may be assigned to the units in an experimental design in such a way that a "balanced" or symmetric configuration is obtained. For example, in a two-factor experiment involving a certain technological process in an accelerated or a steady production, the two factors may be A and B. If the same number of responses are obtainedfor each $A - B$ combination, we say that the design is *balanced*. Otherwise, it is unbalanced, or the data are missing.

See also BLOCKS, BALANCED INCOMPLETE; DESIGN OF EXPERIMENTS; and GENERAL BALANCE.

BALDUCCI HYPOTHESIS. See LIFE TABLES, BALDUCCI HYPOTHESIS

BALLOT PROBLEMS

The first ballot theorem was discovered by J. Bertrand in 1887 [5]. He found the following result:

Theorem 1. If in a ballot candidate A scores a votes and candidate B scores b votes where $a \geqslant b$, then the probability that throughout the counting the number of votes registered for A is always greater than the number of votes registered for B is given by

$$P(a,b) = (a - b)/(a + b) \qquad (1)$$

provided that all the possible voting records are equally probable.

$P(a, b)$ can be expressed as $N(a,b)/\binom{a+b}{a}$, where $N(a, b)$ is the number of favorable voting records and $\binom{a+b}{a}$ is the number of possible voting records. Bertrand observed that

$$N(a,b) = N(a - 1,b) + N(a,b - 1)$$

for $a > b$, that $N(a,b) = 0$ for $a \leqslant b$ and deduced (1) from these equations. Also in 1887, D. André [2] proved that

$$N(a,b) = \binom{a + b}{a} - 2\binom{a + b - 1}{a} \qquad (2)$$

by using an ingenious method based on the reflection principle*.

Bertrand's result (1) can be traced back to 1708, when A. De Moivre* (1667–1754) solved the following problem of games of chance* [9].

Two players, A and B, agree to play a series of games. In each game, independently of the others, either A wins a coin from B with probability p or B wins a coin from A with probability q, where $p > 0, q > 0$ and $p + q = 1$. Let us suppose that A has an unlimited number of coins, B has only k coins, and the series ends when B is ruined, i.e., when B loses his last coin. Denote by $\rho(k)$ the duration of games, i.e., the number of games played until B is ruined. The problem is to determine the distribution of $\rho(k)$. De Moivre discovered that

$$\mathbb{P}\{\rho(k) = k + 2j\} = \frac{k}{k + 2j} \binom{k + 2j}{j} p^{k+j} q^j \quad (3)$$

for $k \geqslant 1$ and $j \geqslant 0$. De Moivre stated (3) without proof; it was proved only in 1773 by P. S. Laplace* (1749–1827) and in 1776 by J. L. Lagrange (1736–1813). In 1802, A. M. Ampère expressed his view that formula (3) is remarkable for its simplicity and elegance.

It is convenient to write

$$L(j,k) = \frac{k}{k + 2j} \binom{k + 2j}{j} \quad (4)$$

for $j \geqslant 0, k \geqslant 1, L(0,0) = 1$ and $L(j,0) = 0$ for $j \geqslant 1$. The numbers $L(j, k)$ might appropriately be called De Moivre numbers. They can also be expressed as

$$L(j,k) = \binom{k + 2j - 1}{j} - \binom{k + 2j - 1}{j - 1} \quad (5)$$

for $j \geqslant 1, k \geqslant 0$ and $L(0,k) = 1$ for $k \geqslant 0$. See Table 1.

By De Moivre's result, the numbers $L(j, k)$ can be interpreted in the following way: *One can arrange $k + j$ letters A and j letters B in $L(j, k)$ ways so that for every $r = 1, 2, \ldots, k + 2j$ among the first r letters there are more A than B.*

Since $N(a,b) = L(b, a - b)$, the result of Bertrand follows from that of De Moivre.

The numbers $L(j,k)$ have appeared in various forms in diverse fields of mathematics.

In 1751, L. Euler (1707–1783) encountered the numbers $L(j, 1)$. In 1751, in a letter to Chr. Goldbach, Euler wrote that the number of different ways of dissecting a convex polygon of n sides into $n - 2$ triangles by $n - 3$ nonintersecting diagonals is

$$D_n = \frac{2 \cdot 6 \cdot 10 \cdots (4n - 10)}{2 \cdot 3 \cdot 4 \cdots (n - 1)} \quad (6)$$

for $n \geqslant 3$. From De Moivre's result one can deduce that $D_n = L(n - 2, 1)$. Euler noticed that $D_n = D_{n-1}(4n - 10)/(n - 1)$ for $n \geqslant 3$ where $D_2 = 1$; however, apparently, he did not succeed in proving it. In his reply to Euler, Goldbach observed that the numbers $D_n(n \geqslant 2)$ satisfy the equation

$$D_n = \sum_{r=2}^{n-1} D_r D_{n-1-r} \quad (7)$$

for $n \geqslant 3$. It seems that it escaped the attention of both Goldbach and Euler that (7) is easy to prove and that (7) implies (6). Formula (6) was proved only in 1758 by J. A. de Segner. Other proofs and generalizations were given by N. v. Fuss (1793), G. Lamé (1838), E. Catalan (1838; ref. 7), O. Rodrigues (1838), J. Binet (1839), J. A. Grunert (1841), J. Liouville (1843), E. Schröder (1870), T. P. Kirkman (1860), A. Cayley (1890), and others.

In 1838, E. Catalan [7] gave an interesting interpretation of the numbers $L(j, 1)$. He proved that the number of ways a product of n factors can be calculated by pairs is

$$P_n = \binom{2n - 1}{n - 1} \frac{1}{2n - 1} \quad (8)$$

for $n \geqslant 2$. By the result of De Moivre, we have $P_n = L(n - 1, 1)$.

In 1859, A. Cayley (1821–1895) encountered the numbers $L(j, 1)$ in the theory of graphs* [8]. He defined a tree as a connected

Table 1. $L(j,k)$

$k \backslash j$	0	1	2	3	4	5
0	1	0	0	0	0	0
1	1	1	2	5	14	42
2	1	2	5	14	42	132
3	1	3	9	28	90	297
4	1	4	14	48	165	572
5	1	5	20	75	275	1001

graph containing no circuits, and a rooted tree as a tree in which one vertex is distinguished as the root of the tree. A planted tree is a tree rooted at an end vertex. The number of edges having an end point in a vertex is called the valence of the vertex. A tree is called trivalent if every vertex has valence 3 except the end vertices, which have valence 1. Cayley proved that the number of trivalent planted trees with $2n$ vertices is given by (8) for $n \geqslant 2$. In 1964, F. Harary, G. Prins, and W. Tutte proved that the number of planted trees with $n + 1$ vertices is also given by (8) for $n \geqslant 1$. Other proofs were given by N. G. DeBruijn and B. J. M. Morselt (1967) and D. A. Klarner (1969).

In 1879, W. A. Whitworth demonstrated that one can arrange $j + k$ letters A and j letters B in

$$M(j,k) = \binom{2j + k}{j} \frac{k + 1}{j + k + 1} \qquad (9)$$

ways such that for every $r = 1, 2, \ldots, 2j + k$ among the first r letters there are at least as many A as B. Since, evidently, $L(j, k + 1) = (j + k + 1)M(j, k)$, (9) follows from (4).

It was already mentioned that in Bertrand's ballot theorem, $N(a, b) = L(b, a - b)$.

In 1897, W. A. Whitworth [32] proved that the number of ways $2n$ points on a circle can be joined in pairs so that the n chords do not intersect each other is $L(n, 1)$. This problem has also been studied by A. Errera (1931), J. Touchard (1950), and J. Riordan (1975).

In 1887, É. Barbier [3] generalized Bertrand's Lemma (1) in the following way.

Theorem 2. If in a ballot candidate A scores a votes and candidate B scores b votes, and $a \geqslant \mu b$, where μ is a nonnegative integer, then the probability that throughout the counting the number of votes registered for A is always greater than μ times the number of votes registered for B is given by

$$P(a, b; \mu) = (a - b\mu)/(a + b) \qquad (10)$$

provided that all the possible voting records are equally probable.

É. Barbier did not prove (10). This was proved only in 1924 by A. Aeppli [1]. Other proofs

were given by A. Dvoretzky and Th. Motzkin (1947), H. D. Grossman (1950), S. G. Mohanty and T. V. Narayana [21], and others.

In 1960, L. Takács [26] generalized Theorem 2 in the following way.

Theorem 3. Let us suppose that a box contains n cards marked $a_1, a_2, \ldots, a_n$, where $a_1, a_2, \ldots, a_n$ are nonnegative integers with sum $a_1 + a_2 + \cdots + a_n = k$, where $k \leqslant n$. We draw all the n cards without replacement. Let us assume that every arrangement has the same probability. The probability that the sum of the first r numbers drawn is less than r for every $r = 1, 2, \ldots, n$ is given by

$$P(n, k) = (n - k)/n. \qquad (11)$$

In 1977, L. Takács [30] pointed out that Theorem 3 can also be deduced from a theorem of G. Hajós [14] which he discovered in 1949 and proposed as a problem. Now there are several proofs for (11), those of J. C. Tanner (1961), M. Dwass (1962), H. Dinges (1963), R. L. Graham (1963), J. L. Mott (1963), S. G. Mohanty (1966), J. G. Wendel (1975), and M. Folledo and I. Vincze (1976).

Theorem 3 has many possible applications in various fields of mathematics, and, in particular, in probability theory and in mathematical statistics. Here are a few examples.

First, an application in mathematical logic. Let $x_1, x_2, x_3, \ldots$ be variables and $f_1, f_2, f_3, \ldots$ be unary, binary, ternary, $\ldots$ operations. A finite expression such as $f_2(x_1, f_3(x_2, x_3, f_1(x_1)))$ is called a "word" if it represents a meaningful mathematical formula. In 1930, K. Menger and in 1948, D. C. Gerneth gave a necessary and sufficient condition for a finite expression to be a word, or, more generally, a sequence of m finite expressions to be a list of m words. Let us replace in each expression every variable by 0 and every $f_i (i = 1, 2, \ldots)$ by i. Denote by a_i the number of symbols i in the m expressions. A sequence of m expressions is a list of m words if and only if $a_0 = m + \sum_{i=1}^{\infty} (i - 1)a_i$ and the sum of the last r numbers is less than r for every $r = 1, 2, \ldots$. In 1960, G. N. Raney found that the number of lists of m words of type $(a_0, a_1, a_2, \ldots)$ with a finite length

$a_0 + a_1 + a_2 + \cdots$ is

$$L(m; a_1, a_2, \ldots) = \frac{m}{(m + \sum_{i=1}^{\infty} i a_i)}$$
$$\cdot \frac{(a_0 + a_1 + a_2 + \cdots)!}{a_0! a_1! a_2! \cdots}.$$

(12)

Formula (12) is an immediate consequence of (11).

As a second example, let us consider a single-server queue* in the time interval $(0, \infty)$. It is supposed that the server starts working at time $t = 0$ and at this time k ($k \geqslant 0$) customers are already waiting for service. Denote by $\nu_r (r = 1, 2, \ldots)$ the number of customers arriving during the rth service and by $\sigma(k)$ the number of customers served in the initial busy period. ($\sigma(0) = 0$.) Set $N_r = \nu_1 + \nu_2 + \cdots + \nu_r$ for $r \geqslant 1$ and $N_0 = 0$. If $\nu_1, \nu_2, \ldots, \nu_r$ are mutually independent, identically distributed (i.i.d.) random variables, then

$$\mathbb{P}\{\sigma(k) = n\} = \mathbb{P}\{N_r < r \text{ for } 1 \leqslant r \leqslant n \text{ and}$$
$$N_n = n - k\}$$ (13)

for $0 \leqslant k \leqslant n$. By Theorem 3 one can conclude that

$$\mathbb{P}\{\sigma(k) = n\} = (k/n)\mathbb{P}\{N_n = n - k\}$$ (14)

for $0 \leqslant k \leqslant n$. Formula (14) was found by É. Borel [6] in 1942 in the particular case where $\mathbb{P}\{\nu_r = j\} = e^{-\lambda} \lambda^j / j!$ (i.e., Poisson* distributed) for $j \geqslant 0$ and $k = 1$. In the general case the distribution of $\sigma(k)$ was determined by D. G. Kendall [18] in 1951 and L. Takács [25] in 1952. The distribution of $\sigma(k)$ can also be obtained from a result of R. Otter [23] for branching processes* and from a result of J. H. B. Kemperman [17] for random walks*.

Theorem 3 has numerous possible applications in the theory of order statistics*. See Takács [27,28] for more details. Here only a few results are mentioned. Let

$$\delta_n^+ = \sup_{-\infty < x < \infty} [F_n(x) - F(x)],$$ (15)

where $F_n(x)$ is the empirical distribution function* of a sample of size n whose elements are mutually independent random variables

each having the same continuous distribution function $F(x)$. By Theorem 3 we have

$$\mathbb{P}\left\{\delta_n^+ \leqslant \frac{k}{n}\right\} = 1 - \sum_{j=1}^{n-k} \frac{k}{(n-j)} \binom{n}{j+k}$$
$$\times \left(\frac{j}{n}\right)^{j+k} \left(1 - \frac{j}{n}\right)^{n-j-k}$$

(16)

for $k = 1, 2, \ldots, n$. The probability $\mathbb{P}\{\delta_n^+ \leqslant x\}$ was obtained by N. V. Smirnov [24] in 1944.

Let $\xi_1, \xi_2, \ldots, \xi_m, \eta_1, \ldots, \eta_n$ be mutually independent random variables each having the same continuous distribution function. Denote by $F_m(x)$ and $G_n(x)$ the empirical distribution functions of the samples $(\xi_1, \xi_2, \ldots, \xi_m)$ and $(\eta_1, \eta_2, \ldots, \eta_n)$, respectively. The distribution of the random variable

$$\delta^+(m, n) = \sup_{-\infty < x < \infty} [F_m(x) - G_n(x)]$$ (17)

was found by B. V. Gnedenko and V. S. Koroljuk [12] for $n = m$ and V. S. Koroljuk [19] for $n = mp$, where p is a positive integer. If $n = mp$, where p is a positive integer and k is a nonnegative integer, then by Theorem 3 we obtain that

$$\mathbb{P}\left\{\delta^+(m, n) \leqslant \frac{k}{n}\right\} = 1 - \sum_{(k+1)/p \leqslant s \leqslant m}$$
$$\times \frac{k+1}{n+k+1-sp} \cdot \binom{sp+s-k-1}{s}$$
$$\times \binom{m+n+k-sp-s}{m-s} \bigg/ \binom{m+n}{m}$$ (18)

Ballot theorems have been used by S. G. Mohanty and T. V. Narayana [21] in characterizing various sampling plans*. Takács [27] used Theorem 3 in studying the urn model of F. Eggenberger and G. Pólya [10] (see Johnson and Kotz [15].)

The following generalization of Bertrand's ballot problem has attracted a great deal of interests.

Let us assume again that in a ballot candidate A scores a votes and candidate B scores b votes, and all the possible voting records are equally probable. Denote by α_r and β_r the number of votes registered for A and B, respectively, among the first r votes counted. The problem is to determine $P_j(a, b; \mu)$, the probability that the inequality $\alpha_r > \mu\beta_r$ holds for exactly j subscripts $r = 1, 2, \ldots, a + b$.

In the case where $\mu = a/b$ and $(a, b) = 1$, H. D. Grossman (1954) proved that $P_j(a, b; \mu) = 1/(a + b)$ for $j = 0, 1, \ldots, a + b - 1$.

In the case where μ is a positive integer $P_j(a, b; \mu)$ was found in various cases by L. Takács (1962 [26], 1963, 1964), J. Riordan (1964), and O. Engelberg (1964, 1965).

In 1966, M. T. L. Bizley made a conjecture concerning the general form of $P_j(a, b; \mu)$. Bizley's formula was proved in 1969 by Takács [29].

In 1909, P. A. MacMahon [20] proved a ballot theorem for n candidates. In his theorem it is supposed that in a ballot candidates $A_1, A_2, \ldots, A_n$ score $a_1, a_2, \ldots, a_n$ votes, respectively. Let $a_1 \geqslant a_2 \geqslant \cdots \geqslant a_n$. Denote by $\alpha_1(r), \alpha_2(r), \ldots, \alpha_n(r)$ the number of votes registered for $A_1, A_2, \ldots, A_n$, respectively, among the first r votes counted. If all the possible voting records are equally probable, then

$$\mathbb{P}\{\alpha_1(r) \geqslant \alpha_2(r) \geqslant \cdots \geqslant \alpha_n(r) \text{ for } r = 1, 2, \ldots,$$

$$a_1 + \cdots + a_n\} = \prod_{l \leqslant i < j \leqslant n} \left(\frac{a_i - a_j + j - i}{a_i + j - i} \right).$$

(19)

We have also

$$\mathbb{P}\{\alpha_1(r) > \alpha_2(r) > \cdots > \alpha_n(r) \text{ for } r = 1, 2, \ldots,$$

$$a_1 + \cdots + a_n\} = \prod_{1 \leqslant i < j \leqslant n} \left(\frac{a_i - a_j}{a_i + a_j} \right). \quad (20)$$

This formula has been proved [13] for $n = 3$ and in another context by R. M. Thrall [31] and R. Srinivasan (1963) for $n \geqslant 3$.

By using a result of S. Karlin and J. McGregor [16], D. E. Barton and C. L. Mallows [4] proved that if $c_1 > c_2 > \cdots > c_n$ are integers, then

$$\mathbb{P}\{\alpha_1(r) + c_1 > \alpha_2(r) + c_2 > \cdots > \alpha_n(r) + c_n$$

$$\text{for } r = 1, 2, \ldots, a_1 + \cdots + a_n\}$$

$$= \det \left[\frac{a_i!}{(a_i + c_i - c_j)!} \right], \quad i, j = 1, 2, \ldots, n,$$

(21)

where in the determinant the (i, j)-entry is 0 for $a_i + c_i - c_j < 0$. See also T. V. Narayana [22].

REFERENCES

1. Aeppli, A. (1924). *Zur Theorie verketteter Wahrscheinlichkeiten*. Markoffsche Ketten höherer Ordnung. Dissertation, Die Eidgenössische Technische Hochschule, Zurich, Switzerland.

2. André, D. (1887). *C. R. Acad. Sci. Paris*, **105**, 436–437.

3. Barbier, É. (1887). *C. R. Acad. Sci. Paris*, **105**, 407; 440 (errata).

4. Barton, D. E. and Mallows, C. L. (1965). *Ann. Math. Statist.*, **36**, 236–260.

5. Bertrand, J. (1887). *C. R. Acad. Sci. Paris*, **105**, 369.

6. Borel, É. (1942). *C. R. Acad. Sci. Paris*, **214**, 452–456.

7. Catalan, E. (1838). *J. Math. Pures Appl.*, **3**, 508–516.

8. Cayley, A. (1859). *Philos. Mag.*, **13**, 172–176; *ibid.*, **18**, 374–378 (1859).

9. De Moivre, A. (1711). *Philos. Trans. R. Soc.*, **27**, 213–264.

10. Eggenberger, F. and Pólya, G. (1923). *Zeit. angew. Math. Mech.*, **3**, 279–289.

11. Fuss, P.-H. (1843). *Correspondance mathématique et physique de quelques célèbres géomètres du XVIIIème siècle*, Vols. 1, 2. St. Petersburg. (Reprinted: *The Sources of Science No. 35.* Johnson Reprint Corporation, New York, 1968.)

12. Gnedenko, B. V. and Koroljuk, V. S. (1951). *Dokl. Akad. Nauk SSSR*, **80**, 525–528. (in Russian). [English translation: *Select. Transl. Math. Statist. Prob.*, **1**, 13–16 (1961).]

13. Grossman, H. D. (1952). *Scr. Math.*, **18**, 298–300.

14. Hajós, G. (1949). *Mat. Lapok*, **1**, 72; *ibid.*, **1**, 152 (1950). [Russian and French translations: *Ibid.*, **7**, 115, 153 (1956).]

15. Johnson, N. L. and Kotz, S. (1977). *Urn Models and Their Application: An Approach to Modern Discrete Probability Theory.* Wiley, New York.

16. Karlin, S. and McGregor, J. (1959). *Pacific J. Math.*, **9**, 1141–1164.

17. Kemperman, J. H. B. (1950). *The General One-Dimensional Random Walk with Absorbing Barriers with Applications to Sequential Analysis.* Thesis, Excelsior, The Hague.

18. Kendall, D. G. (1951). *J. R. Statist. Soc. B*, **13**, 151–185.

19. Koroljuk, V. S. (1955). *Izv. Akad. Nauk SSSR. Ser. Mat.*, **19**, 81–96. [English translation: Select. Transl. Math. Statist. Prob., **4**, 105–121 (1963).]

20. MacMahon, P. A. (1909). *Philos. Trans. R. Soc. Lond. A*, **209**, 153–175. [Reprinted in *Combinatorics*, G. E. Andrews, ed. MIT Press, Cambridge, Mass., 1978, pp. 1292–1314.]

21. Mohanty, S. G. and Narayana, T. V. (1961). *Biom. Zeit.*, **3**, 252–258; *Ibid.*, **5**, 8–18 (1963).

22. Narayana, T. V. (1955). *J. Indian Soc. Agric. Statist.*, **7**, 169–178.

23. Otter, R. (1949). *Ann. Math. Statist.*, **20**, 206–224.

24. Smirnov, N. V. (1944). *Uspekhi Mat. Nauk*, **10**, 179–206 (in Russian).

25. Takács, L. (1955). *Acta Math. Acad. Sci. Hung.*, **6**, 101–129.

26. Takács, L. (1962). *J. Amer. Statist. Ass.*, **57**, 327–337.

27. Takács, L. (1967). *Combinatorial Methods in the Theory of Stochastic Processes*. Wiley, New York.

28. Takács, L. (1970). In *Nonparametric Techniques in Statistical Inference*, M. L. Puri, ed. Cambridge University Press, Cambridge, pp. 359–384.

29. Takács, L. (1970). *J. Appl. Prob.*, **7**, 114–123.

30. Takács, L. (1980). *Periodica Math. Hung.*, **11**, 159–160.

31. Thrall, R. M. (1952). *Mich. Math. J.*, **1**, 81–88.

32. Whitworth, W. A. (1897). *DCC Exercises Including Hints for the Solution of All the Questions in Choice and Chance*. Deighton Bell, Cambridge. [Reprinted by Hafner, New York, 1965.]

See also COMBINATORICS; GRAPH THEORY and QUEUEING THEORY.

LAJOS TAKÁCS

BANACH'S MATCH-BOX PROBLEM

Banach's match-box problem is a classical problem, related to "waiting time" (negative binomial*), discrete distributions, and combinatorial probability theory. (For its history, see Feller [1, p. 166].)

A man carries two boxes of matches, one in his left and one in his right pocket. Initially, they contain N matches each. They are emptied one match at a time—the box to have a match taken from it being chosen each time at random, with each equally likely. Denote by R the random variable representing the number of matches remaining in the other box when, for the first time, one of the boxes is found to be empty. The probability distribution of R is given by

$$\Pr[R = r] = \binom{2N - r}{N} 2^{-2N + r}$$

and

$$E(R) = 2^{-2N}(2N + 1)\binom{2N}{N - 1} - 1$$

$$\approx 2\sqrt{N/\pi} - 1$$

(for N sufficiently large, using Stirling's formula*). The value of $E(R)$ for $N = 50$ is approximately 7.04.

REFERENCES

1. Feller, W. (1968). *An Introduction to Probability Theory and Its Applications*, 3rd. ed., Vol. 1. Wiley, New York, pp. 166, 170, 238.

2. Moran, P. A. P. (1968). *An Introduction to Probability Theory*. Clarendon Press, Oxford, p. 59.

See also COMBINATORICS.

BANDWIDTH SELECTION

The bandwidth controls the amount of smoothness of kernel estimators* and related methods, such as local polynomials, for nonparametric curve estimation including density estimation* and nonparametric regression*. Kernel estimators are weighted moving averages*, and the bandwidth is the window width over which local averaging is done. If the bandwidth is too large, important features of the data can be smoothed away; if it is too small, spurious artifacts of the sampling process appear. Introduction to these ideas and further discussion may be found in refs. 5, 19, 20.

A useful approach to bandwidth selection is interactive trial and error by an experienced data analyst. This results in an effective choice, and different insights can be available from different levels of smoothing. A drawback is that the process is time-consuming and requires some expertise in determining which features are important and which are sampling artifacts.

For situations where interactive choice is impractical (e.g., when many curve estimates are required), or simply for use as a starting point for interactive choice, data-based methods of bandwidth selection are useful. The first generation of proposed methods includes rules of thumb and cross validation* methods [19,13]. These do not perform well, but effective performance is available from second-generation methods, including plug-in and smoothed bootstrap* approaches. These points have been amply demonstrated through real data examples, simulation studies, and asymptotic analysis.

A good collection of real data examples may be found in Sheather [17]. Figure 1 shows one example from that paper, the Hidalgo stamp data (see also ref. 9). The bandwidth used here is essentially that of Sheather and Jones [18]. The amount of smoothing by this method (and by other

second-generation methods as well) is correct, as verified by auxiliary historical information in [9]. In particular, the largest five modes are known to be important features that are not artifactual. First-generation methods either break down completely or else grossly oversmooth, showing only two modes and thus missing the other features in the data, as shown in Fig. 4.2 in ref. 18. Extensive simulation studies that compare first- and second-generation methods have been performed in refs. 1, 2, 12, 15.

Most mathematical analysis of bandwidth selectors has measured effectiveness through squared-error-type criteria, such as the mean integrated squared error,

$$\mathrm{MISE} = E\left[\int (f_h - f)^2 dx\right],$$

where f_h is the kernel estimate with bandwidth h, f is the curve being estimated (regression or density), the integral is over some domain such as $(-\infty, \infty)$, and the expectation is with respect to random sampling. The optimal bandwidth with respect to this measure

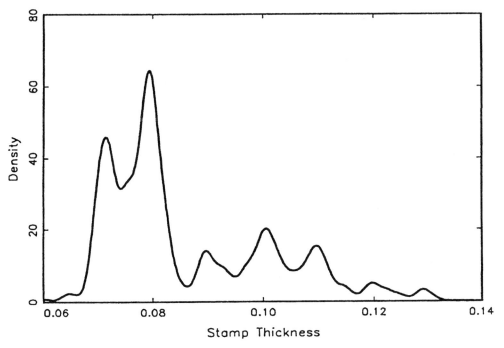

Figure 1. Gaussian kernel density estimate, with bandwidth selected by the Sheather–Jones method, for the Hidalgo stamp data.

of error is not always the most useful for data analysis; Marron and Tsybakov [14] provide further discussion and propose some alternatives. However, the asymptotic tractability of the MISE has allowed particularly deep analysis.

Detailed discussion of the asymptotic results on bandwidth selection may be found in ref. 12. An informative tool for comparison of bandwidth selectors has been the *bandwidth rate of convergence*. These are results of the form

$$\frac{\hat{h} - h_{\text{MISE}}}{h_{\text{MISE}}} \sim n^{-p}$$

for some power p, where $\hat{h}$ is the bandwidth chosen by some data-based method, and h_{MISE} is a minimizer of the mean integrated squared error. The poor performance of first-generation methods is made clear by $p \leqslant \frac{1}{10}$ (an excruciatingly slow rate of convergence); see refs. 6, 16. But much better rates, with $p \in \left[\frac{4}{17}, \frac{1}{2}\right]$, are available for second-generation methods under some smoothness conditions [3,18,10,2,8]. The rate of $p = 1/2$ was shown to be best possible by Hall and Marron [7], and the best possible constant coefficient was obtained by Fan and Marron [4].

See Wand and Jones [20] and Jones et al. [11] for an overview of second-generation bandwidth selection methods and more detailed references.

The preceding remarks all apply to the simplest form of bandwidth selection, where the same amount of smoothing, i.e., the same window width, is used at all locations. In some situations, it is desirable to use different amounts of smoothing in different locations. For example, in Fig. 1 it may be desirable to smooth out the two minor modes at thicknesses 0.12 and 0.13. If this is done with a larger global bandwidth, some of the important modes will also disappear. However, local bandwidth selection entails a much more difficult choice, because the need to choose a single parameter is now replaced by the need to choose an entire function. Methods for choosing a bandwidth function are not yet so well developed or understood as methods for choosing a single global parameter. The term "adaptive" has been used to describe methods which allow different amounts of smoothing in different locations, although "location-adaptive" is better terminology, since "adaptive" has also been used to describe techniques for using the data to choose a global smoothing parameter.

REFERENCES

1. Berlinet, A. and Devroye, L. (1994). *Publ. Inst. Statist. Univ. Paris*, **38**, 3–59.

2. Cao, R., Cuevas, A., and González-Manteiga, W. (1994). A comparative study of several smoothing methods in density estimation. *Comput. Statist. Data Anal.*, **17**, 153–176.

3. Chiu, S. -T. (1991). Bandwidth selection for kernel density estimation. *Ann. Statist.*, **19**, 1528–1546.

4. Fan, J. and Marron, J. S. (1992). Best possible constant in bandwidth selection. *Ann. Statist.*, **20**, 2057–2070.

5. Härdle, W. (1990). *Applied Nonparametric Regression*. Cambridge University Press.

6. Hall, P. and Marron, J. S. (1987). Extent to which least-squares cross-validation minimizes integrated squared error in nonparametric density estimation. *Probab. Theory and Related Fields*, **74**, 567–581.

7. Hall, P. and Marron, J. S. (1991). Lower bounds for bandwidth selection in density estimation. *Probab. Theory and Related Fields*, **90**, 149–173.

8. Hall, P., Marron, J. S., and Park, B. -U. (1992). Smoothed cross-validation. *Probab. Theory and Related Fields*, **92**, 1–20.

9. Izenman, A. J. and Sommer, C. J. (1988). Philatelic mixtures and multimodal densities. *J. Amer. Statist. Ass.*, **83**, 941–953.

10. Jones, M. C., Marron, J. S., and Park, B. -U. (1991). A simple root n bandwidth selector. *Ann. Statist.*, **19**, 1919–1932.

11. Jones, M. C., Marron, J. S., and Sheather, S. J. (1996). A brief survey of bandwidth selection for density estimation. *J. Amer. Statist. Ass.*, **91**, 401–407.

12. Jones, M. C., Marron, J. S., and Sheather, S. J. (1996). Progress in data-based bandwidth selection for kernel density estimation. *Comput. Statist.*, **11**, 337–381.

13. Marron, J. S. (1988). Automatic smoothing parameter selection: a survey. *Empirical Econ.*, **13**, 187–208.

14. Marron, J. S. and Tsybakov, A. B. (1995). Visual error criteria for qualitative smoothing. *J. Amer. Statist. Ass.*, **90**, 499–507.

15. Park, B. -U. and Turlach, B. A. (1992). Practical performance of several data driven bandwidth selectors. *Comput. Statist.*, **7**, 251–285.

16. Scott, D. W. and Terrell, G. R. (1987). Biased and unbiased cross-validation in density estimation. *J. Amer. Statist. Ass.*, **82**, 1131–1146.

17. Sheather, S. J. (1992). The performance of six popular bandwidth selection methods on some real data sets. *Comput. Statist.*, **7**, 225–250.

18. Sheather, S. J. and Jones, M. C. (1991). A reliable data-based bandwidth selection method for kernel density estimation. *J. R. Statist. Soc. B*, **53**, 683–690.

19. Silverman, B. W. (1986). *Density Estimation for Statistics and Data Analysis*. Chapman and Hall, London, England.

20. Wand, M. P. and Jones, M. C. (1995). *Kernel Smoothing*. Chapman and Hall, London, England.

See also Density Estimation—I; Kernel Estimators; Local Regression; Moving Averages; Nonparametric Regression and Window Width.

J. S. Marron

BANKING, STATISTICS IN

Although the large-scale application of statistical methods in banking is a relatively new phenomenon, at least compared to applications in agriculture, medicine, psychology, and other areas, such tools are already beginning to have a major impact. They have now been widely adopted, to the extent that modern banking at all levels would be impossible without them.

There are various types of banks, characterized by their main business. One distinction is between primary banks (chiefly concerned with transfers of money) and secondary banks (providing other services). We can also make a distinction between central banks, commercial retail banks, cooperative banks, merchant banks, and other specialist financial organizations. Central banks are typically government owned. They do not serve individuals, but play a role in monetary policy. Commercial retail banks are the ones we are familiar with in the High Street (and, now, via the telephone and internet). Cooperative banks are established by a small group of individuals with common concerns, and are important in the United States of America

and Europe. Merchant banks are concerned with mediating between those with large sums of money to invest and those who wish to borrow large sums, as well as with offering financial services to large organizations. Other types of financial body are important in particular countries—for example, discount houses and building societies in the United Kingdom, and savings and loan associations in the United States of America.

Given the broad range of activities carried out by these different types of banking organization, it is not surprising that they make use of different kinds of statistical tools and models. All, at some level, are concerned with modeling and minimizing risk. At a high level, this might be corporate default risk, and at a low level, predicting the riskiness and creditworthiness of individual customers.

Before discussing some of the tools employed, it is appropriate to say something about mathematical finance. Although its genesis can be traced back to the doctoral dissertation of Louis Bachelier in 1900, entitled *Théorie de la Speculation*, it is only in the last thirty years or so of the twentieth century that the discipline has really come into its own. This has been possible because of the developments in computer power over that period. The discipline is built on a variety of mathematical developments, including Kolmogorov's axiomatization of probability theory*, the modeling of Brownian motion* by Wiener, the introduction of martingales*, and the development of Ito's stochastic calculus*. At base, the discipline is concerned with putting a value on risk: how valuable is the right, but not the obligation, to buy a given number of shares at a time T for a specified price P? There are innumerable variants on this basic "option," and mathematical finance provides a formalism for valuing them. The ideas are based on the principle of constructing a portfolio, for example, of a mixture of the underlying asset and cash, and being able to trade these so that the portfolio's value is equal to the value that the option will have should it be exercised at time T. The earliest and simplest such model is the Black-Scholes model, described in 1973, but a huge amount of research into such models has been carried out and some highly sophisticated tools have resulted. Mathematical finance has been described as almost unique in that it involves advanced mathematics that is immediately applicable and that front rank mathematicians and physicists are employed by investment banks to develop new models.

However, as is implicit throughout the preceding paragraph, mathematical finance is mathematics. Despite being constructed upon a foundation of probability, it is not statistics. In particular, this is evident in the lack of a formalism for handling uncertainty in the values of the parameters used in the models. However, statistical ideas are beginning to be integrated into such models [4].

Turning to default risk, various kinds of statistical approaches have been used, and some of these are described below. At one level, phenomenological models of customer behavior have been built, while at another level, purely descriptive models have been used.

There are two main phenomenological default risk models used in the industry, yielding distributions for the loss a lender will experience, given that they have made loans to a set of customers. The basic model consists of a set of loans having been made to n borrowers, each of whom will either repay the loan or default (i.e. not repay their loan). If all borrowers had the same probability q of defaulting, the distribution of the number r defaulting would be binomial:

$$P(r) = \binom{n}{r} q^r (1-q)^{n-r} \quad r = 0, 1, \ldots, n.$$

$$(1)$$

In one approach [16], borrowers will default if a measure of their "riskiness" drops below a threshold. So, define the riskiness of the ith borrower as

$$R_i = \sqrt{\rho}c + (\sqrt{1-\rho})a_i, \quad (2)$$

where c is a common factor such as a general economic variable, thought to influence all borrowers equally, and a_i is a random effect specific to the ith borrower. Assuming that c and a_i are independent standard normal variables, the R_i will also be standard normal variables, with correlation $corr(R_i, R_j) = \rho$ when $i \neq j$. Then, the ith borrower will

default if R_i drops to a threshold v. That is, given c, the probability of the ith borrower defaulting is

$$q_c = P(R_i \leqslant v|c) = \Phi\left(\frac{v - c\sqrt{\rho}}{\sqrt{1-\rho}}\right), \quad (3)$$

where $\Phi(.)$ is the standard normal distribution function. From this, the limiting conditional probability that the proportion defaulting is no greater than x is $I(q_c \leqslant x)$, $I(.)$ being the indicator function. Integrating over c yields

$$P(r/n \leqslant x) \to \Phi\left(\frac{\sqrt{1-\rho}\Phi^{-1}(x) - v}{\sqrt{\rho}}\right) \quad (4)$$

as $n \to \infty$. The second approach [5] assumes a gamma mixture of Poisson distributions. Take the Poisson limit of Equation 1 as $n \to \infty$ and $q \to 0$, with $nq \to \lambda$, yielding

$$P(r) = e^{-\lambda}\lambda^r/r!, \quad r = 0, 1, 2, \ldots,. \quad (5)$$

Assuming λ to follow a gamma distribution with density

$$f(\lambda) = \frac{\lambda^{v-1}e^{-\lambda/\xi}}{\xi^v \Gamma(v)}, \lambda > 0; \xi > 0, v > 0, \quad (6)$$

it follows that

$$P(r) = \frac{\Gamma(v+r)\xi^r}{r!\Gamma(v)(1+\xi)^{v+r}}. \quad (7)$$

Once again, various extensions and sophistications of these basic models have been developed.

Naturally, since one hopes that major financial crises or default conditions will be rare events, the statistics of extreme values* also plays a part in many financial problems.

Other models are based on Markov chain approaches, in which the borrower is regarded as moving between states such as up-to-date with repayments, one month in arrears, two months in arrears, in default, and so on. An extension of the basic Markov chain model, the so-called *mover-stayer model**, treats the borrowers as heterogeneous, falling into two classes: those who tend to stay in a good state and those who tend to move between states.

While the application of statistical tools in retail banking are also generally concerned with risk assessment, very different classes of models have been explored. In particular, while those in investment banking have tended to be more phenomenological, those in retail banking have been more descriptive. They have some similarities to those used in actuarial work (*see* ACTUARIAL SCIENCE). However, whereas the distributions arising in actuarial work (for example, mortality statistics) might justifiably be assumed usually to change only slowly over the course of time, those arising in retail banking are subject to the vagaries of marketing departments and changing economic conditions. These can lead to very sudden and dramatic changes in the risk characteristics of customer populations. There are also other differences between the data used in investment and retail banking operations. For example, the corporate sector generally deals with a small volume of large loans, with a one-by-one approval process, based on extensive information such as from audited financial statements but without payment history, and manages these on a loan-by-loan basis. In contrast, the retail sector generally has a large volume of small loans, with a volume approval process, based on limited information such as application characteristics but also including repayment history, and the retail sector manages entire portfolios of customers. It will be obvious from this description that retail banking, with large numbers of relatively homogeneous customers, is a natural application domain for statistical methods. The surprising thing is that awareness of this has been so slow to develop. As Thomas et al. [14] say: "The tools [of credit scoring] are based on statistical and operational research techniques and are some of the most successful and profitable applications of statistical theory in the last 20 years. Yet the area is almost completely neglected in modern statistical textbooks and academic courses."

Retail banking data sets are characterized by being large (with billions of credit card transactions per year, for example), typically using categorical data* (recoded if necessary), and often, as with any large multivariate data set relating to human beings, problems of missing and distorted data. Sometimes the problems are deeper than merely measurement, recording, or data collection* errors. For example, the available data will typically only refer to those customers who were offered and accepted a financial product in the past, but one might want to use these data to build a model that is predictive for the entire population of potential customers. Thus, in deciding to whom to give a loan, one will have available outcome measures only for those previously granted a loan, but one will want to apply any model to all applicants. This problem of selectivity bias is a common one in the retail banking context, with the term reject inference often being used to describe attempts to infer the unknown good/bad risk class of customers previously rejected.

Loan decisions are just one type of decision for which a statistical model may be built. There are many others. For example, decisions are also needed about churn (who is likely to switch to another supplier—especially relevant in today's mortgage and credit card markets), who should be offered a top-up loan, who will turn down a loan offer, whether fraud is being committed, whether a good risk customer is likely to be profitable, how to segment the population of customers for marketing strategies, and so on. The range of questions requiring statistical models is unlimited but we can notice several distinct classes of question.

Many of the models are predictive; the aim is to predict some likely future characteristic of a customer, on the basis of information currently available. Such models are often called *scorecards* in the industry. The predictor variables in such models might be static, including application form data and additional data from a credit bureau describing repayment behavior in the past, the aim being to make a decision about whether or not to offer a product to a customer. These are termed *application scorecards*. In contrast, *behavioral scorecards* are used for predicting the behavior of existing customers on the basis of their continuing record. This would be appropriate, for example, in monitoring payments on credit cards or loans. Clearly, tools such as dynamic linear models have a role here.

Another distinction can be made between *front end* and *back end* statistical models. The term "front end" refers to models in which the result is communicated to the customer. This would be the case, for example, in an application scorecard, where the applicant will be told the outcome of the application. In such situations, there are often legal obligations to tell the customer on what basis a rejection has been made. This means that such models must be interpretable, so that sophisticated neural network* or support vector machine models will not be appropriate. In contrast, back end models are used for making internal decisions. An example would be fraud detection, where an unusual pattern of credit card purchases might trigger a warning, which can then be investigated by a phone call to the customer. In such situations, there are no constraints on the form of the model that might be used, and highly sophisticated tools can be, and are, applied.

It will be obvious from the above that rapid processing and decision making is often required, sometimes in real time. Customer segmentation, using cluster analysis*, may be carried out over a period of weeks, but the decision on whether to give a loan must be made in minutes and the detection of possible fraud must be made in real time. This again contrasts with investment banking, where fraud detection may occur years after the event. Statistical methods of fraud detection in general are reviewed in Reference 3.

There are often stringent legal restrictions on the variables that may be used as predictors; for example, many countries do not allow gender to be used. The enforced exclusion of variables as a way of combating discrimination is not uncontroversial. In insurance, a very similar application, there are no such restrictions. Moreover, by preventing scorecards using the objective predictive information that is available in such variables, one is necessarily biasing the result against some subgroups of the population.

The most popular tool for scorecard construction is logistic regression*, having taken over from the earlier discriminant analysis*. In many applications, however, linear regression is effectively just as accurate as logistic regression. Classification trees (*see* CLASSIFICATION—I) are also fairly widely used, especially in the form of *derogatory trees*—small trees that serve as a way to introduce interactions between predictor variables into larger linear models. For back end problems, the type of models that have been used is unlimited and includes projection pursuit* regression, nearest neighbor methods*, linear programming*, artificial neural networks, support vector machines, and rule-based methods derived from machine learning rather than statistics.

It is clear that the role of statistical tools in banking applications, both investment and retail, will continue to grow dramatically. Apart from drivers such as the accumulation of large data sets in banking databases, increased competition in tougher economic conditions, and the simple possibility of building better and more predictive models, increased legislation [2] is also beginning to have an effect. It may be that the last of these will have the same effect on statistics as legislation requiring statistical input into new drug development had via the pharmaceutical industry.

FURTHER READING

A large number of books on mathematical finance have appeared of late, some more abstract than others. Good introductions are given by references 1, 8, and 17. Several review papers on retail banking and credit scoring have appeared, including reference 6, 7, 11, 12, and 13, but there are far fewer books on statistics in retail banking and credit scoring than on mathematical finance. Three that can be recommended are reference 9, a collection of articles on different aspects of credit scoring, reference 10, which describes the problems from the practitioner's perspective, and reference 15, which describes the technical statistical aspects.

REFERENCES

1. Avellaneda, M. and Laurence, P. (2000). *Quantitative Modeling of Derivative Securities: from Theory to Practice*. Chapman & Hall/CRC Press, Boca Raton, Fla.

2. Basel (2001). *Overview of the New Basel Capital Accord*. http://www.bis.org.

3. Bolton, R. J. and Hand, D. J. (2002). Statistical fraud detection. *Stat. Sci.*

4. Bunnin, F. O., Guo, Y., and Ren, Y. (2002). Option pricing under model and parameter uncertainty using predictive densities. *Stat. Comput.*, **12**, 37–44.

5. Credit Suisse Financial Products (1997). *CreditRisk +—a Credit Risk Management Framework.* http://www.csfb.com/creditrisk.

6. Hand, D. J. (2001). Modelling consumer credit risk. *IMA J. Manage. Math.*, **12**, 139–155.

7. Hand, D. J. and Henley, W. E. (1997). Statistical classification methods in consumer credit scoring: a review. *J. R. Stat. Soc. A*, **160**, 523–541.

8. Hunt, P. J. and Kennedy, J. E. (2000). *Financial Derivatives in Theory and Practice.* Wiley, Chichester, UK.

9. Mays, E., ed. (2001). *Handbook of Credit Scoring.* Glenlake, Chicago, Ill.

10. McNab, H. and Wynn, A. (2000). *Principles and Practice of Consumer Credit Risk Management.* CIB, Kent, Ohio.

11. Rosenberg, E. and Gleit, A. (1994). Quantitative methods in credit management: a survey. *Operations Res.*, **42**, 589–613.

12. Thomas, L. C. (1998). "Methodologies for Classifying Applicants for Credit". In *Statistics in Finance*, D. J. Hand and S. D. Jacka, eds. Arnold, London, pp. 83–103.

13. Thomas, L. C. (2000). A survey of credit and behavioral scoring: forecasting financial risk of lending to consumers. *Int. J. Forecasting*, **16**, 149–172.

14. Thomas, L. C., Crook, J. N., and Eedelman, D. B. (1992). *Credit Scoring and Credit Control.* Clarendon Press, Oxford.

15. Thomas, L. C., Edelman, D. B., and Crook, J. N. (2002). *Credit Scoring and its Applications.* SIAM, Pa.

16. Vasicek, O. A. (1987). *Probability of Loss on Loan Portfolio.* KMV Corporation, San Francisco, Calif.

17. Wilmott, P., Howison, S., and Dewynne, J. (1995). *The Mathematics of Financial Derivatives.* Cambridge University Press, Cambridge, UK.

See also FINANCE, STATISTICS IN; RISK MANAGEMENT, STATISTICS OF; RISK MEASUREMENT, FOUNDATIONS OF; and RISK THEORY.

DAVID J. HAND

BAR CHART

A form of graphical representation applicable to data classified in a number of (usually nonordered) categories. Equal-width rectangular bars are constructed over each category with *height* equal to the frequency or other measurement associated with the category. (This is a widespread representation method in newspapers, magazines, and general publications.)

Example. Energy consumption per capita in millions BTUs in the United States in 1977 was:

Gas	Coal	Petroleum	Other	Total
120	60	170	25	375

The corresponding bar chart is shown in Fig. 1.

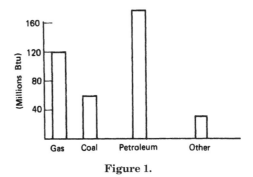

Figure 1.

See also GRAPHICAL REPRESENTATION OF DATA; HISTOGRAMS; and PIE CHART.

BARLOW–SCHEUER RELIABILITY GROWTH MODEL. See RELIABILITY GROWTH MODEL, BARLOW–SCHEUER

BARTLETT ADJUSTMENT—I

One of the most widely used methods for deriving tests in parametric problems is the (maximized) likelihood-ratio test* introduced by Neyman and Pearson [16]. Suppose that we observe a vector random variable Y assumed to have a density $f(y; \theta)$ depending on an unknown parameter θ which can be partitioned as (ψ, χ). To test the hypothesis $\psi = \psi_0$ with χ a nuisance parameter* we introduce the log-likelihood function $l(\theta) = \log f(y; \theta)$ and consider the test statistic

$$w(\psi_0) = 2[l(\hat{\psi}, \hat{\chi}) - l(\psi_0, \hat{\chi}_{\psi_0})],$$

where $(\hat{\psi}, \hat{\chi})$ is the overall maximum-likelihood estimate and $\hat{\chi}_{\psi_0}$ is the maximum-likelihood estimate* of χ for given $\psi = \psi_0$. Under the null hypothesis the test statistic has approximately a chi-squared distribution* with d_ψ degrees of freedom, where d_ψ denotes the dimensionality of ψ. The proof hinges on the usual asymptotic arguments of maximum-likelihood theory, suggesting that the distributional result will be a useful approximation when errors of estimation are in some sense relatively small and when sufficient independence is present for a central-limit effect to operate.

The adequacy of the distributional approximation is an obvious source of concern, and this, among other aspects of the procedure, was considered by Bartlett [4,5]. He noted in two specific contexts that for large numbers n of observations an expansion of the form

$$E\{w(\psi_0)\} = d_\psi(1 + b/n + \cdots)$$

is possible. Now if the chi-squared distribution applied exactly, the expectation would be exactly d_ψ, suggesting the modification of the test statistic to

$$w_B(\psi_0) = \frac{w(\psi_0)}{1 + b/n}.$$

If b depends on unknown parameters, it is replaced by a good estimate.

If confidence regions* for ψ are of interest rather than significance tests*, then such regions are formed from values of ψ such that $w_B(\psi)$ is not significant at the appropriate level.

Correcting the mean to be closer to that for chi-squared does not necessarily improve the approximation in the upper tail, which is what is typically required in applications. Bartlett showed, however, by the use

of characteristic functions*, that in his two examples, remarkably, the second and third cumulants were also corrected to agree with those of chi-squared to the same order as the mean. This was confirmed in a heroic calculation by Lawley [14], who showed for a quite general class of problems that cumulants of all orders are modified in the correct way, suggesting that the distribution function of w_B agrees with that of chi-squared with error $0(n^{-1})$. Hayakawa [12] obtained a similar result by use of characteristic functions, although agreement of distribution functions required the vanishing of a certain constant. Some intricate direct calculations by Cordeiro [8,9] showed that indeed this extra condition was satisfied. Barndorff-Nielsen and Cox [1,2], first in a very special case and then in some generality, gave a fairly simple derivation stemming from Laplace expansions associated with the exponential family of distributions. Bickel and Ghosh [7] gave a different derivation emphasizing the connection with Bayesian theory.

Bartlett [6] gave the factors $1 + b/n$ for a number of standard problems in multivariate analysis (*see* HOMOGENEITY OF VARIANCES, BARTLETT'S TEST FOR; WILKS'S LAMBDA CRITERION); numerical comparisons with exact and simulation results have shown that the adjustment usually produces a very substantial improvement in the adequacy of the chi-squared approximation. Usually, although not always, $b > 0$. McCullagh and Cox [15] gave a differential geometric interpretation of b relating its magnitude to a combination of the departure of the model from simple exponential family structure and the departure of the distribution of the score statistic from Gaussian or inverse Gaussian form. Jensen and Frydenberg [13] showed that the asymptotic distribution theory does not apply to discrete problems, although in their numerical examples there was typically an improvement in using the correction.

If ψ is one-dimensional and it is required to obtain one-sided tests or separate upper and lower confidence limits, the likelihood-ratio statistic is replaced by its signed square root, i.e.

$$r(\psi_0) = \text{sgn}(\hat{\psi} - \psi_0)\sqrt{w(\psi_0)},$$

with approximately a standard normal distribution. Adjustment to improve the approximation is now more complicated, requiring a correction for skewness as well as a scale change. For an account of asymptotic statistical theory putting some emphasis on r, see Barndorff-Nielsen and Cox [1].

Analogous distributional adjustments to other parametric test statistics, such as the score test, are typically more complicated than simple scaling; see Cordeiro et al. [10]. When there are many nuisance parameters, modification of the likelihood otherwise than by simple maximization may be desirable, an idea also stemming from Bartlett [4], and adjustment of the resulting statistic for distributional accuracy is considered by DiCiccio and Stern [11].

Study of the Bartlett adjustment is part of the currently active field of higher-order asymptotics. Two open issues are the clarification of the problems to which the adjustment in its simple form is theoretically appropriate and (not quite the same thing) practically helpful, and the setting out of a simple explanation of why an adjustment merely by rescaling works so powerfully.

REFERENCES

1. Barndorff-Nielsen, O. E. and Cox, D. R. (1979). Edgeworth and saddle-point approximations with statistical applications (with discussion). *J. R. Statist. Soc. B*, **41**, 279–312.

2. Barndorff-Nielsen, O. E. and Cox, D. R. (1984). Bartlett adjustments to the likelihood ratio statistic and the distribution of the maximum likelihood estimator. *J. R. Statist. Soc. B*, **46**, 483–495.

3. Barndorff-Nielsen, O. E. and Cox, D. R. (1994). *Inference and Asymptotics*. Chapman & Hall, London.

4. Bartlett, M. S. (1937). Properties of sufficiency and statistical tests. *Proc. R. Soc. (London) A*, **160**, 268–282.

5. Barlett, M. S. (1938). Further aspects of the theory of multiple regression. *Proc. Cambridge Phil. Soc.*, **34**, 33–40.

6. Bartlett, M. S. (1954). A note on multiplying factors for various χ^2 approximations. *J. R. Statist. Soc. B*, **16**, 296–298.

7. Bickel, P. J. and Ghosh, J. K. (1990). A decomposition for the likelihood ratio statistic and the Bartlett correction—a Bayesian argument. *Ann. Statist.*, **18**, 1070–1090.

8. Cordeiro, G. M. (1983). Improved likelihood-ratio tests for generalized linear models. *J. R. Statist. Soc. B*, **45**, 404–413.

9. Cordeiro, G. M. (1987). On the corrections to the likelihood ratio statistics. *Biometrika*, **74**, 265–274.

10. Cordeiro, G. M., Ferrari, S. L. P., and Paula, G. A. (1993). Improved score tests for generalized linear models. *J. R. Statist. Soc. B*, **55**, 661–674.

11. DiCiccio, T. J. and Stern, S. E. (1994). Frequentist and Bayesian Bartlett correction of test statistics based on adjusted profile likelihoods. *J. R. Statist. Soc. B*, **56**, 397–408.

12. Hayakawa, T. (1977). The likelihood ratio criterion and the asymptotic expansion of its distribution. *Ann. Inst. Statist. Math.*, **29**, 359–378.

13. Jensen, J. L. and Frydenberg, M. (1989). Is the 'improved likelihood ratio statistic' really improved in the discrete case? *Biometrika*, **76**, 655–661.

14. Lawley, D. N. (1956). A general method for approximating to the distribution of likelihood ratio criteria. *Biometrika*, **43**, 295–303.

15. McCullagh, P. and Cox, D. R. (1986). Invariants and likelihood ratio statistics. *Ann. Statist.*, **14**, 1419–1430.

16. Neyman, J. and Pearson, E. S. (1928). On the use and interpretation of certain test criteria for purposes of statistical inference. *Biometrika*, **20A**, 175–240, 263–294.

See also ASYMPTOTICS, HIGHER ORDER; HOMOGENEITY OF VARIANCES, BARTLETT'S TEST FOR; LIKELIHOOD RATIO TESTS; and WILKS'S LAMBDA CRITERION.

D. R. COX

BARTLETT ADJUSTMENT—II

Bartlett adjustments for 10 multivariate normal testing problems related to structured covariance matrices were obtained by Møller [15]. His results apply to real, complex, and quaternion Wishart distributions, and he examines a number of tests.

Papers dealing with closed-form Bartlett corrections include Reference 16 for logistic regressions and Reference 7 for a class of Jørgensen's dispersion* models [12]. Corrections for likelihood ratio tests for heteroscedastic regression models are discussed in References 2 and 4. For Bartlett corrections in testing slopes in regression models when the independent variables are subject to error and for testing the equality of normal variances against an increasing alternative, see References 19 and 20. Corrections to the log-likelihood ratio statistic in a regression model with Student t-errors [10], for heteroscedastic linear models [8] and for multivariate regression [9], have been proposed. For Bartlett adjustments in log-likelihood tests of the unit root* hypothesis, see References 13 and 17. A valuable historical sketch and new algorithms are due to Jensen [11]. Algorithms for computing Bartlett corrections using symbolic computations are provided in References 1 and 18. General matrix formulas are provided by Cordeiro [5]. For exponential families, in addition to McCullagh and Cox [14], Bartlett corrections are discussed in Reference 6.

A general set of results due to Aubin and Cordeiro [3] derive from heteroscedastic regression models of the form

$$E(\boldsymbol{Y}) = \boldsymbol{\mu} = \boldsymbol{X\beta},$$

$$\boldsymbol{Y'} = (y_1, \ldots, y_n)',$$

$$\sigma_\ell^2 = \mathrm{Var}(Y_\ell) = H(\boldsymbol{\omega}_\ell' \boldsymbol{\gamma}),$$

for $\ell = 1, \ldots, n$. Here $\boldsymbol{X}$ is a known $n \times p$ design matrix of full rank $p(p < n), \boldsymbol{x}_\ell' = (x_{\ell 1}, \ldots, x_{\ell p})$ is the ℓth row of $\boldsymbol{X}$, $\boldsymbol{\beta}$ is an unknown p-dimensional vector of regression parameters, $\boldsymbol{w}_\ell'(1 \times q)$ is a vector of known constants, the variance function $H(\cdot)$ is a known smooth differentiable function, and $\gamma(q \times 1)$ is an unknown vector. The components of $\boldsymbol{Y}$ are uncorrelated with mean μ and diagonal covariance matrix $\Gamma = \mathrm{diag}\{\sigma_\ell^2\}$.

REFERENCES

1. Andrews, D. F. and Stafford, J. E. (1993). Tools for the symbolic computation of asymptotic expansions. *J. R. Stat. Soc. B*, **55**, 613–627.

2. Attfield, C. L. F. (1991). A Bartlett adjustment to the likelihood ratio test for homoskedasticity in the linear model. *Econ. Lett.*, **37**, 119–123.

3. Aubin, E. C. Q. and Cordeiro, G. M. (1998). Some adjusted likelihood ratio tests for heteroscedastic regression models. *Braz. J. Probab. Stat.*, **2**, 131–148.

4. Cordeiro, G. M. (1993a). Bartlett corrections and bias correction for two heteroscedastic regression models. *Commun. Stat.—Theor. Methods*, **22**, 169–188.

5. Cordeiro, G. M. (1993b). General matrix formulae for computing Bartlett corrections. *Stat. Probab. Lett.*, **16**, 11–18.

6. Cordeiro, G. M., Cribari-Neto, F., Augin, E. C. Q., and Ferrari, S. L. P. (1995). Bartlett corrections for one-parameter exponential family models. *J. Stat. Comp. Simul.*, **53**, 211–231.

7. Cordiero, G. M., Paula, G. A., and Botter, D. A. (1994). Improved likelihood ratio tests for dispersion models. *Int. Stat. Rev.*, **62**, 257–276.

8. Cribari-Neto, F. and Ferrari, S. L. P. (1995). Bartlett-corrected tests for heteroskedastic linear models. *Econ. Lett.*, **48**, 113–118.

9. Cribari-Neto, F. and Zarkos, S. (1995). Improved test statistics for multivariate regression. *Econ. Lett.*, **49**, 113–120.

10. Ferrari, S. L. P. and Arellano-Valle, R. B. (1996). Modified likelihood ratio and score tests in linear regression models using the t distribution. *Braz. J. Probab. Statist.*, **10**, 15–33.

11. Jensen, J. L. (1993). A historical sketch and some new results on the improved likelihood ratio statistic. *Scand. J. Stat.*, **20**, 1–15.

12. Jørgensen, B. (1987). Exponential dispersion models. *J. R. Stat. Soc. B*, **49**, 127–162.

13. Larsson, R. (1994). *Bartlett Corrections for Unit Root Test Statistics*. Unpublished paper, Institute of Mathematical Statistics, University of Copenhagen.

14. McCullagh, P. and Cox, D. R. (1986). Invariants and the likelihood ratio statistic. *Ann. Stat.*, **14**, 1419–1430.

15. Møller, J. (1986). Bartlett adjustments for structured covariances. *Scand. J. Stat.*, **13**, 1–15.

16. Moulton, L. H., Weissfeld, L. A., and St. Laurent, R. T. (1993). Bartlett corrections factors in logistic regression models. *Comput. Stat. Data Anal.*, **15**, 1–11.

17. Nielsen, B. (1995). *Bartlett Correction for the Unit Root Test in Autoregressive Models*. Unpublished paper, Nuffield College, University of Oxford.

18. Stafford, J. E. and Andrews, D. F. (1993). A symbolic algorithm for studying adjustments to the profile likelihood. *Biometrika*, **80**, 715–730.

19. Wang, Y. (1994). A Bartlett-type adjustment for the likelihood ratio statistic with an ordered alternative. *Stat. Probab. Lett.*, **20**, 347–352.

20. Wong, M. Y. (1991). Bartlett adjustment to the likelihood ratio statistic for testing several slopes. *Biometrika*, **78**, 221–224.

BARTLETT, MAURICE STEVENSON

Born: March 18, 1910 in Chiswick, London, England

Died: January 8, 2002 in Exmouth, Devon, England

Contributed to: statistical theory and methods, time series analysis, stochastic processes, population and epidemic models, spatial statistics

Maurice Bartlett, one of the outstanding statisticians of the twentieth century, was the third child of a family in modest circumstances. After elementary school, he won a scholarship to the famous Latymer Upper School in London; he later wrote that his lifelong interest in probability began at this school with the chapter on that topic in Hall and Knight's *Algebra*.

In 1929, he took up a scholarship at Queens' College, Cambridge, to study mathematics. Three years later, he published his first paper, jointly with John Wishart [1], on second order moment statistics in a normal system. He graduated with the rank of Wrangler, obtaining a Distinction in his Tripos Examination (Schedule B). After graduation, he continued at Cambridge for a fourth year as Wishart's first mathematical postgraduate student, published a second paper [2] with him, and was awarded the Rayleigh Prize for 1933. His scientific interests were very broad; he attended several courses "for fun," among them Eddington's lectures on relativity, and Dirac's on quantum theory. He retained his fascination with statistical physics throughout his life. 'While at Cambridge, he also found time for extracurricular activities such as rowing for his College, and designing the front cover of the College journal.

In 1933, Egon Pearson* offered Bartlett his first job as Assistant Lecturer in the new Statistics Department at University College London. R.A. Fisher* was the newly appointed Galton Professor and J.B.S. Haldane was a frequent visitor to the Galton Laboratory. Jerzy Neyman* had also joined Pearson, and Bartlett was able to hear their lectures on the newly developed Neyman–Pearson theory of statistical hypothesis testing*. This exciting environment was a great stimulus to Bartlett's research in statistical theory, but he felt the need to get to closer grips with statistical methods.

In 1934, he accepted the post of statistician with the Imperial Chemical Industries (ICI) Agricultural Research Station at Jealott's Hill. He spent four years at ICI, and later wrote of them [21] as "not only the happiest period (professionally) of my life, but also the most creative." During his time at ICI, he wrote some two dozen papers on a wide range of topics: on inbreeding (with J.B.S. Haldane) [3], the estimation of general intelligence [4], the effect of nonnormality on the *t*-distribution [5], contingency tables [6], field and laboratory sampling errors (with A.W. Greenhill) [7], homogeneity of variances in the paper in which he developed his famous test [8], mental factors [9], correlation among genetic components of ability [10], cotton experiments in Egypt [11], and the nutritive value of "summer" milk (with J.C. Drummond et al.) [12].

In October 1938, he took up a lectureship at the University of Cambridge, and soon after the beginning of World War II was allocated to a Ministry of Supply establishment, initially in Kent, where research on rockets was being conducted. He met Frank Anscombe*, David Kendall, and Pat Moran at one time or another at this establishment. Shortly after the fall of France in 1940, he began to correspond with Jo Moyal, who had escaped to Britain through Bordeaux. Moyal's knowledge of continental and Russian research on stochastic processes proved to be of great help to Bartlett.

On returning to Cambridge in 1946, Bartlett concentrated on the study of time series and diffusion processes. Harold Hotelling* invited him to spend four months at the University of North Carolina at

Chapel Hill, where Bartlett lectured on stochastic processes. He produced a set of mimeographed notes that were to form the nucleus of his book [14] on the subject 9 years later. During his time in North Carolina, and on his later visits to Harvard in 1958, Chicago in 1964 and Santa Barbara in 1987, Bartlett met many of the important statisticians in the United States of America, among them Anderson, Doob, Karlin, Kruskal, Meier, Mosteller, Robbins, Savage, Stein, and Wilks.

Meanwhile, Bartlett pursued his research on stochastic processes, and in 1947 accepted the Chair in Mathematical Statistics at the University of Manchester, where Jo Moyal soon joined him. He collaborated with David Kendall and Jo Moyal in the Royal Statistical Society's ground-breaking 1949 *Symposium on Stochastic Processes*. His paper on evolutionary stochastic processes [13], together with Kendall's and Moyal's contributions, opened new vistas of research to students of the field.

During his 13 years in the Statistical Laboratory of the Department of Mathematics at the University of Manchester, Bartlett developed modern courses in mathematical statistics, inaugurated a Diploma in Statistics and started a postgraduate course in operational research. In 1955, his most important book *An Introduction to Stochastic Processes* [14], based on his earlier North Carolina notes, was published. He was also extending his study of epidemic models, and wrote his well-known Berkeley Symposium paper [15] on recurrent epidemics, and his study of the periodicity of measles and community size [16] during the following two years. These were followed in 1960 by his second book *Stochastic Population Models in Ecology and Epidemiology* [17].

In 1960, Bartlett succeeded Egon Pearson* in the Chair of Statistics at University College London, where he continued his research on epidemic processes, stochastic path integrals, statistical physics, the spectral analysis of point processes, spatial patterns and multivariate statistics. His *Essays in Probability and Statistics* [18] appeared in 1962. In 1967, he was offered the newly created Chair of Biomathematics at the University of Oxford, from which he eventually retired

in 1975. His fourth book *Probability, Statistics and Time* [19] was published in the same year.

Bartlett found his position at Oxford challenging: he wrote in [21] that his Department of Biomathematics was "viewed with suspicion by the biologists as being too mathematical." Nevertheless, he persevered in his efforts to "foster biological statistics and mathematics" and met with a considerable measure of success; he studied spatial systems, nearest neighbour and population models, and published several papers on these topics. His fifth book *The Statistical Analysis of Spatial Pattern* [20] appeared in 1976.

In 1973–1974, he spent a year in his old friend Pat Moran's Department of Statistics at the Australian National University, Canberra. After his retirement, he returned to Australia for shorter visits in 1977 and 1980; he had remained active, and had continued to publish steadily on Ising models, random walks, spatial patterns, population processes and birth–death and catastrophe models. A full list of his 167 publications between 1932 and 1989 as well as a reproduction of his autobiographical article [21] are available in Volume 1 of the three volume *Selected Papers of M.S. Bartlett* [23]. While his papers always reflected his breadth of interest and his originality of thought, they were not always easy reading, partly because of "the high density of ideas," in David Cox's words in the Foreword to [23].

Bartlett received wide recognition for his scientific achievements. He was elected a Member of the International Statistical Institute* in 1949, and an Honorary Member in 1980. He became a Fellow of the Royal Society of London (FRS) in 1961, and was President of the Royal Statistical Society* for 1966–1967. He was awarded an honorary DSc from the University of Chicago in 1966, and another from the University of Hull in 1976. The Royal Statistical Society awarded him its Guy Silver Medal in 1952 and its Guy Gold Medal in 1969, while the University of Oxford presented him with the Weldon Prize and Medal in 1971. He became a Foreign Associate of the US National Academy of Sciences in 1993.

Maurice Bartlett was a quiet and reserved man of great sensitivity. He was scrupulously fair to his colleagues and students, all of whom remember him with great affection. He was a gifted draughtsman with considerable artistic talent; he used to collect nineteenth-century water colors which he often tracked down in dusty sales rooms. He married Sheila Chapman in 1957, and they had a daughter Penny Robinson, and three granddaughters, all of them great favourites with Sheila and Maurice. Bartlett died at his home in Exmouth at the age of 91. The opinion of many statisticians is that the subject of statistics would not be quite the same today without Bartlett's major contributions to probability, stochastic processes, and statistical theory. Further details may be found in [22] and [24].

REFERENCES

1. Bartlett, M. S. (with Wishart, J.) (1932). Distribution of second-order moment statistics in a normal system. *Proc. Camb. Phil. Soc.*, **28**, 455–459.

2. Bartlett, M. S. (with Wishart, J.) (1933). Generalised product moment distribution in a normal system. *Proc. Camb. Phil. Soc.*, **29**, 260–270.

3. Bartlett, M. S. (with Haldane, J. B. S.) (1935). Theory of inbreeding with forced heterozygosis. *J. Genetics.*, **41**, 327–340.

4. Bartlett, M. S. (1935a). The statistical estimation of 'G'. *British. J. Psychol.*, **26**, 199–206.

5. Bartlett, M. S. (1935b). The effect of non-normality on the t-distribution. *Proc. Camb. Phil. Soc.*, **31**, 223–231.

6. Bartlett, M. S. (1935c). Contingency table interactions. *J. R. Statist. Soc. Suppl.*, **2**, 248–252.

7. Bartlett, M. S. (with Greenhill, A. W.) (1936). Relative importance of field and laboratory sampling errors in small plot pasture productivity experiments. *J. Agric. Sci.*, **26**, 258–262.

8. Bartlett, M. S. (1937). Properties of sufficiency and statistical tests. *Proc. R. Soc. Lond. A*, **160**, 268–282.

9. Bartlett, M. S. (1937a). The statistical conception of mental factors. *British J. Psychol.*, **28**, 97–104.

10. Bartlett, M. S. (1937b). Note on the development of correlation among genetic components of ability. *Ann. Eug.*, **7**, 299–302.

11. Bartlett, M. S. (with Crowther, F.) (1938). Experimental and statistical testing of some complex cotton experiments in Egypt. *Emp. J. Exp. Agric.*, **6**, 53–68.

12. Bartlett, M. S. (with Drummond, J. C., Singer, E., Watson, S.J., and Ferguson, W.S.) (1938). An experiment on the nutritive value of winter-produced 'summer' milk. *J. Hygiene*, **38**, 25–39.

13. Bartlett, M. S. (1949). Some evolutionary stochastic processes. *J. R. Statist. Soc. B*, **11**, 211–229.

14. Bartlett M. S. (1955). *An Introduction to Stochastic Processes*, (2nd ed. 1966). Cambridge University Press, Cambridge.

15. Bartlett, M. S. (1956) Deterministic and stochastic models for recurrent epidemics. *Proc. 3rd Berkeley Symp.Statist.Prob.*, **4**, 81–109.

16. Bartlett, M. S. (1957). Measles periodicity and community size. *J. R. Stat. Soc. A*, **120**, 48–70.

17. Bartlett, M. S. (1960). *Statistical Population Models in Ecology and Epidemiology*, Methuen, London.

18. Bartlett, M. S. (1962). *Essays in Probability and Statistics*. Methuen, London.

19. Bartlett, M. S. (1975). *Probability, Statistics and Time*. Chapman and Hall, London.

20. Bartlett, M. S. (1976). *The Statistical Analysis of Spatial Pattern*. Chapman and Hall, London.

21. Bartlett, M. S. (1982). "Chance and Change". Contribution to *The Making of Statisticians*, Edited by J. Gani, ed. Springer-Verlag, New York, 41–60.

22. Gani, J. (2002). Obituary: Maurice Stevenson Bartlett. *J. Appl. Prob.*, **39**, 664–670.

23. Stanton, R.G., Johnson, E. D., and Meek, D. S. eds. (1989). *Selected Papers of M.S. Bartlett*. in 3 Volumes. Charles Babbage Research Centre, Winnipeg.

24. Whittle, P. (2004). Maurice Stevenson Bartlett. *Biogr. Mems Fell. R. Soc. Lond.*, **50**, 15–33.

J. GANI

BARTLETT'S TEST OF HOMOGENEITY OF VARIANCES. See HOMOGENEITY OF VARIANCES, BARTLETT'S TEST FOR

BARTON–DAVID TEST. See SCALE TESTS, BARTON–DAVID

BARYCENTRIC COORDINATES

Introduced by A. F. Möbius in 1827 [2], they are a special case of homogeneous coordinates.

PHYSICAL MOTIVATION IN TWO DIMENSIONS

Let masses m_1, m_2, and m_3 be concentrated on three noncollinear points A_1, A_2, and A_3. Then in the plane generated by these points a unique point A exists (the centroid or center of gravity of these points) containing mass m such that $m = m_1 + m_2 + m_3$, and the weighted sums of projections of AA_j ($j = 1$, 2, 3; weights m_j/m). Conversely, for any point A—located in the plane of three fixed points A_1, A_2, and A_3—possessing mass $m(m \neq 0)$, there correspond three specific masses m_1, m_2, $m_3(m_1 + m_2 + m_3 = m)$ which, being located at points A_1, A_2, and A_3, respectively, have point A as their center of gravity. The "masses" m_1, m_2, and m_3 may take on positive or negative values. These masses m_1, m_2, m_3 can be viewed as coordinates of the point A possessing mass m and are called barycentric coordinates of point A with respect to A_1, A_2, and A_3.

If the mass m increases k-fold, the coordinates m_1, m_2, and m_3 also increase k-fold. The same point A may have different barycentric coordinates. The point A with mass m possessing barycentric coordinates m_1, m_2, and m_3 may be located inside, outside, or on the boundary of the triangle A_1, A_2, A_3. To normalize, we usually take $m_1 + m_2 + m_3 = 1$.

GENERAL DEFINITION

Let $\mathbf{x}_1, \mathbf{x}_2, \ldots, \mathbf{x}_n$ be linearly independent points (or vectors) in n-dimensional Euclidean space. For any vector (point) $\mathbf{x}$ belonging to E_n, there is only one set of real numbers $b_1, b_2, \ldots, b_n$ which satisfy $\sum_{i=1}^{n} b_i = 1$ and are such that $\mathbf{x} = b_1\mathbf{x}_1 + \cdots + b_n\mathbf{x}_n$. These numbers $b_1, \ldots, b_n$ are called the barycentric coordinates of the vector $\mathbf{x}$ with respect to simplex determined by $\mathbf{x}_1, \ldots, \mathbf{x}_n$. If $\mathbf{x} = b_1\mathbf{x}_1 + b_2\mathbf{x}_2 + b_3\mathbf{x}_3 + \cdots + b_n\mathbf{x}_n$ with the additional requirements $b_1 \geqslant 0, b_2 \geqslant 0, \ldots, b_n \geqslant 0$, it is called a *convex linear combination* of $\mathbf{x}_1, \mathbf{x}_2, \ldots, \mathbf{x}_n$. It includes all points belonging to the boundary and the interior of the simplex determined by $\mathbf{x}_1, \mathbf{x}_2, \ldots, \mathbf{x}_n$. Barycentric coordinates are utilized in noncooperative game theory* (see Vorob'yev [4]) and in problems relating to definitions of measures of distances between categories: e.g., in multinomial distributions* (see Gibbons et al. [1]).

REFERENCES

1. Gibbons, J. D., Olkin, I. and Sobel, M. (1977). *Selecting and Ordering Populations*. Wiley, New York.
2. Möbius, A. F. (1827). *Der baryzentrische Kalkül*. Gesammelte Werke, Vol. 1. Leipzig.
3. Skibinsky, M. (1976). *Ann. Statist.*, **4**, 187–213.
4. Vorob'yev, N. N. (1977). *Game Theory for System Scientists*, S. Kotz, trans./ed. Springer-Verlag, New York.

BASIS

Any set of linearly independent vectors $\mathbf{v}_1$, $\mathbf{v}_2, \ldots, \mathbf{v}_n$ such that every vector in a vector space can be written as a linear combination of $\mathbf{v}_1, \ldots, \mathbf{v}_n$ is a *basis* of the space (provided that $v_1, \ldots, v_n$ belong to the space).

In general, there are many different possible bases for a given vector space. The minimum number of members needed for a basis is the *dimension* of the vector space.

Example. The vectors $b_1 = \binom{1}{1}$ and $b_2 = \binom{1}{-1}$ form a basis in the two-dimensional Euclidean space R^2. We can express the vector $\binom{2}{1}$ as

$$\binom{2}{1} = \frac{3}{2}\binom{1}{1} + \frac{1}{2}\binom{1}{-1}.$$

A basis consisting of *unit* vectors δ_i is called a *standard* or *canonical* basis. A canonical basis of R^n consists of vectors

$$\delta_1 = \begin{bmatrix} 1 \\ 0 \\ \vdots \\ 0 \end{bmatrix}, \delta_2 = \begin{bmatrix} 0 \\ 1 \\ \vdots \\ 0 \end{bmatrix}, \ldots, \delta_n = \begin{bmatrix} 0 \\ 0 \\ 0 \\ \vdots \\ 1 \end{bmatrix}.$$

See also Orthogonal Transformation.

BASU, DEBABRATA

Born: July 5, 1924, in Dhaka, now the capital of Bangladesh

Died: March 24, 2001, in Kolkata, India

Contributed to: Sample surveys, foundation of statistics, Bayesian analysis, mathematical statistics.

Basu got a Master's degree in Mathematics from Dhaka University around 1945. He wanted to be a mathematician like his father Nalini Mohan Basu, who was the head of the department of Mathematics, and his favorite teacher T. Vijayraghaban. Basu has described in Reference 4 how he learned modern analysis on his own under the inspiring guidance of Vijayraghaban. He also loved studying number theory from Landau's book and modeled his own style of exposition on Landau's.

He taught mathematics briefly at Dhaka University, but political events in the Indian subcontinent, namely, India's independence and partition in 1947, changed his life completely. He ended up at the Indian Statistical Institute in Kolkata in 1950 and began studying statistics as C. R. Rao's first doctoral student. In 1950, Rao was still working on minimum variance unbiased estimates, a subject that he himself had introduced in 1945. As a Ph.D. student of Fisher at Cambridge, Rao was also interested in maximum likelihood estimates, ancillary statistics, and conditional inference. Moreover, owing to Wald's tragic visit to India around this time—Wald died in a plane crash in India—there was a lot of interest in the emerging subject of statistical decision theory. Because of Mahalanobis, survey sampling was a popular topic too. Basu worked on all these topics as well as characterization problems. He finished his thesis in 2 years and spent 2 years in Berkeley as a postdoctoral fellow. He also listened to and interacted with Fisher, who was a regular visitor to the Indian Statistical Institute. This led to a keen interest in foundational questions arising from both the Neyman-Pearson–Wald theory and Fisher's idea on using conditional inference to avoid conditionality paradoxes as in Welch's example. Throughout the decade, Basu worked on many topics, all at the cutting edge, and, on each, had something important to contribute.

One of his most interesting results is a characterization of the normal distribution through independence of two linear functions $\sum_{i=1}^{n} a_i X_i$, $\sum_{i=1}^{n} b_i X_i$, $a_i \neq 0$, $b_i \neq 0$, X_i's independent. Basu proved this under moment assumptions. A little later, the result was proved independently without this assumption by Darmois and Skitovich. It is usually called the Darmois–Skitovich theorem.

Basu's most well known theorem in mathematical statistics is his result on the independence of a boundedly complete sufficient statistic and an ancillary statistic. There are two beautiful discussions of these results and their many applications in proving independence, calculating moments and finding a sampling distribution, by Lehmann [6] and Ghosh [2]. Basu [1] points to a similar, somewhat weaker, result obtained by Hogg and Craig.

During this early period—1950 to 1960—Basu was at work on other seminal ideas. Some of his famous counterexamples appeared in this period. He presented a new idea of relative efficiency of two estimates based on a comparison of tail probabilities rather than variance. He showed that a small change in the squared error loss can lead to the nonexistence of a best unbiased estimate even for $N(\theta, 1)$. In this period, he also had one of the first examples of an estimatable parametric function without a best unbiased estimate as well as the unexpected effects of truncating the parameter space. In reference 6, he reminisces on problems with improper priors and his first encounter with improper priors and his first encounter with Wald. He recalls an early counterexample on a maximum likelihood estimate in reference 4 [pp. 1,2].

Though Basu had been interested in ancillaries right from the beginning, they became his main interest from the late-fifties to mid-sixties. In 1965, Basu was still trying to find a middle ground in inference, between Fisher and Neyman, by finding an unambiguous definition of an appropriate ancillary statistic. In his paper at the Fifth Berkeley Symposium, he invokes Zorn's lemma to show that in many examples there are many maximal ancillary statistics but none that is largest.

Sufficiency was also a major interest. He rediscovered a result known to ergodic theorists but not in the language of sufficient statistics used by him. In a paper with the present writer, it was shown that a minimal sufficient statistic always exists for families of discrete distributions and this was used to show the minimal sufficiency of distinct units in survey sampling. Basu's interest in survey sampling goes back to the earlier period when he showed how one can improve estimates by using distinct units only.

In 1968, Basu became a passionate Bayesian while preparing for an invited lecture on the Bayesian paradigm. His subsequent polemical papers on foundations, specially the likelihood principle and its violation in survey sampling, as well as a critical analysis of the Horvitz–Thompson estimate and Fisher's exact test for a 2×2 table, reflect his complete disenchantment with both classical and Fisherian statistics. Basu's delightful circus example illustrates how devastating the Horvitz–Thompson estimate can be. He was among the few who pioneered a critical look at both randomization and design-based inference. It was his hope that an informal Bayesian approach, rather than a rigid Bayesian analysis with thousands of parameters, was appropriate. He expected this would lead to methods of analyzing survey data that have "the hard to define property of face validity."

Basu was a Professor at the Indian Statistical Institute until the mid-seventies. In 1976, he joined Florida State University as Professor and worked there until 1986. After that, he was often at the Indian Statistical Institute. Both these institutions made him Professor Emeritus. He also visited many of the leading universities in the United States, Britain, and Japan, and spent some time at the American University at Beirut.

He was a great teacher. Many of his students, both at ISI and FSU, remember him with affection and respect. He also thought a great deal about what it is that makes for good teaching and reliable exams. As the first Dean of Studies at ISI, he joined Haldane, Mahalanobis, and C. R. Rao in making the new B.Stat. course one of the most creative and interactive programs in statistics in the world in the sixties.

Basu was married to Kalyani, a gifted singer, who like most Indian women of her generation, chose to be a dedicated wife and mother rather than have a career of her own. Both husband and wife were very devoted to each other. A stroke in 1984 had left Kalyani wheelchair bound but she continued to take good care of her husband until the end. She died within a couple of months of her husband's death. Their daughter, Monimala, is a journalist and son, Shantanu, is an astrophysicist. Both live in North America but were frequently with their parents in Kolkata.

Basu loved trees, flowers, and children. All his life, he took great interest in gardens and gardening. He cared for all vulnerable people. That is how his family wants him to be remembered.

Most of Basu's papers, along with some biographical material is available in reference 3. More material on Basu can be found in References 5 and 4.

REFERENCES

1. Basu, D. (1982). "Basu Theorems". In *Encyclopedia of Statistical Sciences*, Vol. 1, N. L. Johnson and S. Kotz, eds. John Wiley & Sons, New York, pp. 193–196.

2. Ghosh, M. (2002). Basu's theorem with applications: a personalistic review. *Sankhya Ser. A*, **64**, 509–531.

3. Ghosh, J. K. (1988). *Statistical Information and Likelihood—A Collection of Critical Essays by Dr. D. Basu*. Springer-Verlag, New York. (Most of Basu's work appears in this volume. Most of his papers are listed.)

4. Ghosh, J. K., Mitra, S. K., and Parthasarathy, K. R. (1992). *Glimpses into India's Statistical Heritage*. Wiley Eastern, New Delhi, India. (Basu's autobiographical article describes his early training in mathematics in Dhaka University and intellectual encounters with some of the giants in statistics when he switched to statistics at the Indian Statistical Institute in the early fifties.)

5. Ghosh, M. and Pathak, P. K. (1992). *Current Issues in Statistical Inference: Essays in Honor of D. Basu*, IMS Lecture Notes and Monograph Series, Hayward, Calif.

6. Lehmann, E. L. (1981). An interpretation of completeness and Basu's theorem. *J. Am. Stat. Assoc.*, **76**, 335–340.

JAYANTA GHOSH

BASU THEOREMS

The theorems are related to the notions of sufficiency*, ancillarity, and conditional independence*. Let X denote the sample and θ the parameter that completely specifies the sampling distribution P_θ of X. An event E is ancillary* if $P_\theta(E)$ is θ-free, i.e., $P_\theta(E) = P_{\theta'}(E)$ for all $\theta, \theta' \in \Theta$, the parameter space. A statistic $Y = Y(X)$ is ancillary if every Y-event (i.e., a measurable set defined in terms of Y) is ancillary (*see* ANCILLARY STATISTICS—I). A statistic T is sufficient* if, for every event E, there exists a θ-free version of the conditional probability* function $P_\theta(E|T)$ (*see* SUFFICIENT STATISTICS). The event E is (conditionally) independent of T if, for each $\theta \in \Theta$, the conditional probability function $P_\theta(E|T)$ is P_θ-essentially equal to the constant $P_\theta(E)$. The statistic Y is independent of T if every Y-event is independent of T. (Independence is a symmetric relationship between two statistics.)

The theorems originated in the following query. Let X_1, X_2 be independent, identically distributed (i.i.d.) $N(\theta, 1)$ and let $Y = X_1 - X_2$. Clearly, Y is ancillary and, therefore, so also is every measurable function $h(Y)$ of Y. The statistic Y is shift invariant* in the sense that $Y(X_1 + a, X_2 + a) = Y(X_1, X_2)$ for all X_1, X_2, and a. It is easy to see that every shift-invariant statistic is a function of Y, and vice versa. Therefore, every shift invariant statistic is ancillary. Is the converse true?

That the answer has to be in the negative is seen as follows. The statistic $T = X_1 + X_2$ is sufficient and is independent of the ancillary statistic $Y = X_1 - X_2$. Let A be an arbitrary T-event and B_1, B_2 be two distinct Y-events such that $P_\theta(B_1) = P_\theta(B_2) = \alpha$, where $0 < \alpha < 1$ is a constant chosen and fixed. Consider the event $E = AB_1 \cup A^c B_2$. The T-events, A, A^c are independent of the Y-events B_1, B_2; therefore,

$$P_\theta(E) = P_\theta(AB_1) + P_\theta(A^c B_2)$$
$$= P_\theta(A)P_\theta(B_1) + P_\theta(A^c)P_\theta(B_2)$$
$$= \alpha[P_\theta(A) + P_\theta(A^c)]$$
$$= \alpha \quad \text{for all } \theta.$$

Thus, E is ancillary even though it is not shift-invariant (not an Y-event). How do we characterize the class of ancillary events in this case?

Consider an arbitrary ancillary event E with $P_\theta(E) \equiv \alpha$. Since $T = X_1 + X_2$ is sufficient, there exists a θ-free version $f(T)$ of the conditional probability function $P_\theta(E|T)$ (*see* SUFFICIENT STATISTICS). Now, $E_\theta f(T) = P_\theta(E) \equiv \alpha$, so $f(T) - \alpha$ is a bounded function of T that has zero mean for each $\theta \in \Theta$. But the statistic T is complete in the sense that no nontrivial function of T can have identically zero mean. Therefore, the event $f(T) \neq \alpha$ is P_θ-null for each $\theta \in \Theta$. In other words, $P_\theta(E|T) = P_\theta(E)$ a.s., $[P_\theta]$ for each θ. That is, every ancillary E is independent of $T = X_1 + X_2$. Is the converse true?

Let T be an arbitrary sufficient statistic and let E be independent of T. Let $f(T)$ be a θ-free version of $P_\theta(E|T)$. Then, for each $\theta \in \Theta, f(T) = P_\theta(E)$ a.s. $[P_\theta]$. If $\mathcal{X}_\theta$ is the set of all sample points for which $f(T) = P_\theta(E)$, then $P_\theta(\mathcal{X}_\theta) = 1$ for all $\theta \in \Theta$. If $P_{\theta_1}(E) \neq P_{\theta_2}(E)$, then the two sets $\mathcal{X}_{\theta_1}$ and $\mathcal{X}_{\theta_2}$ are disjoint and so P_{θ_1} and P_{θ_2} have disjoint supports, which is a contradiction in the present case. Thus the class of ancillary events may be characterized as the class of events that are independent of $X_1 + X_2$.

The "Basu theorems" are direct generalizations of the foregoing results and may be stated as follows.

Theorem 1. Let T be sufficient and boundedly complete*. Then a statistic Y is ancillary only if it is (conditionally) independent of T for each θ.

The measures P_θ and $P_{\theta'}$ are said to overlap if they do not have disjoint supports. The family $\mathcal{P} = \{P_\theta\}$ of measures on a space $\mathcal{X}$ is said to be connected if for all θ, θ' there exists a finite sequence $\theta_1, \theta_2, \ldots, \theta_k$ such that every two consecutive members of the sequence $P_\theta, P_{\theta_1}, P_{\theta_2}, \ldots, P_{\theta_k}, P_{\theta'}$ overlap. For example, if under $P_\theta, -\infty < \theta < \infty$, the random variables $X_1, X_2, \ldots, X_n$ are i.i.d. with a common uniform distribution concentrated on the interval $(\theta, \theta + 1)$, then the family $\mathcal{P} = \{P_\theta\}$ is connected even though P_θ and $P_{\theta'}$ do not overlap whenever $|\theta - \theta'| \geqslant 1$.

Theorem 2. Let $\mathcal{P} = \{P_\theta\}$ be connected and T be sufficient. Then Y is ancillary if it is (conditionally) independent of T for each θ.

Neither the condition of bounded completeness (in Theorem 1) nor that of connectedness (in Theorem 2) can be entirely dispensed with. If $T = X$, the whole sample, then it is sufficient. Consider, therefore, a case where a nontrivial ancillary statistic $Y = Y(x)$ exists. Such an Y cannot be independent of $T = X$, because, if it were, then Y has to be independent of itself, which it cannot be unless it is a constant (a trivial ancillary). On the other hand, if $\mathcal{P}$ is not connected, then it is typically true that there exists nonempty proper subsets $\mathcal{X}_0 \subset \mathcal{X}$ and $\Theta_0 \subset \Theta$ of the sample space $\mathcal{X}$ and the parameter space Θ, respectively, such that

$$P_\theta(\mathcal{X}_0) = \begin{cases} 1 & \text{for all } \theta \in \Theta_0 \\ 0 & \text{for all } \theta \in \Theta - \Theta_0 \end{cases}.$$

Koehn and Thomas [7] called such a set $\mathcal{X}_0$ a splitting set. A splitting set (event) $\mathcal{X}_0$ is clearly not ancillary; however, for every $\theta \in \Theta$, it is P_θ-equivalent either to the whole space $\mathcal{X}$ (the sure event) or to the empty set ϕ (the impossible event). Therefore, the nonancillary event $\mathcal{X}_0$ is independent of every other event E; that is, $\mathcal{X}_0$ is independent of the sufficient statistic $T = X$. Basu [2] gave a pathological example of a statistical model, with a disconnected $\mathcal{P}$, where we have two independent sufficient statistics.

Consider the following three propositions:

(a) T is sufficient

(b) Y is ancillary.

(c) T and Y are (conditionally) independent for each $\theta \in \Theta$.

Under suitable conditions (a) and (b) together imply (c) (Theorem 1) and (a) and (c) together imply (b) (Theorem 2). The following theorem completes the set.

Theorem 3. If (T, Y) is jointly sufficient, then (b) and (c) imply (a).

Theorem 1 is often used to solve diverse problems involving sampling distributions. The following example illustrates this.

Example. Let $X_1, X_2, \cdots$ be a sequence of mutually independent gamma variables with shape parameters $\alpha_1, \alpha_2, \ldots$, (i.e., the PDF of X_n is $Cx^{\alpha_n - 1}e^{-x}, x > 0, n = 1, 2, \ldots$.) Let $T_n = X_1 + X_2 + \cdots + X_n$ and $Y_n = T_n/T_{n+1}$, $n = 1, 2, \ldots$. It is easy to show that T_n has a gamma distribution* with shape parameter $\alpha_1 + \alpha_2 + \cdots + \alpha_n$ and that Y_n has a beta distribution* with parameters $\alpha_1 + \alpha_2 + \cdots + \alpha_n$ and α_{n+1}. But it is not clear why $Y_1, Y_2, \ldots, Y_{n-1}, T_n$ have to be mutually independent. This is seen as follows.

Introduce a scale parameter θ into the joint distribution of $X_1, X_2, \ldots$. That is, suppose that the X_n's are mutually independent and that the PDF of X_n is $c(\theta) \cdot x^{\alpha_n - 1}e^{-x/\theta}, x > 0, \theta > 0, n = 1, 2, \ldots$. Regard the α_n's as known positive constants and θ as the unknown parameter. With $\boldsymbol{X}^{(n)} = (X_1, X_2, \ldots, X_n)$ as the sample, $T_n = X_1 + X_2 + \cdots + X_n$ is a complete sufficient statistic*. The vector-valued statistic $\boldsymbol{Y}^{(n-1)} = (Y_1, Y_2, \ldots, Y_{n-1})$ is scale-invariant. Since θ is a scale parameter, it follows that $\boldsymbol{Y}^{(n-1)}$ is an ancillary statistic. From Theorem 1 it then follows that, for each $n, \boldsymbol{Y}^{(n-1)}$ is independent of T_n. Since $(\boldsymbol{Y}^{(n-1)}, T_n)$ is a function of the first nX_i's, the pair $(\boldsymbol{Y}^{(n-1)}, T_n)$ is independent of X_{n+1}. Thus $\boldsymbol{Y}^{(n-1)}, T_n$, and X_{n+1} are mutually independent. It follows at once that $\boldsymbol{Y}^{(n-1)}$ is independent of $Y_n = T_n/(T_n + X_{n+1})$. Therefore, for each $n \geqslant 2$, the vector $(Y_1, Y_2, \ldots, Y_{n-1})$ is independent of Y_n and this means that the Y_i's are mutually independent.

(Refer to Hogg and Craig [6] for Several interesting Uses of Theorem 1 in Proving Results in Distribution Theory.)

Basu [1] stated and proved Theorem 1 in the generality stated here. At about the same time, Hogg and Craig proved a particular case of the theorem. Basu [1] stated Theorem 2 without the condition of connectedness for $\mathcal{P}$. This has resulted in the theorem being incorrectly stated in several statistical texts. Basu [2] stated and proved Theorem 2 in the general form stated here. Koehn and Thomas [7] noted that the proposition "every Y that is conditionally independent of a sufficient T is ancillary" is true if and only if there do not exist a splitting set as specified above. It turns out, however, that the two notions of connectedness and nonexistence of splitting sets coincide for all typical statistical models.

The Basu theorems are of historical interest because they established a connection between the three apparently unrelated notions of sufficiency, ancillarity, and independence. That the three notions really hang together is not easy to see through if we adopt an orthodox Neyman-Pearson* point of view. However, if we take a Bayesian* view of the matter and regard θ as a random variable with a (prior) probability distribution ξ and the model $\mathcal{P}$ as a specification of the set of conditional distributions of X given θ, then the notions of ancillarity and sufficiency will appear to be manifestations of the notion of conditional independence.

For a model $\mathcal{P}$ and for each prior ξ, consider the joint probability distribution Q_ξ of the pair (θ, X). A statistic $Y = Y(X)$ is ancillary if its conditional distribution, given θ, is θ-free. In other words, Y is ancillary if, for each joint distribution Q_ξ of (θ, X), the two random variables Y and θ are stochastically independent. A statistic $T = T(X)$ is sufficient if the conditional distribution of X, given θ and T, depends only on T (i.e., the conditional distribution is θ-free). Sufficiency of T may, therefore, be characterized as follows:

Definition. The statistic T is sufficient if, for each Q_ξ, X and θ are conditionally independent given T.

Thus a neo-Bayesian* version of Theorem 1 may be stated as:

Theorem 1(a). Suppose that, for each Q_ξ, the variables Y and θ are stochastically independent and also X and θ are conditionally independent given T. Then Y and T are conditionally independent given θ provided that the statistic T is boundedly complete in the sense described earlier.

Refer to Florens and Mouchart [5] for more on the Bayesian insight on the theorems and also to Dawid [4] for a clear exposition on conditional independence as a language of statistics.

REFERENCES

1. Basu, D. (1955). *Sankhyā*, **15**, 377.
2. Basu, D. (1958). *Sankhyā*, **20**, 223.
3. Basu, D. (1959). *Sankhyā*, **21**, 247.
4. Dawid, A. P. (1979). *J. R. Statist. Soc. B*, **41**, 1.
5. Florens, J. P. and Mouchart, M. (1977). *Reduction of Bayesian Experiments. CORE Discuss. Paper 1737.*
6. Hogg, R. V. and Craig, A. T. (1956). *Sankhyā*, **37**, 209.
7. Koehn, U. and Thomas D. L. (1975). *Amer. Statist.*, **39**, 40.

See also ANCILLARY STATISTICS—I; CONDITIONAL INFERENCE; STRUCTURAL INFERENCE; and SUFFICIENT STATISTICS.

D. BASU

BASU'S ELEPHANT

This is an example that purports to demonstrate the absurdity of the Horvitz—Thompson estimator* for the total of some quantity in a finite population. A circus owner has 50 elephants and proposes estimating their total weight by selecting and weighing one elephant. One of the elephants, Sambo, is known to be about average weight. The rest do not differ much from Sambo except that one, Jumbo, is very heavy. A selection procedure is then proposed which selects Sambo with probability 99/100 and each other elephant, including Jumbo, with probability 1/4900. If, as is overwhelmingly most likely, Sambo is selected and found to have weight y, the Horvitz—Thompson estimate is $100y/99$, only a little more than y. Had, most unusually, Jumbo been selected and found to have weight y, the estimate is $4900y$, grossly in excess of Jumbo's already large weight. The natural procedure is to weigh Sambo and estimate the total weight as $50y$.

REFERENCE

1. Basu, D. (1971). An essay on the logical foundations of survey sampling, part 1 (with discussion). In *Foundations of Statistical Inference*, V. P. Godambe and D. A. Sprott, eds. Holt, Rinehart & Winston, Toronto, pp. 203–242.

DENNIS V. LINDLEY

BATHTUB CURVE

For many observed lifetime distributions, the hazard function starts (at age 0) at a fairly high level, decreases fairly soon to a minimum, and then starts to increase again, at first slowly but then more and more rapidly. The resultant curve is as shown in Fig. 1.

The name "bathtub curve" is given to curves of this type, although most of them would represent rather uncomfortable hathtubs! Such curves are typical of force of mortality* for human populations. They also apply to some mechanical and electrical products for which there are early failures of substandard specimens, whereas the standard product wears out more slowly but at an increasing rate as time progresses. Of the periods shown in Fig. 1, period I is often called the "infant mortality" period or "break-in" period, and period III is the "wear-out" period.

BIBLIOGRAPHY

Bury, A. K. (1975). *Statistical Methods in Applied Science*. Wiley, New York.

See also FORCE OF MORTALITY and LIFE TABLES.

BAULE'S EQUATION. See CROP YIELD ESTIMATION, MITSCHERLICH'S LAW IN

BAYES FACTORS

Bayes factors are the primary tool used in Bayesian inference* for hypothesis testing*

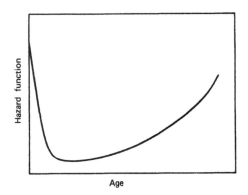

Figure 1.

and model selection. They also are used by non-Bayesians in the construction of test statistics*. For instance, in the testing of simple hypotheses, the Bayes factor equals the ordinary likelihood ratio*. BAYESIAN MODEL SELECTION discusses the use of Bayes factors in model selection. STATISTICAL EVIDENCE discusses the general use of Bayes factors in quantifying statistical evidence. To avoid overlap with these entries, we concentrate here on the motivation for using Bayes factors, particularly in hypothesis testing.

Suppose we are interested in testing two hypotheses:

$$H_1 : X \text{ has density } f_1(x|\theta_1) \text{ vs.}$$

$$H_2 : X \text{ has density } f_2(x|\theta_2).$$

If the parameters θ_1 and θ_2 are unknown, a Bayesian generally specifies prior densities $\pi_1(\theta_1)$ and $\pi_2(\theta_2)$ for these parameters, and then computes the Bayes factor of H_1 to H_2 as the ratio of the marginal densities of x under H_1 and H_2,

$$B = m_1(x)/m_2(x),$$

where

$$m_i(x) = \int f_i(x|\theta_i)\pi_i(\theta_i)\, d\theta_i.$$

The Bayes factor is typically interpreted as the "odds provided by the data for H_1 to H_2," although this interpretation is strictly valid only if the hypotheses are simple; otherwise, B will also depend on the prior distributions*. Note, however, that B does not depend on the prior probabilities of the hypotheses. If one wants a full Bayesian analysis, these prior probabilities*, $P(H_1)$ and $P(H_2)$ (which, of course, sum to one), must also be specified. Then the posterior probability* of H_1 is given by

$$P(H_1|x) = \frac{B}{B + [P(H_2)/P(H_1)]}. \tag{1}$$

Clearly, the posterior probability of H_2 is just $1 - P(H_1|x)$.

The attraction of using a Bayes factor to communicate evidence about hypotheses is precisely that it does not depend on $P(H_1)$ and

$P(H_2)$, which can vary considerably among consumers of a study. Any such consumer can, if they wish, determine their own $P(H_1)$ and convert the reported Bayes factor to a posterior probability using (1). Although B will still depend on $\pi_1(\theta_1)$ and/or $\pi_2(\theta_2)$, the influence of these priors is typically less significant than the influence of $P(H_1)$ and $P(H_2)$. Also, default choices of $\pi_1(\theta_1)$ and $\pi_2(\theta_2)$ are sometimes available (see below), in which case B can be used as a default measure of evidence. General discussion of Bayes factors can be found in Jeffreys [29], Berger [2], Kass and Raftery [30], and Berger and Pericchi [11].

Example 1. Suppose we observe an i.i.d. normal* sample, $X_1, X_2, \ldots, X_n$, from an $N(\theta, \sigma^2)$ distribution, with σ^2 known. It is desired to test $H_1 : \theta = \theta_0$ vs. $H_2 : \theta \neq \theta_0$. For the prior distribution of θ under H_2, it is common to choose an $N(\theta_0, \tau^2)$ distribution, where the standard deviation τ is chosen to reflect the believed plausible range of θ if H_2 were true. (One could, of course, also choose prior means other than θ_0.)

A simple computation then shows that the Bayes factor of H_1 to H_2 is

$$B = \left(1 + \frac{n\tau^2}{\sigma^2}\right)^{1/2} \exp\left(\frac{-\frac{1}{2}z^2}{1 + (\sigma^2/n\tau^2)}\right), \quad (2)$$

where $z = \sqrt{n}(\bar{x} - \theta_0)/\sigma$ is the usual standardized test statistic. A frequently used default choice of τ^2 is $\tau^2 = 2\sigma^2$ (the quartiles of the prior on θ are then roughly $\pm\sigma$), in which case

$$B = \sqrt{1 + 2n} \exp\left(\frac{-z^2}{2 + (1/n)}\right).$$

For instance, if $n = 20$ and $|z| = 1.96$, then $B = 0.983$. As this very nearly equals 1, which would correspond to equal odds for H_1 and H_2, the conclusion would be that the data provide essentially equal evidence for H_1 and H_2.

Note that $|z| = 1.96$ corresponds to a P-value* of 0.05, which is typically considered to be significant evidence against H_1, in contradiction to the message conveyed by $B = 0.983$. This conflict between P-values and Bayes factors is discussed further below. For now, it is interesting to note that the conflict magnifies in severity as $n \to \infty$ (or $\tau^2 \to \infty$) in (2). Indeed, for *any* fixed z, (2) then converges to ∞, so that B would indicate overwhelming evidence for H_1 even though z was any (fixed) large value (and the P-value was, correspondingly, *any* fixed small value).

Various versions of this phenomenon have become known as Jeffreys' paradox [29], Lindley's paradox [33], and Bartlett's paradox [1] (*see* SHARP NULL HYPOTHESES). The "paradox" depends crucially on H_1 being a believable exact point null hypothesis. While this may sometimes be true, a point null hypothesis is more typically an approximation to a small interval null, and the validity of the approximation disappears as $n \to \infty$ (but not as $\tau^2 \to \infty$); see Berger and Delampady [7] for discussion of this and its effect on the "paradox."

MOTIVATION FOR USING BAYES FACTORS

Since posterior probabilities are an integral component of Bayesian hypothesis testing and model selection, and since Bayes factors are directly related to posterior probabilities, their role in Bayesian analysis is indisputable. We concentrate here, therefore, on reasons why their use should be seriously considered by all statisticians.

Reason 1. Classical P-values* can be highly misleading when testing precise hypotheses. This has been extensively discussed in Edwards et al. [20], Berger and Sellke [13], Berger and Delampady [7], and Berger and Mortera [8]. In Example 1, for instance, we saw that the P-value and the posterior probability of the null hypothesis could differ very substantially. To understand that the problem here is with the P-value, imagine that one faces a long series of tests of new drugs for AIDS. To fix our thinking, let us suppose that 50% of the drugs that will be tested have an effect on AIDS, and that 50% are ineffective. (One could make essentially the same point with any particular fraction of effective drugs.) Each drug is tested in an independent experiment, corresponding to a normal test of no effect, as in Example 1. (The experiments could all have different sample sizes and variances, however.) For each drug, the P-value is computed, and those with P-values smaller than 0.05 are deemed to be effective. (This is perhaps an unfair caricature of standard practice, but that is not relevant to the point we are trying to make about P-values.)

Suppose a doctor reads the results of the published studies, but feels confused about the meaning of P-values. (Let us even assume here that all studies are published, whether they obtain statistical significance or not; the real situation of publication selection bias only worsens the situation.) So the doctor asks the resident statistician to answer a simple question: "A number of these published studies have P-values that are between 0.04 and 0.05; of these, what fraction of the corresponding drugs are ineffective?"

The statistician cannot provide a firm answer to this question, but can provide useful bounds if the doctor is willing to postulate a prior opinion that a certain percentage of the drugs being originally tested (say, 50% as above) were ineffective. In particular, it is then the case that at least 23% of the drugs having P-values between 0.04 and 0.05 are ineffective, and in practice typically 50% or more will be ineffective (see Berger and Sellke [13]). Relating to this last number, the doctor concludes: "So if I start out believing that a certain percentage of the drugs will be ineffective, say 50%, then a P-value near 0.05 does not change my opinion much at all; I should still think that about 50% are ineffective."

This is essentially right, and this is essentially what the Bayes factor conveys. In Example 1 we saw that the Bayes factor is approximately one when the P-value is near 0.05 (for moderate sample sizes). And a Bayes factor of one roughly means that the data are equally supportive of the null and alternative hypotheses, so that posterior beliefs about the hypotheses will essentially equal the prior beliefs.

We cast the above discussion in a frequentist framework to emphasize that this is a fundamental fact about P-values; in situations such as that above, a P-value of 0.05 essentially does not provide any evidence against the null hypothesis. (Note, however, that the situation is quite different in situations where there is not a precise null hypothesis; then P-values and posterior probabilities often happen to be reasonably similar—see Casella and Berger [16].) That the meaning of P-values is commonly misinterpreted is hardly the fault of consumers of statistics. It is the fault of statisticians for providing a concept so ambiguous in meaning. The real point here is that the Bayes factor essentially conveys the right message easily and immediately.

Reason 2. Bayes factors are consistent* for hypothesis testing and model selection. Consistency is a very basic property. Its meaning is that, if one of the entertained hypotheses (or entertained models) is actually true, then a statistical procedure should guarantee selection of the true hypothesis (or true model) if enough data are observed. Use of Bayes factors guarantees consistency (under very mild conditions), while use of most classical selection tools, such as P-values, C_p, and AIC (see MODEL SELECTION: AKAIKE'S INFORMATION CRITERION), does not guarantee consistency (cf. Gelfand and Dey [25]).

In model selection it is sometimes argued that consistency is not a very relevant concept because no models being considered are likely to be exactly true. There are several possible counterarguments. The first is that, even though it is indeed typically important to recognize that entertained models are merely approximations, one should not use a procedure that fails the most basic property of consistency when you do happen to have the correct model under consideration. A second counterargument is based on the results of Berk [14] and Dmochowski [18]; they show that asymptotically (under mild conditions), use of the Bayes factor for model selection will choose the model that is closest to the true model in terms of Kullback-Leibler divergence (see INFORMATION, KULLBACK). This is a remarkable and compelling property of use of Bayes factors. However, not all criteria support Bayes factors as optimal when the true model is not among those being considered; see Shibata [37] and Findley [22] for examples.

Reason 3. Bayes factors behave as automatic Ockham's razors (see PARSIMONY, PRINCIPLE OF), favoring simple models over more complex models, if the data provide roughly comparable fits for the models. Overfitting is a continual problem in model selection, since more complex models will always provide a somewhat better fit to the data than will simple models. In classical statistics overfitting is avoided by introduction of an ad hoc penalty term (as in AIC), which increases as the complexity (i.e., the number of unknown parameters) of the model increases. Bayes factors act naturally to penalize model complexity, and hence need no ad hoc penalty terms. For an interesting historical example and general discussion and references, see Jefferys and Berger [28].

Reason 4. The Bayesian approach can easily be used for multiple hypotheses or models. Whereas classical testing has difficulty with more than two hypotheses, consideration of such poses no additional difficulty in the Bayesian approach. For instance, one can easily extend the Bayesian argument in Example 1 to test between $H_0 : \theta = 0, H_1 : \theta < 0$, and $H_2 : \theta > 0$.

Reason 5. The Bayesian approach does not require nested hypotheses or models, standard distributions, or regular asymptotics. Classical hypothesis testing has difficulty if the hypotheses are not nested or if the distributions are not standard. There are general classical approaches based on asymptotics, but the Bayesian approach does not require any of the assumptions under which an asymptotic analysis can be justified. The following example illustrates some of these notions.

Example 2. Suppose we observe an i.i.d. sample, $X_1, X_2, \ldots, X_n$, from either a normal or a Cauchy distribution f, and wish to test the hypotheses

$$H_1 : f \text{ is } N(\mu, \sigma^2) \text{ vs. } H_2 : f \text{ is } C(\mu, \sigma^2).$$

This is awkward to do classically, as there is no natural test statistic and even no natural null hypothesis. (One can obtain very different answers depending on which test statistic is used and which hypothesis is considered to be the null hypothesis.) Also, computations of error probabilities are difficult, essentially requiring expensive simulations.

In contrast, there is a natural and standard automatic Bayesian test for such hypotheses. In fact, for comparison of any location-scale* distributions, it is shown in Berger et al. [12] that one can legitimately compute Bayes factors using the standard noninformative prior density $\pi(\mu, \sigma^2) = 1/\sigma^2$ (*see* JEFFREYS' PRIOR DISTRIBUTION). For testing H_1 vs. H_2 above, the resulting Bayes factor is available in closed form (see Franck [23] and Spiegelhalter [38]).

Reason 6. The Bayesian approach can allow for model uncertainty and is often predictively optimal. Selecting a hypothesis or model on the basis of data, and then using the same data to estimate model parameters or make predictions based upon the model, is well known to yield (often seriously) overoptimistic estimates of accuracy. In the classical approach it is often thus recommended to use part of the data to select a model and the remaining part for estimation and prediction. When only limited data are available, this can be difficult.

The Bayesian approach takes a different tack: ideally, all models are left in the analysis, and (say), prediction is done using a weighted average of the predictive distributions from each model, the weights being determined from the posterior probabilities (or Bayes factors) of each model. See Geisser [24] and Draper [19] for discussion and references.

Although keeping all models in the analysis is an ideal, it can be cumbersome for communication and descriptive purposes. If only one or two models receive substantial posterior probability, it would not be an egregious sin to eliminate the other models from consideration. Even if one must report only one model, the fact mentioned above, that Bayes factors act as a strong Ockham's razor, means that at least the selected model will not be an overly complex one, and so estimates and predictions based on this model will not be quite so overoptimistic. Indeed, one can even establish certain formal optimality properties of selecting models on the basis of Bayes factors. Here is one such:

Result 1. Suppose it is of interest to predict a future observation Y under, say, a symmetric loss $L(|Y - \hat{Y}|)$ (*see* DECISION THEORY), where L is nondecreasing. Assume that two models, M_1 and M_2, are under consideration for the data (present and future), that any unknown parameters are assigned proper prior distributions, and that the prior probabilities of M_1 and M_2 are both equal to 1/2. Then the optimal model to use for predicting Y is M_1, or M_2, as the Bayes factor exceeds, or is less than, one.

Reason 7. Bayes factors seem to yield optimal conditional frequentist tests. The standard frequentist testing procedure, based on the Neyman-Pearson lemma*, has the disadvantage of requiring the reporting of fixed error probabilities, no matter what the data. (The data-adaptive versions of such testing, namely P-values, are not true frequentist procedures and suffer from the rather severe interpretational problems discussed earlier.) In a recent surprising development (based on ideas of Kiefer [32]), Berger et al. [5,6] show for simple versus simple testing and for testing a precise hypothesis, respectively, that tests based on Bayes factors (with, say, equal prior probabilities of the hypotheses) yield posterior probabilities that have direct interpretations as conditional frequentist error probabilities. Indeed, the posterior probability of H_1 is the conditional Type I frequentist error probability, and the posterior probability of H_2 is a type of average conditional Type II error probability (when the alternative is a composite hypothesis). The reported error probabilities thus vary with the data, exactly as do the posterior probabilities. Another benefit that accrues is the fact that one can accept H_1 with a specified error probability (again data-dependent).

The necessary technical detail to make this work is the defining of suitable conditioning sets upon which to compute the conditional error probabilities. These sets necessarily include data in both the acceptance region* and the rejection region, and can roughly be described as the sets which include data points providing equivalent strength of evidence (in terms of Bayes factors) for and against H_1. Computation of these sets is, however, irrelevant to practical implementation of the procedure.

The primary limitation of this Bayesian-frequentist equivalence is that there will

typically be a region, which is called the *no-decision region*, in which frequentist and Bayesian interpretations are incompatible. Hence this region is excluded from the decision space. In Example 1, for instance, if the default $N(0, 2\sigma^2)$ prior is used and $n = 20$, than the no-decision region is the set of all points where the usual z-statistic is between 1.18 and 1.95. In all examples we have studied, the no-decision region is a region where both frequentists and Bayesians would feel indecisive, and hence its presence in the procedure is not detrimental from a practical perspective.

More surprises arise from this equivalence of Bayesian and conditional frequentist testing. One is that, in sequential analysis* using these tests, the stopping rule (*see* OPTIMAL STOPPING RULES) is largely irrelevant to the stated error probabilities. In contrast, with classical sequential testing, the error probabilities depend very much on the stopping rule. For instance, consider a sequential clinical trial* which is to involve up to 1000 patients. If one allows interim looks at the data (after, say, each 100 patients), with the possibility of stopping the experiment if the evidence appears to be conclusive at an interim stage, then the classical error probability will be substantially large than if one had not allowed such interim analysis. Furthermore, computations of classical sequential error probabilities can be very formidable. It is thus a considerable surprise that the conditional frequentist tests mentioned above not only provide the freedom to perform interim analysis without penalty, but also are much simpler than the classical tests. The final surprise is that these conditional frequentist tests provide frequentist support for the stopping-rule principle (see Berger and Berry [4]).

CARE IN SPECIFICATION OF HYPOTHESES

In classical statistics, whether one formulates a test as a one-sided or a two-sided test makes little practical difference; the α-level or P-value changes by at most a factor of two. In Bayesian hypothesis testing, however, the difference between the formulations can lead to strikingly different answers, and so considerable care must be taken in formulation of the hypotheses. Let us begin with two examples.

Example 3. Suppose one is comparing a standard chemotherapy treatment with a new radiation treatment for cancer. There is little reason to suspect that the two treatments have the same effect, so the correct test will be a one-sided test comparing the two treatments.

Example 4. Suppose two completely new treatments for AIDS are being compared. One should now be concerned with equality of treatment effects, because both treatments could easily have no (and hence equal) effect. Hence one should test the null hypothesis of no treatment difference against the alternative that there is a difference. (One might well actually formulate three hypotheses here, the null hypothesis of no difference, and the two one-sided hypotheses of each treatment being better; this is perfectly permissible and adds no real complications to the Bayesian analysis.)

The difference in Example 3 is that the standard chemotherapy treatment is presumably known to have a nonzero effect, and there is no reason to think that a radiation treatment would have (nearly) the same nonzero effect. Hence possible equality of treatment effects is not a real concern in Example 3. (In Bayesian terms, this event would be assigned a prior probability of essentially zero.) Note, however, that if the second treatment had, instead, been the same chemotherapy treatment, but now with (say) steroids added, then equality of treatments would have been a real possibility since the steroids might well have no effect on the cancer.

Deciding whether or not to formulate the test as testing a precise hypothesis or as a one-sided test thus centers on the issue of deciding if there is a believable precise hypothesis. Sometimes this is easy, as in testing for the presence of extrasensory perception, or testing that a proposed law of physics holds. Often it is less clear; for instance, in medical testing scenarios it is often argued that any treatment will have some effect, even if only a very small effect, so that exact equality of treatment effects will never occur.

While perhaps true, it will still typically be reasonable to formulate the test as a test of no treatment difference, since such a test can be shown to be a good approximation to the "optimal" test unless the sample size is very large (cf. Berger and Delampady [7]).

Another aspect of this issue is that Bayesians cannot test precise hypotheses using confidence intervals*. In classical statistics one frequently sees testing done by forming a confidence region for the parameter, and then rejecting a null value of the parameter if it does not lie in the confidence region. This is simply wrong if done in a Bayesian formulation (and if the null value of the parameter is believable as a hypothesis).

USING BAYES FACTORS

Several important issues that arise in using Bayes factors for statistical inference are discussed here.

Computation

Computation of Bayes factors can be difficult. A useful simple approximation is the Laplace approximation*; see BAYESIAN MODEL SELECTION for its application to Bayes factors. The standard method of numerically computing Bayes factors has long been Monte Carlo importance sampling*. There have recently been a large variety of other proposals for computing Bayes factors; see BAYESIAN MODEL SELECTION and Kass and Raftery [30] for discussion and references.

Multiple Hypotheses Or Models

When considering k (greater than two) hypotheses or models, Bayes factors are rather cumbersome as a communication device, since they involve only pairwise comparisons. There are two obvious solutions. One is just to report the marginal densities $m_i(x)$ for all hypotheses and models. But since the scaling of these is arbitrary, it is typically more reasonable to report a scaled version, such as

$$P_i^* = \frac{m_i(x)}{\sum_{j=1}^{k} m_j(x)}.$$

The P_i^* have the additional benefit of being interpretable as the posterior probabilities of the hypotheses or models, if one were to assume equal prior probabilities. Note that a consumer of such a report who has differing prior probabilities $P(H_i)$ can compute his or her posterior probabilities as

$$P(H_i|x) = \frac{P_i^* P(H_i)}{\sum_{j=1}^{k} P_j^* P(H_j)}.$$

Minimal Statistical Reports and Decision Theory

While Bayes factors do summarize the evidence for various hypotheses or models, they are obviously not complete summaries of the information from an experiment. In Example 1, for instance, along with the Bayes factor one would typically want to know the location of θ given that H_2 were true. Providing the posterior distribution of θ, conditional on the data and H_2 being true, would clearly suffice in this regard. A "minimal" report would, perhaps, be a credible set for θ based on this posterior—along with the Bayes factor, of course. (It is worth emphasizing that the credible set alone would not suffice as a report; the Bayes factor is needed to measure the strength of evidence against H_1.)

We should also emphasize that, often, it is best to approach hypothesis testing and model selection from the perspective of decision analysis (Bernardo and Smith [15] discuss a variety of decision and utility-based approaches to testing and model selection). It should be noted that Bayes factors do not necessarily arise as components of such analyses. Frequently, however, the statistician's goal is not to perform a formal decision analysis, but to summarize information from a study in such a way that others can perform decision analyses (perhaps informally) based on this information. In this regard, the "minimal" type of report discussed above will often suffice as the statistical summary needed for the decision maker.

Updating Bayes Factors

Full posterior distributions have the pleasant property of summarizing all available information, in the sense that if new, independent information becomes available, one can simply update the posterior with the new information through Bayes' rule. The same is not generally true with Bayes factors. Updating Bayes factors in the presence of new information typically also requires knowledge of the full posterior distributions (or, at least, the original likelihoods). This should be kept in mind when reporting results.

DEFAULT BAYES FACTORS

Ideally, $\pi_1(\theta_1)$ and/or $\pi_2(\theta_2)$ are derived as subjective prior distributions. In hypothesis testing, especially nested hypothesis testing, there are strong reasons to do this. In Example 1, for instance, the Bayes factor clearly depends strongly on the prior variance, τ^2. Thus, at a minimum, one should typically specify this prior quantity (roughly the square of the prior guess as to the possible spread of θ if H_2 is true) to compute the Bayes factor. An attractive alternative for statistical communication is to present, say, a graph of the Bayes factor as a function of such key prior inputs, allowing consumers of a study easily to determine the Bayes factor corresponding to their personal prior beliefs (cf. Dickey [17] and Fan and Berger [21]).

Another possibility is to use Bayesian robustness* methods, presenting conclusions that are valid simultaneously for a large class of prior inputs. In Example 1, for instance, one can show that the lower bound on the Bayes factor over all possible τ^2, when $z = 1.96$, is 0.473. While this is a useful bound here, indicating that the evidence against H_1 is no more than 1 to 2, such bounds will not always answer the question. The problem is that the upper bounds on the Bayes factor, in situations such as this, tend to be infinite, so that one may well be left with an indeterminate conclusion. (Of course, one might very reasonably apply robust Bayesian methods to a reduced class of possible prior inputs, and hope to obtain sensible upper and lower bounds on the Bayes factor.) Discussions of robust Bayesian methods in testing can be found in refs. 20, 2, 13, 7, 3.

Sometimes use of noninformative priors is reasonable for computing Bayes factors. One-sided testing provides one such example, where taking limits of symmetric proper priors as they become increasingly vague is reasonable and can be shown to give the same answer as use of noninformative priors (cf. Casella and Berger [16]). Another is nonnested testing, when the models are of essentially the same type and dimension. Example 2 above was of this type. See Berger et al. [12] for discussion of other problems of this kind.

In general, however, use of noninformative priors is not legitimate in hypothesis testing and model selection. In Example 1, for instance, the typical noninformative prior for θ (under H_2) is the constant density, but any constant could be used (since the prior is improper regardless) and the resulting Bayes factor would vary with the arbitrary choice of the constant. This is unfortunate—especially in model selection, because at the initial stages of model development and comparison it is often not feasible to develop full subjective proper prior distributions. This has led to a variety of alternative proposals for the development of default Bayes factors. A few of these methods are briefly discussed below. For discussion of other methods and comparisons, see Berger and Pericchi [11] and Iwaki [27].

The most commonly used default procedure is the Bayes information criterion (BIC) of Schwarz [36], which arises from the Laplace approximation to Bayes factors (*see* BAYESIAN MODEL SELECTION for discussion). The BIC is often a quite satisfactory approximation (cf. Kass and Wasserman [31]), but it avoids the problem of prior specification only by simply ignoring that term of the expansion.

Another common approach is to simply choose default proper prior distributions. Thus, in Example 1, we commented that the $N(0, 2\sigma^2)$ distribution is a standard default prior for this testing problem. Jeffreys [29] pioneered this approach (although he actually recommended a $C(0, \sigma^2)$ default prior in Example 1); see also Zellner and Siow [40] and the many references to this approach in Berger and Pericchi [11].

An attempt to directly use noninformative priors, but with a plausible argument for choosing particular constant multiples of them when they are improper, was proposed for linear models in Spiegelhalter and Smith [39].

Two recent default approaches are the intrinsic Bayes factor approach of Berger and Pericchi [9,10,11] and the fractional Bayes factor approach of O'Hagan [34,35]. These use, respectively, parts of the data ("training samples") or a fraction of the likelihood to create, in a sense, a default proper prior distribution. These approaches operate essentially automatically, and apply in great generality to hypothesis testing and model selection problems. And the better versions of these approaches can be shown to correspond to use of actual reasonable default proper prior distributions. They thus provide the best general default methods of testing and model selection that are currently available.

REFERENCES

1. Bartlett, M. S. (1957). A comment on D. V. Lindley's statistical paradox. *Biometrika*, **44**, 533–534.

2. Berger, J. (1985). *Statistical Decision Theory and Bayesian Analysis*, 2nd edition. Springer-Verlag, New York.

3. Berger, J. (1994). An overview of robust Bayesian analysis. *Test*, **3**, 5–124.

4. Berger, J. and Berry, D. (1988). The relevance of stopping rules in statistical inference. In *Statistical Decision Theory and Related Topics IV*, S. S. Gupta and J. Berger, eds. Springer-Verlag, New York, pp. 29–48.

5. Berger, J., Brown, L., and Wolpert, R. (1994). A unified conditional frequentist and Bayesian test for fixed and sequential hypothesis testing. *Ann. Statist.*, **22**, 1787–1807.

6. Berger, J., Boukai, B., and Wang, Y. (1994). Unified frequentist and Bayesian testing of a precise hypothesis. *Statist. Sci.*, **12**, 133–156.

7. Berger, J. and Delampady, M. (1987). Testing precise hypotheses. *Statist. Sci.*, **3**, 317–352.

8. Berger, J. and Mortera, J. (1991). Interpreting the stars in precise hypothesis testing. *Int. Statist. Rev.*, **59**, 337–353.

9. Berger, J. and Pericchi, L. R. (1995). The intrinsic Bayes factor for linear models. In *Bayesian Statistics 5*, J. M. Bernardo et al., eds. Oxford University Press, London, pp. 23–42.

10. Berger, J. and Pericchi, L. R. (1996). On the justification of default and intrinsic Bayes factors. In *Modeling and Prediction*, J. C. Lee et al., eds. Springer-Verlag, New York, pp. 276–293.

11. Berger, J. and Pericchi, L. R. (1996). The intrinsic Bayes factor for model selection and prediction. *J. Amer. Statist. Ass.*, **91**, 109–122.

12. Berger, J., Pericchi, L., and Varshavsky, J. (1998). *Bayes factors and marginal distributions in invariant situations, Sankhya.*

13. Berger, J. and Sellke, T. (1987). Testing a point null hypothesis: the irreconcilability of *P*-values and evidence. *J. Amer. Statist. Ass.*, **82**, 112–122.

14. Berk, R. (1966). Limiting behavior of posterior distributions when the model is incorrect. *Ann. Math. Statist.*, **37**, 51–58.

15. Bernardo, J. M. and Smith, A. F. M. (1994). *Bayesian Theory.* Wiley, New York.

16. Casella, G. and Berger, R. (1987). Reconciling Bayesian and frequentist evidence in the onesided testing problem. *J. Amer. Statist. Ass.*, **82**, 106–111.

17. Dickey, J. (1973). Scientific reporting. *J. R. Statist. Soc. B*, **35**, 285–305.

18. Dmochowski, J. (1996). Intrinsic Bayes factors via Kullback—Leibler geometry. In *Bayesian Statistics 5*, J. M. Bernardo et al., eds. Oxford University Press, London, pp. 543–550.

19. Draper, D. (1995). Assessment and propagation for model uncertainty (with discussion). *J. R. Statist. Soc. B*, **57**, 45–98.

20. Edwards, W., Lindman, H., and Savage, L. J. (1963). Bayesian statistical inference for psychological research. *Psych. Rev.*, **70**, 193–242.

21. Fan, T. H. and Berger, J. (1995). *Robust Bayesian Displays for Standard Inferences Concerning a Normal Mean. Technical Report*, Purdue University, West Lafayette, Ind.

22. Findley, D. (1991). Counterexamples to parsimony and BIC. *Ann. Inst. Statist. Math.*, **43**, 505–514.

23. Franck, W. E. (1981). The most powerful invariant test of normal versus Cauchy with applications to stable alternatives. *J. Amer. Statist. Ass.*, **76**, 1002–1005.

24. Geisser, S. (1993). *Predictive Inference: An Introduction.* Chapman and Hall, London.

25. Gelfand, A. and Dey, D. (1994). Bayesian model choice: asymptotics and exact calculations. *J. R. Statist. Soc. B*, **56**, 501–514.

26. George, E. I. and McCulloch, R. E. (1993). Variable selection via Gibbs sampling. *J. Amer. Statist. Ass.*, **88**, 881–889.

27. Iwaki, K. (1995). Posterior expected marginal likelihood for testing hypotheses. *J. Econ. (Asia Univ.)*, **21**, 105–134.

28. Jefferys, W. and Berger, J. (1992). Ockham's razor and Bayesian analysis. *Amer. Sci.*, **80**, 64–72.

29. Jeffreys, H. (1961). *Theory of Probability*, 3rd ed. Clarendon, Oxford.

30. Kass, R. E. and Raftery, A. (1995). Bayes factors and model uncertainty. *J. Amer. Statist. Ass.*, **90**, 773–795.

31. Kass, R. E. and Wasserman, L. (1995). A reference Bayesian test for nested hypotheses and its relationship to the Schwarz criterion. *J. Amer. Statist. Ass.*, **90**, 928–934.

32. Kiefer, J. (1977). Conditional confidence statements and confidence estimators. *J. Amer. Statist. Ass.*, **72**, 789–827.

33. Lindley, D. V. (1957). A statistical paradox. *Biometrika*, **44**, 187–192.

34. O'Hagan, A. (1994). *Bayesian Inference.* Edward Arnold, London.

35. O'Hagan, A. (1995). Fractional Bayes factors for model comparisons. *J. R. Statist. Soc. B*, **57**, 99–138.

36. Schwarz, G. (1978). Estimating the dimension of a model. *Ann. Statist.*, **6**, 461–464.

37. Shibata, R. (1981). An optimal selection of regression variables. *Biometrika*, **68**, 45–54.

38. Spiegelhalter, D. J. (1985). Exact Bayesian inference on the parameters of a Cauchy distribution with vague prior information. In *Bayesian Statistics 2*. J. M. Bernardo et al., eds. North-Holland, Amsterdam, pp. 743–750.

39. Spiegelhalter, D. J. and Smith, A. F. M. (1982). Bayes factors for linear and log-linear models with vague prior information. *J. R. Statist. Soc. B*, **44**, 377–387.

40. Zellner, A. and Siow, A. (1980). Posterior odds for selected regression hypotheses. In *Bayesian Statistics 1*, J. M. Bernardo et al., eds. pubValencia University Press, Valencia, pp. 585–603.

See also Bayesian Model Selection; Decision Theory; Hypothesis Testing; Model Selection: Bayesian Information Criterion; Parsimony, Principle of; *P*-Values; Scientific Method and Statistics; Sharp Null Hypotheses; and Statistical Evidence.

James O. Berger

BAYES LINEAR ANALYSIS

The Bayes linear approach is concerned with problems in which we want to combine prior judgements of uncertainty with observational data, and we use expected value* rather than probability as the primitive for expressing these judgements. This distinction is of particular relevance in complex problems with too many sources of information for us to be comfortable in making a meaningful full joint prior probability specification of the type required for Bayesian inference*. Therefore, we seek methods of prior specification and analysis that do not require this extreme level of detail. For such problems, expectation may be considered as a more natural primitive than probability (*see* PREVISION for a summary of de Finetti's treatment of expectation as a primitive, and, from a rather different viewpoint, Whittle [15]). Thus the Bayes linear approach is similar in spirit to a full Bayesian analysis, but offers a practical methodology for analyzing partially specified beliefs for large problems.

ADJUSTED MEANS AND VARIANCES

In the Bayes linear approach, we make direct prior specifications for that collection of means, variances, and covariances which we are both willing and able to assess, and update these prior assessments by linear fitting. Suppose that we have two collections of random quantities, namely, vectors $\boldsymbol{B} = (B_1, \ldots, B_r), \boldsymbol{D} = (D_0, D_1, \ldots, D_s)$, where $D_0 = 1$, and we intend to observe $\boldsymbol{D}$ in order to improve our assessments of belief about $\boldsymbol{B}$. The *adjusted* or *Bayes linear* expectation for B_i given $\boldsymbol{D}$ is the linear combination $\boldsymbol{a}_i^T\boldsymbol{D}$ minimizing $\mathrm{E}[(B_i - \boldsymbol{a}_i{}^T\boldsymbol{D})^2]$ over choices of $\boldsymbol{a}_i$. To do so, we must specify prior mean vectors and covariance matrices for $\boldsymbol{B}$ and $\boldsymbol{D}$ and a covariance matrix between $\boldsymbol{B}$ and $\boldsymbol{D}$. The adjusted expectation vector, $\mathrm{E}_{\boldsymbol{D}}(\boldsymbol{B})$, for $\boldsymbol{B}$ given $\boldsymbol{D}$ is evaluated as

$$\mathrm{E}_{\boldsymbol{D}}(\boldsymbol{B}) = \mathrm{E}(\boldsymbol{B}) + \mathrm{cov}(\boldsymbol{B},\boldsymbol{D})[\mathrm{var}(\boldsymbol{D})]^{-1}$$
$$\times \, [\boldsymbol{D} - \mathrm{E}(\boldsymbol{D})].$$

If $\mathrm{var}(\boldsymbol{D})$ is not invertible, then we may use an appropriate generalized inverse* in the preceding, and following, equations. The *adjusted variance matrix* for $\boldsymbol{B}$ given $\boldsymbol{D}$, denoted by $\mathrm{var}_{\boldsymbol{D}}(\boldsymbol{B})$, is evaluated as

$$\mathrm{var}_{\boldsymbol{D}}(\boldsymbol{B}) = \mathrm{var}[\boldsymbol{B} - \mathrm{E}_{\boldsymbol{D}}(\boldsymbol{B})]$$
$$= \mathrm{var}(\boldsymbol{B}) - \mathrm{cov}(\boldsymbol{B},\boldsymbol{D})[\mathrm{var}(\boldsymbol{D})]^{-1}$$
$$\times \, \mathrm{cov}(\boldsymbol{D},\boldsymbol{B}).$$

Stone [14] and Hartigan [9] were among the first to discuss the role of such assessments in Bayes analysis with partial prior specification. For examples of papers with practical

details of assessing and generalizing Bayes linear estimators in particular problems, see refs. 1, 10, 11.

We may write $\boldsymbol{B}$ as the sum of the two uncorrelated quantities $\boldsymbol{B} - \mathrm{E}_{\boldsymbol{D}}(\boldsymbol{B})$ and $\mathrm{E}_{\boldsymbol{D}}(\boldsymbol{B})$, so that $\mathrm{var}(\boldsymbol{B}) = \mathrm{var}[\boldsymbol{B} - \mathrm{E}_{\boldsymbol{D}}(\boldsymbol{B})] + \mathrm{var}[\mathrm{E}_{\boldsymbol{D}}(\boldsymbol{B})]$. We term $\mathrm{rvar}_{\boldsymbol{D}}(\boldsymbol{B}) = \mathrm{var}[\mathrm{E}_{\boldsymbol{D}}(\boldsymbol{B})]$ the *resolved variance matrix*, so that $\mathrm{var}(\boldsymbol{B}) = \mathrm{var}_{\boldsymbol{D}}(\boldsymbol{B}) + \mathrm{rvar}_{\boldsymbol{D}}(\boldsymbol{B})$, and, informally, the resolved variance matrix expresses the uncertainty about $\boldsymbol{B}$ removed by the adjustment.

INTERPRETATIONS OF BELIEF ADJUSTMENT

Belief adjustment is a general and powerful concept, which can be described in diverse ways.

1. Within the usual Bayesian view, adjusted expectation offers a simple, tractable approximation to conditional expectation which is useful in complex problems, while adjusted variance is a strict upper bound to expected posterior variance over all prior specifications consistent with the given moment structure. The approximations are exact in certain important special cases, and in particular if the joint probability distribution of $\boldsymbol{B},\boldsymbol{D}$ is multivariate normal. Therefore, there are strong formal relationships between Bayes linear calculations and the analysis of Gaussian structures, so that the linear adjustments arise in contexts as diverse as dynamic linear models and kriging*.

2. $\mathrm{E}_{\boldsymbol{D}}(\boldsymbol{B})$ may be viewed as an estimator of $\boldsymbol{B}$, combining the data with simple aspects of prior beliefs in an intuitively plausible manner, so that $\mathrm{var}_{\boldsymbol{D}}(\boldsymbol{B})$ is the expected mean squared error of $\mathrm{E}_{\boldsymbol{D}}(\boldsymbol{B})$. As a class of estimators, the Bayes linear rules have certain important admissibility* properties; *see* LINEAR ESTIMATORS, BAYES.

3. Adjusted expectation is numerically equivalent to conditional expectation in the particular case where $\boldsymbol{D}$ comprises the indicator functions for the elements of a partition, i.e., where each D_i takes value one or zero and precisely one element D_i will equal one. We may view adjusted expectation as a

natural generalization of the approach to conditional expectation based on "conditional" quadratic penalties (*see* PREVISION) where we drop the restriction that we may only condition on the indicator functions for a partition. Here, adjusted variance may be interpreted as a primitive quantity, analogous to prior variance but applied to the residual variation when we have extracted the variation in B accounted for by adjustment on D.

4. A more fundamental interpretation, which subsumes each of the above views, is based on foundational considerations concerning the implications of a partial collection of prior belief statements about B, D for the posterior assessment that we may make for the expectation of B having observed D. Any linkage between belief statements at different times requires some form of temporal coherence* condition. The *temporal sure preference condition* says, informally, that if it is logically necessary that you will prefer a certain small random penalty A to C at some given future time, then you should not now have a strict preference for penalty C over A. This condition is weak enough to be acceptable as a temporal coherence* condition for many situations, and has the consequence that your actual posterior expectation, $E_T(B)$, at time T when you have observed D, satisfies the relation

$$E_T(B) = E_D(B) + R, \qquad (1)$$

where R has, *a priori*, zero expectation and is uncorrelated with D. Therefore, adjusted expectation may be viewed as a prior inference for your actual posterior judgements, which resolves a portion of your current variance for B, and whose difference from the posterior judgement is not linearly predictable. The larger this resolved variance, the more useful is the prior analysis, and this is determined by the choice of D. In the special case where D represents a partition, $E_D(B)$ is equal to the conditional expectation given D,

and R has conditional expectation zero for each member of the partition. In this view, relationships between actual belief revisions and formal analysis based on partial prior specifications are entirely derived through stochastic relations such as (1); see the discussion in Goldstein [7].

5. The geometric interpretation of adjusted beliefs is as follows. For any collection $C = (C_1, C_2, \ldots)$ of random quantities, we denote by $\langle C \rangle$ the collection of (finite) linear combinations $\sum_i r_i C_i$ of the elements of C. Adjusted expectation is linear, that is, $E_D(X + Y) = E_D(X) + E_D(Y)$, so that defining adjusted expectation over the elements of C is equivalent to defining adjusted expectation over $\langle C \rangle$. We view $\langle C \rangle$ as a vector space. Prior covariance acts as an inner product on this space. If we choose C to be the union of the collection of elements of the vectors B and D, then the adjusted expectation of an element $Y \in \langle B \rangle$ given D is the orthogonal projection of Y into the linear subspace $\langle D \rangle$, and adjusted variance is the squared distance between Y and $\langle D \rangle$. Each of the finite-dimensional analyses described in this article has an infinite-dimensional counterpart within this construction. For example, the usual Bayes formalism is represented as follows. Corresponding to the vector space $\langle C \rangle$ is the collection of all random variables defined over the outcome space, and the inner-product space may therefore be identified with the Hilbert space of square-integrable functions over the outcome space, with respect to the joint prior measure over the outcomes, with squared norm the expected squared distance between functions, so that conditional expectation given a sample is equivalent to orthogonal projection into the subspace of all functions defined over the sample space. This Hilbert space, for many problems, is large and hard to specify. Bayes linear analysis may be viewed as restricting prior specification and subsequent projection into the largest subspace of the full space that we are prepared to specify prior

beliefs over. The geometric formulation extends the Bayes linear approach to general analyses of uncertainty over linear spaces; see, for example, Wilkinson and Goldstein [16], in which a Bayes linear approach is developed for adjusting beliefs over covariance matrices considered as elements of an appropriate inner-product space.

INTERPRETIVE AND DIAGNOSTIC MEASURES

Much of Bayes linear methodology is built around the following interpretive and diagnostic cycle: (1) we interpret the expected effects of the adjustment, *a priori*; (2) given observations, we interpret the outcome of the actual adjustment; (3) we make diagnostic comparisons between observed and expected beliefs. These comparisons may be carried out as follows.

In the first step of the cycle, the expected effects of the adjustment of vector B by D may be examined through the eigenstructure of the *resolution transform*, defined to be $T_D = [\text{var}(B)]^{-1}\text{rvar}_D(B)$. Denote the eigenvectors of T_D as $c_1, \ldots, c_r$ corresponding to eigenvalues $1 \geqslant \lambda_1 \geqslant \lambda_2 \geqslant \cdots \geqslant \lambda_r \geqslant 0$, and let $Z_i = c_i^T B$, where each c_i is scaled so that $\text{var}(Z_i) = 1$. By analogy with canonical analysis*, we call the collection $Z_1, \ldots, Z_r$ the *canonical variables* for the belief adjustment. The eigenvalues are termed the *canonical resolutions*, since for each i, one has $\lambda_i = \text{rvar}_D(Z_i)$.

The canonical variables form a grid of uncorrelated directions over $\langle B \rangle$ which summarize the effects of the adjustment in the following sense. For any $Y = a^T B$, we have $\text{rvar}_D(Y) = \text{cov}(Y, T_D Y)$, from which we may deduce that, for any $Y \in \langle B \rangle$ with prior variance 1, $\text{rvar}_D(Y) = \sum_i \lambda_i \text{cov}(Y, Z_i)$, and, in particular, for each i, $\text{rvar}_D(Z_i)$ Maximizes $\text{rvar}_D(a^T B)$ over all choices a for which $a^T B$ is uncorrelated with $Z_1, \ldots, Z_{i-1}$. Therefore, adjustment by D is mainly informative about those directions in $\langle B \rangle$ with large correlations with canonical variables having large canonical resolutions [3]. Comparisons of canonical structures therefore provide guidance, for example in choosing between alternative choices of sampling frame or experimental design.

In the second step of the cycle, we summarize the actual adjustment in beliefs as follows. Observing the vector $D = d$, gives, for any $Y \in \langle B \rangle$, the observed value $E_d(Y)$ for the adjusted expectation $E_D(Y)$. We construct the *bearing* of the adjustment from the canonical variables as $Z_d = \sum_i E_d(Z_i)Z_i$. We have $\text{cov}(Y, Z_d) = E_d(Y) - E(Y)$. Therefore

1. for any Y uncorrelated with Z_d, the prior and adjusted expectations are the same;
2. $\text{var}(Z_d)$ maximizes over $Y \in \langle B \rangle$ the value of $S_d(Y) = [E_d(Y) - E(Y)]^2 / \text{var}(Y)$, this maximum occurring for $Y = Z_d$;
3. multiplying Z_d by a constant multiplies each adjustment $E_d(Y) - E(Y)$ by that constant.

Thus, all changes in belief are in the "direction" of Z_d, and $\text{var}(Z_d)$ represents the "magnitude" of the adjustment in belief; for discussion and applications, see Goldstein [5]. In the infinite-dimensional version of the analysis, the bearing is constructed as the Riesz representation of the adjusted expectation functional on the inner-product space over $\langle B \rangle$, and in a full Bayes analysis, the bearing is usually equivalent to a normalized version of the likelihood.

For the third step, a natural diagnostic comparison between observed and expected adjustments is to compare the maximum value of $S_d(Y)$, namely $\text{var}(Z_d)$, with the prior expectation for this maximum. This prior expectation is equal to the trace of the resolution transform, T_D, for the adjustment. Thus if $\text{var}(Z_d)$ is much larger than the trace, this may suggest that we have formed new beliefs which are surprisingly discordant with our prior judgements. It is important to identify such discrepancies, but whether this causes us to reexamine the prior specification, or the adjusted assessments, or the data, or to accept the analysis, depends entirely on the context.

Each of the above interpretive and diagnostic quantities has a corresponding partial form to assess the effect on adjusted beliefs of individual aspects of a collection of pieces of evidence. In the simplest case, suppose that we split data D into portions E and

F, based on some criterion such as time or place of measurement. The eigenstructure of T_E summarizes the usefulness of E for the assessment of B. Similarly, the eigenstructure of the partial resolution matrix $T_{[F/E]} = T_D - T_E$ summarizes the additional effect on B of adding the adjustment by F to that of E, and the trace of $T_{[F/E]}$ is equal to the prior expectation for the maximum value of $[E_d(Y) - E_e(Y)]^2 / \mathrm{var}(Y)$, for $Y \in \langle B \rangle$. The partial bearing $Z_{[f/e]} = Z_d - Z_e$ summarizes the additional changes in adjusted expectation from observing $F = f$. We term $\mathrm{corr}(Z_e, Z_{[f/e]})$ the *path correlation*; this quantity is a measure of the degree of support, if positive, or conflict, if negative, between the two collections of evidence in determining the overall adjustment of beliefs.

Each feature of the preceding analysis may be usefully displayed using *Bayes linear influence diagrams*. These are constructed in a similar manner to standard influence diagrams*, but under the relation that the vectors B and C are *separated* by the vector D, provided that $\mathrm{corr}[B - E_D(B), C - E_D(C)] = 0$. Separation acts as a generalized conditional independence property [13], and diagrams defined over vector nodes based upon such a definition may be manipulated by the usual rules governing influence diagrams. Because these diagrams are defined through the covariance structure, they share many formal properties with Gaussian diagrams. They may be used firstly to build up the qualitative form of the covariance structure between the various components of the problem, and secondly to give a simple graphical representation of the interpretive and diagnostic features that we have described. For example, we may shade the outer ring of a node to express the amount of variation that is reduced by partial adjustment by each parent node, and we may shade within a node to show diagnostic warnings of differences between observed and expected outcomes; for details and computational implementation see Goldstein and Wooff [8].

SECOND-ORDER EXCHANGEABILITY

One of the principle motivations for the Bayes linear approach is to develop a methodology based strictly on the combination of meaningful prior judgements with observational data. Central to this aim is the need for an approach to statistical modeling in which models may be constructed directly from simple collections of judgments over observable quantities. We achieve this direct construction of statistical models using the representation theorem for second-order exchangeable random vectors.

We say that an infinite sequence of random vectors X_i is *second-order exchangeable* if the mean, variance, and covariance structure is invariant under permutation, namely $E(X_i) = \mu$, $\mathrm{var}(X_i) = \Sigma$, $\mathrm{cov}(X_i, X_j) = \Gamma, \forall i \neq j$. We may represent each X_i as the uncorrelated sum of an underlying "population mean" M and individual variation R_i, according to the following representation theorem: for each i, we may express X_i via $X_i = M + R_i$, where $E(M) = \mu$, $\mathrm{var}(M) = \Gamma$, $E(R_i) = 0$, $\mathrm{var}(R_i) = \Sigma - \Gamma \forall i$, and the vectors $M, R_1, R_2, \ldots$ are mutually uncorrelated [4]. This representation is similar in spirit to de Finetti's representation theorem for fully exchangeable sequences; *see* EXCHANGEABILITY. However, while second-order exchangeability is usually straightforward to specify over the observable quantities, the specification of beliefs for a fully exchangeable sequence requires such an extreme level of detail of specification that, even for the simplest of problems, it is impossible to apply the full representation theorem in practice to create statistical models from beliefs specified over observables.

Suppose that we want to adjust predictive beliefs about a future observation X_j given a sample of n previous observations $X_{[n]} = (X_1, \ldots, X_n)$. We obtain the same results if we directly adjust X_j by $X_{[n]}$ or if we adjust M by $X_{[n]}$ and derive adjusted beliefs for X_j via the representation theorem [6]. This adjustment is of a simple form in the following sense: The canonical variables for the adjustment of M by $X_{[n]}$ are the same for each sample size n, and if the canonical resolution for a particular canonical variable is λ for $n = 1$, then the corresponding canonical resolution for a sample of size n is $n\lambda / [1 + (n - 1)\lambda]$. Therefore, the qualitative features of the adjustment are the same for all sample sizes, and it is simple and natural to compare choices of possible sample sizes based on analysis of the eigenstructure of the resolution transform.

CONCLUDING COMMENTS

We have discussed general features that characterize the Bayes linear approach. For examples of its application, see Craig et al. [2] and O'Hagan et al. [12], which illustrate how the Bayes linear approach may be used to combine expert assessments with observational information for large and complex problems in which it would be extremely difficult to develop full Bayes solutions.

REFERENCES

1. Cocchi, D. and Mouchart, M. (1996). Quasi-linear Bayes estimation in stratified finite populations. *J. R. Statist. Soc. B*, **58**, 293–300.

2. Craig, P. S., Goldstein, M., Seheult, A. H., and Smith, J. A. (1996). Bayes linear strategies for history matching of hydrocarbon reservoirs. In *Bayesian Statistics 5*, J. Bernardo et al., eds. Oxford University Press, pp. 69–98.

3. Goldstein, M. (1981). Revising previsions: a geometric interpretation (with discussion). *J. R. Statist. Soc. B*, **43**, 105–130.

4. Goldstein, M. (1986). Exchangeable belief structures. *J. Amer. Statist. Ass.*, **81**, 971–976.

5. Goldstein, M. (1988). The data trajectory. In *Bayesian Statistics 3*, J. Bernardo et al., eds. Oxford University Press, pp. 189–209.

6. Goldstein, M. (1994). Revising exchangeable beliefs: subjectivist foundations for the inductive argument. In *Aspects of Uncertainty: A Tribute to D. V. Lindley*. P. Freeman and A. F. M. Smith, eds. Wiley, New York, pp. 201–222.

7. Goldstein, M. (1996). Prior inferences for posterior judgements. *Proc. 10th Int. Congress of Logic, Methodology and Philosophy of Science*. M. L. D. Chiara et al., eds. Kluwer, Amsterdam, pp. 55–71.

8. Goldstein, M. and Wooff, D. A. (1995). Bayes linear computation: concepts, implementation and programming environment. *Statist. Comput.*, **5**, 327–341.

9. Hartigan, J. A. (1969). Linear Bayes methods. *J. R. Statist. Soc. B*, **31**, 446–454.

10. Mouchart, M. and Simar, L. (1980). Least squares approximation in Bayesian analysis. In *Bayesian Statistics*, J. Bernardo et al., eds. Valencia University Press, Valencia, pp. 207–222.

11. O'Hagan, A. (1987). Bayes linear estimators of randomized response models. *J. Amer. Statist. Ass.*, **82**, 580–585.

12. O'Hagan, A., Glennie, E. B., and Beardsall, R. E. (1992). Subjective modelling and Bayes linear estimation in the UK water industry. *Appl. Statist.*, **41**, 563–577.

13. Smith, J. Q. (1990). Statistical principles on graphs. In *Influence Diagrams, Belief Nets and Decision Analysis*, R. M. Oliver and J. Q. Smith, eds. Wiley, New York, pp. 89–120.

14. Stone, M. (1963). Robustness of non-ideal decision procedures. *J. Amer. Statist. Ass.*, **58**, 480–486.

15. Whittle, P. (1992). *Probability via Expectation*. Springer-Verlag, New York.

16. Wilkinson, D. J. and Goldstein, M. (1996). Bayes linear adjustment for variance matrices. In *Bayesian Statistics 5*, J. Bernardo et al., eds. Oxford University Press, pp. 791–800.

See also BAYESIAN INFERENCE; CANONICAL ANALYSIS; COHERENCE—I; EXCHANGEABILITY; LINEAR ESTIMATORS, BAYES; PREVISION; and REPRODUCING KERNEL HILBERT SPACES.

MICHAEL GOLDSTEIN

BAYES *p*-VALUES

When we have a simple null hypothesis*$H_0 : \theta = \theta_0$, a test statistic T will order the possible observed data values from least extreme to most extreme (relative to H_0). The p-value* associated with T is then P_{θ_0} (we would get data at least as extreme as the observed data).

In some testing stituations, there is only one sensible way to rank the possible data values. For example, if the sample mean $\overline{X} \sim$ Normal$(\theta, 1/n)$ and we are testing $H_0 : \theta = 0$ vs. $H_1 : \theta > 0$, the data become more extreme (relative to H_0) as $\overline{X}$ increases. The only reasonable p-value is $p = P_{\theta=0} (\overline{X} \geqslant$ observed $\bar{x})$.

In other testing situations, there will be many reasonable ways of ranking the possible data values and generating a p-value. Some well-known techniques include rankings involving a likelihood ratio test* statistic, a most powerful test statistic, or a locally most powerful test statistic [1]. In some two-sided tests, a common method for generating a p-value is to double a corresponding one-tailed p-value [2, p. 365]. This method is somewhat controversial (*see* P-VALUES).

Another method of generating a ranking that has many desirable properties [3] is to rank according to a test statistic of the form

$$T = \frac{\sum_{i=1}^{k} \pi(\theta_i) f_{\theta_i}(x_1, x_2, \dots, x_n)}{f_{\theta_0}(x_1, x_2, \dots, x_n)}$$

or

$$T = \frac{\int_{\theta} \pi(\theta) f_{\theta}(x_1, x_2, \dots, x_n) d\theta}{f_{\theta_0}(x_1, x_2, \dots, x_n)}.$$

Here f_{θ} is the joint pdf of the data values $x_1, x_2, \dots, x_n$, and π is a probability on the θ's corresponding to H_1. The p-value generated is called a *Bayes p-value* with respect to the prior π.

Our interpretation of π here is as a weighting function. When H_0 is false, we would like to get a p-value as small as possible. For each θ corresponding to H_1 and each ranking of the possible observed data values, an expected p-value can be found. The Bayes p-value approach generates a p-value that minimizes the weighted average of these expected p-values (using weighting function π).

Example. Suppose we have a two-sided test on the variance of a normal distribution, $\sigma^2, H_0 : \sigma^2 = 4$ vs. $H_1 : \sigma^2 \neq 4$. We have a sample of size 9, and will base our analysis on the sample variance $Y = \sum (X_i - \overline{X})^2 / 8$. Suppose further that we decide that a procedure which is optimal, on average, when $\sigma^2 = 2$ or 6 would suit our needs. We take $\pi(\sigma^2 = 2) = 0.5 = \pi(\sigma^2 = 6)$ and get

$$
\begin{aligned}
T(y) &= \frac{0.5 f_{\sigma^2=2}(y) + 0.5 f_{\sigma^2=6}(y)}{f_{\sigma^2=4}(y)} \\
&= \frac{(0.5)\frac{8}{3} y^3 e^{-2y} + (0.5)\frac{8}{243} y^3 e^{-2y/3}}{\frac{1}{6} y^3 e^{-y}} \\
&= 8 e^{-y} + \frac{8}{81} e^{y/3}.
\end{aligned}
$$

Now suppose we get data and observe $Y = 1$. What is the p-value? $T(1) = 3.08 = T(10.32)$ also. (Thus, T views 10.32 as just as

extreme as 1.) Since T is concave up in y, the data values at least as extreme as $Y = 1$ are $Y \leqslant 1$ and $Y \geqslant 10.32$. The p-value is

$$
\begin{aligned}
p &= P_{\sigma^2=4}(Y \leqslant 1 \text{ or } Y \geqslant 10.32) \\
&= 0.0190 + 0.0082 \\
&= 0.0272.
\end{aligned}
$$

It is also possible to interpret π as representing the researcher's probability views on θ (given H_1 is true) prior to collecting data. However, in this setup a Bayesian analysis is typically performed rather than a p-value calculation [1].

REFERENCES

1. Casella, G. and Berger, R. L. (1990). *Statistical Inference*. Wadsworth, Pacific Grove, Calif.
2. Hogg, R. V. and Tanis, E. A. (1997). *Probability and Statistical Inference*, 5th ed. Prentice-Hall, Englewood Cliffs, NJ.
3. Thompson, P. (1997). Bayes *P*-values. *Statist. Probab. Lett.*, **31**, 267–274.

See also BAYESIAN INFERENCE; HYPOTHESIS TESTING; P-VALUES; and SIGNIFICANCE TESTS, HISTORY AND LOGIC OF.

PETER THOMPSON

BAYES' RISK. See DECISION THEORY

BAYES' THEOREM

This theorem is obtained from the relationships

$$\Pr[E|H_1] \Pr[H_1] = \Pr[E \cap H_1]$$
$$= \Pr[H_1|E] \Pr[E]$$

and (if $\sum_{i=1}^{k} \Pr[H_i] = 1$)

$$\Pr[E] = \sum_{i=1}^{k} \Pr[H_i] \Pr[E|H_i].$$

It is usually stated in the form

$$\Pr[H_j|E] = \frac{\Pr[H_j] \Pr[E|H_j]}{\sum_{i=1}^{k} \Pr[H_i] \Pr[E|H_i]}.$$

In applications, $H_1, \ldots, H_k$ usually represent possible "states of nature" ("hypotheses") and E an observed event. The formula then provides a value for the probability of H_j given that E has been observed—the posterior probability*. Calculation of this value requires not only a knowledge of the probabilities $\Pr[E|H_i]$, but also of the prior (a priori) probabilities* $\Pr[H_i]$.

Another form of Bayes's theorem relates the posterior distribution of a parameter θ, given observed values $\mathbf{x}$ to its prior distribution*. When $\mathbf{x}$ has a PDF $f(\mathbf{x}|\theta)$, and $p(\theta)$ is the prior PDF of θ, the posterior PDF of θ is

$$p(\theta|\mathbf{x}) = \frac{p(\theta)f(\mathbf{x}|\theta)}{\int \cdots \int p(\theta)f(\mathbf{x}|\theta)\,d\theta}$$

(θ and/or $\mathbf{x}$ may be vectors; the integral is over the whole range of variation of θ).

Considerable controversy has existed, and still continues, on the possibility of using Bayes' theorem when—as is often the case—the values to be used for the prior probabilities are not clearly established.

More detailed discussion of these matters will be found in the entries listed below.

See also BAYES FACTORS; BAYESIAN INFERENCE; DECISION THEORY; FIDUCIAL DISTRIBUTIONS; and FIDUCIAL PROBABILITY.

BAYES, THOMAS

Born: 1701, in London, England.
Died: April 7, 1761, in Tunbridge Wells, England.
Contributed to: statistical inference, probability.

Relatively little is known about Thomas Bayes, for whom the Bayesian school of inference and decision is named. He was a Nonconformist minister, a Fellow of the Royal Society, and according to the certificate proposing him for election to that society, was "well skilled in Geometry and all parts of Mathematical and Philosophical Learning" (see ref. 6, p. 357).

The Nonconformist faith played a major role in the scientific ideas of the eighteenth century, a time when religion and philosophy

of science were inextricably linked. The growth of Nonconformism was influenced by the rise of Natural Philosophy, which encouraged a scientific examination of the works of the Deity so that men could come to understand His character. The Nonconformists included Deists and Arians (forerunners of modern Unitarians) among them, all sharing a rejection of the Trinity and a skepticism about the divinity of Christ. The Royal Society is known to have had a strong antitrinitarian component [6, p. 356].

Thomas was born to Joshua and Ann Bayes in 1702 in London. Joshua was the first Nonconformist minister to be publicly ordained (1694) after the Act of Uniformity was passed. He became minister of the Presbyterian meeting house at Leather Lane in 1723; Thomas served as his assistant.

Thomas was educated privately, as was the custom with Nonconformists at that time. After serving as an assistant to his father, he went to Tunbridge Wells as minister of a dissenting Presbyterian congregation. It is not known precisely when he began that post, but he was there by 1731 when his religious tract, *Divine Benevolence, or an attempt to prove that the Principal End of the Divine Providence and Government is the Happiness of his Creatures*, was printed by John Noon at the White Hart in Cheapside, London [1].

Bayes was elected a Fellow of the Royal Society in 1742. The signers of his certificate for election included officers of the Royal Society, John Eames, Mathematician and personal friend of Newton, and Dr. Isaac Watts, Nonconformist minister and wellknown composer of hymns. Unfortunately, there is little indication of published work by Bayes meriting his election to the Royal Society. In fact, there is no concrete evidence of any published paper, mathematical or theological by Bayes from his election in 1742 until his death in 1761 [6, p. 358]. There is however, an anonymous 1736 tract, published by John Noon, which has been ascribed to Bayes, entitled *An Introduction to the Doctrine of Fluxions, and a Defence of the Mathematicians against the objections of the Author of the Analyst, in so far as they are designed to affect the general method of reasoning*. This work was a defense of the logical foundations of Newtonian calculus against attacks made by Bishop Berkeley, the noted philosopher. Augustus DeMorgan, in an 1860 published query about Bayes, noted the practice of Bayes' contemporaries of writing in the author's name on an anonymous tract, and DeMorgan (and the British Museum) accepted the ascription.

DeMorgan gives some sense of Bayes' early importance by writing, "In the last century there were three Unitarian divines, each of whom has established himself firmly among the foremost promoters of a branch of science. Of Dr. Price and Dr. Priestley ... there is no occasion to speak: their results are well known, and their biographies are sufficiently accessible. The third is Thomas Bayes ..." [5]. De Morgan recognized Bayes as one of the major leaders in the development of the mathematical theory of probability, claiming that he was "of the calibre of DeMoivre and Laplace in his power over the subject," and noting in addition that Laplace* was highly indebted to Bayes, although Laplace made only slight mention of him [5].

Bayes is also known to have contributed a paper in 1761, published in the *Philosophical Transactions* in 1763 (see ref. 2), on semiconvergent asymptotic series. Although a relatively minor piece, it may have been the first published work to deal with semiconvergence [6, p. 358].

Bayes was succeeded in the ministry at Tunbridge Wells in 1752, but continued to live there until his death on April 17, 1761. He was buried in the Bayes family vault in Bunhill Fields, a Nonconformist burial ground. The bulk of his considerable estate was divided among his family, but small bequests were made to friends, including a $200 bequest to Richard Price.

Bayes' greatest legacy, however, was intellectual. His famous work, "An Essay towards Solving a Problem in the Doctrine of Chance," was communicated by Richard Price to the Royal Society in a letter dated November 10, 1763, more than two years after Bayes' death, and it was read at the December 23 meeting of the society. Although Bayes' original introduction was apparently lost, Price claims in his own introduction (see ref. 3) that Bayes' purpose was "to find out a method by which we might judge concerning the probability that an event has to happen in given circumstances, upon supposition that we know

nothing concerning it but that under the same circumstances it has happened a certain number of times and failed a certain other number of times." Price wrote that the solution of this problem was necessary to lay a foundation for reasoning about the past, forecasting the future, and understanding the importance of inductive reasoning. Price also believed that the existence of the Deity could be deduced from the statistical regularity governing the recurrency of events.

Bayes' work begins with a brief discussion of the elementary laws of chance. It is the second portion of the paper in which Bayes actually addresses the problem described by Price, which may be restated as: Find $\Pr[a < \theta < b|X = x]$, where $X \sim$ Binomial (n, θ). His solution relies on clever geometric arguments (using quadrature* of curves instead of the more modern integral calculus), based on a physical analog of the binomial model—relative placements of balls thrown upon a flat and levelled table. Bayes' findings can be summarized as follows [7]:

$$\Pr[a < \theta < b, X = x]$$
$$= \int_a^b \binom{n}{x} \theta^x (1 - \theta)^{n-x} \, d\theta,$$

$$\Pr[X = x] = \int_0^1 \binom{n}{x} \theta^x (1 - \theta)^{n-x} \, d\theta,$$

and finally,

$$\Pr[a < \theta < b|X = x]$$
$$= \frac{\int_a^b \binom{n}{x} \theta^x (1 - \theta)^{n-x} \, d\theta}{\int_0^1 \binom{n}{x} \theta^x (1 - \theta)^{n-x} \, d\theta}.$$

The uniform prior distribution* for the unknown θ followed naturally by construction in Bayes' analogy. According to Price, Bayes originally obtained his results by assuming that θ was uniformly distributed, but later decided that the assumption may not always be tenable and resorted to the example of the table. Bayes added a *Scholium* to his paper in which he argued for an a priori uniform distribution for unknown probabilities.

The use of a uniform prior distribution to represent relative lack of information (sometimes called the *Principle of Insufficient Reason*) has long been controversial, partly because such a distribution is not invariant under monotone transformation of the unknown parameter. Many statisticians and philosophers have interpreted the *Scholium* in terms of the principle of insufficient reason, but more recently, Stigler [7] has argued that Bayes was describing a uniform predictive distribution for the unknown X, and *not* a prior distribution for the parameter θ. Such an interpretation seems plausible, particularly due to its emphasis on probabilities for ultimately observable quantities and due to Bayes' view of probability as an expectation (see also ref. 4 for a view of probability as expectation). The assumption of a uniform predictive distribution is more restrictive than that of a uniform prior distribution; it does not follow, in general, that the implied distribution of θ would be uniform. One advantage of the uniform predictive distribution is that the problem of invariance no longer arises: If $\Pr[X = x]$ is constant, then so is $\Pr[f(X) = f(x)]$, for monotone functions f.

Thomas Bayes did not extend his results beyond the binomial model, but his views of probability and inductive inference have been widely adopted and applied to a variety of problems in statistical inference and decision theory.

REFERENCES

1. Barnard, G. A. (1958). *Biometrika*, **45**, 293–295. (A biographical note that introduces a "reprinting" of Bayes' famous paper.)

2. Bayes, T. (1763). *Philos. Trans. R. Soc.*, London, **53**, 269–271.

3. Bayes, T. (1763). *Philos. Trans. R. Soc.*, London, **53**, 370–418. (Also reprinted in *Biometrika*, **45**, 296–315. The reprinted version has been edited and modern notation has been introduced.)

4. De Finetti, B. (1974–1975). *Theory of Probability*, Vols. 1 and 2. Wiley, New York.

5. DeMorgan, A. (1860). *Notes and Queries*, January 7, 9–10.

6. Pearson, K. (1978). *The History of Statistics in the 17th and 18th Centuries*, E. S. Pearson, ed. Macmillan, New York, pp. 355–370.

7. Stigler, S. M. (1982). *J. R. Statist. Soc. A*, **145**, 250–258. (Argues that uniform distribution is

intended for the predictive or marginal distribution of the observable events; discusses reasons behind common "misinterpretations" of Bayes' *Scholium*.)

BIBLIOGRAPHY

Holland, J. D. (1962). *J. R. Statist. Soc. A*, **125**, 451–461. (Contains many references to primary sources concerning Bayes' life.)

Laplace, P. S. (1951). *A Philosophical Essay on Probabilities*. Dover, New York. (Laplace's indebtedness to Bayes is apparent in his General Principles.)

Stigler, S. M. (1983). *Amer. Statist.*, **37**, 290–296. (Evidence is presented that Nicholas Saunderson, an eighteenth century mathematician, may have first discovered the result attributed to Bayes.)

Todhunter, I. (1865). *A History of the Mathematical Theory of Probability*. Chelsea, New York (reprint 1965).

See also BAYESIAN INFERENCE; LAPLACE, PIERRE SIMON; LOGIC OF STATISTICAL REASONING; and PRIOR DISTRIBUTIONS.

R. L. TRADER

BAYESIAN CATCHALL FACTOR

This term was presumably introduced by L. J. Savage* in connection with Bayes' theorem*. In assessing the posterior probability* of an event (or hypothesis) H, given evidence e, we have

$$P(H|e) = \frac{P(e|H)P(H)}{P(e|H)P(H) + p(e|\text{not } H)P(\text{not } H)}.$$

In assessing $P(H|e)$, the probability $P(e|\text{not } H)$ is referred to as the *Bayesian catchall factor* (with evidence e) and, in the opinion of many, has caused substantial difficulties in the logical interpretation of Bayes theorem.

The concept has been criticized by Salmon [2] and examined by Mayo [1].

REFERENCES

1. Mayo, D. G. (1996). *Error and the Growth of Experimental Knowledge*. University of Chicago Press, Chicago, London.

2. Salmon, W. C. (1991). The Appraisal of Theories: Kuhn Meets Bayes. In *PSA (Philosophy of Science Association) 1990*, Vol. 2, A. Fine, M. Forbes, and L. Wessels, eds. East Lansing, Mich., pp. 325–332.

BAYESIAN EXPERIMENTAL DESIGN

Consider a one-factor experiment with a single continuous response variable Y, where the usual assumptions of the analysis of variance hold. This can be represented by a mathematical model $Y \sim N(X\tau, \sigma^2 I)$, where X is the design matrix, τ is the vector of means of the response variable at the different levels of the factor, and I is the identity matrix. Given a sample size of n homogeneous experimental units, the usual experimental design allocates equal number of units to each level of the factor, commonly referred to as *treatments*. Generally, the units are randomly allocated. Such a design is optimal according to most criteria based on the covariance matrix of the estimates of the treatment effects, that is, the matrix $\sigma^2(X'X)^{-1}$. This includes the *A-optimality* criterion, which minimizes the trace of the covariance matrix, the *D-optimality* criterion, which minimizes the determinant of this matrix, and also a criterion based on the power of the usual hypothesis test of no treatment effect. An unequal allocation can be optimal if the experimenter's interest is in the estimation of a function of the treatment effects which is not invariant to permutation of the treatment labels. (*see also* OPTIMAL DESIGN OF EXPERIMENTS).

In many experimental situations, it is quite likely that the researcher has prior knowledge of some aspect of the planned experiment. For example, it is quite possible that one or more of the treatments have been previously studied. This is especially true for experiments where one of the treatments is a control. In these cases, it is sensible to use such prior information in the planning of the experiment. This process is known as Bayesian experimental design.

The first task is generally to convert the prior information into a probability distribution of the parameters of the mathematical model that is being used. Known as prior elicitation*, this is a crucial step in the design process. There is quite extensive literature on prior elicitation in the normal linear model; some examples are; DuMouchel [16], Garthwaite and Dickey [20], and Kadane et al. [23]. In the simplest case of one-way analysis of variance*, the prior information is usually converted into a normal distribution of the treatment effects, that is, $\tau \sim N(\theta, \sigma^2 R)$, where θ is the prior mean of τ, R is the prior covariance matrix, and σ^2 is considered known or has a prior distribution that is independent of the design and of τ. The most common Bayesian experimental design is the optimal allocation* according to criteria based on the posterior covariance matrix of the treatment effects, that is, on functions of $\sigma^2(R^{-1} + X'X)^{-1}$.

Bayesian experimental design is a relatively new statistical discipline, originating with the work of Raiffa and Schlaifer [30] and Lindley [25, pp. 19–20]. Their approach was based on decision theory*, that is, the design was chosen to maximize the expected utility* of the experiment. Since then, the field has advanced in many directions. The most extensively studied is the area of design for linear models. Some of the early contributions are those in refs. 1, 21, 28, 29. Experimental design for nonlinear models has also received considerable attention. The earliest work is due to Tsutakawa [39] and Zacks [42]. Even though most of the work is theoretical, Bayesian methods have recently been applied to design problems in clinical trials*. The following sections give a brief review of this interesting and active area of statistics; see also Chaloner and Verdinelli [5] and DasGupta [9].

BAYESIAN DESIGN FOR THE NORMAL LINEAR MODEL

In the normal linear model the response variable Y, conditional on the mean τ and covariance C, has the normal distribution $N(X\tau, C)$. In most of the Bayesian design literature it is further assumed that $C = \sigma^2 I$, where σ^2 is known, and that the prior information on τ can be converted into a normal distribution with a given mean θ and a given covariance matrix $\sigma^2 R$. Applying Bayes' theorem*, the posterior distribution* of τ is normal with mean $\mu = \overline{Y} - N^{-1}(N^{-1} + R)^{-1}(\overline{Y} - \theta)$, where $N = X'X$, and covariance matrix $V = \sigma^2(R^{-1} + N)^{-1}$. If n is the total sample size and n_i observations are taken at x_i, then N can be written as $n \sum (n_i/n) x_i x_i' = nM$.

Most of the early work on Bayesian optimal design was in the tradition of optimum design of experiments* as pioneered by Kiefer [24]. That is, the designs were selected according to Bayesian analogues of criteria such as A-, D-, and ψ-optimality, and further, the designs were approximate in the sense that they were obtained by rounding off the optimal noninteger allocations n_i.

Even though the roots of the alphabetic criteria lie in classical optimal design, the criteria can usually be given a Bayesian or decision-theoretical interpretation. For example, Bayesian ψ-optimal designs, which minimize trace$(\psi V) =$ trace$[\psi \sigma^2(R^{-1} + nM)^{-1}]$, arise naturally as designs minimizing the expected posterior loss of $L(\hat{\tau}) = (\tau - \hat{\tau})' \psi (\tau - \hat{\tau})$.

In cases when interest is in the estimation of a contrast $c'\tau$, the matrix $\psi = cc'$. There are many applications of this criterion, for example, numerous theoretical results in the context of regression [2,18]. In factorial models, the criterion was used to find the optimal allocation of several new treatments and a control treatment under one-way blocking [28], and two-way blocking [38]. Another application was to find optimal blocking and fractionation schemes for 2^k and 3^k factorial experiments* [35].

Bayesian D-optimal designs arise naturally as designs which minimize the expected loss in Shannon information [31], that is,

$$- \int \log \frac{p(\tau|y)}{p(\tau)} p(y, \tau) \, dy \, d\tau.$$

After some simplification the criterion becomes $-\det V = -\det[\sigma^2(R^{-1} + nM)^{-1}]$. Bayesian D-optimality was first discussed in the context of the normal hierarchical model [33]. It was later applied to the treatment-control problem [21,22]. DuMouchel and Jones [17] used it for fractionation of the 2^k design.

A notable characteristic of Bayesian ψ- and D-optimal designs is their dependence on the sample size n. For large sample sizes the

Bayes designs are very similar to their classical counterparts. The mathematical tools used to obtain them are also very similar to those used in classical optimal design. Most often they are in the form of an equivalence theorem such as those described in Silvey [32]. Another similarity is that an upper bound on the number of support points can be obtained [2].

Recently, the focus of Bayesian experimental design for linear models has shifted back to decision-theoretic origins. The approach is to select an appropriate loss function for the experimental situation and then choose the design which minimizes its expected value averaged over both the posterior distribution of the parameters and the marginal distribution of the data. One example is Verdinelli [40], who suggested a loss function with two components, one appropriate for inference on the parameters, one for prediction of a future observation. Verdinelli and Kadane [41] used a loss function that guarantees precise inference as well as a large value of the output. Toman [36] used a loss function appropriate for testing multiple hypotheses.

While the conjugate normal prior distribution for the means of the linear model has generally been used, there are some exceptions. The hierarchical prior distribution* has been used [33,22]. Other studies [11,34,37] used classes of prior distributions and found designs which performed well for the entire class. This type of optimal design is called a *robust Bayes design*. Other nonconjugate prior distributions have not been widely used, mostly because the required integrals do not have a closed-form expression.

Until very recently, the integration of the loss function over the sample space was carried out as a matter of course. This non-Bayesian formulation was considered justifiable because in a design problem the data are not yet available. Recently some authors [12,10] proposed using a conditional Bayes formulation in the selection of the optimal sample size in the normal linear model. Their optimal sample size is a solution to the minimax problem min max trace $V(y|n)$, where the minimum is over n and the maximum is over a set of samples y with a predictive probability $1 - \varepsilon$.

BAYESIAN DESIGN FOR NONLINEAR MODELS

When the response variable is a nonlinear function of the parameters, or when one is interested in nonlinear functions of the parameters of the linear model, Bayesian optimal experimental design is more difficult. The main problem is that functions of the posterior distribution used in design, such as the posterior variance, are usually very complicated integrals. The most common approach has been to use a normal approximation to the posterior distribution of the parameters τ, that is, $\tau|y \sim N(\hat{\tau}, [nI(\hat{\tau})]^{-1})$, where $I(\tau)$ is the expected Fisher information* matrix and $\hat{\tau}$ is the maximum likelihood estimate of τ. Optimal designs are found by minimizing criteria such as $\int \text{tr}\{\psi[nI(\hat{\tau})]^{-1}\}p(\tau)\,d\tau$, the ψ-optimality criterion, where the integration over the prior distribution $p(\tau)$ is a further approximation. Another common criterion is Bayes D-optimality, that is, $\int \log \det[nI(\hat{\tau})]\,p(\tau)\,d\tau$.

Variations on these two criteria were used in some of the earliest work in this area [39,42]. Tsutakawa [39] used a version of ψ-optimality to find optimal designs for one-parameter logistic regression with a known slope coefficient and unknown parameter LD50. Zacks [42] used a variation on D-optimality in quantal response problems. Another interesting application of these criteria was Lohr [26]. She found optimal designs for the estimation of the variance component in the one-way random-effects model.

The mathematical properties of these two criteria were studied in some detail by Chaloner and Larntz [4]. One of their major contributions is an equivalence theorem that can be used to verify that a candidate design is optimal. They also pointed out that a bound on the number of support points is not generally available in nonlinear models, but that the number of support points increases as the prior distribution becomes more dispersed. Others have recently found, for specific models, characterizations of situations when an optimal design with a given number of support points exists [3,13,14,15].

It is clear that the optimality of the designs obtained using criteria based on the Fisher information matrix depends on the degree to which approximations to normality are appropriate. This point was addressed by Clyde [6], who used constraints on the optimization to ensure that normality was obtained.

More often than for linear models, Bayesian optimal designs for nonlinear models have to be found numerically. A software system has been developed [7] that can be used to obtain optimal designs for a multitude of problems. Another example of numerical optimization [27] used Markov chain Monte Carlo* methods to optimize the exact expected posterior loss, thus avoiding the need for approximations of the posterior distribution.

APPLICATIONS

Although Bayesian methods are naturally suitable for experimental design, since prior information is often quite plentiful, most practitioners have yet be convinced of their advantage. Partly this is due to the conceptual difficulty of prior elicitation, partly due to the fact that computer software is necessary and not widely available. Thus unfortunately much of the work in this area remains purely theoretical. The most promising area of potential application appears to be in clinical trials. Here, Bayesian methods have been used in at least two real experiments [19,8].

REFERENCES

1. Chaloner, K. (1982). Optimal Bayesian experimental design for linear models, Department of Statistics Technical Report 238, Carnegie-Mellon University, Pittsburgh.

2. Chaloner, K. (1984). Optimal Bayesian experimental designs for linear models. *Ann. Statist.*, **12**, 283–300.

3. Chaloner, K. (1993). A note on optimal Bayesian design for nonlinear problems. *J. Statist. Plann. Inference*, **37**, 229–235.

4. Chaloner, K. and Larntz, K. (1989). Optimal Bayesian designs applied to logistic regression experiments. *J. Statist. Plann. Inference*, **21**, 191–208.

5. Chaloner, K. and Verdinelli, I. (1994). Bayesian experimental design: a review. *Statist. Sci.*, **10**, 273–304.

6. Clyde, M. A. (1993). Bayesian optimal designs for approximate optimality. Ph.D. thesis, University of Minnesota, Minneapolis.

7. Clyde, M. A. (1993). An object-oriented system for Bayesian nonlinear design using Xlisp-Stat. *Technical Report 587*, School of Statistics, University of Minnesota, Minneapolis.

8. Clyde, M. A., Müller, P., and Parmigiani, G. (1996). Optimal design for heart defibrillators, In *Case Studies in Bayesian Statistics II*, C. Gatsonis et al., eds. Springer-Verlag, New York, pp. 278–292.

9. DasGupta, A. (1995). *Review of optimal Bayes designs*, Technical Report 95-4, Department of Statistics, Purdue University, West Lafayette, IN.

10. DasGupta, A. and Mukhopadhyay, S. (1994). Uniform and subuniform posterior robustness: the sample size problem. *J. Statist. Plann. Inference*, **40**, 189–204.

11. DasGupta, A. and Studden, W. J. (1991). Robust Bayesian experimental designs in normal linear models. *Ann. Statist.*, **19**, 1244–1256.

12. DasGupta, A. and Vidakovic, B. (1994). Sample sizes in ANOVA: the Bayesian point of view, Department of Statistics, Technical Report, Purdue University, West Lafayette, IN.

13. Dette, H. and Neugebauer, H. M. (1996). Bayesian *D*-optimal designs for exponential regression models. *J. Statist. Plann. Inference*, **60**, 331–345.

14. Dette, H. and Sperlich, S. (1994). Some applications of continued fractions in the construction of optimal designs for nonlinear regression models. *J. Comput. Statist. & Data Anal.*, **21**, 273–292.

15. Dette, H. and Sperlich, S. (1994). A note on Bayesian *D*-optimal designs for general exponential growth models. *S. Afr. Statist. J.*, **28**, 103–117.

16. DuMouchel, W. H. (1988). A Bayesian Model and a graphical elicitation procedure for multiple comparisons. *Bayesian Statist.*, **3**, 127–146.

17. DuMouchel, W. H. and Jones, B. (1994). A Simple Bayesian modification of *D*-optimal designs to reduce dependence on an assumed model. *Technometrics*, **36**, 37–47.

18. El-Kruntz, S. and Studden, W. J. (1991). Bayesian optimal designs for linear regression models. *Ann. Statist.*, **19**, 2183–2208.

19. Flournoy, N. (1993). A clinical experiment in bone marrow transplantation: estimating a percentage point of a quantal response curve. In *Case Studies in Bayesian Statistics*, C. Gatsonis et al., eds., Springer-Verlag, pp. 234–336.

20. Garthwaite, P. H. and Dickey, J. M. (1988). Quantifying expert opinion in linear regression problems. *J. R. Statist. Soc. B*, **50**, 462–474.

21. Giovagnoli, A. and Verdinelli, I. (1983). Bayes *D*-optimal and *E*-optimal block designs. *Biometrika*, **70**, 695–706.

22. Giovagnoli, A. and Verdinelli, I. (1985). Optimal block designs under a hierarchical linear model. *Bayesian Statist.*, **2**, 655–662.

23. Kadane, J. B., Dickey, J. M., Winkler, R. L., Smith, W. S., and Peters, S. C. (1980). Interactive elicitation of opinion for a normal linear model. *J. Amer. Statist. Ass.*, **75**, 845–854.

24. Kiefer, J. (1959). Optimum experimental designs. *J. R. Statist. Soc.*, **21**, 272–319.

25. Lindley, D. V. (1972). *Bayesian Statistics—A Review*. SIAM, Philadelphia, pp. 19–20.

26. Lohr, S. L. (1995). Optimal Bayesian design of experiments for the one-way random effects model. *Biometrika*, **82**, 175–186.

27. Müller, P. and Parmigiani, G. (1995). Optimal design via curve fitting of Monte Carlo experiments. *J. Amer. Statist. Ass.*, **90**, 1322–1330.

28. Owen, R. J. (1970). The optimum design of a two-factor experiment using prior information. *Ann. Math. Statist.*, **41**, 1917–1934.

29. Pilz, J. (1983). *Bayesian Estimation and Experimental Design in Linear Regression*, Teubner-Texte zur Mathematik, Teubner-Verlag, Leipzig.

30. Raiffa, H. and Schlaifer, R. (1961). *Applied Statistical Decision Theory*. Division of Research, Harvard Business School, Boston.

31. Shannon, C. E. (1948). A mathematical theory of communication. *Bell System Tech. J.*, **27**, 379–423–623–656.

32. Silvey, S. D. (1980) *Optimal Design*. Chapman and Hall, London.

33. Smith, A. F. M. and Verdinelli, I. (1980). A note on Bayes designs for inference using a hierarchical linear model. *Biometrika*, **67**, 613–619.

34. Toman, B. (1992). Bayesian robust experimental designs for the one-way analysis of variance. *Statist. Probab. Lett.*, **15**, 395–400.

35. Toman, B. (1994). Bayes optimal designs for two-and three-level factorial experiments. *J. Amer. Statist. Ass.*, **89**, 937–946.

36. Toman, B. (1995). Bayesian experimental design for multiple hypothesis testing. *J. Amer. Statist. Ass.*, **91**, 185–190.

37. Toman, B. and Gastwirth, J. (1993). Robust Bayesian experimental design and estimation for analysis of variance models using a class of

normal mixtures. *J. Statist. Plann. Inference*, **35**, 383–398.

38. Toman, B. and Notz, W. (1991). Bayesian optimal experimental design for treatment-control comparisons in the presence of two-way heterogeneity. *J. Statist. Plann. Inference*, **27**, 51–63.

39. Tsutakawa, R. K. (1972). Design of an experiment for bioassay. *J. Amer. Statist. Ass.*, **67**, 584–590.

40. Verdinelli, I. (1992). Advances in Bayesian experimental design. *Bayesian Statist.*, **4**, 467–481.

41. Verdinelli, I. and Kadane, J. B. (1992). Bayesian designs for maximizing information and outcome. *J. Amer. Statist. Ass.*, **87**, 510–515.

42. Zacks, S. (1977). Problems and approaches in design of experiments for estimation and testing in non-linear models. In *Multivariate Analysis IV*, P. R. Krishnaiah, ed. North-Holland, pp. 209–223.

BIBLIOGRAPHY

Palmer, J. and Muller, P. (1998). Bayesian optimal design in population models of hematologic data. *Statist. in Medicine*, **17**, 1613–1622.

See also BAYESIAN INFERENCE; DESIGN OF EXPERIMENTS; ELICITATION; and OPTIMAL DESIGN OF EXPERIMENTS.

BLAZA TOMAN

BAYESIAN FORECASTING

Bayesian forecasting encompasses statistical theory and methods in time series* analysis and time series forecasting*, particularly approaches using dynamic and state-space models, though the underlying concepts and theoretical foundation relate to probability modeling and inference more generally. This entry focuses specifically on time series and dynamic modeling, with mention of related areas.

BACKGROUND

Bayesian forecasting and dynamic modeling has a history that can be traced back to the late 1950s in short-term forecasting and time series monitoring in commercial environments [19,20], and many of the developments since then have retained firm links with roots in applied modeling and forecasting problems in industrial and socioeconomic areas [21,22,23,24,46,71]. Almost in parallel, technically similar developments arose in the systems and control engineering areas [29,30], with a central focus on adaptive estimation and filtering theory for automatic control. Central to the field are state-space models for time-varying systems, and related statistical methodology. These methods are most naturally viewed from a Bayesian inference* perspective, and indeed, the developments in both forecasting and control environments can be seen as one of the important practical successes of Bayesian methods.

In the time series analysis and forecasting area, Bayesian methods based on dynamic linear models were initiated largely as outgrowths of simple and widely used smoothing and extrapolation* techniques, especially exponential smoothing* and exponentially weighted moving average* methods [20,71]. Developments of smoothing and discounting techniques in stock control and production planning areas led to formulations in terms of linear state-space models for time series with time-varying trends and seasonality*, and eventually to the associated Bayesian formalism of methods of inference and prediction. From the early 1960s, practical Bayesian forecasting systems in this context involved the combination of formal time series models and historical data analysis together with methods for subjective intervention and forecast monitoring, so that complete forecasting systems, rather than just routine and automatic data analysis and extrapolation, were in use at that time [19,22]. Methods developed in those early days are still in use in some companies, in sales forecasting and stock control. There have been major developments in models and methods since then, especially promulgated in the seminal article by Harrison and Stevens [23], though the conceptual basis remains essentially unchanged. The more recent history of the field is highlighted in ref. 69, in various articles in ref. 61, in books [46,71], and in commentary below.

Parallel developments in control engineering highlighted the use of sequential statistical learning and optimization in linear state-space models. From the early foundational work of Wiener, this field developed significantly in the 1960s, and was hallmarked by the key contributions of Kalman [29,30]. The approach to statistical inference based on linear least squares* was central, and the associated sequences of updating equations for state-parameter estimation are now widely referred to as the *Kalmanfilter*equations*. Variants of Kalman filtering techniques now permeate various engineering areas, and the relationships with Bayesian forecasting in dynamic linear models are obviously intricate [2,3,39,48,61,71; §1.4]. The same may be said of the related developments of structural modeling* in econometric time series analysis [26]. In recent times, Bayesian concepts and methods have become more apparent in these related fields.

During the last two decades, the range of developments and applications of Bayesian forecasting and dynamic models has grown enormously. Much of the development involves theoretical and methodological extensions mentioned below. It is simply impossible to do justice to the field here with a comprehensive review of application areas and contexts, even ignoring the vast engineering control arena. To give some of the flavor of the breadth and scope of the field, the results of a scan of the *Current Index to Statistics* database in early 1995 reveal application areas running through the traditional fields of commercial and industrial time-series forecasting, including sales and demand forecasting (e.g., refs. 17, 28, 50, 71), energy demand and consumption, advertising market research (e.g., ref. 42), inventory/stock management (e.g., ref. 21), and others. This is complemented by ranges of applications in macroeconomics and econometrics*, especially in connection with problems in financial statistics*, structural change, aggregation*, and combined forecasts/models (e.g., refs. 51, 52, 53, 49). There are similar ranges of applications in biomedical monitoring contexts (e.g., refs. 16, 59), and specific kinds of applications in focused areas, including reliability* analysis (e.g., refs. 6, 40), survival analysis* in biostatistics and economics [12,13,64], point processes* [14], geological statistics* [67], stochastic hydrology* and river dam management [54,55], competitive

bidding [4], spectral analysis* (e.g., refs. 32, 66), and many others, including traffic accident forecasting, marketing*, finite population* inference, prediction in queueing theory*, small area estimation*, optimum design* and optimal stochastic control*, and novel directions in spatial statistics* and imaging. This is a vibrant and rapidly developing field, and one likely to continue to grow well into the twenty-first century.

DYNAMIC BAYESIAN MODELS

Dynamic models provide the central technical components of Bayesian forecasting methods and systems [46,71]. In by far the most widely used incarnation as dynamic linear models, and with associated assumptions of normally distributed components, they also feature centrally in various areas of classical time series analysis and control areas (as state-space models [2,3]), and econometrics (as state-space or structural time-series models [26]).

The approach of Bayesian forecasting and dynamic modeling comprises, fundamentally,

> sequential model definitions for series of observations observed over time,
> structuring using parametric models with meaningful parametrizations
> probabilistic representation of information about all parameters and observables, and hence
> inference and forecasting derived by summarizing appropriate posterior* and predictive probability distributions.

The probabilistic view is inherent in the foundation in the Bayesian paradigm and has important technical and practical consequences. First, the routine analysis of time series data, and the entire process of sequential learning and forecasting, is directly determined by the laws of probability. Joint distributions for model parameters and future values of a time series are manipulated to compute parameter estimates and point forecasts, together with a full range of associated summary measures of uncertainties. Second, and critical in the context of practical forecasting systems that are open to external information sources, subjective interventions to incorporate additional information into data analysis and inference are both explicitly permissible and technically feasible within the Bayesian framework. All probability distributions represent the views of a forecaster, and these may be modified/updated/adjusted in the light of any sources of additional information deemed relevant to the development of the time series. The only constraint is that the ways in which such additional information is combined with an existing model form and the historical information must abide by the laws of probability.

The sequential approach focuses attention on statements about the future development of a time series conditional on existing information. Thus, if interest lies in the scalar series Y_t, statements made at time $t - 1$ are based on the existing information set D_{t-1}, whatever this may be. These statements are derived from a representation of the relevant information obtained by structuring the beliefs of the forecaster in terms of a parametric model defining the probability density $f_Y(Y_t|\theta_t, D_{t-1})$, where θ_t is the defining parameter vector at time t. Through conditioning arguments, the notation explicitly recognizes the dependence of y_t on the model parameters and on the available historical information. Parameters represent constructs meaningful in the context of the forecasting problem. For example, θ_t may involve terms representing the current level of the Y_t-series, regression parameters on independent variables, seasonal factors, and so forth. Indexing θ_t by t indicates that the parametrization may be dynamic, i.e., varying over time through both deterministic and stochastic mechanisms. In some cases, the parameter may even change in dimension and interpretation.

NORMAL DYNAMIC LINEAR MODELS

The class of normal *dynamic linear models* (DLMs) is central to Bayesian forecasting and time-series analysis. The basic model over all time t is defined by the observation and evolution equations

$$\mathbf{Y}_t = \mathbf{F}'_t\theta_t + v_t,$$

$$\theta_t = \mathbf{G}_t\theta_{t-1} + \omega_t,$$

with components as follows:

θ_t is the state vector at time t;

$\mathbf{F}_t$ is a known vector of regression variables and constants;

v_t is an observation noise term, representing measurement and sampling errors corrupting the observation of Y_t, assumed normally distributed with zero mean and known variance v_t, i.e., $N(v_t|0, v_t)$;

$\mathbf{G}_t$ is the state evolution matrix, defining the deterministic map of state vectors between times $t - 1$ and t; and

ω_t is the evolution noise term, or innovation, representing stochastic changes in the state vector and assumed normally distributed with zero mean and known variance matrix $\mathbf{W}_t$, i.e., $N(\omega_t|0, \mathbf{W}_t)$.

Additionally, the noise sequences v_t and ω_t, over time t, are assumed independent and mutually independent. Variations on this basic framework allow for correlated noise sequences, nonzero-mean noise terms, and other minor modifications.

The model structure is Markovian, with state vectors varying over time according to a linear Markov evolution equation (see MARKOV PROCESSES). The class of DLMs includes many kinds of time-series models, such as models with time-varying, "smooth" trends, time-varying seasonal/periodic behavior, regression effects in which the regression parameters may change (usually slowly) over time, transfer responses, and others, including stationary and nonstationary variants of autoregressive moving-average (ARMA) models[*] [71].

Routine analysis of the sequentially received data series involves sequentially updating summary statistics that characterize sequences of posterior/predictive distributions for inference about subsets of the θ_t over all time t and for future values of the series $\mathbf{Y}_t$. Assuming the only information used in updating is the set of observed values $y_1, y_2, \ldots$, we have a closed model in which information updates via $D_t = \{D_{t-1}, y_t\}$ at each time point, given an initial information set D_0 (at an arbitrary time origin $t = 0$). Assuming a normal distribution for the initial state vector $\boldsymbol{\theta}_0$, we have complete joint normality of all the $\boldsymbol{\theta}_t$ and Y_s, and the sequential updating of distributions is based, essentially, on the Kalman filter equations. At time t, we have a current posterior for the current state vector defined by $(\boldsymbol{\theta}_t|D_t) \sim N(\boldsymbol{\theta}_t|\boldsymbol{m}_t, \boldsymbol{C}_t)$ with the updating defined via

$$\boldsymbol{m}_t = \boldsymbol{a}_t + \boldsymbol{A}_t e_t, \quad \boldsymbol{C}_t = \boldsymbol{R}_t - \boldsymbol{A}_t \boldsymbol{A}_t' q_t,$$

where $\boldsymbol{a}_t = \boldsymbol{G}_t \boldsymbol{m}_{t-1}$, $\boldsymbol{R}_t = \boldsymbol{G}_t \boldsymbol{C}_{t-1} \boldsymbol{G}_t' + \boldsymbol{W}_t$, $q_t = \boldsymbol{F}_t' \boldsymbol{R}_t \boldsymbol{F}_t + v_t$, and $e_t = y_t - \boldsymbol{F}_t' \boldsymbol{a}_t$. By way of interpretation, $\boldsymbol{a}_t$ and $\boldsymbol{R}_t$ are the defining moments of the state prior, or state prediction, distribution at time $t-1$, $f_{\boldsymbol{\theta}}(\boldsymbol{\theta}_t|\mathbf{D}_{t-1}) = N(\boldsymbol{\theta}_t|\boldsymbol{a}_t, \boldsymbol{R}_t)$, q_t is the variance of the one-step ahead predictive distribution $f_Y(Y_t|\mathbf{D}_{t-1}) = N(Y_t|\boldsymbol{F}_t' \boldsymbol{a}_t, q_t)$, and e_t is the observed forecast error.

This set of equations provides a computationally efficient algorithm for sequentially updating the distributional summaries. Related forecasting and smoothing algorithms [71, Chap. 4] provide for computation of the practically relevant distributions

$f_Y(Y_{t+k}|D_t)$, for $k > 0$, the k-step-ahead forecast distribution at time t, and

$f_{\boldsymbol{\theta}}(\theta_{t-k}|D_t)$, for $k > 0$, the k-step-filtered distribution at time t.

Forecasting is based on the former, and retrospective time series analysis and decomposition, especially in connection with the evaluation, of changes over time in elements of the state vector, are based on the latter.

Various practically important extensions of this basic model are discussed in [71]. Two that are of key practical relevance are estimation of the observation variances v_t—possibly constant though often slowly varying over time, perhaps with additional weights—and the use of uninformative or reference initial prior distributions for the initial state vector [44].

Practical forecasting systems based on dynamic models require consideration of issues of model specification and structuring, forecast monitoring, and intervention. These kinds of developments, discussed below, are covered in detail in refs. [71] and [46].

COMPONENT DYNAMIC LINEAR MODELS

Model building and structuring relies heavily on the development of overall DLMs from simpler and specific components. Refer to the DLM described earlier by the model quadruple $\{\boldsymbol{F}_t, \boldsymbol{G}_t, v_t, \boldsymbol{W}_t\}$, defined for all t. DLMs are built from individual components describing features such as slowly varying trends, usually local polynomials, seasonal components, regression components, and possibly others, including residual AR or ARMA components. Each component can be viewed as a sub-DLM, and the linear combination of components provides an overall DLM for the series. Suppose a set of $m > 1$ independent DLMs is defined by individual elements model quadruples $\{\boldsymbol{F}_{i,t}, \boldsymbol{G}_{i,t}, v_{i,t} \boldsymbol{W}_{i,t}\}$, for all t and $i = 1, \ldots, m$; write $Y_{i,t}$ for the observation on the ith series at time t. Add the series following these models, to obtain a series $Y_t = \sum_{i=1}^m Y_{i,t}$. Then Y_t follows a DLM $\{\boldsymbol{F}_t, \boldsymbol{G}_t, v_t, \boldsymbol{W}_t\}$, in which (1) the state vector, regression vector, and evolution innovation vector are obtained via the catenation of the corresponding elements of the individual models, so that $\boldsymbol{\theta}_t = (\theta_{1,t}'; \ldots; \theta_{m,t}')'$, with similar forms for $\boldsymbol{F}_t$ and $\boldsymbol{\omega}_t$ in the overall model; (2) the observation error variances are added, $v_t = \sum_{i=1}^m v_{i,t}$; and (3) the evolution matrices are built from the components in block-diagonal form, namely $\boldsymbol{G}_t =$ block diag$(\boldsymbol{G}_1; \ldots; \boldsymbol{G}_m)$ and $\boldsymbol{W}_t =$ block diag $(\boldsymbol{W}_1; \ldots; \boldsymbol{W}_m)$. The model for Y_t is said to be formed by the superposition of the component DLMs.

Key model components for many applications are those for smooth trends and seasonal patterns, in which the $\boldsymbol{F}$ and $\boldsymbol{G}$ elements are time-independent. For example, a second-order polynomial (or locally linear) trend is defined via

$$\boldsymbol{F} = \begin{pmatrix} 1 \\ 0 \end{pmatrix} \quad \text{and} \quad \boldsymbol{G} = \begin{pmatrix} 1 & 1 \\ 1 & 0 \end{pmatrix},$$

with higher-order local polynomial forms by extension [71, Chap. 7].

Fourier representations of seasonal or cyclical components are based on the time-varying sinusoidal DLM with

$$\boldsymbol{F} = \begin{pmatrix} 1 \\ 0 \end{pmatrix} \quad \text{and} \quad \boldsymbol{G} = \begin{pmatrix} \cos\alpha & \sin\alpha \\ \sin\alpha & \cos\alpha \end{pmatrix},$$

where $\alpha = 2\pi/p$ is the frequency and p the period or wavelength. In typical cases, p is an integer, e.g. $p = 12$ for a monthly seasonal pattern, and the above cyclical DLM describes a sine wave of varying amplitude and phase but fixed period p. More complicated seasonal patterns require additional periodic components at harmonic frequencies, thus adding components like that above but with frequencies 2α, 3α, etc. Often a complete description of an arbitrary seasonal pattern of integer period p requires the full set of harmonic components [24,71; Chap. 8].

There are often alternative DLM representations of specific component forms, choice among which is something of a matter of taste, convenience, and interpretation. For example, the above cyclical component is alternatively modeled via

$$\boldsymbol{F} = \begin{pmatrix} 1 \\ 0 \end{pmatrix} \quad \text{and} \quad \boldsymbol{G} = \begin{pmatrix} 2\cos\alpha & -1 \\ 1 & 0 \end{pmatrix},$$

a special case of an autoregressive DLM, here defining an AR(2) component with a unit root (e.g., refs. [71, Chap. 5] and [66]). Extensive discussion of component representation and this issue of similar models with differing $\boldsymbol{F}$ and $\boldsymbol{G}$ appears in ref. [71, Chaps. 5, 6]. This reference also includes wider development of the connections with linear systems theory.

Other useful components include regression effects, in which $F_{i,t}$ is a regressor variable and $G_{i,t} = 1$ and more general regression transfer function models* [47,71]. ARMA components may also be developed, in various forms [3,26,33,34,66], and the use, in particular, of AR component DLMs is becoming more widespread with the advent of appropriate computational methods (see below).

DISCOUNTING

In dynamic modeling generally, and with DLMs in particular, the notion of information discounting, and its use in structuring models, has been usefully exploited. Given a DLM form defined by $\boldsymbol{F}_t$ and $\boldsymbol{G}_t$, model structuring is completed by specification of the evolution variance matrix $\boldsymbol{W}_t$, a key ingredient that controls and determines the extent and nature of stochastic changes in state vectors over time. In practical models, the dimension of $\boldsymbol{\theta}_t$ may range up to 12–15 or more, so that this is a challenging specification task unless $\boldsymbol{W}_t$ is structured in terms of just a few basic parameters. Traditional concepts of information discounting in time series led to the development of methods of DLM component discounting that rather neatly and naturally provide practitioners with interpretable ways of specifying evolution variance matrices [71, Chap. 6].

For example, the various DLMs implemented in the BATS package [46] associate a single discount with each of the polynomial trend, seasonal, and regression components of an overall model. These determine individual evolution variances $\boldsymbol{W}_{i,t}$ in the block-diagonal form of $\boldsymbol{W}_t$, and provide opportunity to model varying degrees of smoothness of (or, reciprocally, variation in) the individual components over time. This can be absolutely critical in terms of both short-term forecasting performance and in adapting to changes in component structure over time in practical time series analysis.

INTERVENTION

In practice, statistical models are only components of forecasting systems, and interactions between forecasters and models is necessary to adequately allow for events and changes that go beyond the existing model form. One of the major features of Bayesian forecasting is the facility for integrating intervention information into existing DLMs and other models [70]. Interventions can be classified as either feedforward or feedback (*see* FEEDFORWARD-FEEDBACK CONTROL SCHEMES). The former is anticipatory, providing opportunity to anticipate changes in the environment and context generating the series. This is very common in commercial environments, for example. Feedback interventions are, on the other hand, corrective actions taken to adjust models for recent events that had not been foreseen or adequately catered for within the existing model.

Various modes of intervention in forecasting models are described in refs. [70] and [71], including a range of practical contexts and exploration of technical aspects of intervening in specified models with existing probabilistic summaries. Key concepts involved include the deletion of observations viewed as suspect or possible outliers*; the imposition of constraints on particular model components to protect against possible spurious adjustments in updating the distributions of state vectors; the extension of an existing DLM form to include additional components representing perhaps transient intervention effects; and, perhaps most importantly, the simple input of additional uncertainty, about state-vector components, usually by momentarily inflating the relevant elements of the evolution variance matrix. Ranges of applications and illustrations are given in refs. 46, 70, 71.

MONITORING AND ADAPTATION

The principle of "management by exception" has been a guide to the development of some forecasting systems. This suggests that, ideally, an adopted model, or set of models, is used routinely to process data and information, providing forecasts and inferences that are used unless exceptional circumstances arise. In exceptional cases, identified either in advance or by the detection of deterioration in model/forecasting performance, some kind of intervention is called for. This raises the need for monitoring of forecast performance and model fit in order to signal the need for feedback, or retrospective, interventions in attempting to cater for exceptional events that occur at unforeseen times.

Typical applications of dynamic models involve various sources of exceptional events. Outliers in the data may occur singly or in groups. Elements of the state vector θ_t, at a single time point t, may change in ways that are greater than predicted under the existing evolution model, being consistent with larger variance elements and perhaps different covariance structure for the matrix $\mathbf{W}_t$. These changes may be quite abrupt, or they may be more subtle and difficult to detect, with parameters "drifting" in unpredicted ways over the course of a few observation periods. Further departures from model form may be due to new external effects that are best explained by additional model components. Whatever the exception, monitoring

of forecast performance to "flag" deteriorating forecast accuracy is the starting point for model adaptation via feedback interventions. Various monitoring techniques exist and are in routine application, including variants of traditional cusum methods (see CUMULATIVE SUM CONTROL CHARTS), and more general Bayesian sequential monitoring methods that are sensitive to forecast deterioration both from single observations, which may be outliers, and from groups of a few recent observations, which may represent more subtle changes in state-vector values [46,63,69,71]. These kinds of techniques are central to the recent development of applied Bayesian forecasting systems. Aspects of model monitoring and assessment related to traditional residual and regression diagnostics, including influence analysis, have also been developed [25,34], though they are of more interest and utility in retrospective time-series analysis than in progressive, sequential analysis with a forecasting and updating focus.

MIXTURES OF DYNAMIC MODELS

Probabilistic mixtures of sets of normal DLMs provide practically useful classes of (explicitly nonnormal) forecasting models, and have been in use at least since the early 1970s in both Bayesian forecasting and engineering [2,22,60]. There are many variants, and much recent work related to model mixing in time series; the kinds of applications can be broadly classified into two groups, sometimes referred to as multiprocess models of classes one and two [23; 71, Chap. 12]. In the first case, an assumed DLM form is accepted as appropriate for a time series, but there are defining parameters (such as variance components*, discount factors, elements of the state evolution matrix $\mathbf{G}$, and so forth) that are uncertain and assumed to belong to a specified, fixed, and finite set of values. Sequential analysis* then formally includes these parameters; the DLM mixture structure can be viewed as induced by marginalization with respect to them. This is precisely the traditional approach to Bayesian model selection* and model averaging.

The second class of mixtures models assumes, in contrast, that, at any time point

t, the model generating the observation is drawn from a set of DLMs, with the choice of model at any time determined by a probability distribution across the set. In general, the probabilities over models at any time may depend on the past selection of models and on past data, so this is a rather general framework for model switching [58]. The early work with this approach in connection with the modeling of time-series outliers and abrupt changes in time-series structure [23,59] is precursor to more recent work in related areas and with other model classes (e.g., refs. 7, 11 and 71, Chap. 12).

Historically, computational constraints have limited wider development and application of mixtures of models, particularly in this second class. Sequential and retrospective analysis becomes quite challenging computationally in even rather modest mixture models, though various analytic approximation methods have been very useful. More recently, the advent of efficient simulation methods has provided potential to overcome these constraints [10,11] and much more can be done in terms of realistic modeling of complex problems.

NONNORMAL NONLINEAR MODELS

From the 1960s, analytic techniques of approximation for analyses of dynamic nonlinear time-series* models have been very widely used in engineering areas, typically under the names of extended Kalman filtering and its variants [2]. Similar methods have been used in Bayesian forecasting, though here wider generalization of DLMs has focused on nonnormal (non-Gaussian) models. A variety of areas have generated interest in dynamic models with nonnormal distributions for either the time-series observations [nonnormality of observation models $f_Y(Y_t|\theta_t, D_{t-1})$], or the evolution equation [nonnormality of state evolution models $f_\theta(\theta_{t-1}, D_{t-1})$], or both.

In addition to the developments using mixtures of DLMs, there has been some interest in other approaches to modeling outliers and changepoints* using heavy-tailed* error distributions, such as t-distribution*, in place of normal forms [31,32,41,43,62].

Perhaps the major developments of nonnormal models relate to the synthesis of dynamic and Bayesian modeling concepts with the widely used class of generalized linear models, for problems of time-series analysis and forecasting with discrete data, such as binary sequences* or Poisson counts, binomial response data, and other error models in the exponential family* class [42,43,68]. Related models, concerning multivariate extensions for compositional data series, appear in refs. [18] and [8].

A rather novel field of development of Bayesian dynamic models is the survival analysis arena [12,13,14,64]. Here the data arise in the traditional survival context, as observed for censored data* on failure times for collections of units, such as diseased and treated individuals, over time. The development of dynamic models for time-varying hazard functions and effects of explanatory variables is a direct extension of dynamic generalized linear modeling*. The utility of these models as tools for exploring nonproportional-hazards structure and time variation in the effects of explanatory variables is illustrated in biostatistical and economic applications in some of the references just noted.

Analysis with nonnormal nonlinear models has historically involved ranges of creative analytic/numerical approximation to overcome computational problems arising in implementation. However, during the last few years the revolution in Bayesian statistics through the development of simulation methods of analysis has provided impetus to develop nonstandard models, as such methods are very widely applicable. Some early examples of nonnormal nonlinear dynamic modeling, with analysis made feasible via simulation methods, appear in refs. 9, 27.

MULTIVARIATE MODELS

Various theoretical classes of dynamic linear models for multivariate time series* exist, though the kinds of multivariate applications of note have tended to inspire models quite specific to the application context (e.g., refs. 50,71). Probably the most widely used models are those based on matrix-variate extensions (*see* MATRIX-VALUED DISTRIBUTIONS) of basic DLMs [51,52]. These are particularly appropriate in contexts of modeling several or many univariate series that are viewed as structurally similar, or perhaps exchangeable*, in the sense of sharing common defining elements F_t, G_t and W_t. These models have been, and continue to be, usefully applied in macroeconomics, especially finance (e.g., ref. [53]), among other areas, where the multivariate structure of most importance is contemporaneous, rather than predictive. Modifications of these multivariate models, as well as quite distinct, nonnormal dynamic models, have been developed for problems in which data series represent proportions or compositions [8,18,52].

One area that has seen some exploration to data, and is likely to be a growth area in future, is the development of multivariate dynamic models using traditional Bayesian hierarchical modeling concepts and methods [72,73]. Though not explicitly based in dynamic linear models, the Bayesian forecasting approaches using vector autoregression and structured prior distributions [35,36] represent a significant area of development and an important branch of the wider Bayesian forecasting arena.

COMPUTATION AND SIMULATION

Early work on numerical methods of Bayesian computation for dynamic models includes methods using mixture distributions* [22,23,60] and references in [2] and [71, Chap. 12]), developments based on numerical quadrature* [31,45,45], and traditional Monte Carlo* simulation approaches [65].

Since the early 1990s, the field has been given dramatic impetus by the development of iterative simulation methods, particularly Markov-chain Monte Carlo* using Gibbs sampling*, Metropolis—Hastings methods, and variants. The entire field of dynamic modeling (state-space and structural time series modeling) is likely to experience much wider use of such methods in coming years. As in other areas of applied Bayesian statistics, researchers can now focus very much more on developing realistic models of specific contexts, in the knowledge that very general

and powerful computational techniques are at hand to implement resulting analyses of some complexity.

Basic work in this area stems from refs. 9, 10, 11, with recent extensions and applications in refs. 57, 66 and elsewhere. In the context of a rather generic dynamic model, suppose a sequence of defining observation and evolution equations are described in terms of conditional densities, i.e.,

$$f_Y(Y_t|\theta_t, \phi, D_{t-1}) \text{ and } f_\theta(\theta_t|\theta_{t-1}, \phi, D_{t-1}),$$

where now additional parameters ϕ are made explicit in conditionings; ϕ may contain, for example, observation and evolution variance components in a normal model, elements of state evolution matrices G in a linear model, and other parameters. These distributions may involve nonnormal and nonlinear components, as in [9,57], for example. Inference is desired for collections of state vectors (current and past), future observations on the Y_t-series, and the defining model parameters ϕ. Iterative simulation methods are focused on the evaluation of posterior and predictive distribution inferences by sampling representative values of the state vectors, parameters, and observations, from the relevant distributions. The utility of this approach is illustrated in various interesting contexts in the above references, indicating that this novel area is one with promise for the future.

Related methodological and computational issues, with interesting complications, arise in chaotic time series modeling [5,56].

RELATED AREAS

The foregoing discussion is rather specifically focused on Bayesian dynamic modeling. Bayesian time-series analysis and forecasting is a much wider field than so far represented, and, from the outset, no attempt has been made to cover the field in any generality. To offset this somewhat, some references are made here to provide access to areas of time-series modeling and Bayesian analysis more widely. As noted, there are strong connections with (largely non-Bayesian) structural time series modeling in econometrics (e.g., ref. [26]), with the use of state-space models and Kalman filtering techniques in ARMA modeling and other areas (e.g., refs. 3, 33, 34), and with the interface of statistics and the control engineering fields (e.g., refs. 3, 61). In connection, in particular, with multivariate time series* forecasting, important Bayesian work has been developing in econometrics and finance. Notable areas include the applications in multivariate forecasting using vector autoregressions and related models (e.g., refs. 35, 36), and ranges of applications of models combining dynamic linear model components with hierarchical/shrinkage methods (e.g., refs. 15, 72, 73). Further, based on the advent of simulation methods for Bayesian analysis, there has recently been much activity in developments of computation Bayesian methods in other, closely related areas of time series (e.g., refs. 1, 37, 38).

REFERENCES

1. Albert, J. H. and Chib, S. (1993). Bayesian inference via Gibbs sampling of autoregressive time series subject to Markov mean and variance shifts. *J. Bus. and Econ. Statist.*, **11**, 1–15.

2. Anderson, B. D. O. and Moore, J. B. (1979). *Optimal Filtering.* Prentice-Hall, Englewood Cliffs, N.J.

3. Aoki, M. (1990). *State Space Modelling of Time Series.* Springer-Verlag, New York.

4. Attwell, D. N. and Smith, J. Q. (1991). A Bayesian forecasting model for sequential bidding. *J. Forecasting*, **10**, 565–577.

5. Berliner, L. M. (1992). Statistics, probability, and chaos (with discussion). *Statist. Sci.* **7**, 69–122.

6. Blackwell, L. M. and Singpurwalla, N. D. (1988). Inference from accelerated life tests using filtering in coloured noise. *J. Roy. Statist. Soc. B*, **50**, 281–292.

7. Bolstad, W. M. (1988). Estimation in the multiprocess dynamic generalized linear model. *Commun. Statist. A—Theory and Methods*, **17**, 4179–4204.

8. Cargnoni, C., Müller P., and West, M. (1997). Bayesian forecasting of multinomial time series through conditionally Gaussian dynamic models. *J. Amer. Statist. Ass.*, **92**, 587–606.

9. Carlin, B. P., Polson, N. G., and Stoffer, D. S. (1992). A Monte Carlo approach to nonnormal and nonlinear state-space modelling. *J. Amer. Statist. Ass.*, **87**, 493–500.

10. Carter, C. K. and Kohn, R. (1994). On Gibbs sampling for state space models. *Biometrika*, **81**, 541–553.

11. Frühwirth-Schnatter, S. (1994). Data augmentation and dynamic linear models. *J. Time Ser. Anal.*, **15**, 183–102.

12. Gamerman, D. and West, M. (1987). A time series application of dynamic survival models in unemployment studies. *Statistician*, **36**, 269–174.

13. Gamerman, D. (1991). Dynamic Bayesian models for survival data. *Appl. Statist.*, **40**, 63–79.

14. Gamerman, D. (1992). A dynamic approach to the statistical analysis of point processes. *Biometrika*, **79**, 39–50.

15. Garcia-Ferrer, A., Highfield, R. A., Palm, F., and Zellner, A. (1987). Macroeconomic forecasting using pooled international data. *J. Bus. and Econ. Statist.* **5**, 53–67.

16. Gordon, K. and Smith, A. F. M. (1990). Modeling and monitoring biomedical time series. *J. Amer. Statist. Ass.*, **85**, 328–337.

17. Green, M. and Harrison, P. J. (1973). Fashion forecasting for a mail order company. *Oper. Res. Quart.*, **24**, 193–205.

18. Grunwald, G. K., Raftery, A. E., and Guttorp, P. (1993). Time series of continuous proportions. *J. R. Statist. Soc. B*, **55**, 103–116.

19. Harrison, P. J. (1965). Short-term sales forecasting. *Appl. Statist.*, **15**, 102–139.

20. Harrison, P. J. (1967). Exponential smoothing and short-term forecasting. *Management Sci.*, **13**, 821–842.

21. Harrison, P. J. (1988). Bayesian forecasting in O.R. In *Developments in Operational Research 1988*, N. B. Cook and A. M. Johnson, eds. Pergamon Press, Oxford.

22. Harrison, P. J. and Stevens, C. (1971). A Bayesian approach to short-term forecasting. *Oper. Res. Quart.*, **22**, 341–362.

23. Harrison, P. J. and Stevens, C. (1976). Bayesian forecasting (with discussion). *J. R. Statist. Soc. B*, **38**, 205–247.

24. Harrison, P. J. and West, M. (1987). Practical Bayesian forecasting. *Statistician*, **36**, 115–125.

25. Harrison, P. J. and West, M. (1991). Dynamic linear model diagnostics. *Biometrika*, **78**, 797–808.

26. Harvey, A. C. (1990). *Forecasting, Structural Time Series Models, and the Kalman Filter.* Cambridge University Press.

27. Jaquier, E., Polson, N. G., and Rossi, P. E. (1994). Bayesian analysis of stochastic volatility. *J. Bus. and Econ. Statist.*, **12**, 371–417.

28. Johnston, F. R. and Harrison, P. J. (1980). An application of forecasting in the alcoholic drinks industry. *J. Oper. Res. Soc.*, **31**, 699–709.

29. Kalman, R. E. (1960). A new approach to linear filtering and prediction problems. *J. Basic Eng.*, **82**, 35–45.

30. Kalman, R. E. (1963). New methods in Wiener filtering theory. *Proc. First Symp. Engineering Applications of Random Function Theory and Probability.* J. L. Bogdanoff and F. Kozin, eds. Wiley, New York.

31. Kitagawa, G. (1987). Non-Gaussian state-space modelling of nonstationiary time series (with discussion). *J. Amer. Statist. Ass.*, **82**, 1032–1063.

32. Kleiner, B., Martin, R. D., and Thompson, D. J. (1979). Robust estimation of power spectra (with discussion). *J. R. Statist. Soc. B*, **41**, 313–351.

33. Kohn, R. and Ansley, C. F. (1986). Estimation, prediction and interpolation for ARIMA models with missing data. *J. Amer. Statist. Ass.*, **81**, 751–761.

34. Kohn, R. and Ansley, C. F. (1993). A fast algorithm for signal extraction, influence and cross-validation in state-space models. *Biometrika*, **76**, 65–79.

35. Litterman, R. B. (1986). Specifying vector autoregressions for macro-economic forecasting. In *Bayesian Inference and Decision Techniques: Essays in Honor of Bruno de Finetti*, P. K. Goel and A. Zellner, eds. North-Holland, Amsterdam.

36. Litterman, R. B. (1986). Forecasting with Bayesian vector autoregressions: five years of experience. *J. Bus. and Econ. Statist.*, **4**, 25–38.

37. McCulloch, R. and Tsay, R. (1993). Bayesian inference and prediction for mean and variance shifts in autoregressive time series. *J. Amer. Statist. Ass.*, **88**, 968–978.

38. McCulloch, R. and Tsay, R. (1994). Bayesian analysis of autoregressive time series via the Gibbs sampler. *J. Time Ser. Anal.*, **15**, 235–250.

39. Meinhold, R. J. and Singpurwalla, N. D. (1983). Understanding the Kalman filter. *Amer. Statist.*, **37**, 123–127.

40. Meinhold, R. J. and Singpurwalla, N. D. (1985). A Kalman filter approach to accelerated life testing. In *Theory of Reliability*, A.

Serra and R. E. Barlow, eds. North-Holland, Amsterdam.

41. Meinhold, R. J. and Singpurwalla, N. D. (1989). Robustification of Kalman filter models. *J. Amer. Statist. Ass.*, **84**, 479–486.

42. Migon, H. S. and Harrison, P. J. (1985). An application of non-linear Bayesian forecasting to television advertising. In *Bayesian Statistics 2*, J. M. Bernardo, M. H. DeGroot, D. V. Lindley, and A. F. M. Smith, eds. North-Holland, Amsterdam, and Valencia University Press, pp. 681–696.

43. Pole, A., West, M., and Harrison, P. J. (1988). Non-normal and non-linear dynamic Bayesian modelling. In *Bayesian Analysis of Time Series and Dynamic Models*, J. C. Spall, ed. Marcel Dekker, New York, pp. 167–198.

44. Pole, A. and West, M. (1989). Reference analysis of the DLM. *J. Time Ser. Anal.*, **10**, 131–147.

45. Pole, A. and West, M. (1990). Efficient Bayesian learning in non-linear dynamic models. *J. Forecasting*, **9**, 119–136.

46. Pole, A., West, M., and Harrison, P. J. (1994). *Applied Bayesian Forecasting and Time Series Analysis.* Chapman and Hall, New York. (An applied treatment of the central classes of dynamic models with many examples and applications, together with the BATS software, manual, and user guide.)

47. Pole, A. (1995). Forecasting and assessing structural change: the case of promotional campaigns. In *Advances in Econometrics (vol. 11B): Bayesian Methods Applied to Time Series Data*, R. Carter Hill and T. Fomby, eds. JAI Press.

48. Priestley, M. B. (1980). System identification, Kalman filtering, and stochastic control. In *Directions in Time Series*, D. R. Brillinger and G. C. Tiao, eds. Institute of Mathematical Statistics.

49. Putnam, B. H. and Quintana, J. M. (1994). New Bayesian statistical approaches to estimating and evaluating models of exchange rate determination. In *1994 Proceedings of the Section on Bayesian Statistical Science.* American Statistical Association, Alexandria, Va., pp. 110–118.

50. Queen, C. M. and Smith, J. Q. (1992). Dynamic graphical models. In *Bayesian Statistics 4*, J. O. Berger, J. M. Bernardo, A. P. Dawid, and A. F. M. Smith, eds. Oxford University Press, pp. 741–752.

51. Quintana, J. M. and West, M. (1987). Multivariate time series analysis: new techniques

applied to international exchange rate data. *Statistician*, **36**, 275–281.

52. Quintana, J. M. and West, M. (1988). Time series analysis of compositional data. In *Bayesian Statistics 3*, J. M. Bernardo, M. H. DeGroot, D. V. Lindley and A. F. M. Smith, eds. Oxford University Press, pp. 747–756.

53. Quintana, J. M. (1992). Optimal portfolios of forward currency contracts. In *Bayesian Statistics 4*, J. O. Berger, J. M. Bernardo, A. P. Dawid, and A. F. M. Smith, eds. Oxford University Press, pp. 753–762.

54. Rios Insua, D. and Salewicz, K. A. (1995). The op-eration of Kariba Lake: a multiobjective analysis. *J. Multicriteria Decision Anal.* **4**, 202–222.

55. Rios Insua, D., Salewicz, K. A., Müller, P., and Bielza, C. (1997). Bayesian methods in reservoir operations: The Zambzi River Case. In *The Practice of Bayesian Analysis*, S. French and J. Q. Smith, eds. Arnold, London, pp. 107–130.

56. Scipione, C. M. and Berliner, L. M. (1993). Bayesian statistical inference in nonlinear dynamical systems. *1993 Proc. Bayesian Statistical Science Section of the ASA*. American Statistical Association, Alexandria, Va., pp. 73–78.

57. Shephard, N. (1994). Partial non-Gaussian state space models. *Biometrika*, **81**, 115–131.

58. Shumway, R. H. and Stoffer, D. S. (1991). Dynamic linear models with switching. *J. Amer. Statist. Ass.*, **86**, 763–769.

59. Smith, A. F. M. and West, M. (1983). Monitoring renal transplants: an application of the multiprocess Kalman filter. *Biometrics*, **39**, 867–878.

60. Sorenson, H. W. and Alspach, D. L. (1971). Recursive Bayesian estimation using Gaussian sums. *Automatica*, **7**, 465–479.

61. Spall, J. C., ed. (1988). *Bayesian Analysis of Time Series and Dynamic Models*. Marcel Dekker, New York.

62. West, M. (1981). Robust sequential approximate Bayesian estimation. *J. R. Statist. Soc. B*, **43**, 157–166.

63. West, M. (1986). Bayesian model monitoring. *J. R. Statist. Soc. B*, **48**, 70–78.

64. West, M. (1992). Modelling time-varying hazards and covariate effects (with discussion). In *Survival Analysis: State of the Art*, J. P. Klein and P. K. Goel, eds. Kluwer, pp. 47–64.

65. West, M. (1993). Mixture models, Monte Carlo, Bayesian updating and dynamic models. *Comput. Sci. and Statist.*, **24**, 325–333.

66. West, M. (1995). Bayesian inference in cyclical component dynamic linear models. *J. Amer. Statist. Ass.*, **90**, 1301–1312.

67. West, M. (1996). Some statistical issues in palæoclimatology (with discussion). In *Bayesian Statistics 5*, J. O. Berger, J. M. Bernardo, A. P. Dawid, and A. F. M. Smith, eds. Oxford University Press, pp. 461–486.

68. West, M., Harrison, P. J., and Migon, H. S. (1985). Dynamic generalised linear models and Bayesian forecasting (with discussion). *J. Amer. Statist. Ass.*, **80**, 73–97.

69. West, M. and Harrison, P. J. (1986). Monitoring and adaptation in Bayesian forecasting models. *J. Amer. Statist. Ass.*, **81**, 741–750.

70. West, M. and Harrison, P. J. (1989). Subjective intervention informal models. *J. Forecasting*, **8**, 33–53.

71. West, M. and Harrison, P. J. (1997). *Bayesian Forecasting and Dynamic Models* (2nd edition). Springer-Verlag, New York. (A full treatment of theory and methods of Bayesian time-series analysis and forecasting with dynamic models.)

72. Zellner, A. and Hong, C. (1989). Forecasting international growth rates using Bayesian shrinkage and other procedures. *J. Econometrics*, **40**, 183–202.

73. Zellner, A., Hong, C., and Min, C. (1991). Forecasting turning points in international output growth rates using Bayesian exponentially weighted autoregression, time-varying parameter, and pooling techniques. *J. Econometrics*, **49**, 275–304.

See also Exponential Smoothing; Forecasting; Kalman Filtering; Nonlinear Time Series; Prediction and Filtering, Linear; Prediction and Forecasting; Predictive Analysis; Stationary Processes; Structural Models; Structural Prediction; Time Series; and Wiener–Kolmogorov Prediction Theory.

Mike West

BAYESIAN INFERENCE

According to the Bayesian view, all quantities are of two kinds: those known to the person making the inference and those unknown to the person; the former are described by their known values, the uncertainty surrounding the latter being described by a joint probability distribution* for them all. In the context of the usual statistical model with a random variable X having possible distributions indexed by a parameter θ, the data x becomes known to the statistician and the object is to make inferences concerning the unknown parameter; so in the Bayesian approach, the statistician will wish to calculate the probability distribution of θ given $X = x$. Once this is done, point estimates* of θ can be calculated; e.g., the mean or mode of this distribution: perform significance tests* of a null value θ_0 by considering the probability that θ equals θ_0: and determine interval estimates* by finding intervals of θ-values containing an assigned probability for θ. If θ is vector-valued, say $\theta = (\theta_1, \theta_2, \ldots, \theta_p)$, and only θ_1 is of interest, then the marginal distribution of θ_1 can be found from the distribution of θ, the remaining values $\theta_2, \ldots, \theta_p$ being regarded as nuisance parameters*. Within the Bayesian framework *all* calculations are performed by the probability calculus, using the probabilities for the unknown quantities.

The statistical model admits, for each θ, a distribution of X. Describe this by a density $p_X(x|\theta)$ for $X = x$, given θ, the suffix being omitted when it is clear which random variable is involved. The Bayesian interpretation of this is that it describes the uncertainty about the data, necessarily before it is known, were the parameter value known. The parameter will have its density $p(\theta)$, so that the required distribution of the first paragraph, again described by a density, can be calculated by Bayes' theorem*:

$$p(\theta|x) = p(x|\theta)p(\theta) \bigg/ \int p(x|\theta)p(\theta)\,d\theta. \quad (1)$$

(The denominator is the unconditional density of X, describing the real uncertainty over the data before it is observed.) It is this basic use of Bayes theorem that gives the method its name. Notice that in addition to the density $p(x|\theta)$ used in most statistical situations, the Bayesian view introduces $p(\theta)$. This is called the prior (or a priori) distribution* of θ, prior, that is, to the data being available. Similarly, $p(\theta|x)$ is termed the posterior distribution*. The data density and the prior together provide a complete probability

specification for the situation from which other uncertainties can be calculated.

As an example, consider a situation in reliability* engineering where n items are put on test, the lifetime x_i of each being supposed exponential* with unknown hazard rate* θ, so that $p(x_i|\theta) = \theta e^{-\theta x_i}$ for $x_i \geqslant 0$ and $\theta > 0$, and given θ, the lifetimes of the n items being independent. Suppose that $r(\leqslant n)$ of the items fail at times $x_1, x_2, \ldots, x_r$ and that the remainder of them are withdrawn at times $x_{r+1}, \ldots, x_n$ without having failed (*see* CENSORING). With $X = (x_1, \ldots, x_r; x_{r+1}, \ldots, x_n)$, $p(x|\theta)$ is $\theta^r e^{=\theta T}$, where $T = \sum_{i=1}^n x_i$, the total time on test. It is technically convenient to suppose that θ has a gamma* density, $p(\theta) = \theta^a e^{-\theta b} b^{a+1}/a!$ for some $b > 0$ and $a > -1$. Then easy calculations in (1) show that

$$p(\theta|x) = \theta^{a+r} e^{-\theta(b+T)}$$
$$\times (b+T)^{a+r+1}/(a+r)!, \quad (2)$$

also gamma. The mean is $(a+r+1)/(b+T)$, providing an estimate of θ. The probability that the hazard rate exceeds an assigned value θ_1 may be found as the integral of (2) for $\theta \geqslant \theta_1$. If values of the hazard rate exceeding θ_1 are considered dangerous, this probability says how reasonable it is, in the light of the data, that a dangerous situation obtains. Notice that this is a direct probability statement about the unknown quantity of interest (unlike the circumlocutory statements provided by confidence intervals*).

An important deduction can be made from (1). Once the data x are known, θ is the only unknown quantity and $p(\theta|x)$ completely describes the uncertainty. But the right-hand side of (1) immediately shows that this depends on the data only through $p(x|\theta)$; or, more exactly, through the function $p(x|\cdot)$ with argument θ. This is called the likelihood function* (of θ, for $X = x$) and we have the likelihood principle*, which says, in its Bayesian form, that, given the data, all inferences about θ require only the likelihood $p_X(x|\cdot)$ for $X = x$. Consider the implications of this for the reliability problem of the preceding paragraph. From (2) we see that r and T, the number of failures and the total time on test, are sufficient* for θ, and that it does not matter how r and T were obtained within the conditions under which (2) was derived. Thus we could have tested for a fixed time t, so that $x_{r+1} = \cdots = x_n = t$, or we could have tested until a preassigned number, r, of failures had been observed; and there are many other possibilities. In all these cases the final inference is the same, and provided by (2), because the likelihood is the same in every one.

Almost all standard, nonBayesian statistical procedures violate the likelihood principle; i.e., they do depend on other aspects of the data besides the likelihood function. This dependence is caused by using integration over x-values to make the inference. Thus the concept of an unbiased estimate*, $t(x)$, demands $\int t(x)p(x|\theta)\,dx = \theta$ for all θ; a test of a hypothesis* (*see* HYPOTHESIS TESTING) uses a significance level* $\int_R p(x|\theta_0)\,dx$, where R is the region of rejection (critical region*) and θ_0 the "null" value. This integration uses $p_X(\cdot|\theta)$ for values other than the observed value, x, thus violating the principle. In many cases it is not obvious which values of X to use for the integration. In the reliability problem we would have one region of integration if T were fixed, and a different one were r to be preselected. The Bayesian argument is often criticized for introducing the prior distribution. It is not often realized that orthodox arguments for inference introduce a set of X-values not needed by the Bayesian. Satisfactory inference scarcely seems possible without using something in addition to the likelihood function, such as other data values or a prior; adherents of the likelihood school having to introduce some device, such as marginal likelihoods*, in order to eliminate nuisance parameters.

There are two fundamental reasons for adopting the Bayesian position. The first is that once the basic step of describing uncertainty through probability is admitted, we have a formal procedure for solving all inference problems; and it is a procedure that works. The procedure, or recipe, runs as follows. First, say what is known; that is, identify the data X. Second, specify the model for the data generation, a family of distributions for x indexed by parameter θ. [This includes nonparametric (distribution-free*) ideas where θ may, e.g., with X a real

number, index all continuous distributions on the real line.] Third, specify the uncertainty concerning θ. Fourth, consider what quantity is of interest; usually, this will be a part of θ, so that if $\theta = (\theta_1, \theta_2)$, it will be θ_1, θ_2 being nuisance. Fifth, calculate $p(\theta_1|x)$, the uncertainty of the quantity of interest given the observed value x of the data; this calculation proceeding by Bayes' theorem and subsequent integration of θ_2. This provides a complete inference for θ_1. It is important to recognize that this is a genuine recipe for inference, saying just what has to be done and how to do it; so that there is no need for the introduction of any other principles. For example, a principle that selects a good estimator is not needed, since $p(\theta_1|x)$ provides everything. Equally, significance levels play virtually no role in Bayesian analysis, although some writers feel they have some use for a Bayesian. The Bayesian recipe is at present the only recipe for inference of wide applicability. It is also important to recognize that the procedure works; there are no examples known to me where the result is unsatisfactory. Indeed, since the procedure is well defined and therefore testable, a single counterexample is all that is needed to make it unacceptable.

The second reason for describing uncertainty in probability terms is mathematical. Using some reasonable requirements of uncertainty as axioms in a mathematical system, it can be proved that such a description exists. The axioms are called those of coherence*, and the Bayesian system is the only coherent one. It therefore follows that other systems are necessarily incoherent and, in some respect, violate simple, axiomatic requirements of uncertainty. It is possible to take any statistical procedure that is non-Bayesian and produce an example where it performs unsatisfactorily. For example, the best, unbiased estimate of θ^2, based on the observation of r successes in n Bernoulli trials* of constant probability θ of success, is $r(r-1)/\{n(n-1)\}$, which is unsound whenever $r = 1$, for to estimate the chance of two future successes as zero when one has been observed is surprising, to say the least.

The concept of probability used in Bayesian inference is not based on frequency considerations; rather, it describes the uncertainty felt by the person making the inference. Some hold that it is objective in the sense that all persons having the same information about θ will have the same probability; others hold a subjective view (see SUBJECTIVE PROBABILITIES), that $p(\theta)$ reflects the subject's own uncertainty; but neither admits any frequency interpretation. At first sight this seems in conflict with the frequency view that dominates most statistical thinking (see PROBABILITY, FOUNDATIONS OF—I), e.g., in the concept of a random sample $x = (x_1, x_2, \ldots, x_n)$ from a population described by θ. There is, however, a simple connection that we illustrate in the context of a Bernoulli sequence of trials*, although the argument is general. The judgment of uncertainty that says a sample is random can be expressed by saying that the order of $x_1, x_2, \ldots, x_n$ is irrelevant. This is captured by saying that $p(x_1, x_2, \ldots, x_n)$ is exchangeable*—the probability here expressing one's uncertainty about the x's. A fundamental theorem due to de Finetti says that if this is true for all n, then this probability must, for the Bernoulli case, be equal to $\int_0^1 \theta^r (1 - \theta)^{n-r} \, dP(\theta)$, where $P(\cdot)$ is a distribution function on [0, 1] and r is the number of successes in n trials; furthermore, the limit of r/n as $n \to \infty$ exists and has the distribution described by $P(\cdot)$. Consequently, the exchangeable model has basically the binomial* structure in $\theta^r (1 - \theta)^{n-r}$ for fixed θ, this last having the "prior" distribution $P(\cdot)$.

In the case where $P(\cdot)$ has a density $p(\cdot)$, we have the same structure as described earlier, P or p, expressing one's belief about the limiting frequency ratio. Ordinary statistical language would refer to θ as the probability of success for a single trial. In the Bayesian view θ is only a probability in a restricted sense; its proper interpretation is a quantity that is unknown to one and is therefore assessed probabilistically by $p(\theta)$. It is referred to as a chance*, or propensity. It appears, almost incidentally, as a probability since $p(x_i = 1|\theta) = \theta$; that is, were θ known it would be equal to the probability, or the belief (see BELIEF FUNCTIONS; BELIEF, DEGREES OF), that a trial will result in a success. In the realistic case where θ is unknown, $p(x_i = 1) = \int \theta dP(\theta)$, the expectation of θ. Thus the concept of exchangeability provides an important link between Bayesian

ideas and the notion of a random sample, a link that is of vital importance if Bayesian ideas are to be applied to the situations of practical importance studied in statistics. Outside the Bernoulli case, a judgment of exchangeability is conveniently described by a random sample, given a parameter θ and a distribution over θ.

There is some advantage in using uncertain quantities that can eventually become known, rather than those, such as, θ, whose value is never known, because they have a reality not possessed by others that makes their uncertainty easier to assess. For this reason it is convenient to express inference rather differently from the manner already mentioned as passing from x to θ, using $p(\theta|x)$. Thus when dealing with a random sample $\mathbf{x}^{(n)} = (x_1, x_2, \ldots, x_n)$, it is preferable to consider x_{n+1}, a further value judged exchangeable with the x_i, and to express one's inference through $p(x_{n+1}|\mathbf{x}^{(n)})$. When x_i is the yield of a variety on the ith plot of an agricultural trial with n plots in all, interest truly centers on what would happen were the variety to be planted on a further plot, $n + 1$. This is a better model of the practical situation than that which introduces parameters such as the "true" yield. In many cases the connection between x_{n+1} and $\mathbf{x}^{(n)}$ is more complicated than simple exchangeability, but for that case the de Finetti representation can simplify the calculation. For then $p(x_{n+1}|\mathbf{x}^{(n)}) = p(\mathbf{x}^{(n+1)})/p(\mathbf{x}^{(n)})$; and the denominator is $\int \prod_{i=1}^n p(x_i|\theta) \, dP(\theta)$, the x's, given θ, being a random sample, i.e., independent and identically distributed. A similar expression with $n + 1$ for n holds for the numerator. Alternatively,

$$p(x_{n+1}|\mathbf{x}^{(n)}) = \int p(x_{n+1}|\theta) p(\theta|\mathbf{x}^{(n)}) \, d\theta$$

since, given θ, x_{n+1} and $\mathbf{x}^{(n)}$ are independent, whence the inference about x_{n+1} is expressed in terms of that about θ. In the Bernoulli case, $p(x_{n+1} = 1|\theta) = \theta$ and $p(x_{n+1} = 1|x^{(n)})$ is simply the mean of the posterior distribution of θ, $E(\theta|x^{(n)})$. In the reliability case above, interests often center on a further item not failing before time s, when it is due for routine inspection. This is $p(x_{n+1} \geq s|x^{(n)})$ and, since $p(x_{n+1} > s|\theta) = e^{-\theta s}$, is $\int e^{-\theta s} p(\theta|r, T) \, d\theta$, which, using (2), is easily evaluated to give $[(b + T)/(b + T + s)]^{a+r+1}$. Inferences concerning x_{n+1} are not easily available within a non-Bayesian framework, although tolerance intervals* can be used.

Bayesian inference is closely connected with (Bayesian) decision theory*. By extending the axioms of uncertainty to include further simple postulates concerning one's attitude toward decisions taken under uncertainty, it is possible to prove that the consequences of decisions that are undertaken can be described numerically, just as uncertainty can, the value being termed utility*; and that the best decision is that of maximum expected utility. The extended axioms are also called coherent and the result can be described by saying that the only coherent method of reaching a decision is through maximizing expected utility, the expectation being with respect to the inference distribution previously considered. Other decision methods, like minimax*, are demonstrably incoherent. These ideas can be used in standard statistical problems such as point estimation. These can be modeled by referring to a decision to treat the parameter as having value t. The consequence of so doing when its true value is θ has some utility, $u(t, \theta)$ say, and the best decision (estimate) is that which maximizes $E[u(t, \theta)|x] = \int u(t, \theta)p(\theta|x)d\theta$. It is common to use a loss $l(t, \theta) = u(\theta, \theta) - u(t, \theta)$, equal to the loss in utility through using t rather than θ, and to minimize the expected loss (see EXPECTED VALUE). If θ is real and quadratic loss $l(t, \theta) = (t - \theta)^2$ is used, then the best estimate is clearly the posterior mean, having smallest mean square error* for the posterior distribution. (Actually, point estimation rarely occurs as a decision problem and its introduction into statistics is, as we saw above, really unnecessary.) A better, because more practical, example is the decision of whether to reject or accept a batch of items on the basis of a test of n of the items in which r fail and $n - r$ pass. If θ describes the quality of the untested items in the batch (an example of a parameter that can become known), then we have $u_A(\theta)$ and $u_R(\theta)$ as utilities for accepting and rejecting, respectively, a batch of quality θ. With density $p(\theta|r, n)$ as our inference concerning θ, the batch is accepted if $\int u_A(\theta)p(\theta|r, n)d\theta$

exceeds the corresponding quantity for rejection. For large batches a binomial approximation might be adequate and $p(\theta|r, n) \propto \theta^r(1 - \theta)^{n-r}p(\theta)$, where $p(\theta)$ is the prior distribution. It is surprising that relatively little use has been made of Bayesian methods in quality control*, where $p(\theta)$ is not too difficult to obtain. It is important to notice that *every* decision problem concerning θ and data x requires only $p(\theta|x)$ as the contribution from the data. Thus the Bayes inference is available for any decision about θ, and similar remarks apply in the alternative inferential form, $p(x_{n+1}|x^{(n)})$. This is why inference is conveniently treated as a separate topic, its merit being irrespective of the decision involved.

The quality of a decision depends on the amount of uncertainty present. If decision d has to be chosen from a set D, then without data the best decision is that which maximizes $\int u(d, \theta)p(\theta)d\theta$. With data x the criterion becomes $\int u(d, \theta)p(\theta|x)d\theta$. The difference between these two values is the gain from observing $X = x$. Actually, the gain can be negative; i.e., one can decrease expected utility by having extra data: e.g., when we obtain "surprising" data. Nevertheless, a simple, but important, result is that the *expected* gain is nonnegative: where the expectation is taken over the values of x, having density $p_X(x) = \int p(x|\theta)p(\theta)d\theta$, the denominator in (1). All relevant data are expected to be of value. The question then arises of whether this gain outweighs the cost, expressed as a loss of utility, in observing X. This leads to the Bayesian solution to the problem of experimental design (see DESIGN OF EXPERIMENTS).

Suppose that we are interested in a parameter θ; a similar treatment is available for an observable quantity such as x_{n+1}. Suppose that we can choose an experiment e, from among a class E of experiments, which will lead to data X having density $p(x|\theta, e)$. Let the purpose of the inference be to make a choice d of a decision from an available class D. Let $u(d, \theta; x, e)$ be the utility of choosing d when the true value is θ after having obtained data x from experiment e. Often this is of the form $u(d, \theta) - c(x, e)$, where $c(\cdot)$ describes the experimental cost. The best decision, with e performed and x observed, maximizes $\int u(d, \theta; x, e)p(\theta|x, e) \cdot d\theta$. Write this maximum $U(x, e)$. Then the best experiment is that which maximizes its expected value $\int U(x,e)p(x|e)dx$. Notice that the device of maximizing an expected value has been used twice, illustrating the point that all decisions can be solved with the aid of this single principle. This method of experimental design requires decisions d in D. Scientists often feel that experimental design should be separated from the decision aspect, just as inference is separated from it when considering a fixed experiment. This can be done here by using the idea that the purpose of an experiment is to "decide" on the distribution of θ. Formally, one identifies d with the posterior distribution $p(\theta|x,e)$ and the utility refers to that of adopting such a distribution when the true value of the parameter is θ. Reasonable restrictions on this utility lead to the concept of information* due to Shannon (see INFORMATION THEORY AND CODING THEORY), using $\int p(\theta|x,e)\log p(\theta|x,e)\,d\theta$, as a criterion for experimental design; and the design chosen maximizes expected information.

The ideas of design explained in the preceding paragraph extend without difficulty to sequential experimentation*. After having observed $x^{(n-1)}$, let the choice be among experiments e_n from a class E_n, consisting in observing X_n. A simple case is where E_n contains only two elements: to stop experimentation or to take a further observation. The same two applications of the principle of maximizing expected utility as used above leads to a solution to the sequential problem. Unfortunately, a serious computational difficulty arises that usually hinders the full operation of this program. It turns out that the choice of e_n cannot be made until e_{n+1} has been selected, which in turns depends on e_{n+2}, and so on. We cannot decide what to do today until we have decided what to do with the tomorrows that today's decisions might bring. Consequently, the calculation has to proceed backwards: from $n+1$ to n, to $n-1$, and back to the start. It often happens that a procedure that looks only one stage ahead at a time gives a reasonable answer, but this is not always so. Only in the case of sequential testing of one simple hypothesis against another, leading to Wald's sequential probability ratio test*, does a complete solution, ignoring boundary effects, exist.

We have seen that in addition to the usual probability $p(x|\theta)$ used in a statistical argument, the coherent approach introduces a density for θ. Before Bayesian ideas can be implemented, a way has to be found for determining $p(\theta)$ as the person's initial opinion about θ. [Notice that $p(x|\theta)$ is also the person's opinion about x were he or she to know θ, but this is only a difference of interpretation from the frequency view, whereas $p(\theta)$ does not have a meaning in that approach.] In the case of the exponential family*, a conjugate distribution* is often used. In the univariate case the density for the exponential family takes the form

$$p(x|\theta) = e^{x\theta}f(x)g(\theta).$$

A member of the conjugate family has the structure

$$p(\theta) = e^{a\theta}g(\theta)^b K(a,b)$$

for some values of a and b that make this a density. The advantage of the conjugate form is that $p(\theta|x)$ has the same structure as $p(\theta)$ (see CONJUGATE FAMILIES OF DISTRIBUTIONS). For by Bayes' theorem, $p(\theta|x) \propto p(x|\theta)p(\theta) \propto e^{(a+x)\theta}g(\theta)^{b+1}$, as before but with $a+x$ for a and b increased by 1. Equation (2) illustrates this. In the Bernoulli case where $p(x|\theta) = \theta^r(1-\theta)^{n-r}$, the conjugate family is that of beta distributions*. The idea, then, is for the values of a and b to be selected and calculations to proceed from there. Bayes' original approach did this in the Bernoulli situation, choosing the uniform distribution* in the unit interval, a special beta distribution. These ideas work well if a member of the conjugate family does adequately describe one's opinion, when the mathematics is tractable and results easily obtained, but is otherwise less satisfactory and, in any case, rarely applies outside the exponential family. There are also difficulties in the multivariate case where the conjugate family is too restricted. Another approach tries to replace $p(\theta)$ by a distribution that describes very little initial knowledge of θ so that the major contribution to the inferential $p(\theta|x)$ comes from the likelihood $p(x|\theta)$, and the data "speak for themselves." For example, small values of b above are clearly indicated since b increases with each observation. Such distributions may be described as "objective," as explained above, and might be used by common agreement among all scientists. Unfortunately, many of these distributions lead to curious paradoxes, usually caused by the resulting objective distributions being improper*, i.e., not having a finite integral, although the work on eliminating these continues. Another approach, using hyper-parameters, will be mentioned later.

The methods just mentioned do not address themselves to the fundamental problem of describing someone's uncertainty about a set of quantities, a description which, according to the coherent argument, must be probabilistic. For simplicity, take the case of an uncertain quantity which takes only one of two values, conventionally 1 and 0, such as an event that can either occur or not. How can one assess the probability that the event will occur? One way of doing this is through a scoring rule which assigns a score $\phi(u)$ if the event occurs after having been assigned probability u, and $\phi(1-u)$ if it does not, since the nonoccurrence will have probability $(1-u)$. If p is the probability, then the expected score, if u is assigned, is $\phi(u)p + \phi(1-u)(1-p)$. If this is identified with the individual's expected utility, or more usually, its negative when the score is a penalty, this should be a minimum when u is p, since otherwise one will be motivated to assign u a value other than p. For this to happen $\phi'(p)p = \phi'(1-p)(1-p)$ and a rule with this property is called proper. $\phi(u) = (1-u)^2$ is an example of such a rule; $\phi(u) = u$ is improper and leads always to stating 0 or 1. Another proper rule is $\phi(u) = \log u$ and this leads again to Shannon's information. Proper scoring rules have been used in weather forecasting. They should be of value in the forecasting of other time series.

We consider how Bayesian ideas apply to the popular statistical situations. Easily the most widely used one in statistics is the general linear model*, in which the data x have expectations that are known linear functions of unknown parameters θ, with the associated concepts of least squares* and the analysis of variance*. The linear relation $E(x) = A\theta$ describes one aspect of $p(x|\theta)$. It is usual to complete the model by supposing $p(x|\theta)$ to be normal* with a known, or partially known, variance–covariance

matrix, although there is a purely linear Bayesian theory which assumes that $E(\theta|x)$ is also linear and parallels the classical Gauss–Markov theory*. To proceed with the coherent analysis it is necessary to assess $p(\theta)$. Let us consider the case where $x = x^{(n)}$ and $E(x_i|\theta) = \theta_i$, where, for given θ and σ^2, the x_i are uncorrelated with equal variance σ^2. This is a canonical form for the linear model. Now it will typically happen that the θ's are related; for example, θ_i may be the true yield of the ith variety, and all yields will increase under favorable conditions. Generally, we may regard the θ's as exchangeable and, more restrictively but in the spirit of the linear model, suppose them to be a random sample from $N(\mu, \tau^2)$ with τ^2 known. Then μ, corresponding to some average yield over varieties, can itself be assigned a distribution. Hence the distribution of θ is described in two stages: for fixed μ, and then over μ. μ is an example of a hyperparameter, describing the distribution of θ just as θ describes that of x, its role here being to exhibit correlation between the θ_i. It turns out that the inference $p(\theta|x)$ is insensitive to the distribution of μ provided that its variance is large and that then the inference is normal with means $E(\theta_i|x) = (\tau^2 x_i + \sigma^2 x.)/(\tau^2 + \sigma^2)$, where $x. = n^{-1} \sum x_i$, the overall mean. The usual least-squares estimate* is x_i, obtained here only as $\tau \to \infty$. The effect of the prior distribution, which recognizes the similarity between the θ's, is to "shrink" the means of the posterior distribution toward the overall mean $x.$, having most effect on the extreme values of x_i. The argument can be extended to the case where σ^2 and τ^2 are themselves unknown. Notice that this canonical form is essentially the problem of estimating the mean θ of a multivariate normal distribution*. The Bayesian approach suggests that the sample mean may not often be a sensible estimate. Similar results apply to the general, linear regression* problem where the largest and the smallest of the sample regression coefficients are typically reduced in value in a procedure closely akin to ridge regression*. One interesting feature of these results is that the least-squares estimate is replaced by another. Even in the sampling-theory view, the least-squares estimate is inadmissible* except when only a few parameters are involved, so that the least-squares procedure needs a complete rethink. Closely related to least squares is the procedure of the analysis of variance with its series of significance tests and various estimates of error. We have remarked above that significance tests, with their tail-area probabilities obtained by integration, play no role in coherent analyses, although it is always possible to calculate the probability of a null hypothesis* given the data. This can be done in the case of the linear model, with the result that one obtains procedures that, for any sample size, can closely parallel the routine of significance tests but which behave quite differently as the sample size changes. For example, let $x^{(n)}$ be a random sample from $N(\theta, 1)$: then a conventional significance test rejects the hypothesis that $\theta = 0$ if $\bar{x}_n^2 > c_1 n^{-1}$, where $\bar{x}_n$ is the sample mean and c_1 is a constant, depending on the chosen level. The Bayes procedure shows that under reasonable assumptions, which include a nonzero prior value for $p(\theta = 0), p(\theta = 0|x^{(n)}) \leqslant \alpha$ when $\bar{x}^2 > c_2 n^{-1} + n^{-1} \log n$, where $c_2 = c_2(\alpha)$. Clearly, for any n, these two inequalities can be made to match for suitable choices of c_1 and c_2, but the agreement cannot be complete for all n. This is an example of coherence between results for different sample sizes, a concept omitted from orthodox statistical theory.

Bayesian analysis of variance does not need significance tests. We illustrate with a two-way, balanced design* (see BALANCING IN EXPERIMENTAL DESIGN) with m rows, n columns, r replicates per cell, and observations $x_{ijk}(i = 1, \ldots, m; j = 1, \ldots, n; k = 1, \ldots, r)$ having $E(x_{ijk}) = \theta_{ij}$. Then using ideas of exchangeability between rows and between columns, in extension of the ideas just discussed, the posterior mean $E(\theta_{ij}|x)$ is of the form, using the usual dot notation for sample means,

$$w_I(x_{ij\cdot} - x_{i\cdot\cdot} - x_{\cdot j\cdot} + x_{\cdot\cdot\cdot})$$
$$+ w_R(x_{i\cdot\cdot} - x_{\cdot\cdot\cdot}) + w_C(x_{\cdot j\cdot} - x_{\cdot\cdot\cdot}) + w x_{\cdot\cdot\cdot},$$

where the w's are weights that can be calculated from the data. [The earlier mean was of this form. $w(x_i - x_\cdot) + (1 - w)x_\cdot$.] If the interaction* is thought, in the light of the data, to be small, then w_I will be small and the interaction, $(x_{ij\cdot} - x_{i\cdot\cdot} - x_{ij\cdot} + x_{\cdot\cdot\cdot})$'s, contribution to the mean will be small. In this way, tests are avoided and a main effect* or interaction included by an amount that is dictated by the weights.

The Bayesian method, or recipe, or paradigm, is complete and provides a tool for all inference (and decision) problems. Bayesian inference is not a branch of statistics: it is a new way of looking at the whole of statistics. Its potential is enormous, but its implementation requires extensive calculations, principally in the performance of numerical integration (to remove nuisance parameters and to find expectations) and maximization (to select optimal decisions). Suitable computer packages to do these calculations, and to interrogate the scientist on his or her knowledge about the situation and convert it into probability forms, have just begun to be written. Until they are more widely available, applications to elaborate sets of data are bound to be limited.

BIBLIOGRAPHY

Box, George E. P. and Tiao, George C. (1973). *Bayesian Inference in Statistical Analysis.* Addison-Wesley, Reading, Mass. (An excellent account of standard statistical models from a Bayesian viewpoint.)

de Finetti, Bruno (1974–1975). *Theory of Probability. A Critical Introductory Treatment* (translated from the Italian by Antonio Machi and Adrian Smith), 2 vols. Wiley, London. (An outstanding book on subjective probability; no one could claim to understand this topic without having read it.)

DeGroot, Morris H. (1970). *Optimal Statistical Decisions.* McGraw-Hill, New York. (In addition to a Bayesian treatment of statistical models, particularly sequential experimentation, this book contains a fine account of the axiomatic treatment. Excellent bibliography.)

Jeffreys, Harold (1961). *Theory of Probability*, 3rd ed. Clarendon Press, Oxford. (An eminent geophysicist writing originally about the theory and practice of statistics. The first edition appeared in 1939, but it is still fresher than many later books.)

Lindley, D. V. (1971). *Bayesian Statistics: A Review.* SIAM, Philadelphia. (A review in outline of the whole field, with an extensive bibliography.)

Novick, Melvin R. and Jackson, Paul H. (1974). *Statistical Methods for Educational and Psychological Research*. McGraw-Hill, New York. (Perhaps the best book on statistical methods from a Bayesian view.)

Raiffa, Howard and Schlaifer, Robert (1961). Applied Statistical Decision Theory. Harvard University, Boston. (Bayesian decision theory with extensive technical developments.)

Savage, Leonard J. (1972). *The Foundations of Statistics*, 2nd ed. Dover, New York. [The first edition (1954) started the modern Bayesian movement.]

Zellner, Arnold (1971). *An Introduction to Bayesian Inference in Econometrics*. Wiley. New York. (A comprehensive account of linear econometric models using objective priors.)

See also BAYES FACTORS; BAYESIAN FORECASTING; BAYES *p*-VALUES; BAYES' THEOREM; BELIEF, DEGREES OF; COHERENCE—I; CONJUGATE FAMILIES OF DISTRIBUTIONS; DECISION THEORY; FIDUCIAL INFERENCE; PROBABILITY, FOUNDATIONS OF—I; and INFERENCE, STATISTICAL—I.

D. V. LINDLEY

BAYESIAN MODEL SELECTION

The Bayesian approach to model selection employs prior probability* distributions to describe the uncertainty surrounding all unknowns in a problem, including the set of models under consideration. The posterior distribution* on the set of models is then used to select a model or make predictions, for example, by choosing that model which has highest posterior probability*.

More precisely, suppose K models $M_1, \ldots, M_K$ are candidates for describing the data $X = (X_1, \ldots, X_n)$. Under model M_k, for $k = 1, \ldots, K$, the data X has density $p(x|\theta_k, M_k)$ where θ_k is a vector of unknown parameters. The Bayesian approach proceeds by assigning a prior distribution* $\pi(\theta_k|M_k)$ to the parameters of each model, and a prior probability $\Pr(M_k)$ to each model. In this context, the prior $\pi(\theta_k|M_k)$ is often considered to be part of the model. Presumably the probabilities $\Pr(M_1), \ldots, \Pr(M_K)$ reflect the analyst's views about the truth of the various models before having observed the data, although this is not strictly necessary.

Under the full prior specification, the posterior probability of model M_k when $X = x$ is

obtained by Bayes' theorem* as

$$\Pr(M_k|x) = \frac{p(x|M_k)\Pr(M_k)}{\sum_{j=1}^{K} p(x|M_j)\Pr(M_j)}, \quad (1)$$

where

$$p(x|M_k) = \int p(x|\theta_k, M_k)\pi(\theta_k|M_k)d\theta_k \quad (2)$$

is the marginal or predictive density (*see* PREDICTIVE ANALYSIS) of X under M_k. Since $\pi(x|\theta_k, M_k)$, regarded as a function of θ_k, is the likelihood* of θ_k under $M_k, p(x|M_k)$ may also be regarded as a marginal or integrated likelihood.

The posterior distribution $\Pr(M_1|x), \ldots, \Pr(M_K|x)$ provides an explicit representation of model uncertainty after having observed the data $X = x$. By treating the posterior probability $\Pr(M_k|x)$ as a measure of the "truth" of the model M_k, a natural strategy for model selection is to choose the M_k for which $\Pr(M_k|x)$ is largest. However, as discussed below, alternative strategies may also be considered.

Pairwise comparisons of model probabilities are obtained by posterior odds ratios (*see* ODDS RATIO ESTIMATORS), which, by (1), can be expressed as

$$\frac{\Pr(M_k|x)}{\Pr(M_j|x)} = \frac{p(x|M_k)}{p(x|M_j)} \times \frac{\Pr(M_k)}{\Pr(M_j)}. \quad (3)$$

The integrated likelihood ratio in (3), namely

$$B_{kj}(x) = \frac{p(x|M_k)}{p(x|M_j)},$$

is the *Bayes factor** for M_k against M_j, and serves to update the prior odds ratio* $\Pr(M_k)/\Pr(M_j)$ to yield the posterior odds ratio. When $B_{kj}(x) > 1$, the data X increase the plausibility of M_k over M_j. Because it does not depend on the prior odds, the Bayes factor is sometimes used as a summary of the evidence provided by the data for M_k against M_j; of course, the Bayes factor still depends on the choice of priors $\pi(\theta_k|M_k)$ and $\pi(\theta_j|M_j)$ for the parameters within each model. Note that the posterior probabilities can be expressed in terms of Bayes factors as

$$\Pr(M_k|x) = \left(\sum_{j=1}^{K} \frac{\Pr(M_j)}{\Pr(M_k)} \times B_{jk}(x) \right)^{-1}.$$

Further discussion and references on Bayes factors are given in refs. [5] and [24]; *see also* BAYES FACTORS.

SOME ILLUSTRATIVE EXAMPLES

The formulation for Bayesian model selection is very general and encompasses a wide range of problems. As a first example, consider the problem of choosing between two competing parametric families for the data X. Suppose that under model M_1, X is a random sample from the geometric distribution* with probability mass function

$$p(\boldsymbol{x}|\theta_1, M_1) = \theta_1^n(1-\theta_1)^s,$$

where $s = \sum_i^n x_i$, but under M_2, X is a random sample from the Poisson distribution* and

$$p(\boldsymbol{x}|\theta_2, M_2) = \frac{\exp(-n\theta_2)\theta_2^s}{\prod_{i=1}^n x_i!}.$$

Suppose further that the uncertainty about θ_1 under M_1 is described by the uniform prior for θ_1 on [0, 1], and that the uncertainty about θ_2 under M_2 is described by the exponential prior $\pi(\theta_2|M_2) = \exp(-\theta_2)$ for $0 \leqslant \theta_2 < \infty$. These model specifications yield the marginal distributions

$$p(\boldsymbol{x}|M_1) = \int_0^1 \theta_1^n(1-\theta_1)^s d\theta_1 = \frac{n!s!}{(n+s+1)!}$$

and

$$p(\boldsymbol{x}|M_2) = \int_0^\infty \frac{\exp[-(n+1)\theta_2]\theta_2^s}{\prod_{i=1}^n x_i!} d\theta_2$$
$$= \frac{s!}{(n+1)^{s+1}\prod_{i=1}^n x_i!}.$$

For a particular choice of prior probabilities $\Pr(M_1)$ and $\Pr(M_2)$, the posterior probability of the model M_1 is then

$$\Pr(M_1|\boldsymbol{x}) =$$

$$\frac{\frac{n!}{(n+s+1)!}\Pr(M_1)}{\frac{n!}{(n+s+1)!}\Pr(M_1) + \frac{1}{(n+1)^{s+1}\prod_{i=1}^n x_i!}\Pr(M_2)},$$

and $\Pr(M_2|\boldsymbol{x}) = 1 - \Pr(M_1|\boldsymbol{x})$. The Bayes factor for M_1 against M_2 is

$$B_{12}(\boldsymbol{x}) = \frac{n!(n+1)^{s+1}\prod_{i=1}^n x_i!}{(n+s+1)!}.$$

Although the Bayesian approach provides a complete package for comparing nonnested models, the choice of parameter priors on θ_1 and θ_2 is not innocuous, and can have a strong influence on the comparison. Bernardo and Smith [7] give a more detailed exploration of prior influence in this example.

Next consider the problem of choosing among nested parametric hypotheses. This problem can be subsumed in the model selection framework by using parameter priors $\pi(\theta_k|M_k)$ to distinguish between hypotheses. As a simple example, suppose $X = (X_1, \ldots, X_n)$ is modeled as a random sample from a normal distribution with known variance σ^2, and it is desired to test the simple hypothesis that $\mu = 0$ against the composite hypothesis* that $\mu \neq 0$ with uncertainty about μ described by a normal distribution with large variance and centered at 0.

This problem may be formulated as a Bayesian model selection problem by using the same parametric form for the models M_1 and M_2, namely

$$p(\boldsymbol{x}|\theta_j, M_j) = (2\pi\sigma^2)^{-n/2}$$
$$\times \exp\left(-\frac{1}{2\sigma^2}\sum_{i=1}^n (x_i - \theta_j)^2\right)$$

and assigning different priors to θ_1 and θ_2, specifically $\pi(\theta_1 = 0|M_1) = 1$ and

$$\pi(\theta_2|M_2) = (2\pi\tau)^{-1/2}\exp\left(-\frac{\theta_2^2}{2\tau^2}\right) \quad (4)$$

with τ^2 chosen to be large. Letting $\bar{x} = n^{-1}\sum x_i$, these priors yield marginal distributions of the form

$$p(\boldsymbol{x}|M_1) = h(\boldsymbol{x})\exp\left(-\frac{\bar{x}^2}{2\sigma^2/n}\right),$$

$$p(\boldsymbol{x}|M_2) = h(\boldsymbol{x})\left(\frac{\sigma^2}{n} + \tau^2\right)^{-n/2}$$
$$\times \exp\left(-\frac{\bar{x}^2}{2(\sigma^2/n + \tau^2)}\right).$$

The posterior odds of M_1 against M_2 are obtained from the prior odds $\Pr(M_1)/\Pr(M_2)$ as

$$\frac{\Pr(M_1|\boldsymbol{x})}{\Pr(M_2|\boldsymbol{x})} = B_{12}(\boldsymbol{x}) \times \frac{\Pr(M_1)}{\Pr(M_2)},$$

where

$$B_{12}(\boldsymbol{x}) = \left(\frac{\sigma^2}{n + \tau^2}\right)^{n/2}$$

$$\times \exp\left(-\frac{\tau^2 \bar{x}^2}{2(\sigma^2/n + \tau^2)\sigma^2/n}\right) \quad (5)$$

is the Bayes factor for M_1 against M_2.

The dependence on τ^2 in (5) may seem troubling in that $B_{12}(\boldsymbol{x}) \to \infty$ as $\tau^2 \to \infty$. Thus, for any fixed value of $\bar{x}$, no matter how extreme, the posterior odds in favor of M_1 can be made arbitrarily large by increasing τ. This phenomenon, which can conflict with the result of a conventional significance test, is sometimes referred to as *Bartlett's paradox* [4] or *Lindley's paradox** [26]; see refs. 7, 15 for further discussion and references.

As a third example, we consider an application of Bayesian model selection to a highly structured problem: variable selection in multiple linear regression*. Suppose $\boldsymbol{Y}$ is an $n \times 1$ vector of observations, thought to depend linearly on some subset of covariate vectors $\boldsymbol{z}_1, \ldots, \boldsymbol{z}_p$. The problem is to select from among the $K = 2^p$ models of the form

$$p(\boldsymbol{y}|\theta_k, M_k) = (2\pi\sigma^2)^{-n/2}$$

$$\times \exp\left(-\frac{1}{2\sigma^2}(\boldsymbol{y} - \boldsymbol{Z}_k\boldsymbol{\beta}_k)^{\mathrm{T}}(\boldsymbol{y} - \boldsymbol{Z}_k\boldsymbol{\beta}_k)\right),$$

where $\theta_k = (\boldsymbol{\beta}_k, \sigma)$, $\boldsymbol{Z}_k$ is an $n \times q_k$ matrix whose columns are a subset of $\boldsymbol{z}_1, \ldots, \boldsymbol{z}_p$, and $\boldsymbol{\beta}_k$ is a $q_k \times 1$ vector of regression coefficients.

A convenient prior for the parameter θ_k of each model is the standard normal-gamma form

$$\pi(\boldsymbol{\beta}_k, \sigma^2|M_k) = \pi(\boldsymbol{\beta}_k|\sigma^2, M_k)\pi(\sigma^2|M_k)$$

with

$$\pi(\boldsymbol{\beta}_k|\sigma^2, M_k) = (2\pi\sigma^2)^{-qk/2}|\boldsymbol{\Sigma}_k|^{-1/2}$$

$$\times \exp\left(-\frac{1}{2\sigma^2}\boldsymbol{\beta}_k^{\mathrm{T}}\boldsymbol{\Sigma}_k^{-1}\boldsymbol{\beta}_k\right),$$

$$\pi(\sigma^2|M_k) = \frac{(\nu\lambda/2)^{\nu/2}\sigma^{-(\nu+2)}}{\Gamma(\nu/2)}$$

$$\times \exp\left(-\frac{\nu\lambda}{2\sigma^2}\right),$$

where ν, λ, and $\boldsymbol{\Sigma}_k$ are hyperparameters chosen by the investigator. This yields

$$p(\boldsymbol{y}|M_k)$$

$$= \frac{\Gamma\left(\frac{\nu+n}{2}\right)(\nu\lambda)^{\nu/2}}{\pi^{n/2}\Gamma\left(\frac{\nu}{2}\right)}|\boldsymbol{I} + \boldsymbol{Z}_k^{\mathrm{T}}\boldsymbol{\Sigma}_k\boldsymbol{Z}_k|^{-1/2}$$

$$\times [\lambda\nu + \boldsymbol{y}^{\mathrm{T}}(\boldsymbol{I} + \boldsymbol{Z}_k^{\mathrm{T}}\boldsymbol{\Sigma}_k\boldsymbol{Z}_k)^{-1}\boldsymbol{y}]^{-(\nu+n)/2}.$$

If all models were considered equally likely, so that $\Pr(M_1) = \cdots = \Pr(M_K) = 2^{-p}$, the posterior probability of model M_k after observing $\boldsymbol{Y} = \boldsymbol{y}$ would then be obtained as

$$\Pr(M_k|\boldsymbol{y}) =$$

$$\frac{|\boldsymbol{I} + \boldsymbol{Z}_k^{\mathrm{T}}\boldsymbol{\Sigma}_k\boldsymbol{Z}_k|^{-1/2}[\lambda\nu + \boldsymbol{y}^{\mathrm{T}}(\boldsymbol{I} + \boldsymbol{Z}_k^{\mathrm{T}}\boldsymbol{\Sigma}_k\boldsymbol{Z}_k)^{-1}\boldsymbol{y}]^{-(\nu+n)/2}}{\sum_{j=1}^{2p}|\boldsymbol{I} + \boldsymbol{Z}_j^{\mathrm{T}}\boldsymbol{\Sigma}_j\boldsymbol{Z}_j|^{-1/2}}.$$

$$\times [\lambda\nu + \boldsymbol{y}^{\mathrm{T}}(\boldsymbol{I} + \boldsymbol{Z}_j^{\mathrm{T}}\boldsymbol{\Sigma}_j\boldsymbol{Z}_j)^{-1}\boldsymbol{y}]^{-(\nu+n)/2}$$

As in the previous example, this application could also be treated directly as a problem of choosing among parametric hypotheses. To see this, the formulation above can be expressed using the same saturated form for each model, i.e.,

$$p(\boldsymbol{y}|\theta_k, M_k) = (2\pi\sigma^2)^{-n/2}$$

$$\times \exp\left(-\frac{1}{2\sigma^2}(\boldsymbol{y} - \boldsymbol{Z}\boldsymbol{\beta})^{\mathrm{T}}(\boldsymbol{y} - \boldsymbol{Z}\boldsymbol{\beta})\right),$$

where $\boldsymbol{Z} = [z_1, \ldots, z_p]$ and $\theta_k = \boldsymbol{\beta} = (\beta_1, \ldots, \beta_p)^{\mathrm{T}}$; one then distinguishes the models by using different priors $\pi(\boldsymbol{\beta}, \sigma^2|M_k)$ on the common parameter space. George and McCulloch [17,18] discuss related approaches and give further references; see also ref. [40].

The potential for the Bayesian approach to model selection goes well beyond these three examples. Many other examples and much useful discussion is given in refs. 3, 7, 8, 12, 19, 21, 27, 29, 35, 41, 44.

DECISION-THEORETIC CONSIDERATIONS

Although choosing the model with maximum posterior probability is a seemingly natural strategy in Bayesian model selection, it is not necessarily the best one. The question of how the posterior is to be used to select a model can be answered by formalizing the problem in the framework of decision theory*, where the goal is to maximize expected utility* [15,7]. More precisely, let m_k represent the action of selecting M_k, and suppose the utility of m_k is given by a utility function $u(m_k, \omega)$, where ω is some unknown of interest. Then the choice which maximizes expected utility is that m_{k^*} for which

$$\overline{u}(m_{k^*}|\boldsymbol{x}) = \max_k \overline{u}(m_k|\boldsymbol{x}),$$

where

$$\overline{u}(m_k|\boldsymbol{x}) = \int u(m_k, \boldsymbol{w})f(\omega|\boldsymbol{x})d\omega \qquad (6)$$

is the expected utility of m_k, and $f(\omega|\boldsymbol{x})$ is the posterior distribution of ω given the observation $\boldsymbol{X} = \boldsymbol{x}$. Based on the posterior (1), this predictive distribution is

$$f(\omega|\boldsymbol{x}) = \sum_{k=1}^{K} f(\omega|M_k, \boldsymbol{x}) \Pr(M_k|\boldsymbol{x}), \qquad (7)$$

where

$$f(\omega|M_k, \boldsymbol{x}) = \int f(\omega|\boldsymbol{\theta}_k, M_k)\pi(\boldsymbol{\theta}_k|M_k, \boldsymbol{x})d\boldsymbol{\theta}_k$$

is the predictive distribution of ω conditional on M_k and $\boldsymbol{x}$.

If ω identifies one of $M_1, \ldots, M_K$ as the true state of nature, and $u(m_k, \omega)$ is 1 or 0 according as a correct selection has or has not been made, then the highest posterior probability model maximizes the expected utility. However, different utility functions can lead to other selections. For example, suppose m_k entails choosing $f(\cdot|M_k, \boldsymbol{x})$ as a predictive distribution for a future observation ω, and that this selection is to be evaluated by the logarithmic score function $u(m_k, \omega) = \ln f(\omega|M_k, \boldsymbol{x})$. Then the selection which maximizes expected utility is that m_{k^*} which maximizes the posterior weighted logarithmic divergence

$$\sum_{k=1}^{K} \Pr[M_k|\boldsymbol{x}] \int f(\omega|M_k, \boldsymbol{x}) \ln \frac{f(\omega|M_k, \boldsymbol{x})}{f(\omega|M_{k^*}, \boldsymbol{x})} d\omega,$$

as shown in ref. 42. Also, if the goal is strictly prediction and not model selection, then expected logarithmic utility is maximized by

using the posterior weighted mixture of $f(\omega|M_k, \boldsymbol{x})$, namely $f(\omega|\boldsymbol{x})$, in (7). This model averaging or mixing procedure incorporates model uncertainty into prediction and has been advocated by many researchers [10,11,12,40,41]. However, if a cost for model complexity is introduced into the utility, then model selection procedures may dominate $f(\omega|\boldsymbol{x})$.

Another useful modification of the decision-theory setup is to allow for the possibility that the true model is not one of $M_1, \ldots, M_K$, a commonly held perspective in many applications. This aspect can be incorporated into a utility analysis by using the actual predictive density in place of $f(\omega|\boldsymbol{x})$ in (6). Where the form of the true model is completely unknown, this approach serves to motivate cross-validation* methods that employ training-sample approaches; Bernardo and Smith [7] and Berger and Pericchi [6] provide a discussion of these.

CHOOSING PRIOR INPUTS

Bayesian model selection procedures can be especially sensitive to the prior inputs. From almost any data set, one can obtain an arbitrary posterior inference by manipulating the choice of the model prior $\Pr(M_1), \ldots, \Pr(M_K)$. Bayes factors can depend strongly on the choice of parameter priors $\pi(\boldsymbol{\theta}_1|M_1), \ldots, p(\boldsymbol{\theta}_K|M_K)$, as demonstrated by Lindley's paradox in previous discussion. In contrast to usual Bayesian estimation, the effect of all these prior inputs does not disappear as the sample size goes to infinity.

A simple and seemingly neutral choice of the model prior is to assign uniform probabilities to all models, so that $\Pr(M_1) = \cdots = \Pr(M_K) = 1/K$. However, this choice should be exercised with care, since it may not be uniform on other characteristics of interest. For instance, in the variable selection example above, the uniform prior puts smaller weight on the class of very parsimonious models, since there are fewer of them. A reasonable alternative might be priors that are decreasing functions of the parameter dimension of the model [5,17,23].

Ideally, the choice of parameter priors can be based on problem-specific prior information. For example, in the variable selection problem, George and McCulloch [17,18] develop Bayesian selection procedures that use structured informative priors to select variables on the basis of practical significance. In many model selection problems, however, the amount of prior information will be limited, especially when there is substantial uncertainty about the form of the model. In such cases, it will often be desirable to use relatively non-informative, traditional priors, as in ref. 5. For example, with the normal mean example discussed before, a conventional choice recommended by Jeffreys [23] is the Cauchy prior $\pi(\mu|M_2) = [\pi\sigma(1 + \mu^2/\sigma^2)]^{-1}$ instead of the normal prior in (4). Examples of such prior choices can be found in refs. 2, 25, 31, 27, 28, 37, 40, 36, 45, 47, 51, 53, 54.

APPROXIMATE BAYESIAN METHODS

Although the use of noninformative parameter priors would be especially desirable for Bayesian model selection, many of the standard choices are improper and so destroy the interpretation of $\Pr(M_k|x)$ as a genuine posterior probability. Even a formal approach can be problematic, because improper priors are only defined up to constant multiples and so lead to Bayes factors that are multiples of those arbitrary constants. This is especially troublesome when comparing distributions of quantities having different dimensions.

A variety of approximate Bayesian methods have been developed to overcome such difficulties. In the context of normal data model selection with improper priors, Smith and Spiegelhalter [45] have proposed a multiplicative correction for the Bayes factor based on using imaginary training samples. Perrichi [34] proposed a multiplicative correction based on information adjustments.

Another route that avoids the pitfalls of improper priors is to directly use part of the data as a training sample in order to obtain an alternative marginal density as follows. Let S_1 and S_2 be two subsets of X. Define the formal predictive density of S_1 given S_2 under M_k by

$$p(S_1|S_2, M_k)$$
$$= \int p(S_1|S_2, \theta_k, M_k)\pi(\theta_k|S_2, M_k)d\theta_k.$$

Note that training samples S_2 that convert an improper prior $\pi(\theta_k|M_k)$ into a proper $\pi(\theta_k|S_2, M_k)$ will yield a proper $p(S_1|S_2, M_k)$ distribution. Bayes factor approximations using $p(S_1|S_2, M_k)$ instead of $p(x|M_k)$ have been proposed in refs. 13, 5, 6. See also refs. 1, 14, 33 for related procedures and further references.

The choice of a prior can also be avoided completely by comparing models on the basis of the Bayes information criterion (BIC) or Schwarz criterion* [43]; here

$$\text{BIC} = -2\ln\frac{p(x|\hat{\theta}_k, M_k)}{p(x|\hat{\theta}_j, M_j)} + (q_k - q_j)\ln n,$$

where $\hat{\theta}_k$ and $\hat{\theta}_j$ are the maximum likelihood estimates of θ_k and θ_j, q_k and q_j are the dimensions of θ_k and θ_j, and n is the sample size. In general exponential families*, the BIC is an asymptotic approximation to minus two times the logarithm of the Bayes factor, in the sense that $-2\ln B_{kj}(x) = \text{BIC} + O(1)$ as $n \to \infty$ [22]. Although the influence of the priors, which is absorbed in the $O(1)$ term (see O, o NOTATION), disappears asymptotically, this influence can be substantial in finite samples. Kass and Wasserman [25] show that in certain nested model comparison problems, the BIC corresponds to using a reference prior.

COMPUTATIONAL CONSIDERATIONS

The calculations required for evaluation of the posterior $\Pr(M_k|x)$ in (1) and (2) can be formidable, especially in large problems. To begin with, analytical evaluation of the marginal density $p(x|M_k)$ in (2) is only possible in special cases such as exponential family distributions with conjugate priors. Furthermore, when the number of models under consideration is very large, which can happen in highly structured problems such as the previous variable selection example, exhaustive evaluation of all K probabilities $\Pr(M_1|x), \ldots, \Pr(M_K|x)$ can be prohibitively expensive, even when closed-form expressions for the marginal densities are available. For example, even in conjugate prior variable selection problems with p covariates and 2^p models, George and McCulloch [18] find that exhaustive evaluation with specialized algorithms is only practical for p less than about 25.

Markov chain Monte Carlo* methods such as Gibbs sampling* and the Metropolis-Hastings algorithm provide a general approach for over-coming some of the obstacles to posterior evaluation. Indeed, in many problems the need for analytical simplification can be entirely avoided by using these methods to obtain an approximate sample from the full posterior $\pi(\boldsymbol{\theta}_k, M_k | \boldsymbol{x})$ [8,17,35,21]. Alternatively, when closed-form expressions for $p(\boldsymbol{x}|M_k)$ are available, these methods can obtain an approximate sample directly from $\Pr(M_k|\boldsymbol{x})$ [18,27,40]. Such samples can be used to estimate a wide variety of posterior characteristics, and although they cannot realistically be used to estimate the full posterior in huge problems, they can still be useful for posterior exploration and for stochastic search of the probability models with high posterior probability that are of principal interest. When the number of models is small or moderate, as in hypothesis testing* problems, evaluation of marginal densities and Bayes factors may also be accomplished by Monte Carlo methods based on importance sampling* [13], sampling from the priors [28], and sampling from the posteriors [9,14,24,30,32,38,52].

Nonstochastic methods can also play a valuable role in posterior evaluation. For example, when expressions for $p(\boldsymbol{x}|M_k)$ are available, an alternative to simulation sampling is the heuristic Occam's Window algorithm of Madigan and Raftery [27], which can identify promising models in large problems. Finally, in problems where analytical evaluation of $p(\boldsymbol{x}|M_k)$ is intractable, a frequently used asymptotic method based on the Laplace approximation* is

$$p(\boldsymbol{x}|M_k) \approx (2\pi)^{d/2} |\boldsymbol{A}_k|^{1/2} p(\boldsymbol{x}|\tilde{\boldsymbol{\theta}}_k, M_k) p(\tilde{\boldsymbol{\theta}}_k|M_k),$$

where $\tilde{\boldsymbol{\theta}}_k$ is the posterior mode of $\boldsymbol{\theta}_k$ under M_k and $\boldsymbol{A}$ is the negative inverse Hessian of $h(\boldsymbol{\theta}_k) = \ln[p(x|\boldsymbol{\theta}_k, M_k)\pi(\boldsymbol{\theta}_k|M_k)]$ evaluated at $\tilde{\boldsymbol{\theta}}_k$ [49,50,24]. This approximation is often fast and accurate, and can also be used to facilitate the use of other methods, as in refs. 3, 38, 41.

REFERENCES

1. Aitkin, M. (1991). Posterior Bayes factors. *J. R. Statist. Soc. B*, **53**, 111–142.

2. Albert, J. H. (1990). A Bayesian test for a two-way contingency table using independence priors. *Can. J. Statist.*, **14**, 1583–1590.

3. Albert, J. A. (1996). Bayesian selection of log-linear models. *Can. J. Statist.*, **24**, 327–347.

4. Bartlett, M. (1957). A comment on D. V. Lindley's statistical paradox. *Biometrika*, **44**, 533–534.

5. Berger, J. O. and Perrichi, L. R. (1996). The intrinsic Bayes factor for model selection and prediction. *J. Amer. Statist. Ass.*, **91**, 109–122.

6. Berger, J. O. and Pericchi, L. R. (1995). The intrinsic Bayes factor for linear models. In *Bayesian Statistics 5*, J. M. Bernardo et al., eds. Oxford University Press, pp. 23–42.

7. Bernardo, J. M. and Smith, A. F. M. (1994). *Bayesian Theory*. Wiley, New York.

8. Carlin, B. P. and Chib, S. (1995). Bayesian model choice via Markov chain Monte Carlo. *J. R. Statist. Soc. B*, **57**, 473–484.

9. Chib, S. (1995). Marginal likelihood from the Gibbs output. *J. Amer. Statist. Ass.*, **90**, 1313–1321.

10. Clyde, M., Desimone, H., and Parmigiani, G. (1996). Prediction via orthogonalized model mixing. *J. Amer. Statist. Ass.*, **91**, 1197–1208.

11. Draper, D. (1995). Assessment and propagation of model uncertainty (with discussion). *J. R. Statist. Soc. B*, **57**, 45–98.

12. Geisser, S. (1993). *Predictive Inference: An Introduction*. Chapman and Hall, London.

13. Geisser, S. and Eddy, W. F. (1979). A predictive approach to model selection. *J. Amer. Statist. Ass.*, **74**, 153–160.

14. Gelfand, A. E. and Dey, D. K. (1994). Bayesian model choice: Asymptotics and exact calculations. *J. R. Statist. Soc. B.* **56**, 501–514.

15. Gelfand, A. E., Dey, D. K., and Chang, H. (1992). Model determination using predictive distributions with implementations via sampling-based methods. In *Bayesian Statistics 4*, J. M. Bernardo et al., eds. Oxford University Press, pp. 147–167.

16. Gelfand, A. and Smith, A. F. M. (1990). Sampling-based approaches to calculating marginal densities. *J. Amer. Statist. Ass.*, **85**, 398–409.

17. George, E. I. and McCulloch, R. E. (1993). Variable selection via Gibbs sampling. *J. Amer. Statist. Ass.*, **88**, 881–889.

18. George, E. I. and McCulloch, R. E. (1997). Approaches to Bayesian variable selection. *Statist. Sinica*, **7**, 339–373.

19. George, E. I., McCulloch, R. E., and Tsay, R. (1996). Two approaches to Bayesian model selection with applications. In *Bayesian Statistics and Econometrics: Essays in Honor of Arnold Zellner*, D. Berry, K. Chaloner, and J. Geweke, eds. Wiley, New York, pp. 339–348.

20. Geweke, J. (1989). Bayesian inference in econometric models using Monte Carlo integration. *Econometrica*, **57**, 1317–1339.

21. Green, P. J. (1995). Reversible jump MCMC computation and Bayesian model determination. *Biometrika*, **82**, 711–732.

22. Haughton, D. (1988). On the choice of a model to fit data from an exponential family. *Ann. Statist.*, **16**, 342–355.

23. Jeffreys, H. (1961). *Theory of Probability*, 3rd ed. Clarendon Press, Oxford.

24. Kass, R. E. and Raftery, A. (1995). Bayes factors and model uncertainty. *J. Amer. Statist. Ass.*, **90**, 773–795.

25. Kass, R. E. and Wasserman, L. (1995). A reference Bayesian test for nested hypothesis and its relationship to the Schwarz criterion. *J. Amer. Statist. Ass.*, **90**, 928–934.

26. Lindley, D. V. (1957). A statistical paradox. *Biometrika*, **44**, 187–192.

27. Madigan, D. and Raftery, A. E. (1994). Model selection and accounting for model uncertainty in graphical models using Occam's window. *J. Amer. Statist. Ass.*, **80**, 1535–1546.

28. McCulloch, R. E. and Rossi, P. E. (1992). Bayes factors for nonlinear hypotheses and likelihood distributions. *Biometrika*, **70**, 663–676.

29. McCulloch, R. E. and Tsay, R. (1994). Bayesian inference of trend and difference stationarity. *Econometric Theory*, **10**, 596–608.

30. Meng, X. -L. and Wong, W. H. (1996). Simulating ratios of normalizing constants via a simple identity: a theoretical exploration. *Statist. Sinica*, **6**, 831–860.

31. Mitchell, T. J. and Beauchamp, J. J. (1988). Bayesian variable selection in regression. *J. Amer. Statist. Ass.*, **83**, 1023–1032.

32. Newton, M. A. and Raftery, A. E. (1994). Approximate Bayesian inference with the weighted likelihood bootstrap. *J. R. Statist. Soc. B*, **56**, 3–48.

33. O'Hagan, A. (1995). Fractional Bayes factors for model comparisons. *J. R. Statist. Soc. B*, **57**, 99–138.

34. Pericchi, L. (1984). An alternative to the standard Bayesian procedure for discrimination between normal linear models. *Biometrika*, **71**, 576–586.

35. Phillips, D. B. and Smith, A. F. M. (1995). Bayesian model comparison via jump diffusions. In *Markov Chain Monte Carlo in Practice*, W. R. Gilks, S. Richardson and D. J. Spiegelhalter, eds. Chapman and Hall, London, pp. 215–240.

36. Poirier, D. J. (1985). Bayesian hypothesis testing in linear models with continuously induced conjugate priors across hypotheses. In *Bayesian Statistics 2*, J. M. Bernardo et al., eds. Elsevier, New York, pp. 711–722.

37. Raftery, A. E. (1996). Approximate Bayes factors and accounting for model uncertainty in generalized linear models. *Biometrika*, **83**, 251–266.

38. Raftery, A. E. (1996). Hypothesis testing and model selection via posterior simulation. In *Markov Chain Monte Carlo in Practice*, W. R. Gilks, S. Richardson, and D. J. Spiegelhalter, eds. Chapman and Hall, London, pp. 163–188.

39. Raftery, A. E. (1995). Bayesian model selection in social research (with discussion). In *Sociological Methodology 1995*, P. V. Marden, ed. Black-well, Cambridge, Mass., pp. 111–196.

40. Raftery, A. E., Madigan, D. M., and Hoeting, J. (1997). Bayesian model averaging for linear regression models. *J. Amer. Statist. Ass.*, **92**, 179–191.

41. Raftery, A. E., Madigan, D., and Volinsky, C. T. (1995). Accounting for model uncertainty in survival analysis improves predictive performance (with discussion). In *Bayesian Statistics 5*, J. M. Bernardo et al., eds. Oxford University Press, pp. 323–350.

42. San Martini, A. and Spezzaferri, F. (1984). A predictive model selection criterion. *J. R. Statist. Soc. B*, **46**, 296–303.

43. Schwarz, G. (1978). Estimating the dimension of a model. *Ann. Statist.*, **6**, 461–464.

44. Smith, M. and Kohn, R. (1996). Nonparametric regression using Bayesian variable selection. *J. Econometrics*, **75**, 317–344.

45. Smith, A. F. M. and Spiegelhalter, D. J. (1980). Bayes factors and choice criteria for linear models. *J. R. Statist. Soc. B*, **42**, 213–220.

46. Spiegelhalter, D. J. and Smith, A. F. M. (1982). Bayes factors for linear and log-linear models with vague prior information. *J. R. Statist. Soc. B*, **44**, 377–387.

47. Stewart, L. (1987). Hierarchical Bayesian analysis using Monte Carlo integration: Computing posterior distributions when there are many possible models. *Statistician*, **36**, 211–219.

48. Tierney, L. (1991). Exploring posterior distributions using Markov chains. *Computer Sci. Statist.: Proc. 23rd Symp. Interface*, E. M. Keramidas, ed. Interface Foundation, Fairfax Station, Va., pp. 563–570.

49. Tierney, L. and Kadane, J. B. (1986). Accurate approximations for posterior moments and marginal densities. *J. Amer. Statist. Ass.*, **81**, 82–86.

50. Tierney, L., Kass, R. E., and Kadane, J. B. (1989). Fully exponential Laplace approximations to expectations and variances of nonpositive functions. *Amer. Statist. Ass.*, **84**, 710–716.

51. Verdinelli, E. and Wasserman, L. (1995). Bayes factors, nuisance parameters, and imprecise tests. In *Bayesian Statistics 5*, J. M. Bernardo et al., eds. Oxford University Press, pp. 765–772.

52. Verdinelli, E. and Wasserman, L. (1995). Computing Bayes factors using a generalization of the Savage-Dickey density ratio. *J. Amer. Statist. Ass.*, **90**, 614–618.

53. Zellner, A. and Min, C. (1993). Bayesian analysis, model selection and prediction. In *Physics and Probability: Essays in Honor of Edwin T. Jaynes*, W. T. Grandy, Jr., and P. W. Milonni, eds. Cambridge University Press, pp. 195–206.

54. Zellner, A. and Siow, A. (1980). Posterior odds for selected regression hypotheses. In *Bayesian Statistics 1*, J. M. Bernardo et al., eds. Valencia University Press, pp. 585–603. Reply, pp. 638–643.

See also Bayes Factors; Gibbs Sampling; Markov Chain Monte Carlo Algorithms; Model Construction: Selection of Distributions; Model Selection: Bayesian Information Criterion; and Parsimony, Principle of.

Edward I. George

BAYESIAN NETWORKS

EXAMPLES AND DEFINITIONS

Numerous real-life applications involve some elements of uncertainty. Probabilistic reasoning is a common approach for dealing with such uncertainty. Probabilistic reasoning starts with the identification of a set of relevant random variables and then the specification of the relationship among them using a joint probability density (JPD) function. The JPD contains a complete (qualitative and quantitative) information about the relationship among the variables.

Let $X = \{X_1, X_2, \ldots, X_n\}$ be a set of jointly distributed random variables with a JPD function, $p(x) = p(x_1, x_2, \ldots, x_n)$. The variables can be discrete, continuous, or mixed. An example of such random variables is the set of diseases and symptoms. The JPD, when known, contains all the information about the relationships among the variables. For example, it can be used to answer queries such as: What is the probability of one variable (e.g., a disease) given that we have observed a set of other variables (e.g., a set of symptoms)? In practice, the problem of specifying the JPD and of answering such queries can be extremely difficult especially when n is large.

The task of specifying the JPD can be made simpler, however, by using graphical models. Graphical models have the advantages that they display the relationships among the variables explicitly and preserve them qualitatively (i.e., for any numerical specification of the parameters). Graphs are also intuitive and easy to explain. Our objective here is to represent the JPD of a set of random variables by a graph as accurately as possible. *Bayesian networks* and *Markov networks* are two graphically based models for representing the JPD of a set of random variables. This entry discusses Bayesian networks, also known as *Belief networks*. Markov networks are discussed in another entry.

Let $Y = \{Y_1, Y_2, \ldots, Y_n\}$ be a permutation of $X = \{X_1, X_2, \ldots, X_n\}$. By virtue of the *chain rule factorization*, any JPD of a set of variables X can be expressed as a product of n *conditional probability distributions* (CPDs), one for each variable in X, as follows:

$$p(x) = p(x_1, x_2, \ldots, x_n) = \prod_{i=1}^{n} p(x_i | b_i), \quad (1)$$

where $B_i = \{Y_1, Y_2, \ldots, Y_{i-1}\}$ is the set of variables before Y_i. One consequence of this factorization is that only CPDs of single variables are needed to specify the JPD of the entire set of variables. Furthermore, these CPDs become even simpler if the JPD contains several subsets of variables that are independent and/or conditionally independent. To clarify this point, let A, B, and C be three disjoint sets of variables, then A is said to be conditionally independent of C given B, if and only if $p(a|b,c) = p(a|b)$, for all possible values a, b, and c of A, B, and C; otherwise, A and C are said to be conditionally dependent given B. When a set of variables A is conditionally independent of another set C given a third set B, we write $I(A, C|B)$. Such a statement is called a *conditional independence statement* (CIS). For example, if a set B_i in Equation 1 is partitioned as $B_i = \{B_{i1}, B_{i2}\}$, where $I(X_i, B_{i2}|B_{i1})$, then the ith CPD can be simplified to $p(x_i|b_i) = p(x_i|\{b_{i1}, b_{i2}\}) = p(x_i|b_{i1})$, that is, we can specify the ith CPD independently of the variables in the set B_{i2}.

A Bayesian network is a model that makes the specification of the JPD and the propagation of evidence both simple and efficient because it makes use of the chain rule factorization and it also takes into consideration the conditional independence information that may be present in the JPD.

Before we formally define Bayesian networks, let us give an example.

Example 1. A Bayesian network. Consider a set $X = \{X_1, X_2, \ldots, X_7\}$ consisting of seven variables. Let us represent each variable by a node and the relationships among the variables by directed links or arrows as shown in Fig. 1. When a link goes from X_i to X_j, we say that X_i is a *parent* of X_j or that X_j is a child of X_i. For example, the parents of X_4 are X_1 and X_2. Let us denote the set of parents of X_i by Π_i. The graph in Fig. 1 is called a *directed acyclic graph* (DAG) because all the links are arrows and it contains no cycles.

In a DAG, the nodes (variables) can always be numbered in such a way that every node comes before all of its children. Such ordering of the nodes is called an *ancestral ordering* or *numbering*. The nodes in Fig. 1 are given in an ancestral ordering. If the variables are arranged in an ancestral ordering, then Π_i is a subset of the variables preceding X_i. Furthermore, using certain graph separation criteria (to be defined later), one can show

that $I(X_i, \{B_i \setminus \Pi_i\}|\Pi_i)$, for every variable X_i, where B_i is the set of nodes preceding X_i and $B_i \setminus \Pi_i$ is B_i minus the set Π_i. This implies that $p(x_i|b_i) = p(x_i|\pi_i)$, that is, we can specify the ith CPD independently of the variables in the set $B_i \setminus \Pi_i$. Accordingly, if the *structural relationship* among the variables can be represented by a DAG, then the JPD of X can be factorized as

$$p(x) = \prod_{i=1}^{n} p(x_i|\pi_i). \tag{2}$$

That is, the JPD can be expressed as a product of n conditional probability distributions (CPDs), one for each variable in X given its set of parents. If X_i has no parents, then $p(x_i|\pi_i) = p(x_i)$, which is the marginal JPD of X_i. When Π_i is smaller than B_i, the factorization in Equation 2 is simpler than that in Equation (1) because it requires a smaller number of parameters. For example, the factorization in Equation 2 and the DAG in Fig. 1 indicate that the JPD of the seven variables can be specified by the product of seven conditional/marginal probability density functions as follows:

$$p(x_1, x_2, x_3, x_4, x_5, x_6, x_7)$$
$$= p(x_1)p(x_2)p(x_3)p(x_4|x_1, x_2)p(x_5|x_2)$$
$$\times p(x_6|x_3, x_4)p(x_7|x_4). \tag{3}$$

These probability density functions can be read directly from the DAG in Fig. 1, which shows the parents of each node. The DAG in Fig. 1 and the CPDs in Equation 2 constitute a Bayesian network, which defines the JPD of the seven variables.

More formally, a *Bayesian network* is a pair (D, P), where D is a DAG, $P = \{p(x_1|\pi_1), \ldots, p(x_n|\pi_n)\}$ is a set of n conditional probability distributions (CPDs), one for each variable, Π_i is the set of parents of node X_i in D, and the set P defines the associated JPD as

$$p(x) = \prod_{i=1}^{n} p(x_i|\pi_i). \tag{4}$$

Thus, a Bayesian network defines the *dependence structure* or the *structural relationship* among the variables. Note, however, that the CPDs $p(x_i|\pi_i)$ may depend on a set of known

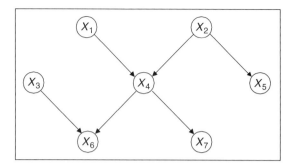

Figure 1. An example of a directed acyclic graph.

or unknown parameters. Bayesian networks are also known as *Belief Networks.*

Given a JPD, we can find the list of all CISs by generating all possible combinations of three disjoint subsets of variables and then checking to see whether they are conditionally dependent or independent. If the JPD implies A and C are conditionally independent given B, we write $I(A, C|B)_M$ to indicate that this CIS is implied by the JPD (i.e., the probabilistic model).

On the other hand, given a DAG, D, we can derive the set of CISs implied by D using a criterion known as *D-separation* [27], [16]. To explain D-separation, we need a few definitions. In a DAG, D, a node X_i is said to be an *ancestor* of node X_j if there is a path from X_i to X_j. The set containing the ancestors of all its nodes is called an *ancestral set.* Finally, the graph obtained from D by first adding links between every pair of nodes with a common child and then dropping the directionality of the links is called a *moral graph.*

To determine whether D implies a CIS such as $I(A, C|B)$, for any three disjoint subsets of nodes A, B, and C in D, we first identify the smallest subgraph containing A, B, and C and their ancestral subsets and then obtain the corresponding moralized graph. Then, we say that B separates A and C if and only if, in the above moralized graph, every path between each node in A and each node in C contains at least one node in B. This is what is referred to as *D-separation.* If B separates A and C in D, we write $I(A, C|B)_D$ to indicate that this CIS is implied by D. For example, given the DAG, D, in Fig. 1, one can show that X_4 D-separates $\{X_1, X_2, X_5\}$ and $\{X_3, X_6, X_7\}$ but X_6 does not D-separate X_3 and X_4.

Accordingly, given a DAG, D, we can find the list of all CISs by generating all possible combinations of three disjoint subsets of variables and checking to see whether they are conditionally dependent or independent using the D-separation criterion. In a sense, the graph can be represented by an equivalent set of CISs.

Therefore, one can generate two sets of CISs, one derived from the JPD, $p(x)$, using the definition of conditional independence and another derived from the DAG, D, using the definition of D-separation. These sets of CISs are known as *input lists.* If the two input lists coincide, we say that D is a *perfect map* of $p(x)$. This raises the following question: Can any JPD be represented perfectly by a DAG? The answer is, unfortunately, no. For examples of JPD that cannot be represented by DAGs see, for example, Reference 9, chapter 6. In cases where a JPD cannot be perfectly represented by a DAG, we should make sure that *(i)* every CIS implied by D is also implied by the JPD, that is, $I(A, C|B)_D \Rightarrow I(A, C|B)_M$ and *(ii)* the number of CISs found in the JPD but not implied by D is kept to a minimum. Every DAG D that satisfies these two requirements is called a *minimal I-map* of the JPD.

From the above discussion, one can think of a Bayesian-network model as an I-map of the probabilistic model given by the JPD because every CIS in D is found in the JPD, but not every CIS in the JPD can be found in D.

There are several types of Bayesian networks depending on whether the variables are discrete, continuous, or mixed, and on the assumed type of distribution for the variables. Two of the most commonly used Bayesian networks are the multinomial Bayesian networks and the Gaussian Bayesian networks. An example of each of these types is given below.

Example 2. Multinomial Bayesian Network.

In a multinomial Bayesian network, all variables in X are discrete, that is, each variable has a finite set of possible values. Without loss of generality, suppose that all variables in Fig. 1 are binary, that is, each can take only one of two possible values (e.g., 0 or 1). In this case, the CPDs are tables of probabilities for the different combinations of values for the variables. An example of the numerical values needed to define the set of CPDs in Equation 3 is shown in Table 1. Thus, the DAG in Fig. 1 and the CPDs in Table 1 constitute what we call a Multinomial Bayesian network, which defines the JPD of the seven variables in Fig. 1. Note that in general the JPD of seven binary variables requires $2^7 - 1 = 127$ free parameters, but the representation of the JPD as in Equation 3 implies that the number of free parameters required to specify the JPD of the seven variables is reduced to $1 + 1 + 1 + 2 + 2 + 4 + 4 = 15$, as can be seen from Table 1. Note that half of the parameters are redundant because they add up to one by groups.

Example 3. Gaussian Bayesian Network.

In a Gaussian (normal) Bayesian network, the variables in X are assumed to have a multivariate normal distribution, $N(\mu, \Sigma)$, where μ is an $n \times 1$ vector of means and Σ is an $n \times n$ variance-covariance matrix. The joint probability density function (PDF) of the set $\{X_i, \Pi_i\}$ is normal with mean vector and variance-covariance matrix given by

$$\begin{pmatrix} \mu_{X_i} \\ \mu_{\Pi_i} \end{pmatrix} \quad \text{and} \quad \begin{pmatrix} \Sigma_{ii} & \Sigma_{\Pi_i, X_i} \\ \Sigma_{X_i, \Pi_i} & \Sigma_{\Pi_i} \end{pmatrix}.$$

Table 1. CPDs Associated with the Multinomial Bayesian Network

x_1	$p(x_1)$
0	0.4
1	0.6

x_2	$p(x_2)$
0	0.2
1	0.8

x_3	$p(x_3)$
0	0.9
1	0.1

x_2	x_5	$p(x_5 \mid x_2)$
0	0	0.5
0	1	0.5
1	0	0.4
1	1	0.6

x_4	x_7	$p(x_7 \mid x_4)$
0	0	0.7
0	1	0.3
1	0	0.3
1	1	0.7

x_1	x_2	x_4	$p(x_4 \mid x_1, x_2)$
0	0	0	0.4
0	0	1	0.6
0	1	0	0.3
0	1	1	0.7
1	0	0	0.6
1	0	1	0.4
1	1	0	0.4
1	1	1	0.6

x_3	x_4	x_6	$p(x_6 \mid x_3, x_4)$
0	0	0	0.8
0	0	1	0.2
0	1	0	0.6
0	1	1	0.4
1	0	0	0.3
1	0	1	0.7
1	1	0	0.2
1	1	1	0.8

A well-known property of the multivariate normal distribution (see, for example, Ref. 31) is that the CPD of X_i given that its parents $\Pi_i = \pi_i$ is normal with (conditional) mean

$$\mu_{X_i} + \Sigma_{X_i,\Pi_i}\Sigma_{\Pi_i}^{-1}(\pi_i - \mu_{\Pi_i}) \qquad (5)$$

and (conditional) variance

$$\Sigma_{ii} - \Sigma_{X_i,\Pi_i}\Sigma_{\Pi_i}^{-1}\Sigma_{\Pi_i,X_i}, \qquad (6)$$

where Σ_{ii} is the variance of X_i.

The JPD of the variables in a Gaussian Bayesian network is specified as in Equation 4 by the product of a set of CPDs, one CPD for each variable. As an example, consider again the DAG given in Fig. 1. Suppose that the seven variables are normally distributed, that is, $f(x_1, x_2, \ldots, x_7) \sim N(\mu, \Sigma)$, where μ is the mean vector of order 7×1 and Σ is the variance-covariance matrix of order 7×7. A normal Bayesian network is defined by specifying the set of CPDs appearing in the factorization Equation 3, where the CPD for X_i given that its parents $\Pi_i = \pi_i$ is normal with mean and variance as given in Equations 5 and 6.

It can be seen from the above discussion and examples that Bayesian-network models have important advantages such as the following:

1. Any CIS derived from the graph D is satisfied by any probabilistic model such that its joint probability can be factorized as in Equation 4.

2. The DAG is minimal in the sense that no link can be removed to keep the validity of the above property.

3. Specification of the JPD of the variables reduces to specification of the conditional probabilities in Equation 4, which is a much simpler task and requires a substantially smaller number of free parameters.

4. No compatibility problem arises because the conditional probabilities in Equation 4 are always compatible due to the chain rule factorization of the JPD.

For these and other reasons, Bayesian networks have been used to solve problems that involve a large number of variables. These applications are frequently encountered in *artificial intelligence*, especially in the building of *expert systems*, see, for example, References 9, 12, 19, 27, and 28.

CONSTRUCTING BAYESIAN-NETWORK MODELS

Although Bayesian networks simplify the construction of the JPD of a set of variables through a factorization such as the one given in Equation 4, in practice, however, the graph D and the corresponding set of CPDs P are often unknown. Consequently, when D and/or P are unknown, an important but very difficult task is how to construct a Bayesian network (D, P) for a given practical problem.

In practice, Bayesian networks can be constructed using information provided by human experts in the domain of knowledge. Building a Bayesian network requires the specification of a DAG D and the corresponding set of CPDs required for the factorization (4) of the JPD. The CPDs may depend on sets of parameters $\Theta = \{\theta_1, \theta_2, \ldots, \theta_n\}$, where θ_i is the set of parameters associated with the CPD of X_i in Equation 4. The graph D describes qualitatively the relationship among the variables and the parameters are the quantitative description of the relationship. The information needed to construct D and to specify the parameters can be either provided by the human experts or learnt from a set of observed data. Below we mention four ways of constructing Bayesian network models:

Models Specified by Input Lists

The human experts in the domain of application are asked to provide an initial set of CISs that they believe to describe the structural relationship among the set of variables. Each CIS indicates which variables contain relevant information about the others and which are conditionally independent of the others. The initial input list are then completed by additional CISs that are derived from the initial list using certain conditional independence properties such as symmetry, decomposition, contraction, and weak and strong union. For details about how additional CISs are derived using these properties, see, for example Reference 9. All initial and derived CISs are then checked for consistency. A single minimal I-map is then constructed to describe the structural relationship among the entire set of variables as contained in the final input list. The JPD is then written as the product of the CPDs suggested by the graph as in Equation 4. The associated sets of parameters Θ can be either specified by the human experts or learnt from a set of observed data as explained later.

Graphically Specified Models

The human experts are asked to provide a DAG or a set of DAGs that they believe to describe the structural relationship among the set of variables. All possible CISs are then derived from the graphs using the D-separation criterion. This initial input list is then augmented by the derived CISs. The final input list is checked for consistency, a single minimal I-map is then constructed to describe the structural relationship among the entire set of variables, and the JPD is then written as the product of the CPDs suggested by the graph as in Equation 4. The associated sets of parameters Θ can be either specified by the human experts or learnt from a set of observed data.

Conditionally Specified Models

The human experts may provide a set

$$P = \{p(y_1|z_1), p(y_2|z_2), \ldots, p(y_m|z_m)\}$$

of CPDs from which the JPD of the entire set of variables can be constructed. This set has to be checked for consistency, compatibility, and uniqueness. For details, see, Reference 2.

Learning Models From Data

The above ways of constructing a Bayesian network assume that the human experts can provide the information needed to construct the Bayesian network. In many practical applications, this may not be possible. Also, different experts can give different and/or even conflicting information because of the subjective nature of the problem at hand or because of the uncertainty surrounding the problem. In fact, in some situations, constructing Bayesian networks from the available data may be the only viable option. This is known as *learning*.

Here we assume that we have available a data set S consisting of N observations on each of the variables $\{X_1, X_2, \ldots, X_n\}$. The data set can be used to construct D, which is known as *structural learning*, and to estimate the associated sets of parameters Θ, which is known as *parametric learning*. In the statistical language, one may think of structural and parametric learning as *model selection* and *parameter estimation* respectively.

There are several methods for learning Bayesian networks from data, see, for example, References 9, 10, 13, 21, and the references therein. Learning methods have two elements in common: a *quality measure* and a *search algorithm*.

Quality measures are used to select the best Bayesian network from among a set of competing Bayesian networks. These measures assess the quality of the Bayesian network as a whole, that is, both the quality of the graphical structure and the quality of the estimated parameters. They are usually functions of prior information (if available), the available data, and the network complexity. A good quality measure should assign high quality to networks that are (i) indicated by the human experts to be more likely than others, (ii) in agreement with the available data (e.g., networks that fit the data well), and (iii) simple (e.g., networks that include a small number of links and a small number of parameters). Examples of quality measures that have been suggested include Bayesian measures (see, e.g., Refs. 11 and 15), information measures (see, e.g., Refs. 1 and 32), and the minimum description length measure (see, e.g., Ref. 3).

Because the number of all possible networks is extremely large for even a small number of variables, trying all possible networks is virtually impossible. A search algorithm is therefore needed to select a small set of high quality Bayesian networks from which the best is selected. Examples of search algorithms are the $K2$-algorithm [11] and the B-algorithm [5].

Good search algorithms should take into account the important fact that different DAGs can lead to the same JPD. In this sense, two or more different Bayesian networks can be *independence equivalent*. To identify independence equivalent graphs, we need two concepts: *(i)* the *v-structure* and *(ii)* the undirected graph associated with a DAG. An ordered triplet of nodes (X_1, X_2, X_3) in a Bayesian network is said to be a *v*-structure iff X_2 has converging arrows from X_1 and X_3 and there is no link between X_1 and X_3. For example, there are two *v*-structures in the DAG in Fig. 1 and these are (X_1, X_4, X_2) and (X_3, X_6, X_4). Every DAG has an associated undirected graph, which is obtained from the DAG after replacing all directed links by undirected links. These two concepts lead to a way of identifying *independence equivalent* Bayesian networks. Two Bayesian networks are said to be independence equivalent iff they have *(i)* the same associated undirected graphs and *(ii)* the same *v*-structures.

Example 4. Independence equivalent Bayesian networks. Consider the six different directed graphs in Fig. 2. The graphs in (a)–(c) lead to the same dependency model, because their associated undirected graphs are the same and there are no *v*-structures in any of them. The graphs in (a)–(d) have the same associated undirected graphs, but there is a *v*-structure, (A, B, C), in graph (d). Therefore, the graph (d) is not independence equivalent to any of the graphs in (a)–(c). The graphs in (e) and (f) are independence equivalent since they have the same associated undirected graph and there are no *v*-structures in any of them. Note that (A, B, C) in graph (e) is not

a *v*-structure because nodes A and C are connected by a link. Thus, the six graphs given in Fig. 2 define only three different dependency models associated with three classes of equivalent graphs: $\{(a), (b), (c)\}$, $\{(d)\}$, $\{(e),$ and $(f)\}$.

PROPAGATION IN BAYESIAN NETWORKS

One of the most important tasks of expert systems is the propagation of evidence, which allows queries to be answered when new evidence is observed. Propagation of evidence consists of calculating the conditional probabilities $p(x_i|e)$ for each nonevidential variable X_i given the evidence $E = e$, where E is a subset of the variables in X for which evidences are available. Propagation of evidence can be carried out once a Bayesian-network model has been constructed using any of the above methods of constructing Bayesian networks. The following example illustrates the complexity of calculating marginal and conditional probabilities and how a Bayesian-network structure can be exploited to reduce the number of required calculations.

Example 5. Exploiting independence structure. Consider a set of binary variables $X = \{X_1, X_2, \ldots, X_7\}$ whose JPD $p(x)$ can be factorized according to the DAG in Fig. 1 as in Equation 3:

$$p(x) = \prod_{i=1}^{7} p(x_i|\pi_i)$$
$$= p(x_1)p(x_2)p(x_3)p(x_4|x_1, x_2)p(x_5|x_2)$$
$$\times p(x_6|x_3, x_4)p(x_7|x_4). \tag{7}$$

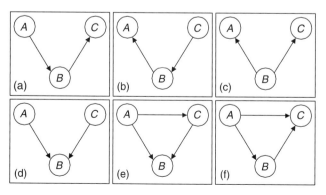

Figure 2. Six different DAGs of three nodes.

Suppose that we wish to calculate the marginal probability density (MPD) function of X_4 (that is, the MPD of X_4 when no evidence is available). We have the following alternatives:

$$p(x_4) = \sum_{x \setminus x_4} p(x)$$

$$= \sum_{x_1, x_2, x_3, x_5, x_6, x_7} p(x_1)p(x_2)p(x_3)p(x_4|x_1, x_2)$$

$$\times p(x_5|x_2)p(x_6|x_3, x_4)p(x_7|x_4),$$

which requires 63 sums and 384 multiplications;

$$p(x_4) = \left(\sum_{x_1, x_2, x_5} p(x_1)p(x_2)p(x_5|x_2)p(x_4|x_1, x_2) \right)$$

$$\times \left(\sum_{x_3, x_6, x_7} p(x_3)p(x_6|x_3, x_4)p(x_7|x_4) \right),$$

which requires 14 sums and 41 multiplications; or

$$p(x_4) = \sum_{x_2} \left[p(x_2) \times \left[\sum_{x_5} p(x_5|x_2) \right] \right.$$

$$\times \sum_{x_1} p(x_1) \times p(x_4|x_1, x_2) \right] \sum_{x_3} \left[p(x_3) \right.$$

$$\times \left[\sum_{x_7} p(x_7|x_4) \right] \times \sum_{x_6} p(x_6|x_3, x_4,) \right],$$

which requires 10 sums and 13 multiplications. Clearly, among the three options, the last one is the most computationally efficient. This example illustrates the savings that can be obtained by exploiting the independence structure of probabilistic models using suitable propagation algorithms.

There are three types of algorithms for propagating evidence:

1. *Exact propagation algorithms:* Apart from precision or round-off errors, these methods compute the conditional probability distribution of the nodes exactly. They include *(i)* propagation methods for polytrees (see Refs. 22, 26, and 29), *(ii)* methods based on conditioning (see Refs. 25 and 37), *(iii)* clustering methods (see Refs. 23, 20, 24, 25, 29, and 33), and *(iv)* join-tree methods (see 6, 20, 35). Exact methods depend on the type of network structure and can be computationally inefficient. In these cases, approximate propagation methods provide attractive alternatives.

2. *Approximate propagation algorithms:* These methods compute the probabilities approximately (mainly by simulation). They include uniform sampling method, acceptance–rejection method [17], likelihood weighing method ([14,36]), systematic sampling method [4], and maximum probability search method ([18,30]). The accuracy of approximation methods can be improved by increasing the number of simulation runs.

3. *Symbolic propagation algorithms:* These methods deal with numerical as well as symbolic parameters. They compute the probabilities in symbolic form. Symbolic propagation methods were developed by [7], [8], and [34]. Symbolic methods are particularly useful for sensitivity analysis and in cases where the parameters cannot be specified exactly.

Example 6. Propagation in Bayesian networks. Consider the Bayesian network defined by the graph in Fig. 1 and the set of CPDs in Table 1. First, suppose we wish to compute the marginal probability densities of all seven variables. The results of various propagation methods are shown in Table 2. The approximation methods are based on 1000 simulation runs. The table shows only $p(X_i = 0)$ because $p(X_i = 1) = 1 - p(X_i = 0)$.

Next, suppose that we have some evidence available, say $X_5 = X_6 = 0$, and we wish to compute the CPD of the other nodes given the available evidence. That is, we wish to compute $p(x_i|X_5 = X_6 = 0)$ for $i = 1, 2, 3, 4, 7$. The results of various propagation methods are shown in Table 3. The table shows only $p(X_i = 0|X_5 = X_6 = 0)$ because $p(X_i = 1|X_5 = X_6 = 0) = 1 - p(X_i = 0|X_5 = X_6 = 0)$. From the comparison with

Table 2. Marginal Probabilities $p(X_i = 0)$ for the Case of No Evidence.
Methods: 1 = Uniform Sampling, 2 = Acceptance–Rejection, 3 = Likelihood
Weighing, 4 = Systematic Sampling, 5 = Maximum Probability

X_i	Exact Methods	Approximate Methods				
		1	2	3	4	5
X_1	0.400	0.501	0.401	0.427	0.400	0.395
X_2	0.200	0.203	0.207	0.203	0.200	0.185
X_3	0.900	0.902	0.897	0.909	0.900	0.936
X_4	0.392	0.386	0.393	0.407	0.392	0.383
X_5	0.420	0.434	0.413	0.407	0.424	0.414
X_6	0.634	0.646	0.635	0.662	0.631	0.641
X_7	0.456	0.468	0.432	0.475	0.456	0.456

Table 3. Conditional Probabilities $p(X_i = 0|X_5 = X_6 = 0)$. Methods:
1 = Uniform Sampling, 2 = Acceptance-Rejection, 3 = Likelihood Weighing,
4 = Systematic Sampling, 5 = Maximum Probability

X_i	Exact Methods	Approximate Methods				
		1	2	3	4	5
X_1	0.391	0.379	0.442	0.387	0.391	0.391
X_2	0.246	0.243	0.200	0.281	0.246	0.246
X_3	0.962	0.965	0.954	0.959	0.962	1.000
X_4	0.469	0.443	0.475	0.476	0.469	0.468
X_7	0.487	0.524	0.500	0.498	0.486	0.487

exact methods, it seems from Tables 2 and 3 that the systematic sampling method is the most accurate of all approximate methods.

AVAILABLE SOFTWARE

There are many free and commercial software packages available for manipulating Bayesian and Markov networks. There are also other resources available on the Internet. The following link is a place to start:
http.cs.berkeley.edu/~murphyk/Bayes/bnsoft.html.

REFERENCES

1. Akaike, H. (1974). A new look at the statistical model identification. *IEEE Trans. Automat. Control*, **19**, 716–723.

2. Arnold, B., Castillo, E., and Sarabia, J. M. (1999). *Conditional Specification of Statistical Models*. Springer-Verlag, New York.

3. Bouckaert, R. R. (1995). Bayesian Belief Networks: From Construction to Inference. Ph.D. Thesis, Department of Computer Science, Utrecht University, The Netherlands.

4. Bouckaert, R. R., Castillo, E., and Gutiérrez, J. M. (1996). A modified simulation scheme for inference in Bayesian networks. *Int. J. Approximate Reason.*, **14**, 55–80.

5. Buntine, W. (1991). "Theory Refinement on Bayesian Networks". *Proceedings of the Seventh Conference on Uncertainty in Artificial Intelligence*. Morgan Kaufmann Publishers, San Mateo, Calif., pp. 52–60.

6. Cano, A., Moral, S., and Salmerón, A. (2000). Penniless propagation in Join Trees. *Int. J. Intell. Syst.*, **15**, 1027–1059.

7. Castillo, E., Gutiérrez, J. M., and Hadi, A. S. (1995). Parametric structure of probabilities in Bayesian networks. *Lecture Notes Artif. Intell.*, **946**, 89–98.

8. Castillo, E., Gutiérrez, J. M., and Hadi, A. S. (1996). A new method for efficient symbolic propagation in discrete Bayesian networks. *Networks*, **28**, 31–43.

9. Castillo, E., Gutiérrez, J. M., and Hadi, A. S. (1997). *Expert Systems and Probabilistic Network Models*. Springer-Verlag, New York.

10. Checkering, D. M. (2002). Learning equivalent classes of Bayesian-network structures. *J. Machine Learn. Res.*, **2**, 445–498.

11. Cooper, G. F. and Herskovits, E. (1992). A Bayesian method for the induction of probabilistic networks from data. *Machine Learn.*, **9**, 309–347.

12. Cowell, R. G., Dawid, A. P., Lauritzen, S. L., and Spiegelhalter, D. J. (1999). *Probabilistic Networks and Expert Systems*. Springer-Verlag, New York.

13. Fisher, D. and Lenz, H. J., eds. (1996). *Learning from Data: Artificial Intelligence and Statistics V*. Springer-Verlag, New York.

14. Fung, R. and Chang, K. (1990). "Weighing and Integrating Evidence for Stochastic Simulation in Bayesian Networks". In *Uncertainty in Artificial Intelligence 5*, M. Henrion, R. D. Shachter, L. N. Kanal, and J. F. Lemmer, eds. North Holland, Amsterdam, pp. 209–219.

15. Geiger, D. and Heckerman, D. (1995). "A Characterization of the Dirichlet Distribution with Application to Learning Bayesian Networks". *Proceedings of the Eleventh Conference on Uncertainty in Artificial Intelligence*. Morgan Kaufmann Publishers, San Mateo, Calif., pp. 196–207.

16. Geiger, D., Verma, T., and Pearl, J. (1990). "D-Separation: From Theorems to Algorithms. In *Uncertainty in Artificial Intelligence 5*, M. Henrion, R. D. Shachter, L. N. Kanal, and J. F. Lemmer, eds. North Holland, Amsterdam, pp. 139–148.

17. Henrion, M. (1988). "Propagation of Uncertainty by Logic Sampling in Bayes' Networks". In *Uncertainty in Artificial Intelligence 2*, J. F. Lemmer and L. N. Kanal, eds. North Holland, Amsterdam, pp. 149–164.

18. Henrion, M. (1991). "Search-Based Methods to Bound Diagnostic Probabilities in Very Large Belief Nets. *Proceedings of the Seventh Conference on Uncertainty in Artificial Intelligence*. Morgan Kaufmann Publishers, San Mateo, Calif., pp. 142–150.

19. Jensen, F. V. (2001). *Bayesian Networks and Decision Diagrams*. Springer-Verlag, New York.

20. Jensen, F. V., Olesen, K. G., and Andersen, S. K. (1990). An algebra of Bayesian belief universes for knowledge-based systems. *Networks*, **20**, 637–660.

21. Jordan, M., ed. (1998). *Learning in Graphical Models*, Vol. 89. Kluwer, Boston, Mass. NATO ASI Series D: Behavioral and Social Sciences Edition.

22. Kim, J. H. and Pearl, J. (1983). "A Computational Method for Combined Causal and Diagnostic Reasoning in Inference Systems". *Proceedings of the Eighth International Joint Conference on Artificial Intelligence (JCAI-83)*. Morgan Kaufmann Publishers, San Mateo, Calif., pp. 190–193.

23. Lauritzen, S. L. and Spiegelhalter, D. J. (1988). Local computations with probabilities on graphical structures and their application to expert systems. *J. R. Stat. Soc., Ser. B*, **50**, 157–224.

24. Lauritzen, S. L. and Wermuth, N. (1989). Graphical models for association between variables, some of which are qualitative and some quantitative. *Ann. Stat.*, **17**, 31–54.

25. Pearl, J. (1986a). "A Constraint-Propagation Approach to Probabilistic Reasoning". In *Uncertainty in Artificial Intelligence*, L. N. Kanal and J. F. Lemmer, eds. North Holland, Amsterdam, pp. 357–369.

26. Pearl, J. (1986b). Fusion, propagation and structuring in belief networks. *Artif. Intell.*, **29**, 241–288.

27. Pearl, J. (1988). *Probabilistic Reasoning in Intelligent Systems: Networks of Plausible Inference*. Morgan Kaufmann Publishers, San Mateo, Calif.

28. Pearl, J. (2000). *Causality*. Cambridge University Press, Cambridge.

29. Peot, M. A. and Shachter, R. D. (1991). Fusion and propagation with multiple observations in belief networks. *Artif. Intell.*, **48**, 299–318.

30. Poole, D. (1993). "Average-Case Analysis of a Search Algorithm for Estimating Prior and Posterior Probabilities in Bayesian Networks with Extreme Probabilities". *Proceedings of the 13th International Joint Conference on Artificial Intelligence (IJCAI-93)*. Morgan Kaufmann Publishers, San Mateo, Calif., pp. 606–612.

31. Rencher, A. C. (1995). *Methods of Multivariate Analysis*. John Wiley & Sons, New York.

32. Schwarz, G. (1978). Estimation of the dimension of a model. *Ann. Stat.*, **17**, 461–464.

33. Shachter, R. D. (1988). Probabilistic inference and influence diagrams. *Operat. Res.*, **36**, 589–605.

34. Shachter, R., D'Ambrosio, B., and Del-Favero, B. (1990). "Symbolic Probabilistic Inference in Belief Networks". *Proceedings of the 8th National Conference on AI (AAAI-90)*. AAAI Press/MIT Press, Menlo Park, Calif, pp. 126–131.

35. Shachter, R., Anderson, B., and Szolovits, P. (1994). "Global Conditioning for Probabilistic Inference in Belief Networks". *Proceedings of the Tenth Conference on Uncertainty in Artificial Intelligence*. Morgan Kaufmann Publishers, San Mateo, Calif., pp. 514–522.

36. Shachter, R. and Peot, M. (1990). "Simulation Approaches to General Probabilistic Inference on Belief Networks. In *Uncertainty in Artificial Intelligence 5*, M. Henrion, R. D. Shachter, L. N. Kanal, and J. F. Lemmer, eds. North Holland, Amsterdam, pp. 221–231.

37. Suermondt, H. J. and Cooper, G. F. (1991). Initialization for the method of conditioning in Bayesian belief networks. *Artif. Intell.*, **50**, 83–94.

FURTHER READING

Almond, R. (1995). *Graphical Belief Modeling*. Chapman & Hall, London.

Borgelt, C. and Kruse, R. (2002). *Graphical Models: Methods for Data Analysis and Mining*. John Wiley & Sons, New York.

Edwards, D. (2000). *Introduction to Graphical Modeling*, 2nd ed. Springer-Verlag, New York.

Geiger, D., Verma, T., and Pearl, J. (1990). Identifying independence in Bayesian networks. *Networks*, **20**, 507–534.

Larrañaga, P., Poza, M., Yurramendi, Y., Murga, R. H., and Kuijpers, C. M. H. (1996). Structure learning of Bayesian networks by genetic algorithms: a performance analysis of control parameters. *IEEE Trans. Pattern Anal. Machine Intell.*, **18**, 912–926.

Lauritzen, S. L. (1996). *Graphical Models*. Oxford University Press, Oxford.

Neapolitan, R. E. (1990). *Probabilistic Reasoning in Expert Systems: Theory and Algorithms*. John Wiley & Sons, New York.

Verma, T. and Pearl, J. (1991). "Equivalence and Synthesis of Causal Models". In *Uncertainty in Artificial Intelligence 6*, P. P. Bonissone, M. Henrion, L. N. Kanal, and J. F. Lemmer, eds. North Holland, Amsterdam, pp. 255–268.

See also NETWORK ANALYSIS.

ENRIQUE CASTILLO

ALI S. HADI

BAYESIAN ROBUSTNESS

A statistical method is robust if the conclusions derived from that method are insensitive to the assumptions that underlie that method. In Bayesian inference* the underlying assumptions include: the family of sampling distributions, the prior distribution, and (often) a loss function. "Bayesian robustness" refers to the analysis of the effect of these assumptions on the posterior inferences. The prior is usually the most contentious part of a Bayesian analysis, and hence robustness to the prior has received the most attention. However, there is more recent work that examines the sensitivity of Bayesian inferences to the other assumptions.

This contrasts with frequentist inference, in which the theory of robustness* has been mostly concerned with finding procedures that are inherently robust to deviations from the sampling model. Nonetheless, finding Bayesian procedures that are robust in the frequentist sense has received some attention, usually by using heavy-tailed distributions* as priors or sampling models; see Berger [2, Section 4.7; 4, Section 2] for references, and also the section on "Sensitivity to the Sampling Model and Outliers" below.

In discussing Bayesian robustness, special mention should be made of Berger [1], which is a milestone in this area. In addition to inspiring a considerable body of research, it is a rich source of references and explains many of the historical antecedents of the subject. A more recent overview is given in Berger [4].

Let Θ be an unknown parameter taking values in a set Ω. Consider i.i.d. observations $X = (X_1, \ldots, X_n)$, and suppose that the distribution for X_i given $\Theta = \theta$ is $F(x|\theta)$ with density $f(x|\theta)$. Let P be a prior distribution for the parameter Θ. Under suitable conditions, the posterior for Θ given $X = x$ satisfies

$$P(A|x) = \frac{\int_A L(\theta)P(d\theta)}{m_P(x)},$$

where A is a measurable subset of Ω, $L(\theta) = \prod_i f(x_i|\theta)$ is the likelihood function, and $m_P(x) = \int_\Omega L(\theta)P(d\theta)$ is the marginal density for $x = (x_1, \ldots, x_n)$.

ASSESSING SENSITIVITY TO THE PRIOR

It is widely known that the effect of the prior diminishes as the sample size increases, a phenomenon called *stable estimation* by Edwards et al. [17]. But for a given data set one needs to investigate whether stable estimation obtains, that is, whether the prior has a substantial effect on the posterior. The usual method for examining the sensitivity of the posterior $P(\cdot|x)$ to the prior P is to embed P into a class of priors Γ. This class might be a parametric class or a nonparametric class. Then one computes the bounds

$$\underline{P}(A|x) = \inf_{P\in\Gamma} P(A|x) \text{ and}$$

$$\overline{P}(A|x) = \sup_{P\in\Gamma} P(A|x). \qquad (1)$$

If $\tau = h(\theta)$ is a function of interest, then bounds on the posterior expectation of τ are denoted by $\underline{E}(\tau|x)$ and $\overline{E}(\tau|x)$.

Perhaps the simplest nonparametric class is the ϵ-contaminated class borrowed from classical robustness (Huber [31]). The class is defined by

$$\Gamma_\epsilon(P) = \{(1-\epsilon)P + \epsilon Q; Q \in \mathcal{Q}\}, \qquad (2)$$

where $\mathcal{Q}$ is some set of alternative priors. Here, $\epsilon \in [0,1]$ and may be loosely interpreted as the degree of doubt in the prior P. If $\mathcal{Q}$ is the set of all priors, then it turns out that

$$\overline{E}(\tau|x) = \sup_\theta \frac{(1-\epsilon)m_P(x)E_P(\tau|x) + \epsilon h(\theta)L(\theta)}{(1-\epsilon)m_P(x) + \epsilon L(\theta)}$$

and

$$\overline{P}(A|x) = \frac{(1-\epsilon)m_P(x)P(A|x) + \epsilon \sup_{\theta\in A}L(\theta)}{(1-\epsilon)m_P(x) + \epsilon \sup_{\theta\in A}L(\theta)}.$$

Robustness can be assessed by plotting $\overline{E}(\tau|x)$ and $\underline{E}(\tau|x)$ as functions of ϵ. Often, the bounds are pessimistically wide. In that case, one might take $\mathcal{Q}$ in (2) to be a more restricted class, such as the set of all priors that are symmetric around some prespecified point θ_0 and unimodal (Sivaganesan and Berger [53]), or by using other restrictions, such as fixing some quantiles. There is a substantial literature on variations of the class (2). Furthermore, many classes of priors not of this form have been considered. Reviews describing the various classes may be found

in Berger [1,3,4] and Wasserman [60]. Parametric classes are discussed in many places; see Kass et al. [38] for example.

Difficult issues arise in the computation of $\underline{E}(\tau|x)$ and $\overline{E}(\tau|x)$. Since $\underline{E}(\tau|x) = -\overline{E}(-\tau|x)$, it suffices to consider only the upper bounds. Recall that

$$E(\tau|x) = \int h(\theta)L(\theta)P(d\theta) \bigg/ \int L\theta P(d\theta).$$

Maximizing this expression over P is a nonlinear problem. The bound may be found by the "linearization technique," apparently used first in this context by DeRobertis [13] and DeRobertis and Hartigan [14], and then developed in more generality by Lavine [39]. The technique works by making the simple observation that $\overline{E}(\tau|x)$ is the solution λ_0 to the equation $s(\lambda) = 0$, where $s(\lambda) = \sup_{P\in\Gamma} E_P\{L(\theta)[h(\theta) - \lambda]\}$. This reduces the nonlinear optimization problem of maximizing $E_P(\tau|x)$ to a two-step process: (1) find $s(\lambda)$ for each λ, and then (2) find the root $s(\lambda) = 0$. The maximizations in step (1) are linear in P. Often, the extreme points of Γ can be identified explicitly, making the computation of $s(\lambda)$ tractable.

SENSITIVITY TO THE SAMPLING MODEL AND OUTLIERS

The assessment of robustness to the sampling model can be performed in a way similar to the assessment of sensitivity to the prior, by embedding $F(x|\theta)$ into a class Γ_θ (Lavine [39], Huber [31]). This approach creates some problems. For one thing, the meaning of the parameter may become obscure. For example, if one places a nonparametric class Γ_θ around an $N(\theta, 1)$ distribution, then θ can no longer be interpreted as the mean of the distribution. Furthermore, the posterior bounds are often hopelessly wide. Alternatively, and perhaps more commonly, one might stick to parametric models but insist that the models be inherently robust. This parametric approach to robustness was used, for example, by Box [7] and Box and Tiao [8]. Generally, parametric robustness is obtained by using heavy-tailed densities such as the t-distribution or finite mixtures of normals (Dawid [10], Freeman [20], Goldstein [23], Guttman et al. [27], Jeffreys [34, Chap. 4], O'Hagan [41–43], Smith [54], Pettit and Smith [44], Verdinelli and Wasserman [57], West [63], Zellner [64]). These heavy-tailed distributions have the effect of protecting the posterior from being unduly affected by outliers. In spirit, this is similar to the M-estimators* of classical statistics (Huber [32]); indeed, Section 4.41 of Jeffreys [34] bears much similarity to a description of M-estimation.

SENSITIVITY IN DECISION PROBLEMS

In a Bayesian decision theory* problem, there is a set of actions $\mathscr{A}$ and a utility* function $U : \Theta \times \mathscr{A} \to \mathbb{R}$ where $U(\theta, a)$ is the utility of choosing action a when the truth state is θ. A decision function δ maps the sample space χ into $\mathscr{A}$; the Bayes decision $\delta(x)$ is satisfied by

$$\int U(\theta, \delta(x))P(d\theta|x) = \sup_{a \in \mathscr{A}} \int U(\theta, a)P(d\theta|x)$$

for each x. It is possible to examine changes in the Bayes risk $\sup_{a \in \mathscr{A}} \int U(\theta, a)P(d\theta|x)$ and in a as the utility, prior, or sampling model changes (Kadane and Chuang [36], Ramsay and Novick [46], Ríos-Insúa [47,48], Srinivasan and Truszczynska [56]).

HYPOTHESIS TESTING

A place where robustness to the prior is of particular concern is in testing precise hypotheses. Consider testing $H_0 : \theta = \theta_0$ versus $H_1 : \theta \neq \theta_0$. In the Bayesian framework the favored approach is due to Jeffreys [34, Chaps. 5, 6]; see also Kass and Raftery [37]. For this one uses a prior of the form $P = a\delta_0 + (1 - a)Q$, where $0 < a < 1$ is the prior probability that H_0 is true, δ_0 is a point mass at θ_0, and Q is a prior for θ under H_1. The Bayes factor* B, defined to be the ratio of the posterior odds to the prior odds, is

$$B = \frac{P(H_0|x)/P(H_1|x)}{P(H_0)/P(H_1)} = \frac{L(\theta_0)}{\int L(\theta)Q(d\theta)}. \quad (3)$$

Values of B larger than 1 support H_0, and values of B smaller than 1 support H_1. It is

well known that B is sensitive to the choice of Q; the methods discussed earlier are helpful for assessing sensitivity of B to Q. For example, one might take Q to vary in the class defined by (2) with $\mathcal{Q}$ the set of unimodal symmetric densities centered at θ_0. The behavior of Bayes factors on this and similar classes is studied in Edwards et al. [17], Berger and Delampady [5], and Berger and Sellke [6]. Interestingly, even for large classes of priors, the bounds are often substantially different than the corresponding significance probability.

UPPER AND LOWER PROBABILITIES

As noted, in Bayesian robustness one is often concerned with bounding probabilities. The tacit assumption is that there is a true probability (prior or sampling distribution) for which one must find bounds. A more radical approach is to forego the notion that there is a true probability. Instead, one might replace probabilities with upper and lower probabilities (see NONADDITIVE PROBABILITY), an approach considered by Fine [19], Good [24], Smith [55], and Walley [58], among others.

In this formulation, every event is assumed to have a lower and upper probability denoted by $\underline{P}(A)$ and $\overline{P}(A)$. But now there is no assumption of a true probability $P(A)$ between these bounds. Suppose one requires that the upper and lower probabilities be coherent*, in the sense that they do not lead to inconsistencies when interpreted as lower and upper betting rates; $\underline{P}$ and $\overline{P}$ are coherent if and only if there exists a nonempty set of probability measures M such that, for all measurable A,

$$\underline{P}(A) = \inf_{P \in M} P(A) \quad \text{and} \quad \overline{P}(A) = \sup_{P \in M} P(A).$$

This approach is detailed in Walley [58]. Manipulating upper and lower probabilities can be numerically demanding. Some simplifications occur if $\overline{P}$ is 2-alternating, meaning that

$$\overline{P}(A \cup B) \leqslant \overline{P}(A) + \overline{P}(B) - \overline{P}(A \cap B)$$

for every pair of events A and B; see Huber [31], Huber and Strassen [33], and

Wasserman and Kadane [61]. Further simplifications occur if $\underline{P}$ is a Dempster—Shafer belief function* (Dempster [12], Shafer [51], Wasserman [59]).

An odd feature of upper and lower probabilities is that they may get uniformly wider after conditioning. Specifically, let $\underline{P}(A|B) = \inf_{P \in M} P(A|B)$ and similarly for $\overline{P}(A|B)$ (assume these are well defined). There may exist an event A and a partition $\mathscr{B} = \{B_1, \ldots, B_m\}$ such that

$$\underline{P}(A|B) < \underline{P}(A) \leqslant \overline{P}(A) < \overline{P}(A|B)$$

for all $B \in \mathscr{B}$; see Walley [58, p. 298]. Seidenfeld and Wasserman [50] refer to this phenomenon as *dilation* and show that it is not at all pathological.

INFLUENCE AND LOCAL SENSITIVITY

In some cases one may be interested in studying the influence of specific observations or specific changes to the prior or model. Such calculations are considered in Geisser [21], Guttman and Peña [28,29], Johnson and Geisser [35], Kass et al. [38], McCulloch [40], Pettit and Smith [44], and Weiss [62]. In many of these papers, the approach is to measure the Kullback information* (or Kullback—Leibler information, divergence) between the posterior obtained using the assumed model, prior, and data and the posterior obtained using some modified version of the prior, model, or data. Geisser has emphasized using predictive distributions (*see* PREDICTIVE ANALYSIS) instead of posteriors. McCulloch [40] takes a local approach by analyzing the effects of small changes to the assumptions. The deviations from the assumptions are taken to lie in a parametric model.

The effect of small changes to the assumptions can also be assessed nonparametrically using appropriate derivatives. Diaconis and Freedman [16] make this precise by computing the Fréchet derivative (*see* STATISTICAL FUNCTIONALS) of the posterior with respect to the prior; see also Cuevas and Sanz [9]. Without any attention to rigor, a simple explanation is as follows. Let $d(P, Q)$ be a measure of distance between P and Q; $d(\cdot, \cdot)$ need not be a metric. Denote the posterior corresponding

to a prior P by P^X. Let $Q_\epsilon = (1 - \epsilon)P + \epsilon Q$, and define

$$S_P(Q) = \lim_{\epsilon \to 0} \frac{d(P^X, Q_\epsilon^X)}{\epsilon},$$

which measures the rate at which P^X changes as P moves in the direction of Q. If $d(P, Q) = \sup_A |P(A) - Q(A)|$, then it can be shown that

$$S_P(Q) = \frac{m_P(x)}{m_Q(x)} d(P^X, Q^X).$$

One potential sensitivity diagnostic is $\sup_{Q \in \mathscr{F}} S_P(Q)$ for some class of distributions $\mathscr{F}$. The class $\mathscr{F}$ cannot be taken to be too large, or $\sup_{Q \in \mathscr{F}} S_P(Q)$ will diverge to infinity as the sample size increases.

The major references on this approach include Delampady and Dey [11], Dey and Birmiwal [15], Gelfand and Dey [22], Gustafson [25], Gustafson et al. [26], Ruggeri and Wasserman [49], Sivaganesan [52], and Srinivasan and Truszczynska [56]. These ideas are intimately related to similar techniques in classical statistics, such as those described in Fernholz [18], Hampel et al. [30], and Huber [32].

REFERENCES

1. Berger, J. O. (1984). The robust Bayesian viewpoint (with discussion). In *Robustness in Bayesian Statistics*, J. Kadane, ed. North-Holland, Amsterdam, pp. 63–124. (Important review article on Bayesian robustness.)

2. Berger, J. O. (1985). *Statistical Decision Theory and Bayesian Analysis*, 2nd ed. Springer-Verlag, New York.

3. Berger, J. O. (1990). Robust Bayesian analysis: sensitivity to the prior. *J. Statist. Plann. Inference*, **25**, 303–328.

4. Berger, J. O. (1994). An overview of robust Bayesian analysis. *Test*, **3**, 5–124. (Overview of the current status of Bayesian robustness.)

5. Berger, J. O. and Delampady, M (1987). Testing precise hypotheses. *Statist. Sci.*, **2**, 317–352.

6. Berger, J. O. and Sellke, T. (1987). Testing a point null hypothesis: the irreconcilability of *P*-values and evidence. *J. Amer. Statist. Assoc.*, **82**, 112–139. (This paper and [5] deal with robustness in testing situations. The emphasis is on comparing Bayesian and robust Bayesian methods with frequentist methods.)

7. Box, G. E. P. (1980). Sampling and Bayes' inference in scientific modelling and robustness. *J. Roy. Statist. Soc. Ser. A*, **143**, 383–430. (Parametric approach to Bayesian robustness.)

8. Box, G. E. P. and Tiao, G. C. (1973). *Bayesian Inference in Statistical Analysis*. Addison-Wesley, Reading, Mass.

9. Cuevas, A. and Sanz, P. (1988). On differentiability properties of Bayes operators. In *Bayesian Statistics 3*, J. M. Bernardo et al., eds. Oxford University Press, pp. 569–577. (Bayes' theorem is regarded as an operator taking priors and likelihoods to posteriors. This paper studies the differentiability properties of this operator.)

10. Dawid, A. P. (1973). Posterior expectations for large observations. *Biometrika*, **60**, 664–667. (Studies the conflict between the tails of the prior and the likelihood in location problems.)

11. Delampady, M. and Dey, D. (1994). Bayesian robustness for multiparameter problems. *J. Statist. Plann. Inference*, **40**, 375–382. (Considers variations of divergence between posteriors when the contamination varies over symmetric and symmetric unimodal classes of priors.)

12. Dempster, A. P. (1967). Upper and lower probabilities induced by a multivalued mapping. *Ann. Math. Statist.*, **38**, 325–339. (One of the earliest papers where belief functions were proposed, though they were not called belief functions at the time.)

13. DeRobertis, L. (1978). The Use of Partial Prior Knowledge in Bayesian Inference. Ph.D. dissertation, Yale University.

14. DeRobertis, L. and Hartigan, J. (1981). Bayesian inference using intervals of measures. *Ann. Statist.*, **9**, 235–244. (Robust Bayesian inference based on intervals of measures.)

15. Dey, D. and Birmiwal, L. R. (1994). Robust Bayesian analysis using divergence measures. *Statist. Probab. Lett.*, **20**, 287–294. (Obtains divergences between posteriors and the curvature for a general phi-divergence measure. The local curvature gives a measure of local sensitivity.)

16. Diaconis, P. and Freedman, D. (1986). On the consistency of Bayes estimates. *Ann. Statist.*, **14**, 1–67. (As an aside to some consistency questions, the authors compute the Fréchet derivative of the posterior with respect to the prior.)

17. Edwards, W., Lindman, H., and Savage, L. J. (1963). Bayesian statistical inference for psychological research. *Psychol. Rev.*, **70**, 193–242. (Introduced the notion of "stable estimation.")

18. Fernholz, L. T. (1983). *Von Mises Calculus for Statistical Functionals*. Springer-Verlag, New York.

19. Fine, T. L. (1988). Lower probability models for uncertainty and nondeterministic processes. *J. Statist. Plann. Inference*, **20**, 389–411. (Summarizes the theory of upper and lower probabilities.)

20. Freeman, P. R. (1979). On the number of outliers in data from a linear model. In *Bayesian Statistics*, J. M. Bernardo et al., eds. University Press, Valencia, Spain, pp. 349–365. (A Bayesian approach to outliers.)

21. Geisser, S. (1992). Bayesian perturbation diagnostics and robustness. In *Bayesian Analysis in Statistics and Econometrics*, P. Goel and S. Iyengar, eds. Springer-Verlag, New York, pp. 289–301. (Considers a variety of perturbation diagnostics in Bayesian inference.)

22. Gelfand, A. and Dey, D. (1991). On Bayesian robustness of contaminated classes of priors. *Statist. and Decisions*, **9**, 63–80. (Introduces geometric contaminations. Variation of the Kullback-Leibler divergence between posteriors is studied for parametric classes of priors.)

23. Goldstein, M. (1982). Contamination distributions. *Ann. Statist.*, **10**, 174–183. (Finds conditions on distributions so that extreme observations have a limited effect on the posterior.)

24. Good, I. J. (1962). Subjective probability as the measure of a non-measurable set. In *Logic, Methodology, and Philosophy of Science*, E. Nagel et al., eds. Stanford University Press, pp. 319–329. (Deals with upper and lower probabilities.)

25. Gustafson, P. (1994). Local Sensitivity in Bayesian Statistics. Ph. D. dissertation, Carnegie Mellon University. (Develops the notion of a derivative of the posterior with respect to the prior in great detail, including asymptotic properties.)

26. Gustafson, P., Srinivasan, C., and Wasserman, L. (1995). Local sensitivity analysis. In *Bayesian Statistics 5*, J. M. Bernardo et al., eds. Oxford University Press, pp. 631–640. (Reviews some approaches to measuring the effect of small perturbations to priors.)

27. Guttman, I., Dutter, R., and Freeman, P. R. (1978). Care and handling of univariate outliers in the general linear model to detect

spuriosity—a Bayesian approach. *Technometrics*, **20**, 187–193. (Studies outliers in linear models.)

28. Guttman, I. and Peña, D. (1988). Outliers and influence: evaluation by posteriors of parameters in the linear model. In *Bayesian Statistics 3*, J. M. Bernardo et al., eds. Oxford University Press, pp. 631–640. (Attempts to deal with outliers and influential points in a unified framework.)

29. Guttman, I. and Peña, D. (1993). A Bayesian look at diagnostics in the univariate linear model. *Statist. Sinica*, **3**, 367–390.

30. Hampel, F. R., Ronchetti, E. M., Rousseeuw, P. J., and Stahel, W. A. (1986). *Robust Statistics: The Approach Based on Influence Functions*, Wiley, New York.

31. Huber, P. (1973). The use of Choquet capacities in statistics. *Bull. Int. Statist. Inst. Proc. 39th Session*, **45**, 181–191. (Shows that Choquet capacities play a role in frequentist and Bayesian robustness.)

32. Huber, P. J. (1981). *Robust Statistics*. Wiley, New York.

33. Huber, P. J. and Strassen, V. (1973). Minimax tests and the Neyman—Pearson lemma for capacities. *Ann. Statist.*, **1**, 251–263. (Extends the Neyman—Pearson lemma for Choquet capacities.)

34. Jeffreys, H. (1961). *Theory of Probability*, 3rd ed. Clarendon Press, Oxford.

35. Johnson, W. and Geisser, S. (1983). A predictive view of the detection and characterization of influential observations in regression analysis. *J. Amer. Statist. Ass.*, **78**, 137–144. (Predictive approach to influential observations.)

36. Kadane, J. B. and Chuang, D. T. (1978). Stable decision problems. *Ann. Statist.*, **6**, 1095–1110. (Deals with robustness in decision problems.)

37. Kass, R. E. and Raftery, A. (1995). Bayes factors. *J. Amer. Statist. Ass.*, **90**, 773–795.

38. Kass, R. E., Tierney, L., and Kadane, J. B. (1989). Approximate methods for assessing influence and sensitivity in Bayesian analysis. *Biometrika*, **76**, 663–674. (Derives diagnostics for influence using asymptotic approximations.)

39. Lavine, M. (1991). Sensitivity in Bayesian analysis: the likelihood and the prior. *J. Amer. Statist. Assoc.*, **86**, 396–399. (Studies robustness to the prior and likelihood using nonparametric classes of distributions.)

40. McCulloch, R. E. (1989). Local model influence. *J. Amer. Statist. Ass.*, **84**, 473–478.

(Derives local perturbation diagnostics based on Kullback-Leibler divergence.)

41. O'Hagan, A. (1979). On outlier rejection phenomena in Bayes inference. *J. R. Statist. Soc. Ser. B*, **41**, 358–367. (Studies the conflict arising from discrepant observations in location problems.)

42. O'Hagan, A. (1988). Modelling with heavy tails. In *Bayesian Statistics 3*, J. M. Bernardo et al., eds. Oxford University Press, pp. 345–360. (Considers the use of heavy-tailed distributions in Bayesian inference.)

43. O'Hagan, A. (1990). Outliers and credence for location parameter inference. *J. Amer. Statist. Ass.*, **85**, 172–176. (Formalizes the idea that tail thickness determines how conflicting sources of observations—such as outliers—behave when combined.)

44. Pettit, L. I. and Smith, A. F. M. (1985). Outliers and influential observations in linear models. In *Bayesian Statistics 2*, J. M. Bernardo et al., eds. North Holland, Amsterdam, pp. 473–494. (Compares Bayesian and frequentist approaches to outliers and influential points. Also proposes some computational simplifications.)

45. Polasek, W. (1984). Multivariate regression systems—estimation and sensitivity analysis of two-dimensional data. In *Robustness of Bayesian Analyses*, J. Kadane, ed. North-Holland, Amsterdam, pp. 229–309.

46. Ramsay, J. O. and Novick, M. R. (1980). PLU robust Bayesian decision theory: point estimation. *J. Amer. Statist. Ass.*, **75**, 901–907. (Deals with robustness of priors, likelihoods and loss functions.)

47. Ríos-Insúa, D. (1990). *Sensitivity Analysis in Multiobjective Decision Making*. Springer-Verlag, Berlin.

48. Ríos-Insúa, D. (1992). Foundations for a robust theory of decision making: the simple case. *Test*, **1**, 69–78.

49. Ruggeri, F. and Wasserman, L. (1993). Infinitesimal sensitivity of posterior distributions. *Canad. J. Statist.*, **21**, 195–203.

50. Seidenfeld, T. and Wasserman, L. (1993). Dilation for sets of probabilities. *Ann. Statist.*, **21**, 1139–1154. (Studies the problem of "dilation," i. e., when imprecise probabilities become uniformly more imprecise by conditioning.)

51. Shafer, G. (1976). *A Mathematical Theory of Evidence*. Princeton University Press.

52. Sivaganesan, S. (1993). Robust Bayesian diagnostics. *J. Statist. Plann. Inference*, **35**,

171–188. (Computes diagnostics based on derivatives of the posterior.)

53. Sivaganesan, S. and Berger, J. O. (1989). Ranges of posterior measures for priors with unimodal constraints. *Ann. Statist.*, **17**, 868–889. (Shows how large classes of priors with unimodality constraints can be used for Bayesian robustness.)

54. Smith, A. F. M. (1983). Bayesian approaches to outliers and robustness. In *Specifying Statistical Models*, J. P. Florens et al., eds. Springer-Verlag, Berlin, pp. 204–221.

55. Smith, C. A. B. (1961). Consistency in statistical inference and decision. *J. R. Statist. Soc. Ser. B*, **23**, 1–37. (Deals with upper and lower probabilities.)

56. Srinivasan, C. and Truszczynska, H. (1990). On the Ranges of Posterior Quantities. *Tech. Rep. 294*, Department of Statistics, University of Kentucky.

57. Verdinelli, I. and Wasserman, L. (1991). Bayesian analysis of outlier problems using the Gibbs sampler. *Statist. Comput.*, **1**, 105–117. (Shows how Markov-chain Monte Carlo can be used for outlier problems.)

58. Walley, P. (1991). *Statistical Reasoning with Imprecise Probabilities*. Chapman and Hall, New York.

59. Wasserman, L. (1990). Prior envelopes based on belief functions. *Ann. Statist.*, **18**, 454–464. (Shows that belief functions can be useful in robust Bayesian inference.)

60. Wasserman, L. (1992). Recent methodological advances in robust Bayesian inference. In *Bayesian Statistics 4*, J. M. Bernardo et al., eds. Oxford University Press, pp. 483–502. (Review article on Bayesian robustness.)

61. Wasserman, L. and Kadane, J. B. (1990). Bayes' theorem for Choquet capacities. *Ann. Statist.*, **18**, 1328–1339. (Gives a version of Bayes' theorem when a class of priors is used instead of a single prior.)

62. Weiss, R. (1993). Bayesian Sensitivity Analysis Using Divergence Measures. *Tech. Rep.*, Department of Biostatistics, UCLA School of Public Health.

63. West, M. (1992). Modeling with mixtures. In *Bayesian Statistics 4*, J. M. Bernardo et al., eds. Oxford University Press, pp. 503–524. (Uses mixture distributions to achieve flexibility in modeling.)

64. Zellner, A. (1976). Bayesian and non-Bayesian analysis of regression models with multivariate student-*t* error terms. *J. Amer. Statist. Ass.*, **71**, 400–405. (Uses *t*-distributions for regression models.)

See also BAYESIAN INFERENCE; POSTERIOR DISTRIBUTIONS; PRIOR DISTRIBUTIONS; and ROBUST ESTIMATION.

LARRY WASSERMAN

BEAMLETS AND MULTISCALE MODELING

INTRODUCTION

A beamlet can be viewed as an extension of a wavelet because a beamlet also is a multi-resolution analysis. However, it is dramatically different from the traditional wavelets* in both construction and applications. Wavelets are good for one-dimensional piecewise smooth functions. Beamlets are good for 2-dimensional (2-D) linear or curvilinear features. The beamlets are defined in continuum. In digital images, beamlets can be discretized. The discretized beamlets are called *digital beamlets*. Digital beamlets are built on a quad-tree. The ways of utilizing digital beamlets are very different from techniques such as wavelet denoising, wavelet coding, etc. They require careful analysis of connectivities among beamlets.

The construction of beamlets is motivated by the ubiquitous presence of linear and curvilinear features in images (such as edges). It is also rooted in geometry [1]. For 2-D beamlets, they can be viewed as a multiscale analysis of 2-D lines.

Beamlets are defined in the following three steps:

1. Recursive Dyadic Partitioning (RDP). We assume that an image resides on a unit square $[0, 1] \times [0, 1]$. The unit square is divided into two by two subsquares that have equal and dyadic sidelengths. The subsquares are then divided into two by two smaller subsquares, each of them has equal and dyadic sidelength. This process is repeated until the finest scale is reached.

2. Vertex Marking. On the boundary (four sides) of each square, starting from the upper left corner, going clockwise (or anti-clockwise) on the boundary, vertices are marked at an equal distance. The interdistance is fixed in advance, and it does not vary according to the sidelength of a square.

3. Connecting. On the boundary of each square, any pair of vertices, which are determined in the previous step, determines a line segment. This line segment is a *beamlet*.

The set of all the beamlets forms a *beamlet dictionary*. Note that a beamlet divides its residing square (or subsquare) into two regions. Each of these two regions is a *wedgelet*. An illustration of some beamlets and wedgelets are given in Figure 1. In [2], the above data structure is called *edgelets*. The renaming reflects a shift of emphasis of developing applications that are based on this data structure.

There are two important properties for beamlets.

1. For any beam, which is a line segment that starts from one pixel corner and ends at another pixel corner, it takes a chain that is made of no more than $O(\log(n))$ beamlets to approximate it.

2. For an n by n image, there are $O(n^2 \log(n))$ beamlets, or wedgelets. At the same time, there are $O(n^4)$ beams.

Beamlet coefficients are defined based on the beamlet data structure. For each beamlet, the line integral of a 2-D function along this beamlet is called a *beamlet coefficient* for this function. The *beamlet transform* is to compute the beamlet coefficients for all beamlets.

Wedgelets

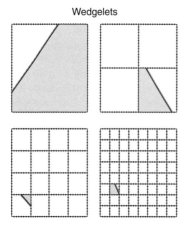

Figure 1. Beamlets and corresponding wedgelets.

The above beamlet transform is defined for *continuous* 2-D functions. For digital images, people can first consider an imagery as a piecewise constant 2-D function: for an n by n image, each pixel is associated with a $1/n$ by $1/n$ square in the unit square; the values of the 2-D continuous functions inside these squares are equal to the intensity of the image at the corresponding pixels. Note that the above scheme is designed for gray scale images. The beamlet coefficients are the line integrals of the interpolated function along beamlets.

Similar to the beamlet transform, a *wedgelet transform* can be defined. A *wedgelet coefficient* is an integral of the interpolated function on a wedgelet. The *wedgelet transform* is utilized to compute all the wedgelet coefficients.

For digital images, it can be shown that the wedgelet and beamlet coefficients are essentially linear combinations of pixel values.

Due to the construction of beamlets, a hierarchy of statistical models for linear and curvilinear features is rendered. The detailed description of such a hierarchy is going to be lengthy. Here we simply refer to [3]. In brief, based on the framework of Recursive Dyadic Partitioning and beamlets, people can create a collection of models. Each model is made by linear features. One model can be refined (coarsened) to get another model. All the models form a hierarchy.

After constructing a hierarchy of models, a statistical principle (such as minimizing the complexity penalized likelihood function) is used to get the statistically optimal estimate. Here each estimate is made by a set of linear features. Note that a curvilinear feature may be approximated well by a few linear features. Hence the above strategy also applies to curvilinear features.

The above methodology can be realized in applications such as denoising, feature extraction, estimation, data reduction. Examples can be found in [4–8]. Figure 2 illustrates on how beamlets can be used in denoising a badly distorted black-and-white image.

There are other methods that are developed to realize the multiscale analysis on different features in images. For example, to emphasize the directional features, at the

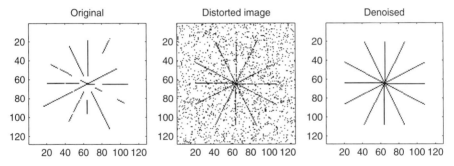

Figure 2. Beamlet based denoising.

same time to retain the function as 2-D continuous functions instead of lines or curves, ridgelets and curvelets [9,10] are developed. Another interesting concept is that of brushlet [11,12], which intends to capture directional features by studying signals in the Fourier domain. Beamlets have also been extended to 3-D [13].

Acknowledgments

The author would like to thank Professor Vidakovic for his invitation.

REFERENCES

1. Jones, P. W. Rectifiable sets and the traveling salesman problem. *Inventiones Mathematicae*, **102**, 1–15, 1990.

2. Donoho, D. L. (1999). Wedgelets: Nearly minimax estimation of edges. *Annals of Statistics*, **27**(3), 859–897.

3. Donoho, D. L. and Huo, X. (2001). Beamlets and multiscale image analysis. In *Multiscale and Multiresolution Methods*, volume 20 of *Lecture Notes in Computational Science and Engineering*, Springer.

4. Huo, X. (1999). *Sparse Image Representation via Combined Transforms*. PhD thesis, Stanford University, August.

5. Donoho, D. and Huo, X. (1999). Combined Image Representation using Edgelets and Wavelets. *Wavelet Applications in Signal and Image Processing VII*.

6. Donoho, D. and Huo, X. (2000). Beamlet pyramids: A new form of multiresolution analysis, suited for extracting lines, curves, and objects from very noisy image data. In *Proceedings of SPIE*, volume 4119, July.

7. Donoho, D. and Huo, X. (2001). Applications of Beamlets to detection and extraction of lines,

curves, and objects in very noisy images. In *Proceedings of Nonlinear Signal and Image Processing*, June.

8. Huo, X. and Donoho, D. (2002). Recovering Filamentary Objects in Severely Degraded Binary Images using Beamlet-Decorated Partitioning. *International Conference on Acoustic Speech and Signal Processing (ICASSP)*, May, Orlando, FL.

9. Candès, E. and Donoho, D. (1999). Ridgelets: a key to higher-dimensional intermittency? *R. Soc. Lond. Philos. Trans. Ser. A Math. Phys. Eng. Sci.* **357** no. 1760, 2495–2509.

10. Candès, E. and Donoho, D. (2001). Curvelets and curvilinear integrals. *J. Approx. Theory* **113** no. 1, 59–90.

11. Meyer, F. G. and Coifman, R. R. (1997). Brushlets: a tool for directional image analysis and image compression. *Applied and Computational Harmonic Analysis* (Academic Press), Vol 4, pages 147-187.

12. Meyer, F. G. and Coifman, R. R. (2001). Brushlets: Steerable wavelet packets. In *Beyond Wavelets*, J. Stoeckler and G. Welland Editors, Academic Press.

13. Donoho, D. L. and Levi, O. (2002). Fast X-Ray and beamlet transform for three-dimensional data. Technical report no. 2002-7, Stanford University, March.

See also WAVELETS.

XIAOMING HUO

BECHHOFER TEST. See FIXED-EFFECTS MODEL, BECHHOFER'S TABLES

BEHRENS' PROBLEM. See BEHRENS–FISHER PROBLEM

BEHRENS–FISHER PROBLEM—I

The Behrens–Fisher problem is that of testing whether the means of two normal populations are the same, without making any assumption about the variances. It is also referred to as the two-means problems and as Behrens' problem. An essentially equivalent problem is that of finding an interval estimate* for the difference between the population means.

Suppose that the first of the normal populations has mean μ_1 and variance σ_1^2 and that a sample of size n_1 yields sample mean $\overline{X}_1$ and sample variance s_1^2. We will use corresponding notation for the second sample, but with subscript 2.

If the population variances are assumed to be identical (at least under the null hypothesis* of no difference between means), then the quantity

$$\frac{(\overline{X}_1 - \overline{X}_2) - (\mu_1 - \mu_2)}{[s_1^2(n_1 - 1) + s_2^2(n_2 - 1)]^{1/2}}$$
$$\times \left[\frac{(n_1 + n_2 - 2)n_1 n_2}{n_1 + n_2}\right]^{1/2} \quad (1)$$

has a t-distribution* with $n_1 + n_2 - 2$ degrees of freedom. Confidence intervals* for $\mu_1 - \mu_2$ and tests for the hypothesis $\mu_1 - \mu_2 = 0$ based on this fact are widely used. The Behren-Fisher problem allows the assumption of equal population variances to be relaxed.

The Behrens–Fisher problem has considerable theoretical importance as a situation where fiducial (*see* FIDUCIAL INFERENCE) and Neyman–Pearson* schools of thought give different results. Kendall and Stuart [12] discuss it extensively in this role. Lee and Gurland [13] and Scheffé [21] provide useful reviews of the various solutions that have been proposed.

BEHRENS' SOLUTION

Behrens [3] was the first to offer a solution to the Behrens–Fisher problem. However, the basis of his solution was not made completely clear. Fisher [6] showed that Behrens' solution could be derived using his theory of statistical inference called fiducial probability*.

Both Fisher and Yates [9] and Isaacs et al. [10] give tables of significance points. Patil [16] gives a useful approximation.

A short fiducial derivation of the test starts by noting that, for fixed μ_1, μ_2, σ_1^2, and $\sigma_2^2, t_1 = (\overline{X}_1 - \mu_1)n_1^{1/2}/s_1$ has a t-distribution with $n_1 - 1$ degrees of freedom and $t_2 = (\overline{X}_2 - \mu_2)n_2^{1/2}/s_2$ has an independent t-distribution with $n_2 - 1$ degrees of freedom. If we treat these distributions as probability statements about μ_1 and μ_2 given $\overline{X}_1, \overline{X}_2, s_1^2$, and s_2^2, then a little manipulation shows that

$$d = \frac{(\overline{X}_1 - \overline{X}_2) - (\mu_1 - \mu_2)}{(s_1^2/n_1 + s_2^2/n_2)^{1/2}}$$
$$= t_1 \sin R - t_2 \cos R, \quad (2)$$

where $\tan R = (s_1/s_2)(n_2/n_1)^{1/2}$. Thus the distribution of d is that of a mixture* of two t-distributions.

The controversial step in this derivation is what Fisher called logical inversion. Here we know the distribution of t_1 and t_2 and it describes the distribution of $\overline{X}_1, s_1, \overline{X}_2$, and s_2 for given μ_1, σ_1^2, μ_2, and σ_2^2 and somehow proceed to the conclusion that the same distribution of t_1 and t_2 describes a distribution of the parameters for given values of the observable random variables.

Jeffreys [11] pointed out that a Bayesian calculation based on a $(\sigma_1^2 \sigma_2^2)^{-1}$ improper prior density for μ_1, μ_2, σ_1^2, and σ_2^2 yields a solution that is numerically the same, although the interpretation is different. This derivation proceeds by finding the posterior distribution* of μ_1, σ_1^2, μ_2, and σ_2^2 for given $\overline{X}_1, s_1^2, \overline{X}_2$, and s_2^2 for a $(\sigma_1^2 \sigma_2^2)^{-1}$ improper prior density. Integrating out σ_1^2 and σ_2^2 shows that the marginal posterior distribution of μ_1 and μ_2 may be described by the t-distributions of the statistics t_1 and t_2 given above. Hence the marginal* posterior distribution of d is that of a mixture* of t_1 and t_2 as in (2). The controversial step in this derivation is the assumption of the prior density.

WELCH–ASPIN SOLUTION

Welch [22] derived a test for which the probability of error was very nearly the nominal value throughout the parameter space. It is based on the same statistic, d, as in Behrens' test but uses different significance points. Asymptotic calculations of Welch [22] were extended by Aspin [1] and the significance points are available in the *Biometrika Tables* [17, Table 11, p. 136].

PRACTICAL COMPARISON OF SOLUTIONS

For sample sizes greater than about 10, the differences between the various proposed solutions are generally much less their differences from the t-test of (1), so the use of any one of them is better than using the t-test, unless the assumption of equal variances is warranted. The tests of McCullough–Banerjee and Cochran and Cox may be regarded as *approximations* to the Behrens–Fisher solution. Most of the other tests discussed by Lee and Gurland [13] may be regarded as approximations to the Welch–Aspin test.

Example. For $n_1 = 13$, $n_2 = 25$, suppose that $s_1^2 = 2.0$, $s_2^2 = 1.0$, and $\overline{X}_1 - \overline{X}_2 = 0.9$. From (2), $d = 2.044$. The 5% significance point is 2.15 for the Behrens–Fisher test, 2.16 for both the McCullough-Banerjee and Cochran and Cox tests, 2.11 for the Welch–Aspin test, and 2.10 for Welch's approximate t-solution. We would not reject the hypothesis of equal population means at the 5% level.

The t-statistic from (1) takes the value 2.27, which is significant at the 5% level, the significance point being 2.03.

If the sample variances were $s_1^2 = 1$ and $s_2^2 = 1.5$, then the t-statistic would still be 2.27. However, the d-statistic would be changed to 2.43, which would be considered significant by the various tests for the Behrens–Fisher problem. (The significance points for all these tests vary with the ratio s_1^2/s_2^2, but are similar to their previously quoted values for this particular case.)

It should be noted that the hypothesis of equal population variances would not be rejected by the usual F-test for either set of sample variances.

CRITICISM OF BEHRENS' SOLUTION

Bartlett [2] noted that the probability of error* for Behrens' test was not always the same as the nominal significance level*. This set the scene for a major controversy between Neymen's theory of confidence intervals and Fisher's theory of fiducial inference. Fisher [7] argued that the probability of including the true value is not an important property, while the Neyman–Pearson school of thought has maintained that it is important, and supported the search for similar*, or approximately similar tests*.

SIMILAR TESTS

A test is said to be similar if its error probability is constant throughout the parameter space. Wilks [23] suggested that similar tests for the Behrens–Fisher problem would be difficult to find. Linnik [14,15] has proved that no similar tests having certain desirable properties exist. The Welch–Aspin solution is approximately similar for moderate sample sizes (see Lee and Gurland [13] for references). Behrens' solution seems to have error rate not greater than the normal value (see Robinson [20]).

The solutions to the Behrens–Fisher problem not discussed here are generally attempts to find approximately similar tests.

CRITICISM OF THE WELCH–ASPIN SOLUTION

Fisher [8] criticized Welch's test in a way that amounts to showing the existence of a negatively biased relevant subset in the sense of Buehler [4]. In our notation, independently of s_1/s_2, $U = (n_1 - 1)(s_1^2/\sigma_1^2) + (n_2 - 1)(s_2^2/\sigma_2^2)$ is distributed as chi-square* with $n_1 + n_2 - 2$ degrees of freedom and $(\overline{X}_1 - \mu_1) + (\overline{X}_2 - \mu_2)$ is distributed normally with mean zero and variance $\sigma_1^2/n_1 + \sigma_2^2/n_2$. It follows that

$$d\sqrt{\frac{s_1^2/n_1 + s_2^2/n_2}{\sigma_1^2/n_1 + \sigma_2^2/n_2}} \frac{n_1 + n_2 - 2}{U} \tag{3}$$

has a t-distribution with $n_1 + n_2 - 2$ degrees of freedom given s_1/s_2, μ_1, μ_2, σ_1^2, and

σ_2^2. When $s_1^2/s_2^2 = 1$, $n_1 = n_2$, and $\sigma_1^2/\sigma_2^2 = w$, expression (3) simplifies to

$$d\sqrt{\frac{4}{2 + 1/w + w}}.$$

Now $1/w + w \geqslant 2$, so

$$\sqrt{\frac{4}{2 + 1/w + w}} \leqslant 1$$

and so

$$\Pr[|d| > a | s_1/s_2 = 1]$$

$$\geqslant \Pr[|t_{(n_1+n_2-2)}| > a]$$

for all $a > 0$, where $t_{(n_1+n_2-2)}$ denotes a variable having a t-distribution with $n_1 + n_2 - 2$ degrees of freedom. In particular, taking $n_1 = n_2 = 7$ and $a = 1.74$ (from the *Biometrika Tables*),

$$\Pr[|d| > 1.74 | s_1/s_2 = 1]$$

$$\geqslant \Pr[t_{(12)} > 1.74] > 0.1.$$

Thus the set where $s_1/s_2 = 1$ is a relevant subset where confidence intervals based on the Welch–Aspin test cover the true value of $\mu_1 - \mu_2$ less often than the nominal confidence level suggests.

Robinson [20] has shown that negatively biased relevant subsets do not exist for Behrens' solution. Positively biased relevant subset similar to that of Buehler and Fedderson [5] for the t-distribution presumably exist, but Robinson [20] argues that this is less important.

CURRENT STATUS OF THE BEHRENS–FISHER PROBLEM

Within the Neyman–Pearson school of thought, the Welch–Aspin test is regarded as satisfactory except that its derivation using asymptotic series makes calculation of significance points for low numbers of degrees of freedom quite difficult. There is still some interest in other approximately similar tests. Other schools of thought tend to regard the Behrens–Fisher problem as an example when the Neyman–Pearson school's quest for a similar test has led it astray.

Fisher's [8] criticism of the Welch–Aspin test and Robinson's [20] support of Behrens'

test have made the Behrens–Fisher problem a focal point for discussion of the importance of relevant subsets.

Pedersen [18] and Fisher [8] suggest that the choice of a reference set—the set relative to which a quoted level of confidence applies—is the major issue illustrated by the Behrens–Fisher problem.

It seems likely that the Behrens–Fisher problem will be quoted as an important example in the discussion of rival theories as long as disagreements about the foundations of statistics persist.

Further Reading

The reviews of Scheffé [21] and Lee and Gurland [13] provide an introduction to practical solutions to the Behrens–Fisher problem. Readers interested in the role of the Behrens–Fisher problem in the foundations of statistics should become familiar with the theories of confidence intervals and fiducial inference and the concepts of relevant subsets and similar regions, then look at Bartlett [2], Fisher [8], Linnik [15], Robinson [20], and Pedersen [18].

In comparing various calculations concerning Behrens' solution, it is important to note whether exact significance points or asymptotic approximations have been used.

REFERENCES

1. Aspin, A. A. (1948). *Biometrika*, **35**, 88–96.
2. Bartlett, M. S. (1936). *Proc. Camb. Philos. Soc.*, **32**, 560–566.
3. Behrens, B. V. (1929). *Landwirtsch. Jb.*, **68**, 807–837.
4. Buehler, R. J. (1959). *Ann. Math. Statist.*, **30**, 845–867.
5. Buehler, R. J. and Fedderson, A. P. (1963). *Ann. Math. Statist.*, **34**, 1098–1100.
6. Fisher, R. A. (1935). *Ann. Eugen. (Lond.)*, **6**, 391–398.
7. Fisher, R. A. (1939). *Ann. Math. Statist.*, **10**, 383–388.
8. Fisher, R. A. (1956). *J. R. Statist. Soc. B*, **18**, 56–60. (There are several replies to this article in the same volume.)
9. Fisher, R. A. and Yates, F. (1957). *Statistical Tables for Biological, Agricultural and*

Medical Research, 5th ed. Oliver & Boyd, Edinburgh.

10. Isaacs, G. L., Christ, D. E., Novick, M. R., and Jackson, P. H. (1974). *Tables for Bayesian Statisticians*. Iowa State University Press, Ames, Iowa.

11. Jeffreys, H. (1940). *Ann. Eugen.* (*Lond.*), **10**, 48–51.

12. Kendall, M. G. and Stuart, A. (1967). *The Advanced Theory of Statistics*, Vol. 2: *Inference and Relationship*, 2nd ed. Charles Griffin, London.

13. Lee, A. F. S. and Gurland, J. (1975). *J. Amer. Statist. Ass.*, **70**, 933–941.

14. Linnik, Yu. V. (1966). *Sankhyā A*, **28**, 15–24.

15. Linnik, Yu. V. (1968). *Statistical Problems with Nuisance Parameters*, Vol. 20. (Amer. Math. Soc. Transl. Math. Monogr.; translated from the Russian.) American Mathematical Society, Providence, R.I.

16. Patil, V. H. (1965). *Biometrika*, **52**, 267–271.

17. Pearson, E. S. and Hartley, H. O., eds. (1954). *Biometrika Tables for Statisticians*, Vol. 1. Cambridge University Press, Cambridge.

18. Pedersen, J. G. (1978). *Int. Statist. Rev.*, **46**, 147–170.

19. Rahman, M. and Saleh, A. K. M. E. (1974). *J. R. Statist. Soc. B*, **36**, 54–60. (Not recommended; note corrigendum on p. 466 of same volume.)

20. Robinson, G. K. (1976). *Ann. Statist.*, **4**, 963–971.

21. Scheffé, H. (1970). *J. Amer. Statist. Ass.*, **65**, 1501–1508.

22. Welch, B. L. (1947). *Biometrika*, **34**, 28–35.

23. Wilks, S. S. (1940). *Ann. Math. Statist.*, **11**, 475–476.

See also FIDUCIAL DISTRIBUTIONS; FIDUCIAL INFERENCE; SIMILAR REGIONS AND TESTS; and STUDENT'S *t*-TESTS.

G. K. ROBINSON

BEHRENS-FISHER PROBLEM—II

Procedures commonly used to test null hypotheses about means for G groups are the analysis of variance (ANOVA)* when the hypotheses concern $p = 1$ variables, and the multivariate analysis of variance (MANOVA)* when the hypotheses concern $p > 1$ variables. These methods are based on several assumptions:

1. the data used to estimate the means for one group are statistically independent of the data used to estimate the means for the other $G - 1$ groups,

2. within each of the G groups the data contributed by different sampling units are statistically independent of one another,

3. the data in each group are drawn from a normal distribution, and

4. the G populations from which the samples are drawn have equal dispersion (equal variances when $p = 1$ and equal covariance matrices when $p > 1$).

Because of the last assumption, statisticians have been interested in the Behrens-Fisher problem*: testing hypotheses about means without assuming equal dispersion. Moreover, a large body of theoretical and empirical evidence demonstrates that if the null hypothesis is true but the G populations have unequal dispersion, the probability of incorrectly rejecting the null hypothesis may be smaller or larger than the nominal Type I error rate α. In addition, the power of the test can be reduced when the population variances are not equal.

Several solutions to the Behrens-Fisher problem have been presented in the ESS (*see* BEHRENS–FISHER PROBLEM—I; WELCH TESTS). The solutions due to Welch are briefly reviewed here. See the former article for the solution due to Behrens.

The earliest solutions were developed for the situation in which two means are to be compared. If the assumptions listed in the first paragraph of this article are met, the independent samples *t*-test should be used. An alternative test statistic to that used in the latter test and designed for use when the inequality of variance assumption is violated is

$$t' = \frac{\bar{x}_1 - \bar{x}_2}{\sqrt{\frac{s_1^2}{n_1} + \frac{s_2^2}{n_2}}}.$$

Welch [24] proposed using t' with the approximate degrees of freedom* (APDF) v and critical value $t_{1-\alpha}(v)$, where α is the nominal Type I error rate and

$$v = \frac{\left(\frac{\sigma_1^2}{n_1} + \frac{\sigma_2^2}{n_2}\right)^2}{\frac{\sigma_1^4}{n_1^2(n_1-1)} + \frac{\sigma_2^4}{n_2^2(n_2-1)}}.$$

In practice v is estimated by substituting sample variances for population variances.

Welch [25] also proposed approximating the critical value for t' by a power series in $1/f_i = 1/(n_i - 1)$, and presented the solution to order $(1/f_i)^2$:

$$z\left(1 + \frac{(1+z^2)V_{21}}{4} - \frac{(1+z^2)V_{22}}{2}\right.$$
$$+ \frac{(3 + 5z^2 + z^4)V_{32}}{3}$$
$$\left. - \frac{(15 + 32z^2 + 9z^4)V_{21}^2}{32}\right),$$

where

$$V_{ru} = \frac{\sum_{i=1}^2 \frac{(S_i^2/n_i)^r}{f_i^u}}{\left(\sum_{i=1}^2 \frac{S_i^2}{n_i}\right)^r}$$

and $z = z_{1-\alpha}$, resulting in the *second-order test*. To order $1/f_i$ the critical value is

$$z\left(1 + \frac{(1+z^2)V_{21}}{4}\right),$$

resulting in the *first-order test*. Aspin* [5] presented the solution to order $(1/f_i)^4$ and in ref. [6] presented an abbreviated table of fourth-order critical values.

Results in Lee and Gurland [17] indicate that with small sample sizes ($\leqslant 9$) the actual Type I error probability (τ) is nearer to α when the fourth-order critical value is used than it is when the second-order or APDF critical values are used. However, all three critical values controlled τ fairly well. Algina et al. [2] investigated larger sample sizes and reported that the APDF critical value and the second-order critical value result in estimates

of τ that typically agree to the third or fourth decimal place.

Welch [26] extended the APDF approach to testing equality of three or more means in a one-way layout. Let n_i denote the number of participants in the ith group and N the total number of participants. Let $w_i = n_i/S_i^2, f_i = n_i - 1, w = \sum_{i=1}^G w_i$, and $\tilde{x} = (1/w)\sum_{i=1}^G \bar{x}_i$. The test statistic is

$$WJ = \sum_{i=1}^G w_i(\bar{x}_i - \tilde{x})^2.$$

The critical value for the APDF test is $cF_{1-\alpha}(G-1, v)$, where

$$c = G - 1 + \frac{2(G-2)}{G+1}\sum_{i=1}^G f_i^{-1}\left(1 - \frac{w_i}{w}\right)^2,$$

$$v = \left[\frac{3}{G^2-1}\sum_{i=1}^G f_i^{-1}\left(1 - \frac{w_i}{w}\right)^2\right]^{-1}.$$

THE POWER SERIES APPROACH FOR $G \geqslant 3$

James* [13] extended the power series approach to testing equality of three or more means in a one-way layout. Consequently the first- and second-order tests are often referred to as *James's first-* and *second-order tests*. The test statistic for the power-series approach is the statistic WJ presented earlier. The critical value for the first-order test is

$$\chi^2\left[1 + \frac{3\chi^2 + G + 1}{2(G^2-1)}\sum_{i=1}^G \frac{1}{f_i}\left(1 - \frac{w_i}{w}\right)^2\right],$$

where $\chi^2 = \chi_{1-\alpha}^2(G-1)$. Let

$$R_{st} = \sum_{i=1}^G \left(\frac{w_i}{w}\right)^t f_i^{-s},$$

and

$$\chi_{2s} = \frac{[\chi_{1-\alpha}^2(G-1)]^s}{(G-1)(G+1)\cdots(G+2s-3)}.$$

The critical value for the second-order test is

$$\chi^2 + \tfrac{1}{2}(3\chi_4 + \chi_2) \sum_{i=1}^{G} \frac{(1-w_i/w)^2}{f_i}$$

$$+ \left[\tfrac{1}{16}(3\chi_4 + \chi_2)^2 \left(1 - \frac{G-3}{\chi^2}\right) \right.$$

$$\times \left(\sum_{i=1}^{G} \frac{(1-w_i/w)^2}{f_i} \right)^2 + \tfrac{1}{2}(3\chi_4 + \chi_2)$$

$$\times [(8R_{23} - 10R_{22} + 4R_{21} - 6R_{12}^2$$

$$+ 8R_{12}R_{11} - 4R_{11}^2) + (2R_{23} - 4R_{22} + 2R_{21}$$

$$- 2R_{12}^2 + 4R_{12}R_{11} - 2R_{11}^2)(\chi_2 - 1)$$

$$+ \tfrac{1}{4}(-R_{12}^2 + 4R_{12}R_{11} - 2R_{12}R_{10} - 4R_{11}^2$$

$$+ 4R_{11}R_{10} - R_{10}^2)(3\chi_4 - 2\chi_2 - 1)]$$

$$+ (R_{23} - 3R_{22} + 3R_{21} - R_{20})(5\chi_6 + 2\chi_4 + \chi_2)$$

$$+ \tfrac{3}{16}(R_{12}^2 - 4R_{23} + 6R_{22} - 4R_{21} + R_{20})$$

$$\times (35\chi_8 + 15\chi_6 + 9\chi_4 + 5\chi_2) + \tfrac{1}{16}(-2R_{22}$$

$$+ 4R_{21} - R_{20} + 2R_{12}R_{10} - 4R_{11}R_{10} + R_{10}^2)$$

$$\times (9\chi_8 - 3\chi_6 - 5\chi_4 - \chi_2) + \tfrac{1}{4}[(-R_{22} + R_{11}^2)$$

$$\times (27\chi_8 + 3\chi_6 + \chi_4 + \chi_2) + (R_{23} - R_{12}R_{11})$$

$$\times (45\chi_8 + 9\chi_6 + 7\chi_4 + 3\chi_2)] \Big].$$

ALTERNATIVES TO THE WELCH-JAMES TESTS

Brown and Forsythe [7] proposed an alternative APDF solution to the Behrens-Fisher problem for a design with $G \geqslant 3$ groups. The test statistic in the Brown-Forsythe test is

$$F^* = \frac{\sum_{i=1}^{G} n_i(\bar{x}_i - \bar{x})^2}{\sum_{i=1}^{G} c_i S_i^2},$$

where $\bar{x} = (1/N) \sum_{i=1}^{G} n_i \bar{x}_i$ is the grand mean and $c_i = 1 - n_i/N$. The numerator of F^* is the numerator sum of squares for a one-way ANOVA. Brown and Forsythe selected the denominator of F^* so that the numerator and the denominator of F^* both estimate the same quantity when $H_0 : \mu_1 = \cdots = \mu_G$ is true. Therefore the value of F^* is expected to be near 1.0 when $H_0 : \mu_1 = \cdots = \mu_G$ is true, regardless of whether the population variances are equal. If H_0 is false, F^* is expected to be larger than 1.0. The critical value is

$F_{1-\alpha}(G - 1, v_2)$, where

$$v_2 = \frac{\left(\sum_{i=1}^{G} c_i S_i^2 \right)^2}{\sum_{i=1}^{G} \frac{(c_i S_i^2)^2}{n_i - 1}}.$$

If $G = 2$ the Brown-Forsythe test and the Welch APDF test are equivalent.

Rubin [21] showed that $F_{1-\alpha}(G - 1, v_2)$ is not the correct critical value for the F^*. The correct critical value is $bF_{1-\alpha}(v_1, v_2)$ where b and v_1 are calculated from the data. Use of the new critical value improves the Type I error rate of the test, particularly when one variance is substantially larger than the others.

Another alternative is the test due to Alexander and Govern [1]. Define

$$t_i = w_i(\bar{x}_i - \tilde{x}).$$

The test statistic is

$$A = \sum_{i=1}^{G} z_i^2,$$

where

$$z_i = c_i + \frac{c_i^3 + 3c_i}{b_i}$$

$$- \frac{4c_i^7 + 33c_i^5 + 240c_i^3 + 855c_i}{10b_i^2 + 8b_i c_i^4 + 1000b_i},$$

$a_i = n_i - 1.5, b_i = 48a_i^2$, and $c_i = \{a_i \ln(1 + t_i^2/f_i)\}^{1/2}$. The critical value is $\chi_{1-\alpha}^2(G - 1)$.

COMPARISON OF THE APPROACHES FOR $G \geqslant 3$

Taken together, simulation results in refs. 1, 7, 27 indicate that the Alexander-Govern and second-order tests control the Type I error rate better than do the APDF or the Brown-Forsythe test. These two tests in turn control the Type I error rate better than does the first-order test. When all tests control the Type I error rate, the Alexander-Govern and second-order tests tend to be more powerful than the other tests.

A drawback to the Alexander-Govern and second-order tests is that they have not been generalized to higher-way layouts. Johansen

[15] generalized the APDF approach to permit testing hypotheses about means in multiway layouts. His results can also be used to construct a first-order critical value for testing hypotheses about means in multiway layouts. The Brown-Forsythe test has also been generalized to multiway layouts.

Using t', its generalization WJ, the Brown-Forsythe test, or A to test equality of means has two shortcomings. First, these tests lose power when the data are long-tailed, [29,33]. However, a Welch APDF test on trimmed means [31,33] can be used to address this problem, as can tests on one-step M-estimates* of location [29,30]. Second, when the data are skewed, the tests may no longer control the Type I error rate [2,20,28]. If one is willing to use trimmed means or one-step M-estimates when the data are skewed, the second problem can be addressed by using the tests cited earlier.

Example 1. Hand and Taylor [12] presented data collected from people who sought treatment for alcoholism or heavy drinking. The participants were separated into three groups defined by the number of self-reported relapses after trying to give up heavy drinking; (1) none, (2) between one and three, and (3) more than three. Each participant completed a survey on factors that might be related to vulnerability to relapse. One variable obtained from the survey was unpleasant mood states, such as depression. Means and variances for the three groups are presented in Table 1. Because of the advantages of James's second-order test and the Alexander-Govern test, these two tests were used to analyze the data. In addition, ANOVA was used.

In Table 1, the number of participants decreases as the sample variability increases. If the population variability decreases as the number of participants decreases, actual

Table 1. Means and Variances for Three Relapse Groups

Relapses	n	$\bar{x}$	S^2
0	125	21.6	106.2
1–3	89	18.2	133.1
> 3	37	17.7	160.1

Type I error rates will be greater than the nominal level. For the second-order test, the test statistic and .05 critical values are 6.66 and 6.23, respectively, indicating that the means for the three groups differ on unpleasant mood states. The test statistic for the Alexander-Govern test is $\chi^2(2) = 6.52, p = .038$, and also indicates that the means for the three groups differ on unpleasant mood states. The test statistic for the ANOVA is $F(2, 248) = 3.38, p = .035$. The p-values for the Alexander-Govern test and the ANOVA suggest that the F-test has not been degraded by the variance differences.

MULTIVARIATE EXTENSIONS

Several authors have generalized univariate solutions to the Behrens-Fisher problem to tests on mean vectors for multivariate data. Johansen [15] generalized the APDF approach to permit testing hypotheses about means in one-way and multiway layouts with multivariate data. His results can also be used to construct a first-order critical value for testing hypotheses about means in multiway layouts with multivariate data. The first-order critical value for the one-way layout is due to James [14]; James also generalized the second-order test approach. Coombs and Algina [9] generalized the Brown-Forsythe approach to test equality of mean vectors for the one-way layout. *See also* MULTIVARIATE BEHRENS-FISHER PROBLEM; WELCH'S v-CRITERION.

The APDF approach can be applied to multiway layouts, but because the other approaches are restricted to one-way layouts, the three methods will be presented for that design. Lix and Keselman [18] have presented a wide variety of examples of the use of Johansen's test in analyzing both univariate and multivariate data in one-way and multiway layouts.

The test statistic for the APDF, first-order and second-order tests is

$$WJ = \sum_{i=1}^{G} (\bar{\boldsymbol{x}}_i - \tilde{\boldsymbol{x}})' \boldsymbol{W}_i (\bar{\boldsymbol{x}}_i - \tilde{\boldsymbol{x}}),$$

where $\bar{\boldsymbol{x}}_i$ is mean vector for the ith sample,

$$\boldsymbol{W}_i = \left(\frac{\boldsymbol{S}_i}{n_i}\right)^{-1},$$

S_i is the dispersion matrix for the ith sample, and

$$W = \sum_{i=1}^{G} W_i, \qquad \tilde{x} = W^{-1} \sum_{i=1}^{G} W_i \bar{x}_i.$$

The critical value for the APDF test is $cF_{1-\alpha}(p(G-1), v)$, where

$$c = p(G-1) + 2A - \frac{6A}{p(G-1)+2},$$

$$A = \sum_{i=1}^{G} \frac{tr(I - W^{-1}W_i)^2 + tr^2(I - W^{-1}W_i)}{2(n_i - 1)},$$

$$v = \frac{p(G-1)[p(G-1)+2]}{3A},$$

and the hypotheses concern p variables. The first-order critical value is $\chi^2(B + \chi^2 C)$, where $\chi^2 = \chi^2_{1-\alpha}[p(G-1)]$ and

$$B = 1 + \frac{1}{2p(G-1)} \sum_{i=1}^{G} \frac{tr^2(I - W^{-1}W_i)}{n_i - 1},$$

$$C = \frac{1}{p(G-1)[p(G-1)+2]}$$
$$\times \left(\sum_{i=1}^{G} \frac{tr(I - W^{-1}W_i)^2}{n_i - 1} \right.$$
$$\left. + \frac{1}{2} \sum_{i=1}^{G} \frac{tr^2(I - W^{-1}W_i)}{n_i - 1} \right).$$

The second-order critical value is substantially more complicated [14].

The multivariate Brown-Forsythe test employs the usual MANOVA criteria applied to multivariate generalizations of the numerator and denominator of the Brown-Forsythe F^*. The numerator from the Brown-Forsythe test is replaced by H, the hypothesis sums-of-squares and cross-products (SSCP) matrix from a one-way MANOVA. Coombs and Algina [9] use

$$E^* = \frac{v_3}{G-1} \sum_{i=1}^{G} c_i S_i = \frac{v_3}{G-1} M$$

in place of E, the within-groups SSCP. They use the sum-of-Wisharts distribution [19] to

approximate the distribution of M. The quantity v_3 is the APDF for M, given by

$$v_3 = \frac{\left(tr \sum_{i=1}^{G} c_i S_i \right)^2 + tr \left(\sum_{i=1}^{G} c_i S_i \right)^2}{\sum_{i=1}^{G} (n_i - 1)^{-1} \{ [tr(c_i S_i)]^2 + [tr(c_i S_i)^2] \}}.$$

Coombs and Algina proposed as test statistics the usual MANOVA functions of HE^{-1} applied to $H(E^*)^{-1}$. Rubin [21] approximated the distribution of the numerator of the Brown-Forsythe test by a sum-of-chi-squares distribution, and improved the Type I error rate of the test. Approximating the distribution of H in the multivariate Brown-Forsythe test by the sum-of-Wisharts distribution may similarly improve control of the Type I error rate for this multivariate generalization of the Brown-Forsythe test.

COMPARISON OF MULTIVARIATE TESTS

Kim [16], Nel and van der Merwe [19], and Yao [32] have developed alternatives to James' first-order and second-order tests and Johansen's APDF test for use when $G = 2$. The Nel-van der Merwe test is equivalent to the multivariate Brown-Forsythe test when $G = 2$. The Nel-van der Merwe and Yao tests use WJ as a test statistic. The critical values for the Nel-van der Merwe and Yao tests are alternative APDF critical values. Kim's test uses a confidence ellipsoid approach. Coombs et al. [11] present test statistics and critical value formulas for all of the competitors as well as for their univariate specializations.

Simulation studies [4,3,22] indicate that when $G = 2$, James' first-order test, James' second-order test, Johansen's test, and Yao's test require n_{min}/p to be sufficiently large in order to control the actual Type I error probability τ, where n_{min} denotes the sample size in the smallest group. The required n_{min}/p is larger with asymmetric distributions than with symmetric distributions. James' second-order test, Johansen's test, and Yao's test, which have similar Type I error rates, allow better control of τ than does James' first-order test. Kim's simulation [16] indicates that when n_{min}/p is small, his test controls τ better than does Yao's test. Kim's test and the Nel-van der Merwe test have similar Type I error rates when n_{min}/p is small, and both control τ better than do Johansen's or James' first-order test [8].

For $G \geqslant 3$, simulation results in Coombs and Algina [10] and in Tang and Algina [23] indicate that James' first-order test and Johansen's test require $n_{\min}/p$ to be sufficiently large in order to control τ. The required sample size is larger with asymmetric distributions than with symmetric distributions and increases as the number of groups increases. James' second-order test, however, tends to have $\tau < \alpha$. Simulation results in ref. [9] indicate that when $G \geqslant 3$ and $n_{\min}/p$ is small, the Coombs-Algina test controls τ better than does Johansen's test.

Christensen and Rencher [8] compared the power of James' first-order, Johansen's Kim's, Nel and van der Merwe's, and Yao's tests for comparing mean vectors for $G = 2$ groups; Kim's test has the best power when all tests are adjusted to the same Type I error rate. However, the unadjusted Type I error rate for Kim's test was often less than α, and in practice its power advantage may not be realized. Basing their recommendation on α-adjusted power, Christensen and Rencher argued for Nel and van der Merwe's test among the four alternatives to Kim's test. Little or nothing is known about how the tests that can be used when $G \geqslant 3$ compare in terms of power.

Example 2. The data from Hand and Taylor [12] used in Example 1 included scores on euphoric states (ES), such as might be experienced at parties or celebrations, and lessened vigilance (LV) to the likelihood or relapse, in addition to unpleasant mood states (UMS), such as depression. Means and dispersion matrices for the three groups are presented in Table 2. Inspection of the dispersion matrices suggest increasing variability as the number of relapses increases. For example, the variance for UMS increases from 106.21 for the first group to 160.14 for the third group. Increases in variance are also observed for ES and LV. The covariance between UMS and ES increases from 36.7 for the first group to 49.91 for the third. Increases in covariance are also observed for UMS and LV and for ES and LV. The number of participants decreases as the variability increases, suggesting that the MANOVA criteria may result in actual Type I error rates above the nominal level.

Table 2. Mean Vectors and Dispersion Matrices for the Hand–Taylor Data

Variable	M	Dispersion Matrix		
		Group 1: No relapses		
UMS	21.6	106.21		
ES	12.0	36.70	40.21	
LV	6.4	12.52	7.50	5.99
		Group 2: 1 to 3 relapses		
UMS	18.2	133.05		
ES	10.7	43.26	43.56	
LV	5.4	18.25	10.36	7.23
		Group 3: More than 3 relapses		
UMS	17.5	160.14		
ES	9.9	49.91	45.00	
LV	4.5	20.51	11.42	9.03

Note. The sample sizes are $n_1, n_2, n_3 = 125, 89, 37$.

Analysis of the data using the Pillai-Bartlett trace applied to the Coombs-Algina generalizations of the Brown-Forsythe numerator and denominator yields $F(6, 248.1) = 2.63$, $p = .017$. The APDF test yields $F(6, 95.2) = 2.77, p = .015$. Thus, using a .05 level of significance and either the APDF test or the multivariate Brown-Forsythe test, the evidence is sufficient to conclude that the three relapse groups differ on one or more of the variables. The Pillai-Bartlett trace from MANOVA yields $F(6, 494) = 3.02$, $p = .0066$; again, with a .05 level of significance there is sufficient evidence to conclude that the three relapse groups differ on one or more of the variables. However, with a .01 level of significance the three tests lead to different decisions about equality of means for the three groups. Only the MANOVA Pillai-Bartlett test results in the conclusion that the three relapse groups differ on one or more of the variables. The fact that the conclusion based on the MANOVA Pillai-Bartlett test is different from those based on the other two tests when $\alpha = .01$ and the substantially smaller p-value for the MANOVA Pillai-Bartlett test are consistent with the possibility that the Type I error rate for the Pillai-Bartlett MANOVA criterion is too large.

REFERENCES

1. Alexander, R. A. and Govern, D. M. (1994). A new and simpler approximation for ANOVA

under variance heterogeneity. *J. Educ. Statist.*, **19**, 91–102. (Develops an alternative to the APDF and power series tests when $G \geqslant 2$.)

2. Algina, J., Oshima, T. C., and Lin, W.-Y. (1994). Type I error rates for Welch's test and James's second-order test under nonnormality and inequality of variance when $G = 2$. *J. Educ. and Behav. Statist.*, **19**, 275–292. (A simulation study of Type I error rates for the univariate APDF and second-order tests when $G = 2$.)

3. Algina, J., Oshima, T. C., and Tang, K. L. (1991). Robustness of Yao's, James', and Johansen's tests under variance-covariance heteroscedasticity and nonnormality. *J. Educ. Statist.*, **16**, 125–139. (A simulation study of Type I error rates for several multivariate tests when $G = 2$.)

4. Algina, J. and Tang, K. L. (1988). Type I error rates for Yao's and James' tests of equality of mean vectors under variance-covariance heteroscedasticity. *J. Educ. Statist.*, **13**, 281–290. (A simulation study of Type I error rates for several multivariate tests when $G = 2$.)

5. Aspin, A. A. (1947). An examination and further development of a formula arising in the problem of comparing two mean values. *Biometrika*, **35**, 88–96. (Provides the fourth-order critical value for the power-series test when $G = 2$.)

6. Aspin, A. A. (1949). Tables for use in comparisons whose accuracy involves two variances separately estimated. *Biometrika*, **36**, 290–293. (Provides tables of the fourth-order critical value for the power-series test when $G = 2$.)

7. Brown, M. B. and Forsythe, A. B. (1974). The small sample behavior of some statistics which test the equality of several means. *Technometrics*, **16**, 129–132. (Develops the univariate Brown-Forsythe test and reports estimates of Type I error rates for the Brown-Forsythe, APDF, and first-order tests.)

8. Christensen, W. F. and Rencher, A. C. (1997). A comparison of Type I error rates and power levels for seven solutions to the multivariate Behrens-Fisher problem. *Commun. Statist. Simul. and Comput.*, **26**, 1251–1273. (A simulation study of Type I error rates and power for several multivariate tests when $G = 2$.)

9. Coombs, W. T. and Algina, J. (1996). New test statistics for MANOVA/descriptive discriminant analysis. *Educ. and Psych. Meas.*, **56**, 382–402. (Develops a multivariate extension of the Brown-Forsythe test.)

10. Coombs, W. T. and Algina, J. (1996). On sample size requirements for Johansen's test. *J. Educ. and Behav. Statist.*, **21**, 169–178. (A simulation study of Type I error rates for the Johansen APDF test when $G \geqslant 3$.)

11. Coombs, W. T., Algina, J., and Oltman, D. O. (1996). Univariate and multivariate omnibus hypothesis tests selected to control Type I error rates when variances are not necessarily equal. *Rev. Educ. Res.*, **66**, 137–180. (Review of the literature on selected solutions to the univariate and multivariate Behrens-Fisher problem.)

12. Hand, D. J. and Taylor, C. C. (1987). *Multivariate Analysis of Variance and Repeated Measures: A Practical Approach for Behavioural Scientists*. Chapman and Hall, London.

13. James, G. S. (1951). The comparison of several groups of observations when the ratios of population variances are unknown. *Biometrika*, **38**, 324–329. (Develops the power series test for the univariate case when $G \geqslant 3$.)

14. James, G. S. (1954). Tests of linear hypotheses in univariate and multivariate analysis when the ratios of the population variances are unknown. *Biometrika*, **41**, 19–43. (Develops power series critical values for the multivariate one-way layout.)

15. Johansen, S. (1980). The Welch–James approximation to the distribution of the residual sum of squares in a weighted linear regression. *Biometrika*, **67**, 85–92. (Develops the APDF test for the multivariate case.)

16. Kim, S.-J. (1993). A practical solution to the multivariate Behrens–Fisher problem. *Biometrika*, **79**, 171–176. (Develops a confidence ellipsoid procedure for comparing two mean vectors when population dispersion matrices are unequal.)

17. Lee, A. F. S. and Gurland, J. (1975). Size and power of tests for equality of means of two normal populations with unequal variances. *J. Amer. Statist. Ass.*, **70**, 933–941. (A study of Type I error rates for the various solutions to the univariate Behrens–Fisher problem when $G = 2$.)

18. Lix, L. M. and Keselman, H. J. (1995). Approximate degrees of freedom tests: a unified perspective on testing for mean equality. *Psych. Bull.*, **117**, 547–560. (Demonstrates the application of Johansen's APDF test in a wide variety of univariate and multivariate examples and presents a SAS PROC IML module for implementing the test.)

19. Nel, D. G. and van der Merwe, C. A. (1986). A solution to the multivariate Behrens–Fisher problem. *Commun. Statist. Theory and Methods*, **15**, 3719–3735. (Develops the APDF for the sum of Wisharts distribution and a multivariate APDF test for $G = 2$.)

20. Oshima, T. C. and Algina, J. (1992). Type I error rates for James's second-order test and Wilcox's H_m test under heteroscedasticity and non-normality. *Brit. J. Math. and Statist. Psych.*, **45**, 255–263. (A simulation study of Type I error rates for the several solutions to the univariate Behrens–Fisher problem when $G \geqslant 3$.)

21. Rubin, S. R. (1982). The use of weighted contrasts in analysis of models with heterogeneity of variance. *Amer. Statist. Ass.: Proc. Bus. and Econ. Statist. Sect.* (Corrects an error in the critical value for the univariate Brown–Forsythe test.)

22. Subrahamian, K. and Subrahamian, K. (1973). On the multivariate Behrens–Fisher problem. *Biometrika*, **60**, 107–111. (A simulation study of Type I error rates for several multivariate tests when $G = 2$.)

23. Tang, K. L. and Algina, J. (1993). Performance of four multivariate tests under variance–covariance heteroscedasticity. *Multivariate Behav. Res.*, **28**, 391–405. (A simulation study of Type I error rates for several multivariate tests when $G = 3$.)

24. Welch, B. L. (1938). The significance of the difference between two means when the population variances are equal. *Biometrika*, **29**, 350–362. (Develops the APDF test for $G = 2$.)

25. Welch, B. L. (1947). The generalization of 'Student's' problem when several different population variances are involved. *Biometrika*, **34**, 23–35. (Develops the power series test for $G = 2$.)

26. Welch, B. L. (1951). On the comparison of several mean values: an alternative approach. *Biometrika*, **38**, 330–336. (Develops the APDF test for $G \geqslant 3$.)

27. Wilcox, R. R. (1988). A new alternative to the ANOVA F and new results on James' second-order method. *Brit. J. Math. and Statist. Psych.*, **41**, 109–117. (A simulation study of Type I error rates for several solutions to the univariate Behrens–Fisher problem when $G \geqslant 3$.)

28. Wilcox, R. R. (1990). Comparing the means of two independent groups. *Biometrical J.*, **32**, 771–780. (A study of Type I error rates for several solutions to the univariate Behrens–Fisher problem when $G = 2$.)

29. Wilcox, R. R. (1992). Comparing one-step m-estimators of location corresponding to two independent groups. *Psychometrika*, **57**, 141–154. (Develops a procedure for comparing one-step estimators of location when $G = 2$.)

30. Wilcox, R. R. (1993). Comparing one-step m estimators of location when there are more than two groups. *Psychometrika*, **58**, 71–78. (Develops a procedure for comparing one-step estimators of location when $G \geqslant 3$.)

31. Wilcox, R. R. (1995). The practical importance of heteroscedastic methods, using trimmed means versus means, and designing simulation studies. *British J. Math. and Statist. Psych.*, **46**, 99–114. (Develops an APDF test for trimmed means when $G \geqslant 3$.)

32. Yao, Y. (1965). An approximate degrees of freedom solution to the multivariate Behrens–Fisher problem. *Biometrika*, **52**, 139–147. (Develops a multivariate APDF test for $G = 2$.)

33. Yuen, K. K. (1974). The two-sample trimmed t for unequal population variances. *Biometrika*, **61**, 165–176. (Develops an APDF test for trimmed means and $G = 2$.)

See also BEHRENS–FISHER PROBLEM—I; HETEROSCEDASTICITY; MULTIVARIATE BEHRENS-FISHER PROBLEM; WELCH'S v-CRITERION; and WELCH TESTS.

JAMES ALGINA

BEHRENS—FISHER PROBLEM, LEE—GURLAND TEST

BACKGROUND

In testing the equality of means of two normal populations, the usual procedure is to use Student's t-test as long as the population variances are known to be equal. However, if the ratio of the variances is unknown, the actual level of significance may err greatly from the nominal value α, especially when the sample sizes are small or moderately different. For example, if the sample sizes are 6 and 11, the actual level of significance for a one-sided test may range from 0.0184 to 0.1319 when the nominal level of significance $\alpha = 0.05$, and from 0.0012 to 0.0440 when $\alpha = 0.005$. This is generally recognized as the Behrens-Fisher* problem. In an effort to narrow down the range of the error, several alternatives to the t-test* have been proposed in statistical literature. These include the Welch—Aspin test [1,9,10], referred to as the AWT*, and Welch's approximate t-test [2,6,8,9,10] (See WELCH TESTS). The test discussed here is the Lee—Gurland test [3,4,5].

LEE—GURLAND TEST

Suppose that $x_{11}, x_{12}, \ldots, x_{1n_1}$ and $x_{21}, x_{22}, \ldots,$ x_{2n_2} are two random samples drawn independently from two normal populations with means μ_1 and μ_2 and variances σ_1^2 and σ_2^2, respectively. The variances are unknown and may be unequal. Let

$$R = \frac{\sigma_1^2}{\sigma_2^2},$$

$$C = \frac{\sigma_1^2/n_1}{\sigma_1^2/n_1 + \sigma_2^2/n_2} = \frac{n_2 R}{n_1 + n_2 R},$$

$$\overline{x}_i = \frac{1}{n_i} \sum_{j=1}^{n_i} x_{ij},$$

$$s_i^2 = \frac{1}{f_i} \sum_{j=1}^{n_i} (x_{ij} - \overline{x}_i)^2,$$

where $f_i = n_i - 1$ for $i = 1, 2$. We are interested in testing the null hypothesis $H_0 : \mu_1 = \mu_2$. against one-sided alternatives $H_1 : \mu_1 > \mu_2$. The usual critical region for this test is

$$v > V(\hat{C}), \qquad (1)$$

where

$$v = \frac{\overline{x}_1 - \overline{x}_2}{(s_1^2/n_1 + s_2^2/n_2)^{1/2}},$$

and

$$\hat{C} = \frac{s_1^2/n_1}{s_1^2/n_1 + s_2^2/n_2}.$$

Furthermore, V is a function of $\hat{C}$. For example, Student's t-test, Welch's approximate t-test, and the Welch—Aspin test can all be rewritten in this form. In fact, the Welch—Aspin test was obtained by applying a series expansion to the equation

$$P[v > V(\hat{C})|H_0] = \alpha. \qquad (2)$$

Welch [9] obtained an asymptotic solution $V(\hat{C})$ up to terms of order $f_i^{-2}, i = 1, 2$. Aspin

[1] extended it to terms of order f_i^{-4}. The *extended* version of the test yields the best control of the actual size as long as $n_1 \geqslant 5$ and $n_2 \geqslant 5$. The form of the function $V(\hat{C})$, however, is too complicated for any practical use. To simplify it, Lee and Gurland [5] suggested the following form of V instead:

$$V(\hat{C}) = \frac{a_1 + a_2\hat{C} + a_3\hat{C}^2}{1 + a_4\hat{C} + a_5\hat{C}^2}, \qquad (3)$$

and solved (2) numerically. The values of the coefficients a_i can be found in Lee [3] for $6 \leqslant n_1 \leqslant n_2 \leqslant 10$ when $\alpha = 0.005$, and for $5 \leqslant n_1 \leqslant n_2 \leqslant 10$ when $\alpha = 0.01, 0.025$, and 0.05. The values of the coefficients for sample sizes up to 31 have also been calculated. To avoid extensive listing of the values, Lee and Fineberg [4] and Lee [3] provided approximations for a_i:

$$a_1 = t_{f_2;\alpha},$$

$$a_i = b_{i0} + \frac{b_{i1}}{f_1} + \frac{b_{i2}}{f_2} + \frac{b_{i3}}{f_1 f_2} + \frac{b_{i4}}{f_1^2} + \frac{b_{i5}}{f_2^2}, \qquad (4)$$

$$i = 2, 3, 4, \qquad (5)$$

$$a_5 = \frac{a_1 + a_2 + a_3}{t_{f_1;\alpha}} - 1 - a_4.$$

Here, $t_{f;\alpha}$ is the upper $100\alpha\%$ point of the t-distribution* with f degrees of freedom*. The values of b are provided in Table 1. Lee and Fineberg [4] detailed the process of obtaining the b's. The Lee—Gurland test or its fitted version is almost identical to the extended Welch—Aspin test in both size and power. For two-sided alternatives, the statistic $|v|$ is used, and the nominal level of significance is 2α.

Example. For $n_1 = 6, n_2 = 11$, suppose $s_1^2 = 3, s_2^2 = 1$, and $\overline{x}_1 - \overline{x}_2 = 2$; then from (1), $v = 2.602$. For the one-sided 2.5% level of significance (5% for two-sided), the coefficients a_i are $a_1 = 2.228, a_2 = -4.815, a_3 = 4.447,$ $a_4 = -1.966, a_5 = 1.690$. These are calculated from (5) using values of b in Table 1. From (3), the critical point is $V(\hat{C}) = 2.450$. Since $v > V(\hat{C})$, we would reject the hypothesis of equal population means at the 2.5% level of significance. No assumption of $\sigma_1^2 = \sigma_2^2$ is needed. The actual level of significance of the test only varies slightly from $\alpha = 0.0250$,

Table 1. Fitted Coefficients for Lee—Gurland Test $a_i = b_{i0} + \dfrac{b_{i1}}{f_1} + \dfrac{b_{i2}}{f_2} + \dfrac{b_{i3}}{f_1 f_2} + \dfrac{b_{i4}}{f_1^2} + \dfrac{b_{i5}}{f_2^2}$

α	i	b_0	b_1	b_2	b_3	b_4	b_5
0.05	2	0.0313	−13.62383	−10.18247	−29.15488	25.12074	−5.62334
	3	−0.0306	10.23573	13.49515	−9.82230	−12.20725	32.39616
	4	0.0242	−8.11294	−4.70330	−12.56419	14.80986	3.52759
0.025	2	−0.0263	−17.62067	−15.44939	−58.30456	38.63035	−9.89857
	3	0.0233	13.54019	19.65693	−20.45043	−11.31067	61.17970
	4	−0.0131	−8.87175	−5.62154	−23.34027	19.81329	5.74252
0.01	2	0.1113	−23.84503	−31.63815	−46.09397	53.07642	−16.50011
	3	−0.0781	17.47112	37.47892	−112.38413	12.72769	112.37644
	4	0.0374	−10.12487	−10.17342	−13.69522	24.17132	10.81035
0.005	2	0.0032	−29.27162	−36.47674	−131.65988	90.04305	−13.88870
	3	0.0364	22.24399	42.73189	−115.38847	13.14024	163.36405
	4	0.0017	−11.23428	−10.53649	−38.73799	36.11356	19.05668

ranging from 0.0247 to 0.0251. The relative error ranges from −1.2% to 0.4%. By "relative error" is meant $100 \times$ (actual size − α)/α. If Student's t-test were used instead, the hypothesis would still be rejected; however, the actual level of significance could vary from 0.0075 to 0.0933, with relative error ranging from −70% to 273%! For Welch's approximate t-test, to be discussed later, the actual level of significance ranges from 0.0239 to 0.0268; the corresponding relative errors are −4.4% and 7.2%.

SIZE AND POWER CALCULATION

The power function Q of the test can be reduced to a form involving only one integration:

$$Q(\delta, C) = P[v > V(\hat{C})|\delta, C]$$

$$= 0.5 - \left\{ B(\tfrac{1}{2}f_1, \tfrac{1}{2}f_2) \right\}^{-1}$$

$$\times \int_0^1 w^{(f_1/2)-1}(1 - w)^{(f_2/2)-1}$$

$$\times g(w; \delta, C)dw, \qquad (6)$$

where $\delta = (\mu_1 - \mu_2)/(\sigma_1^2/n_1 + \sigma_2^2/n_2)^{1/2}$ is the noncentrality parameter, and $g(w; \delta, C)$ is a series depending on the parameters δ and C. For the details and a way to evaluate the integral numerically, see Lee and Gurland [5].

Lee [3] showed that the Lee—Gurland test possesses reasonably high power. For

example, for $n_1 = 6, n_2 = 11$, and $\alpha = 0.05$, its power is in excess of 88% relative to that of the twosample Student t-test, assuming R is known. The relative powers are above 79%, 66%, and 57% if $\alpha = 0.025, 0.01$, and 0.005, respectively. The lower values occur when $R > 1$, i.e., when the smaller sample comes from a population with a larger variance. If the opposite is the case, the relative powers are all above 97%, 93%, 88%, and 81% for $\alpha = 0.05, 0.025, 0.01$, and 0.005, respectively.

COMPARISON WITH WELCH'S APPROXIMATE T-TEST

Welch's approximate t-test, obtained by the method of moments*, has long been regarded as a satisfactory practical solution to the Behrens—Fisher problem [7]. However, Lee [3] found that for a small α, say 0.005 for a one-sided test (or 0.01 for a two-sided test), $n_1 = 6$ or 7, and n_2 from 6 to 31, the relative error of the actual level of significance can be 20% to 40% above α. In view of these results, the Lee—Gurland test or its fitted version is recommended for use if more accurate results are desired. The recommendation is summarized in Table 2. For illustration, consider one-sided $\alpha = 0.01$. The Lee—Gurland test is recommended for $n_1 = 5$ to 8 and $n_1 \leqslant n_2$, if one desires to maintain the actual level of significance within ±10% margin of error from the nominal level of significance. If the

Table 2. Cases when Lee—Gurland Test is Recommended ($n_1 \leqslant n_2$)

Margin of Errors (%)	n_1			
	$\alpha = 0.05$ $2\alpha = 0.1$	$\alpha = 0.025$ $2\alpha = 0.05$	$\alpha = 0.01$ $2\alpha = 0.02$	$\alpha = 0.005$ $2\alpha = 0.01$
±10	–	5	5–8	6–10
±5	5	5–7	5–11	6–14

margin of error is to be kept within ±5%, then the recommendation includes n_1 from 5 to 11.

REFERENCES

1. Aspin, A. A. (1948). An examination and further development of a formula arising in the problem of comparing two mean values. *Biometrika*, **35**, 88–96. (Extended Welch's series expansion solution to the fourth degree in the reciprocal of the degrees of freedom.)

2. Lee, A. F. S. (1992). Optimal sample sizes determined by two-sample Welch's t test. *Commun. Statist. Simulation and Comput.*, **21**, 689–696. (Provides tables of optimal sample sizes based on the Welch's approximate t-test.)

3. Lee, A. F. S. (1995). Coefficients of Lee—Gurland two sample test on normal means. *Commun. Statist. Theory and Methods*, **24**, 1743–1768. (Derives the coefficients for the Lee—Gurland test to cover more $\alpha's$. Also summarizes three categories of solutions to the Behrens—Fisher problem that appeared in the literature.)

4. Lee, A. F. S. and Fineberg, N. S. (1991). A fitted test for the Behrens—Fisher problem. *Commun. Statist. Theory and Methods*, **20**, 653–666. (Explains how to fit the coefficients needed for the Lee—Gurland test in order to avoid a substantial number of tables.)

5. Lee, A. F. S. and Gurland, J. (1975). Size and power of tests for equality of means of two normal populations with unequal variances. *J. Amer. Statist. Ass.*, **70**, 933–941. [Suggests a numerical solution, instead of the series expansion, to the probability equation (2) and obtains the Lee—Gurland test. Provides an extensive review of the literature on practical solutions of the Behrens—Fisher problem.]

6. Satterthwaite, F. E. (1946). An approximate distribution of estimates of variance components. *Biometrics Bull.*, **2**, 110–114. (Suggests the same test as Welch's approximate t-test. This is why the test is also referred to as the Satterthwaite test.)

7. Scheffé, H. (1970). Practical solutions of the Behrens—Fisher problem. *J. Amer. Statist. Ass.*, **65**, 1501–1508. (Concludes that Welch's approximate t-test is the most practical solution to the Behrens—Fisher problem.)

8. Welch, B. L. (1937). The significance of the difference between two means when the population variances are unequal. *Biometrika*, **29**, 350–362. (The first in a series of articles about Welch's approximate t-test. It is sometimes called the approximated degrees-of-freedom t-test, the approximation being obtained by the method of moments.)

9. Welch, B. L. (1947). The generalization of "student's" problem when several different population variances are involved, *Biometrika*, **24**, 28–35. [Welch proposed his second solution to the Behrens—Fisher problem by making a series expansion of the probability equation (2). The expansion was carried out to the second degree in the reciprocal of the degrees of freedom, and later extended to the fourth degree by A. A. Aspin.]

10. Welch, B. L. (1949). Further note on Mrs. Aspin's tables and on certain approximations to the tabled function. *Biometrika*, **36**, 293–296.

See also Behrens–Fisher Problem; Student's t-Tests; and Welch Tests.

Austin F. S. Lee

BELIEF, DEGREES OF

The main controversy in statistics is between "Bayesian" or neo-Bayesian methods and non-Bayesian or sampling theory methods. If you, as a statistician, wish to attach a probability to a hypothesis or to a prediction, or if you wish to use statistics for decision making, you are usually forced to use Bayesian methods. (*See* Bayesian Inference; Decision Theory.) The kinds of probability that are basic to Bayesian methods are known as

(1) subjective (personal) probabilities, here taken as a synonym for "degrees of belief," and (2) logical probabilities (credibilities) or the degrees of belief of a hypothetical perfectly rational being. Sometimes the expressions "intuitive probability" or "epistemic probability" are used to cover both subjective probability and credibility. Although "Each man's belief is right in his own eyes" (William Cowper: Hope), you would presumably accept "credibilities" as your own probabilities if you knew what they were. Unfortunately, the question of the existence of credibilities is controversial even within the Bayesian camp.

Sometimes "degree of belief" is interpreted as based on a snap judgment of a probability, but it will here be assumed to be based on a well-considered mature judgment. This perhaps justifies dignifying a degree of belief with the alternative name "subjective probability." We shall see that a theory of subjective probability provides help in arriving at mature judgments of degrees of belief. It does this by providing criteria by which inconsistencies in a body of beliefs can be detected. We shall return to this matter in more detail.

The simplest way to give an operational meaning to degrees of belief is by combining them with payoffs as in a bet. For example, suppose that your subjective probability of an event is $\frac{3}{4}$, and that you know that your opponent has the same judgment. If you are forced to bet, the unit stake being $\pm\$100$ and the sign being chosen by your opponent, then you would offer odds of 3 to 1. Any other odds will give you a negative expected gain. More generally, the basic recommendation of Bayesian decision theory is "maximize the mathematical expectation of the utility" where "utility" is a measure of value, not necessarily monetary or even material. This "principle of rationality" is used informally even in matters of ordinary life, as when we decide whether to carry an umbrella. In fact, it is a characteristic of the Bayesian philosophy that its basic principles are the same in all circumstances, in ordinary life, in business decisions, in scientific research, and in professional statistics. For other applications of the principle of rationality, see Good [12] and DECISION THEORY. There are those who

believe that Bayesian decision theory is the true basis for the whole of statistics, and others who consider that utilities should be avoided in purely scientific reporting.

Because of the subjectivity of degrees of belief, most statisticians still try to avoid their formal use though much less so than in the second quarter of the twentieth century. For this avoidance the concept of a confidence interval was introduced for the estimation of a parameter. (Confidence intervals have a long history; see Wilks [44, p. 366].) It is possible that clients of statisticians usually incorrectly interpret confidence intervals as if they were Bayesian estimation intervals. As far as the present writer knows, a survey has not been made to determine whether this is so.

Although subjective probabilities* vary from one person to another, and even from time to time for a single person, they are not arbitrary because they are influenced by common sense, by observations and experiments, and by theories of probability. Sampling is just as important in Bayesian statistics as in non-Bayesian or sampling theory methodology, and the staunch Bayesian would claim that whatever is of value in the definition of probability by long-run proportional frequency can be captured by theorems involving subjective probabilities. For the sake of emphasis we repeat here that "subjective probability" is used in this article interchangeably with "degree of belief." But in the article BELIEF FUNCTIONS a "degree of belief" is for some reason identified with "lower probability" in a sense that will be explained later.

According to the *Oxford English Dictionary*, "belief" is an all-or-none affair: you either believe something or you do not. It was this sense of belief that Bernard Shaw had in mind when he said: "It is not disbelief that is dangerous to our society; it is belief" (preface to *Androcles and the Lion*). But the extension of language that is captured by the expression "degrees of belief" is justified because we all know that beliefs can be more or less strong; in fact, the O.E.D. itself refers to "strong belief" in one definition of "conviction"! Clearly, Shaw was attacking dogmatic belief, not degrees of belief.

Beliefs, then, are quantitative, but this does not mean that they are necessarily numerical; indeed, it would be absurd to say that your degree of belief that it will rain tomorrow is 0.491336230. Although this is obvious, Keynes [26, p. 28] pointed out that "there are very few writers on probability who have explicitly admitted that probabilities, although in some sense quantitative, may be incapable of numerical comparison," and what he said was still true 40 years later, but is appreciably less true in 1980. He quotes Edgeworth as saying in 1884 that "there may well be important quantitative, although not numerical, estimates." In fact, sometimes, but not always, one can conscientiously say that one degree of belief exceeds or "subceeds" another one. In other words, degrees of belief are only partially ordered, as emphasized by Keynes [26], Koopman [27,28], and in multitudinous publications by Good, beginning with Good [11]. Keynes's emphasis was on the degrees of belief of a perfectly rational person, in other words on logical probabilities or credibilities, whereas Good's is on those of a real person, in other words on subjective or personal probabilities. (For "kinds of probability" see, e.g. Fréchet [10], Kemble [25], Good [13], and Fine [9].)

One reason why degrees of belief are not entirely numerical is that statements often do not have precise meanings. For example, it is often unclear whether a man has a beard or whether he has not shaved for a few days. But in spite of the fuzziness of faces it is often adequate to ignore the fuzziness of language; *see* FUZZY SET THEORY. Even when language is regarded as precise, it must still be recognized that degrees of belief are only partially ordered. They may be called comparative probabilities.

Some degrees of belief are very sharp, such as those occurring in many games of chance, where well-balanced roulette wheels and well-shuffled packs of cards are used. These sharp degrees of belief are fairly uncontroversial and are sometimes called "chances." Every rational number (the ratio of two integers, such as p/q) between 0 and 1 can thus be an adequately uncontroversial sharp degree of belief because a pack of cards could contain just q cards of which p are of a specialized class. With only a little idealization, of the kind that is conventional in all applied mathematics, we can include the irrational numbers by imagining a sharp rotating pointer instead of a roulette wheel. Such degrees of belief provide sharp landmarks. Every judgment of a subjective probability is in the form of an inequality of the form $P(A|B) \leqslant (P(C|D)$. By allowing for the sharp landmark probabilities provided by games of chance, each subjective probability can be enclosed in an interval of values; so we may regard subjective probabilities as "interval valued." This leads one to talk about "upper" and "lower" degrees of belief, also often called upper and lower (subjective) probabilities. These are defined as the right-hand and left-hand end points of the shortest interval that is regarded as definitely containing a specified degree of belief. (The deduction of upper and lower probabilities from other sharp probabilities, as was done by Boole [2, Chap. XIX], and by Dempster [8] by means of multivalued mappings, is logically distinct from recognizing that degrees of belief can only be *judged* to be "interval valued." The terminology of upper and lower probabilities has caused some people to confuse a "philosophy" with a "technique.") Thus a theory of partially ordered probabilities is essentially the same as a theory of probabilities that are "interval valued." An extreme anti-Bayesian would regard these intervals for subjective probabilities to be (0, 1), whereas they reduce to points for "sharp" Bayesians. Otherwise, the partially ordered or comparative theory constitutes a form of Bayes/non-Bayes compromise.

In a theory of comparative degrees of belief the upper and lower degrees of belief are themselves liable to be fuzzy, but such a theory is at any rate more realistic than a theory of sharp (precise) degrees of belief if no allowance is made for the greater simplicity of the latter.

This greater simplicity is important and much of the practical formal applications and theory of subjective probability have so far depended on sharp probabilities [3,30,37,38]. This is true also for writings on logical probability [24,29].

Without loss of generality, a degree of belief can be regarded as depending on propositions that might describe events or hypotheses. A proposition can be defined as the meaning of a statement, and a subjective probability always depends upon *two* propositions, say A and B. We then denote the "subjective probability of A given or assuming B" by $P(A|B)$. Upper and lower probabilities are denoted by P^* and P_*. The axioms are all expressed in terms of these notations in the article on axiomatics. (*See* AXIOMS OF PROBABILITY.)

Theories of degrees of belief can be either descriptive or prescriptive (= normative). Descriptive theories belong strictly to psychology; but they are useful also in the application of a normative theory because the understanding of your own psychology enables you to introspect more effectively. See Hogarth [23] for a review of the psychological experiments. It is perhaps fair to say that these experiments show that in some respects people are good at judging probabilities and in other respects are not good even when they are uninfluenced by gambler's superstitions and wishful thinking.

The function of a normative theory of degrees of belief is similar to that of logic; it removes some of your freedom because it constrains your beliefs and tends to make them more rational, in other words, it encourages your subjective judgments to be more objective. We then have the apparent paradox that those people who refuse to use a good theory of subjective probability might be more subjective than those who do use the theory. The freedom to be irrational is a freedom that some of us can do without. It is because the beliefs of real people are not entirely rational that it is useful to have a normative theory of degrees of belief: at least this is the opinion of those people who make use of such theories.

It is convenient to distinguish between (1) psychological probabilities which depend on your snap judgments, and (2) subjective probabilities that depend on your mature judgments made in an unemotional disinterested manner and with some attempt to make your body of beliefs more consistent with a good theory. There is, of course, no sharp line of demarcation between psychological and subjective probabilities. Research on the nature of psychological probability is more a problem for psychology and artificial intelligence than for statistics.

It is not known in detail how people make judgments; if it were we could build an intelligent but possibly not entirely rational machine. Obviously, the neural circuitry of the brain gradually adapts, both by conscious and unconscious thought, and not necessarily by formal Bayesian methods, to some degree of rationality. One difficulty in the analysis of judgments is that thought is not purely linguistic; as Balth van der Pol commented at the First London Symposium on Information Theory in 1951, you usually know in some sense what you are going to say before you know what words you are going to use. It is as if thought depends on propositions more than on statements. This might give some justification for expressing the axioms of probability in terms of propositions. (*See* AXIOMS OF PROBABILITY.)

The two primary methods for measuring subjective probabilities are linguistic and behavioral. In the linguistic method a person is asked to estimate a probability, and in the behavioral method he has to make bets, although the stakes need not be monetary. He has to put his decisions "where his mouth is." Or in the words of James 2:26: "Faith without works is dead." This principle constitutes a threat to the use of confidence intervals. A statistician who provides confidence intervals would not necessarily in the long run "break even" if his client, taking independent advice from a Bayesian of good judgment, had the choice of whether or not to bet. A Bayesian of bad judgment might suffer even more.

The approach in terms of betting can be used to arrive at the usual axioms of sharp subjective probability: see, e.g., Ramsey [36], de Finetti [6], and Savage [38]. (*See* AXIOMS OF PROBABILITY.) De Finetti's definition of $P(A|B)$ is the price you are just willing to pay for a small unit amount of money conditional on A's being true, B being regarded as certain all along (e.g., de Finetti [7, p. 747]). By means of a modification of Savage's approach, axioms for partially ordered subjective probability can be derived [43]. See also Shimony [42] and AXIOMS OF PROBABILITY.

The simplest form for a theory of subjective probability can be regarded as a "black box" theory. The black box or mathematical theory contains the axioms of a theory of *sharp* probabilities. Its input consists of judgments of inequalities between subjective probabilities, and its output consists of *discernments* of the same kind. The purposes of the theory are to increase your body of beliefs and to detect inconsistencies in it [11]. This description of a theory applies *mutatis mutandis* to other scientific theories, not just to a theory of subjective probability. For example, it applies to the theory of rational decisions [12]. In a more complete description of the black box theory of probability, the probabilities within the black box are represented by a different notation from the ones outside. A typical judgment, $P(A|B) < P(C|D)$, can be regarded as represented by $P'(A|B) < P'(C|D)$ in the black box, where the prime indicates that the probability is a real number and is not interval valued. This double notation, P and P', can be avoided only by using somewhat complicated axioms relating to upper and lower probabilities, such as are mentioned in the article AXIOMS OF PROBABILITY.

The plugging in of the input judgments and the reading out of the discernments can be regarded as the simplest rules of application of the theory. But for formulating your subjective probabilities you need more than just the axioms and rules of application. You also need a variety of "suggestions," and these are perhaps almost as important as the axioms, although many mathematical statisticians might not agree. In an attempt to codify the suggestions, 27 of them were collected together by Good [15] and called "priggish principles," with citations of earlier writings. For example, number 5 was: "The input and output to the abstract theories of probability and rationality are judgments (and discernments) of inequalities of probabilities, odds, Bayesian factors (ratios of final to initial odds), log-factors or weights of evidence, information, surprise indices, utilities, *and any other functions of probabilities and utilities*.... It is often convenient to forget about the inequalities for the sake of simplicity and to use precise estimates (see Principle 6)."

Principle 6 was: "When the expected time and effort taken to think and to do calculations is allowed for in the costs, then one is using the principle of *rationality of type*

II. This is more important than the ordinary principle of rationality, but is seldom mentioned because it contains a veiled threat to conventional logic by incorporating a time element. It often justifies *ad hoc* procedures such as confidence methods and this helps to decrease controversy." (For an extended account of this "time element" effect, see Good [19].)

Principle 12 dealt with the Device of Imaginary Results, which is the recommendation that you can derive information about an initial or prior distribution by an imaginary (*Gedanken*) experiment. Perhaps this principle is overlooked so often because of the "tyranny of words," that is, because a "prior distribution" sounds as if it must be thought of before the posterior distribution. (The expression "tyranny of words" is apparently due to Stuart Chase, but the concept was emphasized by Francis Bacon, who called it the Idol of the Market Place.) To take a simple example, suppose that you would be convinced that a man had extrasensory perception if he could correctly "guess" 40 consecutive random decimal digits, fraud having been ruled out. Then your prior probability must, for consistency, be at least 10^{-40} and not strictly zero. This can be shown by using Bayes' theorem backwards. The Device of Imaginary Results can be used in more standard statistical problems, such as in Bayesian testing for equiprobability of the cells of a multinomial distribution, or for "independence" in contingency tables [14,17,21]. Consider, for example, the estimation of a binomial parameter p. A class of priors that seems broad enough in some cases is that of symmetric beta priors, proportional to $p^{k-1}(1-p)^{k-1}$. If you have difficulty in selecting a value for k, you can consider implications of a few values of k when used in connection with various imaginary samples of the form "r successes in n trials." You can ask yourself, for example, whether you would in a bet give odds of 3 to 1 that $p > 0.6$ if $n = 100$ and $r = 80$. As an aid to making such judgments you can even make use of non-Bayesian methods. See also the comment below concerning hyperpriors.

Other priggish principles dealt with the Bayes/non-Bayes compromise, compromises between subjective and logical probability, quasi-utilities, such as amounts of information or weights of evidence whose expectations can be maximized in the design of experiments, the sharpening of the Ockham—Duns razor (expounded more fully by Good [18]), and the hierarchical Bayes technique. The latter is related to the fuzziness of upper and lower probabilities, and attempts, with some success, to cope with it by distributions of higher types. These deal with subjective probabilities concerning the values of rational subjective probabilities. For example, a binomial parameter p might be assumed provisionally to have a beta prior proportional to $p^{k-1}(1-p)^{k-1}$, where the "hyperparameter" k is not fixed, but is assumed to have a distribution or "hyperprior." This comes to the same as assuming a prior that is a *mixture* of beta distributions and the extension to mixtures of Dirichlet distributions is also useful. Some of the history and applications of the hierarchical Bayes technique is reviewed by Good [20]. The technique may well be necessary for the sound application of neo-Bayesian theory to multiparameter problems. Experience so far seems to support the following comments from Good [12]: "It might be objected that the higher the type the woollier (fuzzier) the probabilities. It will be found, however, that the higher the type the less the woolliness matters, provided the calculations do not become too complicated."

In short, "suggestions" for handling degrees of belief are probably essential for the practical implementation of neo-Bayesian theory.

Introspection can be aided by working with appropriate interactive computer programs. See, e.g., Novick and Jackson [33], Novick et al. [34], and Schlaifer [40]. These programs are designed to return deductions from your judgments and to allow you to change them. In other words, the computer is more or less the embodiment of the abstract black box mentioned before. The programs provide an excellent example of human—machine synergy.

A problem that has received much attention since 1950 is that of eliciting accurate degrees of belief by means of suitable rewards. The assessor specifies sharp degrees of belief $p_1, p_2, \ldots, p_n$ to n mutually exclusive events, and the reward depends on these probabilities and on the event that occurs. The reward scheme is known to the assessor in advance and is chosen in such a manner that the assessor's expected payoff is maximized, in his own opinion, if he is honest. The concept was independently suggested by Brier [4], without explicit reference to expected payoffs, and by Good [12]. Marschak [31, p. 95] says that the problem "opens up a new field in the economics of information."

According to McArthy [32], it was pointed out by Andrew Gleason (see also Aczél and Daróczy, [1, p. 115]) that if the reward is a function only of the asserted probability p of the event that occurs, and if $n > 2$, then the reward must be of the form $a + b \log p$, a formula proposed by Good [12]. This is therefore an especially pertinent form for eliciting degrees of belief, provided that the assessor has a good "Bayesian" (or neo-Bayesian) judgment. Such judgment, like any other, can presumably be improved by education and experience. The logarithmic payoff formula has an intimate relationship with weight of evidence in the sense of Peirce [35] and with the communication theory of Shannon [41] and these relationships do not depend on the condition $n > 2$.

The topic of eliciting honest degrees of belief has generated a large literature, and a very good review of the topic was written by Savage [39]. For some later work, see Buehler [5], Hendrickson and Buehler [22], and Good [16].

REFERENCES

1. Aczél, J. and Daróczy, Z. (1975). *On Measures of Information and Their Characterizations.* Academic Press, New York.

2. Boole, G. (1854). *An Investigation of the Laws of Thought.* (Reprinted by Dover, New York, undated.)

3. Box, G. E. P. and Tiao, G. C. (1973). *Bayesian Inference in Statistical Analysis.* Addison-Wesley, Reading, Mass.

4. Brier, G. W. (1950). *Monthly Weather Rev.*, **78**, 1–3.

5. Buehler, R. J. (1971). In *Foundations of Statistical Inference*, V. P. Godambe and D. A. Sprott, eds. Holt, Rinehart and Winston, Toronto, pp. 330–341. (With discussion.)

6. de Finetti, B. (1937). *Ann. Inst. Henri Poincaré*, 7, 1–68. English translation in *Studies in Subjective Probability*, H. E. Kyburg and H. E. Smokler, eds. Wiley, New York.

7. de Finetti, B. (1978). *International Encyclopedia of Statistics*, Vol. 2, W. H. Kruskal and J. M. Tanur, eds. Free Press, New York, pp. 744–754.

8. Dempster, A. P. (1967). *Ann. Math. Statist.*, **38**, 325–329. [*Math. Rev.*, **34** (1967), Rev. No. 6817.]

9. Fine, T. (1973). *Theories of Probability*. Academic Press, New York.

10. Fréchet, M. (1938). *J. Unified Sci.*, **8**, 7–23.

11. Good, I. J. (1950). *Probability and the Weighing of Evidence*. Charles Griffin, London/Hafner, New York.

12. Good, I. J. (1952). *J.R. Statist. Soc. B*, **14**, 107–114.

13. Good, I. J. (1959). *Science*, **129**, 443–447.

14. Good, I. J. (1967). *J.R. Statist. Soc. B*, **29**, 399–431. (With discussion.)

15. Good, I. J. (1971). In *Foundations of Statistical Inference*, V. P. Godambe and D. A. Sprott, eds. Holt, Rinehart and Winston, Toronto, pp. 124–127.

16. Good, I. J. (1973). In *Science, Decision, and Value*, J. J. Leach, R. Butts, and G. Pearce, eds. D. Reidel, Dordrecht, Holland, pp. 115–127.

17. Good, I. J. (1976). *Ann. Statist.* **4**, 1159–1189.

18. Good, I. J. (1977). *Proc. R. Soc. Lond. A*, **354**, 303–330.

19. Good, I. J. (1977). In *Machine Intelligence*, Vol. 8, E. W. Elcock and D. Michie, eds. Ellis Horwood, /Wiley, New York, 1976.

20. Good, I. J. (1980). International Meeting on Bayesian Statistics, 1979, Valencia, Spain. *Trab. Estadist. Invest. Oper.* (in press). Also in *Bayesian Statistics*, J. M. Bernardo, M. H. DeGroot, D. V. Lindley, and A. F. M. Smith, eds. Univ. of Valencia Press, 1981, pp. 489–519 (with discussion).

21. Good, I. J. and Crook, J. F. (1974). *J. Amer. Statist. Ass.*, **69**, 711–720.

22. Hendrickson, A. and Buehler, R. J. (1972). *J. Amer. Statist. Ass.*, **67**, 880–883.

23. Hogarth, R. M. (1975). *J. Amer. Statist. Ass.*, **70**, 271–294. (With discussion.)

24. Jeffreys, H. (1939/1961). *Theory of Probability*. Clarendon Press, Oxford.

25. Kemble, E. C. (1942). *Amer. J. Phys.*, **10**, 6–16.

26. Keynes, J. M. (1921). *A Treatise on Probability*. Macmillan, London.

27. Koopman, B. O. (1940). *Bull. Amer. Math. Soc.*, **46**, 763–774.

28. Koopman, B. O. (1940). *Ann. Math.*, **41**, 269–292.

29. Lindley, D. V. (1965). *Introduction to Probability and Statistics from a Bayesian Viewpoint*, 2 vols. Cambridge University Press, Cambridge.

30. Lindley, D. V. (1971). *Bayesian Statistics: A Review*. SIAM, Philadelphia.

31. Marschak, J. (1959). In *Contributions to Scientific Research in Management*, Western Data Processing Center, University of California at Los Angeles, ed. University of California, Berkeley, Calif., pp. 79–98.

32. McArthy, J. (1956). *Proc. Natl. Acad. Sci. USA*, **42**, 654–655.

33. Novick, M. R. and Jackson, P. H. (1974). *Statistical Methods for Educational and Psychological Research*. McGraw-Hill, New York.

34. Novick, M. R. Isaacs, G. L., and DeKeyrel, D. F. (1977). *In Manual for the CADA Monitor*. University of Iowa, Iowa City, Iowa.

35. Peirce, C. S. (1878). *Popular Sci. Monthly*; reprinted in *The World of Mathematics*, Vol. 2, J. R. Newman, ed. Simon and Schuster, New York, 1956, pp. 1341–1354.

36. Ramsey, F. P. (1931). *The Foundations of Mathematics and Other Logical Essays*. Kegan Paul, London.

37. Rosenkrantz, R. (1977). *Inference, Method and Decision*. D. Reidel, Dordrecht, Holland.

38. Savage, L. J. (1954). *The Foundations of Statistics*. Wiley, New York.

39. Savage, L. J. (1971). *J. Amer. Statist. Ass.*, **66**, 783–801.

40. Schlaifer, R. (1971). *Computer Programs for Elementary Decision Analysis*. Graduate School of Business Administration, Harvard University.

41. Shannon, C. E. (1948). *Bell Syst. Tech. J.*, **27**, 379–423, 623656.

42. Shimony, A. (1955). *J. Symb. Logic*, **20**, 1–28.

43. Smith, C. A. B. (1961). *J.R. Statist. Soc. B*, **23**, 1–37. (With discussion).

44. Wilks, S. S. (1962). *Mathematical Statistics*, Wiley, New York.

See also BAYESIAN INFERENCE and SUBJECTIVE PROBABILITIES.

I. J. GOOD

BELIEF FUNCTIONS

The theory of belief functions is a generalization of the Bayesian theory of subjective probability*. It generalizes the Bayesian theory in two ways. First, it permits nonadditive degrees of belief*. If, for example, there is no evidence for a proposition A and yet only inconclusive evidence for its negation $\overline{A}$, then the theory permits us to assign the degree of belief zero to A while assigning a degree of belief less than one to A. Secondly, it generalizes the Bayesian rule of conditioning to "Dempster's rule of combination," a rule for combining degrees of belief based on one body of evidence with degrees of belief based on another body of evidence.

If Ω denotes the set of possible answers to some question, then a "belief function over Ω" assigns to each subset A of Ω a number Bel[A] that is interpreted as one's degree of belief that the correct answer is in A. Although belief functions are not required to be additive (they may fail to satisfy Bel[A] + Bel[$\overline{A}$] = 1), they are required to have a weaker property called "monotonicity of order ∞." This property was first studied in the early 1950s by Gustave Choquet.

In addition to their general interest as a means for quantifying the weight of evidence*, belief functions are also of interest because of the statistical methods that use them. Some of these methods can be understood as generalizations of Bayesian methods (*see* NONADDITIVE PROBABILITY).

BIBLIOGRAPHY

Choquet, G. (1954). *Ann. Inst. Fourier*, **5**, 131–295.

Dempster, A. P. (1968). *J. R. Statist. Soc. B*, **30**, 205–247.

Shafer, G. (1976). *A Mathematical Theory of Evidence*. Princeton University Press, Princeton, N.J.

See also BELIEF, DEGREES OF; NONADDITIVE PROBABILITY; and SUBJECTIVE PROBABILITIES.

G. SHAFER

BELL-DOKSUM TESTS.   See DISTRIBUTION-FREE TESTS, BELL-DOKSUM

BELLMAN–HARRIS PROCESS.   See BRANCHING PROCESSES

BELL POLYNOMIALS

These polynomials represent expressions for derivatives of composite functions. The equations

$$D\{f(g(x))\} = f'(g(x)) \cdot g'(x),$$

$$D^2\{f(g(x))\} = f''(g(x)) \cdot \{g'(x)\}^2 + f'(g(x)) \cdot g''(x),$$

etc., can be written formally as

$$Df(g(x)) = f_1 g_1,$$
$$D^2 f(g(x)) = f_2 g_1^2 + f_1 g_2,$$

etc.

The quantities on the right-hand side are called (by Riordan [2,3]) *Bell polynomials* (see Bell [1]) of order $1, 2, \ldots$. The Bell polynomial of order r is

$$Y_r = r! \sum \cdots \sum f_k \prod_{j=1}^{n} \left(\frac{g_j^{k_j}}{j! k_j!} \right),$$

where the summation is over all sets of nonnegative integer values $k_1, k_2, \ldots, k_n$ such that $k_1 + 2k_2 + \cdots + nk_n = n$.

Interpreting f^k as f_k, and Y^k by Y_k, the Bell polynomials are generated by the identity

$$\exp(sY) = \exp\left(f \sum_{j=1}^{\infty} \frac{s^j g_j}{j!} \right).$$

Bell polynomials are useful in the analysis of Lagrange distributions*.

For further details, refer to Riordan [2,3].

REFERENCES

1. Bell, E. T. (1927). *Ann. Math.*, **29**, 38–46.
2. Riordan, J. (1958). *An Introduction to Combinatorial Analysis*. Wiley, New York.
3. Riordan, J. (1968). *Combinatorial Identities*. Krieger, Huntington, N.Y.

See also LAGRANGE AND RELATED PROBABILITY DISTRIBUTIONS.

BENFORD'S   LAW.   See   FIRST-DIGIT PROBLEM

BENINI DISTRIBUTION

This distribution is one of the earliest modifications of the Pareto Type I distribution. It was proposed by Benini [1,2], early in the twentieth century, and has since then been rediscovered many times. Among these it has been termed "approximate lognormal" [3] and "quasi-lognormal" [4].

The cdf is given by

$$F(x) = 1 - (x/x_0)^{-\alpha + \beta \log x}, \quad x \geqslant x_0 > 0,$$

$$\alpha > 0.$$

When $x_0 = 1$, the density is

$$f(x) = x^{-\alpha - 1 - \beta \log x}(\alpha + 2\beta \log x).$$

This reduces to Pareto Type I* for $\beta = 0$; for $\alpha = 0$, it can be viewed as a log-Rayleigh distribution*.

The distribution is "Stieltjes-indeterminate" in the sense (analogous to the lognormal distribution) that the moments do not determine the distribution. Benini [1] and subsequent Italian statisticians dealing with income distribution problems made extensive use of this distribution for fitting personal income data. Benini [2] uses a regression technique for estimators of the parameters.

REFERENCES

1. Benini, R. (1905). I diagrammi a scala logarithmica. *Giornale degli Economisti, Ser. 2*, **30**, 222–231.

2. Benini, R. (1906). *Principii di Statistica Metodologica*, Torino, UTET (Unione Tipogratico-Editrice Torines.)

3. DuMouchel, W. H. and Olshen, R. A. (1975). On the distribution of claims costs. In: Kahn (ed.) *Credibility: Theory and Applications* Proc. Berkeley Actuar. Res. Conference on Credibility, Sept. 19–21, 1974, 23–46.

4. Shpilberg, D. C. (1977). The probability distribution of fire loss amount. *J. Risk Insurance*, **44**, 103–115.

BENJAMINI–HOCHBERG PROCEDURE. See FALSE DISCOVERY RATE

BERGE INEQUALITY

A two-dimensional probability inequality of the Chebyshev* type. Given a two-dimensional random variable $X = (X_1, X_2)$ with

$$\mu_i = E[X_i], \quad \text{var}(X_i) = \sigma_i^2 \quad (i = 1, 2)$$

$$\text{cov}(X_1, X_2) = \sigma_{12} \quad \text{and } \rho = \frac{\sigma_{12}}{\sigma_1 \sigma_2},$$

then

$$\Pr\left(\max\left\{\frac{|X_1 - \mu_1|}{\sigma_1}, \frac{|X_2 - \mu_2|}{\sigma_2}\right\} \geqslant \lambda\right)$$

$$\leqslant \frac{1 + \sqrt{1 - \rho^2}}{\lambda^2};$$

also,

$$\Pr\left(\text{either} \frac{|X_1 - \mu_1|}{\sigma_1} \geqslant \lambda \text{ or } \frac{|X_2 - \mu_2|}{\sigma_2} \geqslant \lambda\right)$$

$$\leqslant \frac{1 + \sqrt{1 - \rho^2}}{\lambda^2}$$

for any $\lambda > 0$.

This inequality bounds the probability of falling outside a rectangle centered at the means of a bivariate distribution; it uses a measure of the dependence* between the random variables.

BIBLIOGRAPHY

Berge, P. O. (1938). *Biometrika*, **29**, 405–406.

Karlin, S. and Studden, W. J. (1966). *Tchebycheff Systems with Applications in Analysis and Statistics*. Wiley, New York.

Leser, C. E. V. (1941). *Biometrika*, **32**, 284–293.

Savage, I. R. (1961). *J. Res. Natl. Bur. Stand.*, **65B**(3), 211–222.

See also BERNSTEIN'S INEQUALITY; BIRNBAUM-RAYMOND-ZUCKERMAN INEQUALITY; and CAMP–MEIDELL INEQUALITY.

BERKSON ERROR MODEL. See LINEAR REGRESSION, BERKSON ERROR MODEL IN

BERKSON'S MINIMUM LOGIT χ^2 PROCEDURE. See MINIMUM CHI-SQUARE

BERKSON'S MINIMUM NORMIT χ^2 PROCEDURE.   See MINIMUM CHI-SQUARE

BERKSON'S $2n$ RULE.   See LOGIT ESTIMATOR, BERKSON'S $2n$ RULE FOR

BERNOULLI DISTRIBUTION

A random variable X possesses a Bernoulli distribution with parameter $p(0 < p < 1)$ if $\Pr\{X = 1\} = p$ and $\Pr\{X = 0\} = 1 - p$. The corresponding CDF is

$$F(x) = \begin{cases} 0 & (x < 0) \\ 1 - p & (0 \leqslant x < 1) \\ 1 & (x \geqslant 1) \end{cases}$$

and the characteristic function* is

$$\phi_x(t) = 1 + p(e^{it} - 1).$$

The moments of this distribution are $\mu'_k = E(X^k) = p \ (k = 1, 2, \ldots)$, $\mathrm{var}(X) = p(1 - p)$.

The Bernoulli distribution is basic in probability theory and statistics, being a model of any random experiment with outcomes belonging to two mutually disjoint classes.

See also BERNOULLI NUMBERS; BERNOULLIS, THE; BINOMIAL DISTRIBUTION; and LAWS OF LARGE NUMBERS.

BERNOULLI DISTRIBUTIONS, MULTIVARIATE

The joint distribution of k Bernoulli variables $X_1, \ldots, X_k$

$$P\left[\cap_{k=1}^{k}(X_i = x_i)\right]$$
$$= P_{X_1, X_2, \ldots, X_k}, x_i = 0, 1; i = 1, \ldots, k \quad (1)$$

is referred to as the *multivariate Bernoulli distribution*.

Teugels notes the one-to-one correspondence between the probabilities (1) and the integers $N = 1, 2, \ldots, 2^k$, with

$$N(\boldsymbol{x}) = N(x_1, x_2, \ldots, x_k) = 1 + \sum_{i=1}^{k} 2^{i-1} x_i.$$

Multivariate Bernoulli variables are of importance [2] in connection with relations between

classification (clustering) and statistical modeling of binary data*, in particular modeling by latent classes [1].

Gyllenberg and Koski [3] make use of this distribution extensively in probabilistic models for bacterial taxonomy. They derive the posterior distribution of (1) utilizing Jeffreys' priors.

More details on the distribution are given in Reference 4. In the case of independent $(X_1, \ldots, X_k)$,

$$p(\boldsymbol{x}|p_{ij}) = \Pi_{i=1}^{k}(1 - p_{ij})^{1-x_i} p_{ij}^{x_i}, \quad \boldsymbol{x} \epsilon R^k,$$

where $\boldsymbol{p}_j = (p_{1j}, \ldots, p_{kj}), 0 \leqslant p_{ij} \leqslant 1$ and $0 \equiv 1$.

REFERENCES

1. Bock, H.-H. (1996). Probability models and hypothesis testing in partitioning cluster analysis. In *Clustering and Classification*, P. Arabil, L. J. Huber, and G. DeSoete, eds. World Scientific Publishers, Singapore, pp. 377–453.

2. Govaert, G. (1990). Classification binaire et modéles. *Rev. Stat. Appl.*, **38**, 67–81.

3. Gyllenberg, M. and Koski, T. (2001). Probabilistic models for bacterial taxonomy. *Int. Stat. Rev.*, **69**, 249–276.

4. Johnson, N. L., Kotz, S., and Balakrishnan, N. (1997). *Discrete Multivariate Distributions*. Wiley, New York.

BERNOULLI NUMBERS

Bernollui numbers $B_r(r = 0, 1, \ldots)$ (first introduced by J. Bernoulli [2]) are used in numerical integration formulas, in the calculus of finite differences*, and also occur in combinatorial-probabilistic problems. They can be defined in various ways.

1. The original definition given by Bernoulli is

$$\sum_{\mu=0}^{m-1} \mu^k = \frac{1}{k+1} \sum_{r=0}^{k} \binom{k+1}{r} B_r m^{k+1-r}$$

for $k = 0, 1, 2$ and $m = 1, 2, \ldots$.

2. For even subscripts,

$$B_{2n} = (-1)^{n-1} \frac{2(2n)!}{(2\pi)^{2n}} \sum_{m=1}^{\infty} \frac{1}{m^{2n}} \quad (n \geqslant 1).$$

For odd subscripts (except for B_1) the Bernoulli numbers are all zero. The values of the first several Bernoulli numbers are $B_0 = 1, B_1 = -\frac{1}{2}, B_2 = \frac{1}{6}, B_4 = -\frac{1}{30}, B_6 = \frac{1}{42}, B_8 = -\frac{1}{30}, B_{10} = \frac{5}{66}, B_{12} = -\frac{691}{2730}$, and $B_{14} = \frac{7}{6}$.

3. Alternative definitions are

$$B_{2n} = (-1)^{n-1} 4n \int_0^\infty \frac{x^{2n-1}}{e^{2\pi x} - 1} \, dx \qquad (n \geqslant 1)$$

or indirectly via the expansion

$$u/(e^u - 1) = \sum_{n=0}^\infty B_n u^n / n! \qquad (0 < |u| < 2\pi),$$

which is often used.

Among the useful properties of these numbers we mention

$$\sum_{v=0}^{n-1} \binom{n}{v} B_v = 0 \quad (n \geqslant 2).$$

These numbers also occur in the Euler-Maclaurin formula:

$$\int_{x_0}^{x_m} f(x)dx = h(f_0/2 + f_1 + f_2 + \cdots$$
$$+ f_{n-1} + f_m/2)$$
$$- \sum_{2}^\infty B_v h^v (f_m^{(v-1)} - f_0^{(v-1)})/v,$$

where $f_i = f(x_i) = f(x_0 + ih)$, as well as in the expansion of the complex-valued trigonometric functions $\cot Z$ and $\tan Z$.

Useful bounds on these numbers are

$$\frac{2 \cdot (2n)!}{(2\pi)^{2n}} < (-1)^{n-1} B_{2n}$$
$$\leqslant \frac{(\pi^2/3)(2n)!}{(2\pi)^{2n}} (n \geqslant 1),$$

which shows that $(-1)^{n-1} B_{2n} \to \infty$ as $n \uparrow \infty$. Tables of Bernoulli numbers are presented in Abramowitz and Stegun [1].

REFERENCES

1. Abramowitz, M. and Stegun, I. A. (1964). Handbook of Mathematical Functions. *Natl. Bur. Stand. U. S. Appl. Math. Ser.* 55 (Washington, D.C.).

2. Bernoulli, J. (1713). *Ars. Conjectandi.*

3. Davis, H. T. (1935). *Tables of the Higher Mathematical Functions.* Principia Press, Bloomington, Ind.

See also BERNOULLI POLYNOMIALS; COMBINATORICS; FINITE DIFFERENCES, CALCULUS OF; and NUMERICAL INTEGRATION.

BERNOULLI POLYNOMIALS

The rth Bernoulli polynomial $B_r(y)$ is defined by

$$xe^{yx}(e^x - 1)^{-1} = \sum_{r=0}^\infty B_r(y) \frac{x^r}{r!}.$$

$B_r(0)$ is called the rth Bernoulli number* and written B_r.

The Bernoulli polynomials of order n are defined by

$$x^n e^{yx}(e^x - 1)^{-n} = \sum_{r=0}^\infty B_r^{(n)}(y) \frac{x^r}{r!}.$$

The following relations are of interest:

$$B_n^{(n+1)}(x) = x^{(n)},$$
$$B_r^{(n)}(x) \simeq (B^{(n)} + x)^r,$$

where, after expansion, $\{B^{(n)}\}^j$ is replaced by $B_j^{(n)}$.

See also BERNOULLI NUMBERS.

BERNOULLI SOCIETY

The Bernoulli Society for Mathematical Statistics and Probability is the international learned society in these fields, and it is the only one. Its objective is the advancement of probability and mathematical statistics in the widest sense through international contacts and international cooperation. Thus theoretical research, applications, dissemination of knowledge inside and outside the circle of the workers in the field, and development of teaching methods in the subject on all levels are all covered.

The Bernoulli Society has only individual members. Membership is open to anybody interested in its fields. In order to emphasize interaction with other branches of statistics such as official (governmental) statistics*, survey sampling*, and statistical computing, the Bernoulli Society is organized as a Section of the International Statistical Institute* (ISI). It is using the facilities of the ISI office in The Hague (the Netherlands) for its administration, and its ordinary meetings are being held biennially conjointly with the ordinary sessions of the ISI.

The statutes foresee, in addition to the general assembly composed of all members, an executive committee (President, elected for 2 years; president-elect; scientific secretary; treasurer; executive secretary of the ISI) and a council, consisting of 12 members elected for 4 years, and the members of the executive committee.

The Bernoulli Society was founded in its present form in 1975 and succeeds the International Association for Statistics in the Physical Sciences (IASPS), which originated in 1963. The following were the presidents of IASPS and the Bernoulli Society: Kitagawa (Japan), Bartlett (United Kingdom), Kolmogorov* (USSR), Neyman* (United States), Schmetterer (Austria), D. G. Kendall (United Kingdom), Blackwell (United States), Krickeberg (France), D. R. Cox (United Kingdom); the president-elect during 1979–1981 is P. Révész (Hungary). Some of the activities of the Bernoulli Society are being conducted by standing committees devoted to a particular subject area or geographical region.

To reach its objective, the Bernoulli Society is active in the following ways: first by organizing its own international meetings and joint meetings with other bodies, or by sponsoring such meetings. Thus the Committee on Statistics in the Physical Sciences, which continues the work of the former IASPS, is involved in meetings in its field of application. The Committee for Conferences on Stochastic Processes*, which is now also a standing committee of the Society, has been organizing annual meetings in its own domain on four continents since 1971. The European Regional Committee of the Society and its forerunner has been arranging the European Meetings of Statisticians since 1962. Similar regional committees in Latin America, in East Asia, and in the Pacific

were recently formed. The Bernoulli Society also attends to relevant parts of the program for the biennial sessions of the ISI.

Second, the Bernoulli Society supports and sponsors relevant publications. In particular, it takes part in the "International Statistical Review"*. All members receive the "Review" and the newsletter of the ISI, "International Statistical Information," which contains, among other items, complete and early announcements of all meetings of interest to statisticians and probabilists. Members also receive the ISI *International Statistical Education Newsletter* and *Short Book Reviews*, which provides a rapid reviewing service. Many professional journals grant reduced rates to members of the Society.

The Bernoulli Society actively seeks contacts with the users of mathematical statistics and probability on an international basis. It is trying to eliminate existing barriers between statisticians or probabilists working in academic institutions and those employed by agricultural, medical, meteorological, industrial, and similar institutions.

One of the main concerns of the Bernoulli Society has been to overcome obstacles posed by geographical and economical conditions, currency problems, or political restrictions. For example, the membership dues may in principle be paid in national currency in all countries whose currency is not convertible for this purpose. The Society feels a very strong responsibility toward advancing its science in the wide sense defined by the opening paragraph, especially in those countries where mathematical statistics and probability is still relatively weak. It is doing this by intensifying its work on a regional basis and by offering the services of its members in various forms: lectures, courses, documentation, and specialized consulting. Its participation in the work of the ISI on statistical education* is seen to a large extent from this angle.

On June 1, 1981, the Bernoulli Society had 936 members in 57 countries. The membership dues are at present 35 Swiss francs per year, and half of this for members of age less than 30 years.

See also INTERNATIONAL STATISTICAL INSTITUTE (ISI).

K. KRICKEBERG

BERNOULLI TRIALS

Named after the Swiss mathematician J. Bernoulli (*see* BERNOULLIS, THE), Bernoulli trials are repeated trials of an experiment that obey the following conditions:

1. Each trial yields one of two outcomes [often called *success* (*S*) and *failure* (*F*)].
2. For each trial the probability of success is the same [usually denoted by $p = P(S)$, and the probability of failure is denoted by $q = 1 - p$].
3. The trials are independent; the probability of success in a trial does not change given any amount of information about the outcomes of other trials.

The simplest example of Bernoulli trials is the model for tossing a fair coin, where the occurrences head and tail are labeled S and F, respectively (here $p = q = \frac{1}{2}$).

See also BERNOULLI DISTRIBUTION and BINOMIAL DISTRIBUTION.

BERNOULLI'S LIMIT THEOREM

James Bernoulli's limit theorem [1] is a simple form of the weak law of large numbers* that follows from Chebyshev's inequality*. If S_n represents the number of successes (or occurrences) in n independent trials of an event with probability p, and if $\epsilon > 0$, then

$$\Pr[|n^{-1}S_n - p| \geqslant \epsilon] \to 0 \quad \text{as } n \to \infty$$

or

$$\Pr[|n^{-1}S_n - p| < \epsilon] \to 1 \quad \text{as } n \to \infty.$$

Equivalently, $n^{-1}S_n \overset{p}{\to} p$ as $n \to \infty$.

The last equation means that the relative frequency of successes approaches *in probability* the probability of success as the number of independent trials increases indefinitely.

REFERENCE

1. Bernoulli, J. (1713). *Ars Conjectandi*. Basle, Switzerland.

See also BERNOULLIS, THE and LAWS OF LARGE NUMBERS.

BERNOULLIS, THE

Bernoulli is one of the most illustrious names in the history of mathematics. At least eight distinguished mathematicians bore the name (see Fig. 1), and three of these, the brothers James and John and John's son Daniel, were in the first rank of the mathematicians of their time. All the mathematicians Bernoulli were descendants of James' and John's father Nicholas Bernoulli, a prominent merchant in Basel, Switzerland. And all considered Basel their home.

Five of the Bernoullis made contributions to the theory of probability: James and John, their nephew Nicholas, Daniel, and John's grandson John III. The major contributions were those of James, Nicholas, and Daniel.

JAMES AND JOHN

James Bernoulli, the eldest of four brothers, was meant by his father to be a theologian. John, the third of the brothers, was meant to be a merchant. Both studied mathematics secretly and against their father's will.

James completed a degree in theology at the University of Basel in 1676. He then left Basel for several years of travel, during which he tutored and taught in several parts of France and Switzerland and further pursued his studies, especially Cartesian philosophy. After his return to Basel in 1680, he turned more to mathematics. In 1681, at a time when his fellow theologians were proclaiming a recent spectacular comet a sign of God's anger, he published a mathematical theory of the motion of comets. And he devoted a second educational journey, to Holland and England in 1681–1682, to visits with mathematicians. He returned to the University of Basel to lecture in experimental physics in 1683, and he obtained the university's chair of mathematics in 1687.

John, too, disappointed his father. He proved unsuited for a business career and obtained permission to enroll at the university when James began teaching there. His father now wanted him to study medicine, and he did so after completing his master of arts degree in 1685. But James secretly tutored him in mathematics, and about the

Figure 1. The mathematicians Bernoulli. The Roman numerals have been added to their names by historians. *James, John,* and *Nicholas* are English names. They become *Jakob, Johann,* and *Nikolaus* in German, and *Jacques, Jean,* and *Nicolas* in French.

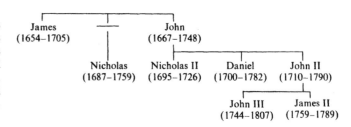

time that James obtained his chair in mathematics, the two began to puzzle out the cryptic version of the infinitesimal calculus that G. W. Leibniz (1646–1716) published in 1684. They were the first to master Leibniz' method, and it was through the brilliant papers that they and Leibniz published from 1690 on that the calculus became known to the learned world.

As their mathematical powers developed, the two pugnacious brothers became bitter rivals. By the late 1690s, after John had obtained his own position in Groningen, the Netherlands, they were publicly trading insults and challenging each other with mathematical problems. It is unfortunate, perhaps, that these problems were so interesting, for memory of them has kept the brothers' quarrels alive even after their own lifetimes.

It is difficult to compare the accomplishments of the two brothers. The slower and more careful James bested his brother in their public exchanges, but John ultimately made the greater contribution to the calculus. John inherited his brother's chair at Basel after his death in 1705, and remained active for 30 years thereafter. After Newton's death in 1727, he was unchallenged as the leading mathematician of all Europe.

In the field of probability, it is certainly James who was the giant. His philosophical training, together with his deep mathematical intuition, uniquely fitted him to attack the conceptual problems involved in applying the theory of games of chance*, as it had been developed by his predecessors Pascal* (1623–1662), Fermat* (1601–1665), and Huygens* (1629–1695), to problems of probability and evidence. And his struggle with these conceptual problems led him to the first limit theorem* of probability—the law of large numbers*.

James' great treatise on probability, *Ars Conjectandi* or the "Art of Conjecture," has four parts. The first three parts are in the tradition of his predecessors; he studies Huygens' pioneering book (*De ratiociniis in ludo aleae,* 1657), develops new combinatorial tools, and applies these to new problems. But the fourth part develops a new theory of probability—a theory that uses Huygens' ideas together with the law of large numbers to go beyond games of chance* to problems of evidence and practical decision. (See below for a brief explanation of James' theory.)

James' mathematical diary shows that he had begun to work on probability in the 1680s and that he had proven the weak law of large numbers* by 1689. Thus his completion of *Ars conjectandi* was impeded by ill health, and after his death its publication was delayed by his wife and son's fears that an editor might treat it unfairly. It was finally published in 1713; *see* BERNOULLI'S LIMIT THEOREM.

NICHOLAS

Nicholas Bernoulli studied mathematics under both his uncles. In 1709 he obtained a law degree with a dissertation that applied some of the ideas of *Ars conjectandi* to problems of law. Later, he was the editor who finally published *Ars conjectandi.* His first academic position was at Padua, Italy, but from 1722 on he held chairs at Basel, first in logic and then in law.

Nicholas was a gifted mathematician, but he published very little. Most of his achievements are hidden in the some 500 letters he wrote to other mathematicians. His correspondents included Leibniz, Euler (1707–1783), De Moivre* (1667–1754), and Montmort (1678–1719).

He is especially remembered for his letters to Montmort*, where he generalized many of the results of Montmort's 1708 book *Essay d'analyse sur les jeux de hazard*. Montmort published most of his correspondence in the second edition of the book, in 1713; he and Nicholas had become so close that Nicholas stayed in Montmort's home for several months helping him prepare that edition.

One of the topics in the correspondence between Nicholas and Montmort was the excess of boys over girls in the christening records in London. This excess had been pointed out around 1711 by John Arbuthnot*, who observed that it was too persistent to attribute to chance*. Arbuthnot saw this as evidence of divine intervention, but Nicholas, mindful of James' law of large numbers, explained the excess by saying that the true odds for a boy are closer to 18 : 17 than to 50 : 50.

DANIEL

Daniel Bernoulli was born during his father John's stay in Groningen, but he was educated at Basel. John made Daniel repeat his own youthful experience—first he tried to make him a merchant, and then he directed him toward medicine. Daniel's doctoral dissertation, completed in 1721, was on the mechanics of breathing.

In 1725, Daniel and his beloved older brother Nicholas II, who had tutored Daniel in mathematics, both obtained academic posts in St. Petersburg, Russia. Daniel was saddened by his brother's sudden death the following year, but he remained in St. Petersburg for eight years, during which he was very productive. In 1733, he returned to Basel, where he successively held chairs in anatomy and botany, in physiology, and in physics.

Daniel was a prolific and original mathematician, especially in physics. His greatest work was his treatise on hydrodynamics, written while he was in St. Petersburg. He also made major contributions to the mechanics of flexible and elastic bodies. It is an indication of his stature, perhaps, that his competitive father was often antagonistic toward him.

Daniel made several contributions to probability and statistics. The first and most influential was his *Specimen theoriae novae de mensura sortis* (1738), in which he developed the idea of utility*. His inspiration for this paper was a problem posed to him by his cousin Nicholas: how to determine the fair price of a game with infinite mathematical expectation*. We usually think of the expectation as the game's fair price, but as Nicholas pointed out, it is hardly fair to pay an infinite amount when one is certain to get only a finite amount back. Daniel's solution was to replace the monetary value of the payoff by its utility to the player, which may have a finite expectation. Since Daniel's paper was published in the St. Petersburg academy's journal, Nicholas' problem has become known as the St. Petersburg Paradox*.

Daniel is also remembered for his spirited statistical defense of inoculation for smallpox and for his statistical test of randomness* for the planetary orbits. And in several papers written around 1770, he participated in the effort to develop a probabilitistic theory of errors* that could be applied to astronomical observations. One of these papers is notable, in retrospect, for its use of what we now call the method of maximum likelihood*. Daniel did not develop this idea extensively, and his work was soon superseded by Laplace's* Bayesian theory.

JAMES' THEORY OF PROBABILITY

In the long tradition of philosophical and theological thought to which James Bernoulli was heir, the idea of probability was not closely tied to the idea of chance. Pascal, Fermat, and Huygens did not even use the word *probability* in their writings on chance [see CHANCE—I; CHANCE—II]; probability, as these scholars knew, was an attribute of opinion, a product of argument or authority. The theory that James set forth in Part IV of *Ars conjectandi* was an attempt to bridge this gap. It was an attempt to apply the new theory of games of chance to probability while preserving the idea that probability is based on argument.

James' theory had two parts: his law of large numbers, which he thought could be used to assess the strength of an argument, and his rules for combining arguments.

Here is one of the examples of practical reasoning that made James see the need for his law of large numbers. Titius is marrying Caja, and an agreement is to be made concerning the division of her estate between him and their children in case she dies before him. But the size of the estate will depend on whether their own fathers are still living. Titius considers two possible agreements: one specifies that two-thirds of the estate will fall to him in any case; the other varies the proportion according to which fathers are still living. Which agreement is most advantageous to Titius? As James saw, the answer depends on how probable it is that Caja will outlive one or both of the fathers. But we cannot evaluate the probabilities a priori*, for we cannot count the equally possible ways in which Caja might or might not outlive the fathers. James thought there must be some number r of ways in which she might outlive them and some number s in which she might not. But since we do not fully understand the possible causes of death for Caja or the fathers, we cannot find r and s so as to calculate the probability $r/r + s$.

But it should be possible, James reasoned, to evaluate this probability *a posteriori*, from outcomes observed in many similar cases. The numbers r and s should be the same for other young women in similar circumstances. So in large number n of such cases, there will be a total of $(r + s)^n$ equally possible ways things might turn out: r^n ways for all the young women to outlive their fathers and fathers-in-law, $nr^{n-1}s$ ways for all but one to, and so on. Studying this binomial expansion, James realized that most of the $(r + s)^n$ equally possible ways involve a number k of the n young women outliving their fathers and fathers-in-law such that k/n is close to $r/(r + s)$. So if we actually observe n such young women, the observed frequency k/n will probably be close to $r/(r + s)$. As James proved, we can always choose n large enough for it to be morally certain (he thought we should decide on some high value of probability to call "moral certainty") that k/n will be close enough to $r/(r + s)$ that the two are identical for our practical purposes. This is the theorem we now call "the weak law of large numbers"*.

There remains the problem of combination. In the preceding example there may be no need to combine the probabilities based on mortality* statistics with probabilities based on other arguments. But in general, James believed, probability will depend on the combination of many arguments. Among the examples he used to make his point was the following very traditional one, which goes back to Cicero (106-43 B.C.) and Quintilian (ca A.D. 35-96). Titius is found dead on the road and Maevius is accused of committing the murder. There are several arguments against him. Maevius hated Titius, and they had been quarreling the day before; Maevius traveled over the road that day; a blood-stained sword is found in his house; and he turns pale and answers apprehensively when he is questioned. Perhaps none of these arguments alone would make Maevius' guilt probable. But together they make it almost certain.

As James saw it, his theorem should enable us to assess probabilities based on each argument. By observing, for example, other instances where a blood-stained weapon is found in a suspect's house, we should be able to assess the probability of Maevius' guilt based on that argument alone. But after thus assessing probabilities based on each argument alone, we should use his rules, which were derived from the theory of games of chance, to combine these probabilities. Only then do we have a probability based on the total evidence. The rules of combination James gave were imperfect, but they are interesting even today. Some of them explicitly recognize the possible non-additivity of probability (*see* NONADDITIVE PROBABILITY).

James' theory was not, of course, an adequate theory of statistical inference*. He suggested we find out how many observations are required for moral certainty and then make that many. He did not recognize, perhaps because he had no opportunity to apply his theory to actual numerical data, that we cannot always make as many observations as we would like. Statistical methods for dealing with limited observations did not emerge until more than 50 years after James' death, in response to the need to combine discrepant astronomical observations. *See* LEAST SQUARES.

Literature

The entries on the Bernoullis in the *Dictionary of Scientific Biography* include excellent descriptions of their contributions to mathematics and mechanics and extensive bibliographies. Detailed accounts of their contributions to probability can be found in Issac Todhunter's *History of the Theory of Probability* (1865; reprinted by Chelsea, New York, in 1965) and in L. E. Maistrov's *Probability Theory: A Historical Sketch* (1974, Academic Press, New York). James' work is also discussed by F. N. David in *Games, Gods and Gambling* (1962, Hafner, New York) and by Ian Hacking in *The Emergence of Probability* (1975, Cambridge University Press, Cambridge).

In 1935, the Bernoulli Commission was created in Basel for the purpose of publishing a new edition of the works of the Bernoullis. The plan is to publish everything left by the three great Bernoullis and a selection of the work of the five lesser ones. So far three volumes have appeared, one volume of John's letters and two volumes of James' work (*Der Briefwechsel von Johann Bernoulli*, Vol. 1, 1955, Otto Spiess, ed.; *Die Werke von Jakob Bernoulli*, Vol. 1, 1969, J. O. Fleckenstein, ed., and Vol. 3, 1975, B. L. van der Waerden, ed.; all published in Basel by Birkhäuser Verlag). The volume edited by van der Waerden includes most of James' work on probability, including the parts of his previously unpublished diary bearing on probability. (It is in this diary that he discusses the problem of Caja's marriage contract.) It also includes Nicholas' dissertation, John de Witt's treatise on annuities, and extensive commentary, in German, on James' work and on most of the Bernoullis' correspondence on probability. [For further information on the new Bernoulli edition, see Truesdell's reviews in *ISIS*: 1958 (**49**:54–62) and 1973 (**64**:112–114).]

James' *Ars Conjectandi* was reprinted in the original Latin in 1968 by the Belgian publishing house Culture et Civilisation. A German translation was published in 1899: *Wahrscheinlichkeitsrechnung* (2 vols., Engelman, Leipzig). Unfortunately, there is still no definitive and widely available English translation, but a rough translation of Part IV is available from the National Technical Information Service, 5285 Port Royal Road, Springfield, VA 22161. (*Translations from James Bernoulli*, by Bing Sung, Harvard University Technical Report, 1966; the order number is AD-631 452.) For a faithful account in English of James' proof of his theorem, see pp. 96–101 of J. V. Uspensky's *Introduction to Mathematical Probability* (1937, McGraw-Hill, New York). For a discussion of his ideas on the combination of arguments, see G. Shafer's "Non-additive Probabilities in the Work of Bernoulli and Lambert," *Archive for History of Exact Sciences*, 1978 (**19**:309–370), and for a discussion of the relation of his ideas to classical rhetoric, see D. Garber and S. Zabell's "On the Emergence of Probability" in the same journal, 1979 (**21**:33–54).

An English translation of Daniel's *De mensura sortis* was published in *Econometrica* in 1954 (**22**:23–36). In 1967, Gregg Press (Farnborough, Hampshire, England) reprinted this translation, together with an earlier German translation and commentary, under the original Latin title. L. J. Savage discussed the relation between Daniel's concept of utility and contemporary concepts in Sec. 5.6 of *The Foundations of Statistics* (1972, Dover, New York). Daniel's essays on inoculation appeared in French in the *Mémoires* of the Paris academy (1760, pp. 1–45) and in *Mercure de France* (June 1760). An English translation of the paper in which he used maximum likelihood* was published in *Biometrika* in 1958 (**45**:293–315) and reprinted in Vol. 1 of *Studies in the History of Statistics and Probability* (E. S. Pearson and M. G. Kendall, eds.; Charles Griffin, London, 1970). Articles by O. B. Sheynin on his work are included in Vol. 2 of these *Studies* (Kendall and Plackett, eds.; 1977).

See also BERNOULLI'S LIMIT THEOREM; BINOMIAL DISTRIBUTION; COMPETING RISKS; GAMES OF CHANCE; LAWS OF LARGE NUMBERS; MAXIMUM LIKELIHOOD ESTIMATION; NONADDITIVE PROBABILITY; and RANDOMNESS, TESTS OF.

G. SHAFER

BERNOULLI'S THEOREM.   See BERNOULLI'S LIMIT THEOREM

BERNSTEIN POLYNOMIAL DISTRIBUTIONS

The Bernstein polynomial distribution was introduced as a prior distribution [1] and extended later to posterior distributions* [2].

For a given integer k and a function F on $[0, 1]$, the Bernstein polynomial is defined as

$$B(x; k, F)$$
$$\equiv \begin{cases} 0, & x < 0, \\ \sum_{j=0}^{k} F\left(\frac{j}{k}\right) \binom{k}{j} x^j (1-x)^{k-j}, & 0 \leqslant x \leqslant 1, \\ 1, & x > 1. \end{cases}$$

If F is a *probability distribution function* (pdf) with support $[0, 1]$, then $B(x; k, F)$ is also a distribution function, comprising a mixture of a point mass $F(0)$ at 0 and an absolutely continuous distribution function, with density function

$$b(x; k, w_k) = \sum_{j=1}^{k} w_{j,k} \beta(x; j, k-j+1),$$

where $\beta(\cdot, a, b)$ is the pdf of a beta distribution with parameters (a, b).

Let N be the set of positive integers, $\mathcal{P}(N)$ the set of all subsets of N, and Δ the class of all probability functions on the unit interval (equipped with appropriate σ-field S). Let K be a random variable with positive-integer values and $\tilde{F}$ a random distribution function with support $[0, 1]$. Let $(K, \tilde{F})$ possess a joint probability P on $(N \times \Delta, \mathcal{P}(N) \times S)$. Then the Bernstein polynomial $B(\cdot; K, \tilde{F})$ is a random distribution function on $[0, 1]$, whose probability law is induced by $\mathcal{P}$ and which is referred to as a *Bernstein prior* with parameter P.

Further details, properties, applications, and construction of Bernstein polynomial posteriors are given in References 1 and 2.

REFERENCES

1. Petrone, S. (1999). Random Bernstein polynomials. *Scand. J. Stat.*, **26**, 373–393.

2. Petrone, S. and Wasserman, L. (2002). Consistency of Bernstein polynomial posteriors. *J. R. Stat. Soc.*, **64**, 79–110.

See also BERNSTEIN POLYNOMIALS; BETA DISTRIBUTIONS; POSTERIOR DISTRIBUTIONS; and PRIOR DISTRIBUTIONS.

BERNSTEIN POLYNOMIALS

These polynomials are useful in approximation and interpolation theory. For a function $f(x)$ defined on the closed interval $[0, 1]$, the nth Bernstein polynomial is

$$B_n(x) = \sum_{k=0}^{\infty} f\left(\frac{k}{n}\right) \binom{n}{k} x^k (1-x)^{n-k}$$

as $n \to \infty$. The polynomial converges *uniformly* to the function $f(x)$. More precisely, [2],

$$|B_n(x) - f(x)| \leqslant \tfrac{5}{4} \omega(f, 1/\sqrt{n}),$$

where the modulus of continuity, $\omega(f; \delta)$, is defined by

$$\omega(f; \delta) = \sup_{|x-y| \leqslant \delta} |f(x) - f(y)|$$
$$= \sup_{|h| \leqslant \delta} |f(x+h) - f(x)|$$

$(x, y; x + h \in [0, 1])$. Also if the derivative is finite, then $\lim_{n \to \infty} B_n'(x) = f'(x)$ and convergence is uniform for a continuously differentiable $f(x)$. Applications of Bernstein's polynomials are for proving Weierstrass' approximation theorem* using the weak law of large numbers and for density function estimation [4].

REFERENCES

1. Lorentz, G. G. (1953). Bernstein Polynomials. *Mathematical Expositions No. 8*. University of Toronto Press, Toronto.

2. Popoviciu, T. (1935). *Mathematica (Cluj)*, **10**, 49–54.

3. Rényi, A. (1970). *Probability Theory*. North-Holland, Amsterdam, p. 165.

4. Vitale, R. A. (1975). *Stochastic Processes*, **2**, 87–100.

BERNSTEIN, SERGEI NATANOVICH

Born: March 5 (n.s.), 1880, in Odessa (a port on the Black Sea in the Ukraine).

Died: October 26, 1968, in Moscow, USSR.

Contributed to: theory and application of differential equations, function approximation theory, probability.

Bernstein received his mathematical education in Paris on completing high school in Odessa (where his father was a doctor and university lecturer) in 1898, and also studied at Göttingen. He defended a doctoral dissertation at the Sorbonne at age 24 in 1904, and another in Kharkov in 1913. From 1908 to 1933 he taught at Kharkov University, and from 1933 he worked at the Mathematical Institute of the USSR Academy of Sciences, also teaching at the University and Polytechnic Institute, in Leningrad, and continued to work at the Mathematical Institute in Moscow from 1943. He was much honored, including foreign membership of the Paris Academy of Sciences, and membership of the USSR and Ukrainian Academies. In his general mathematical work he united the traditions of the St. Petersburg "school", founded by P. L. Chebyshev* with those of modern western European thinking. The scope of his probabilistic work was in general ahead of its time, and his writings, including his textbook [4], which first appeared in 1927, were largely responsible for determining the course of development of this subject area in the USSR.

Bernstein's early publications of a probabilistic nature have a heavily analytical character: a constructive proof, using the sequence of what came to be known as Bernstein polynomials* and Bernoulli's law of large numbers*, of Weierstrass's uniform approximation theorem, and a consideration of the accuracy of approximation of the binomial distribution* by the normal distribution*, a problem to which he was to return several times. These early writings include an interesting attempt (1917) at the axiomatization* of probability theory (see ref. 5, pp. 10–60, and refs. 6 and 8).

One of the most significant areas of Bernstein's creativity, however, was a reexamination in a new light of the main existing theorems of probability theory of his times. For example, if $X_1, X_2, \ldots$ are independent random variables with finite variance, adjusted to have zero mean, and $S_n = \sum_{i=1}^{n} X_i$, then Chebyshev's inequality* may be written

$$P\{|S_n| < tB_n\} > 1 - t^{-2},$$

where $B_n = \operatorname{var} S_n$, and Kolmogorov's inequality* strengthens this result by replacing $|S_n|$ by $\sup_{1 \leqslant k \leqslant n} |S_k|$. Bernstein succeeded in raising the lower bound $1 - t^{-2}$ to $1 - 2e^{-t^2}$ in both, under additional assumptions on the X_k, this variant assuming the name Bernstein's inequality* and even more significantly showed the refinements to hold where the X_k are no longer necessarily independent but form what is now termed a martingale-difference sequence, so that $\{S_n\}$ is a martingale*. Another direction taken by Bernstein within this area was to generalize the conditions of Liapunov* for the applicability of the central limit theorem* for sums of random variables. In 1992, Bernstein proved this under conditions which, when specialized to the same setting, are equivalent to those of Lindeberg, whose paper appeared in the same year; a fundamental and perhaps his best-known paper [3] extended this work to sums of "weakly dependent" random variables X_k and Markov chains*; he later proved it for martingales $\{S_n\}$. (He speaks of "sums of dependent variables, having mutually almost null regression," since for a martingale difference sequence $E[X_{k+1}|X_k, X_{k-1}, \ldots, X_1] = 0$.) A group of papers, and Appendix 6 of [4], deal in essence with the weak convergence of a discrete-time stochastic process* to one in continuous time whose probability distribution satisfies a diffusion equation, together with an examination of the boundary theory of such an equation, anticipating later extensive development of this subject matter. Little known also are surprisingly advanced mathematical investigations in population genetics (e.g., ref 2), including a synthesis of Mendelian inheritance and Galtonian "laws" of inheritance. In addition, Bernstein took a keen interest in the methodology of the teaching of mathematics at secondary and tertiary levels.

In the years 1952–1964 he devoted much time to the editing and publication of the four-volume collection of his works [5], which contains commentaries by his students and experts in various fields.

REFERENCES

1. Alexandrov, P. S., Akhiezer, N. I., Gnedenko, B. V., and Kolmogorov, A. N. (1969). *Uspekhi Mat. Nauk*, **24**, 211–218 (in Russian). (Obituary; emphasis is mathematical rather than probabilistic, but interesting commentary on Bernstein's population mathematics.)

2. Bernstein, S. N. (1924). *Uch. Zap. Nauchno-Issled. Kafedr Ukr., Otd. Mat.*, **1**, 83–115 (in Russian). [Also in ref. 5, pp. 80–107; in part (Chap. 1) in English translation (by Emma Lehner) in *Ann. Math. Statist.*, **13**, 53–61 (1942).]

3. Bernstein, S. N. (1926). *Math. Ann.*, **97**, 1–59. (Published in 1944 in Russian translation, in *Uspekhi Mat. Nauk*, **10**, 65–114. Also in ref. 5, pp. 121–176.)

4. Bernstein, S. N. (1946). *Teoriya veroiatnostei*, 4th ed. Gostehizdat, Moscow-Leningrad. (1st ed.: 1927; 2nd and 3rd eds.: 1934. Portions on nonhomogeneous Markov chains—pp. 203–213, 465–484 reprinted in ref. 5, pp. 455–483.)

5. Bernstein, S. N. (1964). *Sobranie sochineniy*, Vol. 4: *Teoriya veroiatnostei i matematicheskaia statistika* [1911–1946]. Izd. Nauka, Moscow. [This fourth volume of Bernstein's collected works contains Russian language versions of most of his probabilistic work. Volume 1 (1952) and Vol. 2 (1954) deal with constructive function theory, and Vol. 3 (1960) with differential equations.]

6. Kolmogorov, A. N. and Sarmanov, O. V. (1960). *Teor. Veroyatn. ee primen.*, **5**, 215–221 (in Russian). (Division of Bernstein's writings into groups, with commentary; complete listing of his probabilistic writings up to ref. 5; full-page portrait.)

7. Kolmogorov, A. N., Linnik, Yu. V., Prokhorov, Yu. V., and Sarmanov, O. V. (1969). Sergei Natanovich Bernstein. *Teor. veroyatn. ee primen.*, **14**, 113–121 (in Russian). (Obituary, with good detail on biography and probabilistic work, and a photograph on p. 112.)

8. Maistrov, L. E. (1974). *Probability Theory. A Historical Sketch*. Academic Press, New York, pp. 250–252. [Translated and edited from the Russian edition (1967) by S. Kotz.]

See also Axioms of Probability; Bernstein Polynomials; Bernstein's Inequality; Chebyshev's

Inequality; Convergence of Sequences of Random Variables; Diffusion Processes; Kolmogorov's Inequality; Laws of Large Numbers; Limit Theorems; Markov Processes; Martingales; Statistical Genetics; and St. Petersburg School of Probability.

E. Seneta

BERNSTEIN'S INEQUALITY

A refinement of Chebyshev's inequality* published in 1926.

If $X_1, X_2, \ldots, X_n$ are independent random variables with $E(X_i) = 0$, then

$$\Pr[|X_1 + X_2 + \cdots + X_n| \leqslant \theta]$$
$$> 1 - 2\exp[-\theta^2(2B_n + c\theta)^{-1}],$$

where B_n is the variance of the sum $S = X_1 + \cdots + X_n$ and c is a constant. (Chebyshev's inequality states: $\Pr[|X_1 + \cdots + X_n| \leqslant t\sqrt{B_n}] > 1 - 1/t^2$.) A useful form of Bernstein's inequality, quoted by Uspensky [3], is

$$\Pr[|X - \mu| \geqslant \lambda] \leqslant 2\exp[-\lambda^2(2\sigma^2 + \tfrac{2}{3}m\lambda)^{-1}]$$

for any $\lambda > 0$, where $\mu_i = EX_i$, $\mu = \sum_{i=1}^{n} \mu_i$; $\sigma_i^2 = \mathrm{var}(X_i)$, $\sigma^2 = \sum_1^n \sigma_i^2$, $X = \sum_1^n X_i$, X_i is independent of X_j for $i \neq j = 1, \ldots, n$, and m is a constant such that $\Pr[|X_i - \mu_i| > m] = 0$. That is, with probability 1, the maximal deviation of a random variable X_i from its mean μ_i does not exceed m. This condition also bounds all the central moments.

For additional information, see refs. 1 and 2.

REFERENCES

1. Renyi, A. (1970). *Probability Theory*. North-Holland, Amsterdam, pp. 284–386.

2. Savage, I. R. (1961). *J. Res. Natl. Bur. Stand.*, **65B**, (3), 211–222.

3. Uspensky, J. V. (1937). *Introduction to Mathematical Probability*. McGraw-Hill, New York, p. 205.

See also Chebyshev's Inequality.

BERNSTEIN LEMMA. See Convergence in Distribution, Bernstein's Lemma

BERNSTEIN'S THEOREM. See DAMAGE MODELS

BERRY–ESSÉEN INEQUALITY. See ASYMPTOTIC NORMALITY

BERRY-ESSÉEN THEOREMS. See CENTRAL LIMIT THEOREMS, CONVERGENCE RATES FOR

BERTRAND'S LEMMA. See BALLOT PROBLEMS

BERTRAND'S THEORY OF HYPOTHESIS TESTING. See FREQUENCY INTERPRETATION IN PROBABILITY AND STATISTICAL INFERENCE; HYPOTHESIS TESTING; INFERENCE, STATISTICAL

BESSEL FUNCTION DISTRIBUTIONS

A system of distributions developing from the work of McKay [3]. Bhattacharyya [1] showed that these could be obtained as the distributions of linear functions of independent chi-squared* variables X_1, X_2 with common degrees of freedom v.

The density function of $Y = a_1 X_1 + a_2 X_2$ (with $a_1 > a_2 > 0$) is

$$\frac{(c^2-1)^{m+1/2}}{\pi^{1/2} 2^m b^{m+1} \Gamma(m+\frac{1}{2})} y^m e^{-cy/b} \, I_m(y/b)$$

$$(y > 0),$$

where $b = 4a_1 a_2 (a_1 - a_2)^{-1}$; $c = (a_1 + a_2)/(a_1 - a_2)$; $m = 2v + 1$ and

$$I_m(x) = \left(\frac{1}{2}x\right)^m \sum_{j=0}^{\infty} \frac{(x/2)^{2j}}{j! \Gamma(m+j+1)}$$

is a modified Bessel function of the first kind.

The density function of $Y = a_1 X_1 - a_2 X_2$ $(a_1 > 0, a_2 > 0)$ is

$$\frac{(1-c'^2)^{m+1/2}}{\pi^{1/2} 2^m b'^{m+1} \Gamma(m+\frac{1}{2})} \cdot |g|^m e^{-cz/b} K_m(|z/b|),$$

where $b' = 4a_1 a_2/(a_1 + a_2)^{-1}$, $c' = (a_2 - a_1)/(a_1 + a_2)$ and

$$K_m(x) = \frac{1}{2} \frac{I_{-m}(x) - I_m(x)}{\sin v\pi}$$

is a modified Bessel function of the second kind.

The curves of this system are rather troublesome to handle numerically and mathematically, mainly because of the functions $I_m(\cdot)$ and (especially) $K_m(\cdot)$, and they are not at all widely used.

The function $I_m(\cdot)$ also appears in the distribution of the difference between two independent Poisson* variables W_1, W_2 with expected values θ_1, θ_2, respectively. We have

$$\Pr[W_1 - W_2 = w] = e^{-(\theta_1 - \theta_2)}(\theta_1/\theta_2)^{w/2}$$

$$\times I_{w/2}(2\sqrt{\theta_1 \theta_2}).$$

REFERENCES

1. Bhattacharyya, B. C. (1942). *Sankhyā*, **6**, 175–182.

2. Johnson, N. L. and Kotz, S. (1970). *Distributions in Statistics: Continuous Univariate Distributions*, Vol. 1. Wiley, New York, Chap. 12.

3. McKay, A. T. (1932). *Biometrika*, **24**, 39–44.

See also BESSEL FUNCTIONS.

BESSEL FUNCTIONS

Introduced by F.W. Bessel (1784–1846), these functions are also called cylinder functions.

Bessel functions of various kinds appear in several statistical distribution formulas. For example: distribution product of normal variables* [4], extreme values* in samples from a Cauchy distribution* [7], distribution of quadratic forms [5], asymptotic distribution of range* [6], noncentral χ^2 distribution* [11], moments of mean differences in samples from a Poisson distribution* [8], distribution of the Cramér-von Mises statistic* [2], distribution of the covariance of two normal correlated variables, directional distributions* of various kinds [9], and distributions of random variables on a cylinder [10] involve Bessel functions. (*See* BESSEL FUNCTION DISTRIBUTIONS.)

BESSEL FUNCTIONS OF THE FIRST AND SECOND KINDS

Bessel functions of the *first* kind of order v and argument z are defined by

$$J_v(z) = \sum_{r=0}^{\infty} \frac{(-1)^r (z/2)^{v+2r}}{r! \Gamma(v+r+1)}.$$

Similarly,

$$J_{-v}(z) = \sum_{r=0}^{\infty} \frac{(-1)^r (z/2)^{-v+2r}}{r! \Gamma(-v + r + 1)}.$$

These are independent solutions for the *Bessel equation*. If n is an integer, we have

$$J_{-n}(z) = (-1)^n J_n(z).$$

The identity

$$\exp\left[\frac{1}{2}z(t - t^{-1})\right] = \sum_{n=-\infty}^{\infty} t^n J_n(z)$$

gives the *generating function** of the Bessel functions $J_n(z)$, which are also called the Bessel coefficients.

Bessel functions of the *second* kind of order v and argument z are defined by

$$Y_v(z) = \frac{(\cos v\pi)J_v(z) - J_{-v}(z)}{\sin v\pi}.$$

If $v = n$ is an integer, we also have

$$Y_{-n}(z) = (-1)^n Y_n(z).$$

SPECIAL CASES OF BESSEL FUNCTIONS OF THE FIRST AND SECOND KINDS

(a) $J_{1/2}(x) = (\pi x/2)^{-1/2} \sin x;$

$J_{-1/2}(x) = (\pi x/2)^{-1/2} \cos x;$

$J_{3/2}(x) = (\pi x/2)^{-1/2}[(\sin x/x) - \cos x],$

and in general

$$J_{\pm n+1/2}(x) = \left(\frac{2}{\pi}\right)^{1/2} x^{n+1/2}$$

$$\times \left(\mp \frac{d}{x\,dx}\right)^n \left(\frac{\sin x}{x}\right)$$

are expressible in a finite number of terms involving sines, cosines, and powers of z and are sometimes referred to as *spherical Bessel functions*.

(b) $J_0(x) = 1 - \dfrac{x^2}{2^2(1!)^2} + \dfrac{x^4}{2^4(2!)^2}$

$$- \frac{x^6}{2^6(3!)^2} + \cdots$$

$J_1(x) = \dfrac{x}{2} - \dfrac{x^3}{2^3 1! 2!} + \dfrac{x^5}{2^5 2! 3!} - \cdots.$

For small x, these series show that $J_0(x) \simeq 1$ and $J_1(x) \simeq x/2$. Similarly, it can be shown that $Y_0(x) = (2/\pi)\{\gamma + \log(x/2)\}$ and $Y_1(x) \simeq -2/(\pi x)$, where $\gamma = 0.5772$ is *Euler's constant**.

Recurrence formulas:

$$J_{v+1}(z) = \frac{v}{z}J_v(z) - J_v'(z),$$

$$J_{v-1}(z) = \frac{v}{z}J_v(z) + J_v'(z),$$

$$J_{v-1}(z) + J_{v+1}(z) = \frac{2v}{z}J_v(z),$$

$$J_{v-1}(z) - J_{v+1}(z) = 2J_v'(z),$$

and the special case

$$J_1(z) = -J_0'(z),$$

which is an identity of great utility.

BESSEL FUNCTIONS OF THE THIRD KIND

Bessel functions of the *third* kind, also known as *Hankel functions*, are defined

$$H_v^{(1)}(z) = J_v(z) + iY_v(z)$$

and

$$H_v^{(2)}(z) = J_v(z) - iY_v(z).$$

These functions are *also* independent solutions of the Bessel equation. The identity

$$H_{n+(1/2)}^{(1)}(z) = J_{n+(1/2)}(z)$$

$$+ i(-1)^{n+1}J_{-n-(1/2)}(z)$$

is sometimes useful.

MODIFIED BESSEL FUNCTIONS

The function $e^{-v\pi i/2}J_v(iz)$ is a real function of z which is a solution of the equation

$$z^2\frac{d^2w}{dz^2} + z\frac{dw}{dz} - (z^2 + v^2)w = 0$$

called the *modified Bessel function* of the *first* kind and denoted by $I_v(z)$. The infinite series representation of $I_v(z)$ is

$$I_v(z) = \sum_{r=0}^{\infty} \frac{(z/2)^{v+2r}}{r! \Gamma(v + r + 1)}.$$

Observe that $I_{-n}(z) = I_n(z)$ for a *positive* integer n. The function

$$K_v(z) = \frac{\pi}{2\sin(v\pi)}\{I_{-v}(z) - I_v(z)\}$$

when v is not an integer is called the *modified Bessel function* of the *second* kind. [Note that $K_v(z) = K_v(-z)$.] Some particular cases of modified Bessel functions are of importance:

1.

$$I_0(x) = 1 + \frac{x^2}{2^2(1!)^2} + \frac{x^4}{2^4(2!)^2}$$

$$+ \frac{x^6}{2^6(3!)^2} + \cdots$$

$$K_0(x) = -\left(\gamma + \log_e \frac{x}{2}\right)I_0(x)$$

$$+ \sum_{r=1}^{\infty} \frac{(x/2)^{2r}}{r!r!}$$

$$\times \left(1 + \frac{1}{2} + \frac{1}{3} + \cdots + \frac{1}{r}\right).$$

Also,

$$I_1(x) = \frac{x}{2} + \frac{x^3}{2^3 1!2!} + \frac{x^5}{2^5 2!3!} + \cdots$$

and

$$K_1(x) = \left(\gamma + \log_e \frac{x}{2}\right)I_1(x) + \frac{1}{x}$$

$$- \frac{1}{2}\sum_{r=0}^{\infty} \frac{(x/2)^{2r+1}}{r!(r+1)!}$$

$$\times \left\{2\left(1 + \frac{1}{2} + \cdots + \frac{1}{r}\right) + \frac{1}{r+1}\right\},$$

which shows that

$$I_0(x) \cong 1, \quad K_0(x) \cong -\gamma - \log_e(x/2),$$

$$I_1(x) \cong (x/2), \quad K_1(x) = 1/x.$$

2. Recurrence relations:

$$-I_{v+1}(z) = \frac{v}{z}I_v(z) - I_v'(z)$$

$$I_{v-1}(z) = \frac{v}{z}I_v(z) + I_v'(z)$$

$$I_{v-1}(z) - I_{v+1}(z) = \frac{2v}{z}I_v(z)$$

$$I_{v-1}(z) + I_{v+1}(z) = 2I_v'(z)$$

$$-K_{v+1}(z) = -\frac{v}{z}K_v(z) + K_v'(z)$$

$$-K_{v-1}(z) = \frac{v}{z}K_v(z) + K_v'(z)$$

$$K_{v-1}(z) - K_{v+1}(z) = -\frac{2v}{z}K_v(z)$$

$$K_{v-1}(z) + K_{v+1}(z) = 2K_v'(z)$$

are worth noting.

Modified functions $I_n(x), K_v(x)$ bear to the exponential functions similar relations to those which the functions $J_v(x), Y_v(x)$ bear to the trigonometric functions; modified functions have *no* zeros for real values of x. [Note that $I_1(z) = I_0'(z)$ and $K_1(z) = -K_0'(z)$; the latter is similar to the corresponding relation for $J_n(z)$.]

Graphs of Bessel functions $J_i(x)$, $Y_i(x)$, $I_i(x)$, and $K_i(x)$ for $i = 0, 1$ are presented in Fig. 1.

Tabulated values of $J_n(x), Y_n(x), I_n(x)$, and $K_n(x)$ for $n = 0(1)20$ are presented in the *British Association Mathematical Tables* [3] and in Abramowitz and Stegun [1].

REFERENCES

1. Abramowitz, M. and Stegun, I. A. (1964). Handbook of Mathematical Functions. *Natl. Bur. Stand. Appl. Math Ser. 55* (Washington, D.C.).

2. Anderson, T. W. and Darling, D. A. (1952). *Ann. Math. Statist.*, **23**, 193–212.

3. *British Association Mathematical Tables (1952). Vol. 10.*

4. Craig, C. C. (1936). *Ann. Math. Statist.*, **7**, 1–15.

5. Grad, A. and Solomon, H. (1955). *Ann. Math. Statist.*, **26**, 464–477.

6. Gumbel, E. J. (1947). *Ann. Math. Statist.*, **18**, 384–412.

7. Gumbel, E. J. and Keeney, R. D. (1950). *Ann. Math. Statist.*, **21**, 523–538.

8. Katti, S. K. (1960). *Ann. Math. Statist.*, **31**, 78–85.

9. Mardia, K. V. (1975). *J. R. Statist. Soc. B*, **37**, 349–393.

10. Mardia, K. V. and Sutton, T. W. (1978). *J. R. Statist. Soc. B*, **40**, 229–233.

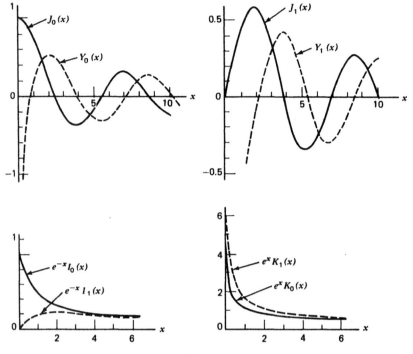

Figure 1.

11. Ruben, H. (1960). *Ann. Math. Statist.*, **31**, 598–618.

BIBLIOGRAPHY

See the following works, as well as the references just given, for more information on the topic of Bessel functions.

Erdelyi, A., ed. (1953). *Higher Transcendental Functions*, Vol. 2. McGraw-Hill, New York.

Luke, Y. L. (1976). *Mathematical Functions and Their Approximations*. Academic Press, New York.

McLachlan, N. W. (1955). *Bessel Functions for Engineers*. Oxford University Press, New York.

Relton, F. E. (1946). *Applied Bessel Functions*, Blackie, London.

Traner, C. J. (1968). *Bessel Functions with Some Physical Applications*. English University Press, London.

Watson, G. N. (1966). *Theory of Bessel Functions. Cambridge University Press, Cambridge.* (*Paperback edition; reprint of the 1944 edition.*)

BESSEL PROCESS. See Brownian Motion

BETA COEFFICIENTS

In writing the model for multiple linear regression*, it is customary to use the Greek symbol β for the coefficients of the independent (predictor) variables. This notation is so widely used that the term "beta coefficients" causes little confusion.

See also Multiple Linear Regression.

BETA DISTRIBUTIONS

The *standard* beta distribution [Beta(a, b)] form of two-parameter family of continuous distributions with density functions

$$\{B(a,b)\}^{-1}x^{a-1}(1-x)^{b-1} \quad (0 < x < 1),$$

where $B(a, b)$ is a beta function ($B(a,b) = \int_0^1 t^{a-1}(1-t)^{b-1}dt$). The parameters are $a(> 0)$ and $b(> 0)$. If X has a standard beta distribution, then $Y = \alpha + \beta X$, with $\beta > 0$, has a beta distribution over the range α to $(\alpha + \beta)$, with the same parameter values as X.

The rth moment of X is

$$E[X] = B(a + r, b)/B(a, b)$$
$$= a^{[r]}/(a + b)^{[r]}$$

if r is an integer $[a^{[r]} = a(a + 1) \cdots (a + r - 1)]$. The expected value of X is $a/(a + b)$; the variance is $ab(a + b)^{-2}(a + b + 1)^{-1}$.

The distribution of Beta(b, a) is the same as that of $\{1 - \text{Beta}(a, b)\}$. This is sometimes referred to as quasi-symmetry*.

Beta distributions arise in statistical theory as the distribution (Beta (α_1, α_2)) of the ratio $Y_1/(Y_1 + Y_2)$, where Y_1, Y_2 are independent standard gamma variables with parameters α_1, α_2. As a particular case of this, if $X_1, \ldots, X_n$ are independent normal* $N(\mu, \sigma^2)$ variables, then

$$\frac{1}{n - 1}(X_i - \overline{X})^2 \bigg/ \sum_{j=1}^{n}(X_j - X)^2$$

(with $\overline{X} = n^{-1}\sum_{j=1}^{n} X_j$) is distributed as Beta $(\frac{1}{2}, \frac{1}{2}(n - 1))$. It is also used as a prior density* in Bayesian inference*. Beta distributions are members (Types I and II) of the Pearson system* of frequency curves.

The four-parameter beta distribution with density of the form

$$\frac{1}{B(p, q)} \frac{(y - a)^{p-1}(b - y)^{q-1}}{(b - a)^{p+q-1}}$$
$$(a \leqslant y \leqslant b)$$

with $p > 0, q > 0$, was studied in detail by Whitby [2].

For additional information, see ref. 1.

REFERENCES

1. Johnson, N. L. and Kotz, S. (1970). *Continuous Univariate Distributions*, Vol. 2. Wiley, New York, Chap. 23.
2. Whitby, O. (1972). Ph.D. dissertation, Stanford University.

See also APPROXIMATIONS TO DISTRIBUTIONS; ARC-SINE DISTRIBUTION; *F*-DISTRIBUTION; PRIOR DISTRIBUTIONS; and UNIFORM DISTRIBUTIONS.

BETTING, LABOUCHÈRE SYSTEMS

The betting system for the even chances (red/black, odd/even, high/low) in roulette, named after the English member of parliament and journalist Henry Du Pre Labouchère (1831–1912), is designed to provide the player with a sequence of bets at the end of which he will have won a fixed amount. In this it is similar in intention to the martingale* (or doubling-up) system, but gives s slower growth in bet size and longer sequences of bets. It suffers from the same flaw as the martingale, in that sooner or later it may require a bet which exceeds either the house limit or the gambler's capital. The gambler then suffers a loss out-weighing earlier gains. The system is sometimes called the cross-out, cancellation, or top-and-bottom system, or in one form the split martingale.

The *standard Labouchère system* involves writing down a line of k integers. The size of the first bet (in suitable units) is the sum of the two extreme integers in the line. If the bet wins, these two numbers are deleted; if it loses, the number corresponding to the amount lost is added at the right of the line. The next bet is then the new sum of the extremes, and this process continues until all the numbers, both the original ones and those added, have been deleted. Thus a win results in the deletion of two numbers, while a loss adds only one; analogy with a random walk on the positive integers in which there is unit displacement to the right in the event of a loss and a displacement of two units to the left in the event of a win shows that the process, if unbounded above, is certain to terminate with all numbers deleted unless the probability of a loss exceeds $\frac{2}{3}$. Any completed sequence of bets results in a win equal to the total of the k initial numbers. The system is commonly used with the four numbers 1, 2, 3, and 4 as the starting values; Labouchère himself used the five numbers 3, 4, 5, 6, and 7 and attributed the system to Condorcet (Thorold [2, pp. 60–61] quotes from an article by Labouchère in the newspaper *Truth* for February 15, 1877).

The way in which sequences develop is illustrated by an eight-bet sequence (four wins and four losses) in Table 1. In the split martingale system the initial bet would have

Table 1. Typical Labouchère Betting Sequence

Bet Number	Amount Bet	Won or Lost	Total Won or Lost	Labouchère Numbers
0	—	—	—	1 2 3 4
1	5	L	−5	1 2 3 4 5
2	6	W	+1	1̶ 2 3 4 5̶
3	6	W	+7	1̶ 2̶ 3 4 5̶
4	3	L	+4	1̶ 2̶ 3 4 5̶ 3
5	6	L	−2	1̶ 2̶ 3 4 5̶ 3 6
6	9	L	−11	1̶ 2̶ 3 4 5̶ 3 6 9
7	12	W	+1	1̶ 2̶ 3̶ 4 5̶ 3 6 9̶
8	9	W	+10	1̶ 2̶ 3̶ 4 5̶ 3̶ 6̶ 9̶

Table 2. Moments of Lengths of Four-Number Labouchère Sequences

	Mean	Variance
Standard (double-zero wheel)	10.28	131.2
Standard (single-zero wheel)	9.42	101.4
Standard or reverse (fair wheel)	8.65	78.9
Reverse (single-zero wheel)	8.00	62.5
Reverse (double-zero wheel)	7.47	50.9

been one unit. If it had won, the bet would be repeated; if it had lost, the Labouchère numbers would become 1, 2, 3, 4, and 1 and the standard Labouchère system would then operate.

The *reverse Labouchère system* adds to the line of numbers after a win and deletes the two extreme numbers after a loss, on the fallacious argument that if the Labouchère system gives a large loss to counterbalance a series of small wins, its mirror image will give an occasional larger win to counterbalance a series of small losses. Although this argument ignores the effect of the zero (or zeros) on a roulette wheel, the reverse Labouchère system has the advantage that it can be operated with relatively small capital since large bets are made only when wins from the bank have provided the stake money.

A *parallel Labouchère system* involves operating separate Labouchère processes (either both standard or both reverse) on a pair of opposing even chances (e.g., red and black). These two processes are not independent and no analysis of their joint behavior is known. Although no system can affect the bank's advantage from the zero on a true wheel, a parallel Labouchère system will reduce that advantage if there is a bias on the wheel; it does not require knowledge of the direction of the bias.

There has been little mathematical analysis of Labouchère systems and few results in closed form are known. Downton [1] used the difference equations for the number of sequences of n games containing m losses which do not contain a completed sequence to compute the probability distributions of the length of completed standard and reverse

Labouchère sequences starting with four numbers for true roulette wheels both with a single and with a double zero. These distributions, which have very long tails, do not provide information about the way in which bet sizes build up, which remains an unsolved problem.

Explicit expressions may be obtained for the moments of the distributions using the random walk analogy. For example, if E_k is the expected number of bets in a Labouchère sequence starting with k numbers, and if p and $1 - p$ are the probabilities of winning and losing a bet, respectively, then

$$E_k = 1 + pE_{k-2} + (1 - p)E_{k+1}, \qquad k \geqslant 1,$$

with $E_k = 0, k \leqslant 0$. These equations give

$$E_k = \frac{1 + (1 + \alpha)k - (-\alpha)^{-k}}{(3p - 1)(1 + \alpha)},$$

where $\alpha = \sqrt{1/p - 3/4} + 1/2$. Higher moments may similarly be expressed as functions of powers of α. The means and variances of the length of standard and reverse Labouchère bet sequences starting with four numbers for true wheels with a single and double zero, and for a fair wheel with no zero are given in Table 2. In practice these means and variances would be reduced by the truncation of the sequences because of stake limits.

REFERENCES

1. Downton, F. (1980). *J. R. Statist. Soc. A*, **143**, 363–366.

2. Thorold, A. L. (1913). *The Life of Henry Labouchère*. Constable, London. (A comprehensive account of Labouchère's eventful life, probably of more interest to historians than to practicing statisticians.)

See also BLACKJACK; GAIN RATIO; GAMBLING, STATISTICS IN; GAMES OF CHANCE; MARTINGALES; RANDOM WALKS; and RUNS.

F. DOWNTON

BIAS

BIAS OF AN ESTIMATOR

The statistical meaning of bias is quite precise. It is defined as the difference between the expected value* of a statistic and the value of a population parameter. It thus depends on both the statistic and the parameter. It is relevant when the statistic is to be used as an estimator of the parameter. (See below for a formal definition.)

Sources of bias fall into three main categories. The most prominently studied in statistical literature is the statistical (technical) bias due to the specific algebraic form of an estimator. The other two categories are selection bias and measurement error. Williams [2] provides an overview of this topic.

Statistical Bias

Consider a statistic T which estimates a parameter θ. If $E_\theta(T) = \theta + b(\theta)$, where E_θ indicates that the expectation of T is computed under the assumption that the true parameter value is θ, the quantity $b(\theta) = E_\theta(T) - \theta$ is the *bias* of the statistic T, regarded as an estimator of θ. If $b(\theta) \equiv 0$ for all values of θ, T is called an *unbiased* estimator of θ (*see* MINIMUM VARIANCE UNBIASED ESTIMATION).

Selection Bias*

This kind of bias can arise when a sample is not drawn according to the prearranged specifications. Population units are included in the sample with probabilities other than those specified in the sampling design*. As a typical example, a design might specify an equal probability of selection for each of a number of households, but in fact, households with children may appear more often than do those without children. If no adjustment is applied, estimates based on the sample are likely to be *biased*. In other words, selection bias occurs if the real probabilities of selection are different from the intended probabilities, and the real probabilities are not known while the incorrect design probabilities are used in the estimator. An extreme case of selection bias occurs when the chance of selection is zero for some subgroup of the population. Then the sampled population is not the same as the target population*.

Measurement Error*

Conceptually, measurement error is easy to understand, but it is often difficult to assess in practice because it can arise in many subtle ways.

A measuring device can malfunction; the device can be mechanical or human and the errors may be generated completely randomly, mechanically, systematically, or even deliberately. Imprecise definitions also cause measurement problems; another major potential source of measurement bias occurs in the handling of the data. The effect of a single mispunched card may not be trivial. Various suggestions available in the literature to combat measurement bias include: (1) to subsample the data and remeasure the subset of observations; (2) to assume that the errors made are randomly distributed (averaging to zero); and (3) to utilize the randomized response* techniques introduced by Warner [1]; [4] to take "parallel" (similar) samples.

BIAS OF A TEST

A statistical test is said to be *biased* if the probability of rejecting the hypothesis tested (H_0), when H_0 is valid—the significance level—is greater than the probability of rejection when some other hypothesis (H, say) is valid. The property of *biasedness* is relative to the set of alternative hypotheses considered (i.e., the set from which H may be chosen). In fact, it is always possible to construct an artificial hypothesis, $\tilde{H}$ say, different from H_0, such that the probability of rejection is zero—simply by excluding the possibility, under $\tilde{H}$, of getting a sample point in the critical region*. (This does not constitute bias unless H is included in the set of alternative hypotheses for which the test is intended to be appropriate.)

ORDER OF BIAS

Let $\hat{\theta}_1$ and $\hat{\theta}_2$ be two estimators of θ defined on a sample of size n with nonnegative biases

$$B_1(n, \theta) = E[\hat{\theta}_1 - \theta] \neq 0$$

and

$$B_2(n, \theta) = E[\hat{\theta}_2 - \theta] \neq 0.$$

If

$$\left| \lim_{n \to \infty} \frac{B_1(n, \theta)}{B_2(n, \theta)} \right| = 1, \tag{1}$$

$\hat{\theta}_1$ and $\hat{\theta}_2$ are said to be *same order* bias estimators of θ. This concept is sometimes abbreviated to $\hat{\theta}_1$ S.O.B.E. $\hat{\theta}_2$. If the absolute value of the limit in (1) is greater than zero and less than 1, $\hat{\theta}_1$ is said to be a *better same order bias estimator* than $\hat{\theta}_2$ (abbreviated as $\hat{\theta}_1$ B.S.O.B.E. $\hat{\theta}_2$). If, without restricting $B_1(n, \theta) \neq 0$,

$$\lim_{n \to \infty} \frac{B_1(n, \theta)}{B_2(n, \theta)} = 0,$$

we then say that $\hat{\theta}_1$ is a lower-order bias estimator than $\hat{\theta}_2$ ($\hat{\theta}_1$ L.O.B.E. $\hat{\theta}_2$). These concepts are useful in bias reduction techniques of estimators.

REFERENCES

1. Warner, S. L. (1965). *J. Amer. Statist. Ass.*, **60**, 63–69.

2. Williams, B. (1978). *A Sampler on Sampling.* Wiley, New York, Chap. 6.

FURTHER READING

Gray, H. L. and Schucany, W. R. (1972). *The Generalized Jackknife Statistic.* Dekker, New York.

See also ESTIMATION, CLASSICAL; HYPOTHESIS TESTING; JACKKNIFE METHODS; MINIMUM VARIANCE UNBIASED ESTIMATION; POWER; and SELECTION BIAS.

BIAS REDUCTION, QUENOUILLE'S METHOD FOR

Quenouille proposed [6] and developed [7] a method to reduce estimation bias*. This work can be considered classical in that it stimulated considerable useful work on unbiased and almost-unbiased point estimation, some of which is discussed here. It also led to new variance estimation techniques, such as jackknifing* and bootstrapping*, for complex sampling designs [3]. If, for a given series of observations $x_1, x_2, \ldots, x_n$, we have a function $t_n(x_1, x_2, \ldots, x_n)$ which provides an estimate of an unknown parameter θ, then for a wide variety of statistics it is true that

$$E(t_n - \theta) = \frac{a_1}{n} + \frac{a_2}{n^2} + \frac{a_3}{n^3} + \cdots$$

and if $t'_n = n t_n - (n-1) t_{n-1}$, then

$$E(t'_n) = \theta - \frac{a_2}{n^2} - \frac{a_2 + a_3}{n^3} - \cdots$$

Similarly,

$$t''_n = [n^2 t'_n - (n-1)^2 t'_{n-1}] / [n^2 - (n-1)^2]$$

is biased to order $1/n^3$, and so on.

It is possible to calculate similar statistics from any subset of the observations to achieve corrections for bias. Durbin [2] studies the idea above in terms of ratio estimators* of the form $r = y/x$, where the regression of y on x is linear ($y = \alpha + \beta x + u$) and where x is normally distributed with $E(x) = 1, V(x) = h$ of order $O(n^{-1})$, and $V(u) = \delta$ of $O(n^{-1})$. He splits the sample n into two halves which yield the ratio estimators $r_1 = y_1/x_1$ and $r_2 = y_2/x_2$, where $y = \frac{1}{2}(y_1 + y_2)$ and $x = \frac{1}{2}(x_1 + x_2)$. Quenouille's estimator is

$$t = 2r - \tfrac{1}{2}(r_1 + r_2).$$

Durbin shows that

$$V(r) = \alpha^2(h + 8h^2 + 69h^3)$$
$$+ \delta(1 + 3h + 15h^2 + 105h^3).$$

and

$$V(t) = \alpha^2(h + 4h^2 + 12h^3)$$
$$+ \delta(1 + 2h + 8h^2 + 108h^3).$$

For sufficiently large n, t has both a smaller bias and variance than r.

If the regression of y on x is as before, but now x has the gamma distribution* with density $x^{m-1}\exp(-x)/\Gamma(m)$, the difference between the mean square errors* of t and r is

$$M(r) - M(t) = \frac{\alpha^2(m-16)}{m(m-1)(m-2)^2(m-4)}$$
$$+ \frac{\delta(m-10)}{(m-1)(m-2)^2(m-4)}.$$

The following conclusions can be drawn from this: If $m > 16$, then $\mathrm{MSE}(t) < \mathrm{MSE}(r)$; if $m \leqslant 10$, then $\mathrm{MSE}(r) < \mathrm{MSE}(t)$; if $V(x) < \frac{1}{4}$, then $\mathrm{MSE}(t) < \mathrm{MSE}(r)$.

Rao [8] derives the variance of t for general g to order $O(n^{-3})$ for a linear relationship between y and x and for x normally distributed. This variance of t is

$$V(t)$$
$$= \alpha^2 \left[h + \frac{2g}{g-1}h^2 - 6g\frac{4g^2 - 14g + 11}{(g-1)^3}h^3 \right]$$
$$+ \delta \left[1 + \frac{g}{g-1}h - 2g\frac{2g^2 + 9g - 8}{(g-1)^3}h^2 - 3g \right.$$
$$\left. \times \frac{28g^4 - 149g^3 + 260g^2 - 173g + 32}{(g-1)^5} \right].$$

Both the bias and variance of t are decreasing functions of g.

Rao and Webster [11] showed that the bias and variance of the generalization t_Q of t,

$$t_Q = gr - (g-1)\bar{r}_g,$$

where

$$\bar{r}_g = g^{-1}\sum_{j=1}^{g} r_j,$$

are decreasing functions of g. Using $g = n$, they showed that there is little difference in efficiency between t_Q and the Tin [12] estimator

$$t_{T_1} = r\left[1 + \theta_1\left(\frac{s_{xy}}{\bar{x}\bar{y}} - \frac{s_x^2}{\bar{x}^2} \right) \right],$$

where $\theta_1 = (1/n - 1/N)$.

Hutchison [4] conducted Monte Carlo comparisons of some ratio estimators. He compared the version of the Quenouille estimator

$$t_Q = wr - (w-1)\bar{r}_g$$

where

$$w = g\{1 - (n-m)/N\},$$

$g = mn$, and $g = n$ so that $m = 1$, with the Tin estimator t_{T_1}, the Beale estimator t_B, the Hartley–Ross estimator t_{HR}, and with Mickey's estimator t_M.

The comparisons are made for two models. The first is $E(y|x) = \beta x$, where x has a lognormal distribution with parameters μ and σ^2. The second is $y = \beta x + e$, where β is a constant and $\epsilon \sim N(0, kx^\gamma)$ for given x, where k and γ are constants. All estimators considered are unbiased for this model. The second model is basically the same except that $x = zx_p$, where z has a Poisson distribution with mean μ and $x_p \sim \chi_{(m)}^2/m$ (Table 1). He concludes that t_{B_1} and t_{T_1} are generally most efficient, closely followed by Quenouille's estimator t_Q.

Rao and Beegle [10] compare eight ratio estimators: $t_Q, t_B, t_M, t_{HR}, r, t_{T_1}, t_{T_3}$ (see RATIO ESTIMATORS, TIN'S), and the Pascual estimator* t_P. They assume two models.

The first model is $y_i = 5(x_i + e_i)$, where $x_i \sim N(10, 4)$ and $e_i \sim N(0, 1)$, so that the correlation* between y_i and x_i is $\rho = 0.89$ and the coefficient of variation* of x is $C_x = 0.2$. They call this the *Lauh and Williams model*. The second model is $y_i = \alpha + \beta x_i + e_i$, where x_i and y_i have a bivariate normal distribution* with $\mu_x = 5, \sigma_x^2 = 45, \bar{y} = 15, \sigma_y^2 = 500$ and $\rho = 0.4, 0.6$, or 0.8, and where $e_i \sim N[0, \sigma_y^2(1 - \rho^2)]$, independent of x_i. They call this *Tin's model*.

Results for the two models are shown in Tables 2 and 3. The authors consider variance comparisons meaningful for the Lauh and Williams model but favor the interquartile range* for the Tin model. This range is the distance between the upper and lower quartile points and hence contains one-half of the 1000 computed values of an estimator. The authors conclude that T_Q, t_{T_1}, and t_B perform quite efficiently, especially if almost

Table 1. Ratio of Mean Square Errors of Estimators for s Strata to that of r from 1000 Replicates of n Pairs (x,y), where $x = zx_p$, z is Poisson with Mean μ, x_p is $\chi^2_{(m)}/m$, $m = 20$, and $y = \chi^2_{(z)}$, given x

| | | $s = 1$ | | | $s = 30$ | | |
| | | μ | | | μ | | |
n	Estimator	10.0	5.0	2.5	10.0	5.0	2.5
6	t_{HR}	1.052	1.090	1.088	0.817	0.818	1.335
	t_{M}	0.989	0.997	1.077	0.763	0.760	1.167
	t_{Q}	0.978	0.981	0.996	0.753	0.752	1.131
	t_{B}	0.976	0.976	0.989	0.767	0.777	1.073
	t_{T1}	0.976	0.976	0.990	0.762	0.768	1.086
12	t_{HR}	1.113	1.079	1.051	0.977	1.065	1.085
	t_{M}	0.993	0.998	0.993	0.873	0.979	0.972
	t_{Q}	0.987	0.993	0.989	0.865	0.973	0.967
	t_{B}	0.987	0.991	0.989	0.868	0.968	0.965
	t_{T1}	0.987	0.991	0.989	0.867	0.969	0.965

Table 2. Ratio of Variance Estimator to That of r for 1000 Replicates using Lauh and Williams Model [10]

Estimator	$n = 6$	$n = 12$
t_{HR}	0.990	0.995
t_{M}	0.990	0.995
t_{Q}	0.990	0.995
t_{B}	0.990	0.995
t_{T1}	0.990	0.995
t_{T3}	0.990	0.995
t_{P}	0.990	0.995

Table 3. Interquartile Ranges of Estimators for 1000 Replicates using Tin's Model [10]

| | | | n | | |
Estimator	4	6	10	20	50
r	2.8	2.3	1.8	1.07	0.72
t_{HR}	4.0	3.2	2.5	1.63	1.05
t_{M}	2.6	1.7	1.4	0.96	0.68
t_{Q}	2.7	1.9	1.4	0.93	0.67
t_{B}	1.7	1.6	1.4	0.96	0.68
t_{T1}	2.1	1.7	1.4	0.94	0.68
t_{T3}	4.0	3.0	2.2	1.24	0.77
t_{P}	4.0	3.1	2.2	1.26	0.77

unbiased rather than unbiased estimators are satisfactory.

DeGraft-Johnson and Sedransk [1] and Rao [9] compare two-phase versions of the Quenouille estimate to other two-phase ratio estimators. The Quenouille estimator with $g = 2$ did not perform as well generally as Tin and Beale estimators in the first study. Two Quenouille-type estimators with $g = n$ performed as well as other ratio estimators in the second study.

Miller [5] concludes that the Quenouille estimator is usually one of the most efficient ratio estimators, requires more computation than competitors, but has an easily computed estimate of its variability.

REFERENCES

1. de Graft-Johnson, K. T. and Sedransk, J. (1974). *Ann. Inst. Statist. Math.*, **26**, 339–350.
2. Durbin, J. (1959). *Biometrika*, **46**, 477–480.
3. Efron, B. (1982). CBMS-NSF Regional Conference Series, *Appl. Math. No. 38*, SIAM, Philadelphia.
4. Hutchison, M. C. (1971). *Biometrika*, **58**, 313–321.
5. Miller, R. G. (1974). *Biometrika*, **61**, 1–15.
6. Quenouille, M. H. (1949). *J. R. Statist. Soc. B*, **11**, 68–84.
7. Quenouille, M. H. (1956). *Biometrika*, **43**, 353–360.
8. Rao, J. N. K. (1965). *Biometrika*, **52**, 647–649.
9. Rao, P. S. R. S. (1981). *J. Amer. Statist. Ass.*, **76**, 434–442.
10. Rao, J. N. K. and Beegle, L. D. (1967). *Sankhyā B*, **29**, 47–56.
11. Rao, J. N. K. and Webster, J. T. (1966). *Biometrika*, **53**, 571–577.

12. Tin, M. (1965). *J. Amer. Statist. Ass.*, **60**, 294–307.

See also RATIO AND REGRESSION ESTIMATORS, MICKEY'S; RATIO ESTIMATOR, PASCUAL'S; RATIO ESTIMATORS—I; RATIO ESTIMATORS, TIN'S; and UNBIASEDNESS.

H. T. SCHREUDER

BIENAYMÉ–CHEBYSHEV INEQUALITY. See CHEBYSHEV'S INEQUALITY

BIENAYMÉ, IRENÉE-JULES

Born: August 28, 1796, in Paris, France.
Died: October 19, 1878, in Paris, France.
Contributed to: probability, mathematical statistics, demography, social statistics.

Bienaymé's life initially followed a somewhat erratic course with regard to scientific pursuits, partly because of the times through which he lived. The École Polytechnique, in which he enrolled in 1815, was dissolved in 1816, because of the Napoleonic sympathies of its students. Subsequently, he worked as a translator for journals, and in 1818 became lecturer in mathematics at the military academy at St. Cyr, leaving in 1820 to enter the Administration of Finances. Here he became Inspector, and, in 1834, Inspector General, but because of the revolution of 1848, retired to devote all his energies to scientific work.

Up to this time he had been active in the affairs of the Société Philomatique de Paris and his contributions to its meetings were reported in the now-obscure scientific newspaper-journal *L'Institut, Paris*, being reprinted at the end of the year of their appearance in the collections *Procés-Verbaux de la Société Philomatique de Paris—Extraits*. The most startling of Bienaymé's contributions to probability occurs in this context, where he gives a completely correct statement of the criticality theorem* for simple branching processes*. His communication [2], which may have been stimulated by work of L. F. B. de Châteauneuf (1776–1856), precedes the partly correct statement of F. Galton* and H. W. Watson

by some 30 years, and the first subsequent completely correct one by over 80 years [5].

Bienaymé began to publish only in 1829, and his early (Civil Service period) writings lean to demography and actuarial matters. For example, a major one [1] discusses the then used *life tables** of A. Deparcieux and E. E. Duvillard, with the object of presenting overwhelming evidence against continued use of the Duvillard table, since its predicted mortality rates were heavier than appropriate. These interests persisted in Bienaymé's work, for despite his retirement, Bienaymé had considerable influence as a statistical expert in the government of Napoleon III, being praised in a report to the Senate in 1864 for his actuarial work in connection with the creation of a retirement fund. He was elected to the Academy of Sciences in 1852.

Even those of his papers with primarily sociological or demographic intent sometimes involve significant methodological contributions to probability and mathematical statistics. Bienaymé was active in three general areas: the stability and dispersion theory* of statistical trials (before his retirement), theory associated with linear least squares* (after his retirement), and limit theorems*.

In dispersion theory (which later came to be associated with the names of W. Lexis*, L. Bortkiewicz*, and A. A. Chuprov*) Bienaymé introduced a physically motivated principle of *durée des causes*, under the operation of which the proportion of successes in a sequence of trials exhibits more variability than in the homogeneous case of all Bernoulli trials*; and which might therefore be used to explain such observed variability. Bienaymé also manifested an understanding of the concept of a sufficient statistic*.

Laplace's treatise [9] acted as a basis for some of Bienaymé's best work. In the area of least squares in particular, he is concerned with generalizing and defending Laplacian positions [3,4]. Reference 4 contains, *interalia*, the general Bienaymé–Chebyshev inequality (*see* CHEBYSHEV'S INEQUALITY). $\Pr[|\overline{X} - EX| \geqslant \varepsilon] \leqslant \operatorname{var} X/(\varepsilon^2 n)$ (proved by the simple reasoning still used today), which is used to deduce a weak law of large numbers*. P. L. Chebyshev* obtained the inequality in 1867 (in a much more restricted setting and with a more difficult proof), in a paper published simultaneously in Russian, and in French juxtaposed to a reprinting of Bienaymé's paper. Later, Chebyshev gave Bienaymé credit for arriving at the inequality via the "method of moments"*, whose discovery he ascribes to Bienaymé. The slightly earlier paper [3], partly on the basis of which Bienaymé was elected to the basis of which Bienaymé was elected to the Academy, contains, again as an incidental result, the deduction of an almost final form of the continuous chi-square* density, with n degrees of freedom, for the sum of squares of n independently and identically distributed normal $N(0, 1)$ random variables.

A limit theorem reobtained in 1919 by R. von Mises [10] (who regarded it as a *Fundamentalsatz* of the same significance as the central limit theorem*), was actually proved by Bienaymé in 1838. It asserts the following: If the random variables W_i, $i = 1, \ldots, m$ ($\sum W_i = 1$) have a joint Dirichlet distribution* with parameters $x_1, x_2, \ldots, x_m$ then, as $n = \sum x_i \to \infty$ with $r = x_i/n = \text{constant}(i) > 0$, the limiting standardized distribution of $V = \sum \gamma_i W_i$, is $N(0, 1)$. Late in his life, he constructed a simple combinatorial "runs up and down" test* of randomness of a series of observations on a continuously varying quantity, based on the number of local maxima and minima in the series. In particular he stated that the number of intervals between extrema ("runs up and down") in a sequence of n observations is, under assumption of randomness* of sample from a continuous distribution, approximately normally distributed (for large n) about a mean of $(2n - 1)/3$ with variance $(16n - 29)/90$.

Bienaymé was far ahead of his time in the depth of his statistical ideas and, as a result, has been largely ignored in the literature, although he was an important figure in nineteenth-century statistics. He was a correspondent of L. A. J. Quetelet* and a friend of A. A. Cournot* and of Chebyshev. However, his papers were verbose, his mathematics was laconic, and he had a penchant for controversy. He invalidly criticized the law of large numbers of S. D. Poisson*, and no sooner was he in the Academy than he engaged in a furious controversy with A. L. Cauchy*.

Bienaymé's life and scientific contributions, within a general framework of nineteenth century statistics, have been extensively described and documented by Heyde and Seneta [6]. Reference 8 provides information on Bienaymé's contributions to linear least squares, and ref. 7 expands on ref. 5 in relation to the discovery of the criticality theorem.

REFERENCES

1. Bienaymé, I. J. (1837). *Ann. Hyg. Paris*, **18**, 177–218.

2. Bienaymé, I. J. (1845). *Soc. Philom. Paris, Extraits, Ser.* 5, 37–39. (Also in *L'Institut, Paris*, **589**, 131–132; and reprinted in ref. 7.)

3. Bienaymé, I. J. (1852). *Liouville's J. Math. Pures Appl.*, **17** (1), 33–78. [Also in *Mém. Pres. Acad. Sci. Inst. Fr.*, **15** (2), 615–663 (1858).]

4. Bienaymé, I. J. (1853). *C. R. Acad. Sci. Paris*, **37**, 309–324. [Also in *Liouville's J. Math. Pures Appl.*, **12** (2) 158–176 (1867).]

5. Heyde, C. C. and Seneta, E. (1972). *Biometrika*, **59**, 680–683.

6. Heyde, C. C. and Seneta, E. (1977). *I. J. Bienaymé: Statistical Theory Anticipated.* Springer-Verlag, New York. (The basic modern source on Bienaymé.)

7. Kendall, D. G. (1975). *Bull. Lond. Math. Soc.*, **7**, 225–253.

8. Lancaster, H. O. (1966). *Aust. J. Statist.*, **8**, 117–126.

9. Laplace, P. S. de (1812). *Théorie analytique des probabilités.* V. Courcier, Paris.

10. Mises, R. von (1964). *Mathematical Theory of Probability and Statistics*, Hilda Geiringer, ed. Academic Press, New York, pp. 352–357.

See also BRANCHING PROCESSES; CHEBYSHEV (or TCHÉBICHEF), PAFNUTY LVOVICH; CHEBYSHEV'S INEQUALITY; CHI-SQUARE DISTRIBUTION; CRITICALITY THEOREM; DEMOGRAPHY; DISPERSION THEORY, HISTORICAL DEVELOPMENT OF; ESTIMATION: METHOD OF MOMENTS; LAWS OF LARGE NUMBERS; LEAST SQUARES; LIFE TABLES; LIMIT THEOREMS; RANDOMNESS, BDS TEST FOR; and SUFFICIENT STATISTICS.

E. SENETA

BIGEOMETRY. See BIPLOTS

BIMODAL. See BIPLOTS

BINARY DATA

These are data arising when measurement can give only one of two values. Conventionally these are represented by 0 and 1, but they need not even be numerical. Survival for a specified period, success or failure in a specified task, sex, are all examples of sources of binary data. The *number* of 1's obtained from a set of n measurements, of course, is a numerical variable, and appropriate statistical analysis can be applied to it. In particular, it is often supposed, as a first approximation, that successive observations are independent and the probability of a 1 (or a 0) is constant (*see* BERNOULLI TRIALS). The number of 1's would then have a binomial distribution*. More elaborate models are, of course, available (*see* LEXIS, WILHELM).

If measurements on concomitant variables* $X_1,\ldots,X_k$ are available, dependence of $\Pr[X = 1]$ on values of the X's is often represented in one of the following ways:

1. *Linear model*: $\Pr[X = 1] =$ linear function of X's
2. *Log-linear model*: $\log\{\Pr[X = 1]\} =$ linear function of X's
3. *Log-logistic model*:

$$\log\left\{\frac{\Pr[X = 1]}{\Pr[X = 0]}\right\} = \text{linear function of } X\text{'s}$$

Although models 1 and 2 (especially Model 1) may have greater immediate appeal, being rather simpler to understand, model 3 has the advantage that as $\Pr[X = 1]$ increases from 0 to 1, the left-hand side increases from $-\infty$ to $+\infty$, and so covers all possible values of the linear function on the right-hand side. However, models 1 and 2 may give good approximations for those ranges of values of the X's which are of practical importance. Provided that risks attendant on extrapolation* are recognized and avoided, such models can be very useful.

Of course, it is possible that some (or all) of the concomitant ("explanatory") variables may be binary. Analysis of multivariate binary data has attracted increasing attention recently (e.g., Cox [3], Ott and Kronmal [5]).

It is common practice to analyze observed proportions of 1's as if the residual variation in the models were normal. For sufficiently large samples this is not likely to be misleading, although some attention should be paid to using appropriate weights (e.g., in least-squares* equations). Lindsey [4] explains how to use direct likelihood* approaches (for model 3 and by analogy for other models).

Binary data can present themselves in many different forms—*see* e.g., CATEGORICAL DATA, LOG-LINEAR MODELS IN CONTINGENCY TABLES, DICHOTOMY, PAIRED COMPARISONS, PROBIT ANALYSIS, and entries on various discrete distributions. There is a very useful collection of methods of analysis appropriate to such data in Cox [2]. There are further interesting examples, and technical developments in Bishop et al. [1].

REFERENCES

1. Bishop, Y. M. M., Fienberg, S. E., and Holland, P. W. (1975). *Discrete Multivariate Analysis: Theory and Practice*. MIT Press, Cambridge, Mass.
2. Cox, D. R. (1970). *Analysis of Binary Data*. Methuen, London.
3. Cox, D. R. (1972). *Appl. Statist.*, **21**, 113–120.
4. Lindsey, J. K. (1975). *Appl. Statist.*, **24**, 1–16.
5. Ott, J. and Kronmal, R. A. (1976). *J. Amer. Statist. Ass.*, **71**, 391–399.

See also BINARY DATA, MULTIVARIATE; CATEGORICAL DATA; LOG-LINEAR MODELS IN CONTINGENCY TABLES; and *PATTERN RECOGNITION*.

BINARY DATA, MULTIVARIATE

Multivariate binary data arise in many contexts. A sample generally consists of n independent binary vectors of possibly varying length and where the vectors are composed of dependent Bernoulli random variables (RVs). For example, they occur commonly as repeated measures on a single dichotomous response for an individual over successive time periods, e.g., when checking for symptoms of heart disease (yes–no) at each successive time period, or when noting unhealthful smog days (yes–no) on a daily basis. They also occur as repeated measures on a single individual that are unlinked to time, e.g., one may be interested in hoof or eye infections in horses and thus examine four hooves or two eyes. Another common situation is when families or clusters of individuals are scrutinized with regard to a single dichotomy, e.g., if families in Amsterdam are sampled and all persons are asked whether or not they favor the death penalty. A single individual can be checked with regard to multiple issues, as in checking for the presence or absence of a collection of symptoms associated with a particular disease. This information can be used to classify the individual as diseased or not in situations where error-free diagnosis is difficult and/or expensive.

According to Cox [12], "the oldest approach to multivariate binary data is to define indices of association following essentially Yule." Goodman and Kruskal [25,26,27] discuss this work and some extensions (see ASSOCIATION, MEASURES OF). Early models for multivariate binary data were introduced and discussed by Bahadur [5] and Cox [12], and early discussions of classification* based on dichotomous response vectors were given by Bahadur [6] and Solomon [42]. Historical comments and a general discussion about the logistic model for marginal distributions were given by Mantel [36] and Cox's monograph [11] catalogues early work concerning independent binary data*. Latent class models for multivariate binary data, which allow for dependence structure, were evidently first discussed by Lazarsfeld [32] (see LATENT STRUCTURE ANALYSIS). Bishop et al. [8] catalog and extend early work in categorical data* analysis. Cox [13] gives an account of binary data up to that time.

The character of multivariate binary data analysis changed in the 1970–1980s, due in part to a combination of the development of fast computers and advances such as computer implementation of methods for obtaining maximum likelihood* estimates such as iterative proportional fitting* (Deming and Stephan [17]) and the EM algorithm* (cf. Dempster et al. [18]). Other crucial developments are the generalized estimation equation (GEE) approach (Zeger and Liang [48]) and Gibbs sampling*, which

was popularized by Gelfand and Smith [24]. The main outcome is that it is now possible to efficiently model dependence and make inferences.

Consider the vector $\mathbf{Y}' = (Y_1, \dots, Y_k)$, with $Y_i \sim \text{Bernoulli}(p_i)$, and assume a joint probability density for $\mathbf{Y}$, $f(\mathbf{y})$. Under the special case of independence,

$$f(\mathbf{y}) = \prod_{i=1}^{k} p_i^{y_i}. \tag{1}$$

The vector $\mathbf{Y}$ is a multivariate Bernoulli vector. There is no simple generalization of the model (1) in the case of dependence, as there is in the case of multivariate normal data wherein the joint distribution is completely characterized by a mean vector and covariance matrix. While pairwise uncorrelatedness for the Bernoulli RVs implies pairwise independence, it does not imply mutual independence as in the normal case.

LOG-LINEAR MODELS

Due to the plethora of results available for log-linear models for categorical data (Bishop et al. [8], Fienberg [19], Plackett [38], Agresti [1], Andersen [3]; Christensen [14]), it is natural to think about dependence among k Bernoulli RVs by constructing a 2^k contingency table and by specifying a saturated log-linear model for the data (see LOG-LINEAR MODELS IN CONTINGENCY TABLES). This is equivalent to modeling the log of the probability of seeing a response vector $\mathbf{y}$ as a standard hierarchical analysis-of-variance model which includes a k-factor interaction and no error, or equivalently by writing

$$\ln \Pr(Y = \mathbf{y}) = \mu + \sum_{i=1}^{k} \gamma_i y_i + \sum_{i,j} \gamma_{ij} y_i y_j$$
$$+ \cdots + \gamma_{12\dots k} y_1 y_2 \dots y_k, \tag{2}$$

where the γ's are unknown parameters. When the γ_{ij}'s are equal and the higher-order γ's are zero, this model is analogous to the multivariate normal with equal correlations and was suggested by Cox [12]. The reduced model with all two-factor and higher γ's set to zero corresponds to the model of mutual independence (1) of the k Bernoulli RVs. There are many other models between the extremes of independence (1) and saturated (2), including many with various independence and conditional independence* interpretations. The above literature discusses the selection and interpretation of such models.

The special case of a sample of n independent k-vectors, $(\mathbf{y}_1, \ldots, \mathbf{y}_n)$, results in a multinomial distribution* with probabilities specified according to the log-linear model (2) with $2^k - 1$ free parameters, provided the ith marginal Bernoulli RVs, obtained from the collection of n sampled vectors, correspond to one another. The standard approach to inference for log-linear models relies on such distributional structure and thus is reasonable in this special case. However, vectors can easily be of varying size. When sampling households, for example, the family or cluster sizes will generally vary, and when taking repeated measurements on the same individual, there will often be missing measurements. In the literature, this model is termed to be not "reproducible" (Fitzmaurice et al. [22, §2.2]), which ultimately translates to mean that the likelihood function in these situations is unwieldy.

Additionally, one must be concerned with whether the family of models (2) is sensible for the data that have been collected. For example, consider $k = 3$ repeated measures in time with no covariates, and define the factors A, B, and C to correspond to the yes–no responses for times 1, 2, and 3 respectively. Suppose standard model-fitting procedures suggest that the factor B is independent of C given A, thus knowing the outcome at time 2 is noninformative for predicting the outcome at time 3 given the outcome at time 1. This model may be difficult to interpret. Alternatively, suppose that households of equal size have been sampled and that all individuals are asked whether or not they favor the death penalty. In order to utilize the log-linear model structure, some sort of ordering must be placed on the individuals in the households to make the ith Bernoulli in each household comparable. So the Bernoulli RVs could be ordered in each household vector according to the age of

the respondents. But under the same scenario just described for repeated measures there is perhaps even greater difficulty with interpreting the model. Such difficulties of interpretation will increase with k. Finally, it is not immediately obvious how one should incorporate covariate information into this model. This topic is considered below.

LATENT VARIABLES

A second approach to modeling dependence arises naturally by consideration of mixtures of distributions. This can be accomplished from a Bayesian perspective (Laird [31]), or through modeling latent variables as discussed by Bartholomew [7] (see LATENT-VARIABLE MODELING). From the purely technical stand-point of generating distributions for dependent data, there is a great deal of similarity in the approaches, so this discussion focuses on the latent-variable approach. Consider a variable or vector, say X, and assume that the components of the k-dimensional vector $\mathbf{Y}$ are independent Bernoulli RVs upon knowing the value of X. Furthermore assume that X has its own density, say $g(x)$. Then the marginal density of $\mathbf{Y}$ is

$$f(\mathbf{y}) = \int f(\mathbf{y}|x)g(x)v(dx). \qquad (3)$$

As the simplest nontrivial version of this model, let $X \sim \text{Bernoulli}(\theta)$, and let the vector $\mathbf{Y}|X = j$ be composed of independent components with $Y_i|X \sim \text{Bernoulli}(p_{ij})$. This results in the model for $\mathbf{Y}$

$$f(\mathbf{y}) = \theta \prod_{i=1}^{k} p_{i1}^{y_i}(1 - p_{i1})^{1-y_i}$$

$$+ (1-\theta) \prod_{i=1}^{k} p_{i2}^{y_i}(1 - p_{i2})^{1-y_i},$$

where a clear dependence has been induced. The idea here is that two populations really are being sampled, but that it is unknown which individuals come from which population. The proportion from the first population is the unknown value θ. Cusick and

Babiker [16] give a specific example of this model with covariates used in modeling the p_{ij}'s. Parameters are estimated by using the EM algorithm [18]. Bartholomew (7) extends this to let X be polytomous, and discusses model selection including the number of categories for X.

Along these same lines one can let the vector of marginal probabilities $\mathbf{P}' \equiv (p_1, \ldots, p_k)$ depend on a parameter vector, say $\lambda = (\lambda_1', \lambda_2')'$, and set $X = \lambda_2$. Then select a continuous distribution for X and apply (3) to obtain a family of models indexed by λ_1 that allows for dependence among the components of $\mathbf{Y}$, e.g., let $\mathrm{logit}(p_j) = \lambda_1 + \lambda_2$ for scalar λ's.

Zeger and Karim [49] have studied such models extended to the following situation. There are n distinct data vectors, $\mathbf{y}_i$, with corresponding probability vectors $\mathbf{p}_i$, where the jth component is defined to be p_{ij}. Corresponding to each p_{ij} is a p-dimensional vector of covariates, say $\mathbf{x}_{ij}$. For example, consider households in Amsterdam with binary responses pertaining to a yes–no question about the death penalty. The covariate information for the jth individual in the ith household could be the vector consisting of the individual specific variables age, highest achieved level of education, and sex. In order to model the multivariate binary dependence structure, define the random q-variate variance component vector $\mathbf{b}_i$. With a logit* link (namely, the log odds of "success" are modeled linearly in the covariates), the Zeger–Karim model would be

$$\mathrm{logit}(p_{ij}) = \mathbf{x}_{ij}'\boldsymbol{\beta} + \mathbf{z}_{ij}'\mathbf{b}_i, \qquad (4)$$

where $\boldsymbol{\beta}$ is a $p \times 1$ vector of fixed regression coefficients and $\mathbf{z}_{ij}$ is a vector composed of a subset of the components of $\mathbf{x}_{ij}$. For the above example, a natural choice for $\mathbf{z}_{ij}$ would simply be $z_{ij} = 1$ with $q = 1$. In general, the $\mathbf{b}_i$'s are assumed to be independent and identically distributed according to some common distribution, usually normal with mean zero and unknown diagonal covariance, say $\boldsymbol{\Theta}$. The likelihood based on the entire sample is

$$\mathrm{Lik}(\boldsymbol{\beta}, \boldsymbol{\Theta}) \propto \prod_{i=1}^{n} \int \prod_{j=1}^{k} f(y_{ij}|\mathbf{b}_i) g(\mathbf{b}_i|\boldsymbol{\Theta}) d\mathbf{b}_i,$$

where $f(y_{ij}|\mathbf{b}_i) = p_{ij}^{y_{ij}}(1 - p_{ij})^{1-y_{ij}}$ and p_{ij} is defined at (4). These integrals may not be evaluated in closed form, so Zeger and Karim [49] formulate a fully Bayesian approach whereby uncertainty about $\boldsymbol{\beta}$ and $\boldsymbol{\Theta}$ are also modeled with prior probability distributions. They then suggest a Gibbs sampling approach [24] in order to perform the calculations, which results in numerical approximations to the posterior distributions of the regression coefficients $\boldsymbol{\beta}$ and the variance components $\boldsymbol{\Theta}$.

Zeger and Karim present an illustration involving the examination of Indonesian preschool children in which they look for the presence or absence of respiratory infection over successive periods of time, and with covariates age, gender, height, and other variables. Their model includes a single variance component corresponding to $z_{ij} = 1$. They find among other things that infection rate decreases with age, that the prevalence of infection is lower in better nourished children, and also that the variance component θ is nonzero, indicating that there is indeed correlation among the observations for the same individual and also that there is heterogeneity among individuals.

THE CLASSIFICATION PROBLEM

An example of the classification problem based on binary data arises in Anderson et al. [4], where the condition keratoconjunctivitis sicca, or dry eyes, is studied. The study refers to 10 symptoms (redness, itchiness, soreness or pain, burning, etc.) that are associated with this condition. Each symptom is either present or absent in each individual, and they are expected to be related to one another. For a given vector it is of interest to make a diagnosis (yes–no) for the disease. A training sample of 40 diseased patients and 37 nondiseased patients was available for use in diagnosis.

Aitchison and Aitken [2] adapted the kernel method in order to estimate the density function $f(\mathbf{y})$ for each group. Define $\hat{g}(\mathbf{x})$ to be the proportion of individuals in the sample of diseased individuals that have symptoms corresponding to the vector $\mathbf{x} \in \{0, 1\}^{10}$ for all 2^{10} such $\mathbf{x}$'s. Then Aitchison and Aitken [2],

Hall [28], Brown and Rundell [10], Tutz [44], and Johnson and Kokolakis [29] gave density estimates of the form

$$\hat{f}(\mathbf{y}) = \sum_x p(\mathbf{y}|\mathbf{x}, \hat{\lambda})\hat{g}(\mathbf{x}), \qquad (5)$$

where the kernel $p(\mathbf{y}|\mathbf{x}, \lambda) \equiv \lambda^{d(\mathbf{x},\mathbf{y})}(1 - \lambda)^{d-d(\mathbf{x},\mathbf{y})}$, and where $d(\mathbf{x},\mathbf{y})$ is the number of agreements between $\mathbf{x}$ and $\mathbf{y}$; $\hat{\lambda}$ is a smoothing-parameter estimate which is determined in various ways by the above authors. These estimates are of the same form as (3) and improve on the usual nonparametric estimates $\hat{g}(\mathbf{y})$. This is so because $\hat{g}(\mathbf{y})$ is nearly always zero due to the large number of cells, $2^{10} = 1024$, and the small number of observations, $n = 40$. The estimate (5) is a weighted average of observed proportions for all vectors, giving highest weights to vectors that are near $\mathbf{y}$. An estimate of the same form is constructed for the nondiseased individuals, and classification of a new vector, say $\mathbf{y}_f$, is based on the "estimated" posterior odds, which is the prior odds times the ratio of these two density estimates evaluated at $\mathbf{y}_f$.

Johnson and Kokolakis [29] showed that the above procedure and methods could be set in the context of a very simple model whereby the errors in recording a one as a zero or a zero as a one for all k responses would be assumed to occur independently with unknown probability $1 - \lambda$. Then letting $\mathbf{X}$ be the actual vector of indicated symptoms, and letting $\mathbf{Y}$ be the observed vector, this assumption gives rise to the transition probability $\Pr(\mathbf{Y} = \mathbf{y}|\mathbf{X} = \mathbf{x}, \lambda) = p(\mathbf{y}|\mathbf{x}, \lambda)$, the kernel defined by Aitchison and Aitken [2]. Defining $\Pr(\mathbf{X} = \mathbf{x}) = g(\mathbf{x})$, the model for $\mathbf{Y}$ is precisely of the form (3) by the total probability law. Johnson and Kokolakis [29] utilized this model to develop a fully Bayesian approach and noted connections with previous approaches.

Ott and Kronmal [30] also introduced a nonparametric method of density estimation* for multivariate binary data which is based on orthogonal expansion of the density in terms of a discrete Fourier series. Liang and Krishnaiah [35] used the same approach, only with different coefficients. Both papers discuss the application of these procedures to the classification problem. Chen et al. [15] further extended this work, and Stoffer [43]

expanded the discussion to binary time-series data.

REGRESSIVE LOGISTIC MODELS

Let $\mathbf{Y}$ be a k-vector as above, and let $\mathbf{x}_i$ denote a p-vector of covariate values corresponding to $\mathbf{Y}_i$ for $i = 1, \dots, k$. Define $\mathbf{X}$ to be the $k \times p$ covariate matrix $(\mathbf{x}_1, \dots, \mathbf{x}_k)'$. In those cases where there is a natural ordering of the responses, e.g., with repeated measures data, Bonney [9] introduced a class of "regressive logistic" models (see LOGISTIC REGRESSION). Assume

$$\Pr(\mathbf{Y} = \mathbf{y}|\mathbf{X})$$
$$= \prod_{i=1}^{k} \Pr(Y_i = y_i|y_1, \dots, y_{i-1}, \mathbf{x}_i), \quad (6)$$

namely, that the distributions for $Y_i|Y_1, \dots, Y_{i-1}, \mathbf{X}$ and $Y_i|Y_1, \dots, Y_{i-1}, \mathbf{x}_i$ are the same. Furthermore define the log conditional odds for the ith Bernoulli as

$$\theta_i = \ln\left(\frac{\Pr(Y_i = 1|Y_1, \dots, Y_{i-1}, \mathbf{x}_i)}{\Pr(Y_i = 0|Y_1, \dots, Y_{i-1}, \mathbf{x}_i)}\right)$$
$$= \mu + \sum_{j=1}^{i-1} \gamma_j Y_j + \mathbf{x}_i'\boldsymbol{\beta}, \qquad (7)$$

where μ, the γ_j's, and $\boldsymbol{\beta}$ are unknown parameters. Then combining (6) and (7) as in Bonney [9], the likelihood function is obtained as

$$\Pr(\mathbf{Y} = \mathbf{y}|\mathbf{X}) = \prod_{i=1}^{k} \frac{e^{\theta_i y_i}}{1 + e^{\theta_i}}.$$

If the $\gamma_j's$ are equal, the log odds of a "success" for the ith Bernoulli given the previous values is a simple linear combination of the number of previous successes and the covariates for the ith individual. Serially dependent observations are modeled by assuming that $\theta_i = \mu + \gamma Y_{i-1} + \mathbf{x}_i'\boldsymbol{\beta}$. Mutual independence results if the γ_j's are all zero. Bonney [9] generalizes (7) to include all possible cross-product terms for the Y_j's, $j = 1, \dots, i-1$. The advantages of this model are the simplicity in modeling, its utility for prediction, and that standard software can be used to obtain estimates of all parameters and their asymptotic standard errors; cf. Bonney [9]. The disadvantages are that it is limited to situations where there is a natural ordering of the responses.

Bonney considers an example abstracted from Wilcox and Gladen [47], where instances of spontaneous abortion are recorded for the same individuals over four pregnancies. The covariates are the pregnancy number and location, of which there are three. One of the well-fitted models was of the form $\theta_i = \mu + \gamma$(#previous abortions $-$ # successful pregnancies) $+\beta_1$(pregnancy number) $+ \beta_2$(indicator for location 1) $+\beta_3$(indicator for location 2). Bonney's analysis indicates that the risk of a spontaneous abortion is positively associated with pregnancy number and with location 1. For a given location and pregnancy number, the risk of a spontaneous abortion is also positively associated with the excess of previous abortions relative to successful pregnancies.

MARGINAL MEANS MODELS

Models that incorporate covariate information and result in specified marginal distributions (logistic regression* for example) are currently under intensive study. Liang and Zeger [33], and Zeger and Liang [48], introduced a multivariate analogue to the quasilikelihood estimation method promoted by Wedderburn [46] and McCullagh [37]. Their method is the *generalized estimation equation* (GEE) approach, and it applies broadly to generalized linear models* and consequently to binary data. The idea is to take the usual likelihood equations for estimating regression coefficients under the assumption of independence and under the assumption that the relationship between the marginal probability p_i and $\mathbf{x}_i'\boldsymbol{\beta}$ is known, and to generalize them in a way that results in estimates of the coefficients that are robust to the possibility of dependence structure among the observations. The equations actually solved generally do not correspond to any known dependence model.

The generalized equations depend upon what is termed a "working" parametric correlation matrix, specified by the analyst. The parameters of the correlation matrix may or may not be specified by the user.

If they are left unspecified, then they can also be estimated. Prentice [39] and Prentice and Zhao [41] indicated how the GEE approach could be generalized to get simultaneous estimates of all parameters. Standard large-sample normal-theory results hold for the regression coefficient estimates with an asymptotic covariance matrix that generalizes the usual inverse of the information matrix obtained under independence. Liang and Zeger [33] established that the naive estimates, assuming independence, are in fact still consistent even when the data are actually dependent, but that the inverse information tends to give the wrong variance estimates. They also established that their estimates are asymptotically normal, gave an appropriate estimate of their covariance, and argued by Monte Carlo methods that GEE estimates are reasonably efficient.

Zhao and Prentice [51] and Prentice and Zhao [41] introduced a quadratic exponential model which is essentially a log-linear model parametrized with unknown main effects and two-factor interactions, and with known higher-order interactions, which are usually set to zero. They reparametrized by defining $\boldsymbol{\mu}_k$ to be the vector of probabilities or means for the k Bernoulli RVs and by defining the vector $\boldsymbol{\sigma}_k$ of $\binom{k}{2}$ pairwise covariances for the k Bernoulli RVs. They assumed a restricted model whereby $\boldsymbol{\mu}_k$ is modeled to be a known function $\boldsymbol{\beta}$ for given covariates, and $\boldsymbol{\sigma}_k$ is modeled to be a known function of $(\boldsymbol{\beta}, \boldsymbol{\alpha})$, where $\boldsymbol{\alpha}$ is generally a new set of nuisance parameters. The standard logistic model for the marginal means corresponds to the specification that p_i is the logistic transform of $\mathbf{x}_i'\boldsymbol{\beta}$ and that the vector $\boldsymbol{\mu}_k$ is the corresponding k-vector. Zhao and Prentice [51] established that the corresponding estimating equations for $(\boldsymbol{\beta}, \boldsymbol{\alpha})$ are of precisely the same form as the GEEs proposed by Liang and Zeger [33], thus providing a specific example of modeled dependence that results in GEEs. However, the two approaches are distinguished by the fact that the Zhao–Prentice (ZP) model is computationally burdensome for large cluster sizes compared with the GEE approach, while on the other hand, if the quadratic exponential model actually holds, the ZP estimates will necessarily be more efficient than those based on GEEs. Zhao and Prentice [51] thus propose a GEE version of their estimating equations which involves the selection of "working" correlation matrices as in the GEE approach and, as mentioned earlier, provides simultaneous GEE estimates for both β and α.

Liang et al. [34], with a slightly different parametrization than ZP, introduced separate estimating equations for β and for "covariance-type" parameters α, essentially treating them as if they were orthogonal. Fitzmaurice and Laird [23] proposed yet another model formulation whereby logistic marginals are fitted, but instead of modeling covariances as in the ZP approach, they modeled two- and higher-way association parameters. While ZP essentially set three- and higher-way parameters to zero, Fitzmaurice and Laird (FL) left them in, and furthermore, while ZP modeled the pairwise covariances, σ_k, as functions of both β and α, FL modeled the higher-order interactions as a function of a single parameter vector α. Fitzmaurice and Laird then developed a maximum likelihood approach to inference, found that α and β are orthogonal, and introduced a robust covariance estimate for $\hat{\beta}$ so that, even when their dependence model is incorrect, they will still get proper estimates of covariances. They calculate these via iterative proportional fitting*, discussed in Fienberg and Meyer [20]. All of the other approaches discussed above employ Fisher scoring or Newton–Raphson or modifications thereof.

Fitzmaurice et al. [22] summarized and reviewed the preceding marginal means models. The relative advantages and disadvantages of the "mixed model" approach taken by FL and the "quadratic exponential" model advocated by ZP are discussed by Prentice and Mancl [40] and in the rejoinder by Fitzmaurice et al [22]. A major advantage of the ZP model is that it is reproducible, whereas the FL model requires equal sample sizes for each vector of correlated observations, thus precluding its use with cluster data and causing difficulties when some longitudinal responses are missing. On the other hand, the ZP model can result in inconsistent estimates of the regression parameters even when the model for the marginal means is correctly specified, unless the model for the covariances is correctly specified; the FL model does not suffer from this potential flaw.

The GEE approach results in consistent estimates even when the working covariance is misspecified, and it does not require equal sample sizes for vectors and thus applies very generally to correlated binary data. However, there is some loss in efficiency by using GEE rather than a parametric approach when the presumed parametric model holds. The GEE approach may thus be preferable unless one can adequately assess to what extent particular parametric alternatives fit observed data. Fitzmaurice and Laird [23] discuss and use the standard likelihood ratio goodness-of-fit statistic G^2 to justify their model choices.

As an illustration of the marginal means approach, Fitzmaurice and Laird analyze a data set from a longitudinal study* of the health effects of air pollution, originally discussed in Ware et al. [45] and also analyzed by Zeger et al. [50]. The study consists of individuals examined annually at ages 7 through 10. The repeated binary response was wheezing status (yes–no) of a child at each time period, and an additional covariate indicating maternal smoking was included. Logistic regression was used to model marginal probabilities of wheezing as a function of age, smoking status, and the interaction term age $\times$ smoking status. Their selected model asserts a negative association between wheezing and age, a positive association between wheezing and maternal smoking, and that the older the child, the greater the effect of maternal smoking. They also found that a model that resulted in exchangeability* of the Bernoulli RVs fitted the data quite well, and additionally that the correlation structure could reasonably be assumed to be the same for all individuals in the sample. Thus there was only a single parameter characterizing the dependence for these data.

REFERENCES

1. Agresti, A. (1990). *Categorical Data Analysis*. Wiley, New York.

2. Aitchison, J. and Aitken, C. G. G. (1976). Multivariate binary discrimination by the kernel method. *Biometrika*, **63**, 413–420. (Seminal paper introducing kernel density estimation into multivariate binary discrimination.)

3. Andersen, E. B. (1990). *The Statistical Analysis of Categorical Data.* Springer-Verlag, Berlin.

4. Anderson, J. A., Whaley, K., Williamson, J. and Buchanan, W. W. (1972). A statistical aid to the diagnosis of keratoconjunctivitis sicca. *Quart. J. Med.*, **162**, 175–189.

5. Bahadur, R. R. (1961). A representation of the joint distribution of responses to n dichotomous items. In *Studies in Item Analysis and Prediction, Stanford Mathematical Studies in the Social Sciences* VI, H. Solomon, ed., Stanford University Press, pp. 158–168.

6. Bahadur, R. R. (1961). On classification based on responses to *n* dichotomous items. In *Studies in Item Analysis and Prediction*, H. Solomon, ed., *Stanford Mathematical Studies in the Social Sciences* VI, Stanford University Press, pp. 169–176.

7. Bartholomew, D. J. (1987). *Latent Variable Models and Factor Analysis.* Oxford University Press, London. (Very nice, easy-to-read book on the theory and application of latent-variable methods.)

8. Bishop, Y. M. M., Fienberg, S. E. and Holland, P. W. (1975). *Discrete Multivariate Analysis: Theory and Practice.* MIT Press, Cambridge, Mass. (Very important reference for theory and application of categorical data analysis.)

9. Bonney, G. E. (1987). Logistic regression for dependent binary observations. *Biometrics*, **43**, 951–973. (Very nice; easy to read; introduces easy-to-use methodology for the case when the regression data have a natural ordering.)

10. Brown, P. J. and Rundell, P. W. K. (1985). Kernel estimates for categorical data. *Technometrics*, **27**, 293–299.

11. Cox, D. R. (1970). *Analysis of Binary Data.* Methuen, London. (Fundamental reference.)

12. Cox, D. R. (1972). The analysis of multivariate binary data. *Appl. Statist*, **21**, 113–120. (A must-read on the early development of the topic.)

13. Cox, D. R. (1982). Binary data. In *Encyclopedia of Statistical Sciences*, vol. 1, S. Kotz, N. L. Johnson and C. B. Read, eds. Wiley, New York, pp. 233–234.

14. Christensen, R. (1990). *Log-linear Models.* Springer-Verlag, New York. (Very nice, up-to-date theory and application of categorical data techniques.)

15. Chen, X. R., Krishnaiah, P. R. and Liang, W. W. (1989). Estimation of multivariate binary density using orthogonal functions. *J. Multivariate Anal.*, **31**, 178–186.

16. Cusick, J. and Babiker, A. (1992). Discussion of "Multivariate regression analyses for categorical data" by Liang, Zeger and Qaqish. *J. R. Statist. Soc. B*, **54**, 30–31. (see ref. 34.)

17. Deming, W. E. and Stephan, F. F. (1940). On the least squares adjustment of a sampled frequency table when th expected marginal totals are known. *Ann. Statist*, **11**, 427–444. (Early, very important work.)

18. Dempster, A. P., Laird, N. M. and Rubin, D. B. (1977). Maximum likelihood from incomplete data via the EM algorithm (with discussion). *J. Amer. Statist. Ass.*, **72**, 77–104. (Seminal paper on EM algorithm.)

19. Fienberg, S. E. (1980). *The Analysis of Cross-Classified Categorical Data.* MIT Press, Cambridge, Mass. (Very nice applications of categorical techniques; low prerequisites.)

20. Fienberg, S. E. and Meyer, M. (1983). Iterative proportional fitting algorithm. In *Encyclopedia of Statistical Sciences*, vol. 4, S. Kotz, N. L. Johnson and C. B. Read, eds. Wiley, New York, pp. 275–279.

21. Fitzmaurice, G. M. and Laird, N. M. (1993). A likelihood-based method for analyzing longitudinal binary responses. *Biometrika*, **80**, 141–151. (They introduce a strong competitor to previous methods for marginal means regression.)

22. Fitzmaurice, G. M., Laird, N. M. and Rotnitzky, A. G. (1993). Regression models for discrete longitudinal responses, with discussion. *Statist. Sci.*, **8**, 284–309. (Very nice in-depth discussion of GEE versus quadratic exponential versus mixed model approaches, but including an overview of other models and methods for marginal means regression.)

23. Fitzmaurice, G. M. and Laird, N. M. (1993). A likelihood-based method for analyzing longitudinal binary responses. *Biometrika*, **80**, 141–151.

24. Gelfand, A. E. and Smith, A. F. M. (1990). Sampling based approaches to calculating marginal densities. *J. Amer. Statist. Ass.*, **8f**, 398–409. (Shines light on the Gibbs sampling approach and makes connections with data augmentation; revitalizes Bayesian computation.)

25. Goodman, L. A. and Kruskal, W. H. (1954). Measures of association for cross-classification. *J. Amer. Statist. Ass.*, **49**, 732–764.

26. Goodman, L. A. and Kruskal, W. H. (1959). Measures of association for cross-classifications, II. Further discussion and references. *J. Amer. Statist. Ass.*, **54**, 123–163.

27. Goodman, L. A. and Kruskal, W. H. (1963). Measures of association for cross-classifications, III. Approximate sampling theory. *J. Amer. Statist. Ass.*, **58**, 310–364.

28. Hall, P. (1981). On nonparametric multivariate binary discrimination. *Biometrika*, **68**, 287–294. (Discusses optimality properties of smoothing parameter estimates for kernel method introduced by Aitchison and Aitken.)

29. Johnson, W. O. and Kokolakis, G. E. (1994). Bayesian classification based on multivariate binary data. *J. Statist. Plann. Inference*, **41**, 21–35. (Gives a Bayesian justification for and alternative to kernel methods introduced by Aitchison and Aitken.)

30. Kronmal, R. A. and Ott, J. (1976). Some classification procedures for multivariate binary data using orthogonal functions. *J. Am. Statist. Ass.*, **71**, 391–399.

31. Laird, N. M. (1991). Topics in likelihood-based methods for longitudinal data analysis. *Statist. Sinica*, **1**, 33–50. (Nice overview of longitudinal data analysis which includes discussion of binary and nonbinary data.)

32. Lazarsfeld, P. W. (1950). Logical and mathematical foundation of latent structure analysis. In *Measurement and Prediction*, S. A. Stouffer et al., eds. Princeton, Princeton University Press, pp. 362–412.

33. Liang, K. Y. and Zeger, S. L. (1986). Longitudinal data analysis using generalized linear models. *Biometrika*, **73**, 13–22. (Seminal paper on generalized estimation equations for marginal means regression.)

34. Liang, K. Y., Zeger, S. L. and Qaqish, B. (1992). Multivariate regression analyses for categorical data. *J. R. Statist. Soc. B*, **54**, 3–40. (Compare and contrast their own GEE approach with the Zhao–Prentice GEE approach to marginal means models.)

35. Liang, W. Q. and Krishnaiah, P. R. (1985). Nonparametric iterative estimation of multivariate binary density. *J. Multivariate Anal.*, **16**, 162–172.

36. Mantel, N. (1966). Models for complex contingency tables and polychotomous dosage response curves. *Biometrics*, **22**, 83–95.

37. McCullagh, P. (1983). Quasi-likelihood functions. *Ann. Statist.*, **11**, 59–67.

38. Plackett, R. L. (1981). *The Analysis of Categorical Data, Griffin's Statistical Monographs and Courses* 35. Griffin, London.

39. Prentice, R. L. (1988). Correlated binary regression with covariates specific to each binary observation. *Biometrics*, **44**, 1033–1048. (Discusses advantages and disadvantages of regressive versus marginal means models and introduces GEE approach to estimate working covariance parameters.)

40. Prentice, R. L. and Mancl, L. A. (1993). Discussion of "Regression models for discrete longitudinal responses" by Fitzmaurice, Laird and Rotnitzky. *Statist. Sci.*, **8**, 302–304.

41. Prentice, R. L. and Zhao, L. P. (1991). Estimating equations for parameters in means and covariances of multivariate discrete and continuous responses. *Biometrics*, **47**, 825–839. (Elaboration of the GEE aspects introduced in their 1990 paper.)

42. Solomon, H. (1961). Classification procedures based on dichotomous response vectors. In *Studies in Item Analysis and Prediction*, H. Solomon, ed., *Stanford Mathematical Studies in the Social Sciences* VI. Stanford University Press, pp. 177–186.

43. Stoffer, D. S. (1991). Walsh–Fourier analysis and its statistical applications (with discussion). *J. Amer. Statist. Ass.*, **86**, 461–485.

44. Tutz, G. (1986). An alternative choice of smoothing for kernel-based density estimates in discrete discriminant analysis. *Biometrika*, **73**, 405–411.

45. Ware, J. H., Dockery, D. W., Spiro, A., III, Speizer, F. E. and Ferris, B. G. (1984). Passive smoking, gas cooking and respiratory health in children living in six cities. *Amer. Rev. Respir. Dis.*, **129**, 366–374.

46. Wedderburn, R. W. M. (1974). Quasi-likelihood functions, generalized linear models and Gauss–Newton method. *Biometrika*, **61**, 439–447.

47. Wilcox, A. J. and Gladen, B. C. (1982). Spontaneous abortion: the role of heterogeneous risk and selective fertility. *Early human. Dev.*, **7**, 165–178.

48. Zeger, S. L. and Liang, K. -Y. (1986). Longitudinal data analysis for discrete and continuous outcomes. *Biometrics*, **42**, 121–130.

49. Zeger, S. L. and Karim, M. R. (1991). Generalized linear models with random effects: a Gibbs sampling approach. *J. Amer. Statist. Ass.*, **86**, 79–86. (Very nice Bayesian technology and methods for mixed generalized linear model.)

50. Zeger, S. L., Liang, K. Y. and Albert, P. S. (1988). Models for longitudinal data: a generalized estimating equation approach. *Biometrics*, **44**, 1049–1060.

51. Zhao, L. P. and Prentice, R. L. (1990). Correlated binary regression using a quadratic exponential model. *Biometrika*, **77**, 642–648. (Introduces a very nice parametric alternative to Liang–Zeger GEE approach to marginal means regression, which gives rise to a new GEE approach and gives partial justification for GEE.)

BIBLIOGRAPHY

Azzalini, A. (1994). Logistic regression for auto-correlated data with application to repeated measures. *Biometrika*, **81**, 767–775.

Bahadur, R. R. (1961). A representation of the joint distribution of responses to n dichotomous items. In *Studies in Item Analysis and Prediction*, H. Solomon, ed. *Stanford Mathematical Studies in the Social Sciences* VI. Stanford University Press.

Carey, V., Zeger, S. L. and Diggle, P. (1993). Modelling multivariate binary data with alternating logistic regressions. *Biometrika*, **80**, 517–526.

le Cessie, S. and van Houwelingen, J. C. (1991). A goodness of fit test for binary regression models based on smoothing methods. *Biometrics*, **47**, 1267–1282.

le Cessie, S. and van Houwelingen, J. C. (1994). Logistic regression for correlated binary data. *Appl. Statist.*, **43**, 95–108.

Cologne, J. B., Carter, R. L., Fujita, S., and Ban, S. (1993). Application of generalized estimation equations to a study of in vitro radiation sensitivity. *Biometrics*, **49**, 927–934.

Collett, D. (1991). *Modelling Binary Data*. Chapman and Hall, London. (Very nice treatment of independent binary data with diagnostics.)

Connolly, M. A. and Liang, K. -Y. (1988). Conditional logistic regression models for correlated binary data. *Biometrika*, **75**, 501–506.

Copas, J. B. (1988). Binary regression models for contaminated data (with discussion). *J. R. Statist. Soc. B*, **50**, 225–265.

Emrich, L. J. and Piedmonte, M. R. (1991). A method for generating high-dimensional multivariate binary variates. *Amer. Statist.*, **45**, 302–303.

Fitzmaurice, G. M. and Lipsitz, S. R. (1995). A model for binary time series data with serial odds ratio patterns. *Appl. Statist.*, **44**, 51–61.

Kokolakis, G. E. (1983). A new look at the problem of classification with binary variables. *Statistician*, **32**, 144–152.

Kokolakis, G. E. (1984). A Bayesian criterion for the selection of binary features in classification problems. *Bayesian Statist.*, **2**, 673–680.

Kokolakis, G. E. and Dellaportas, P. Hierarchical modelling for classifying binary data. *Bayesian Statist.*, **5**, J. M. Bernardo, J. O. Berger, A. P. Dawid and A. F. M. Smith, eds. Oxford Science, pp. 647–652.

Laird, N. M. and Ware, J. H. (1982). Random-effects models for longitudinal data. *Biometrics*, **38**, 963–974.

Lefkopoulou, M. and Ryan, L. (1993). Global tests for multiple binary outcomes. *Biometrics*, **49**, 975–988.

Liang, K. Y. and Zeger, S. L. (1989). A class of logistic regression models for multivariate binary time series. *J. Amer. Statist. Ass.*, **84**, 447–451.

Lipsitz, S. R., Laird, N. M., and Harrington, D. P. (1991). Generalized estimating equations for correlated binary data: using the odds ratio as a measure of association. *Biometrika*, **78**, 153–160.

Lipsitz, S. R., Fitzmaurice, G. M., Orav, E. J., and Laird, N. M. (1994). Performance of generalized estimating equations in practical situations. *Biometrics*, **50**, 270–278.

Neuhaus, J. M. (1993). Estimation efficiency and tests of covariate effects with clustered binary data. *Biometrics*, **49**, 989–996.

Neuhaus, J. M., Kalbfleisch, J. D. and Hauck, W. W. (1991). A comparison of cluster-specific and population-averaged approaches for analyzing correlated binary data. *Statist. Rev.*, **59**, 25–36.

Prentice, R. L. (1986). A case-cohort design for epidemiologic cohort studies and disease prevention trials. *Biometrika*, **73**, 1–11.

Rao, J. N. K. and Scott, A. J. (1992). A simple method for the analysis of clustered binary data. *Biometrics*, **48**, 577–585.

Rosner, B. (1984). Multivariate methods in ophthalmology with application to other paired-data situations. *Biometrics*, **40**, 1025–1035.

Rotnitzky, A. and Jewell, N. P. (1990). Hypothesis testing of regression parameters in semiparametric generalized linear models for cluster correlated data. *Biometrika*, **77**, 485–489.

Strauss, D. (1992). The many faces of logistic regression. *Amer. Statist.*, **46**, 321–327.

Stiratelli, L. N. and Ware, J. H. (1984). Random-effects model for serial observations with binary response. *Biometrics*, **40**, 961–971.

Titterington, D. M. (1993). A contamination model and resistant estimation within the sampling paradigm. *J. R. Statist. Soc. B*, **55**, 817–827.

Walter, S. D. and Irwig, L. M. (1988). Estimation of test error rates, disease prevalence and relative risk from misclassified data: a review. *J. Clin. Epidemiol.*, **41**, 923–937.

Zeger, S. L., Liang, K. Y., and Self, S. G. (1985). The analysis of binary longitudinal data with time-independent covariates. *Biometrika*, **72**, 31–38.

Zeger, S. L. and Qaqish, B. (1988). Markov regression models for time series: a quasi-likelihood approach. *Biometrics*, **44**, 1019–1031.

Zhao, L. Ping, Prentice, R. L. and Self, S. G. (1992). Multivariate mean parameter estimation by using a partly exponential model. *J. R. Statist. Soc. B*, **54**, 805–811.

See also Association, Measures of; Binary Data; Latent Class Analysis; Logistic Regression; and Log-linear Models in Contingency Tables.

WESLEY O. JOHNSON

BINARY SEQUENCE

In a binary sequence, each element has one of only two possible values—usually 0 or 1. Such a sequence can represent the results of a series of trials at each of which the occurrence or nonoccurrence of an event—*E*, say—is observed, with 0 representing nonoccurrence and 1 representing occurrence. The resulting sequence is an example of binary data*. Binary sequences also arise in nonstatistical contexts: e.g., in the representation of a number of the binary scale.

See also Bernoulli Trials and Binary Data.

BINARY STRING. See Algorithmic Information Theory

BINARY STRING TECHNIQUE. See Editing Statistical Data

BINARY TIME SERIES, KEENAN MODELS FOR

Let $\{X_t\}$ be an unobserved stationary process with state space the real line R and let the response function $F : R \rightarrow [0, 1]$ be monotone. Conditional on $\{X_t\}$, the random variables $\{S_t\}$ are independent, with the distribution

$$P(S_t = 1) = 1 - P(S_t = 0) = F(X_t).$$

A special case is when $\{X_t\}$ is a Gaussian process* with mean 0 and F is the normal cdf with mean $\mu = 0$ and variance σ^2.

For this case, explicit expressions for the distribution of two and three consecutive observations in $\{S_t\}$ are derived by Keenan [1]. The joint distribution of more than three consecutive observations is not available in closed form, but Keenan provides six different approximations.

These models are examples of parameter-driven processes.

REFERENCE

1. Keenan, D. M. (1982). A time series analysis of binary data. *J. Amer. Statist. Ass.*, **77**, 816–821.

BINGHAM DISTRIBUTION. See Directional Distributions

BINNING

The statistical uses of binning are generally associated with *smoothing*. They are of at least three types: (1) smoothing techniques in their own right, (2) computational devices for increasing the performance of other smoothing methods, and (3) ways of interpreting rounding errors in otherwise continuous data. In the case of the latter application, binning usually represents unwanted statistical smoothing, and the aim is generally to remove its effects as much as possible, for example by using *Sheppard's correction**. The remainder of this entry will focus solely on the first two uses of binning, where it is deliberately introduced.

Arguably the simplest approach to smoothing uses *histograms**. In their application to *nonparametric density estimation**, for example, the data are allocated to connected, usually convex regions, called *bins*. The proportion of data falling into a given bin, divided by the content of the bin (i.e., its length if it is an interval, its area if estimation is in two dimensions, and so on), is taken as an estimator of the probability density associated with that region. If the density f of the sampling distribution is continuous, then by letting the bin size shrink to zero in an appropriate way as the sample size increases, consistent estimation of f may be ensured.

Binning can be used to simplify implementation of other approaches to smoothing. In particular, consider the *kernel estimator** of a univariate probability density f, based on a random sample $X_1, \ldots, X_n$:

$$\hat{f}(x) = \frac{1}{nh} \sum_{j=1}^{n} K\left(\frac{x - X_j}{h}\right).$$

Here, K is a kernel function and h is a bandwidth*. The amount of computation involved in calculating $\hat{f}(x)$ is of course $O(n)$ with respect to sample size, but can be reduced by binning, as follows. Let b_j, $-\infty < j < \infty$, denote an increasing sequence of equally spaced points (the bin centers), and let $\mathscr{B}_j$ (the jth bin) be the set of all real numbers that are closer to b_j than to any of the other b_k's. (Ties may be broken in any systematic way.) Write N_j for the number of elements of $\mathscr{B}_j$. Then, to a first approximation, the value of $\hat{f}$ at b_j is given by

$$\bar{f}_j = \frac{1}{nh} \sum_{k} N_j K\left(\frac{b_j - b_k}{h}\right),$$

where the sum is taken over all indices of all bins. If the bin width is of larger order than n^{-1}, then the number B of bins will be of smaller order than n, and so the level of computation, $O(n + mB)$, required to calculate $\bar{f}_j$ for m values of j will be of smaller order than the level $O(mn)$ needed for the same values of $\hat{f}(b_j)$.

Of course, $\hat{f}_j$ is only an approximation to $\hat{f}(x)$, for those values of x lying in the jth bin. Nevertheless, the approximation can be surprisingly accurate, and the error only becomes noticeable in a first-order sense when the width of the bin is so large that it approaches the bandwidth. (This is assuming that the kernel is scaled so that all or most of its support is contained in the interval $[-1, 1]$.)

Histograms and binned approximations have many generalizations, for use in both *density estimation** and *nonparametric regression**. The averaged shifted histogram* (ASH) method was introduced by Scott [4] and is based on averaging histograms over a range of different choices of origin. The operation of averaging removes the sharp discontinuities of a regular histogram, and produces a result that is equivalent to using the triangular kernel in the definition of $\bar{f}_j$ above. This similarity was exploited by Härdle and Scott [2] in their development of WARPing (the acronym for weighted average of rounded points), where the biweight kernel* is used. (The triangular kernel is $K(x) = 1 - |x|$ for $|x| \leqslant 1$, and 0 elsewhere; the biweight kernel is $K(x) = \frac{15}{16}(1 - x^2)^2$ for $|x| \leqslant 1$, and 0 elsewhere.)

The *fast Fourier transform* approach to computing kernel estimators also employs binning methods, and requires $O(n + B \log B)$ computations to calculate $\bar{f}_j$ for $1 \leqslant j \leqslant B$ if there are B bins. Indeed, the binning step is arguably the principal source of its computational speed. Details of the algorithm are discussed by Silverman [5] and Wand and Jones [6], for example. More sophisticated binning rules, including those for bivariate data, are addressed by Jones and Lotwick [3] and Hall and Wand [1].

REFERENCES

1. Hall, P. and Wand, M. P. (1996). On the accuracy of binned kernel density estimators. *J. Multivar. Anal.*, **56**, 165–184. (Discusses generalizations of standard binning rules.)

2. Härdle, W. and Scott, D. W. (1992). Smoothing by weighted averaging of rounded points. *Comput. Statist.*, **7**, 97–128. (Introduces weighted averaging of rounded points.)

3. Jones, M. C. and Lotwick, H. W. (1984). A remark on algorithm AS176: kernel density estimation using the fast Fourier transform

(remark ASR50). *Appl. Statist.*, **33**, 120–122. (Discusses a generalization of standard binning rules.)

4. Scott, D. W. (1985). Average shifted histograms: effective nonparametric density estimators in several dimensions. *Ann. Statist.*, **13**, 1024–1040. (Introduces averaged shifted histograms.)

5. Silverman, B. W. (1986). *Density Estimation for Statistics and Data Analysis*. Chapman and Hall, London. (Section 3.5 discusses computational methods for kernel density estimators, including the fast Fourier transform.)

6. Wand, M. P. and Jones, M. C. (1995). *Kernel Smoothing*, Chapman and Hall, London. (Binning methods are discussed in a number of places; see for example pp. 5–7 and 183–188. The fast Fourier transform is described on pp. 184–187.)

See also AVERAGED SHIFTED HISTOGRAM; DENSITY ESTIMATION—I; GRADUATION; HISTOGRAMS; and KERNEL ESTIMATORS.

PETER HALL

BINOMIAL AND MULTINOMIAL PARAMETERS, INFERENCE ON

The binomial distribution* is important in many statistical applications. Assume that an experiment consists of a sequence of n i.i.d. trials, each trial resulting in one of two outcomes, denoted by "success" or "failure", with probability p and $q = 1 - p$, $0 < p < 1$, respectively, where p is the *success probability* and q is the *failure probability*. Let Y be the total number of "successes". Then, Y follows (or is distributed as) a *binomial distribution* with a total number of trials n, $n = 1, 2\ldots$, and a success probability p (or with parameters n and p for short), $0 < p < 1$, and written as $Y \sim Bin(n, p)$. The parameter p is also called the *binomial proportion*. The probability mass function (pmf) $f_{n,p}(y)$ of Y is

$$f_{n,p}(y) = P(Y = y) = \binom{n}{y} p^y (1 - p)^{n-y},$$

$$y = 0, 1, \ldots, n. \tag{1}$$

A binomial distribution has two parameters n and p. One or both of them could be unknown in applications. The mean and

variance of a random variable Y distributed as $Bin(n, p)$ are np and $np(1 - p)$ respectively. If Y_i, $i = 1, \ldots, k$, are independent $Bin(n_i, p)$, then $\sum_{i=1}^{k} Y_i \sim Bin(\sum_{i=1}^{k} n_i, p)$. Putting $\theta = \log(p/(1 - p))$ when n is known, the binomial PMF in Equation 1 can be written in the form of an exponential family as

$$f_{n,p}(y) = \binom{n}{y}(1 + e^\theta)^{-n} e^{y\theta}, \ y = 0, 1, \ldots, n. \tag{2}$$

A binomial distribution can be approximated by either a normal distribution or a Poisson distribution when n is large. If n is large, p is not small, both np and nq are not small (e.g., $np, nq > 5$ [36, p. 211]), then a $Bin(n, p)$ distribution can be approximated by a normal $N(np, npq)$ distribution. Since a binomial distribution is discrete, a continuity correction* of 0.5 is recommended in the normal approximation to compute a binomial cdf, that is,

$$F_{n,p}(y) \approx \Phi\left(\frac{y + 0.5 - np}{\sqrt{npq}}\right),$$

where Φ is the standard normal cdf and $F_{n,p}$ is the $Bin(n, p)$ cdf.

If n is large and p is small (e.g., $n \geqslant 20$ and $p \leqslant 0.05$ or $n \geqslant 100$ and $np \leqslant 10$, [36, p. 204]), then a $Bin(n, p)$ distribution can be approximated by a Poisson(np) distribution. By using the *delta method* (see STATISTICAL DIFFERENTIALS, METHOD OF), one can find the variance stabilizing transformation [30, p. 54] in the normal approximation of the $Bin(n, p)$ distribution to be the arcsine square-root transformation

$$\sqrt{n}(\arcsin\sqrt{y/n} - \arcsin\sqrt{p}) \xrightarrow{L} N(0, 1/4), \tag{3}$$

as $n \to \infty$. The Anscombe [5] transformation $\arcsin[\sqrt{(Y + 3/8)/(n + 3/4)}]$ provides a better variance stabilization* for moderate n.

POINT ESTIMATION OF P

In this section, we assume that n is known but p is not, and denote $f_{n,p}$ by f_p. The maximum likelihood* estimator (m.l.e.), the moment estimator (ME) and the uniformly minimum variance unbiased* estimator (UMVUE) are all given by $\hat{p} = Y/n$. In fact, the estimator $\hat{p}$ is also the only unbiased estimator of p based on Y [51, p. 312]. When n is large and neither p nor $1 - p$ is small, $\hat{p}$ is approximately distributed as $N(p, pq/n)$.

Under squared error loss $L(p, \delta) = (p - \delta)^2$, the risk function of an estimator $\delta = \delta(Y)$ is

$$R(p, \delta) = E_p(\delta - p)^2 = \sum_{y=0}^{n} [\delta(y) - p]^2 f_p(y).$$

The Bayes risk under a general prior measure Π and squared error loss is

$$r(\Pi, \delta) = \int_0^1 R(p, \delta)\Pi(dp)$$

(see DECISION THEORY). Usually, one uses a prior that has a density, say π. Then, the posterior density $g_\pi(p|y)$ is the conditional density of p given $Y = y$. Sometimes, priors that do not yield $\int_0^1 \Pi(dp) < \infty$ are nevertheless used; these are called *improper priors*; (see PRIOR DISTRIBUTIONS).

One conjugate prior for p is the Beta prior, say $Beta(a, b)$, given by the density

$$\pi_{a,b}(p) = \frac{\Gamma(a + b)}{\Gamma(a)\Gamma(b)} p^{a-1} q^{b-1}, \qquad (4)$$

where a, b are positive constants. The posterior density $g_{a,b}(p|y)$ of p under the prior $\pi_{a,b}(p)$ (see POSTERIOR DISTRIBUTIONS) is the density of the $Beta(y + a, n - y + b)$ distribution,

$$g_{a,b}(p|y)$$
$$= \frac{\Gamma(n + a + b)}{\Gamma(y + a)\Gamma(n - y + b)} p^{y+a-1} q^{n-y+b-1}$$
$$\qquad (5)$$

(See CONJUGATE FAMILIES OF DISTRIBUTIONS). If either $a = 0$ or $b = 0$, then $\pi_{a,b}$ is improper. However, formally, the posterior density still exists unless $a = 0$ and $y = 0$ or $b = 0$ and $y = n$.

An estimator δ of p is *admissible* if there is no estimator δ' of p such that $R(p, \delta') \leqslant R(p, \delta)$ for all p and $R(p, \delta') < R(p, \delta)$ for some p (see ADMISSIBILITY); δ is *minimax* if it minimizes $\sup_p R(p, \delta')$. A *Bayes estimator* $\hat{p}_\Pi$ of p under

the prior Π minimizes the Bayes risk $r(\Pi, \delta)$. If there is a unique Bayes estimator $\hat{p}_\Pi$, then it is admissible.

When the prior has a density π, the Bayes estimator $\hat{p}_\pi$ can be uniquely found as

$$\hat{p}_\pi = \int_0^1 pg_\pi(p|y)dp.$$

If $\pi = \pi_{a,b}$ for $a, b > 0$, then

$$\hat{p}_\pi = \hat{p}_{a,b} = \frac{a + y}{a + b + n}. \qquad (6)$$

Even though formally $\hat{p} = \hat{p}_{0,0}$, the m.l.e. $\hat{p}$ cannot be legitimately written as $E(p|y)$ since when $y = 0$ or $y = n$, $g_{0,0}$ is not a probability density. Since the m.l.e. $\hat{p}$ is unbiased, it cannot be Bayes with respect to any prior Π unless its Bayes risk with respect to Π is 0 [29, p. 53]. It follows that if we restrict the parameter space to $0 < p < 1$, then $\hat{p}$ cannot be Bayes.

The Jeffreys' prior for p is the $Beta(1/2, 1/2)$ density, which is proper. Although it is common among Bayesians to use this as a noninformative prior (see JEFFREYS' PRIOR DISTRIBUTION), there are many other definitions for a noninformative prior. Consequently, a variety of priors have been suggested as noninformative for the binomial parameter p. A selection of such priors [9, p. 89] are:

$$\pi_1(p) = \frac{1}{\sqrt{p(1 - p)}},$$

$$\pi_2(p) = 1,$$

$$\pi_3(p) = \frac{1}{p(1 - p)}, \qquad (7)$$

$$\pi_4(p) = p^p(1 - p)^{1-p},$$

$$\pi_5(p) = \frac{1}{\sqrt{1 - p^2}}.$$

The first two are the most ubiquitous; all except π_3 are proper.

Empirical and robust Bayes estimation of the binomial parameter p have also been considered in the literature. In the empirical Bayes* formulation, m conditionally independent experiments with their individual parameters are performed, and interest is in estimating the parameter of the next experiment, usually called the *current experiment*. Precisely, for some given m, $p_1, p_2, \ldots, p_{m+1}$ are i.i.d. from some (unknown) prior Q and $Y_i | p_i$ are independent $Bin(n_i, p_i)$. The Bayes estimate $E(p | Y_{m+1})$ depends on the unknown prior Q. An empirical Bayes estimate replaces the unknown Q by an estimate $\hat{Q}$ on the basis of data $Y_1, Y_2, \ldots, Y_m$ from the past experiments. Various methods of estimation of Q result in different empirical Bayes estimates for p. There is no clearly superior one among them. Interestingly, estimates that incorporate only the current observation Y_{m+1} are often quite satisfactory, and one does not gain much by using more complicated empirical Bayes estimates that use all of the past data. Explicit empirical Bayes estimates for p are presented in References 9, 40, and 52.

In the robust Bayesian viewpoint (*see* BAYESIAN ROBUSTNESS), the parameter p is assumed to have a prior distribution belonging to an appropriate family of priors, say, Γ. A question of central interest is the sensitivity of the posterior expectation to the choice of a specific prior from Γ. Usually, one computes the supremum and the infimum of the posterior expectation over all the priors in Γ for a given Y. If the sup and the inf are approximately equal, then the posterior action is insensitive with respect to the choice of the prior. Zen and DasGupta [84] consider five different families of priors, some being parametric and others nonparametric. For example, they consider the family

$$\Gamma = \text{set of all priors symmetric}$$
$$\text{and unimodal about } 1/2.$$

They derive the supremum and infimum of the posterior expectation for general n and show that the decision rule defined as the average of the sup and the inf is Bayes and admissible for every n. This addresses the question of recommending a specific decision rule when the statistician does not have a unique prior. Similar results are derived in Reference 84 for other families of priors considered. See Reference 10 for additional discussions and exposition.

Under squared error loss, $\hat{p}_{a,b}$ is admissible for all positive a and b. Indeed, a sufficient condition for an estimator of the form $\delta =$ $\alpha \hat{p} + \beta$, $(\alpha, \beta \in R)$ to be admissible is $\alpha \geqslant 0$, $\beta \geqslant 0$, and $\alpha + \beta \leqslant 1$ [51, p. 335]. Thus, the m.l.e. $\hat{p}$ is admissible, a constant between 0 and 1 is also admissible, and a nonconstant estimator of the form $\delta = \alpha \hat{p} + \beta$ of p is admissible if and only if there exist $a \geqslant 0$ and $b \geqslant 0$ such that $\delta = \hat{p}_{a,b}$.

In order to find a minimax estimator of p, one can search for a and b so that $\hat{p}_{a,b}$ has a constant risk function. Here

$$R(p, \hat{p}_{a,b}) = \frac{1}{(a+b+n)^2}$$
$$\{np(1-p) + [a(1-p) - bp]^2\},$$

which equals a constant only when $a = b = \sqrt{n}/2$. Hence, a minimax admissible estimator of p is $\hat{p}_{\sqrt{n}/2, \sqrt{n}/2}$. This minimax estimator is unique, but it is Bayes with respect to infinitely many priors.

It is clear that $\hat{p}_{\sqrt{n}/2, \sqrt{n}/2}$ is biased. A comparison of the risks of $\hat{p}_{\sqrt{n}/2, \sqrt{n}/2}$ and $\hat{p}$ is obtained from

$$\Delta(p) = {}_*R(p, \hat{p}_{\sqrt{n}/2, \sqrt{n}/2}) - R(p, \hat{p})$$
$$= \frac{1}{4(1+\sqrt{n})^2} - \frac{p(1-p)}{n},$$

and $\Delta(p) \leqslant 0$ if and only if $1/2 - c_n \leqslant p \leqslant 1/2 + c_n$, where

$$c_n = \frac{\sqrt{1 + 2\sqrt{n}}}{2(1 + \sqrt{n})}. \tag{8}$$

The minimax estimator is better for most of the range of p when n is small. Even if n is not small, it is better if p is not far away from $1/2$. In fact, the minimax estimator is better than the m.l.e. for most values of p when $n \leqslant 40$.

For a general discussion of the admissibility of any estimator of a function of p, say $g(p)$, see References 47, 48, and 68. Johnson [47] concluded that an estimator δ of $g(p)$ is admissible under squared error loss if and only if there exist integers r, s, and probability measure Π on $[0, 1]$ that satisfy $-1 \leqslant r < s \leqslant n + 1$ and $\Pi(\{0\} \cup \{1\}) < 1)$, such that $\delta(y) = g(0)$ when $y \leqslant r$, $\delta(y) = g(1)$ when $y \geqslant s$, and

$$\delta(y) = \frac{\int_0^1 g(p) p^{y-r-1} (1-p)^{s-y-1} \Pi(dp)}{\int_0^1 p^{y-r-1} (1-p)^{s-y-1} \Pi(dp)}$$

when $r + 1 \leqslant y \leqslant s + 1$. Interestingly, admissibility in this set up is closed under pointwise convergence.

Suppose $Y_1, \ldots, Y_k$ are independently distributed as $Bin(n_i, p_i)$, where $p_i \in (0, 1)$ for $i = 1, \ldots, k$, and g_i, $i = 1, \ldots, k$, are continuous real functions on $[0, 1]$. For the problem of estimating each $g_i(p_i)$, if $\delta_i(Y_i)$ is an admissible estimator of $g_i(p_i)$ under the squared loss, then $\delta = (\delta_1(Y_1), \ldots, \delta_k(Y_k))$ is an admissible estimator of $(g_1(p_1), \ldots, g_k(p_k))$ under the sum of squared error loss, that is, there is no Stein effect* [69]. See References 13 and 42 for very general results on this. A couple of interesting examples are given by Johnson, for example, $\delta(Y_1, Y_2) = \max(Y_1/n_1, Y_2/n_2)$ is an admissible estimator of $\max(p_1, p_2)$ and $\delta(Y_1, Y_2) = (Y_1/n_1)(Y_2/n_2)$ is an admissible estimator of $p_1 p_2$ under squared error loss.

On the basis of the above criterion given by Johnson for the case when $g(p) = p$, Skibinsky and Rukhin [68] obtained an explicit form for an inverse Bayes rule map, which they used to derive another admissibility criterion for an estimator of p. Some interesting results can be obtained from this criterion. Let $\delta = (\delta(0), \delta(1), \delta(2), \delta(3))$ be an estimator of p when $n = 3$ such that $0 < \delta(0) \leq \delta(1) \leq \delta(2) \leq \delta(3) < 1$. If $\delta(1) = \delta(2)$ or if $\delta(2) > \delta(1)$ and

$$\delta(2)[1 - \delta(1)][\delta(1) - \delta(0)][\delta(3) - \delta(2)]$$
$$= \delta(0)[\delta(2) - \delta(1)]^2[1 - \delta(3)],$$

then δ is the Bayes estimator of p with respect to exactly one prior distribution for p and this prior distribution has finite support. If $\delta(2) > \delta(1)$ and

$$\delta(2)[1 - \delta(1)][\delta(1) - \delta(0)][\delta(3) - \delta(2)]$$
$$> \delta(0)[\delta(2) - \delta(1)]^2[1 - \delta(3)],$$

then δ is the Bayes estimator of p with respect to uncountably many prior distributions for p. If δ does not satisfy any of the above conditions, then it is not a Bayes estimator of p.

Here are some examples from Skibinsky and Rukhin. Let $\delta = (a, b, b, b)$, $0 < a < b < 1$. Then δ is admissible. It is Bayes only with respect to the prior distribution whose support is at 0 and b with mass $a/[1 - (b - a)(b^2 + 3(1 - b))]$ at the latter point. Let

$\delta = (\epsilon, \epsilon, 1 - \epsilon, 1 - \epsilon)$, $0 < \epsilon < 1/2$. Then δ is inadmissible. Let $\delta = (\epsilon, 2\epsilon, 3\epsilon, 4\epsilon)$, $0 < \epsilon < 1/4$. Then, δ is admissible and Bayes with respect to uncountably many prior distributions for p.

Absolute error loss is considered by many to be more natural than squared error loss. Unfortunately, it is a very difficult loss function to work with in the binomial case due to the lack of a useful group structure. For example, it is not known whether the m.l.e. Y/n is admissible under this loss. It would be remarkably surprising if it is a generalized Bayes estimate at all under absolute error loss. The form of a minimax estimate for general n is also not known, and it seems virtually impossible to write down a general form. It would be technically interesting to investigate these basic questions.

ESTIMATION OF $1/P$ AND SEQUENTIAL METHODS

The problem of estimation of $1/p$ arises in many situations. For example, if we know an unbiased estimator of $1/p$, then we are able to obtain an unbiased estimator of the mean of the binomial waiting time distribution, which is also called the negative binomial or Pascal distribution. Estimation of $1/p$ can also help in estimating fish in a tank or wild population in a forest by the capture-recapture method.

More generally, sequential estimation* of a function of p, say $g(p)$, has received a great deal of attention in the literature [27,41,67, 77,80]. The problem is motivated by the fact that only polynomials of degree $\leqslant n$ can be estimated unbiasedly on the basis of an i.i.d. Bernoulli sample of size n. Thus, $1/p$ cannot be estimated unbiasedly by using a sample of a fixed size. So far as the problem of unbiased estimation of $1/p$ is concerned, the relevant work of DeGroot [27] put forward a remarkable solution. He introduced the notion of *efficient sampling plans*, when observations are independently taken on the random variable X with $P[X = 1] = p$ and $P[X = 0] = q = 1 - p$, $0 < p < 1$. A sequential estimation of $g(p)$ is obtained under a sampling plan S and an estimator f.

Let us introduce some useful definitions given by DeGroot [27]: A *sampling plan* is a function S defined on the points γ whose horizontal coordinate u_γ and vertical coordinate v_γ are nonnegative integers, taking only the values 0 and 1, and such that $S(\gamma) = 1$ if $u_\gamma = v_\gamma = 0$. A *path* to γ is a sequence of points $(0,0) = \gamma_0, \gamma_1, \ldots, \gamma_n = \gamma$ such that $S(\gamma_k) = 1$ for $k = 0, 1, \ldots, n-1$, and either $u_{\gamma_{k+1}} = u_{\gamma_k} + 1$, $v_{\gamma_{k+1}} = v_{\gamma_k}$ or $u_{\gamma_{k+1}} = u_{\gamma_k}$, $v_{\gamma_{k+1}} = v_{\gamma_k} + 1$. A point γ is called a *continuation point* if there is a path to γ and $S(\gamma) = 1$, a *boundary point* if there is a path to γ and $S(\gamma) = 0$, and an *inaccessible point* if there is no path to γ. Under a given sequential plan S, a point mass $K(\gamma)p^{v_\gamma}q^{u_\gamma}$ is assigned at every $\gamma \in B$, where B is the set of all boundary points and $K(\gamma)$ is the number of distinct paths to γ. A sampling plan is *closed* if $\sum_{\gamma \in B} K(\gamma)p^{v_\gamma}q^{u_\gamma} = 1$. An *estimator* f is a real function defined on B. The functions defined on B and taking values u_γ, v_γ, and $n_\gamma = u_\gamma + v_\gamma$ are denoted by U_γ, V_γ, and N_γ respectively.

The above sampling plan S can also be explained as a stopping time in a random walk, in which a path can be explained as a record of successive steps of the random walk. In this random walk, one starts from $(0,0)$ and stops at a boundary point γ. It moves one horizontal unit for failure and one vertical unit for success. The point mass $K(\gamma)p^{v_\gamma}q^{u_\gamma}$ that has been assigned to a boundary point γ in the sampling plan S can be explained as the probability for the stopping time to terminate at γ. Any closed stopping time will be finite with probability 1.

DeGroot [27] proved that for any estimator f,

$$Var(f|p) \geqslant pq[g'(p)]^2/E(N|p), \quad (9)$$

where $g(p) = E(f|p)$. This is essentially a sequential Cramér-Rao* inequality. A nonconstant estimator f under a given sampling plan S is said to be *efficient* at p_0 if equality holds when $p = p_0$. If equality holds for all $0 < p < 1$, then the pair (S, f) is called *efficient*. DeGroot proved that an estimator f is efficient if and only if there exist two functions of p, say $a(p)$ and $b(p)$, with $a(p) \neq 0$, such that $f(\gamma) = a(p)(qV - pU) + b(p)$.

DeGroot [27] also showed that the only efficient sampling plans are the single sample plans, in which $B = \{\gamma : N_\gamma = n\}$ for a positive integer n, and the inverse binomial plans, in which either $B = \{\gamma : U_\gamma = c\}$ or $B = \{\gamma : V_\gamma = c\}$ for some positive integer c. He also showed that $f = a + bV$ as efficient estimators of $a + bnp$ are the only efficient estimators in single sample plans, and that $f = a + bN$ as efficient estimators of $a + bc$ $(1/p)$ are the only efficient estimators in the inverse binomial sampling plans by taking $B = \{\gamma : V_\gamma = c\}$. So N/c is the only efficient estimator of $1/p$ in the inverse binomial sampling plans by the same choice of B.

For the single sample plan with $B = \{\gamma : N_\gamma = n\}$, only polynomials in p of degree at most n are unbiasedly estimable. For the inverse binomial plan with $B = \{\gamma : V_\gamma = c\}$, a function $h(q)$ is unbiasedly estimable if and only if it can be expanded in Taylor's series in the interval $|q| < 1$. If $h(q)$ is unbiasedly estimable, then its unique unbiased estimator is given by

$$f(\gamma_k) = \frac{(c-1)!}{(k+c-1)!} \frac{d^k}{dq^k}\left[\frac{h(q)}{(1-q)^c}\right]_{q=0}, \quad (10)$$

for $k = 0, 1, \ldots$, where γ_k is the unique point in B such that $N(\gamma_k) = c + k$.

If a closed plan with $B = \{\gamma = (u_\gamma, v_\gamma)\}$ is such that by changing its points from γ to $\gamma' = (u_\gamma, v_\gamma + 1)$ one gets another closed plan $B' = \{\gamma' = (u_\gamma, v_\gamma + 1)\}$, then $1/p$ is unbiasedly estimable for plan B [41,67]. If $1/p$ is unbiasedly estimable under a sampling plan B, then B must be horizontally unbounded, that is, for any finite u_0, there is a boundary point $\gamma = (u, v)$ so that $u > u_0$. More generally [51, p. 103], if (a, b) is an accessible point, the quantity $p^b q^a$ is unbiasedly estimable and an unbiased estimator depending only on (U_γ, V_γ) is given by $K'(\gamma)/K(\gamma)$, where $K'(\gamma)$ denotes the number of paths passing through (a, b) and terminating in γ and $K(\gamma)$ is the number of paths terminating in γ as before.

INTERVAL ESTIMATION OF P

In many applications, a confidence interval is more useful than a point estimate. Let us denote the confidence level by $1 - \alpha, \alpha \in (0, 1)$. The coverage probability of a nonrandomized confidence interval cannot be a constant since a binomial distribution is a discrete distribution. Neyman's classical confidence interval [54] has exactly $1 - \alpha$ coverage probability, but it is achieved by a randomized procedure.

A textbook standard (confidence) interval for p is given by $\hat{p} \pm z_{\frac{\alpha}{2}} \sqrt{\hat{p}\hat{q}/n}$, where $\hat{q} = 1 - \hat{p}$ and $z_{\frac{\alpha}{2}} = \Phi^{-1}(1 - \alpha/2)$. It is obtained from the normal approximation of the m.l.e. $\hat{p}$ as $\hat{p} \sim N(p, pq/n)$ in which the asymptotic variance pq/n is estimated by $\hat{p}\hat{q}/n$. The standard interval is also known as the *Wald interval*, because it comes from the Wald large-sample test of the binomial proportion. It is well known that the standard interval has a low coverage probability when p is close to 0 or 1, since the normal approximation is poor in these cases. Many people believe that the coverage probabilities of the standard interval would be much closer to the nominal level $1 - \alpha$ when p is not close to 0 or 1. However, this has been shown to be incorrect [14,15]. Since 1990, many articles have pointed out that the coverage probability of the standard interval can be erratically poor even if p is not near the boundaries [2,53,65,76]. This situation has been studied in more detail by Brown et al. [14] in a sequence of articles. The coverage probability could decrease from a value which is close to $1 - \alpha$ to a much lower value than $1 - \alpha$ when p only changes by a small value (see Fig. 1). Even when p is fixed, the coverage probability is not strictly increasing in n. It could drop dramatically to a very low level when n increases (see Fig. 2).

An alternative to the standard interval is the confidence interval based on the solutions to the quadratic equations $(\hat{p} - p)/\sqrt{pq/n} = \pm z_{\alpha/2}$. This interval, the *Wilson (confidence) interval* [79], has the form

$$\frac{Y + z_{\frac{\alpha}{2}}^2/2}{n + z_{\frac{\alpha}{2}}^2} \pm \frac{z_{\frac{\alpha}{2}}\sqrt{n}}{n + z_{\frac{\alpha}{2}}^2} \sqrt{\hat{p}\hat{q} + \frac{z_{\frac{\alpha}{2}}^2}{4n}}; \quad (11)$$

it is derived by inverting the score test [60] for p; so it is also called the *score (confidence) interval*.

Some statistics textbooks recommend the "exact" confidence interval for p, called the *Clopper-Pearson interval* [18], based on inverting an equal-tailed test for the binomial proportion (*see* BINOMIAL PARAMETER,

CLOPPER-PEARSON INTERVAL ESTIMATION). Its lower bound ℓ and upper bound u are the solutions to the equations $1 - F_{n,\ell}(y) = \alpha/2$ and $F_{n,u}(y) = \alpha/2$ when y is not 0 or n. It suggests using the one-sided confidence interval when $y = 0$ or $y = n$. The Clopper-Pearson interval has at least $1 - \alpha$ coverage probability for all p and is easy to calculate by statistical software. The Wilson interval tends to perform much better than the Clopper-Pearson or the Wald intervals.

Blyth and Still [12] considered the connections between a confidence region for p and a family of tests for the simple hypothesis $p = p_0$. They studied the nonrandomized procedures so that the acceptance region is an interval $[A_n(p_0), B_n(p_0)]$. It is required that both $A_n(p_0)$ and $B_n(p_0)$ are monotonely increasing in n for fixed p_0 and in p_0 for fixed n so that a confidence interval can be solved from $\{\ell_n(Y) \leqslant p_0 \leqslant u_n(Y)\} = \{A_n(p_0) \leqslant Y \leqslant B_n(p_0)\}$. The Clopper-Pearson confidence interval can be obtained from the inequalities

$$P_{p_0}[Y < A_n(p_0)] \leqslant \alpha/2,$$
$$P_{p_0}[Y > B_n(p_0)] \leqslant \alpha/2.$$

Sterne [70] suggested making up the acceptance region for $p = p_0$ by including the most probable value of Y, then the next most probable one and so on, until the total probability is greater than $1 - \alpha$. Crow [23] pointed out that although Sterne's acceptance region is always an interval, it does not always give an interval-valued confidence region.

The *Agresti-Coull interval* [2] is a revised version of the standard interval, replacing $\hat{p}$ by $\tilde{p} = (Y + z_{\frac{\alpha}{2}}^2/2)/(n + z_{\frac{\alpha}{2}}^2)$ in the form of the standard interval and leading to the interval $\tilde{p} \pm z_{\frac{\alpha}{2}} \sqrt{\tilde{p}\tilde{q}/n}$. Note that the Agresti-Coull intervals are never shorter than the Wilson intervals. Since both of them are centered at $\tilde{p}$, the Agresti-Coull interval always covers the Wilson interval.

The *Jeffreys interval* is the Bayesian credible interval for p under the noninformative *Jeffreys prior Beta(1/2, 1/2)* distribution for p. Historically, Bayes procedures under noninformative priors have a track record of good frequentist properties. A $(1 - \alpha)$-level Bayesian credible interval has an exact $1 - \alpha$

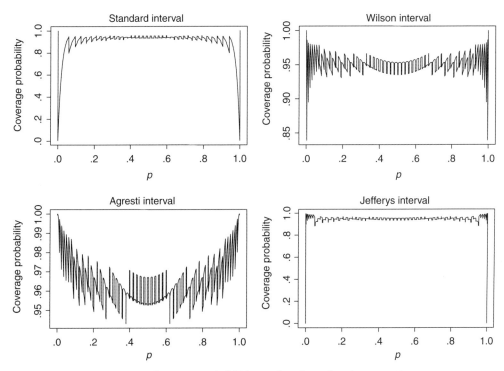

Figure 1. Coverage probabilities as functions of p when $n = 50$.

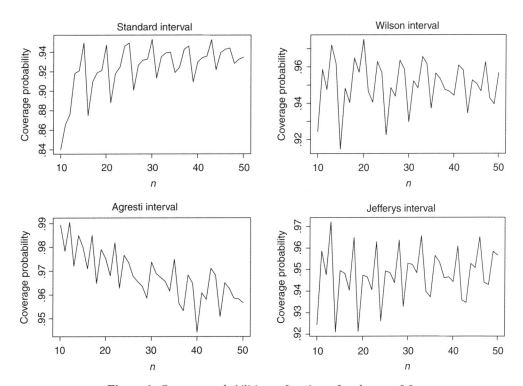

Figure 2. Coverage probabilities as functions of n when $p = 0.3$.

posterior coverage probability. A $100(1 - \alpha)\%$ equal-tailed Jeffreys interval is given by

$$[B(\alpha/2; Y + 1/2, n - Y + 1/2),$$
$$B(1 - \alpha/2; Y + 1/2, n - Y + 1/2)],$$

where $B(\alpha; m_1, m_2)$ denotes the α-quantile of a $Beta(m_1, m_2)$ distribution. A $100(1 - \alpha)\%$ level highest posterior density (HPD) Jeffreys interval, say $[\ell, u]$, can be solved from

$$
\begin{aligned}
B(u; Y &+ 1/2, n - Y + 1/2) \\
-B(\ell; Y &+ 1/2, n - Y + 1/2) = 1 - \alpha, \\
b(u; Y &+ 1/2, n - Y + 1/2) \\
&= b(\ell; Y + 1/2, n - Y + 1/2),
\end{aligned}
\tag{12}
$$

where $b(x; m_1, m_2)$ is the density of a $Beta(m_1, m_2)$ distribution. The equal-tailed Jeffreys credible interval is easy to compute but it has a positive lower bound when $Y = 0$, an upper bound less than 1 when $Y = n$, and a coverage probability approaching 0 when p goes to 0 or 1. Brown, Cai and DasGupta [14] modified the end points of the Jeffreys interval so that its lower bound is 0 when $Y = 0$ and its upper bound is 1 when $Y = n$. The HPD Jeffreys interval is one-sided when $Y = 0$ or n and it automatically transforms from a one-sided interval to a two-sided interval, but it is not easy to compute.

The *likelihood ratio interval* is constructed by the inversion of the likelihood ratio test criterion, which accepts the null hypothesis $H_0 : p = p_0$ if $-2\log(\Lambda_n) \leqslant z_{\alpha/2}^2$, the likelihood ratio being given by

$$
\Lambda_n = \frac{L(p_0)}{\sup_p L(p)} = \frac{p_0^Y (1 - p_0)^{n-Y}}{(Y/n)^Y (1 - Y/n)^{n-Y}},
$$

and L denotes the likelihood function [60]. This interval has nice properties, but it is hard to compute [15].

The *arcsine interval* is based on a widely used variance stabilizing transformation for the binomial distribution [5] but it replaces the m.l.e. $\hat{p}$ by a new estimator $\check{p} = (Y + 3/8)/(n + 3/4)$. This method leads to an approximate $100(1 - \alpha)\%$ confidence interval for p as

$$\sin^2\left(\arcsin\sqrt{\check{p}} \pm \frac{z_{\alpha/2}}{2\sqrt{n}}\right). \tag{13}$$

This interval performs much worse if $\hat{p}$ is used as an estimator of p [14].

The *logit interval* is obtained by inverting a Wald-type interval for the log odds $\theta = \log(p/(1-p))$ [71, p. 667]. So $\hat{\theta} = \log(\hat{p}/(1-\hat{p}))$, leading to an approximate $100(1 - \alpha)\%$ confidence interval for p,

$$\left[\frac{e^{\theta_\ell}}{1 + e^{\theta_\ell}}, \frac{e^{\theta_u}}{1 + e^{\theta_u}}\right], \tag{14}$$

where $\hat{\theta}_\ell = \hat{\theta} - z_{\alpha/2}\sqrt{\hat{V}}$, $\hat{\theta}_u = \hat{\theta} + z_{\alpha/2}\sqrt{\hat{V}}$ and $\hat{V} = n/[Y(n - Y)]$. The logit interval has no solutions when $Y = 0$ or $Y = n$, but it can be revised by adding positive values to both the numerator Y and the denominator $n - Y$ to complete its definition.

The coverage probabilities of the Wilson interval and the Jeffreys interval are close to $1 - \alpha$ except for p near 0 or 1. The coverage probabilities are still spiky; the spikes exist for all n and α, but they can be removed by using a one-sided Poisson approximation for y close to 0 or n [14]. The resulting confidence intervals are called the *modified Wilson interval* and the *modified Jeffreys interval*, respectively. The coverage probabilities of the modified Wilson and modified Jeffreys interval are somewhat better when p is close to 0 or 1. The coverage probability of the arcsine interval could be close to 0 when p is close to 0 or 1. The logit interval performs quite well when p is away from 0 and 1, but it is unnecessarily long and its expected length is larger than even that of the Clopper-Pearson exact interval [14].

The coverage probabilities of all of these intervals oscillate erratically even for large n, and the actual coverage can be very significantly smaller than the claimed nominal value. Brown, Cai, and DasGupta [15] report (see Table 1) the smallest n when $\alpha = 0.05$ after which the coverage probability stays at 0.93 or above for selected values of p, for the standard interval, the Jeffreys interval, the Wilson interval, and the Agresti-Coull interval, denoted by n_s, n_J, n_W and n_{AC} respectively. From Table 1, one may think that the Agresti-Coull interval is the obvious interval of choice. However, since it tends to be longer than the other intervals, the Agresti-Coull interval may not be the most desirable.

Table 1. Smallest n After which the Coverage Stays at 0.93 or Above

p	0.01	0.025	0.05	0.1	0.15	0.2	0.25	0.3	0.35	0.4	0.45	0.5
n_s	2757	1005	526	286	141	118	89	82	56	64	54	71
n_J	956	384	134	47	62	32	36	25	29	22	22	32
n_W	407	42	111	36	24	41	13	31	18	7	13	32
n_{AC}	1	5	10	11	8	4	1	1	10	1	1	32

Interestingly, however, analytical lower bounds on the coverage probability can be obtained for some of these intervals, including the Wilson interval. For example (see PROBABILITY INEQUALITIES FOR SUMS OF BOUNDED RANDOM VARIABLES, using the Bennett-Hoeffding inequality [8,46] given by

$$P(\sqrt{n}|\frac{Y}{n} - p| \geq \lambda) \leqslant 2\exp\left[-\frac{\lambda^2}{2p}\psi\left(\frac{\lambda}{p\sqrt{n}}\right)\right],$$

$$0 < \lambda,\ 0 < p < 1, n \geqslant 1,$$

where $\psi(x) = 2[(1+x)\log(1+x) - x]/x^2$, Das-Gupta and Zhang [26] show that the coverage probability of the nominal 99% Wilson interval is greater than or equal to 0.8140 for any $n \geqslant 100$ if $p = 0.05$. Likewise, if $p = 0.1$, an analytical lower bound is 0.8225 for any $n \geqslant 100$. They also obtain alternative analytical lower bounds by using martingale inequalities (see Refs. 4, 11). It would be useful but probably difficult to derive better analytical lower bounds on the coverage probability of the Wilson and other important confidence intervals for p.

To summarize, the detailed results in Brown, Cai, and DasGupta [15] show that the pure bias in the coverage is the smallest for the Jeffreys interval, the average oscillation in the coverage the smallest for the Agresti-Coull interval, and the average length the smallest for the Jeffreys and the likelihood ratio interval. The Wilson interval ranks second in both bias and oscillation. It is easy to compute. A general recommendation is the Wilson interval for small n (such as $n \leqslant 40$). When $n > 40$, the Agresti-Coull interval is recommended, because of computational ease and because the benefits of the Wilson interval become marginal.

In certain applications, one-sided confidence bounds are more relevant. Similar complexities arise there too, but Hall [45] gives a method to obtain more accurate one-sided confidence bounds for p, using a continuity correction.

HYPOTHESIS TESTING

In some applications, it is of interest to test $H_0 : p = p_0$ versus $H_1 : p \neq p_0$ for $p_0 \in (0, 1)$ (see HYPOTHESIS TESTING). In this case, a uniformly most powerful (UMP) test does not exist since the alternative is two-sided [50, p. 138], but an α-level UMP unbiased test exists, the test function being

$$\phi(y) = \begin{cases} 1, & y < C_1 \text{ or } y > C_2 \\ \gamma_i, & y = C_i, i = 1, 2 \\ 0, & C_1 < y < C_2, \end{cases}$$

where $C_1, C_2, \gamma_1, \gamma_2$ are uniquely determined by

$$\begin{aligned} &F_{n,p_0}(C_2 - 1) - F_{n,p_0}(C_1) + \\ &\sum_{i=1}^{2}(1 - \gamma_i)f_{n,p_0}(C_i) = 1 - \alpha, \\ &F_{n-1,p_0}(C_2 - 2) - F_{n-1,p_0}(C_1 - 1) \\ &+ \sum_{i=1}^{2}(1 - \gamma_i)f_{n-1,p_0}(C_i - 1) = 1 - \alpha. \end{aligned} \quad (15)$$

The two-sided UMP unbiased test has randomized components at $y = C_1$ and $y = C_2$, and its power function $E_p[\Phi(Y)]$ is globally minimized at $p = p_0$ with minimum value α.

Sometimes it is also of interest to test $H_0 : p \leqslant p_0$ versus $H_1 : p > p_0$ or $H_0 : p \geqslant p_0$ versus $H_1 : p < p_0$. UMP tests for both problems exist [50, p. 93]. By symmetry, it is enough to exhibit the first case, where the test function of the α-level UMP test is given by

$$\phi(y) = \begin{cases} 1, & y > C, \\ \gamma, & y = C, \\ 0, & y < C, \end{cases}$$

and C and γ are uniquely determined by

$$\gamma f_{n,p_0}(C) + [1 - F_{n,p_0}(C)] = \alpha. \quad (16)$$

This test also has a randomized component at $y = C$. Its power function is less than or equal to α when $p \leqslant p_0$ and greater than α when $p > p_0$.

A standard test statistic,

$$Z = \frac{\hat{p} - p_0}{\sqrt{p_0 q_0 / n}},$$

is derived from the normal approximation of $\hat{p}$ if $p = p_0$. The α-level two-sided standard test rejects the null hypothesis $H_0 : p = p_0$ if $|Z| > z_{\frac{\alpha}{2}}$. The α-level one-sided standard test rejects the null hypothesis $H_0 : p \leqslant p_0$ if $Z > z_\alpha$ and rejects the null $H_0 : p \geqslant p_0$ if $Z < -z_\alpha$. The power function of the standard as well as the UMPU test can be poor when n is not large or the alternative nears the null values. Their power curves cross. Also, their type I errors could be seriously biased from α (see Fig. 3 for details).

Two-sample tests for binomial distributions are very useful in many applications, in which the observations Y_1 and Y_2 are independently observed and distributed as $Bin(m, p_1)$ and $Bin(n, p_2)$, respectively, where m and n are known but p_1 and p_2 are not. It is of interest to test either $H_0 : p_1 = p_2$ versus $H_1 : p_1 \neq p_2$ or $H_0 : p_1 \leqslant p_2$ versus $H_1 : p_1 > p_2$. The solutions are extensively useful in medical sciences, political science, and social sciences.

Under $p_1 = p_2 = p$, an exact distribution that is free of the unknown parameter p results from conditioning on the marginal frequencies $m, n, y_1 + y_2$, and $m + n - y_1 - y_2$ in both samples. Then, we obtain the PMF of the hypergeometric distribution*

$$f(y_1 | y_1 + y_2, m, n) = \binom{m}{y_1}\binom{n}{y_2} / \binom{m+n}{y_1 + y_2},$$

for $\max(0, y_1 + y_2 - n) \leqslant y_1 \leqslant \min(m, y_1 + y_2)$. To test $H_0 : p_1 = p_2$, the P-value is the sum of $f(y_1 | y_1 + y_2, m, n)$ over sample values

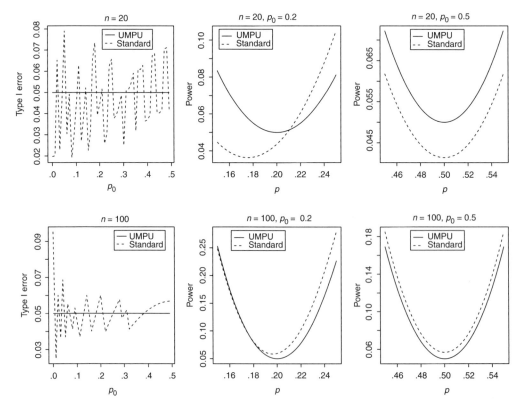

Figure 3. Type I error $E_{p_0}[\phi(Y)]$ as functions of p_0 for selected n and test power $E_p[\phi(Y)]$ as functions of p for selected p_0 and n when $H_0 : p = p_0$ versus $H_1 : p \neq p_0$ and $\alpha = 0.05$.

at least as favorable to the alternative hypothesis as the observed y_1. This test of H_0 is *Fisher's exact test** [32,33].

A textbook large-sample test procedure is obtained from the central limit theorem for $\hat{p}_1 - \hat{p}_2$:

$$\hat{p}_1 - \hat{p}_2 \overset{L}{\to} N\left(p_1 - p_2, \frac{p_1 q_1}{m} + \frac{p_2 q_2}{n}\right),$$

as $\min\{m, n\} \to \infty$, where $\hat{p}_1 = Y_1/m$, $\hat{p}_2 = Y_2/n$, $q_1 = 1 - p_1$, and $q_2 = 1 - p_2$. The asymptotic variance can be estimated either by $\hat{p}_1\hat{q}_1/m + \hat{p}_2\hat{q}_2/n$ or by $\hat{p}\hat{q}(1/m + 1/n)$, where $\hat{p} = (Y_1 + Y_2)/(m + n)$, $\hat{q} = 1 - \hat{p}$, which give the test statistic

$$Z_1 = \frac{\hat{p}_1 - \hat{p}_2}{\sqrt{\hat{p}\hat{q}(1/m + 1/n)}},$$

or

$$Z_2 = \frac{\hat{p}_1 - \hat{p}_2}{\sqrt{\hat{p}_1\hat{q}_1/m + \hat{p}_2\hat{q}_2/n}},$$

respectively. Then, the α-level textbook large-sample test rejects $H_0 : p_1 = p_2$ if $|Z_1| \leqslant z_{\frac{\alpha}{2}}$ or $|Z_2| \leqslant z_{\frac{\alpha}{2}}$ and rejects $H_0 : p_1 \leqslant p_2$ if $Z_1 > z_\alpha$ or $Z_2 > z_\alpha$ respectively.

For $H_0 : p_1 = p_2$, Fisher's exact test is the UMPU test and for $H_0 : p_1 \leqslant p_2$ it is the UMP test. Even though the power of the test based on Z_1 is less than that of the test based on Z_2 [28], the Type I error of Z_2 is higher than that of Z_1 [21]. Cressie [21] compares the tests based on Z_1 and Z_2 and concludes that in some cases Z_1 is to be preferred and in other cases Z_2 is to be preferred.

Another large-sample test is constructed by using the logarithm of the odds ratio*, say $\log\theta = \log[(p_1/q_1)/(p_2/q_2)]$, where θ is the odds ratio. Here $p_1 = p_2$ is equivalent to $\log\theta = 0$, and the m.l.e. of θ is

$$\hat{\theta} = \frac{Y_1/(m - Y_1)}{Y_2/(n - Y_2)}.$$

The asymptotic variance of $\log\hat{\theta}$ is estimated [1, p. 54] by

$$\hat{\sigma}_{\log\hat{\theta}}^2 = \widehat{Var}(\log\hat{\theta})$$

$$= \frac{1}{Y_1} + \frac{1}{m - Y_1} + \frac{1}{Y_2} + \frac{1}{n - Y_2}.$$

Then, the Wald large-sample test rejects $H_0 : p_1 = p_2$ if $|\log\hat{\theta}/\hat{\sigma}_{\log\hat{\theta}}| > z_{\alpha/2}$ and rejects $H_0 : p_1 < p_2$ if $\log\hat{\theta}/\hat{\sigma}_{\log\hat{\theta}} > z_\alpha$. Gart and Zweigel [38] compared many modifications of θ in their paper and concluded that adding 0.5 to all cell frequencies is useful in the event that some of the frequencies are 0.

A Pearson χ^2 statistic defined by

$$X^2 = (m + n)\frac{[Y_1(n - Y_2) - Y_2(m - Y_1)]^2}{(Y_1 + Y_2)(m + n - Y_1 - Y_2)mn} \tag{17}$$

as well as several corrections of it, such as Pearson's adjusted X^2 [59], Yates's correction [82], and Cook's correction [74]), are studied by Upton [74]. In studying the power and size of rejection region for each of these tests, Upton concluded that if the set of data being analyzed cannot be regarded as a random sample from the population(s) of interest, then only the exact test or Yates' corrected X^2 approximation to the exact test are appropriate. In general, a reasonably good test suggested by Upton is a slight improvement on X^2, via

$$X_U^2 = (m + n - 1)$$
$$\times \frac{[Y_1(n - Y_2) - Y_2(m - Y_1)]^2}{(Y_1 + Y_2)(m + n - Y_1 - Y_2)mn}. \tag{18}$$

Comparisons of some corrections form the χ^2 test for this problem in literature can be found in References 43 and 83. After a few steps of trivial computations, one can see that the Pearson chi-square statistic X^2 is in fact equal to Z_1^2.

The problem of testing the homogeneity of several binomial proportions was first addressed by Pearson [58]. In this problem, we assume that $Y_i, i = 1, \ldots, k$, are independently observed and distributed as $Bin(n_i, p_i)$, respectively, $n_1, \ldots, n_k$ being known but $p_1, \ldots, p_k$ being unknown. It is of interest to test $H_0 : p_1 = \ldots = p_k$ versus H_1 : not all of $p_1, \ldots, p_k$ are equal.

Under H_0, a traditional Pearson χ^2 (chi-square) statistic defined by

$$X^2 = \sum_{i=1}^{k}\left[\frac{(Y_i - n_i\hat{p})^2}{n_i\hat{p}} + \frac{(n - Y_i - n_i\hat{q})^2}{n_i\hat{q}}\right], \tag{19}$$

and a likelihood-ratio statistic defined by

$$G^2 = \sum_{i=1}^{k}$$

$$= 2 \sum_{i=1}^{k} \left[(n_i - Y_i) \log \frac{\hat{q}_i}{\hat{q}} + Y_i \log \frac{\hat{p}_i}{\hat{p}} \right], \quad (20)$$

where $\hat{p}_i = Y_i/n_i$, $\hat{q}_i = 1 - \hat{p}_i$, $\hat{p} = \sum_{i=1}^{k} Y_i / \sum_{i=1}^{k} n_i$ and $\hat{q} = 1 - \hat{p}$, are approximately distributed as χ_{k-1}^2 if $\min_i(n_1, n_2, \ldots, n_k)$ is large. So, H_0 is rejected in the corresponding test if X^2 or G^2 is greater than $\chi_{\alpha,k-1}^2$, where $\chi_{\alpha,m}^2$ denotes the $(1 - \alpha)$-quantile of the χ^2 distribution with m degrees of freedom.

For practical validity of the χ^2 as the limiting distribution of X^2 in large samples, it is customary to recommend, in applications of the test, that the smallest expected frequency in any class should be 10 or (with some writers) 5. If this requirement is not met in the original data, combination of classes is recommended [19]. If X^2 has less than 30 degrees of freedom and the minimum expectation is 2 or more, use of the ordinary χ^2 tables is usually adequate. If X^2 has more than 30 degrees of freedom, it tends to become normally distributed, but when the expectations are low, the mean and variance are different from those of the χ^2 distribution [20]. Yarnold [81] studied an Edgeworth expansion of $P(X^2 \geqslant c)$ and concluded that one should never use the incomplete gamma or continuous Edgeworth approximations since the χ^2 approximation is as accurate or better. Agresti [1, p. 49] points out that for fixed k, X^2 usually converges more quickly than G^2. The usual χ^2 approximation is poor for G^2 if $\sum_{i=1}^{k} n_i/2k < 5$. When k is large, it can be usable for X^2 for $\sum_{i=1}^{k} n_i/2k$ as small as 1.

For I independent random variables $Y_1, \ldots, Y_I$ distributed as binomial with parameters n_i and p_i, if Y_i is observed with covariates $X_{i,1}, \ldots, X_{i,p-1}$, $p \leqslant I$, then a regular logistic regression* model could be fitted as

$$\log \frac{p_i}{1 - p_i} = \beta_0 + \sum_{j=1}^{p-1} X_{i,j}\beta_j, \quad (21)$$

where $i = 1, \ldots, I$. The loglikelihood function for the logistic regression model is strictly concave, and the m.l.e. $\hat{\beta}$ of $\beta = (\beta_0, \ldots, \beta_{p-1})$ exists and is unique except in certain boundary cases [66, p. 120, 78]. Since the likelihood equations of a logistic model are nonlinear functions of $\hat{\beta}$, computation requires a numerical iterative method to obtain the solutions [1, p. 114]. The Wald large-sample test, the score test, and the likelihood ratio test can be constructed for the logistic regression model as well.

POINT ESTIMATION OF n

Estimating the parameter n on the basis of independent success counts $Y_1, \ldots, Y_k$ from a binomial distribution with unknown n and p was discussed many years ago by Fisher [34] and Haldane [44]. Fisher's argument for using $Y_{(k)}$ as an estimator of n was that $Y_{(k)} \overset{a.s.}{\to} n$ as $k \to \infty$, where $Y_{(k)}$ is the sample maximum, but the speed of convergence is very slow when p is small even when n is not large. $Y_{(k)}$ is atrociously downward biased and essentially worthless without a serious bias correction (see Fig. 4 for details). Using inequalities in van Zwet [75], Das-Gupta and Rubin [25] give a very effective method for bias correction of $Y_{(k)}$, without requiring the jackknife*. They also show that no functions of just n or just p are unbiasedly estimable. The maximum likelihood equations when both n and p are unknown are presented by Haldane [44], but the question of the uniqueness of the solution was not answered. Sensitive dependence on a small number of observations is characteristic of the problem. Thus, although the problem sounds innocuous, it is a notoriously hard one.

Since $\mu = np$ and $\sigma^2 = np(1 - p)$ are the mean and variance of the $Bin(n, p)$ distribution, the method of moments* gives $\hat{p} = \hat{\mu}/\hat{n}$ and $\hat{n} = \hat{\mu}^2/(\hat{\mu} - \hat{\sigma}^2)$, where $\hat{\mu} = \overline{Y}$ and $\hat{\sigma}^2 = \sum_{i=1}^{k}(Y_i - \overline{Y})^2/k$. So this method estimates n via $\hat{n}_{MME} = \hat{n}$.

The moment estimator $\hat{n}_{MME}$ could be very unstable when the sample mean $\hat{\mu}$ is close to the sample variance $\hat{\sigma}^2$ and is even negative when $\hat{\mu} < \hat{\sigma}^2$. A moment-stabilized estimator has been proposed [56] as

$$\hat{n}_{MME:S} = \max\{\hat{\sigma}^2 \phi^2/(\phi - 1), Y_{(k)}\},$$

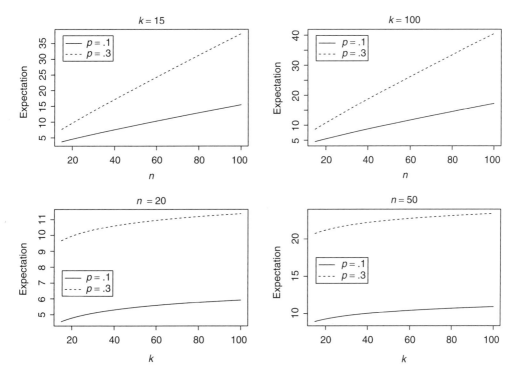

Figure 4. Expected values of $Y_{(k)}$ as functions of n (upper panel) and as functions of k (lower panel) for selected p.

where

$$\phi = \begin{cases} \hat{\mu}/\hat{\sigma}^2, & \text{if } \hat{\mu}/\sigma^2 \geqslant 1 + 1/\sqrt{2}, \\ \max\{(Y_{(k)} - \hat{\mu})/ & \\ \quad \hat{\sigma}^2, 1 + \sqrt{2}\}, & \text{if } \hat{\mu}/\hat{\sigma}^2 < 1 + 1/\sqrt{2}. \end{cases}$$

(22)

The loglikelihood function

$$\ell(n, p) = \sum_{i=1}^{k} \log \left(\frac{n!}{Y_i!(n - Y_i)!} \right)$$

$$+ \sum_{i=1}^{k} Y_i \log \left(\frac{p}{1-p} \right) + kn \log(1 - p)$$

is maximized at $n = \infty$ when $\hat{\mu} \leq \hat{\sigma}^2$ and at a finite $n \geqslant Y_{(k)}$ when $\hat{\mu} > \hat{\sigma}^2$. On the basis of this fact, a stabilized maximum likelihood estimator of n is proposed [56] as

$$\hat{n}_{MLE:S}$$

$$= \begin{cases} MLE, & \text{if } \hat{\mu}/\hat{\sigma}^2 \geqslant 1 + 1/\sqrt{2}, \\ Y_{(k)} + \frac{k-1}{k}(Y_{(k)} & \\ \quad - Y_{(k-1)}), & \text{if } \hat{\mu}/\hat{\sigma}^2 < 1 + 1/\sqrt{2}. \end{cases}$$

(23)

Carroll and Lombard [16] proposed an estimator of n, called the *C-L estimator*, by maximizing an integrated likelihood function of n restricted to $n \geqslant Y_{(k)}$, obtained by integrating p out with respect to a $Beta(a, b)$ prior for p. The integrated likelihood function then is only a function of n; it is given by

$$L(n) = \left\{ \prod_{i=1}^{k} \binom{n}{Y_i} \right\}$$

$$\times \left[(kn + a + b - 1) \right.$$

$$\left. \times \binom{kn + a + b - 2}{a - 1 + \sum_{i=1}^{k} Y_i}^{-1} \right].$$

(24)

Since $\lim_{n \to \infty} L(n) = 0$ if $a \geqslant 1$ and $b \geqslant 1$, $L(n)$ is maximized at a finite value of n, but Carroll and Lombard did not prove the uniqueness of the maximum. They suggested either using $a = b = 1$ or $a = b = 2$. The estimate requires a lot of care to compute.

DasGupta and Rubin [25] proposed two new estimators of n. One is a new moment estimator, given by

$$\hat{n}_1 = \max\{\frac{Y_{(k)}^{\alpha+1}(\hat{\sigma}^2)^{\alpha}}{\hat{\mu}^{\alpha}(Y_{(k)} - \hat{\mu})^{\alpha}}, Y_{(k)}\}, \qquad (25)$$

when $\hat{\sigma}^2 > 0$ and $\hat{n}_1 = Y_{(k)}$ when $\hat{\sigma}^2 = 0$. The estimator $\hat{n}_1$ is derived from the identity

$$n = n^{\alpha+1}(npq)^{\alpha}/[(np)^{\alpha}(nq)^{\alpha}]$$

via estimating np by $\hat{\mu}$, nq by $Y_{(k)} - \hat{\mu}$ and n by $Y_{(k)}$ on the right side of the identity. The other is a *bias-corrected estimator* given by

$$\hat{n}_2 = Y_{(k)} + \sum_{i=1}^{[\hat{n}_1]-2} B(\frac{1}{k}; i+1, [\hat{n}_1] - i), \quad (26)$$

and is derived from a bias correction of $Y_{(k)}$.

The C-L estimator is at least competitive with, and in some cases superior to, the stabilized moment and the stabilized maximum likelihood estimators [16]. DasGupta and Rubin [25] compared the biases and the mean-squared errors (MSEs) of $Y_{(k)}$, the C-L estimator, $\hat{n}_1$ and $\hat{n}_2$, for small and moderate p with respect to small n, moderate n and large n. They concluded that $Y_{(k)}$ is seriously downward biased if p is not very large. The C-L estimator is also downward biased if p is small and the bias is reduced significantly if p is not small. The estimators $\hat{n}_1$ and $\hat{n}_2$ are downward biased for small p, but they could be biased upwards for moderate or large p. On the basis of their study on the MSEs, Das-Gupta and Rubin [25] concluded that when p is small, the new moment estimator $\hat{n}_1$ is the best unless n is small, in which case the C-L estimator is the best. When p is not small, the bias-corrected estimator $\hat{n}_2$ is the best. Further research is needed in this problem. The software and capture-recapture literature has also addressed this problem. It is usually not assumed in that context that $Y_1, \ldots, Y_n$ are i.i.d. See [7,6] for a selection of references and certain computational Bayesian approaches, and Reference 24 for a discussion of empirical Bayes methods.

For the problem of finding an admissible estimator of n when p is known see Reference 39, and Reference 63 for practical applications to animal counting.

Again, let $Y \sim Bin(n,p)$ be the counts, where p is known but n is not. If the parameter space is $N = \{0, 1, 2, \ldots\}$, then $T^0 = Y/p$ is the m.l.e. and the only unbiased estimator of n [63]; further [39], T^0 is admissible and minimax under quadratic loss. If $n = 0$ is excluded, T^0 is no longer admissible, but an adjusted estimator T^* defined by

$$T^*(y) = \begin{cases} -q/(p \log p), & y = 0, \\ y/p, & y > 0, \end{cases}$$

is admissible under quadratic loss [64], and is the only admissible adjustment when $p \geqslant 1/2$.

MULTINOMIAL PARAMETERS

In n independent trials each with I possible outcomes, let the probability of the ith outcome be p_i. If Y_i denotes the number of trials resulting in outcome i, $i = 1, \ldots, I$, then the joint distribution of $\mathbf{Y} = (Y_1, \ldots, Y_I)$ follows the *multinomial distribution** with parameter n and probabilities $\mathbf{p} = (p_1, p_2, \ldots, p_I)$, denoted by $\mathbf{Y} \sim M(n; p_1, \ldots, p_I)$, where $p_i \geqslant 0$, $i = 1, \ldots, I$, and $\sum_{i=1}^{I} p_i = 1$. The PMF of Y is

$$L(\mathbf{p}|\mathbf{y}) = P(Y_1 = y_1, \ldots, Y_I = y_I)$$
$$= \frac{n!}{y_1! \cdots y_I!} p_1^{y_1} \cdots p_I^{y_I}, \qquad (27)$$

where $\mathbf{y} = (y_1, \ldots, y_I)$, y_i's are nonnegative integers and satisfy $\sum_{i=1}^{I} y_i = n$. Let $\theta_i = \log(p_1/p_I)$, $i = 1, \ldots, I - 1$. Then the PMF of Y can be written in the form of an exponential family as

$$P(Y_1 = y_1, \ldots, Y_I = y_I)$$
$$= \frac{n!}{\prod_{i=1}^{I} y_i!} \left(1 + \sum_{i=1}^{I-1} e^{\theta_i}\right)^{-n} \exp\left(\sum_{i=1}^{I-1} y_i \theta_i\right).$$
$$(28)$$

The maximum likelihood estimator of $\mathbf{p}$ is $\hat{\mathbf{p}} = (\hat{p}_1, \ldots, \hat{p}_I) = (Y_1/n, \ldots, Y_I/n)$.

Consider the Bayes estimator of the vector $\mathbf{p}$ when the prior for $\mathbf{p}$ is the Dirichlet distribution** with density

$$\pi_{a_1, \ldots, a_I}(p_1, \ldots, p_I) = \frac{\Gamma(\sum_{i=1}^{I} a_i)}{\prod_{i=1}^{I} \Gamma(a_i)} \prod_{i=1}^{I} p_i^{a_i - 1},$$
$$(29)$$

for $a_i > 0$, $0 \leqslant p_i \leqslant 1$ and $\sum_{i=1}^{I} p_i = 1$. The Bayes estimator of p_i under squared error loss $L(\mathbf{p}, \delta) = \sum_{i=1}^{I}(p_i - \delta_i)^2$, where δ_i is the ith component of an estimator δ, is $\hat{\mathbf{p}}_{a_1,\dots,a_I}$ where its i-th component is given [51, p. 349] by

$$\hat{p}_{i;a_1,\dots,a_I}(\mathbf{Y}) = \frac{a_i + Y_i}{n + \sum_{i=1}^{I} a_i}. \tag{30}$$

The Bayes estimator given by Equation 30 can be written in the form

$$\hat{p}_{i;a_1,\dots,a_I} = \alpha \hat{p}_i + \beta_i,$$

where $\alpha = 1/(1 + \sum_{i=1}^{I} a_i/n)$ and $\beta_i = a_i/(n + \sum_{i=1}^{I} a_i)$. On the other hand, if an estimator $\delta = (\delta_1, \dots, \delta_I)$ of $\mathbf{p}$ can be written via the form $\delta_i = \alpha_i \hat{p}_i + \beta_i$, then the α_i must be all equal, so that $\sum_{i=1}^{I} \delta_i = 1$ for all combinations of $\hat{p}_i$ that satisfy $\sum_{i=1}^{I} \hat{p}_i = 1$. Suppose that $\alpha_1 = \dots = \alpha_I = \alpha$. Then, if $0 \leqslant \alpha < 1$, $0 \leqslant \beta_i \leqslant 1$, and $\alpha + \sum_{i=1}^{I} \beta_i = 1$, there exist $a_1, \dots, a_I > 0$, such that δ is the Bayes estimator under squared-error loss and prior $\pi_{a_1,\dots,a_I}$ for $\mathbf{p}$. So, δ is an admissible estimator of $\mathbf{p}$ under squared-error loss. Further, a linear estimator δ of $\hat{\mathbf{p}}$ is admissible under squared-error loss if and only if it can be written in the form $\delta_i = \alpha \hat{p}_i + \beta_i$, where $0 \leqslant \alpha \leqslant 1, 0 \leqslant \beta_i \leqslant 1$ and $\alpha + \sum_{i=1}^{I} \beta_i = 1$. So, $\hat{\mathbf{p}}$ is admissible. Characterization of all admissible estimates by a stepwise Bayesian argument is given in Reference 13. Earlier literature includes References 3, 57.

Curious admissibility phenomena are plentiful, as always. Consider the admissibility of an estimator of $\max_i p_i$ under squared-error loss. For $I = 2$, the estimator $\max_i(Y_i/n)$ is admissible if $n = 1, 2, 3, 4, 5, 7$. For $n \geqslant 6$ and n even, curiously, it is inadmissible. Likewise, if $I = 3$ and $n \geqslant 12$ and a multiple of 3, then $\max_i(Y_i/n)$ is inadmissible [47].

Let $a_1 = \dots = a_I = a$ in Equation 30. Then, the estimator reduces to

$$\hat{p}_{i;a}(\mathbf{Y}) := \hat{p}_{i;a,\dots,a} = a + Y_i/(n + aI). \tag{31}$$

Since the estimator with components as in Equation 31 has a constant risk if $a = \sqrt{n}/I$, the estimator $\hat{\mathbf{p}}_{a_1,\dots,a_I}$ is minimax.

Consider next the simple testing problem $H_0 : \mathbf{p} = \mathbf{p}_0$ versus $H_1 : \mathbf{p} \neq \mathbf{p}_0$, where $\mathbf{p} = (p_1, \dots, p_I)$, $\mathbf{p}_0 = (p_{10}, \dots, p_{I0})$ are probability vectors and $\mathbf{p}_0$ is known. The problem

is related to the problem of goodness of fit. A special case is $H_0 : p_1 = \dots = p_I$ with $\mathbf{p}_0 = (1/I, \dots, 1/I)$. An α-level *likelihood ratio test* rejects H_0 if $G^2 \leqslant \chi^2_{\alpha, I-1}$, where the likelihood-ratio statistic G^2 is defined as

$$G^2 = 2[\log(\mathbf{p}; \mathbf{y}) - \log(\mathbf{p}_0; \mathbf{y})]$$

$$= 2 \sum_{i=1}^{I}[y_i \log y_i - y_i \log(np_{i0})]. \tag{32}$$

The most well known test for the problem is the Pearson χ^2 test. The great virtue of the Pearson χ^2 test is its versatility, in the context of goodness of fit*. An α-level *Pearson χ^2 test* rejects H_0 if $X^2 < \chi^2_{\alpha, I-1}$, where X^2 could be either the *usual Pearson χ^2* statistic defined by

$$X^2 = \sum_{i=1}^{I} \frac{(Y_i - np_{i0})^2}{np_{i0}}, \tag{33}$$

or the *Neyman modified Pearson χ^2* statistic [55] defined by

$$X^2 = \sum_{i=1}^{I} \frac{(Y_i - np_{i0})^2}{Y_i} \tag{34}$$

(*see* CHI-SQUARE TEST—I). Yet another well known alternative is the Freeman-Tukey test* [35]. It rejects H_0 if $T^2 \leqslant \chi^2_{\alpha, I-1}$, where

$$T^2 = \sum_{i=1}^{I} \left[\sqrt{Y_i} + \sqrt{Y_i + 1} - \sqrt{1 + 4np_{i0}} \right]^2. \tag{35}$$

Those tests are constructed from the properties that G^2, X^2, and T^2 are approximately distributed as a χ^2 distribution with $I - 1$ degrees of freedom under H_0 if n is large. The χ^2 approximations for X^2 and G^2 are usually not bad when $\min(np_{10}, \dots, np_{I0}) \geqslant 5$ [19]. Cochran [20] also concluded that with more than 1 degree of freedom, if relatively few np_{i0} are less than 5, $\min(np_{i0}) = 1$ is not very bad. For small expected values, the Pearson statistic does not follow χ^2 well [19,73,81], and simulation indicates that the likelihood ratio statistic would be better in such situations. For certain models, T^2 may have true significance levels closer to the nominal level than does either X^2 or G^2 [72]. On the basis of a criterion of the closeness of the small sample distribution to the asymptotic χ^2 approximation, X^2 is the most desirable [49]. Both G^2 and T^2 yield too many rejections under the null distribution, that is, their type I error rates are too high.

Cressie and Read [22] propose a class of goodness-of-fit statistics called *Power Divergence (PD)*. The PD class is defined via

$$PD(\lambda) = \frac{2}{\lambda(\lambda+1)}$$
$$\times \sum_{i=1}^{I} \left\{ y_i \left[\left(\frac{y_i}{np_{i0}} \right)^{\lambda} - 1 \right] \right\}, \quad \lambda \neq 0,$$
(36)

and $PD(0) = \lim_{\lambda \to 0} PD(\lambda)$. It can be seen from Equation 36 that $PD(0) = G^2$ given by Equation 32, $PD(1) = X^2$ given by Equation 33, $PD(-1/2) = T^2$ given by Equation 35, and $PD(-2) = X^2$ given by Equation 34. They show that the $PD(\lambda)$ class is approximately distributed as χ^2_{k-1} for a given λ under H_0. Cressie and Read explore how well the moments of $PD(\lambda)$ statistics fit those of χ^2 and propose a new goodness-of-fit statistic as $PD(2/3)$, which appears as an excellent and compromising alternative to G^2 and X^2. Read [62] proposes another class of goodness-of-fit statistics by extending the Freeman-Tukey T^2. He compares it to many of the test statistics introduced above and shows by simulation that it performs favorably when $\mathbf{p}_0 = (1/I, \ldots, 1/I)$. Reference [66, p. 64–70] gives comparative information about most of these tests.

Testing problems with nuisance parameters* are of more interest than the simple testing problem. In such problems, $\mathbf{p}_0$ is a function of certain parameters $\theta_1, \ldots, \theta_s$, $s < I - 1$, which can be estimated from samples from a distribution of a given form with parameters $\theta_1, \ldots, \theta_s$. The estimates can be obtained from either: (i) the frequencies of specified cells or (ii) the original observations. Under suitable regularity conditions, the asymptotic distribution of X^2 when $\mathbf{p}_0$ is estimated from just the cell frequencies is χ^2_{I-s-1} [55]. The asymptotic distribution of X^2 when $\mathbf{p}_0$ is estimated using all the data is not a χ^2 distribution. The asymptotic distribution depends, in general, on the population

from which the samples are obtained, on the true value of the underlying parameters $\theta_1, \ldots, \theta_s$, and (to the surprise of many) even the exact method used to estimate $\theta_1, \ldots, \theta_s$. See Reference 31. If m.l.e.s are used to estimate $\theta_1, \ldots, \theta_s$, then the asymptotic distribution lies stochastically between χ^2_{I-s-1} and χ^2_{I-1} and has the same distribution as

$$\sum_{i=1}^{I-s-1} X_i^2 + \sum_{i=I-s}^{I-1} \lambda_i X_i^2, \qquad (37)$$

where X_i^2 are χ^2_1 i.i.d. distributed for $1 \leqslant i \leqslant I - 1$ and $0 \leqslant \lambda_i < 1$ for $I - s \leqslant i \leqslant I - 1$, and they depend on the true values of $\theta_1, \ldots, \theta_s$, [17]. Thus, the use of χ^2_{I-s-1} percentiles as critical values is definitely incorrect, but the seriousness of the error depends on the specific problem. Chernoff and Lehmann [17] show numerically that using the χ^2_{I-s-1} approximation is probably not a serious error in the case of fitting a Poisson distribution, but may be very serious for the fitting of a normal. Versatility and power being conflicting demands, the χ^2 tests generally do not provide much power. Case-by-case power simulations are available but necessarily to a limited extent. See Reference 61.

REFERENCES

1. Agresti, A. (1990). *Categorical Data Analysis*. Wiley, New York.

2. Agresti, A. and Coull, B. A. (1998). Approximate is better than "exact" for interval estimation of binomial proportions. *Am. Stat.*, **52**, 119–126.

3. Alam, K. (1979). Estimation of multinomial probabilities. *Ann. Stat.*, **7**, 282–283.

4. Alon, N., Spencer, J., and Erdös, P. (1992). *The Probabilistic Method*. Wiley, New York.

5. Anscombe, F. J. (1948). The transformation of Poisson, binomial and negative-binomial data. *Biometrika*, **35**, 246–254.

6. Basu, S. (2003). Bayesian inference for the number of undetected errors. *Handbook Stat.*, **22**, 1131–1150.

7. Basu, S. and Ebrahimi, N. (2001). Bayesian capture-recapture methods for error detection and estimation of population size: heterogeneity and dependence. *Biometrika*, **88**, 269–279.

8. Bennett, G. (1962). Probability inequalities for the sum of independent random variables. *J. Am. Stat. Assoc.*, **57**, 33–45.

9. Berger, J. O. (1985). *Statistical Decision Theory and Bayesian Analysis*, 2nd ed. Springer, New York.

10. Berger, J. O. (1994). An overview of robust Bayesian analysis. *Test*, **3**, 5–124. (With comments and a rejoinder by the author).

11. Blackwell, D. (1997). "Large Deviations for Martingales". *Festschrift for Lucien Le Cam: Research Papers in Probability and Statistics.* Springer, New York, pp. 89–91.

12. Blyth, C. and Still, H. (1983). Binomial confidence intervals. *J. Am. Stat. Assoc.*, **78**, 108–116.

13. Brown, L. (1981). A complete class theorem for statistical problems with finite sample spaces. *Ann. Stat.*, **9**, 1289–1300.

14. Brown, L., Cai, T., and DasGupta, A. (2001). Interval estimation for a binomial proportion (with discussions). *Stat. Sci.*, **2**, 101–133.

15. Brown, L., Cai, T., and DasGupta, A. (2002). Confidence interval for a binomial proportion and asymptotic expansions. *Ann. Stat.*, **30**, 160–201.

16. Carroll, R. and Lombard, F. (1985). A note on N estimators for the binomial distribution. *J. Am. Stat. Assoc.*, **80**, 423–426.

17. Chernoff, H. and Lehmann, E. L. (1954). The use of maximum likelihood estimates in χ^2 tests for goodness of fit. *Ann. Math. Stat.*, **25**, 579–586.

18. Clopper, C. J. and Pearson, E. S. (1934). The use of confidence or fiducial limits illustrated in the case of the binomial. *Biometrika*, **26**, 404–413.

19. Cochran, W. G. (1952). The χ^2 test of goodness of fit. *Ann. Math. Stat.*, **23**, 315–346.

20. Cochran, W. G. (1954). Some methods of strengthening the common χ^2 test. *Biometrics*, **10**, 417–451.

21. Cressie, N. (1978). Testing for the equality of two binomial proportions. *Ann. Inst. Stat. Math.*, **30**, 421–427.

22. Cressie, N. A. C. and Read, T. R. C. (1984). Multinomial goodness-of-fit tests. *J. R. Stat. Soc. Ser. B*, **46**, 440–464.

23. Crow, E. L. (1956). Confidence intervals for a proportion. *Biometrika*, **45**, 275–279.

24. DasGupta, A., Haff, L. R., and Strawderman, W. (1996). Estimation problems in the binomial distribution when n and p are both unknown, Technical Report #96-28. Purdue University, Purdue, Ind.

25. DasGupta, A. and Rubin, H. (2004). Estimation of binomial parameters when both n, p are unknown. To appear.

26. DasGupta, A. and Zhang, T. (2004). Analytical lower bounds on coverage probabilities of binomial confidence intervals. Preprint.

27. DeGroot, M. (1959). Unbiased sequential estimation for binomial populations. *Ann. Math. Stat.*, **30**, 80–101.

28. Eberhardt, K. R. and Fligner, M. A. (1977). Comparison of two tests for equality of two proportions. *Am. Stat.*, **4**, 151–155.

29. Ferguson, T. (1973). *Mathematical Statistics: A Decision Theoretic Approach*. Academic Press, New York.

30. Ferguson, T. (1996). *A Course in Large Sample Theory*. Chapman & Hall, New York.

31. Fisher, R. A. (1924). The conditions under which χ^2 measures the discrepancy between observation and hypothesis. *J. R. Stat. Soc.*, **87**, 442–450.

32. Fisher, R. A. (1934). *Statistical Methods for Research Workers*. Oliver and Boyd, Edinburgh, Scotland.

33. Fisher, R. A. (1935). The logic of inductive inference (with discussion). *J. R. Stat. Soc. A*, **98**, 39–82.

34. Fisher, R. A. (1941). The negative binomial distribution. *Ann. Eugenics*, **11**, 182–187.

35. Freeman, M. F., and Tukey J. W. (1950). Transformations related to the angular and the squared root. *Ann. Math. Stat.*, **27**, 607–611.

36. Freund, J. E. (1992). *Mathematical Statistics*, 5th ed. Prentice-Hall, Englewood Cliffs, N.J.

37. Garside, G. R. and Mack, C. (1976). Actual type 1 error probabilities for various tests in the homogeneity case of the 2×2 contingency table. *Am. Stat.*, **30**, 18–21.

38. Gart, J. and Zweifel, J. (1967). On the bias of various estimators of the logit and its variance with application to quantal bioassay. *Biometrika*, **54**, 181–187.

39. Ghosh, M. and Meeden, G. (1975). How many tosses of the coin? *Sankhyã A*, **37**, 523–529.

40. Griffin, B. S. and Krutchkoff, R. G. (1971). An empirical Bayes estimator for P [success] in the binomial distribution. *Sankhyã B*, **33**, 217–224.

41. Gupta, M. K. (1967). Unbiased estimate for $1/p$. *Ann. Inst. Stat. Math.*, **19**, 413–416.

42. Gutmann, S. (1984). Decisions immune to Stein's effect. *Sankhyã A*, **46**, 186–194.

43. Haber, M. (1980). A comparison of some continuity corrections for the Chi-squared test

on 2×2 tables. *J. Am. Stat. Assoc.*, **75**, 510–515.

44. Haldane, J. B. S. (1941, 1942). The negative binomial distributions. *Ann. Eugenics*, **11**, 182–187.

45. Hall, P. (1982). Improving the normal approximation when constructing one-sided confidence intervals for binomial or Poisson parameters. *Biometrika*, **69**, 647–652.

46. Hoeffding, W. (1963). Probability inequalities for sums of bounded random variables. *J. Am. Stat. Assoc.*, **58**, 13–30.

47. Johnson, B. M. (1971). On the admissible estimators for certain fixed sample binomial problems. *Ann. Math. Stat.*, **42**, 1579–1587.

48. Kozek, A. (1982). Towards a calculus of admissibility. *Ann. Stat.*, **10**, 825–837.

49. Larntz, K. (1978). Small-sample comparison of exact levels for chi-square goodness-of-fit. *J. Am. Stat. Assoc.*, **73**, 253–263.

50. Lehmann, E. L. (1986). *Testing Statistical Hypotheses*, 2nd ed. Wiley, New York.

51. Lehmann, E. L. and Casella, G. (1998). *Theory of Point Estimation*, 2nd ed. Springer, New York.

52. Martz, H. F. and Lian, M. G. (1974). Empirical Bayes estimation of the binomial parameter. *Biometrika*, **61**, 517–523.

53. Newcombe, R. G. (1998). Two-sided confidence intervals for the single proportions; comparison of several methods. *Stat. Med.*, **17**, 857–872.

54. Neyman, J. (1937). Outline of a theory of statistical estimation based on the classical theory of probability. *Philos. Trans. R. Soc. Lond. A*, **236**, 333–380.

55. Neyman, J. (1949). Contribution to the theory of the χ^2 test, *Proceedings of the First Berkeley Symposium on Mathematical Statistics and Probability*, Berkeley, CA, pp. 239–273.

56. Olkin, I., Petkau, A. J. and Zidek, J. V. (1981). A comparison of n estimators for binomial distribution. *J. Am. Stat. Assoc.*, **76**, 637–642.

57. Olkin, I. and Sobel, M. (1979). Admissible and minimax estimation for the multinomial distribution and for K independent binomial distributions. *Ann. Stat.*, **7**, 284–290.

58. Pearson, K. (1900). On the criterion that a given system of deviations from the probable in the case of a correlated system of variables is such that it can be reasonably supported to have arisen from random sampling. *Philos. Mag.*, **50**, 157–175.

59. Pearson, K. (1947). The choices of statistical tests illustrated on the interpretation of

data classed in a 2×2 table. *Biometrika*, **34**, 139–167.

60. Rao, C. R. (1973). *Linear Statistical Inference and Its Applications*, 2nd ed. Wiley, New York.

61. Rao, K. C. and Robson, D. S. (1974). A chi-square statistic for goodness-of-fit tests within the exponential family. *Commun. Stat.*, **3**, 1139–1153.

62. Read, C. B. (1993). Freeman-Tukey chi-squared goodness-of-fit statistics. *Stat. Probab. Lett.*, **18**, 271–278.

63. Rukhin, A. L. (1975). Statistical decision about the total number of observable objects. *Sankhā A*, **37**, 514–522.

64. Sadooghi-Alvandi, S. M. (1986). Admissible estimation of the binomial parameter n. *Ann. Stat.*, **14**, 1634–1641.

65. Santner, T. J. (1998). Teaching large sample binomial confidence Intervals. *Teach. Stat.*, **20**, 20–23.

66. Santner, T. and Duffy, D. (1989). *The Statistical Analysis of Discrete Data*. Springer, New York.

67. Sinha, B. K. and Sinha, B. K. (1972). Some problems of unbiased sequential binomial estimation. *Ann. Inst. Stat. Math.*, **24**, 245–258.

68. Skibinsky, M. and Rukhin, A. L. (1989). Admissible estimators of binomial probability and the inverse Bayes rule map. *Ann. Inst. Stat. Math.*, **41**, 699–716.

69. Stein, C. (1956). "Inadmissibility of the Usual Estimator for the Mean of a Multivariate Normal Distribution". *Proceedings of the Third Berkeley Symposium on Mathematical Statistics and Probability*, Vol. 1, University of California Press, Berkeley, Calif., pp. 197–206.

70. Sterne, T. E. (1954). Some remarks on confidence of fiducial limits. *Biometrika*, **41**, 275–278.

71. Stone, C. J. (1995). *A Course in Probability and Statistics*. Duxbury, Belmont, Calif.

72. Sylwester, D. (1974). Abstract: a Monte Carlo study of multidimensional contingency table analysis. *Biometrics*, **30**, 386.

73. Tate, M. W. and Hyer, L. A. (1973). Inaccuracy of the χ^2 test of goodness of fit when expected frequencies are small. *J. Am. Stat. Assoc.*, **68**, 836–841.

74. Upton, G. J. G. (1982). A comparison of alternative tests for the 2×2 comparative trial. *J. R. Stat. Soc. A*, **145**, 86–105.

75. van Zwet, W. R. (1967). An inequality for expected values of sample quantiles. *Ann. Math. Stat.*, **38**, 1817–1821.

76. Vollset, S. E. (1993). Confidence interval for a binomial proportion. *Stat. Med.*, **12**, 809–824.

77. Wasan, M. T. (1964). Sequential optimum procedures for unbiased estimation of binomial parameter. *Technometrics*, **6**, 259–272.

78. Wedderburn, R. W. M. (1976). On the existence and uniqueness of the maximum likelihood estimates for certain generalized linear models. *Biometrika*, **63**, 27–32.

79. Wilson, E. B. (1927). Probable inference, the law of succession, and statistical inference. *J. Am. Stat. Assoc.*, **22**, 209–212.

80. Wolfowitz, J. (1946). On sequential binomial estimation. *Ann. Math. Stat.*, **17**, 489–493.

81. Yarnold, J. K. (1970). The minimum expectation in χ^2 goodness of fit tests and the accuracy of approximations for the null distribution. *J. Am. Stat. Assoc.*, **65**, 864–886.

82. Yates, F. (1934). Contingency tables involving small numbers and the χ^2 test. *J. R. Stat. Soc. Suppl.*, **1**, 217–235.

83. Yates, F. (1984). Tests of significance for 2×2 contingency tables. *J. R. Stat. Soc. Ser. A*, **147**, 426–463.

84. Zen, M. M. and DasGupta, A. (1993). Estimating a binomial parameter: is robust Bayes real Bayes? *Stat. Decis.*, **11**, 37–60.

See also BINOMIAL DISTRIBUTION; BINOMIAL PARAMETER, CLOPPER-PEARSON INTERVAL ESTIMATION; CHI-SQUARE DISTRIBUTION; CONTINUITY CORRECTIONS; CRESSIE–READ STATISTIC; FISHER'S EXACT TEST; INVERSE SAMPLING; JEFFREYS' PRIOR DISTRIBUTION; MULTINOMIAL DISTRIBUTIONS; and TWO-BY-TWO (2×2) TABLES.

ANIRBAN DASGUPTA

TONGLIN ZHANG

BINOMIAL COEFFICIENT, GAUSSIAN

This is a generalization of the ordinary binomial coefficient and is defined as

$$\binom{n}{m}_q : (q;q)_{n+m} | (q;q)_n (q;q)_m ,$$

where $(q;q)_n = (1-q)(1-q^2) \ldots (1-q^n)$. As $q \to 1$, the limit tends to the ordinary binomial coefficient.

The Gamma binomial coefficient is a polynomial of degree mn in q. Applications in statistics include the distribution of the Mann-Whitney statistic [1]. See Reference 2 for earlier applications.

REFERENCES

1. Di Bucchianico, A. (1999). Combinatorics, computer algebra and the Wilcoxon-Mann-Whitney test. *J. Stat. Plann. Inference*, **79**, 349–364.

2. Handa, B. R. and Mohanty, S. C. (1980). On q-binomial coefficients and some statistical applications. *SIAM J. Math. Anal.*, **11**, 1027–1035.

BINOMIAL DISTRIBUTION

This important discrete distribution is commonly used to model observation counts of events. It is based on the assumption that the counts X can be represented as the results of a sequence of independent Bernoulli trials*. If the probability of observing an event E is p for each of n such trials, the probability that E will be observed in exactly x of the trials is

$$P_x = \binom{n}{x} p^x (1-p)^{n-x} \quad (x = 0, 1, \ldots, n).$$

The distribution defined by

$$\Pr[X = x] = P_x \quad (x = 0, 1, \ldots, n)$$

a *binomial distribution with parameters n and p*. The name comes from the fact that the binomial expansion of $(p + \overline{1-p})^n$ is $P_0 + P_1 + \cdots + P_n$.

It is customary to use the symbol q to represent the quantity $(1-p)$. This leads to some simplification of formulae; it does not, of course, increase the number of parameters beyond two. The parameter p is sometimes called the *population binomial proportion*, to distinguish it from X/n, the *sample binomial proportion*.

An extensive discussion of the binomial distribution is given in [2], Chapter 3. It belongs to the exponential family, and to the families of power series distributions*, of factorial series distributions*, of generalized hypergeometric distributions*, and of distributions having an increasing hazard rate.* Among tables of individual and cumulative probabilities of X is ref. [1]. When $p = q = \frac{1}{2}$, the distribution is symmetrical.

The identity

$$\Pr[X \geq x] = I_p(x, n - x + 1),$$

where $I_p(a, b)$ is the incomplete beta function* ratio, is useful in constructing confidence intervals and tests for p (see BINOMIAL AND MULTINOMIAL PARAMETERS, INFERENCE ON), as well as approximations to binomial probabilities.

The probability generating function* for the variable X is given by

$$G(z) = (q + pz)^n, \quad 0 < p < 1.$$

MOMENTS

The expected value of X is np; the standard deviation is $\sqrt{npq}$. The third and fourth central moments μ_3 and μ_4, and the moment ratios* $\sqrt{\beta_1}$ and β_2 are given by

$$\mu_3 = npq(q - p),$$
$$> 0 \quad \text{if } p < 1/2,$$
$$< 0 \quad \text{if } p > 1/2;$$
$$\mu_4 = 3(npq)^2 + npq(1 - 6pq),$$
$$\sqrt{\beta_1} = (q - p)/\sqrt{npq},$$
$$\beta_2 = 3 + (1 - 6pq)/(npq).$$

The moment generating function of the distribution is given by

$$M(t) = (q + pe^t)^n.$$

Two recursion relations of interest are for the central moments

$$\mu_{r+1} = pq\left(nr\mu_{r-1} + \frac{d\mu_r}{dp}\right)$$

and for the cumulants:

$$\kappa_{r+1} = pq\frac{\partial \kappa_r}{\partial p}, \quad r = 1, 2, \ldots.$$

Further discussion is given in [1].

The rth factorial moment* is

$$E[X^{(r)}] = E[X(X - 1) \cdots (X - r + 1)]$$
$$= n^{(r)}p^r.$$

The mean deviation* is

$$n\binom{n - 1}{[np]}p^{[np]+1}q^{n-[np]},$$

where $[np]$ = integer part of np.

FURTHER PROPERTIES

The ratio of successive P_x's is

$$\frac{P_{x+1}}{P_x} = \frac{n - x}{x + 1}\frac{p}{q},$$

from which it can be seen that as x increases from zero, P_x increases until $x = [(n + 1)p]$. Thereafter P_x decreases, so that there is a mode at $x = [(n + 1)p]$. If $(n + 1)p < 1$, the mode is at zero, and $P_{x+1}/P_x < 1$ for all $x = 0, 1, \ldots$. If $(n + 1)p$ is an integer, there are two equal maximum values of P_x at $x = (n + 1)p - 1$ and $x = (n + 1)p$.

Although direct calculation of P_x is feasible, calculations become very heavy as n increases. There are very good approximations which may be used when n is large.

APPROXIMATIONS

Several approximations to cumulative probabilities and to individual terms P_x are given in [2, 3, 4, 5]; see also NORMAL APPROXIMATIONS TO SOME DISCRETE DISTRIBUTIONS.

If $\min(np, n(1 - p)) > 10$, the normal approximation*

$$\Pr[X \leq x] = \Phi\left(\frac{x + \frac{1}{2} - np}{(npq)^{1/2}}\right), \quad (1)$$

where $\Phi(\cdot)$ is the unit normal distribution function, incorporating the continuity correction* $\frac{1}{2}$, performs well (see [2], Sec. 6.1). Better approximations can usually be obtained from formulas developed by Molenaar [3].

Many of these appear in [2], Sec. 6.1, and in [5], Sec. 7.1. For example, the approximation

$$\Pr[X \le x] = \Phi[(4x + 3)q]^{1/2}$$
$$- [(4n - 4x - 1)p]^{1/2}), \quad (2)$$

$0.05 < p < 0.93$, combines accuracy wth simplicity.

When $np < 10$ and p is small (less than 0.1, say) the Poisson* approximation

$$\Pr[X \le x] = e^{-np} \sum_{j=0}^{x} \frac{(np)^j}{j!}$$

can be used. If $n(1 - p) < 10$ and p is large (greater than 0.9, say) the Poisson approximation may be applied to the distribution of $Y = n - X$.

Littlewood has published an exhaustive study of approximations to sums of binomial probabilities [2].

REFERENCES

1. Johnson, N. L. and Kotz, S. and Kemp, A. W. (1992). *Univariate Discrete Distributions* (2nd ed.). Wiley, New York. (Chap. 3.)

2. Littlewood, J. E. (1969). *Adv. Appl. Prob.*, **1**, 43–72.

3. Molenaar, W. (1970). *Approximations to the Poisson, Binomial and Hypergeometric Distribution Functions*. Mathematical Centre, Tracts 31, Amsterdam: Mathematisch Centrum.

4. Patel, J. K. and Read, C. B. (1996). *Handbook of the Normal Distribution* (2nd ed.). Dekker, New York. (Sec. 7.1)

5. Pearson, E. S. and Hartley, H. O. (1976). Biometrika Tables for Statisticians, Vol. 1 (3rd ed.). Biometrika Trust, London.

FURTHER READING

von Collani, E. and Dräger, K. (2001). *Binomial Distribution Handbook for Scientists and Engineers*. Birkhäuser, Cambridge, MA.

See also BERNOULLI DISTRIBUTION; BERNOULLI'S LIMIT THEOREM; BERNOULLI TRIALS; CLASSICAL DISCRETE DISTRIBUTIONS, GENERALIZATIONS OF; LAWS OF LARGE NUMBERS; LEXIAN DISTRIBUTION; LIMIT THEOREMS; NEGATIVE BINOMIAL DISTRIBUTION; NORMAL APPROXIMATIONS TO SOME DISCRETE DISTRIBUTIONS; and POISSON DISTRIBUTION.

BINOMIAL DISTRIBUTION, GENERALIZED (COMPOUND)

Numerous generalizations of the binomial distribution have been attempted during the past five decades (see, e.g., ref. 1 for a comprehensive discussion of results up to 1992). The following model seems to be a most natural one.

A random variable X has a *generalized (compound) binomial distribution* with parameters $n, p_1, p_2, \ldots, p_n (0 < p_i < 1, i = 1, 2, \ldots, n)$ if

$$\Pr[X = k]$$

$$= \begin{cases} \prod_{i=1}^{n}(1 - p_i) & \text{for} \quad k = 0, \\[2mm] \sum_{\ell_k=1}^{n} \cdots \sum_{\ell_1=1}^{n} \prod_{i \neq \ell_l}^{n}(1 - p_i) \prod_{j=1}^{k} p_{\ell_j} \\[2mm] \vdots \\[1mm] \quad\quad i \neq \ell_k \\[1mm] \quad\quad \text{for} \quad k = 1, \ldots, n-1, \\[2mm] \prod_{i=1}^{n} p_i, & k = n. \end{cases}$$

The characteristic function* of this random variable is

$$\phi_X(t) = \prod_{k=1}^{n}[1 + p_k(e^{it} - 1)],$$

and the first two moments are:

$$E(X) = \sum_{k=1}^{n} p_k, \quad E(X^2) = \sum_{k=1}^{n} p_k + \sum_{k \neq \ell} p_k p_l,$$

so that

$$\text{var}(X) = \sum_{k=1}^{n} p_k(1 - p_k)$$

(compare with the Lexian* and ordinary binomial distributions*).

If $X_1, X_2, \ldots, X_n$ are independent Bernoulli random variables with parameters $p_1, p_2, \ldots, p_n$, respectively, then $X = \sum_{k=1}^{n} X_k$ possesses the generalized binomial distribution. Moreover, if $\bar{p} = (1/n) \sum_{k=1}^{n} p_k$ and $\sigma_{\bar{p}} = (1/n) \sum_{k=1}^{n} p_k^2 - \bar{p}^2$, then

$$\text{var}(X) = n\bar{p}(1 - \bar{p}) - \bar{n}\sigma_{\bar{p}}\bar{p}.$$

Thus, the variance of a binomial random variable with parameters n and $\bar{p}$ is larger than the variance of the random variable X with a corresponding generalized binomial distribution.

REFERENCE

1. Johnson, N. L., Kotz, S., and Kemp, A. W. (1993). *Distributions in Statistics, Discrete Univariate Distributions*, 2nd ed. Wiley, New York.

See also BINOMIAL DISTRIBUTION; CLASSICAL DISCRETE DISTRIBUTIONS, GENERALIZATIONS OF; and LEXIAN DISTRIBUTION.

BINOMIAL DISTRIBUTION: SAMPLE SIZE ESTIMATION

The main focus of this entry is on the point estimation of n when p is unknown and observations $\mathbf{x} = (x_1, x_2, x_3, \ldots, x_k)$ are from the binomial distribution with parameters n and p

$$P(\mathbf{x}|n,p) = \prod_{i=1}^{k} \binom{n}{x_i} p^{x_i} q^{n-x_i}$$

for $x_i \in \{0, 1, 2, \ldots, n\}$ and $q = 1 - p$.

This problem arises in many practical situations, some examples of which are given below. Originally, standard method of moments and then maximum likelihood estimators were proposed, to be denoted MME and MLE, but the problems associated with these estimators were eventually realized, and some of these problems are discussed below. Many attempts were made to modify these estimators to alleviate some of their problems, and many attempts at Bayesian estimation have been made as well. Thus in this entry, estimators are classified as either method of moments related, maximum likelihood related, or Bayesian. These classifications encompass nearly all of the estimators that have been proposed, and the few that do not fit in these categories can be found in the references listed.

In each case, the estimator is given, and an attempt is made to comment on some its associated properties. It should be noted that this problem is a special case of a more general problem addressed in [6].

Where appropriate, comments are made on the known p case, but unless stated otherwise, both n and p will be assumed unknown.

EXAMPLES

There are many situations in which the need to estimate n when p is unknown arises in practice. In this section, some examples of such situations are presented.

One of the best known is found in [13]. Here an appliance company is interested in estimating the total number n of a certain type of appliance in use in a given service area. The number of such appliances $(x_1, x_2, \ldots, x_k)$ brought in for repair on k consecutive weeks can then be modeled as independent binomial(n, p) observations, where p is the probability an appliance is brought in for repair.

Another well known example is in [26]. Here the data $(x_1, x_2, \ldots, x_k)$ represent counts of reported crimes, where not all crimes are reported, and interest is in estimating to total numbers of crimes $(n_1, n_2, \ldots, n_k)$ committed in successive time periods. Under the assumption that each crime has the same probability p of being reported, independently of the other crimes, and if the n_i's are assumed to be equal to a constant n, then the data can be modeled as binomial(n, p) observations. An argument for the constant n_i's assumption is given in [26].

Two more examples are given in [8]. In these examples $(x_1, x_2, \ldots, x_k)$ represent counts from light aircraft on k successive days, and n the total number, of individual waterbucks or herds of impala in a national park in South Africa. The assumption of course is that each waterbuck or impala herd has probability p of being counted, independently of the others, and that counts on successive days are independent.

Other examples appear in [25] (counts of animals caught by traps, with n representing the animal population size), [25] and [30] (traffic counts with n representing the maximum possible count).

An example of a situation where p is known is found in [5]. Here $(x_1, x_2, \ldots, x_k)$

represent counts of a certain species in sub-samples of plankton, each subsample representing a known proportion p of the original sample.

HISTORY

Interest in the problem of fitting a binomial distribution when both n and p are unknown dates back at least as far as 1914, when the binomial, negative binomial, and Poisson distributions were often all considered to be equally plausible candidates for modeling a given set of count data, each being representable through the coefficients in the expansion of $(p + q)^n$ where $p + q = 1$, with $p \in [0, 1]$ and $n \in \{1, 2, 3, \ldots\}$ representing the binomial model, $p < 0$ (so that $q > 1$) and $n < 0$ representing the negative binomial, and $p = \lambda/n$ and $n \to \infty$, with λ constant, representing the Poisson.

Thus early papers [31], [32], [19] mainly addressed ways of deciding between these three models for a given set of data. For example, Whittaker [32] reanalyzes datasets to which previous authors had fit Poisson models, and assesses whether a binomial or negative binomial model might be more appropriate. Many of these early papers, however, contain mistakes, as noted in [6].

Detailed studies of the properties of point estimators of n when p is unknown were first carried out by Fisher [15], Haldane [19], and Binet [4], all of whom addressed the issue of deciding between the MME (which is easy to compute) and the MLE (which is efficient). Hoel [23] addresses both point and interval estimation of n as well as hypothesis testing.

Olkin et al. [26] provide more comments on the history of this estimation problem.

SOME DIFFICULTIES IN ESTIMATION

The problem of estimating n when p is unknown is a notoriously difficult one. In this section, some of the difficulties are outlined. These include various problems with the performance of both the MME and the MLE, and so various attempts have been made to modify these estimators to improve performance. Bayesian estimators are not without problems either, as will be seen.

Perhaps the most obvious difficulty is the fact that n and p are so confounded as to make it very hard to estimate each individually. The likelihood function (7) has a sharp ridge, with large values of n and small p being nearly as likely as small n and large p for a given set of data. Contrast this with the much easier problem of estimating np and npq, the mean and variance of the binomial distribution. Kahn [24] suggests a reparameterization for alleviating this problem in a Bayesian setting.

Many of the problems associated with the MLE and the MME arise when the sample mean $\bar{x} = k^{-1} \sum_{i=1}^{k} x_i$ is not much bigger than the sample variance $s^2 = k^{-1} \sum_{i=1}^{k} (x_i - \bar{x})^2$. The MME (1) is negative when $s^2 > \bar{x}$ and infinite when $s^2 = \bar{x}$, both of which happen with positive probability, and the MLE is infinite when $s^2 \geq \bar{x}$ (see [11]).

Even when $s^2 < \bar{x}$, both the MME and the MLE can be highly unstable, i.e. small perturbations in the data can have drastic effects on the estimates when $\bar{x}$ and s^2 are nearly equal [26]. This behavior is often associated with samples from populations in which n is large and p small [26], [20]. Hall [20] gives results, discussed in more detail below, showing that under certain configurations of n, p and k both the MME and MLE converge to random variables with heavy tailed distributions, thereby explaining some of their erratic behavior. Also, Aitkin and Stasinopoulos [2,1] explain some of the erratic behavior of the MLE by the flatness of the profile likelihood for n near it's maximum. Carroll and Lombard [8] too mention the flatness of the profile likelihood function, pointing out that even as $n \to \infty$, it does not necessarily follow that this function decreases to 0. Blumenthal and Dahiya explain in more detail why s^2 nearly equal to $\bar{x}$ can be a problem. Casella [9] suggested a method for determining whether the MLE is too unstable.

Gupta et al. [18] show that no unbiased estimators of n exist when both n and p are unknown.

They also show in the same paper that there are no nonconstant ancillary statistics for n and that there is no complete sufficient statistic for n.

Difficulties arise with the Bayesian approach to estimation of n too. Raftery [28] points out the lack of tractable priors for n due to it's integer nature. Kahn [24] points out that choice of the prior can have a strong effect on the inference about n. More specifically, he shows that for the commonly used beta prior on p, i.e. when p has the prior density proportional to $p^a(1-p)^b$, and n has prior $f(n)$, independent of p, such that $f(n) > 0$ for all sufficiently large n, then there is some positive constant C such that

$$\lim_{n\to\infty} \frac{P(n|\mathbf{x})}{f(n)/n^{a+1}} = C$$

where $P(n|\mathbf{x})$ is the marginal posterior distribution of n given the data. It follows that the existence of the posterior mean (and higher moments) and median are determined in advance by the particular choice of the prior $f(n)$, independently of the data.

When p is known, estimating n is less problematic. For example, in this case, the MLE is finite for every sample [5]. Still though, some problems exist. For example, even when p is known, a UMVUE for n does not exist [29], and the prior for n can still have a strong influence on the inference [21].

All these problems point to the need for alternative estimators, a subject to which we now turn.

METHOD OF MOMENTS AND RELATED ESTIMATORS

The method of moments estimator $\tilde{n}$ for n has the advantage over other estimators that it's easy to compute. It is obtained by equating the sample mean $\bar{x}$ and sample variance s^2 to their population counterparts np and npq, and solving for n in terms of $\bar{x}$ and s^2, and is given by

$$\tilde{n} = \frac{\bar{x}^2}{(\bar{x} - s^2)} \tag{1}$$

Blumenthal and Dahiya [5] define a method of moments estimator to be the nearest integer $\tilde{n}^*$ to $\tilde{n}$. They then establish consistency of $\tilde{n}^*$ in the sense that for fixed n,

$$\lim_{k\to\infty} P(\tilde{n}^* = n) = 1. \tag{2}$$

On the other hand, they also show that $\tilde{n}^*$ is not consistent when k is fixed but $n \to \infty$,

and they give an expression for a random variable whose distribution is the same as the limiting distribution of $\tilde{n}^*/n$. Finally, they also show that when $n \to \infty$ and $k \to \infty$ with $\sqrt{k}/n \to 0$,

$$\sqrt{k}/n(\tilde{n}^* - n) \stackrel{\mathcal{D}}{\to} N(0, 2q^2/p^2). \tag{3}$$

where $\stackrel{\mathcal{D}}{\to}$ denotes convergence in distribution.

Fisher [15] and Binet [4] compute the asymptotic efficiency of the MME, showing that it's not fully efficient.

An explanation of the erratic behavior of the MME when n is large and p small is provided by Hall [20]. He describes the asymptotic behavior of $\tilde{n}$ under certain conditions on the behavior of n, p, and k, among them that $n \to \infty$, $p \to 0$, and $np \to \lambda$, where $0 < \lambda \le \infty$. If the conditions hold, then under the additional condition $kp^2 \to \infty$,

$$(kp^2/2)^{\frac{1}{2}}(\tilde{n} - n)/n \stackrel{\mathcal{D}}{\to} Z, \tag{4}$$

where $Z \sim N(0,1)$. On the other hand, if $kp^2 \to \rho$, where $0 \le \rho < \infty$, then there exists a continuous random variable $X(\rho)$ such that

$$(kp^2/2)^{-\frac{1}{2}}\tilde{n}/n \stackrel{\mathcal{D}}{\to} X(\rho), \tag{5}$$

where $X(\rho)$ has no finite moments, i.e. it's distribution has very large tails.

To curtail some of the erratic behavior of the MME, Olkin, Petkau and Zidek [26] propose a stabilized version of the estimator, denoted by MME:S. They label the MME 'stable' whenever

$$\bar{x}/s^2 \ge 1 + 1/\sqrt{2} \tag{6}$$

and 'unstable' otherwise. The proposed estimator is then

$MME : S$

$$= \begin{cases} \max(\tilde{n}, x_{(k)}) & \text{if (6) holds} \\ \max(s^2 w^2 & \text{if (6) does not} \\ \quad /(w-1), x_{(k)})2 & \text{hold} \end{cases}$$

where $w = \max(1 + \sqrt{2}, (x_{(k)} - \bar{x})/s^2)$. This estimator is shown in [26] to be an improvement over $\tilde{n}$ in terms of stability.

Feldman and Fox [14] and Blumenthal and Dahiya [5] examine asymptotic properties of the MME when p is known.

MAXIMUM LIKELIHOOD AND RELATED ESTIMATORS

The maximum likelihood estimator $\hat{n}$ is obtained by maximizing the likelihood function

$$L(n,p) = \prod_{i=1}^{k} \binom{n}{x_i} p^{x_i} q^{n-x_i} \qquad (7)$$

simultaneously over integers $n \geq x_{(k)}$, where $x_{(k)} = \max(x_1, x_2, x_3, \ldots, x_k)$, and $p \in [0,1]$. For any fixed n, (7) is maximized over p by $\hat{p} = \bar{x}/n$. Thus $\hat{n}$ is found by maximizing the profile likelihood $L(n,\hat{p})$ over n. The MLE $\hat{n}$ exists and is finite if and only if

$$\bar{x} > s^2, \qquad (8)$$

otherwise it is infinite (see [26]).

DeRiggi [11] shows that when (8) holds, the likelihood function is unimodal whenever $\bar{x}$ is an integer and claims incorrectly that this is the case in general. He later corrects himself [12], giving a counterexample to unimodality for the noninteger $\bar{x}$ case, but shows that a sufficient condition for unimodality even when $\bar{x}$ is a noninteger is

$$\int_{[\bar{x}]}^{x_{(k)}} (F - G)(y) dy \geq 0$$

where

$$F(y) = \min(y, \bar{x})$$

$$G(y) = k^{-1}\left(M_y([y] + 1) + \sum_{x_i \leq y} x_i\right),$$

M_y denotes the number of x_i's strictly greater than y, and $[x]$ denotes integer part of x. An upper bound on $\hat{n}$ when (8) holds is given in [11], facilitating the search for a global maximum when the likelihood is multimodal.

Various procedures for finding the MLE have been suggested [5], [10]. For example, one procedure [10] is to find the (possibly noninteger valued) $\hat{v}$ satisfying

$$L(\hat{v} - 1, \hat{p}) = L(\hat{v}, \hat{p})$$

and then set $\hat{n} = [\hat{v}]$. Another [5] is to set $\hat{n}$ equal to the greatest integer n satisfying $L(n - 1, \hat{p}) < L(n, \hat{p})$. Still another [5] is to

momentarily regard n as a continous parameter and find the roots of $d\mathcal{L}(n,\hat{p})/dn$, the derivative of the log of the likelihood function. The resulting equation to be solved is

$$-\log(\hat{q}) = \frac{1}{k}\sum_{i=1}^{k}\left\{\sum_{j=0}^{x_i - 1}(n - j)^{-1}\right\} I_{\{x_i \geq 1\}}$$

where $\hat{q} = 1 - \hat{p}$ and I_A is the indicator of the event A, giving a solution n_c'. Note that n_c' may be less than $x_{(k)}$. Then $\hat{n}$ is found by comparing $L(n,\hat{p})$ at $n = [\hat{n}_c]$ and $n = [\hat{n}_c + 1]$, where $\hat{n}_c = \max(x_{(k)}, n_c')$.

Asymptotic properties of $\hat{n}$ have been studied by Blumenthal and Dahiya [5] and by Hall [20]. Blumenthal and Dahiya [5] show that both (2) and (3) hold with $\hat{n}$ replacing $\tilde{n}^*$ (under the same conditions on n and k), and that $\hat{n}$ is also not consistent when k is fixed but $n \to \infty$, in which case $\hat{n}/n$ converges to the same random variable as $\tilde{n}^*/n$ does.

Hall's results explaining the erratic behavior of the MLE are similar those for the MME. Specifically, under the same conditions necessary for (4) to hold for $\tilde{n}$, (4) holds with $\hat{n}$ replacing $\tilde{n}$. Also, under the same conditions necessary for (5) to hold for $\tilde{n}$,

$$P(\hat{n} = \infty) \to P(X(\rho) < 0) > 0$$

and

$$(kp^2/2)^{-\frac{1}{2}}\hat{n}/n|(\hat{n} < \infty) \xrightarrow{\mathcal{D}} X(\rho)|(X(\rho) > 0),$$

where | denotes conditioning and $X(\rho)|(X(\rho) > 0)$ has no finite moments, i.e. its distribution has very large tails.

To avoid some of the erratic behavior associated with the MLE, Olkin, Petkau and Zidek [26] propose a stabilized version, which they denote MLE:S, with

$$MLE:S = \begin{cases} \hat{n} & \text{if (6) holds} \\ JK & \text{if (6) does not hold} \end{cases}$$

where JK is a bias-reduced version of $x_{(k)}$ using the jackknife technique. It's shown in [26] that this estimator is an improvement over $\hat{n}$, but not over MME:S, in terms of stability.

Another variant of the MLE, proposed by Carroll and Lombard [8] is obtained by integrating (7) with respect to a beta prior distribution for p, and maximizing the result.

More formally, suppose p has a density proportional to $p^a(1-p)^b$. Then the Carroll and Lombard estimator $\hat{n}_{CL}$ of n is obtained by maximizing

$$\prod_{i=1}^{k} \binom{n}{x_i} \int_0^1 p^{a+T}(1-p)^{b+kn-T} dp$$

$$= \left\{\prod_{i=1}^{k} \binom{n}{x_i}\right\} \left[(kn+a+b+1)\right.$$

$$\left. \times \binom{kn+a+b}{a+\sum_{i=1}^{k} x_i}\right]^{-1} \qquad (9)$$

over integer values of $n \geq x_{(k)}$. If $a \geq 0$ and $b \geq 0$, then (9) is maximized at some finite n (see [8]). Gupta, Nguyen and Wang [18] give an example to show that the likelihood function (9) may not be unimodal, but they also give an upper bound on $\hat{n}_{CL}$, facilitating the search for a global maximum.

Carroll and Lombard [8] examined the efficiencies of their estimator numerically for the cases $a = b = 0$ (corresponding to a uniform prior on p) and $a = b = 1$ in terms of mean squared error, finding them to be generally superior to MME:S and MLE:S. A plot of (9) is provided in [2] for a dataset known to produce unstable estimates, and shows a much sharper peak than the corresponding plot for the profile likelihood function for n.

Another explanation for the improved stability properties of the $\hat{n}_{CL}$ is given by Hall [20]. He finds (4) holds with $\hat{n}_{CL}$ replacing $\tilde{n}$ under the same conditions necessary for (4) to hold for $\tilde{n}$. However, under the same conditions necessary for (5) to hold for $\tilde{n}$, there exists a continuous random variable $Y(\rho)$ such that

$$(kp^2/2)^{-\frac{1}{2}} \hat{n}_{CL}/n \xrightarrow{\mathcal{D}} Y(\rho), \qquad (10)$$

where $Y(\rho)$ is strictly positive and has all moments finite, in contrast to the asymptotic behavior of the MLE and MME, which both converge to very heavy tailed distributions.

Blumenthal and Dahaya [5] suggest a modified version of the MLE, denoted $\hat{n}_m$, obtained by maximizing

$$p^a(1-p)^b L(n,p)$$

over both n and p rather than integrating p out. They show that $\hat{n}_m$ has the same asymptotic properties they showed for $\hat{n}$ and $\tilde{n}^*$ discussed above, but also that their new estimator is always finite whenever $a > 0$. Recall that both $\hat{n}$ and $\tilde{n}^*$ can be infinite with positive probability.

Another variant of the MLE for n obtained by conditioning on $T = \sum_{i=1}^{k} x_i$ and maximizing the resulting conditional likelihood function is discussed in [3], [1], [8], [2], and [18] . Carroll and Lombard [8] noted that this conditional likelihood estimator is a special case of their estimator $\hat{n}_{CL}$ with $a = -1$ and $b = 0$. Some asymptotic properties of this estimator are discussed [20], and plots of the conditional likelihood function are shown in [1] and [2], showing it to be very flat as $n \to \infty$. Although conditioning on T eliminates the nuisance parameter p from the likelihood function, since the distribution of T depends strongly on n, some information about n is lost in the conditioning [2] (recall that there are no nonconstant ancillary statistics for n). This estimator is discussed in some detail in [3].

In contrast to the unknown p case, when p is known the likelihood function always has a unique root [14]. Asymptotic properties of the MLE in the known p case are discussed in [14] and [5].

BAYES AND RELATED ESTIMATORS

An advantage of the Bayesian approach to estimating n is that the posterior distribution for n gives not only a point estimate of n (eg the posterior mean or posterior mode), but also an assessment of the precision of that estimate (eg highest posterior density sets). Various choices of prior distributions on n and p have been studied, with varying degrees of success in the resulting estimators.

Draper and Guttman [13] treat n and p as independent, with n distributed uniformly on a finite range of positive integers $\{1, 2, 3, \ldots, N\}$, with N chosen to be large, and p distributed as a beta random variable. Then p is integrated out to get the marginal posterior distribution of n, and the mode of this distribution is used as the estimate. Although the authors say that the choice of the upper limit N of the domain of the prior on n does not have a strong impact on the inference about n, Kahn [24] points out that this is not the case, with both the marginal posterior mean and median becoming arbitrarily large as N grows (see also [21]).

Both Raftery [28] and Hamedani and Walter [21] use a Poisson(μ) prior distribution for n. Then conditional on p, each x_i has a Poisson(λ) distribution, where $\lambda = \mu p$. Hamedani and Walter then treat n and p as indepenent, with a beta prior assumed for p, and use the posterior mean to estimate n. Raftery assumes a uniform prior for p. But whereas Hamedani and Walter [21] appear to have chosen a value for the hyperparameter μ somewhat arbitrarily in their examples, Raftery adopts a hierarchical Bayes approach, assuming that λ and p are independent (so that μ and p are negatively correlated), with λ given a standard vague prior $p(\lambda) \propto \lambda^{-1}$. This leads to the posterior distribution of n given the data

$$p(n|\mathbf{x}) \propto \left\{ \frac{(kn - T)!}{(kn + 1)!n} \right\} \left\{ \prod_{i=1}^{k} \binom{n}{x_i} \right\} \quad (11)$$

for $n \geq x_{(k)}$. The estimator $\hat{n}_{MRE}$ of n is then defined to be the Bayes estimator associated with (11) and the relative squared error loss $(\bar{n}/n - 1)^2$ of an estimator $\bar{n}$. Simulations indicate $\hat{n}_{MRE}$ performs favorably to both $\hat{n}_{CL}$ with $a = b = 1$ and MME:S in terms of relative mean squared error.

The decision to treat n and p as independent is called into question in [7]. An empirical Bayes approach to estimating n is presented in [22]. In [17], n is treated as a continuous random variable, and assigned a gamma prior distribution.

For the p known case, Draper and Guttman [13] again place a uniform prior on n positive integers $\{1, 2, 3, \ldots, N\}$, and estimate n with the the posterior mode, which is just the maximum likelihood estimate $\hat{n}$ of n provided N is large. Hamedani and Walter [21] give an simple expression for the posterior mean of n in terms of the marginal distribution of the x_i's for a general prior on n. Ghosh and Meeden [16] and Sadooghi-Alvandi [29] address admissible estimation of n. Wiper and Petit [33] and O'Quigley [27] address the problem of estimating estimating

n when p is known and the sample proportions $(\hat{p}_1, \hat{p}_2, \hat{p}_3, \ldots, \hat{p}_k)$ are observed, where $\hat{p}_i = x_i/n$.

REFERENCES

1. Aitkin, M. (1991). Posterior Bayes factors (with discussion). *J. Roy. Statist. Soc. Ser. B* **53**, 111–142.

2. Aitkin, M. and Stasinopoulos, M. (1989). Likelihood analysis of a binomial sample size problem. In *Contributions to Probability and Statistics*, Gleser, L. J., Perlman, M. D., Press, S. J. and Sampson, A. R., eds. Springer, New York, 1989, 399–411.

3. Bain, L., Engelhardt, M. and Williams, D. (1990). Confidence bounds for the binomial n parameter: a classical approach. *Comm. Statist. Simulation Comput.* **19**, 335–348.

4. Binet, F. E. (1954). The fitting of the positive binomial distribution when both parameters are estimated from the sample. *Ann. Eugenics* **18**, 117–119.

5. Blumenthal, S. and Dahiya, R. C. (1981). Estimating the binomial parameter n. *J. Amer. Statist. Assoc.* **76**, 903–909.

6. Blumenthal, S. and Dahiya, R. C. (1982). Population or sample size estimation. In *Encyclopedia of statistical sciences*, Kotz, S., Johnson, N. L., Read, C. B., eds. Wiley, New York, 1982, 100–110.

7. Burnham, K. P. (1987). Comment on Kahn. *Amer. Statist.* **41**, 338.

8. Carroll, R. J. and Lombard, F. (1985). A note on n estimators for the binomial distribution. *J. Amer. Statist. Assoc.* **80**, 423–426.

9. Casella, G. (1986). Stabilizing binomial n estimators. *J. Amer. Statist. Assoc.* **81**, 172–175.

10. Dahiya, R. C. (1981). An improved method of estimating an integer-parameter by maximum likelihood. *Amer. Statist.* **35**, 34–37.

11. DeRiggi, D. F. (1983). Unimodality of likelihood functions for the binomial distribution. *J. Amer. Statist. Assoc.* **78**, 181–183.

12. DeRiggi, D. (1994). Sufficient conditions for unimodality of the positive binomial likelihood function. *Statist. and Prob. Letters* **19**, 1–4.

13. Draper, N. and Guttman, I. (1971). Baysian estimation of the binomial parameter. *Technometrics* **13**, 667–673.

14. Feldman, D. and Fox, M. (1968). Estimation of the parameter n in the binomial distribution. *J. Amer. Statist. Assoc.* **63**, 150–158.

15. Fisher, R. A. (1942). The negative binomial distribution. *Ann. Eugenics* **11**, 182–187.

16. Ghosh, M. and Meeden, G. (1975). How many tosses of the coin? *Sankhyā Ser. A* **37**, pt. 4, 523–529.

17. Günel, E. and Chilko, D. (1989). Estimation of parameter n of the binomial distribution. *Comm. Statist. Simulation* **18**, 537–551.

18. Gupta, A. K., Nguyen, T. T., and Wang, Y. (1999). On maximum likelihood estimation of the binomial parameter n. *Canad. J. Statist.* **27**, no. 3, 599–606.

19. Haldane, J.B.S. (1942). The fitting of binomial distributions. *Ann. Eugenics* **11**, 179–181.

20. Hall, P. (1994). On the erratic behavior of estimators of n in the binomial n,p distribution. *J. Amer. Statist. Assoc.* **89**, 344–352.

21. Hamedani, G. G. and Walter, G. G. (1988). Bayes estimation of the binomial parameter n. *Comm. Statist. Theory Methods* **17**, 1829–1843.

22. Hamedani, G. G. and Walter, G. G. (1990). Empirical Bayes estimation of the binomial parameter n. *Comm. Statist. Theory Methods* **19**, 2065–2084.

23. Hoel, P. (1947). Discriminating between binomial distributions. *Ann. Math. Statist.* **18**, 556–564.

24. Kahn, W. D. (1987). A cautionary note for Bayesian estimation of the binomial parameter n. *Amer. Statist.* **41**, 38–40.

25. Moran, P. A. (1951). A mathematical theory of animal trapping. *Biometrika* **38**, 307–311.

26. Olkin, I., Petkau, A. J., and Zidek, J. V. (1981). A comparison of n estimators for the binomial distribution. *J. Amer. Statist. Assoc.* **37**, pt. 4, 523–529.

27. O'Quigley, J. (1992). Estimating the binomial parameter n on the basis of pairs of known and observed proportions. *Appl. Statist.* **41**, 173–180.

28. Raftery, A. (1988). Inference for the binomial n parameter: A hierarchical Bayes approach. *Biometrika* **75**, 223–228.

29. Sadooghi-Alvandi, S. M. (1986). Admissible estimation of the binomial parameter n. *Ann. Statist.* **14**, 1634–1641.

30. Skellam, J. G. (1948). A probability distribution derived from the binomial distribution by regarding the probability of success as variable between the sets of trials. *J. Roy. Statist. Soc. B* **10**, 257–261.

31. Student. (1918). An explanation of deviations from Poisson's law in practice. *Biometrika* **12**, 211–215.

32. Whittaker, L. (1914). On the Poisson law of small numbers. *Biometrika* **10**, 36–71.

33. Wiper, M. P. and Pettit, L. I. (1994). Bayesian estimation of the binomial parameter n. *Appl. Statist.* **43**, 233–236.

34. Johnson, R. W. (1987). Simultaneous estimation of binomial n's. *Sankhyā Ser. A* **37**, pt. 2, 264–266.

See also BINOMIAL AND MULTINOMIAL PARAMETERS, INFERENCE ON.

NELS GREVSTAD

BINOMIAL, DISTRIBUTIONS RELATED TO THE. See CLASSICAL DISCRETE DISTRIBUTIONS, GENERALIZATIONS OF

BINOMIAL MOMENT. See FACTORIAL MOMENTS

BINOMIAL PARAMETER, CLOPPER-PEARSON INTERVAL ESTIMATION

Confidence intervals for the parameter p of a *binomial distribution** were proposed by Clopper and Pearson [1]. From an observed value of X successes in n trials, the 100. $(1 - \alpha)\%$ confidence interval is obtained by including all values of p such that

$$\sum_{x=0}^{X-1} \binom{n}{x} p^x (1-p)^{n-X} \geqslant 1 - \frac{1}{2}\alpha \qquad (1)$$

and

$$\sum_{x=X+1}^{n} \binom{n}{x} p^x (1-p)^{n-x} \geqslant 1 - \frac{1}{2}\alpha, \qquad (2)$$

that is, all values of p such that neither "tail" of the distribution (either below X or above X) is less then $\frac{1}{2}\alpha$. This leads to an interval with upper and lower limits given by solution of the equations (in p) obtained by replacing (1) and (2) by corresponding equalities.

The actual *confidence coefficient** varies with p; it cannot be less than $100(1 - \alpha)\%$ for any p. The same would be true if the right-hand sides of (1) and (2) were replaced by $1 - \alpha'$ and $1 - \alpha''$ with $\alpha' + \alpha'' = \alpha$. The symmetrical choice ($\alpha' = \alpha'' = \frac{1}{2}\alpha$) used by Clopper and Pearson is sometimes termed a *central* confidence interval.

REFERENCE

1. Clopper, C. J. and Pearson, E. S. (1934). *Biometrika*, **26**, 404–413.

See also BINOMIAL AND MULTINOMIAL PARAMETERS, INFERENCE ON; BINOMIAL DISTRIBUTION; and CONFIDENCE INTERVALS AND REGIONS.

BINOMIAL PROBABILITY PAPER

This is a graph paper with both scales proportional to square roots. It was introduced by Mosteller and Tukey [2]. The paper is used by plotting the nominal sample point $(n - x, x)$ (where x is the number of occurrences of an event in n binomial trials). The actual coordinate distances are $(\sqrt{n - x}, \sqrt{x})$, so the point lies somewhere on a quarter-circle of radius $\sqrt{n}$ with center at the origin. The angle between the line connecting the origin with the point and the horizontal axis is arcsin $\sqrt{x/n}$. Repeated random samples of size n will result in sample points at different points on the circumference of the circle. The standard error*, measured around the circumference, will be about $\frac{1}{2}$. It is a graphical representation of the *angular transformation* (*see* VARIANCE STABILIZATION).

For additional information, see ref. 1.

REFERENCES

1. Hald, A. (1952). *Statistical Theory with Engineering Applications*, Wiley, New York.
2. Mosteller, F. and Tukey, J. W. (1949). *J. Amer. Statist. Ass.*, **44**, 174–212.

See also BINOMIAL DISTRIBUTION and VARIANCE STABILIZATION

BINOMIAL PROPORTION, INFERENCE ON. See BINOMIAL AND MULTINOMIAL PARAMETERS, INFERENCE ON

BINOMIAL PROPORTION. See BINOMIAL DISTRIBUTION

BINOMIAL TEST

A test of the value of a (population) binomial proportion*. It is supposed that we have an observed value which can be represented by a random variable R that has a binomial distribution*.

$$\Pr[R = r] = \binom{n}{r} p^r (1 - p)^{n-r}$$

$$(r = 0, 1, \dots, n)$$

n is supposed known; the hypothesis tested relates to the value of p.

The sample size (n) corresponds to the number of trials. For example, if $H_0 : p = \frac{1}{2}$ and the alternative hypotheses* are $H_1 : p > \frac{1}{2}$ with $n = 25$, and the significance level* is $\alpha = 0.05$, the test rejects H_0 when $R > r_0$, where r_0 is usually chosen as the least integer for which the cumulative probability $\Pr[R \leqslant r_0] = \sum_{r=0}^{r_0} \binom{n}{r} p^r (1 - p)^{n-r}$ exceeds $1 - \alpha$. Sometimes r_0 is chosen to make this quantity as close to α as possible.

In the example above, when the hypothesis H_0 is valid,

$$\Pr[R > 16] = \sum_{r=17}^{25} \binom{n}{r} p^r (1 - p)^{n-r}$$

$$= \sum_{r=17}^{25} \binom{25}{r} \left(\frac{1}{2}\right)^r \left(\frac{1}{2}\right)^{25-r}$$

$$= 0.0530$$

and $\Pr[R > 17] = 0.0217$. The critical region* $R > 16$ gives the level of significance $\alpha = 0.0530$.

See also BINOMIAL AND MULTINOMIAL PARAMETERS, INFERENCE ON; BINOMIAL DISTRIBUTION; and HYPOTHESIS TESTING.

BIOASSAY, STATISTICAL METHODS IN

Bioassay refers to the process of evaluating the potency of a stimulus by analyzing the responses it produces in biological organisms. Examples of a stimulus are a drug, a hormone, radiation, an environmental effect, and various forms of toxicants. Examples of biological organisms are experimental animals, human volunteers, living tissues, and bacteria.

When a new drug is introduced, we are often interested in how it compares with a standard drug. One means of approaching this problem is to use the two drugs on living organisms from a common population. On the other hand, when an insecticide is being considered for use on an unfamiliar insect population, we may be more interested in the tolerance* of the insect population to the insecticide than in the relative strength of the insecticide compared to another. These are examples of two basic problems in bioassay. One is the evaluation of the relative potency of an unknown drug to a standard. The other is the estimation of the stimulus response.

The response to a stimulus may often be classified as quantitative or quantal. If the response can be measured, as would be the case if we were studying weight changes following the use of a vitamin, it is quantitative. On the other hand, if the response is all or nothing, as would be the case of death or survival, it is quantal (*see* BINARY DATA).

The development of statistical methods in bioassay has paralleled the development of statistics in general. Not only are the familiar techniques of regression analysis* and maximum likelihood* extensively used in bioassay, but a number of statistical problems have originated in bioassay and have stimulated the general development of statistics. The companion books by Finney [14,16], which are now in their third edition, have contributed greatly toward unifying the field and introducing many biomedical researchers, as well as seasoned statisticians, to current practices in bioassay. Reference 14 deals primarily with the probit model* for quantal responses, and ref. 16 with bioassay more generally, including recent developments in the field.

One of the earliest applications of the normal distribution* and normal deviate* to quantal responses was Fechner's [12] psychophysical experiment on human sensitivity to various physical stimuli. Probit analysis*, which is based on normal deviates, was widely used for estimating quantal response curves in the late 1920s and early 1930s and put into its present form by R. A. Fisher* [17]. During this period the need for standardizing various drugs, hormones, and toxicants brought about statistical methods for

estimating relative potencies [22]. The multitude of competing methods for estimating parameters of quantal response curves were unified under the theory of RBAN estimation* by Neyman [25]. A number of sequential designs* for quantal responses were proposed in the 1950s and 1960s [9] and a limited number of Bayesian methods were introduced in the 1970s [18]. The extensive use of animals for measuring concentrations of hormones and enzymes have been recently replaced by radioimmunoassay and related procedures known for their high precision [27].

ESTIMATION OF RELATIVE POTENCY FROM QUANTITATIVE RESPONSES

Consider a test drug T being compared to a standard drug S. T and S are said to be similar if there is a constant $\rho > 0$ such that the distribution of the response to dose z_2 of the test drug is equivalent to that of dose $z_1 = \rho z_2$ of the standard drug. When the drugs are similar, ρ is called the relative potency of T to S.

In many dose-response experiments the quantitative response Y is linearly related to the log dose, $x = \log z$, and we can assume the model

$$E[Y|x] = \alpha + \beta x, \qquad (1)$$

where α and β are unknown real-valued parameters (*see* REGRESSION (Various)). If S and T are similar, then the responses Y_1 and Y_2 to S and T, respectively, are related by

$$E[Y_1|x] = \alpha + \beta x$$
$$E[Y_2|x] = \alpha + \beta \log \rho + \beta x \qquad (2)$$

for some α, β, and ρ. This follows from the assumption that dose z of T is equivalent to dose ρz of S. The estimation of ρ based on this model is called a parallel-line assay, in contrast to a slope-ratio assay [16], which is based on a model where the regression is linear with respect to $x = z^\lambda$, for some $\lambda \neq 0$.

Given the intercepts $\alpha_1 = \alpha$ and $\alpha_2 = \alpha + \beta \log \rho$ and the common slope β, the horizontal distance between the regression lines (2) is $\log \rho = (\alpha_2 - \alpha_1)/\beta$. Thus if one has estimates a_1, a_2, and b of α_1, α_2, and β, he or she may estimate $\log \rho$ by $M = (a_2 - a_1)/b$ and ρ by $\log^{-1}(M)$.

A typical experiment for a parallel-line assay consists of a series of n_1 observations $(Y_{11}, \ldots, Y_{1n_1})$ at log doses $(x_{11}, \ldots, x_{1n_1})$ of the standard and n_2 observations $(Y_{21}, \ldots, Y_{2n_2})$ at log doses $(x_{21}, \ldots, x_{2n_2})$ of the unknown. In addition to the linearity assumption, suppose that the observations are independent and normally distributed with common variance σ^2. Then α_1, α_2, β and σ^2 may be estimated by the standard least-squares* method, and a γ-level confidence interval* for $\log \rho$ is given by

$$
(\bar{x}_1 - \bar{x}_2) + (1 - g)^{-1}
$$
$$
\times \left\{ M - \bar{x}_1 + \bar{x}_2 \pm \frac{st}{b} \right.
$$
$$
\times \left[(1 - g)\left(\frac{1}{n_1} + \frac{1}{n_2} \right) \right.
$$
$$
\left. \left. + \frac{(M - \bar{x}_1 + \bar{x}_2)^2}{\sum_{i=1}^{2} \sum_{j=1}^{n_i} (x_{ij} - \bar{x}_i)^2} \right]^{1/2} \right\}, \quad (3)
$$

where

$$
\bar{x}_i = \sum_{j=1}^{n_i} x_{ij}/n_i,
$$

$$
s^2 = \sum_{i=1}^{2} \sum_{j=1}^{n_i} (Y_{ij} - a_i - b_i x_{ij})^2/(n_1 + n_2 - 3),
$$

$$
g = \frac{t^2 s^2}{b^2 \sum_{i=1}^{2} \sum_{j=1}^{n_j} (x_{ij} - \bar{x}_i)^2},
$$

and t is the $(1 + \gamma)/2$ quantile of the t-distribution* with $n_1 + n_2 - 3$ d.f.

The expression (3) follows from the celebrated Fieller's theorem* [13], which has been widely used in deriving confidence intervals for ratios of two parameters whose estimators have a bivariate normal distribution*. We note that small values of g indicate a significant departure of β from 0. However, large values of g, (i.e., $g > 1$) indicate that β is not significantly different from 0 and the confidence region is the complement of the interval defined by (3) and is of little practical value.

Table 1. Uterine Weights in Coded Units

Standard Dose (mg)			Unknown Dose (mg)	
0.2	0.3	0.4	1.0	2.5
73	77	118	79	101
69	93	85	87	86
71	116	105	71	105
91	78	76	78	111
80	87	101	92	102
110	86		92	107
	101			102
	104			112

Source: Brownlee [7].

As an example, consider an estrogen hormone assay illustrated by Brownlee [7] using data derived from Emmons [11]. A total of 33 rats were assigned to three levels of the standard and two levels of the unknown. The response variable Y is some linear function of the logarithm of the weights of the rats' uteri and the independent variable x the logarithm of 10 times the dose. (See Table 1.)

In this example the estimated variance about the regression lines is $s^2 = 153.877$ and the estimated regression coefficients* are $a_1 = 67.7144$, $a_2 = 32.8817$, and $b = 50.3199$. Thus the estimated $\log \rho$ is $M = -0.6722$. Moreover, $g = 0.307316$ and the 95% confidence interval for $\log \rho$ based on (3) is $(-0.8756, -0.4372)$. Thus the estimate of ρ is 0.213 and its 95% confidence interval is $(0.133, 0.365)$. For computational details and tests for the validity of assumptions, see Brownlee [7, pp. 352–358].

QUANTAL RESPONSE MODELS

When the response to a drug (or other stimulus) is quantal, it is often reasonable and convenient to assume that each member of the population has a tolerance level such that the member responds to any dose greater than this level and does not to any lesser dose. This gives rise to the concept of a continuous tolerance distribution*, F, which has the characteristics of a CDF. In particular, the probability of a response from a randomly chosen member at dose x (usually measured in log units) is $F(x)$ and the probability of nonresponse is $(1 - F(x))$. In practice, this model is sometimes generalized to include an unknown proportion C of the population consisting of those who will respond in the absence of any stimulus and an unknown proportion D consisting of those that are immune to the stimulus and will not respond to any finite dose. When the responses are deaths and survivals, the dose which is lethal to 50% of the population is called LD50. More generally ED05, ED50, and ED90 denote doses that affect 5, 50 and 90% of the population.

Relative to a suitable transformation of the dose, such as the logarithm, the tolerance distribution is often approximated by a normal* or logistic* distribution, i.e.,

$$F(x) = \Phi(\alpha + \beta x) \qquad (4)$$

or

$$F(X) = \psi(\alpha + \beta x), \qquad (5)$$

where

$$\Phi(t) = \int_{-\infty}^{t} \frac{1}{\sqrt{2\pi}} \exp(-w^2/2)\,dw$$

$$\psi(t) = [1 + \exp(-t)]^{-1},$$

$-\infty < t < \infty$, and α and β are unknown parameters $-\infty < \alpha < \infty$, $0 < \beta < \infty$. Given the probability of response P, the probit* is defined for the normal (probit model) by the value of y such that $P = \Phi(y - 5)$ and the logit* for the logistic (logit model) by $y = \ln[P/(1 - P)]$. Berkson [4] promoted the use of the logit model because of its similarity to the normal and numerical tractability. In view of the recent development of computational facilities, the numerical convenience is usually negligible and the choice between the two is generally not crucial. Prentice [26] has introduced other distributions, including those that are skewed and may be better suited for modeling extreme percent points such as the ED99.

ESTIMATION OF QUANTAL RESPONSE CURVES

A typical experiment for quantal responses consists of taking n_i independent observations at log dose x_i and observing the frequency of responses $r_i, i = 1, \ldots, k$, where the number, k, of dose levels is generally at least two, and more than two when the model is being tested. If we assume that the tolerance distribution $F(x|\theta)$ belongs to a family of distributions with parameter θ, the log likelihood* function is given by

$$\sum_{i=1}^{k} \{r_i \log F(x_i|\theta) + (n_i - r_i)\log[1 - F(x_i|\theta)]\}.$$

For most of the commonly used models, including the probit and logit, the maximum likelihood estimators of θ do not have explicit expressions and estimates must be found numerically by iterative schemes such as the Newton-Raphson*. The probit method originally proposed by Fisher [17] is still a widely used method for finding such estimates and has been described in detail by Finney [14].

There are a number of other estimators, such as the minimum chi-square* and weighted least squares* which belong to the RBAN family [25] and hence have the same asymptotic efficiency* as the maximum likelihood estimator*. Their relative merits for small samples have not been studied extensively and the choice among them is usually a matter of convenience.

For the probit model, with x the log dose, $\log$ ED50 is estimated by $\hat{\mu} = a/b$, where (a, b) is the maximum likelihood estimate of (α, β). The γ-level confidence interval for log ED50, valid for large samples, is given by

$$\hat{\mu} + \frac{g}{1-g}(\hat{\mu} - \bar{x}) \pm \frac{K}{b(1-g)}$$

$$\times \left[\frac{1-g}{\sum_{i=1}^{k} n_i w_i} + \frac{(\hat{\mu} - \bar{x})^2}{\sum n_i w_i (x_i - \bar{x})^2}\right]^{1/2},$$

where

$$\bar{x} = \sum_{i=1}^{k} n_i w_i x_i \Big/ \sum_{i=1}^{k} n_i w_i,$$

$$w_i = z_i^2 / P_i Q_i,$$

$$z_i = (2\pi)^{1/2} \exp[-\tfrac{1}{2}(a + bx_i)^2],$$

$$g = \frac{K^2}{b^2 \sum_{i=1} n_i w_i (x_i - \bar{x})^2},$$

and K is the $(1 + \gamma)/2$ quantile of the standard normal distribution. As in the case of relative potency, the interval should be used only when $g < 1$.

Among the several nonparametric (distribution-free*) estimators of the ED50 that have been used, the Spearman–Kärber* estimator [23,28] is not only quite simple to use but is considered quite efficient [5]. If the levels are ordered such that $x_1 < \cdots < x_k$, this estimator is defined by

$$\tilde{\mu} = \sum_{i=1}^{k-1} (p_{i+1} - p_i)(x_i + x_{i+1})/2,$$

provided that $p_1 = 0$ and $p_k = 1$, where $p_i = r_i/n_i, i = 1, \ldots, k$. If $p_1 > 0$, then an extra level is added below x_1, where no responses are assumed to occur. Similarly, if $p_k < 1$, an extra level is added above x_k, where responses only are assumed to occur. When the levels are equally spaced with interval d, the variance of $\tilde{\mu}$ is estimated by

$$\text{var}(\tilde{\mu}) = d^2 \sum p_i(1 - p_i)/(n_i - 1),$$

provided that $n_i \geqslant 2, i = 1, \ldots, k$.

The maximum likelihood and Spearman–Kärber methods may be illustrated using data from Finney [14] on the toxicity of quinidine to frogs, given in Table 2. The dose was measured in units of 10^{-2} ml per gram of body weight. If the probit model with respect to log dose is assumed, maximum likelihood estimates may be computed using a SAS program [3]. The resulting maximum likelihood estimates are $a = -1.313$, $b = 4.318$, and log LD50 $= 1.462$. The 95% confidence interval for log LD50 is $(1.398, 1.532)$. On the other hand, the Spearman–Kärber estimates are $\tilde{\mu} = 1.449$ and $\text{var}(\tilde{\mu}) = 9.334 \times 10^{-4}$. (The estimate here uses an additional level at log dose $= 1.879$, where $p_i = 1$ is assumed.)

Table 2. Toxicity of Quinidine to Frogs

Log Dose	n	r	p
1.0000	26	0	0.0000
1.1761	24	2	0.0833
1.3522	23	9	0.3913
1.5276	24	17	0.7083
1.7033	26	20	0.7692

Source: Finney [14].

There are a few nonparametric procedures for estimating the entire tolerance distribution F. One approach is to use the isotone regression* method for estimating ordered binomial parameters [2]. Under this method the estimates of F at the points $x_1 < \cdots < x_k$ are given by

$$\hat{F}(x_i) = \min_{i \leqslant v \leqslant k} \max_{1 \leqslant u \leqslant i} \left(\sum_{v=u}^{v} r_v \Big/ \sum_{v=u}^{v} n_v \right),$$

$i = 1, \ldots, k$. Between these points and outside the interval $[x_1, x_k]$ the estimate $\hat{F}$ may be defined arbitrarily, subject to the constraint that $\hat{F}$ be nondecreasing.

Another parametric approach, which is Bayesian*, assumes that F is a random distribution function, whose distribution is defined by a Dirichlet process*, and uses the Bayes estimate of F with respect to a suitable loss function. Antoniak [1] gives a theoretical discussion of the Dirichlet prior and resulting posterior, as well as references to related works.

ESTIMATION OF RELATIVE POTENCY FROM QUANTAL RESPONSES

When quantal responses satisfy the probit or logit model with respect to log dose and the drugs are similar, a parallel-line assay may be performed to estimate the relative potency of an unknown to a standard. Under the probit model, for example, the condition of similarity requires that the probabilities of response at log dose x must satisfy

$$F_1(x) = \Phi(\alpha_1 + \beta x),$$
$$F_2(x) = \Phi(\alpha_2 + \beta x),$$

for the standard and unknown, respectively, for some parameters α_1, α_2, and $\beta > 0$. In this case, $\alpha_2 = \alpha_1 + \beta \log \rho$ or $\log \rho = (\alpha_2 - \alpha_1)/\beta$.

If we have two independent series of independent quantal response observations, one for the standard and the other for the unknown, the joint likelihood function may be formed and the maximum likelihood estimates a_1, a_2, and b of α_1, α_2, and β may be computed by the Newton-Raphson or similar method. These estimates may be used to estimate $\log \rho$ by $M = (a_2 - a_1)/b$. Use of

Fieller's theorem gives an approximate confidence interval for $\log \rho$ and ρ. Computational details and numerical illustrations are given in Finney [16].

DESIGN OF THE EXPERIMENT

The design of the experiment* for bioassay involves the selection of dose levels and the allocation of the living organisms to these levels in order to obtain experimental results that can be used to answer some predetermined question.

In estimating the relative potency, the objective may be to select the design that minimizes the width of the confidence interval* for a given confidence level and total sample size. If a parallel-line assay is being considered, it is important to have a rough estimate of ρ so that doses chosen for the unknown will give results comparable to those of the standard. One recommended approach is to use the same number of dose levels, usually between 3 and 6, for both the unknown and standard and the same number of organisms at each level. For each level of the standard, the level of the unknown should be chosen so that the predicted outcome will be like that of the standard. Although there are certain advantages to using dose levels that cover a wide range for qualitative responses, this does not hold for quantal responses since responses at extreme levels are quite predictable (being all responses or all nonresponses) and little information is gained. (See Finney [16] for more detailed instructions.)

In estimating the ED50 from quantal responses, again the objective may be to minimize the width of the confidence interval. For both the probit and Spearman–Kärber methods, it is generally recommended that an equal number of observations be taken at three to six equally spaced levels which are between 0.5 and 2 standard deviations (of F) apart. The specific recommendations vary according to the number of levels to be used, total sample size available, and previous information. (See Finney [14] for the probit method and Brown [6] for the Spearman–Kärber.)

A number of sequential* designs have been proposed as a means of allocating N

experimental units to different levels as information becomes available. The Robbins–Monro* process [9] and up-and-down method* [10], together with their variations, are the most intensively studied sequential procedures for estimating the ED50. For the up-and-down method, one observation at a time is taken, starting at some initial level x_0 and successively at levels $x_{i+1} = x_i \pm d$, where d is the step size chosen to be close to the standard deviation of F, a $+$ used if the ith observation is a nonresponse and a $-$ if it is a response. Under this scheme, the experimental dose levels tend to fluctuate about the ED50 and the average, $\Sigma_{i=2}^{N+1} x_i / N$, is used to estimate the ED50. For the probit model this estimator approximates the maximum likelihood estimator and has better efficiency than do the fixed sample designs [8,10]. The up-and-down method, when modified by using several observations at a time, not only reduces the number of trials but has been shown to provide even greater efficiency [29].

The selection of dose levels for estimating ED50 and extreme percent points, such as the ED90, of F may also be based on Bayesian principles. Freeman [18] has proposed a sequential method for estimating the ED50 of a one-parameter logit model, where the slope β is assumed known, and Tsutakawa [30] has proposed nonsequential designs for logit models, with unknown slope. These methods depend on the explicit and formal use of a prior distribution of the parameters and are aimed at minimizing the posterior variance of the percent point of interest.

RELATED AREAS

Some of the techniques that have been described above have been modified and extended to more complex problems. We will briefly describe some of these where further work can be expected.

The analysis of quantal responses to combinations or mixtures of two or more drugs introduces many additional problems, since we must not only consider the relative proportion of the drugs but also the interaction of the drugs. When there are two drugs and one is effectively a dilution of the other or they act independently, it is not difficult to extend the models for single drugs to those for a mixture. However, when the drugs interact, either antagonistically or synergistically, the model building becomes considerably more complex. Hewlett and Plackett [21] have discussed different modes in which the effect of the mixture of drugs may depend on the amount of drugs reaching the site of the action.

In many quantal response studies the time of response is an important variable. For example, in a carcinogenic experiment animals are often exposed to different doses of a carcinogen and kept under observation until they develop a tumor, die, or the experiment is terminated. For a discussion of such experiments and related references, see Hartley and Sielken [20].

Statistical techniques for bioassay have been used in radioimmunoassay and related techniques for measuring minute concentrations of hormones and enzymes. Radioactivity counts, resulting from antigen-antibody reactions, are observed at different doses of a ligand in order to estimate a dose-response curve. Under appropriate transformations, the counts are often related to the dose by a logistic model with unknown asymptotes (generally different from 0 and 1). For transformations, weighted least-squares methods*, and references, see Rodbard and Hutt [27] and Finney [15].

The probit and logit models have also been applied to mental testing and latent trait analysis, where human subjects with different mental abilities respond to questions of different degrees of difficulty [24]. When the responses are classified correct or incorrect, the probability of a correct answer to a particular question usually depends on the ability of the subject and can often be approximated by one of these models. In such cases, the ability of each subject can be estimated by using the joint response to several questions. See Hambleton and Cook [19] for a review of this area and related references.

REFERENCES

1. Antoniak, C. E. (1974). *Ann. Statist.*, **2**, 1152–1174.

2. Ayer, M., Brunk, H. D., Ewing, G. M., Reid, W. T., and Silverman, E. (1955). *Ann. Math. Statist.*, **26**, 641–647.

3. Barr, A. J., Goodnight, J. H., Sall, J. P., and Helwig, J. T. (1976). *A User's Guide to SAS76*. SAS Institute, Raleigh, N.C.

4. Berkson, J. (1944). *J. Amer. Statist. Ass.*, **39**, 357–365.

5. Brown, B. W., Jr. (1961). *Biometrika*, **48**, 293–302.

6. Brown, B. W., Jr. (1966). *Biometrics*, **22**, 322–329.

7. Brownlee, K. A. (1965). *Statistical Theory and Methodology in Science and Engineering*, 2nd ed. Wiley, New York.

8. Brownlee, K. A., Hodges, J. L., Jr., and Rosenblatt, M. (1953). *J. Amer. Statist. Ass.*, **48**, 262–277.

9. Cochran, W. G. and Davis, M. (1965). *J. R. Statist. Soc. B*, **27**, 28–44.

10. Dixon, W. J. and Mood, A. M. (1948). *J Amer. Statist. Ass.*, **43**, 109–126.

11. Emmons, C. W. (1948). *Principles of Biological Assay*. Chapman & Hall, London.

12. Fechner, G. T. (1860). *Elemente der Psychophysik*, Breitkopf und Hartel, Leipzig. (Translated into English in 1966 by H. E. Adler, Holt, Rinehart and Winston, New York.)

13. Fieller, E. C. (1940). *J. R. Statist. Soc. Suppl.*, **7**, 1–64.

14. Finney, D. J. (1971). *Probit Analysis*, 3rd ed. Cambridge University Press, Cambridge.

15. Finney, D. J. (1976). *Biometrics*, **32**, 721–740.

16. Finney, D. J. (1978). *Statistical Methods in Biological Assay*, 3rd ed. Macmillan, New York.

17. Fisher, R. A. (1935). *Ann. Appl. Biol.*, **22**, 134–167.

18. Freeman, P. R. (1970). *Biometrika*, **57**, 79–89.

19. Hambleton, R. K. and Cook, L. L. (1977). *J. Educ. Meas.*, **14**, 75–96.

20. Hartley, H. O. and Sielken, R. L., Jr. (1977). *Biometrics*, **33**, 1–30.

21. Hewlett, P. S. and Plackett, R. L. (1964). *Biometrics*, **20**, 566–575.

22. Irwin, J. O. (1937). *J. R. Statist. Soc. Suppl.*, **4**, 1–48.

23. Kärber, G. (1931). *Arch. exp. Pathol. Pharmakol.*, **162**, 480–487.

24. Lord, F. M. and Novick, M. R. (1968). *Statistical Theories of Mental Test Scores*. Addison-Wesley, Reading, Mass.

25. Neyman, J. (1949). *Proc. Berkeley Symp. Math. Statist. Prob.*, University of California Press, Berkeley, Calif., pp. 239–273.

26. Prentice, R. L. (1976). *Biometrics*, **32**, 761–768.

27. Rodbard, D. and Hutt, D. M. (1974). *Radioimmunoassay and Related Procedures in Medicine*, Vol. 1. International Atomic Energy Agency, Vienna, pp. 165–192.

28. Spearman, C. (1908). *Br. J. Psychol.*, **2**, 227–242.

29. Tsutakawa, R. K. (1967). *J. Amer. Statist. Ass.*, **62**, 842–856.

30. Tsutakawa, R. K. (1980). *Appl. Statist.*, **29**, 25–33.

See also Bivariate Normal Distribution, Fieller's Theorem; Isotonic Inference; Logit; Probit Analysis; Rankit; and Regression Analysis (Various Entries).

Robert K. Tsutakawa

BIOAVAILABILITY AND BIOEQUIVALENCE

DEFINITION

The bioavailability of a drug product is the rate and extent to which the active drug ingredient or therapeutic moiety is absorbed and becomes available at the site of drug action. A comparative bioavailability study refers to the comparison of bioavailabilities of different formulations of the same drug or different drug products. When two formulations of the same drug or different drug products are claimed to be bioequivalent, it is meant that they will provide the same therapeutic effect and that they are therapeutically equivalent. This assumption is usually referred to as the *fundamental bioequivalence assumption*. Under this assumption, the United States Food and Drug Administration* (FDA) does not require a complete new drug application (NDA) submission for approval of a generic drug product if the sponsor can provide evidence of bioequivalence between the generic drug product and the innovator through bioavailability studies.

HISTORY OF BIOAVAILABILITY

The study of drug absorption (e.g., sodium iodide) can be traced back to 1912 [2].

However, the concept of bioavailability and bioequivalence did not become a public issue until the late 1960s, when concern was raised that a generic drug product might not be as bioavailable as that manufactured by the innovator. These concerns rose from clinical observations in humans. The investigation was facilitated by techniques permitting the measurement of minute quantities of drugs and their metabolites in biologic fluids. In 1970, the FDA* began to ask for evidence of biological availability in applications submitted for approval of certain new drugs. In 1974, a Drug Bioequivalence Study Panel was formed by the Office of Technology Assessment (OTA) to examine the relationship between the chemical and therapeutic equivalence of drug products. Based on the recommendations in the OTA report, the FDA published a set of regulations for submission of bioavailability data in certain new drug applications. These regulations became effective on July 1, 1977 and are currently codified in 21 Code of Federal Regulation (CFR) Part 320. In 1984, the FDA was authorized to approve generic drug products under the Drug Price Competition and Patent Term Restoration Act. In recent years, as more generic drug products become available, there is a concern that generic drug products may not be comparable in identity, strength, quality, or purity to the innovator drug product. To address this concern, the FDA conducted a hearing on Bioequivalence of Solid Oral Dosage Forms in Washington, D.C. in 1986. As a consequence of the hearing, a Bioequivalence Task Force (BTF) was formed to evaluate the current procedures adopted by the FDA for the assessment of bioequivalence between immediate solid oral dosage forms. A report from the BTF was released in January 1988. Based on the recommendations of the report by the BTF, a guidance on statistical procedures for bioequivalence studies was issued by the FDA Division of Bioequivalence, Office of Generic Drugs in 1992 [6].

BIOEQUIVALENCE MEASURES

In bioavailability/bioequivalence studies, following the administration of a drug, the blood or plasma concentration—time curve is often used to study the absorption of the drug. The curve can be characterized by taking blood samples immediately prior to and at various time points after drug administration. The profile of the curve is then studied by means of several pharmacokinetic parameters such as the area under it (AUC), the maximum concentration (C_{max}), the time to achieve maximum concentration (t_{max}), the elimination half-life $(t_{1/2})$, and the rate constant (k_e). Further discussion of the blood or plasma concentration—time curve can be found in Gibaldi and Perrier [9]. The AUC, which is one of the primary pharmacokinetic parameters, is often used to measure the extent of absorption or total amount of drug absorbed in the body. Based on these pharmacokinetic parameters, bioequivalence may be assessed by means of the difference in means (or medians), the ratio of means (or medians), and the mean (or median) of individual subject ratios of the primary pharmacokinetic variables. The commonly used bioequivalence measures are the difference in means and the ratio of means.

DECISION RULES

Between 1977 and 1980, the FDA proposed a number of decision rules for assessing bioequivalence in average bioavailability [14]. These decision rules include the 75/75 rule, the 80/20 rule, and the ±20 rule. The 75/75 rule claims bioequivalence if at least 75% of individual subject ratios [i.e., individual bioavailability of the generic (test) product relative to the innovator (reference) product] are within (75%, 125%) limits. The 80/20 rule concludes there is bioequivalence if the test average is not statistically significantly different from the reference average and if there is at least 80% power for detection of a 20% difference of the reference average. The BTF does not recommend these two decision rules, because of their undesirable statistical properties. The ±20 rule suggests that two drug products are bioequivalent if the average bioavailability of the test product is within ±20% of that of the reference product with a certain assurance (say 90%). Most recently, the FDA guidance recommends a 80/125 rule for log-transformed

data. The 80/125 rule claims bioequivalence if the ratio of the averages between the test product and the reference product falls within (80%, 125%) with 90% assurance. The ±20 rule for raw data and the 80/125 rule for log-transformed data are currently acceptable to the FDA for assessment of bioequivalence in average bioavailability. Based on the current practice of bioequivalence assessment, it is suggested that the 80/125 rule be applied to AUC and C_{max} for all drug products across all therapeutic areas.

DESIGNS OF BIOAVAILABILITY STUDIES

The *Federal Register* [7] indicated that a bioavailability study (single-dose or multiple-dose) should be crossover in design. A crossover design (*see* CHANGEOVER DESIGNS) is a modified randomized block design in which each block (i.e., subject) receives more than one formulation of a drug at different time periods. The most commonly used study design for assessment of bioequivalence is the two-sequence, two-period crossover* design (Table 1), which is also known as the standard crossover design. For the standard crossover design, each subject is randomly assigned to either sequence 1 (R-T) or sequence 2 (T-R). In other words, subjects within sequence R-T (T-R) receive formulation R (T) during the first dosing period and formulation T (R) during the second dosing period. Usually the dosing periods are separated by a washout period of sufficient length for the drug received in the first period to be completely metabolized and/or excreted by the body.

In practice, when differential carryover effects are present, the standard crossover design may not be useful because the formulation effect is confounded with the carryover effect. In addition, the standard crossover design does not provide independent estimates of intrasubject variability for each formulation, because each subject only receives each formulation once. To overcome these drawbacks, Chow and Liu [3] recommend a higher-order crossover design be used. Table 1 lists some commonly used higher-order crossover designs, which include Balaam's design, the two-sequence dual design,

Table 1. Crossover Designs for Two Formulations

Standard Crossover Design		
	Period	
Sequence	I	II
1	R	T
2	T	R

Balaam Design		
	Period	
Sequence	I	II
1	T	T
2	R	R
3	R	T
4	T	R

Two-Sequence Dual Design			
	Period		
Sequence	I	II	III
1	T	R	R
2	R	T	T

Four-Sequence Optimal Design				
	Period			
Sequence	I	II	III	IV
1	T	T	R	R
2	R	R	T	T
3	T	R	R	T
4	R	T	T	R

and the optimal four-sequence design. As an alternative, Westlake [16] suggested the use of an incomplete block design*. Note that crossover designs for comparing more than two formulations are much more complicated than those for comparing two formulations. In the interest of the balance property, Jones and Kenward [10] recommend a Williams design be used.

STATISTICAL METHODS FOR AVERAGE BIOAVAILABILITY

Without loss of generality, consider the standard two-sequence, two-period crossover experiment. Let Y_{ijk} be the response [e.g., log(AUC)] of the ith subject in the kth sequence at the jth period, where $i = 1,\ldots,n_k$, $k = 1,2$, and $j = 1,2$. Under the assumption that there are no carryover effects, Y_{ijk} can be described by the statistical

model

$$Y_{ijk} = \mu + S_{ik} + F_{(j,k)} + P_j + e_{ijk},$$

where μ is the overall mean, S_{ik} is the random effect of the ith subject in the kth sequence, P_j is the fixed effect of the jth period, $F_{(j,k)}$ is the direct fixed effect of the formulation in the kth sequence which is administered at the jth period, and e_{ijk} is the within-subject random error in observing Y_{ijk}.

The commonly used approach for assessing bioequivalence in average bioavailability is the method of the classical (shortest) confidence interval. Let μ_T and μ_R be the means of the test and reference formulations, respectively. Then, under normality assumptions, the classical $(1 - 2\alpha) \times 100\%$ confidence interval for $\mu_T - \mu_R$ is as follows:

$$L = (\overline{Y}_T - \overline{Y}_R) - t(\alpha, n_1 + n_2 - 2)$$
$$\times \hat{\sigma}_d \sqrt{\frac{1}{n_1} + \frac{1}{n_2}},$$
$$U = (\overline{Y}_T - \overline{Y}_R) + t(\alpha, n_1 + n_2 - 2)$$
$$\times \hat{\sigma}_d \sqrt{\frac{1}{n_1} + \frac{1}{n_2}},$$

where $\overline{Y}_T$ and $\overline{Y}_R$ are the least-squares means for the test and reference formulations, $t(\alpha, n_1 + n_2 - 2)$ is the upper αth critical value of a t-distribution* with $n_1 + n_2 - 2$ degrees of freedom, and $\hat{\sigma}_d^2$ is given by

$$\hat{\sigma}_d^2 = \frac{1}{n_1 + n_2 - 2} \sum_{k=1}^{2} \sum_{i=1}^{n_k} (d_{ik} - \overline{d}_{.k})^2,$$

in which

$$d_{ik} = \tfrac{1}{2}(Y_{i2k} - Y_{i1k}) \text{ and}$$
$$\overline{d}_{.k} = \frac{1}{n_k} \sum_{i=1}^{n_k} d_{ik}.$$

According to the 80/125 rule, if the exponentiations of L and U are within (80%, 125%), then we conclude the two formulations are bioequivalent.

As an alternative, Schuirmann [15] proposed a procedure with two one-sided tests to evaluate whether the bioavailability of the test formulation is too high (safety) for one side and is too low (efficacy) for the other side. In it, we conclude that the two formulations are bioequivalent if

$$T_L = \frac{(\overline{Y}_T - \overline{Y}_R) - \theta_L}{\hat{\sigma}_d \sqrt{\frac{1}{n_1} + \frac{1}{n_2}}} > t(\alpha, n_1 + n_2 - 2)$$

and

$$T_U = \frac{(\overline{Y}_T - \overline{Y}_R) - \theta_U}{\hat{\sigma}_d \sqrt{\frac{1}{n_1} + \frac{1}{n_2}}}$$
$$< -t(\alpha, n_1 + n_2 - 2),$$

where $\theta_L = \log(0.8) = -0.2331$ and $\theta_U = \log(1.25) = 0.2331$ are bioequivalence limits.

In addition to the ± 20 rule (and the recent 80/125 rule), several other methods have been proposed. These include the Westlake symmetric confidence interval [17], Chow and Shao's joint-confidence-region approach [5], an exact confidence interval based on Fieller's theorem* [8,12], Anderson and Hauck's test for interval hypotheses [1], a Bayesian approach for the highest posterior density (HPD) interval, and nonparametric methods. Some of these methods are actually operationally equivalent in the sense that they will reach the same decision on bioequivalence. More details can be found in Chow and Liu [2].

SAMPLE-SIZE DETERMINATION

For bioequivalence trials, a traditional approach for sample size determination is to conduct a power analysis based on the 80/20 decision rule. This approach, however, is based on point hypotheses rather than interval hypotheses and therefore may not be statistically valid. Phillips [13] provided a table of sample sizes based on power calculations of Schuirmann's two one-sided tests procedure using the bivariate noncentral t-distribution*. However, no formulas are provided. An approximate formula for sample-size calculations was provided in Liu and Chow [11]. Table 2 gives the total sample sizes needed to achieve a desired power

Table 2. Sample Sizes for Schuirmann's Procedures with Two One-Sided Tests

Power %	CV (%)	$\theta = \mu_T - \mu_R$ 0%	5%	10%	15%
80	10	8	8	16	52
	12	8	10	20	74
	14	10	14	26	100
	16	14	16	34	126
	18	16	20	42	162
	20	20	24	52	200
	22	24	28	62	242
	24	28	34	74	288
	26	32	40	86	336
	28	36	46	100	390
	30	40	52	114	448
	32	46	58	128	508
	34	52	66	146	574
	36	58	74	162	644
	38	64	82	180	716
	40	70	90	200	794
90	10	10	10	20	70
	12	10	14	28	100
	14	14	18	36	136
	16	16	22	46	178
	18	20	28	58	224
	20	24	32	70	276
	22	28	40	86	334
	24	34	46	100	396
	26	40	54	118	466
	28	44	62	136	540
	30	52	70	156	618
	32	58	80	178	704
	34	66	90	200	794
	36	72	100	224	890
	38	80	112	250	992
	40	90	124	276	1098

Source: Liu and Chow [11].

for a standard crossover design for various combinations of $\theta = \mu_T - \mu_R$ and CV, where

$$\text{CV} = 100 \times \frac{\sqrt{2\hat{\sigma}_d^2}}{\mu_R}.$$

CURRENT ISSUES

As more generic drug products become available, current regulatory requirements and unresolved scientific issues that may affect the identity, strength, quality, and purity of generic drug products have attracted much attention in recent years. For example, for regulatory requirements, the FDA has adopted the analysis for log-transformed data and the 80/125 rule. To allow for variability of bioavailability and to ensure drug exchangeability, the FDA is seeking alternative pharmacokinetic parameters, decision rules, and statistical methods for population and individual bioequivalence. Some unresolved scientific issues of particular interest include the impact of add-on subjects for dropouts, the use of female subjects in bioequivalence trials, in vitro dissolution as a surrogate for in vivo bioequivalence, postapproval bioequivalence, and international harmonization for bioequivalence requirements among the European Community, Japan, and the United States. A comprehensive overview of these issues can be found in Chow and Liu [4].

REFERENCES

1. Anderson, S. and Hauck, W. W. (1983). A new procedure for testing equivalence in comparative bioavailability and other clinical trials. *Commun. Statist. Theory Methods*, **12**, 2663–2692.

2. Chow, S. C. and Liu, J. P. (1992). *Design and Analysis of Bioavailability and Bioequivalence Studies*. Marcel Dekker, New York.

3. Chow, S. C. and Liu, J. P. (1992). On assessment of bioequivalence under a higher-order crossover design. *J. Biopharm. Statist.*, **2**, 239–256.

4. Chow, S. C. and Liu, J. P. (1995). Current issues in bioequivalence trials. *Drug. Inform. J.*, **29**, 795–804.

5. Chow, S. C. and Shao, J. (1990). An alternative approach for the assessment of bioequivalence between two formulations of a drug. *Biometrics J.*, **32**, 969–976.

6. FDA (1992). Guidance on Statistical Procedures for Bioequivalence Studies Using a Standard Two-Treatment Crossover Design. Division of Bioequivalence, Office of Generic Drugs, Food and Drug Administration, Rockville, Md., July 1.

7. Federal Register (1977). Vol. 42, No. 5, Section 320.26(b).

8. Fieller, E. (1954). Some problems in interval estimation. *J. R. Statist. Soc. B*, **16**, 175–185.

9. Gibaldi, M. and Perrier, D. (1982). *Pharmacokinetics*. Marcel Dekker, New York.

10. Jones, B. and Kenward, M. G. (1989). *Design and Analysis of Crossover Trials*. Chapman & Hall, London.

11. Liu, J. P. and Chow, S. C. (1992). Sample size determination for the two one-sided tests procedure in bioequivalence. *J. Pharmacokin. Biopharm.*, **20**, 101–104.

12. Locke, C. S. (1984). An exact confidence interval for untransformed data for the ratio of two formulation means. *J. Pharmacokin. Biopharm.*, **12**, 649–655.

13. Phillips, K. F. (1990). Power of the two one-sided tests procedure in bioequivalence. *J. Pharmacokin. Biopharm.*, **18**, 137–144.

14. Purich, E. (1980). Bioavailability/bioequivalency regulations: an FDA perspective. In *Drug Absorption and Disposition: Statistical Considerations*, K. S. Albert, ed. American Pharmaceutical Association, Academy of Pharmaceutical Sciences, Washington, pp. 115–137.

15. Schuirmann, D. J. (1987). A comparison of the two one-sided tests procedure and the power approach for assessing the equivalence of average bioavailability. *J. Pharmacokin. Biopharm.*, **15**, 657–680.

16. Westlake, W. J. (1973). The design and analysis of comparative blood-level trials. In *Current Concepts in the Pharmaceutical Sciences*, J. Swarbrick, ed. Lea & Febiger, Philadelphia.

17. Westlake, W. J. (1976). Symmetrical confidence intervals for bioequivalence trials. *Biometrics*, **32**, 741–744.

See also BIOEQUIVALENCE CONFIDENCE INTERVALS; CHANGEOVER DESIGNS; FDA STATISTICAL PROGRAMS: HUMAN DRUGS; and PHARMACEUTICAL INDUSTRY, STATISTICS IN.

SHEIN-CHUNG CHOW

BIOEQUIVALENCE CONFIDENCE INTERVALS

As described in the entry BIOAVAILABILITY AND BIOEQUIVALENCE [1], the objective of *bioequivalence* testing is to see whether there is sufficient evidence that a generic drug is equivalent in practice to that manufactured by the innovator, in terms of various pharmacokinetic parameters. This entry clarifies the relationship between tests of hypotheses* and confidence intervals* for bioequivalence.

Let μ_T and μ_R denote the means—for the test drug and the reference drug respectively—of a specific pharmacokinetic parameter. For simplicity, our definition will consider two drugs to be *bioequivalent* in that parameter if $|\mu_T - \mu_R| < \delta$ for some prespecified $\delta > 0$ (cf. the 80/125 rule described in Chow [1]). See Hsu et al. [4] for an analogous development when the definition is $\delta_1 < \mu_T/\mu_R < \delta_2$ (cf. the ±20% rule in ref. [1]).

Referring to the two-sequence two-period crossover design in ref. [1], denote by $\overline{Y}_T, \overline{Y}_R$, and $\hat{\sigma}^2$ the usual estimates of μ_T, μ_R, and σ^2, and let n_1 and n_2 denote the numbers of observations obtained under the test-followed-by-reference drug sequence and the reference-followed-by-test drug sequence, respectively. Under the usual assumptions, $\overline{Y}_T - \overline{Y}_R$ is an unbiased estimate* of $\mu_T - \mu_R$, normally distributed with mean $\mu_T - \mu_R$ and standard deviation $\sigma\sqrt{n_1^{-1} + n_2^{-1}}$; $\hat{\sigma}^2$ is independent of $\overline{Y}_T - \overline{Y}_R$, and if $\nu = n_1 + n_2 - 2$, then $\nu\hat{\sigma}^2/\sigma^2$ has a χ^2distribution* with ν degrees of freedom. Let $t_{q,\nu}$ denote the qth quantile* of the t-distribution* with ν degrees of freedom.

THE TWO ONE-SIDED TESTS PROCEDURE

To establish bioequivalence, both the FDA bioequivalence guideline [3] and the European Community guideline [2] specify that a *two one-sided tests* procedure should be performed at level 5%. The α-level two one-sided tests of Westlake [8] (cf. Schurimann [6]) tests

$$H : \mu_T - \mu_R \leqslant -\delta \text{ or } \mu_T - \mu_R \geqslant \delta$$
vs.
$$K : -\delta < \mu_T - \mu_R < \delta,$$

using a size-α t-test for

$$H_- : \mu_T - \mu_R \leqslant -\delta \qquad (1)$$

vs.

$$K_- : \mu_T - \mu_R > -\delta, \qquad (2)$$

and a size-α t-test for

$$H_+ : \mu_T - \mu_R \geqslant \delta \qquad (3)$$

vs.

$$K_+ : \mu_T - \mu_R < \delta. \qquad (4)$$

If both H_- and H_+ are rejected, then H is rejected in favor of K, establishing *bioequivalence* between the test treatment and the reference treatment.

One would expect the two one-sided tests procedure to be identical to some *confidence interval* procedure: For some appropriate $100(1 - \alpha)\%$ confidence interval $[C_T^-, C_T^+]$ for $\mu_T - \mu_R$, declare the test treatment to be bioequivalent to the reference treatment if and only if $[C_T^-, C_T^+] \subset (-\delta, \delta)$. It has been noted (e.g., Westlake [8, p. 593]; Schuirmann [6, p. 661]) that the two one-sided tests procedure is operationally identical to the procedure of declaring equivalence only if the ordinary $100(1 - 2\alpha)\%$, not $100(1 - \alpha)\%$, confidence interval for $\mu_T - \mu_R$, i.e.,

$$\overline{Y}_T - \overline{Y}_R \pm t_{1-\alpha,\nu}\hat{\sigma}\sqrt{n_1^{-1} + n_2^{-1}}, \qquad (5)$$

is completely contained in the interval $(-\delta, \delta)$. Indeed, both the FDA bioequivalence guideline [3] and the European Community guideline [2] specify that the two one-sided tests should be executed in this fashion.

WESTLAKE'S SYMMETRIC CONFIDENCE INTERVAL

The fact that the two one-sided tests procedure seemingly corresponds to a $100(1 - 2\alpha)\%$, not a $100(1 - \alpha)\%$, confidence interval perhaps caused some concern (Westlake [8, p. 593]). As an alternative, Westlake [7] proposed his *symmetric* confidence interval procedure, which declares μ_R to be practically equivalent to μ_T when the symmetric confidence interval

$$\left[\overline{Y}_T - \overline{Y}_R - k_2\hat{\sigma}\sqrt{n_1^{-1}, + n_2^{-1}}, \right.$$
$$\left. \overline{Y}_T - \overline{Y}_R + k_1\hat{\sigma}\sqrt{n_1^{-1} + n_2^{-1}} \right] \qquad (6)$$

is contained in $(-\delta, \delta)$. Here k_1 and k_2 are determined by the pair of equations

$$P\left\{ -k_2 \leqslant \frac{\overline{Y}_T - \overline{Y}_R - (\mu_T - \mu_R)}{\hat{\sigma}\sqrt{n_1^{-1} + n_2^{-1}}} \leqslant k_1 \right\}$$
$$= 1 - \alpha, \qquad (7)$$

$$2(\overline{Y}_T - \overline{Y}_R) + (k_1 - k_2)\hat{\sigma}\sqrt{n_1^{-1} + n_2^{-1}} = 0, \quad (8)$$

of which the first states that the probability of a t random variable with ν degrees of freedom being between $-k_2$ and k_1 must be $1 - \alpha$, and the second states that the interval (6) must be symmetric about 0. Then (6) is a conservative confidence interval for $\mu_T - \mu_R$ [7].

A BIOEQUIVALENCE CONFIDENCE INTERVAL

Invoking the connection between tests and confidence sets (cf. Lehmann [5, p. 90]), Hsu et al. [4] showed that the $100(1 - \alpha)\%$ confidence interval for $\mu_T - \mu_R$ corresponding to the family of tests such that, for $\mu_T - \mu_R \geqslant 0$,

$$\phi_{\mu_T - \mu_R}(\overline{Y}_T - \overline{Y}_R)$$

$$= \begin{cases} 1 & \text{if } \overline{Y}_T - \overline{Y}_R - (\mu_T - \mu_R) \\ & < -t_{1-\alpha,\nu}\hat{\sigma}\sqrt{n_1^{-1} + n_2^{-1}}, \quad (9) \\ 0 & \text{otherwise}, \end{cases}$$

and for $\mu_T - \mu_R < 0$,

$$\phi_{\mu_T - \mu_R}(\overline{Y}_T - \overline{Y}_R)$$

$$= \begin{cases} 1 & \text{if } \overline{Y}_T - \overline{Y}_R - (\mu_T - \mu_R) \\ & > t_{1-\alpha,\nu}\hat{\sigma}\sqrt{n_1^{-1} + n_2^{-1}}, \quad (10) \\ 0 & \text{otherwise}, \end{cases}$$

is

$$\left[-\left(\overline{Y}_T - \overline{Y}_R - t_{1-\alpha,\nu}\hat{\sigma}\sqrt{n_1^{-1} + n_2^{-1}}\right)^-, \right.$$

$$\left. \left(\overline{Y}_T - \overline{Y}_R + t_{1-\alpha,\nu}\hat{\sigma}\sqrt{n_1^{-1} + n_2^{-1}}\right)^+ \right], \quad (11)$$

where $-x^- = \min(0, x)$ and $x^+ = \max(0, x)$. (Technically the upper limit is open, but this point is inconsequential in bioequivalence trials, since the inference $\mu_T \neq \mu_R$ is not of interest.) While the two one-sided tests (1)–(4) are in the family of tests (9) and (10), the same cannot be said of the family of tests corresponding to the ordinary $100(1 - 2\alpha)\%$ confidence interval (5). Thus, it is more appropriate to say that the $100(1 - \alpha)\%$ equivalence confidence interval (11), rather than the ordinary $100(1 - 2\alpha)\%$

confidence interval (5), corresponds to the two level-α one-sided tests.

The equivalence confidence interval (11) is always contained in Westlake's symmetric confidence interval (6): Without loss of generality, suppose $\overline{Y}_T > \overline{Y}_R$. Then $k_1 > t_{1-\alpha,\nu}$ by (7) (with equality only if $k_2 = \infty$). Therefore,

$$\overline{Y}_T - \overline{Y}_R + t_{1-\alpha,\nu}\hat{\sigma}\sqrt{n_1^{-1} + n_2^{-1}}$$

$$< \overline{Y}_T - \overline{Y}_R + k_1\hat{\sigma}\sqrt{n_1^{-1} + n_2^{-1}},$$

establishing that the equivalence confidence upper bound is smaller than Westlake's symmetric confidence upper bound. By (8), $k_2 > k_1$ when $\overline{Y}_T > \overline{Y}_R$. Therefore, noting $k_2 > 0$, we have

$$-\left(\overline{Y}_T - \overline{Y}_R - t_{1-\alpha,\nu}\hat{\sigma}\sqrt{n_1^{-1} + n_2^{-1}}\right)^-$$

$$> \overline{Y}_T - \overline{Y}_R - k_2\hat{\sigma}\sqrt{n_1^{-1} + n_2^{-1}},$$

establishing that the equivalence confidence lower bound is larger than Westlake's symmetric lower confidence bound.

REFERENCES

1. Chow, S.-C. (1997). Bioavailability and bioequivalence. In *Encyclopedia of Statistical Sciences*, Update Vol. 1, Wiley, New York. S. Kotz, C. B. Read, and D. L. Banks, eds., pp. 51–56.

2. CPMP Working Party on Efficacy of Medical Products (1993). *Biostatistical Methodology in Clinical Trials in Applications for Marketing Authorization for Medical Products*, draft guideline ed. Commission of the European Communities, Brussels.

3. FDA (1992). Bioavailability and bioequivalence requirements. In *U.S. Code of Federal Regulations*, vol. 21. U.S. Government Printing Office, Chap. 320.

4. Hsu, J. C., Hwang, J. T. G., Liu, H. K., and Ruberg, S. J. (1994). Confidence intervals associated with bioequivalence trials. *Biometrika*, **81**, 103–114.

5. Lehmann, E. L. (1986). *Testing Statistical Hypotheses*, 2nd ed. Wiley, New York.

6. Schuirmann, D. J. (1987). A comparison of the two one-sided test procedure and the power approach for assessing the equivalence of average bioavailability. *J. Pharmacokinet. and Biopharmaceut.*, **15**(6), 657–680.

7. Westlake, W. J. (1976). Symmetric confidence intervals for bioequivalence trials. *Biometrics*, **32**, 741–744.

8. Westlake, W. J. (1981). Response to T. B. L. Kirkwood: bioequivalence testing—a need to rethink. *Biometrics*, **37**, 589–594.

See also BIOAVAILABILITY AND BIOEQUIVALENCE and CONFIDENCE INTERVALS AND REGIONS.

JASON C. HSU

BIOGEOGRAPHY, STATISTICS IN

E. C. Pielou has defined the work of a biogeographer to consist in observing, recording, and explaining the geographic ranges of all living things [15]. Biogeographers have introduced a variety of mathematical models to explain the observed phenomena. Many of these models are stochastic by nature, and hence the use of statistical methods is inevitable. Furthermore, biogeographical data are usually incomplete, consisting of a set of measurements or a sample of observations. For this reason we would like to add the word "estimating" to Pielou's definition to emphasize the need for statistical treatment also in the assessment of the range.

Many of the problems in biogeography and the methods to solve them are similar to those in other fields of geography*. Conversely, the ideas presented here are applicable in several fields involving spatial data.

MAPPING GEOGRAPHICAL RANGE

The geographical range of a wildlife population at some point of time can be defined as the collection of the locations of the individuals in the population at that moment [7]. Naturally, obtaining such complete observations is often practically impossible. For this reason it is typically necessary in mapping of geographical ranges to coarsen either the spatial or the temporal resolution, or both. By temporal resolution we mean the duration of the study, and by spatial resolution the geographic accuracy which is used in the recording. Choosing the resolutions means balancing the wished-for accuracy against the resources available. For a detailed consideration of the scale problems, see ref. 19.

ATLAS METHOD

A common way to fix the spatial resolution is to apply a grid mapping technique known as the *atlas method*. In an atlas survey the study area is divided into a grid of sites, which are typically squares of equal size. The aim is to confirm in each site whether it is inhabited by the target species during the survey or not. Thus the size of the site determines the spatial resolution. Atlas surveys are usually vast, often nationwide, projects; a typical size of an atlas square is 10×10 km. For practical reasons such surveys are based on fieldwork of voluntary observers. Sometimes the atlas data are collected more systematically, yielding typically a sample of thoroughly explored sites.

The most notable problem in large atlas surveys utilizing voluntary naturalists is the insufficient and heterogeneous coverage of fieldwork. As a result the observed atlas maps underestimate the true range because some of the sites inhabited by the target population remain uncovered. The coverage problems are most serious when a survey involves either remote areas having scarce human population, or species that are hard to detect.

The advantages of the atlas method are its comprehensiveness, objectivity, and methodological simplicity, which is important in vast surveys exploiting large numbers of voluntary observers. Prior to atlas surveys, distribution maps were typically subjective and based on rough limits of ranges produced by joining known "extreme" locations of occurrence. More detailed discussions about the atlas method can be found in Sharrock [17] and Högmander [10].

In the following sections we review how atlas data have been applied to estimate and explain wildlife distributions. First, we consider the sitewise presence or absence of the target population. Furthermore, atlas-type data can be utilized in estimating geographical variation in the abundance and the "commonness" of the population. Commonness can be roughly defined as the probability of encountering an individual of the population at a random location. Hence it depends not only on the density of the population but also on how regularly (as opposed to clustered) the individuals are located.

One has to be careful with the popular term "probability of occurrence," since its definition is not unique in biogeography. It has been used for the estimated probability of presence (at a site) in prediction of range, or for the commonness of a species on some scale, sometimes neither exactly.

MODELING GEOGRAPHICAL RANGE

Several more or less statistical procedures have been used in attempts to explain observed wildlife distributions by environmental factors; examples include discriminant analysis* [5], correspondence analysis* [9], classification trees [20], and generalized additive models* [21]. Here we shall restrict attention to *logistic regression* models, which seem to be most natural and popular in this context.

Ordinary logistic regression* has been applied in refs. 4, 9, 13, 14, 20. The presence or absence at all sites of the study area or at a random sample of them is regressed against environmental or geographical covariates, or both. In order to remove correlation* between the explanatory variables and to reduce their number, Osborne and Tigar [14] and Buckland and Elston [4] performed a preliminary principal-component analysis*. Canonical correspondence analysis* was applied in Hill [9] for this purpose. Buckland and Elston [4] also explain how to handle data with different resolutions for different variables.

It should be noticed that this approach models the observed rather than the actual distribution. Highly variable coverage or low detection probability makes the interpretation of the model in terms of the true distribution somewhat questionable. Osborne and Tigar include a measure of the observational effort as a covariate to allow for uneven coverage.

It is typical that neighboring sites have similar environments. If available covariates do not fully reflect this, then the residuals at neighboring sites tend to be positively correlated. Another obvious source of spatial autocorrelation is the spread of the species. Smith [18] and Augustin et al. [1] used a generalized version of the logistic model that allows for spatial correlation. This autologistic model includes counts or proportions of occupied sites in a suitably chosen neighborhood as an additional explanatory variable (autocovariate). In the same spirit Le Duc et al. [12] combine spatial smoothing with logistic regression.

Due to the introduced dependence structure, maximum-likelihood estimation* becomes complicated. An alternative maximum-pseudo-likelihood* method of Besag [2] corresponds to acting as if the autocovariate were just another independent variable. Better approximations are typically highly computer-intensive*. Another complication occurs if only a sample of squares is surveyed. Then the autocovariate cannot be calculated, and data augmentation* techniques are required.

PREDICTING GEOGRAPHICAL RANGE

For each site i where the covariates are available, a fitted value $\hat{p}_i \in (0, 1)$ can be calculated from the estimated logistic model. This is often interpreted as a "probability of occurrence," although that is justified only if the sample sites can be assumed thoroughly covered. If a considerable proportion of observed "absences" are actually nondetected presences, then the $\hat{p}_i$ tend to give an underestimate of the true distribution. Hence they can only be considered as relative measures. If observational effort is used as a covariate, then the $\hat{p}_i$ clearly estimate the probability of finding the species (given the effort) rather than that of the actual occurrence. Again, the relative likelihood of occurrence can be obtained by looking at the fits at a constant level of effort, but absolute probabilities for prediction cannot be derived.

Högmander and Møller [11] and Heikkinen and Högmander [8] used separate models for the "truth" and the observation process to enable prediction in the case where no site can be assumed completely covered or the coverage varies. They interpreted the task of predicting the range as equivalent to restoring binary pixel images, and applied the methodology, of Bayesian image analysis (see, e.g., refs. 3 and [16]). Despite the different line of approach the models in [11] and [8] are actually of the same autologistic form as those of the previous section. In the model of the actual range they use only the autocovariate, but environmental predictors can be included, as demonstrated by Frigessi and Stander [6] in a similar context.

The use of Bayesian methodology seems quite natural here. The role of the prior distribution* is played by the model of the actual range, and the likelihood is defined by the sitewise probabilities of detecting the species if it were present. The posterior probabilities* of occurrence combine the information from the model and from the observations, and can be directly used in prediction. The Bayesian approach also allows for taking the uncertainty of model estimation into account when calculating the predictive probabilities [8].

Example. For illustration we briefly review the application in Heikkinen and Högmander [8]. The aim there was to estimate the range of the common toad *Bufo bufo* in Finland from the data of an atlas survey based on voluntary fieldwork (see Fig. 1).

Let X_i denote the true status in site i (1 for presence, 0 for absence) and Y_i the observation there (1 if detected, 0 otherwise). The model for the true distribution was purely spatial:

$$\Pr[X_i = 1|\text{rest}] = \frac{\exp(\beta k_i)}{1 + \exp(\beta k_i)},$$

where $k_i =$ (number of inhabited neighbor squares) $-$ (number of uninhabited ones). Rough classification of the effort into three classes $E_i \in \{0, 1, 2\}$ (see Fig. 2) was used to define the observation model. The probabilities

$$g_e = \Pr[Y_i = 0|X_i = 1, E_i = e]$$

of not detecting present toads were assumed equal within each effort class; other conditional probabilities were defined in an obvious way. In addition some plausible independence assumptions were made. The quantities β and g_e ($e = 0, 1, 2$) were treated as random variables with uniform prior distributions over appropriate intervals.

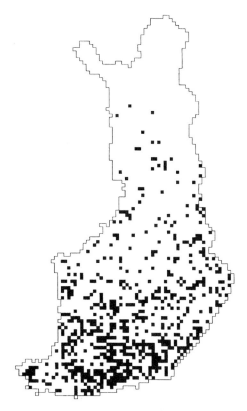

Figure 1. The 10 × 10-km squares of Finland with at least one record of the common toad in the 1980s. (Reproduced by kind permission of the Royal Statistical Society, *Appl. Statist.*, **43**, 569–582. The data are from the Finnish Museum of Natural History.)

From the resulting joint posterior distribution of X, β, and g a sample of 15,000 realizations was simulated using the Markov-chain Monte Carlo* methodology. For each site i the predictive probability of occurrence (see Fig. 3) was obtained simply as the proportion of realizations with $X_i = 1$ in the sample.

GEOGRAPHICAL VARIATION OF ABUNDANCE AND COMMONNESS

The geographical variation in abundance can also be assessed via spatial regression models. Whereas the logistic model is appropriate for binary presence–absence data, here other forms of the generalized linear model* should be applied. Other possibilities include generalized additive models and kriging*.

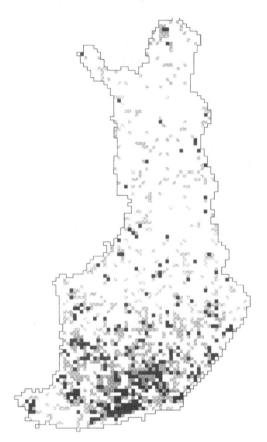

Figure 2. The map of research activity: $e = 2$ in the black squares, $e = 1$ in the gray squares, and $e = 0$ in the white area. (Reproduced by kind permission of the Royal Statistical Society, *Appl. Statist.*, **43**, 569–582.)

Figure 3. Posterior probability estimates $\hat{p}_i$ for the occurrence of the common toad. The eight gray levels refer to probabilities $0 - \frac{1}{8}$ (white), $\frac{1}{8} - \frac{2}{8}, \ldots, \frac{7}{8} - 1$ (black). The contour distinguishes between the shades below and above $\frac{1}{2}$. (Reproduced by kind permission of the Royal Statistical Society, *Appl. Statist.*, **43**, 569–582.)

Le Duc et al. [12] can be seen as one attempt to assess local commonness. Their data are at two resolutions: 2-km tetrads within 10-km squares. They measure the commonness by the proportion of occupied tetrads within a 10-km square.

Application of image restoration* methods along the lines of refs. 8 and 11 would seem like another appropriate approach for estimating local abundance or commonness as well as the range.

SPATIO-TEMPORAL MODELING

Buckland and Elston [4] presented a stochastic model for temporal changes of a range. The colonization parameters in their model can

be estimated if atlas data regarding the same population are available from two points of time. Using the observed range of the latter survey as the starting point, the model enables predicting future development of the distribution. Furthermore, temporal variation of the covariates can be included. Hence, for instance, estimation of the probabilities for local extinction of the species of interest under different land use schemes is possible.

In addition to monitoring of geographical ranges and predicting their changes there are many other tasks in several branches of biogeography which require spatiotemporal modeling. Obviously, a lot of statistical effort is needed in this direction.

OTHER TOPICS

This review has concentrated on the statistical analysis of data from atlas surveys. However, biogeography contains also many other branches where statistical methods play important roles. For the mathematical models applied in biogeographic classification, island biogeography, interspecific competition, mapping of intraspecific variation, partitioning of a geographically variable species population, or gene flow along a geographic gradient, see ref. [15] and the references therein. In addition to the basic mathematical and statistical methods, these models apply more advanced tools such as fitting of trend surfaces in mapping of intraspecific variation and tessellations* with Gabriel graphs in the partitioning of a geographically variable species population. The *Journal of Biogeography* is a good source for more recent applications in the field.

REFERENCES

1. Augustin, N. H., Mugglestone, M. A., and Buckland, S. T. (1996). An autologistic model for the spatial distribution of wildlife. *J. Appl. Ecol.*, **33**, 339–347.

2. Besag, J. (1975). Statistical analysis of non-lattice data. *Statistician*, **24**, 179–195.

3. Besag, J. (1986). On the statistical analysis of dirty pictures (with discussion). *J. R. Statist. Soc. B*, **48**, 259–302.

4. Buckland, S. T. and Elston, D. A. (1993). Empirical models for the spatial distribution of wildlife. *J. Appl. Ecol.*, **30**, 478–495.

5. Caughley, G., Short, J., Grigg, G. C., and Nix, H. (1987). Kangaroos and climate: an analysis of distributions. *J. Anim. Ecol.*, **56**, 751–761.

6. Frigessi, A. and Stander, J. (1994). Informative priors for the Bayesian classification of satellite images. *J. Amer. Statist. Ass.*, **89**, 703–709.

7. Gaston, K. J. (1994). *Rarity*. Chapman & Hall, London. (A good discussion of basic concepts of abundance and range.)

8. Heikkinen, J. and Högmander, H. (1994). Fully Bayesian approach to image restoration with an application in biogeography. *Appl. Statist.*, **43**, 569–582.

9. Hill, M. O. (1991). Pattern of species distribution in Britain elucidated by canonical correspondence analysis. *J. Biogeogr.*, **18**, 247–255.

10. Högmander, H. (1995). Methods of Spatial Statistics in Monitoring of Wildlife Populations. Dissertation. Jyväskylä Studies in Computer Science, Economics and Statistics, 25.

11. Högmander, H. and Møller, J. (1995). Estimating distribution maps from atlas data using methods of statistical image analysis. *Biometrics*, **51**, 393–404.

12. Le Duc, M. G., Hill, M. O., and Sparks, T. H. (1992). A method for predicting the probability of species occurrence using data from systematic surveys. *Watsonia*, **19**, 97–105.

13. Nicholls, A. O. (1989). How to make biological surveys go further with generalised linear models. *Biol. Conserv.*, **50**, 51–75. (A good general discussion. Easy to read for nonstatisticians.)

14. Osborne, P. E. and Tigar, B. J. (1992). Interpreting bird atlas data using logistic models: an example from Lesotho, Southern Africa. *J. Appl. Ecol.*, **29**, 55–62.

15. Pielou, E. C. (1979). *Biogeography*. Wiley, New York. (A classic.)

16. Ripley, B. D. (1988). *Statistical Inference for Spatial Processes*. Cambridge University Press, Cambridge, England.

17. Sharrock, J. T. R. (1976). *The Atlas of Breeding Birds in Britain and Ireland*. Poyser, Calton. (Includes history, background, and practice of the atlas method.)

18. Smith, P. A. (1994). Autocorrelation is logistic regression modelling of species' distributions. *Global Geol. and Biogeogr. Lett.*, **4**, 47–61.

19. Turner, M. G., Dale, V. H., and Gardner, R. H. (1989). Predicting across scales: theory development and testing. *Landscape Ecol.*, **3**, 245–252.

20. Walker, P. A. (1990). Modelling wildlife distributions using a geographic information system: kangaroos in relation to climate. *J. Biogeogr.*, **17**, 279–289.

21. Yee, T. W. and Mitchell, N. D. (1991). Generalized additive models in plant ecology. *J. Vegetation Sci.*, **2**, 587–602.

BIBLIOGRAPHY

Cressie, N. (1991). *Statistics for Spatial Data*. Wiley, New York. (A comprehensive handbook for analysis of spatial data.)

Gaston, K. J. (1991). How large is a species' geographic range? *Oikos*, **61**, 434–438.

Hengeveld, R. (1990). *Dynamic Biogeography*. Cambridge University Press, Cambridge, England. (A discussion of classification and ordination procedures in biogeography and of analysis of ranges.)

Miller, R. I., ed. (1994). *Mapping the Diversity of Nature*. Chapman & Hall, London. (Introduction to modern technical aids such as GIS in range mapping.)

See also ANIMAL SCIENCE, STATISTICS IN; FORESTRY, STATISTICS IN; GEOGRAPHY, STATISTICS IN; GEOSTATISTICS; KRIGING; LINE INTERCEPT SAMPLING; PROFILES OF DIVERSITY; QUADRAT SAMPLING; SPATIAL DATA ANALYSIS; and SPATIAL PROCESSES.

JUHA HEIKKINEN
HARRI HÖGMANDER

BIOMETRIC FUNCTIONS

In biometrics, actuarial science*, and demography*, one wishes to study a nonnegative random variable X, the lifetime* of a subject selected at random from a population of interest.

The natural quantities of interest in the statistical analysis of lifetime data are $S(x)$, the probability that an individual survives longer than age x, and $e(x)$, the expected remaining lifetime of an individual who has attained age x. Consistent with the definition of lifetime, the age of an individual is either the chronological age or the time since entering the study as the case may be. Mathematically, these are defined in terms of $F(x)$, the cumulative distribution function* of the lifetime X, by the equations

$$S(x) = \Pr[X > x] = 1 - F(x) \qquad (1)$$

$$e(x) = E[X - x | X > x]$$
$$= \int_x^\infty S(v)\,dv / S(x) \qquad (2)$$

for any x such that $S(x) > 0$. Note that

$$S(x) = e(0)e^{-1}(x)\exp\left[-\int_0^x e^{-1}(v)\,dv\right]. \qquad (3)$$

The functions $S(x)$ and $e(x)$ are termed *biometric functions* (BF). Generally speaking, any function arising in probability modeling of lifetimes may be regarded as a BF. Other examples include $\mu(x)$, the force of mortality*

(or intensity function), and $q(x,y)$, the conditional probability of dying in age interval (x,y) given that the subject is alive at age x. The defining equations are

$$\mu(x) = F'(x)/S(x) = -(d/dx)\log S(x), \qquad (4)$$

$$q(x,y) = [S(x) - S(y)]/S(x), \qquad (x \leqslant y). \qquad (5)$$

The quantities $S(x)$, $e(x)$, $\mu(x)$, and $q(x,y)$ are mathematically equivalent [9] in the sense that each expression can be derived from any of the others as illustrated by (1) to (5). In actuarial science $e(x)$ is denoted by e_x and is termed the life expectancy at age x.

Use of the term "biometric function" can be traced back to a paper by Lotka [7], who called the birthrate B_t of a population at time t a BF. B_t and $S(x)$ are related by

$$m_t = \int_0^\infty B_{t-x} S(x)\,dx,$$

where m_t is the population size at time t. Chiang [2] refers to the statistical estimators of $e(x)$, $S(x)$, and other life-table functions as BF.

In the engineering, physical, and management sciences, the study of lifetime distributions is part of reliability theory*. In this context the functions $\mu(x)$ and $e(x)$ and $S(x)$ are usually called the hazard function*, the mean residual life*, and the survival function*, respectively.

STATISTICAL INFERENCE

A general estimation technique for these biometric functions is the construction of life tables*. This method produces nonparametric estimates of $S(x)$, $e(x)$, and $q(x, y)$ for a sequence of ages $0, x_1, x_2, \ldots, x_w$ [3]. For nonparametric estimates of $\mu(x)$, see HAZARD RATE ESTIMATION.

In the parametric approach, Gompertz in 1825 modeled the force of mortality by $\mu(x) = \theta c^x$ for appropriate values of θ and c. Other models considered in the literature include the exponential*, Weibull*, gamma*, log normal*, and extreme value* distributions.

CENSORED DATA

In practice, some lifetime observations may not be completed. Such data are called censored*. The censored lifetimes complicate statistical analyses, but they cannot be ignored without introducing bias into the estimates of the BF.

The statistical treatment of the censored data varies with types of censorship. For example, in the case of right censorship* (i.e., instead of the lifetime X, one observes a random variable Y and the fact that $X > Y$), the survival function can be estimated using life-table methods or the Kaplan–Meier product limit (PL) estimator*. The PL estimator is nonparametric and generalizes the usual empirical distribution function* for uncensored data. Some asymptotic properties of the PL estimator have been rigorously established by Breslow and Crowley [1]. Nonparametric estimation of $e(x)$ has been investigated by Yang [8,9]. Various parametric models for censored data have also been analyzed [5,6]. Cox [4] has proposed a regression model for inference on biometric functions in the presence of covariates. See MULTIVARIATE COX REGRESSION MODEL.

COMPETING RISKS

In a competing risk model*, an individual is exposed to several possible causes of death (called risks). One wishes to study, in the presence of all risks, the force of mortality, the life expectancy, and the survival function (the so-called crude survival probability). One also wishes to study these BF pertaining to a particular risk when some or all other risks have been eliminated. The survival function corresponding to the elimination of all but one risk is termed the net survival function.

REFERENCES

1. Breslow, N. and Crowley, J. (1974). *Ann. Statist.*, **2**(3), 437–453.
2. Chiang, C. L. (1960). *Biometrics*, **16**, 618–635.
3. Chiang, C. L. (1968). *Introduction of Stochastic Processes in Biostatistics*. Wiley, New York.
4. Cox, D. R. (1972). *J. R. Statist. Soc. B*, **34**, 187–220.
5. David, H. A. and Moeschberger, M. L. (1978). *The Theory of Competing Risks*. Charles Griffin, London.
6. Kalbfleisch, J. D. and Prentice, R. L. (1980). *The Statistical Analysis of Failure Time Data*. Wiley, New York.
7. Lotka, A. J. (1929). *Proc. Natl. Acad. Sci. USA*, **15**, 793–798.
8. Yang, G. L. (1977). *Stoch. Processes Appl.*, **6**, 33–39.
9. Yang, G. L. (1978). *Ann. Statist.*, **6**, 112–116.

See also COMPETING RISKS; KAPLAN–MEIER ESTIMATOR—I; LIFE TABLES; RELIABILITY, PROBABILISTIC; and SURVIVAL ANALYSIS.

GRACE L. YANG

BIOMETRIC SOCIETY. See INTERNATIONAL BIOMETRIC SOCIETY

BIOMETRICAL JOURNAL

[This entry has been updated by the Editors.]

The *Biometrical Journal* was founded as *Biometrische Zeitschrift* by Ottokar Heinisch and Maria Pia Geppert, with four issues per year. The first volume was published in what was then the German Democratic Republic in 1959, with an Editorial Board drawn both from Western Europe and the Eastern (Soviet) bloc. In 1973, the publication frequency stabilized with eight issues per year, each year constituting one volume.

The current title *Biometrical Journal* was introduced in 1977 (Vol. 19). The *Biometrical Journal* is an independent international publication, under two editors and a team of 37 associate editors from across the world. It is published by Wiley; its website is www3.interscience.wiley.com/cgi-bin/jhome/5007702.

As the website indicates, the scope of the journal covers papers on the development of statistical and related methodology and its application to problems in all areas of the life sciences (including biology, agriculture, and forestry, but particularly medicine), or on the application of known statistical methods to these areas that are motivated by relevant problems ideally worked out as illustrative examples. Papers containing solely extensive mathematical theory are not included.

A section is devoted to case studies and regulatory aspects in medicine as well as reviews of current research activities in this field.

Preference is given to papers submitted electronically.

H. Ahrens

BIOMETRICS

[This entry has been updated by the Editors.]

Biometrics is the journal of The International Biometric Society*. Publication was initiated in 1945 as the *Biometric Bulletin*, under the auspices of the Biometrics Section of the American Statistical Association*. The founding editor of the journal was Gertrude M. Cox*, who served in this capacity until 1955. The name was changed to *Biometrics* in 1947 and it became the official publication of the Biometric Society in 1950. The journal appears annually in four issues. There are three Co-Editors and more than 70 Associate Editors from across the world.

The general objectives of *Biometrics* are to promote and extend the use of mathematical and statistical methods in pure and applied biological sciences, by describing and exemplifying developments in these methods and their applications in a form readily usable by experimental scientists. Published papers generally focus on the development of new methods and on results that are of use in the biological sciences. They may deal with (1) statistical methodology applied to specific biological contexts, (2) topics in mathematical biology, and (3) statistical principles and methodology of general applicability in these sciences. Typically, the journal seeks to publish illustrations of new methods with real data wherever possible.

In addition to regular articles, the journal has a Consultant's Forum, a Reader Reaction section referring directly to articles previously published in the journal, and a Book Reviews section. Invited papers are published from time to time.

The website for *Biometrics* can be accessed via that of the parent society, www.tibs.org.

See also International Biometric Society.

Foster B. Cady

BIOMETRIKA

[This entry has been updated by the Editors.]

Primarily a journal of statistics, *Biometrika* places an emphasis on papers containing original theoretical or methodological contributions of direct or potential value in applications. From time to time, papers in bordering fields are published.

Biometrika is published quarterly by the Biometrika Trust, and is distributed by Oxford University Press. The journal has a long-established international reputation; as an illustration Table 1 shows the countries of residence of authors of papers published in 1979.

Some review papers have been published and also a series on the history of probability and statistics. There is a Miscellanea section for shorter contributions. Book reviews are no longer included.

Currently, all papers submitted for publication are reviewed under the supervision of the Editor and 13 Associate Editors.

HISTORY

Biometrika was founded by Karl Pearson* and W. F. R. Weldon* in consultation with

Table 1. Country of Residence of Authors of the 102 Papers Published in 1979

Country	Number of Authors	Country	Number of Authors
United States	62	Israel	2
United Kingdom	34	Brazil	1
Australia	20	Denmark	1
Canada	10	Italy	1
Hong Kong	4	Norway	1
Japan	4	Pakistan	1
India	3	Total	144

Francis Galton*; the first issue appeared in October 1901. From 1906, Pearson assumed entire editorial responsibility, and since then the editorship has changed only infrequently; E. S. Pearson (Karl Pearson's son) was Editor from 1936 to 1965, D. R. Cox from 1966 to 1991, D. V. Hinkley from 1991 to 1992, A. P. Dawid from 1993 to 1996, and D. M. Titterington from 1997 to the present time (2004).

The origin of the journal was due in part to a request by the Royal Society that, in papers submitted for publication, mathematics be kept apart from biological applications.

The editorial in 1901 stated that the journal would include:

(a) memoirs on variation, inheritance, and selection in Animals and Plants, based on the examination of statistically large numbers of specimens ...;

(b) those developments of statistical theory which are applicable to biological problems;

(c) numerical tables and graphical solutions tending to reduce the labour of statistical arithmetic.

The early volumes contained many diagrams, photographs, and tables of measurements of parts of the human body, animals, and plants. Over the years the amount of biological and anthropological work has become less, and applications of theory have not been restricted to biological problems. Publication of tables continued on a reduced scale.

There were in general two journal issues per year up to 1967 and three per year from 1968 to 1986. World Wars I and II resulted not only in fewer papers and pages per year but also in the frequency of publications falling to one issue per year. Since 1987 there have been four journal issues per year.

In 2001, a special volume was published [1] to celebrate *Biometrika*'s one-hundredth year.

MANAGEMENT

In the early days, a guarantee fund was provided by the founders and their friends.

After Weldon's death in 1906, the practical management was in Karl Pearson's hands. Circulation dropped during World War I, and Pearson and Mrs. Weldon contributed money to help keep the journal going. *Biometrika* has never been run for financial profit. Auxiliary publications, such as books of tables, have been issued since 1914, notably *Biometrika Tables for Statisticians*, two volumes, edited by E. S. Pearson and H. O. Hartley. In 1935, the Biometrika Trust was set up by Karl Pearson with at least five trustees, responsible for financial matters and the issuance of back numbers, auxiliary publications, and so on. The Trust also holds the copyright of *Biometrika* and its associated publications. Since 1979, the Business Manager of the Trust has been R. F. Galbraith.

The *Biometrika* website is www3.oup.co.cuk/biomet. Enquiries concerning the journal or the Trust may also be made to the business office: *Biometrika* Office, University College London, Gower Street, London WC1E 6BT, United Kingdom. Email correspondence goes to jnl.info@oup.co.uk.

REFERENCE

1. Titterington, D. M. and Cox, D. R., eds. (2001). *Biometrika: One Hundred Years*. Oxford University Press, New York. [Several articles are seminal papers from Biometrika's first 100 years; others review important contributions by the journal to areas of statistical activity.]

BRENDA SOWAN

BIOMETRISCHE ZEITSCHRIFT. See BIOMETRICAL JOURNAL

BIOMETRY. See BIOSTATISTICS, CLASSICAL

BIOSTATISTICS, CLASSICAL

Biostatistics is that branch of science which applies statistical methods to biological problems, the common prefix being derived from the Greek word *bios*, meaning life.

The first major applications started in the middle of the seventeenth century when Sir William Petty and John Graunt conceived new and creative methods to analyze the London Bills of Mortality. Petty and Graunt essentially invented the field of *vital statistics** by studying the reported christenings and causes of death, and proposing measures of what they called "political arithmetick." Graunt recognized problems of inference when there has been inaccurate reporting of causes of death; he created methods of estimating mortality rates by age when age was not even recorded on the death certificate; and he devised estimates of birthrates, as well as a method to estimate the population from birthrates and other ingenious techniques for interpreting the data in the records of christenings and burials. Sir William Petty developed an enumeration schedule for a population census, proposed a centralized statistical department in the government, conceived the idea of life expectancy before Halley developed the first actual life table*, and proposed clever and original ideas on how to estimate population sizes. For further details, see the monograph by Greenwood [29].

Today, vital statistics is generally restricted by definition to the statistics of births, deaths, marriages, and divorces, and thus the term has a current connotation considerably more limited than "biostatistics," despite its derivation from the same root in its Latin form *vita*. Biometry or biometrics is another term closely identified with biostatistics but also more restricted in scope. The biostatistician must deal not only with biometrical techniques used in the design and analysis of experiments but also with some sociometric and psychometric procedures plus most of the methods used by demographers. Thus, the biostatistician works closely not only with the biological researcher but also with the epidemiologist, survey researcher, local community planner, state and national health policy analyst, and those government officials concerned with developing procedures for registering births, deaths, marriages, divorces, abortions, morbidity reports, the description of populations by sample surveys and census enumeration, and with health regulatory agencies.

FIELDS OF APPLICATION OR AREAS OF CONCERN

The biostatistician differs from the traditional statistician in that he or she is confronted by a wider range of problems dealing with all the phenomena that affect people's physical, social, and mental wellbeing. These phenomena consist of our relationship to other human beings, to animals and microbes, to plants, and to the physical and chemical elements in the environment. In dealing with these problems the biostatistician encounters theoretical difficulties, such as analyzing autocorrelated data in time series, in addition to practical and applied problems, such as working with accountants and economists to calculate costs versus benefits in evaluating the efficiency of a health program.

This means that the biostatistician must have familiarity with the concepts, goals, and specialized techniques of numerous fields beyond what might be considered a standard knowledge of statistics and probability. Some of these fields and areas of concern are mentioned here briefly and the remainder of this article will comment on a few of them at greater length.

Statistical Genetics

After the early developments in vital statistics, the field of *statistical genetics** was the next area that benefited most from the new ideas emerging in statistics. Any discussion of biostatistics and biometry would be incomplete without the names of Charles Darwin (1809–1882), Francis Galton* (1822–1911), Karl Pearson* (1857–1936), and Ronald A. Fisher* (1890–1962).

Galton is responsible for the use of the term "regression"* when he observed that sons regressed linearly on their fathers with respect to stature. His thesis was to call the phenomenon a "regression to mediocrity" because children deviated less from the mean height of all children than the amount their fathers deviated from the mean height of all fathers. This bivariate normal distribution* gave rise to the measurement of the association by the coefficient of (product-moment) correlation* in 1897 by Karl Pearson* and to many other contributions by him. He is also generally credited with the creation of the new discipline of biometry and established with Walter F. R. Weldon and C. B. Davenport, in consultation with Galton, a new journal called *Biometrika** to provide for study of these problems. The journal has been in continuous publication since 1901, and after an unsigned editorial presumably written by Pearson, had as its first paper an article entitled "Biometry" by Francis Galton. The journal is still a highly regarded source for communications in biometry. Fisher's major contributions were to genetics and statistical theory, and he published the genetical theory of natural selection in 1930. This landmark book, plus earlier and later publications, represented attempts by Fisher to give quantitative form to Darwin's views and a statistical theory of evolution.

For more detailed discussion of current statistical problems in genetics, readers are referred to STATISTICAL GENETICS. For a history of early developments in statistical genetics, see Norton [48]. For biographical accounts of the statistical geneticists, see the appropriate entries in Kruskal and Tanur [35], and for a stimulating account of the life of Fisher, the biography by his daughter, Joan Fisher Box [4], is unrivaled (*see also* FISHER, RONALD AYLMER).

Bioassay

*Bioassay** techniques cover the use of special transformations such as probits and logits, as well as the application of regression to the estimation of dosages that are p percent effective within stated confidence limits. There are also problems in measuring relative potency, slope-ratio assays, and quantal responses* vis-à-vis tolerance distributions*. The reader interested in this subject is well advised to consult bioassay and a standard textbook such as Finney [18].

Demography

A knowledge of *demography** which includes traditional vital statistics, rates and ratios, life tables, competing risks, actuarial statistics, and census enumeration techniques, is necessary in biostatistics. In this category, many tabulations of data will consist of a time series of events or rates classified by age. For the appropriate analysis of such data, reference should be made to cohort analysis techniques collected in a monograph by Hastings and Berry [31]. For further details in this broad area, *see* ACTUARIAL SCIENCE, LIFE TABLES, COMPETING RISKS, and DEMOGRAPHY, as well as Linder and Grove [37] and the book by Spiegelman [52].

Epidemiology

Some knowledge is required about the measurement of disease, including false-negative and false-positive results, so that sensitivity and specificity of a diagnostic test can be estimated, as well as survey results used to estimate the true incidence and prevalence of disease. It is necessary to have knowledge of epidemic theory and the use of deterministic and stochastic models [1]. Fundamental to this whole field of application is an understanding of causality and association [3,24,36].

In the case where clinical trials can be conducted, two groups of persons, one "treated" and the other "untreated," are observed over a period of time with respect to attack by or relief from the disease that is the object of the study. Here the biostatistician must know how to develop a protocol [17], how to randomize (*see* RANDOMIZATION), use double-blind techniques, and combine multiple-response variables into a multivariate analysis*. If several medical centers are involved, it is important to know how to operate a statistical coordinating center for collaborative clinical trials (see refs. 20 and 23, and CLINICAL TRIALS—II).

In situations where moral concerns prohibit a comparative experiment, such as in the study of whether exposure of a woman during pregnancy to infection by German measles (rubella) causes congenital malformations, it is necessary to know how to conduct retrospective case-control studies and measure the relative risk caused by exposure. In fact, with the sole exception of clinical trials, almost all the statistical research in epidemiology is retrospective in nature. That is, the research is ex post facto because the investigators seek to describe and analyze a series of events that are customarily a rather sudden, unusual, and significant increase in the incidence of disease.

The so-called case-control study is the most common procedure used to investigate an epidemic or unusual increase in disease. By this approach, a special group, frequently a 100% sample of available cases, is studied in detail to ascertain whether there were one or more common factors to which the members of the group were exposed. The exposure might be a drug, a food, or an environmental factor. A comparable group of non-cases, frequently called controls or compeers or referents, is also selected at random in order to determine whether its members had the same, less, or more exposure to the suspected factor(s).

In the typical design of such studies, the data are presented in a 2×2 contingency table* of the following form, wherein a, b, c, and d are category frequencies.

Factor F	Cases	Compeers	Total
Exposed	a	b	$a + b$
Nonexposed	c	d	$c + d$
Total	$a + c$	$b + d$	N

If the proportion $a/(a + c)$ is significantly greater than $b/(b + d)$, one can safely assume that factor F is associated in some way with the occurrence of the event. The test of significance* to validate this may be the common χ^2 with 1 degree of freedom.

Owing to the fact that the design is retrospective, the comparable groups are cases and compeers, *not* exposed and nonexposed. Thus one cannot calculate the rates of disease as simply $a/(a + b)$ and $c/(c + d)$ in order to divide the former by the latter to derive a measure of relative risk associated with factor F. Although other researchers in genetics had previously used a solution similar to his, it was Cornfield [10] who demonstrated clearly that an estimate of relative risk is obtainable from the ratio of cross-products, ad/bc. [If $a/(a + c)$ is designated as p_1, and $b/(b + d)$ is designated as p_2, the relative risk is equivalently estimated as $p_1(1 - p_2)/p_2(1 - p_1)$.] The ratio of cross-products is commonly referred to as the *odds ratio**, motivated by the comparison of exposed-to-nonexposed "odds" in the two groups, $a : c$ and $b : d$.

Cornfield clearly emphasized that the validity of such estimation is contingent upon the fulfillment of three assumptions:

1. The rate of disease in the community must be comparatively small, say in the order of magnitude of 0.001 or less, relative to both the proportion of exposed cases and the proportion of nonexposed persons in the nonattacked population.

2. The $(a + c)$ cases must represent a random, unbiased sample* of all cases of the disease.

3. The $(b + d)$ controls must represent a random, unbiased sample of all non-cases of the disease.

In actual practice, fulfillment of the first assumption is usually easily attainable, and any minor deviation from it causes no serious distortion in the results. The remaining two assumptions, however, are extremely difficult, if not actually impossible, to satisfy. Failure can cause considerable bias in the results, and is the basis of disagreement among both biostatisticians and epidemiologists. For instance, the detection of cases, referred to as ascertainment by R. A. Fisher, may be biased because cases are selected in a large medical referral center which is not representative of all cases in the community. In addition to being certain that the cases have all been diagnosed properly, the biostatistician must check to be sure that the cases were not selected because of distinguishing attributes such as socioeconomic status, location with respect to the center, race, sex, medical care previously received, or even whether the cases had close relatives with a similar disease and sought special diagnostic attention. The controls are sometimes chosen to be persons in the same hospital with a different diagnosis, or neighbors in the community. The biostatistician has to determine whether they are comparable in such factors as age, race, sex, severity (stage or grade) of the disease, and many other variables that tend to confound a just and fair comparison on factor F alone.

Three statistical problems are mentioned as a result of this type of study.

1. How to select the cases and controls? The literature on this is voluminous, but a few references may be mentioned [9,32,33].

2. In selecting the compeers, is it worthwhile to try to pair one or more controls to each case based upon certain characteristics which influence the probability of disease so that factor F will be the primary residual influence? References to this are found in Cochran [8] and McKinlay [38,39], among others.

3. After selecting cases and controls, it is difficult to estimate the influence of all factors other than F. These other variables are referred to as confounding variables, and they need to be adjusted or accounted for so as to enable a valid comparison to be made on factor F. Historically, the procedure of adjusted or standardized rates is one way of achieving this goal. Another common statistical procedure is the use of covariance analysis [see ANALYSIS OF COVARIANCE] [22]. More elaborate statistical procedures for dealing with confounding* may be found in Miettinen [42–45], Rothman [51], and many others.

An excellent introductory reference to this class of problems is the monograph of Fleiss [19] and a review paper by Walter [54].

Clinical Trials

In the case of *clinical trials**, a whole host of special problems arise for which the biostatistician has had to develop special techniques. One of these is the detection of unexpected and untoward rare effects of drugs with the consequent need to terminate a trial early. Moreover, when data are demonstrating a trend earlier than expected, there is also a need to end the accession of patients and to stop further treatment with what may be an inferior regimen. This means the biostatistician must be familiar with the problems of multiple examinations of data [40,41], multiple comparisons* [46,53], and other adjustment procedures made necessary by the ex post facto dredging of data. An excellent pair of references on randomized clinical trials is the set of articles by

Peto et al. [49,50] and one by Byar et al. [5]. *See also* CLINICAL TRIALS—II. For the ethical problems involved in conducting clinical trials, the reader is urged to read Gilbert et al. [21] and Courand [12].

Confidentiality and Privacy of Records

The biostatistician is continually confronted with the demands of *confidentiality* and *privacy* of records in dealing with health and medical records. This subject begins with the controversial area of what constitutes informed consent when a patient is a minor, ill, comatose, or otherwise incompetent. The use of informed consent and randomization may present a conflict as far as a patient is concerned, and other design techniques may have to be used to overcome this problem [55].

Safeguarding of Computer Data

Related to the problem of privacy and confidentiality of data is how to safeguard these attributes when data are stored in *computers*, or records are to be linked from multiple sources. The linkage of records regarding an individual or an event requires the knowledge of computers, computer languages, and programming, as well as means for protecting identification of information stored in computers [6,14,15]. Of course, knowledge of computers is necessary in general because the tabulation of large volumes of data and analysis by current techniques could not be carried out if the methods in vogue 25 years ago, such as punch-card machines and desk calculators, were still relied upon to carry out the necessary mechanical procedures. Reference should be made to COMPUTERS AND STATISTICS*.

Nature of Human Surveys

The nature of the field of inquiry in *human surveys* involving personal knowledge, attitudes, and behavior confront the biostatistician with a challenge akin to that of the sociologist. The goal is to obtain cooperation and a truthful response when the question(s) to be asked may be highly personal, sensitive, or stigmatizing in nature. Biostatisticians and sociologists have developed techniques to maximize cooperation with incentives of various kinds, including even monetary payment, as well as devices to assure anonymity. The use of response cards, telephone surveys, and a technique known as randomized response* have all shown promise in this field [27,28].

Censoring of Observations

The *censoring* of observations in statistics is not a matter related to the sensitivity of highly classified data but rather is the purposeful or accidental blurring of an actual value for other reasons. For instance, in studies of life expectancy subsequent to an event such as exposure to a carcinogenic agent, the limited amount of time available for observation may require the experimenter to terminate the experiment after 1 or 2 years. Those individuals who have survived up to that point have had their *actual* life expectancy "censored," and the appropriate analysis must consider such observations in estimating parameter values. Similarly, the speed of some reactions may be so rapid, or the recording of data so crude at the beginning of the period of observation, that a certain number of early observations are censored initially. Thus, in deaths of infants under 1 day old, some countries do not record the actual number of hours lived under 24. For most of these problems the use of order statistics* has been the preferred solution to the problem of calculating unbiased estimates of the distribution parameters. (See refs. 16 and 26, and ORDER STATISTICS.)

Community Diagnosis

The biostatistician collaborates with health planning personnel to establish bench marks that describe the health status of a community so as to earmark places where greater attention and/or funds should be directed. The first step in community diagnosis is to study the population in terms of its magnitude and distribution by attributes such as age, sex, ethnicity, occupation, residence, and other factors that are related either to health or to the ability to obtain needed health services. Special studies may also be made of the community itself with respect to special environmental factors (industries and occupations, climate, pollution, etc.) and availability of health facilities, resources, and personnel. This information is combined with vital statistics and morbidity data to ascertain health problems characteristic of the community. For example, an unusually high birth rate may signify the need for a program of family planning services or it may simply be a fact caused by the peculiar age and sex distribution of the population. An excessive rate of lung cancer may suggest the need for an antismoking campaign or an investigation as to whether air pollution or a special industry, like the manufacture of products containing asbestos, may be involved.

In making judgments regarding the health status of a community, the biostatistician must be aware of the many possible comparisons that might be drawn. The age-adjusted death rate in a county, for example, might be compared to that of an adjacent county, to the rate of the entire state or nation, to the lowest county rate in the area, to the rate of a county or counties similar in the composition of the population of the study county, or simply to the trend of the rates for that county over the past 10 to 20 years.

Finally, community diagnosis would be incomplete without an attempt to study the effectiveness of treatment efforts. The biostatistician collaborates with health service providers to evaluate the effectiveness of any program instituted to improve the status of the community. With the assistance of cost-accounting specialists, the benefits (and any undesirable effects) of the program are compared with the costs in terms of funds and personnel so as to balance the relative weights of these items from a societal point of view. There are many references for studying this aspect in greater detail, and two helpful ones are Caro [7] and Greenberg [25].

FUTURE TRENDS

There are two areas in which the biostatistician has been playing a leading role lately, and these cut across many kinds of applications and problems. It is highly likely that considerable research in methodology will continue to be devoted to these two special areas of concern. The first of these areas might be called modeling.

Mathematical Models

The relationship between a set of independent variables and the dependent or response variable(s) is usually referred to as the mathematical model. This model may take the form of a standard multiple regression* analysis with a single response variable, a surface, or multiple response variables as in multivariate analysis.

It is generally assumed that the technical specialist with substantive knowledge of the field of application (epidemiology, toxicology, pharmacology, radiology, genetics, etc.) will play a crucial role in determining the model or relationship between a set of independent variables and the response variable(s). In actual practice, however, the biostatistician is the one who finally selects the specific model that establishes this functional relationship and then attempts to measure the strength or influence of the independent variables therein. Moreover, the biostatistician is often expected to contribute strongly to the decision as to whether the relationship is a causal one or merely one of association and correlation.

For example, in the measurement of the carcinogenicity of a food additive or drug, questions may arise as to whether a substance can be judged harmful if it "accelerates" the appearance of a tumor even though it does not increase the incidence of the abnormal growth. In general, the answer to this question is in the affirmative provided that there can be unequivocally demonstrated a dosage-response relationship between the substance and the tumor—i.e., the more of the suspected compound that is given or exposed to the test animal, the greater is the probability or likelihood that the tumor will occur earlier.

In the case of bioassay procedures, the biostatistician may be called upon to decide which sigmoid or S-shaped curve to use in order to relate the dosage to the response. This problem is more than simply a decision between the integrated normal curve (or probit*) vis-à-vis the logit function [i.e., $\log_e(p/(1-p))$, where $0 < p < 1$]. In the case of harmful or toxic substances or low-level irradiation (as explained in the subsequent section), it is a question as to whether even

a regression relationship can be assumed for purposes of extrapolation.

In many data sets that arise in biostatistics the data are in the form of contingency tables* arising from frequency counts in cells created as a result of an n-way cross-classification of the population. The variable being measured is frequently on a nominal scale, i.e., the categories are simply the set of names corresponding to an attribute such as occupation, place of residence, cause of death, religion, or sex. In a few instances the categories of classification may be ordered or ranked on a scale whose orientation is clear but whose spacings are not known. For example, socioeconomic class may be classified as low, medium, or high; or birth order may be determined as first, second, third, and so on. In those special cases where the number of possible categories for an attribute is limited to two, the data are referred to as *dichotomous**. Thus one can have dichotomous information on sex disaggregated by male, female; or data where simply the presence or absence of a factor is noted and a dummy variable is created such that 1 = yes and 0 = no.

If we consider one kind of contingency table encountered frequently in surveys, we may have as cell entries the annual mean number of doctor visits, illnesses, or days of disability. The number of respondents in each cell may be known but the relationship of the attributes in the classification scheme to the dependent variable is not. The aim of the analysis is to study how the different subclasses of the categories relate to the response variable.

For instance, let us consider that the number of doctor visits for a given age group has been averaged in each cell, and the number of classification variables is three: location (L = urban, rural), sex (S = male, female), and highest level of education (E = no high school, high school, college, postcollege). This means that there are $2 \times 2 \times 4 = 16$ cells, and the response variable, R, can be a function of all 16 parameters.

Instead of having to contend with 16 parameters, a general linear model* might consist of

$$R = \lambda + \lambda_L + \lambda_S + \lambda_E + \lambda_{LS} + \lambda_{LE}$$

$$+ \lambda_{SE} + \lambda_{LSE},$$

where λ = general mean,

$\lambda_L, \lambda_S, \lambda_E$ = main effects of location, sex, and education, respectively

$\lambda_{LS}, \lambda_{LE}, \lambda_{SE}$ = the first-order interactions, each the simultaneous effect of two factors shown

λ_{LSE} = a second-order interaction, or the simultaneous effect of all three factors

In fitting the parameters in this model, it is highly unlikely that the general mean will itself provide an adequate fit. By adding the main effects, one can ascertain whether an adequate fit has been obtained. If it doesn't, one proceeds to the next level and introduces whichever of the first-order interactions seem necessary. Finally, and only if required, the highest-order interaction would be brought into the final model. This hierarchical approach is similar to that of a stepwise regression* analysis and can be carried out either forward or backward.

Now, if the model is restructured so that one may consider the effects as a linear sum for the *logarithm* of R, then the model is referred to as multiplicative, because effects are being multiplied. Thus, if

$$\log_e R = \lambda + \lambda_L + \lambda_S + \lambda_E + \lambda_{LS}$$
$$+ \lambda_{LE} + \lambda_{SE} + \lambda_{LSE},$$

we can define a new set of parameters such that $\lambda = \log_e \lambda'$, whence

$$\log_e R = \log \lambda' + \log \lambda'_L + \log \lambda'_S + \log \lambda'_E$$
$$+ \log \lambda'_{LS} + \log \lambda'_{LE} + \log \lambda'_{SE}$$
$$+ \log \lambda'_{LSE}$$

or

$$\log_e R = \log(\lambda' \cdot \lambda'_L \cdot \lambda'_S \cdot \lambda'_E \cdot \lambda'_{LS}$$
$$\cdot \lambda'_{LE} \cdot \lambda'_{SE} \cdot \lambda'_{LSE}).$$

Taking antilogarithms results in a form that shows why the model is referred to as a multiplicative model:

$$R = \lambda' \cdot \lambda'_L \cdot \lambda'_S \cdot \lambda'_E \cdot \lambda'_{LS} \cdot \lambda'_{LE} \cdot \lambda'_{SE} \cdot \lambda'_{LSE}.$$

This form of relationship is referred to as a *log-linear model**, and the predicted response value must always be positive.

In the case where the response variable is dichotomous yes–no, the mean values in the cell entries are really proportions. In the case of proportions, one may also use a logit-linear model not only to assure that the predicted values of p will lie between zero and 1 but also to help obtain a better fit.

A frequently encountered multiway contingency table consists of s samples from s multinomial distributions having r categories of response. There are then counterparts to the linear models and log-linear models discussed earlier, but the problem of estimation of the parameters involves choices between ordinary least squares*, weighted least squares*, maximum likelihood*, and minimum χ^2*. The complexity of these considerations is beyond the scope of this section, but the reader will find that refs. 2, 13, 30, and 47 fairly well summarize the state of the art.

Detection of Hazardous Substances

With the successful conquest of most of the infectious diseases that have plagued mankind throughout history, health authorities have been concentrating recently upon two chronic diseases whose etiology is yet to be determined: cardiovascular disease and cancer. In both cases, there is no disagreement with the thesis that heredity exercises a determining influence, but the role of the environment in causing many cases is also unquestioned. Attempts to measure the harm that may be caused by potentially hazardous substances in the environment, principally with respect to these two diseases, represent the greatest challenge to the biostatistician today. The benefit to society when successes are obtained make this area of research rewarding emotionally and scientifically despite the exceptional complexities involved.

The number of factors included under the rubric of environment which have a human impact is probably infinite. In addition to food, air, and water, environment includes everything that is not included under the genetics label. Thus in addition to known or unknown chemical and physical substances, there are such variables as exercise, noise, tension, and stress, plus all the psychosocial elements that affect people.

When one starts to limit study to hazardous substances by themselves, say cigarette smoking and exposure to asbestos, the interaction of these two factors and of these two with stress, sleep, use of alcohol, and psychosocial elements soon points to the impracticality of drawing too constraining a line around the hazardous substances. Thus if one accepts the assertion that an overwhelming majority of cancers are environmentally induced, and that of these perhaps 5 to 10% are occupation-related, any good inquiry into the incriminating substance(s) in the workplace cannot overlook the personal characteristics and habits of the workers themselves.

This highlights the first two difficulties in measuring the health importance of hazardous substances: that the list of substances and important factors is substantially great if not infinite, and that these factors have interactions or synergistic reactions that may be more important than the main effects themselves. (The effect of cigarette smoking and asbestos exposure referred to a moment ago is a perfect example of how the two factors in combination are much more important than the addition of the two by themselves in promoting lung cancer.)

Some of the other difficulties in studying the hazardous substances is that they are frequently available only in low doses and administered over a long period of time. The low dosage creates many problems of its own. For example, there is the question of how reliable the measurement is, especially when many of the estimates of exposure have to be retrospective or ex post facto in nature. Even if prospective in time, a sampling of the air in an environment requires a suitable model of air circulation so as to know whether to collect samples at different locations, at different times of the day, indoors or outdoors, with or without sunlight and wind, and so on. Furthermore, total exposure over an n-hour period may be more or less important than a peak period of exposure during a short period of time.

Determination of the impact of low doses of hazardous substances is especially complex because of other peculiarities. Experiments on human beings are, of course, out of the question, so reliance must be placed on accidental exposure of human beings, long-term exposure retrospectively, or the effect upon animals. Since low doses are likely to cause small effects, the results are extremely difficult to detect and usually require large numbers of animals plus extensive and expensive examinations of many organs. Since the number of animals required might be prohibitive anyway, reliance is often placed upon artificially high doses. This confuses the picture further because one needs to know what model to use in projecting by extrapolation from high doses the effect at low dose levels, as well as what rationale for judging how valuable it is to measure the possible effects on human beings from data on the health of animals, usually mice and other rodents. The models used to extrapolate from high dose to low dose, especially in problems involving radiation, include the one-hit, two-hit, probit*, and a variety of empirical models. This subject gets still more complicated when assumptions are made about threshold levels and rates of neutralization or detoxification [11].

There is here also the effect of confounding variables, which were referred to earlier, when discussing the selection of epidemiological cases and controls, as a special field of application.

There is also the entire question of risk versus benefit, which is an important consideration in determining public policy. As an example of how legislation in the United States which fails to consider both risk and benefit can adversely affect public policy, the Delaney Amendment to the Food Additive Amendments of 1958 (P.L. 85–959) can be cited. The Delaney clause stated that any food additive that is capable of producing cancer in animals or human beings is assumed harmful to human beings and must be banned by the Food and Drug Administration regardless of any benefit. Although the methodology for studying risks vis-à-vis benefits is in a primitive state, the present law does not encompass the possibility that the risk might be miniscule and the benefit substantial. Such has probably been the situation regarding the use of saccharin. Another drawback with the Delaney clause is that certain food additives that have a low carcinogenic effect, such as nitrites, might be found naturally in other foods. They have to be banned as a food additive, even though the benefits might outweigh the disadvantages and even though other foods already possess the particular substance.

This has been a brief overview of a most important area. Readers who are interested in this subject are urged to examine the technical report by Hunter and Crowley [34].

EDUCATION AND EMPLOYMENT OF BIOSTATISTICIANS

What does it take to make a biostatistician? To this question there are a host of responses but no firm, positive answer. This is not surprising in view of the fact that two other quite different occupations that have been studied probably more than any other during the past 40 years—airplane piloting and medicine—still have a long way to go before most of the answers are known regarding the optimal selection of applicants and the most effective method of education.

Undoubtedly, one of the most important characteristics in the outstanding biostatistician is the individual himself or herself. Essential attributes are an inquisitive curiosity or "burning yearning for learning," a constancy of purpose, an ability to think quantitatively, an interest in applying statistical methods to biological problems, and probably a personality or mental disposition that encourages close working relationships with collaborators from many fields.

The field of biostatistics has many avenues of access for entry purposes. Although most persons probably enter from a traditional mathematical or mathematical–statistical background, many others have come from an original interest in biomedical fields, sociology, psychology, engineering, and computer sciences. There is no unique or assured pathway to biostatistics; the right person can approach it from whatever direction maximizes his or her own potential.

The most frequently used institutions for educating biostatisticians are the departments of biostatistics and statistics. These departments should be located in a university setting where there are, first, an academic health center or, at the very least, a medical school and hospital. There must also be strong units in the remainder of the university concerned with the teaching of graduate students in mathematics, probability and statistics, and computer sciences. More than the structural units in the university, however, there must be a pattern or tradition of close working relationships between the biostatistics faculty and those from the other entities. Training at the doctoral level will not be truly meaningful unless the biostatistical faculty are actively engaged in and publish work on applied as well as theoretical research concerning statistical methods.

Merely because the university has, say, an accredited school of public health with a department of biostatistics is no guarantee that it is a good one. (Moreover, biostatisticians can be, and have been, educated in statistical departments other than those found in schools of public health.) Students seeking training in biostatistics would be well advised to be certain that the teaching faculty are engaged in both applied and methodological research, and most important, that there is a close affiliation or working relationship with a medical unit of some kind. Unless such affiliation exists, it will be more difficult to get exposure to good experience involving clinical data, clinical trials, epidemiology, hospital and clinic studies, and health services research.

The doctoral training might consist of either the program for the traditional academic Ph.D. degree or a program that is professionally oriented, such as one directed to the Doctor of Hygiene or Doctor of Public Health degree. The Ph.D. is usually intended for those persons planning careers in an academic or research setting where emphasis is on developing statistical methodology to solve important biological and public health problems. The related doctoral dissertations are most often published in statistical journals.

The professional doctoral degree is usually intended for persons who plan careers in government or industry and whose emphasis is on service to persons seeking statistical advice. The nature of the doctoral dissertation is frequently the new application of statistical concepts to important public health or biological problems. What is novel in the dissertation is the application of a known statistical technique to solve an important health problem. This type of dissertation is usually published in biological and public health journals.

Training at the master's degree level for persons interested in beginning and intermediate-level positions in biostatistics is determined and evaluated by approximately the same guidelines as the foregoing. The main difference, perhaps, is that the criterion of an active ongoing research program is not as stringent. Instead, emphasis should be placed on good teaching and other pedagogical processes that will enable students to learn practical techniques at the same time that a few of them are stimulated to pursue advanced training at the doctoral level.

Employers of biostatisticians have ranged from local, state, and federal government to industry and academic institutions. Each employer will require a different set of areas of knowledge over and above the general field of statistics and probability. For example, one type of government biostatistician may be required to be an expert in registration of vital statistics, demography, survey research, and the special problems associated with confidentiality. The person working as a biostatistician in a pharmaceutical firm may require special training in bioassay techniques, mathematical modeling, and those aspects of clinical trials related to the study of new drugs and their toxicity, dosage, and effectiveness, in order to collaborate in preparing applications for a new drug to be approved by a regulatory agency such as the Food and Drug Administration, U.S. Public Health Service.

It is difficult to know even approximately how many persons there are in the United States who would classify themselves as biostatisticians. There is no one professional organization designed for affiliation of persons who are primarily biostatisticians. A rough guess would be that one-fifth of the 15,000 statisticians listed in the 1978 Directory published by the American Statistical Association* are strongly interested in biostatistical problems, and, of these, about one-half would be individuals who classify themselves primarily as biostatisticians. An international Biometric Society* was established in 1947, and in 1950 it assumed responsibility for publication of the journal *Biometrics**, which had had its inception as the *Biometrics Bulletin* in 1945 under the aegis of the American Statistical Association.

The reader interested in further readings about biostatistics is recommended to consult especially the journals *Biometrika** and *Biometrics**.

REFERENCES

The citations referred to in the article are presented in alphabetical order. For convenience, they have been classified into one of seven categories shown at the end of the reference and coded according to the following scheme.

A: historical and biographical
B: epidemiology: models and causality
C: epidemiology: relative risk
D: epidemiology: clinical trials
E: demography and community diagnosis
F: surveys, privacy and confidentiality
G: general biometry

1. Bailey, N. T. J. (1957). *The Mathematical Theory of Epidemics*. Charles Griffin, London/Hafner, New York. (B)
2. Bishop, Y. M. M., Fienberg, S. E., and Holland, P. W. (1975). *Discrete Multivariate Analysis*. MIT Press, Cambridge, Mass. (G)
3. Blalock, H. C., Jr. (1964). *Causal Inference in Nonexperimental Research*. University of North Carolina Press, Chapel Hill, N.C. (B)
4. Box, J. F. (1978). *R. A. Fisher: The Life of a Scientist*. Wiley, New York. (A)
5. Byar, D. P., Simon, R. M., Friedewald, W. T., Schlesselman, J. J., DeMets, D. L., Ellenberg, J. H., Gail, M. H., and Ware, J. H. (1976). *N. Engl. J. Med.*, **295**, 74–80. (D)
6. Campbell, D. T., Baruch, R. F., Schwartz, R. D., and Steinberg, J. (1974). *Confidentiality—Preserving Modes of Access to Files and to Interfile Exchange for Useful Statistical Analysis*. Report of the National Research Council Committee on Federal Agency Evaluation Research. (F)
7. Caro, F. G., ed. (1971). *Readings in Evaluation Research*. Russell Sage Foundation, New York. (E)

8. Cochran, W. G. (1953). *Amer. J. Public Health*, **43**, 684–691. (C)

9. Cochran, W. G. (1965). *J. R. Statist. Soc. A*, **128**, 234–255. (C)

10. Cornfield, J. (1951). *J. Natl. Cancer Inst.*, **11**, 1269–1275. (C)

11. Cornfield, J. (1977). *Science*, **198**, 693–699. (G)

12. Courand, A. (1977). *Science*, **198**, 699–705. (D)

13. Cox, D. R. (1970). *The Analysis of Binary Data*. Methuen, London. (G)

14. Dalenius, T. (1974). *Statist. Tidskr.*, **3**, 213–225. (F)

15. Dalenius, T. (1977). *J. Statist. Plan. Infer.*, **1**, 73–86. (F)

16. David, H. A. (1981). *Order Statistics*. (2nd ed.) Wiley, New York. (G)

17. Ederer, F. (1979). *Amer. Statist.*, **33**, 116–119. (D)

18. Finney, D. J. (1964). *Statistical Method in Biological Assay*, 2nd ed. Hafner, New York. (G)

19. Fleiss, J. L. (1973). *Statistical Methods for Rates and Proportions*. Wiley, New York. (C)

20. George, S. L. (1976). *Proc. 9th Int. Biom. Conf., (Boston)*, **1**, 227–244. (D)

21. Gilbert, J. P., McPeek, B., and Mosteller, F. (1977). *Science*, **198**, 684–689. (D)

22. Greenberg, B. G. (1953). *Amer. J. Public Health*, **43**, 692–699. (C)

23. Greenberg, B. G. (1959). *Amer. Statist.*, **13**(3), 13–17, 28. (D)

24. Greenberg, B. G. (1969). *J. Amer. Statist. Ass.*, **64**, 739–758. (B)

25. Greenberg, B. G. (1974). *Medikon*, **6/7**, 32–35. (E)

26. Greenberg, B. G. and Sarhan, A. E. (1959). *Amer. J. Public Health*, **49**, 634–643. (G)

27. Greenberg, B. G. and Sirken, M. (1977). *Validity Problems, Advances in Health Survey Research Methods*. National Center of Health Services Research, Research Proceedings Series, DHEW Publication No. (HRA) 77-3154, pp. 24–31. (F)

28. Greenberg, B. G. and Abernathy, J. R., and Horvitz, D. G. (1970). *Milbank Mem. Fund Quart.*, **48**, 39–55. (F)

29. Greenwood, M. (1948). *Medical Statistics from Graunt to Farr*. Cambridge University Press, Cambridge. (A)

30. Grizzle, J. E., Starmer, C. F., and Koch, G. G. (1969). *Biometrics*, **25**, 489–503. (G)

31. Hastings, D. W. and Berry, L. G., eds. (1979). *Cohort Analysis: A Collection of Interdisciplinary Readings*. Scripps Foundation for Research in Population Problems, Oxford, Ohio. (E)

32. Horwitz, R. I. and Feinstein, A. R. (1978). *N. Engl. J. Med.*, **299**, 1089–1094. (C)

33. Hulka, B. S., Hogue, C. J. R., and Greenberg, B. G. (1978). *Amer. J. Epidemiol.*, **107**, 267–276. (C)

34. Hunter, W. G., and Crowley, J. J. (1979). *Hazardous Substances, the Environment and Public Health: A Statistical Overview*. *Wisconsin Clinical Cancer Center Tech. Rep. No. 4*, University of Wisconsin, Madison, Wis. (G)

35. Kruskal, W. and Tanur, J. M., eds. (1978). *International Encyclopedia of Statistics*. Free Press, New York. (A)

36. Lave, L. B. and Seskin, E. P. (1979). *Amer. Sci.*, **67**, 178–186. (B)

37. Linder, F. E. and Grove, R. D. (1947). *Vital Statistics Rates in the United States, 1900–1940*. U.S. Government Printing Office, Washington, D.C. (See especially, Chaps. 3 and 4.) (E)

38. McKinlay, S. M. (1975). *J. Amer. Statist. Ass.*, **70**, 859–864. (C)

39. McKinlay, S. M. (1977). *Biometrics*, **33**, 725–735. (C)

40. McPherson, K. (1974). *N. Engl. J. Med.*, **290**, 501–502. (D)

41. McPherson, C. K. and Armitage, P. (1971). *J. R. Statist. Soc. A*, **134**, 15–25. (D)

42. Miettinen, O. S. (1970). *Biometrics*, **26**, 75–86. (C)

43. Miettinen, O. S. (1970). *Amer. J. Epidemiol.*, **91**, 111–118. (C)

44. Miettinen, O. S. (1972). *Amer. J. Epidemiol.*, **96**, 168–172. (C)

45. Miettinen, O. S. (1974). *Amer. J. Epidemiol.*, **100**, 350–353. (C)

46. Miller, R. G. (1966). *Simultaneous Statistical Inference*. McGraw-Hill, New York. (G)

47. Nelder, J. A. and Wedderburn, R. W. M. (1972). *J. R. Statist. Soc. A*, **135**, 370–384. (G)

48. Norton, B. J. (1976). *Proc. 9th Int. Biom. Conf. (Boston)*, **1**, 357–376. (A)

49. Peto, R., Pike, M. C., Armitage, P., Breslow, N. E., Cox, D. R., Howard, S. V., Mantel, N., McPherson, K., Peto, J., and Smith, P. G. (1976). *Br. J. Cancer*, **34**, 585–612. (D)

50. Peto, R., Pike, M. C., Armitage, P., Breslow, N. E., Cox, D. R., Howard, S. V., Mantel, N.,

McPherson, K., Peto, J., and Smith, P. G. (1977). *Br. J. Cancer*, **35**, 1–39. (D)

51. Rothman, K. J. (1976). *Amer. J. Epidemiol.*, **103**, 506–511. (C)

52. Spiegelman, M. (1968). *Introduction to Demography*, Harvard University Press, Cambridge, Mass. (E)

53. Tukey, J. W. (1977). *Science*, **198**, 679–684. (D)

54. Walter, S. D. (1976). *Biometrics*, **32**, 829–849. (C)

55. Zelen, M. (1979). *N. Engl. J. Med.*, **300**, 1242–1245. (D)

FURTHER READING

Armitage, P. and David, H. A. (1996). *Advances in Biometry: 50 Years of the International Biometric Society*. Wiley, New York.

Dunn, G. and Everitt, B. (1995). *Clinical Biostatistics: An Introduction to Evidence-Based Medicine*. Wiley, New York/Chichester.

Redmond, C. K. and Colton, T., eds. (2001). *Biostatistics in Clinical Trials*. Wiley, New York.

Wassertheil-Smoller, S. (1995). *Biostatistics and Epidemiology: A Primer for Health Professionals*. Springer, New York/Berlin.

See also CLINICAL TRIALS—I; EPIDEMICS; and MEDICINE, STATISTICS IN.

BERNARD G. GREENBERG

BIPLOTS

DEFINITION

A *biplot* is a graphical display of a matrix $Y_{n \times m}$ by means of markers $\mathbf{a}_1, \ldots, \mathbf{a}_n$ for its rows and $\mathbf{b}_1, \ldots, \mathbf{b}_m$ for its columns, such that inner product $\mathbf{a}_i'\mathbf{b}_v$ represents element $y_{i,v}$ ($i = 1, \ldots, n; v = 1, \ldots, m$).[1] The prefix "bi" indicates that this is a joint display of the rows and columns; in that, biplots differ from most other plots which display only rows or only columns.

(A Geometric Discussion for The Mathematical Reader Is Given in The Concluding Section.)

USES

Biplots are useful for visual inspection of data matrices, allowing the eye to pick up patterns, regularities, and outliers. Multivariate data on single batches can be biplotted in a manner analogous to principal component analysis*, multiple sample data in analogy with MANOVA*, and discriminant analysis*. Biplots are also available for canonical correlations* and contingency tables* and can incorporate approximate graphical significance tests. They can be used to diagnose models, such as Tukey's degree of freedom for nonadditivity* or a harmonic series*.

A SIMPLE EXAMPLE

A biplot of the matrix Y of Table 1 is shown in Fig. 1, with row (tribe) markers $\mathbf{a}_1, \mathbf{a}_2, \ldots, \mathbf{a}_5$ displayed as points and column (characteristics) markers $\mathbf{b}_1, \mathbf{b}_2, \mathbf{b}_3$ displayed as arrows from the origin.

The biplot represents each element $y_{i,v}$ geometrically. This is illustrated as follows for element $y_{1,2}$: A perpendicular is dropped from point $\mathbf{a}_1$ onto arrow $\mathbf{b}_2$ (or onto the straight line through $\mathbf{b}_2$). The distance from the origin to the foot P of this perpendicular is measured as 2.3 and this is multiplied by 12.6, the length of arrow $\mathbf{b}_2$. The product $2.3 \times 12.6 = 29.0$ corresponds to value $y_{1,2}$. This construction is the *inner product* $\mathbf{a}_1'\mathbf{b}_2 \Rightarrow y_{1,2}$.

The $\mathbf{a}$-scatter reflects the tribes. The cluster of three points shows that Apaches, Sioux, and Hopis were quite similar in the characteristics displayed here, whereas Shoshones and Navajos differed from them in opposite directions.

Linear combinations of rows of Y can be represented on the biplot. Thus the average of rows $\mathbf{y}_1'$ and $\mathbf{y}_4'$ can be represented by a point midway between $\mathbf{a}_1$ and $\mathbf{a}_4$. This would be close to the cluster of the other three points. Evidently, the Shoshone–Navajo average was similar to the other three tribes. Similarly, the Navajo–Apache difference $\mathbf{a}_4 - \mathbf{a}_2$ can be constructed as a point in the same direction and distance from the origin as $\mathbf{a}_4$ is from $\mathbf{a}_2$.

The configuration of arrows reflects the magnitude and relations of the characteristics. The longer arrow $\mathbf{b}_2$ shows that the numbers in the second column are larger than those in the other columns. The near collinearity of arrows $\mathbf{b}_1$ and $\mathbf{b}_3$ reflects the rough proportionality of the first and third columns.

Table 1. Demographic Characteristics of Some American Indian Tribes

Tribe	Median Years of Schooling	Percentage below Poverty Line	Economic Index[a]
Shoshones	10.3	29.0	9.08
Apaches	8.9	46.8	10.02
Sioux	10.2	46.3	10.75
Navajos	5.4	60.2	9.26
Hopis	11.3	44.7	11.25

Source. Kunitz [8].

[a] 0.6 median school year + 0.1% poverty.

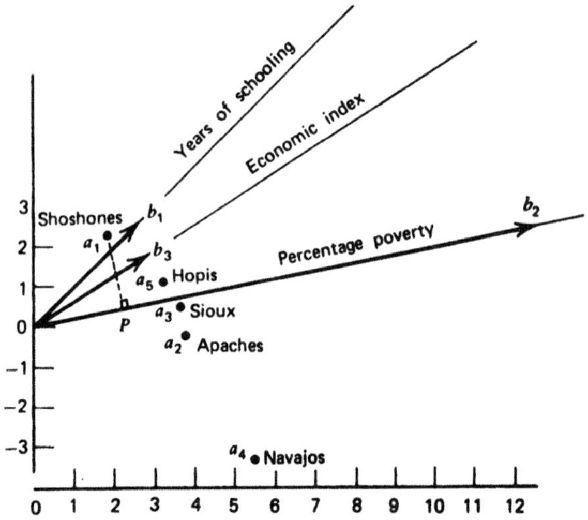

Figure 1. Biplots of demographic characteristics of Indian tribes.

Combinations of columns may be represented just like combinations of rows. As an example, a new variable defined as (economic index)−(median years of schooling) can be represented by an arrow in the direction from $\mathbf{b}_1$ to $\mathbf{b}_3$. This is very much the same direction as the Navajo−Apache difference $\mathbf{a}_4 - \mathbf{a}_2$. Evidently, these two tribes differed most strongly on that variable. The special feature of the biplot is that it displays such interrelations between the rows and columns.

APPROXIMATION AND COMPUTATION

In practical applications, biplot display is approximative because matrices are usually of rank greater than 2. Inner-product representation relation $\mathbf{a}_i'\mathbf{b}_v \Rightarrow y_{i,v}$ can be written $AB' \Rightarrow Y$, where the $\mathbf{a}$'s and $\mathbf{b}$'s are rows of A and B, respectively. But this AB'-biplot can represent Y well only if the rank 2 product AB' is close to Y. One may obtain AB' by least squares* through the Householder=-Young theorem (see below) [1]. Algorithms are also available for weighted least-squares* fitting [6] and for adaptive fits [9]. (If either A or B is given from some prior considerations, the other factor may be obtained by linear least squares.)

Adding another dimension to the biplot display may yield a closer approximation. Such a three-dimensional bimodel can be partly represented by a number of planar projections but is most useful when displayed on a CRT with a fast rotation-projection algorithm [12].

PRINCIPAL COMPONENT* BIPLOTS AND AN EXAMPLE

A special type of biplot is suitable for multivariate data matrices in which i indexes the n units of a batch and v the m variables, the observations $y_{i,v}$ being measured from the batch means. These *principal component biplots* [3] are fitted by least squares as follows. Solve $Y'Y\mathbf{q}_\alpha = \lambda_\alpha^2 \mathbf{q}_\alpha$ subject to $\mathbf{q}_\alpha'\mathbf{q}_\alpha = 1$ for the largest two roots $\lambda_1^2 \geqslant \lambda_2^2$; compute $\mathbf{p}_\alpha = Y\mathbf{q}_\alpha\lambda_\alpha^{-1}$ and define $G = (\mathbf{p}_1, \mathbf{p}_2)$, $H = (\lambda_1\mathbf{q}_1, \lambda_2\mathbf{q}_2)$, $J = (\lambda_1\mathbf{p}_1, \lambda_2\mathbf{p}_2)$, and $K = (\mathbf{q}_1, \mathbf{q}_2)$. This gives the GH'- and the JK'-biplots, both of which approximate Y to the extent of $(\lambda_1^2 + \lambda_2^2)/\sum_{\alpha=1}^y \lambda_\alpha^2$. The GH'-biplot gives an even better fit to the variables (columns) *configuration* (i.e., the inner products $Y'Y$) in that $h'_v h_w \Rightarrow y'_{(v)}y_{(w)}$, which is n times the (v, w)-covariance; but its representation $\|\mathbf{g}_i - \mathbf{g}_e\| \Rightarrow \sqrt{(\mathbf{y}_i - \mathbf{y}_e)' \cdot (Y'Y)^{-1}(\mathbf{y}_i - \mathbf{y}_e)}$ of the standardized interunit distances is poorer. Conversely, the JK'-biplot represents the interunit *distances* very well, but not the configuration of the variables. (See Table 2.)

Figure 2 displays a GH'-biplot of 27 European countries' per capita consumption of protein from nine sources, measured from the European means [11].

1. Lengths of **h**-arrows are roughly proportional to the standard deviations* of the variables (sources) represented. Cosines of angles between **h**-vectors approximate the correlations* between the sources represented.

Cereal protein consumption had by far the highest variability, followed by milk. Eggs and starches had little variability. Cereal protein consumption was negatively correlated with meat proteins, $(180°$ angle $\Rightarrow -1$ correlation). Vegetables and fruit appeared uncorrelated with either of the above, but somewhat with pulses, nuts, and oilseeds and with fish.

2. Distances between **g**-points represent standardized intercountry differences

Table 2. Goodness of Fit of Biplot Approximations

Biplot Approximation		Goodness of
GH'	JK'	Fit[a]
$\mathbf{g}_i'\mathbf{h}_v \Rightarrow y_{i,v}$	$\mathbf{j}'ik_v \Rightarrow y_{i,v}$	$(\lambda_1^2 + \lambda_2^2)/\sum_{\alpha=1}^y \lambda_2^2$
$\mathbf{g}_i'\mathbf{g}_e \Rightarrow \mathbf{y}_i'(Y'Y)^{-1}y_e$	$\mathbf{k}_v'\mathbf{k}_w \Rightarrow y'_{(v)}(YY')^{-1}y_{(w)}$	$2/y$
$\mathbf{h}_v'\mathbf{h}_w \Rightarrow \mathbf{y}'_{(v)}y_{(w)}$	$\mathbf{j}_i'\mathbf{j}_e \Rightarrow \mathbf{y}_i'\mathbf{y}_e$	$(\lambda_1^4 + \lambda_2^4)/\sum_{\alpha=1}^y \lambda_\alpha^4$

[a]Goodness of fit
$$= 1 - \frac{\text{sum of squares of residuals from approximation}}{\text{sum of squares of data}}$$

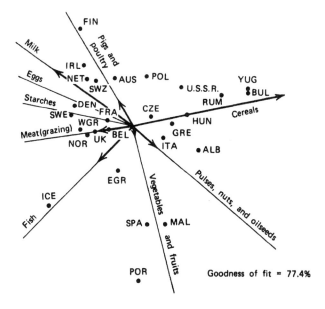

Goodness of fit = 77.4%

Figure 2. GH-biplot of European countries' protein consumption.

in protein consumption. Points close together represent countries with similar patterns.

There was a clear geographical division—the eastern countries to the right, the northwestern ones to the left, and the southwestern ones at the bottom. Geographical contiguity appears to have been related to similarity of protein consumption. (Only Iceland and Finland were exceptional.)

3. Inner products $\mathbf{g}_i'\mathbf{h}_v$ represent the actual v-source protein consumption in country i (as measured from the European mean). If $\mathbf{g}_i$ is in (opposite to) the direction of $\mathbf{h}_v$, country i consumes more (less) of protein v than the European average.

Eastern European countries are clustered along the arrow for cereals and opposite those for animal proteins. Evidently, Eastern Europeans consumed more cereals and fewer animal proteins than average Europeans did. Northwestern Europeans have the reverse pattern. Southwestern Europeans are seen to have consumed large amounts of vegetables and fruit, of pulses, nuts, and oilseeds, and of fish. (Iceland and Finland showed unusually

large consumption of fish and milk, respectively.)

The biplot's special feature is that it shows the relation of the variables' configuration—$\mathbf{h}$-arrows—to the units' scatter—$\mathbf{g}$-points—and thus indicates on what variables the clusters of units differ.

A DIAGNOSTIC EXAMPLE

A biplot can display data from a three-way and higher-order layout and can be used to diagnose a model. Figure 3 displays data from a three-factor, two-replication experiment (*see* DESIGN OF EXPERIMENTS) on absorption of gamma radiation. The factors metal, distance, and replication were cross-classified within the rows of Y and are jointly represented by $\mathbf{a}$'s; $\mathbf{b}$'s represent number of plates, the column factor.

Figure 3 shows collinearity of the $\mathbf{b}$ markers as well as of the $\mathbf{a}$ markers for each metal. One may apply the diagnostic rules from Table 3. When these patterns hold for only a part of the row markers or the column markers, the model may be diagnosed for the appropriate submatrix. These rules suggest two separate *additive* models, one for lead and another for aluminum. Both replicates

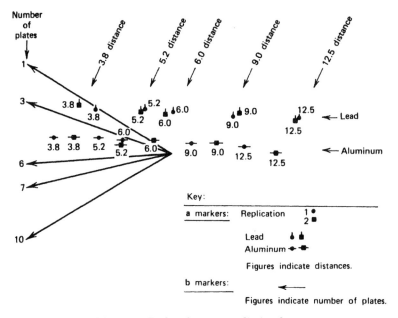

Figure 3. Biplot of gamma-radiation data.

Table 3. Biplot Diagnostic Rules

Row Markers	Column Markers	The Model is:	
Collinear	—	$y_{i,v} = \beta_v + \alpha_i \beta_v^*$	Columns' Regression
—	Collinear	$y_{i,v} = \alpha_i + \alpha_i^* \beta_v$	Rows' Regression
Collinear	Collinear	$y_{i,v} = \mu + \alpha_i \beta_v$	Concurrent
Collinear at 90° to	Collinear	$y_{i,v} = \alpha_i + \beta_v$	Additive

Source. Bradu and Gabriel [2].

fit the models, but whereas lead replicates hardly differ, those for aluminum do.

The order of markers allows one to detail the diagnosis further. The **b** markers show a trend proportional to the number of plates on which radiation is seen to depend linearly. For each metal, the **a** markers have a trend that indicates quadratic dependence on distance. The two metals' **a**'s are close to two parallel lines. This indicates null metal × distance interaction*. Hence the biplot suggests diagnosis of a model

$$\text{radiation}_{(\text{plts,dist,metal})} = \mu + \alpha(\text{dist})$$
$$+ \beta(\text{dist})^2$$
$$+ \gamma_{(\text{metal})}(\text{plts}).$$

Least-squares fitting confirmed this as a good model [2].

Nonlinear models may also be revealed. Thus an elliptic structure of **b**'s in three dimensions corresponds to a harmonic model for the rows of Y [10].

A MULTIVARIATE ANALYSIS-OF-VARIANCE* (MANOVA) EXAMPLE

Data of the first randomized Israeli rainmaking experiment [4] are displayed in Fig. 4. The days were classified into four samples by the area allocated to seeding (north or center) and by whether seeding actually took place (suitable clouds available). There also was a fifth sample of preexperimental days. The variables were daily precipitation in the following four areas: north, center, a buffer zone between these two alternate target areas, and south.

A JK'-biplot was fitted to deviations from the overall mean, weighted by sample sizes n_i and by the inverse of the "within" variance–covariance* matrix. Around each

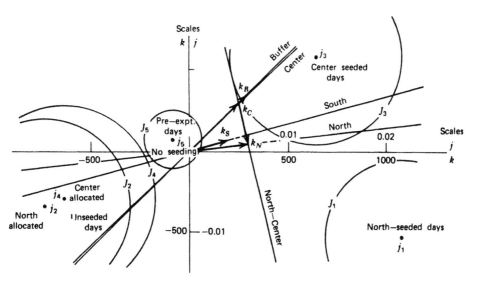

Figure 4. MONOVA biplot of results of rainmaking experiment, Israel, 1961–1967 (rain amounts in millimeters).

sample marker $\mathbf{j}_i$, a circle J_i was drawn with radius $[n_i\theta(1-\theta)/2]^{1/2}$, where θ is the upper 5% point of the maximum characteristic root distribution*.

1. Each $\mathbf{k}$-vector represents "within" variation of one area's rainfall and the cosines approximate the correlations.
   $\mathbf{k}$-vectors for linear combinations of variables can be constructed. Thus $\mathbf{k}_{N-C} = \mathbf{k}_N - \mathbf{k}_C$ is shown; it displays the "within" variation of the north–center rainfall differences.

2. The distance between $\mathbf{j}$ markers displays the dissimilarity of the samples with respect to rainfall in the four areas. (It is the Mahalanobis distance* of the planar approximation.)
   The two unseeded samples cluster together, with the nonseeded preexperimental sample between them and the two seeded samples. This reflects the division of days into those suitable for seeding and those not suitable, and the preexperimental days, which included both types. The seeded samples differed more than the unseeded; this may have been due to seeding.

3. Approximate tests are: "Declare any two samples significantly different if their J circles do not overlap."
   Only circles J_2 and J_4 overlap. All samples are therefore declared "significantly" different except the two unseeded samples.

4. For a rough test of whether two samples differ significantly on a particular variable v, one may project the corresponding two circles onto the $\mathbf{k}_v$-vector.
   North-seeded days are seen to have had a "significantly" larger north-center rainfall difference than center-seeded days, although these two samples of days did not differ "significantly" on the rainfall in any single one of the areas.

This MANOVA biplot allowed graphic approximation of complex data and simultaneous significance testing*, and it provides a picture of the multivariate structure of the data. It would have been difficult to obtain these by purely analytical methods.

We must caution against interpreting "significance" literally in this particular example, as rainfall data are far from Gaussian* and observations are dependent from day to day.

RELATION TO OTHER TECHNIQUES

GH'- and JK'-biplots are complementary to principal component analysis* in that they display two (or three, for bimodels) dimensions simultaneously. The MANOVA* biplot displays samples along discriminant function*, or canonical variate* axes. Biplots of canonical correlations* or contingency tables* also complement rather than supplant standard analyses.

Multidimensional scaling* methods differ from classical multivariate analyses and the related biplots in using nonmetric methods of fitting [7]. This makes them more flexible for the purpose they are designed for, but less amenable to joint representation of rows and columns. However, a multidimensional scaling model for the variables (or units) can be supplemented by a superimposed biplot display of the units (or variables) [5].

The relation to factor analysis* is less clear, although rotational techniques may be applied to the biplot.

RELATION TO MULTIVARIATE GEOMETRY

Any matrix $Y_{(n \times m)}$ can be factored as $Y = A_{(n \times r)}(B_{(m \times r)})'$ for any $r \geqslant y$. A display of $\mathbf{a}_1, \ldots, \mathbf{a}_n$ and $\mathbf{b}_1, \ldots, \mathbf{b}_m$ in E_r is called the AB'-bigeometry of Y and represents the matrix elementwise by inner products $y_{i,v} = \mathbf{a}_i'\mathbf{b}_v$. The factorization is not unique and neither is the bigeometry of Y [3].

Two bigeometries of particular interest are the columns geometry—much used for representing the configuration $Y'Y$ of a general linear model*—and the rows geometry—commonly known, for $m = 2$, as the scatter plot. The former is used for representation of the variance-covariance configuration of the columns of Y, whereas the latter displays interrow distances.

The columns geometry corresponds to the IY-bigeometry with $r = n$, in which the columns of Y are displayed as $\mathbf{b}_v = \mathbf{y}_{(v)}$, and the rows as $\mathbf{a}_i' = \mathbf{u}_i'$,—unit-length vectors along the n orthogonal axes of reference of E_n.

The inner products are $\mathbf{a}_i'\mathbf{b}_v = \mathbf{u}_i'\mathbf{y}_{(v)} = y_{i,v}$, and the column configuration is preserved by the $\mathbf{b}$'s, i.e., $\mathbf{b}_v'\mathbf{b}_w = \mathbf{y}_{(v)}'\mathbf{y}_{(w)}$.

The rows geometry corresponds to the YI-bigeometry, with $r = m$, in which the rows are displayed as $\mathbf{a}_i' = \mathbf{y}_i'$, and the columns as $\mathbf{b}_v = \mathbf{u}_v$—unit-length vectors along the orthogonal axes of reference of E_m. Again, the inner products are $\mathbf{a}_i'\mathbf{b}_b = \mathbf{y}_i'\mathbf{u}_v = y_{i,v}$, but here interrow distances are preserved by the $\mathbf{a}$'s display—$\|\mathbf{a}_i - \mathbf{a}_e\| = \|\mathbf{y}_i - \mathbf{y}_e\|$.

The columns' configuration does not change if it is rotated into another set of orthogonal axes*. Thus if $\tilde{P}_{(n \times m)}$ satisfies $\tilde{P}\tilde{P}' = \tilde{P}'\tilde{P} = I_n$, then the rotated $(\tilde{P}')(\tilde{P}Y)$-bigeometry has the same configuration $(\tilde{P}Y)' \cdot (\tilde{P}Y) = Y'Y$ as the IY-bigeometry. One such rotation is that obtained from the singular-value decomposition:

$$Y = \tilde{P}_{(n \times m)}'\tilde{\Lambda}_{(n \times m)}\tilde{Q}_{(m \times m)}$$

$$= (P_{(n \times y)}', P^{*'}{}_{(n \times (n-y))}) \begin{bmatrix} \Lambda_{(y \times y)} & 0 \\ 0 & 0 \end{bmatrix}$$

$$\times \begin{pmatrix} Q_{(y \times m)} \\ Q^*{}_{((m-y) \times m)} \end{pmatrix}$$

$$= P'\Lambda Q$$

$$= \sum_{\alpha=1}^{y} \lambda_\alpha \mathbf{p}_\alpha \mathbf{q}_\alpha',$$

which would yield the $(\tilde{P}')(\tilde{\Lambda}\tilde{Q})$-bigeometry. Since only the first y rows of Q are nonzero, this may be represented in E_y by the GH'-bigeometry with $G = P'$ and $H' = \Lambda Q$.

Omission of the last $(n - y)$-axes is equivalent to projection onto what are known as the first y principal axes* for columns.

The vectors $\mathbf{h} = (\lambda_1 q_{1,v}, \ldots, \lambda_y q_{y,v})'$ have the same configuration $\mathbf{h}_v'\mathbf{h}_w = \mathbf{y}_{(v)}'\mathbf{y}_{(w)}$, as the columns of Y. Their elements $h_{v,\alpha} = \lambda_\alpha q_{\alpha,v}$ are known as the *loadings** of column v onto principal axis α. The vectors $\mathbf{g}_i' = (p_{1,i}, \ldots, p_{y,i})$, are projections of unit length $\tilde{\mathbf{p}}_{(i)}$'s onto these *principal axes*. Since $\mathbf{g}_i'\mathbf{g}_e = \mathbf{y}_i'(Y'Y)^{-1}\mathbf{y}_e$, the $\mathbf{g}$'s display the scatter of the rows of Y in a standardized form.

The rows scatter will similarly be preserved if the YI-bigeometry is rotated into another set of axes. Choice of rotation $\tilde{Q}$ from the singular-value decomposition, and projection into the y-dimensional subspace spanned

by what are known as the *principal axes for rows*, yields the JK'-bigeometry with $J = P'\Lambda$ and $K' = Q$. Vectors $\mathbf{k}_v = (q_{1,v}, \ldots, q_{y,v})$, are projections of unit length $\tilde{q}_{(v)}$'s onto these axes. They display the column configuration of Y in standardized form $\mathbf{k}_v'\mathbf{k}_w = \mathbf{y}_{(v)}'(YY')^{-}\mathbf{y}_{(w)}$. The vectors $\mathbf{j}_i' = (\lambda_1 p_{1,i}, \ldots, \lambda_y p_{y,i})$ preserve the row scatter in that $\mathbf{j}_i'\mathbf{j}_e = \mathbf{y}_i'\mathbf{y}_e$. Their elements $j_{i,\alpha} = \lambda_\alpha p_{\alpha,i}$ are known as row i's *principal coordinates** with respect to axis α, or αth *principal component scores**.

The rows geometry and the columns geometry, regarded, respectively, as the JK' and the GH' bigeometries, are seen to differ only by reciprocal scalings along their principal axes, i.e., $j_{i,\alpha}\lambda_\alpha^{-1} = g_{i,\alpha}$ and $k_{v,\alpha}\lambda_\alpha = h_{v,\alpha}$. The Y rows' scatter, as displayed by the $\mathbf{j}$'s, is scaled down to the standardized scatter displayed by the $\mathbf{g}$'s, whereas the standardized Y columns' configuration displayed by the $\mathbf{k}$'s is scaled up to the original configuration displayed by the $\mathbf{h}$'s. The relation between the two geometries is evident when they are regarded as bigeometries.

For $y = 2$ or 3 the bigeometry reduces to a *biplot* or *bimodel*, in the plane or in three dimensions, respectively. When y is larger, the bigeometry may be projected onto a plane to obtain an *approximate biplot* (and similarly for an approximate bimodel). Projection can be obtained by omitting all but the first two (or three) coordinates of the $\mathbf{a}$'s and $\mathbf{b}$'s. In case of the GH' and JK' bigeometries, this leads to least-squares approximation of rank 2 (or 3). (This is the Householder-Young theorem.) The GH'-biplot thus displays vectors $\mathbf{g}_i' = (p_{1,i}, p_{2,i})$ and $\mathbf{h}_v = (\lambda_1 q_{1,v}, \lambda_2 q_{2,v})$, and the JK'-biplot displays $\mathbf{j}_i' = (\lambda_1 p_{1,i}, \lambda_2 p_{2,i})$ and $\mathbf{k}_v = (q_{1,v}, q_{2,v})'$. (For brevity, we use the same notation for the y-dimensional $\mathbf{g}$, $\mathbf{h}$, $\mathbf{j}$, and $\mathbf{k}$ for their two-dimensional biplot projections.) These biplots serve as approximate displays of Y with properties similar to those of the corresponding bigeometries.

NOTE

1. A matrix is denoted by an uppercase letter; its (i, v)th element is subscripted i, v; its ith row and vth column are set in boldface type and subscripted

i and (v), respectively; $(\;)^{-1}$ indicates a generalized inverse. $\|\mathbf{x}\|$ denotes the length of $\mathbf{x}$. 8The row vectors of identity matrix I_m are denoted $\mathbf{u}_i (i = 1, \ldots, m)$. The notation $\mathbf{a}'_i\mathbf{b}_v \Rightarrow y_{i,v}$ means "$\mathbf{a}'_i\mathbf{b}_v$ represents $y_{i,v}$."

REFERENCES

[References 2 and 3 provide useful introductions to the biplot and its diagnostic uses. References 4, 5, and 6 give additional applications.]

1. BIPLOT computer program available from Division of Biostatistics, University of Rochester, Rochester, NY 14642.

2. Bradu, D. and Gabriel, K. R. (1978). *Technometrics*, **20**, 47–68.

3. Gabriel, K. R. (1971). *Biometrika*, **58**, 453–467.

4. Gabriel, K. R. (1971). *J. Appl. Meteorol.*, **11**, 1071–1077.

5. Gabriel, K. R. (1978). In *Theory Construction and Data Analysis in the Behavioral Sciences*, S. Shye, ed. Jossey-Bass, San Francisco, pp. 350–370.

6. Gabriel, K. R. and Zamir, S. (1979). *Technometrics*, **21**, 489–498.

7. Kruskal, J. B. (1978). *International Encyclopedia of Statistics*, W. H. Kruskal and J. M. Tanur, eds. Free Press, New York. pp. 307–330.

8. Kunitz, S. J. (1976). *Hum. Biol.*, **48**, 361–377.

9. McNeil, D. R. and Tukey, J. W. (1975). *Biometrics*, **31**, 487–510.

10. Tsianco, M. and Gabriel, K. R. (1981). Biplot diagnosis of harmonic models for meteorological data. (To be published.)

11. Weber, A. (1973). Agrarpolitik in Spannungsfeld der internationalen Ernährungspolitik. Institut für Agrarpolitik und Marktlehre, Kiel (mimeographed).

12. Weber, E. and Berger, J. (1978). *COMSTAT 78–Proceedings*, L. C. A. Corsten and J. Hermans, eds. Physica Verlag, Vienna. pp. 331–337.

See also ANDREWS FUNCTION PLOTS; FACTOR ANALYSIS; GRAPHICAL REPRESENTATION OF DATA; and PRINCIPAL COMPONENT ANALYSIS, GENERALIZED.

K. RUBEN GABRIEL

BIRNBAUM, ALLAN

Born: May 27, 1923, in San Francisco, California, USA.

Died: July 1, 1976, in London, England.

Contributed to: statistical inference, foundations of statistics, statistical genetics, statistical psychology, history of science.

Allan Birnbaum was one of the most profound thinkers in the field of foundations of statistics. In spite of his short life span, he made a significant impact on the development of statistical ideas in the twentieth century. His tragic, and apparently self-inflicted, death in 1976 was a severe blow to the progress of research into foundations for statistical science and the philosophy of science in general.

Born in San Francisco in 1923 of Russian-Jewish orthodox parents, he studied mathematics as an undergraduate in the University of California at Berkeley and at Los Angeles, completing simultaneously a premedical program in 1942, and a first degree in 1945. He remained at Berkeley for the next two years, taking graduate courses in science, philosophy, and mathematics. He was influenced by the work of Ernest Nagel and Hans Reichenbach. Following the latter's advice, he moved to Columbia University, New York, in 1947. Here he obtained a Ph. D. degree in mathematical statistics in 1954 (partly under the guidance of E. L. Lehmann, who was spending a semester at Columbia University in 1952). He had already been a faculty member for three years and continued at Columbia until 1959. During this time he was active in research on classification and discrimination, and development of mathematical models and statistical techniques for the social sciences. In the latter, he interacted with such luminaries as Duncan, Luce, Nagel, and Lazarsfeld. In 1959, Birnbaum moved to the Courant Institute of Mathematical Sciences, where he remained (officially) until 1974. During this period he spent considerable periods as a visiting professor in several places—notably Stanford University, Imperial College (London), for most of 1971–1974 as a visiting Fellow at Peterhouse, Cambridge, and also (in 1974) as a visiting professor at Cornell University; and he was also associated with the Sloan-Kettering Cancer Research Center. He finally left the Courant Institute in 1974, reputedly, in part, due to an apparent lack of appreciation for statistics by mathematicians at the Institute. Birnbaum was appointed Professor of Statistics at City University, London, in 1975, less than a year before his death in 1976.

Birnbaum combined a forceful personality and great intellectual honesty and power with a quiet and unassuming nature, and was a sensitive and kind person. He had a great variety of interests, and his subtle sense of humor made him especially effective on informal occasions and in private conversations. The author of this note remembers well how Birnbaum arranged for him to visit the Courant Institute in 1965 to learn about the status of statistics in the Soviet Union at that time. Allan was very much attached to his only son, Michael, born late in his life.

He was elected a Fellow of the Institute of Mathematical Statistics, the American Statistical Association, and the American Association for the Advancement of Science. His scientific output includes 41 papers on statistical theory and applications listed in a memorial article by G. A. Barnard and V. P. Godambe in the *Annals of Statistics* (1982), **10**, 1033–1039. His early work was in the spirit of E. L. Lehmann's famous textbook *Testing of Statistical Hypotheses*, and is, indeed, referred to in that book. His most notable contribution is generally considered to be a paper originally published in the *Journal of the American Statistical Association* (1962), **57**, 269–306 (discussion, pp. 307–326), which was reprinted in vol. I of *Breakthroughs in Statistics*, Samuel Kotz and Norman L. Johnson, eds., Springer-Verlag, New York (1992), pp. 478–518, with a perceptive introduction by J. F. Bjornstad (pp. 461–477). This paper studies the likelihood principle (LP) and its application in measuring the evidence about unknown parameter values contained in data. Birnbaum showed that the LP is a consequence of principles of sufficiency and conditionality and claimed that this implies, inter alia, the irrelevance, at the inference stage, of stopping rules. (Birnbaum interpreted the LP as asserting that if two experiments produce proportional likelihood functions, the same inference should be made in each case.) Birnbaum's result is regarded by many as one

of the deeper theorems of theoretical statistics (with a remarkably simple proof). The result has given rise to continuing controversy, and claims have been made (notably by the prominent statistician S. W. Joshi) that it may be false. L. J. Savage* remarked, on the other hand, that appearance of the paper was an event "really momentous in the history of statistics. It would be hard to point to even a handful of comparable events" (obituary notice in *The Times* of London). Indeed, the paper has had substantial effects on estimation in the presence of nuisance parameters, on prediction in missing-data problems, and in metaanalysis (combination of observations).

Another groundbreaking initiative by Birnbaum, in the field of foundations of statistics, is his paper on Concepts of Statistical Evidence, published in *Philosophy, Science and Method: Essays in Honor of Ernest Nagel* (St. Martin's, New York). His ideas on statistical scholarship were expressed in an article published in 1971 in the *American Statistician*, **25**, pp. 14–17. His final views on the Neyman–Pearson theory are discussed in a paper which appeared posthumously (with a discussion) in *Synthèse*, 36, 19–49 (1978), having previously been rejected by the Royal Statistical Society.

Further important contributions include an article on conceptual issues in experimental genetics [*Genetics*, **72**, 734–758 (1972)] and historical articles on John Arbuthnot* [*American Statistician*, **21**, 23–25 and 27–29 (1961)] and on statistical biographies [*Journal of the Royal Statistical Society, Series A*, **133**, 1265–1282 (1970)].

REFERENCES

1. Barnard, G. A. and Godambe, V. P. (1982). Allan Birnbaum 1923–1976, *Ann. Statist.*, **10**, 1033–1039.

2. Bjørnstad, J. F. (1992). Introduction to Birnbaum. In *Breakthroughs in Statistics, 1*, S. Kotz and N. L. Johnson, eds. Springer-Verlag, New York, pp. 461–467.

3. Lindley, D. V. (1978). Birnbaum, Allan. In *International Encyclopedia of Statistics*, W. H. Kruskal and J. Tanur, eds. Free Press, New York, vol. 1, pp. 22–23.

4. Norton, B. (1977). Obituaries: *J. R. Statist. Soc. A*, **140**, 564–565; *The Times*.

BIRNBAUM–HALL TEST.   See
KOLMOGOROV-SMIRNOV TEST, THREE-SAMPLE

BIRNBAUM-RAYMOND-ZUCKERMAN INEQUALITY

A probability inequality of the Chebyshev* type. Let X_i be independent random variables with common variance, $\mu_i = EX_i$ and $\sigma^2 = \mathrm{var}(\dot{X}_i)$, $i = 1, \ldots, n$. Then for even n

$$
\Pr\left(\sum_{i=1}^{n}(X_i - \mu_i)^2 \geqslant \lambda^2\right)
$$

$$
\leqslant \begin{cases}
1 & \text{if } \lambda^2 \leqslant n\sigma^2 \\
\frac{n\sigma^2}{2\lambda^2 - n\sigma^2} & \text{if } n\sigma^2 \leqslant \lambda^2 \\
& \quad \leqslant \frac{n\sigma^2}{4}(3 + \sqrt{5}) \\
\frac{n\sigma^2}{\lambda^2}\left(1 - \frac{n\sigma^2}{4\lambda^2}\right) & \text{if } \frac{n\sigma^2}{4}(3 + \sqrt{5}) \leqslant \lambda^2,
\end{cases}
$$

and for odd n

$$
\Pr\left(\sum_{i=1}^{n}(X_i - \mu_i)^2 \geqslant \lambda^2\right)
$$

$$
\leqslant \begin{cases}
1 & \text{if } \lambda^2 \leqslant n\sigma^2 \\
\frac{(n+1)\sigma^2}{2\lambda^2 - (n-1)\sigma^2} & \text{if } n\sigma^2 \leqslant \lambda^2 \\
& \quad \leqslant \frac{\sigma^2}{4}(3n + 1 \\
& \qquad + \sqrt{5n^2 + 6n + 5}) \\
\frac{n\sigma^2}{\lambda^2} - \frac{(n^2-1)}{4}\frac{\sigma^4}{\lambda^4} & \text{if } \frac{\sigma^2}{4}(3n + 1 \\
& \qquad + \sqrt{5n^2 + 6n + 5}) \\
& \quad \leqslant \lambda^2.
\end{cases}
$$

This inequality provides an upper bound on the probability of the sample point falling outside a hypersphere centered at the population mean. It has applications to bombing and other aiming problems. Although the random variables are assumed to be independent, it is a *multidimensional* inequality in the sense that it gives a bound on the probability of a multidimensional set. (Compare with Berge's inequality*.) As n becomes large, the results for odd and even integers n approach each other.

BIBLIOGRAPHY

Birnbaum, Z. W., Raymond, J., and Zuckerman, H. S. (1947). *Ann. Math. Statist.*, **18**, 70–79.

Savage, I. R. (1961). *J. Res. Natl. Bur. Stand.*, **65B**(3), 211–222.

See also BERGE INEQUALITY; CAMP–MEIDELL INEQUALITY; and PROBABILITY INEQUALITIES FOR SUMS OF BOUNDED RANDOM VARIABLES.

BIRNBAUM'S THEOREM.   See
ANCILLARY STATISTICS—I

BIRTH-AND-DEATH PROCESSES

It would seem that the name "birth-and-death process" should apply to any stochastic process* that models the development of a population, the individual members of which are subject to the vicissitudes of births and deaths. It is usual, however, for the name to be used with reference to a very specific class of stationary Markov* chains in continuous time; this article is restricted to a discussion of this class.

We begin by describing what we shall call a *generalized* birth-and-death process; this terminology is convenient but not universal, and some authors refer to these "generalized" processes simply as "birth-and-death" processes.

Let $N(t)$ be the size of the population at time t, i.e., the number of *live* individuals at that time; assume that $N(T)$ is, as mentioned above, a Markov chain with stationary transition probabilities*:

$$
P_{ij}(s) = \Pr[N(t + s) = j | N(t) = i], \quad (1)
$$

where, for $s > 0$, $i = 0, 1, 2, \ldots$,

$$
P_{i,i+1}(s) = \alpha_i s + o(s),
$$
$$
P_{i,i}(s) = 1 - (\alpha_i + \beta_i)s + o(s),
$$

and, provided that $i \neq 0$,

$$
P_{i,i-1}(s) = \beta_i s + o(s).
$$

Here the numbers α_i and β_i are nonnegative constants that may be referred to as the birthrates and deathrates, respectively.

Unfortunately, there are technical mathematical difficulties if the $\{\alpha_i\}$ sequence grows too rapidly with increasing i [causing the $N(t)$-process to "explode" to arbitrarily large values]; the remarks of this article assume that the birthrates, $\{\alpha_i\}$, grow no more rapidly than a polynomial in i. There are few applications where such an assumption is violated; indeed, a large part of the interesting ones need the birthrates to grow no faster than a linear function of i.

If we set $P_j(t) = \Pr[N(t) = j]$, then it is possible to show these $P_j(t)$ satisfy the system of differential equations

$$\frac{dP_j(t)}{dt} = \alpha_{j-1}P_{j-1}(t) - (\alpha_j + \beta_j)P_j(t)$$
$$+ \beta_{j+1}P_{j+1}(t) \qquad (2)$$

($j = 0, 1, 2, \ldots$), where we introduce a constant $\beta_0 = 0$, for ease of writing.

Thus, in theory, if we know the probability distribution of $N(0)$, say, we can derive the probability distribution of $N(t)$ for any $t > 0$. Unfortunately, it is impossible to obtain usable solutions of the system (2) for quite arbitrary birth and death rates and, indeed, the class of processes for which some sort of explicit solution to (2) may be exhibited is decidedly circumscribed; we shall discuss specific examples of such tractable processes below.

Nonetheless, if the birth and death rates are such that a steady-state (stationary) solution* to (2) exists, then that stationary solution is easily calculated for quite general processes. If we set

$$S = 1 + \frac{\alpha_0}{\beta_1} + \frac{\alpha_0\alpha_1}{\beta_1\beta_2} + \frac{\alpha_0\alpha_1\alpha_2}{\beta_1\beta_2\beta_3} + \cdots \qquad (3)$$

then the finiteness of S ensures the existence of a steady-state solution $P_j(t) = \pi_j$, say ($j = 0, 1, \ldots$), where every $\pi_j \geqslant 0$ and $\sum_0^\infty \pi_j = 1$. Indeed, it is easy to discover that

$$\pi_j = S^{-1}\left(\frac{\alpha_0\alpha_1\cdots\alpha_{j-1}}{\beta_1\beta_2\cdots\beta_j}\right) \qquad (j \geqslant 1) \qquad (4)$$

and $\pi_0 = S^{-1}$.

A consequence of (4) is the revelation that a very large number of different mechanisms can lead to the same steady-state solution to (2). Thus the practical observation of a certain probability distribution (e.g., the Poisson*) in, shall we say, a biological context, can offer, at best, only partial confirmation of some hypothesis about the mechanism of that biological process.

When $S < \infty$, one can infer from the general theory of Markov chains* in continuous time that, whatever the initial probability distribution $\{P_j(0)\}_{j=0}^\infty, P_j(t) \to \pi_j$ as $t \to \infty$. This result demonstrates the supreme importance, from the viewpoint of applications, of solution (4) to equations (2) of the generalized birth-and-death process.

If specific assumptions are made about the $\{\alpha_i\}$ and $\{\beta_i\}$ as functions of i, then it is sometimes possible to determine useful information about the "transient" probabilities $P_j(t)$. The most useful tool here is the probability generating function* (PGF):

$$\Pi(z, t) = \sum_{j=0}^\infty z^j P_j(t), \qquad (5)$$

whose use in the present context was championed (although not invented) by M. S. Bartlett (see, e.g., Bartlett [1]). In (5) the dummy variable z may be supposed to be real, $0 \leqslant z \leqslant 1$.

Let us examine briefly certain interesting examples where this generating function technique is successful.

"PURE" BIRTH-AND-DEATH PROCESS

In this example we imagine that for all $j \geqslant 0$, $\alpha_j = \lambda j$, and $\beta_j = \mu j$, where $\lambda > 0$ and $\mu > 0$ are two constants. The intuitive idea is that if $N(t) = j$, then, in the short time interval $(t, t + \delta t)$, each individual has a small probability $\lambda\delta t + o(\delta t)$ of splitting into two individuals (a "birth") and a small independent probability $\mu\delta t + o(\delta t)$ of dying. It can be shown that the infinite set of equations (2) reduce, with the introduction of the PGF, to the single equation

$$\frac{\partial}{\partial t}\Pi(z, t) = (\lambda z - \mu)(z - 1)\frac{\partial}{\partial z}\Pi(z, t). \qquad (6)$$

This partial differential equation is a special example of what is often, in the literature devoted to applied probability, called the Lagrange equation; fortunately, a method of attacking such equations has long been known. It yields (if $\lambda \neq \mu$) the solution

$$\Pi(t, z) = \Psi\left[\frac{\mu(z - 1) - (\lambda z - \mu)e^{-(\lambda - \mu)t}}{\lambda(z - 1) - (\lambda z - \mu)e^{-(\lambda - \mu)t}}\right], \qquad (7)$$

where we have written $\Psi(z)$ for the initial PGF $\Pi(z, 0)$, which we assume to be known. The case $\lambda = \mu$ presents no special difficulty, but must be treated specially.

In many cases formula (7) can be expanded as a power series in z to yield the desired probabilities $P_j(t)$.

Even in intractable problems where such expansion is impracticable, one can still obtain useful information by falling back on the fact that repeated differentiations, followed by the substitution $z = 1$, yield factorial moments* of $N(t)$.

Notice that if $\lambda > \mu$, then, as $t \to \infty$,

$$\Pi(z, t) \to \Psi(\mu/\lambda);$$

while if $\lambda < \mu$,

$$\Pi(z, t) \to \Psi(1) = 1.$$

From the first result, by examining coefficients of z^n, we can infer that, as $t \to \infty$, $P_j(t) \to 0$ if $j \geqslant 1$ and $P_0(t) \to \Psi(\mu/\lambda)$. Thus we may deduce that if the birthrate exceeds the deathrate, there is a probability $\Psi(\mu/\lambda)$ that the population will ultimately become extinct; but if that extinction does not occur, the population will become arbitrarily large. The latter deduction follows by observing that if K is any large number, then $P\{1 \leqslant N(t) \leqslant K\}$ tends to zero as $t \to \infty$. In a similar way we see that the second result implies that if the deathrate exceeds the birthrate then the ultimate extinction has a probability 1 of occurring.

A possible modification of this pure birth-and-death process is to suppose that at a birth, instead of there being exactly one "off-spring," there appears a random number ξ, say, of offspring; one may assume that the probabilities

$$\Pr[\xi = r] = g_r(t) \qquad (r = 0, 1, 2, \ldots)$$

depend on the time t at which the births occur. If one introduces the PGF

$$\Gamma(z, t) = \sum_{r=0}^{\infty} z^r g_r(t),$$

it can be shown that (6) is to be replaced by

$$\frac{\partial \pi}{\partial t} = \{\lambda[\Gamma(z, t) - z] + \mu[z - 1]\}\frac{\partial \pi}{\partial z}.$$

The tractability of this equation depends very much on assumptions made about $\Gamma(z, t)$. We refer to the references cited at the end of this article for further details.

Another possible, and interesting, modification of the pure birth-and-death model arises if we suppose λ and μ to be dependent on t, the time. This yields the so-called *non-homogeneous* pure birth-and-death process. It is governed by an equation identical to (6) except that λ and μ should now become $\lambda(t)$ and $\mu(t)$. This model was fully explored by Kendall [4], who showed, for instance, that if $\Pi(z, 0) = z$ (corresponding to exactly *one* initial member of the population), then

$$\Pi(z, t) = 1 + \frac{1}{e^\rho(z - 1)^{-1} - \int_0^t \lambda e^\rho \, dt}, \qquad (8)$$

where

$$\rho \equiv \rho(t) \equiv \int_0^t (\mu - \lambda) \, dt.$$

BIRTH–DEATH–IMMIGRATION PROCESS

This process is similar to (a) except that (in addition to the births and deaths) in every small time interval there is an independent probability $v\delta t + 0(\delta t)$, where $v > 0$ is the constant "immigration" rate, of an addition to the population from outside the system being studied. The differential equations (2) can again be summed up in a single equation, such as (6), involving the PGF:

$$\frac{\partial \Pi}{\partial t} = (\lambda z - \mu)(z - 1)\frac{\partial \Pi}{\partial z} + v(z - 1)\Pi. \qquad (9)$$

This is another partial differential equation of the Lagrange kind mentioned earlier and, once again, textbook methods are adequate and yield the solution (for $\lambda \neq \mu$):

$$\Pi(z, t) = A(z, t)B(z, t), \qquad (10)$$

where $B(z, t)$ is the function displayed in (7) as the PGF associated with the pure-birth process and involves $\Psi(z)$, the assumed initial PGF, while $A(z, t)$ is given by

$$A(z, t) = \left[1 - \frac{\lambda(z - 1)(1 - e^{(\lambda - \mu)t})}{(\mu - \lambda)}\right]^{-v/\lambda}. \qquad (11)$$

Thus $A(z, t)$ is the PGF of $N_I(t)$, say, that part of the total population $N(t)$ that has descended from immigrants. As is clear from the considerable amount of independence assumed in our model, if we write $N_0(t)$ for those in the population at time t who have descended from the original population, then $N(t) = N_0(t) + N_I(t)$, where, as (10) shows, N_0 and N_I are independent.

If $\lambda < \mu$, we see that

$$\Pi(z,t) \to \left(\frac{\mu}{\mu - \lambda} - \frac{\lambda z}{\mu - \lambda} \right)^{-\nu/\lambda}.$$

The PGF on the right-hand side of this result is that of a negative-binomial distribution*. Thus we see that, with the passage of time, the distribution of $N(t)$ approaches more and more closely a particular negative-binomial distribution, whose parameters depend on μ, λ, and ν.

IMMIGRATION–DEATH PROCESS

This model arises in several fields of application; it is really a special case of the birth-death–immigration process with $\lambda = 0$. It is possible to deduce the appropriate PGF by letting $\lambda \to 0$ in (10), (7), and (11). One finds that

$$\Pi(z,t) = e^{(\nu/\mu)(z-1)(1-e^{\mu t})}$$
$$\cdot \Psi(1 + [z - 1]e^{-\mu t}), \qquad (12)$$

where, as before, we write $\Psi(z)$ for the initial $\Pi(z, 0)$, assumed known. Notice that if the population contains, initially, zero particles [so that $\Psi(z) \equiv 1$], then $\Pi(z, t)$ has the form of the PGF of a Poisson* distribution; thus, in this case, $N(t)$ always has a familiar Poisson distribution with mean value

$$\mathcal{E}N(t) = \frac{\nu}{\mu}(1 - e^{\mu t}). \qquad (13)$$

In any case, whatever $\Psi(z)$ may be, as $t \to \infty$

$$\Pi(z,t) \to e^{(\nu/\mu)(z-1)}.$$

The limit function is the PGF of a Poisson distribution with mean ν/μ. Thus for large t, the distribution of $N(t)$ will always be close to a Poisson distribution (with mean value ν/μ).

SIMPLE QUEUE PROCESS

This model would seem, from its specifications, to be the simplest of all possible birth-and-death processes. It stems from the basic assumptions

$$\alpha_n = \lambda > 0 \qquad \text{for } n = 0, 1, 2, \ldots$$
$$\beta_n = \mu > 0 \qquad \text{for } n = 1, 2, 3, \ldots.$$

The corresponding equation for $\Pi(z, t)$ here becomes

$$\frac{\partial \Pi}{\partial t} = \left(\lambda z - \lambda - \mu + \frac{\mu}{z} \right) \Pi$$
$$+ \mu \left(1 - \frac{1}{z} \right) P_0(t). \qquad (14)$$

The partial differential equations previously encountered for $\Pi(z, t)$ have involved no unknown function other than $\Pi(z, t)$ itself and they have thus been amenable to conventional textbook attack. The present equation (14) raises an awkward difficulty in that it involves the unknown $P_0(t)$ as well as $\Pi(z, t)$. Special methods are needed to solve this problem; we cannot go into them here (but see the article on queueing theory*, which will also explain the choice of name for the present special birth-and-death process).

We here end our discussion of certain special birth-and-death processes and turn to some related matters. The useful notion of the *characteristic functional* has been discussed by several authors, but the most useful results are due to Bartlett and Kendall [2]. For a population initially of unit size, we may define it as the expectation

$$\mathcal{E}[\omega, \theta(x); t] = \mathcal{E} \left\{ \omega^l \exp \left[i \sum_t \theta(x) \right] \right\},$$

Where l is a "marker" variable equal to unity throughout the life of the initial ancestor and zero after his or her death, and the summation $\sum_t$ is over the ages of all individuals (other than the initial ancestor) alive at t (*see* CHARACTERISTIC FUNCTIONS). The function $\theta(x)$ is purely arbitrary and by skillful choices of $\theta(x)$ one can learn a great deal about the development of the birth-and-death process. It is gratifying that for many specific birth-and-death processes the characteristic functional can be explicitly determined. For the pure birth-and-death process, Kendall [5] showed that

$$
\mathcal{E}[\omega, \theta(x); t] = 1 + \left\{ (\omega - 1)e^{-\mu t} \right.
$$

$$
+ \int_0^t (e^{i\theta(x)} - 1)g(x,t)dx \Big\}
$$

$$
\times \left\{ 1 - \int_0^t (e^{i\theta(x)} - 1)h(x,t)dx \right\}^{-1}
$$

where
$$
g(x,t) = \lambda e^{-\mu x} e^{(\lambda - \mu)(t - x)}
$$

and
$$
h(x,t) = \frac{\lambda}{\lambda - \mu} e^{-\mu x} [\lambda e^{(\lambda - \mu)(t - x)} - \mu].
$$

No discussion, however short, of birth-and-death processes would be complete without some reference to the original and profound approach to the general problem adopted by Karlin and McGregor [3], who relate the task of solving equations [2] to the solution of a Stieltjes moment problem*. Their methods require extensive discussion for a complete exposition and we must settle here for the slightest of sketches. They develop a sequence of polynomials $Q_n(x), n = 0, 1, \ldots$, by taking $Q_0(x) \equiv 1$ and then requiring

$$
-xQ_0(x) = -(\alpha_0 + \beta_0)Q_0(x) + \alpha_0 Q_1(x),
$$

while, for $n \geqslant 1$,

$$
-xQ_n(x) = \alpha_n Q_{n-1}(x) - (\alpha_n + \beta_n)Q_n(x)
$$
$$
+ \alpha_n Q_{n+1}(x).
$$

They define $\pi_0 = 1$ and, for $n \geqslant 1$,

$$
\pi_n = \frac{\alpha_0 \alpha_1 \cdots \alpha_{n-1}}{\beta_1 \beta_2 \cdots \beta_n}.
$$

They are then able to show the existence of a weight function $\Psi(x)$ on $[0, \infty)$ such that the transition probabilities $P_{ij}(t)$ of the general birth-and-death process are given by equations

$$
P_{ij}(t) = \pi_j \int_0^\infty e^{-xt} Q_i(x) Q_j(x) d\Psi(x), \quad (15)
$$

where

$$
\int_0^\infty \{Q_j(x)\}^2 d\Psi(x) = \frac{1}{\pi_j}.
$$

Remarkably, for many specific birth-and-death processes of practical interest, the weight function $\Psi(x)$ has been explicitly determined and general formulas derived for the Q-polynomials. The integral representations (15) then give considerable insight into the temporal development of the process.

There are many possible generalizations of birth-and-death processes that we have no room to discuss in this, necessarily brief, general article. One important inquiry allows the particles comprising the population to be of various "sexes" and the birthrate to depend in some way on the relative numbers of particles of different sexes. Another inquiry is concerned with "variable generation time"; a parent particle's ability to spawn an offspring depends on the age of the parent. A birth-and-death process in which over-crowding has a deterrent effect on births is the *logistic process**; this is notoriously difficult to treat analytically and merits a separate article.

REFERENCES

1. Bartlett, M. S. (1955). *An Introduction to Stochastic Processes*. Cambridge University Press, Cambridge. (This book gives a brief and authoritative discussion of, particularly, those aspects of birth-and-death processes on which Bartlett did pioneering work. However, the style of the book is taciturn and some readers may find it difficult.)

2. Bartlett, M. S. and Kendall, D. G. (1951). *Proc. Camb. Philos. Soc.*, **47**, 65–76.

3. Karlin, S. and McGregor, J. L. (1957). *Trans. Amer. Math. Soc.*, **85**, 489–546.

4. Kendall, D. G. (1948). *Ann. Math. Statist.*, **19**, 1–15.

5. Kendall, D. G. (1950). *J. R. Statist. Soc. B*, **12**, 278–285.

BIBLIOGRAPHY

See the following works, as well as the references just given, for more information on the topic of birth and death processes.

Bharucha-Reid, A. T. (1960). *Elements of the Theory of Markov Processes and Their Applications.* McGraw-Hill, New York.

Bhat, U. N. (1972). *Elements of Applied Stochastic Processes.* Wiley, New York. (See, especially, Chap. 7.)

Cox, D. R. and Miller, M. D. (1965). *The Theory of Stochastic Processes.* Methuen, London.

Hoel, P. G., Port, S. C., and Stone, C. J. (1972). *An Introduction to Stochastic Processes.* Houghton Mifflin, Boston. (Chapter 3 is an accessible and clear introduction to the sort of random processes discussed in this article).

Karlin, S. and Taylor, H. M. (1975). *A First Course in Stochastic Processes*, 2nd ed. Academic Press, New York. (Chapter 4 is a good introductory account containing a wealth of examples in applied fields.)

Moyal, J. E., Bartlett, M. S., and Kendall, D. G. (1949). *J. R. Statist. Soc. B*, **11**, 150–282. (A most valuable source of information touching on many of the points raised in this entry.)

See also APPLIED PROBABILITY; BRANCHING PROCESSES; GALTON–WATSON PROCESS; MARKOV PROCESSES; POPULATION GROWTH MODELS; and QUEUEING THEORY.

WALTER L. SMITH

BISERIAL CORRELATION

Biserial correlation refers to an association (*see* ASSOCIATION, MEASURES OF) between a random variable X which takes on only two values (for convenience 0 and 1), and a random variable Y measured on a continuum. Choice of a parameter to measure such association, and a statistic to estimate and test the parameter, depend on the conceptualization of the nature of the (X, Y)-population. Common forms are the point biserial, the biserial, forms that here are termed the modified biserial and the rank biserial correlation coefficient.

For example, in the Appendix are data representing a sample used to test the hypothesis that expectant mothers with extreme lengths of labor (either very short or very long) are more likely to be administered analgesia during labor. In this case X refers to use or nonuse of analgesia, Y is the absolute deviation* of length of labor from 6 hours (median labor).

POINT BISERIAL CORRELATION

The point biserial correlation coefficient is probably the earliest statistical approach to this problem because of its close relationship both to the product-moment correlation coefficient* and to the two-sample t-test*.

If it is assumed that the distributions of Y, conditional on $X = 0$ and 1, are normal with different means but with a common variance, the product moment correlation coefficient* between X and $Y(\rho_{pb})$ is estimated by the *point* biserial correlation coefficient

$$r_{pb} = (pq)^{1/2}(\bar{y}_1 - \bar{y}_0)/s_y,$$

where $(x_1, y_1), (x_2, y_2), \ldots, (x_n, y_n)$ is a sample from the (X, Y) population, $\bar{y}_1$ and $\bar{y}_0$ are the mean y-values of observations having $x_i = 1$ and $x_i = 0$, respectively; s_y^2 is the sample variance of Y; and p is the proportion of the X-sample with $x_i = 1$, $(q = 1 - p)$.

The statistic t may be used to test the null hypothesis* that $\rho_{pb} = 0$, where

$$t = (n - 2)^{1/2} r_{pb} (1 - r_{pb}^2)^{-1/2},$$

and t is distributed as Student's t^* with $n - 2$ degrees of freedom. This test is equivalent to a two-sample t-test of the null hypothesis that the mean of Y values with $X = 1$ equals that with $X = 0$. When $\rho_{pb} \neq 0$, the statistic t, where

$$t = (n - 2)^{1/2}(r_{pb} - \rho_{pb})$$
$$\cdot [(1 - r_{pb}^2)(1 - \rho_{pb}^2)]^{-1/2}$$

is asymptotically normally distributed (*see* ASYMPTOTIC NORMALITY) with mean zero and variance

$$1 + \rho_{pb}^2 \lambda/4,$$

where λ is the kurtosis* of the X-distribution:

$$\lambda = (1 - 6PQ)/4PQ$$

and $P = 1 - Q = \Pr[X = 1]$.

This distribution theory is generally robust with respect to departures from the assumption of the normal conditional distributions* of Y, but sensitive to departures from the assumption of equal variances.

In the numerical example presented in the Appendix, the mean Y-value for subjects with $X = 1$ was 5.31 hours, with $X = 0$ was 2.48 hours, and $s_y = 3.86$ hours. Of the 34 subjects, a proportion $p = 0.382$ had $X = 1$. Hence

$$r_{pb} = (0.382 \times 0.618)^{1/2}(5.31 - 2.48)/3.86$$

$$= 0.36$$

$$t_{32} = (32)^{1/2} \times 0.36(1 - 0.36^2)^{1/2}$$

$$= 2.2 \quad (p < 0.05).$$

Statistical papers related to the use of r_{pb} appear in the same era (1900–1920) as those introducing the biserial correlation. Development of the latter approach was motivated by the fact that underlying the dichotomy* X there often exists an unobservable random variable Z measured on a continuum, with $X = 1$ if Z exceeds some unknown threshold value and $X = 0$ otherwise. In the illustration, for example, Z might represent the physician's perception of the difficulty of labor. There is no loss in generality entailed in assuming that $E(Z) = 0$ and $\text{var}(Z) = 1$. The parameter of interest in this approach is some correlation coefficient between Z and Y, the estimation of which is based on the sample from the (X, Y)-population.

BISERIAL CORRELATION

If (Z, Y) is assumed to have a bivariate normal distribution* with correlation coefficient ρ_b, a consistent estimator* of ρ_b based on the sample from the (X, Y) population is the biserial correlation coefficient r_b:

$$r_b = pq(\bar{y}_1 - \bar{y}_0)/(s_y u),$$

where

$$u = (2\pi)^{-1/2}e^{-h^2/2}$$

and h is defined by

$$\Pr[Z \geqslant h] = p,$$

with Z as a standard normal variate.

The statistic $z(2r_b/\sqrt{5})$, where z is Fisher's z-transformation* $(z(r) = 0.5\ln[(1+r)/(1-r)])$ is asymptotically normally distributed with mean $z(2\rho_b/\sqrt{5})$ and variance $5/(4n)$ when ρ_b is near zero.

The distribution theory of r_b is sensitive to departures from the assumption of bivariate normality* of (Z, Y). Under such circumstances, the magnitude of $|r_b|$ is often found to exceed 1, although by definition ρ_b cannot exceed 1. Even when the assumptions are met, the asymptotic efficiency of $|r_b|$ as an estimator of ρ_b approaches zero as $|\rho_b|$ approaches 1.

In the illustrative data, since $p = 0.382$, $h = 0.300$, and $u = 0.381$. Thus

$$r_b = (0.382 \times 0.618)$$

$$\times (5.31 - 2.48)/\{(3.86)(0.381)\}$$

$$= 0.45.$$

Although both point biserial and biserial correlation coefficients were known and used in 1920, knowledge of the nonnull distribution theory and general statistical properties of these statistics were not developed until the 1950s with Robert F. Tate's work [4,5]. In the long interim period (1920–1950), there were a number of papers dealing with computation methods. This signaled that, even lacking theoretical base, these approaches were being utilized. The awakening of interest in theoretical development in 1950–1960, however, is marked by an unfortunate series of papers that confused the conditional distribution* theory of r_{pb} with its unconditional distribution. On this basis procedures were evolved and published for estimating and testing point biserial correlation coefficients.

Tate's study corrected these problems with r_{pb}, and in addition suggested the lack of robustness* of r_b and stimulated interest in more robust approaches. The first of these was actually introduced by Brogden in 1949 [1], but further interest does not appear until the late 1950s.

MODIFIED BISERIAL CORRELATION

If one weakens the foregoing assumptions by assuming only a monotonic relationship between Y and Z, then the product moment correlation coefficient between Y and Z is ρ_{mb}, where

$$\rho_{mb} = \frac{E(\bar{y}_1) - E(\bar{y}_0)}{\sigma_Y [E(Z|Z \geqslant h) - E(Z|Z < h)]},$$

where σ_Y^2 is the population variance of Y. A consistent* estimator would be of the form

$$r_{mb} = (\bar{y}_1 - \bar{y}_0)/D.$$

There have been several suggestions as to the choice of the statistic D. The one suggested by Brogden [1] requires that the Y-sample be rank-ordered:

$$y_1 > y_2 > y_3 > \cdots > y_n,$$

with

$$D_m = \left(\sum_{i=1}^{m} y_i/m\right) - \left(\sum_{i=m+1}^{n} y_i/(n-m)\right),$$

where m is an integer between 1 and the sample size n. If $\bar{y}_1 \geqslant \bar{y}_0$, let $D = D_{np}$, and if $\bar{y}_1 < \bar{y}_0$, let $D = D_{nq}$. Such a choice guarantees that if $|\rho_{mb}| = 1$, its estimator r_{mb} will equal ρ_{mb} with probability 1. Empirical study indicates that r_{mb} is preferable to r_b for estimation of ρ_b when $|\rho_b|$ is large, even if the more stringent assumptions underlying the definition of the biserial correlation coefficient, ρ_b, are met.

In the illustrative data, $np = 13$ and

$$D_m = \text{(average of 13 most extreme labors)}$$

$$- \text{(average of 21 least extreme labors)}$$

$$= 7.08 - 1.38 = 5.70.$$

Thus $r_{mb} = (5.31 - 2.48)/5.70 = 0.50$.

RANK BISERIAL CORRELATION

Another modification of the biserial approach is to focus attention on the *rank* correlation coefficient* between Y and Z, rather than the product moment correlation coefficient

between Y and Z (as for r_b) or the product moment correlation coefficient between X and Y (as for r_{pb}). This association would then include not just linear relationships, but any monotonic relationship between Y and Z. The rank correlation coefficient (either Spearman's* or Kendall's* forms) between Y and Z, as estimated from observation of X and Y, is simply

$$r_{rb} = 2(\bar{R}_1 - \bar{R}_0)/n,$$

where $\bar{R}_1$ and $\bar{R}_0$ are the average ranks of the y-values associated with $x_i = 1$ and $x_i = 0$, respectively.

The test of the null hypothesis that $\rho_{rb} = 0$ is equivalent to a Mann–Whitney test* applied to the groups with $x_i = 1$ and $x_i = 0$, respectively.

In the illustrative data, $\bar{R}_1 = 21.15, \bar{R}_0 = 13.76$, and

$$r_{rb} = 2(21.15 - 13.76)/34 = 0.43.$$

At present, although several forms of robust or nonparametric (distribution-free*) biserial correlation coefficients have been proposed, little is known of the distribution theory or any of their statistical properties. In particular, little is known of the advantages or disadvantages these forms may have over the classical forms.

In the illustration, the fact that the conditional distributions of Y when $X = 0$ and 1 appear nonnormal, and that the sample variances of these distributions differ widely, would militate against use of r_{pb}. The fact that the marginal distribution of Y is nonnormal warns against the use of r_b. Whether r_{mb} or r_{rb} is preferred remains an open question.

REFERENCES

1. Brogden, H. E. (1949). *Psychometrika*, **14**, 169–182.

2. Cureton, E. E. (1956). *Psychometrika*, **21**, 287–290. (A brief and simple definition with an illustration of its use.)

APPENDIX .

X (Analgesia)[a]	Rank	Y |Labor-6|	X (Analgesia)[a]	Rank	Y |Labor-6|
Y	33	14.8	Y	16	2.3
N	32	13.8	Y	15	2.1
Y	31	12.4	N	13.5	1.7
Y	30	10.1	N	13.5	1.7
Y	29	7.1	N	12	1.5
Y	28	6.1	N	10.5	1.3
N	27	5.8	N	10.5	1.3
Y	26	4.6	N	8.5	1.2
N	25	4.3	N	8.5	1.2
N	24	3.5	N	7	1.1
N	23	3.3	Y	6	0.8
Y	22	3.2	N	5	0.7
Y	21	3.0	N	4	0.6
N	19.5	2.8	N	3	0.5
N	19.5	2.8	N	2.5	0.2
N	18	2.5	N	2.5	0.2
Y	17	2.4	Y	1	0.1

[a]Y, yes; N, no.

3. Lord, F. M. (1963). *Psychometrika*, **28**, 81–85. (Introduces a new form of modified biserial, reviews earlier suggested forms, and presents a brief empirical comparison of one such form with the biserial.)

4. Tate, R. F. (1954). *Ann. Math. Statist.*, **25**, 603–607.

5. Tate, R. F. (1955). *Biometrika*, **42**, 205–216. (The level of Tate's two papers requires a degree of mathematical facility. However, the exposition is clear and the results are basic to an understanding of the biserial and point biserial coefficients.)

See also ASSOCIATION, MEASURES OF; KENDALL'S TAU—I; SPEARMAN RANK CORRELATION COEFFICIENT; and TETRACHORIC CORRELATION COEFFICIENT.

HELENA CHMURA KRAEMER

residual. These weights are assigned by the value of a weight function called *bisquare*. Residuals* close to zero have a weight close to unity. In the next step the parameters are reestimated using a weighted least-squares* procedure with weight r_i.

The iterative procedure continues until the fit converges. Cleveland [1] provides details and several numerical examples.

REFERENCE

1. Cleveland, W. S. (1993). *Visualizing Data*. Hobart Press, Summit, N.J.

See also LINEAR REGRESSION; ROBUST REGRESSION; and WEIGHTED LEAST SQUARES.

BISQUARE FITTING

Bisquare fitting is a method used for bivariate data (x_i, y_i) to obtain robust estimates. The first step is to fit a line l_y by the method of least squares* (*see* LINEAR REGRESSION). The outliers*—if any—are marked, usually by the plotting symbol +. Next, a robustness weight r_i is assigned to each observation (x_i, y_i), based on the size of the corresponding

BIT (BINARY DIGIT). See ENTROPY

BIVARIATE DISCRETE DISTRIBUTIONS

The random variables (X, Y) are said to have a *bivariate discrete distribution* over the subset T of the Cartesian product of the set of non-negative integers if the joint probability mass function is $f(x,y)$ for $(x,y) \in T$. We define the probability generating function* (pgf) of (X,Y) by the equation $\Pi(t_1,t_2) = E\{t_1^X t_2^Y\} = \sum_{(x,y) \in T} t_1^x t_2^y f(x,y)$. The series defining the pgf is absolutely convergent over the rectangle $|t_1| \leqslant 1, |t_2| \leqslant 1$. This enables us to differentiate the pgf any number of times with respect to the arguments t_1 and t_2 at $t_1 = t_2 = 0$. This yields the probability function (pf) of the random variables (X,Y) at (x,y) as $\Pi^{(x,y)}(0,0)/(x!y!)$, the numerator being the mixed partial derivative of the pgf of order (x,y) evaluated at $(0,0)$.

An alternative way of relating the pf with the pgf is to expand the pgf in powers of t_1 and t_2. The coefficient of $t_1^x t_2^y$ in this expansion will be the pf at the point (x,y).

An important property of the pgf is its one-to-one relationship to the pf. The marginal distributions of the random variables X and Y are $\Pi_x(t) = \Pi(t,1)$ and $\Pi_y(t) = \Pi(1,t)$. This gives a necessary and sufficient condition for the independence of X and Y as $\Pi(t_1,t_2) = \Pi(t_1,1)\Pi(1,t_2)$.

MOMENTS

The factorial moment of order (r,s) of (X,Y) is $\mu_{(r,s)} = E\{X^{(r)} Y^{(s)}\}$, where $x^{(r)} = x!/(x-r)!$. The factorial moment generating function (fmgf) is $G(t_1,t_2)$ which is the series $\sum_r \sum_s (t_1^r t_2^s / r!s!) \mu_{(r,s)}$. Then $G(t_1,t_2) = \Pi(1+t_1, 1+t_2)$. The factorial moment of order (r,s) can also be found from the pgf as $\mu_{(r,s)} = \Pi^{(r,s)}(1,1)$. If the joint moment generating function is $M(t_1,t_2) = E\{e^{t_1 X + t_2 Y}\}$, provided this expectation exists for $|t_1| \leqslant h_1$ and $|t_2| \leqslant h_2$ (Hogg and Craig [15, p. 77]), then $M(t_1,t_2) = \Pi(e^{t_1}, e^{t_2})$. This is helpful in establishing the relationships between the moments and the factorial moments of the random variables (X,Y). Finally, the factorial cumulant generating function (fcgf) and the joint factorial cumulants $k_{(r,s)}$ of (X,Y) are given by the expansion $H(t_1,t_2) = \log G(t_1,t_2) = \sum_r \sum_s [(t_1^r, t_2^s)/r!s!]k_{(r,s)}, (r,s) \neq (0,0)$. The relationships between the various types of moments in the bivariate case have been extensively discussed in Kocherlakota [25] and in the monograph by Kocherlakota and Kocherlakota [28, p. 5–8].

CONDITIONAL PGF

The conditional pf of Y given $X = x$ is $f(y|x) = f(x,y)/g(x)$ where $g(x)$ is the marginal pf of x at x. However, in practice it is preferable to determine the pgf of the conditional distribution, as in most instances this gives a better insight into the form of the conditional distribution.

Theorem for Conditional Distributions (TCD) (Subrahmaniam [39]). The conditions pgf of Y given $X = x$ is given by $\Pi_y(t|x) = \Pi^{(x,0)}(0,t)/\Pi^{(x,0)}(0,1)$. Similarly, the regression of Y on X can also be expressed in this case as $E[Y|X=x] = \Pi^{(x,1)}(0,1)/\Pi^{(x,0)}(0,1)$.

It is possible to express the joint pgf of X and Y via the conditional pgfs. Thus $\Pi(t_1,t_2) = \sum_x t_1^x \Pi_y(t_2|x) f(x)$, where $\Pi_y(t_2|x)$ is the conditional pgf of Y given $X = x$. Subrahmaniam [40] considers the special case when the conditional distribution is the convolution of the conditional distributions of Y_1 and Y_2 given $X = x$; then $\Pi(t_1,t_2) = \sum_x \Pi_1(t_2|x)\Pi_2(t_2|x)t_1^x f(x)$.

Special cases involve *homogeneous* and *nonhomogeneous* pgfs. Kemp [19] defines a bivariate discrete distribution to be of the *homogeneous type* if and only if the pgf is of the form $\Pi(t_1,t_2) = H(at_1 + bt_2)$, where a and b are nonnegative constants with $H(a+b) = 1$. The random variables X and Y are of the homogeneous type if and only if the conditional distribution of X given $X + Y = z$ is binomial* with index parameter z and probability $p = a/(a+b)$. While no general results are available for nonhomogeneous distributions, Subrahmaniam [40] has established a characterization of the binomial distribution based on the conditional distribution of Y given X in particular instances:

1. Let the marginal distribution of X be Poisson*, and the conditional distribution of Y given X be the convolution of Y_1, having a Poisson distribution, and Y_2. The joint distribution of X and Y is a bivariate Poisson distribution if and only if Y_2 has the binomial distribution.

2. Let the marginal distribution of X be binomial, and the conditional distribution of Y given X be the convolution of Y_1, having a binomial distribution, and

Y_2. The joint distribution of X and Y is a bivariate binomial distribution if and only if Y_2 has a binomial distribution.

3. Let the marginal distribution of X be negative binomial*, and the conditional distribution of Y given X be the convolution of Y_1, having a negative binomial distribution, and Y_2. The joint distribution of X and Y is a bivariate negative binomial distribution if and only if Y_2 has the binomial distribution.

We use the following notation: $B(n; \theta) =$ bin-omial distribution with index parameter n and probability θ; $P(\lambda) =$ Poisson distribution with parameter λ; $NB(r, \theta) =$ negative binomial distribution with index parameter r and probability θ.

BIVARIATE BINOMIAL DISTRIBUTIONS

In the univariate situation where one characteristic is under study, the binomial distribution is developed through Bernoulli trials. Bivariate binomial distributions will be based on Bernoulli trials in which two characteristics are studied simultaneously on each individual.

Bivariate Bernoulli Trials

Consider a sequence of trials in which two characteristics are studied simultaneously on each individual. We can then generalize the notion of Bernoulli trials to include the study of two charteristics as follows: (1) Each trial has four possible outcomes: $AB, A\overline{B}, \overline{A}B, \overline{A}\,\overline{B}$. (2) The probabilities of the outcomes, $P(AB) = p_{11}, P(A\overline{B}) = p_{10}, P(\overline{A}B) = p_{01}, P(\overline{A}\,\overline{B}) = p_{00}$, remain constant over the trials. (3) The trials are independent. As usual, the indicator random variables I_1 and I_2 will be defined in association with the events A and B, respectively.

Type I Bivariate Binomial Distribution

Genesis. The number of times the outcome $(I_1 = r, I_2 = s)$, for $r = 0, 1$ and $s = 0, 1$, occurs in n trials is now of interest. Define

$X = \Sigma_{i=1}^{n} I_{1i}$ and $Y = \Sigma_{i=1}^{n} I_{2i}$. The joint pgf of (X, Y) is $[1 + p_1 + (t_1 - 1) + p_{+1}(t_2 - 1) + p_{11}(t_1 - 1)(t_2 - 1)]^n$; hence, the pf of (X, Y) is

$g(x, y)$

$$= \sum_{n_{11}} \frac{n! p_{11}^{n_{11}} (p_{1+} - p_{11})^{x - n_{11}} (p_{+1} - p_{11})^{y - n_{11}}}{n_{11}!(x - n_{11})!(y - n_{11})!(n - x - y + n_{11})!}$$

$$\times (1 - p_{+1} - p_{1+} + p_{11})^{n - x - y + n_{11}},$$

where $p_{1+} = P(A), p_{+1} = P(B)$, and the sum extends over $\max(0, x + y - n) \leqslant n_{11} \leqslant \min(x, y)$; see Wicksell [43]. Alternatively, Aitken and Gonin [1] developed this distribution in conjunction with a fourfold sampling scheme.

Properties. Using the pgf of (X, Y), the marginal pgfs of X and of Y can be determined, each of which is a binomial distribution with the same index, but with probability of success p_{1+} for X and p_{+1} for Y. Using the fmgf $G(t_1, t_2) = (1 + p_{1+}t_1 + p_{+1}t_2 + p_{11}t_1t_2)^n$, the (r, s) th factorial moment is found as

$$\mu_{(r,s)} = (r + s)!$$

$$\times \left(\sum_{i=0}^{\min(r,s)} \frac{\binom{r}{i}\binom{s}{i}\binom{n}{r+s-i}}{\binom{r+s}{i}} \right.$$

$$\left. \times \left(\frac{p_{11}}{p_1 + p_{+1}} \right)^i \right) p_{1+}^r p_{+1}^s.$$

The correlation coefficient is $\rho = (p_{11} - p_{1+} \times p_{+1})/[p_{1+}(1 - p_{1+})p_{+1}(1 - p_{+1})]^{1/2}$. Hence $\rho = 0$ if and only if $p_{11} = p_{1+}$; i.e., $\rho = 0$ is a necessary and sufficient condition for independence.

The pgf of the conditional distribution of Y given $X = x$ can be developed using the TCD: $\Pi_y(t|x) = (Q + Pt)^{n-x}(q + pt)^x$, where $Q = 1 - P, P = (p_{+1} - p_{11})/(1 - p_{1+}), q = 1 - p$, and $p = p_{11}/p_{1+}$. The conditional distribution of Y given $X = x$ is the convolution of $B(n - x, P)$ with $B(x, p)$. Then the regression of Y given $X = x$ is $E[Y|x] = nP + x(p - P)$, which is linear in x with the regression coefficient $p - P = (p_{11} - p_{1+}p_{+1})/p_{1+}(1 - p_{1+})$. The conditional variance is $\text{var}[Y|x] = n(p_{01}p_{00})/p_{0+}^2 - x(p - P)(p - Q)$, also linear in x.

Estimation of Parameters. The estimation of the parameters for the various types of bivariate binomial distribution is mainly carried out using the maximum likelihood[*] principle. Estimation of the parameters of the Type I distribution will be discussed here. In practice data from the 2×2 contingency table can be of two types: (i) only the marginal totals n_{1+} and n_{+1} are known or (ii) $\{n_{rs}\}$ for all the entries of the 2×2 table are available.

Case (i): Estimation of the parameters is not possible if information is available from only one table (i.e., a single sample of size n). In order to estimate the parameters p_{1+}, p_{+1}, p_{11}, we will assume that marginal information is available from k independent samples; see Hamdan and Nasro [14]. Now the data consist only of pairs of observations (n_{1+}, n_{+1}) rather than the entire 2×2 table. Consider a sample of k such independent pairs. Writing $x_1 = n_{1+}$ and $y_i = n_{+1}$ for the ith sample with $i = 1, 2, \ldots, k$, we have the likelihood function $L = \Pi_{i=1}^{k} g(x_i, y_i; n)$ with $g(x_i, y_i; n)$ given by the pf of the Type I distribution. Differentiating L with respect to p_{1+}, p_{+1}, and p_{11} and solving, we have $\hat{p}_{1+} = x_+/nk, \hat{p}_{+1} = y_+/nk$, where x_+ and y_+ are summations over the sample values. Here $\hat{p}_{11}$ is the solution to $\overline{R} = 1$, where $\overline{R} = (1/k) \sum_{i=1}^{k} R_i$ with $R_i = g(x_i - 1, y_i - 1; n - 1)/g(x_i, y_i; n)$. The equation $\overline{R} = 1$ has to be solved iteratively. A preliminary estimate for p_{11} can be obtained using the method of moments: $\tilde{p}_{11} = (1/nk) \sum_{i=1}^{k} x_i y_i + \overline{xy} - n\overline{x}\overline{y}$ with $\overline{x} = x_+/(nk)$ and $\overline{y} = y_+/(nk)$.

Case (ii): Consider a 2×2 table in which neither of the margins is fixed. For a sample of size n, the data $\{n_{rs}\}$ are available from the table. The joint distribution is quadrinomial with corresponding likelihood $L \propto p_{11}^{n_{11}} p_{10}^{n_{10}} p_{01}^{n_{01}} p_{00}^{n_{00}}$. Here we are interested in the Type I distribution, which arises as the joint distribution of the margins (n_{1+}, n_{+1}). Since the pf is a function of p_{1+}, p_{+1}, and p_{11}, we have to reparametrize L in terms of the parameters of interest p_{1+}, p_{+1} and the covariance $\delta = p_{11} - p_{1+}p_{+1}$: thus

$$L \propto (\delta + p_{1+}p_{+1})^{n_{11}}(p_{1+} - \delta - p_{1+}p_{+1})^{n_{10}}$$
$$\times (p_{+1} - \delta - p_{1+}p_{+1})^{n_{01}}$$
$$\times (1 - p_{1+} - p_{+1} + p_{1+}p_{+1} + \delta)^{n_{00}},$$

resulting in estimates $\hat{p}_{1+} = n_{1+}/n, \hat{p}_{+1} = n_{+1}/n$, and $\hat{\delta} = (n_{00}n_{11} - n_{01}n_{10})/n^2$.

Type II Bivariate Binomial Distribution

Genesis. A more general version of the Type I distribution was introduced by Hamdan [11] and Hamdan and Jensen [13], permitting the index parameters as well as the probability parameters in the marginal pfs to be *unequal*. We will refer to this model as the *Type II bivariate binomial distribution*; it can be developed using the following modified sampling scheme. Three independent samples are drawn with replacement from a population in which each individual can be classified according to one of two characteristics A and B. In the first sample k individuals are classified according to both characteristics; let W_1 be the number of times that A occurs and W_2 the number of times that B occurs. In a second sample of size $n_1 - k$, only characteristic A is studied. Let S be the number of times that A occurs in this sample. In a third sample of size $n_2 - k, B$ alone is studied and it occurs T times. Then the total number of times A occurs is $X = W_1 + S$, while the total number of times B occurs is $Y = W_2 + T$. Suppose that the proportions of A, B, and AB in the population are p_{1+}, p_{+1}, and p_{11}. Then marginally

$$S \sim B(n_1 - k, p_{1+}),$$
$$T \sim B(n_2 - k, p_{+1}),$$

and (W_1, W_2) is a Type I bivariate binomial with parameters $(k; p_{00}, p_{10}, p_{01}, p_{11})$ with S, T, and (W_1, W_2) independent due to the sampling scheme. The pgf of (X, Y) (see Kocherlakota [24]) is

$$\prod (t_1, t_2)$$
$$= (p_{0+} + p_{1+}t_1)^{n_1-k}(p_{+0} + p_{+1}t_2)^{n_2-k}$$
$$\times (p_{00} + p_{10}t_1 + p_{01}t_2 + p_{11}t_1t_2)^k.$$

Properties. The probability function of the distribution is quite complicated, as are the factorial moments. The marginal distributions are binomial with $X \sim B(n_1, p_{1+})$ and $Y \sim B(n_2, p_{+1})$. Hamdan [11] considers the case when $p_{10} = p_{01} = 0$, giving $p_{1+} = p_{11} =$

$p_{+1} = p, p_{0+} = p_{00} = p_{+0} = 1 - p = q$, and the joint pgf of (X, Y):

$$\prod (t_1, t_2)$$
$$= (q + pt_1)^{n_1-k}(q + pt_2)^{n_2-k}(q + pt_1t_2)^k.$$

In this case the marginals are each binomial with the indices n_1 and n_2, respectively, but with the same probability of success p.

As in the Type I distribution, the pgf of X given $Y = y$ can be found using the TCD:

$$\prod{}_X(t|y) = (p_{0+} + p_{1+}t)^{n_1-k}$$

$$\times \sum_{i=0}^{y} \frac{\binom{n_2-k}{i}\binom{k}{y-i}}{\binom{n_2}{y}} \left(\frac{p_{01}}{p_{+1}} + \frac{p_{11}}{p_{+1}}t\right)^{y-i}$$

$$\times \left(\frac{p_{00}}{p_{+0}} + \frac{p_{10}}{p_{+0}}t\right)^{k-(y-i)}.$$

For the conditional distribution of X given $Y = y$, see Kocherlakota [24]; the regression of X on Y can then be determined by viewing the conditional distribution as a convolution:

$$E(X|y) = n_1 p_{1+}$$
$$+ k\left[\left(\frac{p_{11}}{p_{+1}} - \frac{p_{10}}{p_{+0}}\right)\frac{y}{n_2} + \left(\frac{p_{10}}{p_{+0}} - p_{1+}\right)\right],$$

which is linear in y.

Polynomial Representations. The rth Krawtchouk polynomial* is defined as

$$G_r(x; n, p) = \sum_{j=0}^{r} (-1)^j \binom{r}{j} \frac{(n-r+j)!}{(n-r)!} p^j x^{(r-j)}$$

Krawtchouk polynomials are orthogonal with respect to the binomial probability mass function $b(x; n, p)$. Using this definition, Aitken and Gonin [1] give the normalized version

$$G_r^*(x; n, p) = \frac{G_r(x; n, p)}{\left[\binom{n}{r}p^r(1-p)^r\right]^{1/2}}.$$

Then the Type II bivariate discrete distribution can be expressed as

$$f(x, y) = b(x; n_1, p_{1+})b(y; n_2, p_{+1})$$
$$\times \left(1 + \sum_{i=1}^{k} \eta_i G_i^*(x; n_1, p_{1+})G_i^*(y; n_2, p_{+1})\right),$$

which is a function of its marginal pfs, the normalized Krawtchouk polynomials, and η_i, a function of the correlation coefficient.

Canonical expansions can also be obtained for the trinomial distribution and for the Type I distribution. Hamdan [11] has studied the difference between these two distributions in terms of their canonical representations, given in Kocherlakota and Kocherlakota [28, pp. 72–75].

BIVARIATE POISSON DISTRIBUTION

The familiar genesis of this distribution is as the limit of the Type I bivariate binomial distribution. Thus, setting $p_{1+} = \lambda_1^*/n, p_{+1} = \lambda_2^*/n, p_{11} = \lambda_{11}/n$ in the joint pgf and letting $n \to \infty$ leads to

$$\prod(t_1, t_2) = \exp[\lambda_1^*(t_1 - 1) + \lambda_2^*(t_2 - 1) + \lambda_{11}(t_1 - 1)(t_2 - 1)]$$

[28, Chap. 4]. The pgf can also be written after reparametrization as

$$\prod(t_1, t_2) = \exp[\lambda_1(t_1 - 1) + \lambda_2(t_2 - 1) + \lambda_3(t_1 t_2 - 1)],$$

which is sometimes a more tractable form.

Properties. The marginal distributions of X and Y can be obtained from the joint pgf and are each univariate Poisson distributions with the parameters $\lambda_1 + \lambda_3$ and $\lambda_2 + \lambda_3$, respectively. Expanding the pgf in powers of t_1 and t_2, the pf is

$$f(r, s) = e^{-(\lambda_1+\lambda_2+\lambda_3)} \sum_{i=0}^{\min(r,s)} \frac{\lambda_1^{r-i}\lambda_2^{s-i}\lambda_3^i}{(r-i)!(s-i)!i!}$$

$$= e^{-(\lambda_1^*+\lambda_2^*-\lambda_3)}$$
$$\times \sum_{i=0}^{\min(r,s)} \frac{(\lambda_1^*-\lambda_3)^{r-i}(\lambda_2^*-\lambda_3)^{s-i}\lambda_3^i}{(r-i)!(s-i)!i!}.$$

Holgate [16] has given the pf in the second form. For $r \geq 1, s \geq 1$, recurrence relations for the pf can be given as

$$rf(r, s) = \lambda_1 f(r-1, s) + \lambda_3 f(r-1, s-1),$$
$$sf(r, s) = \lambda_2 f(r, s-1) + \lambda_3 f(r-1, s-1).$$

These, combined with the probability functions for the first row and column which are determined from the univariate Poisson distribution, permit the fitting of the distribution in general. The factorial moments can be determined from either the pgf or the fmgf to yield

$$\mu_{(r,s)} = \sum_{i=0}^{\min(r,s)} \binom{r}{i} \binom{s}{i}$$
$$\times (\lambda_1 + \lambda_3)^{r-i}(\lambda_2 + \lambda_3)^{s-i} i! \lambda_3^i$$
$$= (\lambda_1 + \lambda_3)^r (\lambda_2 + \lambda_3)^s \sum_{i=0}^{\min(r,s)} \binom{r}{i}\binom{s}{i} i! \tau^i,$$

where $\tau = \lambda_3/[(\lambda_1 + \lambda_3)(\lambda_2 + \lambda_3)]$. The coefficient of correlation for the distribution is $\rho = \lambda_3/[(\lambda_1 + \lambda_3)(\lambda_2 + \lambda_3)]^{1/2}$. Since $\lambda_3 \geqslant 0$, the correlation is nonnegative; $\lambda_3 = 0$ is a necessary and sufficient condition for the independence of X and Y.

Conditional Distribution and Regression. The conditional pgf of X given $Y = y$ is

$$\prod_X(t|y) = e^{\lambda_1(t-1)} \left(\frac{\lambda_2 + \lambda_3 t}{\lambda_2 + \lambda_3}\right)^y,$$

which is the convolution of $P(\lambda_1)$ with $B[y, \lambda_3/(\lambda_2 + \lambda_3)]$. This gives the regression of X on Y as $E[X|y] = \lambda_1 + \lambda_3 y/(\lambda_2 + \lambda_3)$, while the conditional variance is $\lambda_1 + \lambda_2\lambda_3 y/[(\lambda_2 + \lambda_3)^2]$; both are linear in y.

Polynomial Representation. The pf can be written in terms of the normalized Charlier* polynomials as

$$f(x,y) = f_1(x)f_2(y)$$
$$\times \left(1 + \sum_{r=1}^{\infty} \rho^r K_r^*(x; \lambda_1 + \lambda_3) K_r^*(y; \lambda_2 + \lambda_3)\right)$$

(Campbell [4]), where the polynomials are of the form $K_r^*(x; \lambda) = K_r(x; \lambda)(\lambda^r/r!)^{1/2}$ with

$$K_r(x; \lambda) = \frac{1}{\lambda^r}\left[x^{[r]} - r\lambda x^{[r-1]}\right.$$
$$\left. + \binom{r}{2}\lambda^2 x^{[r-2]} - \cdots + (-1)^r \lambda^r\right],$$

and are orthonormal with respect to the mass function $p(x; \lambda)$, the Poisson pf. In this representation $f_1(x)$ and $f_2(y)$ are the Poisson pfs and ρ is the correlation of (X, Y).

Estimation. Holgate [16] examined the problem of estimation using various methods. Let $(x_i, y_i), i = 1, 2, \ldots, n$, be a random sample of size n from the bivariate Poisson. We can consider the data to be grouped, the frequency of the pair (r, s) being n_{rs}, for $r = 0, 1, 2, \ldots, s = 0, 1, 2, \ldots$. Note that $\sum_{r,s} n_{rs} = n$. Defining the sample moments as usual, since the population moments are $E[X] = \lambda_1^*, E[Y] = \lambda_2^*$, and $\text{Cov}(X, Y) = \lambda_3$, the method-of-moments estimators are $\tilde{\lambda}_1^* = \bar{x}, \tilde{\lambda}_2^* = \bar{y}, \tilde{\lambda}_3 = m_{1,1}$. If zero-cell frequency is used in place of the sample covariance, the estimator for the parameter λ_3 is $\tilde{\lambda}_3 = \log(n_{00}/n) + \tilde{\lambda}_1^* + \tilde{\lambda}_2^*$.

The method of maximum likelihood can be used to determine the estimators for the parameters in Holgate's form of the pf $f(r, s)$. The likelihood equations yield $\hat{\lambda}_1^* = \bar{x}, \hat{\lambda}_2^* = \bar{y}$, while λ_3 has to be determined from the equation $\overline{R} = 1$, where

$$\overline{R} = \frac{1}{n}\sum_{i=1}^{n} \frac{f(x_i - 1, y_i - 1)}{f(x_i, y_i)}$$
$$= \frac{1}{n}\sum_{r,s} n_{rs} \frac{f(r-1, s-1)}{f(r,s)}.$$

This equation can be solved iteratively. The covariance matrices of the estimators are discussed in ref. 28, p. 104. Some special methods of estimation have been developed [29,38].

BIVARIATE NEGATIVE BINOMIAL DISTRIBUTION

Like the univariate negative binomial distribution*, the bivariate negative binomial can be described in terms of a variety of chance mechanisms:

1. *Inverse Sampling*: Here the random variables of interest are (X, Y), where X is the number of times that A occurs and Y is the number of times that B occurs before C occurs r times. The pgf of (X, Y) is

$$(1 - p_1 - p_2)^r(1 - p_1 t_1 - p_2 t_2)^{-r},$$

yielding the pf

$$f(x, y)$$
$$= \frac{(r + x + y - 1)!}{(r - 1)!x!y!} p_1^x p_2^y (1 - p_1 - p_2)^r$$

for $x = 0, 1, 2, \ldots, y = 0, 1, 2, \ldots$. Strictly speaking, this type should be referred to as the negative trinomial distribution.

2. *Shock Model*[*]: The random variables under study are X, the number of shocks to component 1, and Y, the number of shocks to component 2, *prior* to the rth failure of the system. Assuming that the probabilities of a shock suffered without failure by the components are p_1 and p_2, while the probability of failure of both components is $1 - p_1 - p_2$, and assuming independence, the pgf and the pf of X and Y are of the form given for inverse sampling (Downton [6]).

3. *Compounding*: This technique was developed by Greenwood and Yule [9] for the univariate case in the study of accidents. They assumed that the number of accidents sustained by individuals in a time interval has a Poisson distribution with parameter λ. To allow for the heterogeneity, or variation from individual to individual, they allowed λ to be a random variable with a gamma distribution. The resulting distribution is univariate negative binomial,

$$f(x) = \frac{\Gamma(\nu + x)}{\Gamma(\nu)x!} \left(\frac{\beta}{\alpha + \beta}\right)^\nu \left(\frac{\alpha}{\alpha + \beta}\right)^x,$$

$$x = 0, 1, 2, \ldots.$$

The index parameter ν in this case can be any real number such that $\Gamma(\nu)$ is defined. On the other hand, in mechanisms 1 and 2, ν has to be a positive integer. The univariate model has been extended to the bivariate case by Arbous and Kerrich [2] and to the multivariate case by Bates and Neyman [3]. The latter authors credit Greenwood and Yule [9] and Newbold [33] with the bivariate extension using the following assumptions:

(i) Let X and Y represent the numbers of accidents incurred by the same individual within the same period of time or in two nonoverlapping consecutive periods. The X and Y are assumed to have Poisson distributions with parameters λ and μ, respectively. Conditional on λ and μ, X and Y are *independent*. Here λ and μ characterize the proneness of the individual to the two types of accidents.

(ii) The parameter μ is taken to be proportional to λ; that is, $\mu = a\lambda$, where a is a constant.

(iii) λ has the gamma distribution with parameters ν and β. Under these assumptions, the pgf of X and Y is $\prod_{X,Y}(t_1, t_2) = \{1 - \beta^{-1}[(t_1 - 1) + a(t_2 - 1)]\}^{-\nu}$ and the pf is

$$g(x, y)$$
$$= \frac{\beta^\nu a^y \Gamma(\nu + x + y)}{\Gamma(x + 1)\Gamma(y + 1)\Gamma(\nu)[\beta + a + 1]^{\nu+x+y}}.$$

The assumption of independence in (i) above was relaxed by Edwards and Gurland [7] and Subrahmaniam [39]; for a given individual the two types of accidents are assumed to be *positively correlated*. Let (X, Y) have the bivariate Poisson distribution with joint pgf, given λ,

$$\prod_{X,Y}(t_1, t_2 | \lambda) = \exp\{\lambda[\alpha_1(t_1 - 1) + \alpha_2(t_2 - 1)$$
$$+ \alpha_3(t_1 t_2 - 1)]\},$$

and let λ have a Pearson Type III (or Gamma) distribution with parameters ν and β. Then the unconditional distribution of X and Y has joint pgf

$$\prod_{X,Y}(t_1, t_2) = q^\nu(1 - p_1 t_1 - p_2 t_2 - p_3 t_1 t_2)^{-\nu},$$

where $p_i = \alpha_i/(\alpha_1 + \alpha_2 + \alpha_3 + \beta), i = 1, 2, 3$, and $q = 1 - (p_1 + p_2 + p_3)$. This general formulation was first given by Edwards and Gurland [7], who unfortunately referred to the distribution as the compound correlated bivariate Poisson (CCBP) distribution. Independently Subrahmaniam [39] developed the distribution, giving it the more generally accepted name *bivariate negative binomial distribution*. Expanding the pgf in powers of t_1 and t_2 and identifying the term involving $t_1^r t_2^s$, we obtain the joint pf $P\{X = r, Y = s\}$ as

$$g(r, s)$$
$$= q^\nu \sum_{i=0}^{\min(r,s)} \frac{\Gamma(\nu + r + s - i)}{\Gamma(\nu)i!(r - i)!(s - i)!} p_1^{r-i} p_2^{s-i} p_3^i.$$

(Subrahmaniam [39, equation 2.4]). The marginal distributions of X and Y are each univariate negative binomial: $X \sim \text{NB}(n, 1 - P_1)$ and $Y \sim \text{NB}(n, 1 - P_2)$ with $P_1 = (p_1 + p_3)/(1 - p_2)$ and $P_2 = (p_2 + p_3)/(1 - p_1)$.

The (r, s)th factorial moment is given by

$$\mu_{(r,s)} = \mu_{(r)}^{(x)} \mu_{(s)}^{(y)} \sum_{i=0}^{\min(r,s)} \frac{\begin{bmatrix} v+r+s-i \\ v+r \end{bmatrix} \binom{r}{i}}{\begin{bmatrix} v+s \\ s \end{bmatrix}}$$

$$\times \left\{ \frac{qp_3}{(p_1 + p_3)(p_2 + p_3)} \right\}^i,$$

where $\begin{bmatrix} v \\ i \end{bmatrix}$ is the ratio $\Gamma(v)/[i!\Gamma(v-i)]$ and $\mu_{(r)}^{(x)}$ is the rth factorial moment of X.

Using the TCD, the conditional distribution of X given $Y = y$ is

$$\prod X(t|y)$$

$$= \left(\frac{p_2 + p_3 t}{p_2 + p_3} \right)^y \left(1 - \frac{p_1}{1 - p_1}(t - 1) \right)^{-(v+y)};$$

hence the conditional distribution of X given $Y = y$ is a convolution of $X_1 \sim B(y, p_3/(p_2 + p_3))$ and $X_2 \sim \text{NB}(v + y, 1 - p_1)$. Hence, $E(X|y) = vp_1/(1 - p_1) + y[p_3/(p_2 + p_3) + p_1/1 - p_1)]$ and $\text{Var}(X|y) = v/(1-p_1)^2 + y[p_2 p_3/(p_2 + p_3)^2 + p_1/(1 - p_1)^2]$, both of which are linear in y.

The bivariate binomial distribution can be expressed in a canonical form in terms of Krawtchouk polynomials. A similar canonical expansion can be developed for the bivariate negative binomial distribution in terms of the canonical correlations and a set of orthogonal polynomials which are orthonormal with respect to the negative binomial probability function; see Hamdan and Al-Bayyati [12].

Estimation

Consider the general model in which $p_3 \neq 0$ (or $\alpha_3 \neq 0$). The marginal moments and the mixed moment $\mu_{1,1}$ suggest a reparameterization as $p_1 = (\gamma_0 - \gamma_2)/(1 + \gamma_0 + \gamma_1 + \gamma_2)$, $p_2 = (\gamma_1 - \gamma_2)/(1 + \gamma_0 + \gamma_1 + \gamma_2)$, $p_3 = \gamma_2/(1 + \gamma_0 + \gamma_1 + \gamma_2)$, and $q = 1/(1 + \gamma_0 + \gamma_1 + \gamma_2)$. With this parametrization the pf becomes

$$g(x, y)$$

$$= \frac{(\gamma_0 - \gamma_2)^x (\gamma_1 - \gamma_2)^y \Gamma(v + x + y)}{x! y! \Gamma(v)(1 + \gamma_0 + \gamma_1 - \gamma_2)^{v+x+y}} S(x, y),$$

$$S(x, y) = \sum_{i=0}^{\min(x,y)} \frac{\binom{x}{i}\binom{y}{i} \tau^i}{\binom{v+x+y-1}{i}},$$

$$\tau = \frac{\gamma_2(1 + \gamma_0 + \gamma_1 - \gamma_2)}{(\gamma_0 - \gamma_2)(\gamma_1 - \gamma_2)}.$$

For methods of estimation using this parametrization see ref. 41:

1. *Method of Moments*: Using the sample marginal moments $\bar{x}, \bar{y}$ and the mixed moment $m_{1,1}$, the moment estimators are $\tilde{\gamma}_0 = \bar{x}/v, \tilde{\gamma}_1 = \bar{y}/v, \tilde{\gamma}_2 = (m_{1,1}/v) - (\overline{xy}/v^2)$.

2. *Zero-Zero Cell Frequency*: The joint pf evaluated at $(0,0)$ is $g(0,0) = (1 + \gamma_0 + \gamma_1 - \gamma_2)^{-v}$. One can estimate the parameters γ_0, γ_1, and γ_2 via $\tilde{\tilde{\gamma}}_0 = \bar{x}/v, \tilde{\tilde{\gamma}}_1 = \bar{y}/v, \tilde{\tilde{\gamma}}_2 = 1 + \bar{x}/v + \bar{y}/v - (n_{00}/n)^{-1/v}$.

3. *Maximum Likelihood Estimation*: The ML estimators of the first two parameters are $\tilde{\gamma}_0 = \bar{x}/v, \tilde{\gamma}_1 = \bar{y}/v$. The third parameter is the iterative solution to the equation $v\hat{\gamma}_2 = \overline{U}$, where $U(x, y) = S'(x, y)/S(x, y)$ with

$$S'(x, y)$$

$$= \sum_{i=0}^{\min(x,y)} i\tau^i \binom{x}{i}\binom{y}{i} \bigg/ \binom{v+x+y-1}{i},$$

and $\overline{U}$ is the sample mean of $U(x, y)$.

Now consider the case in which $p_3 = 0$ (or $\alpha_3 = 0$) and the pgf is of the form $\prod X,Y(t_1, t_2) = \{1 - (1/\beta)[(t_1 - 1) + \alpha(t_2 - 1)]\}^{-v}$. If the index parameter v is known, as is usually the case in inverse sampling, then the ML estimators for β and α are $\hat{\beta} = v/\bar{x}$ and $\hat{\alpha} = \bar{y}/\bar{x}$. In the case in which v is not necessarily an integer and is unknown, Bates and Neyman [3] have derived the ML estimators $\hat{\beta} = \hat{v}/\bar{x}, \hat{\alpha} = \bar{y}/\bar{x}$, and $\hat{v}$ is the solution to the equation

$$\log \left(1 + \frac{\bar{x} + \bar{y}}{v} \right) = \sum_{t=0}^{\infty} \frac{1 - \sum_{i=0}^{t} q_i/n}{v + t},$$

where q_t represents the frequency of pairs (x, y) such that $x + y = t$. This latter equation has to be solved iteratively.

Testing the Model

Two types of bivariate negative binomial distributions were developed by compounding bivariate Poisson random variables that are (1) independent or (2) correlated. The correlation coefficient in the case (1) is zero, since X and Y are independent. In the correlated case (Subrahmaniam [39]), the correlation coefficient is $\rho_I = \alpha_3/[(\alpha_2 + \alpha_3)(\alpha_1 + \alpha_3)]^{1/2}$. Here ρ_I is the correlation between X and Y for a given individual in the population; it is not a function of the bivariate Poisson parameter λ. Subrahmaniam refers to ρ_I as the coefficient of "intrinsic correlation" as distinct from that arising from the unconditional distribution.

The intrinsic correlation coefficient ρ_I can be used for distinguishing between the two models. Testing for $\rho_I = 0$ is equivalent to testing for $\alpha_3 = 0$; thus, testing for the significance of $\rho_I = 0$ is actually testing for the adequacy of the model based on independent bivariate Poisson random variables. Subrahmaniam developed a test of this hypothesis using a locally asymptotic most powerful test introduced by Neyman [32] for composite hypotheses.

DISTRIBUTIONS ARISING FROM THE BIVARIATE POISSON

A rich class of distributions is generated from the bivariate Poisson distribution by the process of compounding [23]; see also Kocherlakota and Kocherlakota [28, pp. 23–25]. Following Gurland [10], compounding is denoted by the symbol $\wedge$. The fundamental result of compounding with the bivariate Poisson distributions is:

Theorem 1. Let the conditional distribution of (X,Y) given τ have the pgf $\prod(t_1, t_2|\tau) = \exp\{\tau[\lambda_1(t_1 - 1) + \lambda_2(t_2 - 1) + \lambda_3(t_1 t_2 - 1)]\}$. Let τ be a random variable with the moment generating function $M(t)$. Then the unconditional distribution of (X,Y) has the pgf $M[\lambda_1(t_1 - 1) + \lambda_2(t_2 - 1) + \lambda_3(t_1 t_2 - 1)]$.

In the case when, conditional on τ, the (X,Y) are independent, then the joint pgf of (X,Y) is $M[\lambda_1(t_1 - 1) + \lambda_2(t_2 - 1)]$, showing

that, in general, X and Y are not unconditionally independent. In this case a necessary and sufficient condition for X and Y to be independent is that the distribution of τ is degenerate [23].

Teicher [42] has related convolution with compounding. A generalization in ref. 28, Chap. 8 is:

Theorem 2. Let X_i be independent bivariate Poisson random variables, for $i = 1, 2$, with the pgfs $\prod_i(t_1, t_2|\tau_i) = \exp\{\tau_i[\lambda_1(t_1 - 1) + \lambda_2(t_2 - 1) + \lambda_3(t_1 t_2 - 1)]\}$. Then the convolution of the compound random variables $X_i \wedge \tau_i, i = 1, 2$, is the compound of the bivariate Poisson random variable with the pgf $\prod(t_1, t_2|\tau) = \exp\{\tau[\lambda_1(t_1 - 1) + \lambda_2(t_2 - 1) + \lambda_3(t_1 t_2 - 1)]\}$ and the random variable τ which is the convolution of τ_1 and τ_2.

Properties. If $Z = X \wedge \tau$ with τ having the mgf $M(t)$, then the pgf of Z is $M(u)$, where $u = \lambda_1(t_1 - 1) + \lambda_2(t_2 - 1) + \lambda_3(t_1 t_2 - 1)$. By differentiating $M(u)$ with respect to the arguments we obtain

1. the probability function

$$f(r,s)$$

$$= \frac{\lambda_1^r \lambda_2^s}{r!s!} \sum_{k=0}^{\min(r,s)} \binom{r}{k}\binom{s}{k} k! M^{(t-k)}(\gamma)\delta^k,$$

where $\gamma = -(\lambda_1 + \lambda_2 + \lambda_3), \delta = \lambda_3/\lambda_1\lambda_2$, $t = r + s$, and $M^{(j)}$ is the jth derivative of M;

2. the factorial moments

$$\mu_{(r,s)} = (\lambda_1 + \lambda_3)^r (\lambda_2 + \lambda_3)^s$$

$$\times \sum_{k=0}^{\min(r,s)} \frac{r!s!\delta^{*k}\mu'_{r+s-k}}{(r-k)!(s-k)!k!},$$

where $\delta^* = \lambda_3/[(\lambda_1 + \lambda_3)(\lambda_1 + \lambda_2)]$ and μ'_k is the kth moment around zero in the distribution with the mgf $M(t)$. When $\lambda_3 = 0$ these expressions reduce to

$$f(r,s) = \frac{\lambda_1^r \lambda_2^s}{r!s!} M^{(t)}(-\lambda_1 - \lambda_2) \quad \text{and}$$

$$\mu_{(r,s)} = \lambda_1^r \lambda_2^s \mu'_{r+s}.$$

Kocherlakota [23] obtains the recurrence relations for the pf:

$$f(r,s)$$

$$= \begin{cases} \dfrac{(r+1)\lambda_2}{s\lambda_1}f(r+1,s-1) + (r-s+1) \\ \qquad \times \dfrac{\lambda_3}{s\lambda_1}f(r,s-1), \qquad r \geqslant s, \\[2mm] \dfrac{(s+1)\lambda_1}{r\lambda_2}f(r-1,s+1) + (s-r+1) \\ \qquad \times \dfrac{\lambda_3}{r\lambda_2}f(r-1.s), \qquad r \leqslant s. \end{cases}$$

Conditional Distribution. The conditional pgf of X, given $Y = y$, is found using the TCD:

$$\prod{}_{X}(t|y) = \frac{M^{(y)}[-(\lambda_2 + \lambda_3) + \lambda_1(t-1)]}{M^{(y)}[-(\lambda_2 + \lambda_3)]}$$
$$\times \left(\frac{\lambda_2}{\lambda_2 + \lambda_3} + \frac{\lambda_3 t}{\lambda_2 + \lambda_3} \right)^y,$$

showing that the conditional distribution of X, given $Y = y$, is the convolution of Y_1, which is $B[y, \lambda_3/(\lambda_2 + \lambda_3)]$, and Y_2, with the pgf

$$\frac{M^{(y)}[-(\lambda_2 + \lambda_3) + \lambda_1(t-1)]}{M^{(y)}[-(\lambda_2 + \lambda_3)]}$$
$$= \sum_{r=0}^{\infty} \frac{M^{[y+r]}[-(\lambda_2 + \lambda_3)]}{r! M^{[y]}[-(\lambda_2 + \lambda_3)]}[\lambda_1(t-1)]^r.$$

Some special distributions generated by this process are discussed below. Each one is a distribution of importance and hence is placed in a separate subsection.

Bivariate Neyman Type A Distribution (NTA)

Neyman [31] obtained the univariate version of the distribution by compounding the Poisson with a Poisson. That is, the random variable X has the Poisson distribution $P(\theta\phi)$, while ϕ is itself $P(\lambda)$. Then the unconditional distribution of X is $P(\theta\phi) \wedge_\phi P(\lambda)$ which gives the pgf of X as $\exp\{\lambda[\exp \theta(t-1) - 1]\}$. Holgate [17] generalized this to the bivariate situation by considering random variables (X,Y), representing the numbers of two types of individuals arising in clusters in a population. It is assumed that the numbers in a cluster have the bivariate Poisson distribution with the pgf

$$\prod{}(t_1, t_2|\tau) = \exp\{\tau[\lambda_1(t_1 - 1) + \lambda_2(t_2 - 1) + \lambda_3(t_1 t_2 - 1)]\}$$

with τ, characterizing the cluster size, itself having the Poisson distribution $P(\lambda)$ with mgf $M(t) = \exp\{\lambda[\exp(t) - 1]\}$. Using the general result given above, the pgf of (X,Y) is

$$\prod{}(t_1, t_2) = \exp(\lambda\{\exp[\lambda_1(t_1 - 1) + \lambda_2(t_2 - 1)$$
$$+ \lambda_3(t_1 t_2 - 1)] - 1\}).$$

Properties. The marginal distributions of X and Y are NTA$(\lambda, \lambda_i + \lambda_3), i = 1, 2$, respectively. The probability function can be obtained from the series representation. Differentiating $M(t)$ with respect to t we have $M^{(r)}(t) = \{\exp[\lambda(e^t - 1)]\}\eta_r$, where η_r is the rth raw moment of $P[\exp \lambda(t)]$. Hence the pf is

$$f(r,s) = \frac{\lambda_1^r \lambda_2^s}{r!s!} \exp[\lambda(e^\gamma - 1)]$$
$$\times \sum_{k=0}^{\min(r,s)} \binom{r}{k}\binom{s}{k} k! \delta^k \psi_{r+s-k},$$

with ψ_r the rth raw moment of $P[\lambda \exp(\gamma)]$, $\delta = \lambda_3/(\lambda_1\lambda_2)$, and $\gamma = -(\lambda_1 + \lambda_2 + \lambda_3)$; see also Gillings [8]. The factorial moments are

$$\mu_{(r,s)} = (\lambda_1 + \lambda_3)^r(\lambda_2 + \lambda_3)^s$$
$$\times \sum_{k=0}^{\min(r,s)} \frac{r!s!\tau^{*k}\mu'_{r+s-k}}{(r-k)!(s-k)!k!},$$

where μ'_r is the rth raw moment of $P(\lambda)$, $\tau^* = \lambda_3/[(\lambda_1 + \lambda_3)(\lambda_2 + \lambda_3)]$.

Conditional Distribution. Referring to the general conditional distribution given above and recalling that $M^{(r)}(t) = M(t)\eta_r(\lambda e^t)$, we have the conditional distribution of X given $Y = y$ as the convolution of $Z \sim B[y, \lambda_3/(\lambda_2 + \lambda_3)]$ and W, which has the pgf

$$\exp(\lambda\{\exp[-(\lambda_2 + \lambda_3)]\}\{\exp[\lambda_1(t-1)] - 1\})$$
$$\times \frac{\eta_y\{\lambda \exp[-(\lambda_2 + \lambda_3) + \lambda_1(t-1)]\}}{\eta_y\{\lambda \exp[-(\lambda_2 + \lambda_3)]\}}$$

Here the pgf shows that the random variable W is the convolution of $W_1 \sim$ NTA$\{\lambda \exp[-(\lambda_2 + \lambda_3)], \lambda_1\}$ and W_2, which has the pgf

$$\frac{\eta_y\{\lambda \exp[-(\lambda_2 + \lambda_3) + \lambda_1(t-1)]\}}{\eta_y\{\lambda \exp[-(\lambda_2 + \lambda_3)]\}},$$

$\eta_y[\cdot]$ being the yth moment of the Poisson distribution with the parameter in the arguments. This is the pgf of the mixture of y Poisson random variables $P(m\lambda_1), m = 1, 2, \ldots, y$, with the weight function

$$\omega_m(y) = \frac{\lambda^m \exp[-m(\lambda_2 + \lambda_3)]S_{m:y}}{\sum_{m=1}^{y} \lambda^m \exp[-m(\lambda_2 + \lambda_3)]S_{m:y}}$$

[28, p. 237], where

$$S_{m;r} = \frac{1}{m!} \sum_{k=0}^{m} (-1)^{m-k} \binom{m}{k} k^r$$

are the Stirling numbers*. Thus the conditional distribution of X given $Y = y$ is the convolution of (1) $Z \sim B[y, p]$ with $p = \lambda_3/(\lambda_2 + \lambda_3)$; (2) $W_1 \sim \text{NTA}\{\lambda \exp[-(\lambda_2 + \lambda_3)], \lambda_1\}$ and (3) $W_2 \sim$ (mixture of y Poisson random variables as given above).

Estimation. Holgate [17] considered moment estimation, whereas Gillings [8] examined the maximum likelihood estimation procedure. If λ is known, the moment equations can be set up in terms of marginal moments and the sample covariance to yield

$$\tilde{\lambda}_1 = \frac{\bar{x}}{\lambda}\left(1 + \frac{\bar{y}}{\lambda}\right) - \frac{m_{1,1}}{\lambda},$$

$$\tilde{\lambda}_2 = \frac{\bar{y}}{\lambda}\left(1 + \frac{\bar{x}}{\lambda}\right)\frac{m_{1,1}}{\lambda},$$

$$\tilde{\lambda}_3 = \frac{m_{1,1}}{\lambda} - \frac{\overline{xy}}{\lambda^2}.$$

On the other hand, if λ is unknown, Holgate [17] considers a reparametrization via $\alpha = \lambda_1 + \lambda_3$ and $\beta = \lambda_2 + \lambda_3$, and solves the resulting equations for λ, α, β, and λ_3 as

$$\tilde{\lambda} = \frac{m_{2,0} + m_{0,2} - \bar{x} - \bar{y}}{\bar{x}^2 + \bar{y}^2}, \quad \tilde{\alpha} = \frac{\bar{x}}{\tilde{\lambda}},$$

$$\tilde{\beta} = \frac{\bar{y}}{\tilde{\lambda}}, \quad \text{and} \quad \tilde{\lambda}_3 = \frac{m_{1,1}}{\tilde{\lambda} - \tilde{\alpha}\tilde{\beta}}.$$

For the maximum likelihood equations Gillings [8] suggests the use of difference relationships among the derivatives with respect to the parameters. An iterative solution of the resulting equations using the information matrix can obtained; see Kocherlakota and Kocherlakota [28, p. 241].

Table 1.

| x | No. of quadrats | | | | | Total |
	$y = 0$	1	2	3	4	
0	34	8	3	1	0	46
1	12	13	6	1	0	32
2	4	3	1	0	0	8
3	5	3	2	1	0	11
4	2	0	0	0	0	2
5	0	0	0	0	1	1
Total	57	27	12	3	1	100

Example. Gillings [8] suggested the use of this distribution for the data in Table 1, in which $x = \#\{$plants of the species *Lacistema aggregatum*$\}$ and $y = \#\{$plants of the species *Protium guianense*$\}$ in each of 100 systematically laid and contiguous quadrats of a tropical forest of Trinidad.

The moment estimates are $\tilde{\alpha} = 0.4051, \tilde{\beta} = 0.2559, \tilde{\lambda} = 2.3450$, and $\tilde{\lambda}_3 = 0.000$. The actual value of $\tilde{\lambda}_3$ is negative and hence is set equal to zero. The maximum likelihood estimates are found to be $\hat{\lambda} = 2.169, \hat{\lambda}_1 = 0.414, \hat{\lambda}_2 = 0.279$, and $\hat{\lambda}_3 = 0.000$, which are quite close to the moment estimates.

Bivariate Hermite Distribution

Several types of this distribution have been presented in the literature. Using the approach given above with X having the bivariate Poisson $[\tau\lambda_1, \tau\lambda_2, \tau\lambda_3]$ and $\tau \sim N(\mu, \sigma^2)$, Kocherlakota [23] obtains the unconditional distribution of X with the pgf $\prod(t_1, t_2) = \exp[\mu u + \frac{1}{2}\sigma^2 u^2]$, where $u = \lambda_1(t_1 - 1) + \lambda_2(t_2 - 1) + \lambda_3(t_1 t_2 - 1)$. Kemp and Papageorgiou [21] give the pgf in two forms:

$$\prod(t_1, t_2) = \exp[a_1(t_1 - 1) + a_2(t_1^2 - 1)$$
$$+ a_3(t_2 - 1) + a_4(t_2^2 - 1)$$
$$+ a_5(t_1 t_2 - 1) + a_6(t_1^2 t_2 - 1)$$
$$+ a_7(t_1 t_2^2 - 1) + a_8(t_1^2 t_2^2 - 1)];$$

alternatively, writing $t_i^* = t_i - 1$,

$$\prod(t_1, t_2) = \exp[b_1 t_1^* + b_2 t_1^{*2} + b_3 t_2^*$$
$$+ b_4 t_2^{*2} + b_5 t_1^* t_2^* + b_6 t_1^{*2} t_2^*$$
$$+ b_7 t_1^* t_2^{*2} + b_8 t_1^{*2} t_2^{*2}].$$

In these pgfs the values of the coefficients (a's and b's) can be expressed in terms of λ_1, λ_2, and λ_3. For other models for the development of the distribution see [21].

Properties. The marginals of X_i are univariate Hermite. The probability function is obtained in terms of the rth derivative of the normal mgf $M(t) = \exp(\mu t + \frac{1}{2}\sigma^2 t^2)$. Writing $M^{(r)}(t) = M(t)P_r(t)$, with $P_r(t)$ a polynomial of degree r in t, we have the pf

$$f(r,s)$$
$$= \frac{\lambda_1^r \lambda_2^s}{r!s!} M(\gamma) \sum_{k=0}^{\min(r,s)} \binom{r}{k}\binom{s}{k} k! P_{r+s-k}(\gamma)\delta^k,$$

where $\gamma = -(\lambda_1 + \lambda_2 + \lambda_3)$ and $\delta = \lambda_3/(\lambda_1\lambda_2)$. Recurrence relations for the pf are given in ref. 28, p. 249. From the factorial cumulant generating function, the factorial cumulants are obtained as $\kappa_{[1,0]} = b_1$, $\kappa_{[2,0]} = 2b_2$, $\kappa_{[0,1]} = b_3$, $\kappa_{[0,2]} = 2b_4$, $\kappa_{[1,1]} = b_5$, $\kappa_{[2,1]} = 2b_6$, $\kappa_{[1,2]} = 2b_7$, $\kappa_{[2,2]} = 4b_8$. The higher-order factorial cumulants are zero.

Conditional Distribution and Regression. The conditional distribution of X given $Y = y$ [23] has the pgf

$$\prod_X(t|y) = \frac{M^{(y)}[-(\lambda_2 + \lambda_3) + \lambda_1(t-1)]}{M^{(y)}[-(\lambda_2 + \lambda_3)]}$$
$$\times \left(\frac{\lambda_2}{\lambda_2 + \lambda_3} + \frac{\lambda_3 t}{\lambda_2 + \lambda_3}\right)^y$$
$$= \exp\{[\mu - \sigma^2(\lambda_2 + \lambda_3)]\lambda_1(t-1)$$
$$+ \frac{1}{2}(\sigma\lambda_1)^2(t-1)^2\}$$
$$\times \left(\frac{\lambda_2}{\lambda_2 + \lambda_3} + \frac{\lambda_3 t}{\lambda_2 + \lambda_3}\right)^y$$
$$\times \sum_{r=0}^{[x/2]} \omega_r(y)(Q + Pt)^{y-2r},$$

showing that the conditional distribution of X given $Y = y$ is the convolution of the random variables $Z_1 \sim B[y, \lambda_3/(\lambda_2 + \lambda_3)]$, $Z_2 \sim P(\lambda_1 \tau) \wedge_\tau N[\mu - \sigma^2(\lambda_2 + \lambda_3), \sigma^2]$, and Z_3, which is the mixture $B[(y-2r), P], r = 0, 1, 2, \ldots, [y/2]$. With $\alpha = 2[\mu/\sigma - \sigma(\lambda_2 + \lambda_3)]^2$, the weight function of the rth component is given by

$$\omega_r(x) = \frac{x!}{(x-2r)!r!}\alpha^{-r} \Big/ \sum_{s=0}^{[x/2]} \frac{x!}{(x-2s)!s!}\alpha^{-s}.$$

Here $P = \sigma^2\lambda_1/[\mu - \sigma^2(\lambda_2 + \lambda_3)]$. Kemp and Papageorgiou [21] derived expressions for the case $\lambda_3 = 0$. Then the conditional distribution becomes Hermite $\{[\mu - \sigma^2(\lambda_2 + \lambda_3)]\lambda_1, \frac{1}{2}(\sigma\lambda_1)^2\}$.

Estimation. The moment estimators for the a's are given by Kemp and Papageorgiou [21] under the assumption $a_i = 0$, for $i = 6, 7$, and 8. These are $\tilde{a}_1 = 2\bar{x} - m_{2,0} - m_{1,1}$, $\tilde{a}_2 = (m_{2,0} - \bar{x})/2$, $\tilde{a}_3 = 2\bar{y} - m_{0,2} - m_{1,1}$, $\tilde{a}_4 = (m_{0,2} - \bar{y})/2$, $\tilde{a}_5 = m_{1,1}$. They also give ML estimators using recurrence relations between the partial derivatives of the pf with respect to the parameters. For details see ref. 28, p. 254.

Example. The data in Table 2 are quoted by them from Cresswell and Frogatt [5]: x and y refer to the numbers of accidents incurred by bus drivers in first and second periods, respectively.

The moment estimators are given by the authors as $\tilde{a}_1 = 0.4813, \tilde{a}_2 = 0.09688, \tilde{a}_3 = 0.6572$, $\tilde{a}_4 = 0.15372, \tilde{a}_5 = 0.3263$. On the other hand, the iterative solution to the ML equations yields $\hat{a}_1 = 0.5053, \hat{a}_2 = 0.1109, \hat{a}_3 = 0.7728, \hat{a}_4 = 0.1219, \hat{a}_5 = 0.2744$.

BIVARIATE LOGARITHMIC SERIES DISTRIBUTION

Subrahmaniam [39] constructed the bivariate version of the logarithmic series distribution (bvlsd) by taking the Fisher limit of the bivariate negative binomial in the form

$$\prod(t_1, t_2) = \lim_{n\to\infty} \theta^n (1 - \theta_1 t_1 - \theta_2 t_2 - \theta_3 t_1 t_2)^{-n},$$

which yields the pgf

$$\prod(t_1, t_2) = \frac{-\log(1 - \theta_1 t_1 - \theta_2 t_2 - \theta_3 t_1 t_2)}{-\log\theta},$$

where $\theta = 1 - \theta_1 - \theta_2 - \theta_3$. When $\theta_3 = 0$ it simplifies to the form

$$\prod(t_1, t_2) = \frac{-\log(1 - \theta_1 t_1 - \theta_2 t_2)}{-\log(1 - \theta_1 - \theta_2)}.$$

A multivariate analogue of the latter case has been studied by Patil and Bildikar [35]. The general form of the distribution defined here has been extensively studied by Kocherlakota and Kocherlakota [27].

Table 2.

| | No. of Periods | | | | | | | | |
x	$y = 0$	1	2	3	4	5	6	7	Total
0	117	96	55	19	2	2	0	0	291
1	61	69	47	27	8	5	1	0	218
2	34	42	31	13	7	2	3	0	132
3	7	15	16	7	3	1	0	0	49
4	3	3	1	1	2	1	1	1	13
5	2	1	0	0	0	0	0	0	3
6	0	0	0	0	1	0	0	0	1
7	0	0	0	1	0	0	0	0	1
Total	224	226	150	68	23	11	5	1	708

Properties. Differentiating the pgf (r, s) times with respect to its arguments, we have

$$\prod{}^{(r,s)}(t_1, t_2)$$

$$= \frac{r!s!T_1^s T_2^r}{\delta T^{r+s}} \sum_{i=0}^{\min(r,s)} \frac{\Gamma(r + s - i)(\tau\theta_3)^i}{(r - i)!(s - i)!i!},$$

for $\tau = (1 - \theta_1 t_1 - \theta_2 t_2 - \theta_3 t_1 t_2)/(\theta_2 + \theta_3 t_1) \times (\theta_1 + \theta_3 t_2)$, $\delta = -\ln\theta$, $T_1 = \theta_2 + \theta_3 t_1$, $T_2 = \theta_1 + \theta_3 t_2$, and $T = 1 - \theta_1 t_1 - \theta_2 t_2 - \theta_3 t_1 t_2$. In the above, and in what follows, $(r, s) \neq (0, 0)$. From this,

1. the probability function is

$$f(r, s)$$

$$= \frac{\theta_1^r \theta_2^s}{\delta} \sum_{i=0}^{\min(r,s)} \frac{\Gamma(r + s - i)}{(r - i)!(s - i)!i!} \left(\frac{\theta_3}{\theta_1\theta_2}\right)^i,$$

where (r, s) is a member of the set $I \times I$ with $(r, s) \neq (0, 0)$ and $I = \{0, 1, 2, \ldots\}$;

2. the factorial moments are obtained by putting the arguments of $\prod{}^{(r,s)}(t_1, t_2)$ each

equal to 1:

$$\mu_{(r,s)} = \frac{r!s!}{\delta}(\theta_1 + \theta_3)^r(\theta_2 + \theta_3)^s\theta^{-(r+s)}$$

$$\times \sum_{i=0}^{\min(r,s)} \frac{\Gamma(r + s - i)(\gamma\theta)^i}{(r - i)!(s - i)!i!},$$

where $\gamma = \theta_3/[(\theta_1 + \theta_3)(\theta_2 + \theta_3)]$.

3. The marginal of X has the pgf $\prod(t) = -\log[1 - \theta_2 - (\theta_1 + \theta_3)t]/\delta$, which yields, when expanded in powers of t, the pf

$$f_X(0) = \frac{-\log(1 - \theta_2)}{\delta},$$

$$f_X(r) = \left(\frac{\theta_1 + \theta_3}{1 - \theta_2}\right)^r \frac{1}{r\delta}, \quad r \geqslant 1;$$

thus the pf is not that of a univariate logarithmic series distribution (lsd), since there is a mass at $x = 0$. Patil and Bildikar, who discovered this phenomenon, refer to this distribution as the *modified logarithmic series distribution*.

Conditional Distributions. Since the marginal distribution of Y has a mass at $y = 0$, we find the conditional pgf of X given $Y = y$ as follows:

1. For $y = 0$, the pgf is $\prod_X(t|y = 0) = \prod(t, 0)/\prod(1, 0) = [-\log(1 - \theta_1 t)]/[-\log(1 - \theta_1)]$, giving a lsd with the parameter θ_1.

2. For $y > 0$, the conditional pgf is

$$\prod_X(t|y > 0) = \left(\frac{\theta_2 + \theta_3 t}{\theta_2 + \theta_3}\right)^y \left(\frac{1 - \theta_1 t}{1 - \theta_1}\right)^{-y},$$

the convolution of $B[y, \theta_3/(\theta_2 + \theta_3)]$ with $NB(y, 1 - \theta_1)$.

Kemp [20] considered the conditional distributions of X and Y given their sum and difference in the case when $\theta_3 = 0$. Kocherlakota and Kocherlakota [27] extended these results. See ref. 28, pp. 199–208 for a detailed study of the distributions of the sum and difference and the conditional distributions.

Estimation. The estimation depends on assumptions made about the presence of the parameter θ_3 in the distribution. If $\theta_3 = 0$, the pf of (X, Y) is

$$f(r,s) = \frac{\Gamma(r+s)}{r!s!} \frac{\theta_1^r \theta_2^s}{-\log(1-\theta_1-\theta_2)},$$

$$(r,s) \in I \times I \quad \text{with } (r,s) \neq (0,0).$$

Here the parameters have the restrictions $0 < \theta_1 < 1, 0 < \theta_2 < 1, 0 < \theta_1 + \theta_2 < 1$. Let (x_i, y_i) for $i = 1, 2, \ldots, n$ be n pairs of independent observations from the distribution, or equivalently, let the frequency of the pair (r,s) be n_{rs} with $\sum_{r,s} n_{rs} = n$. In this case the maximum likelihood estimators are $\hat{\theta}_1 = (\bar{x}/T)\hat{\Phi}, \hat{\theta}_2 = (\bar{y}/T)\hat{\Phi}$, where T is the sum of the two sample means and $\hat{\Phi}$ is the maximum likelihood estimator of $\theta_1 + \theta_2$, the parameter of a univariate lsd for the random variable $X + Y$; see Patil and Bildikar [35], who also give the uniformly minimum variance unbiased estimators* of the parameters as $\tilde{\theta}_i = z_i|S_{z-1}^n|/|S_z^n|$ for $z > n, i = 1, 2$, and $= 0$ for $z = n$. Here S_x^n is Stirling's number of the first kind with arguments n and x.

In the case when $\theta_3 \neq 0$, Kocherlakota and Kocherlakota [27] have given the ML estimators by iterative techniques.

Models Giving Rise to LSD. Several models other than the Fisher limit to the negative binomial distribution give rise to the bvlsd. Kemp [20] points out that any model giving rise to the negative binomial also leads to the bvlsd. See ref. 27 for details of these derivations.

MISCELLANEOUS DISTRIBUTIONS

Several other distributions of interest have been presented in the literature with possible applications in the area of accident theory.

Weighted Distributions

These were introduced by Mahfoud and Patil [30]. Let the random variables (X, Y) have the joint pf $f(x,y)$. Then the pf of the weighted distribution with the weight function $w(x,y)$ is

$f^w(x,y) = f(x,y)w(x,y)/E[w(X,Y)]$. The marginal distribution of X is given by the weighted univariate pf, where $w_1(x) = E[w(x,Y)|x]$, the conditional expectation of the weight function $w(x,y)$. The conditional pf of X given $Y = y$ has the weighted distribution with the pf $f_1^w(x|y) = f_1(x|y)w(x,y)/E[w(X,y)|y]$, where $f_1(x|y)$ is the unweighted conditional pf of X given $Y = y$. Kocherlakota [26] has studied the weighted distribution with the *multiplicative weight function* $w(x,y) = x^{(\alpha)}y^{(\beta)}$, where $x^{(\alpha)} = x!/(x-\alpha)!$ and a similar ratio for $y^{(\beta)}$. For this case the pgfs of (X^w, Y^w) and of $(X^w - \alpha, Y^w - \beta)$ are

$$\prod{}^w(t_1,t_2) = t_1^\alpha t_2^\beta \prod{}^{(\alpha,\beta)}(t_1,t_2)/\prod{}^{(\alpha,\beta)}(1,1),$$

$$\prod{}^{*w}(t_1,t_2) = \prod{}^{(\alpha,\beta)}(t_1,t_2)/\prod{}^{(\alpha,\beta)}(1,1),$$

respectively. The TCD can be used to find the pgf of the conditional distribution of X^w given $Y^w = y^w$ as $\prod_x^{*w}(t|y^*) = \prod{}^{(\alpha,\beta+y^*)}(t,0)/\prod{}^{(\alpha,\beta+y^*)}(1,0)$, where $y^* = y^w - \beta$. If X and Y are independent random variables, then for the special type of weight function under consideration, the weighted random variables are also independent. Kocherlakota has studied the weighted forms of the bivariate Poisson, binomial, negative binomial, and logarithmic series distributions, and the relationships among the weighted versions. The problem of statistical inference in connection with all these distributions is quite complicated and is an open problem.

Waring Distributions*

Irwin [18] introduced univariate versions of these distributions for the study of accidents. His results were generalized by Xekalaki [44,45] to the bivariate situation. These distributions arise as a result of assuming that each individual in the population is subject to two factors, *liability* and *proneness*. Under this setup, it is assumed that the conditional distribution of the number of accidents sustained by an individual over nonoverlapping periods is a double Poisson with the parameters λ_1/ν and λ_2/ν. Here the liability parameters λ_1 and λ_2 each have the two-parameter gamma distribution, while the proneness parameter ν has the beta distribution. Combining all these assumptions, the joint pgf of (X, Y) is $\prod(t_1, t_2) = (\beta_{[\tau+\kappa]}/(\alpha + \beta)_{[\tau+\kappa]})F(\alpha; \tau, \kappa; \alpha + \beta + \tau + \kappa; t_1, t_2)$, with an appropriate representation for these parameters in terms of λ_1, λ_2, and ν. Here $x_{[r]}$ is the ascending factorial ration, and F is Gauss' confluent hypergeometric function* of two arguments; see ref. 28, p. 275.

The marginal and conditional distributions are all univariate Waring distributions, as is the sum of X and Y. Xekalaki [45] has suggested several models leading to the bivariate Waring distribution and has also studied the estimation of the parameters in the pf.

Short Distributions

Here the distribution arises as a convolution of the Neyman Type A with the a Poisson distribution. Papageorgiou [36] and Kocherlakota and Kocherlakota [28, pp. 289–291] have further generalized this to the bivariate situation. The term *short* is used to describe these distributions. Papageorgiou has studied the corresponding estimation problems.

Generalized Power-Series Distributions*

These were introduced in the multidimensional setting by Khatri [22] and by Patil [34]. In the two-dimensional case the following definition has been given [28, p. 287]: Let T be a subset of the twofold Cartesian product of the set of nonnegative integers and $(r, s) > 0$ over T. Also, for $(r, s) \in T$ let $u(\theta_1, \theta_2) = \sum a(r, s)\theta_1^r\theta_2^s$ with $\theta_1 \geq 0, \theta_2 \geq 0$ and $(\theta_1, \theta_2) \in \Omega$, a two-dimensional space such that $u(\theta_1, \theta_2)$ is finite and differentiable for all such (θ_1, θ_2). Under these assumptions, the random variables (X, Y) with the probability function $f(r, s) = a(r, s)\theta_1^r\theta_2^s/u(\theta_1, \theta_2), (r, s) \in T, (\theta_1, \theta_2) \in \Omega$, are said to have the *bivariate generalized power-series distribution*. A necessary and sufficient condition for the random variables X and Y to be independent is that the series function $u(\theta_1, \theta_2)$ is equal to $u_1(\theta_1)u_2(\theta_2)$, where u_1 and u_2 are the series functions of the marginal distributions. An alternative necessary and sufficient condition is that the coefficient $a(r, s)$ in the series function be representable as $a_1(r)a_2(s)$ for all $(r, s) \in T$. Here a_1 and a_2 are the coefficients in the series functions of the marginal distributions.

Bivariate Compounded Distributions

Papageorgiou and David [37] have studied a family of bivariate discrete distributions whose marginal and conditional distributions are convolved Poisson distributions. They consider applications to accident data.

REFERENCES

1. Aitken, A. C. and Gonin, H. T. (1935). On fourfold sampling with and without replacement *Proc. R. Soc. Edinburgh*, **55**, 114–125.

2. Arbous, A. G. and Kerrich, J. E. (1951). Accident statistics and the concept of accident proneness. *Biometrics*, **7**, 340–432.

3. Bates, G. E. and Neyman, J. (1952). Contributions to the theory of accident proneness, I: An optimistic model of correlation between light and severe accidents. *Univ. Calif. Publi. Statist.*, **1**, 215–254.

4. Campbell, J. T. (1934). The Poisson correlation function. *Proc. Edinburgh Math. Soc. 2*, **4**, 18–26.

5. Cresswell, W. L. and Froggatt, P. (1963). *The Causation of Bus Driver Accidents. An Epidemiological Study*. Oxford University Press, London.

6. Downtown, F. (1970). Bivariate exponential distributions in reliability theory. *J. R. Statist. Soc. B*, **32**, 408–417.

7. Edwards, C. B. and Gurland, J. (1961). A class of distributions applicable to accidents. *J. Amer. Statist. Ass.*, **56**, 503–517.

8. Gillings, D. B. (1974). Some further results for bivariate generalizations of the Neyman Type A distribution. *Biometrics*, **30**, 619–628.

9. Greenwood, M. and Yule, G. U. (1920). An inquiry into the nature of frequency distribution representations of multiple happenings, with particular references to the occurrence of multiple attacks of disease of repeated accidents. *J. R. Statist. Soc.*, **83**, 255–279.

10. Gurland, J. (1957). Some interrelations among compound and generalized distributions. *Biometrika*, **44**, 265–268.

11. Hamdan, M. A. (1972). Canonical expansion of the bivariate binomial distribution with unequal marginal indices. *Int. Statist. Rev.*, **40**, 277–280.

12. Hamdan, M. A. and Al-Bayyati, H. A. (1971). Canonical expansion of the compound correlated bivariate Poisson distribution. *J. Amer. Statist. Ass.*, **66**, 390–393.

13. Hamdan, M. A. and Jensen, D. R. (1976). A bivariate binomial distribution and some applications. *Austral. J. Statist.*, **18**, 163–169.

14. Hamdan, M. A. and Nasro, M. O. (1986). Maximum likelihood estimation of the parameters of the bivariate binomial distribution. *Commun. Statist. Theory and Methods*, **15**, 747–754.

15. Hogg, R. V. and Craig, A. T. (1978). *Introduction to Mathematical Statistics*, 4th ed. Macmillan, New York.

16. Holgate, P. (1964). Estimation for the bivariate Poisson distribution. *Biometrika*, **51**, 241–245.

17. Holgate, P. (1966). Bivariate generalizations of Neyman's Type A distribution. *Biometrika*, **53**, 241–244.

18. Irwin, J. O. (1968). The generalized Waring distribution applied to accident theory. *J. R. Statist. Soc. A*, **131**, 205–225.

19. Kemp, A. W. (1981). Computer sampling from homogeneous bivariate discrete distributions. *ASA Proc. Statist. Comput. Section*, pp. 173–175.

20. Kemp, A. W. (1981). Conditionality properties for the bivariate logarithmic distribution with an application to goodness of fit. In *Statistical Distributions in Scientific Work* vol. 5, C. Taillie, G. P. Patil, and B. A. Baldessari, eds. Reidel, Dordrecht, pp. 57–73.

21. Kemp, C. D. and Papageorgiou, H. (1982). Bivariate Hermite distributions. *Sankhyā A*, **44**, 269–280.

22. Khatri, C. G. (1959). On certain properties of power-series distributions. *Biometrika*, **46**, 486–490.

23. Kocherlakota, S. (1988). On the compounded bivariate Poisson distribution; a unified approach. *Ann. Inst. Statist. Math.*, **40**, 61–76.

24. Kocherlakota, S. (1989). A note on the bivariate binomial distribution. *Statist. Probab. Lett.*, **7**, 21–24.

25. Kocherlakota, S. (1990). Factorial cumulants: bivariate discrete distributions. *Gujarat Statist. J.*, Khatri Memorial Volume, 117–124.

26. Kocherlakota, S. (1995). Discrete bivariate weighted distributions under multiplicative weight function. *Commun. Statist. Theory and Methods*, **24**, 533–551.

27. Kocherlakota, S. and Kocherlakota, K. (1990). The bivariate logarithmic series distribution. *Commun. Statist. Theory and Methods*, **19**, 3387–3432.

28. Kocherlakota, S. and Kocherlakota, K. (1992). *Bivariate Discrete Distributions*. Marcel Dekker, New York.

29. Loukas, S., Kemp, C. D., and Papageorgiou, H. (1986). Even point estimation for the bivariate Poisson distribution. *Biometrika*, **73**, 222–223.

30. Mahfoud, M. and Patil, G. P. (1982). On weighted distributions. In *Statistics and Probability: Essays in Honor of C. R. Rao*, G. Kallianpur et al., eds., North Holland, Amsterdam, 479–492.

31. Neyman, J. (1939). On a new class of "contagious" distributions applicable in entomology and bacteriology. *Ann. Math. Statist.*, **10**, 35–57.

32. Neyman, J. (1959). Optimal asymptotic tests of composite statistical hypothesis. In *Probability and Statistics: H. Cramér Volume*, U. Grenander, ed. Wiley, New York, pp. 213–234.

33. Newbold, E. M. (1926). *A Contribution to the Study of the Human Factor in the Causation of Accidents. Rep. 34*, Industrial Fatigue Research Board, London.

34. Patil, G. P. (1965). On multivariate generalized power series distribution and its application to the multinomial and negative multinomial. In *Classical and Contagious Discrete Distributions*, G. P. Patil, ed. Pergamon, Oxford, and Statistical Publishing Society, Calcutta, India, pp. 225–238.

35. Patil, G. P. and Bildikar, S. (1967). Multivariate logarithmic series distribution as a probability model in population and community ecology and some of its statistical properties, *J. Amer. Statist. Ass.*, **62**, 655–674.

36. Papageorgiou, H. (1986). Bivariate "short" distributions. *Commun. Statist. Theory and Methods*, **15**, 893–906.

37. Papageorgiou, H. and David, K. M. (1995). On a class of bivariate compounded Poisson distributions. *Statist. Probab. Lett.*, **23**, 93–104.

38. Papageorgiou, H. and Loukas, S. (1988). Conditional even point estimation for bivariate discrete distributions. *Commun. Statist. Theory and Methods*, **17**, 3403–3412.

39. Subrahmaniam, K. (1966). A test for "intrinsic correlation" in the theory of accident proneness. *J. Roy. Statist. Soc. B*, **35**, 131–146.

40. Subrahmaniam, K. (1967). On a property of the binomial distribution. *Trab. Estadist.*, **18**, 89–103.

41. Subrahmaniam, K. and Subrahmaniam, K. (1973). On the estimation of the parameters in the bivariate negative binomial distribution. *J. R. Statist. Soc. B*, **35**, 131–146.

42. Teicher, H. (1960). On the mixture of distributions. *Ann. Math. Statist.*, **31**, 55–73.

43. Wicksell, S. D. (1916). Some theorems in the theory of probability, with special reference to their importance in the theory of homograde correlations. *Svenska Aktuarieforeningens Tidskrift*, 165–213.

44. Xekalaki, E. (1984). Models leading to the bivariate generalized Waring distribution. *Utilitas Math.*, **25**, 263–290.

45. Xekalaki, E. (1984). The bivariate generalized Waring distribution and its application to accident theory. *J. R. Statist. Soc. A*, **147**, 488–498.

See also BIVARIATE DISTRIBUTIONS, SPECIFICATION OF and DISCRETE MULTIVARIATE DISTRIBUTIONS.

SUBRAHMANIAM KOCHERLAKOTA

KATHLEEN KOCHERLAKOTA

BIVARIATE DISTRIBUTIONS, SPECIFICATION OF

Visualization of a joint density function is not trivial. Knowledge of the marginal distributions is clearly inadequate to specify a distribution. Almost inevitably we must deal with cross sections of the density in order to describe it. Such cross sections, taken parallel to the axes, are, save for appropriate normalization, recognizable as conditional densities. Some role for conditional densities in the specification of joint densities seems thus to be assured.

To specify the joint density of a random vector (X_1, X_2), one common approach involves postulating the form of one marginal density, say $f_{X_1}(x_1)$, and the family of conditional densities of the second variable given the first, i.e., $f_{X_2|X_1}(x_2|x_1)$. In cases where visualization of features of marginal distributions is difficult, recourse may be taken to the technique of conditional specification. In this framework both families of conditional densities $f_{X_1|X_2}(x_1|x_2)$ and $f_{X_2|X_1}(x_2|x_1)$ are given. In such a setting one must concern oneself with questions of uniqueness and compatibility. If we are given two functions $a(x_1,x_2)$ and $b(x_1,x_2)$ designated as putative families of conditional densities, i.e., $f_{X_1|X_2}(x_1|x_2) = a(x_1,x_2)$ and $f_{X_2|X_1}(x_2|x_1) = b(x_1,x_2)$, certain conditions must be satisfied to ensure that a joint density $f_{X_1,X_2}(x_1,x_2)$

exists with those given functions as it conditional densities. One must verify that the ratio $a(x_1,x_2)/b(x_1,x_2)$ factors into a form $g_1(x_1)/g_2(x_2)$, where the g_i's are integrable functions. To ensure that a compatible pair of functions $a(x_1,x_2)$ and $b(x_1,x_2)$ should determine a unique joint density, further restrictions are required. The simplest is a positivity condition corresponding to the assumption that the joint density be strictly positive on a support set which is a Cartesian product $A \times B$. More generally one can visualize a Markov process using $a(x_1,x_2)$ and $b(x_1,x_2)$ alternatively as kernels. If this chain has an indecomposable state space, then the given $a(x_1,x_2)$ and $b(x_1,x_2)$ will determine a unique joint distribution.

Upon introspection it is evident that the above development involves a degree of overspecification. Instead of being given $f_{X_1|X_2}(x_1|x_2)$ and $f_{X_2|X_1}(x_2|x_1)$ for every x_1 and x_2, it is enough to be given $f_{X_1|X_2}(x_1|x_2) = a(x_1,x_2)$ for every x_1 and x_2 and $f_{X_2|X_1}(x_2|x_{10}) = b(x_{10},x_2)$ for just one value x_{10} [provided that $f_{X_2|X_1}(x_2|x_{10}) > 0$ for every x_2]. With this information one can obtain the marginal density of X_2 by normalizing the ratio $f_{X_2|X_1}(x_2|x_{10})/f_{X_1|X_2}(x_{10}|x_2) = b(x_{10},x_2)/a(x_{10},x_2)$, provided only that it is integrable. Compatibility is guaranteed and no checking is required.

A simple discrete example may help visualize these consideration. Consider a situation where X_1 and X_2 each have possible values 0, 1, 2 and the arrays of conditional discrete densities are as in Table 1.

For these arrays, $a(x_1,x_2)/b(x_1,x_2)$ factors into a ratio of the form $g_1(x_1)/g_2(x_2)$ (or equivalently, in the finite case, the two arrays have identical cross-product ratios). Consequently, there is a unique joint density for (X_1,X_2) determined by the given arrays $a(x_1,x_2)$ and $b(x_1,x_2)$. If instead we are given the full array $a(x_1,x_2)$ and the first column of $b(x_1,x_2)$, then the joint density is still uniquely determined (note the first column of $b(x_1,x_2)$ contains no zeros). The vector

$$\left(\frac{b(0,0)}{a(0,0)}, \frac{b(0,1)}{a(0,1)}, \frac{b(0,2)}{a(0,2)} \right) = \left(\tfrac{8}{11}, \tfrac{9}{11}, \tfrac{21}{11} \right),$$

normalized to sum to one, will yield the marginal density of X_2, i.e. $\left(\tfrac{1}{6}, \tfrac{1}{4}, \tfrac{7}{12} \right)$. This

Table 1. (a)

$f_{X_1|X_2}(x_1|x_2) =$
$a(x_1, x_2)$

x_1 \ x_2	0	1	2
0	$\frac{1}{6}$	$\frac{2}{6}$	$\frac{3}{6}$
1	$\frac{4}{9}$	0	$\frac{5}{9}$
2	$\frac{6}{21}$	$\frac{7}{21}$	$\frac{8}{21}$

Table 1. (b)

$f_{X_2|X_1}(x_2|x_1) =$
$b(x_1, x_2)$

x_1 \ x_2	0	1	2
0	$\frac{1}{11}$	$\frac{2}{9}$	$\frac{3}{16}$
1	$\frac{4}{11}$	0	$\frac{5}{16}$
2	$\frac{6}{11}$	$\frac{7}{9}$	$\frac{8}{16}$

combined with the array $a(x_1, x_2)$ provides us with the joint density of (X_1, X_2). If instead we had been provided with the full array $a(x_1, x_2)$ and only the *second* column of $b(x_1, x_2)$ (a column which contains a zero), it would not be possible uniquely to determine the marginal density of X_2, and consequently the joint density of (X_1, X_2) would not uniquely be determined. Note that any density for X_2 of the form $\left(\frac{2}{9}\alpha, 1 - \alpha, \frac{7}{9}\alpha\right)$ for $\alpha \in (0, 1)$ will be compatible with the full array $a(x_1, x_2)$ and the second column of $b(x_1, x_2)$.

Instead of postulating the precise form of the conditional densities, for example, it may be only possible to argue that they belong to specified parametric families. Thus one might ask that for each x_2, the conditional distribution of X_1 given $X_2 = x_2$ should be normal with mean and variance depending on x_2 and also that for each x_1, the conditional distribution of X_2 given $X_1 = x_1$ should be normal with mean and variance depending on x_1. The class of such "conditionally normal" distributions is readily determined (an eight-parameter exponential family that of course includes and extends the classical bivariate normal model). A comprehensive reference for general discussions of models specified by requiring that conditional distributions be members of specified parametric families is Arnold et al. [1].

If we wish to consider related questions in higher dimensions, we quickly recognize that there is an enormous variety of ways in which one could envision conditional specification paradyms. Gelman and Speed [2] provide an excellent discussion of what combinations of marginal and/or conditional densities are adequate to uniquely specify a multivariate distribution. Their presentation deals with the finite discrete case under a positivity assumption, but careful extensions to more general settings are not difficult (as in the two-dimensional case, attention must be paid to integrability and indecomposability issues).

REFERENCES

1. Arnold, B. C., Castillo, E., and Sarabia, J. M. (1992). *Conditionally Specified Distributions*, Lecture Notes in Statistics, Vol. 73. Springer-Verlag, Berlin.

2. Gelman, A. and Speed, T. P. (1993). Characterizing a joint probability distribution by conditionals, *J. R. Statist. Soc. B*, **55**, 185–188.

See also EXPONENTIAL FAMILIES and MULTIVARIATE DISTRIBUTIONS.

BARRY C. ARNOLD

BIVARIATE NORMAL DISTRIBUTION

This is a special case of the multinormal distribution*, for two variables. It has a position of considerable historical importance (*see* F. Galton*).

The *standard* bivariate normal distribution is a joint distribution of two standardized normal variables* X_1, X_2 with correlation coefficient* ρ. It has density function

$$\{2\pi \sqrt{(1 - \rho^2)}\}^{-1} \exp[-\tfrac{1}{2}(1 - \rho^2)^{-1}$$
$$\times (x_1^2 - 2\rho x_1 x_2 + x_2^2)]. \quad (1)$$

The regression* of either variable on the other is linear; in fact, the distribution of X_2, given $X_1 = x_1$, is normal with expected value ρx_1 and standard deviation $\sqrt{(1 - \rho^2)}$—a similar result holds for the distribution of X_1, given $X_2 = x_2$.

The density is constant on the ellipse

$$x_1^2 - 2\rho x_1 x_2 + x_2^2 = K \quad \text{(a constant)}. \quad (2)$$

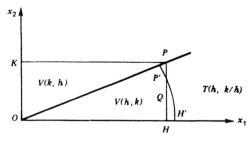

Figure 1.

The distribution of $(X_1^2 - 2\rho X_1 X_2 + X_2^2)(1 - \rho^2)^{-1}$ is chi-square* with 2 degrees of freedom (which is a special case of the exponential* distribution), so the integral of (1) over the interior of the ellipse (2) is

$$\Pr[\chi_2^2 \leqslant K(1 - \rho^2)]$$
$$= 1 - \exp[-\tfrac{1}{2}K(1 - \rho^2)].$$

Extensive tables of integrals of (1) over various regions of the (x_1, x_2) space have been published [1]. These include values of

(a) $\Pr[(X_1 > h) \cap (X_2 > k)] = L(h, k; \rho)$;

(b) $V(h, k) = \Pr[(0 < X_1 < h) \cap (0 < X_2 < k) \cap (X_1 > X_2)]$.

[This is the integral of (1) over triangle OPH in Fig. 1.]

(c) $\Pr[\sigma_1^2 X_1^2 + \sigma_2^2 X_2^2 < K]$;

(d) $\Pr[(X_1 + \xi_2)^2 + (X_2 + \xi_2)^2 < K]$.

The *general* bivariate normal distribution is obtained by the transformation

$$Y_i = \eta_i + \sigma_i X_i \qquad (i = 1, 2; \quad \sigma_i > 0).$$

The variables Y_1, Y_2 are each normally distributed; their correlation coefficient is ρ, and $E[Y_i] = \eta_i$; $\text{var}(Y_i) = \sigma_i^2 (i = 1, 2)$.

Relations such as

$$\Pr[(Y_1 > a_1) \cap (Y_2 > a_2)]$$
$$= \Pr\left[\left(X_1 > \frac{a_1 - \eta_1}{\sigma_1}\right) \cap \left(X_2 > \frac{a_2 - \eta_2}{\sigma_2}\right)\right]$$
$$= L\left(\frac{a_1 - \eta_1}{\sigma_1}, \frac{a_2 - \eta_2}{\sigma_2}; \rho\right)$$

make it possible to use tables of the standard bivariate normal in evaluating probabilities for general bivariate normal distributions.

Two normally distributed variables do not necessarily have a joint bivariate normal distribution (*see*, e.g., FARLIE–GUMBEL–MORGENSTERN DISTRIBUTIONS). However, there are several other sets of conditions that do have this consequence (*see* CHARACTERIZATIONS OF DISTRIBUTIONS).

REFERENCE

1. Kotz, S., Balakrishnan, N. and Johnson, N. L. (2000). *Continuous Multivariate Distributions–1.* Wiley, New York.

FURTHER READING

Patel, J. K. and Read, C. B. (1996). *Handbook of the Normal Distribution.* Dekker, New York. (Chapter 9.)

See also BISERIAL CORRELATION; BIVARIATE NORMAL DISTRIBUTION, FIELLER'S THEOREM; CONTINUOUS MULTIVARIATE DISTRIBUTIONS; LIMIT THEOREM, CENTRAL; LINEAR REGRESSION; MULTIVARIATE NORMAL DISTRIBUTIONS; and TETRACHORIC CORRELATION COEFFICIENT.

BIVARIATE NORMAL DISTRIBUTION, FIELLER'S THEOREM

This result provides confidence intervals for the ratio of mean values of two random variables having a bivariate normal distribution*. Originally developed by Fieller [2], it is more clearly explained by Fieller [3].

Let x and y be observed estimates of the means μ_x and μ_y, respectively, and $\gamma = \mu_x / \mu_y$. Let s_{xx}, s_{yy}, and s_{xy} be estimates of the variances and covariance, respectively; in practice, x and y might be observed means of a joint random sample from the underlying bivariate normal distribution. Then the pivotal quantity*

$$\frac{x - \gamma y}{s_{xx} - 2\gamma s_{xy} + \gamma^2 s_{yy}} \tag{1}$$

results from a Student t-distribution* with an appropriate number of degrees of freedom. If the $100(1 - \tfrac{1}{2}\alpha)$ percentile of this distribution is denoted by t, then Fieller [3] showed that the corresponding $100(1 - \alpha)$ percent confidence limits for γ are given by

$$\frac{xy - t^2 s_{xy} \pm [f(x, y, s_{xx}, s_{yy}, s_{xy})]^{1/2}}{y^2 - t^2 s_{yy}},$$

$$f(x, y, s_{xx}, s_{yy}, s_{xy})$$
$$= (xy - t^2 s_{xy})^2 - (x^2 - t^2 s_{xx})(y^2 - t^2 s_{yy}). \tag{2}$$

If one thinks of x/y as a point estimate of γ, a more natural but lengthier equivalent expression for (2) is

$$\left\{ \frac{x}{y} - \frac{g s_{xy}}{s_{yy}} \pm \frac{t}{y} \left[s_{xx} - 2\frac{x}{y} s_{xy} + \frac{x^2}{y^2} s_{yy} \right. \right.$$
$$\left. \left. - g \left(s_{xx} - \frac{s_{xy}^2}{s_{yy}} \right) \right]^{1/2} \right\} \Big/ (1 - g), \tag{3}$$

where $g = t^2 s_{yy}/y^2$.

Fieller [3] describes the values in (2) as fiducial limits, as does Finney [4], but Stone points out elsewhere in this encyclopedia (see FIDUCIAL PROBABILITY) that "Fieller's confidence level is clearly not interpretable as fiducial probability since, when less than unity, it is inapplicable to an interval consisting of all possible values of the ratio." Fieller demonstrates this difficulty with Cushny and Peebles' data for the effect of two drugs on hours of sleep gained by 10 patients (a set of data used incidentally by Gosset* in the original paper [6] which developed the t-distribution). Figure 1 shows in Fieller's notation the curve leading to the values in (2) as solutions. For this data set, $1 - \alpha > 0.9986$ leads to all values of γ, but this would logically correspond to a fiducial probability of 1.0. Figure 1 also shows that small values of t (or of g) give finite confidence intervals, but for $0.995 \leqslant 1 - \alpha < 0.9986$, the corresponding percentiles t for this set of data lead to confidence regions that *exclude* only a finite interval. These anomalies arise because the function of γ involved is not monotonic. Except when g is reasonably small, the data will not in general be useful for interval estimation of γ.

The Fieller-Creasy paradox arises from a related paper by Creasy [1], who obtains her fiducial distributions of γ from the separate fiducial distributions of μ_x and μ_y. The resulting distribution differs from that of Fieller and so leads to different intervals. Neyman [5] related the paradox to the question of whether or not a meaningful definition of a fiducial distribution exists. "Miss Creasy's solution differs from that of Mr. Fieller and in the past repeated assertions were made that in a given set of conditions a parameter may have only one fiducial distribution."

The principal applications of Fieller's theorem have been in bioassay*, in estimating ratios of regression coefficients and distances between regression lines. Finney [4, pp. 32–35] developed two analogs of the

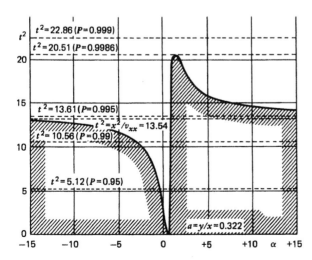

Figure 1. Cushny-Peebles Data: Fiducial Limits for The Ratio $\alpha =$ (Effect of Drug A)/(Effect of Drug B). (Reproduced by Permission of The Royal Statistical Society.)

theorem, and gave applications for several types of assay problems [4, p. 663].

REFERENCES

1. Creasy, M. (1954). *J. R. Statist. Soc. B*, **16**, 186–194.

2. Fieller, E. C. (1940). *J. R. Statist. Soc. Suppl.*, **7**, 1–64. (See pp. 45–63.)

3. Fieller, E. C. (1954). *J. R. Statist. Soc. B*, **16**, 175–185.

4. Finney, D. J. (1962). *Statistical Method in Biological Assay*. Hafner, New York.

5. Neyman, J. (1954). *J. R. Statist. Soc. B*, **16**, 216–218.

6. "Student" (1908). *Biometrika*, **6**, 1–25.

See also CONFIDENCE INTERVALS AND REGIONS; FIDUCIAL PROBABILITY; and PIVOTAL QUANTITIES.

CAMPBELL B. READ

BIVARIATE SIGN TEST, BENNETT'S

For the problem of testing whether a univariate distribution has a specified median*, the sign test* holds a prominent role in regard to historical importance, simplicity, versatility, and minimal assumptions about the shape of the distribution. Bennett [1] proposed an extension of the sign test for testing the null hypothesis* that a bivariate distribution has specified marginal medians. Let $(X_i, Y_i), i = 1, \ldots, n$ be a random sample from a bivariate population with an unspecified continuous distribution function $F(x - \mu, y - v)$ whose marginal medians are denoted by μ and v and are assumed to be unique. Bennett's test is concerned with testing H_0 : $(\mu, v) = (0, 0)$ vs. $H_1 : (\mu, v) \neq (0, 0)$, with no assumption made on the shape or association of the bivariate distribution. A typical context of application is a matched-pair* experiment where two, possibly associated, characteristics are observed simultaneously, and (X_i, Y_i) is identified as the vector of paired differences for the ith pair.

Consider the Euclidean plane divided in four quadrants with the origin taken at the median point specified under H_0. Denoting the respective quadrant probabilities by p_1, p_2, p_3, and p_4, an equivalent formulation of the null hypothesis is then $H_0 : p_1 = p_3, p_2 =$

p_4. Bennett's test is basically structured in the frame of Pearson's chi-square statistic* for goodness of fit* based on the quadrant frequency counts. Specifically, the test statistic has the form $B = \sum_{j=1}^{4} (n_j - n\hat{p}_j)^2 / n\hat{p}_j$, where n_j denotes the number of (X_i, Y_i)'s in the jth quadrant, and $\hat{p}_j, j = 1, \ldots, 4$, are the maximum likelihood* estimates of the p_j's based on the multinomial distribution* of $(n_1, \ldots, n_4)$ with cell probabilities constrained by $p_1 = p_3$ and $p_2 = p_4$. These estimates are $\hat{p}_1 = \hat{p}_3 = (n_1 + n_3)/2n$, $\hat{p}_2 = \hat{p}_4 = (n_2 + n_4)/2n$, and, after substitution, the expression for B simplifies to

$$ B = \frac{(n_1 - n_3)^2}{n_1 + n_3} + \frac{(n_2 - n_4)^2}{n_2 + n_4}, $$

which is the test statistic of Bennett's bivariate sign test. Noting that the asymptotic null distribution of B is central chi-squared* with 2 degrees of freedom, the level α critical region* is set as $B \geqslant \chi_{2,\alpha}^2$, where $\chi_{2,\alpha}^2$ is the upper α-point of the chi-squared distribution with 2 degrees of freedom. Evidently, usefulness of the test is limited to large samples. In fact, the exact distribution of B involves the unknown parameter p_1. While the univariate sign test is strictly distribution-free*, Bennett's bivariate extension is only asymptotically distribution-free.

Bhattacharyya [2] studied the Pitman efficiency* of B under local translation alternatives $F(x - \delta_1 n^{-1/2}, y - \delta_2 n^{-1/2})$ and some smoothness conditions on F. In particular, when $F(x, y)$ is diagonally symmetric with a covariance matrix Σ, the asymptotic efficiency of B relative to the optimum normal theory (Hotelling's) T^{2*}-test is given by $e(B : T^2) = \delta' \tau^{-1} \delta / \delta' \Sigma^{-1} \delta$, where $\delta' = (\delta_1, \delta_2)$,

$$ \tau = \frac{1}{4\alpha_1^2 \alpha_2^2} \begin{bmatrix} \alpha_2^2 & \alpha_1 \alpha_2 (4p_1 - 1) \\ \alpha_1 \alpha_2 (4p_1 - 1) & \alpha_1^2 \end{bmatrix} $$

and $\alpha_1 = f_1(0)$, $\alpha_2 = f_2(0)$ are the respective marginal pdf's evaluated at 0. When X and Y are independent normal, $e(B : T^2) = 2/\pi$, which is also the Pitman efficiency of the univariate sign test relative to the t-test. With a scale contamination* model of the form $F(x, y) = (1 - \epsilon)G_1(x, y) + \epsilon G_2(x/\sigma_1, y/\sigma_2)$, the efficiency can be arbitrarily large with increasing σ_1 or σ_2 for any fixed $0 < \epsilon < 1$. On the other hand, the efficiency could approach 0 when the bivariate distribution tends to be singular.

A strictly distribution-free bivariate sign test is proposed by Chatterjee [4] through a conditioning approach to remove the nuisance parameter*p_1. His test statistic is equivalent to B, but the rejection region is determined from the fact that conditionally, given $(n_1 + n_3)$, the counts n_1 and n_2 are independent binomial* variables. This conditional version of Bennett's test is unbiased* and consistent* under a wide class of alternatives and has the same Pitman efficiency as the unconditional test.

Two other bivariate extensions of the sign test are due to Hodges [5] and Blumen [3]. These tests are exact and share a property of Hotelling's T^{2*} in that they are invariant under nonsingular linear transformations of the observations (X_i, Y_i). On the other hand, Bennett's test is invariant under transformations of the form $X_i' = h_1(X_i), Y_i' = h_2(Y_i)$, where h_1 and h_2 are monotone increasing and zero-preserving functions. Bennett [1] also discusses a general multivariate extension of the sign test along the same lines as the bivariate test. Such extensions of Hodges' and Blumen's tests do not appear to be tractable. For additional discussions and relevant references on multivariate sign tests, see Puri and Sen [6].

REFERENCES

1. Bennett, B. M. (1962). *J. R. Statist. Soc. B*, **24**, 159–161.

2. Bhattacharyya, G. K. (1966). *J. R. Statist. Soc. B*, **28**, 146–149.

3. Blumen, I. (1958). *J. Amer. Statist. Ass.*, **53**, 448–456.

4. Chatterjee, S. K. (1966). *Ann. Math. Statist.*, **37**, 1771–1782.

5. Hodges, J. L., Jr. (1955). *Ann. Math. Statist.*, **26**, 523–527.

6. Puri, M. L. and Sen, P. K. (1971). *Nonparametric Methods in Multivariate Analysis*. Wiley, New York.

See also DISTRIBUTION-FREE METHODS and SIGN TESTS

G. K. BHATTACHARYYA

BLACK BOX

The term "Black Box" was borrowed from physics, and introduced into statistical theory by D. A. S. Fraser [1] to describe features common to most techniques of statistical inference*. The following quotations from Fraser [2] describe the concept. "The black box has an input variable θ receiving any value in Ω, it has an output variable x producing values in X, and it has behavioral characteristics $f(x; \theta)$ describing the frequency distribution* of output values x for any input θ...." "The black box has *complete behavioral characteristics*—provided θ is treated as output." But "an investigator ... needs the black box with the opposite direction of throughput; in by means of x, out by means of θ."

"The decision theory model introduces an additional black box ... The Bayesian method also introduces an additional black box."

In Fraser [2] the distinction between the two additional black boxes is explained.

The following earlier uses of the term 'Black Box' have come to our attention:

BIBLIOGRAPHY

McArthur, D. S. and Heigh, J. J. (1957), *Trans. Amer. Soc. Qual. Control*, 1–18.

Good, I. J. (1954), *J. Inst. Actu., 80*, 19–20 (Contribution to discussion of a paper by R.D. Clarke).

REFERENCES

1. Fraser, D. A. S. (1968). *The Structure of Inference*. Wiley, New York.

2. Fraser, D. A. S. (1968). *Technometrics*, **10**, 219–229. (A concise description of the concept.)

See also BAYESIAN INFERENCE; DECISION THEORY; PRIOR PROBABILITIES; and STRUCTURAL INFERENCE.

BLACKJACK

Blackjack or "twenty-one" is a popular casino card game attracting some 24 million gamblers wagering over \$15 billion annually. Its origin is unknown; however, the most recent version can be traced to the French game vingt-et-un. The game was introduced to the United States as "twenty one" in the early 1900s, but the name "blackjack" did not evolve until 1912, when a gambling house in Evansville, Indiana, offered a bonus if the first two cards dealt were the ace of spades and either of the black jacks. The name persisted as the game spread throughout the United States in the 1920s. (See Scarne [7] for a complete historical background.) The game continues to gain in popularity. Las Vegas casinos in the fiscal year 1977–1978 had at least 3,670 tables in operation. These tables generated a median win per unit ranging from $106,720 for casinos with $2 to 5 million total gaming revenue to $321,720 for casinos with an excess of $20 million. These and similar statistics are available in the *Nevada Gaming Abstract*, which may be obtained by writing to the State Gaming Control Board, Las Vegas, Nevada.

Under casino rules, the dealer has a rigid line of play. With 16 or fewer points, the dealer *must* draw. With 17 or more, the dealer *must* stand (not draw). An ace counts 11 points unless the total count exceeds 21, in which case the ace counts 1.

Bets are made, then two cards are dealt to each player, including the dealer. Players exercise their options, basically to draw or stand, starting with the player to the immediate left of the dealer. If a player goes bust (exceeds 21), he or she loses regardless of the dealer's actions. This rule puts the odds in the casino's favor because both player and dealer simultaneously go bust about 7% of the time.

The player's odds are somewhat improved because the dealer pays out a 3 : 2 ratio when the player has "blackjack," an ace and a 10-point card as the first two cards dealt. An exception occurs when the dealer also has "blackjack"; in this case no money exchanges hands. Baldwin et al. [1] were the first to publish an analytical strategy for the player. The use of their method increases the player's probability of winning to 49.5%. It was E. O. Thorp's best-selling book *Beat the Dealer* [8] that terrified casino owners along the Las Vegas strip and brought joy to studious gamblers. As a result of Thorp's classic, many

casinos shifted from a single deck to a four-deck "shoe" or resorted to shuffling a single deck after each hand. Many players used Thorp's counting methods to improve their play. In addition, Thorp's work stimulated the creation of new computer evaluations to give more accurate estimates of probabilities associated with the "correct" line of play. Analytical computations are too cumbersome to use because of the myriad of combinations involved. A more formal development, including some analytical computations, can be found in Griffin [3]. Julian Braun [6,8] has devised some useful computer algorithms, and Manson et al. [4] have made successful computer applications for the four-deck game to provide further sources of favorable strategy. On the other hand, the literature abounds with misleading schemes. (See Wilson [10] for an informative list.)

Casino operators turn away customers who use Thorp's methods and consider "counters" to be "cheaters." Ken Uston, a gambler who has managed to outwit casino owners many times by such means as disguise, had made blackjack a profitable vocation for the past 10 years. According to Uston, casinos are depriving him of a living. He has introduced several court cases [9], which are still pending. The courts will decide whether counting cards in blackjack is legal.

An important concept used by counters is that of a favorable or "rich" deck. A rich deck contains a large number of high-value cards (aces and 10-point cards). When the deck is rich, the counter gains an advantage because he or she can increase the size of a bet and optimally exercise several options with the standard playing procedures. The options include:

1. Doubling down; after two cards are drawn, a bet is doubled and only one more card is dealt.

2. Pair splitting; if the first two cards make a pair whose indices show the same value, the cards may be split into two separate hands.

3. Insurance; if the dealer has an ace showing, the player may wager up to half his or her bet that the dealer has blackjack.

Because few players are familiar with scientific strategies, dealers should expect an approximate 3% advantage; however, returns typically run over 20%. This hefty percentage may be accounted for by both amateur players and the casino atmosphere—in which player concentration is diminished by free drinks from scantily dressed waitresses and by continuous long hours at uninterrupted play. Another possible reason for returns over 20% is the wrong stopping policy used by the majority of "average" players. They stop playing only when they are broke or nearly broke; just a few have the will-power to stop when winning.

REFERENCES

1. Baldwin, R. R., Cantey, W. E., Maisel, H., and McDermott, J. P. (1956). *J. Amer. Stat. Ass.*, **51**, 429–439. (For the researcher or the serious player.)

2. Epstein, R. A. (1977). *The Theory of Gambling and Statistical Logic*. Academic Press, New York. (Excellent at all levels; contains techniques for computing probabilities in blackjack and other games as well.)

3. Griffin, P. (1979). *The Theory of Blackjack (The Complete Card Counter's Guide to the Casino Game of "21")*. GBC Press (Gambler's Book Club), Las Vegas, Nev. (Thorp's book [8] should be read as a prerequisite for complete understanding. Some new results that can benefit all levels of players.)

4. Manson, A. R., Barr, A. J., and Goodnight, J. A. (1975). *Amer. Statist.*, **29**(2), 84–88. (Interesting tables that can be used by all level players.)

5. Morehead, A., Frey, R., and Mott Smith, G. (1964). *The New Complete Hoyle*. Doubleday (Garden City Books), Garden City, NY. (For the beginner who knows little about cards.)

6. Revere, L. (1973). *Playing Blackjack as a Business*. Paul & Mann, Las Vegas, Nev. (For those interested in computer simulations and for serious players. Contains Braun's computer calculations of exact strategies.)

7. Scarne, J. (1978). *Scarne's Guide to Casino Gambling*. Simon and Schuster, New York. (Some misleading information concerning player strategy; good for those interested in historical background and basic rules.)

8. Thorp, E. O. (1966). *Beat the Dealer*. Random House, New York. (Excellent; a classic; readable at all levels. The novice should read ref. 2 first. Contains Braun's computer calculations of exact strategies.)

9. Uston, K. and Rapoport, R. (1977). *The Big Player*. Holt, Rinehart and Winston, New York. (Strategy; for the serious player.)

10. Wilson, A. N. (1965). *The Casino Gambler's Guide*. Harper & Row, New York. (Well written and entertaining; a must for all players as well as those interested in computer simulations.)

See also GAMBLING, STATISTICS IN; GAMES OF CHANCE; and SAMPLING WITHOUT REPLACEMENT.

A. GOLDMAN

BLOCK BOOTSTRAP

The classical form of the bootstrap* involves resampling *independently* and with replacement, using the operation of "sampling from the sample" to mimic the manner in which the original random data set was drawn from the population. In the context of dependent data, however, independently resampling individual data values is no longer an option.

One alternative procedure, the *block bootstrap*, involves dividing the data into blocks and resampling these, rather than individual data, at random. The dependence structure of the process generating the data is preserved within blocks in the resampled sequence, but is corrupted at the ends of blocks. Thus, good performance relies on appropriate choice of block length, which in this problem plays the role of a smoothing parameter. Increasing the length of blocks tends to reduce the bias but increase the variance of the bootstrap estimator.

The block bootstrap was first suggested by Hall [1] in the context of spatial* statistics. Two methods, emloying nonoverlapping or overlapping blocks—sometimes called "fixed blocks" and "moving blocks," respectively—were proposed there. Versions of these methods in the context of time-series* analysis were addressed, respectively, by Carlstein [2] and Künsch [3]. Alternative bootstrap methods for dependent data include those based on structural models for the process that generated the data [4], on the so-called "stationary bootstrap" [5], and on moving-window methods [6].

REFERENCES

1. Hall, P. (1985). *Stoch. Processes Appl.*, **20**, 231–246. (Suggests both fixed- and moving-block bootstrap methods in the setting of spatial data.)

2. Carlstein, E. (1986). *Ann. Statist.*, **14**, 1171–1179. (Suggests a form of the fixed-block bootstrap for time series data.)

3. Künsch, H. R. (1989). *Ann. Statist.*, **17**, 1217–1241. (Suggests the moving-block bootstrap for time series data.)

4. Freedman, D. (1984). *Ann. Statist.*, **12**, 827–842. (An early contribution to bootstrap methods for dependent data in the setting of structural models.)

5. Politis, D. N. and Romano, J. P. (1994). *J. Amer. Statist. Assoc.*, **89**, 1303–1313. (Suggests the stationary bootstrap for time series data.)

6. Hall, P. and Jing, B. (1996). *J. R. Statist. Soc. B*, **58**. (Suggests moving-window methods for the bootstrap with dependent data.)

See also BOOTSTRAP; SPATIAL DATA ANALYSIS; SPATIAL PROCESSES; and TIME SERIES.

PETER HALL

BLOCKING. See BLOCKS, RANDOMIZED COMPLETE

BLOCKS, BALANCED INCOMPLETE

When R. A. Fisher* introduced randomized complete block designs (*see* BLOCKS, RANDOMIZED COMPLETE, which should be read in conjunction with this discussion), he had in mind experiments such as agricultural field trials, in which block sizes were moderately flexible. In these studies, the requirement that the common block size equal the number of treatments under study created no difficulty for the experimenter. In many areas of research, however, the choice of block size is severely limited. Yates [14] mentions the restrictions on block size in studies where animal litters or monozygotic human twins form the blocks. Scheffé [13, p. 160], citing examples from consumer product testing, alludes to the limitations on block size imposed by an automobile's having only four wheels, or by an individual's wearing only two shoes at a time. He also comments that in certain cases the block size could theoretically be set large enough to accommodate all t treatments of interest, but one might then lose both precision and accuracy in assessing treatment differences because a large block size would contribute to measurement instability. To illustrate this point, consider the taste testing and rating of wines and spirits. Expert tasters find it increasingly difficult to evaluate treatments in a meaningful fashion as the number of tastings per session increases ("palate paralysis," or simply "paralysis" is sometimes used to describe the phenomenon associated in these studies with block sizes that are too large).

Yates [14] perceived the need in many studies for a common block size k that was smaller than the number of treatments t, so he relaxed the randomized complete block (RCB) design requirement that k equal t. Instead, he imposed the weaker requirement that each *pair* of treatments in a blocked experiment occur together in λ blocks. Designs satisfying this restriction, originally labeled by Yates symmetrical incomplete randomized block arrangements, are more commonly known today as *balanced incomplete block (BIB) designs*, the need for randomization* being understood. Yates illustrated his new class of designs with a design for six treatments, say a, b, c, d, e, and f, to be studied in 10 blocks of size 3:

$$
\begin{array}{llll}
1: & a\,b\,c & 6: & b\,c\,f \\
2: & a\,b\,d & 7: & b\,d\,e \\
3: & a\,c\,e & 8: & b\,e\,f \\
4: & a\,d\,f & 9: & c\,d\,e \\
5: & a\,e\,f & 10: & c\,d\,f
\end{array}
$$

For this design, it is easily verified that $\lambda = 2$.

For the implementation in practice of a BIB design, Cochran and Cox [4] list the following steps that require randomization:

1. Randomly assign the blocks of a design plan as given in a reference work to the groupings of experimental units.

2. Randomly assign the letters (or numbers, which some authors substitute for letters) of each block to the experimental units associated with that block in step 1.

3. Randomly assign treatments to the letters or numbers.

The number of blocks b and the number of replications r per treatment, together with t, k, and λ, serve as a set of five integers to index the class of BIB designs. Two easily verified restrictions on the class are:

$$rt = bk \quad \text{and} \quad r(k-1) = \lambda(t-1).$$

Not so easily seen is the further restriction (Fisher's inequality) that $b \geqslant t$ or, equivalently, $k \leqslant r$ [8, p. 220].

For a given k and t, with $k < t$, there always exists a BIB design; it may be formed by listing all subsets of the numbers 1 to t that are of size k. This subclass of BIB designs is referred to as the unreduced class, and is described further by the following equations:

$$b = \binom{t}{k}, \quad r = \binom{t-1}{k-1}, \quad \text{and}$$

$$\lambda = \binom{t-2}{k-2}.$$

Unfortunately, even for moderate values of k and t, these unreduced designs frequently require too many blocks for most studies; e.g., for $t = 8$ and $k = 4$, one would need 70 blocks and a total of 280 experimental units.

The combinatorial problems encountered in studying the class of BIB designs have long challenged both statisticians and mathematicians. A detailed discussion of this research on design existence and construction for given (t, k, b, r, λ) is found in John [8, Sec. 13]. From the standpoint of practice, the important results of this research are extensive catalogs of BIB designs, such as those found in Fisher and Yates [7, Tables XVIII and XIX] and Cochran and Cox [4, pp. 469–482].

The parametric model commonly adopted for the analysis of data from a BIB design is identical in form to the model assumed for a RCB design (*see* BLOCKS, RANDOMIZED COMPLETE):

$$y_{ij} = \mu + \tau_i + \beta_j + \epsilon_{ij}. \quad (1)$$

Here y_{ij} is the response observed if treatment i has been applied to a randomly chosen experimental unit in block j, μ is an overall mean, τ_i is the contribution of the ith treatment, β_j is the contribution of the jth block and ϵ_{ij} is an observational error. The $\{\epsilon_{ij}\}$ are assumed to be independent, each with mean zero and variance σ^2.

In contrast to the analysis of randomized complete block (RCB) designs, there is a difference in analysis for balanced incomplete block (BIB) designs depending upon whether blocks are considered to be fixed or random. Cochran and Cox [4, p. 383] comment that the random assumptions made for block contributions are very reasonable in certain contexts, but that these "assumptions cannot be taken for granted, and with certain types of data we might not wish to make them." Whether block contributions are assumed to be fixed or random, the corresponding analysis of a BIB design is more complex than is the analysis for a RCB design. This increased complexity is attributable to the lack of orthogonality of treatments and blocks.

For the case of fixed block effects, the least-squares* normal equations* are of the form

$$rt\hat{\mu} + r\sum_{i=1}^{t}\hat{\tau}_i + k\sum_{j=1}^{b}\hat{\beta}_j = S,$$

$$r\hat{\mu} + r\hat{\tau}_i + \sum_{j=1}^{b}n_{ij}\hat{\beta}_j = T_i,$$

$$k\hat{\mu} + \sum_{i=1}^{t}n_{ij}\hat{\tau}_i + k\hat{\beta}_j = B_j,$$

where n_{ij} takes the value 1 if treatment i occurs in block j and is otherwise zero, S is the sum of all observed responses, T_i is the sum of the responses for experimental units administered treatment i, and B_j is the sum of responses for experimental units in block j. As with randomized complete block designs, the $\{\tau_i\}$ in (1) are not estimable (*see* ESTIMABILITY); the symmetry requirement for BIB designs, however, results in simplified normal equations. Indeed, the desire for simplicity of analysis in the precomputer era was one argument in favor of balanced incomplete block designs vis-á-vis alternatives lacking in balance. Today, it is known [10] that a balanced incomplete block design is optimal under a broad range of optimality criteria.

For any contrast in the $\{\tau_i\}$, say

$$L = \sum_{i=1}^{t} c_i \tau_i, \tag{2}$$

with $\sum c_i = 0$, its least-squares estimator is given by

$$T = \sum_{i=1}^{t} c_i Q_i/(\lambda t), \tag{3}$$

where

$$Q_i = kT_i - \sum_{j=1}^{b} n_{ij} B_j. \tag{4}$$

If I_i is the set of block indices for blocks containing treatment i, then (4) may be rewritten as

$$Q_i = \sum_{j \in I_i} (k y_{ij} - B_j). \tag{5}$$

The quantity in parentheses in (5) is a contrast of the responses within the jth block. Consequently, Q_i is obtained from *intrablock information**, and T is an intrablock estimator. Being a least-squares estimator, T is unbiased for L; its variance is

$$\mathrm{var}(T) = \{k/(\lambda t)\} \left(\sum_{i=1}^{t} c_i^2 \right) \sigma^2.$$

For example, the variance of the estimate of a simple difference $(\tau_i - \tau_j), i \neq j$, is $2k\sigma^2/(\lambda t)$, irrespective of i and j.

It is easy to verify that

$$\sum_{i=1}^{t} Q_i = 0.$$

Hence the least-squares estimate in (3) of the ith treatment effect (relative to the average treatment contribution), i.e., of

$$\tau_i - \left(\sum_{j=1}^{t} \tau_j/t \right),$$

reduces simply to $Q_i/(\lambda t)$ and has variance $k(t-1)\sigma^2/(\lambda t^2)$.

Discussions of BIB designs appear in the five texts cited as general references for RCB designs: [2,4,5,8,9]. Where tables of these designs are presented, an "estimation efficiency factor" E is frequently included. Yates [14] formulated this factor by comparing the variance of a simple treatment difference estimate, i.e., of $\tau_i - \tau_j, i \neq j$, in a RCB design to the variance of an estimate of such a difference in a BIB design when the two designs have an equal number of treatments and an equal number of observations.

The ratio of these variances is

$$(2\sigma_{\mathrm{RCB}}^2/r)/(2k\sigma_{\mathrm{BIB}}^2/(\lambda t)), \tag{6}$$

where σ_{RCB}^2 and σ_{BIB}^2 are the intrablock variances that would obtain in the RCB and BIB designs, respectively. The efficiency factor E is the value of (6) if $\sigma_{\mathrm{RCB}}^2 = \sigma_{\mathrm{BIB}}^2$, i.e.,

$$E = \lambda t/(kr) = (1 - k^{-1})/(1 - t^{-1}). \tag{7}$$

Its reciprocal represents the proportionate increase in variance of a simple treatment difference *if no reduction in intrablock variability is achieved* when a RCB design experiment is replaced by a BIB design experiment. Since it is intuitively clear that in almost all cases $\sigma_{\mathrm{BIB}}^2 \leqslant \sigma_{\mathrm{RCB}}^2$ because of the smaller block size, E represents a lower bound to the efficiency of BIB designs compared to RCB designs. Of course, for cases where one cannot employ blocks of size t, this comparison is meaningless. Nevertheless, one can still use (7) to compare alternative BIB designs.

As with point estimation, the partition of observational variability into block components, treatment differences, and residual variability is made more complex than for RCB designs by the lack of orthogonality of blocks and treatments. Formulas for the constituents of the analysis of variance* are presented in Table 1, where the notation

$$\bar{y} = S/rt = \sum_{i=1}^{t} \sum_{j \in I_i} y_{ij}/rt$$

is employed.

If, in addition to the earlier error assumptions, the observational errors are assumed to

Table 1. Analysis of Variance for a Balanced Incomplete Block Design

Source	Degrees of Freedom	Sum of Squares	Mean Square
Blocks (unadjusted)	$b - 1$	$k^{-1} \sum_{j=1}^{b}(B_j - k\bar{y})^2$	$[k(b-1)] \sum_{j=1}^{b}(B_j - k\bar{y})^2$
Treatments (adjusted for blocks)	$t - 1$	$(\lambda tk)^{-1} \sum_{i=1}^{t} Q_i^2$	$[\lambda tk(t-1)]^{-1} \sum_{i=1}^{t} Q_i^2$
Residual or error	$rt - t - b + 1$	Residual SS obtained by subtraction	(Residual SS) / $(rt - t - b + 1)$
Corrected total	$rt - 1$	$\sum_i \sum_{j \in I_i}(y_{ij} - \bar{y})^{-2}$	

be normally distributed, then the hypothesis of no differences among treatments, i.e.,

$$H_0 : \tau_i = \tau(\text{ all } i)$$

can be tested (*see* F-TESTS) by the ratio

$$F = \frac{(\text{treatment mean square})}{(\text{error mean square})}.$$

Under H_0, this statistic has an F-distribution* with $(t - 1)$ and $(rt - t - b + 1)$ degrees of freedom. Confidence limits* or tests of hypotheses for a contrast* in (2) are based on the result that

$$(T - L)/$$

$$\times \left[(\text{error mean square}) \cdot (k/(\lambda t)) \cdot \sum c_i^2\right]^{\frac{1}{2}}$$

has a Student's t-distribution* with $(rt - t - b + 1)$ degrees of freedom. As with RCBs, inferences about more than one contrast raise the issue of multiple comparisons*.

The discussion of an analysis of a BIB design, to this point, has been predicated on the assumption that block contributions are fixed unknowns. As mentioned earlier, there are diverse research studies for which it is reasonable to assume that the block contributions are normal random variables, independent of each other and of the observational errors, and that each has mean zero and variance σ_B^2. All other aspects of (1) remain unchanged. Yates [15] formulated this analysis for BIB designs with $b > t$, and labeled it an *analysis with recovery of interblock information*. One consequence of

the new assumption regarding block contributions is that the variance of a difference between two observations from different blocks is $2(\sigma^2 + \sigma_B^2)$, whereas the variance of a difference between two observations in the same block is smaller, equaling $2\sigma^2$. A second and most important consequence of the randomness of block contributions is that

$$E(B_j) = k\mu + \sum_{i=1}^{t} n_{ij}\tau_i. \tag{8}$$

The block totals $\{B_j\}$, which are independent of the intrablock estimates (3) obtained earlier, can be used to obtain an unbiased estimate T^* of L in (2) independent of T:

$$T^* = \sum_{i=1}^{t} c_i \sum_{j=1}^{b} n_{ij}B_j/(r - \lambda). \tag{9}$$

It is these estimates that represent the recovered interblock information. The variance of T^* is

$$\text{var}(T^*) = \left(\sum_{i=1}^{t} c_i^2\right)(k\sigma^2 + k^2\sigma_B^2)/(r - \lambda).$$

Were σ^2 and σ_B^2 known, one would combine T and T^* by weighting them proportionally to the reciprocals of these variances to obtain

$$\tilde{T} = (w_T T + w_{T^*}T^*)/(w_T + w_{T^*}) \tag{10}$$

for $w_T = \lambda t/(k\sigma^2)$ and $w_{T^*} = (r - \lambda)/(k\sigma^2 + k^2\sigma_B^2)$. Whereas (10) was originally proposed by Yates [15] simply as a weighted estimate, Rao [12] later showed that (10) is the maximum likelihood* estimate of L for known σ^2 and σ_B^2.

It is informative to consider the behavior of (10) under two extremes. If σ_B^2 is very large compared to σ^2, i.e., the variability among block contributions is great compared to observational error variability, then $\tilde{T}$ differs little from T. In this case, there is little interblock information to recover. At the other extreme, if σ_B^2 is essentially equal to 0, then $\tilde{T}$ reduces to a contrast in the unadjusted means of the observations from units administered treatment i, $i = 1, \ldots, t$, i.e.,

$$\sum_{i=1}^{t} c_i T_i / r.$$

This is the estimator of L that would be obtained in a completely randomized design or one-way layout, in which blocking is entirely ignored (see ONE-WAY ANALYSIS OF VARIANCE).

In reality, σ^2 and σ_B^2 are unknown. Standard practice has been to estimate these parameters from the data (see John [8, p. 238] or Kempthorne [9, p. 536], and to replace the parameters in (10) by estimates, without worrying about this substitution. Brown and Cohen [3] have shown that certain weighted combinations of T and T^*, with weights estimated from the data, always have smaller variance than T alone for any BIB design, unless $b = 3$.

Two final remarks are in order. First, as with RCB designs, nonparametric (distribution-free*) methods of analysis are available for BIB designs (see Durbin [6] or Noether [11, p. 54]. Second, there are numerous cases where no BIB design exists that satisfies the research project's constraints on the number of blocks. In that event, it is necessary to sacrifice the symmetry of BIB designs in order to reduce the number of blocks. The *partially balanced incomplete block designs* of Bose and Nair [1] represent the next level of compromise (see [4,5,8,9]; *see also* PARTIALLY BALANCED DESIGNS).

REFERENCES

1. Bose, R. C., and Nair, K. R. (1939). *Sankhyā*, **4**, 337–372.

2. Box, G. E. P., Hunter, W. G., and Hunter, J. S. (1978). *Statistics for Experimenters*. Wiley, New York.

3. Brown, L. D. and Cohen, A. (1974). *Ann. Statist.*, **2**, 963–976.

4. Cochran, W. G. and Cox, G. M. (1957). *Experimental Designs*, 2nd ed. Wiley, New York.

5. Cox, D. R. (1958). *Planning of Experiments*. Wiley, New York.

6. Durbin, J. (1951). *Br. J. Statist. Psychol.*, **4**, 85–90.

7. Fisher, R. A. and Yates, F. (1953). *Statistical Tables for Biological, Agricultural and Medical Research*, 4th. ed. Oliver & Boyd, Edinburgh.

8. John, P. W. M. (1971). *Statistical Design and Analysis of Experiments*, Macmillan, New York.

9. Kempthorne, O. (1952). *The Design and Analysis of Experiments*. Wiley, New York.

10. Kiefer, J. (1959). *J. R. Statist. Soc. B*, **21**, 272–319.

11. Noether, G. E. (1967). *Elements of Nonparametric Statistics*, Wiley, New York.

12. Rao, C. R. (1947). *J. Amer. Statist. Ass.*, **42**, 541–561.

13. Scheffé, H. (1959). *The Analysis of Variance*, Wiley, New York.

14. Yates, F. (1936). *Ann. Eugen. (Lond.)*, **7**, 121–140.

15. Yates, F. (1940). *Ann. Eugen. (Lond.)*, **10**, 317–325.

See also ANALYSIS OF VARIANCE; BLOCKS, RANDOMIZED COMPLETE; DESIGN OF EXPERIMENTS; FIXED-, RANDOM-, AND MIXED-EFFECTS MODELS; F-TESTS; INTERBLOCK INFORMATION; INTRABLOCK INFORMATION; INCOMPLETE BLOCK DESIGNS; ONE-WAY ANALYSIS OF VARIANCE; and PARTIALLY BALANCED DESIGNS.

BARRY H. MARGOLIN

BLOCKS, RANDOMIZED COMPLETE

The concept of *blocking* in statistically designed experiments originated during the intensive agricultural field trial research conducted at Rothamsted Experimental Station during Sir R. A. Fisher's* tenure as Chief Statistician. It was during this period that Fisher [4, p. 269] perceived the "peculiarity of agricultural field experiments"—"that the area of ground chosen for the experimental plots may be assumed to be markedly heterogeneous, in that its fertility varies in a systematic, and often a complicated manner from point to point."

Fisher first concluded that complete randomization* of the allocation of treatments to experimental units, here the plots of land, would overcome the potential bias in assessing treatment differences due to systematic variation in fertility. He then realized that far more was achievable; it was possible to insulate the estimates of treatment differences from much of the effect of soil heterogeneity, and thereby to increase the sensitivity of the experiment, by first grouping the experimental plots so that plots within a common group or *block* were nearly homogeneous, and then randomly applying all treatments to different plots within each block. This blocking permitted Fisher to partition the variability inherent in the plot yields after treatment into components attributable to (1) treatment differences, (2) within-block variability, and (3) between-block variability. With this conception, the *randomized complete block design* was born.

Agricultural research is not unique in having potentially heterogeneous experimental units that invite blocking; today, randomized complete block designs may involve blocking experimental units on a diversity of physical, chemical, genetic, socioeconomic, psychological, or temporal characteristics. Reports of only those studies that have used human subjects or litters of mice as blocks would fill this encyclopedia. Discussion of this topic may be found in most books on the design and analysis of experiments (*see* DESIGN OF EXPERIMENTS), including Box et al. [1], Cochran and Cox [2], Cox [3], John [6], and Kempthorne [7].

The term *block* is used with great generality to refer to any group of experimental units that share a set of characteristics thought to possibly affect the response to be observed after treatment. A randomized complete block (RCB) design is a design in which each of t treatments under study is applied

to a distinct experimental unit once in each of b blocks. For such a design, the number of experimental units per block is constant, say k, and $k = t$; moreover, the number of replications per treatment is constant, say r, and $r = b$. Allocation of the t treatments to the individual experimental units within each block is determined by randomization. The reader should note that certain authors, e.g., Cochran and Cox [2, p. 105], adopt a broader definition of a RCB design permitting certain of the treatments to occur more than once within each block.

In the following example, the blocking characteristic is temporal. Margolin et al. [8] reported on what was conceivably the first published on-line experiment upon a computer system servicing its normal user population. Two algorithms to manage computer free storage, labeled Old and New, represented the treatments of interest. Blocking on both weeks and days of the week was employed because it was thought that the user workload might vary substantially over time. For present purposes the first blocking characteristic, weeks, will be ignored. The experiment ran on two Mondays, two Tuesdays, two Wednesdays, and two Thursdays, so that $k = 2 = t$, and $b = 4 = r$. At 5:00 A.M. each morning of a designated test day, a computer operator powered up a computer system utilizing either the New or Old free storage algorithm, and allowed the system to run for 24 hours. The data in Table 1 represent the average supervisor time in microseconds needed to obtain required free storage for each day of the experiment.

Without employing formal statistical inference, it is clear that the New algorithm substantially reduced the average supervisor time needed to obtain free storage.

Although the blocking characteristic clearly must be specified before experimentation, it need not represent a proven source

Table 1. Free Storage Experiment Average Timings (microseconds)

Treatment	Monday	Tuesday	Wednesday	Thursday
New	37	50	45	45
Old	325	346	313	426

of heterogeneity for the response of interest. In this respect, blocking, at times, may be likened to an insurance policy taken to guard against a possible eventuality, here heterogeneity of experimental units; such was the case with the blocking in the free storage experiment. As with all insurance, there is a cost. To better appreciate this cost, it is necessary to understand the model implicitly assumed in the analysis of a randomized complete block design. To achieve this understanding, one must first arbitrarily number the treatments and blocks 1 to t and 1 to b, respectively.

The statistical model underlying the common parametric analysis of a RCB design assumes that the response y_{ij} observed when the ith treatment is applied to a randomly chosen experimental unit in the jth block is simply a sum of an overall mean μ, a treatment contribution τ_i, a block contribution β_j, and an observational error ϵ_{ij}:

$$y_{ij} = \mu + \tau_i + \beta_j + \epsilon_{ij}. \tag{1}$$

Block contributions may be considered as fixed or random quantities; this distinction in no way alters inferences concerning the treatment contributions in (1). The observational errors are assumed to be independent, each with mean zero and variance σ^2; normality of the errors is sometimes assumed as well.

Treatment and block contributions enter (1) *additively*. This implies that the difference between two observations in any one block, say y_{ij} and y_{hj}, is unaltered by the block contribution. Therefore, up to observational errors, $(y_{ij} - y_{hj})$ measures solely the difference between the corresponding treatment contributions. This is clear mathematically from (1), for the model implies that

$$y_{ij} - y_{hj} = (\tau_i - \tau_h) + (\epsilon_{ij} - \epsilon_{hj}). \tag{2}$$

The model in (1) does not uniquely specify the treatment contributions, since a constant c may be added to each τ_i and subtracted from μ to produce a model of the form in (1) but with different parameter values:

$$y_{ij} = (\mu - c) + (\tau_i + c) + \beta_j + \epsilon_{ij}$$
$$= \mu^* + \tau_i^* + \beta_j + \epsilon_{ij}. \tag{3}$$

This lack of uniqueness or identifiability implies that the treatment contributions are not estimable (*see* ESTIMABILITY). Any linear combination of the $\{\tau_i\}$ of the form

$$L = \sum_{i=1}^{t} c_i \tau_i \tag{4}$$

for which $\sum c_i = 0$, however, is estimable; such a linear combination is labeled a contrast*. The least-squares* estimator of L is given by

$$T = \sum_{i=1}^{t} c_i \bar{y}_{i\cdot}, \tag{5}$$

where $\bar{y}_{i\cdot} = \sum_{j=1}^{b} y_{ij}/b$.

Particular contrasts of the $\{\tau_i\}$ that are frequently of interest to researchers are:

(1) Simple differences: $\tau_i - \tau_j, i \neq j$;
(2) The ith treatment effect (relative to the average treatment contribution): $\tau_i - \bar{\tau}$, for $\bar{\tau} = \Sigma_{i=1}^{t} \tau_i/t$.

The estimator in (5) has mean and variance given by

$$E(T) = \sum_{i=1}^{t} c_i \tau_i \quad \text{(unbiased)}$$

$$\text{var}(T) = \left(\sum_{i=1}^{t} c_i^2/b \right) \sigma^2,$$

and is normally distributed if the observational errors are assumed normally distributed. The estimator T may be written equivalently in the form

$$T = b^{-1} \sum_{j=1}^{b} \left(\sum_{i=1}^{t} c_i y_{ij} \right). \tag{6}$$

The quantity in parentheses is a contrast of the responses within the jth block, and is unchanged by the addition of a common constant to each observation in that block. Consequently, under (1), the behavior of T is unaffected by the block contributions β_j, so T is insulated from heterogeneity of the experimental units to the extent that (1) holds. This should improve the detection of treatment differences.

The partition of observational variability into the three components Fisher envisioned—those attributable to (a) block differences, (b) treatment differences, and (c) within-block or residual variability—is formalized in an analysis-of-variance* table. Table 2 presents a prototype of such an analysis for a RCB design, where $\bar{y}_{\cdot j}$ and $\bar{y}_{\cdot \cdot}$ denote the average in the jth block and the overall average, respectively.

Under the normal error* assumption, the partitioned components are independent. The hypothesis of no differences among treatments, i.e.,

$$H_0 : \tau_i = \tau \quad \text{(all } i),$$

can then be tested against an all-inclusive alternative hypothesis by the ratio

$$F = \frac{\text{(treatment mean square)}}{\text{(error mean square)}}.$$

Under H_0, this statistic has an F-distribution* with $(t-1)$ and $(b-1)(t-1)$ degrees of freedom*. In addition, the error mean square is an unbiased estimator of σ^2, and can be used together with (5) to set confidence limits on or test hypotheses for contrasts* in (4). The main result employed in either computation is that

$$(T-L)/\left(\text{error mean square} \cdot \sum c_i^2/b\right)^{1/2}$$

has a Student's t-distribution* on $(b-1)(t-1)$ degrees of freedom. If inferences are to be made concerning more than one contrast, the issue of *multiple comparisons** must be faced.

A researcher may wish to determine whether his or her blocking has isolated an important component of variability; this can be done only after the fact. An F-test* can be formed with $(b-1)$ and $(b-1)(t-1)$ degrees of freedom, based on the ratio

$$R = \frac{\text{(block mean square)}}{\text{(error mean square)}}, \quad (7)$$

to assess whether the blocking, together with any other extraneous contributions that may be confounded with blocks, represented a source of substantial variability.

Recall that the insurance afforded by blocking comes at a cost; this cost is the reduction in the degrees of freedom available for estimation of σ^2, which makes the detection of treatment differences more difficult. Again, after the fact, this cost may be compared with the intended gain of reduced variability for contrasts achieved through the isolation of the block component of variability. Various measures have been proposed (e.g., Cochran and Cox [2, p. 112]) to assess the efficiency of a RCB design relative to a comparable completely randomized design, i.e., a design with the same number of experimental units but no blocking. As far as is known, all such measures are monotonically increasing functions of (7).

The computation of the analysis of variance* in Table 2 is frequently available

Table 2. Analysis of Variance for a Randomized Complete Block Design

Source	Degrees of Freedom	Sum of Squares	Mean Square
Blocks	$b-1$	$t\sum_{j=1}^{b}(\bar{y}_{\cdot j}-\bar{y}_{\cdot\cdot})^2$	$\frac{t}{b-1}\sum_{j=1}^{b}(\bar{y}_{\cdot j}-\bar{y}_{\cdot\cdot})^2$
Treatments	$t-1$	$b\sum_{i=1}^{t}(\bar{y}_{i\cdot}-\bar{y}_{\cdot\cdot})^2$	$\frac{b}{t-1}\sum_{i=1}^{t}(\bar{y}_{i\cdot}-\bar{y}_{\cdot\cdot})^2$
Residual or error	$(t-1)(b-1)$	$\sum_i\sum_j(y_{ij}-\bar{y}_{i\cdot}-\bar{y}_{\cdot j}+\bar{y}_{\cdot\cdot})^2$	$\frac{1}{(t-1)(b-1)}\sum_i\sum_j(y_{ij}-\bar{y}_{i\cdot}-\bar{y}_{\cdot j}+\bar{y}_{\cdot\cdot})^2$
Corrected total	$bt-1$	$\sum_i\sum_j(y_{ij}-\bar{y}_{\cdot\cdot})^2$	

in packages of statistical computer programs, although it may not be identified as the analysis of a randomized complete block design. This is so because model (1) is also standardly adopted for the more common *two-way layout* with one observation per cell. The analysis-of-variance table is the same in both cases, as are normal probability plots* to assess the normality assumption; and checks for outliers* and lack of additivity. Box et al. [1, Sec. 6.5] discuss diagnostic checking of the basic model in (1). The major distinction between the two designs is that in a two-way layout, there are two sets of treatments of roughly equal interest, both of whose allocation to the experimental units is determined by randomization; whereas in a randomized complete block design, the blocks represent some intrinsic property of the experimental units, and randomization is employed to allocate the single set of treatments to experimental units so that each treatment occurs once per block. Thus since the blocks themselves are in no sense randomized, the test in (7) is not a valid test for block effects, but rather, only for the combined effects of blocks as labeled, plus any other properties of the experimental units confounded with blocks.

A last comment concerns situations where the response of interest is unlikely to satisfy the normality assumption, even after transformation. In the extreme, for example, the observed responses for an entire block may only be a ranking of the t treatments within that block. In these situations, one may turn to nonparametric (distribution-free*) methods to analyze a RCB design [5, Chap. 7, Sec. 1].

REFERENCES

1. Box, G. E. P., Hunter, W. G., and Hunter, J. S. (1978). *Statistics for Experimenters*. Wiley, New York.

2. Cochran, W. G. and Cox, G. M. (1957). *Experimental Designs*, 2nd ed. Wiley, New York.

3. Cox, D. R. (1958). *Planning of Experiments*. Wiley, New York.

4. Fisher, R. A. (1938). *Statistical Methods for Research Workers*. Oliver & Boyd, Edinburgh.

5. Hollander, M. and Wolfe, D. A. (1973). *Nonparametric Statistical Methods*. Wiley, New York.

6. John, P. W. M. (1971). *Statistical Design and Analysis of Experiments*. Macmillan, New York.

7. Kempthorne, O. (1952). *The Design and Analysis of Experiments*. Wiley, New York.

8. Margolin, B. H., Parmelee, R. P., and Schatzoff, M. (1971). *IBM Syst. J.*, **10**, 283–304.

See also ANALYSIS OF VARIANCE; BLOCKS, BALANCED INCOMPLETE; DESIGN OF EXPERIMENTS; FACTORIAL EXPERIMENTS; F-TESTS; INTERACTION; KRUSKAL–WALLIS TEST; and MULTIPLE COMPARISONS—I.

BARRY H. MARGOLIN

BLOCKS, STATISTICALLY EQUIVALENT. See TOLERANCE REGIONS; SEQUENTIAL PROCEDURES, JIŘINA

BLUMEN'S BIVARIATE SIGN TEST. See BIVARIATE SIGN TEST, BENNETT'S

BMDP. See STATISTICAL SOFTWARE

BODE'S LAW

Bode's Law is an example of a model suggested by existing data, or of the construction of a model after gathering data. The Polish-German astronomer J. D. Titius (1729–1796) discovered in 1766 that the mean distances of planets from the sun, in order of increasing distance, are approximately in ratios

$$d_n = 4 + 3(2^n) \qquad (n = -\infty, 0, 1, 2, \ldots). \quad (1)$$

For the first eight planets this formula implies mean distances proportional to 4, 7, 10, 16, 28, 52, 100, and 196. The seven planets known up to 1800 had mean distances of 3.9, 7.2, 10 (the Earth), 15.2, 52, 95, and 192 and fit the theoretical sequence remarkably well. The exception was the missing planet located 28 units from the sun. Remarkably, J. E. Bode (1747–1826) and five other German astronomers discovered a small planet, Ceres, on January 1, 1801, in the vicinity of 28 units from the sun. Since then several other small planets (the "minor planets") have been located in this area, which are conjectured to be fragments of a single large planet. Consequently, the "law" given in (1) is now referred to as Bode's law (or the Titius–Bode law). (According to Good [3], the actual discovery was made by Piazzi and Bode merely publicized it.)

In the 1970s there was a heated controversy as to the extent to which Bode's discovery adds significantly to the believability of the "law". Moreover, it would seem that it is necessary to assess the degree of believability it had prior to 1801 and that the real test should be based on future observations. A detailed analysis of controversy surrounding Bode's law, taking into account that the data were "instigated", has been given by Leamer [5, pp. 300–305]. This work and refs. 1,4 and 6 present fascinating reading and touch on the outstanding points of difficulty in the contemporary theory and methods of data analysis.

REFERENCES

1. Efron, B. (1971). *J. Amer. Statist. Ass.*, **66**, 552–568. (Has comments by Good, Bross, Stuart, Danby, Blyth, and Pratt.)

2. Good, I. J. (1969). *J. Amer. Statist. Ass.*, **64**, 23–66. (With discussion.)

3. Good, I. J. (1972). Letter to the Editor, *Amer. Statist.*, **26**, 48–49.

4. Leamer, E. E. (1974). *J. Amer. Statist. Ass.*, **69**, 122–131.

5. Leamer, E. E. (1978). *Specification Searches.* Wiley, New York, Sect. 9.5.

6. Polanyi, M. (1964). *Personal Knowledge.* Harper & Row, New York.

See also ASTRONOMY, STATISTICS IN; BELIEF, DEGREES OF; and STATISTICAL MODELING.

BOL'SHEV, LOGIN NIKOLAEVICH

Born: March 6, 1922 in Moscow, USSR.

Died: August 29, 1978 in Moscow, USSR.

Contributed to: transformations of random variables; asymptotic expansions for probability distributions; tables of mathematical statistics; mathematical epidemiology and demography.

L. N. Bol'shev was one of the leading Soviet experts in the field of mathematical statistics during the period 1950–1978.

He was born in 1922 in the family of a white-collar worker. His father was a military man who distinguished himself in World War I, and his grandfather a well-known military topographer of his time.

Bol'shev graduated in 1951 from the Mechanical–Mathematical Faculty of the Moscow State University. Based on A. N Kolmogorov's* recommendations and initiative, he started his tenure at the Probability Theory and Mathematical Statistics Branch of the prestigious Steklov Mathematical Institute in Moscow, where he worked throughout his whole life and where A. Ya. Khinchine*, N. V. Smirnov*, and Yu. V. Prohorov were his colleagues. Later, Bol'shev combined his work at the Institute with lecturing at the Moscow State University.

In 1955 Bol'shev defended his Master's dissertation entitled "On the problem of testing composite statistical hypotheses" at the Institute, and in 1966 he was awarded the Doctor of Science degree for his research on "Transformation of random variables" by the Scientific Council of the Institute.

The theory of transformation of random variables and the methods of asymptotic expansions for the basic probability distributions utilized in statistics (in particular Pearson, Poisson, and generalized hypergeometric) which Bol'shev developed in his thesis are widely used in statistical theory and practice [1,4,5,6,7].

L. N. Bol'shev was highly successful in developing and implementing the construction of tables of mathematical statistics which are required for probabilistic and statistical calculations [12–15]. These activities were originated in the USSR by E. E. Slutsky and N. V. Smirnov. Jointly with Smirnov, he complied in 1965 the well-known *Tables of Mathematical Statistics* [13], which are the most advanced and authoritative international contribution to this area.

Additional contributions of L. N. Bol'shev which have attracted international attention are in the areas of statistical parameter estimation [2,8,11], testing for outliers [3,9], and epidermiology and demography [10].

An excellent and popular teacher, Bol'shev attracted many students who later became well-known Soviet statisticians. He was often called as an expert in various scientific discussions connected with the application of statistical methodology in industry, geology, biology, medicine, and sports refereeing.

In 1974 he was elected a corresponding member of the Academy of Sciences of the USSR, and the following year a member of the International Statistical Institute*. He served with great distinction in the Red Army during World War II and was awarded the Order of the Red Banner of Labor and other medals.

REFERENCES

1. Bol'shev, L. N. (1959). On transformation of random variables. *Theory Probab. Appl.*, **4**(2), 136–149.

2. Bol'shev, L. N. (1961). A refinement of the Cramér–Rao inequality. *Theory Probab. Appl.*, **6**(3), 319–326.

3. Bol'shev, L. N. (1961). On elimination of outliers. *Theory Probab. Appl.*, **6**(4), 482–484.

4. Bol'shev, L. N. (1963). Asymptotic Pearsonian transformations. *Theory Probab. Appl.*, **8**(2), 129–155; Corr., 473.

5. Bol'shev, L. N. (1964). Distributions related to hypergeometric. *Theory Probab. Appl.*, **9**(4), 687–692.

6. Bol'shev, L. N. (1964). Some applications of Pearson transformations. *Rev. Int. Statist. Inst.*, **32**, 14–16.

7. Bol'shev, L. N. (1965). On characterization of Poisson distribution and its statistical applications. *Theory Probab. Appl.*, **10**(3), 488–499.

8. Bol'shev, L. N. (1965). On construction of confidence intervals. *Theory Probab. Appl.*, **10**(1), 187–192.

9. Bol'shev, L. N. (1966). Testing of outliers in the case of least squares estimation. In *Abstracts of Scientific Papers at the International Congress of Mathematicians, Section 11*, Moscow, pp. 30–31. (In Russian.)

10. Bol'shev, L. N. (1966). A comparison of intensities of simple streams. *Theory Probab. Appl.*, **11**(3), 353–355.

11. Bol'shev, L. N. and Loginov, E. A. (1966). Interval estimation in the presence of nuisance parameters. *Theory Probab. Appl.*, **11**(1), 94–107.

12. Bol'shev, L. N. and Smirnov, N. V. (1962). *Tables for Calculation of Bivariate Normal Distribution Functions*. Fizmatgiz, Moscow.

13. Bol'shev, L. N. and Smirnov, N. V. (1965). *Tables of Mathematical Statistics*. Fizmatgiz, Moscow. (Translated into English.)

14. Bol'shev, L. N., Gladkov, B. V., and Shcheglova, V. (1961). Tables for calculation of functions of B- and Z-distributions. *Theory Probab. Appl.*, **6**(4), 446–455.

15. Bol'shev, L. N., Bark, L. S., Kuznetsov, P. I., and Cherenkov, A. P. (1964). *Tables of the Rayleigh–Rice Distribution*. Fizmatgiz, Moscow.

SERGEI A. AIVAZIAN

(Translated and edited by S. KOTZ)

BOLTZMANN, LUDWIG EDWARD

Born: February 20, 1844, in Vienna, Austria.

Died: September 5, 1906, in Duino, near Trieste (now Italy).

Contributed to: physics (especially statistical mechanics).

Boltmann's statistical contributions are associated with his efforts, following on from J. C. Maxwell*, to explain the thermodynamics of gases on the basis of kinetic theory: i.e., to view gases as particles undergoing movement at different velocities and collisions according to the principles of mechanics. Maxwell, considering an equilibrium situation, had arrived at the probability density function $f(v) = 4\alpha^{-3}\pi^{-1/2}v^2 \exp(-v^2/\alpha^2)$, $v \geqslant 0$, for the (root-mean-square) velocity V of a particle in three dimensions, with $V = (X^2 + Y^2 + Z^2)^{1/2}$, where X, Y, Z are independent, identically distributed (i.i.d.) zero-mean normal random variables, corresponding to velocity components along the coordinate axes. The density of kinetic energy, U (of a randomly chosen particle), since it is proportional to V^2, is in this situation of the form const. $u^{1/2} \exp(-hu)$, $u \geqslant 0$, which is essentially the density of a chi-square* distribution with 3 degrees of freedom (*see* MAXWELL DISTRIBUTION). Boltzmann [1] in 1881 actually wrote down the density of the equilibrium kinetic energy distribution in the case of movement in n dimensions, apparently unaware of the work of earlier authors (Abbe*, Bienaymé*, and Helmert*), who had obtained expressions for the chi-square density of the sum of squares of n i.i.d. $N(0, 1)$ random variables.

His major contribution in the kinetic theory of gases, however, is in connection not with the equilibrium state itself, but with approach to equilibrium, and, consequently, with the relation between the evolution of thermodynamic entropy and the probability distribution of velocities. If $f(x, y, z; t)$ is the joint density of velocity components X, Y, Z at time t, he asserted that

$$E(t)$$
$$= -\iiiint f(x, y, z; t) \log f(x, y, z; t)\, dx\, dy\, dz$$

is nondecreasing to the value given by $f(x, y, z; t) = \text{const.} \exp[-h(x^2 + y^2 + z^2)]$: i.e., the equilibrium-state velocity distribution. This is essentially Boltzmann's H-theorem, with $E(t)$ corresponding to modern probabilistic usage of entropy* in connection with information*. Boltzmann's derivation was not rigorous and the fact that in classical mechanics collisions between particles are reversible, anticipating recurrence of any overall configuration, seemed inconsistent with the overall irreversibility predicted by increasing entropy. As a result of the ensuing controversy, the celebrated paper [2], which introduced the Ehrenfest urn model* as an illustrative example, suggested a probabilistic formulation in Markov chain* terms. The appropriate Markov chain $\{\mathbf{X}_n\}$, $n \geqslant 0$, describes the evolution of the relative-frequency table of velocities assumed by N particles at a specific time, where the states of $\mathbf{X}_n$ are vectors describing all such possible frequency tables, it being assumed that each particle must have one of r velocity values (such subdivision into quanta, leading to combinatorial treatment, is characteristic of Boltzmann's work). The chain $\{\mathbf{X}_n\}$ is such that each state will recur with probability 1, but a suitable scalar function $\phi(\mathcal{E}\mathbf{X}_n)$ ("entropy") is nondecreasing with n. The probability distribution, $\mathcal{E}\mathbf{X}_n$, reflects the density $f(v, t)$ of root-mean-square velocity at time t. The limiting-stationary distribution of

$N\mathbf{X}_n$ is the multinomial with general term

$$\frac{N!}{x_1! \cdots x_r!} r^{-N},$$

now identified with "Maxwell-Boltzmann statistics" in physics, and manifests the ultimate (equilibrium) tendency of any one particle to have any of the possible velocities equiprobably. (See FERMI-DIRAC, MAXWELL-BOLTZMANN, AND BOSE-EINSTEIN STATISTICS.)

We owe to Boltzmann also the word "ergodic"*, although its meaning has evolved considerably up to the present.

Additional information on Boltzmann's contributions to statistics can be found in refs. 3 and 6.

REFERENCES

1. Boltzmann, L. (1909). *Wissenschaftliche Abhandlungen*, F. Hasenöhrl, ed., 3 vols. Leipzig. (Boltzmann's collected technical papers; the expression appears in Vol. 2, p. 576.)

2. Ehrenfest, P. and T. (1907). *Phys. Zeit.*, **8**, 311–314.

3. Gnedenko, B. V. and Sheynin, O. B. (1978). In *Matematika XIX veka* [*Mathematics of the 19th Century*]. Izd. Nauka, Moscow, pp. 184–240. (Pages 229–232 contain a picture of Boltzmann and an assessment of his statistical contributions.)

4. Kac, M. (1959). *Probability and Related Topics in Physical Sciences*. Interscience, London.

5. Moran, P. A. P. (1961). *Proc. Camb. Philos. Soc.*, **57**, 833–842. (A modern exploration of the theorem in a Markovian setting.)

6. Sheynin, O. B. (1971). *Biometrika*, **58**, 234–236. (Historical manifestations of the chi-square distribution.)

See also CHI-SQUARE DISTRIBUTION; ENTROPY; ERGODIC THEOREMS; EXPECTED VALUE; FERMI-DIRAC, MAXWELL-BOLTZMANN, AND BOSE-EINSTEIN STATISTICS; MARKOV PROCESSES; MAXWELL, JAMES CLERK; and MULTINOMIAL DISTRIBUTIONS.

E. SENETA

BONFERRONI, CARLO EMILIO

Born: January 28, 1892, in Bergamo, Italy.

Died: August 18, 1960, in Florence, Italy.

Contributed to: probability theory, actuarial mathematics, linear algebra, geometry.

Carlo Emilio Bonferroni attended the University of Torino, where he was exposed to some very outstanding mathematicians as instructors. He studied analysis with Giuseppe Peano and geometry with Corrado Segre. After receiving the degree he was appointed an assistant to Filadelfo Insolera, holder of the chair of mathematics of finance. Soon he became an instructor in courses in classical mechanics and geometry at the Polytechnic Institute of Torino.

His first professorial appointment was in 1923 to the chair of mathematics of finance at the University of Bari, where he stayed for ten years. For seven out of those years he was the president of the University of Bari. In 1933 he was appointed to the chair of general and financial mathematics of the University of Florence, and he remained in Florence for twenty-seven years until his death on August 18, 1960. During this period he was the dean of the Faculty of Economics and Commerce in 1945–1949. Simultaneously he also taught courses in statistical methods at the Bocconi University in Milan and analysis and geometry for the Faculty of Architecture in Florence. The Tuscan Academy "La Colombaria" honored him with the presidency of the Class of Physical, Natural, and Mathematical Sciences. He was also a member of the Italian Institute of Actuaries.

He was married to Mrs. Jolenda Bonferroni. He had a strong interest in music and was an excellent pianist and composer. Among his hobbies was also mountain climbing.

While teaching courses in different subjects he was an active writer and researcher. He wrote about 65 research papers and books. Among statisticians he is mostly known as the author of Bonferroni's inequalities [1]. However, he had several original ideas in other fields of mathematics. His work can be grouped in three broad categories: (1) analysis, geometry, and mechanics, (2) actuarial mathematics and economics, and (3) statistics and probability. In the first category there are 14 entries in the bibliography, including three textbooks (or lecture notes). In the second category there are 19 entries (including one textbook), and finally his list of publications contains 33 entries in the general area of statistics and probability.

In the first group we shall quote here two of his theorems. One concerns an extension of the so-called "Napoleon" theorem. Napoleon observed and asked Laplace to prove that "if one constructs three equilateral triangles, each on one of three sides of an arbitrary triangle, their centers form an equilateral triangle." Bonferroni extended this theorem to the following: "Consider three triangles with angles, α, β, γ $(\alpha + \beta + \gamma = \pi)$ constructed internally or externally on three sides of an arbitrary triangle respectively. Their centroids form a triangle with angles α, β, γ."

The following theorem in linear algebra with application to analysis was also established by Bonferroni: "The determinant obtained by horizontal multiplication of two conjugate matrices does not exceed the product of an element of the main diagonal times its minor, and if this product is nonzero, the determinant is equal to it only when the lines that cross at that element have every other element nonzero." Hadamard's theorem and other results on Hermitian bilinear forms follow as simple consequences.

An article by Carlo Benedetti in *Metron* contains an extensive review of Bonferroni's books. His textbook *Elements of Mathematical Analysis* had its fourth edition (1957) shortly before the author died. This reviewer has very high praise for its style, elegance, and level of generality.

The book on *Foundations of Actuarial Mathematics* includes several of his earlier ideas on the subject.

In the area of statistics and probability most of Bonferroni's research is concentrated in the period 1917–1942. There are only three papers published in the 1950s, all concerning the same subject of medians in continuous distributions.

The last revision of the text *Elements of General Statistics* (1941–1942) contains many of his ideas on the subject and some original proofs of results known at the time. He introduces and discusses various indices and coefficients. His coefficient of parabolic correlation of order k is revisited. Contingency indices of Pearson* and Chuprov* get unified treatment, together with other indices which he proposes.

Bonferroni was also interested in measures of concentration. He proposed a concentration index, unaware of earlier work of Dalton [18]. Another concentration index proposed by him could be considered a variation of Gini's* concentration ratio. This could have created some friction with Gini, in addition to that arising from Gini's defense of his assistant, Cisbani, who had a dispute with Bonferroni.

REFERENCES

1. Alt, F. B. (1982). Bonferroni inequalities and intervals. In *Encyclopedia of Statistical Sciences*, S. Kotz and N. L. Johnson, eds. Wiley, New York, vol. 1, p. 294.

2. Benedetti, C. (1982). Carlo Emilio Bonferroni (1892–1960). *Metron*, **40**(3–4), 3–36.

3. Bonferroni, C. E. (1919). *Principi di Geometria (ad Uso degli Studenti del Politecnico di Torino)*. Gnocchi, Turin, Italy.

4. Bonferroni, C. E. (1920). *Corso di Meccanica Raxionale (ad Uso degli Studenti del Politecnico di Torino)*. Gili, Turin, Italy.

5. Bonferroni, C. E. (1921). Probabilita assolute, relative, di precessione, *Gior. Mat. Finanz.*

6. Bonferroni, C. E. (1927–1928). *Elementi di Statistica Generale*, 1st ed. Bocconi, Milan, Italy. [Last (expanded) edition 1940.]

7. Bonferroni, C. E. (1928). On the measures of the variations of a collectivity. *J. Inst. Actuaries Lond.* **59**, 62–65.

8. Bonferroni, C. E. (1936). Teoria statistica delle classi e calcolo delle probabilità. In *Volume in Honor of Riccardo Della Volta*. University of Florence, Italy.

9. Bonferroni, C. E. (1939). Di una estensione del coefficiente di correlazione. *Giorn. Econ.* (Expanded version in vol. 1, Institute of Statistics, Bocconi, 1941.)

10. Bonferroni, C. E. (1940). Le condizione d'equilibrio per operazionè finanziarie finite ed infinite, *Giorn. Istituto. Ital. Attuar.* **11**, 190–213.

11. Bonferroni, C. E. (1940). Di un coefficiente del correlazione. *Cong. Unione Mat. Ital.* (Also in vol. 1, Institute of Statistics, Bucconi, 1941.)

12. Bonferroni, C. E. (1940). *Un indice quadratico di concentrazione.* Ibid.

13. Bonferroni, C. E. (1941). Nuovi indici di connessione fra variabili statistiche. *Cong. Unione Mat. Ital.*

14. Bonferroni, C. E. (1942). *Fondamenti di Matematica Attuariale*, 2nd ed. Gili, Turin, Italy. (1st ed. 1935–1936.)

15. Bonferroni, C. E. (1950). Di una disuguaglianza sui determinanti ed il teorema di Hadamard. *Boll. Unione Mat. Ital.*

16. Bonferroni, C. E. (1950). *Una teorema sul triangolo ed id teorema di Napoleone*, Ibid.

17. Bonferroni, C. E. (1957). *Elemente di Analisi Matematica*, 6th ed. Cedam, Padua, Italy. (1st edn. 1933–1934.)

18. Dalton, H. (1920). *The Inequalities of Income*, Routledge, London; Dulton, New York.

PIOTR W. MIKULSKI

BONFERRONI INEQUALITIES AND INTERVALS

BONFERRONI INEQUALITIES

If $E_1, E_2, \ldots, E_n$ are n random events, the probability that *exactly* r of them occur is

$$P_{[r]} = \sum_{j=r}^{n} (-1)^{j-r} \binom{j}{j-r} S_j, \qquad (1)$$

where

$$S_j = \sum_{1 \leqslant \alpha_1 < \cdots < \alpha_j \leqslant n} \Pr\left[\bigcap_{h=1}^{j} E_{\alpha_h} \right]$$

and $S_0 = 1$. The probability that *at least* r of them occur is

$$P_r = \sum_{i=r}^{n} P_{[i]} = \sum_{j=r}^{n} (-1)^{j-r} \binom{j-1}{j-r} S_j. \qquad (2)$$

(*See* INCLUSION-EXCLUSION METHOD; BOOLE'S INEQUALITY, and, for an example of application, MATCHING PROBLEM.)

Bonferroni [5,6] showed that

$$0 \leqslant P_{[0]} \leqslant 1,$$
$$1 - S_1 \leqslant P_{[0]} \leqslant 1 - S_1 + S_2,$$

and generally

$$1 - S_1 + S_2 - \cdots - S_{2i-1}$$
$$\leqslant P_{[0]} \leqslant 1 - S_1 + S_2 - \cdots + S_{2i} \tag{3}$$

for $i = 1, 2, \ldots, n/2$.

Galambos [12,13] gives an overview of methods of proving these inequalities, and also some improved bounds. The inequality

$$1 - S_1 \leqslant P_{[0]}$$

is equivalent to

$$1 - \sum_{i=1}^{n} \Pr[E_i] \leqslant \Pr\left[\bigcap_{i=1}^{n} \bar{E}_i\right].$$

Replacing the complement $\bar{E}_i$ by A_i (and so E_i by $\bar{A}_i$), we have

$$1 - \sum_{i=1}^{n} \Pr[\bar{A}_i] \leqslant \Pr\left[\bigcap_{i=1}^{n} A_i\right]. \tag{4}$$

This result has been the most frequently used of Bonferroni's inequalities and is often referred to as "*the* Bonferroni inequality."

The set of inequalities (3) generalize to

$$\left.\begin{array}{l} S_r - \binom{r+1}{1} S_{r+1} \leqslant P_{[r]} \leqslant S_r \\[2mm] S_r - \binom{r+1}{1} S_{r+1} + \binom{r+2}{2} S_{r+2} - \binom{r+3}{3} S_{r+3} \\[2mm] \leqslant P_{[r]} \leqslant S_r - \binom{r+1}{1} S_{r+1} + \binom{r+2}{2} S_{r+2}, \end{array}\right\} \tag{5}$$

and so on.

Similar sets of inequalities can be obtained for P_r. Specifically,

$$\left.\begin{array}{l} S_r - \binom{r}{1} S_{r+1} \leqslant P_r \leqslant S_r \\[2mm] S_r - \binom{r}{1} S_{r+1} + \binom{r+1}{2} S_{r+2} - \binom{r+2}{3} S_{r+3} \\[2mm] \leqslant P_r \leqslant S_r - \binom{r}{1} S_{r+1} + \binom{r+1}{2} S_{r+2}, \end{array}\right\} \tag{6}$$

and so on.

Proofs of these inequalities are given by Fréchet [11] and Feller [10].

Meyer [16] has extended Bonferroni's inequalities to classes of events $\{E_{ij}, i = 1, \ldots, N_j\}$, $j = 1, 2, \ldots, k$. For nonnegative integers r_j, $0 \leqslant r_j \leqslant N_j$, he has obtained Bonferroni inequalities for $P_{[r_1, r_2, \ldots, r_k]}$ and $P_{r_1, r_2, \ldots, r_k}$, where $P_{[r_1, r_2, \ldots, r_k]}$ denotes the probability that exactly r_1 of the E_{i1}'s, exactly r_2 of the E_{i2}'s, $\ldots$, and exactly r_k of the E_{ik}'s will occur; $P_{r_1, r_2, \ldots, r_k}$ is defined analogously with "at least" replacing "exactly."

From a statistical point of view, it is the Bonferroni inequality stated in (4) that is of paramount importance because of its use in simultaneous inference to maintain the family confidence level, at least in a bounded sense. *See* MULTIPLE COMPARISONS—I. An extensive treatment of simultaneous inference is given in Miller [17] with updated references in ref. 18.

BONFERRONI INTERVALS

Let $\theta = (\theta_1, \theta_2, \ldots, \theta_k)'$ be a $(k \times 1)$ vector of parameters—e.g., multinomial* proportions, elements in an expected value vector or a variance-covariance matrix* or partial regression coefficients in a general linear model*. If separate two-sided confidence intervals are constructed for each of the k parameters, each with confidence coefficient $100(1 - \alpha)\%$, and if A_i denotes the event that the interval for θ_i includes the actual value of θ_i, then, from the Bonferroni inequality (4), it follows that the probability $(\Pr[\cap_{i=1}^{k} A_i])$ that every interval includes the value of the parameter it estimates is at least $(1 - k\alpha)$. Formally,

$$\Pr\left[\bigcap_{i=1}^{k} A_i\right] \geqslant 1 - k\alpha. \tag{7}$$

Thus the "family confidence coefficient" is at least $100(1 - k\alpha)\%$, whatever be the dependence among the statistics used in constructing the confidence intervals. If the confidence level for each separate interval is increased to $100(1 - \alpha k^{-1})\%$, then the family confidence coefficient is at least $100(1 - \alpha)\%$. The resulting confidence intervals are called *Bonferroni intervals*.

In obtaining Bonferroni intervals, it is not necessary that all the separate confidence coefficients $[100(1 - \alpha_i)\%, i = 1, \ldots, k]$ be equal, only that $\sum_{i=1}^{k} \alpha_i = \alpha$. Thus, if a few of the parameters warrant greater interest than the others, then the confidence coefficients for these parameters could be larger. Regardless of the allocation of the α_i's, the conservative Bonferroni intervals provide a viable alternative for achieving a family confidence coefficient of at least $(1 - \alpha)$.

To illustrate the preceding remarks, consider the normal error, multiple linear regression* model.

$$Y_i = \beta_0 + \beta_1 X_{i1}$$
$$+ \beta_2 X_{i2} + \cdots + \beta_{p-1} X_{i,p-1} + \epsilon_i,$$

$i = 1, \ldots, n$, where the ϵ_i's are mutually independent $N(0, \sigma^2)$.

A $100(1 - \alpha)\%$ confidence interval for β_i, $i = 1, \ldots, p - 1$, is

$$b_i \pm s_i t_{n-p, 1-\alpha/2},$$

where $s_i^2 = $ (residual mean square) $\times$ (($i + 1$, $i + 1$)th diagonal element of $(\mathbf{X}'\mathbf{X})^{-1}$), and $t_{n-p, 1-\alpha}$ is the corresponding t-percentile.

The corresponding set of Bonferroni intervals with family (joint) confidence coefficient at least $100(1 - \alpha)\%$ is given by

$$b_i \pm s_i t_{n-p, 1-\alpha/(2(p-1))}, \qquad i = 1, \ldots, p - 1. \tag{8}$$

This set of intervals may be compared with the ellipsoidal confidence region

$$(\mathbf{Ab} - \mathbf{A}\beta)'(\mathbf{A}(\mathbf{X}'\mathbf{X})^{-1}\mathbf{A}')^{-1}(\mathbf{Ab} - \mathbf{A}\beta)$$

$$< (p - 1)(\text{residual mean square})$$

$$\times F_{p-1, n-p, 1-\alpha}, \tag{9}$$

where $\mathbf{A}$ is a $((p - 1) \times p)$ matrix whose first column is the $\mathbf{0}$ vector and whose last $(p - 1)$ columns constitute the identity matrix of dimension $(p - 1)$, and $F_{p-1, n-p, 1-\alpha}$ is the corresponding F-percentile.

Comparison between the rectangular confidence region formed by intersection of the intervals (8) and the ellipsoidal confidence region (9) depends somewhat on the correlations among the b_i's (see Fig. 1).

Formulas will now be presented for constructing Bonferroni intervals for k linear combinations of the parameters, k mean responses, and the prediction of k new observations.

If the k linear functions of the β's are $\mathbf{C}'\beta$, where $\mathbf{C}'$ is a full (k) rank ($k \times p$) matrix, the Bonferroni intervals with family confidence coefficient at least $100(1 - \alpha)\%$ are

$$\mathbf{c}_i'\mathbf{b} \pm (\text{residual mean square})^{1/2}$$

$$\times [\mathbf{c}_i'(\mathbf{X}'\mathbf{X})^{-1}\mathbf{c}_i]^{1/2} t_{n-p, 1-\alpha/(2k)} \tag{10}$$

($i = 1, \ldots, k$), where $\mathbf{c}_i'$ is the ith row of $\mathbf{C}'$.

To estimate $E(Y_i)$ at k different $\mathbf{x}$ vectors, $\mathbf{x}_1, \mathbf{x}_2, \ldots, \mathbf{x}_k$, use the following Bonferroni intervals:

$$\mathbf{x}_i'\mathbf{b} \pm (\text{residual mean square})^{1/2}$$

$$\times \left[\mathbf{x}_i'(\mathbf{X}'\mathbf{X})^{-1}\mathbf{x}_i\right]^{1/2} t_{n-p, 1-\alpha/(2k)}, \tag{11}$$

$i = 1, 2, \ldots, k$. Bonferroni prediction intervals for the Y_i at k different values of $\mathbf{x}$ are given by

$$\mathbf{x}_i'\mathbf{b} \pm (\text{residual mean square})^{1/2}$$

$$\times \left[1 + \mathbf{x}_i'(\mathbf{X}'\mathbf{X})^{-1}\mathbf{x}_i\right]^{1/2} t_{n-p, 1-\alpha/(2k)}. \tag{12}$$

In all cases, the intervals are of standard form with the exception of the t-percentile.

Neter and Wasserman [21] utilize the foregoing formulas in a detailed, numerical example for two independent variables. Seber [25] empirically compares Bonferroni intervals with maximum modulus t-intervals and intervals generated by Scheffé's S-method (*see* SIMULTANEOUS COMPARISON PROCEDURE, SCHEFFÉ'S). In general, the Scheffé intervals are widest, and the maximum modulus intervals fare only slightly better than Bonferroni intervals.

Dunn and Clark [9] give detailed numerical examples showing how to construct Bonferroni intervals for model parameters and their contrasts when dealing with multifactor, randomized block* and Latin square* designs. Neter and Wasserman [21] also present several numerical examples for these cases. They point out that the superiority of one procedure over the other hinges on the number of comparisons to be investigated. Again, Miller's works [17,18] are excellent references, as is the paper by O'Neill and Wetherill [22].

Because of flexibility, Bonferroni's inequality has received wide usage in almost all aspects of multivariate statistics. For example, suppose that a random sample of size n is drawn from a p-variate normal population, where $\mu' = (\mu_1, \mu_2, \ldots, \mu_p)'$ is the $(p \times 1)$ vector of unknown means. If the $(p \times p)$ positive definite covariance matrix Σ is known completely, then $100(1 - \alpha)\%$ simultaneous confidence intervals* for k distinct

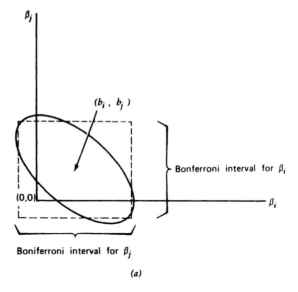

(a)

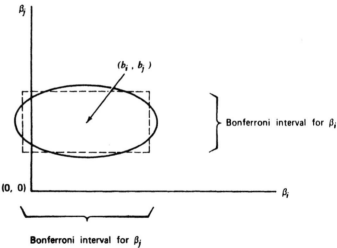

Figure 1. Rectangular Bonferroni confidence region and the elliptical region of the F-statistic in the presence of: (a) high correlation between b_i and b_j; (b) no correlation between b_i and b_j.

(b)

linear combinations $\mathbf{a}'_i\boldsymbol{\mu}$ of $\boldsymbol{\mu}$ are given by

$$\mathbf{a}'_i\bar{\mathbf{x}} \pm \left(\sqrt{\mathbf{a}'_i\boldsymbol{\Sigma}\mathbf{a}_i/n}\right)z_{1-\alpha/(2k)}, \quad i = 1, 2, \ldots, k,$$
(13)

where $\bar{\mathbf{x}}$ denotes the $(p \times 1)$ vector of sample means. If intervals are desired only for the p individual means (μ_i), as would be the case when the $H_0 : \boldsymbol{\mu} = \boldsymbol{\mu}_0$ is rejected, then these are obtained from (13) by successively letting $\mathbf{a}_i$ be the $(p \times 1)$ vector whose entries are 0's except for a 1 in the ith position. Thus $100(1 - \alpha)\%$ simultaneous confidence intervals for $\boldsymbol{\mu}$

are of the form

$$\bar{x}_i \pm (\sigma_i/\sqrt{n})z_{1-\alpha/(2p)} \quad (i = 1, 2, \ldots, p),$$
(14)

whatever be the covariance in $\boldsymbol{\Sigma}$. Alt and Spruill [1] have shown that, for $p \geqslant 2$ and confidence levels above 50%, the Bonferroni intervals in (14) are shorter than the corresponding Scheffé intervals, which are obtained by replacing $z_{1-\alpha/(2p)}$ by $\sqrt{\chi^2_{p,1-\alpha}}$.

In the more frequently occurring situation when the covariance matrix is completely unknown and the only information available is the elements $\mathbf{X}_1, \mathbf{X}_2, \ldots, \mathbf{X}_n$ of a random sample from the population, one can construct Bonferroni intervals for the k linear combinations of the form

$$\mathbf{a}_i' \overline{\mathbf{x}} \pm \left(\sqrt{\mathbf{a}_i^t \mathbf{S} \mathbf{a}_i / n} \right) t_{n-1, 1-\alpha/(2k)},$$

$$i = 1, 2, \ldots, k, \qquad (15)$$

where $\mathbf{S} = (n-1)^{-1} \sum_{i=1}^n (\mathbf{X}_i - \overline{\mathbf{X}})(\mathbf{X}_i - \overline{\mathbf{X}})'$.
Bonferroni intervals for the p means are

$$\overline{x}_i \pm (s_i/\sqrt{n}) t_{n-1, 1-\alpha/(2p)}$$

$$(i = 1, 2, \ldots, p), \qquad (16)$$

where s_i^2 is simply the sample variance for the ith variable. To obtain the Scheffé intervals (in this context sometimes called the Roy-Bose intervals) corresponding to (15) and (16), just replace the t-percentile by $\sqrt{p(n-1)F_{p,n-p,1-\alpha}/(n-p)}$. Determination of which procedure yields shorter intervals is easily accomplished by comparing the t-percentile with the $\sqrt{F}$ − percentile multiplied by $\sqrt{p(n-1)/(n-p)}$. In general, when intervals are obtained only for the p-means, the Bonferroni intervals are shorter. However, when k linear combinations are to be estimated and k is much larger than p, then the Scheffé intervals may be shorter [17,19].

Bonferroni's inequality has also been used when there is more than one p-variate normal population [19].

Occasionally, simultaneous confidence intervals are needed for the variances of a p-variate normal distribution. Bens and Jensen [4] and Jensen and Jones [14] show that using the usual univariate intervals, each with confidence coefficient $1 - (\alpha/p)$, yields satisfactory results.

Bonferroni's inequality plays an important role in the construction of simultaneous confidence intervals for multinomial* proportions. Let $\pi_1, \pi_2, \ldots, \pi_p$ be the multinomial cell probabilities and let $n_1 n_2, \ldots, n_p$ denote the corresponding observed cell frequencies for a sample of size N from this multinomial population. Then asymptotic $(N \to \infty)$ confidence intervals for the π_i are of the form

$$\hat{\pi}_i \pm \left(\sqrt{\hat{\pi}_i(1 - \hat{\pi}_i)/N} \right) z_{1-\alpha/(2p)},$$

$$i = 1, \ldots, p. \qquad (17)$$

Alt and Spruill [1] demonstrate the superiority of the Bonferroni intervals over the Scheffé intervals in this case. Bonferroni intervals for $\pi_i - \pi_j$ are given in Miller [17]. He also treats the case when there are r multinomial populations. Pairwise median tests controlling type I error by using Bonferroni inequalities were studied by Ryan [23]; see MULTIPLE COMPARISONS PROCEDURE, RYAN'S. See also Ryan [24] for historical remarks.

The preceding examples illustrate the wide use of Bonferroni's inequality, but the list is by no means exhaustive. For example, the conservative Bonferroni approach has also been used to obtain prediction* intervals [2,7]. Its use in reliability problems is demonstrated in Kwerel [15]. Most of the applications require percentage points of Student's t-distribution which are not commonly available. Bailey [3] provides tables for $\alpha = 0.05$ and 0.01 and for a range of values of v, the number of degrees of freedom. When the error allocation is unequal, the tables of Dayton and Schafer [8] may prove useful. The charts of Moses [20] should also prove helpful.

For multinormal distributions*, Bonferroni intervals have shown themselves to be strong competitors to intervals generated using other specialized techniques, but slightly shorter intervals are obtained by using Šidák's inequality [26] when it is applicable. If $\mathbf{Y}$ is distributed as a p-variate normal with zero means and arbitrary covariance matrix, then Šidák's inequality states that

$$\left. \begin{array}{l} P(|Y_1| \leqslant c_1, |Y_2| \leqslant c_2, \ldots, |Y_p| \leqslant c_p) \\ \quad \geqslant P(|Y_1| \leqslant c_1) \cdot P(|Y_2| \leqslant c_2) \cdots \\ \quad P(|Y_p| \leqslant c_p), \end{array} \right\} \quad (18)$$

for any positive numbers $c_1, c_2, \ldots, c_p$; see ŠIDÁK—SCOTT—KHATRI INEQUALITY. Simultaneous confidence intervals for the individual means of any p-variate normal distribution with known covariance matrix Σ are obtained by putting $Y_i = \sqrt{n}(\overline{X}_i - \mu_i)/\sigma_i$. The intervals obtained are those presented in (14) but with $z_{1-\alpha/(2p)}$ replaced by z_*, where z_* is such that $P(Z \leqslant z_*) = [1 + (1 - \alpha)^{1/p}]/2$. This form of Šidák's inequality also applies to the intervals for multinomial proportions presented in (17). Since $(1 - \alpha_1) \cdots (1 - \alpha_p) > 1 - \sum_{i=1}^p \alpha_i$, for $p > 1$, the Šidák intervals are slightly shorter than the Bonferroni intervals. For example, if $p = 2$ and $\alpha_1 = \alpha_2 = 0.05$, then $(1 - \alpha_1)(1 - \alpha_2) = 0.9025$ while $1 - \sum_{i=1}^{2} \alpha_i = 0.90$. Šidák also provided a more general inequality which can be used to construct confidence intervals for t-statistics. *See* ŠIDÁK—SCOTT—KHATRI INEQUALITY.

In summary, when the number of intervals (k) is small, the Bonferroni intervals are competitive with intervals obtained using other methods. The Bonferroni intervals use a confidence level of $(1 - \alpha_i)$ per interval, resulting in a family confidence level of at least $1 - \sum_{i=1}^{k} \alpha_i$. When an equal error allocation of α/k per interval is used, the family confidence level is at least $1 - k(\alpha/k) = 1 - \alpha$. When k is large, however, the Bonferroni intervals are unnecessarily long.

REFERENCES

1. Alt, F. and Spruill, C. (1977). *Commun. Statist. A*, **6**, 1503–1510.

2. Angers, C. and McLaughlin, G. (1979). *Technometrics*, **21**, 383–385.

3. Bailey, B. (1977). *J. Amer. Statist. Ass.*, **72**, 469–478.

4. Bens, G. and Jensen, D. (1967). *Percentage Points of the Bonferroni Chi-Square Statistics*. Tech. Rep. No 3, Dept. of Statistics, Virginia Polytechnic Institute, Blacksburg, Va.

5. Bonferroni, C. E. (1936). *Pubbl. Ist Sup. Sci. Econ. Commun. Firenze*, **8**, 1–62.

6. Bonferroni, C. E. (1936). Il Calcolo delle assicurazioni su gruppi di teste. In *Studii in onore del prof. S. O. Carboni*. Roma.

7. Chew, V. (1968). *Technometrics*, **10**, 323–331.

8. Dayton, C. and Schafer, W. (1973). *J. Amer. Statist. Ass.*, **68**, 78–83.

9. Dunn, O. and Clark, V. (1974). *Applied Statistics: Analysis of Variance and Regression*. Wiley, New York. (This introductory text provides a detailed treatment of the use of Bonferroni's inequality in analysis of variance.)

10. Feller, W. (1968). *An Introduction to Probability Theory and its Applications*, 3rd ed., Vol. 1. Wiley, New York. (Chapter IV of this classic text covers combinations of events and states the generalized inequalities of Bonferroni.)

11. Fréchet, M. (1940). *Les Probabilitiés associées à un système d'événements compatibles et dépendants*. (Actualités scientifiques et industrielles, no. 859.) Hermann & Cie, Paris. [This monograph (in French) gives a unified treatment of the inequalities of Bonferroni and others, such as Boole's and Gumbel's.]

12. Galambos, J. (1975). *J. Lond. Math. Soc.*, **9**(2), 561–564.

13. Galambos, J. (1977). *Ann. Prob.*, **5**, 577–581.

14. Jensen, D. and Jones, M. (1969). *J. Amer. Statist. Ass.*, **64**, 324–332.

15. Kwerel, S. (1975). *J. Amer. Statist. Ass.*, **70**, 472–479.

16. Meyer, R. M. (1969). *Ann. Math. Statist.*, **40**, 692–693.

17. Miller, R. (1966). *Simultaneous Statistical Inference*. McGraw-Hill, New York. (This frequently referenced book provides a lucid summary of the theory, methods, and applications of simultaneous inference through 1966. Although numerical examples are not given, specific formulas for using Bonferroni's inequality are presented for numerous scenarios.)

18. Miller, R. (1977). *J. Amer. Statist. Ass.*, **72**, 779–788. (Over 250 references are given.)

19. Morrison, D. (1976). *Multivariate Statistical Methods*, 2nd ed. McGraw-Hill, New York. (Although this book develops the multivariate linear hypothesis through the union-intersection principle of Roy and uses the corresponding simultaneous intervals, emphasis is also given to the Bonferroni intervals. Numerical examples are given.)

20. Moses, L. (1976). *Charts for Finding Upper Percentage Points of Student's t in the Range .01 to .00001*. Tech. Rep. No. 24 (5 R01 GM21215-02), Stanford University, Stanford, Calif.

21. Neter, J. and Wasserman, W. (1974). *Applied Linear Statistical Models*. Richard D. Irwin, Homewood, Ill. (This introductory text contains many numerical examples and provides a good treatment of the importance and use of Bonferroni's inequality.)

22. O'Neill, R. and Wetherill, G. (1971). *J. R. Statist. Soc. B*, **33**, 218–250.

23. Ryan, T. A. (1959). *Psychol. Bull.*, **56**, 26–47.

24. Ryan, T. A. (1980). *Amer. Statist.*, **34**, 122–123.

25. Seber, G. (1977). *Linear Regression Analysis*. Wiley, New York. (Chapter 5 of this theoretical treatment of regression compares the use of Bonferroni t-intervals, maximum modulus t-intervals, and Scheffé's S-method in constructing simultaneous confidence intervals. Numerical examples are not given.)

26. Šidák, Z. (1967). *J. Amer. Statist. Ass.*, **62**, 626–633.

See also CONFIDENCE BANDS,
WORKING–HOTELLING–SCHEFFÉ; MULTIPLE
COMPARISONS—I; MULTIPLE COMPARISONS
PROCEDURE, RYAN'S; MULTIVARIATE MULTIPLE
COMPARISONS; SIMULTANEOUS COMPARISON
PROCEDURE, SCHEFFÉ'S; and SIMULTANEOUS TESTING.

F. B. ALT

BONFERRONI *t*-STATISTIC

A commonly used tool for simultaneous statistical inference when the parent distribution is normal. The technique is based on the following specialization of the Bonferroni inequalities*.

Let $Y_1, \ldots, Y_k$ be normally distributed (not necessarily independent) random variables with means $\mu_1, \ldots, \mu_k$ and variances $\sigma_1^2, \ldots, \sigma_k^2$, respectively. Let $s_i^2, \ldots, s_k^2$ be independent sample estimators of $\sigma_1^2, \ldots, \sigma_k^2$, respectively, with $\nu_i s_i^2/\sigma_i^2$ distributed as χ^2 variables* with $\nu_i = n_i - 1$ degrees of freedom ($i = 1, \ldots, k$). It is assumed that Y_i is independent of $s_i^2, i = 1, \ldots, k$, so that $T_i = (Y_i - \mu_i)/s_i, i = 1, \ldots, k$, are t variables* with $\nu_i = n_i - 1$ degrees of freedom. (s_i^2 may depend on $Y_{i'}, i' \neq i$.)

If $t_{r_i, \alpha/(2k)}, i = 1, \ldots, k$, are the *upper* $\alpha/(2k)$ percentile points (or equivalently $\pm t_{r_i, \alpha/(2k)}$ are two-tailed α/k percentile points) of the t-distribution with r_i degrees of freedom, $i = 1, \ldots, k$, then

$$\Pr\left(\bigcap_{i=1}^{k}\left[\left|\frac{Y_i - \mu_i}{s_i}\right| \leqslant t_{r_i, \alpha/(2k)}\right]\right) \geqslant 1 - \left(k \times \frac{\alpha}{k}\right)$$
$$= 1 - \alpha$$

since $\Pr\{|Y_i - \mu_i|/s_i| \leqslant t_{r_i, \alpha/(2k)}$ is by definition equal to $1 - \alpha/k$. In other words, with probability greater than or equal to $1 - \alpha, \mu_i \in Y_i \pm t_{r_i, \alpha/2k} s_i$ simultaneously for all $i = 1, \ldots, k$. This is valid if the significance levels* for the component intervals are not all the same (α/k), provided only that $\alpha_1 + \alpha_2 + \cdots + \alpha_k = \alpha$.

Tables of $t_{\nu, \alpha/(2k)}$ for $\alpha = 0.05, k = 1(1)10, 15, 20, 50$ and $\nu = 5, 10, 15, 20, 24, 30, 40, 60, 120, +\infty$ are given by Dunn [1]. More

extensive tables are given in Dunn [2], Dunn and Massey [3], and Miller [4].

REFERENCES

1. Dunn, O. J. (1959). *J. Amer. Statist. Ass.*, **54**, 613–621.
2. Dunn, O. J. (1961). *J. Amer. Statist. Ass.*, **56**, 52–64.
3. Dunn, O. J. and Massey, F. J. (1965). *J. Amer. Statist. Ass.*, **60**, 573–583.
4. Miller, P. G. (1966). *Simultaneous Statistical Inference*. McGraw-Hill, New York, pp. 67–70.

See also BONFERRONI INEQUALITIES AND INTERVALS;
CONFIDENCE INTERVALS AND REGIONS; and
STUDENTIZED MAXIMAL DISTRIBUTIONS, CENTERED.

BOOLE'S INEQUALITY

If $A_1, A_2, \ldots, A_n$ belong to an algebra of events, then

$$P(A_1 \cup A_2 \cup \cdots \cup A_n)$$
$$\leqslant P(A_1) + P(A_2) + \cdots + P(A_n).$$

Boole's inequality is widely used in combinatorial applied probability.

See also BONFERRONI INEQUALITIES AND INTERVALS.

BOOTSTRAP

INTRODUCTION

In its present form, the bootstrap was introduced by Efron [18] as a method of computing the sampling distribution of a statistic and its various functionals, such as the bias and the standard error, without going through the large sample theory calculation. This computational procedure usually involves resampling the data values and can be carried out in an almost mechanical way. Among other things, it provides a simpler and often better alternative (see the next section) to the large sample theory-based inference. Consider a statistic $T = T(\underline{X})$, based on an independent and identically distributed sample $\underline{X} = (X_1, \cdots, X_n)$, from a common distribution

F, that estimates a parameter $\theta = \theta(F)$ so that inference could be made on the basis of the sampling distribution of $R = \sqrt{n}(T - \theta)$ or its studentized form $\sqrt{n}(T - \theta)/v$, where $v = v(\underline{X})$ is an estimate of the asymptotic standard deviation of $\sqrt{n}T$. The bootstrap estimates the sampling distribution of R by that of the (conditional) distribution of R^*, given the original data $\underline{X}$, where $R^* = \sqrt{n}(T^* - \widehat{\theta})$, for the normalized statistic, and $= \sqrt{n}(T^* - \widehat{\theta})/v^*$, for the studentized statistic. Here, $\widehat{\theta} = T(\underline{X})$, $T^* = T(\underline{X}^*)$, and $v^* = v(\underline{X}^*)$, where $\underline{X}^* = (X_1^*, \cdots, X_n^*)$ is a random sample from $\widehat{F} = F_n$, the empirical distribution of the original sample $\underline{X}$. In most cases, a Monte Carlo approximation to this estimated distribution is carried out via resampling (which has become synonymous with bootstrap) as follows:

1. Pick a large positive integer B. Given the original data $\underline{X}$, generate B batches of independent random samples, each of size n, denoted $\underline{X}_1^*, \cdots, \underline{X}_B^*$, via simple random sampling with replacement (SRSWR) from the finite population of the original sample values $\{X_1, \cdots, X_n\}$. (Note that replicated values need to be counted multiple times in this representation.)
2. Calculate $R_i^* = \sqrt{n}(T(\underline{X}_i^*) - \widehat{\theta})$ or $\sqrt{n}(T(\underline{X}_i^*) - \widehat{\theta})/v(\underline{X}_i^*)$, for each $i = 1, \cdots, B$.
3. The histogram of the bootstrapped R values $R_1^*, \cdots, R_B^*$ (or a smoothed version of it, if so desired) gives a bootstrap approximation to the sampling distribution of R.

VALIDITY AND ACCURACY

Ignoring the Monte Carlo error in bootstrap resampling, we say that a bootstrap approximation is asymptotically valid if the difference between the sampling distribution and its bootstrap version (usually measured by the supremum norm or the Kolmogorov distance) goes to zero (in probability or almost surely) as the sample size grows to infinity. It can be regarded as a form of consistency property of the bootstrap procedure. The first set of asymptotic validity results for the bootstrap were obtained by Bickel and Freedman [7] and Singh [39] for simple statistics such

as the sample mean and quantiles. Asymptotic validity of the bootstrap for the sample mean is described in the following result. Let P^* denote the conditional distribution of the bootstrap sample $\underline{X}^*$ given the original data $\underline{X}$.

Suppose $0 < Var(X_1) < \infty$, $\mu = E(X_1)$ and $s^2 = \sum_i (X_i - \overline{X})^2/(n - 1)$. Then as $n \to \infty$, almost surely,

$$\sup_x |P\{R \leqslant x\} - P^*\{R^* \leqslant x\}| \to 0, \quad (1)$$

where $R = \sqrt{n}(\overline{X} - \mu)$ or $\sqrt{n}(\overline{X} - \mu)/s$.

For studentized statistics, it turned out that the bootstrap is not only asymptotically valid but the resulting approximation is better than the large sample normal approximation. Pioneering works in this area include References 39 and 4. Reference 25 obtains the following asymptotic accuracy of the bootstrapped student's t-statistic under minimal moment conditions.

Let $E|X_1|^{3+\epsilon} < \infty$, for some $\epsilon > 0$, $Var(X_1) > 0$, and the distribution of X_1 is not concentrated on a lattice. Then as $n \to \infty$, almost surely,

$$\sup_x |P\{R \leqslant x\} - P^*\{R^* \leqslant x\}| = o(n^{-1/2}),$$

where $R = \sqrt{n}(\overline{X} - \mu)/s$.

Note that, in contrast, Edgeworth expansion shows that the error in normal approximation is only $O(n^{-1/2})$. Therefore, bootstrap is said to be second-order correct in such cases.

BOOTSTRAP STANDARD ERROR

One popular use of the bootstrap is in the calculation of the standard error of a statistic T. A Monte Carlo evaluation of this is made by the empirical standard deviation of $T(\underline{X}_i^*)$ calculated over a large number B of independent bootstrap replicates leading to

$$SE_{\text{Boot}}(T) = \sqrt{(B - 1)^{-1} \sum_i \left(T(\underline{X}_i^*) - \overline{T}^*\right)^2},$$

where $\overline{T}^* = B^{-1} \sum_i T(\underline{X}_i^*)$. As $B \to \infty$ and $n \to \infty$, $\sqrt{n}SE_{\text{Boot}}(T)$ consistently estimates the asymptotic standard deviation of $\sqrt{n}T$ for a variety of statistics, including the sample median [24] for which the jackknife is known to produce an incorrect standard error.

BOOTSTRAP CONFIDENCE INTERVALS

A primary use of the bootstrap is to construct a confidence interval for the parameter of interest θ. By inverting the studentized statistic $R = \sqrt{n}(\overline{X} - \mu)/s$, one gets the one-sided confidence interval for θ as $(T - n^{-1/2}\hat{r}_{1-\alpha}s, \infty)$, where $\hat{r}_{1-\alpha}$ is the $(1 - \alpha)$th quantile of the bootstrap distribution of R^*. This method is referred to as the *percentile t-method* and typically produces intervals more accurate than obtained from the normal approximation. Various two-sided versions also exist, notably, the equitailed t-interval and the short confidence interval [27]. In contrast, the simpler percentile method [19,20] bootstraps the raw statistic $T = \hat{\theta}$, without normalization or studentization, and reports the interval $(\hat{\theta}_\alpha, \infty)$, where $\hat{\theta}_\alpha$ is the αth quantile of the bootstrap distribution of $\hat{\theta}^*$. Two-sided versions are constructed similarly. While these intervals are invariant to monotone transformation, they are typically less accurate than the percentile t-intervals. A bootstrap bias-corrected (BC) percentile interval is given in Reference 20 that improves upon the accuracy. Letting $\hat{H}$ and $\hat{H}^{-1}$ denote the bootstrap distribution of $\hat{\theta}^*$ and its quantiles, respectively, a one-sided BC(α) confidence interval for θ is given by $\left(H^{-1}\circ\Phi(z_\alpha + 2\Phi^{-1}\circ\hat{H}(\hat{\theta})), \infty\right)$, where Φ and Φ^{-1} denote the standard normal cumulative distribution function and the quantile function, respectively, and $z_\alpha = \Phi^{-1}(\alpha)$. A correction for the skewness term is provided by the bootstrap accelerated bias-corrected (ABC) percentile interval [21] but its construction is much more complicated and less automatic. Theoretical comparisons of various bootstrap confidence intervals are carried out in Reference 26.

BOOTSTRAP FOR NON I.I.D. DATA

The notion of bootstrap resampling to estimate the sampling distribution of a statistic extends beyond the realm of i.i.d. data, although more caution must be exercised to ensure a valid procedure. The resampling procedures generally fall under the following categories.

Residual-Based Bootstrap

Originally proposed for the regression model by Freedman [22], this concept can be generalized to cover time series models that admit model residuals. For simplicity of illustration, we consider a first-order AR(1) model $X_t = \phi X_{t-1} + \epsilon_t, t \geq 2$, where ϕ is a real parameter and ϵ are i.i.d. errors (innovations) with $E(\epsilon) = 0$, $E(\epsilon^2) < \infty$. A resampling scheme can be formulated as follows. Having observed the data $\underline{X}$, compute the model residuals as $e_t = X_t - \hat{\phi}X_{t-1}, \hat{\phi}$ being a consistent estimator of ϕ (e.g., the least squares estimator), $t = 2, \cdots, n$. The next step is to construct the bootstrapped errors $\epsilon_2^*, \cdots, \epsilon_n^*$ by SRSWR from the finite population of the centered residuals $\{e_2^c, \cdots, e_n^c\}, e_i^c = e_i - (n - 1)^{-1}\sum e_i$. Finally, a bootstrap sample from this model is constructed as $X_1^* = X_1, X_t^* = \hat{\phi}X_{t-1}^* + \epsilon_t^*, 2 \leq t \leq n$. Asymptotic validity and second-order accuracy of this scheme for estimating the sampling distribution of the least squares estimate of the autoregressive parameter of a stationary AR(1) model was established by Bose [9]. Generalizations to nonstationary and higher-order ARMA models were considered by Bose [10], Basawa et al. [6], Datta [14], and others.

An extension of this residual based resampling scheme is called the *sieve bootstrap* [11], where the order p of the underlying autoregressive process used in construction of the bootstrap values can increase with the sample size. The method then applies to more general stationary processes that can be approximated by autoregressive processes.

Bootstrapping Markov Chains

If the data are known to have come from a Markov chain, then it is appealing to consider a model-based resampling scheme that also generates the bootstrap sample from an estimated Markov chain, given the original data. In case of a discrete sample space, the transition probabilities can be estimated by the method of maximum likelihood. In the case of a continuous state space, smoothing techniques could be used to estimate the transition density nonparametrically or maximum likelihood can be used if a parametric model is imposed. In any case, a bootstrap sample can be generated recursively as follows: let $X_1^* = X_1$ and, having observed $X_1^*, \cdots, X_{t-1}^*$, generate X_t^* from the estimated transition kernel $\widehat{p}(X_{t-1}^*, \cdot)$. Asymptotic validity and accuracy of this bootstrap scheme has been demonstrated by Datta and McCormick [17].

Block Bootstrap

The moving block bootstrap (MBB) is the most well known of the block bootstrap schemes. Introduced by Künsch [29] and Liu and Singh [34] as a model-free method, this scheme attempts to capture the general dependence structure of the underlying stationary data.

Suppose one observes data $X_1, \cdots, X_n$ from a stationary process $\{X_t : t \geqslant 1\}$. Consider the overlapping (moving) blocks $B_i = (X_i, \cdots, X_{i+l-1})$ of size l, $1 \leqslant i \leqslant N$, with $N = n - l + 1$. Instead of resampling the individual observations X_i, the idea here is to resample the blocks B_i via SRSWR from the collection of blocks $\{B_1, \cdots, B_N\}$. To make the size of the bootstrap sample the same (or nearly same) as that of the original data, one needs to resample $k = [n/l]$ blocks and concatenate them to obtain a bootstrap sample of size $n' = lk$. The block length l needs to grow with the sample size n in such a way that $l(n) \to \infty$ but $l(n)/n \to 0$. Theoretical and empirical studies suggest that the order of l should be smaller than that of k. Optimal selection of the block length is a difficult problem. See References 28 and 33 for some answers in this direction.

Under appropriate centering and studentization, asymptotic accuracy of this procedure has been established under general weak dependence for the smooth functions of means framework by Lahiri [30,31]. Other block bootstrap schemes include the circular bootstrap and the stationary bootstrap of Politis and Romano [35,36].

BOOTSTRAP INVALIDITY AND REDUCED RESAMPLE SIZE BOOTSTRAP

Asymptotic validity of the bootstrap cannot be taken for granted even for simple statistics

such as the sample mean. Finiteness of the second moment is a necessary condition for the convergence in Equation 1 and, for heavier tailed distributions, the randomness in the bootstrap distribution does not diminish as the sample size grows, leading to invalidity. This was first demonstrated by Athreya [2]. Other instances of bootstrap invalidity include autoregressive processes with unit roots [5,13], U-statistics* at critical values [3], and explosive branching processes [15]. One way of fixing the invalidity of the standard bootstrap is to use a smaller resample size m than the original-sample size n (see, e.g., Refs. 15 and 16) in constructing the bootstrap statistic, such that $m = m(n) \to \infty$ but $m(n)/n \to 0$. More precise control of the growth of $m(n)$ may be necessary depending on the situation, especially, for almost sure validity. See Reference 1 for the results in the sample mean case. This strategy is also known as "m out of n bootstrap" [8].

SUBSAMPLING

Subsampling [12] is another nonparametric technique for estimating the sampling distribution of a statistic. Given the original data $X_1, \cdots, X_n$ and a subsample size $m = m(n)$ (that is typically of smaller order than the original sample size n), the statistic is computed on every possible subset of size m (called a subsample) of $\{X_1, \cdots, X_n\}$. These $\binom{n}{m}$ values of T (or its studentized form) form an approximation to the sampling distribution of T. This procedure is closely related to the 'm out of n bootstrap' and enjoys similar theoretical consistency properties [37]. A detailed account of it can be found in the monograph by Politis et al. [38].

SPATIAL BOOTSTRAP

There appears to be a lot of interest in bootstrapping spatial data. In real applications, this could be a difficult proposition unless the spatial pattern is regular (e.g., on an integer grid), in which case a direct generalization of the moving block bootstrap for time series data is possible. In the case of irregularly spaced data, a resampling plan can be devised using stochastic designs. See Reference 32 for details of these methods. Garcia-Soidan and Hall [23] proposed a different sample reuse* method for spatial data, which they called the "sampling window" method. It can be regarded as a form of spatial subsampling. The sampling window is small relative to the total size of the data region. It is moved to all possible locations within the data set and the spatial patterns are examined repeatedly.

REFERENCES

1. Arcones, M. A. and Giné, E. (1989). The bootstrap of the mean with arbitrary bootstrap sample size. *Ann. Inst. Henri Poincaré*, **25**, 457–481.

2. Athreya, K. B. (1987). Bootstrap of the mean in the infinite variance case. *Ann. Stat.*, **15**, 724–731.

3. Babu, G. J. (1984). Bootstrapping statistics with linear combinations of chi-squares as weak limit. *Sankhya, Ser. A*, **46**, 85–93.

4. Babu, G. J. and Singh, K. (1984). On one term Edgeworth correction by Efron's bootstrap. *Sankhya, Ser. A*, **46**, 219–232.

5. Basawa, I. V., Mallik, A. K., McCormick, W. P., Reeves, J. H., and Taylor, R. L. (1991). Bootstrapping unstable first-order autoregressive processes. *Ann. Stat.*, **19**, 1098–1101.

6. Basawa, I. V., Mallik, A. K., McCormick, W. P., and Taylor, R. L. (1989). Bootstrapping explosive autoregressive processes. *Ann. Stat.*, **17**, 1479–1486.

7. Bickel, P. and Freedman, D. (1981). Some asymptotic theory for the bootstrap. *Ann. Stat.*, **9**, 1196–1217.

8. Bickel, P. J., Götze, F., and van Zwet, W. R. (1997). Resampling fewer than n observations: gains, losses, and remedies for losses. *Stat. Sin.*, **7**, 1–31.

9. Bose, A. (1988). Edgeworth correction by bootstrap in autoregression. *Ann. Stat.*, **16**, 827–842.

10. Bose, A. (1990). Bootstrap in moving average models. *Ann. Inst. Stat. Math.*, **42**, 753–768.

11. Bühlmann, P. (1997). Sieve bootstrap for time series. *Bernoulli*, **3**, 123–148.

12. Carlstein, E. (1986). The use of subseries values for estimating the variance of a general statistic from a stationary sequence. *Ann. Stat.*, **14**, 1171–1179.

13. Datta, S. (1996). On asymptotic properties of bootstrap for AR(1) processes. *J. Stat. Plann. Inference*, **53**, 361–374.

14. Datta, S. (1995). Limit theory and bootstrap for explosive and partially explosive autoregression. *Stochastic Proc. Appl.*, **57**, 285–304.

15. Datta, S. (1998). "Making the Bootstrap Work". In *Frontiers in Probability and Statistics*, S. P. Mukherjee, S. K. Basu, and B. K. Sinha, eds. Narosa Publishing, Narosa, New Delhi, pp. 119–129.

16. Datta, S. and McCormick, W. P. (1995a). Bootstrap inference for a first order autoregression with positive innovations. *J. Am. Stat. Assoc.*, **90**, 1289–1300.

17. Datta, S. and McCormick, W. P. (1995b). Some continuous Edgeworth expansions for Markov chains with applications to bootstrap. *J. Multivariate Anal.*, **52**, 83–106.

18. Efron, B. (1979). Bootstrap methods: another look at the jackknife. *Ann. Stat.*, **7**, 1–26.

19. Efron, B. (1981). Nonparametric estimates of standard errors and confidence intervals. *Can. J. Stat.*, **9**, 139–172.

20. Efron, B. (1982). *The Jackknife, The Bootstrap, and Other Resampling Plans*. CBMS 38, SIAM-NSF.

21. Efron, B. (1987). Better bootstrap confidence intervals (with discussion). *J. Am. Stat. Assoc.*, **82**, 171–200.

22. Freedman, D. A. (1981). Bootstrapping regression models. *Ann. Stat.*, **9**, 1218–1228.

23. Garcia-Soidan, P. H. and Hall, P. (1997). On sample reuse methods for spatial data. *Biometrics*, **53**, 273–281.

24. Ghosh, M., Parr, W. C., Singh, K., and Babu, G. J. (1984). A note on bootstrapping the sample median. *Ann. Stat.*, **12**, 1130–1135.

25. Hall, P. (1987). Edgeworth expansion for Student's statistic under minimal moment conditions. *Ann. Probab.*, **15**, 920–931.

26. Hall, P. (1988). Rate of convergence in bootstrap approximations. *Ann. Stat.*, **16**, 927–953.

27. Hall, P. (1992). *The Bootstrap and Edgeworth Expansion*. Springer, New York.

28. Hall, P., Horowitz, J. L., and Jing, B.-Y. (1995). On blocking rules for the bootstrap with dependent data. *Biometrika*, **82**, 561–574.

29. Künsch, H. R. (1989). The jackknife and the bootstrap for general stationary observations. *Ann. Stat.*, **17**, 1217–1241.

30. Lahiri, S. N. (1991). Second order optimality of stationary bootstrap. *Stat. Probab. Lett.*, **11**, 335–341.

31. Lahiri, S. N. (1992). "Edgeworth Correction by 'moving block bootstrap' for Stationary and Nonstationary Data". In *Exploring the Limits of Bootstrap*, R. LePage and L. Billard, eds. Wiley, New York, pp. 183–214.

32. Lahiri, S. N. (2003). *Resampling Methods for Dependent Data.* Springer, New York.

33. Lahiri, S. N., Furukawa, K., and Lee, Y-D. (2004). A nonparametric plug-in method for selecting the optimal block length. *J. Am. Stat. Assoc.*, **99**; to appear.

34. Liu, R. Y. and Singh, K. (1992). "Moving Blocks Jackknife and Bootstrap Capture Weak Dependence". In *Exploring the Limits of Bootstrap*, R. LePage and L. Billard, eds. Wiley, New York, pp. 225–248.

35. Politis, D. N. and Romano, J. P. (1992). "A Circular Block-Resampling Procedure for Stationary Data". In *Exploring the Limits of Bootstrap*, R. LePage, L., eds. Wiley, New York, pp. 263–270.

36. Politis, D. N. and Romano, J. P. (1994a). The stationary bootstrap. *J. Am. Stat. Assoc.*, **89**, 1303–1313.

37. Politis, D. N. and Romano, J. P. (1994b). Large sample confidence regions based on subsamples under minimal assumptions. *Ann. Stat.*, **22**, 2031–2050.

38. Politis, D. N., Romano, J. P., and Wolf, M. (1999). *Subsampling.* Springer-Verlag, New York.

39. Singh, K. (1981). On asymptotic accuracy of Efron's bootstrap. *Ann. Stat.*, **9**, 1187–1195.

SOMNATH DATTA

BOREL-CANTELLI LEMMA

Let $A_1, A_2, \ldots$ be an infinite sequence of events in a sigma-field of events, and let A* be the event: "infinitely many of the events A_n occur", so that

$$A* = \lim_{n \to \infty} \sup A_n = \cap^{\infty}_{n=1} \cup^{\infty}_{j=m} A_j.$$

The Borel-Cantelli Lemma [1, 2, 3], sometimes called the Borel-Cantelli zero-one law, states that

a) If $\sum_{j=1}^{\infty} P(A_j) < \infty$, then $P(A^*) = 0$.

b) If $\sum_{j=1}^{\infty} P(A_j) = \infty$ and $\{A_n\}$ is a sequence of *independent* events, then $P(A^*) = 1$.

The proof involves an application of a theorem of Toeplitz (*see* TOEPLITZ LEMMA).

The Borel-Cantelli lemma is a useful tool for establishing strong convergence of certain sequences of random variables (the so-called strong law of large numbers*). Extensions to certain types of dependent events are presented by Loève [4]. The lemma is also applied in the derivation of asymptotic distributions of various test statistics. Stoyanov [5] illustrates how the independence condition in b) is essential.

REFERENCES

1. Borel, E. (1909). *Rend. Circ. Math. Palermo*, **27**, 247–271.

2. Cantelli, F. P. (1917). *Atti. Accad. Naz. Lincei*, **26**, 39–45.

3. Feller, W. (1957). *An Introduction to Probability Theory and its Applications* (2nd ed.). Wiley, New York. (Sec. VIII.3).

4. Loève, M. (1977). *Probability Theory* (4th ed.). Springer, New York.

5. Stoyanov, J. M. (1997). *Counterexamples in Probability* (2nd ed.). Wiley, New York.

FURTHER READING

Durbins, E. and Freedman, D. A. (1965). *Ann. Math. Statist.*, **36**, 800–807.

Freedman, D. (1973). *Ann. Prob.*, **1**, 910–925.

See also CONVERGENCE OF SEQUENCES OF RANDOM VARIABLES

BOREL–KOLMOGOROV PARADOX

This paradox is based on a property of conditional densities of a random variable given a condition having probability zero. These densities can be modified arbitrarily on any set of conditions, provided the total probability of conditions is zero.

A classical example is the conditional density of X, given $X = Y$. If we represent the condition as $Y - X = 0$, we shall get a different result when the same condition is given as $\frac{Y}{X} = 1$. A popular elementary example is when the joint density of X and Y is

$$f(x,y) = 4xy \quad 0 \leqslant x \leqslant 1, \quad 0 \leqslant 1 \leqslant 1.$$

In this case, easy calculations show that the density

$$g_X(x|X - Y = 0) = 3x^2, \quad 0 \leqslant x \leqslant 1,$$

while

$$h_X\left(x\left|\frac{X}{Y} = 1\right.\right) = 4x^3, \quad 0 \leqslant x \leqslant 1.$$

An explanation is that for any fixed ϵ (however small), the conditioning event $|X - Y| < \epsilon$ is different from $|Y/X - 1| < \epsilon$. In the first case, the admissible values take the form shown in Fig. 1, uniformly distributed around the diagonal.

In the second case, the admissible values are as shown in Fig. 2, where the ratio Y/X favors large values of x.

Proschan and Presnell [6] refer to this paradox as "the equivalent event fallacy." Singpurwalla and Swift [8] utilize it in problems related to the assessment of the reliability of a network.

Another example of the paradox, discussed in the literature [4,5], is as follows: Let X_1 and X_2 be independent, standard normal random variables with the restriction that $x_2 = 0$ is impossible (which does not change the joint distribution function). Owing to

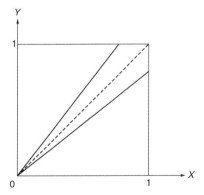

Figure 2. Admissible values, second case

independence, the conditional distribution of X_2 given the event $A = \{X_1 = 0\}$ is standard normal. However, the conditional distribution of X_2 given $B = \{X_1/X_2 = 0\}$ is that of $(-1)^Y \chi_2^2$, where Y takes values 1 and 2 with probability $\frac{1}{2}$ and is independent of χ_2^2. However, the events A and B are identical.

The classical resolution of the paradox is that in the second case we have conditional probability, given the σ-field of events determined by the random variable X_1/X_2, and not given individual events in the σ-field under consideration is criticized by Kadane et al. [5], who recommend application of the Dubins theory of conditional probability spaces [3], which does not involve σ-fields.

REFERENCES

1. Bartoszyiski, R. and Niewiadomska-Buga, J. (1996). *Probability and Statistical Inference*. Wiley, New York.

2. Borel, E. (1925). *Traités Des Probabilities*. Gauthier-Villars, Paris.

3. Dubins, L. (1975). *Ann. Probab.*, **3**, 8999.

4. Hill, B. (1980). "On Some Statistical Paradoxes and Non-Conglomerability". In *Bayesian Statistics*, J. M. Bernardo, M. DeGroot, D. Lindley, and A. Smith, eds. University Press, Valencia, Spain, pp. 39–49.

5. Kadane, J. B., Schervish, M. J. and Seidenfeld, T. I. (1986). "Statistical Implications of Finitely Additive Probability". In *Bayesian Inference and Decision Techniques*, P. K. Goel and A. Zellner, eds. Elsevier, Amsterdam, pp. 59–76.

6. Kolmogorov, A. (1956). *Foundations of Probability Theory*. Chelsea, New York.

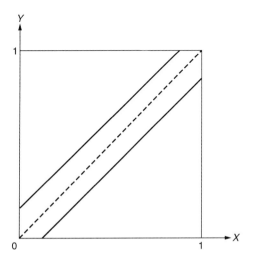

Figure 1. Admissible values, first case

7. Proschan, M. A. and Presnell, B. (1998). *Am. Stat.*, **52**, 248–252.

8. Singpurwalla, N. D. and Swift, A. (2001). *Am. Stat.*, **55**, 213–218.

BOREL–TANNER DISTRIBUTION

A random variable X possesses a Borel–Tanner distribution with parameters r, α, where r is a positive integer and $0 < \alpha < 1$ if

$$\Pr(X = k) = \frac{r}{(k - r)!} k^{k-r-1} e^{-\alpha k} \alpha^{k-r}$$

$$(k = r, r + 1, \ldots).$$

The expected value of this variable is $\mu_1' = E(X) = r/(1 - \alpha)$ and variance $V(X) = \alpha r/(1 - \alpha)^3$.

In queueing theory* the Borel-Tanner distribution arises as the distribution of the number of customers served in a queueing system with Poisson* input with parameter α and a constant service time, given that the length of the queue at the initial time is r.

BIBLIOGRAPHY

Johnson, N. L., Kotz, S. and Kemp, A. W. (1992). *Univariate Discrete Distributions*. (2nd ed.). Wiley, New York, 394–396. (This volume discusses additional properties and contains references to original papers as well as information about tables of this distribution.)

FURTHER READING

Johnson, N. L., Kotz, S., and Kemp, A. W. (1992). *Discrete Distributions* (2nd ed.), Wiley, New York. (Secs 9.11, 11.13).

See also LAGRANGE AND RELATED PROBABILITY DISTRIBUTIONS and POISSON DISTRIBUTION.

BOREL ZERO-ONE LAW.   See ZERO-ONE LAWS

BORTKIEWICZ, LADISLAUS VON

German statistician and economist. Born in St. Petersburg, Russia, August 7, 1868, Bortkiewicz graduated from the University of St. Petersburg Faculty of Law in 1890, and subsequently studied under Lexis* in Göttingen, where he defended his doctoral thesis in 1893. Privatdozent in Strasbourg from 1895 to 1897, Bortkiewicz then returned to Russia until 1901, when he was appointed to a professorial position at the University of Berlin, where he taught statistics and economics until his death on July 15, 1931.

Best known for his modeling of rare-event phenomena by the Poisson distribution*, Bortkiewicz also made numerous other contributions to mathematical statistics, notably the statistical analysis of radioactivity, the theory of runs*, and the distributional properties of extreme values*. His work on population theory, actuarial science, and political economy was also noteworthy; for the latter, see Gumbel [2].

Bortkiewicz's monograph *Das Gesetz der kleinen Zahlen* (The Law of Small Numbers) [1] is unquestionably his best known work and was in large part responsible for the subsequent popularity of the Poisson distribution. Contrary to popular belief, Bortkiewicz meant by the expression "law of small numbers" not the Poisson distribution itself, but the tendency of data in binomial* sampling with small and variable success probability to appear as though the success probability were constant when the sample size was large. The Lexis ratio* was advocated as a means of detecting the presence of variable success probabilities.

A meticulous scholar, Bortkiewicz wrote in a difficult style that may have lessened his influence. In England, one of the few to appreciate his work was John Maynard Keynes*, who devoted a chapter of *A Treatise on Probability* [3] to describing the results of Lexis and Bortkiewicz.

Literature

Thor Andersson's lengthy obituary in the *Nordic Statistical Journal*, 1931 (**3**:9-26), includes an essentially complete bibliography. The biography by E. J. Gumbel in the *International Encyclopedia of Statistics* (1978, Free Press, New York) contains a useful list of secondary literature; see also the entry by O. B. Sheynin in the *Dictionary of Scientific Biography* (1970, Scribner's, New York). Bortkiewicz's contributions to dispersion theory* and the Lexis ratio are discussed within a historical perspective by C. C. Heyde and

E. Seneta in *I. J. Bienaymé: Statistical Theory Anticipated* (1977, Springer-Verlag, New York), pp. 49–58. Other historical comments, including a description of Bortkiewicz's applications of the Poisson distribution and the controversies with Gini and Whittaker, may be found in Frank Haight's *Handbook of the Poisson Distribution* (1967, Wiley, New York), Chap. 9.

Bortkiewicz was a member of the International Statistical Institute and participated at its meetings; William Kruskal and Frederick Mosteller describe his criticism of A. N. Kiaer's concept of "representative sample"* at one such meeting in their "Representative Sampling IV: The History of the Concept in Statistics," *International Statistical Review*, 1980 (**48**: August).

REFERENCES

1. Bortkiewicz, L. von (1898). *Das Gesetz der kleinen Zahlen* (The Law of Small Numbers). Teubner, Leipzig.

2. Gumbel, E. J. (1978). In *International Encyclopedia of Statistics*. Free Press, New York.

3. Keynes, J. M. (1921). *A Treatise on Probability*. Macmillan, London.

See also LAW OF SMALL NUMBERS; POISSON DISTRIBUTION; and POISSON LIMIT.

SANDY L. ZABELL

BOSCOVICH, RUGGIERO GIUSEPPE

Born: May 18, 1711, in Ragusa (now Dubrovnik), Dalmatia (now in Croatia).

Died: February 13, 1787, in Milan, Italy.

Contributed to: astronomy, geodesy, and physics.

Boscovich was a son of a Serb who settled in Ragusa (we use above the Italian version of his name). His early education (in his native city) was at a Jesuit school, and subsequently he attended the Collegium Romanum. Ordained a priest within the Society of Jesus in Rome in 1744, he propounded a very early version of the atomic theory of matter and was recognized by the foremost scientific bodies in Europe; he was made a Fellow of the Royal Society of London in 1760.

After the suppression of the Jesuits in 1773, Boscovich, then in his sixties, went to Paris at the invitation of the King of France and spent 9 years as Director of Optics for the Marine before returning to Italy.

In 1750, Pope Benedict XIV commissioned Boscovich and his fellow Jesuit, C. Le Maire, to carry out several meridian measurements in Italy; part of the report of this journey, reprinted as ref. 1, contains his presently known contribution to statistical methodology. (Several other considerations on the effects of chance* are reported in ref. 3.) Boscovich considered the simple version $Z_i = \alpha + \beta w_i + \epsilon_i i = 1, \ldots, n$ of the linear model*, where the data set is (Z_i, w_i), $i = 1, \ldots, n$, the ϵ_i are residuals* due to errors in measurement, and α and β are to be determined from the conditions

(a) $$\sum_{i=1}^{n} |\epsilon_i| = \min.$$

(b) $$\sum_{i=1}^{n} \epsilon_i = 0.$$

The second of the conditions became familiar in the eighteenth and nineteenth centuries. Noting, as was done by Boscovich, that it implies that $\overline{Z} = \alpha + \beta\overline{w}$, which yields by substitution in the model that $Y_i = Z_i - \overline{Z} = \beta(w_i - \overline{w}) + \epsilon_i, i = 1, \ldots, n$, where it is automatically satisfied, it can be ignored. Putting $x_i = w_i - \overline{w}$, Boscovich therefore proposes to determine β in this model according to criterion (a), i.e., by minimizing the sum of absolute deviations* (the l_1 criterion). Boscovich's method for achieving this was geometric; Laplace* gave an analytical solution, acknowledging Boscovich, and used it in his own astronomical writings as early as 1789. In the second supplement (1818) to the *Théorie Analytique des Probabilités*, he calls it the "method of situation" to distinguish it from the l_2 (least-squares*) procedures, which he calls the "most advantageous method." This robust* method of estimation (in principle, tantamount to the use of the sample median as a measure of location of a sample rather than the least-squares measure, the sample mean), whose efficient general application requires linear

programming* methodology, therefore predates least squares in the linear model framework. After Laplace, it was taken up [3] by
Gauss*, who indicated the form of analytic
solution in the general case of the rank $r \geqslant 1$
linear model.

Additional information on Boscovich's life
and work can be found in refs. 2 and 4.

REFERENCES

1. Bosković, R. J. (1757). In *Geodetski rad R.
Boskovića*, N. Cubranić, ed., Zagreb, 1961.
(Printed with Serbo-Croatian translation. See
especially pp. 90–91.)

2. Eisenhart, C. (1961). In *R. J. Boscovich: Studies
of His Life and Work*, L. L. Whyte, ed. London,
pp. 200–213.

3. Sheynin, O. B. (1973). *Arch. History Exact Sci.*,
9, 306–324. (The most comprehensive overview
of Boscovich's statistical concepts.)

4. Stigler, S. M. (1973). *Biometrika*, **60**, 439–445.
(Section 3 discusses Laplace's treatment of the
"method of situation.")

See also ABSOLUTE DEVIATION; GENERAL LINEAR MODEL;
LEAST SQUARES; LINEAR PROGRAMMING; and ROBUST
ESTIMATION.

E. SENETA

BOSE, RAJ CHANDRA

Born: June 19, 1901, in Hoshangabad,
Madhya Pradesh, India.

Died: October 30, 1987, in Fort Collins,
Colorado.

Contributed to: coding theory, design
and analysis of experiments, geometry, graph theory, multivariate analysis
distribution theory.

Raj Chandra Bose was educated at Punjab
University, Delhi University, and Calcutta
University. He received M.A. degrees from
the latter two institutions, in 1924 (in applied
mathematics) and 1927 (in pure mathematics), respectively. At Calcutta he came under
the influence of the outstanding geometer
S. Mukhopadhyaya.

There followed a lectureship at Asutosh
College, Calcutta. This involved very heavy
teaching duities, but in his limited spare

time, Bose continued to produce research,
including joint publications with Mukhopadhyaya, in multidimensional and non-Euclid-
ean geometry.

In 1932, this came to the attention of
Mahalanobis*, who was in the process of
forming the Indian Statistical Institute*. He
needed a research worker who could apply
geometrical techniques to distributional problems in multivariate analysis*, along the
lines pioneered by Fisher*, and invited Bose
to join the young institute.

Initially, Bose worked, as expected, on
applications of geometrical methods in multivariate analysis, in particular on the distribution of Mahalanobis' D^2-statistic*. Much of
this work was in calibration with Roy (e.g.,
ref. 6). Later, as a consequence of attending
seminars given by Levi (appointed Hardinge
Professor of Pure Mathematics at Calcutta
University in 1936), he developed an interest in the application of geometrical methods
to the construction of experiment designs*.
His monumental paper [2] on the construction of balanced incomplete block designs*
established his reputation in this field. Further development of these ideas constitutes
a major part of Bose's work. It includes the
now well-known concept of partially balanced
incomplete block designs* [see ref. 4], also the
demonstration (with S. S. Shrikhande and
E. T. Parker) of the falsity of Euler's conjecture* (ref. 7, foreshadowed by ref. 1) on the
nonexistence of $n \times n$ Graeco–Latin squares*
for values of n other than a prime or a
power of a prime. Reference 3 contains a
broad summary of this work, including treatment of confounding* and fractional factorial designs*.

Bose took a part-time position in the Department of Mathematics at Calcutta University in 1938, and moved to the Department of
Statistics upon its formation in 1941. He was
head of the department from 1945 to 1949,
and received a D.Litt. degree in 1947. In 1949,
he became a professor in the Department of
Statistics at the University of North Carolina, where he stayed until his "retirement"
in 1971.

During this period, Bose's interests came
to include, with growing emphasis, coding
theory*—in particular, the use of geometry to construct codes. Among the outcomes

of this interest there resulted, from collaboration with Ray-Chaudhuri, the well-known Bose–Chaudhuri (BCH) codes [5].

After his "retirement," Bose accepted a position, jointly with the Departments of Mathematics and Statistics (later, Mathematics alone), at Colorado State University in Fort Collins. His interests continued to change, gradually returning to more purely mathematical topics, including graph theory. He finally retired in 1980, but retained an active interest in research.

REFERENCES

1. Bose, R. C. (1938). *Sankhyā*, **3**, 323–339.
2. Bose, R. C. (1938). *Ann. Eugen. (Lond.)*, **9**, 358–399.
3. Bose, R. C. (1947). *Sankhyā*, **8**, 107–166.
4. Bose, R. C.,   Clatworthy, W. H.,   and Shrikhande, S. S. (1954). *Tables of Partially Balanced Designs with Two Associate Classes. Tech. Bull.* **107**, North Carolina Agricultural Experiment Station, Raleigh, NC. (An extensive revision, by W. H. Clatworthy, with assistance from J. M. Cameron and J. A. Speakman, appeared in 1963 in National Bureau of Standards (U.S.) *Applied Mathematics Series*, **63**.)
5. Bose R. C. and Ray-Chaudhuri, D. K. (1960). *Inform. Control*, **3**, 68–79.
6. Bose, R. C. and Roy, S. N. (1936). *Sankhyā*, **4**, 19–38.
7. Bose, R. C., Shrikhande, S. S., and Parker, E. T. (1960). *Canad. J. Math.*, **12**, 189–203.

See also BLOCKS, BALANCED INCOMPLETE; EULER'S CONJECTURE; GRAPH THEORY; INFORMATION THEORY AND CODING THEORY; and PARTIALLY BALANCED DESIGNS.

BOSE–EINSTEIN STATISTICS. See FERMI-DIRAC, MAXWELL-BOLTZMANN, AND BOSE-EINSTEIN STATISTICS

BOUNDS, COOKE'S METHOD OF ESTIMATION

Cooke [1] proposed the following method of estimating the upper (or lower) bounds of the range of variation of a continuous random variable X, based on a random sample of size n with order statistics* $X_1 \leqslant X_2 \leqslant \cdots \leqslant X_n$.

The central idea of Cooke's suggestion is that if θ is the upper, and ϕ the lower bound, then

$$\theta = E[X_n] + \int_{\phi}^{\theta} \{F(x)\}^n dx, \qquad (1)$$

where $F(x)$ is the CDF of X. On the righthand side of (1), replacing $E[X_n]$ by the greatest order statistic X_n, ϕ by X_1, and $F(x)$ by the empirical distribution function* (EDF),

$$\hat{F}(x) = \begin{cases} 0, & x < X_1, \\ jn^{-1}, & X_j \leqslant x < X_{j+1}, \\ 1, & x \geqslant X_n, \end{cases}$$

we obtain *Cooke's estimator*

$$X_n + \int_{X_1}^{X_n} \{\hat{F}(x)\}^n \, dx$$

$$= 2X_n - \sum_{j=0}^{n-1} \left[(1 - jn^{-1})^n \right.$$

$$\left. -\{1 - (j+1)n^{-1}\} \right] X_{n-j}.$$

Approximately, for n large, the estimator of θ is

$$2X_n - (1 - e^{-1}) \sum_{j=0}^{n-1} e^{-j} X_{n-j}.$$

Similarly, the estimator of ϕ is

$$2X_1 - \sum_{j=1}^{n} \left[\{1 - (j-1)n^{-1}\}^n \right.$$

$$\left. -(1 - jn^{-1})^n \right] X_j$$

$$\doteq 2X_1 - (e-1) \sum_{j=1}^{n} e^{-j} X_j.$$

Cooke compared his estimator of θ with the estimator

$$T = 2X_n - X_{n-1} \qquad (2)$$

and found (2) was slightly (10–17%) less asymptotically efficient when $\nu \ll 1$ and

$$\{F(x)\}^n \sim \exp\left\{ -\left(\frac{\theta - x}{\theta - u_n} \right)^{1/\nu} \right\},$$

where $u_n = F^{-1}(1 - n^{-1})$. If v is known, however, the modified Cooke estimator

$$X_n + \left\{1 - (1 - e^{-1})^{-v}\right\}^{-1}$$
$$\times \left\{X_n - (1 - e^{-1}) \sum_{j=0}^{n-1} e^{-j}X_{n-j}\right\}$$

has about 30% greater efficiency than the appropriate modification of $T, T' = X_n + v^{-1}(X_n - X_{n-1})$.

REFERENCE

1. Cooke, P. (1979). *Biometrika*, **66**, 367–374.

See also EXTREME-VALUE DISTRIBUTIONS; ORDER STATISTICS; and THRESHOLD PARAMETER.

BOWLEY, ARTHUR LYON

Born: November 6, 1869, in Bristol, England.

Died: January 21, 1957, in Haslemere, Cumbria, England.

Contributed to: application of sampling techniques in social surveys, index-number construction, wage and income studies, mathematical economics, statistical education, popularization of statistics.

Sir Arthur Lyon Bowley was one of the most distinguished British statisticians in the first half of the twentieth century. His parents were Rev. J. W. L. Bowley, vicar of St. Philip and St. James in Bristol, and his wife, Maria Johnson. He attended Christ's Hospital School (1879–1888) and Trinity College, Cambridge (1888–1892), reading mathematics (bracketed as Tenth Wrangler in 1891). He gained an M.A. in 1895 and an Sc.D. in 1913, and was made an honorary fellow of Trinity in 1938.

Under the influence of Alfred Marshall and others in Cambridge active at that time in developing social sciences, he applied his mathematical abilities to problems in economic statistics. During this period, there appeared (in 1893) his earliest publication—*A Short Account of England's Foreign Trade in the Nineteenth Century* (3rd ed.; 1922, Allen and Urwin, London)—for which he was awarded the Cobden prize.

After leaving Cambridge, he taught mathematics at St. John's School, Leatherhead, Surrey from 1893 to 1899. During that period, his interest in statistics developed rapidly. His first paper in the *Journal of the Royal Statistical Society (JRSS)*, on "Changes in average wages in the United Kingdom between 1860 and 1891," was published in 1895—the year in which the London School of Economics (LSE) was established. On Marshall's recommendation, Bowley was invited to serve as a lecturer. His Wednesday evening courses in statistics (at 5:45 or 6:00 P.M.) at London School of Economics continued from October, 1895 for some 38 years with only occasional interruption. For 18 years Bowley combined his lectureship at London School of Economics with teaching mathematics and economics at University College, Reading (later Reading University) as a lecturer from 1900 to 1907, and professor from 1907–1913. It was only in 1915 that he was given the title of professor at London School of Economics, having been a reader since 1908. In 1919 he was elected to a newly-established Chair of Statistics—a post he held until his retirement in 1936. He still continued many of his activities, both in London School of Economics and elsewhere, notably acting as director of the Institute of Statistics in Oxford University (1940–1944) during World War Two. He was created a C.B.E. (Companion of the Order or the British Empire) in 1937, and knighted in 1950.

In 1904, Bowley married Julie, daughter of Thomas Williams, a land agent of Spalding, Lincolnshire. They had three daughters, one of whom, Marian, became professor of political economy at University College, London. Bowley was a reserved and shy person, but had a dry sense of humor. He did not readily make friends. However, Joan Box records in her account of the life of her father, R. A. Fisher* [*Life of a Scientist* (1978), Wiley, New York] that in the 1920s Fisher and Bowley were close neighbors in Harpenden, Hertfordshire and frequently enjoyed evenings of bridge together. This did not prevent Bowley from criticizing Fisher at a meeting of the Royal Statistical Society in 1935.

Bowley's famous textbook *Elements of Statistics*, first published in 1910, went through seven editions, the seventh appearing in 1937. His stated goal was to put "in the simplest possible way those formulae and ideas which appear to be most useful in the fields of economic and social investigation and of showing their relationship to the treatment followed in the text." Bowley's influence as a teacher, especially at the postgraduate level, was considerable. [One of his many students was Maria (Mary) Smith-Faulkner (1878–1968), a fiery communist revolutionary who emigrated to the Soviet Union, and was responsible for the development of Soviet statistics in accord with rigid Marxist ideology. She also translated works of D. Ricardo and W. Petty* into Russian.]

In the field of economics, his most influential work seems to be on British incomes. He authored a number of studies on the definition and measurement of national income, which occupied his attention intermittently for more than twenty years before the first official estimates were issued, under the guidance of J.M. Keynes, during World War Two. Bowley's main publication during this period is *Wages and Income in the United Kingdom since 1860*, published in 1937 by Cambridge University Press.

His other pioneering activity in this field was in the London and Cambridge Economic Service, over the years 1923 to 1953. He was highly respected among British official statisticians, exercising his influence through teaching and research on the one hand, and extensive international contacts on the other. The International Statistical Institute (ISI) owes much to him as a member from 1903, as treasurer (1929–1936 and 1947–1949), and as author of several reports sponsored by the Institute. In 1949 he was honored by election as honorary president. He traveled extensively, influencing statisticians and others in many parts of the world. Among these trips was a visit to India in 1934 to review "Indian economic conditions and the present statistical organization of India." Publication of *A Scheme for an Economic Census of India* (with Professor Sir Dennis Robertson) resulted from this journey.

Bowley's most original contributions were in the application of sampling techniques to economic and social surveys. He was forming his ideas on these topics as early as the 1890s, and propagated them through the ISI, though he was at first opposed by many leading official statisticians. He was attracted by the possibilities of the so-called "representative method," and developed an appropriate mathematical formulation of sampling precision and optimal ways of interpreting the results of such sample surveys to laymen. These were later presented in several publications, including *The New Survey of London Life and Labour* (vol. 3, pp. 29–96 and 216–253, in 1932, and vol. 6, pp. 29–117, in 1934). His pioneering work contributed subsequently to the development of market research. This work was closely associated with Bowley's studies on index numbers, reported in *JRSS* (1926), *Economic Journal* (1897, 1928), *Econometrica* (1938), and *London and Cambridge Economic Service* [Memo 5 (1924) and Memo 28 (1929)]. [See also Jazairi (1983).]

Bowley published relatively little in either mathematical statistics or mathematical economics. These subjects developed rapidly during the latter parts of his life, and he quite naturally felt less intense interest in them. Mathematical techniques themselves were incidental to his main purposes, and he used them essentially as a matter of convenience. He was a practitioner in applied statistics *par excellence*, though he cannot be regarded as a pathbreaker (in the sense of R.A. Fisher, K. Pearson,* and other British statisticians) in either statistical methodology or mathematical economics.

Bowley was closely associated with the Royal Economic Society (elected a fellow in 1893 and a member of Council in 1901, and publishing 15 papers in the *Economic Journal*). He was elected a fellow of the Royal Statistical Society in 1894, became a member of Council in 1899, was vice-president in 1907–1909 and 1912–1914, and was president in 1938–1940. He was awarded the Society's Guy Medals in Silver in 1895, and in Gold in 1935. His other scientific honors include a D. Litt. from Oxford University (1944), a D.Sc. from Manchester University (1926), and fellowship of the British Academy (1922).

REFERENCES

1. Allen, R. G. D. and George, R. F. (1957). Obituary of Professor Sir Arthur Lyon Bowley. *J. R. Statist. Soc. A*, **120**, 236–241. (Contains an extensive bibliography.)
2. Jazairi, N. (1983). Index numbers. In *Encyclopedia of Statistical Sciences*, 4, 54–62. Wiley: New York.
3. Roberts, F. C., compiler. *(1979). Obituaries from The Times of London, 1951–1960.* Newspaper Archive Department, Reading, England.
4. Anon. (1973). *Who Was Who in America, Vol. 5 (1969–1973).* Marquis Who's Who, Chicago.
5. Williams, E. T. and Palmer, H. T., eds. (1971). *The Dictionary of National Biography.* Oxford University Press, Oxford, England.

BOX–COX TRANSFORMATION—I

The usual assumptions when one analyzes data are the standard assumptions of the linear model (*see* GENERAL LINEAR MODEL), i.e., the additivity of effects, the constancy of variance, the normality of the observations, and the independence of observations. If these assumptions are not met by the data, Tukey [8] suggested two alternatives: either a new analysis must be devised to meet the assumptions, or the data must be transformed to meet the usual assumptions. If a satisfactory transformation is found, it is almost always easier to use than to develop a new method of analysis.

Tukey suggested a family of transformations with an unknown power parameter and Box and Cox [1] modified it to

$$y^{(\lambda)} = \begin{cases} (y^{(\lambda)} - 1)/\lambda & \text{for } \lambda \neq 0 \\ \log y, & \text{for } \lambda = 0, \end{cases}$$

where y is an original observation, $y^{(\lambda)}$ the "new" observation, and λ a real unknown parameter.

Box and Cox assumed that for some λ the n transformed observations

$$\mathbf{y}^{(\lambda)} = \begin{bmatrix} y_1^{(\lambda)} \\ y_2^{(\lambda)} \\ \vdots \\ y_n^{(\lambda)} \end{bmatrix}$$

are independent and normally distributed with constant variance σ^2 and mean vector

$$E\mathbf{y}^{(\lambda)} = A\boldsymbol{\theta},$$

where A is a known $n \times p$ matrix and $\boldsymbol{\theta}$ is a vector of parameters associated with the transformed observations.

Box and Cox [1] estimate the parameters by maximum likelihood* as follows. First, given λ,

$$\hat{\boldsymbol{\theta}}(\lambda) = (A'A)^{-1}A'\mathbf{y}^{(\lambda)}$$

and

$$\hat{\sigma}^2(\lambda) = \mathbf{y}'^{(\lambda)}[I - A(A'A)^{-1}A']\mathbf{y}^{(\lambda)} \qquad (1)$$

are the estimators of $\boldsymbol{\theta}$ and σ^2, respectively. Second, λ is estimated by maximizing the log likelihood function*

$$l(\lambda) = -n \log \hat{\sigma}^2(\lambda)/2 + \log J(\lambda : y),$$

where

$$J(\lambda : y) = \prod_{i=1}^{n} y_i^{\lambda-1}.$$

The maximum likelihood estimator of λ, say $\hat{\lambda}$, is then substituted into (1), which determines the estimators of the other parameters.

Their ideas are nicely illustrated with the analysis of the survival times* of animals that were exposed to three poisons.

Since the original Box–Cox article, there have been many related papers. For example, Draper and Cox [3] derive the precision of the maximum likelihood estimator of λ for a simple random sample, i.e., $Ey_i^{(\lambda)} = \mu$ for $i = 1, 2, \ldots, n$. They found that the approximate variance is

$$V(\hat{\lambda}) = \frac{2}{3n\Delta^2}\left(1 - \frac{1}{3}\gamma_1^2 + \frac{7}{18}\gamma_2\right)^{-1},$$

where

$$\gamma_1 = \mu_3/\sigma^3,$$
$$\gamma_2 = \mu_4/\sigma^4 - 3,$$
$$\Delta = \lambda\sigma/(1 + \lambda\mu),$$

σ^2 is the variance, and μ_i is the ith central moment of the original observations.

Hinkley [5] generalized the Box–Cox transformation to include power transformations to symmetric distributions and showed his "quick" estimate of λ to be consistent* and have a limiting normal distribution. He discusses the same exponential distribution* as Draper and Cox.

Literature

There are many references related to the Box and Cox transformation. Poirier [6] has extended the Box–Cox work to truncated normal distributions*. Some recent results related to large-sample behavior are given by Harnández and Johnson [4]. For information about other power transformations, *see* TRANSFORMATIONS.

The Box and Cox [1], Draper and Cox [3], Hinkley [5], and Poirier [6] articles are quite informative but highly technical.

At the textbook level, Chapter 10 of Box and Tiao [2] and Chapter VI of Zellner [9] are excellent introductions to the Box–Cox theory of transformations. For computer programs, see Tran Cong Liem [7] and references therein.

REFERENCES

1. Box, G. E. P. and Cox, D. R. (1964). *J. R. Statist. Soc. B*, **26**, 211–252.

2. Box, G. E. P. and Tiao, G. C. (1973). *Bayesian Inference in Statistical Analysis*. Addison-Wesley, Reading, Mass.

3. Draper, N. R. and Cox, D. R. (1969). *J. R. Statist. Soc. B*, **31**, 472–476.

4. Harnández, F. and Johnson, R. A. (1979). *Tech. Rep. No. 545*, Dept. of Statistics, University of Wisconsin, Madison, Wis.

5. Hinkley, D. V. (1975). *Biometrika*, **62**, 101–111.

6. Poirier, D. J. (1978). *J. Amer. Statist. Ass.*, **73**, 284–287.

7. Tran Cong Liem (1980). *Amer. Statist.*, **34**, 121.

8. Tukey, J. W. (1957). *Ann. Math. Statist.*, **28**, 602–632.

9. Zellner, A. (1971). *An Introduction to Bayesian Inference in Econometrics*. Wiley, New York.

See also TRANSFORMATIONS.

LYLE D. BROEMELING

BOX–COX TRANSFORMATIONS—II

The classical general linear model*, which includes linear regression* and analysis of variance*, may be written as

$$\mathbf{y} = \mathbf{X}\boldsymbol{\beta} + \boldsymbol{\epsilon}, \qquad (1)$$

where $\mathbf{y}' = (y_1, y_2, \ldots, y_n)$ is a vector of n observations, $\mathbf{X}$ is a known design matrix, $\boldsymbol{\beta}$ is an unknown vector of constants, and $\boldsymbol{\epsilon}' = (\epsilon_1, \ldots, \epsilon_n)$ is a vector of measurement errors. The value of $\boldsymbol{\beta}$ is usually estimated using least squares*. Statistical properties of the estimate $\hat{\boldsymbol{\beta}}$ can be justified under the so-called normal-theory assumptions, which state that the components of $\boldsymbol{\epsilon}$

1. are independent,
2. are normally distributed, and   (2)
3. have constant variance σ^2.

When these assumptions hold, many statistical procedures have desirable properties. For example, the components of $\hat{\boldsymbol{\beta}}$ have minimum variance among all unbiased estimates (see e.g., Ref. 13 p. 380), and F-tests* of hypotheses about $\boldsymbol{\beta}$ are optimal within certain classes of tests (see e.g., Ref. 18). Normality of the errors can be checked with a normal quantile plot of the residuals* (*see* PROBABILITY PLOTTING) and variance homogeneity can be checked with a plot of the residuals against the fitted values. The independence assumption is, however, often difficult if not impossible to verify. Comprehensive discussions of the consequences of violating the assumptions can be found in References 16 and 18.

When the normal-theory assumptions are not satisfied, we can either *(i)* use non-least squares methods to estimate $\boldsymbol{\beta}$ or *(ii)* transform $\mathbf{y}$ or $\mathbf{X}$ to better satisfy the assumptions. If we adopt the first solution, we lose the computational simplicity of least squares procedures and their statistical optimality properties. If we adopt the second solution, we must decide which variables to transform and how to transform them.

Among the first articles to explore transformations of the $\mathbf{y}$ vector were References 1, 17 and 20. The latter introduced the two-parameter family of power transformations

$$y_i(\lambda_1, \lambda_2) = \begin{cases} \text{sign}(\lambda_1)(y_i + \lambda_2)^{\lambda_1}, & 0 \neq \lambda_1 \leqslant 1, \\ \log(y_i + \lambda_2), & \lambda_1 = 0, \end{cases}$$

$$(3)$$

where λ_2 is such that $y_i + \lambda_2 > 0$. Transformation of the columns of $\mathbf{X}$ was considered in Reference 5. The family that has received the most attention in theory and practice is a one-parameter modification of Equation 3 proposed by Box and Cox [3]:

$$y_i(\lambda) = \begin{cases} (y_i^\lambda - 1)/\lambda, & \lambda \neq 0, \\ \log y_i, & \lambda = 0, \end{cases} \quad (4)$$

which holds for $y_i > 0$. This is the family of *Box–Cox transformations*. Note that $y_i(\lambda)$ is continuous in λ. Since the form of a linear model is unchanged by a linear transformation, it is often more convenient to consider the simpler but equivalent transformation

$$y_i(\lambda) = \begin{cases} y_i^\lambda, & \lambda \neq 0, \\ \log y_i, & \lambda = 0. \end{cases} \quad (5)$$

PARAMETER ESTIMATION

Let $\mathbf{y}(\lambda) = (y_1(\lambda), \ldots, y_n(\lambda))'$. Box and Cox [3] showed how the parameters in the model

$$\mathbf{y}(\lambda) = \mathbf{X}\boldsymbol{\beta} + \boldsymbol{\epsilon} \quad (6)$$

can be estimated by maximum likelihood* and Bayes theory. They assume that there exists a value of λ such that Equation 6 holds, with $\boldsymbol{\epsilon}$ satisfying the normal-theory conditions in (2). Given the values of $\mathbf{X}$ and $\mathbf{y}$, the likelihood function of the transformed data is

$$l(\lambda, \boldsymbol{\beta}, \sigma^2)$$
$$= \frac{\exp\{-(2\sigma^2)^{-1}(\mathbf{y}(\lambda) - \mathbf{X}\boldsymbol{\beta})'(\mathbf{y}(\lambda) - \mathbf{X}\boldsymbol{\beta})\}}{(2\pi\sigma^2)^{n/2}}$$
$$\times J(\lambda, \mathbf{y}),$$

where

$$J(\lambda, \mathbf{y}) = \prod_{i=1}^{n} \left| \frac{dy_i(\lambda)}{dy_i} \right|$$

is the Jacobian of the transformation. Let $\dot{y} = (y_1 y_2 \ldots y_n)^{1/n}$ denote the geometric mean* of the y_i's. For the family (4),

$$J(\lambda, \mathbf{y}) = \begin{cases} \dot{y}^{(\lambda-1)n}, & \lambda \neq 0, \\ \dot{y}^{-n}, & \lambda = 0, \end{cases}$$

and for the simpler family (5),

$$J(\lambda, \mathbf{y}) = \begin{cases} \lambda^n \dot{y}^{(\lambda-1)n}, & \lambda \neq 0, \\ \dot{y}^{-n}, & \lambda = 0. \end{cases}$$

Maximum likelihood estimates of λ, $\boldsymbol{\beta}$, and σ^2 are obtained in two steps. For fixed λ, compute the maximum log-likelihood $L_M(\lambda) = \max_{\boldsymbol{\beta}, \sigma^2} \log l(\lambda, \boldsymbol{\beta}, \sigma^2)$. The maximizing values of $\boldsymbol{\beta}$ and σ^2 are the ordinary least squares estimates and the maximum likelihood estimate of λ is the maximizer of $L_M(\lambda)$. An explicit expression for $L_M(\lambda)$ can be given in terms of the normalized vector $\mathbf{z}^{(\lambda)} = (z_1^{(\lambda)}, z_2^{(\lambda)}, \ldots, z_n^{(\lambda)})'$, where

$$z_i^{(\lambda)} = \frac{y_i^{(\lambda)}}{J(\lambda, \mathbf{y})^{1/n}} = \begin{cases} (y_i^\lambda - 1)/(\lambda \dot{y}^{\lambda-1}), & \lambda \neq 0 \\ \dot{y} \log y_i, & \lambda = 0. \end{cases}$$

Let $S(\lambda, \mathbf{z})$ denote the residual sum of squares of $\mathbf{z}^{(\lambda)}$. Then

$$L_M(\lambda) = -(n/2) \log\{n^{-1} S(\lambda, \mathbf{z})\}, \quad (7)$$

and the maximum likelihood estimate $\hat{\lambda}$ is the value that minimizes $S(\lambda, \mathbf{z})$. An approximate $100(1 - \alpha)\%$ confidence interval for λ may be obtained from the usual chi-squared asymptotics: $L_M(\hat{\lambda}) - L_M(\lambda) < \chi_1^2(\alpha)/2$, where $\chi_1^2(\alpha)$ is the α-quantile of the chi-squared distribution* with one degree of freedom.

Box and Cox showed that the maximum likelihood estimates coincide with the Bayesian estimates if the joint prior density is chosen to be

$$\frac{d\boldsymbol{\beta} \, d(\log \sigma)}{\{J(\lambda, \mathbf{y})\}^{(n - v_r)/n}} p_0(\lambda) \, d\lambda, \quad (8)$$

where $p_0(\lambda)$ denotes the prior density of λ. Then the posterior density $p(\lambda)$ is proportional to $p_0(\lambda)/S(\lambda, \mathbf{z})^{v_r/2}$, and if $p_0(\lambda)$ is taken to be locally uniform, $p(\lambda) \propto S(\lambda, \mathbf{z})^{-v_r/2}$. It follows that the Bayes estimate of λ is the maximizer of $L_B(\lambda) = -(v_r/2) \log\{v_r^{-1} S(\lambda, \mathbf{z})\}$ and the Bayes estimates of $\boldsymbol{\beta}$ and σ^2 are the corresponding least squares estimates. The similarity between $L_B(\lambda)$ and the expression for $L_M(\lambda)$ in Equation 8 implies that if v_r is constant, the Bayes and maximum likelihood estimates of the parameters are the same. The unusual dependence of Equation 7 on $\mathbf{y}$ was noted by some discussants at the end of Reference 3.

EXAMPLES

A 3×4 Experiment

Box and Cox [3] give two examples to illustrate the methodology. The first example is a four-replicate 3×4 factorial experiment on the survival times of animals (see Ref. 3, p. 220 for the data). The factors are three poisons and four treatments. The ANOVA results in the upper part of Table 1 show no significant interaction between the two factors. But the plots of the residuals in the left column of Fig. 1 reveal strong variance heterogeneity and nonnormality of the residuals.

The likelihood function $L_M(\lambda)$ is maximized near $\lambda = -1$. The ANOVA table for the data following a reciprocal transformation is shown in the lower part of Table 1. The residuals now satisfy the normal-theory assumptions much better, as can be seen from the plots in the right column of Fig. 1. Further, the ANOVA results indicate that while the interaction remains insignificant, there is a two- to threefold increase in the F-statistics for the main effects.

A 3^3 Experiment

For their second example, Box and Cox [3] used an unreplicated 3^3 factorial experiment from the textile industry (see Ref. 3, p. 223 for the data). When a full second-order model is fitted to the data, there are highly significant main effects and interactions, as shown in Table 2. A plot of the residuals versus fitted values (Fig. 2a), however, indicates a lack of fit. To demonstrate that a transformation can

not only improve the normal-theory assumptions but also reduce the number of columns in the $\mathbf{X}$ matrix, Box and Cox estimated λ from an *additive model*, that is, one that contains only the main-effect terms. The value of $\hat{\lambda}$ turns out to be close to 0, suggesting a log transformation. When a full second-order model is fitted to the log-transformed data, only the main effects are significant as shown on the right side of Table 2. Further, the F values for the main-effect terms are again increased by two to three times. Finally, no lack of fit is evident from the plot of residuals versus fitted values for the additive model (see Fig. 2b).

A Counter Example

The impressive success of the transformation technique demonstrated in these two examples cannot be taken for granted in practice. Consider, for instance, the data in Table 3, which come from a study on the effects of five factors on the production of fibers when xanthan gum, a milk stabilizer, is added to whey protein. The experiment was an unreplicated 2^5 factorial, the factors and their levels being:

A: Xanthan homogenized (+) or not (−).
B: Xanthan heated (+) or not (−).
C: pH adjusted (+) or not (−).
D: Whey protein heated (+) or not (−).
E: Concentration of xanthan gum at 0.05% (+) or 0.01% (−).

The yield is the percent of solid fibers.

Table 1. ANOVA Tables for the 3 × 4 Experiment

| | | Before Data Transformation | | | |
Source	Df	Sum Sq	Mean Sq	F value	P value
Poison	2	1.033	0.517	23.22	3.33×10^{-7}
Treatment	3	0.921	0.307	13.81	3.78×10^{-6}
Poison × Treatment	6	0.250	0.042	1.87	0.11
Residuals	36	0.801	0.022		

| | | After Reciprocal Transformation | | | |
Source	Df	Sum Sq	Mean Sq	F value	P
Poison	2	34.877	17.439	72.64	2.31×10^{-13}
Treatment	3	20.414	6.805	28.34	1.38×10^{-9}
Poison × Treatment	6	1.571	0.262	1.09	0.39
Residuals	36	8.643	0.240		

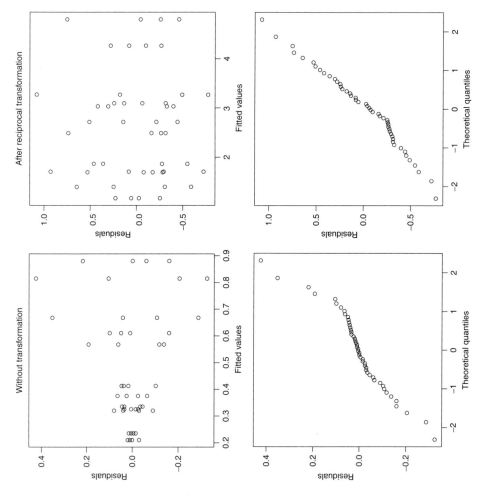

Figure 1. Plots of residuals versus fitted values and normal quantile plots of residuals for the 3×4 example of Box and Cox (1964). The plots in the left column are before transformation and those in the right are after a reciprocal transformation.

Table 2. ANOVA Tables for the 3^3 Experiment

Source	Df	Before Data Transformation			After Log Transformation		
		Mean Sq	F value	P value	Mean Sq	F value	P value
x_1	1	7840800	106.07	1.00×10^{-8}	12.4715	331.94	1.37×10^{-12}
x_2	1	5169184	69.93	1.99×10^{-7}	7.1667	190.75	1.14×10^{-10}
x_3	1	1738491	23.52	1.50×10^{-4}	2.7729	73.80	1.37×10^{-7}
x_1^2	1	341771	4.62	0.05	0.0441	1.17	0.29
x_2^2	1	455953	6.17	0.02	0.0035	0.09	0.76
x_3^2	1	14017	0.19	0.67	0.0273	0.73	0.41
$x_1 x_2$	1	2500707	33.83	2.06×10^{-5}	0.0175	0.47	0.50
$x_1 x_3$	1	666465	9.02	8.01×10^{-3}	0.0562	1.49	0.24
$x_2 x_3$	1	245388	3.32	0.09	0.0052	0.14	0.71
Residuals	17	73922			0.0376		

Table 3. Data from a 2^5 Experiment

Run	A	B	C	D	E	%Solids	Run	A	B	C	D	E	%Solids
23	−	−	−	−	−	14.00	24	−	−	−	+	−	12.30
5	−	−	−	−	+	13.10	26	−	−	−	+	+	10.20
22	+	−	−	−	−	13.70	29	+	−	−	+	−	12.50
19	+	−	−	−	+	14.20	8	+	−	−	+	+	10.70
10	−	+	−	−	−	13.80	1	−	+	−	+	−	9.32
30	−	+	−	−	+	13.00	3	−	+	−	+	+	7.31
7	+	+	−	−	−	13.00	16	+	+	−	+	−	9.89
25	+	+	−	−	+	12.80	9	+	+	−	+	+	7.89
27	−	−	+	−	−	13.40	13	−	−	+	+	−	12.00
2	−	−	+	−	+	11.80	6	−	−	+	+	+	10.20
11	+	−	+	−	−	12.30	14	+	−	+	+	−	12.30
20	+	−	+	−	+	13.00	28	+	−	+	+	+	11.40
21	−	+	+	−	−	11.40	32	−	+	+	+	−	10.40
17	−	+	+	−	+	10.70	31	−	+	+	+	+	8.85
15	+	+	+	−	−	11.80	18	+	+	+	+	−	9.86
12	+	+	+	−	+	10.60	4	+	+	+	+	+	7.94

The upper left panel of Fig. 3 shows a normal quantile plot of the estimated main and interaction effects. Four significant effects are indicated, namely the main effects of B, D, E, and the CD interaction. A model containing only these four effects appears to be inadequate, however, as shown by the residual plot on the upper right panel of Fig. 3.

Following the previous example, we can search for a power λ for which an additive model suffices. The plot of $S(\lambda, \mathbf{z})$ versus λ in Fig. 4 shows that $\hat{\lambda} = 2.5$. However, this transformation does not eliminate all the interaction* effects because a normal quantile plot of estimated effects shows that the main effects* of B, C, D, E, and the CD interaction are significant. Further, a plot of the residuals versus fitted values for a model containing these five effects exhibits more curvature than that before transformation—see

the plots in the bottom row of Fig. 3. It turns out that it is impossible simultaneously to eliminate all the interaction effects *and* satisfy the normal-theory assumptions for this data set. The best transformation appears to be the log. Using the model selection method in Reference 15, the log-transformed data are satisfactorily fitted with a model containing three main effects and three two-factor interactions: B, D, E, DE, BD, and CD.

FURTHER DEVELOPMENTS

In the above examples, statistical inference for $\boldsymbol{\beta}$ and σ in the transformed model was carried out as if $\hat{\lambda}$ were known rather than estimated (see Ref. 3, p. 219). Bickel and Doksum [2] studied the behavior of $\hat{\lambda}$ and $\hat{\boldsymbol{\beta}}$ as σ^2 decreases to zero (also known as "small-σ

(a)

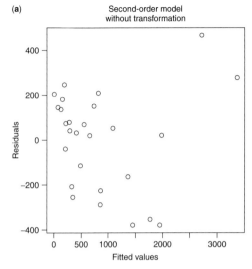

(b)

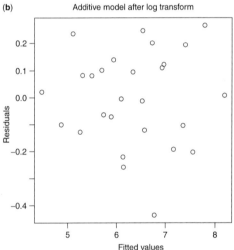

Figure 2. Residual plots for 3^3 example of Box and Cox (1964). The plot in (**a**) is for a full second-order model fitted to the data before transformation. The one in (**b**) is for a first-order model after a log transformation.

asymptotics"). They find that in some problems such as transformed linear regression, $\hat{\lambda}$ and $\hat{\beta}$ can be highly correlated. As a result, estimation of λ can cause the variance of $\hat{\beta}$ to be very much larger than what it would be if λ were known. Box and Cox [4] responded that it is not scientifically meaningful to interpret the value of $\hat{\beta}$ when λ is estimated. A lively debate on this philosophical issue is provided by Hinkley and Runger [14] and discussants.

Conditions for the strong consistency and asymptotic normality of the parameter estimates are derived in Reference 10.

A related and less controversial question [4] is under what circumstances do confidence intervals and hypothesis tests about β, calculated as if $\hat{\lambda}$ were preassigned, remain approximately valid? In the two-sample problem, in which the populations differ by a location shift after transformation, the t-test after transformation has asymptotically the correct level and the same local power as if λ were known [12] (see also Ref. 19). In the opposite situation, in which the populations differ by a location shift before transformation, the asymptotic efficiency of t-test after transformation is at least as large as that before transformation [9]. In the problem of prediction interval estimation in linear regression, under regularity conditions, the intervals calculated from the transformed data have the correct asymptotic coverage probabilities [11] (see also Ref. 6, 7, and 8).

Acknowledgment

The author's work was prepared under U. S. Army Research Office grant DAAD19-01-1-0586.

REFERENCES

1. Bartlett, M. S. (1947). The use of transformatons. *Biometrics*, **3**, 39–52.

2. Bickel, P. J. and Doksum, K. A. (1981). An analysis of transformations revisited. *J. Am. Stat. Assoc.*, **76**, 296–311.

3. Box, G. E. P. and Cox, D. R. (1964). An analysis of transformations. *J. R. Stat. Soc., Ser. B*, **26**, 211–252.

4. Box, G. E. P. and Cox, D. R. (1982). An analysis of transformations revisited, rebutted. *J. Am. Stat. Assoc.*, **77**, 209–210.

5. Box, G. E. P. and Tidwell, P. W. (1982). Transformation of the independent variables. *Technometrics*, **4**, 531–550.

6. Carroll, R. J. (1982). Prediction and power transformation when the choice of power is restricted to a finite set. *J. Am. Stat. Assoc.*, **77**, 908–915.

7. Carroll, R. J. and Ruppert, D. (1981). On prediction and the power transformation family. *Biometrika*, **68**, 609–615.

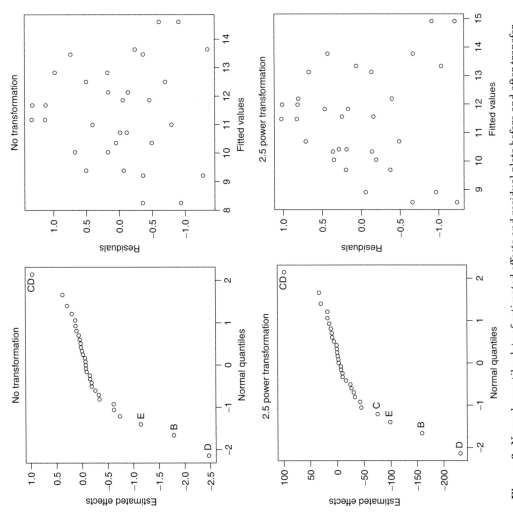

Figure 3. Normal quantile plots of estimated effects and residual plots before and after transformation for the data in Table 3. The residuals are calculated from models fitted with the effects identified in the quantile plots.

643

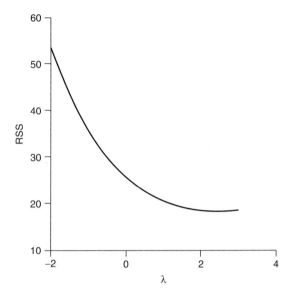

Figure 4. Plot of the residual sum of squares $S(\lambda, \mathbf{z})$ versus λ for the data in Table 3. The residual sum of squares is computed assuming a model without interaction terms.

8. Carroll, R. J. and Ruppert, D. (1991). Prediction and tolerance intervals with transformation and/or weighting. *Technometrics*, **33**, 197–210.

9. Chen, H. and Loh, W.-Y. (1992). Bounds on AREs of tests following Box-Cox transformations. *Ann. Stat.*, **20**, 1485–1500.

10. Cho, K., Yeo, I., Johnson, R. A., and Loh, W.-Y. (2001a). Asymptotic theory for Box-Cox transformations in linear models. *Stat. Probab. Lett.*, **51**, 337–343.

11. Cho, K., Yeo, I., Johnson, R. A., and Loh, W.-Y. (2001b). Prediction interval estimation in transformed linear models. *Stat. Probab. Lett.*, **51**, 345–350.

12. Doksum, K. A. and Wong, C.-W. (1983). Statistical tests based on transformed data. *J. Am. Stat. Assoc.*, **78**, 411–417.

13. Graybill, F. A. (1976). *Theory and Application of the Linear Model*. Duxbury Press, North Scituate, Massachusettes.

14. Hinkley, D. V. and Runger, G. (1984). The analysis of transformed data (with discussion). *J. Am. Stat. Assoc.*, **79**, 302–309.

15. Loh, W.-Y. (1992). Identification of active contrasts in unreplicated factorial experiments. *Comput. Stat. Data Anal.*, **14**, 135–148.

16. Miller, R. G. Jr. (1986). *Beyond ANOVA, Basics of Applied Statistics*. Wiley, New York.

17. Moore, P. G. and Tukey, J. W. (1954). Answer to query 112. *Biometrics*, **10**, 562–568.

18. Scheffé, H. (1959). *The Analysis of Variance*. Wiley, New York.

19. Taylor, J. M. G. (1986). The retransformed mean after a fitted power transformation. *J. Am. Stat. Assoc.*, **81**, 114–118.

20. Tukey, J. W. (1957). On the comparative anatomy of transformations. *Ann. Math. Stat.*, **28**, 602–632.

FURTHER READING

Atkinson, A. C. (1985). *Plots, Transformations and Regression*. Oxford University Press, New York.

Carroll, R. J. and Ruppert, D. (1988). *Transformation and Weighting in Regression*. Chapman and Hall, New York.

See also FACTORIAL EXPERIMENTS; GENERAL LINEAR MODEL; and TRANSFORMATIONS.

WEI-YIN LOH

BOX–COX TRANSFORMATIONS: SELECTING FOR SYMMETRY

Deciding whether a square root, reciprocal or other transformation of (positive-valued) data will make a given distribution of data values more nearly bell-shaped, or at least more nearly symmetric can be a matter of trial and error, or involve the use of computing packages. One simple alternative results from noting that if X_M is the median of the data set, if X_L and X_U are a pair of lower and upper percentiles, such as the two quartiles, of the data, and if λ is the power that results in the transformed values of X_L and X_U being equal distances from the transformed value of X_M, then

$$X_M^\lambda - X_L^\lambda = X_U^\lambda - X_M^\lambda$$

or

$$1 - \left(\frac{X_L}{X_M}\right)^\lambda = \left(\frac{X_M}{X_U}\right)^\lambda - 1$$

or

$$2 = x^\lambda + y^{-\lambda}$$

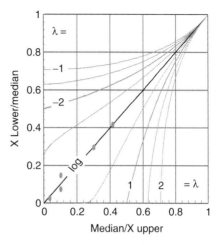

Figure 1.

where $x = X_L/X_M$ and $y = X_M/X_U$. With a graph of curves of the form $2 = x^\lambda + y^{-\lambda}$, and with values of x and y calculated from the data, it is a simple matter to plot the point (x, y) and note the value of λ for a nearby plotted curve. A sensible precaution is to use several pairs of lower and upper percentiles to discover if the indicated value of λ is consistent over a wide range of the data.

For example, Krebs [11, page 469] reports that the Protozoa levels of 12 samples of forest soil were 56, 160, 320, 640, 900, 900, 5,200, 5,200, 7200, 20,800, 20,800 and 59,000 and asks for a recommendation as to whether a transformation seems appropriate, and if so which one? Since the number of observations is even, we need first to determine some median value, X_M, which lies between the sixth and seventh values, 900 and 5,200. Of the two obvious candidates as a possible median, X_M, the geometric mean and the arithmetic mean [i.e., $\sqrt{/(900 \cdot 5200)}$, or 2163.331 and $(900 + 5200)/2$, or 3050, respectively], the geometric mean has the convenient property that it automatically transforms into the corresponding median under any power-law transformation, but either choice of median works well.

Figure 1 consists of a family of curves for which $2 = x^\lambda + y^{-\lambda}$, along with the five informative observation-pair plotted points for the data set discussed above. (The point for the two central observations, from which the median is determined, always falls on

the "log" transformation curve, the diagonal line in the graph if the geometric-mean based determination of the median is used, regardless of the transformation indicated by the remaining plotted points. As such, that point is uninformative, or even potentially misleading. If included, that sixth point would have fallen exactly on the fifth plotted point for this data set, because of repeated values in the data.) The plotted points very strongly suggest a log transformation.

[Transformation selection methods have long been studied, from various perspectives. The method described here and initially presented by Hines and OHara Hines [7] is suitable for symmetrizing a single distribution and resembles procedures proposed by authors such as Emerson and Stoto [1982], Hinkley [1975, 1976, 1977], and Tukey [1957, 1977] in focussing on the essential property that the upper and corresponding lower quartiles, deciles or other quantiles of a symmetric distribution are equidistant from its median. See also discussions by Box and Cox [1964], Andrews [1971], Atkinson [1973, 1985] and Emerson [1983]

REFERENCES

1. Andrews, D.F. (1971) A Note on the Selection of Data Transformations, *Biometrica*, **58**, pp. 249–254.

2. Atkinson, A.C. (1973) Testing Transformations to Normality, *Journal of the Royal Statistical Society*, Ser. B, **35**, pp. 473–479.

3. Atkinson, A.C. (1985) *Plots, Transformations and Regression : An Introduction to Graphical Methods of Diagnostic Regression Analysis*, Oxford, UK, Clarendon Press

4. Box, G.E.P. and Cox, D.R. (1964) An Analysis of Transformations (with discussion), *Journal of the Royal Statistical Society* **26**, pp. 211–252.

5. Emerson, J. D. (1983) Mathematical Aspects of Transformations, in *Understanding Robust and Exploratory Data Analysis*, eds. D.C. Hoaglin, F. Mosteller and J.W. Tukey, New York, Wiley, pp. 247–282.

6. Emerson, J. D. and Stoto, M. A. (1982) Exploratory Methods for Choosing Power Transformations, *Journal of the American Statistical Association*, 77, pp. 103–108

7. Hines, W.G.S. and OHara Hines, R.J. (1987) Quick Graphical Power-Law Transformation Selection, *American Statistician* **41**, pp. 21–24.

8. Hinkley, D. V. (1975) On Power Transformations to Symmetry, *Biometrika*, **62**, pp. 101–111.

9. Hinkley, D. V. (1975) On Estimatiing a Symmetric Distribuiton, *Biometrika*, **63**, pp. 680–681.

10. Hinkley, D. V. (1975) On Quick Choice of Power Transformation, *Applied Statistics*, **26**, pp. 67–69.

11. Krebs, C. J. (1989) Ecological Methodology, Harper and Row, New York.

12. Tukey, J. W. (1957), On the Comparative Anatomy of Transformations, *Annals of Mathematical Statistics*, **28**, 602–632.

13. Tukey, J. W. (1977), Exploratory Data Analysis, Addison Wesley, Reading, Mass.

W. G. S. HINES
R. J. O'HARA HINES

BOX–JENKINS MODEL

Consider a time series* of "length" n denoted by $z_1, \ldots, z_n$. For example, the $z_i (i = 1, \ldots, n)$ might be the weekly wage bills for a large company during a calendar year, with $n = 52$ in that case. Then, if the series $\{z_i : i = 1, \ldots, n\}$ is taken as a realization of some underlying stochastic (time) process* $\{Z_i\}$, we are evidently interested in the statistical structure of this process.

One way of describing such a structure is to obtain a parametric model for the process, and the method of time-series analysis associated with and advocated by Box and Jenkins first postulates a plausible class of models for initial investigation. It then proceeds to tentatively choose or "identify" a promising member from this class, whose particular parameters are next efficiently estimated; and finally, the success of the resulting fit is assessed. The now precisely defined model (identified and estimated) is either accepted by this verification stage or the diagnostic checks carried out will find it wanting in certain respects and should then suggest a sensible modified identification—after which further estimation and checking takes place, and the cycle of identification, estimation, and verification is repeated until a satisfactory fit obtains.

It is important not to overparameterize* the model, since, although this might improve the goodness of fit* for the series history at hand, it is likely to result in the model portraying spurious features of the sampled data, which may detract from the usefulness of the achieved fit. For example, unnecessarily poor forecasts for future observations on the series are a typical outcome of ignoring the principle of parsimonious parameterization.

(Although it is frequently not realized, part of the verification should be to ensure that the fitted model does make sense in the context from which the data were drawn. This model interpretation is sometimes thought of as an extra fourth stage to the Box–Jenkins cycle.)

The univariate models, allowed for by Box and Jenkins, are a very general class of linear process* which may be taken as being driven by some completely random sequence of unknown "shocks," denoted by $\{A_i\}$, say. These A_i are uncorrelated and identically distributed zero-mean random variables, all with the same variance, σ_A^2 say, and $\{A_i\}$ is then referred to as a "white noise"* process. It is frequently convenient to assume that the shocks are in fact normally distributed*, but this assumption then needs to be justified in applications.

The process of interest, $\{Z_i\}$, is considered to be obtained by applying a linear filter* to the shocks $\{A_i\}$, according to

$$Z_i = A_i + \psi_1 A_{i-1} + \psi_2 A_{i-2} + \cdots \quad (1)$$

for suitable choices of $\psi_1, \psi_2, \ldots$. In this representation, each Z_i is taken as being formed from a weighted sum of the current and previous shocks, with psi weights $\psi_0 = 1, \psi_1, \psi_2, \ldots$.

The simplest case is when Z_i is itself completely random, giving for all i,

$$Z_i = A_i,$$

which is just white noise.

Next, we have the so-called moving average models* of general order $q \geqslant 0$, denoted by MA(q) and satisfying

$$Z_i = A_i + \theta_1 A_{i-1} + \cdots + \theta_q A_{i-q}, \qquad (2)$$

which of course reduces to white noise in the special case with $q = 0$. One usually restricts the theta parameters $\theta_1, \ldots, \theta_q$ to values such that the polynomial $\theta_q(\zeta) \equiv 1 + \theta_1 \zeta + \cdots + \theta_q \zeta^q$, in the complex variable ζ, has no zeros within the unit circle. This is the "invertibility" condition.

It is convenient to introduce the backshift operator B such that, for any function $f(\cdot)$ (e.g., z_i or A_i) and all integers i and j, $B^j f(i) \equiv f(i - j)$. Then, for instance, (2) can be rewritten as

$$Z_i = \theta_q(B) A_i, \qquad (3)$$

where $\theta_q(B)$, a polynomial in B of degree q, is an operator obtained by writing B in place of ζ in $\theta_q(\zeta)$.

If θ_j is put at φ_1^j for $j = 1, \ldots, q$ and q is allowed to go to infinity, $\theta_q(B)$ becomes $(1 - \varphi_1 B)^{-1}$ and we can then rewrite (3) as

$$(1 - \varphi_1 B) Z_i = A_i \qquad (4)$$

or, alternatively, as

$$Z_i = \varphi_1 Z_{i-1} + A_i. \qquad (5)$$

Expressions (4) and (5) represent the first-order autoregressive model, or AR(1), in which each Z_i is "autoregressed" on its previous Z_{i-1} value (see AUTOREGRESSIVE–MOVING AVERAGE (ARMA) MODELS).

This type of model generalizes to AR($p + d$), with $p + d \geqslant 0$, the $(p + d)$th-order autoregressive model

$$Z_i = \varphi_1 Z_{i-1} + \cdots + \varphi_{p+d} Z_{i-p-d} + A_i \qquad (6)$$

or, in backshift notation,

$$\varphi_{p+d}(B) Z_i = A_i,$$
$$\varphi_{p+d}(B) = 1 - \varphi_1 B - \cdots - \varphi_{p+d} B^{p+d}$$

and, again, none of the zeros of $\varphi_{p+d}(B)$ must lie within the unit circle. Once more, when $p + d = 0$, (6) reduces to white noise.

Box and Jenkins in fact distinguish between those zeros of $\varphi_{p+d}(B)$ that lie on the unit circle, say d of them, as opposed to those lying outside it, the remaining p zeros. They would then write $\varphi_{p+d}(B)$ in the associated factored form $\phi_p(B) S_d(B)$, where $\phi_p(B)$, the "stationary" autoregressive part, has no zeros on the unit circle. If $d = 0$, this condition ensures the "stationarity"* of the process.

A typical example of $S_d(B)$ is $(1 - B)^d$, where the operator $(1 - B)$ effects a (unit) differencing of the series under study. Thus $(1 - B) z_i = z_i - z_{i-1}$ and $(1 - B)^d$ results in d successive (unit) differences being taken. (It is very rare that a degree of differencing d needs to be greater than 1 or 2.)

Operators of the form $S_d(B)$ are termed "simplifying operators." They represent ways in which the raw series should be transformed initially before detailed statistical analysis is begun. As they yield linear transformations of the data, they do not lead to inefficient analysis in the ways that nonlinear transformations do.

However, in certain instances, notably for the purpose of stabilizing the variance of a series and obtaining more nearly Gaussian behavior, nonlinear transformations of the raw data might be made initially. These are usually of the Box-Cox* [6] form: i.e., $z_i \rightarrow z_i^{(\lambda, m)}$, where

$$z_i^{(\lambda, m)} = \begin{cases} (z_i + m)^\lambda & (\lambda \neq 0) \\ \ln(z_i + m) & (\lambda = 0) \end{cases}$$

and m and λ need to be chosen by the analyst. See BOX − COX TRANSFORMATION—I. But it should be noted that, unless the resulting change in metric appears desirable from other considerations, such transformation tends to be controversial. (For instance, good forecasts might be obtained for the transformed series but the advantage lost on transforming back.) To avoid cumbrous notation, we will moreover suppress the transformation superscripts in what follows.

Note that the general stationary AR(p) model

$$Z_i = \phi_1 Z_{i-1} + \cdots + \phi_p Z_{i-p} + A_i \qquad (7)$$

can also be written as an MA(∞), $Z_i = \phi_p^{-1}(B) A_i$, which is a special case of (1), with $\psi(B) = (1 + \psi_1 B + \psi_2 B^2 + \cdots) = \phi_p^{-1}(B)$. A further generalization, to (7) and (2), is the mixed (stationary) autoregressive moving average model of order (p, q), or ARMA (p, q),

$$Z_i = \phi_1 Z_{i-1} + \cdots + \phi_p Z_{i-p}$$
$$+ A_i + \theta_1 A_{i-1} + \cdots + \theta_q A_{i-q},$$

which, in operator notation, becomes

$$\phi_p(B) Z_i = \theta_q(B) A_i.$$

Introducing unit circle nonstationary zeros into the autoregressive part, we first get models of the form

$$\phi_p(B)(1 - B)^d Z_i = \theta_q(B) A_i, \qquad (8)$$

which are termed autoregressive integrated moving average* models of order (p, d, q), or ARIMA (p, d, q); when $(1 - B)^d$ is replaced by more general $S_d(B)$, the models have been described as ARUMA $(p, d, q)^*$.

Should B be replaced by B^T in (8), where T is some integer greater than unity, we get a purely seasonal model of period T. Such models are usually denoted by

$$\Phi_P(B^T)(1 - B^T)^D Z_i = \Theta_Q(B^T) A_i, \qquad (9)$$

where $(1 - B^T)$ effects a seasonal differencing, according to $(1 - B^T) z_i = z_i - z_{i-T}$, and capital letters help to distinguish (9) from the earlier discussed nonseasonal models. Thus the purely seasonal first-order stationary autoregressive process of period 12, conveniently written as AR$_{12}(1)$, would have the form $Z_i = \Phi_1 Z_{i-12} + A_i$.

Mixed nonseasonal seasonal models can occur. These may be expressed as

$$\phi_p(B)\Phi_P(B^T)(1 - B)^d(1 - B^T)^D Z_i$$
$$= \theta_q(B)\Theta_Q(B) A_i \qquad (10)$$

and indeed, models with more than one seasonal period are possible. For instance, hourly electricity demand over the year would be likely to depend, not only on B and B^{24} (daily), but on B^{168} (weekly) as well. Also, note that multiplicative factors such as

$$(1 - \phi_1 B)(1 - \Phi_1 B^T)$$
$$= 1 - \phi_1 B - \Phi_1 B^T + \phi_1 \Phi_1 B^{1+T}$$

may be generalized to, say,

$$1 - \phi_1 B - \Phi_1 B^T + \alpha B^{1+T}.$$

Finally, univariate models can be written with a deterministic trend $t(i)$ on the right-hand side, although this is frequently removed (at least in part) by the unit and seasonal differencing operators* which are often employed. For instance, if a series contains a linear trend ci, with slope c, simple differencing of the raw data will reduce this to just a constant c, since $(1 - B)ci = c[i - (i - 1)] = c$. The simplest "trend"* occurs when $E[Z_i]$ exists and is a nonzero constant μ. Then $t(i) = \mu$ and to achieve models such as (1), Z_i is replaced by $Z_i - \mu$. So when $\bar{z}$, the series mean, is significantly different from zero (as would be the case in our wages example), the z_i are considered to be replaced by the mean-corrected series, $\{\tilde{z}_i = z_i - \bar{z}\}$, which is a (linear) Box-Cox transform* of the original data.

The family of linear models of the types described above are commonly referred to as Box–Jenkins models. Although they were mostly originally due to earlier workers, Box and Jenkins deserve the credit for bringing together, developing and popularizing an extensive methodology (known as the Box–Jenkins approach), which has been highly successful as a means of analyzing time series met with in a very wide range of application areas. This success is founded on the fact that the various Box–Jenkins models can, among them, mimic the behaviors of diverse types of series—and do so adequately, usually without requiring very many parameters to be estimated in the final choice of model. The disadvantage, however, is that successful analysis generally requires considerable skill—although some quite promising automatic modeling computer packages are beginning to appear.

The formal objective of a Box–Jenkins analysis may be considered as discovering that parsimoniously parameterized filter* which satisfactorily reduces the original series to a residual white noise series $\{a_i\}$, with small variance. What is satisfactory will depend on the context from which the data were drawn, and on the purpose of the analysis, as well as on purely statistical criteria.

The main analytical tool for series identification is the sequence of sample serial correlations*, $\{r_1, \ldots, r_{n-1}\}$, where

$$r_k = \sum_{i=1}^{n-k} \tilde{z}_i \tilde{z}_{i-k} \left/ \sum_{i=1}^{n} \tilde{z}_i^2 \right.,$$

although frequently only about the first quarter of them are computed. These $\{r_k : k = 1, \ldots, n-1\}$ are taken to mimic the theoretical autocorrelations $\{\rho_k\}$, defined as $\text{cov}[Z_i, Z_{i-k}]/\text{var}[Z_i]$, so the task of identification is, given the observed sample correlation pattern for the series, to try to match it with the known population values for some particular process.

For instance, a proper MA(1) model is characterized by $\rho_1 \neq 0, \rho_k = 0(k > 1)$. So a set of serials with r_1 substantial, but later r_k negligible, would suggest that an MA(1) should be tentatively tried. What count as substantial or negligible serial "spikes" depend on the particular model being considered and the length of the observed series. Given these facts, significance tests are available.

Certain structures in the sampled correlations can suggest that a simplifying operator first be applied to the raw data. For example, a slow, roughly linear declining sequence of positive values, for the early r_k, is often taken as an indication that unit differencing is necessary.

Another useful tool for identification is the sequence of partial correlations—sampled $\{p_k\}$, theoretical $\{\pi_k\}$. Thus an AR(1) is characterized by $\pi_1 \neq 0, \pi_k = 0(k > 0)$. So if p_1 is significant, but none of the later p_k are, an AR(1) model would be indicated.

A frequent purpose for analyzing time series is to obtain good forecasts*. Given a series $\{z_1, \ldots, z_n\}$ running up to time $n = $ now, the aim then, typically, is to forecast z_{h+n} at h time intervals hence. If we assume that the generating process has the form (1), it can be shown that the optimal least-squares* forecast, $_h f_n$ say, is the expected value of Z_{h+n} conditional on the information available at time n. Now

$$Z_{h+n} = A_{h+n} + \psi_1 A_{h-1+n} + \cdots$$
$$+ \psi_{h-1} A_{1+n} + \psi_h A_n + \cdots.$$

So $_h f_n = \psi_h a_n + \cdots$ (since the expectations of future shocks are all zero, whereas those for past and present ones take the actual values that have already occurred) and the forecast error, $_h \epsilon_n = z_{h+n} - _h f_n$, is given by

$$_h \epsilon_n = a_{h+n} + \psi_1 a_{h-1+n} + \cdots + \psi_{h-1} a_{1+n},$$

with variance $_h V_n = (1 + \psi_1^2 + \cdots + \psi_{h-1}^2) \cdot \sigma_A^2$, from which probability limits for the forecasts can be obtained, on replacing σ_A^2 by its estimate, the sample variance* of the estimated shock series, $\{\hat{a}_i : i = 1, \ldots, n\}$, which turns out to be just the series of residuals, $\{\hat{a}_i = z_i - \hat{z}_i\}$, where the $\hat{z}_i$ are the estimated values for the original series, as obtained from the fitted model.

Currently, there is much interest in extending the linear models discussed so far, to cater for at least part of the nonlinearity common in some applications areas. One extension gives the bilinear model, which is achieved by introducing additional product terms $Z_{i-u} A_{i-c}$ into the right of the linear model. Other popular generalizations involve substituting time-varying parameters for the constants in (10).

However, Box and Jenkins themselves (in conjunction with their co-workers) have developed certain major extensions to the univariate modeling described above. First, they considered building transfer function* models, which would perhaps improve the forecasts obtained for the series of interest $\{z_i\}$ by extracting relevant information contained in some appropriate leading indicator series $\{y_i\}$, say. This is done by relating the current z to current and previous y, according to a model

$$Z_i = \omega(B) B^\delta Y_i + E_i,$$

where $\omega(B)$ is a linear filter $(\omega_0 + \omega_1 B + \cdots)$, with $\omega_0 \neq 0$, and the B^δ factor indicates that there is a delay of δ units before a y-value can begin to affect the observed z. $\{E_i\}$ is a sequence of error terms, which is assumed to follow some ARMA (p, q) process. In general, parsimonious parameterization can be achieved by writing $\omega(B)$ as the quotient of two finite-length operators, $\alpha(B)$ and $\beta(B)$, say, so that the model fitted has the form

$$Z_i = \frac{\alpha(B)}{\beta(B)} Y_{i-\delta} + \frac{\theta_q(B)}{\phi_p(B)} A_i. \qquad (11)$$

Box and Jenkins [7] provide a well-defined iterative model-building procedure for estimating these transfer function-noise processes, which is analogous to that for the univariate case. Identification of the relationship between the "input" y and "output" z series relies heavily on the cross-correlations between the two series, preferably after a procedure called "prewhitening"* has been effected. Here the filter needed to reduce the y_i to white noise is initially determined and then this prewhitening filter is applied to both the input and the output series before the cross-correlations are computed. (However, not all experts are agreed that this is the best way to "prewhiten.")

Equation (11) can be simply generalized to cater for several input series; again, a deterministic trend can be incorporated on the right, the Z_i and Y_i can be transformed initially, seasonal factors can be introduced into the various filters, and differencing can be employed.

The univariate models can also be generalized to multivariate ones with the basic form

$$\boldsymbol{\phi}(B)\mathbf{Z}_i = \boldsymbol{\theta}(B)\mathbf{A}_i, \qquad (12)$$

where the matrix operators $\boldsymbol{\phi}(B)$ and $\boldsymbol{\theta}(B)$ have elements that are polynomials in B of general finite order, with the restrictions that those along the leading diagonals start with unity, whereas the rest start with powers of B, and the stationarity and (strict) invertibility conditions are, respectively, that all the zeros of the determinants $|\boldsymbol{\phi}(B)|$ and $|\boldsymbol{\theta}(B)|$ lie outside the unit circle.

A further extension is to consider multivariate transfer function models whose basic structure is

$$\mathbf{Z}_i = \boldsymbol{\Omega}(B) \otimes \mathbf{Y}_i + \mathbf{E}_i, \qquad (13)$$

where the transfer function matrix, $\boldsymbol{\Omega}(B)$, has elements of the form $\omega(B)B^\delta$, $\otimes$ denotes the Kronecker product*, and $\mathbf{E}_i$ follows a multivariate model such as (12).

Again, for models (12) and (13), as previously, there is a three-stroke cycle of identification, estimation, and diagnostic checks for obtaining satisfactory fits; and, as well as cross-correlations, partial cross-correlations between individual pairs of residual series are also used. When building any of the models (11), (12), or (13), univariate stochastic modeling—of the sort extensively discussed above—is required for the individual series. So skillful univariate analysis is a prerequisite for all of the more advanced Box–Jenkins methodology.

Finally, in all the processes so far mentioned, it may be necessary to take account of "abnormal" events, such as strikes, changes in the law or freak weather conditions. Box–Jenkins "intervention" models allow such effects to be represented by dummy variables, typically introduced as a filtered pulse on the input side. For instance, to model a step change in level of magnitude Λ, occurring at time I, one needs to include a term $(1 - B)^{-1}\Lambda \prod_i^{(I)}$ on the right, where

$$\prod_i^{(I)} = \begin{cases} 1 & \text{for } i = I \\ 0 & \text{otherwise.} \end{cases}$$

Interventions are thus treated just as particularly simple cases of leading indicator series. (Outliers* and missing observations* can also be dealt with by interpolating realistic values which are estimated from the remaining data.)

Literature

We complete this discussion of the Box–Jenkins model with a brief look at the literature. First, some readable articles: Anderson [2] provides a formula-free introduction, and a more mathematical treatment of the same material is found in Anderson [3]. Newbold [10] gives an excellent treatment at about the same level, and Anderson [4] states a later updated view.

As for books, Anderson [1] and Nelson [9] supply simple introductions, Box and Jenkins [7] is the best work for reference purposes, and Jenkins (in Anderson [5]) provides about the only reputable published account of the more advanced topics, although Granger and Newbold [8] is of related interest for multivariate* modeling, as is Newbold and Reed (in Anderson [5]).

REFERENCES

1. Anderson, O. D. (1975). *Time Series Analysis and Forecasting: The Box–Jenkins Approach*. Butterworths, London. Second Edition, 1982.

2. Anderson, O. D. (1976). *Statist. News*, No. 32 (February), 14–20.

3. Anderson, O. D. (1976). *Math. Sci.*, **1** (January), 27–41.

4. Anderson, O. D. (1977). *The Statistician*, **25** (December), 285–303.

5. Anderson, O. D., ed. (1979). *Forecasting*. North-Holland, Amsterdam.

6. Box, G. E. P. and Cox, D. R. (1964). *J. R. Statist. Soc. B*, **26**, 211–252.

7. Box, G. E. P. and Jenkins, G. M. (1970). *Time Series Analysis: Forecasting and Control*. Holden-Day, San Francisco.

8. Granger, C. W. J. and Newbold, P. (1977). *Forecasting Economic Time Series*. Academic Press, New York.

9. Nelson, C. R. (1973). *Applied Time Series Analysis for Managerial Forecasting*. Holden-Day, San Francisco.

10. Newbold, P. (1975). *Operat. Res. Quart.*, **26**, 397–412.

See also AUTOREGRESSIVE–INTEGRATED MOVING AVERAGE (ARIMA) MODELS; AUTOREGRESSIVE–MOVING AVERAGE (ARMA) MODELS; BUSINESS FORECASTING METHODS; and PREDICTION AND FORECASTING.

O. D. ANDERSON

BOX–MULLER TRANSFORMATION

If U_1 and U_2 are independent standard uniform* variables then

$$X_1 = (-2 \ln U_1)^{1/2} \cos 2\pi U_2$$

$$X_2 = (-2 \ln U_1)^{1/2} \sin 2\pi U_2$$

are independent unit normal* variables [1].

By means of this transformation, random normal deviates can be derived directly from a source of random uniform variates, such as a table of random numbers* or a generator program in a digital computer.

In ref. 1 it is pointed out that simple extensions of the method provide values of random variables distributed as $\chi^{2*}, F^*,$ and t^*. For a more recent discussion, see Golder and Settle [2].

REFERENCES

1. Box, G. E. P. and Muller, M. E. (1958). *Ann. Math. Statist.*, **29**, 610–611.

2. Golder, E. R. and Settle, J. G. (1976). *Appl. Statist.*, **25**, 12–20.

See also GENERATION OF RANDOM VARIABLES, COMPUTER and VERTICAL DENSITY REPRESENTATION.

BOXPLOT, BIVARIATE

The univariate boxplot (or box-and-whiskers plot*) of Tukey [15] is a well-known exploratory tool (*see* EXPLORATORY DATA ANALYSIS). Two examples of univariate boxplots are shown in Fig. 1, which displays the weight and engine displacement of 60 cars [2, pp. 46–47]. The univariate boxplot of the weights is drawn along the x-axis. There are no observations outside the fence, which is indicated by two dashed vertical bars. The whiskers are the horizontal lines going from the box to the most extreme points inside the fence. Along the y-axis we see the univariate boxplot of the engine displacements. The points lying outside of the fence are flagged as outliers*. (Note that the Camaro and the Caprice share the same engine displacement, as do the Mustang and the Victoria.)

The univariate boxplot is based on ranks, since the box goes from the observation with rank $\lfloor n/4 \rfloor$ to that with rank $\lceil 3n/4 \rceil$, and the central bar of the box is drawn at the median. A natural generalization of ranks to multivariate data is Tukey's notion of *halfspace depth* [14,3]. Using this concept, Rousseeuw and Ruts [12] constructed a bivariate version of the boxplot. Its main components are a *bag* that contains 50% of the data points, a *fence* that separates inliers from outliers, and *whiskers* from the bag to points inside the fence. The resulting graph is called a *bagplot*.

Consider again Fig. 1. The Tukey median, i.e. the point with highest halfspace depth, lies at the center and is indicated by an asterisk (*see also* MULTIVARIATE MEDIAN and REGRESSION DEPTH). The bag is the polygon drawn as a full line, and the fence is the larger polygon drawn as a dashed line. The whiskers go to the observations that lie outside the bag but inside the fence. We also see four observations outside the fence. These outliers* are indicated by small black triangles and are labeled. One inlier is labeled because it lies close to the fence, so it is a boundary case.

Like the univariate boxplot, the bagplot also visualizes several characteristics of the data: its location (the median), spread (the size of the bag), skewness (the shape of the bag and the whiskers), and tails (the long whiskers and the outliers).

To illustrate these characteristics, Fig. 2 contains the bagplots of two generated data sets, both with 100 points. Their medians (indicated by asterisks) are far apart. (The data points inside the bags are not shown, to avoid overplotting.) The bags are of roughly the same size (area), so the data sets have a similar spread. But the bags have a different orientation: the left one slopes upward (positive correlation) and the right one slopes downward. We also see that the first data set is very skewed, for the median lies in the lower left part of the bag, where the whiskers are also short, whereas the right part of the

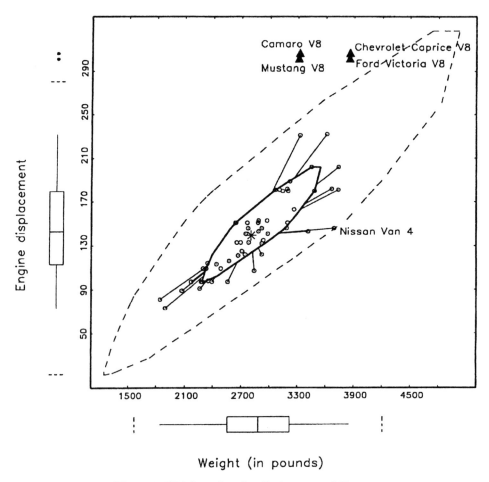

Figure 1. Weight and engine displacement of 60 cars.

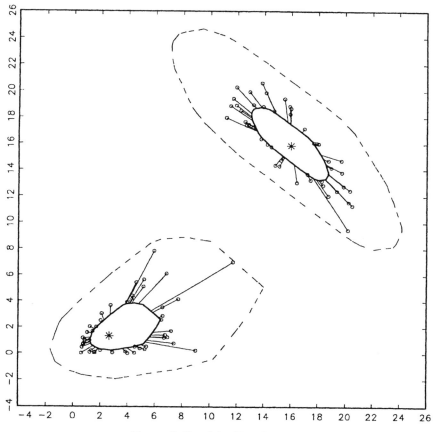

Figure 2. Bagplots of two data sets.

bag is wider and has much longer whiskers. By contrast, the bagplot of the second data set is nicely balanced, and its form suggests an elliptic distribution. Finally, both data sets are medium-tailed, judging from the length of the whiskers and the absence of outliers.

CONSTRUCTION OF THE BAGPLOT

The halfspace location depth ldepth(θ, Z) of a point $\theta \in \mathbb{R}^2$ relative to a bivariate data cloud $Z = \{\mathbf{z}_1, \mathbf{z}_2, \ldots, \mathbf{z}_n\}$ is the smallest number of $\mathbf{z}_i$ contained in any closed halfplane with boundary line through θ. A time-efficient algorithm for ldepth(θ, Z) was provided in ref. [10]. The depth region D_k is the set of all θ with ldepth$(\theta, Z) \geqslant k$, and was constructed in ref. [13]. The depth regions are convex polygons, and $D_{k+1} \subset D_k$.

The *Tukey median* $\mathbf{T}^*$ of Z is defined as the θ with highest ldepth(θ, Z) if there is only one such θ. Otherwise, $\mathbf{T}^*$ is defined as the center of gravity of the deepest region. An algorithm for $\mathbf{T}^*$ was obtained in [11].

The bag B is constructed as follows. Let $\#D_k$ denote the number of data points contained in D_k. One first determines the value k for which $\#D_k \leqslant \lfloor n/2 \rfloor < \#D_{k-1}$, and then one interpolates between D_k and D_{k-1} (relative to the point $\mathbf{T}^*$) to obtain the set B for which $\#B = \lfloor n/2 \rfloor$. The bag B is again a convex polygon.

The fence is obtained by inflating B (relative to $\mathbf{T}^*$) by a factor of 3. A whisker is a line segment connecting an observation between the bag and the fence to the boundary of B, in the direction of $\mathbf{T}^*$. This generalizes the whiskers of the univariate case. The entire bagplot is constructed by the algorithm given in ref. [12].

When the observations $\mathbf{z}_i = (x_i, y_i)$ are subjected to a translation and/or a nonsingular linear transformation, their bagplot is transformed accordingly, because the halfspace depth is invariant under such mappings. Consequently, the points inside the bag remain inside, the outliers remain outliers, and so on.

EXAMPLE

Figure 3 plots the concentration of plasma triglycerides against that of plasma cholesterol for 320 patients with evidence of narrowing arteries [5, p. 221]. Here the Tukey median T^* is represented by the "nucleus" of the bag, obtained by deflating the bag relative to T^*. Because there are many points, the number of whiskers is reduced: from each edge of the bag only the major whisker is drawn. The fence (dashed line) originally extended below the x-axis, which is not possible for the actual data because the y-values are concentrations. Therefore that part of the

fence was truncated. The bagplot highlights three outliers, which are indicated by black triangles.

OTHER BIVARIATE BOXPLOTS

Becketti and Gould [1] considered the univariate boxplot of x (as in Fig. 1 above), from which they keep the median, the quartiles, and the endpoints of both whiskers. They do the same for the y-variable, and then use these numbers to draw vertical and horizontal lines on the scatterplot, thereby forming a cross and a rectangle. Lenth [6] modified this plot to put more emphasis on the univariate quartiles. However, neither version reflects the bivariate shape and correlation of the data.

Goldberg and Iglewicz [4] proposed two generalizations of the boxplot which are truly bivariate. When the data can be assumed to

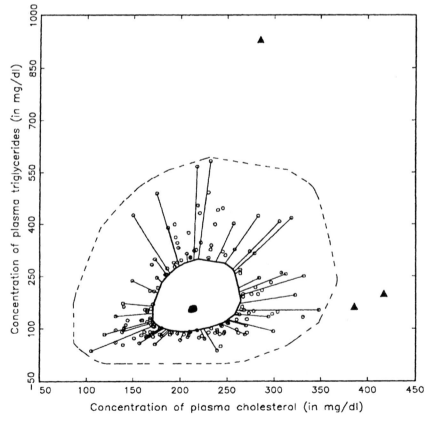

Figure 3. Concentrations of cholesterol and triglycerides in plasma of 320 patients.

be elliptically symmetric, they construct a *robust elliptic plot* (relplot); the "box" is an ellipse, obtained by a robust method such as the *minimum volume ellipsoid estimator* (MVE) proposed in ref. [9]. For asymmetric data they construct a *quarter-elliptic plot* (quelplot), where the "box" consists of four quarter ellipses, computed by a kind of M-estimator*.

The bagplot differs from the relplot and the quelplot in that its shape is more general. Whereas the relplot-quelplot approach estimates parameters of (nearly) elliptical models, the bagplot is model-free because the halfspace depth is. Further variations on the bagplot are currently under development [7,8].

HIGHER DIMENSIONS

The halfspace depth and the Tukey median exist in any dimension, so the bag can still

be defined. In three dimensions the bag is a convex polyhedron and the whiskers stick out in all directions, but in more dimensions this becomes hard to visualize. However, in any dimension one can draw the *bagplot matrix* which contains the bagplot of each pair of variables, as in Fig. 4. Each diagonal cell is the bagplot of a variable against itself; hence all the points lie on the 45° line. By definition, such a bagplot reduces to the usual univariate boxplot.

REFERENCES

1. Becketti, S. and Gould, W. (1987). Range-finder box plots. *Amer. Statist.*, **41**, 149.
2. Chambers, J. M. and Hastie, T. J. (1992). *Statistical Models in S*. Wadsworth and Brooks, Pacific Grove, Calif.
3. Donoho, D. L. and Gasko, M. (1992). Breakdown properties of location estimates based on halfspace depth and projected outlyingness. *Ann. Statist.*, **20**, 1803–1827.

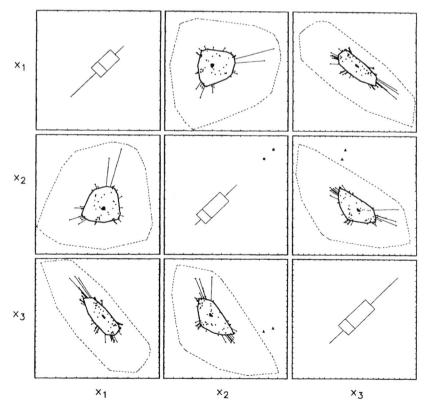

Figure 4. Bagplot matrix of 85 observations in three dimensions.

4. Goldberg, K. M. and Iglewicz, B. (1992). Bivariate extensions of the boxplot. *Technometrics*, **34**, 307–320.

5. Hand, D. J., Daly, F., Lunn, A. D., McConway, K. J., and Ostrowski, E. (1994). *A Handbook of Small Data Sets*. Chapman and Hall, London.

6. Lenth, R. (1988). Comment on Becketti and Gould (with reply by Becketti). *Amer. Statist.*, **42**, 87–88.

7. Liu, R. Y., Parelius, J. M., and Singh, K. (1998). Multivariate analysis by data depth; descriptive statistics, graphics and inference. Tech. Rep., Dept. of Statistics, Rutgers Univ., Rutgers, NJ.

8. Romanazzi, M. (1997). A bivariate extension of the box-plot based on halfspace depth. Presented at the Third Conference on Statistical Data Analysis Based on the L^1 Norm and Related Methods, Neuchâtel, Switzerland.

9. Rousseeuw, P. J. (1984). Least median of squares regression. *J. Amer. Statist. Ass.*, **79**, 871–880.

10. Rousseeuw, P. J. and Ruts, I. (1996). AS 307: Bivariate location depth. *Appl. Statist.*, **45**, 516–526.

11. Rousseeuw, P. J. and Ruts, I. (1998). Constructing the bivariate Tukey median. *Statist. Sinica*, **8**.

12. Rousseeuw, P. J. and Ruts, I. (1997). The bagplot: a bivariate box-and-whiskers plot. Tech. Rep. 97-28, Univ. of Antwerp, Antwerp, Belgium. Abstract: *IMS Bull.*, **26**, 102.

13. Ruts, I. and Rousseeuw, P. J. (1996). Computing depth contours of bivariate point clouds. *Comput. Statist. and Data Anal.*, **23**, 153–168.

14. Tukey, J. W. (1975). Mathematics and the picturing of data. *Proc. Int. Congr. Math.*, **2**, 523–531.

15. Tukey, J. W. (1977). *Exploratory Data Analysis*. Addison-Wesley, Reading, Mass.

See also EXPLORATORY DATA ANALYSIS; GRAPHICAL REPRESENTATION OF DATA; MULTIVARIATE MEDIAN; and NOTCHED BOX-AND-WHISKER PLOT.

PETER J. ROUSSEEUW
IDA RUTS

BRADFORD DISTRIBUTION

The Bradford distribution is a Pearson Type VI distribution, truncated from above, with density function

$$[\theta / \log(1 + \theta)](1 + \theta x)^{-1}$$

$$(0 < x < 1; \theta > -1).$$

It is related to the Yule* and Zipf* distributions. Bradford [1] used it in connection with distributions of frequency of citation of references.

Additional information on the Bradford distribution is provided in [2], Chapter 28, Sec. 8.

REFERENCES

1. Bradford, S. C. (1948). *Documentation*. Crosby Lockwood, London.

2. Johnson, N. L., Kotz, S., and Balakrishnan, N. (1995). *Continuous Univariate Distributions*, Vol. 2. (2nd ed.). Wiley, New York.

FURTHER READING

Asai, I. (1981). A general formulation of Bradford's distribution; the graph-oriented approach. *J. Amer. soc. Information Sci.*, **32**, 113–119.

Brookes, B. C. (1981). A critical commentary on Leimkuhler's "exact" formulation of the Bradford law. *J. Documentation*, **37**, 77–88.

Egghe, L. (1985). Consequences of Lutka's law for the law of Bradford. *J. Documentation*, **41**, 173–189.

Maia, M. J. F. and Maia, M. D. (1984). On the unity of Bradford's law. *J. Documentation*, **40**, 206–216.

Morse, P. M. (1981). Implications of the exact Bradford distribution. *J. Amer. Soc. Information Sci.*, **32**, 43–50.

See also PARETO DISTRIBUTION.

BRADLEY–TERRY MODEL

A model representing the results of experiments in which responses are pairwise rankings of treatments (so-called paired comparison experiments*). It assigns preference probabilities to each of the $\binom{t}{2}$ pairs among t treatments in terms of a set of $t - 1$ parameters. Specifically, it postulates "treatment parameters" $\pi_1, \ldots, \pi_t, \pi_i \geqslant 0, i = 1, \ldots, t$, associated with treatments $T_1, \ldots, T_t$. These parameters represent relative selection probabilities for the treatments, subject to the constraints $\pi_i \geqslant 0 (i = 1, \ldots, t)$ and $\sum_{i=1}^{t} \pi_i = 1$.

The probability that treatment T_i is preferred over treatment T_j in a single comparison is $\pi_i/(\pi_i + \pi_j)$ for all i and j.

Let a_{ij} denote the *number* of times treatment T_i is preferred over treatment T_j in n_{ij} comparisons of T_i and T_j. The likelihood function* is then given by

$$L(\boldsymbol{\pi}, \mathbf{A}) = \prod_{i=1}^{t} \pi_i^{a_i} \prod_{i<j} (\pi_i + \pi_j)^{-n_{ij}},$$

where $a_i = \sum_{j \neq i} a_{ij}$ and $\boldsymbol{\pi}$ and $\mathbf{A}$ are vectors of π's and a's, respectively. The maximum likelihood estimators* $\hat{\pi}_1, \ldots, \hat{\pi}_t$ of $\pi_1, \ldots, \pi_t$ are solutions of the equations

$$\hat{\pi}_i = \frac{a_i}{\sum_{j \neq i} n_{ij}/(\hat{\pi}_i + \hat{\pi}_j)} \quad (i = 1, \ldots, t)$$

subject to $\sum_{i=1}^{t} \hat{\pi}_i = 1$.

For testing the null hypothesis* of equality of treatment selection probabilities,

$$H_0 : \pi_1 = \pi_2 = \cdots = \pi_t = 1/t$$

against the alternative,

$$H_1 : \pi_i \neq \pi_j \quad \text{for some} \quad i, j, i \neq j,$$
$$i, j, = 1, \ldots, t,$$

the likelihood ratio statistic* is

$$-2 \ln \lambda_1 = 2N \ln 2 - 2B,$$

where $N = \sum_{i<j} n_{ij}$ and $B_1 = \sum_{i<j} n_{ij} \ln(\hat{\pi}_i + \hat{\pi}_j) - \sum a_i \ln \pi_i$. For large n_{ij}, the statistic $-2 \ln \lambda_1$ is distributed under H_0 approximately as a χ_{t-1}^2 variable. Tables of values of B_1, together with exact significance levels, are given by Bradley and Terry [3] and Bradley [1]. Factorial experiments* based on this model are discussed by Littell and Boyett [5]. Extensions were also studied by van Baaren [6].

REFERENCES

1. Bradley, R. A. (1954). *Biometrika*, **41**, 502–537.
2. Bradley, R. A. (1976). *Biometrics*, **32**, 213–232. (A comprehensive article covering various extensions as well.)
3. Bradley, R. A. and Terry, M. E. (1952). *Biometrika*, **39**, 324–345.
4. Dykstra, O. (1960). *Biometrics*, **16**, 176–188.
5. Littell, R. C. and Boyett, J. M. (1977). *Biometrika*, **64**, 73–77.
6. van Baaren, A. (1978). *Statist. Neerlandica*, **32**, 57–66.

FURTHER READING

Glickman, M. E. (1999). Parameter estimation in large dynamic paired comparison experiments. *Appl. Statist.*, **48**, 377–394.

Li, L. and Kim, K. (2000). Estiamting driver crash risks based on the extended Bradley-Terry model: An induced exposure method. *J. Roy. Statist. Soc. A*, **163**, 227–240.

Hastie, T. and Tibshirani, R. (1998). Classification by pairwise coupling. *Ann. Statist.*, **26**, 451–471.

Matthews, J. N. S. and Morris, K. P. (1995). An application of Bradley-Terry-type models to the measurement of pain. *Appl. Statist.*, **44**, 243–255.

Simons, G. and Yao, Y.-C. (1999). Asymptotics when the number of parameters tends to infinity in the Bradley-Terry model for paired comparisons. *Ann. Statist.*, **27**, 1041–1060.

Stigler, S. M. (1994). Citation patterns in the journals of statistics and probability. *Statist. Sci.*, **9**, 94–108.

See also PAIRED COMPARISONS.

BRADSTREET INDEX.   See INDEX NUMBERS

BRANCH-AND-BOUND METHOD

The method of branch and bound was originally devised by Land and Doig [7] and independently by Little et al. [8] to solve integer programs*. It has since proved to be so powerful and general that its use has broadened enormously, and now includes a number of statistical applications. Since all but illustrative examples must be run on computers, this extension has also gone hand in hand with the development in computer science of tree-processing algorithms.

Branch and bound is useful for solving problems (usually combinatoric in nature) that can be manipulated into the following form:

1. One wishes to find the minimum of some function over a denumerable but usually large discrete set of possibilities. (Maximization problems can be solved by a trivial modification.)

2. The possibilities can be regarded as the terminal nodes of a tree.

3. The function that one wishes to minimize can be extended to the interior nodes* of a tree* (*see* GRAPH THEORY) in such a way that its value at any node cannot exceed that at any successor node and must equal or exceed that at any predecessor node.

Selection of a subset regression* provides a good example. Suppose that we have a problem with 11 predictors and seek the best subset of 5. We can set up a tree whose first node represents the regression on all 11 predictors. This has 11 successor nodes corresponding respectively to the deletion of predictors $1, 2, \ldots, 11$. Each of these nodes generates a number of successors corresponding to the deletion of a second variable, and so on. (It is obviously sensible, but is not necessary theoretically, to apply additional rules to ensure that each such subregression appears only once in the tree.) Continuing to define additional levels of the tree as corresponding to the deletion of further variables, we end up with a tree whose terminal nodes correspond to the 462 possible subsets of 5 predictors. The function we wish to minimize is the residual* sum of squares for that subset, and this extends at once to the interior nodes and satisfies requirement 3. The problem is then ripe for solution by branch and bound.

The solution of a problem by branch and bound involves traversing the tree (conceptually at least) from top to bottom. From the time the method first reaches a terminal node, it maintains a record of v, the function value at the best node located so far, and this is used to prune the tree. If any interior node has a value exceeding v, then so by property 3 must all its successors, so these need not be looked at. The secret of a successful application of branch and bound lies in pruning so successfully that only a tiny fraction of the branches need be followed down to their terminal nodes.

An example adapted from Narula and Wellington [10] is illustrated in Fig. 1. It shows a four-predictor regression problem; at each node the circled numbers show the variables included in the regression, while the superscript number is the sum of relative errors (SRE) for that subregression. By introducing variables in stepwise regression* order, we get successively the subsets 1, 12, 123, and 1234 with SREs 2.47, 2.45, 2.22, and 2.20, respectively.

Let us suppose that we wish to find the best subset of size 2. The bound v is initially 2.45 (from the set 12 used on the way to the root of the tree). Traversing the tree from the right, we may then prune at the 234 node at once since its SRE of 2.89 exceeds v. The 134 node passes the pruning test and its successor nodes are investigated. Node 14 fails the test, but 13 has an SRE of 2.22, which becomes the new value of v. The 134 node has now been "fathomed" (i.e., analyzed completely). Next, the 124 node is investigated, but as its SRE of 2.44 exceeds the current value of $v = 2.22$, it is pruned. The 123 node may be

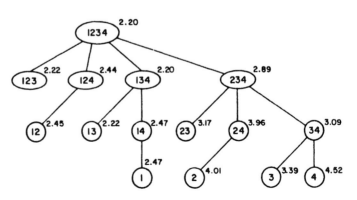

Figure 1.

ignored, as it has no successors corresponding to two-predictor regressions, so the problem is solved.

In this example we used the subset of size 2 found by stepwise regression* to provide what turned out to be an excellent starting bound v, and by putting the best predictor first and the worst last, were able to prune very successfully. If it were not convenient to compute this bound initially, it would be better to reorder the variables from worst to best. This ensures that the 12 node is the first of size 2 investigated by the algorithm, and again leads to effective pruning.

Another important principle is illustrated. If one wants not the best solution, but any one guaranteed to be within say 5% of optimum, this is given by any two-variable solution within 5% of the four-variable SRE or the best three-variable SRE. This possibility of finding not necessarily optimal but good solutions is often useful.

Much of the successful application of branch and bound is an art. First, it is necessary to manipulate the problem into a tree structure and extend the function's definition to the interior nodes, which are usually artificial constructs. Then it is extremely important to arrange the tree so that good terminal nodes are found quickly. Logically equivalent formulations that differ in one or both of these two aspects may lead to solution times differing by a factor of 1000 or more. Unfortunately, although the general principles involved are very simple, each application area of branch and bound has its own special structure and features, and a successful implementation depends on a recognition of these.

Further Reading

There is little published on branch and bound per se, but a welter of information on its applications, especially to integer programming problems: numbers of the latter may be found, for example, in every volume of *Operations Research**, which may be considered the natural home for mathematical writings on the topic. A useful exposition of the mathematical properties of branch and bound has been given by Mitten [9], and a very readable commentary on computer implementation of the method by Baker [2].

A discussion of the solution of subset regressions problems by branch and bound is given in the extremely important paper by Furnival and Wilson [4], and Narula and Wellington [10] extend the ideas to include use of the L_1 norm instead of least squares*. In the examples of the first of these papers, the number of operations in a p predictor regression was 88×1.4^p, a much more slowly varying function that the $O(2^p)$ required for investigating all subsets, and one that implies that as many as 30 predictors may be handled quite comfortably. Both these papers are quite technical and devote much attention to "labeling," which is computationally important because it defines the way in which the tree is traversed. The regression problem in which coefficients are constrained to be nonnegative is set out in a readable account by Armstrong and Frome [1]. Other short and easy-to-follow papers describing the application of branch-and-bound methods to statistical problems are on finding the nearest neighbors* in a set of points A to the points in set B [3], cluster analysis* [6], and determination of stochastic rankings [11], and Hawkins [5] points out that the Furnival and Wilson code may also be used for subset discriminant analysis*.

REFERENCES

1. Armstrong, R. D. and Frome, E. L. (1976). *Technometrics*, **18**, 447–450.

2. Baker, K. R. (1974). *Introduction to Sequencing and Scheduling*. Wiley, New York.

3. Fukunaga, K. and Narendra, P. M. (1975). *IEEE Trans. Computers*, **C-24**, 750–753.

4. Furnival, G. M. and Wilson, R. W. (1974). *Technometrics*, **13**, 403–408.

5. Hawkins, D. M. (1976). *J. R. Statist. Soc. B*, **38**, 132–139.

6. Koontz, W. L. G., Narendra, P. M. and Fukunaga, K. (1975). *IEEE Trans. Computers*, **C-24**, 908–914.

7. Land, A. and Doig, A. (1960). *Econometrica*, **28**, 497–520.

8. Little, J. D. C., Murty, K. G., Sweeney, D. W., and Karel, C. (1963). *Operat. Res.*, **11**, 972–989.

9. Mitten, A. (1970). *Operat. Res.*, **18**, 24–34.

10. Narula, S. C. and Wellington, J. F. (1979). *Technometrics*, **21**, 299–306.

11. Singh, J. (1976). *Ann. Statist.*, **4**, 651–654.

FURTHER READING

Dotzauer, E., Jönsson, H. and Ravn, H. F. (2001). Optimal unit commitment by branch-and-bound exploiting dual optimality conditions. *Theor. Stoch. Proc.*, **7**, 74–87.

Leenen, I. and van Mechelen, I. (2001). An evaluation of two algorithms for hierarchical classes analysis. *J. Classification*, **18**, 57–80.

Niemann, H. and Goppert, R. (1988). An efficient branch-and-bound nearest neighbor classifier. *Pattern Recog. Lett*, **7**, 67–72.

Wah, B. W. and Yu, C. F. (1985). Stochastic modeling of branch-and-bound algorithms with best-first search. *IEEE Software Eng.*, **11**, 922–934.

Welch, W. J. (1982). Branch-and-bound search for experimental designs based on D-optimality and other criteria. *Technometrics*, **24**, 41–48.

See also Graph Theory; Integer Programming; Stepwise Regression; and Traveling-Salesman Problem.

D. M. Hawkins

BRANCHING PROCESSES

Branching processes originate as models for the development of a population, the members of which are termed *individuals* (or *particles*) and which are each capable of reproducing. The basic assumption, the *branching property*, states that different individuals reproduce *independently* of each other and, more generally, that their whole lines of descent are independent.

Most applications arise in biology and physics. Obvious examples in biology are the growth of a population of cells, say in a tumor, or the time evolution of a species within a closed area (in examples of this type, one often considers a single-sex model in order to maintain the branching property). The population modeled by the branching process could also be the individuals in some larger population having some specific characteristic, say carrying a mutant gene inherited by some of the offspring or having an infectious disease that may be transmitted to healthy individuals. A notable example in physics is the neutron chain reaction in nuclear reactors and bombs; and a number of special cascade phenomena, in particular in cosmic rays (electron–photon), have received considerable attention.

THE GALTON–WATSON PROCESS

The prototype of a branching process in discrete time is the Galton–Watson process* $Z_0, Z_1, \ldots$ (historically connected also with the name of Bienaymé)*. The individuals reproduce independently and each according to the *offspring distribution*, specified by the probabilities p_k of getting k children. That is, if $X_{n,i}$ represents the number of children of the ith individual of the nth generation, then

$$Z_{n+1} = \sum_{i=1}^{Z_n} X_{n,i}$$

and the $X_{n,i}$ are independent with $P(X_{n,i} = K) = p_k$. The realizations of the process are often depicted in a *family tree* (Fig. 1). Because of the branching property, there is essentially no loss of generality in assuming that $Z_0 = 1$ since otherwise one just adds Z_0 independent copies. To avoid trivialities, assume also that $p_k < 1$ for all k.

The Galton–Watson process is a time-homogeneous Markov chain with state space $\{0, 1, 2, \ldots\}$. State 0 is absorbing and all other states transient, and almost all realizations belong either to the set $E = \{Z_n = 0$ eventually$\}$ of extinction or to $\{Z_n \to \infty\}$. Incidentally, this instability of the process (typical also of many generalizations) is a severe objection to the model in many applied situations since it predicts either infinite growth (irrespective of capacity of the environment) or ultimate extinction. For this reason, branching models are often appropriate only as approximations for the population development in its early stages.

Exact distribution results are available if the offspring distribution is modified geometric*, $p_k = cb^k, k > 0$, in which case the distribution of Z_n is again of the same type (with parameters depending on n). Simple recurrence formulas for the probability generating functions* (PGF) $f_n(s) = Es^{Z_n}$ follow immediately from the branching property (Fig. 2). In fact, letting $f(s) = f_1(s) = \sum_0^\infty s^k p_k$ be the offspring PGF, one gets $f_{n+1}(s) = f_n(f(s)) = \cdots = f^{0(n+1)}(s)$, the $(n+1)$st functional iterate of f. The extinction probability $q = P(Z_n = 0$ eventually$)$ can be found as the smallest root of the equation $s = f(s)$.

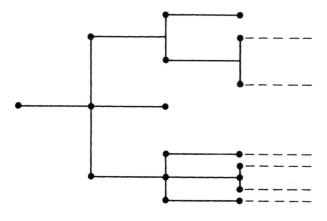

Figure 1. Family tree of a Galton–Watson path.

In a closer study of the limit theory, three main cases arise according to the value of the offspring mean $m = \sum_0^\infty k p_k$.

Case 1. The supercritical case $1 < m < \infty$. Here $q < 1, \{W_n\} = \{Z_n/m^n\}$ is a nonnegative martingale* with unit mean and hence a.s. convergent to a limit $W \in [0, \infty)$. The process indeed grows at rate m^n under weak regularity conditions. In fact, $0 < W < \infty \{=$ $Z_n \to \infty\}$ a.s. if and only if

$$\sum_{k=2}^\infty k \log k \, p_k < \infty \qquad (1)$$

while $W = 0$ a.s. otherwise.

Case 2. The Critical Case $m = 1$. Here $q = 1$. More precisely (with a second moment

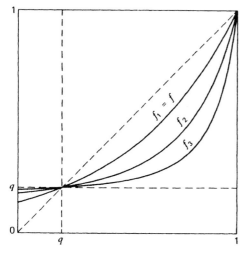

Figure 2. Iterated probability generating functions.

assumption), $nP(Z_n \neq 0)$ has a limit and conditioned on $\{Z_n \neq 0\}$, Z_n/n has a limiting exponential distribution*.

Case 3. The Subcritical Case $0 < m < 1$. Here $q = 1$ and conditional on $Z_n \neq 0, Z_n$ has a limiting distribution (the *Yaglom limit*). Furthermore, $P(Z_n \neq 0) \simeq m^n$ if and only if (1) holds.

Numerous generalizations and refinements of these results have been obtained. The flavor of the subcritical and critical case is analytic (expansions of generating functions*), while in the supercritical case almost sure (a.s.) convergence and martingales* also enter.

OTHER BRANCHING PROCESSES

The continuous-time counterpart of the Galton–Watson process* is the *Markov* branching process* $Z_{t \; t \geq 0}$, where each individual alive at time t reproduces in $[t, t+h]$ with probability $\beta h + o(h)$. A notable example is the linear birth-and-death process*, where $\beta = \lambda + \mu$ with λ the individual birth rate and μ the individual death rate. A Markov branching process has the property that any equidistant discrete skeleton $Z_0, Z_\delta, Z_{2\delta}, \ldots$ is a Galton–Watson process. (The skeletons of birth-and-death processes correspond to Galton–Watson processes with modified geometric offspring distribution.) As applied probability models, the Galton–Watson and Markov branching process are in general very crude approximations, so in most situations, one needs to incorporate features specific to the phenomenon under study. An obvious example is the need to study *age-dependent branching processes*. Examples of such are the *age-dependent birth-and-death process*, with individual birth and death rates $\lambda(a)$ and $\mu(a)$ depending on the age a, and the *Bellman–Harris process*, where reproduction takes place at the end of the life and the lifetime distribution is general (the exponential lifetime distribution corresponds to the Markov branching process). Both are special cases of the very general *Crump–Mode–Jagers–process*, where the distribution of the birth instants is a general point process, not necessarily independent of the life-length distribution. Also, one sometimes needs to introduce a spatial factor; e.g., to study the speed of spread of an epidemic. A simple way to do this is to superimpose on the branching a particle movement. *Branching diffusions* and *branching random walks** are examples of this and have received much attention.

An important way to generalize the simplest (one-type) models is by allowing for a finite number of types, each of which may produce offspring (of any type). This setting is very flexible from the application point of view: age-dependent models can be approximated by letting types be particles in various age groups, and spatial dependence can be treated by letting the types be particles in the elements of a partition of the region; and in epidemics, one can allow for several periods of the illness, such as incubation, latent, and infectious phases.

Passing to a continuum, we may consider a distribution of types and treat factors such as age and location within a unified setting. In some main cases (typical of which are age-dependent models and branching diffusions on bounded domains), the type distribution approaches a stable* one, the process grows from then on exponentially (the rate is the *Malthusian* parameter*), and initial conditions enter through a factor, the *reproductive value*. But the type distribution may also (typically in branching random walks and diffusions on unbounded domains) drift off to infinity.

Also, branching processes with *immigration* have been considered. One of their main features is that in the subcritical case, a stationary distribution will typically exist. From the point of view of many applications, this is intuitively appealing, although the immigration will often be an artifice. Instead, it might be more appropriate to consider *controlled* branching processes, where the branching mechanism depends on the state, but the theory of such processes is so far comparatively incomplete.

Finally, we mention branching processes with *varying environment*, i.e., permitting time inhomogeneity, and with *random environment*, i.e., a varying environment determined as the outcome of a stochastic process*.

BIBLIOGRAPHY

Athreya, K. B. and Ney, P. (1972). *Branching Processes*. Springer-Verlag, Berlin. (The standard monograph since its publication. Not all parts are still up to date.)

Harris, T. E. (1963). *The Theory of Branching Processes*. Springer-Verlag, Berlin. (The first standard reference.)

Heyde, C. C. and Seneta, E. (1977). *I. J. Bienaymé: Statistical Theory Anticipated*. Springer-Verlag, Berlin. (Gives some early history.)

Jagers, P. (1975). *Branching Processes with Biological Applications*. Wiley, New York. (Broad introduction.)

Joffe, A. and Ney, P. (eds.) (1978). *Branching Processes: Advances in Probability and Related Topics*, Vol. 5. Marcel Dekker, New York. (Contains a number of up-to-date surveys of special areas.)

Kendall, D. G. (1966). *J. Lond. Math. Soc.*, **41**, 385–406. (Historical.)

See also BIRTH-AND-DEATH PROCESSES; EPIDEMIOLOGICAL STATISTICS—I; GALTON–WATSON PROCESS; MARKOV PROCESSES; and STATISTICAL GENETICS;

S. ASMUSSEN

BRAZILIAN JOURNAL OF PROBABILITY AND STATISTICS

[This entry has been updated by the Editors.]

The *Brazilian Journal of Probability and Statistics (Revista Brasileira de Probabilidade e Estatística)* is the official publication of the Brazilian Statistical Association (website www.ime.usp.br/~abe/) and is issued twice a year. The first issue was published in 1987.

The journal is supported in part by the Brazilian Statistical Association and also by the Research Council of Brazil.

The aim of the journal is to further advance statistical methods, to promote the applications of probabilistic and statistical methods in other fields, and to provide an international forum for matters of general interest to probabilists and statisticians. It is open to all areas of the theory, methodology, and application of probability and statistics. Research papers, survey articles, technical notes, book reviews, and case studies are welcome.

One of the goals of the journal is to achieve an equilibrium between theoretical and applied articles. The following types of contributions are considered:

1. original articles dealing with methodological developments, comparison of competing techniques or their computational aspects;

2. articles aimed at novel applications of existing methodologies to practical problems;

3. survey articles containing a thorough coverage of topics of broad interest for probabilists and statisticians.

The journal will occasionally publish invited papers and essays on statistics in education, statistical consulting, or computational statistics. Some issues or part issues may be devoted to papers presented in Brazilian meetings on probability and statistics or devoted to a special theme.

The international board of the *Brazilian Journal of Probability and Statistics* consists of an editor-in-chief, two editors (one for theory and methods, the other for applications), a book-review editor, and 20 distinguished scholars from several countries working as associate editors, who have expertise in areas of probability and statistics such as foundations of statistical inference, regression models, Bayesian methods, multivariate analysis, asymptotic theory, stochastic processes, nonparametric methods, time series, and statistical data analysis.

Papers must be written in English. The editorial policy is to publish papers containing original and significant research contributions. Papers appearing in the journal are abstracted in *Statistical Theory and Methods Abstracts** and indexed in *Current Index to Statistics**.

The website for the journal is www.de. ufpe.br/rebrape/bjps.html.

Gauss M. Cordeiro
The Editors

BREAKDOWN POINT

In many statistical estimation situations, the only data available are possibly contaminated by recording errors, miscoding, or some other factors. The *breakdown point* of an estimator is, roughly speaking, the minimum proportion of the data for which contamination can lead to a completely noninformative estimation result. It is a measure of the resistance* of the estimator to data contamination; the higher the breakdown point is, the more resistant the estimator is to data contamination.

HODGES' TOLERANCE

If a location estimate is attracted to a small number of extreme values, the estimate fails to convey location information about the other observations. A natural way of defining the breakdown point in location estimation is thus to consider what happens to the estimate if some of the observations are replaced with positively or negatively extreme values. The *tolerance* proposed by Hodges [4] formally implements this idea.

Let $\hat{\theta}$ be a location estimate of interest, and $(y_1, y_2, \ldots, y_n)$ the data set of arbitrary real numbers sorted in ascending order. Suppose that there exist integers $m_l \geqslant 0$, $m_r \geqslant 0$ such that: (1) $y_{m_l+1} \leqslant \hat{\theta} \leqslant y_{n-m_r}$ for any $(y_1, y_2, \ldots, y_n)$; (2) when $(y_{m_l+2}, y_{m_1+3}, \ldots, y_n)$ are fixed, for any sequence of the replacements of $(y_1, y_2, \ldots, y_{m_l+1})$ along which $y_{m_l+1} \to -\infty$ with the order of $(y_1, y_2, \ldots, y_{m_l+1})$ being preserved, $\hat{\theta} \to -\infty$; and (3) when $(y_1, y_2, \ldots, y_{n-m_r-1})$ are fixed, for any sequence of the replacements of $(y_{n-m_r}, y_{n-m_r+1}, \ldots, y_n)$ along which $y_{n-m_r} \to \infty$ with the order of $(y_{n-m_r}, y_{n-m_r+1}, \ldots, y_n)$ being preserved, $\hat{\theta} \to \infty$. Then it is said that $\hat{\theta}$ can tolerate m_l extreme values on the left and m_r extreme values on the right.

Example 1. Both the sample average and the sample median are location estimators. Let n be the sample size. Then the sample average can tolerate no extreme values either on the left or on the right, while the sample median can tolerate $\lceil n/2 \rceil - 1$ extreme values both on the left and on the right, where $\lceil a \rceil$ is the minimum integer greater than or equal to a for each real number a.

HAMPEL'S BREAKDOWN POINT

Unlike Hodges, Hampel [3] considers the effect of data contamination strictly in the probabilistic framework. Suppose that the "true" distribution is F in a class $\mathscr{G}$ of probability distributions. The data of interest would be generated by random draws from probability distribution F, if there were no data contamination. Actual draws are from the probability distribution $G \in \mathscr{G}$ in some neighborhood of F (in terms of Prokhorov distance), due to data contamination. Let $B_\delta(F)$ be the set of all probability distributions in $\mathscr{G}$ within distance δ of F. Also let $\hat{\theta}_n$ be the estimator of interest with sample size n.

Suppose that G can be any member of $B_\delta(F)$. If the probability limit of $\hat{\theta}_n$ as $n \to \infty$ can be any point in the parameter space Θ depending on G in $B_\delta(F)$, the estimator fails to convey any information about F. On the other hand, if there is a proper subset K of Θ such that $G \in B_\delta(F)$ implies that the probability of the estimator $\hat{\theta}_n$ belonging to K converges to one as the sample size grows to ∞, then $\hat{\theta}_n$ is considered to convey some information about F when the observations are drawn from any G in $B_\delta(F)$. The supremum of δ for which the estimator $\hat{\theta}_n$ is informative about F when the observations are drawn from any probability distribution in $B_\delta(F)$ is the *Hampel breakdown point*.

Example 2. Let $\mathscr{G}$ be the set of all probability distributions on the entire real line with finite first absolute moment. Suppose that $F \in \mathscr{G}$. Pick an arbitrary $\delta \in (0, 1)$. Consider a probability distribution G created by reducing by 100δ percent the probability of each event specified by F, and putting probability δ on some positive point on the real line. By moving the point to which probability δ is added, we can set the expectation of the

probability distribution G to any value we like. Because the sample average converges to the expectation of G by the Kolmogorov law of large numbers, this implies that the Hampel breakdown point is no higher than δ. Because δ is an arbitrary positive number, this implies that the Hampel breakdown point is zero for the sample average. In the same situation, the Hampel breakdown point of the sample median is 50%.

DONOHO-HUBER FINITE-SAMPLE BREAKDOWN POINT

Although Hampel's breakdown point can be applied to general parametric estimation problems, it only gives us insight into the large-sample behavior of the estimators. On the other hand, Hodges' tolerance, which is defined without recourse to a probability model, is a sensible measure for the resistance of the estimator with each fixed sample size, though it is only applicable to location estimators. While preserving its useful properties for finite sample sizes, Donoho and Huber [1] extend Hodges' tolerance to cover a wider range of problems. This notion is called the *finite-sample breakdown point*. Among the three versions of the finite-sample breakdown point proposed in ref. 1, we first explain the one with "ϵ-replacement."

Let $Q \subset \mathbb{R}^v$ be a sample space, where v is a natural number. Let n denote the sample size. For given $m \in \{0, \ldots, n\}$ and a given data set $\bar{z}^n \in Q^n \equiv \times_{t=1}^n Q$, consider choosing m arbitrary observations in $\bar{z}^n$ and replacing each of them with an arbitrary point in Q. Let $D_m^n(\bar{z}^n)$ denote the set of all data sets with m replaced observations and $n - m$ original observations.

Let a given parameter space Θ be a subset of a metric space with metric d (e.g., a Euclidean space, as assumed in Donoho and Huber [1]). The finite-sample breakdown point of an estimator $\hat{\theta}_n : Q^n \to \Theta$ at a data set $\bar{z}^n \in Q^n$ is defined as the minimum element of the following set of fractions:

$$\left\{ \frac{m}{n} : b(m, n; \bar{z}^n, \hat{\theta}_n) = \infty, \right.$$

$$\left. m \in \{1, \ldots, n\} \right\} \cup \{1\},$$

where $b(m, n; \bar{z}^n, \hat{\theta}_n) \equiv \sup_{z^n \in D_m^n(\bar{z}^n)} d(\hat{\theta}_n(z^n),$ $\hat{\theta}_n(\bar{z}^n))$. We call this finite-sample break-down point the *Donoho—Huber breakdown point with ϵ-replacement* (or simply the Donoho—Huber breakdown point) for convenience and clarity in what follows.

The value $b(m, n; \bar{z}^n, \hat{\theta}_n)$ can be considered to be the maximum bias (more accurately, the supremum bias) caused by m observation replacements. By checking this maximum bias for $m = 1, m = 2$, and so on, one may find that the maximum bias becomes ∞ for some m. Then m/n is the Donoho—Huber breakdown point with this m. When there is no $m \leqslant n$ for which the maximum bias is ∞, the breakdown point is one. An example of an estimator whose Donoho—Huber breakdown point is one is the single-parameter estimator whose value is always a fixed value, say zero, regardless of the observations.

The Donoho—Huber breakdown point of an estimator is dependent on the noncontaminated data set $\bar{z}^n$ in general. Nevertheless, it takes the same value at all or at most possible data sets in many interesting applications. For this reason, the Donoho—Huber breakdown point of an estimator is often referred to without specifying what the noncontaminated data set $\bar{z}^n$ is.

Example 3. In the location estimation problem, the Donoho—Huber breakdown point is essentially the same as Hodges' tolerance if the metric of the location parameter space is Euclidean distance. Thus, the Donoho—Huber breakdown points of the sample average and the sample median are $1/n$ and $\lceil n/2 \rceil / n$, respectively, where n is the sample size.

The other two versions of the breakdown points proposed by Donoho and Huber [1] are derived from the ϵ-replacement version by changing the way that the data are corrupted. In some situations, data corruption is caused by m erroneous observations adjoined to the original data set. Donoho and Huber call this type of data corruption ϵ-*contamination*. The breakdown point with ϵ-contamination is the minimum fraction of the contaminated observations, $m/(n + m)$, for which the distance between the estimate based on the contaminated data set can be arbitrarily far from that based on the original data set.

The third version of the Donoho—Huber breakdown point, unlike the previous two versions, does not limit the way in which the data are corrupted. All possible data corruption, including replacement and contamination, is considered. The degree of data corruption is measured by the distance between the empirical distribution of the modified data and that of the original data. The breakdown point is defined as the minimum data corruption level for which the estimate based on the modified data set can be arbitrarily far from that based on the original data set. Donoho and Huber call this breakdown point the finite-sample breakdown point with ϵ-*modification*.

As Donoho and Huber [1] mention, it is necessary to modify the definition of the three Donoho—Huber breakdown points to cover certain problems. Scale estimators* are an example of this. Suppose that a small number of contaminated observations can cause a scale estimate to be arbitrarily close to zero. We can call this behavior of the estimator *implosion*. The implosion of a scale estimator is as problematic as its explosion. The Donoho—Huber breakdown point ignores implosion, because the maximum bias in scale estimation caused in this way is never infinity. To take account of the possibility of implosion, the breakdown point with ϵ-replacement, for instance, needs to be redefined as m/n, where m is the minimum number of replaced observations for which the maximum bias is ∞ or for which the estimate can be arbitrarily close to zero.

EXTENSION OF THE DONOHO—HUBER BREAKDOWN POINT

Although the location estimation problem can be considered as a special case of the regression estimation problem, the Donoho—Huber breakdown point of a regression estimator may be very different from that of the corresponding location estimator. The L_1-estimator (the least-absolute-deviation estimator) is a good example. In the simple location setting L_1-estimation delivers the sample median. Nevertheless, as Ellis and Morgenthaler [2] precisely describe, the L_1-estimator for linear regression can be greatly affected by an observation which is far from the majority of observations in the regressor space. (Such an observation is a *leverage** point.) It thus has the Donoho—Huber breakdown point $1/n$. Many other M-estimators* share the same property. On the other hand, the Donoho—Huber breakdown points of certain other estimators are known to have positive lower bound independent of n. For an excellent treatment of the properties of estimators in the linear regression setting, *see* ROBUST REGRESSION, POSITIVE BREAKDOWN IN.

A natural question that arises from studying the breakdown properties of linear regression estimators is what breakdown properties hold for estimators for other types of models. The Donoho—Huber breakdown point is inconvenient for answering this question, in that it is not invariant with respect to reparametrizations (possibly involving change of the metric of the parameter space). To see this point, consider the regression model

$$Y = X\alpha + \epsilon, \tag{1}$$

where the set of all real numbers is denoted $\mathbb{R}, Y \in \mathbb{R}$ is the dependent variable, $X \in \mathbb{R}$ is the explanatory variable, α is the model parameter belonging to $\mathbb{R}$, and ε is an error term. The Donoho—Huber breakdown point of the least squares (LS) estimator for this model is known to be $1/n$, where n is the sample size. Nevertheless, we can rewrite this model with an alternative parametrization as

$$Y = X \tan^{-1}\theta + \epsilon, \tag{2}$$

where $\theta \in (-\pi/2, \pi/2)$. If we employ Euclidean distance in this new parameter space $(-\pi/2, \pi/2)$, the Donoho—Huber breakdown point for the LS estimator is one with the new parametrization. Note that the LS estimator picks exactly the same regression function in the two estimation problems. This means that essentially the same estimator may have very different breakdown points depending on the parameterization.

The dependence of the breakdown point on the parametrization is not such a serious problem in this simple linear regression example. As long as we restrict the model to be a simple linear regression model, we would

always use (1), because (2) is not linear in the parameter. Nevertheless, there are models for which parameterization may be arbitrary. Nonlinear regression models are an example of this. The dependence of the breakdown point on the parametrization makes the analysis unnecessarily complicated.

Note also that the bias considered by the Donoho—Huber breakdown point may be an inappropriate measure for the effects of data contamination in certain situations. Stromberg and Ruppert [6] give the following example. Consider the nonlinear regression* model known as the Michaelis—Menten model,

$$Y = \frac{VX}{K + X} + \epsilon, \tag{3}$$

where $Y > 0$ is the dependent variable, $X > 0$ is the explanatory variable, ε is the error term, and V and K are parameters of this model, both of which can be any positive real numbers. Suppose that replacing m data points can drive both an estimate $\hat{V}$ for V and an estimate $\hat{K}$ for K to ∞, keeping their ratio $\alpha = \hat{V}/\hat{K}$ constant. This behavior of the parameter estimates does not completely invalidate the estimated regression function, which still retains information on the conditional location of Y given X. This point is made clear by comparing this case with that in which the ratio of the estimated parameters, α, can be arbitrarily close to zero or arbitrarily large. We thus need to modify the definition of the breakdown point in such estimation problems as (3).

Taking into account this limitation of the Donoho—Huber breakdown point, Stromberg and Ruppert [6] propose an alternative finite-sample breakdown point for nonlinear regression estimators. In Stromberg and Ruppert's version of the finite-sample breakdown point, the behavior of the estimated regression function is of concern instead of the estimated parameters. At each point in the explanatory-variable space, consider the possible range of the regression function obtained by letting the parameter range over the parameter space. If the estimated regression function can be arbitrarily close to the upper or lower boundary of its range at a certain contamination level, the estimator is considered to break down at the point in the explanatory-variable space at the given contamination level. The breakdown point at that point in the explanatory-variable space is thus defined. The minimum value of the breakdown point over the points in the explanatory variable space is the *Stromberg—Ruppert breakdown point*.

This idea resolves the problems related to the dependence of the finite-sample breakdown point on the model parameterization, because the definition of the Stromberg—Ruppert breakdown point does not depend on the parameter space. Nevertheless, there are some cases in which it is inappropriate to judge the breakdown of the estimators in the way proposed by Stromberg and Ruppert, as the next example shows.

Example 4. A regression estimator is said to have the *exact fit property* if the estimate of the regression coefficient vector is $\bar\theta$ whenever more than the half of the observations are perfectly fitted by the regression function with $\bar\theta$. Some estimators (e.g., the least-median-of-squares estimator) are known to have the exact fit property.

Let $\Theta = [0, \infty)$, and consider the linear model regressing $Y \in \mathbb{R}$ on $X \in \mathbb{R}$:

$$Y = \theta X + \epsilon,$$

where $\theta \in \Theta$, and ϵ is an error term. Suppose that we have a data set that consists of the following pairs of observations on Y and X: (0.0, −4.0), (0.0, −3.0), (0.0, −2.0), (0.0, −1.0), (0.001,0.1), (0.01, 1.0), (0.02, 2.0), (0.03, 3.0), (0.04, 4.0). Then any estimator that has the exact fit property gives 0.01 as the estimate for θ with the given data set. The estimate, however, becomes 0 if we replace the fifth observation with (0.0, −0.1), i.e., the estimated regression function attains its lower bound at each $x \in (0, \infty)$ and its upper bound at each $x \in (-\infty, 0)$. Thus the Stromberg—Ruppert breakdown point of the estimator at the above dataset is $\frac{1}{9}$. For the dependent variable whose sample space is $\mathbb{R}$, would everyone agree that $\theta = 0$ is a crucially bad choice?

Although Stromberg and Ruppert mention the possibility that their breakdown point is inappropriate and one may instead need to

use either the upper or the lower breakdown point similarly defined, both upper and lower breakdown points agree with the Stromberg—Ruppert breakdown point in the above example. This suggests that we may want to reconsider the criterion for crucially bad behavior of estimators. Also, the Stromberg—Ruppert breakdown point is not easily extended to parametric estimation problems other than nonlinear regression.

To overcome these limitations, Sakata and White [5] propose another version of the finite-sample breakdown point based on the fundamental insights of Donoho and Huber [1] and Stromberg and Ruppert [6]. This version of the breakdown point is like the Stromberg and Ruppert's in that it is based on the behavior of the estimated object of interest, e.g., a regression function, instead of the estimated parameters. For the Sakata—White breakdown point the model user specifies the criterion against which breakdown is to be judged. This criterion (e.g., the negative of goodness of fit) is the *badness measure*; the breakdown point is the *badness-measure-based breakdown point*. It constitutes the minimum proportion of observations for which date contamination leads to the worst value of the badness measure. Because of its flexibility, the Sakata—White breakdown point is generic. It can generate the Donoho—Huber breakdown point and the Stromberg—Ruppert breakdown point for appropriate badness measures. It can also be applied in nonregression contexts.

Table 1 summarizes the five breakdown points explained here.

REFERENCES

1. Donoho, D. L. and Huber, P. J. (1983). The notion of breakdown point. In *A Festschrift for Erich L. Lehmann in Honor of His 65th Birthday*, P. J. Bickel, K. A. Doksum, and J. L. Hodges, Jr., eds. Wadsworth International Group, Belmont, Calif., pp. 157–184.

2. Ellis, S. P. and Morgenthaler, S. (1992). Leverage and breakdown in L_1 regression. *J. Amer. Statist. Ass.*, **87**, 143–148.

3. Hampel, F. R. (1971). A general qualitative definition of robustness. *Ann. Math. Statist.*, **42**, 1887–1896.

4. Hodges, J. L., Jr., (1967). Efficiency in normal samples and tolerance of extreme values for

Table 1. Comparison of Breakdown-Point Concepts

Breakdown Point	Applicable	Notion of Breakdown
Tolerance	Location	The estimate is arbitrarily (negatively or positively) large.
Hampel	Parametric in general	The stochastic limit of the estimator can be anywhere in the parameter space.
Donoho—Huber	Univariate and multivariate location, etc.	The estimate under data corruption can be arbitrarily far from that without data corruption.
Stromberg—Ruppert	Nonlinear regression	The estimated regression function can be arbitrarily close to its possible highest or lowest value at some point in the regressor space.
Sakata—White	Parametric in general	The estimated model can be arbitrarily bad in terms of the model user's criterion.

some estimates of location. In *Proc. Fifth Berkeley Symp. Math. Statist. and Probab.*, vol. 1, pp. 163–168.

5. Sakata, S. and White, H. (1995). An alternative definition of finite sample breakdown point with applications to regression model estimators. *J. Amer. Statist. Ass.*, **90**, 1099–1106.

6. Stromberg, A. J. and Ruppert, D. (1992). Breakdown in nonlinear regression. *J. Amer. Statist. Ass.*, **87**, 991–997.

See also Maxbias Curve; Robust Regression, Positive Breakdown in; and Trimming and Winsorization.

Shinichi Sakata
Halbert White

BREAKDOWN VALUES. See Robust Regression, Positive Breakdown in

BRITISH JOURNAL OF MATHEMATICAL AND STATISTICAL PSYCHOLOGY

[This entry has been updated by the Editors.]

The *British Journal of Mathematical and Statistical Psychology* was founded in 1947 by the British Psychological Society (BPS) with the title *British Journal of Psychology (Statistical Section)* under the joint editorship of Cyril Burt and Godfrey Thomson. It changed its name to *The British Journal of Statistical Psychology* in 1953, taking its broader current title in 1966.

The journal publishes articles relating to all areas of psychology that have a greater mathematical or statistical or other formal aspect to their argument than is usually acceptable to other journals. Articles that have a clear reference to substantive issues in psychology are preferred.

The aims and scope of the journal are stated on the journal website: "The Journal is recognized internationally as publishing high quality articles in quantitative psychology. Contributions are invited on any aspect of mathematics or statistics of relevance to psychology; these include statistical theory and methods, decision making, mathematical psychology, psychometrics, psychophysics and computing". The journal also publishes book reviews, software reviews, expository articles, short notes, and papers of historical or philosophical nature.

The journal is international, most submissions coming from outside the U.K. All submissions are treated equally on their scientific merits within their area. The journal is one of a group of ten psychology journals published by the BPS, which also publishes books. At about the same time as the Society moved into book publishing, it took over responsibility for publishing its own journals.

In addition to the Editor, the *Journal* has a Consultant Editor, a Book Review Editor, a

Statistical Software Editor, and an international Editorial Board of 18 or so members. It is published in May and November of each year, these two issues constituting a single volume.

The website link to the journal is www.bps .org.uk/publications/jMS_1.cfm.

BROUILLON INDEX. See DIVERSITY INDICES

BROWNIAN BRIDGE

INTRODUCTION

A Brownian Bridge process $\{B(t), 0 \leqslant t \leqslant 1\}$ is a Gaussian process derived from Brownian motion* or Wiener process. The name "bridge" comes from the fact that Brownian motion forming the Brownian bridge process is tied to 0 at two time instants, $t = 0$ and $t = 1$. Figure 1 shows two realizations of the process.

Let $X(t), t \geqslant 0$ be a Brownian motion. The Brownian bridge is a conditional process

$$B(t) = \{X(t), 0 \leqslant t \leqslant 1 | X(1) = 0\}.$$

Gaussianity, continuity and irregularity of paths of $B(t)$ is inherited from $X(t)$. Since the conditional distribution of $X(s)$ given $\{X(t) = A\}$ is Gaussian and proportional to $\exp\left\{-\frac{t(x - As/t)^2}{2s(t-s)}\right\}$, it follows that

$$E(X(s)|X(t) = A) = A\frac{s}{t},$$
$$Var(X(s)|X(t) = A) = \frac{s(t-s)}{t}, \quad s < t. \tag{1}$$

Equations (1) yield $EB(s) = 0$, and $Var(B(s)) = s(1-s)$. Also, taking into account that $EB(t) = 0$, the covariance function is $Cov(B(s), B(t)) = s(1-t)$, $s < t$. Indeed, for $s < t < 1$,

$$Cov(B(s), B(t))$$
$$= E(X(s)X(t)|X(1) = 0)$$
$$= E(E(X(s)X(t)|X(t))|X(1) = 0)$$
$$= E(X(t)\frac{s}{t}E(t)|X(1) = 0)$$
$$= \frac{s}{t}Var(B(t))$$
$$= s(1-t),$$

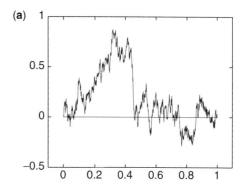

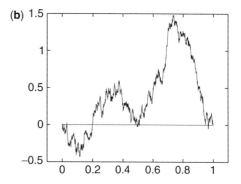

Figure 1. Two realizations of Brownian Bridge Process.

where we used conditioning and (1). Since a Gaussian process is uniquely determined by its second order structure (means and covariances), the definition of Brownian bridge is simple:

Definition 1. The Brownian Bridge is a Gaussian process with mean value 0 and covariance function $s(1-t), s < t$.

Finite-dimensional distributions of Brownian Bridge processes are zero-mean multivariate normal. Let $0 = t_0 < t_1 < \cdots < t_n < t_{n+1} = 1$, $x_0 = 0, x_{n+1} = 0$, and $x_1, x_2, \ldots, x_n \in R$; then

$$P(B(t_1) = x_1, \ldots, B(t_n) = x_n)$$
$$= \prod_{k=1}^{n+1} \phi_{t_k - t_{k-1}}(x_{k-1}, x_k)/\phi_1(0, 0),$$

where $\phi_t(x, y) = (2\pi t)^{-1/2} \exp\{-\frac{1}{2t}(y - x)\}$.

Result. Let $X(t)$ be a Brownian motion process. Consider the two processes defined for

$0 \leqslant t \leqslant 1$: (i) $B_1(t) = X(t) - tX(1)$, and (ii) $B_2(t) = (1-t)X\left(\frac{t}{1-t}\right)$. Each of the processes $B_1(t)$ and $B_2(t)$ is a Brownian Bridge.

Obviously, both processes are Gaussian and zero-mean, since $X(t)$ is. All finite dimensional distributions are fully determined by their covariance functions.

Let $s < t < 1$.

For the process B_1:

$$E[B_1(s)B_1(t)]$$

$$= E[(X(s) - sX(1))(X(t) - tX(1))]$$

$$= E[X(s)X(t) - sX(1)X(t) - tX(s)X(1)$$

$$\quad + stX(1)X(1)]$$

$$= [(s \wedge t) - s(1 \wedge t) - t(s \wedge 1) + st(1 \wedge 1)]$$

$$= s - st - st + st$$

$$= s(1 - t).$$

For the process B_2:

$$E[B_2(s)B_2(t)]$$

$$= (1-s)(1-t)E\left[X\left(\frac{s}{1-s}\right)X\left(\frac{t}{1-t}\right)\right]$$

$$= (1-s)(1-t)\left[\left(\frac{s}{1-s}\right)\right.$$

$$\left. \wedge \left(\frac{t}{1-t}\right)\right]$$

$$= (1-s)(1-t)\left(\frac{s}{1-s}\right)$$

$$= s(1 - t).$$

From the definition of B_2 it is clear that B_2 is a Markov process, but it does not have independent increments and it is not a martingale*;

$$E\left[B_2(t) - B_2(s)|\mathcal{F}_s\right] = -\frac{t-s}{1-s}B_2(s).$$

The Brownian Bridge process can be constructed using the Haar orthonormal system in $L_2([0,1])$. We describe the construction and omit many technical details needed for a strict formal construction. Interested readers can consult [8, 5].

Let

$$s_{jk}(t) = \int_0^t h_{jk}(x)dx,$$

where $h_{jk}(x) = 2^{j/2}h(2^j x - k)$ and $h(x) = 2\mathbf{1}_{[0,1/2]}(x) - 1$, is the Haar function. The Haar system of functions $\{h_{jk}(x), j \geqslant 0, 0 \leqslant k \leqslant 2^j - 1; \mathbf{1}_{[0,1]}(x)\}$ constitutes a complete orthonormal basis of $L_2([0,1])$, while the family of the functions $\{s_{jk}, j \geqslant 0, 0 \leqslant k \leqslant 2^j - 1\}$ is known as the Schauder system.

Define

$$B(t) = \sum_{j=0}^{\infty} V_j(t),$$

where $V_j(t) = \sum_{k=0}^{2^j-1} X_{jk}s_{jk}(t)$ is a random sum and X_{jk} are independent standard normal random variables. Then, $B(t)$ is the Brownian Bridge.

$B(t)$ is a Gaussian, zero mean process. Its covariance function is

$$EB(s)B(t) = E\sum_{j=0}^{\infty}V_j(s)\sum_{j=0}^{\infty}V_j(t)$$

$$\overset{1}{=} \sum_{j=0}^{\infty}E(V_j(s)V_j(t))$$

$$= \sum_{j=0}^{\infty}E\left\{\sum_{k=0}^{2^j-1}X_{jk}\int_0^s h_{jk}(x)dx\sum_{m=0}^{2^j-1}X_{jm}\right.$$

$$\left. \times \int_0^t h_{jm}(x)dx\right\}$$

$$= \sum_{j=0}^{\infty}\sum_{k=0}^{2^j-1}\int_0^s h_{jk}(x)dx\int_0^t h_{jk}(x)dx$$

$$= \sum_{j=0}^{\infty}\sum_{k=0}^{2^j-1}\int_0^1 \mathbf{1}_{[0,s]}(x)h_{jk}(x)dx$$

$$\times \int_0^1 \mathbf{1}_{[0,t]}(x)h_{jk}(x)dx + st - st$$

$$\overset{2}{=} \int_0^1 \mathbf{1}_{[0,s]}(x)\mathbf{1}_{[0,t]}(x)dx - st$$

$$= s \wedge t - st.$$

The equality $\overset{1}{=}$ is justified by Tonelli's theorem while the equality $\overset{2}{=}$ is Parseval's identity.

THE BROWNIAN BRIDGE AND KOLMOGOROV-SMIRNOV STATISTICS

The most significant application of Brownian Bridge Processes is in the distribution of a functional of the empirical distribution function, used in statistics.

Let X_i be independent and identically distributed random variables (a sample) from some unknown distribution μ_X with cumulative distribution function $F(t) = P(X_i < t) = \mu_X((-\infty, t])$. A sample counterpart of $F(t)$ is the empirical distribution function*

$$F_n(t) = \frac{\#[i \leqslant n : X_i \leqslant t]}{n} = \sum_{i=1}^{n} \mathbf{1}_{(-\infty, t]}(X_i),$$

a random function of t that starts at $F_n(-\infty) = 0$ and jumps by $1/n$ at each observation X_i. Kolmogorov and Smirnov studied the probability distribution of the quantity

$$Y_n = \sup_{-\infty < t < \infty} \sqrt{n} |F_n(t) - F(t)|,$$

the normalized largest deviation of the empirical distribution from the true distribution. It turns out that Y_n has the same distribution for any continuous distribution F, and in particular, the same as that for uniformly distributed random variables with $F(t) = t$, $0 \leqslant t \leqslant 1$.

Regarded as a stochastic process, $F_n(t)$ has mean $F(t)$ and if $s < t$,

$$Cov(F_n(s), F_n(t)) = \frac{1}{n}[F(s)(1 - F(t))].$$

Indeed,

$$E[F_n(t)] = \frac{1}{n} \sum_{i=1}^{n} E\mathbf{1}_{(-\infty, t]}(X_i)$$

$$= \frac{1}{n} \sum_{i=1}^{n} P[X_i \leqslant t]$$

$$= F(t).$$

$$E[(F_n(s) - F(s))(F_n(t) - F(t))]$$

$$= E\left\{ \left[\frac{1}{n} \sum_{i=1}^{n} (\mathbf{1}_{(-\infty, s]}(X_i) - F(s)) \right] \right.$$

$$\left. \times \left[\frac{1}{n} \sum_{i=1}^{n} (\mathbf{1}_{(-\infty, t]}(X_i) - F(t)) \right] \right\}$$

$$= n^{-2} \sum_{i=1}^{n} E[(\mathbf{1}_{(-\infty, s]}(X_i) - F(s))$$

$$\times (\mathbf{1}_{(-\infty, t]}(X_i) - F(t))]$$

$$= n^{-1} E[(\mathbf{1}_{(-\infty, s]}(X_1) - F(s))$$

$$\times (\mathbf{1}_{(-\infty, t]}(X_1) - F(t))]$$

$$= n^{-1} E[(\mathbf{1}_{(-\infty, s \wedge t]}(X_1) - F(s)\mathbf{1}_{(-\infty, t]}(X_1)$$

$$- F(t)\mathbf{1}_{(-\infty, s]}(X_1) - F(s)F(t))]$$

$$= n^{-1}[F(s \wedge t) - F(s)F(t)]$$

$$= n^{-1}[F(s) \wedge F(t) - F(s)F(t)]$$

$$= n^{-1}[F(s)(1 - F(t))]$$

Thus, $\sqrt{n}[F_n(t) - F(t)]$ has the same covariance function as $B(F(t))$, for a Brownian Bridge $B(t)$.

In fact, a stronger result is true. Any continuous functional of $\sqrt{n}[F_n(t) - F(t)]$ converges weakly to a similar functional of the Brownian Bridge, and in particular that of the Kolmogorov-Smirnov statistics Y_n converges to $Y = \sup_{0 \leqslant t \leqslant 1} |B(t)|$, in distribution.

Kolmogorov [4] (cf. also Smirnov, [6]) proved that

$$P(Y \leqslant y) = 1 + 2 \sum_{k=1}^{\infty} (-1)^k \exp\left\{ -2k^2 y^2 \right\},$$

$$y > 0. \qquad (2)$$

The graphs of this cumulative distribution function and its density are given in Figure 2. The following results hold:

$$P(\sup_{0 \leqslant t \leqslant 1} B(t) > y) = \exp\left\{ -2y^2 \right\};$$

$$P(\sup_{0 \leqslant t \leqslant 1} B(t) - \inf_{0 \leqslant t \leqslant 1} B(t) \geqslant y) =$$
$$2 \sum_{n=1}^{\infty} (4n^2 y^2 - 1) \exp\left\{ -2n^2 y^2 \right\}$$

For proofs, see [2] and [3].

REFERENCES

1. Anderson, T.W. and Darling, D.A. (1952). Asymptotic theory of certain goodness of fit criteria based on stochastic processes. *Ann. Math. Statist.*, **23**, 193-212.

2. Durrett, R. (1991). *Probability: Theory and Examples,* Duxbury Press, Belmont, Ca.

3. Feller, W. (1971). *An Introduction to Probability Theory and its Applications* Vpl. II, Second Edition, Wiley, NY.

4. Kolmogorov, A.N. (1933). Sulla determinazione empirica di une legge di distribuzione. *Giorn. Inst. Ital. Attuari*, **4**, 83-91.

(a)

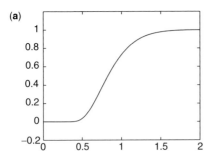

(b)

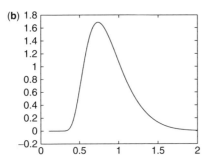

Figure 2. (above) The Kolmogorov-Smirnov distribution from (2); (below) Density for the cdf in (2).

5. Salminen, P. and Borodin, A. N. (2002). Handbook of Brownian Motion: Facts and Formulae (Probability and Its Applications) Second Edition, Birkhauser Boston, 688pp.

6. Smirnov, N.V. (1939). On the estimation of the discrepancy between empirical curves of distribution for two independent samples. *Bull. Math. de l'Université de Moscou*, **2**, 3-14.

7. Smirnov, N.V. (1944). Approximate laws of distribution of random variables from empirical data (in Russian). *Uspehi Mat. Nauk.*, **10**, 179-206.

8. Van Der Vaart, A. and Wellner, J. (2000). Weak Convergence and Empirical Processes: With Applications to Statistics, Second Edition, Springer Series in Statistics, Springer-Verlag, NY.

See also BROWNIAN MOTION; GAUSSIAN PROCESSES; and KOLMOGOROV–SMIRNOV STATISTICS.

BROWNIAN MOTION

Brownian motion with drift parameter μ and variance parameter $\sigma^2 > 0$ is the continuous-time, continuous-state stochastic process* $\{X(t); t \geqslant 0\}$ for which

1. $X(t) - X(0)$ is normally distributed with mean μt and variance $\sigma^2 t$;

2. all nonoverlapping increments $X(t_1) - X(t_0), X(t_2) - X(t_1), \ldots, X(t_n) - X(t_{n-1})$, where $0 \leqslant t_0 \leqslant t_1 \leqslant \cdots \leqslant t_n$, are independent random variables.

3. It is common to assume that $X(0) = 0$ unless explicitly stated otherwise (Brownian motion started at $X(0) = x$). *Standard Brownian motion* $\{B(t); t \geqslant 0\}$ refers to the case $\mu = 0$ and $\sigma^2 = 1$, and is related to the general Brownian motion by $X(t) = \mu t + \sigma B(t)$.

Brownian motion is important for a myriad of reasons, the primary ones being:

a. Numerous functionals of the process, such as distributions of first passage times, can be calculated explicitly.

b. Brownian motion is the analog in continuous time of sums of independent random variables, the statement being made precise by the *invariance principle*[*]: If $S_n = \xi_1 + \cdots + \xi_n$ is a sum of independent identically distributed random variables $\{\xi_k\}$ having zero means and unit variances, then for n tending to infinity, the processes $B_n(t) = n^{-1/2} S_{[nt]}$ converge weakly in the appropriate function space to the standard Brownian motion. Thus functionals* of Brownian motion often serve as good approximations to analogous functionals for sums of independent random variables.

c. Brownian motion is the central example of a continuous-time, continuous-path strong Markov process* (a *diffusion process*[*]). An arbitrary one-dimensional diffusion process can be constructed from Brownian motion by a homeomorphic transformation in the state space plus a random rescaling of time.

We provide a sampling of some of the known quantities concerning Brownian motion. Let $\phi(\xi) = (2\pi)^{-1/2} \exp(-\frac{1}{2}\xi^2)$ be the standard normal density and $\Phi(z) = \int_{-\infty}^{z} \phi(\xi) d\xi$ the corresponding cumulative distribution function.

Table 1.

Functional W	Distribution $\Pr[W \leqslant y]$
$B(t)$	$\Phi(y/\sqrt{t}), -\infty < y < \infty$
$|B(t)|$	$2\Phi(y/\sqrt{t}) - 1, y \geqslant 0$
$B(t) - B(s), 0 < s < t$	$\Phi\left(y/\sqrt{(t-s)}\right), -\infty < y < \infty$
$\sup\{B(s); 0 \leqslant s \leqslant t\}$	$2\Phi(y/\sqrt{t}) - 1, y \geqslant 0$
$B(t) - \inf\{B(s); 0 \leqslant s \leqslant t\}$	$2\Phi(y/\sqrt{t}) - 1, y \geqslant 0$
$\inf\{t > 0; B(t) = a\}, a > 0$	$2[1 - \Phi(a/\sqrt{y})], y > 0$
$t^{-1}m\{s \leqslant t; B(s) > 0\}$	$(2/\pi)\arcsin\sqrt{y}, 0 \leqslant y \leqslant 1$
($m = $ Lebesgue measure)	
$t^{-1}\sup\{s \leqslant t; B(s) = 0\}$	$(2/\pi)\arcsin\sqrt{y}, 0 \leqslant y \leqslant 1$
$\sup\{|B(s)|; 0 \leqslant s \leqslant t\}$	$\sum_{k=-\infty}^{+\infty}(-1)^k\left[\Phi\left(\frac{(2k+1)y}{\sqrt{t}}\right) - \Phi\left(\frac{2(k-1)y}{\sqrt{t}}\right)\right]$

First consider standard Brownian motion $\{B(t)\}$ (Table 1).

Let $T_z = \inf\{t \geqslant 0; B(t) = z\}$ be the hitting time to a point z. Then the probability density function for T_z is

$$f(t) = \frac{z}{\sqrt{2\pi}}t^{-3/2}\exp\left(-\frac{z^2}{2t}\right),$$

sometimes called the *inverse Gaussian distribution**.

Also for $a > 0$ and $b < 0$,

$$\Pr[\text{Hit } a \text{ before } b] = \Pr[T_a < T_b]$$
$$= |b|/(a + |b|),$$
$$E[\min\{T_a, T_b\}] = a|b|.$$

Turning to the general Brownian motion $X(t)$ with parameters μ and $\sigma^2 > 0$, we again let $T_z = \inf\{t > 0; X(t) = z\}$ be the hitting time to z. Of relevance in the *Wald approximation** in the *sequential probability ratio test** is the formula

$$\Pr[T_a < T_b] = \frac{1 - e^{-2\mu b/\sigma^2}}{e^{-2\mu a/\sigma^2} - e^{-2\mu b/\sigma^2}}.$$

When $\mu < 0$, then $M = \sup_{t \geqslant 0} X(t)$ is a finite random variable for which $\Pr[M > y] = e^{-2|\mu|y/\sigma^2}, y \geqslant 0$. When $\mu > 0$, then T_z is a finite random variable for $z > 0$ whose probability density function is

$$f(t) = \frac{z}{\sigma\sqrt{2\pi t^3}}\exp\left[-\frac{(z - \mu t)^2}{2\sigma^2 t}\right](t \geqslant 0).$$

We complete this sampling of Brownian motion formulas with a quantity relevant to the one-sided cumulative sum control chart* method. Let $M(t) = \sup X(s); 0 \leqslant s \leqslant t, Y(t) = M(t) - X(t)$ and $\tau_a = \inf\{t \geqslant 0; Y(t) \geqslant a\}$. Then

$$E[\tau_a] = \frac{1}{\mu}\left[\frac{\sigma^2}{2\mu}(e^{2\mu a/\sigma^2} - 1) - a\right], \quad \mu \neq 0.$$

Some processes closely related to Brownian motion that appear frequently are the following.

Reflected Brownian Motion.. $Y(t) = |B(t)|$. Then $E[Y(t)] = \sqrt{2t/\pi}$ and $\text{var}\{Y(t)\} = (1 - 2/\pi)t$. Also arises as $Y(t) = B(t) - \inf\{B(s); 0 \leqslant s \leqslant t\}$.

Absorbed Brownian Motion.. $Y(t) = B(\min t, T_0)$, where $B(t)$ is standard Brownian motion starting at $y > 0$ and $T_0 = \inf\{t \geqslant 0; B(t) = 0\}$ is the hitting time to zero.

Geometric Brownian Motion.. $Y(t) = e^{X(t)}$. Arises in certain population models and in mathematical economics. $E[Y(t)] = e^{\alpha t}$, where $\alpha = \mu + \frac{1}{2}\sigma^2$ and $\text{var}[Y(t)] = e^{2\alpha t}(e^{\sigma^2 t} - 1)$.

n-Dimensional Brownian Motion.. $\mathbf{B}(t) = (B_1(t), \ldots, B_n(t))$, where $B_1(t), \ldots, B_n(t)$ are independent standard Brownian motions.

Bessel Process.. $Y(t) = ||\mathbf{B}(t)|| = [B_1^2(t) + \cdots + B_n^2(t)]^{1/2}$, the radial distance of n-dimensional Brownian motion from the origin.

*Wiener measure** is the probability distribution on the Borel subsets of $C[0, \infty)$ that corresponds to standard Brownian motion.

By considering Wiener measure one may assume that all trajectories $t \to B(t)$ are continuous. In contrast to this smoothness, however, with probability 1 the following statements hold: The paths are nowhere differentiable, and in fact,

$$\limsup_{h \downarrow 0} h^{-1/2} |B(t+h) - B(t)| = \infty$$

for all t. Consider a partition sequence $\pi_0 \subset \pi_1 \subset \cdots$ of $[0,t]$ where $\pi_k = \{0 = t_{k0} < t_{k1} < \cdots < t_{kn(k)} = t\}$ and for which $\max_{1 \leqslant j \leqslant n(k)} |t_{kj} - t_{kj-1}| \to 0$ as $k \to \infty$. Then

$$\lim_{k \to \infty} \sum_{j=1}^{n(k)} |B(t_{kj}) - B(t_{kj-1})|^2 = t, \quad (1)$$

while $\lim_{k \to \infty} \sum_{j=1}^{n(k)} |B(t_{kj}) - B(t_{kj-1})| = \infty$.

Let K be the class of absolutely continuous functions on $[0, 1]$ with $f(0) = 0$ and $\int_0^1 f'(t)^2 dt \leqslant 1$. Strassen's law of the iterated logarithm* asserts that the family $\{Z_n\} \subset C[0, 1]$ defined by

$$Z_n(t) = (2n \log \log n)^{-1/2} \times B(nt)$$

for $0 \leqslant t \leqslant 1$ is relatively compact with limit set K. From this, the classical law of the iterated logarithm

$$\limsup_{t \to \infty} (2t \log \log t)^{-1/2} B(t) = 1, \quad (2)$$

$$\liminf_{t \to \infty} (2t \log \log t)^{-1/2} B(t) = -1, \quad (3)$$

follows.

Although the Brownian path is nowhere differentiable, a theory of stochastic integration* has been developed, defining such expressions as

$$\mathcal{I}(Y)(t) = \int_0^t Y(s) dB(s), \quad (4)$$

where $\{Y(s)\}$ is a stochastic process progressively measurable with respect to the Brownian motion and locally square integrable. There are two approaches to (2) in common use, both involving approximating sums of the form $\sum Y(s_j^*)[B(s_j) - B(s_{j-1})]$, where $0 = s_0 < s_1 < \cdots < s_n = t$ and $s_{j-1} \leqslant s_j^* \leqslant s_j$. In the Ito integral*, one takes $s_j^* = s_{j-1}$, the

left endpoint. Then $\mathcal{I}(Y)(t)$ is a local martingale with respect to the Brownian motion. In the Stratonovich interpretation, denoted by $\mathcal{I}_S(Y)(t)$, one takes $s_j^* = (s_{j-1} + s_j)/2$, the midpoint of the interval. The Stratonovich calculus is more closely aligned with deterministic calculus and thus better suited for many models of physical systems perturbed by noise. The martingale property that the Ito calculus enjoys makes it the preferred approach in most theoretical work. It is important to recognize the distinction, as the two interpretations may lead to vastly differing conclusions, as exemplified by:

Ito	Stratonovich
$dB(t)^2$ $= dt$[compare with (1)]	$dB(t)^2 = 0$
$\int_0^t 2B(s) dB(s)$ $= B^2(t) - t$	$\int_0^t 2B(s) dB(s)$ $= B^2(t)$

In the *semigroup** approach to Markov processes standard Brownian motion is characterized by the operator $T_t f(x) = \int p(t,x,y) \times f(y) dy$, on the space $BC(-\infty, +\infty)$ of bounded continuous real-valued functions f that corresponds to the *Gauss kernel**

$$p(t,x,y) = (2\pi t)^{-1/2} \exp[-(y-x)^2/(2t)].$$

Given f in $BC(-\infty, +\infty)$, the bounded solution of $\partial u / \partial t = \frac{1}{2} \partial^2 u / \partial x^2, t > 0$ and $u(0+, \cdot) = f$, is $u = u(t,x) = T_t f(x)$. The resolvent $R_\lambda f(x) = \int_0^\infty e^{-\lambda t} T_t f(x) dt, \lambda > 0$, has the kernel representation $R_\lambda f(x) = \int_{-\infty}^{+\infty} r(\lambda, x, y) f(y) dy$, where

$$r(\lambda, x, y) = (1/\sqrt{2\lambda}) \exp(-\sqrt{2\lambda}|y - x|).$$

R_λ maps $BC(-\infty, +\infty)$ one-to-one onto the set Δ of f in $BC(-\infty, +\infty)$ having two bounded continuous derivatives. Let R_λ^{-1} denote the operator on Δ that is inverse to R_λ. Then $\mathscr{A} = \lambda - R_\lambda^{-1}$ together with the domain Δ is called the *infinitesimal generator*, and $\mathscr{A}u(x) = \frac{1}{2} \partial^2 u / \partial x^2$ for u in Δ or, alternatively,

$$\mathscr{A}u(x) = \lim_{\epsilon \downarrow 0} \epsilon^{-1} \left[T_\epsilon u(x) - u(x) \right],$$

where the convergence takes place boundedly and pointwise. Different notions of convergence, e.g., strong and uniform operator topologies, lead to different notions of infinitesimal generator.

The set $\Xi = t \geq 0; B(t) = 0$ of zeros of the Brownian path is a topological Cantor set, i.e., closed, uncountable, of topological dimension 0 and having no isolated points. Its Lebesgue measure is zero. However, its Hausdorff dimension* is $\frac{1}{2}$ and related to this fact is the possibility of assigning a nontrivial measure to the amount of time the process spends at 0, called the *local time** at 0 and denoted $l(0,t)$. For standard Brownian motion, the probability law of $l(0,t)$, is the same as that of $M(t) = \max_{0 \leq s \leq t} B(s)$.

BIBLIOGRAPHY

Freedman, D. (1971). *Brownian Motion and Diffusion*. Holden-Day, San Francisco. (This is an advanced, difficult-to-read, but carefully written text. Contains some topics difficult to find elsewhere, but is suitable only for the mathematically advanced reader.)

Ito, K. and McKean, H. P., Jr. (1965).*Diffusion Processes and Their Sample Paths*. Springer-Verlag, Berlin. (The definitive book on Brownian motion and diffusion processes. A wealth of information, but very terse and mathematically advanced.)

Karlin, S. and Taylor, H. M. (1975). *A First Course in Stochastic Processes*, 2nd ed. Academic Press, New York. (Recommended starting point for an applied statistician having a modest mathematical background. The 40-page chapter on Brownian motion contains many derivations of functionals and examples of their use.)

Karlin, S. and Taylor, H. M. (1980). *A Second Course in Stochastic Processes*. Academic Press, New York. (Contains a complete discussion of Brownian motion as a diffusion process, stochastic calculus, and numerous examples of Brownian motion and related processes as they arise in applications. Intermediate level of presentation.)

The following three references contain results relating Brownian motion to Cusum control charts*:

Reynolds, M. R., (1975). *Ann. Statist.*, **3**, 382–400.
Reynolds, M. R., (1975). *Technometrics*, **17**, 65–72.
Taylor, H. M., (1975). *Ann. Probab.*, **3**, 234–246.

See also DIFFUSION PROCESSES; HAUSDORFF DIMENSION; LOCAL TIME; and MARKOV PROCESSES.

H. TAYLOR

BROWNIAN MOTION, FRACTIONAL.
See FRACTIONAL BROWNIAN MOTIONS AND FRACTIONAL GAUSSIAN NOISES

BRUSHLETS.   See BEAMLETS AND MULTISCALE MODELING

BUFFON'S NEEDLE PROBLEM

' Buffon's needle problem is a celebrated classical problem of geometric probability*. A plane is lined with parallel straight lines 1 unit apart and a thin needle N of length l is thrown "at random" on the plane. ["At random" here means that: (1) the center of the needle is uniformly distributed over the interval of unit length perpendicular to and between two neighboring straight lines, and (2) the probability that the angle ϕ between the needle and the straight lines is $\phi_1 \leq \phi \leq \phi_1 + \Delta\phi$ is proportional to $\Delta\phi$ for any ϕ_1, whatever be the position of the center.] What is the probability p that N intersects (at least) one of these lines? The simple answers are:

$$p = 2l/\pi \qquad \text{if } l \leq 1$$
$$p = 2\left[\cos^{-1}(1/l) + 1 - (1 - l^{-2})^{1/2}\right]/\pi$$
$$\text{if } l \geq 1.$$

For a history of this problem, see Uspensky [4].

(Problems of this kind arise in some applications to ballistics which take the size of the shell into account.)

Additional information on the needle problem is provided by refs. 1–3.

REFERENCES

1. Buffon, L. (1777). *Essai d'arithmétique morale*.
2. Santaló, L. A. (1976). Integral geometry and geometric probability In *Encyclopedia of Mathematics and Its Applications*, Vol. 1, G. C. Rota, ed. Addison-Wesley, Reading, Mass.
3. Schuster, E. F. (1979). *Amer. Math. Monthly*, **81**, 26–29.

4. Uspensky, J. V. (1937). *Introduction to Mathematical Probability*. McGraw-Hill, New York.

See also GEOMETRIC PROBABILITY THEORY.

BULK SAMPLING

Bulk material is sampled by taking "increments" (small portions) of the material, blending these increments into a single composite sample, and then, if necessary, reducing this gross sample to a size suitable for laboratory testing. When dealing with bulk material there comes a point in the sampling procedure (either before or after the creation of the composite sample) when the material cannot be viewed as consisting of discrete constant preexistent and identifiable unique units that may be sampled as such. New units are created by means of some optional sampling device at the time of sampling. Another feature of bulk sampling is the physical mixing of the initial increments of the material. When the initial increments are blended into a single sample it is not possible to measure variation between the increments making up the composite. This is a disadvantage. In the case of bulk sampling, the object of the sampling is usually measurement of the mean quality of material sampled.

A detailed analysis of bulk sampling and formulation of the relevant theoretical models is given in Duncan [2]. For sampling standards, see ref. 1.

REFERENCES

1. American Society for Testing Materials, *Book of ASTM Standards*, 1958 Symposium on Bulk Sampling (ASTM Spec. Publ. No. 242). Philadelphia.
2. Duncan, A. J. (1962). *Technometrics*, **4**, 319–343.

BULLETIN IN APPLIED STATISTICS (BIAS)

SCOPE OF THE JOURNAL

The scope of this journal is very wide and varied. Although it is a journal for descriptive statistics, it overlaps greatly with the *Journal of Applied Statistics** (see JOURNAL OF THE ROYAL STATISTICAL SOCIETY, SERIAL CORRELATION) and *Statistical News and Notes*. This loose demarcation of the journal has provided greater interest and deeper stimulation to the applied statistician and others involved in practicing statistical methods in whatever field.

Applied statisticians in the course of their duties always come across the situation where the existing techniques either do not fit or have to be modified for their work. In other words, they have to engineer the existing statistical techniques to carry on their work. *BIAS* encourages such activities by way of publishing articles on statistical engineering and statistical case studies.

Table 1 should serve to indicate the nature and range of topics covered:

EDITORIAL POLICY

BIAS, although a serious and informative journal, is also meant to be lively and entertaining, and is therefore aimed at all those who are using statistics in any sector of this modern world. The purpose of the *Bulletin* is to create a medium through which news, views, and ideas may be aired and shared and in this way acts as a link between people with a variety of statistical interest. It is endeavoring to produce *readable* articles about what is going on in statistics which will be of interest to practitioners of statistics. It is trying to reveal the reasons for the apparent failure of certain techniques in some areas, and it is also concerned with the problems of communication within the statistical profession and between statisticians and users of statistics.

The editorial policy is to encourage the submission of articles to the *Bulletin* regarding:

1. Statistical activities in schools, colleges, polytechnic institutes, universities, research units, industries, local and central government bodies, and so on.

2. General nontechnical articles of a statistical nature (semitechnical articles may be considered).

3. Research reports, short communications in the nature of notes, or brief

Table 1. Contents of *BIAS*, Vol. 6, No. 1, 1979

Title	Author
Experimental Design from the Time of Fisher to the 1970's	G. H. Freeman
Some Practical Aspects of Canonical Variate Analysis	N. A. Campbell
Age, Period and Cohort Effects—A Confounded Confusion	H. Goldstein
Cluster Analysis of Some Social Survey Data	B. Jones
Probability and Special Relativity	T. F. Cox
A Report on the 1978 IOS Conference on Time-Series Analysis	C. Chatfield
The Census of Production and Its Uses	B. Mitchell and E. Swires-Hennesy
FORUM: Treatment of Myopia	C. I. Phillips
Letter to the Editor	

Table 2. BIAS in brief

First issue	1974
Editor and Founder	G. K. Kanji, Department of Mathematics and Statistics, Sheffield City Polytechnic, Sheffield S1 1WB, United Kingdom
Publisher	Sheffield City Polytechnic
Number of issues per year/volume	Two
Number of pages per issue	100 (approx.)
Description of subscribers	schools; colleges; polytechnic and university teachers; industrial, social, business, and medical statisticians from over 21 countries
Description of contributors	From all spheres of the statistical world
Types of articles	Any aspects of applied statistics
Subscription rates/volume	Personal (UK = £3.00, foreign = £4.00) Library, etc. (UK = £4.00, foreign = £6.00)

accounts of work in progress (prior to more formal publication).

4. Statistical case studies.

5. Reviews of statistical techniques.

6. Book reviews.

The articles are refereed. Two issues of the journal are produced each year. Table 2 is a concise presentation of pertinent facts about *BIAS*.

Figure 1, from the cover page of *BIAS*, wordlessly portrays the editor's feelings and communicates his sentiments about applied statistics.

G. K. KANJI

Figure 1. Picture on the cover page of *BIAS* portrays the editor. (Copyright © *BIAS*.)

BULLETIN ON MATHEMATICAL STATISTICS

A journal published by the Research Association of Statistical Sciences, Kyusha University, Fukuoka, Japan. Started in 1947, it appears somewhat irregularly; in recent years there has been one issue per year. All papers are in English and most (but not all) are authored by Japanese statisticians. The journal covers a wide spectrum of topics in probability theory and mathematical statistics at a rather high theoretical level. The most recent issues contain about 110 to 125 pages and 9 or 10 articles.

The founding and current editor is T. Kitagawa; there are 16 associate editors, including some well-known probabilists such as K. Ito and Y. Ishida.

BURN-IN

Burn-in is a selection procedure that is applicable when a survival* distribution function (SDF) has a decreasing hazard rate*. Items are put on test until a predetermined number fail, or for a predetermined time. The first of these corresponds to censored sampling (or censored data*), the second to truncated sampling, but the purpose here is different. Rather than using the data to make inferences about the SDF, we are simply hoping to be left with items having greater survival probability than the original product.

BIBLIOGRAPHY

Barlow, R. E., Madansky, A., Proschan, F., & Scheuer, E. M. (1968). *Technometrics*, **10**, 51–62. (A useful general presentation of estimation procedures based on data from burn-in processes.)

Marcus, R. & Blumenthal, S. (1974). *Technometrics*, **16**, 229–234. (Discusses a generalization of burn-in called sequential screening*.)

See also CENSORED DATA; CENSORING; and TRUNCATED DATA.

BURR DISTRIBUTIONS

Burr system of distributions was constructed in 1941 by Irving W. Burr. Since the corresponding density functions have a wide variety of shapes, this system is useful for approximating histograms*, particularly when a simple mathematical structure for the fitted cumulative distribution function (CDF) is required. Other applications include simulation*, quantal response, approximation of distributions*, and development of nonnormal control charts*. A number of standard theoretical distributions are limiting forms of Burr distributions.

The original motivation for developing the Burr system was to provide a method for fitting CDFs to frequency data. This approach avoids the analytical difficulties often encountered when a fitted density (such as a Pearson curve*) is integrated to obtain probabilities or percentiles*. Burr [1] chose to work with the CDF $F(x)$ satisfying the differential equation

$$\frac{dy}{dx} = y(1-y)g(x,y) \qquad [y = F(x)],$$

an analog of the differential equation that generates the Pearson system (*see* PEARSON SYSTEM OF DISTRIBUTIONS; FREQUENCY CURVES, SYSTEMS OF). The function $g,(x,y)$ must be positive for $0 \leqslant y \leqslant 1$ and x in the support of $F(x)$. Different choices of $g(x,y)$ generate various solutions $F(x)$; e.g., when $g(x,y) = g(x)$,

$$F(x) = \left\{ \exp\left[-\int_{-\infty}^{x} g(u)du \right] + 1 \right\}^{-1}.$$

The solutions $F(x)$ of Burr's differential equation can be classified by their functional forms, each of which gives rise to a family of CDFs within the Burr system. Burr [1] listed 12 such families:

(I)	x	$(0 < x < 1)$
(II)	$(1 + e^{-x})^{-k}$	$(-\infty < x < \infty)$
(III)	$(1 + x^{-c})^{-k}$	$(0 < x)$
(IV)	$\left[1 + \left(\dfrac{c-x}{x} \right)^{1/c} \right]^{-k}$	$(0 < x < c)$
(V)	$(1 + ce^{-\tan x})^{-k}$	$(-\pi/2 < x < \pi/2)$
(VI)	$(1 + ce^{-r\sinh x})^{-k}$	$(-\infty < x < \infty)$
(VII)	$2^{-k}(1 + \tanh x)^k$	$(-\infty < x < \infty)$
(VIII)	$\left(\dfrac{2}{\pi} \tan^{-1} e^x \right)^k$	$(-\infty < x < \infty)$
(IX)	$1 - \dfrac{2}{c[(1 + e^x)^k - 1] + 2}$	$(-\infty < x < \infty)$
(X)	$(1 + e^{-x^2})^k$	$(0 < x)$
(XI)	$\left(x - \dfrac{1}{2\pi} \sin 2\pi x \right)^k$	$(0 < x < 1)$
(XII)	$1 - (1 + x^c)^{-k}$	$(0 < x).$

(In each case, $c > 0; k > 0$.) The Roman numeral designation for the 12 types was first used by Johnson and Kotz [8]. The parameters c, k, and r are positive; when fitting data, a parameter (ξ) for location* and a parameter (λ) for scale* are introduced by replacing x with $(x - \xi)/\lambda$.

The Burr I family consists of the uniform distribution* whose densities have a rectangular shape. The Burr III, IV, V, IX, and XII families yield a variety of density shapes and involve four parameters, the most that can be efficiently estimated with the method of moments*. Among these families, Types III and XII are the simplest functionally and thus the most attractive for statistical modeling. Only the Type XII family was studied in detail by Burr [1] in his original work.

Burr XII density functions are of the form

$$f(x) = \frac{ckx^{c-1}}{(1 + x^c)^{k+1}} \quad (x > 0),$$

which is unimodal* (with a mode* at $x = [(c-1)/(ck+1)]^{1/c})$ if $c > 1$ and L-shaped if $c \leqslant 1$. Two typical Burr XII densities are illustrated in Fig. 1. The rth moment about the origin*

of a Burr XII distribution is

$$\mu_r' = kB(r/c + 1, k - r/c) \quad (r < ck).$$

The skewness* $(\sqrt{\beta_1})$ and kurtosis* (β_2) are given by

$$\sqrt{\beta_1} = \frac{\Gamma^2(k)\lambda_3 - 3\Gamma(k)\lambda_2\lambda_1 + 2\lambda_1^3}{[\Gamma(k)\lambda_2 - \lambda_1^2]^{3/2}}$$

and

$$\beta_2 = \frac{\Gamma^3(k)\lambda_4 - 4\Gamma^2(k)\lambda_3\lambda_1 + 6\Gamma(k)\lambda_2\lambda_1^2 - 3\lambda_1^4}{[\Gamma(k)\lambda_2 - \lambda_1^2]^2},$$

where $\lambda_j = \Gamma(j/c + 1)\Gamma(k - j/c)(j = 1, 2, 3, 4)$ and $ck > 4$. Here skewness is defined as $\sqrt{\beta_1} = \mu_3/\mu_2^{3/2}$, and kurtosis is defined as $\beta_2 = \mu_4/\mu_2^2$; as usual, μ_2, μ_3, and μ_4 denote moments about the mean μ_1.)

Rodriguez [9] constructed a moment ratio* diagram for the coverage area in the $(\sqrt{\beta_1}, \beta_2)$ plane corresponding to the Burr XII family. As shown in Fig. 2, the boundaries of the Burr XII area can be identified with limiting forms of Burr XII distributions. The northern boundary, labeled "Wei-bull," is generated by the $(\sqrt{\beta_1}, \beta_2)$ points of the Weibull

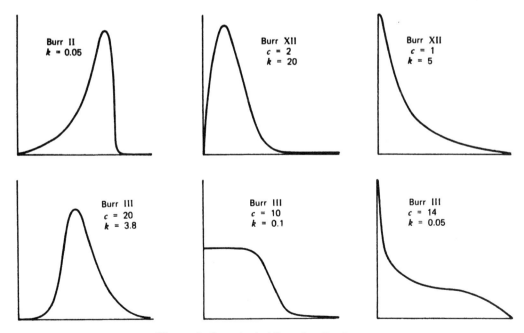

Figure 1. Some typical Burr density shapes.

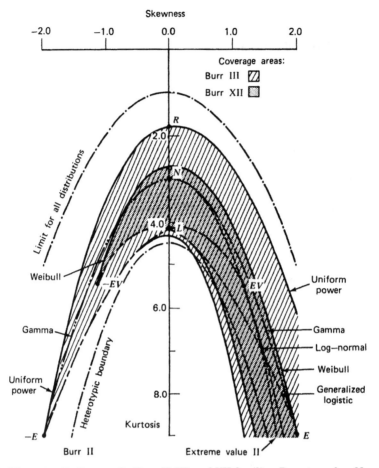

Figure 2. Moment ratio diagram for Burr II, III, and XII families. R, rectangular; N, normal; L, logistic; E, exponential; EV, extreme value. The symbol−denotes reflection.

distributions[*]; if X has a Burr XII distribution, then for fixed k,

$$\lim_{k \to \infty} \Pr\left[X \leqslant (1/k)^{1/c}x\right] = 1 - \exp(-x^c).$$

The southwest boundary is part of a curve labeled "generalized logistic," which corresponds to the generalized logistic distributions[*],

$$1 - (1 + e^x)^{-k}.$$

This family represents the limiting forms of Burr XII distributions as $c \to \infty$ with k fixed. The southeast boundary is generated by the Burr XII distributions with $k = 1$.

The curve labeled "generalized logistic" subdivides the Burr XII coverage area. In the single occupancy region north of this division, each $(\sqrt{\beta_1}, \beta_2)$ point corresponds to a unique (c, k) combination. In the double occupancy region south of this division, each $(\sqrt{\beta_1}, \beta_2)$ point corresponds to exactly two (c, k) combinations.

The Burr XII area occupies portions of the moment ratio diagram also covered by various well known families of distributions represented by points or curves in Fig. 2. These include the normal[*], log-normal[*], gamma[*], logistic[*], and extreme value Type I[*] distributions. Furthermore, the Burr XII region covers areas corresponding to the Type I, Type IV, and Type VI families of the Pearson system[*], as well as the S_U and S_B families of the translation (Johnson) system[*]. *See* JOHNSON'S SYSTEM OF DISTRIBUTIONS.

Although the Burr XII family covers a large portion of the moment ratio diagram, a much greater area is covered by the Burr III family (see Fig. 2). These two families are related, in the sense that if X has a Burr XII distribution with parameters c and k, then X^{-1} has a Burr III distribution with parameters c and k. Burr III density functions have the form

$$g(x) = \frac{ckx^{ck-1}}{(1+x^c)^{k+1}} \quad (x > 0),$$

which is unimodal (with a mode at $x = [(ck-1)/(c+1)]^{1/c}$ if $ck > 1$, and twisted L-shaped if $ck \leqslant 1$. Figure 1 provides examples of these shapes. To obtain the rth moment about the origin, skewness, and kurtosis of a Burr III distribution with parameters c and k, replace c with $-c$ in the expressions for μ'_r, $\sqrt{\beta_1}$, and β_2 given above for the Bur III family.

The northern boundary, labeled "uniform power," of the Burr III moment ratio coverage area in Fig. 2 is generated by the distributions of positive powers of a uniformly distributed random variable; these are limiting forms of Burr III distributions as $k \to 0$ with c fixed. The southwest boundary, labeled "Burr II," corresponds to the Burr II distributions with $k \leqslant 1$, whereas the southeast boundary is an envelope of $(\sqrt{\beta_1}, \beta_2)$ points for the Burr III family.

The Burr III area, like the Burr XII area, is subdivided into a northern single occupancy region and a southern double occupancy region. The division between these regions corresponds to the $(\sqrt{\beta_1}, \beta_2)$ points for two families of distributions: the Burr II distributions with $k > 1$, which are limiting forms as $c \to \infty$ of Burr III distributions, and the extreme value Type II* distributions, which are limiting forms of Burr III distributions as $k \to \infty$ with c fixed (see Rodriguez [10]).

Various methods are available for fitting* Burr distributions to frequency data. Burr [1] developed a cumulative moment technique for estimating the Burr XII parameters c and k from the observed skewness and kurtosis. This approach requires the use of a table (see Burr [4]), which can be difficult to interpolate due to the high nonlinearity of $\sqrt{\beta_1}$ and β_2 as functions of c and k. Other methods that can be used with both Burr XII and Burr III distributions include maximum likelihood estimation* and the method of percentiles*.

In addition to fitting frequency data, Burr distributions are useful for dealing with a number of statistical problems in which a class of distributions with functional simplicity and a variety of density shapes is required. For example, since the Burr XII and Burr III distributions can be inverted in closed form, they can be applied in simulation work, quantal response, and approximation of theoretical distributions whose moments are known, but whose functional forms cannot be expressed directly. See Drane et al. [6] for a discussion of Burr XII distributions as response* functions in analyzing quantal response experiments. This family was also used by Burr [2] to examine the effect of nonnormality on constants used in computing sample averages and ranges for plotting control charts.

It should also be noted that the sample medians* and ranges* for Burr III and Burr XII distributions have convenient distributional properties. For instance, if the sample size is $n = 2m + 1$, then the density function of the median is

$$\frac{(2m+1)!}{(m!)^2}(1+x^{-c})^{-mk}\left[1-(1+x^{-c})^{-k}\right]^m$$
$$\times \frac{ckx^{ck-1}}{(1+x^c)^{k+1}} \quad (x > 0)$$

for Burr XII distributions, and

$$\frac{(2m+1)!}{(m!)^2}\left[1-(1+x^c)^{-k}\right](1+x^c)^{-mk}$$
$$\times \frac{ckx^{c-1}}{(1+x^c)^{k+1}} \quad (x > 0)$$

for Burr XII distributions. Related results were obtained by Rodriguez [10] for the Burr III family and by Burr [3] and Burr and Cislak [5] for the Burr XII family.

A multivariate* Burr XII distribution was constructed by Takahasi [12]; the joint density function for variables $X_1, X_2, \ldots, X_m$ with this distribution is

$$f(x_1, \ldots, x_m)$$

$$= [\Gamma(k+m)/\Gamma(k)]$$

$$\times \left(1 + \sum_{j=1}^{m} \alpha_j x_j^{c_j}\right)^{-(k+m)} \prod_{j=1}^{m}(\alpha_j c_j x_j^{c_j-1})$$

for $x_j > 0, c_j > 0$ $(j = 1, \ldots, m)$, and $k > 0$. Here $\alpha_j^{1/c_j} X_j$ has a univariate Burr XII distribution with $c = c_j$, and any subset of the X_j's has a joint density of the foregoing form with an appropriate change of parameters. The regression* function of X_1 on $X_2, \ldots, X_m$ is

$$E[X_1 | X_2, \ldots, X_m]$$

$$= \left[\alpha_1^{-1} \left(1 + \sum_{j=2}^{m} \alpha_j X_j^{c_j}\right)\right]^{1/c_1} (k+m-1)$$

$$\times B(1 + c_1^{-1}, k + m - 1 - c_1^{-1})$$

and

$$\text{cov}(X_1, X_2)$$

$$= \alpha_1^{-1/c_1} \alpha_2^{-1/c_2} \Gamma(1 + c_1^{-1}) \Gamma(1 + c_2^{-1})$$

$$\times \left[\frac{\Gamma(k - c_1^{-1} - c_2^{-1})}{\Gamma(k)} \right.$$

$$\left. - \left(\frac{\Gamma(k - c_1^{-1})}{\Gamma(k)}\right) \left(\frac{\Gamma(k - c_2^{-1})}{\Gamma(k)}\right) \right].$$

Durling [7] discussed a bivariate Burr XII distribution whose joint cumulative distribution function is

$$F(x_1, x_2) = 1 - (1 + x_1^{c_1})^{-k} - (1 + x_2^{c_2})^{-k}$$

$$+ (1 + x_1^{c_1} + x_2^{c_2} + r x_1^{c_1} x_2^{c_2})^{-k},$$

provided that $x_1 \geqslant 0$, $x_2 \geqslant 0$, and $0 \leqslant r \leqslant k + 1$; otherwise, $F(x_1, x_2) = 0$. If $r = 0, F(x_1, x_2)$ reduces to the bivariate forms of the distribution introduced by Takahasi [12]. If $r = 1, F(x_1, x_2)$ can be written as the product of two independent Burr XII distributions.

The approach followed by Takahasi [12] can be used to derive a multivariate Burr III distribution whose joint density function is

$$g(x_1, \ldots, x_m) = [\Gamma(k+m)/\Gamma(k)]$$

$$\times \left(1 + \sum_{j=1}^{m} \alpha_j x_j^{-c_j}\right)^{-(k+m)}$$

$$\times \prod_{j=1}^{m}(\alpha_j c_j x_j^{-c_j-1})$$

for $x_j > 0, c_j > 0 (j = 1, \ldots, m)$, and $k > 0$. See Rodriguez [11] for further details concerning this distribution and other multivariate Burr III distributions. See FREQUENCY SURFACES, SYSTEMS OF.

REFERENCES

1. Burr, I. W. (1942). *Ann. Math. Statist.*, **13**, 215–232.

2. Burr, I. W. (1967). *Ind. Quality Control*, **23**, 563–569.

3. Burr, I. W. (1968). *J. Amer. Statist. Ass.*, **63**, 636–643.

4. Burr, I. W. (1973). *Commun. Statist.*, **2**, 1–21.

5. Burr, I. W. and Cislak, P. J. (1968). *J. Amer. Statist. Ass.*, **63**, 627–635.

6. Drane, J. W., Owen, D. B., and Seibert, G. B., Jr. (1978). *Statist. Hefte*, **19**, 204–210.

7. Durling, F. C. (1975). In *Statistical Distributions in Scientific Work*, Vol. 1, G. P. Patil, S. Kotz, and J. K. Ord, eds. D. Reidel, Dordrecht, Holland, 329–335.

8. Johnson, N. L. and Kotz, S. (1970). *Continuous Univariate Distributions*, Vol. 1. Houghton Mifflin, Boston.

9. Rodriguez, R. N. (1977). *Biometrika*, **64**, 129–134.

10. Rodriguez, R. N. (1979). *The Moment Ratio Geography of Burr Type III Distributions. Forthcoming research publication*, General Motors Research Laboratories, Warren, Mich.

11. Rodriguez, R. N. (1979). *Multivariate Burr III Distributions*. Forthcoming research publication, General Motors Research Laboratories, Warren, Mich.

12. Takahasi, K. (1965). *Ann. Inst. Statist. Math. Tokyo*, **17**, 257–260.

See also FREQUENCY CURVES, SYSTEMS OF; FREQUENCY SURFACES, SYSTEMS OF; and MOMENT-RATIO DIAGRAMS.

ROBERT N. RODRIGUEZ

BUSINESS FORECASTING METHODS

All businesses plan for an uncertain future, formally or informally. Planning and decision-making processes inevitably incorporate some view of projected developments in the relevant environment. The forecasting activity, then, is not merely an option in business management; it is inevitable. Forecasts are not invariably developed through some formal mechanism—much less through one that embodies principles familiar to statisticians. Indeed, much business forecasting is implicit rather than explicit, informal rather than formal, judgmental rather than objective, and ad hoc rather than based on familiar statistical principles.

It is not appropriate to view "business forecasting" as a coherent entity, whose methodology can be readily described. In practice, businesses face a wide array of forecasting questions. These range from the need to take a view of the long term, as it relates to a large investment project, to the requirement to generate short-term forecasts of demand for many products, as an input to production planning and inventory management. Quite different approaches to forecasting will be appropriate at these extremes; the approaches taken in practice depend greatly on the resources available to, and the inclinations of, management. These resources should include appropriate subject-matter knowledge and experience of the relevant environment, as well as an array of quantitative techniques. The mix of resources applied in any specific case is very much problem-specific.

Two scholarly journals—the *Journal of Forecasting* and the *International Journal of Forecasting*—began publication in the 1980s, and some feel for the very wide range of available forecasting methods can be obtained through scanning these journals. However,

the great majority of the published articles are written by academics for other academics, and do not discuss actual business case studies. They suggest a level of technical sophistication far in excess of current practice. Reported surveys—including refs. 29, 40, 10, 42, and 39—suggest that very often purely judgmental approaches to forecasting are employed. These include the personal judgment of an individual executive, the combined judgment of a group of executives, and, in the case of sales forecasting, an amalgamation of the views of the sales force. When "objective" approaches are employed, they are often very rudimentary, based for example on a simple moving average* of, or trend line fitted to, recent past observations. Two editorial articles in the *International Journal of Forecasting* emphasize the point. Chatfield [7] asks if "Simple is best?" while DeRoeck [11] asks "Is there a gap between forecasting theory and practice?"

In this article we shall concentrate on nontrivial quantitative methods employed in business forecasting. Although these methods are at the sophisticated end of the spectrum of contemporary practice, most are very unsophisticated by the standards of contemporary academic theoretical developments. However, it would be misleading to discuss here such interesting issues as nonlinear time-series* models when there appears to be little chance of their extensive incorporation into business forecasting practice in the foreseeable future. The methods we shall discuss have been mostly applied to short-term forecasting, particularly sales forecasting. Long-term forecasting is far more likely to be based on subjective judgmental considerations.

From the statistician's viewpoint, it is reasonable to ask if "forecasting" is an interesting issue distinct from the elements of model building. After all, formal models are often fitted, through the methods of time-series* analysis, regression*, and econometrics*, to historical records. Such models are often used to understand observed phenomena, but their projection for forecasting is then relatively routine. It might therefore be concluded that forecasting is a fairly straightforward adjunct to statistical model building. To some extent, this is correct. However, some aspects—notably prediction through exponential smoothing and the combination of forecasts*—are somewhat divorced from traditional statistical model-building methodology.

Often in practice, forecasts are based exclusively on the history of a single time series. More generally, information on related time series can also be incorporated, through an amalgamation of the methodologies of regression and time-series analysis. On many occasions the forecaster has available several alternative possible approaches. It might then be useful to consider *combining* forecasts from different sources. In addition, since alternative approaches might be adopted, it is sensible to *evaluate* the performance of a specific approach to forecasting. This might suggest inadequacies and point to directions for possible improvements in forecast quality.

TREND CURVES

Many business forecasting methods project future values of a variable as a function of its own past history. Let X_t denote the observed quantity of interest at time t. Given a time series of observations, one possibility is to fit to these data a deterministic function of time. The equation to be estimated is then of the general form

$$g(X_t) = f(t, \beta) + \varepsilon_t, \tag{1}$$

where g and f are specified functions, the latter involving a vector of unknown parameters, β, and where ε_t is a random error term. For example, f might be a polynomial in time, while g might be the logarithmic function. Once the parameters of the trend function have been estimated, the fitted function is projected forward to derive forecasts.

While the choices of linear, quadratic, and log-linear functions are obvious possibilities, special considerations suggest other functions in some applications. In marketing* applications, the quantity X_t might represent the percentage of households purchasing or owning some product. Then the function f should have an asymptote, representing market saturation. Further, it is believed

that, when a new product is introduced, sales and ownership initially experience a period of rapid growth, followed by a mature period of steady growth, and eventually slow movement to a saturation level. This suggests a trend function that is S-shaped, such as the logistic function, with $g(X_t) = X_t$ in (1), and

$$f(t, \beta) = \alpha[1 + \beta \exp(1 - \gamma t)]^{-1}.$$

Meade [28] provides a much fuller discussion of sales forecasting through fitted trend functions.

There are two main difficulties in this approach. First, the specification of an appropriate trend function is quite arbitrary. Possibly several alternative functions might be tried. Unfortunately, experience suggests that, while different functions might fit observed data almost equally well, these fitted functions can diverge quite rapidly when projected forward (see, e.g., Freedman et al. [14]). In these circumstances, it is difficult to place much confidence in medium- or long-term forecasts. Second, it is commonly assumed, for both parameter estimation and the derivation of forecasts, that the error terms ε_t in (1) are uncorrelated. However, experience suggests that many business time series exhibit very strong autocorrelation patterns. Indeed, for short-term forecasting it is likely to be more important to anticipate departures from any long-run trend function than to specify that function well. Consequently, short-term forecasts derived under the assumption that the error terms in (1) are uncorrelated are likely to be seriously suboptimal. Of course, it is quite feasible to incorporate sophisticated autocorrelated error structures in (1), but this is rarely done in practice.

The forecasting methods we discuss next are primarily designed for short-term prediction, and concentrate on the autocorrelation structure. It is, however, quite straightforward to incorporate deterministic time trends in the framework of ARIMA models.

EXPONENTIAL SMOOTHING

Exponential smoothing is quite widely used in business forecasting (*see also* GEOMETRIC

MOVING AVERAGE and PREDICTION AND FORECASTING). Gardner [16] provides an extensive survey of this approach. The most common application is to routine short-term sales forecasting for inventory management and production planning. Typically, sales forecasts a few months ahead are required for a large number of product lines. The attraction of exponential smoothing is that it is quickly and easily implemented, forecasts being generated routinely, except perhaps in a few cases where there is evidence of unsatisfactory forecast quality. Relatively little attention is paid to each individual time series, so that it is not expected that individual forecasts will be optimal. Rather, the hope is that the prediction algorithms are sufficiently robust that forecasts of adequate quality should be obtained for sales of the great majority of product lines.

Viewed by a statistician, the development of exponential smoothing was quite extraordinary. Although the procedure is entirely quantitative, this development owed very little to fundamental statistical concepts, such as probability and uncertainty, random variables, inference, and model building. Indeed, the rationale behind those exponential smoothing methods now commonly used can be easily understood without ever having encountered these concepts.

Exponential smoothing is neither a single forecast-generating mechanism nor a logically coherent general approach to forecasting. Rather, what is available is a set of forecast-generating *algorithms*, one of which must be selected for any specific task, and all of which can be viewed as developments of *simple exponential smoothing*. The simple exponential smoothing algorithm is appropriate for nonseasonal time series with no predictable trend. The task is to estimate the current *level* of the series. This level estimate is then used as the forecast of all future values. Let L_t denote the level estimate at time t, to be based on current and past observations X_{t-j}, $j \geqslant 0$. This estimate is formed as a *weighted average* of current and past observations, with highest weights associated with the most recent observations. This coincides with the plausible intuition that, in developing forecasts, most weight should be associated with the recent past, and very

little weight should be given to the distant past. If a system of exponentially decreasing weights is used, then, for any $0 < \alpha < 1$, a weighted average is provided by

$$L_t = \alpha X_t + \alpha(1 - \alpha)X_{t-1}$$
$$+ \alpha(1 - \alpha)^2 X_{t-2} + \cdots$$
$$= \alpha X_t + (1 - \alpha)L_{t-1}, \quad 0 < \alpha < 1. \quad (2)$$

Equation (2) is the computational algorithm for simple exponential smoothing. Imposing an initial condition, such as $L_1 = X_1$, simple recursive calculations follow for $t \geqslant 2$. The parameter α in (2) is a *smoothing constant*. Often in practice its value is set judgmentally—either through graphical inspection of the time series, or as a value that has proved successful in the prediction of similar series. A more objective approach is to try a grid of possible values—say $0.1, 0.2, \ldots, 0.9$. That particular value which best "predicts" the sample observations one period ahead—generally in a least-squares* sense—is then retained for future forecasting.

Forecasts of all future values are set at the most recent estimate of level. Then, standing at time $t = n$, the forecast of X_{n+h} is

$$\hat{X}_n(h) = L_n, \quad h = 1, 2, 3, . \quad (3)$$

Often, it is felt that upward or downward trend is predictable. If a linear forecast function, possibly in the logarithms of the original process, is thought to be adequate over the horizon of interest, *Holt's linear trend algorithm* provides a natural extension of simple exponential smoothing. In addition to local level L_t, a local slope T_t is introduced. When a new observation X_t becomes available, these estimates are updated through the pair of equations

$$L_t = \alpha X_t + (1 - \alpha)(L_{t-1} + T_{t-1}),$$
$$0 < \alpha < 1, \quad (4a)$$
$$T_t = \beta(L_t - L_{t-1}) + (1 - \beta)T_{t-1}.$$
$$0 < \beta < 1. \quad (4b)$$

The smoothing constants α and β in these equations can again be chosen subjectively or objectively, and forecasts are obtained from

$$\hat{X}_n(h) = L_n + hT_n, \quad h = 1, 2, \dots \quad (5)$$

Many business time series, including a large number of product sales series, are seasonal. Let s denote the number of periods in the year, so that $s = 4$ for quarterly data and $s = 12$ for monthly data. The *Holt–Winters algorithm* extends Holt's linear trend algorithm through the introduction of s seasonal factors F_t, which may be taken to be additive or multiplicative. In the additive case, the updating equations of the algorithm are

$$L_t = \alpha(X_t - F_{t-s}) + (1 - \alpha)(L_{t-1} + T_{t-1}),$$
$$0 < \alpha < 1, \quad (6a)$$

$$T_t = \beta(L_t - L_{t-1}) + (1 - \beta)T_{t-1},$$
$$0 < \beta < 1, \quad (6b)$$

$$F_t = \gamma(X_t - L_t) + (1 - \gamma)F_{t-s},$$
$$0 < \gamma < 1, \quad (6c)$$

The three smoothing constants α, β, γ may be determined subjectively or objectively, and forecasts are obtained from

$$\hat{X}_n(h) = L_n + hT_n + F_{n+h-s},$$
$$h = 1, 2, \dots, s, \quad (7)$$

and so on. In the case of multiplicative seasonal factors, $(X_t - F_{t-s})$ is replaced by X_t/F_{t-s} in (6a), and $(X_t - L_t)$ by X_t/L_t in (6c). The righthand side of (7) is then $(L_n + hT_n)F_{n+h-s}$.

The three exponential smoothing algorithms introduced here are all commonly used. Chatfield [6] and Chatfield and Yar [8] provide considerably fuller discussion of the most general of them. However [16], there are many possible alternative algorithms that have also been applied in practice. Unfortunately, there does not appear to be a well-defined objective methodology allowing the user to select a particular algorithm for the prediction of a specific time series. Typically, in practice, this selection is made subjectively.

The practical attraction of exponential smoothing is that fully automatic prediction algorithms allow the routine computation of forecasts, without manual intervention. However, it is recognized that forecast quality should be monitored, preferably automatically. Gardner [15] and McClain [26] discuss so-called *tracking signals*, based on prediction errors, that have been employed to monitor forecast quality. Given a signal of unsatisfactory forecast performance—for example, evidence of bias—the prediction algorithm might be manually or automatically adjusted.

Exponential smoothing consists of a collection of numerical algorithms for forecast generation. At least overtly, these do not incorporate the stochastic elements of a statistical model or the ingredients of a statistical-model-building exercise. Of course, the objective determination of smoothing constants through minimizing a sum of squared prediction errors is in effect least-squares parameter estimation. Moreover, tracking signals of forecast quality are assessed through what amount to traditional hypothesis tests. Nevertheless, as Newbold and Bos [34] have shown, the absence of explicit stochastic models from the development of exponential smoothing algorithms has led to a number of practices of questionable validity in the application of exponential smoothing algorithms.

EXPONENTIAL SMOOTHING AND ARIMA MODEL BUILDING

Box and Jenkins [5] introduced an approach to forecasting based on statistical model building (*see also* Box—Jenkins Model and Prediction and Forecasting). The Box—Jenkins approach begins by postulating a class of models as potential generators of observed time series. Given a particular time series, a specific model from this general class is fitted through an iterative cycle of model selection, parameter estimation, and model checking. Then, forecasts are derived through mechanical projection of the fitted model. This approach is used in business forecasting, though perhaps not as widely as one might guess. The available survey evidence suggests for example that its obvious competitor, exponential smoothing, is more widely applied in practice.

For a single nonseasonal time series, Box and Jenkins consider the autoregressive integrated moving average* [ARIMA (p, d, q)]

class of models

$$(1 - \phi_1 B - \cdots - \phi_p B^p)[(1 - B)^d X_t - \mu]$$
$$= (1 - \theta_1 B - \cdots - \theta_q B^q)\varepsilon_t, \qquad (8)$$

where B is a back-shift operator, so $B^j X_t = X_{t-j}$; μ, ϕ_i, and θ_j are fixed parameters; and the ε_t are uncorrelated deviates.

It is well known (see, for example, refs. 32, 24, and 27) that prediction from particular exponential smoothing algorithms is equivalent to forecasting from restricted variants of specific simple ARIMA models. For example, from (2) and (3) a little algebraic manipulation shows that simple exponential smoothing is equivalent to forecasting from an ARIMA (0, 1, 1) model, and from (4) and (5) that Holt's linear trend algorithm is equivalent to forecasting from an ARIMA (0, 2, 2) model. Indeed, virtually all exponential smoothing algorithms in common use for the prediction of nonseasonal time series are equivalent to sparsely parametrized members of the ARIMA class.

Some years ago, it could be argued that computational considerations limited the scope for practical adoption of the Box–Jenkins approach, but that case is now difficult to sustain. Presumably, given a moderately long series of observations, sensible application of the ARIMA model-building methodology should identify those cases where specific exponential smoothing algorithms are appropriate, and also detect cases in which they are clearly inappropriate. In short, there is now little justification for employing exponential smoothing to predict nonseasonal time series.

The seasonal case is rather different; see Abraham and Ledolter [1] and Newbold [33]. The ARIMA class of models (8) was extended by Box and Jenkins to what is generally called the "multiplicative seasonal ARIMA class." The additive seasonal variant of the Holt–Winters algorithm is equivalent to prediction from a model in which $(1 - B)(1 - B^s)X_t$ is moving average of order $q = s + 1$, the moving-average coefficients being specific functions of the three smoothing constants. There is no facility in the standard ARIMA framework for building in such functional restrictions. Moreover, the principle of parsimony* renders it unlikely

that an ARIMA model with $s + 1$ free parameters will be fitted in practice, at least in the case of monthly time series. The more commonly applied multiplicative variant of the Holt–Winters algorithm generates predictors that are nonlinear functions of the observed data. For that reason, it does not correspond to a member of the ARIMA class, though Abraham and Ledolter have demonstrated that in some circumstances it is possible to find a seasonal time-series model that generates similar one-step predictions. As a practical matter, then, in the case of seasonal time series, exponential smoothing offers a genuinely distinct alternative to ARIMA model building. However, exponential smoothing is still relatively inflexible in terms of the alternative predictors available, and little effort is made to assess whether the prediction algorithm applied is appropriate for the specific time series of interest.

Exponential smoothing algorithms are also closely related to statistical models that view a time series as the sum of unobserved components—often designated "trend, cycle, seasonal, and irregular." Many of these components models are special cases of the ARIMA class; see refs. 22 and 33 for further discussion.

REGRESSION AND ECONOMETRIC METHODS

The methods discussed so far are all designed for the prediction of a time series exclusively on the basis of its own history. Often, other relevant information will be available, and it is desirable that such information be incorporated in the forecast-generating mechanism. The methodology of regression* analysis provides an obvious starting point. Let Y denote the series to be predicted, and assume that Y is influenced by K variables $X_1, \ldots, X_K$. The static linear regression model is

$$Y_t = \alpha + \beta_1 X_{1t} + \cdots + \beta_K X_{Kt} + \varepsilon_t, \qquad (9)$$

where ε_t is a stochastic error term. If the error terms are assumed to be uncorrelated, the parameters of this model can be estimated by least squares. The fitted model is then projected forward for forecasting. This procedure will generate *conditional forecasts* of Y, given assumed future values of the X_i. *Unconditional forecasts* of Y can only be obtained by inserting forecasts of future X_i, which must be developed elsewhere. In practice, this is often done judgmentally, though a single-series time-series approach can be applied.

The preceding approach is too crude to be of much value, in practice, for two reasons. First, it does not capture the *dynamics* of typical relationships among business time series. One way to achieve this is to include among the regressors on the right-hand side of (9) past "lagged" values of Y and the X_i. Second, the error terms in a regression equation often are autocorrelated. It is important that any autocorrelation structure in the errors be taken into account in both the estimation of the parameters of (9) and the calculation of the forecasts. One possibility is to specify an ARIMA model for these error terms.

Often theory suggests that a set of "dependent," or *endogenous*, variables $Y_1, \ldots, Y_M$ are jointly determined by a set of "independent," or *exogenous*, variables $X_1, \ldots, X_K$. Theory then postulates a system of M simultaneous stochastic equations linking the endogenous and exogenous variables (*see* ECONOMETRICS). Once such a model has been estimated, it can be projected forward to derive conditional forecasts of the endogenous variables given assumed future values of the exogenous variables.

The most sophisticated practical applications of econometric simultaneous equations models are in macroeconomic forecasting. Klein [23] provides an excellent introduction and a nice taste of the scope of contemporary practice. Econometric models of national economies frequently involve systems of hundreds of equations. The specifications of these equations are based on some economic theory, which often leads to nonlinear relationships, so that the solution for the endogenous variables in terms of the exogenous variables must be accomplished numerically. In addition, some care is taken in specifying appropriate dynamic structures of the individual equations. In spite of the sophistication and complexity of these models, the forecasts they generate are often judgmentally adjusted. Simultaneous-equation econometric models are relatively infrequently used in business forecasting, and most applications involve quite small models. However, business forecasters do use the forecasts from large macroeconomic models as inputs into their own forecasting exercises.

REGRESSION AND TIME-SERIES PRINCIPLES

We have stressed the importance of capturing the dynamics of a relationship and the autocorrelation structure of the errors in a time-series regression model. One way to achieve this, discussed in detail by Box and Jenkins [5], is through a *transfer-function noise model*. In the case of a single independent variable X_t, the model is

$$Y_t = \frac{\omega_0 + \omega_1 B + \cdots + \omega_r B^r}{1 - \delta_1 B - \cdots - \delta_s B^s} X_t + u_t, \quad (10)$$

where ω_i and δ_i are fixed parameters, and the stochastic error term u_t is permitted to follow an ARIMA process.

Because of the structure of typical business and economic time series, further considerations arise in the construction of regression models linking such series. Empirical evidence suggests that many business and economic time series may be generated by models that are *integrated of order one*—that is, the series are not trend-stationary, but first differences are stationary, so that an ARIMA $(p, 1, q)$ model might be appropriate. It is well known (see, for example, refs. 19, 36, and 38) that inference based on regressions involving the levels of such series can be unreliable, leading to a tendency to "discover," through the usual significance tests, relationships where none exist. It is tempting then to consider regression analysis based on first differences of such series. However, as discussed by Engle and Granger [13], this ignores an important and interesting possibility—that stable long-run relationships among the series exist. For example, let Y_t and X_t be integrated of order one. Then there may exist a constant β such that $Y_t - \beta X_t$ is stationary. In that case, the series are said to be *cointegrated* (*see* COINTEGRATION). Economists model relationships among cointegrated time series through *error-correction models*, involving current and past first differences of the time series, and error-correction terms, such as $Y_{t-1} - \hat{\beta}X_{t-1}$, where $\hat{\beta}$ is an estimate of β, in our two-variable example. Although these considerations are potentially extremely important in model building for business forecasting, they take us some considerable distance from contemporary practice; see Banerjee et al. [2] and Hamilton [21].

COMBINATION OF FORECASTS*

Forecasters often have available two or more alternative forecasts or forecast-generating methodologies. One possibility is to select a single forecast, either on subjective grounds, or on the basis of past performance. An alternative strategy, originally proposed by Bates and Granger [3], is to form a composite, or *combined*, forecast. This simple idea has proved to be very effective over a wide range of applications. Some subsequent developments in this area are summarized by Clemen [9], Winkler [43], and Granger [17].

Let $F_i (i = 1, \ldots, K)$ be K forecasts of the quantity X, made in an earlier time period. Bates and Granger proposed employing a combined forecast that is a weighted average of the individual forecasts; that is,

$$C = \sum_{i=1}^{K} W_i F_i, \quad \sum_{i=1}^{K} W_i = 1,$$
$$W_i \geqslant 0 \quad i = 1, \ldots, K.$$

The simplest possibility is the equal-weights case, where the combined forecast is a simple average of the constituent forecasts. This can prove very effective in practice [25].

Given a record of the past performances of the constituent forecasts, it seems more reasonable to use this information to suggest appropriate combination weights. Then, most weight would be given to the forecaster, or forecasting method, that has been most successful in the recent past. It might be reasonable to permit the weighting scheme to adapt over time, allowing for changes in the relative performance of the individual forecasts. This suggests that the weights should be based on just a few past forecasts, or possibly that some discounting scheme should

be employed in determining the weights. Let $F_{it} (i = 1, \ldots, K, t = 1, \ldots, n)$ be forecasts of X_t, made h time periods earlier. Then we will have observed the forecast errors

$$e_{it} = X_t - F_{it},$$
$$i = 1, \ldots, K, \quad t = 1, \ldots, n.$$

One possibility is to employ for the current set of forecasts, weights inversely proportional to the sum of squared errors of previous forecasts; that is,

$$W_i = \left(\sum_{t=1}^{n} e_{it}^2 \right)^{-1} \Big/ \sum_{j=1}^{K} \left(\sum_{t=1}^{n} e_{jt}^2 \right)^{-1}$$
$$i = 1, \ldots, K. \tag{11}$$

A more obvious possibility is to estimate the appropriate weights through regression. A simple linear regression equation is provided by

$$X_t = W_1 F_{1t} + \cdots + W_K F_{Kt} + \varepsilon_t, \tag{12}$$

where ε_t denotes the error of the combined forecast. The requirement that the weights should sum to one can be achieved by estimating W_i $(i = 1, \ldots, K-1)$ through the regression of $X_t - F_{Kt}$ on $F_{it} - F_{Kt}, i = 1, \ldots, K-1$. This does not assure that the resulting weights estimates will be nonnegative. One could drop those forecasts with negative weight estimates, and reestimate the regression equation. For one-step-ahead prediction, it is generally reasonable to assume that the combined forecast errors will not be autocorrelated, so that ordinary least-squares estimation of (12) is appropriate. For prediction h steps ahead, the forecast errors will follow a moving-average process of order $h - 1$, and allowance for this autocorrelated error structure can be made in estimating (12). Diebold [12] provides a fuller discussion on serially correlated errors in forecast combination.

Combination as a weighted average is appealing for two reasons. First, it seems reasonable that nonnegative weights should be applied to the constituent forecasts. Second, if all the constituent forecasts are unbiased, then so will be their weighted average. However [20], often individual forecasts may be biased, yet still provide useful information; the more general multiple regression

$$X_t = \alpha + \beta_1 F_{1t} + \cdots + \beta_K F_{Kt} + \varepsilon_t$$

should also be considered. It is, of course, possible to test the constraints implied by (12) after fitting the more general model.

The intuitive appeal of the combination of forecasts is the same as that of investment in a portfolio of securities rather than in a single security—risk is diversified. In fact, many studies, including Newbold and Granger [37], Winkler and Makridakis [44], and those cited in Clemen [9], have provided empirical evidence on the practical effectiveness of combination. The combined forecast very often outperforms the individual forecasts. The accumulated empirical evidence suggests that most often the weighting scheme (11) is more successful than a regression-based scheme for determining combination weights. In the latter case, when there are two forecasts, the weight given to the first is

$$W_1 = \frac{\sum_{t=1}^n e_{2t}^2 - \sum_{t=1}^n e_{1t}e_{2t}}{\sum_{t=1}^n e_{1t}^2 + \sum_{t=1}^n e_{2t}^2 - 2\sum_{t=1}^n e_{1t}e_{2t}},$$

while the scheme (11) replaces the sum of products by zero in this expression. No completely convincing theoretical explanation of why this should so often be preferable is available. The regression-based approach is most often useful when one or more of the constituent forecasts is very poor and should therefore be given zero weight in the composite.

The theoretical appeal of forecast combination does not match its practical success. After all, in principle at least, it should be possible to find a model that incorporates all constituent forecast-generating models as special cases. Many econometric practitioners would then prefer to develop from sample data a single forecasting model, perhaps employing the encompassing principle of Mizon and Richard [31]. The practical success of the combination of forecasts may result from its relative parsimony and a minimum of data mining* over the sample period.

EVALUATION OF FORECASTS

Since there will not be complete assurance that any forecast-generating mechanism will produce the best possible, or even adequate, forecasts, it is important to monitor performance. Forecast evaluation criteria in common use are surveyed by Granger and Newbold [18], Stekler [41], and Newbold and Bos [35]. Forecasts are produced as an aid to decision making, and occasionally it is possible to associate actual costs with forecasting errors. Boothe and Glassman [4] evaluate exchange-rate forecasts in terms of the profitability of forward market speculation. Generally, however, this is not feasible, and most evaluation criteria in common use adopt squared error as a proxy for costs. Then competing forecasts of the same quantity are compared through postsample mean squared errors.

Assume that we have available a record of h-steps-ahead forecasts F_t ($t = 1, \ldots, n$) and have observed the corresponding outcomes X_t, so that the forecast errors are

$$e_t = X_t - F_t, \quad t = 1, \ldots, n.$$

A number of forecast evaluation criteria in common use attempt to assess in isolation this set of forecasts. For example, Mincer and Zarnowitz [30] proposed estimation of the regression equation

$$X_t = \alpha + \beta F_t + \varepsilon_t. \tag{13}$$

These authors describe as "efficient" forecasts for which $\alpha = 0$ and $\beta = 1$ in (13). In principle, Mincer–Zarnowitz efficiency could be viewed as a null hypothesis to be tested through a standard F-test*. However, some care is needed. For fully optimal h-steps-ahead forecasts, the error terms in (13) will follow a moving-average process of order $h - 1$. This autocorrelated error structure should be allowed for when the regression equation is estimated. Indeed, a further test of "efficiency" would check whether this error structure actually holds. For example, for one-step-ahead forecasts, it is important to test whether the error terms in (13) are uncorrelated.

In fact, relatively little can be learned from viewing in isolation a single set of forecasts, that, for example, could be efficient in the Mincer–Zarnowitz sense, and yet be inferior to alternative forecasts that could have been developed. Often an investigator has available an alternative set of forecasts for comparison. When this is not so, it is common practice to develop such a set. For example, forecasts from sophisticated econometric models are often compared with "naive" forecasts, obtained exclusively from the past history of the time series, possibly through a fitted ARIMA model. Granger and Newbold [18] argue that, when forecasts are evaluated in this way, the analyst should not be completely satisfied by merely outperforming the naive forecasts. A more stringent requirement is that the naive forecasts should contain no additional useful information. In that case, the forecasts of interest are said to be *conditionally efficient* with respect to the naive forecasts.

Let, F_1 and F_2 be competing forecasts of X. Conditional efficiency can be assessed through the combination of forecasts. If F_1 is conditionally efficient with respect to F_2, then F_2 should receive zero weight in the composite forecast. Given a set observations, the regression

$$X_t - F_{1t} = W(F_{2t} - F_{1t}) + \varepsilon_t$$

can be estimated, and the null hypothesis of conditional efficiency–that is, $W = 0$—tested. It is easy to test simultaneously for Mincer–Zarnowitz efficiency and conditional efficiency. In the regression

$$X_t = \alpha + \beta_1 F_{1t} + \beta_2 F_{2t} + \varepsilon_t \qquad (14)$$

this implies $\alpha = 0$, $\beta_1 = 1$, $\beta_2 = 0$. Again, for forecasting h steps ahead, the error term in (14) should be moving average of order $h - 1$.

The evaluation procedures discussed here are employed in practice. Nevertheless, frequently evaluation is ignored, so that too often business forecasters persevere with methods that are unnecessarily suboptimal.

SUMMARY

Practice in business forecasting is quite diverse, partly as a result of the diversity of forecasting problems that are encountered. Much of the diversity can be attributed to the inclinations of business forecasters. Very often those inclinations suggest skepticism about statistical methods that are judged to be complicated, or difficult to understand. Consequently, frequently forecasting is purely judgmental, even when adequate data for statistical model building are available. The quantitative methods that are used most often are the most straightforward, such as trend curve extrapolation, exponential smoothing, and forecast combination.

REFERENCES

1. Abraham, B. and Ledolter, J. (1986). Forecast functions implied by autoregressive integrated moving average models and other related forecasting procedures. *Int. Statist. Rev.*, **54**, 51–66.

2. Banerjee, A., Dolado, J., Galbraith, J. W., and Hendry, D. F. (1993). *Cointegration, Error Correction, and the Econometric Analysis of Nonstationary Data*. Oxford University Press, Oxford, England.

3. Bates, J. M. and Granger, C. W. J. (1969). The combination of forecasts. *Op. Res. Quart.*, **20**, 451–468.

4. Boothe, P. and Glassman, D. (1987). Comparing exchange rate forecasting models: accuracy versus profitability. *Int. J. Forec.*, **3**, 65–79.

5. Box, G. E. P. and Jenkins, G. M. (1970). *Time Series Analysis, Forecasting, and Control*. Holden Day, San Francisco.

6. Chatfield, C. (1978). The Holt–Winters forecasting procedure. *Appl. Statist.*, **27**, 264–279.

7. Chatfield, C. (1986). Simple is best? *Int. J. Forec.*, **2**, 401–402.

8. Chatfield, C. and Yar, M. (1988). Holt–Winters forecasting: some practical issues. *Statistician*, **37**, 129–140.

9. Clemen, R. T. (1989). Combining forecasts: a review and annotated bibliography. *Int. J. Forec.*, **5**, 559–583.

10. Dalrymple, D. J. (1987). Sales forecasting practices: results from a United States survey. *Int. J. Forec.*, **3**, 379–391.

11. DeRoeck, R. (1991). Is there a gap between forecasting theory and practice? *Int. J. Forec.*, **7**, 1–2.

12. Diebold, F. X. (1988). Serial correlation and the combination of forecasts. *J. Bus. Econ. Statist.*, **6**, 105–111.

13. Engle, R. F. and Granger, C. W. J. (1987). Cointegration and error correction: representation, estimation and testing. *Econometrica*, **55**, 251–276.

14. Freedman, D., Rothenberg, T., and Sutch, R. (1983). On energy policy models. *J. Bus. Econ. Statist.*, **1**, 24–32.

15. Gardner, E. S. (1983). Automatic monitoring of forecast errors. *J. Forec.*, **2**, 1–21.

16. Gardner, E. S. (1985). Exponential smoothing: the state of the art. *J. Forec.*, **4**, 1–28.

17. Granger, C. W. J. (1989). Combining forecasts—twenty years later. *J. Forec.*, **8**, 167–173.

18. Granger, C. W. J. and Newbold, P. (1973). Some comments on the evaluation of economic forecasts. *Appl. Econ.*, **5**, 35–47.

19. Granger, C. W. J. and Newbold, P. (1974). Spurious regressions in econometrics. *J. Economet.*, **2**, 111–120.

20. Granger, C. W. J. and Ramanathan, R. (1984). Improved methods of combining forecasts. *J. Forec.*, **3**, 197–204.

21. Hamilton, J. D. (1994). *Time Series Analysis*. Princeton University Press, Princeton, N.J.

22. Harvey, A. C. (1984). A unified view of statistical forecasting procedures. *J. Forec.*, **3**, 245–275.

23. Klein, L. R. (1988). The statistical approach to economics. *J. Economet.*, **37**, 7–26.

24. Ledolter, J. and Box, G. E. P. (1978). Conditions for the optimality of exponential smoothing forecast procedures. *Metrika*, **25**, 77–93.

25. Makridakis, S. and Winkler, R. L. (1983). Averages of forecasts: some empirical results. *Man. Sci.*, **29**, 987–996.

26. McClain, J. O. (1988). Dominant tracking signals. *Int. J. Forec.*, **4**, 563–572.

27. McKenzie, E. (1984). General exponential smoothing and the equivalent ARMA process. *J. Forec.*, **3**, 333–344.

28. Meade, N. (1984). The use of growth curves in forecasting market development–a review and appraisal. *J. Forec.*, **3**, 429–451.

29. Mentzer, J. T. and Cox, J. E. (1984). Familiarity, application, and performance of sales forecasting techniques. *J. Forec.*, **3**, 27–36.

30. Mincer, J. and Zarnowitz, V. (1969). The evaluation of economic forecasts. In *Economic Forecasts and Expectations*, J. Mincer, ed. National Bureau of Economic Research, New York.

31. Mizon, G. E. and Richard, J. F. (1986). The encompassing principle and its application to testing non-nested hypotheses. *Econometrica*, **54**, 657–678.

32. Muth, J. F. (1960). Optimal properties of exponentially weighted forecasts. *J. Am. Statist. Ass.*, **55**, 299–306.

33. Newbold, P. (1988). Predictors projecting linear trend plus seasonal dummies. *Statistician*, **37**, 111–127.

34. Newbold, P. and Bos, T. (1989). On exponential smoothing and the assumption of deterministic trend plus white noise data-generating models. *Int. J. Forec.*, **5**, 523–527.

35. Newbold, P. and Bos, T. (1994). *Introductory Business and Economic Forecasting*, 2nd ed. South-Western. Cincinnati, Ohio.

36. Newbold, P. and Davies, N. (1978). Error misspecification and spurious regression. *Int. Econ. Rev.*, **19**, 513–519.

37. Newbold, P. and Granger, C. W. J. (1974). Experience with forecasting univariate time series and the combination of forecasts. *J. Roy. Statist. Soc. A*, **137**, 131–165.

38. Phillips, P. C. B. (1986). Understanding spurious regression in econometrics. *J. Economet.*, **33**, 311–340.

39. Sanders, N. R. and Manrodt, K. B. (1994). Forecasting practices, in U.S. corporations: Survey results. *Interfaces*, **24**, 92–100.

40. Sparkes, J. R. and McHugh, A. K. (1984). Awareness and use of forecasting techniques in British industry. *J. Forec.*, **3**, 37–42.

41. Stekler, H. O. (1991). Macroeconomic forecast evaluation techniques. *Int. J. Forec.*, **7**, 375–384.

42. West, D. C. (1994). Number of sales forecast methods and marketing management *J. Forec.*, **13**, 395–407.

43. Winkler, R. L. (1989). Combining forecasts: a philosophical basis and some current issues. *Int. J. Forec.*, **5** 605–609.

44. Winkler, R. L. and Makridakis, S. (1983). The combination of forecasts. *J. R. Statist. Soc. A*, **146**, 150–157.

See also Autoregressive–Integrated Moving Average (ARIMA) Models; Box–Jenkins Model; Econometrics; Forecasting; Forecasts, Combination of; Geometric Moving Average; Prediction and Forecasting; Regression (Various); and Time Series.

Paul Newbold

BUYS–BALLOT TABLE.   See Time Series